Dedication

KU-417-162

Once Again, to Sheila
With Love

UNIVERSITY OF STRATHCLYDE
30125 00803021 4

ELECTRONIC DEVICES

Books are to be returned on or before the last date below.
ONE WEEK LOAN
CHECK ON RETURN FOR CD/DISK
DUE 20 DEC 2007
DUE 2 APR 2008
DUE 13 NOV 2008
3 DEC 2009
10 APR 2012
5 JUL 2011
11 AUG 2011
4 MAR 2013
DUE 17 MAR 2009
28 JAN 2010
29 MAR 2012
LIBREX

PEARSON
Prentice Hall

Pearson Education International

ANDERSONIAN LIBRARY
WITHDRAWN FROM LIBRARY STOCK
UNIVERSITY OF STRATHCLYDE

If you purchased this book within the United States or Canada you should be aware that if has been wrongfully imported without the approval of the Publisher or the Author.

Editor-in-Chief: Vernon Anthony
Editorial Assistant: Lara Dimmick
Production Editor: Rex Davidson
Production Manager: Matt Ottenweller
Design Coordinator: Diane Ernsberger
Cover Designer: Linda Sorrells-Smith
Cover Photo: Getty Images
Art Coordinator: Janet Portisch
Director of Marketing: David Gesell
Marketing Assistant: Les Roberts

This book was set in Times Roman by Techbooks and was printed and bound by R. R. Donnelley & Sons Company. The cover was printed by Phoenix Color Corp.

Copyright © 2008, 2005, 2002, 1999, 1996, 1992, 1988, 1984 by Pearson Education, Inc., Upper Saddle River, New Jersey 07458. Pearson Prentice Hall. All rights reserved. Printed in the United States of America. This publication is protected by Copyright and permission should be obtained from the publisher prior to any prohibited reproduction, storage in a retrieval system, or transmission in any form or by any means, electronic, mechanical, photocopying, recording, or likewise. For information regarding permission(s), write to: Rights and Permissions Department.

Pearson Prentice Hall™ is a trademark of Pearson Education, Inc.
Pearson® is a registered trademark of Pearson plc
Prentice Hall® is a registered trademark of Pearson Education, Inc.

Pearson Education Ltd.
Pearson Education Singapore Pte. Ltd.
Pearson Education Canada, Ltd.
Pearson Education—Japan
Pearson Education, Upper Saddle River, New Jersey
Pearson Education Australia Pty. Limited
Pearson Education North Asia Ltd.
Pearson Educación de Mexico, S.A. de C.V.
Pearson Education Malaysia Pte. Ltd.

D
621.3815
FLO

UNIVERSITY OF STRATHCLYDE
23 MAY 2007
UNIVERSITY LIBRARY

PEARSON
Prentice Hall

10 9 8 7 6 5 4 3 2 1

ISBN-13: 978-0-13-615581-2
ISBN-10: 0-13-615581-2

Preface

Electronic Devices has been carefully revised to make this eighth edition a textbook with a smoother flow, improved clarity, and a more comprehensive coverage. Many recommendations from reviewers and current users have been incorporated. Coverage of electronic devices and circuits, including extensive troubleshooting and applications, is provided. Chapters 1 through 11 are essentially devoted to discrete devices and circuits. Chapters 12 through 18 primarily cover linear integrated circuits. Extensive exercises and problems, many using Multisim circuit simulation, are designed to strengthen students' understanding of circuit theory and to develop troubleshooting and measurement skills. The Multisim circuit files in versions 8 and 9 are on the CD packaged with this book. Newer versions of these files will be posted on the website at *www.prenhall.com/floyd* as they become available.

New in This Edition

Application Activity Feature This feature replaces previous System Applications and appears at the end of most chapters. Many new activities are presented in an instructive format that is easy to follow. This feature is closely linked to the lab manual by David Buchla and Steve Wetterling, *Laboratory Exercises for Electronic Devices*. Students can build and test circuits similar to the ones introduced in the Application Activities. Each Application Activity takes the reader through the process of specification, simulation, prototyping, testing, and finalizing a circuit in the context of a practical system.

Programmable Analog Design Feature This new optional feature, following the Application Activity in Chapters 12 through 16, introduces the implementation of analog circuits by programming integrated circuit chips known as field-programmable analog arrays (FPAAs) or dynamically programmable analog signal processors (dpASPs). This feature is also linked to the lab manual.

Class D Amplifiers The coverage of amplifiers now includes the class D nonlinear amplifier. The basics of pulse width modulation are discussed to lay a foundation for understanding this important class of amplifier.

Analog and Digital Switching Circuits Analog switches are widely used in data acquisition systems, in which analog signals are converted to digital form, and in switched capacitor circuits used in programmable ICs. The fundamentals of both analog switching and digital logic gates are introduced.

Expanded Coverage Numerous topics, including high-intensity LEDs used in large color displays, automotive lighting, and traffic signals, have been expanded or revised. Also, zener diodes and transistor biasing are among other topics that have been rewritten or revised. Problems have been added in many chapters and there are a number of new examples, including some on computer analysis.

True/False Quiz This quiz has been added at the end of all chapters.

Margin Features History Notes, Tech Notes, and Safety Notes are now part of this new edition.

PowerPoint® Slides A completely new set of unique PowerPoint slides developed by Dave Buchla is available. These innovative, interactive slides are coordinated with each text chapter and are an excellent tool to supplement classroom presentations.

Standard Features

- Full-color format
- Chapter openers provide a chapter outline, chapter objectives, introduction, key terms list, Application Activity preview, and website reference.
- Brief introduction and objectives for each section within a chapter.
- Abundant worked-out examples, each with a related problem similar to that illustrated in the example. Answers to related problems are at the end of the chapter.
- Multisim circuit files for selected examples, troubleshooting, and problems are on the CD packaged with the book.
- Section checkup questions appear at the end of each section. Answers are at the end of the chapter.
- A section completely devoted to troubleshooting appears in many chapters.
- An Application Activity at the end of most chapters.
- A Programmable Analog Design feature at the end of some chapters provides an optional introduction to programmable analog devices and software.
- A sectionalized chapter summary, key term glossary, key formula list, true/false quiz, circuit-action quiz, sectionalized self-test, and sectionalized problem set at the end of each chapter.
- The end-of-chapter problem set is organized into basic and advanced problems. Additionally, many chapters have categories of datasheet, application activity, and Multisim troubleshooting problems.

Changes from Previous Edition

Some topics have been moved to provide a more logical arrangement and a better flow of material. The chapter on power amplifiers now follows the chapter on BJT amplifiers (Chapter 6). The topic of programmable analog arrays now appears in the new optional feature Programmable Analog Design at the end of certain chapters. Obsolete devices have been updated or omitted from the text.

The accompanying lab manual by Dave Buchla and Steve Wetterling has been reorganized so that it is more closely linked to the textbook. The eighteen experiments correspond to the eighteen chapters of the text, and include a component linked to the text's Application Activity.

Student Resources

Companion Website *(www.prenhall.com/floyd)* This website offers students an online study guide that they can check for conceptual understanding of key topics.

Multisim® CD-ROM Packaged with each textbook, this CD includes simulation circuits in Multisim 8 and 9 for selected examples, troubleshooting sections, and selected problems in the text. These circuits were created for use with Multisim software. A free download of the software is available at *electronicsworkbench.com.*

Multisim is widely regarded as an excellent circuit simulation tool for classroom and laboratory learning. However, no part of your textbook is dependent upon the Multisim software or provided files. These files are provided for those who choose to utilize Multisim software.

Laboratory Exercises for Electronic Devices, Eighth Edition, by Dave Buchla and Steve Wetterling. ISBN: 0-13-242971-3

Instructor Resources

To access supplementary materials online, instructors need to request an instructor access code. Go to **www.prenhall.com,** click the **Instructor Resource Center** link, and then click **Register Today** for an instructor access code. Within 48 hours after registering you will receive a confirmation e-mail including an instructor access code. Once you have received your code, go to the site and log on for full instructions on downloading the materials you wish to use.

Instructor's Resource Manual Includes solutions to chapter problems, Application Activity results, summary of Multisim circuit files, and a test item file. Solutions to the lab manual are also included. ISBN: 0-13-242985-3. Also available online.

Online Course Support If your program is offering your electronics course in a distance learning format, please contact your local Prentice Hall sales representative for a list of product solutions.

PowerPoint® CD-ROM This innovative, interactive PowerPoint slide presentation for each chapter in the book provides a very effective supplement to classroom lectures. Includes all figures from the text. Also available online.

TestGen This is a test bank of over 800 questions on CD-ROM. ISBN: 0-13-242987-X. Also available online.

Chapter Features

Chapter Opener Each chapter begins with an opening page, as shown in Figure P–1. The chapter opener includes a chapter introduction, a list of chapter sections, chapter objectives, key terms, an Application Activity preview, and a website reference for associated study aids.

Section Opener Each section in a chapter begins with a brief introduction and section objectives. An example is shown in Figure P–2.

Section Checkup Each section in a chapter ends with a checkup consisting of questions that highlight the main concepts presented in the section. This feature is also illustrated in Figure P–2. The answers to the Section Checkups are at the end of the chapter.

Worked Examples, Related Problems, and Multisim Exercises Numerous worked-out examples throughout each chapter illustrate and clarify basic concepts or specific procedures. Each example ends with a Related Problem that reinforces or expands on the example by requiring the student to work through a problem similar to the example. Selected examples feature a Multisim exercise keyed to a file on the CD-ROM that contains the circuit illustrated in the example. A typical example with a Related Problem and a Multisim exercise is shown in Figure P–3. Answers to Related Problems are at the end of the chapter.

Troubleshooting Sections Many chapters include a troubleshooting section that relates to the topics covered in the chapter and that illustrates troubleshooting procedures and techniques. The Troubleshooting section also provides Multisim Troubleshooting exercises.

Application Activity This feature follows the last section in most chapters and are identified by a special graphic design. A practical application of devices or circuits covered in the chapter is presented. The student learns how the specific device or circuit is used and is taken through the steps of design specification, simulation, prototyping, circuit board

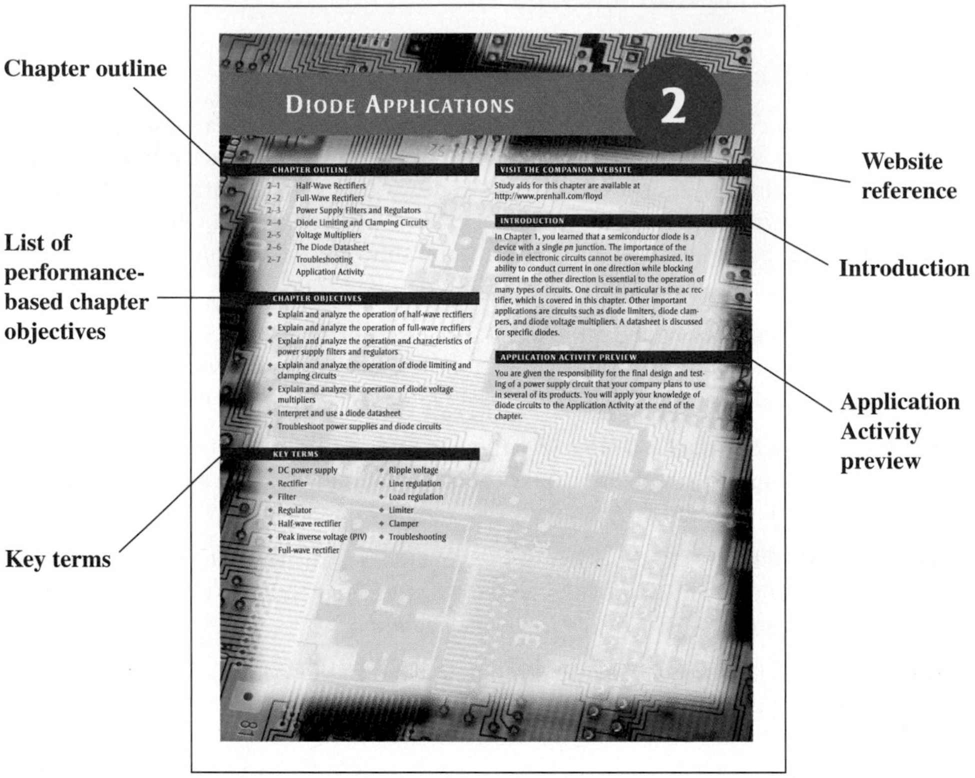

FIGURE P–1

A typical chapter opener.

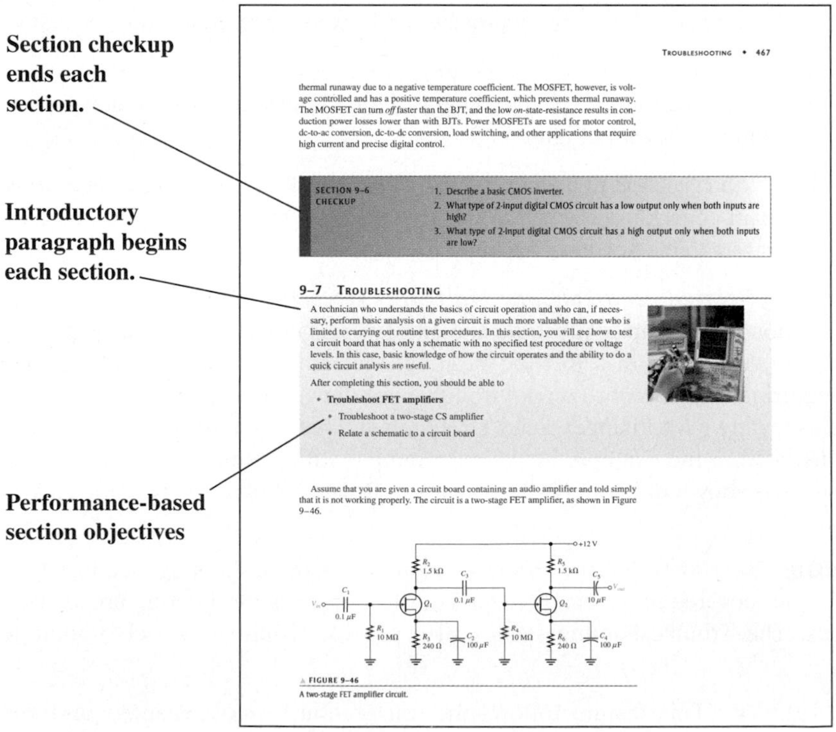

FIGURE P–2

A typical section opener and section review.

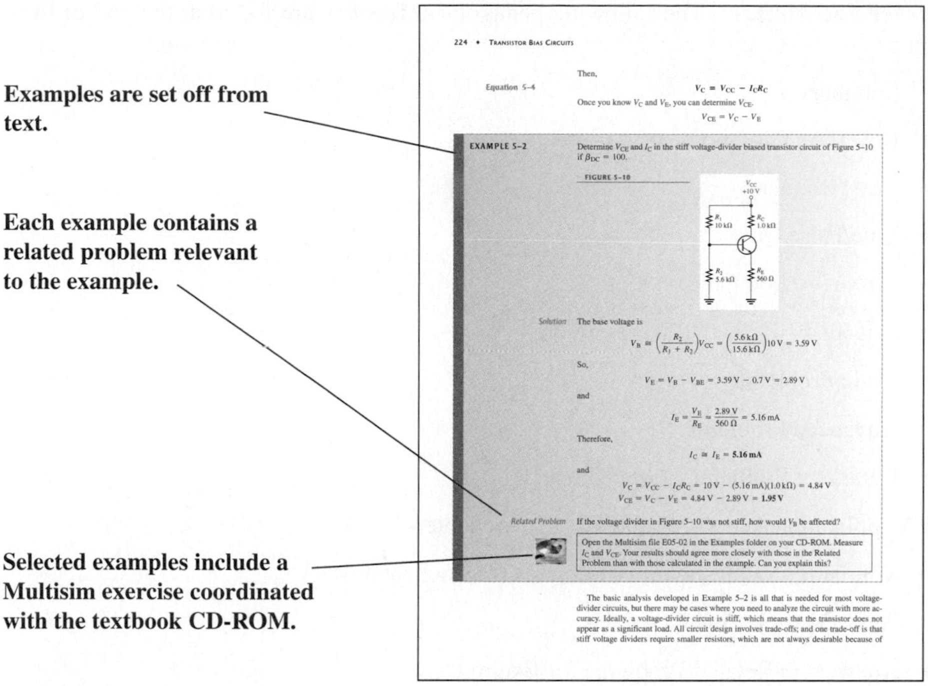

224 • TRANSISTOR BIAS CIRCUITS

Then,

Equation 5–4 $V_C = V_{CC} - I_C R_C$

Once you know V_C and V_E, you can determine V_{CE}.

$V_{CE} = V_C - V_E$

EXAMPLE 5–2

Determine V_{CE} and I_C in the stiff voltage-divider biased transistor circuit of Figure 5–10 if $\beta_{DC} = 100$.

FIGURE 5–10

Solution The base voltage is

$$V_B \cong \left(\frac{R_2}{R_1 + R_2}\right)V_{CC} = \left(\frac{5.6\text{ k}\Omega}{15.6\text{ k}\Omega}\right)10\text{ V} = 3.59\text{ V}$$

So,

$$V_E = V_B - V_{BE} = 3.59\text{ V} - 0.7\text{ V} = 2.89\text{ V}$$

and

$$I_E = \frac{V_E}{R_E} = \frac{2.89\text{ V}}{560\ \Omega} = 5.16\text{ mA}$$

Therefore,

$$I_C \cong I_E = \mathbf{5.16\text{ mA}}$$

and

$$V_C = V_{CC} - I_C R_C = 10\text{ V} - (5.16\text{ mA})(1.0\text{ k}\Omega) = 4.84\text{ V}$$
$$V_{CE} = V_C - V_E = 4.84\text{ V} - 2.89\text{ V} = \mathbf{1.95\text{ V}}$$

Related Problem If the voltage divider in Figure 5–10 was not stiff, how would V_B be affected?

Open the Multisim file E05-02 in the Examples folder on your CD-ROM. Measure I_C and V_{CE}. Your results should agree more closely with those in the Related Problem than with those calculated in the example. Can you explain this?

The basic analysis developed in Example 5–2 is all that is needed for most voltage-divider circuits, but there may be cases where you need to analyze the circuit with more accuracy. Ideally, a voltage-divider circuit is stiff, which means that the transistor does not appear as a significant load. All circuit design involves trade-offs; and one trade-off is that stiff voltage dividers require smaller resistors, which are not always desirable because of

▲ FIGURE P–3

A typical example with a related problem and Multisim exercise.

implementation, and testing. A typical Application Activity is shown in Figure P–4. The Application Activities are optional; skipping any or all of them does not affect comprehension of the chapter content. Results for the Application Activities are provided in the Instructor's Resource Manual.

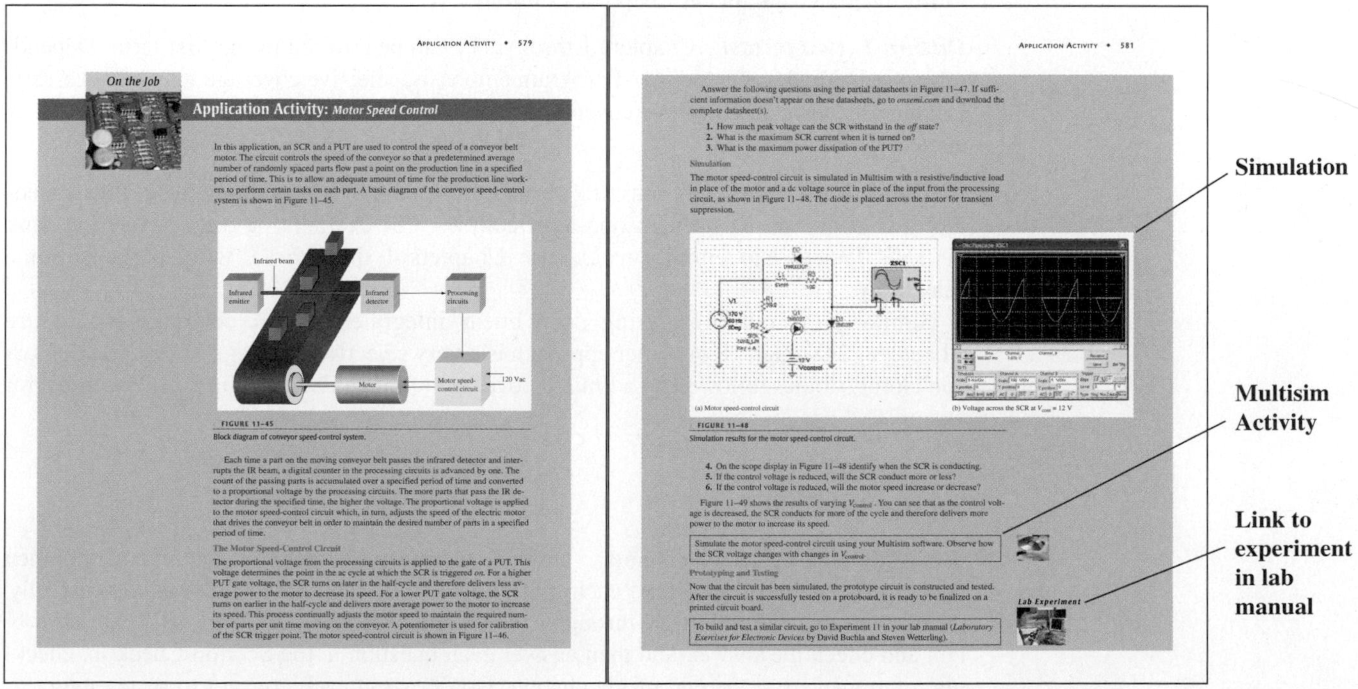

APPLICATION ACTIVITY • 579

On the Job

Application Activity: *Motor Speed Control*

In this application, an SCR and a PUT are used to control the speed of a conveyor belt motor. The circuit controls the speed of the conveyor so that a predetermined average number of randomly spaced parts flow past a point on the production line in a specified period of time. This is to allow an adequate amount of time for the production line workers to perform certain tasks on each part. A basic diagram of the conveyor speed-control system is shown in Figure 11–45.

FIGURE 11–45

Block diagram of conveyor speed-control system.

Each time a part on the moving conveyor belt passes the infrared detector and interrupts the IR beam, a digital counter in the processing circuits is advanced by one. The count of the passing parts is accumulated over a specified period of time and converted to a proportional voltage by the processing circuits. The more parts that pass the IR detector during the specified time, the higher the voltage. The proportional voltage is applied to the motor speed-control circuit which, in turn, adjusts the speed of the electric motor that drives the conveyor belt in order to maintain the desired number of parts in a specified period of time.

The Motor Speed-Control Circuit

The proportional voltage from the processing circuits is applied to the gate of a PUT. This voltage determines the point in the ac cycle at which the SCR is triggered *on*. For a higher PUT gate voltage, the SCR turns on later in the half-cycle and therefore delivers less average power to the motor to decrease its speed. For a lower PUT gate voltage, the SCR turns on earlier in the half-cycle and delivers more average power to the motor to increase its speed. This process continually adjusts the motor speed to maintain the required number of parts per unit time moving on the conveyor. A potentiometer is used for calibration of the SCR trigger point. The motor speed-control circuit is shown in Figure 11–46.

APPLICATION ACTIVITY • 581

Answer the following questions using the partial datasheets in Figure 11–47. If sufficient information doesn't appear on these datasheets, go to *onsemi.com* and download the complete datasheet(s).

1. How much peak voltage can the SCR withstand in the *off* state?
2. What is the maximum SCR current when it is turned on?
3. What is the maximum power dissipation of the PUT?

Simulation

The motor speed-control circuit is simulated in Multisim with a resistive/inductive load in place of the motor and a dc voltage source in place of the input from the processing circuit, as shown in Figure 11–48. The diode is placed across the motor for transient suppression.

(a) Motor speed-control circuit

(b) Voltage across the SCR at $V_{cont} = 12$ V

FIGURE 11–48

Simulation results for the motor speed-control circuit.

4. On the scope display in Figure 11–48 identify when the SCR is conducting.
5. If the control voltage is reduced, will the SCR conduct more or less?
6. If the control voltage is reduced, will the motor speed increase or decrease?

Figure 11–49 shows the results of varying $V_{control}$. You can see that as the control voltage is decreased, the SCR conducts for more of the cycle and therefore delivers more power to the motor to increase its speed.

Simulate the motor speed-control circuit using your Multisim software. Observe how the SCR voltage changes with changes in $V_{control}$.

Prototyping and Testing

Now that the circuit has been simulated, the prototype circuit is constructed and tested. After the circuit is successfully tested on a protoboard, it is ready to be finalized on a printed circuit board.

Lab Experiment

To build and test a similar circuit, go to Experiment 11 in your lab manual (*Laboratory Exercises for Electronic Devices* by David Buchla and Steven Wetterling).

▲ FIGURE P–4

Portion of a typical Application Activity section.

Chapter End Matter The following pedagogical features are found at the end of most chapters:

- Summary
- Key Term Glossary
- Key Formulas
- True/False Quiz
- Circuit-Action Quiz
- Self-Test
- Basic Problems
- Advanced Problems
- Datasheet Problems (selected chapters)
- Application Activity Problems (many chapters)
- Multisim Troubleshooting Problems (most chapters)
- Answers to Section Checkups
- Answers to Related Problems for Examples
- Answers to True/False Quiz
- Answers to Circuit-Action Quiz
- Answers to Self-Test

Suggestions for Using This Textbook

As mentioned, this book covers discrete devices and circuits in Chapters 1 through 11 and linear integrated circuits in Chapters 12 through 18.

Option 1 (two terms) Chapters 1 through 11 can be covered in the first term. Depending on individual preferences and program emphasis, selective coverage may be necessary. Chapters 12 through 18 can be covered in the second term. Again, selective coverage may be necessary.

Option 2 (one term) By omitting certain topics and by maintaining a rigorous schedule, this book can be used in one-term courses. For example, a course covering only discrete devices and circuits would use Chapters 1 through 11 with, perhaps, some selectivity.

Similarly, a course requiring only linear integrated circuit coverage would use Chapters 12 through 18. Another approach is a very selective coverage of discrete devices and circuits topics followed by a limited coverage of integrated circuits (only op-amps, for example).

To the Student

When studying a particular chapter, study one section until you understand it and only then move on to the next one. Read each section and study the related illustrations carefully, think about the material, work through each example step by step, work its Related Problem and check the answer, and then answer each question in the Section Checkup, checking your answers at the end of the chapter. Don't expect each concept to be crystal clear after a single reading; you may have to read the material two or even three times. Once you think that you understand the material, review the chapter summary, key formula list, and key term definitions at the end of the chapter. Take the true/false quiz, the circuit-action quiz, and the self-test. Finally, work the assigned problems at the end of the chapter. Working through these problems is perhaps the most important way to check and reinforce your comprehension of the chapter. By working problems, you acquire an additional level of insight and understanding, and develop logical thinking that reading or classroom lectures alone do not provide.

Generally, you cannot fully understand a concept or procedure by simply watching or listening to someone else. Only hard work and critical thinking will produce the results you expect and deserve.

Milestones in Electronics

Before you begin your study of electronic devices, let's briefly look at some of the important developments that led to the electronics technology we have today. The names of many of the early pioneers in electricity and electromagnetics still live on in terms of familiar units and quantities. Names such as Ohm, Ampere, Volta, Farad, Henry, Coulomb, Oersted, and Hertz are some of the better known examples with which you are already familiar. More widely known names such as Franklin and Edison are also significant in the history of electricity and electronics because of their tremendous contributions. Brief biographies of a few important figures in the history of electronics are shown.

Early experiments with electronics involved electric currents in vacuum tubes. Heinrich Geissler (1814–1879) removed most of the air from a glass tube and found that the tube glowed when there was current through it. Later, Sir William Crookes (1832–1919) found the current in vacuum tubes seemed to consist of particles. Thomas Edison (1847–1931) experimented with carbon filament bulbs with plates and discovered that there was a current from the hot filament to a positively charged plate. He patented the idea but never used it.

Other early experimenters measured the properties of the particles that flowed in vacuum tubes. Sir Joseph Thompson (1856–1940) measured properties of these particles, later called *electrons.*

Although wireless telegraphic communication dates back to 1844, electronics is basically a 20th century concept that began with the invention of the vacuum tube amplifier. An early vacuum tube that allowed current in only one direction was constructed by John A. Fleming in 1904. Called the Fleming valve, it was the forerunner of vacuum tube diodes. In 1907, Lee DeForest added a grid to the vacuum tube. The new device, called the audiotron, could amplify a weak signal. By adding the control element, DeForest ushered in the electronics revolution. It was an improved version of his device that made transcontinental telephone service and radios possible. In 1912, a radio amateur in San Jose, California, was regularly broadcasting music!

In 1921, the secretary of commerce, Herbert Hoover, issued the first license to a broadcast radio station; within two years over 600 licenses were issued. By the end of the 1920s radios were in many homes. A new type of radio, the superheterodyne radio, invented by Edwin Armstrong, solved problems with high-frequency communication. In 1923, Vladimir Zworykin, an American researcher, invented the first television picture tube, and in 1927 Philo T. Farnsworth applied for a patent for a complete television system.

The 1930s saw many developments in radio, including metal tubes, automatic gain control, "midget" radios, and directional antennas. Also started in this decade was the development of the first electronic computers. Modern computers trace their origins to the work of John Atanasoff at Iowa State University. Beginning in 1937, he envisioned a binary machine that could do complex mathematical work. By 1939, he and graduate student Clifford Berry had constructed a binary machine called ABC, (for Atanasoff-Berry Computer) that used vacuum tubes for logic and condensers (capacitors) for memory. In 1939, the magnetron, a microwave oscillator, was invented in Britain by Henry Boot and

BIOGRAPHY

Thomas Alva Edison 1847–1931

Born in Milan, Ohio, Thomas Edison was the most prolific inventor of all time. He is credited with 1093 patents and is the only person to ever have at least one patent every year for 65 consecutive years. Mr. Edison's inventions and enterprises encompass many different areas. One of his most famous inventions, the light bulb, was introduced in 1879. Mr. Edison is credited with discovering the diode effect while working with vacuum tubes for the light bulb. Most of his work was done in his laboratory in West Orange, NJ. He also maintained a laboratory at his winter home in Fort Myers, Florida, which was devoted principally to the development of a synthetic rubber using the goldenrod plant. (Photo credit: Library of Congress)

BIOGRAPHY

Lee DeForest
1873–1961

Born in Iowa, Lee DeForest became an inventor while in college to help defray expenses. He graduated from Yale in 1899 with a PhD. His doctoral thesis, "Reflection of Hertzian Waves from the Ends of Parallel Wires," began his long career in radio. His invention of the vacuum tube triode for use in amplification (audion amplifier) was the most important of his more than 300 inventions. (Photo credit: The National Cyclopedia of American Biography, courtesy AIP Emilio Segrè Visual Archives, T.J.J. See Collection)

John Randall. In the same year, the klystron microwave tube was invented in America by Russell and Sigurd Varian.

The decade of the 1940s opened with World War II. The war spurred rapid advancements in electronics. Radar and very high-frequency communication were made possible by the magnetron and klystron. Cathode ray tubes were improved for use in radar. Computer work continued during the war. By 1946, John von Neumann had developed the first stored program computer, the Eniac, at the University of Pennsylvania. One of the most significant inventions ever occurred in 1947 with the invention of the transistor. The inventors were Walter Brattain, John Bardeen, and William Shockley. All three won Nobel prizes for their invention. PC (printed circuit) boards were also introduced in 1947. Commercial manufacturing of transistors didn't begin until 1951 in Allentown, Pennsylvania.

The most important invention of the 1950s was the integrated circuit. On September 12, 1958, Jack Kilby, at Texas Instruments, made the first integrated circuit (Figure P–5), for which he was awarded a Nobel prize in the fall of 2000. This invention literally created the modern computer age and brought about sweeping changes in medicine, communication, manufacturing, and the entertainment industry. Many billions of "chips"—as integrated circuits came to be called—have since been manufactured.

The 1960s saw the space race begin and spurred work on miniaturization and computers. The space race was the driving force behind the rapid changes in electronics that followed. The first successful "op-amp" was designed by Bob Widlar at Fairchild Semiconductor in 1965. Called the μA709, it was very successful but suffered from "latch-up" and other problems. Later, the most popular op-amp ever, the 741, took shape at Fairchild. This op-amp became the industry standard and influenced design of op-amps for years to come. Precursors to the Internet began in the 1960s with remote

BIOGRAPHY

John Bardeen
1908–1991

An electrical engineer and physicist born in Madison, Wisconsin, Dr. Bardeen was on the faculty of the University of Minnesota from 1938 to 1941 and a physicist at the Naval Ordnance Lab from 1941 to 1945. He then joined Bell Labs and remained there until 1951. Some of his fields of interest were conduction in semiconductors and metals, surface properties of semiconductors, and superconductivity. While at Bell Labs he jointly invented the transistor with colleagues Walter Brattain and William Shockley. After leaving Bell Labs in 1951, Dr. Bardeen joined the faculty at the University of Illinois. (Photo credit: AIP Emilio Segrè Visual Archives, W. F. Meggers Gallery of Nobel Laureates)

BIOGRAPHY

William Shockley
1910–1989

An American born in London, England, Dr. Shockley obtained his PhD in 1936 from M.I.T. He joined Bell Labs upon graduation and remained there until 1955. His research emphasis included areas of energy bands in solids, theory of vacuum tubes, photoelectrons, ferromagnetic domains, and transistor physics. While at Bell Labs, Dr. Shockley joined John Bardeen and Walter Brattain in the invention of the transistor in 1947. After leaving Bell Labs, Dr. Shockley spent time at Beckman Instruments and at Stanford University. (Photo credit: AIP Emilio Segrè Visual Archives, Physics Today Collection)

BIOGRAPHY

Walter H. Brattain
1902–1987

An American born in China, Dr. Brattain joined Bell Telephone Laboratories in 1929. One of his main areas of research was the surface properties of semiconductive materials. His chief contributions were the discovery of the photo effect at the surface of a semiconductor and the invention of the point-contact transistor in 1947, which he jointly invented with John Bardeen and William Shockley. (Photo credit: AIP Emilio Segrè Visual Archives, W. F. Meggers Gallery of Nobel Laureates)

FIGURE P–5

The first integrated circuit. (Courtesy of Texas Instruments)

BIOGRAPHY

Jack S. Kilby
1923–2005

Jack Kilby was born in Missouri and earned degrees in electrical engineering from the University of Illinois and the University of Wisconsin. In 1958, he joined Texas Instruments in Dallas. Within a year after joining TI he developed the monolithic integrated circuit, and the rest is history. At about the same time Robert Noyce of Fairchild Semiconductor independently developed an integrated circuit. Both men are generally credited with the invention. (Photo credit: Courtesy of Texas Instruments)

networked computers. Systems were in place within Lawrence Livermore National Laboratory that connected over 100 terminals to a computer system (colorfully called the "Octopus system").

By 1971, a new company that had been formed by a group from Fairchild introduced the first microprocessor. The company was Intel and the product was the 4004 chip, which had the same processing power as the Eniac computer. Later in that same year, Intel announced the first 8-bit processor, the 8008. In 1975, the first personal computer was introduced by Altair, and *Popular Science* magazine featured it on the cover of the January 1975 issue. The 1970s also saw the introduction of the pocket calculator and new developments in optical integrated circuits.

By the 1980s, half of all U.S. homes were using cable hookups instead of television antennas. The reliability, speed, and miniaturization of electronics continued throughout the 1980s, including automated testing and calibrating of PC boards. The computer became a part of instrumentation and the virtual instrument was created. Computers became a standard tool on the workbench.

The 1990s saw a widespread application of the Internet. In 1993, there were 130 websites; by the start of the new century (in 2001) there were over 24 million. In the 1990s, companies scrambled to establish a home page and many of the early developments of radio broadcasting had parallels with the Internet. The exchange of information and e-commerce fueled the tremendous economic growth of the 1990s. The Internet became especially important to scientists and engineers, becoming one of the most important scientific communication tools ever.

In 1995, the FCC allocated spectrum space for a new service called Digital Audio Radio Service. Digital television standards were adopted in 1996 by the FCC for the nation's next generation of broadcast television. As the 20th century drew toward a close, historians could only breathe a sign of relief. As one person put it, "I'm all for new technologies, but I wish they'd let the old ones wear out first."

The 21st century dawned on January 1, 2001. The major story was the continuing explosive growth of the Internet; shortly thereafter, scientists were planning a new supercomputer system that would make massive amounts of information accessible in a computer network. The new international data grid will be an even greater resource than the World Wide Web, giving people the capability to access enormous amounts of information and the resources to run simulations on a supercomputer. Research in the 21st century continues along lines of faster and smaller circuits using new technologies.

Acknowledgments

Many capable people have contributed to the eighth edition of *Electronic Devices*. It has been thoroughly reviewed and checked for both content and accuracy. Those at Prentice Hall who have contributed greatly to this project throughout the many phases of development and production include Rex Davidson, Lara Dimmick, and Vern Anthony. Lois Porter has once more done an outstanding job editing the manuscript. Dave Buchla did a thorough line-by-line review of the manuscript and contributed extensively to the content of the book, helping to make this edition the best one yet. Gary Snyder created the circuit files for the Multisim features in this edition. A note of appreciation goes to Evan Robinson, Erik Luther, and Shekhar Sharad of Electronics Workbench for their contributions to the Multisim content.

I wish to express my appreciation to those already mentioned as well as the reviewers who provided many valuable suggestions and constructive criticism that greatly influenced this edition. These reviewers are William Dolan, Kennebec Valley Community College; John Duncan, Kent State University; Art Eggers, Community College of Southern Nevada; Paul Garrett, ITT Technical Institute; Mark Hughes, Cleveland Community College; Lisa Jones, Southwest Tennessee Community College; Max Rabiee, University of Cincinnati; and Jim Rhodes, Blue Ridge Community College.

Tom Floyd

Brief Contents

CONTENTS

INTRODUCTION TO SEMICONDUCTORS

1

CHAPTER OUTLINE

CHAPTER OBJECTIVES

- Discuss the basic structure of atoms
- Discuss insulators, conductors, and semiconductors and how they basically differ
- Describe how current is produced in a semiconductor
- Describe the properties of *n*-type and *p*-type semiconductors
- Describe a diode and how a *pn* junction is formed
- Discuss biasing of a diode
- Analyze the voltage–current (*V–I*) characteristic curve of a diode
- Discuss the operation of diodes and explain the three diode models
- Test a diode using a digital multimeter

KEY TERMS

- Atom
- Proton
- Electron
- Shell
- Valence
- Ionization
- Free electron
- Insulator
- Conductor
- Semiconductor
- Silicon
- Crystal
- Hole
- Doping
- Diode
- *PN* junction
- Barrier potential
- Bias
- Forward bias
- Reverse bias
- *V–I* characteristic
- Cathode
- Anode

VISIT THE COMPANION WEBSITE

Study aids for this chapter are available at http://www.prenhall.com/floyd

INTRODUCTION

Electronic devices such as diodes, transistors, and integrated circuits are made of a semiconductive material. To understand how these devices work, you should have a basic knowledge of the structure of atoms and the interaction of atomic particles. An important concept introduced in this chapter is that of the *pn* junction that is formed when two different types of semiconductive material are joined. The *pn* junction is fundamental to the operation of devices such as the diode and certain types of transistors. The operation and characteristics of the diode are covered. Also, three diode models representing three levels of approximation are presented and testing is discussed.

1–1 ATOMIC STRUCTURE

All matter is made of atoms; and all atoms consist of electrons, protons, and neutrons. In this section, you will learn about the structure of the atom, electron orbits and shells, valence electrons, ions, and two semiconductive materials—silicon and germanium. The configuration of certain electrons in an atom is the key factor in determining how a given material conducts electrical current.

After completing this section, you should be able to

- **Discuss the basic structure of atoms**
  - Define *nucleus, proton, neutron,* and *electron*
  - Describe an element's atomic number
  - Explain electron shells
  - Describe a valence electron
  - Describe ionization
  - Describe a free electron

An **atom*** is the smallest particle of an element that retains the characteristics of that element. Each of the known 109 elements has atoms that are different from the atoms of all other elements. This gives each element a unique atomic structure. According to the classical Bohr model, atoms have a planetary type of structure that consists of a central nucleus surrounded by orbiting electrons, as illustrated in Figure 1–1. The **nucleus** consists of positively charged particles called **protons** and uncharged particles called **neutrons.** The basic particles of negative charge are called **electrons**.

FIGURE 1–1

The Bohr model of an atom showing electrons in orbits around the nucleus, which consists of protons and neutrons. The "tails" on the electrons indicate motion.

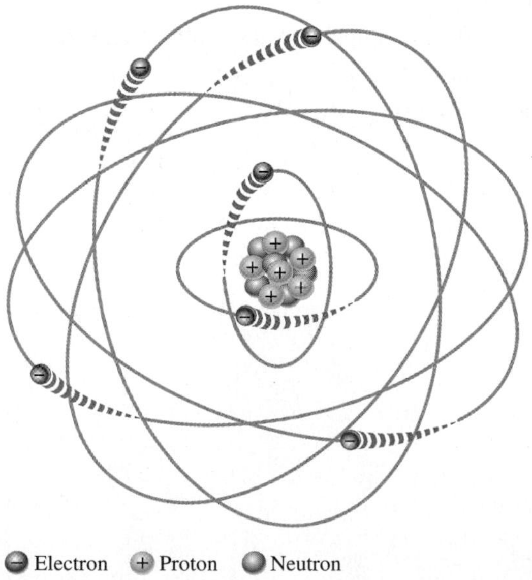

***All bold terms are in the end-of-book glossary. The bold terms in color are key terms and are also defined at the end of the chapter.**

Each type of atom has a certain number of electrons and protons that distinguishes it from the atoms of all other elements. For example, the simplest atom is that of hydrogen, which has one proton and one electron, as shown in Figure 1–2(a). As another example, the helium atom, shown in Figure 1–2(b), has two protons and two neutrons in the nucleus and two electrons orbiting the nucleus.

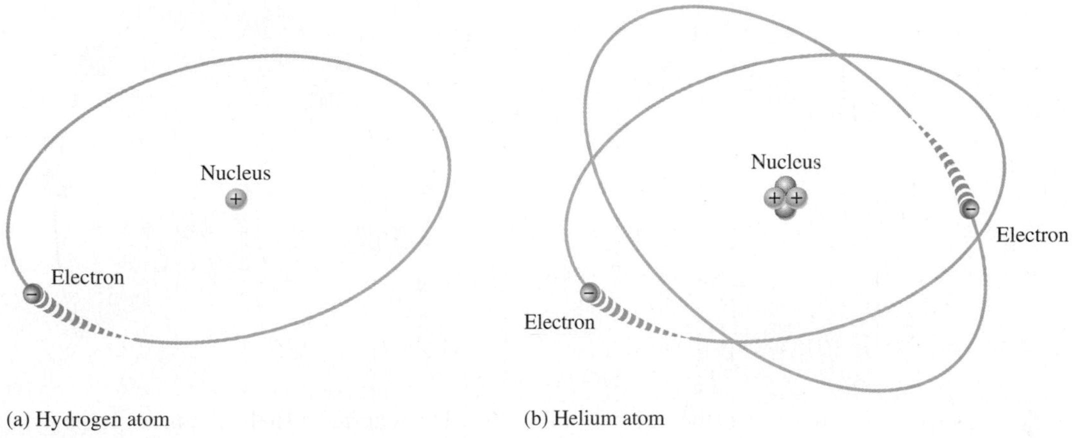

▲ FIGURE 1–2

Two simple atoms, hydrogen and helium.

Atomic Number

All elements are arranged in the periodic table of the elements in order according to their atomic number. The **atomic number** equals the number of protons in the nucleus, which is the same as the number of electrons in an electrically balanced (neutral) atom. For example, hydrogen has an atomic number of 1 and helium has an atomic number of 2. In their normal (or neutral) state, all atoms of a given element have the same number of electrons as protons; the positive charges cancel the negative charges, and the atom has a net charge of zero.

Electron Shells and Orbits

Electrons orbit the nucleus of an atom at certain distances from the nucleus. Electrons near the nucleus have less energy than those in more distant orbits. Only discrete (separate and distinct) values of electron energies exist within atomic structures. Therefore, electrons must orbit only at discrete distances from the nucleus.

Energy Levels Each discrete distance **(orbit)** from the nucleus corresponds to a certain energy level. In an atom, the orbits are grouped into energy bands known as **shells**. A given atom has a fixed number of shells. Each shell has a fixed maximum number of electrons at permissible energy levels. The differences in energy levels within a shell are much smaller than the difference in energy between shells. The shells are designated 1, 2, 3, and so on, with 1 being closest to the nucleus. This energy band concept is illustrated in Figure 1–3, which shows the 1st shell with one energy level and the 2nd shell with two energy levels. Additional shells may exist in other types of atoms, depending on the element.

The Number of Electrons in Each Shell The maximum number of electrons (N_e) that can exist in each shell of an atom is a fact of nature and can be calculated by the formula,

$$N_e = 2n^2 \qquad \text{Equation 1–1}$$

▶ **FIGURE 1–3**

Energy increases as the distance from the nucleus increases.

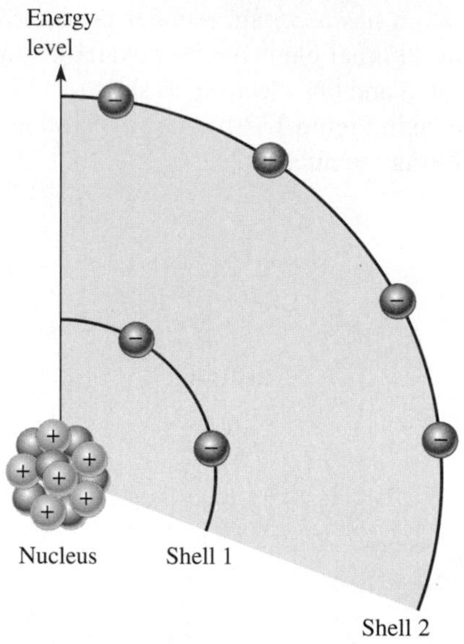

where n is the number of the shell. The innermost shell is number 1, the next shell is number 2, and so on. The maximum number of electrons that can exist in the innermost shell (shell 1) is

$$N_e = 2n^2 = 2(1)^2 = 2$$

The maximum number of electrons that can exist in the second shell is

$$N_e = 2n^2 = 2(2)^2 = 2(4) = 8$$

The maximum number of electrons that can exist in the third shell is

$$N_e = 2n^2 = 2(3)^2 = 2(9) = 18$$

The maximum number of electrons that can exist in the fourth shell is

$$N_e = 2n^2 = 2(4)^2 = 2(16) = 32$$

Valence Electrons

Electrons that are in orbits farther from the nucleus have higher energy and are less tightly bound to the atom than those closer to the nucleus. This is because the force of attraction between the positively charged nucleus and the negatively charged electron decreases with increasing distance from the nucleus. Electrons with the highest energy exist in the outermost shell of an atom and are relatively loosely bound to the atom. This outermost shell is known as the **valence** shell and electrons in this shell are called *valence electrons.* These valence electrons contribute to chemical reactions and bonding within the structure of a material and determine its electrical properties.

Ionization

When an atom absorbs energy from a heat source or from light, for example, the energies of the electrons are raised. The valence electrons possess more energy and are more loosely bound to the atom than inner electrons, so they can easily jump to higher orbits within the valence shell when external energy is absorbed by the atom.

If a valence electron acquires a sufficient amount of energy, it can actually escape from the outer shell and the atom's influence. The departure of a valence electron leaves a previously neutral atom with an excess of positive charge (more protons than electrons). The process of losing a valence electron is known as **ionization**, and the resulting positively charged atom is called a *positive ion.*

For example, the chemical symbol for hydrogen is H. When a neutral hydrogen atom loses its valence electron and becomes a positive ion, it is designated H^+. The escaped valence electron is called a **free electron**.

For some nonmetallic materials such as chlorine, a free electron can be captured by the neutral atom, forming a negative ion. In the case of chlorine, the ion is more stable than the neutral atom because it has a filled outer shell. The chlorine ion is designated as Cl^-.

SECTION 1–1 CHECKUP
Answers are at the end of the chapter.

1. Describe an atom.
2. What is an electron?
3. What is a valence electron?
4. What is a free electron?
5. How are ions formed?

1–2 INSULATORS, CONDUCTORS, AND SEMICONDUCTORS

In terms of their electrical properties, materials can be classified into three groups: conductors, semiconductors, and insulators. When atoms combine to form a solid, crystalline material, they arrange themselves in a symmetrical pattern. The atoms within the crystal structure are held together by covalent bonds, which are created by the interaction of the valence electrons of the atoms. Silicon is a crystalline material.

After completing this section, you should be able to

- **Discuss insulators, conductors, and semiconductors and how they basically differ**
  - Define the core of an atom
  - Describe the atomic structure of copper, silicon, germanium, and carbon
  - List the four best conductors
  - List four semiconductors
  - Discuss the difference between conductors and semiconductors
  - Discuss the difference between silicon and germanium semiconductors
  - Discuss covalent bonding in silicon

All materials are made up of atoms. These atoms contribute to the electrical properties of a material, including its ability to conduct electrical current.

For purposes of discussing electrical properties, an atom can be represented by the valence shell and a **core** that consists of all the inner shells and the nucleus. This concept is illustrated in Figure 1–4 for a carbon atom. Carbon is used in some types of electrical resistors. Notice that the carbon atom has four electrons in the valence shell and two electrons in the inner shell. The nucleus consists of six protons and six neutrons so the $+6$ indicates the positive charge of the six protons. The core has a net charge of $+4$ ($+6$ for the nucleus and -2 for the two inner-shell electrons).

Insulators

An **insulator** is a material that does not conduct electrical current under normal conditions. Most good insulators are compounds rather than single-element materials. Valence

▶ FIGURE 1–4

Diagram of a carbon atom.

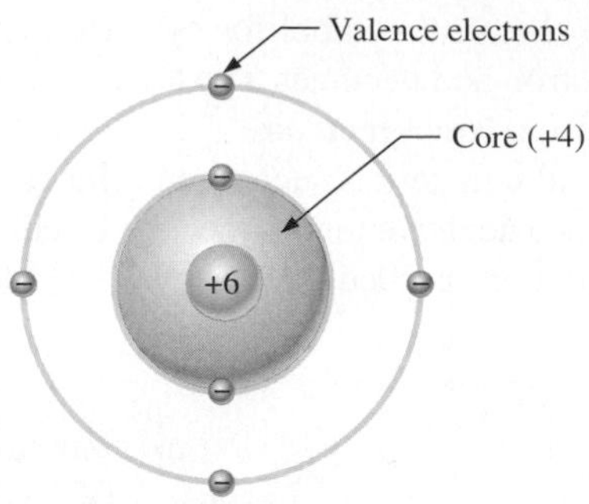

electrons are tightly bound to the atoms; therefore, there are very few free electrons in an insulator. Examples of insulators are rubber, plastics, glass, mica, and quartz.

Conductors

A **conductor** is a material that easily conducts electrical current. Most metals are good conductors. The best conductors are single-element materials, such as copper, silver, gold, and aluminum, which are characterized by atoms with only one valence electron very loosely bound to the atom. These loosely bound valence electrons become free electrons. Therefore, in a conductive material the free electrons are valence electrons.

Semiconductors

A **semiconductor** is a material that is between conductors and insulators in its ability to conduct electrical current. A semiconductor in its pure (intrinsic) state is neither a good conductor nor a good insulator. The most common single-element semiconductors are **silicon, germanium,** and **carbon.** Compound semiconductors such as gallium arsenide and indium phosphide are also commonly used. The single-element semiconductors are characterized by atoms with four valence electrons.

Energy Bands

Recall that the valence shell of an atom represents a band of energy levels and that the valence electrons are confined to that band. When an electron acquires enough additional energy, it can leave the valence shell, become a *free electron,* and exist in what is known as the *conduction band.*

The difference in energy between the valence band and the conduction band is called an *energy gap.* This is the amount of energy that a valence electron must have in order to jump from the valence band to the conduction band. Once in the conduction band, the electron is free to move throughout the material and is not tied to any given atom.

Figure 1–5 shows energy diagrams for insulators, semiconductors, and conductors. Notice in part (a) that insulators have a very wide energy gap. Valence electrons do not jump into the conduction band except under breakdown conditions where extremely high voltages are applied across the material. As you can see in part (b), semiconductors have a much narrower energy gap. This gap permits some valence electrons to jump into the conduction band and become free electrons. By contrast, as part (c) illustrates, the energy bands in conductors overlap. In a metallic conductive material there is always a large number of valence electrons that are free electrons.

Comparison of a Semiconductor Atom to a Conductor Atom

Silicon is a semiconductor and copper is a conductor. Diagrams of the silicon atom and the copper atom are shown in Figure 1–6. Notice that the core of the silicon atom has a net charge of +4 (14 protons − 10 electrons) and the core of the copper atom has a net charge of +1 (29 protons − 28 electrons). The core includes everything except the valence electrons.

TECH NOTE

Next to silicon, the second most common semiconductive material is gallium arsenide, GaAs. This is a crystalline compound, not an element. Its properties can be controlled by varying the relative amount of gallium and arsenic.

GaAs has the advantage of making semiconductor devices that respond very quickly to electrical signals. This makes it better than silicon for applications like amplifying the high frequency (1 GHz to 10 GHz) signals from TV satellites, etc. The main disadvantage of GaAs is that it is more difficult to make and the chemicals involved are quite often poisonous!

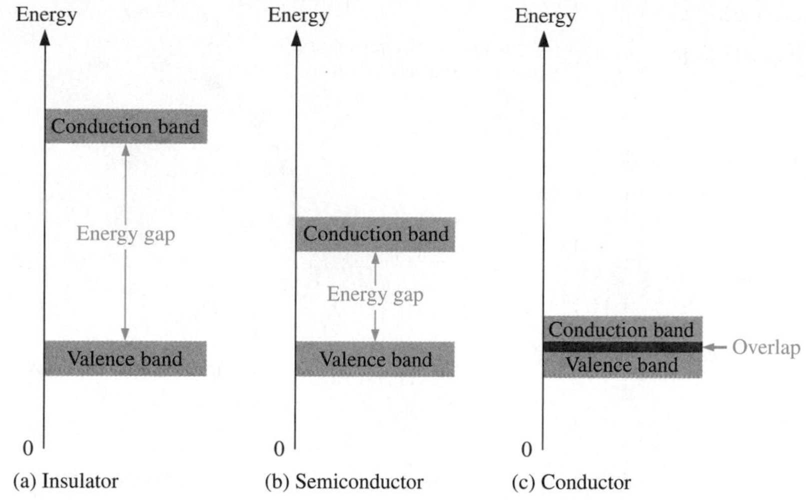

FIGURE 1–5

Energy diagrams for the three types of materials.

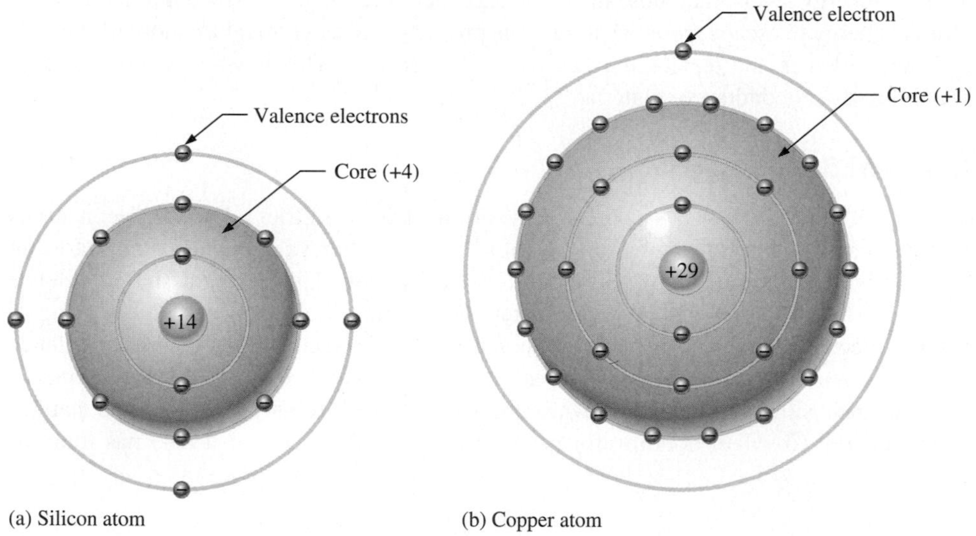

FIGURE 1–6

Diagrams of the silicon and copper atoms.

The valence electron in the copper atom "feels" an attractive force of +1 compared to a valence electron in the silicon atom which "feels" an attractive force of +4. Therefore, there is more force trying to hold a valence electron to the atom in silicon than in copper. The copper's valence electron is in the fourth shell, which is a greater distance from its nucleus than the silicon's valence electron in the third shell. Recall that electrons farthest from the nucleus have the most energy. The valence electron in copper has more energy than the valence electron in silicon. This means that it is easier for valence electrons in copper to acquire enough additional energy to escape from their atoms and become free electrons than it is in silicon. In fact, large numbers of valence electrons in copper already have sufficient energy to be free electrons at normal room temperature.

Silicon and Germanium

The atomic structures of silicon and germanium are compared in Figure 1–7. **Silicon** is by far the most widely used material in diodes, transistors, integrated circuits, and other semiconductor devices. Notice that both silicon and germanium have the characteristic four valence electrons.

The valence electrons in germanium are in the fourth shell while those in silicon are in the third shell, closer to the nucleus. This means that the germanium valence electrons are

FIGURE 1–7

Diagrams of the silicon and germanium atoms.

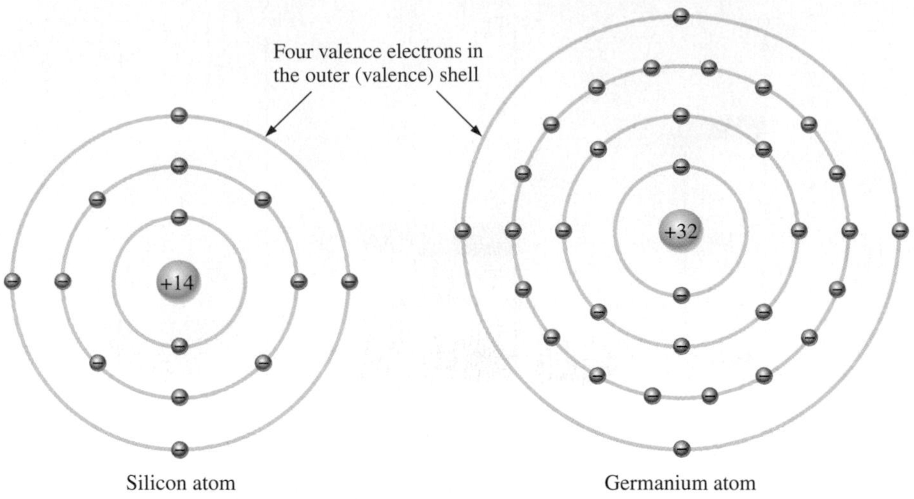

at higher energy levels than those in silicon and, therefore, require a smaller amount of additional energy to escape from the atom. This property makes germanium more unstable at high temperatures and results in excessive reverse current. This is why silicon is a more widely used semiconductive material.

Covalent Bonds

Figure 1–8 shows how each silicon atom positions itself with four adjacent silicon atoms to form a silicon **crystal**. A silicon (Si) atom with its four valence electrons shares an electron with each of its four neighbors. This effectively creates eight shared valence electrons for each atom and produces a state of chemical stability. Also, this sharing of valence electrons produces the **covalent** bonds that hold the atoms together; each valence electron is attracted equally by the two adjacent atoms which share it. Covalent bonding in an intrinsic silicon crystal is shown in Figure 1–9. An **intrinsic** crystal is one that has no impurities. Covalent bonding for germanium is similar because it also has four valence electrons.

FIGURE 1–8

Illustration of covalent bonds in silicon.

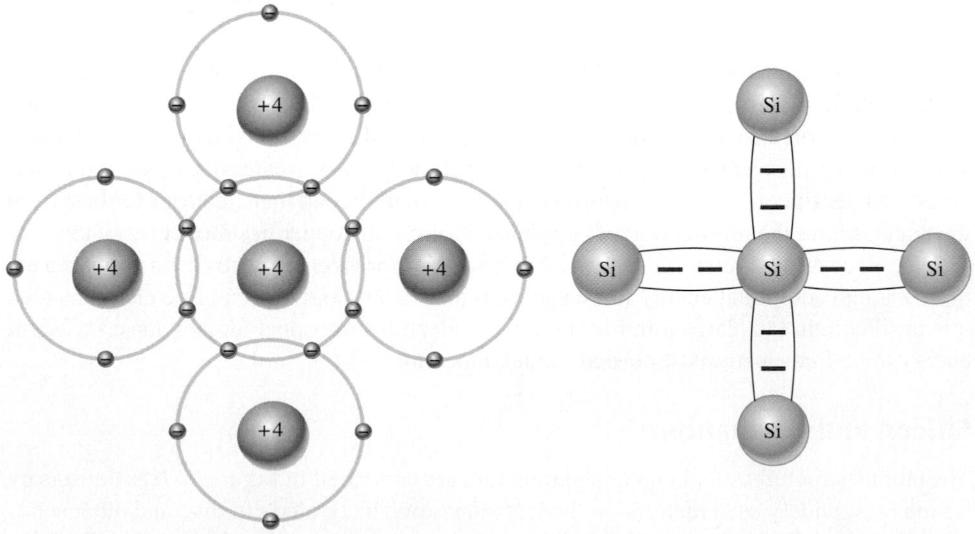

(a) The center silicon atom shares an electron with each of the four surrounding silicon atoms, creating a covalent bond with each. The surrounding atoms are in turn bonded to other atoms, and so on.

(b) Bonding diagram. The red negative signs represent the shared valence electrons.

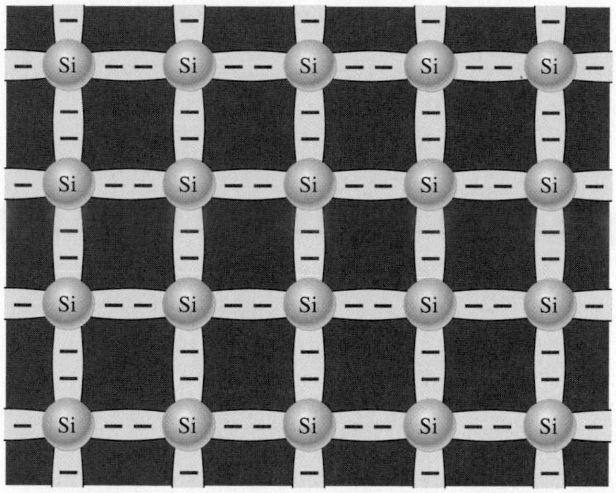

FIGURE 1–9

Covalent bonds in a silicon crystal.

SECTION 1–2 CHECKUP

1. What is the basic difference between conductors and insulators?
2. How do semiconductors differ from conductors and insulators?
3. How many valence electrons does a conductor such as copper have?
4. How many valence electrons does a semiconductor have?
5. Name three of the best conductive materials.
6. What is the most widely used semiconductive material?
7. Why does a semiconductor have fewer free electrons than a conductor?
8. How are covalent bonds formed?
9. What is meant by the term *intrinsic*?
10. What is a crystal?

1–3 CURRENT IN SEMICONDUCTORS

The way a material conducts electrical current is important in understanding how electronic devices operate. You can't really understand the operation of a device such as a diode or transistor without knowing something about current.

After completing this section, you should be able to

- **Describe how current is produced in a semiconductor**
  - Describe a conduction electron
  - Define *hole*
  - Explain what an electron-hole pair is
  - Discuss recombination
  - Explain the difference between electron current and hole current

As you have learned, the electrons of an atom can exist only within prescribed energy bands. Each shell around the nucleus corresponds to a certain energy band and is separated from adjacent shells by energy gaps, in which no electrons can exist. Figure 1–10 shows the energy band diagram for an unexcited (no external energy such as heat) atom in a pure silicon crystal. This condition occurs *only* at a temperature of absolute 0 Kelvin.

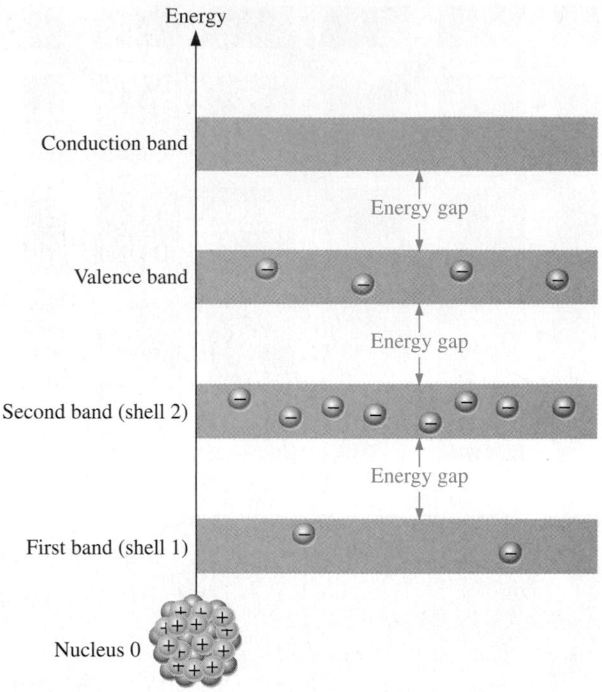

▶ FIGURE 1–10

Energy band diagram for an unexcited atom in a pure (intrinsic) silicon crystal. There are no electrons in the conduction band.

Conduction Electrons and Holes

An intrinsic (pure) silicon crystal at room temperature has sufficient heat (thermal) energy for some valence electrons to jump the gap from the valence band into the conduction band, becoming free electrons. Free electrons are also called **conduction electrons.** This is illustrated in the energy diagram of Figure 1–11(a) and in the bonding diagram of Figure 1–11(b).

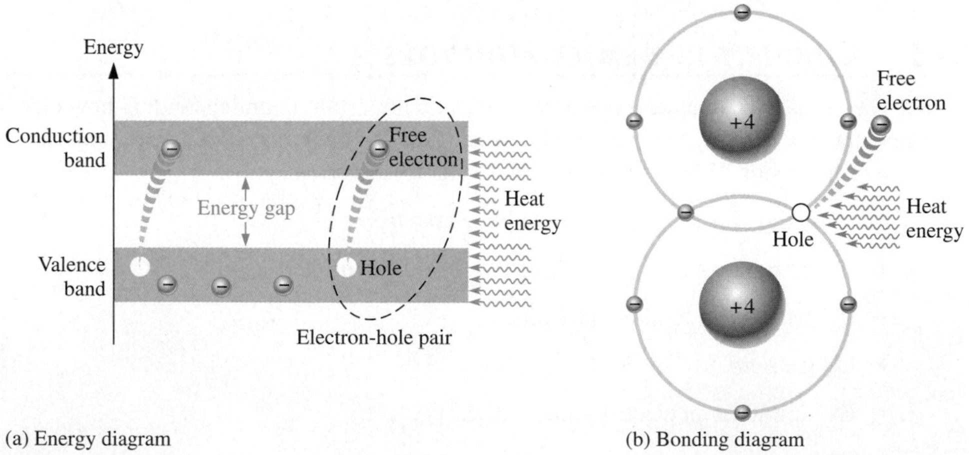

▶ FIGURE 1–11

Creation of electron-hole pairs in a silicon crystal. Electrons in the conduction band are free electrons.

When an electron jumps to the conduction band, a vacancy is left in the valence band within the crystal. This vacancy is called a **hole**. For every electron raised to the conduction band by external energy, there is one hole left in the valence band, creating what is called an **electron-hole pair. Recombination** occurs when a conduction-band electron loses energy and falls back into a hole in the valence band.

To summarize, a piece of intrinsic silicon at room temperature has, at any instant, a number of conduction-band (free) electrons that are unattached to any atom and are essentially drifting randomly throughout the material. There is also an equal number of holes in the valence band created when these electrons jump into the conduction band. This is illustrated in Figure 1–12.

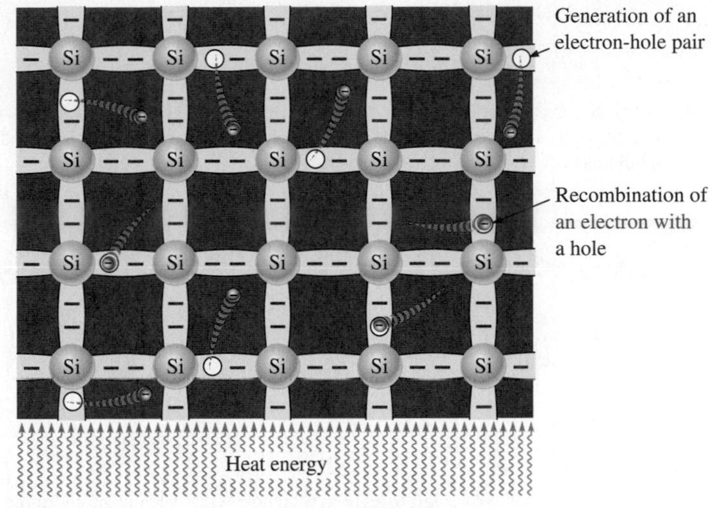

FIGURE 1–12

Electron-hole pairs in a silicon crystal. Free electrons are being generated continuously while some recombine with holes.

Electron and Hole Current

When a voltage is applied across a piece of intrinsic silicon, as shown in Figure 1–13, the thermally generated free electrons in the conduction band, which are free to move randomly in the crystal structure, are now easily attracted toward the positive end. This movement of free electrons is one type of **current** in a semiconductive material and is called *electron current.*

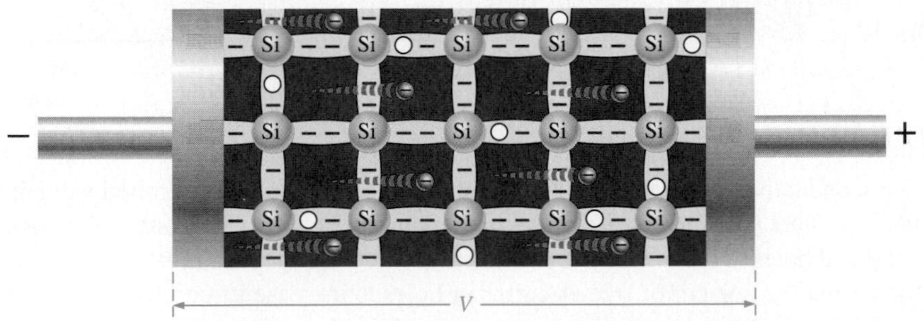

FIGURE 1–13

Electron current in intrinsic silicon is produced by the movement of thermally generated free electrons.

Another type of current occurs in the valence band, where the holes created by the free electrons exist. Electrons remaining in the valence band are still attached to their atoms and are not free to move randomly in the crystal structure as are the free electrons. However, a valence electron can move into a nearby hole with little change in its energy level, thus leaving another hole where it came from. Effectively the hole has moved from one place to another in the crystal structure, as illustrated in Figure 1–14. Although current in the valence band is produced by valence electrons, it is called *hole current* to distinguish it from electron current in the conduction band.

As you have seen, conduction in semiconductors is considered to be either the movement of free electrons in the conduction band or the movement of holes in the valence band, which is actually the movement of valence electrons to nearby atoms, creating hole current in the opposite direction.

It is interesting to contrast the two types of charge movement in a semiconductor with the charge movement in a metallic conductor, such as copper. Copper atoms form a different type of crystal in which the atoms are not covalently bonded to each other but consist of a "sea" of positive ion cores, which are atoms stripped of their valence electrons. The valence electrons are attracted to the positive ions, keeping the positive ions together and forming the metallic bond. The valence electrons do not belong to a given atom, but to the crystal as a whole. Since the valence electrons in copper are free to move, the application of a voltage results in current. There is only one type of current—the movement of free electrons—because there are no "holes" in the metallic crystal structure.

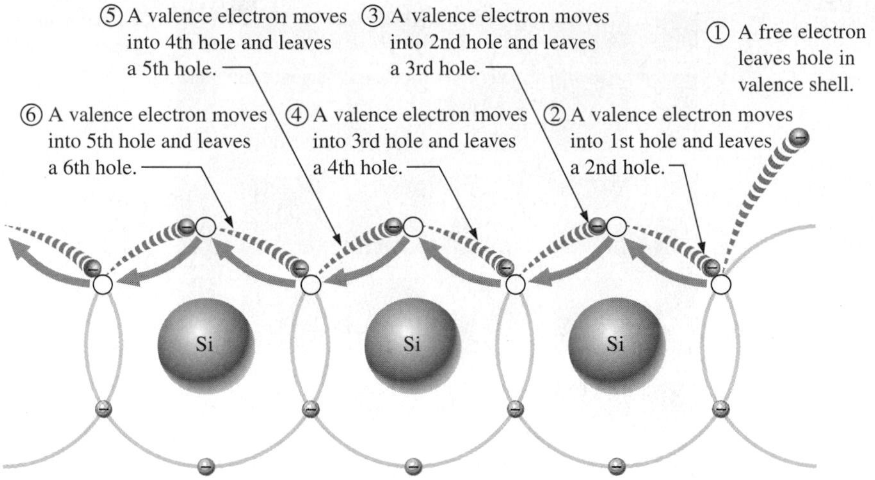

When a valence electron moves left to right to fill a hole while leaving another hole behind, the hole has effectively moved from right to left. Gray arrows indicate effective movement of a hole.

FIGURE 1–14

Hole current in intrinsic silicon.

SECTION 1–3 CHECKUP

1. Are free electrons in the valence band or in the conduction band?
2. Which electrons are responsible for electron current in silicon?
3. What is a hole?
4. At what energy level does hole current occur?

1–4 *N*-TYPE AND *P*-TYPE SEMICONDUCTORS

Semiconductive materials do not conduct current well and are of limited value in their intrinsic state. This is because of the limited number of free electrons in the conduction band and holes in the valence band. Intrinsic silicon (or germanium) must be modified by increasing the number of free electrons or holes to increase its conductivity and make it useful in electronic devices. This is done by adding impurities to the intrinsic material. Two types of extrinsic (impure) semiconductive materials, *n*-type and *p*-type, are the key building blocks for most types of electronic devices.

After completing this section, you should be able to

- **Describe the properties of *n*-type and *p*-type semiconductors**
  - Define *doping*
  - Explain how *n*-type semiconductors are formed
  - Explain how *p*-type semiconductors are formed
  - Describe a majority carrier and minority carrier

Doping

The conductivity of silicon and germanium can be drastically increased by the controlled addition of impurities to the intrinsic (pure) semiconductive material. This process, called **doping**, increases the number of current carriers (electrons or holes). The two categories of impurities are *n*-type and *p*-type.

N-Type Semiconductor

To increase the number of conduction-band electrons in intrinsic silicon, **pentavalent** impurity atoms are added. These are atoms with five valence electrons such as arsenic (As), phosphorus (P), bismuth (Bi), and antimony (Sb).

As illustrated in Figure 1–15, each pentavalent atom (antimony, in this case) forms covalent bonds with four adjacent silicon atoms. Four of the antimony atom's valence electrons are used to form the covalent bonds with silicon atoms, leaving one extra electron. This extra electron becomes a conduction electron because it is not involved in bonding. Because the pentavalent atom gives up an electron, it is often called a *donor atom.* The number of conduction electrons can be carefully controlled by the number of impurity atoms added to the silicon. A conduction electron created by this doping process does not leave a hole in the valence band because it is in excess of the number required to fill the valence band.

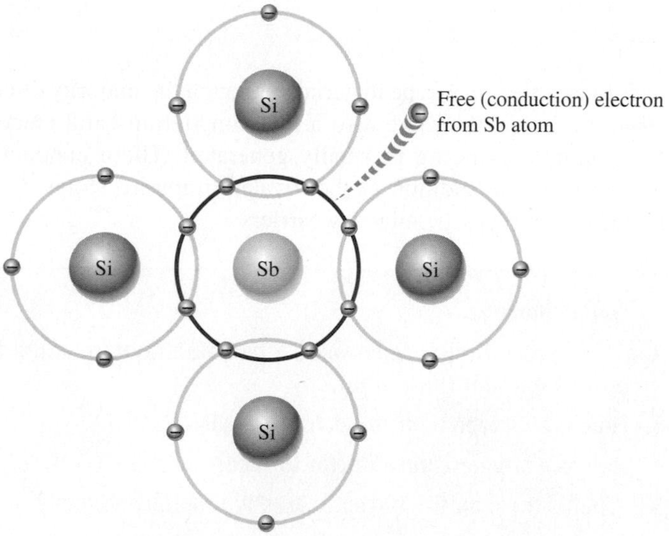

FIGURE 1–15

Pentavalent impurity atom in a silicon crystal structure. An antimony (Sb) impurity atom is shown in the center. The extra electron from the Sb atom becomes a free electron.

Majority and Minority Carriers Since most of the current carriers are electrons, silicon (or germanium) doped with pentavalent atoms is an *n*-type semiconductor (the *n* stands for the negative charge on an electron). The electrons are called the **majority carriers** in *n*-type material. Although the majority of current carriers in *n*-type material are electrons, there are also a few holes that are created when electron-hole pairs are thermally generated. These holes are *not* produced by the addition of the pentavalent impurity atoms. Holes in an *n*-type material are called **minority carriers.**

P-Type Semiconductor

To increase the number of holes in intrinsic silicon, **trivalent** impurity atoms are added. These are atoms with three valence electrons such as boron (B), indium (In), and gallium (Ga). As illustrated in Figure 1–16, each trivalent atom (boron, in this case) forms covalent bonds with four adjacent silicon atoms. All three of the boron atom's valence electrons are used in the covalent bonds; and, since four electrons are required, a hole results when each trivalent atom is added. Because the trivalent atom can take an electron, it is often referred to as an *acceptor atom.* The number of holes can be carefully controlled by the number of trivalent impurity atoms added to the silicon. A hole created by this doping process is *not* accompanied by a conduction (free) electron.

Majority and Minority Carriers Since most of the current carriers are holes, silicon (or germanium) doped with trivalent atoms is called a *p*-type semiconductor. The holes are the

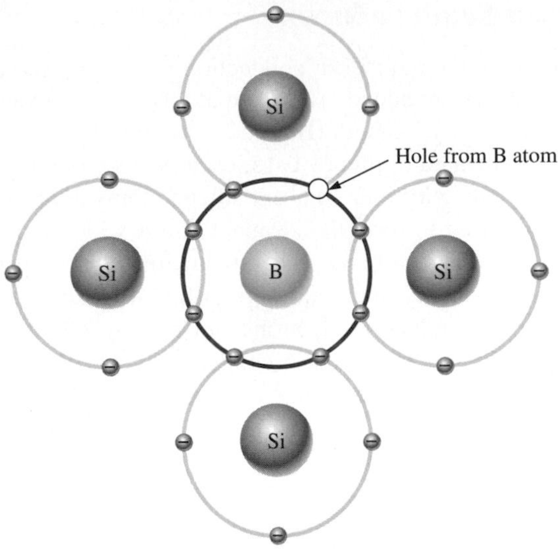

▶ FIGURE 1–16

Trivalent impurity atom in a silicon crystal structure. A boron (B) impurity atom is shown in the center.

majority carriers in *p*-type material. Although the majority of current carriers in *p*-type material are holes, there are also a few conduction-band electrons that are created when electron-hole pairs are thermally generated. These conduction-band electrons are *not* produced by the addition of the trivalent impurity atoms. Conduction-band electrons in *p*-type material are the minority carriers.

SECTION 1–4 CHECKUP

1. Define *doping.*
2. What is the difference between a pentavalent atom and a trivalent atom? What are other names for these atoms?
3. How is an *n*-type semiconductor formed?
4. How is a *p*-type semiconductor formed?
5. What is the majority carrier in an *n*-type semiconductor?
6. What is the majority carrier in a *p*-type semiconductor?
7. By what process are the majority carriers produced?
8. By what process are the minority carriers produced?
9. What is the difference between intrinsic and extrinsic semiconductors?

1–5 THE DIODE

If you take a block of silicon and dope part of it with a trivalent impurity and the other part with a pentavalent impurity, a boundary called the *pn* junction is formed between the resulting *p*-type and *n*-type portions and a basic diode is created. A **diode** is a device that conducts current in only one direction. The *pn* junction is the feature that allows diodes, certain transistors, and other devices to work.

After completing this section, you should be able to

- **Describe a diode and how a *pn* junction is formed**
  - Discuss diffusion across a *pn* junction
  - Explain the formation of the depletion region
  - Define *barrier potential* and discuss its significance
  - State the values of barrier potential in silicon and germanium

A p-type material consists of silicon atoms and trivalent impurity atoms such as boron. The boron atom adds a hole when it bonds with the silicon atoms. However, since the number of protons and the number of electrons are equal throughout the material, there is no net charge in the material and so it is neutral.

An n-type silicon material consists of silicon atoms and pentavalent impurity atoms such as antimony. As you have seen, an impurity atom releases an electron when it bonds with four silicon atoms. Since there is still an equal number of protons and electrons (including the free electrons) throughout the material, there is no net charge in the material and so it is neutral.

If a piece of intrinsic silicon is doped so that part is n-type and the other part is p-type, a ***pn* junction** forms at the boundary between the two regions and a diode is created, as indicated in Figure 1–17(a). The p region has many holes (majority carriers) from the impurity atoms and only a few thermally generated free electrons (minority carriers). The n region has many free electrons (majority carriers) from the impurity atoms and only a few thermally generated holes (minority carriers).

HISTORY NOTE

After the invention of the light bulb, Edison continued to experiment and in 1883 found that he could detect electrons flowing through the vacuum from the lighted filament to a metal plate mounted inside the bulb. This discovery became known as the *Edison effect.*

An English physicist, John Fleming, took up where Edison left off and found that the Edison effect could also be used to detect radio waves and convert them to electrical signals. He went on to develop a two-element vacuum tube called the *Fleming valve,* later known as the *diode.*

Formation of the Depletion Region

The free electrons in the n region are randomly drifting in all directions. At the instant of the pn junction formation, the free electrons near the junction in the n region begin to diffuse across the junction into the p region where they combine with holes near the junction, as shown in Figure 1–17(b).

Before the pn junction is formed, recall that there are as many electrons as protons in the n-type material, making the material neutral in terms of net charge. The same is true for the p-type material.

When the pn junction is formed, the n region loses free electrons as they diffuse across the junction. This creates a layer of positive charges (pentavalent ions) near the junction. As the electrons move across the junction, the p region loses holes as the electrons and holes combine. This creates a layer of negative charges (trivalent ions) near the junction. These two layers of positive and negative charges form the **depletion region,** as shown in Figure 1–17(b). The term *depletion* refers to the fact that the region near the pn junction is depleted of charge carriers (electrons and holes) due to diffusion across the junction. Keep

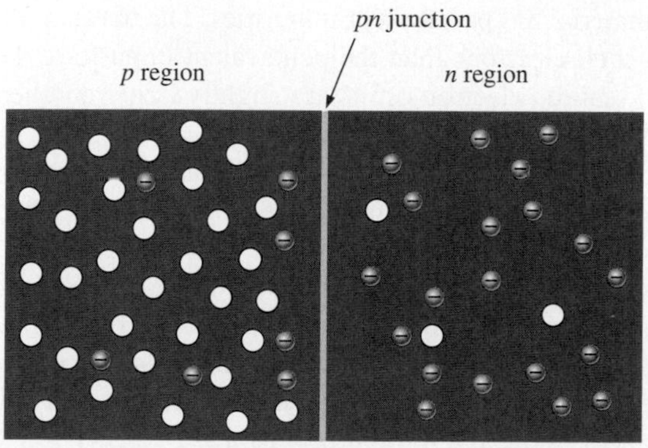

(a) The basic diode structure at the instant of junction formation showing only the majority and minority carriers. Free electrons in the n region near the pn junction begin to diffuse across the junction and fall into holes near the junction in the p region.

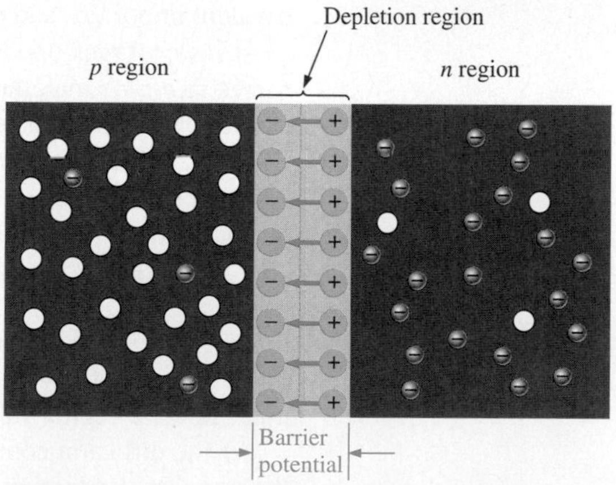

(b) For every electron that diffuses across the junction and combines with a hole, a positive charge is left in the n region and a negative charge is created in the p region, forming a barrier potential. This action continues until the voltage of the barrier repels further diffusion. The blue arrows between the positive and negative charges in the depletion region represent the electric field.

▲ FIGURE 1–17

Formation of the depletion region. The width of the depletion region is exaggerated for illustration purposes.

HISTORY NOTE

Russell Ohl, working at Bell Labs in 1940, stumbled on the semiconductor *pn* junction. Ohl was working with a silicon sample that had an accidental crack down its middle. He was using an ohmmeter to test the electrical resistance of the sample when he noted that when the sample was exposed to light, the current that flowed between the two sides of the crack made a significant jump. This discovery was fundamental to the work of the team that invented the transistor in 1947.

in mind that the depletion region is formed very quickly and is very thin compared to the *n* region and *p* region.

After the initial surge of free electrons across the *pn* junction, the depletion region has expanded to a point where equilibrium is established and there is no further diffusion of electrons across the junction. This occurs as follows. As electrons continue to diffuse across the junction, more and more positive and negative charges are created near the junction as the depletion region is formed. A point is reached where the total negative charge in the depletion region repels any further diffusion of electrons (negatively charged particles) into the *p* region (like charges repel) and the diffusion stops. In other words, the depletion region acts as a barrier to the further movement of electrons across the junction.

Barrier Potential Any time there is a positive charge and a negative charge near each other, there is a force acting on the charges as described by Coulomb's law. In the depletion region there are many positive charges and many negative charges on opposite sides of the *pn* junction. The forces between the opposite charges form an *electric field,* as illustrated in Figure 1–17(b) by the blue arrows between the positive charges and the negative charges. This electric field is a barrier to the free electrons in the *n* region, and energy must be expended to move an electron through the electric field. That is, external energy must be applied to get the electrons to move across the barrier of the electric field in the depletion region.

The potential difference of the electric field across the depletion region is the amount of voltage required to move electrons through the electric field. This potential difference is called the **barrier potential** and is expressed in volts. Stated another way, a certain amount of voltage equal to the barrier potential and with the proper polarity must be applied across a *pn* junction before electrons will begin to flow across the junction. You will learn more about this when we discuss *biasing* in Section 1–6.

The barrier potential of a *pn* junction depends on several factors, including the type of semiconductive material, the amount of doping, and the temperature. The typical barrier potential is approximately 0.7 V for silicon and 0.3 V for germanium at 25°C. Because germanium devices are rare, silicon will be used throughout the rest of the book.

Energy Diagrams of the *PN* Junction and Depletion Region

The valence and conduction bands in an *n*-type material are at slightly lower energy levels than the valence and conduction bands in a *p*-type material. Recall that *p*-type material has trivalent impurities and *n*-type material has pentavalent impurities. The trivalent impurities exert lower forces on the outer-shell electrons than the pentavalent impurities. The lower forces in *p*-type materials mean that the electron orbits are slightly larger and hence have greater energy than the electron orbits in the *n*-type materials.

An energy diagram for a *pn* junction at the instant of formation is shown in Figure 1–18(a). As you can see, the valence and conduction bands in the *n* region are at lower energy levels than those in the *p* region, but there is a significant amount of overlapping.

The free electrons in the *n* region that occupy the upper part of the conduction band in terms of their energy can easily diffuse across the junction (they do not have to gain additional energy) and temporarily become free electrons in the lower part of the *p*-region conduction band. After crossing the junction, the electrons quickly lose energy and fall into the holes in the *p*-region valence band as indicated in Figure 1–18(a).

As the diffusion continues, the depletion region begins to form and the energy level of the *n*-region conduction band decreases. The decrease in the energy level of the conduction band in the *n* region is due to the loss of the higher-energy electrons that have diffused across the junction to the *p* region. Soon, there are no electrons left in the *n*-region conduction band with enough energy to get across the junction to the *p*-region conduction band, as indicated by the alignment of the top of the *n*-region conduction band and the bottom of the *p*-region conduction band in Figure 1–18(b). At this point, the junction is at equilibrium; and the depletion region is complete because diffusion has ceased. There is an energy gradiant across the depletion region which acts as an "energy hill" that an *n*-region electron must climb to get to the *p* region.

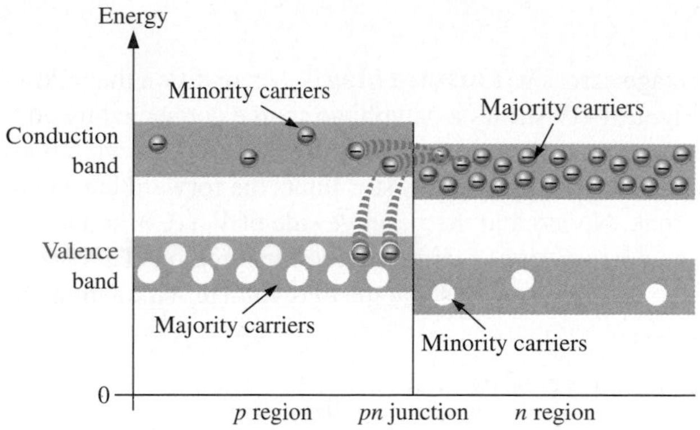

(a) At the instant of junction formation

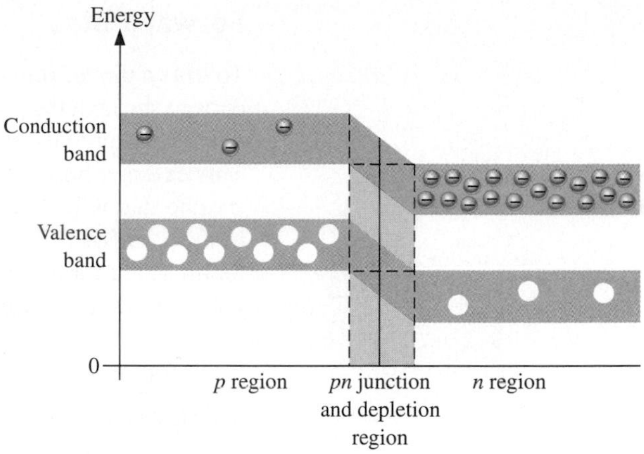

(b) At equilibrium

▲ **FIGURE 1–18**

Energy diagrams illustrating the formation of the *pn* junction and depletion region.

Notice that as the energy level of the *n*-region conduction band has shifted downward, the energy level of the valence band has also shifted downward. It still takes the same amount of energy for a valence electron to become a free electron. In other words, the energy gap between the valence band and the conduction band remains the same.

SECTION 1–5 CHECKUP

1. What is a *pn* junction?
2. Explain what diffusion is.
3. Describe the depletion region.
4. Explain what the barrier potential is and how it is created.
5. What is the typical value of the barrier potential for a silicon diode?
6. What is the typical value of the barrier potential for a germanium diode?

1–6 BIASING A DIODE

As you have learned, no electrons move through the *pn* junction at equilibrium. Generally the term *bias* refers to the use of a dc voltage to establish certain operating conditions for an electronic device. In relation to a diode, there are two bias conditions: forward and reverse. Either of these bias conditions is established by connecting a sufficient dc voltage of the proper polarity across the *pn* junction.

After completing this section, you should be able to

- **Discuss biasing of a diode**
  - Define *forward bias* and state the required conditions
  - Define *reverse bias* and state the required conditions
  - Discuss the effect of barrier potential on forward bias
  - Explain how current is produced in forward bias
  - Explain reverse current
  - Describe reverse breakdown of a diode

Forward Bias

To **bias** a diode, you apply a dc voltage across it. **Forward bias** is the condition that allows current through the *pn* junction. Figure 1–19 shows a dc voltage source connected by conductive material (contacts and wire) across a diode in the direction to produce forward bias. This external bias voltage is designated as V_{BIAS}. The resistor limits the forward current to a value that will not damage the diode. Notice that the negative side of V_{BIAS} is connected to the *n* region of the diode and the positive side is connected to the *p* region. This is one requirement for forward bias. A second requirement is that the bias voltage, V_{BIAS}, must be greater than the barrier potential.

FIGURE 1–19

A diode connected for forward bias.

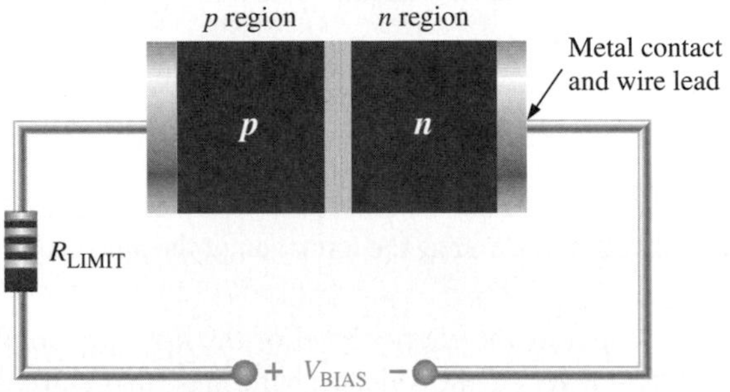

A fundamental picture of what happens when a diode is forward-biased is shown in Figure 1–20. Because like charges repel, the negative side of the bias-voltage source "pushes" the free electrons, which are the majority carriers in the *n* region, toward the *pn* junction. This flow of free electrons is called *electron current.* The negative side of the source also provides a continuous flow of electrons through the external connection (conductor) and into the *n* region as shown.

FIGURE 1–20

A forward-biased diode showing the flow of majority carriers and the voltage due to the barrier potential across the depletion region.

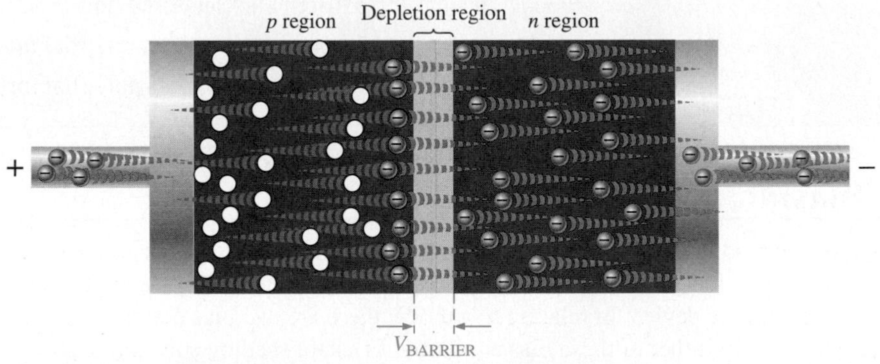

The bias-voltage source imparts sufficient energy to the free electrons for them to overcome the barrier potential of the depletion region and move on through into the *p* region. Once in the *p* region, these conduction electrons have lost enough energy to immediately combine with holes in the valence band.

Now, the electrons are in the valence band in the *p* region, simply because they have lost too much energy overcoming the barrier potential to remain in the conduction band. Since unlike charges attract, the positive side of the bias-voltage source attracts the valence electrons toward the left end of the *p* region. The holes in the *p* region provide the medium or "pathway" for these valence electrons to move through the *p* region. The valence electrons move from one hole to the next toward the left. The holes, which are the majority carriers in the *p* region, effectively (not actually) move to the right toward the junction, as you can see in Figure 1–20. This *effective* flow of holes is the hole current. You can also view the hole current as being created by the flow of valence electrons through the *p* region, with the holes providing the only means for these electrons to flow.

As the electrons flow out of the p region through the external connection (conductor) and to the positive side of the bias-voltage source, they leave holes behind in the p region; at the same time, these electrons become conduction electrons in the metal conductor. Recall that the conduction band in a conductor overlaps the valence band so that it takes much less energy for an electron to be a free electron in a conductor than in a semiconductor and that metallic conductors do not have holes in their structure. There is a continuous availability of holes effectively moving toward the pn junction to combine with the continuous stream of electrons as they come across the junction into the p region.

The Effect of Forward Bias on the Depletion Region As more electrons flow into the depletion region, the number of positive ions is reduced. As more holes effectively flow into the depletion region on the other side of the pn junction, the number of negative ions is reduced. This reduction in positive and negative ions during forward bias causes the depletion region to narrow, as indicated in Figure 1–21.

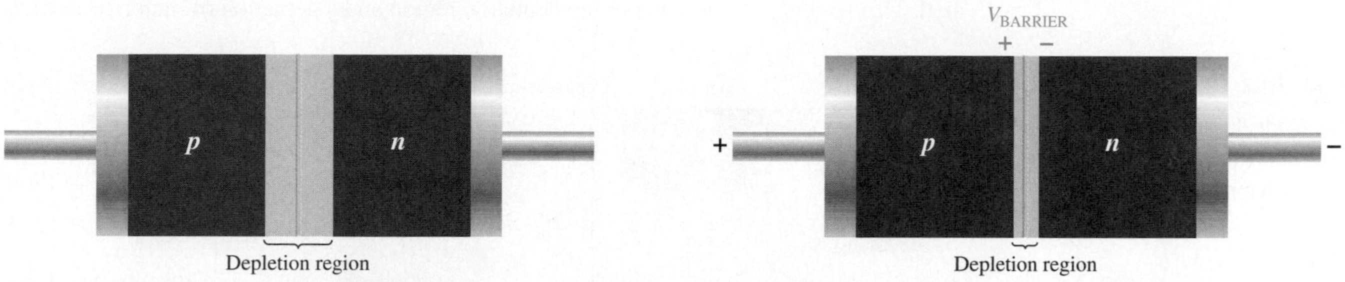

(a) At equilibrium (no bias)

(b) Forward bias narrows the depletion region and produces a voltage drop across the pn junction equal to the barrier potential.

▲ FIGURE 1–21

The depletion region narrows and a voltage drop is produced across the pn junction when the diode is forward-biased.

The Effect of the Barrier Potential During Forward Bias Recall that the electric field between the positive and negative ions in the depletion region on either side of the junction creates an "energy hill" that prevents free electrons from diffusing across the junction at equilibrium (see Figure 1–18(b)). This is known as the *barrier potential.*

When forward bias is applied, the free electrons are provided with enough energy from the bias-voltage source to overcome the barrier potential and effectively "climb the energy hill" and cross the depletion region. The energy that the electrons require in order to pass through the depletion region is equal to the barrier potential. In other words, the electrons give up an amount of energy equivalent to the barrier potential when they cross the depletion region. This energy loss results in a voltage drop across the pn junction equal to the barrier potential (0.7 V), as indicated in Figure 1–21(b). An additional small voltage drop occurs across the p and n regions due to the internal resistance of the material. For doped semiconductive material, this resistance, called the **dynamic resistance,** is very small and can usually be neglected. This is discussed in more detail in Section 1–7.

Reverse Bias

Reverse bias is the condition that essentially prevents current through the diode. Figure 1–22 shows a dc voltage source connected across a diode in the direction to produce reverse bias. This external bias voltage is designated as V_{BIAS} just as it was for forward bias. Notice that the positive side of V_{BIAS} is connected to the n region of the diode and the negative side is connected to the p region. Also note that the depletion region is shown much wider than in forward bias or equilibrium.

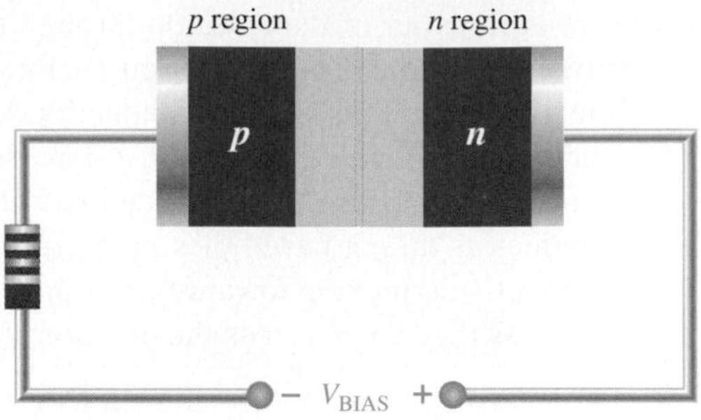

▶ FIGURE 1–22

A diode connected for reverse bias. A limiting resistor is shown although it is not important in reverse bias because there is essentially no current.

An illustration of what happens when a diode is reverse-biased is shown in Figure 1–23. Because unlike charges attract, the positive side of the bias-voltage source "pulls" the free electrons, which are the majority carriers in the *n* region, away from the *pn* junction. As the electrons flow toward the positive side of the voltage source, additional positive ions are created. This results in a widening of the depletion region and a depletion of majority carriers.

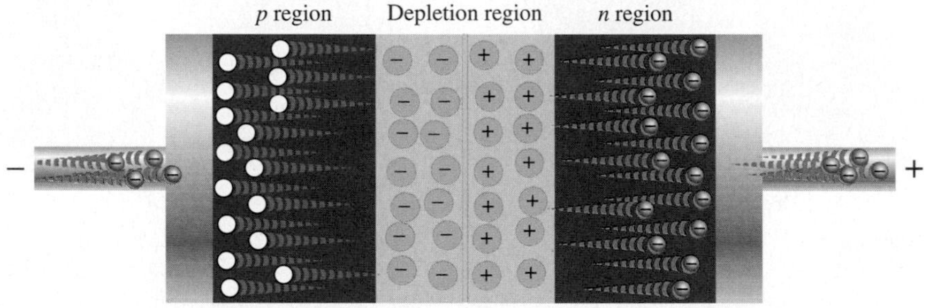

▶ FIGURE 1–23

The diode during the short transition time immediately after reverse-bias voltage is applied.

In the *p* region, electrons from the negative side of the voltage source enter as valence electrons and move from hole to hole toward the depletion region where they create additional negative ions. This results in a widening of the depletion region and a depletion of majority carriers. The flow of valence electrons can be viewed as holes being "pulled" toward the positive side.

The initial flow of charge carriers is transitional and lasts for only a very short time after the reverse-bias voltage is applied. As the depletion region widens, the availability of majority carriers decreases. As more of the *n* and *p* regions become depleted of majority carriers, the electric field between the positive and negative ions increases in strength until the potential across the depletion region equals the bias voltage, V_{BIAS}. At this point, the transition current essentially ceases except for a very small reverse current that can usually be neglected.

Reverse Current The extremely small current that exists in reverse bias after the transition current dies out is caused by the minority carriers in the *n* and *p* regions that are produced by thermally generated electron-hole pairs. The small number of free minority electrons in the *p* region are "pushed" toward the *pn* junction by the negative bias voltage. When these electrons reach the wide depletion region, they "fall down the energy hill" and combine with the minority holes in the *n* region as valence electrons and flow toward the positive bias voltage, creating a small hole current.

The conduction band in the *p* region is at a higher energy level than the conduction band in the *n* region. Therefore, the minority electrons easily pass through the depletion region because they require no additional energy. Reverse current is illustrated in Figure 1–24.

Reverse Breakdown Normally, the reverse current is so small that it can be neglected. However, if the external reverse-bias voltage is increased to a value called the *breakdown voltage,* the reverse current will drastically increase.

This is what happens. The high reverse-bias voltage imparts energy to the free minority electrons so that as they speed through the *p* region, they collide with atoms with enough

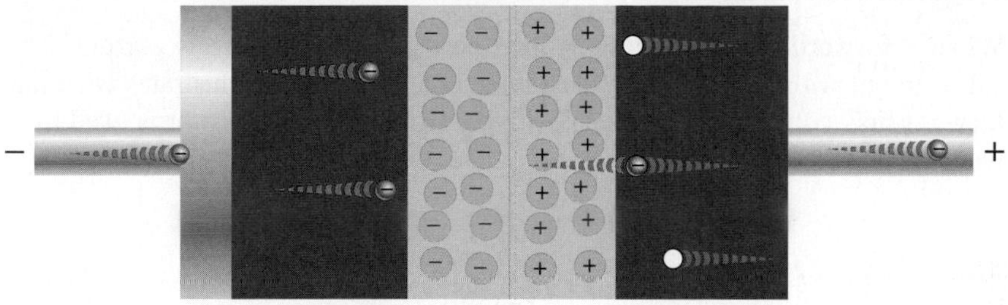

◄ FIGURE 1–24

The extremely small reverse current in a reverse-biased diode is due to the minority carriers from thermally generated electron-hole pairs.

energy to knock valence electrons out of orbit and into the conduction band. The newly created conduction electrons are also high in energy and repeat the process. If one electron knocks only two others out of their valence orbit during its travel through the *p* region, the numbers quickly multiply. As these high-energy electrons go through the depletion region, they have enough energy to go through the *n* region as conduction electrons, rather than combining with holes.

The multiplication of conduction electrons just discussed is known as the **avalanche effect,** and reverse current can increase dramatically if steps are not taken to limit the current. When the reverse current is not limited, the resulting heating will permanently damage the diode. Most diodes are not operated in reverse breakdown, but if the current is limited (by adding a series-limiting resistor for example), there is no permanent damage to the diode.

SECTION 1–6 CHECKUP

1. Describe forward bias of a diode.
2. Explain how to forward-bias a diode.
3. Describe reverse bias of a diode.
4. Explain how to reverse-bias a diode.
5. Compare the depletion regions in forward bias and reverse bias.
6. Which bias condition produces majority carrier current?
7. How is reverse current in a diode produced?
8. When does reverse breakdown occur in a diode?
9. Define *avalanche effect* as applied to diodes.

1–7 VOLTAGE-CURRENT CHARACTERISTIC OF A DIODE

As you have learned, forward bias produces current through a diode and reverse bias essentially prevents current, except for a negligible reverse current. Reverse bias prevents current as long as the reverse-bias voltage does not equal or exceed the breakdown voltage of the junction. In this section, we will examine the relationship between the voltage and the current in a diode on a graphical basis.

After completing this section, you should be able to

- **Analyze the voltage-current (*V-I*) characteristic curve of a diode**
  - Explain the forward-bias portion of the *V-I* characteristic curve
  - Explain the reverse-bias portion of the *V-I* characteristic curve
  - Identify the barrier potential
  - Identify the breakdown voltage
  - Discuss temperature effects on a diode

V-I Characteristic for Forward Bias

When a forward-bias voltage is applied across a diode, there is current. This current is called the *forward current* and is designated I_F. Figure 1–25 illustrates what happens as the forward-bias voltage is increased positively from 0 V. The resistor is used to limit the forward current to a value that will not overheat the diode and cause damage.

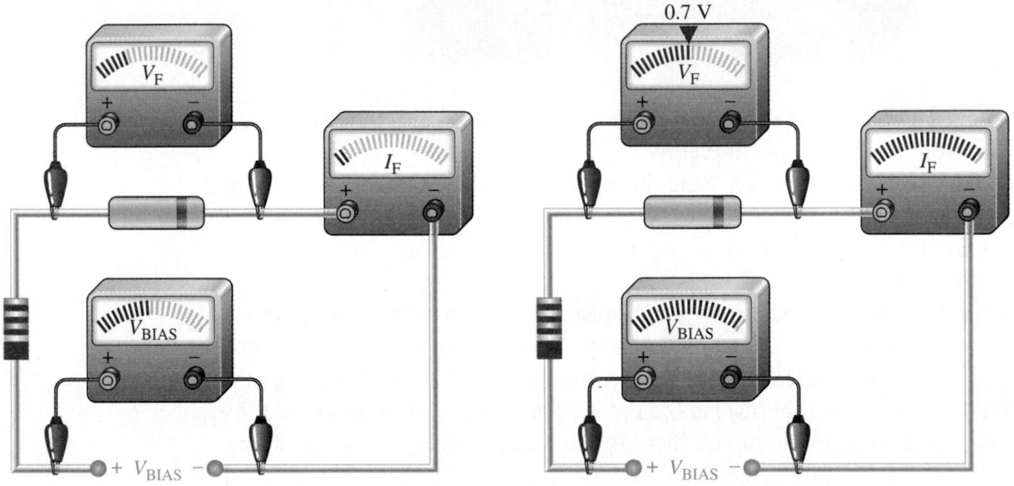

(a) Small forward-bias voltage ($V_F < 0.7$ V), very small forward current.

(b) Forward voltage reaches and remains at approximately 0.7 V. Forward current continues to increase as the bias voltage is increased.

▲ **FIGURE 1–25**

Forward-bias measurements show general changes in V_F and I_F as V_{BIAS} is increased.

With 0 V across the diode, there is no forward current. As you gradually increase the forward-bias voltage, the forward current *and* the voltage across the diode gradually increase, as shown in Figure 1–25(a). A portion of the forward-bias voltage is dropped across the limiting resistor. When the forward-bias voltage is increased to a value where the voltage across the diode reaches approximately 0.7 V (barrier potential), the forward current begins to increase rapidly, as illustrated in Figure 1–25(b).

As you continue to increase the forward-bias voltage, the current continues to increase very rapidly, but the voltage across the diode increases only gradually above 0.7 V. This small increase in the diode voltage above the barrier potential is due to the voltage drop across the internal dynamic resistance of the semiconductive material.

Graphing the V-I Curve If you plot the results of the type of measurements shown in Figure 1–25 on a graph, you get the ***V-I* characteristic** curve for a forward-biased diode, as shown in Figure 1–26(a). The diode forward voltage (V_F) increases to the right along the horizontal axis, and the forward current (I_F) increases upward along the vertical axis.

As you can see in Figure 1–26(a), the forward current increases very little until the forward voltage across the *pn* junction reaches approximately 0.7 V at the knee of the curve. After this point, the forward voltage remains at approximately 0.7 V, but I_F increases rapidly. As previously mentioned, there is a slight increase in V_F above 0.7 V as the current increases due mainly to the voltage drop across the dynamic resistance. The I_F scale is typically in mA, as indicated.

Three points *A*, *B*, and *C* are shown on the curve in Figure 1–26(a). Point *A* corresponds to a zero-bias condition. Point *B* corresponds to Figure 1–25(a) where the forward voltage is less than the barrier potential of 0.7 V. Point *C* corresponds to Figure 1–25(a) where the forward voltage *approximately* equals the barrier potential. As the external bias voltage and forward current continue to increase above the knee, the forward voltage will increase

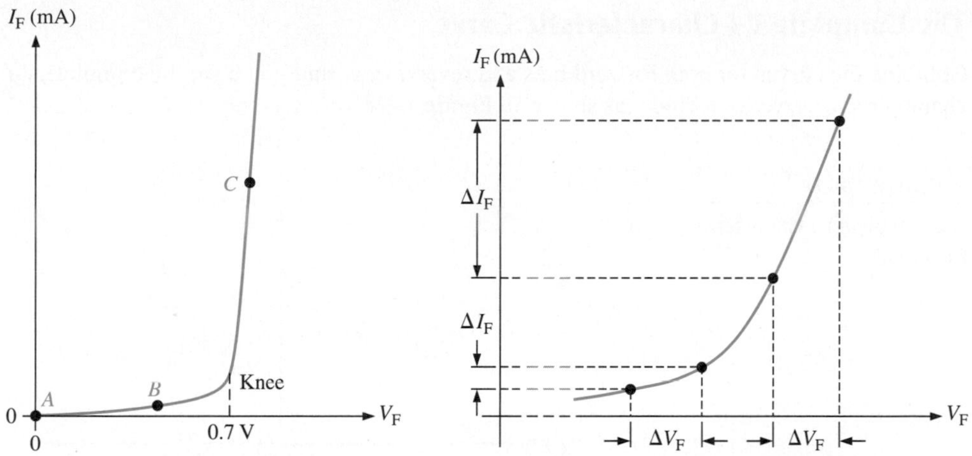

(a) *V-I* characteristic curve for forward bias.

(b) Expanded view of a portion of the curve in part (a). The dynamic resistance r'_d decreases as you move up the curve, as indicated by the decrease in the value of $\Delta V_F / \Delta I_F$.

FIGURE 1–26

Relationship of voltage and current in a forward-biased diode.

slightly above 0.7 V. In reality, the forward voltage can be as much as approximately 1 V, depending on the forward current.

Dynamic Resistance Figure 1–26(b) is an expanded view of the *V-I* characteristic curve in part (a) and illustrates dynamic resistance. Unlike a linear resistance, the resistance of the forward-biased diode is not constant over the entire curve. Because the resistance changes as you move along the *V-I* curve, it is called *dynamic* or *ac resistance.* Internal resistances of electronic devices are usually designated by lowercase italic *r* with a prime, instead of the standard *R*. The dynamic resistance of a diode is designated r'_d.

Below the knee of the curve the resistance is greatest because the current increases very little for a given change in voltage ($r'_d = \Delta V_F / \Delta I_F$). The resistance begins to decrease in the region of the knee of the curve and becomes smallest above the knee where there is a large change in current for a given change in voltage.

V-I Characteristic for Reverse Bias

When a reverse-bias voltage is applied across a diode, there is only an extremely small reverse current (I_R) through the *pn* junction. With 0 V across the diode, there is no reverse current. As you gradually increase the reverse-bias voltage, there is a very small reverse current and the voltage across the diode increases. When the applied bias voltage is increased to a value where the reverse voltage across the diode (V_R) reaches the breakdown value (V_{BR}), the reverse current begins to increase rapidly.

As you continue to increase the bias voltage, the current continues to increase very rapidly, but the voltage across the diode increases very little above V_{BR}. Breakdown, with exceptions, is not a normal mode of operation for most *pn* junction devices.

Graphing the V-I Curve If you plot the results of reverse-bias measurements on a graph, you get the *V-I* characteristic curve for a reverse-biased diode. A typical curve is shown in Figure 1–27. The diode reverse voltage (V_R) increases to the left along the horizontal axis, and the reverse current (I_R) increases downward along the vertical axis.

There is very little reverse current (usually μA or nA) until the reverse voltage across the diode reaches approximately the breakdown value (V_{BR}) at the knee of the curve. After this point, the reverse voltage remains at approximately V_{BR}, but I_R increases very rapidly, resulting in overheating and possible damage if current is not limited to a safe level. The breakdown voltage for a diode depends on the doping level, which the manufacturer sets, depending on the type of diode. A typical rectifier diode (the most widely used type) has a breakdown voltage of greater than 50 V. Some specialized diodes have a breakdown voltage that is only 5 V.

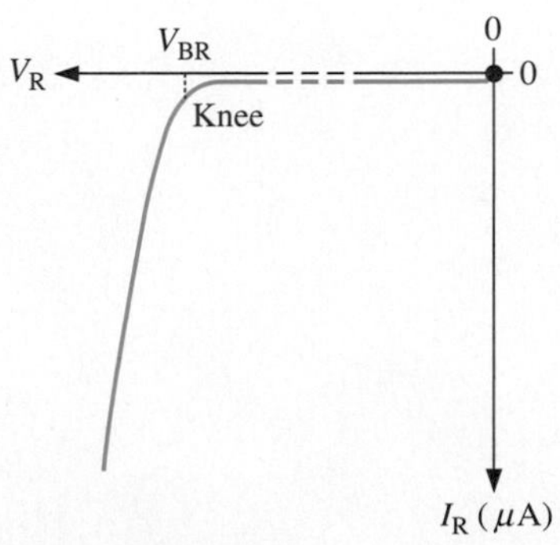

FIGURE 1–27

V-I **characteristic curve for a reverse-biased diode.**

The Complete *V-I* Characteristic Curve

Combine the curves for both forward bias and reverse bias, and you have the complete *V-I* characteristic curve for a diode, as shown in Figure 1–28.

FIGURE 1–28

The complete *V-I* characteristic curve for a diode.

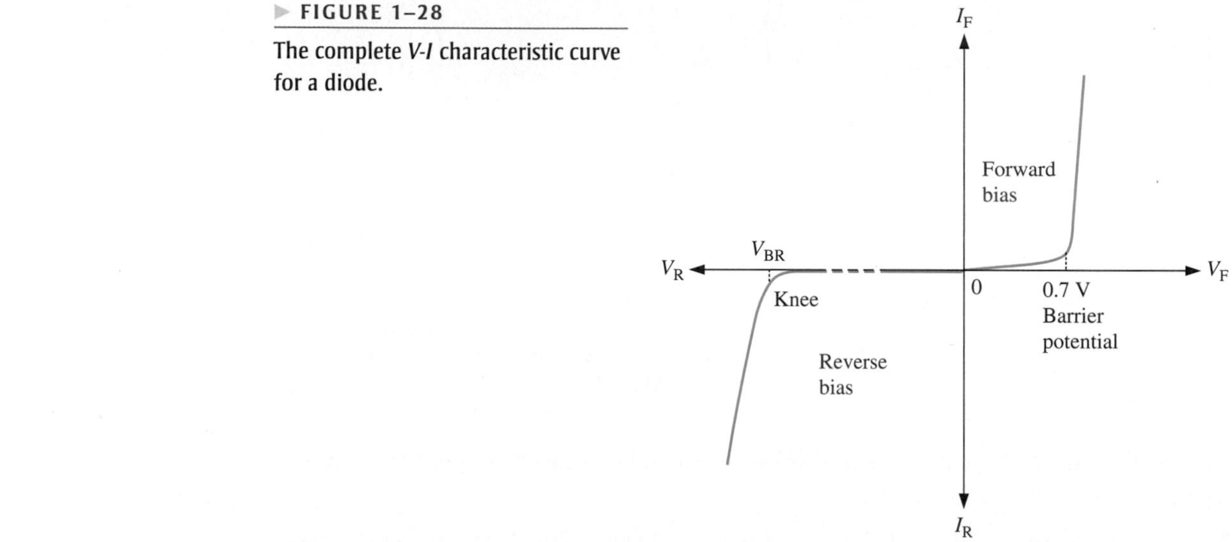

Temperature Effects For a forward-biased diode, as temperature is increased, the forward current increases for a given value of forward voltage. Also, for a given value of forward current, the forward voltage decreases. This is shown with the *V-I* characteristic curves in Figure 1–29. The blue curve is at room temperature (25°C) and the red curve is at an elevated temperature (25°C + ΔT). The barrier potential decreases by 2 mV for each degree increase in temperature.

FIGURE 1–29

Temperature effect on the diode *V-I* characteristic. The 1 mA and 1 μA marks on the vertical axis are given as a basis for a relative comparison of the current scales.

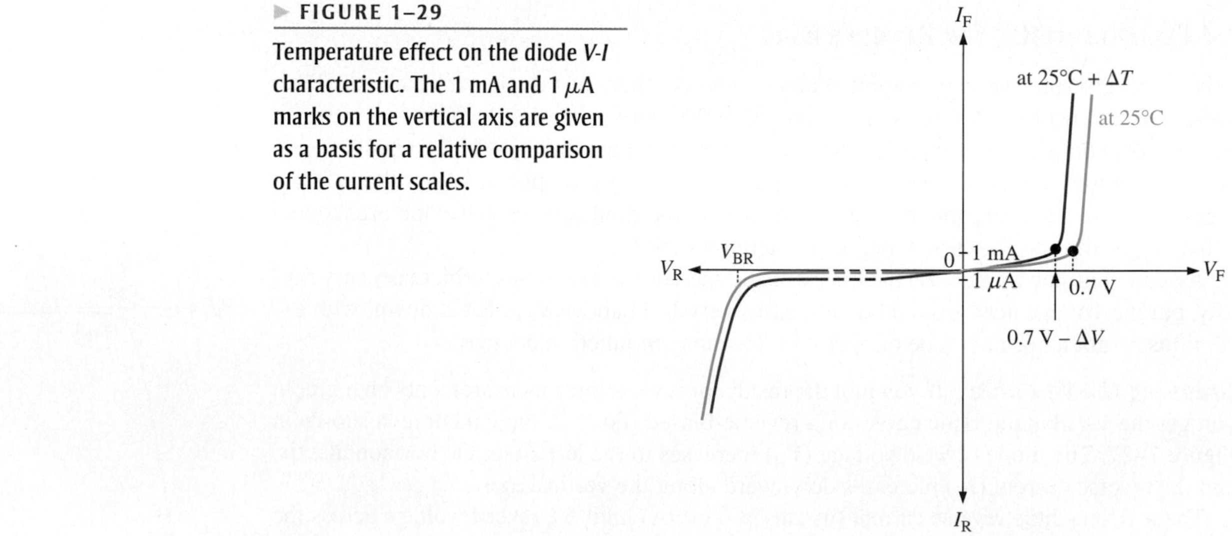

For a reverse-biased diode, as temperature is increased, the reverse current increases. The difference in the two curves is exaggerated on the graph in Figure 1–29 for illustration. Keep in mind that the reverse current below breakdown remains extremely small and can usually be neglected.

SECTION 1–7 CHECKUP

1. Discuss the significance of the knee of the characteristic curve in forward bias.
2. On what part of the curve is a forward-biased diode normally operated?
3. Which is greater, the breakdown voltage or the barrier potential?
4. On what part of the curve is a reverse-biased diode normally operated?
5. What happens to the barrier potential when the temperature increases?

1–8 DIODE MODELS

You have learned that a diode is a *pn* junction device. In this section, you will learn the electrical symbol for a diode and how a diode can be modeled for circuit analysis using any one of three levels of complexity. Also, diode packaging and terminal identification are introduced.

After completing this section, you should be able to

- **Discuss the operation of diodes and explain the three diode models**
  - Recognize a diode symbol and indentify the diode terminals
  - Recognize diodes in various physical configurations

Diode Symbol

There are several types of diodes, but the schematic symbol for a general-purpose or rectifier diode, such as introduced in this chapter, is shown in Figure 1–30. The *n* region is called the **cathode** and the *p* region is called the **anode**. The "arrow" in the symbol points in the direction of conventional current (opposite to electron flow).

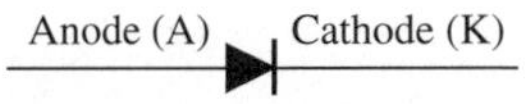

FIGURE 1–30
Diode schematic symbol.

Forward-Bias Connection A diode is forward-biased when a voltage source is connected as shown in Figure 1–31(a). The positive terminal of the source is connected to the anode through a current-limiting resistor. The negative terminal of the source is connected to the cathode. The forward current (I_F) is from anode to cathode as indicated. The forward voltage drop (V_F) due to the barrier potential is from positive at the anode to negative at the cathode.

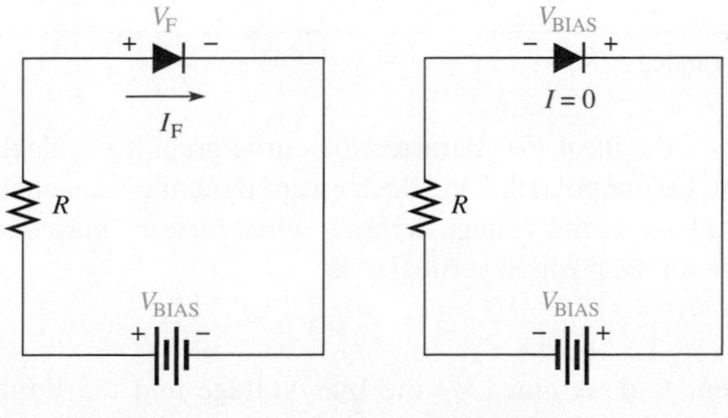

FIGURE 1–31
Forward-bias and reverse-bias connections showing the diode symbol.

Reverse-Bias Connection A diode is reverse-biased when a voltage source is connected as shown in Figure 1–31(b). The negative terminal of the source is connected to the anode side of the circuit, and the positive terminal is connected to the cathode side. A resistor is not necessary in reverse bias but it is shown for circuit consistency. The reverse current is extremely small and can be considered to be zero. Notice that the entire bias voltage (V_{BIAS}) appears across the diode.

Diode Approximations

The Ideal Diode Model The ideal model of a diode is the least accurate approximation and can be represented by a simple switch. When the diode is forward-biased, it ideally acts like a closed (on) switch, as shown in Figure 1–32(a). When the diode is reverse-biased, it ideally acts like an open (off) switch, as shown in part (b). Although the barrier potential, the forward dynamic resistance, and the reverse current are all neglected, this model is adequate for most troubleshooting when you are trying to determine if the diode is working properly.

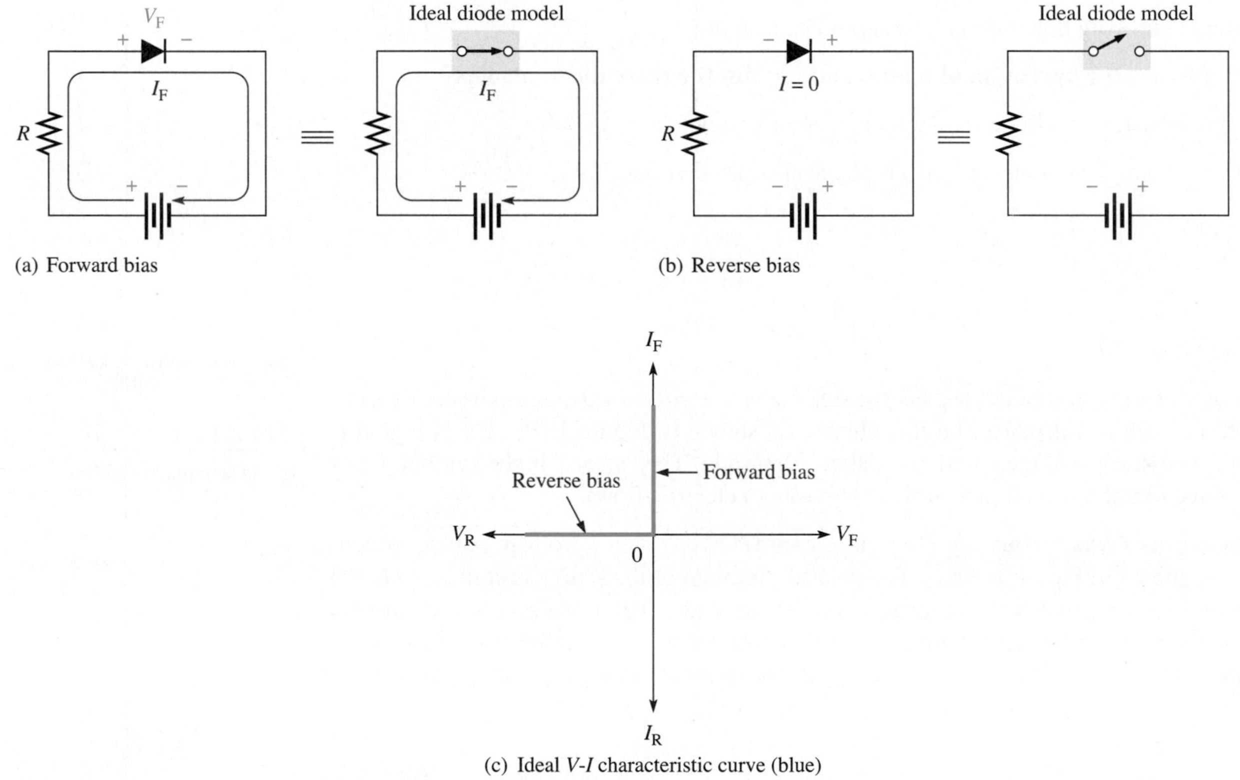

▲ **FIGURE 1–32**

The ideal model of a diode.

In Figure 1–32(c), the ideal *V-I* characteristic curve graphically depicts the ideal diode operation. Since the barrier potential and the forward dynamic resistance are neglected, the diode is assumed to have a zero voltage across it when forward-biased, as indicated by the portion of the curve on the positive vertical axis.

$$V_F = 0\text{ V}$$

The forward current is determined by the bias voltage and the limiting resistor using Ohm's law.

Equation 1–2

$$I_F = \frac{V_{BIAS}}{R_{LIMIT}}$$

Since the reverse current is neglected, its value is assumed to be zero, as indicated in Figure 1–32(c) by the portion of the curve on the negative horizontal axis.

$$I_R = 0\text{ A}$$

The reverse voltage equals the bias voltage.

$$V_R = V_{BIAS}$$

You may want to use the ideal model when you are troubleshooting or trying to figure out the operation of a circuit and are not concerned with more exact values of voltage or current.

The Practical Diode Model The practical model includes the barrier potential. When the diode is forward-biased, it is equivalent to a closed switch in series with a small equivalent voltage source (V_F) equal to the barrier potential (0.7 V) with the positive side toward the anode, as indicated in Figure 1–33(a). This equivalent voltage source represents the barrier potential that must be exceeded by the bias voltage before the diode will conduct and is not an active source of voltage. When conducting, a voltage drop of 0.7 V appears across the diode.

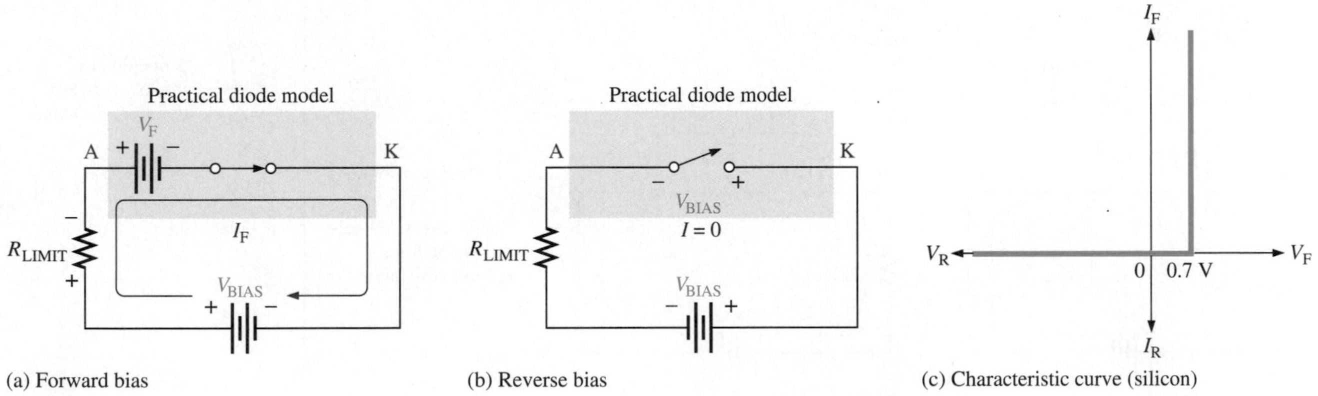

FIGURE 1–33

The practical model of a diode.

When the diode is reverse-biased, it is equivalent to an open switch just as in the ideal model, as shown in Figure 1–33(b). The barrier potential does not affect reverse bias, so it is not a factor.

The characteristic curve for the practical diode model is shown in Figure 1–33(c). Since the barrier potential is included and the dynamic resistance is neglected, the diode is assumed to have a voltage across it when forward-biased, as indicated by the portion of the curve to the right of the origin.

$$V_F = 0.7\text{ V}$$

The forward current is determined as follows by first applying Kirchhoff's voltage law to Figure 1–33(a):

$$V_{BIAS} - V_F - V_{R_{LIMIT}} = 0$$

$$V_{R_{LIMIT}} = I_F R_{LIMIT}$$

Substituting and solving for I_F,

$$I_F = \frac{V_{BIAS} - V_F}{R_{LIMIT}}$$

Equation 1–3

The diode is assumed to have zero reverse current, as indicated by the portion of the curve on the negative horizontal axis.

$$I_R = 0\text{ A}$$
$$V_R = V_{BIAS}$$

The practical model is useful when you are troubleshooting in lower-voltage circuits. In these cases, the 0.7 V drop across the diode may be significant and should be taken into account. The practical model is also useful when you are designing basic diode circuits.

The Complete Diode Model The complete model of a diode is the most accurate approximation and includes the barrier potential, the small forward dynamic resistance (r'_d), and the large internal reverse resistance (r'_R). The reverse resistance is taken into account because it provides a path for the reverse current, which is included in this diode model.

When the diode is forward-biased, it acts as a closed switch in series with the equivalent barrier potential voltage (V_B) and the small forward dynamic resistance (r'_d), as indicated in Figure 1–34(a). When the diode is reverse-biased, it acts as an open switch in parallel with the large internal reverse resistance (r'_R), as shown in Figure 1–34(b). The barrier potential does not affect reverse bias, so it is not a factor.

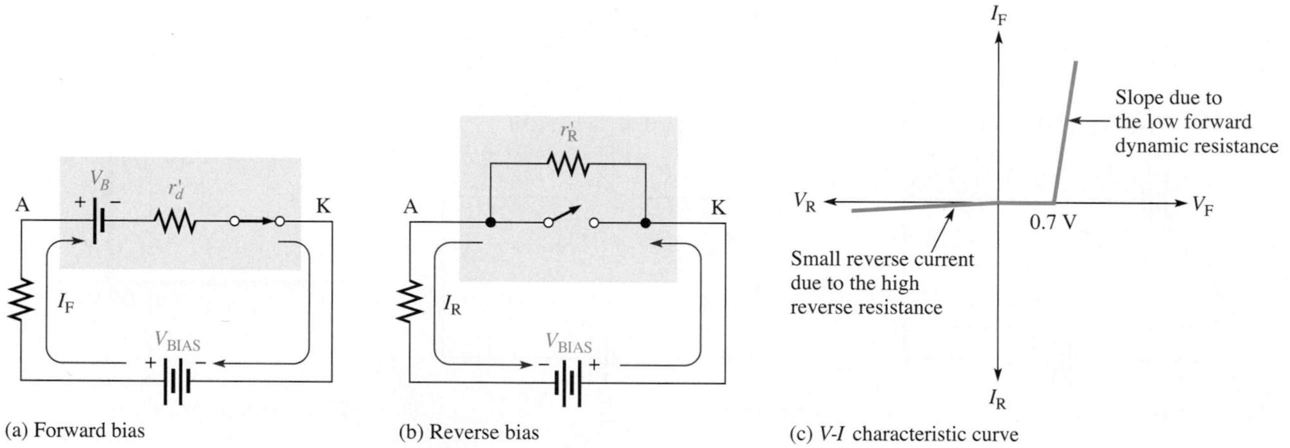

FIGURE 1–34

The complete model of a diode.

The characteristic curve for the complete diode model is shown in Figure 1–34(c). Since the barrier potential and the forward dynamic resistance are included, the diode is assumed to have a voltage across it when forward-biased. This voltage (V_F) consists of the barrier potential voltage plus the small voltage drop across the dynamic resistance, as indicated by the portion of the curve to the right of the origin. The curve slopes because the voltage drop due to dynamic resistance increases as the current increases. For the complete model of a silicon diode, the following formulas apply:

$$V_F = 0.7\text{ V} + I_F r'_d$$

$$I_F = \frac{V_{BIAS} - 0.7\text{ V}}{R_{LIMIT} + r'_d}$$

The reverse current is taken into account with the parallel resistance and is indicated by the portion of the curve to the left of the origin. The breakdown portion of the curve is not shown because breakdown is not a normal mode of operation for most diodes.

For troubleshooting work, it is unnecessary to use the complete model, as it involves complicated calculations. This model is generally suited to design problems using a computer for simulation. The ideal and practical models are used for circuits in this text, except in the following example, which illustrates the differences in the three models.

EXAMPLE 1–1

(a) Determine the forward voltage and forward current for the diode in Figure 1–35(a) for each of the diode models. Also find the voltage across the limiting resistor in each case. Assume $r'_d = 10\ \Omega$ at the determined value of forward current.

(b) Determine the reverse voltage and reverse current for the diode in Figure 1–35(b) for each of the diode models. Also find the voltage across the limiting resistor in each case. Assume $I_R = 1\ \mu A$.

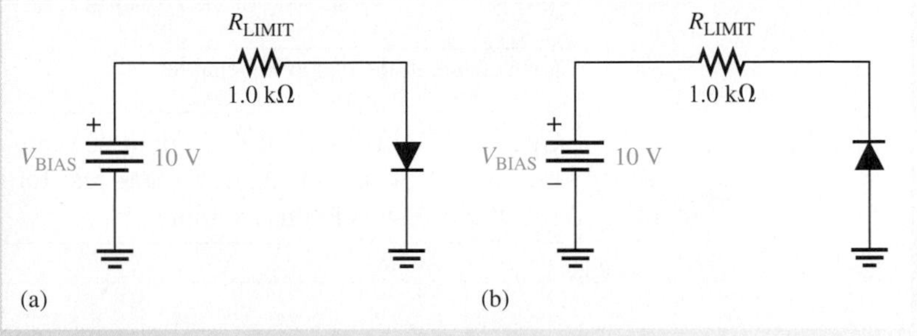

FIGURE 1–35

Solution **(a)** Ideal model:

$$V_F = \mathbf{0\ V}$$

$$I_F = \frac{V_{BIAS}}{R_{LIMIT}} = \frac{10\ V}{1.0\ k\Omega} = \mathbf{10\ mA}$$

$$V_{R_{LIMIT}} = I_F R_{LIMIT} = (10\ mA)(1.0\ k\Omega) = \mathbf{10\ V}$$

Practical model:

$$V_F = \mathbf{0.7\ V}$$

$$I_F = \frac{V_{BIAS} - V_F}{R_{LIMIT}} = \frac{10\ V - 0.7\ V}{1.0\ k\Omega} = \frac{9.3\ V}{1.0\ k\Omega} = \mathbf{9.3\ mA}$$

$$V_{R_{LIMIT}} = I_F R_{LIMIT} = (9.3\ mA)(1.0\ k\Omega) = \mathbf{9.3\ V}$$

Complete model:

$$I_F = \frac{V_{BIAS} - 0.7\ V}{R_{LIMIT} + r'_d} = \frac{10\ V - 0.7\ V}{1.0\ k\Omega + 10\ \Omega} = \frac{9.3\ V}{1010\ \Omega} = \mathbf{9.21\ mA}$$

$$V_F = 0.7\ V + I_F r'_d = 0.7\ V + (9.21\ mA)(10\ \Omega) = \mathbf{792\ mV}$$

$$V_{R_{LIMIT}} = I_F R_{LIMIT} = (9.21\ mA)(1.0\ k\Omega) = \mathbf{9.21\ V}$$

(b) Ideal model:

$$I_R = \mathbf{0\ A}$$

$$V_R = V_{BIAS} = \mathbf{10\ V}$$

$$V_{R_{LIMIT}} = \mathbf{0\ V}$$

Practical model:

$$I_R = \mathbf{0\ A}$$

$$V_R = V_{BIAS} = \mathbf{10\ V}$$

$$V_{R_{LIMIT}} = \mathbf{0\ V}$$

Complete model:

$$I_R = \mathbf{1\ \mu A}$$
$$V_{R_{LIMIT}} = I_R R_{LIMIT} = (1\ \mu A)(1.0\ k\Omega) = \mathbf{1\ mV}$$
$$V_R = V_{BIAS} - V_{R_{LIMIT}} = 10\ V - 1 mV = \mathbf{9.999\ V}$$

*Related Problem** Assume that the diode in Figure 1–35(a) fails open. What is the voltage across the diode and the voltage across the limiting resistor?

*Answers are at the end of the chapter.

Open the Multisim file E01-01 in the Examples folder on your CD-ROM. Measure the voltages across the diode and the resistor in both circuits and compare with the calculated results in this example.

Typical Diode Packages

Several common physical configurations of through-hole mounted diodes are illustrated in Figure 1–36(a). The anode (A) and cathode (K) are indicated on a diode in several ways, depending on the type of package. The cathode is usually marked by a band, a tab, or some other feature. On those packages where one lead is connected to the case, the case is the cathode.

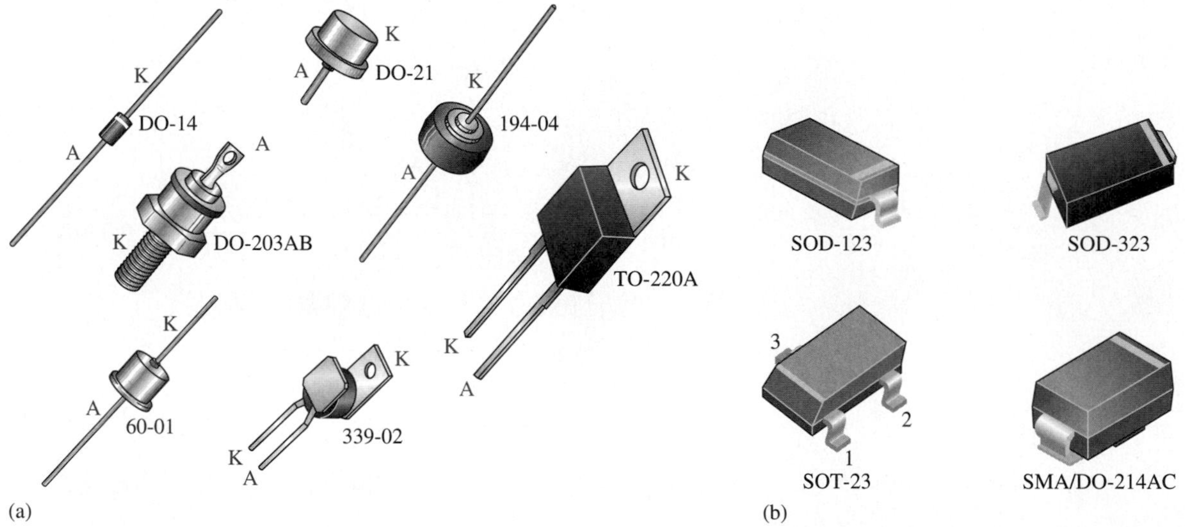

▲ **FIGURE 1–36**

Typical diode packages with terminal identification. The letter K is used for cathode to avoid confusion with certain electrical quantities that are represented by *C*. Case type numbers are indicated for each diode.

Surface-Mount Diode Packages Figure 1–36(b) shows typical diode packages for surface mounting on a printed circuit board. The SOD and SOT packages have gull-wing shaped leads. The SMA package has L-shaped leads that bend under the package. The SOD and SMA types have a band on one end to indicate the cathode. The SOT type is a three-terminal package in which there are either one or two diodes. In a single-diode SOT package, pin 1 is usually the anode and pin 3 is the cathode. In a dual-diode SOT package, pin 3 is the common terminal and can be either the anode or the cathode. Always check the datasheet for the particular diode to verify the pin configurations.

SECTION 1–8 CHECKUP

1. What are the two conditions under which a diode is operated?
2. Under what condition is a diode never intentionally operated?
3. What is the simplest way to visualize a diode?
4. To more accurately represent a diode, what factors must be included?
5. Which diode model represents the most accurate approximation?

1–9 TESTING A DIODE

A multimeter can be used as a fast and simple way to check a diode. A good diode will show an extremely high resistance (ideally an open) with reverse bias and a very low resistance with forward bias. A defective open diode will show an extremely high resistance (or open) for both forward and reverse bias. A defective shorted or resistive diode will show zero or a low resistance for both forward and reverse bias. An open diode is the most common type of failure.

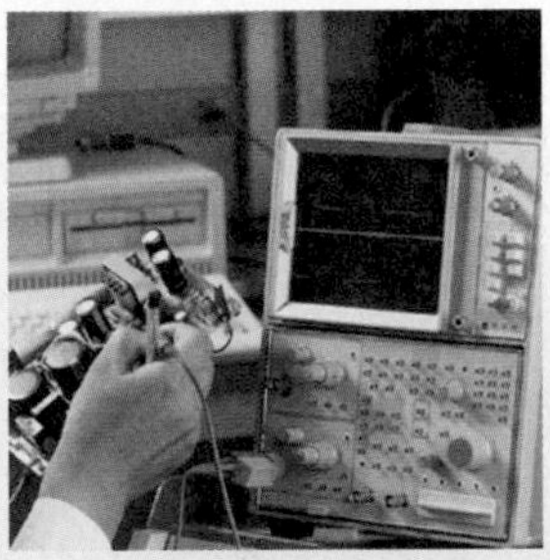

After completing this section, you should be able to

- **Test a diode using a digital multimeter**
  - Identify a properly functioning diode
  - Identify a faulty diode

The DMM Diode Test Position Many digital multimeters (DMMs) have a diode test function that provides a convenient way to test a diode. A typical DMM, as shown in Figure 1–37, has a small diode symbol to mark the position of the function switch. When set to *diode test,* the meter provides an internal voltage sufficient to forward-bias and reverse-bias a diode. This internal voltage may vary among different makes of DMM, but 2.5 V to 3.5 V is a typical range of values. The meter provides a voltage reading or other indication to show the condition of the diode under test.

When the Diode Is Working In Figure 1–37(a), the red (positive) lead of the meter is connected to the anode and the black (negative) lead is connected to the cathode to forward-bias the diode. If the diode is good, you will get a reading of between approximately 0.5 V and 0.9 V, with 0.7 V being typical for forward bias.

In Figure 1–37(b), the diode is turned around for reverse bias as shown. If the diode is working properly, you will typically get a reading of "OL". Some DMMs may display the internal voltage for a reverse-bias condition.

When the Diode Is Defective When a diode has failed open, you get an out-of-range "OL" indication for both the forward-bias and the reverse-bias condition, as illustrated in Figure 1–38(a). If a diode is shorted, the meter reads 0 V in both forward- and reverse-bias tests, as indicated in part (b).

Checking a Diode with the OHMs Function DMMs that do not have a diode test position can be used to check a diode by setting the function switch on an OHMs range. For a forward-bias check of a good diode, you will get a resistance reading that can vary depending on the meter's internal battery. Many meters do not have sufficient voltage on the OHMs setting to fully forward-bias a diode and you may get a reading of from several hundred to several thousand ohms. For the reverse-bias check of a good diode, you will get an out-of-range indication such as "OL" on most DMMs because the reverse resistance is too high for the meter to measure.

▶ **FIGURE 1–37**

DMM diode test on a properly functioning diode.

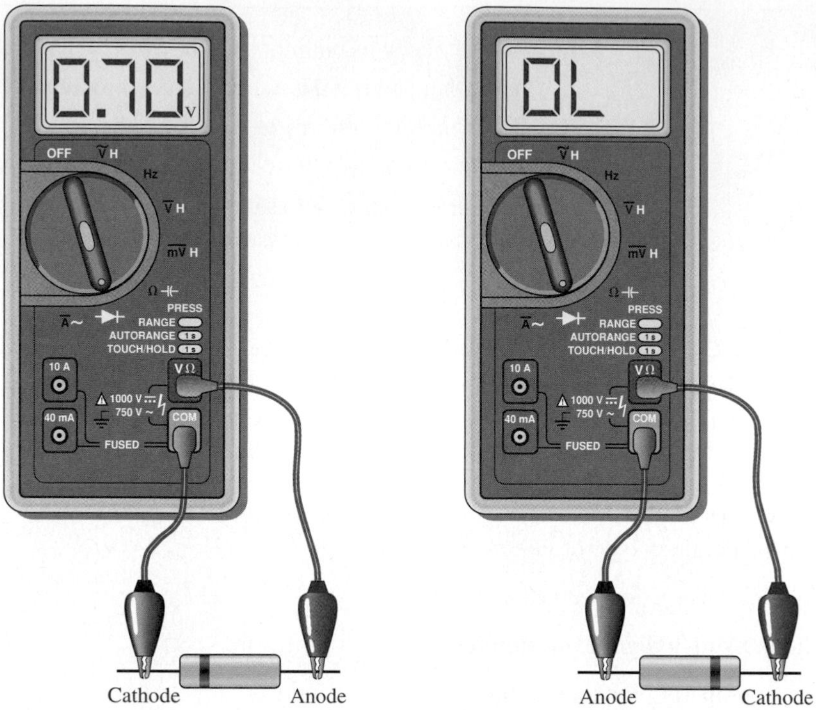

(a) Forward-bias test

(b) Reverse-bias test

▶ **FIGURE 1–38**

Testing a defective diode.

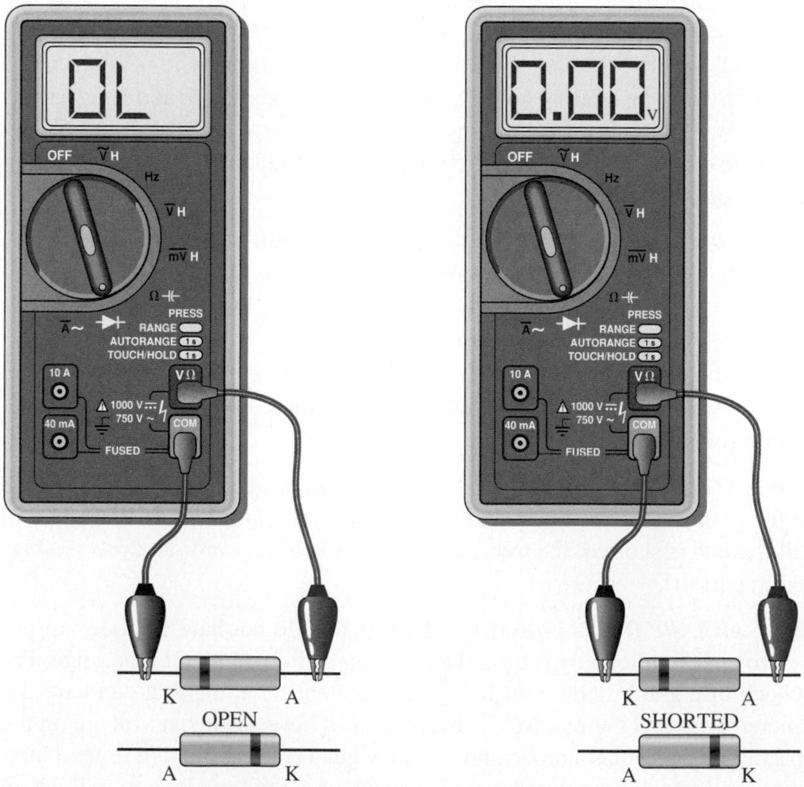

(a) Forward- and reverse-bias tests for an open diode give the same indication.

(b) Forward- and reverse-bias tests for a shorted diode give the same 0 V reading.

Even though you may not get accurate forward- and reverse-resistance readings on a DMM, the relative readings indicate that a diode is functioning properly, and that is usually all you need to know. The out-of-range indication shows that the reverse resistance is extremely high, as you expect. The reading of a few hundred to a few thousand ohms for forward bias is relatively small compared to the reverse resistance, indicating that the diode is working properly. The actual resistance of a forward-biased diode is typically much less than 100 Ω.

SECTION 1–9 CHECKUP

1. A properly functioning diode will produce a reading in what range when forward-biased?
2. What reading might a DMM produce when a diode is reverse-biased?

SUMMARY OF DIODE BIAS

FORWARD BIAS: PERMITS MAJORITY-CARRIER CURRENT

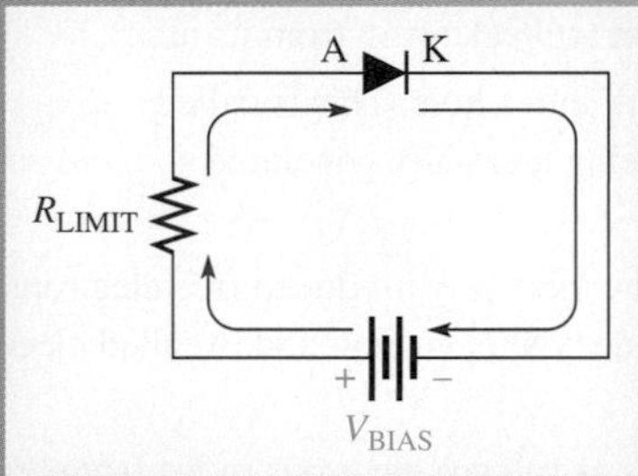

- Bias voltage connections: positive to anode (A); negative to cathode (K).
- The bias voltage must be greater than the barrier potential.
- Barrier potential: 0.7 V for silicon.
- Majority carriers provide the forward current.
- The depletion region narrows.

REVERSE BIAS: PREVENTS MAJORITY-CARRIER CURRENT

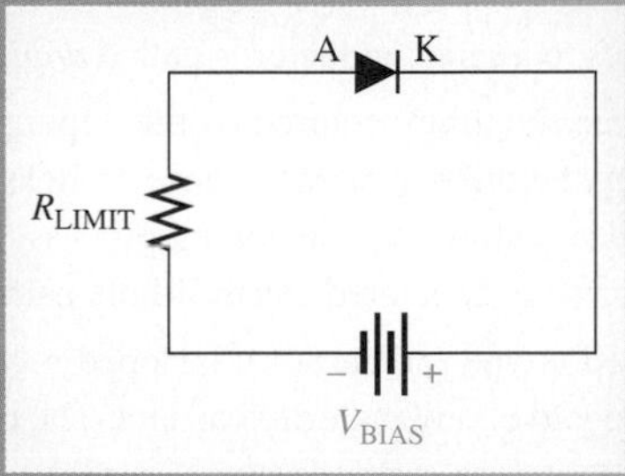

- Bias voltage connections: positive to cathode (K); negative to anode (A).
- The bias voltage must be less than the breakdown voltage.
- There is no majority carrier current after transition time.
- Minority carriers provide a negligibly small reverse current.
- The depletion region widens.

SUMMARY

Section 1–1

- ◆ According to the classical Bohr model, the atom is viewed as having a planetary-type structure with electrons orbiting at various distances around the central nucleus.
- ◆ The nucleus of an atom consists of protons and neutrons. The protons have a positive charge and the neutrons are uncharged. The number of protons is the atomic number of the atom.
- ◆ Electrons have a negative charge and orbit around the nucleus at distances that depend on their energy level. An atom has discrete bands of energy called *shells* in which the electrons orbit. Atomic structure allows a certain maximum number of electrons in each shell. In their natural state, all atoms are neutral because they have an equal number of protons and electrons.

- The outermost shell or band of an atom is called the *valence band,* and electrons that orbit in this band are called *valence electrons.* These electrons have the highest energy of all those in the atom. If a valence electron acquires enough energy from an outside source such as heat, it can jump out of the valence band and break away from its atom.

Section 1–2

- Insulating materials have very few free electrons and do not conduct current at all under normal circumstances.
- Materials that are conductors have a large number of free electrons and conduct current very well.
- Semiconductive materials fall in between conductors and insulators in their ability to conduct current.
- Semiconductor atoms have four valence electrons. Silicon is the most widely used semiconductive material.
- Semiconductor atoms bond together in a symmetrical pattern to form a solid material called a *crystal.* The bonds that hold a crystal together are called *covalent bonds.*

Section 1–3

- The valence electrons that manage to escape from their parent atom are called *conduction electrons* or *free electrons.* They have more energy than the electrons in the valence band and are free to drift throughout the material.
- When an electron breaks away to become free, it leaves a hole in the valence band creating what is called an *electron-hole pair.* These electron-hole pairs are thermally produced because the electron has acquired enough energy from external heat to break away from its atom.
- A free electron will eventually lose energy and fall back into a hole. This is called *recombination.* Electron-hole pairs are continuously being thermally generated so there are always free electrons in the material.
- When a voltage is applied across the semiconductor, the thermally produced free electrons move toward the positive end and form the current. This is one type of current and is called electron current.
- Another type of current is the hole current. This occurs as valence electrons move from hole to hole creating, in effect, a movement of holes in the opposite direction.

Section 1–4

- An *n*-type semiconductive material is created by adding impurity atoms that have five valence electrons. These impurities are *pentavalent atoms.* A *p*-type semiconductor is created by adding impurity atoms with only three valence electrons. These impurities are *trivalent atoms.*
- The process of adding pentavalent or trivalent impurities to a semiconductor is called *doping.*
- The majority carriers in an *n*-type semiconductor are free electrons acquired by the doping process, and the minority carriers are holes produced by thermally generated electron-hole pairs. The majority carriers in a *p*-type semiconductor are holes acquired by the doping process, and the minority carriers are free electrons produced by thermally generated electron-hole pairs.

Section 1–5

- A *pn* junction is formed when part of a material is doped *n*-type and part of it is doped *p*-type. A depletion region forms starting at the junction that is devoid of any majority carriers. The depletion region is formed by ionization.
- The barrier potential is typically 0.7 V for a silicon diode and 0.3 V for germanium.

Section 1–6

- There is current through a diode only when it is forward-biased. Ideally, there is no current when there is no bias nor when there is reverse bias. Actually, there is a very small current in reverse bias due to the thermally generated minority carriers, but this can usually be neglected.
- Avalanche occurs in a reverse-biased diode if the bias voltage equals or exceeds the breakdown voltage.
- A diode conducts current when forward-biased and blocks current when reversed-biased.
- Reverse breakdown voltage for a diode is typically greater than 50 V.

Section 1–7

- The *V-I* characteristic curve shows the diode current as a function of voltage across the diode.
- The resistance of a forward-biased diode is called the *dynamic* or *ac resistance.*
- Reverse current increases rapidly at the reverse breakdown voltage.
- Reverse breakdown should be avoided in most diodes.

Section 1–8

- The ideal model represents the diode as a closed switch in forward bias and as an open switch in reverse bias.
- The practical model represents the diode as a switch in series with the barrier potential.
- The complete model includes the dynamic forward resistance in series with the practical model in forward bias and the reverse resistance in parallel with the open switch in reverse bias.

Section 1–9

- Many DMMs provide a diode test function.
- DMMs display the diode drop when the diode is operating properly in forward bias.
- Most DMMs indicate "OL" when the diode is open.

KEY TERMS

Key terms and other bold terms are defined in the end-of-book glossary.

Anode The *p* region of a diode.

Atom The smallest particle of an element that possesses the unique characteristics of that element.

Barrier potential The amount of energy required to produce full conduction across the *pn* junction in forward bias.

Bias The application of a dc voltage to a diode to make it either conduct or block current.

Cathode The *n* region of a diode.

Conductor A material that easily conducts electrical current.

Crystal A solid material in which the atoms are arranged in a symmetrical pattern.

Diode A semiconductor device with a single *pn* junction that conducts current in only one direction.

Doping The process of imparting impurities to an intrinsic semiconductive material in order to control its conduction characteristics.

Electron The basic particle of negative electrical charge.

Forward bias The condition in which a diode conducts current.

Free electron An electron that has acquired enough energy to break away from the valence band of the parent atom; also called a *conduction electron.*

Hole The absence of an electron in the valence band of an atom.

Insulator A material that does not normally conduct current.

Ionization The removal or addition of an electron from or to a neutral atom so that the resulting atom (called an ion) has a net positive or negative charge.

***PN* junction** The boundary between two different types of semiconductive materials.

Proton The basic particle of positive charge.

Reverse bias The condition in which a diode prevents current.

Semiconductor A material that lies between conductors and insulators in its conductive properties. Silicon, germanium, and carbon are examples.

Shell An energy band in which electrons orbit the nucleus of an atom.

Silicon A semiconductive material.

Valence Related to the outer shell of an atom.

***V-I* characteristic** A curve showing the relationship of diode voltage and current.

KEY FORMULAS

1–1	$N_e = 2n^2$	Maximum number of electrons in any shell
1–2	$I_F = \dfrac{V_{BIAS}}{R_{LIMIT}}$	Forward current, ideal diode model
1–3	$I_F = \dfrac{V_{BIAS} - V_F}{R_{LIMIT}}$	Forward current, practical diode model

TRUE/FALSE QUIZ

Answers are at the end of the chapter.

1. An atom is the smallest particle in an element.
2. An electron is a negatively charged particle.
3. An atom is made up of electrons, protons, and neutrons.
4. Electrons are part of the nucleus of an atom.
5. Valence electrons exist in the outer shell of an atom.
6. Crystals are formed by the bonding of atoms.
7. Silicon is a conductive material.
8. All diodes have one *pn* junction.
9. The *p* and *n* regions in a diode are formed by a process called *ionization*.
10. The two regions of a diode are the anode and the collector.
11. A diode can conduct current in two directions with equal ease.
12. A diode conducts current when forward-biased.
13. When reverse-biased, a diode ideally appears as a short.
14. Two types of current in a diode are electron and hole.

CIRCUIT-ACTION QUIZ

Answers are at the end of the chapter.

1. When a diode is forward-biased and the bias voltage is increased, the forward current will
   (a) increase **(b)** decrease **(c)** not change
2. When a diode is forward-biased and the bias voltage is increased, the voltage across the diode (assuming the practical model) will
   (a) increase **(b)** decrease **(c)** not change
3. When a diode is reverse-biased and the bias voltage is increased, the reverse current (assuming the practical model) will
   (a) increase **(b)** decrease **(c)** not change
4. When a diode is reverse-biased and the bias voltage is increased, the reverse current (assuming the complete model) will
   (a) increase **(b)** decrease **(c)** not change
5. When a diode is forward-biased and the bias voltage is increased, the voltage across the diode (assuming the complete model) will
   (a) increase **(b)** decrease **(c)** not change
6. If the forward current in a diode is increased, the diode voltage (assuming the practical model) will
   (a) increase **(b)** decrease **(c)** not change
7. If the forward current in a diode is decreased, the diode voltage (assuming the complete model) will
   (a) increase **(b)** decrease **(c)** not change
8. If the barrier potential of a diode is exceeded, the forward current will
   (a) increase **(b)** decrease **(c)** not change

SELF-TEST

Answers are at the end of the chapter.

Section 1–1

1. Every known element has
   (a) the same type of atoms **(b)** the same number of atoms
   (c) a unique type of atom **(d)** several different types of atoms

2. An atom consists of
   (a) one nucleus and only one electron (b) one nucleus and one or more electrons
   (c) protons, electrons, and neutrons (d) answers (b) and (c)
3. The nucleus of an atom is made up of
   (a) protons and neutrons (b) electrons
   (c) electrons and protons (d) electrons and neutrons
4. Valence electrons are
   (a) in the closest orbit to the nucleus (b) in the most distant orbit from the nucleus
   (c) in various orbits around the nucleus (d) not associated with a particular atom
5. A positive ion is formed when
   (a) a valence electron breaks away from the atom
   (b) there are more holes than electrons in the outer orbit
   (c) two atoms bond together
   (d) an atom gains an extra valence electron

Section 1–2

6. The most widely used semiconductive material in electronic devices is
   (a) germanium (b) carbon (c) copper (d) silicon
7. The difference between an insulator and a semiconductor is
   (a) a wider energy gap between the valence band and the conduction band
   (b) the number of free electrons
   (c) the atomic structure
   (d) answers (a), (b), and (c)
8. The energy band in which free electrons exist is the
   (a) first band (b) second band (c) conduction band (d) valence band
9. In a semiconductor crystal, the atoms are held together by
   (a) the interaction of valence electrons (b) forces of attraction
   (c) covalent bonds (d) answers (a), (b), and (c)
10. The atomic number of silicon is
    (a) 8 (b) 2 (c) 4 (d) 14
11. The atomic number of germanium is
    (a) 8 (b) 2 (c) 4 (d) 32
12. The valence shell in a silicon atom has the number designation of
    (a) 0 (b) 1 (c) 2 (d) 3
13. Each atom in a silicon crystal has
    (a) four valence electrons
    (b) four conduction electrons
    (c) eight valence electrons, four of its own and four shared
    (d) no valence electrons because all are shared with other atoms

Section 1–3

14. Electron-hole pairs are produced by
    (a) recombination (b) thermal energy (c) ionization (d) doping
15. Recombination is when
    (a) an electron falls into a hole
    (b) a positive and a negative ion bond together
    (c) a valence electron becomes a conduction electron
    (d) a crystal is formed
16. The current in a semiconductor is produced by
    (a) electrons only (b) holes only (c) negative ions (d) both electrons and holes

Section 1–4

17. In an intrinsic semiconductor,
- **(a)** there are no free electrons
- **(b)** the free electrons are thermally produced
- **(c)** there are only holes
- **(d)** there as many electrons as there are holes
- **(e)** answers (b) and (d)

18. The process of adding an impurity to an intrinsic semiconductor is called
(a) doping **(b)** recombination **(c)** atomic modification **(d)** ionization

19. A trivalent impurity is added to silicon to create
(a) germanium **(b)** a *p*-type semiconductor
(c) an *n*-type semiconductor **(d)** a depletion region

20. The purpose of a pentavalent impurity is to
(a) reduce the conductivity of silicon **(b)** increase the number of holes
(c) increase the number of free electrons **(d)** create minority carriers

21. The majority carriers in an *n*-type semiconductor are
(a) holes **(b)** valence electrons
(c) conduction electrons **(d)** protons

22. Holes in an *n*-type semiconductor are
- **(a)** minority carriers that are thermally produced
- **(b)** minority carriers that are produced by doping
- **(c)** majority carriers that are thermally produced
- **(d)** majority carriers that are produced by doping

Section 1–5

23. A *pn* junction is formed by
- **(a)** the recombination of electrons and holes
- **(b)** ionization
- **(c)** the boundary of a *p*-type and an *n*-type material
- **(d)** the collision of a proton and a neutron

24. The depletion region is created by
(a) ionization **(b)** diffusion **(c)** recombination **(d)** answers (a), (b), and (c)

25. The depletion region consists of
(a) nothing but minority carriers **(b)** positive and negative ions
(c) no majority carriers **(d)** answers (b) and (c)

Section 1–6

26. The term *bias* means
- **(a)** the ratio of majority carriers to minority carriers
- **(b)** the amount of current across a diode
- **(c)** a dc voltage is applied to control the operation of a device
- **(d)** neither (a), (b), nor (c)

27. To forward-bias a diode,
- **(a)** an external voltage is applied that is positive at the anode and negative at the cathode
- **(b)** an external voltage is applied that is negative at the anode and positive at the cathode
- **(c)** an external voltage is applied that is positive at the *p* region and negative at the *n* region
- **(d)** answers (a) and (c)

28. When a diode is forward-biased,
- **(a)** the only current is hole current
- **(b)** the only current is electron current
- **(c)** the only current is produced by majority carriers
- **(d)** the current is produced by both holes and electrons

29. Although current is blocked in reverse bias,

(a) there is some current due to majority carriers

(b) there is a very small current due to minority carriers

(c) there is an avalanche current

30. For a silicon diode, the value of the forward-bias voltage typically

(a) must be greater than 0.3 V

(b) must be greater than 0.7 V

(c) depends on the width of the depletion region

(d) depends on the concentration of majority carriers

31. When forward-biased, a diode

(a) blocks current **(b)** conducts current

(c) has a high resistance **(d)** drops a large voltage

Section 1–7

32. A diode is normally operated in

(a) reverse breakdown

(b) the forward-bias region

(c) the reverse-bias region

(d) either (b) or (c)

33. The dynamic resistance can be important when a diode is

(a) reverse-biased

(b) forward-biased

(c) in reverse breakdown

(d) unbiased

34. The *V-I* curve for a diode shows

(a) the voltage across the diode for a given current

(b) the amount of current for a given bias voltage

(c) the power dissipation

(d) none of these

Section 1–8

35. Ideally, a diode can be represented by a

(a) voltage source **(b)** resistance

(c) switch **(d)** all of these

36. In the practical diode model,

(a) the barrier potential is taken into account

(b) the forward dynamic resistance is taken into account

(c) none of these

(d) both (a) and (b)

37. In the complete diode model,

(a) the barrier potential is taken into account

(b) the forward dynamic resistance is taken into account

(c) the reverse resistance is taken into account

(d) all of these

Section 1–9

38. When a silicon diode is working properly in forward bias, a DMM in the diode test position will indicate

(a) 0 V **(b)** OL

(c) approximately 0.7 V **(d)** approximately 0.3 V

39. When a silicon diode is open, a DMM will generally indicate

(a) 0 V **(b)** OL

(c) approximately 0.7 V **(d)** approximately 0.3 V

PROBLEMS

Answers to all odd-numbered problems are at the end of the book.

BASIC PROBLEMS

Section 1–1 Atomic Structure

1. If the atomic number of a neutral atom is 6, how many electrons does the atom have? How many protons?

2. What is the maximum number of electrons that can exist in the 3rd shell of an atom?

Section 1–2 Insulators, Conductors, and Semiconductors

3. For each of the energy diagrams in Figure 1–39, determine the class of material based on relative comparisons.

FIGURE 1–39

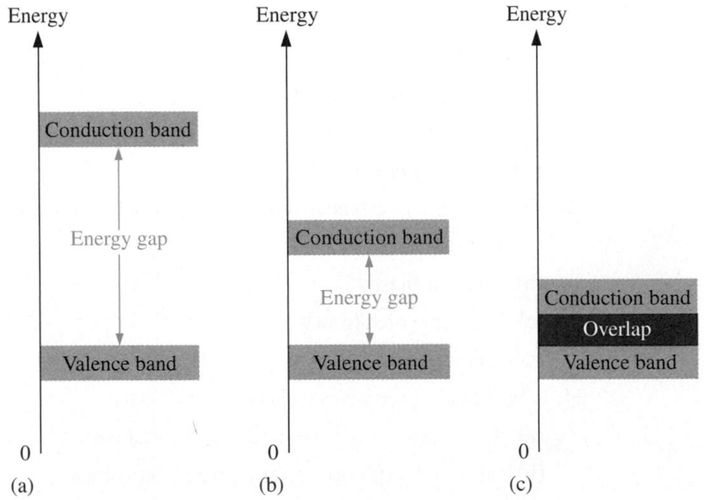

4. A certain atom has four valence electrons. What type of atom is it?

5. In a silicon crystal, how many covalent bonds does a single atom form?

Section 1–3 Current in Semiconductors

6. What happens when heat is added to silicon?

7. Name the two energy bands at which current is produced in silicon.

Section 1–4 *N*-Type and *P*-Type Semiconductors

8. Describe the process of doping and explain how it alters the atomic structure of silicon.

9. What is antimony? What is boron?

Section 1–5 The Diode

10. How is the electric field across the *pn* junction created?

11. Because of its barrier potential, can a diode be used as a voltage source? Explain.

Section 1–6 Biasing a Diode

12. To forward-bias a diode, to which region must the positive terminal of a voltage source be connected?

13. Explain why a series resistor is necessary when a diode is forward-biased.

Section 1–7 Voltage-Current Characteristic of a Diode

14. Explain how to generate the forward-bias portion of the characteristic curve.

15. What would cause the barrier potential of a silicon diode to decrease from 0.7 V to 0.6 V?

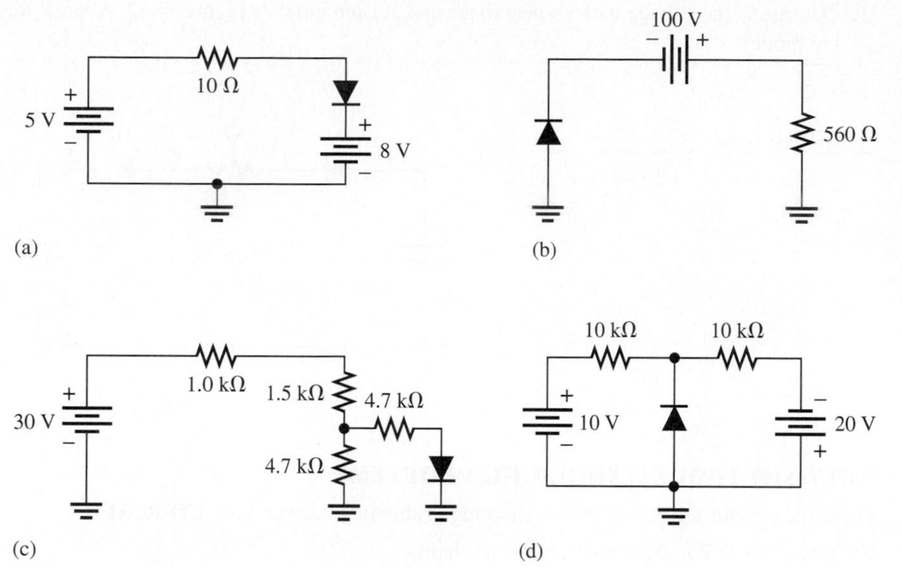

◀ **FIGURE 1–40**

Multisim file circuits are identified with a CD logo and are in the Problems folder on your CD-ROM. Filenames correspond to figure numbers (e.g., F01-40).

Section 1–8 Diode Models

16. Determine whether each silicon diode in Figure 1–40 is forward-biased or reverse-biased.

17. Determine the voltage across each diode in Figure 1–40, assuming the practical model.

18. Determine the voltage across each diode in Figure 1–40, assuming an ideal diode.

19. Determine the voltage across each diode in Figure 1–40, using the complete diode model with $r'_d = 10\ \Omega$ and $r'_R = 100\ \text{M}\Omega$.

Section 1–9 Testing a Diode

20. Consider the meter indications in each circuit of Figure 1–41, and determine whether the diode is functioning properly, or whether it is open or shorted. Assume the ideal model.

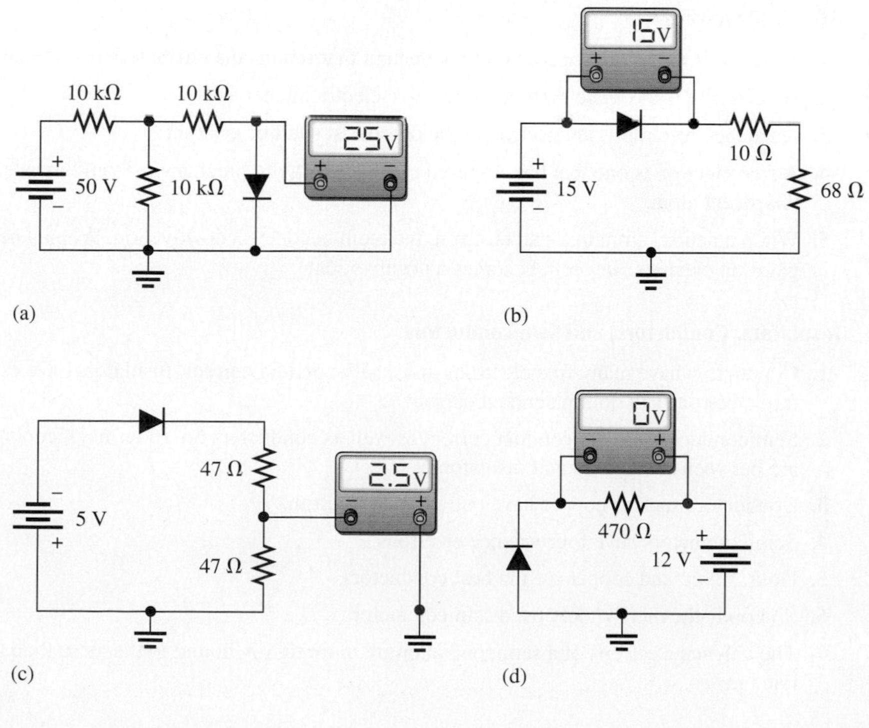

◀ **FIGURE 1–41**

21. Determine the voltage with respect to ground at each point in Figure 1–42. Assume the practical model.

▶ FIGURE 1–42

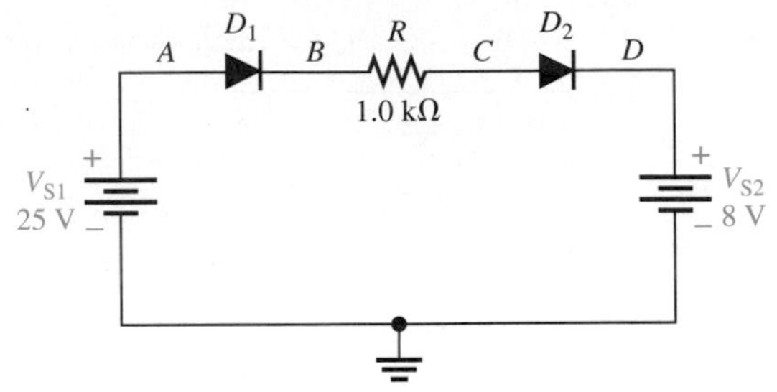

MULTISIM TROUBLESHOOTING PROBLEMS

These file circuits are in the Troubleshooting Problems folder on your CD-ROM.

22. Open file TSP01-22 and determine the fault.
23. Open file TSP01-23 and determine the fault.
24. Open file TSP01-24 and determine the fault.
25. Open file TSP01-25 and determine the fault.
26. Open file TSP01-26 and determine the fault.
27. Open file TSP01-27 and determine the fault.
28. Open file TSP01-28 and determine the fault.
29. Open file TSP01-29 and determine the fault.
30. Open file TSP01-30 and determine the fault.

ANSWERS

SECTION CHECKUPS

Section 1–1 Atomic Structure

1. An atom is the smallest particle of an element that retains the characteristics of that element.
2. An electron is the basic particle of negative electrical charge.
3. A valence electron is an electron in the outermost shell of an atom.
4. A free electron is one that has acquired enough energy to break away from the valence band of the parent atom.
5. When a neutral atom loses an electron, the atom becomes a positive ion. When a neutral atom gains an electron, the atom becomes a negative ion.

Section 1–2 Insulators, Conductors, and Semiconductors

1. Conductors have many free electrons and easily conduct current. Insulators have essentially no free electrons and do not conduct current.
2. Semiconductors do not conduct current as well as conductors do. In terms of conductivity, they are between conductors and insulators.
3. Conductors such as copper have one valence electron.
4. Semiconductors have four valence electrons.
5. Gold, silver, and copper are the best conductors.
6. Silicon is the most widely used semiconductor.
7. The valence electrons of a semiconductor are more tightly bound to the atom than those of conductors.

8. Covalent bonds are formed by the sharing of valence electrons with neighboring atoms.
9. An intrinsic material is one that is in a pure state.
10. A crystal is a solid material formed by atoms bonding together in a symmetrical pattern.

Section 1–3 Current in Semiconductors

1. Free electrons are in the conduction band.
2. Free (conduction) electrons are responsible for electron current in silicon.
3. A hole is the absence of an electron in the valence band.
4. Hole current occurs at the valence level.

Section 1–4 *N*-Type and *P*-Type Semiconductors

1. Doping is the process of adding impurity atoms to a semiconductor in order to modify its conductive properties.
2. A pentavalent atom (donor) has five valence electrons and a trivalent atom (acceptor) has three valence electrons.
3. An *n*-type material is formed by the addition of pentavalent impurity atoms to the intrinsic semiconductive material.
4. A *p*-type material is formed by the addition of trivalent impurity atoms to the intrinsic semiconductive material.
5. The majority carrier in an *n*-type semiconductor is the free electron.
6. The majority carrier in a *p*-type semiconductor is the hole.
7. Majority carriers are produced by doping.
8. Minority carriers are thermally produced when electron-hole pairs are generated.
9. A pure semiconductor is intrinsic. A doped (impure) semiconductor is extrinsic.

Section 1–5 The Diode

1. A *pn* junction is the boundary between *p*-type and *n*-type semiconductors in a diode.
2. Diffusion is the movement of the free electrons (majority carriers) in the *n*-region across the *pn* junction and into the *p* region.
3. The depletion region is the thin layers of positive and negative ions that exist on both sides of the *pn* junction.
4. The barrier potential is the potential difference of the electric field in the depletion region and is the amount of energy required to move electrons through the depletion region.
5. The barrier potential for a silicon diode is approximately 0.7 V.
6. The barrier potential for a germanium diode is approximately 0.3 V.

Section 1–6 Biasing a Diode

1. When forward-biased, a diode conducts current. The free electrons in the *n* region move across the *pn* junction and combine with the holes in the *p* region.
2. To forward-bias a diode, the positive side of an external bias voltage is applied to the *p* region and the negative side to the *n* region.
3. When reverse-biased, a diode does not conduct current except for an extremely small reverse current.
4. To reverse-bias a diode, the positive side of an external bias voltage is applied to the *n* region and the negative side to the *p* region.
5. The depletion region for forward bias is much narrower than for reverse bias.
6. Majority carrier current is produced by forward bias.
7. Reverse current is produced by the minority carriers.
8. Reverse breakdown occurs when the reverse-bias voltage equals or exceeds the breakdown voltage of the *pn* junction of a diode.
9. Avalanche effect is the rapid multiplication of current carriers in reverse breakdown.

Section 1–7 Voltage-Current Characteristic of a Diode

1. The knee of the characteristic curve in forward bias is the point at which the barrier potential is overcome and the current increases drastically.
2. A forward-biased diode is normally operated above the knee of the curve.
3. Breakdown voltage is always much greater than the barrier potential.
4. A reverse-biased diode is normally operated between 0 V and the breakdown voltage.
5. Barrier potential decreases as temperature increases.

Section 1–8 Diode Models

1. A diode is operated in forward bias and reverse bias.
2. A diode should never be operated in reverse breakdown.
3. The diode can be ideally viewed as a switch.
4. A diode includes barrier potential, dynamic resistance, and reverse resistance in the complete model.
5. The complete diode model is the most accurate diode approximation.

Section 1–9 Testing a Diode

1. 0.5 V to 0.9 V
2. OL

RELATED PROBLEM FOR EXAMPLE

1–1 $V_D = 5\text{ V}$; $V_{LIMIT} = 0\text{ V}$

TRUE/FALSE QUIZ

1. F	**2.** T	**3.** T	**4.** F	**5.** T	**6.** T	**7.** F
8. T	**9.** F	**10.** F	**11.** F	**12.** T	**13.** F	**14.** T

CIRCUIT-ACTION QUIZ

1. (a)	**2.** (c)	**3.** (c)	**4.** (a)	**5.** (a)	**6.** (c)	**7.** (b)	**8.** (a)

SELF-TEST

1. (c)	**2.** (d)	**3.** (a)	**4.** (b)	**5.** (a)	**6.** (d)	**7.** (d)	**8.** (c)
9. (d)	**10.** (d)	**11.** (d)	**12.** (d)	**13.** (c)	**14.** (b)	**15.** (a)	**16.** (d)
17. (e)	**18.** (a)	**19.** (b)	**20.** (c)	**21.** (c)	**22.** (a)	**23.** (c)	**24.** (d)
25. (d)	**26.** (c)	**27.** (d)	**28.** (d)	**29.** (b)	**30.** (b)	**31.** (b)	**32.** (d)
33. (b)	**34.** (a)	**35.** (c)	**36.** (a)	**37.** (d)	**38.** (c)	**39.** (b)	

Diode Applications

2

CHAPTER OUTLINE

CHAPTER OBJECTIVES

- Explain and analyze the operation of half-wave rectifiers
- Explain and analyze the operation of full-wave rectifiers
- Explain and analyze the operation and characteristics of power supply filters and regulators
- Explain and analyze the operation of diode limiting and clamping circuits
- Explain and analyze the operation of diode voltage multipliers
- Interpret and use a diode datasheet
- Troubleshoot power supplies and diode circuits

KEY TERMS

- DC power supply
- Rectifier
- Filter
- Regulator
- Half-wave rectifier
- Peak inverse voltage (PIV)
- Full-wave rectifier
- Ripple voltage
- Line regulation
- Load regulation
- Limiter
- Clamper
- Troubleshooting

VISIT THE COMPANION WEBSITE

Study aids for this chapter are available at
http://www.prenhall.com/floyd

INTRODUCTION

In Chapter 1, you learned that a semiconductor diode is a device with a single *pn* junction. The importance of the diode in electronic circuits cannot be overemphasized. Its ability to conduct current in one direction while blocking current in the other direction is essential to the operation of many types of circuits. One circuit in particular is the ac rectifier, which is covered in this chapter. Other important applications are circuits such as diode limiters, diode clampers, and diode voltage multipliers. A datasheet is discussed for specific diodes.

APPLICATION ACTIVITY PREVIEW

You are given the responsibility for the final design and testing of a power supply circuit that your company plans to use in several of its products. You will apply your knowledge of diode circuits to the Application Activity at the end of the chapter.

2–1 HALF-WAVE RECTIFIERS

Because of their ability to conduct current in one direction and block current in the other direction, diodes are used in circuits called rectifiers that convert ac voltage into dc voltage. Rectifiers are found in all dc power supplies that operate from an ac voltage source. A power supply is an essential part of each electronic system from the simplest to the most complex.

After completing this section, you should be able to

- **Explain and analyze the operation of half-wave rectifiers**
  - Describe a basic dc power supply and half-wave rectification
  - Determine the average value of a half-wave rectified voltage
  - Discuss the effect of barrier potential on a half-wave rectifier output
  - Define *peak inverse voltage (PIV)*
  - Describe the transformer-coupled half-wave rectifier

The Basic DC Power Supply

All active electronic devices require a source of constant dc that can be supplied by a battery or a dc power supply. The **dc power supply** converts the standard 120 V, 60 Hz ac voltage available at wall outlets into a constant dc voltage. The dc power supply is one of the most common circuits you will find, so it is important to understand how it works. The voltage produced is used to power all types of electronic circuits including consumer electronics (televisions, DVDs, etc.), computers, industrial controllers, and most laboratory instrumentation systems and equipment. The dc voltage level required depends on the application, but most applications require relatively low voltages.

A basic block diagram of the complete power supply is shown in Figure 2–1(a). Generally the ac input line voltage is stepped down to a lower ac voltage with a transformer (although it may be stepped up when higher voltages are needed or there may be no transformer at all in rare instances). As you learned in your dc/ac course, a **transformer** changes ac voltages based on the turns ratio between the primary and secondary. If the secondary has more turns than the primary, the output voltage across the secondary will be higher and the current will be smaller. If the secondary has fewer turns than the primary, the output voltage across the secondary will be lower and the current will be higher. The rectifier can be either a half-wave rectifier or a full-wave rectifier (covered in Section 2–2). The **rectifier** converts the ac input voltage to a pulsating dc voltage, called a half-wave rectified voltage, as shown in Figure 2–1(b). The **filter** eliminates the fluctuations in the rectified voltage and produces a relatively smooth dc voltage. The power supply filter is covered in Section 2–3. The **regulator** is a circuit that maintains a constant dc voltage for variations in the input line voltage or in the load. Regulators vary from a single semiconductor device to more complex integrated circuits. The load is a circuit or device connected to the output of the power supply and operates from the power supply voltage and current.

TECH NOTE

The standard line voltage in North America is 120 V/240 V at 60 Hz. Most small appliances operate on 120 V and larger appliances such as dryers, ranges, and heaters operate on 240 V. Occasionally, you will see references to 110 V or 115 V, but the standard is 120 V. Some foreign countries do use 110 V or 115 V at either 60 Hz or 50 Hz.

The Half-Wave Rectifier

Figure 2–2 illustrates the process called *half-wave rectification.* A diode is connected to an ac source and to a load resistor, R_L, forming a **half-wave rectifier**. Keep in mind that all ground symbols represent the same point electrically. Let's examine what happens during one cycle of the input voltage using the ideal model for the diode. When the sinusoidal input voltage (V_{in}) goes positive, the diode is forward-biased and conducts current through

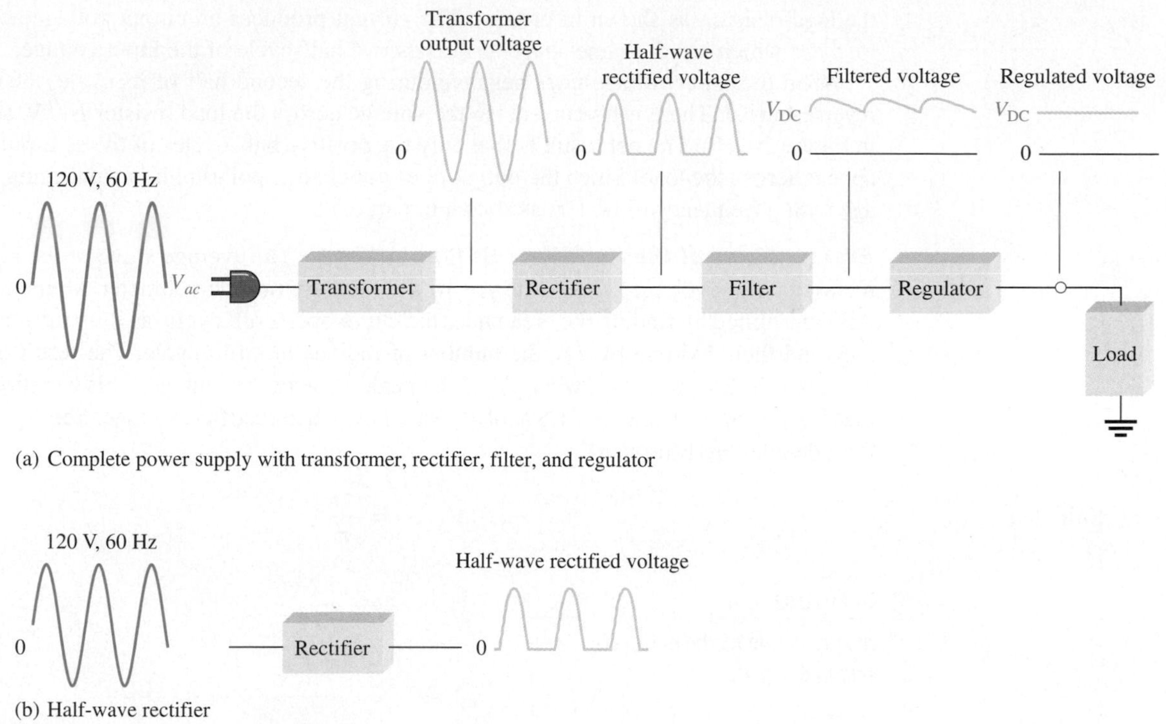

▲ FIGURE 2–1

Block diagram of a dc power supply with a load and a rectifier.

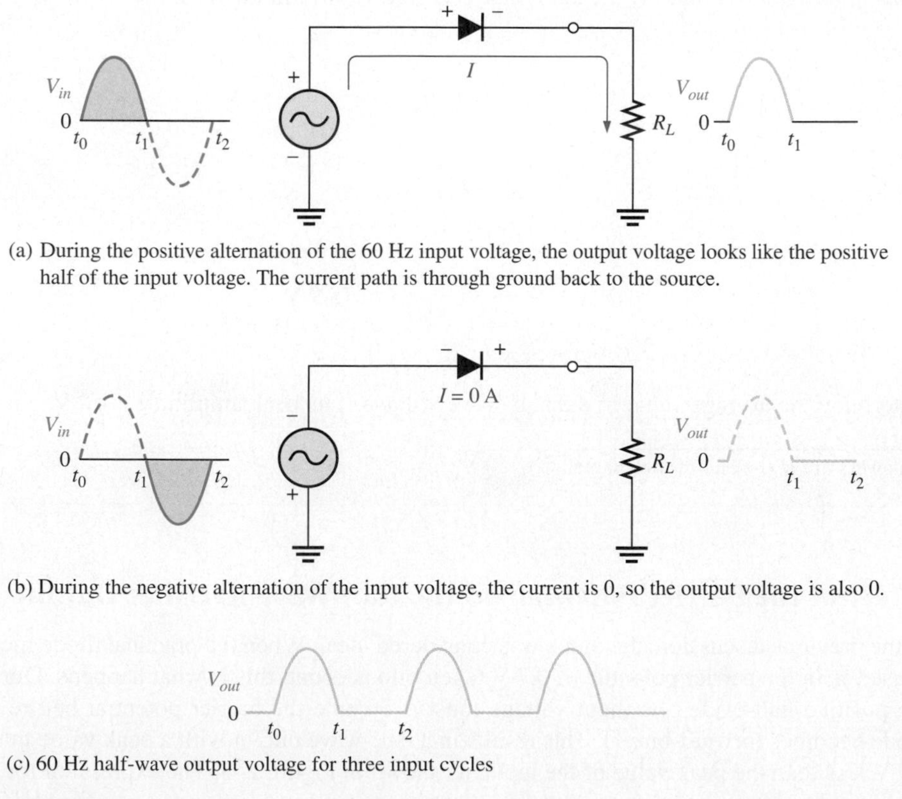

▲ FIGURE 2–2

Half-wave rectifier operation. The diode is considered to be ideal.

the load resistor, as shown in part (a). The current produces an output voltage across the load R_L, which has the same shape as the positive half-cycle of the input voltage.

When the input voltage goes negative during the second half of its cycle, the diode is reverse-biased. There is no current, so the voltage across the load resistor is 0 V, as shown in Figure 2–2(b). The net result is that only the positive half-cycles of the ac input voltage appear across the load. Since the output does not change polarity, it is a pulsating dc voltage with a frequency of 60 Hz, as shown in part (c).

Average Value of the Half-Wave Output Voltage The average value of the half-wave rectified output voltage is the value you would measure on a dc voltmeter. Mathematically, it is determined by finding the area under the curve over a full cycle, as illustrated in Figure 2–3, and then dividing by 2π, the number of radians in a full cycle. The result of this is expressed in Equation 2–1, where V_p is the peak value of the voltage. This equation shows that V_{AVG} is approximately 31.8% of V_p for a half-wave rectified voltage. See Appendix B for a detailed derivation.

Equation 2–1

$$V_{AVG} = \frac{V_p}{\pi}$$

FIGURE 2–3

Average value of the half-wave rectified signal.

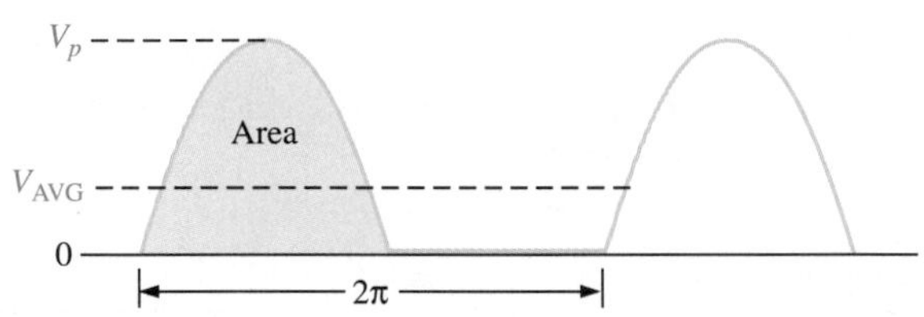

EXAMPLE 2–1

What is the average value of the half-wave rectified voltage in Figure 2–4?

FIGURE 2–4

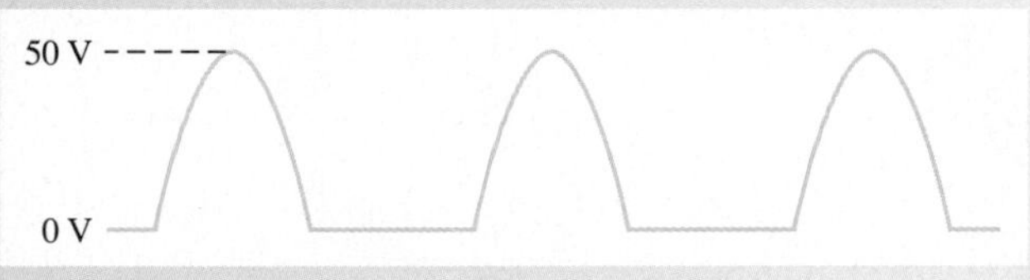

Solution

$$V_{AVG} = \frac{V_p}{\pi} = \frac{50\text{ V}}{\pi} = \mathbf{15.9\ V}$$

Notice that V_{AVG} is 31.8% of V_p.

*Related Problem** Determine the average value of the half-wave voltage if its peak amplitude is 12 V.

*Answers are at the end of the chapter.

Effect of the Barrier Potential on the Half-Wave Rectifier Output

In the previous discussion, the diode was considered ideal. When the practical diode model is used with the barrier potential of 0.7 V taken into account, this is what happens. During the positive half-cycle, the input voltage must overcome the barrier potential before the diode becomes forward-biased. This results in a half-wave output with a peak value that is 0.7 V less than the peak value of the input, as shown in Figure 2–5. The expression for the peak output voltage is

Equation 2–2

$$V_{p(out)} = V_{p(in)} - 0.7\text{ V}$$

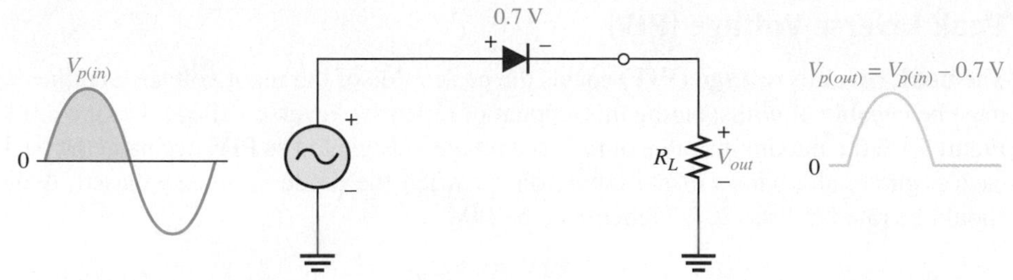

FIGURE 2–5

The effect of the barrier potential on the half-wave rectified output voltage is to reduce the peak value of the input by about 0.7 V.

It is usually acceptable to use the ideal diode model, which neglects the effect of the barrier potential, when the peak value of the applied voltage is much greater than the barrier potential (at least 10 V, as a rule of thumb). However, we will use the practical model of a diode, taking the 0.7 V barrier potential into account unless stated otherwise.

EXAMPLE 2–2

Draw the output voltages of each rectifier for the indicated input voltages, as shown in Figure 2–6. The 1N4001 and 1N4003 are specific rectifier diodes.

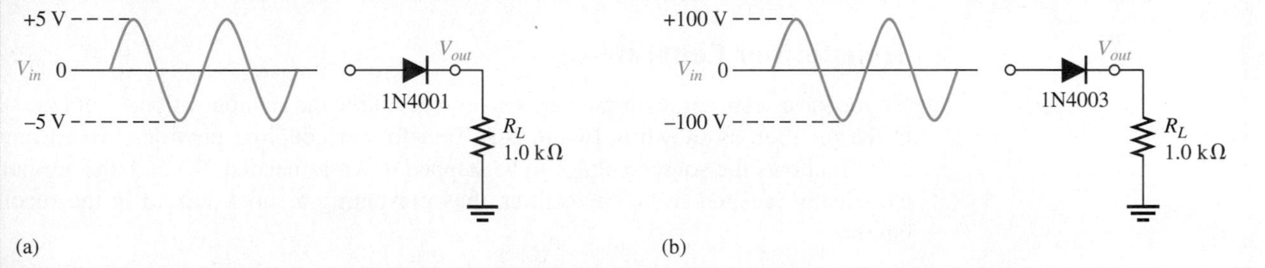

FIGURE 2–6

Solution The peak output voltage for circuit (a) is

$$V_{p(out)} = V_{p(in)} - 0.7\text{ V} = 5\text{ V} - 0.7\text{ V} = \mathbf{4.30\text{ V}}$$

The peak output voltage for circuit (b) is

$$V_{p(out)} = V_{p(in)} - 0.7\text{ V} = 100\text{ V} - 0.7\text{ V} = \mathbf{99.3\text{ V}}$$

The output voltage waveforms are shown in Figure 2–7. Note that the barrier potential could have been neglected in circuit (b) with very little error (0.7 percent); but, if it is neglected in circuit (a), a significant error results (14 percent).

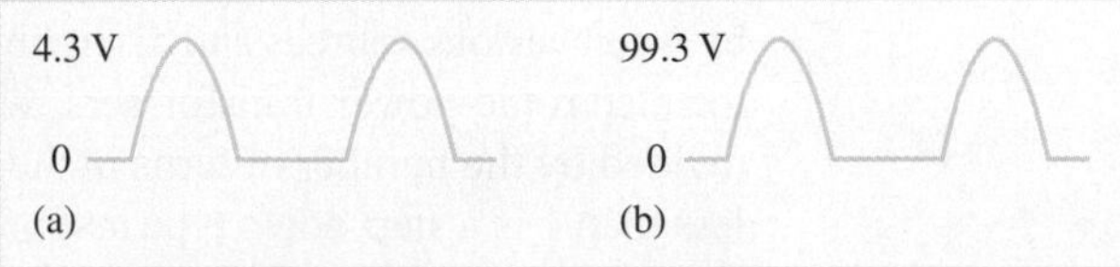

FIGURE 2–7

Output voltages for the circuits in Figure 2–6. They are not shown on the same scale.

Related Problem Determine the peak output voltages for the rectifiers in Figure 2–6 if the peak input in part (a) is 3 V and the peak input in part (b) is 50 V.

Open the Multisim file E02-02 in the Examples folder on your CD-ROM. For the inputs specified in the example, measure the resulting output voltage waveforms. Compare your measured results with those shown in the example.

Peak Inverse Voltage (PIV)

The **peak inverse voltage (PIV)** equals the peak value of the input voltage, and the diode must be capable of withstanding this amount of repetitive reverse voltage. For the diode in Figure 2–8, the maximum value of reverse voltage, designated as PIV, occurs at the peak of each negative alternation of the input voltage when the diode is reverse-biased. A diode should be rated at least 20% higher than the PIV.

Equation 2–3

$$\mathbf{PIV} = V_{p(in)}$$

FIGURE 2–8

The PIV occurs at the peak of each half-cycle of the input voltage when the diode is reverse-biased. In this circuit, the PIV occurs at the peak of each negative half-cycle.

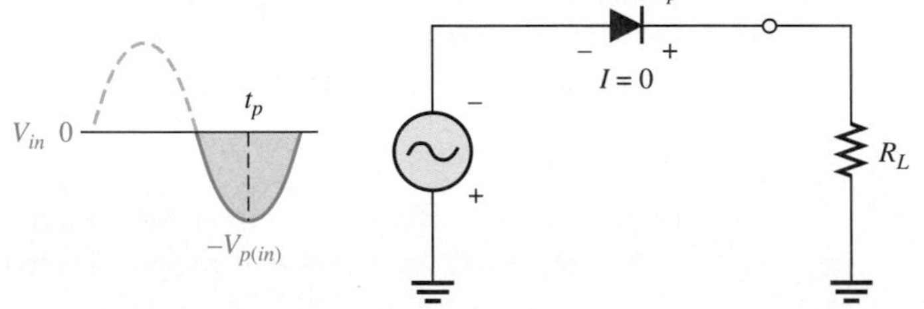

Transformer Coupling

As you have seen, a transformer is often used to couple the ac input voltage from the source to the rectifier, as shown in Figure 2–9. Transformer coupling provides two advantages. First, it allows the source voltage to be stepped down as needed. Second, the ac source is electrically isolated from the rectifier, thus preventing a shock hazard in the secondary circuit.

FIGURE 2–9

Half-wave rectifier with transformer-coupled input voltage.

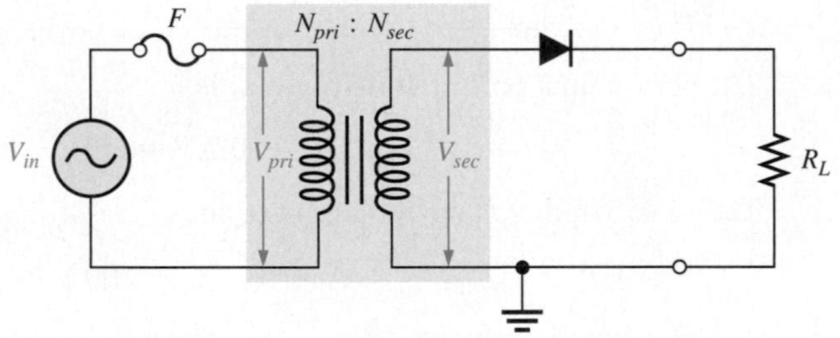

The amount that the voltage is stepped down is determined by the **turns ratio** of the transformer. Unfortunately, the definition of turns ratio for transformers is not consistent between various sources and disciplines. In this text, we use the definition given by the IEEE for electronic power transformers, which is "the number of turns in the secondary (N_{sec}) divided by the number of turns in the primary (N_{pri})." Thus, a transformer with a turns ratio less than 1 is a step-down type and one with a turns ratio greater than 1 is a step-up type. To show the turns ratio on a schematic, it is common practice to show the numerical ratio directly above the windings.

The secondary voltage of a transformer equals the turns ratio, n, times the primary voltage.

$$V_{sec} = nV_{pri}$$

If $n > 1$, the secondary voltage is greater than the primary voltage. If $n < 1$, the secondary voltage is less than the primary voltage. If $n = 1$, then $V_{sec} = V_{pri}$.

The peak secondary voltage, $V_{p(sec)}$, in a transformer-coupled half-wave rectifier is the same as $V_{p(in)}$ in Equation 2–2. Therefore, Equation 2–2 written in terms of $V_{p(sec)}$ is

$$V_{p(out)} = V_{p(sec)} - 0.7\text{ V}$$

and Equation 2–3 in terms of $V_{p(sec)}$ is

$$PIV = V_{p(sec)}$$

Turns ratio is useful for understanding the voltage transfer from primary to secondary. However, transformer datasheets rarely show the turns ratio. A transformer is generally specified based on the secondary voltage rather than the turns ratio.

EXAMPLE 2–3

Determine the peak value of the output voltage for Figure 2–10 if the turns ratio is 0.5.

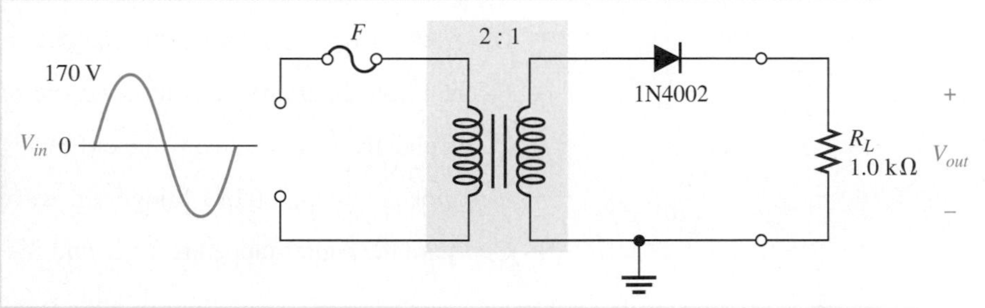

▲ FIGURE 2–10

Solution

$$V_{p(pri)} = V_{p(in)} = 170 \text{ V}$$

The peak secondary voltage is

$$V_{p(sec)} = nV_{p(pri)} = 0.5(170 \text{ V}) = 85 \text{ V}$$

The rectified peak output voltage is

$$V_{p(out)} = V_{p(sec)} - 0.7 \text{ V} = 85 \text{ V} - 0.7 \text{ V} = \mathbf{84.3\ V}$$

where $V_{p(sec)}$ is the input to the rectifier.

Related Problem

(a) Determine the peak value of the output voltage for Figure 2–10 if $n = 2$ and $V_{p(in)} = 312$ V.

(b) What is the PIV across the diode?

(c) Describe the output voltage if the diode is turned around.

Open the Multisim file E02-03 in the Examples folder on your CD-ROM. For the specified input, measure the peak output voltage. Compare your measured result with the calculated value.

SECTION 2–1 CHECKUP

Answers are at the end of the chapter.

1. At what point on the input cycle does the PIV occur?
2. For a half-wave rectifier, there is current through the load for approximately what percentage of the input cycle?
3. What is the average of a half-wave rectified voltage with a peak value of 10 V?
4. What is the peak value of the output voltage of a half-wave rectifier with a peak sine wave input of 25 V?
5. What PIV rating must a diode have to be used in a rectifier with a peak output voltage of 50 V?

2–2 FULL-WAVE RECTIFIERS

Although half-wave rectifiers have some applications, the full-wave rectifier is the most commonly used type in dc power supplies. In this section, you will use what you learned about half-wave rectification and expand it to full-wave rectifiers. You will learn about two types of full-wave rectifiers: center-tapped and bridge.

After completing this section, you should be able to

- **Explain and analyze the operation of full-wave rectifiers**
  - Discuss how full-wave rectification differs from half-wave rectification
  - Determine the average value of a full-wave rectified voltage
  - Describe the operation of a center-tapped full-wave rectifier
  - Explain how the transformer turns ratio affects the rectified output voltage
  - Determine the peak inverse voltage (PIV)
  - Describe the operation of a bridge full-wave rectifier
  - Compare the center-tapped rectifier and the bridge rectifier

A **full-wave rectifier** allows unidirectional (one-way) current through the load during the entire 360° of the input cycle, whereas a half-wave rectifier allows current through the load only during one-half of the cycle. The result of full-wave rectification is an output voltage with a frequency twice the input frequency and that pulsates every half-cycle of the input, as shown in Figure 2–11.

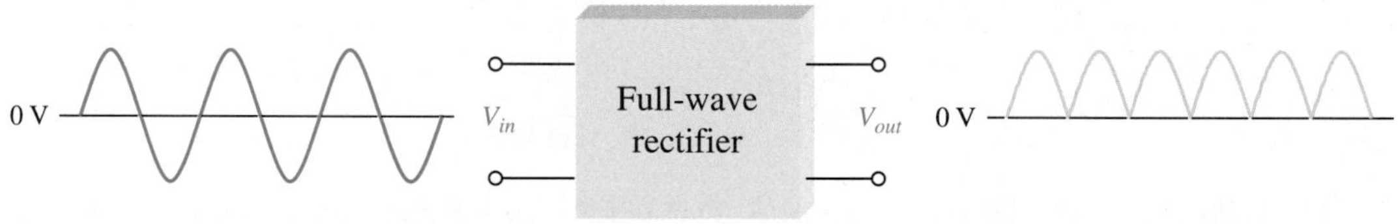

▲ FIGURE 2–11

Full-wave rectification.

The number of positive alternations that make up the full-wave rectified voltage is twice that of the half-wave voltage for the same time interval. The average value, which is the value measured on a dc voltmeter, for a full-wave rectified sinusoidal voltage is twice that of the half-wave, as shown in the following formula:

Equation 2–4

$$V_{AVG} = \frac{2V_p}{\pi}$$

V_{AVG} is approximately 63.7% of V_p for a full-wave rectified voltage.

EXAMPLE 2–4

Find the average value of the full-wave rectified voltage in Figure 2–12.

FIGURE 2–12

Solution

$$V_{AVG} = \frac{2V_p}{\pi} = \frac{2(15\text{ V})}{\pi} = \mathbf{9.55\ V}$$

V_{AVG} is 63.7% of V_p.

Related Problem Find the average value of the full-wave rectified voltage if its peak is 155 V.

The Center-Tapped Full-Wave Rectifier

A **center-tapped rectifier** is a type of full-wave rectifier that uses two diodes connected to the secondary of a center-tapped transformer, as shown in Figure 2–13. The input voltage is coupled through the transformer to the center-tapped secondary. Half of the total secondary voltage appears between the center tap and each end of the secondary winding as shown.

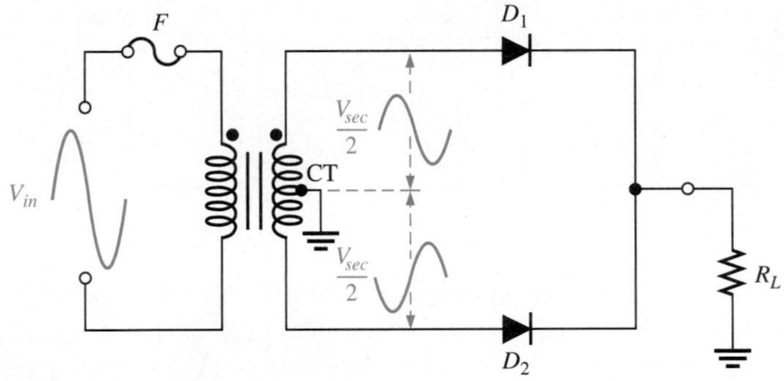

FIGURE 2–13
A center-tapped full-wave rectifier.

For a positive half-cycle of the input voltage, the polarities of the secondary voltages are as shown in Figure 2–14(a). This condition forward-biases diode D_1 and reverse-biases diode D_2. The current path is through D_1 and the load resistor R_L, as indicated. For a negative half-cycle of the input voltage, the voltage polarities on the secondary are as shown in

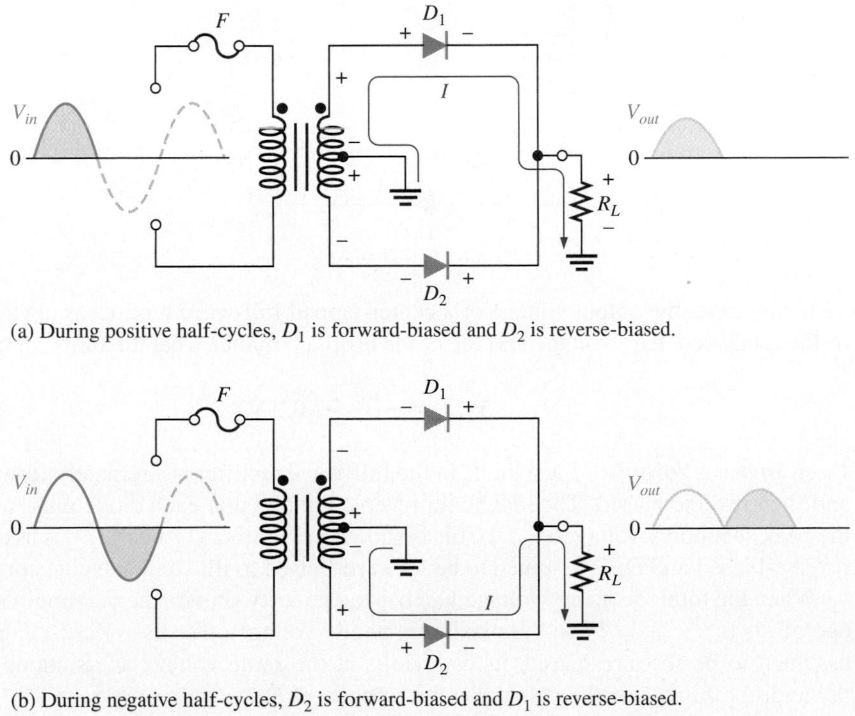

FIGURE 2–14
Basic operation of a center-tapped full-wave rectifier. Note that the current through the load resistor is in the same direction during the entire input cycle, so the output voltage always has the same polarity.

Figure 2–14(b). This condition reverse-biases D_1 and forward-biases D_2. The current path is through D_2 and R_L, as indicated. Because the output current during both the positive and negative portions of the input cycle is in the same direction through the load, the output voltage developed across the load resistor is a full-wave rectified dc voltage, as shown.

Effect of the Turns Ratio on the Output Voltage If the transformer's turns ratio is 1, the peak value of the rectified output voltage equals half the peak value of the primary input voltage less the barrier potential, as illustrated in Figure 2–15. Half of the primary voltage appears across each half of the secondary winding ($V_{p(sec)} = V_{p(pri)}$). We will begin referring to the forward voltage due to the barrier potential as the **diode drop.**

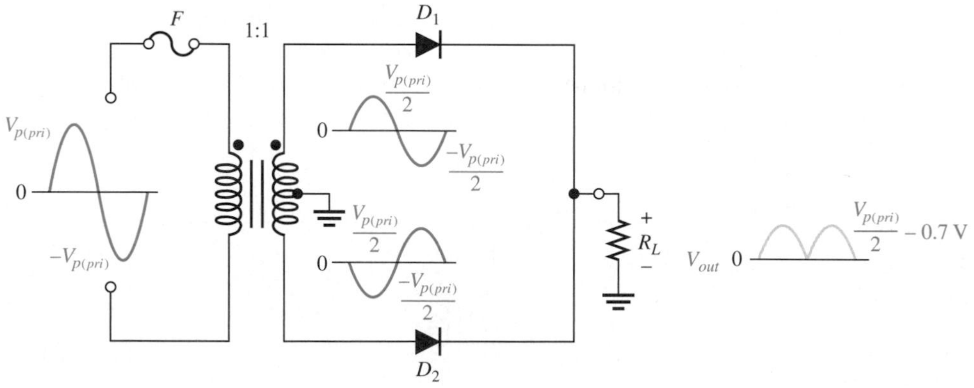

▶ FIGURE 2–15

Center-tapped full-wave rectifier with a transformer turns ratio of 1. $V_{p(pri)}$ is the peak value of the primary voltage.

In order to obtain an output voltage with a peak equal to the input peak (less the diode drop), a step-up transformer with a turns ratio of $n = 2$ must be used, as shown in Figure 2–16. In this case, the total secondary voltage (V_{sec}) is twice the primary voltage ($2V_{pri}$), so the voltage across each half of the secondary is equal to V_{pri}.

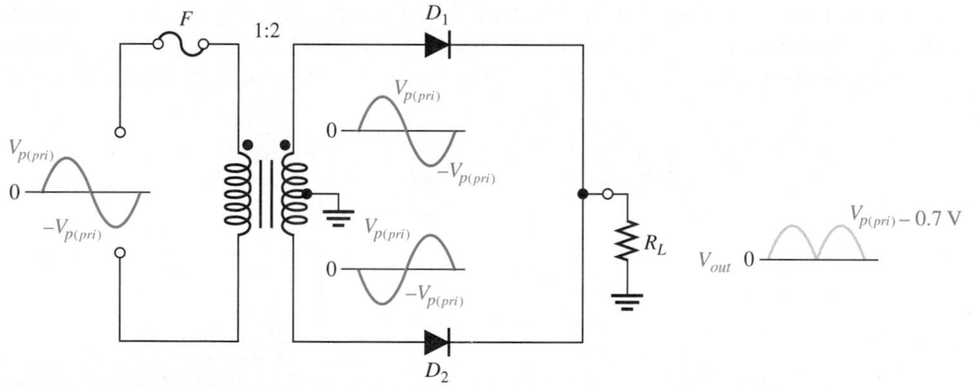

▶ FIGURE 2–16

Center-tapped full-wave rectifier with a transformer turns ratio of 2.

In any case, the output voltage of a center-tapped full-wave rectifier is always one-half of the total secondary voltage less the diode drop, no matter what the turns ratio.

Equation 2–5

$$\mathbf{V_{out} = \frac{V_{sec}}{2} - 0.7\ V}$$

Peak Inverse Voltage Each diode in the full-wave rectifier is alternately forward-biased and then reverse-biased. The maximum reverse voltage that each diode must withstand is the peak secondary voltage $V_{p(sec)}$. This is shown in Figure 2–17 where D_2 is assumed to be reverse-biased and D_1 is assumed to be forward-biased to illustrate the concept.

When the total secondary voltage V_{sec} has the polarity shown, the maximum anode voltage of D_1 is $+V_{p(sec)}/2$ and the maximum anode voltage of D_2 is $-V_{p(sec)}/2$. Since D_1 is assumed to be forward-biased, its cathode is at the same voltage as its anode minus the diode drop; this is also the voltage on the cathode of D_2.

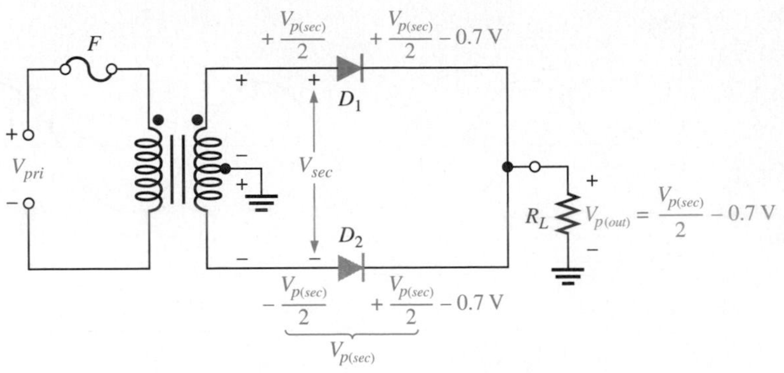

FIGURE 2–17

Diode reverse voltage (D_2 shown reverse-biased and D_1 shown forward-biased).

The peak inverse voltage across D_2 is

$$\text{PIV} = \left(\frac{V_{p(sec)}}{2} - 0.7\text{ V}\right) - \left(-\frac{V_{p(sec)}}{2}\right) = \frac{V_{p(sec)}}{2} + \frac{V_{p(sec)}}{2} - 0.7\text{ V}$$

$$= V_{p(sec)} - 0.7\text{ V}$$

Since $V_{p(out)} = V_{p(sec)}/2 - 0.7$ V, then by multiplying each term by 2 and transposing,

$$V_{p(sec)} = 2V_{p(out)} + 1.4\text{ V}$$

Therefore, by substitution, the peak inverse voltage across either diode in a full-wave center-tapped rectifier is

$$\mathbf{PIV = 2V_{p(out)} + 0.7\ V}$$ **Equation 2–6**

EXAMPLE 2–5

(a) Show the voltage waveforms across each half of the secondary winding and across R_L when a 100 V peak sine wave is applied to the primary winding in Figure 2–18.

(b) What minimum PIV rating must the diodes have?

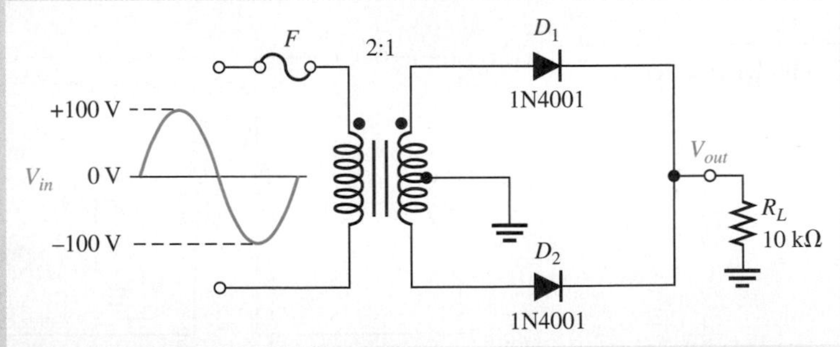

FIGURE 2–18

Solution

(a) The transformer turns ratio $n = 0.5$. The total peak secondary voltage is

$$V_{p(sec)} = nV_{p(pri)} = 0.5(100\text{ V}) = 50\text{ V}$$

There is a 25 V peak across each half of the secondary with respect to ground. The output load voltage has a peak value of 25 V, less the 0.7 V drop across the diode. The waveforms are shown in Figure 2–19.

(b) Each diode must have a minimum PIV rating of

$$\text{PIV} = 2V_{p(out)} + 0.7\text{ V} = 2(24.3\text{ V}) + 0.7\text{ V} = \mathbf{49.3\ V}$$

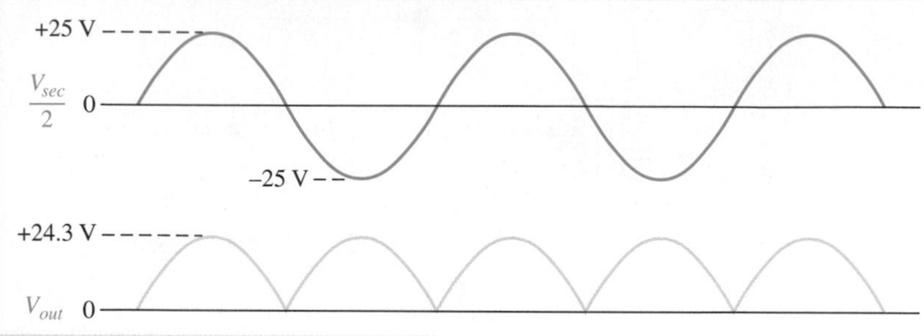

FIGURE 2–19

Related Problem What diode PIV rating is required to handle a peak input of 160 V in Figure 2–18?

Open the Multisim file E02-05 in the Examples folder on your CD-ROM. For the specified input voltage, measure the voltage waveforms across each half of the secondary and across the load resistor. Compare with the results shown in the example.

The Bridge Full-Wave Rectifier

The **bridge rectifier** uses four diodes connected as shown in Figure 2–20. When the input cycle is positive as in part (a), diodes D_1 and D_2 are forward-biased and conduct current in the direction shown. A voltage is developed across R_L that looks like the positive half of the input cycle. During this time, diodes D_3 and D_4 are reverse-biased.

FIGURE 2–20

Operation of a bridge rectifier.

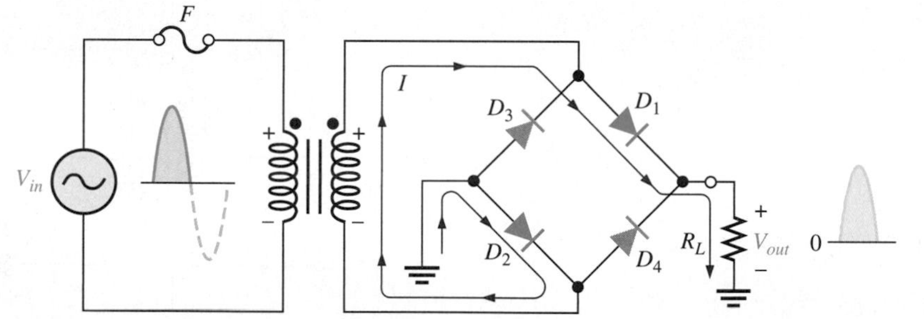

(a) During the positive half-cycle of the input, D_1 and D_2 are forward-biased and conduct current. D_3 and D_4 are reverse-biased.

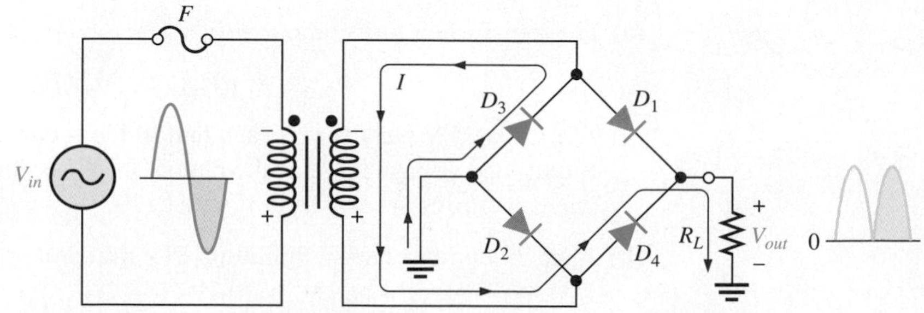

(b) During the negative half-cycle of the input, D_3 and D_4 are forward-biased and conduct current. D_1 and D_2 are reverse-biased.

When the input cycle is negative as in Figure 2–20(b), diodes D_3 and D_4 are forward-biased and conduct current in the same direction through R_L as during the positive half-cycle. During the negative half-cycle, D_1 and D_2 are reverse-biased. A full-wave rectified output voltage appears across R_L as a result of this action.

Bridge Output Voltage A bridge rectifier with a transformer-coupled input is shown in Figure 2–21(a). During the positive half-cycle of the total secondary voltage, diodes D_1 and D_2 are forward-biased. Neglecting the diode drops, the secondary voltage appears across the load resistor. The same is true when D_3 and D_4 are forward-biased during the negative half-cycle.

$$V_{p(out)} = V_{p(sec)}$$

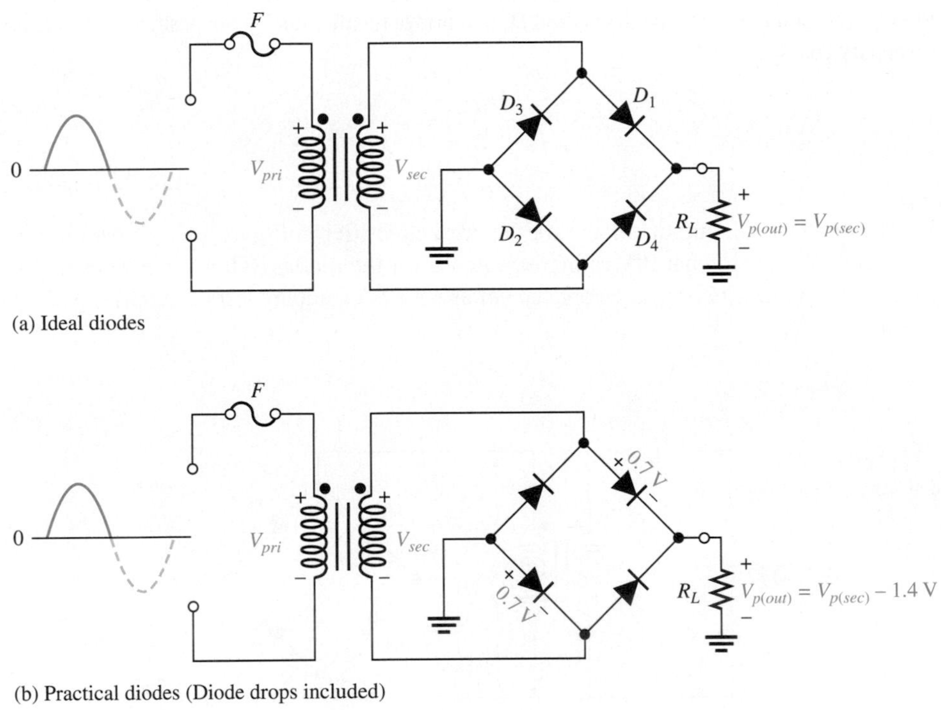

FIGURE 2–21

Bridge operation during a positive half-cycle of the primary and secondary voltages.

As you can see in Figure 2–21(b), two diodes are always in series with the load resistor during both the positive and negative half-cycles. If these diode drops are taken into account, the output voltage is

$$\mathbf{V_{p(out)} = V_{p(sec)} - 1.4\,V}$$ **Equation 2–7**

Peak Inverse Voltage Let's assume that D_1 and D_2 are forward-biased and examine the reverse voltage across D_3 and D_4. Visualizing D_1 and D_2 as shorts (ideal model), as in Figure 2–22(a), you can see that D_3 and D_4 have a peak inverse voltage equal to the peak secondary voltage. Since the output voltage is *ideally* equal to the secondary voltage,

$$\text{PIV} = V_{p(out)}$$

If the diode drops of the forward-biased diodes are included as shown in Figure 2–22(b), the peak inverse voltage across each reverse-biased diode in terms of $V_{p(out)}$ is

$$\mathbf{PIV = V_{p(out)} + 0.7\,V}$$ **Equation 2–8**

The PIV rating of the bridge diodes is less than that required for the center-tapped configuration. If the diode drop is neglected, the bridge rectifier requires diodes with half the PIV rating of those in a center-tapped rectifier for the same output voltage.

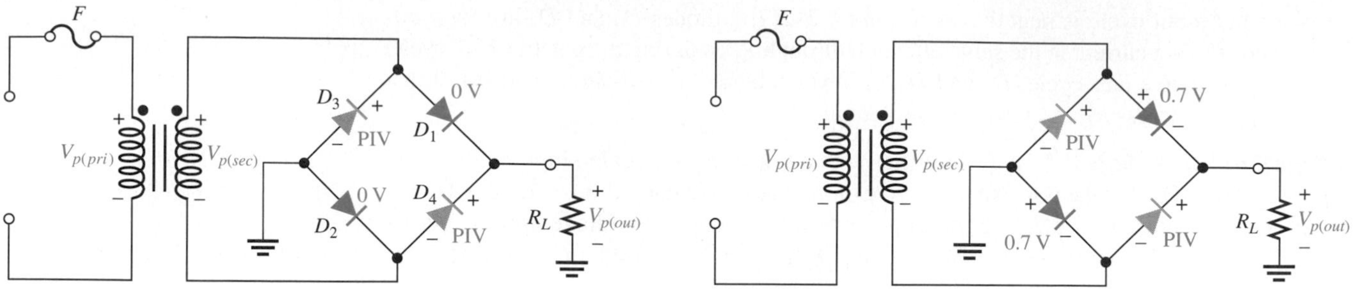

(a) For the ideal diode model (forward-biased diodes D_1 and D_2 are shown in green), PIV = $V_{p(out)}$.

(b) For the practical diode model (forward-biased diodes D_1 and D_2 are shown in green), PIV = $V_{p(out)}$ + 0.7 V.

▲ FIGURE 2–22

Peak inverse voltages across diodes D_3 and D_4 in a bridge rectifier during the positive half-cycle of the secondary voltage.

EXAMPLE 2–6

Determine the peak output voltage for the bridge rectifier in Figure 2–23. Assuming the practical model, what PIV rating is required for the diodes? The transformer is specified to have a 12 V rms secondary voltage for the standard 120 V across the primary.

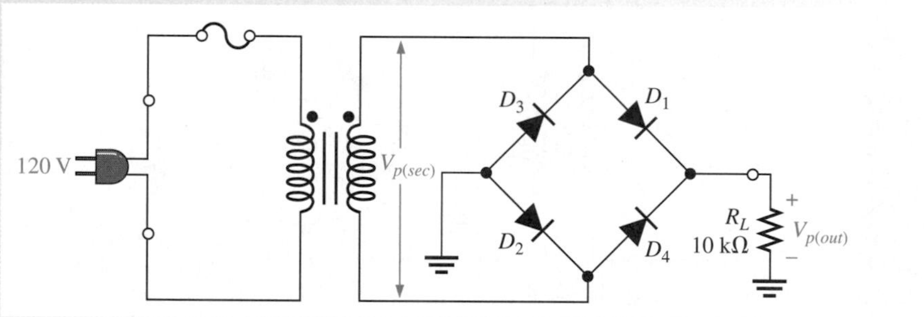

▲ FIGURE 2–23

Solution The peak output voltage (taking into account the two diode drops) is

$$V_{p(sec)} = 1.414V_{rms} = 1.414(12\text{ V}) \cong 17\text{ V}$$

$$V_{p(out)} = V_{p(sec)} - 1.4\text{ V} = 17\text{ V} - 1.4\text{ V} = \mathbf{15.6\text{ V}}$$

The PIV rating for each diode is

$$\text{PIV} = V_{p(out)} + 0.7\text{ V} = 15.6\text{ V} + 0.7\text{ V} = \mathbf{16.3\text{ V}}$$

Related Problem Determine the peak output voltage for the bridge rectifier in Figure 2–23 if the transformer produces an rms secondary voltage of 30 V. What is the PIV rating for the diodes?

Open the Multisim file E02-06 in the Examples folder on your CD-ROM. Measure the output voltage and compare to the calculated value.

SECTION 2–2 CHECKUP

1. How does a full-wave voltage differ from a half-wave voltage?
2. What is the average value of a full-wave rectified voltage with a peak value of 60 V?
3. Which type of full-wave rectifier has the greater output voltage for the same input voltage and transformer turns ratio?
4. For a peak output voltage of 45 V, in which type of rectifier would you use diodes with a PIV rating of 50 V?
5. What PIV rating is required for diodes used in the type of rectifier that was not selected in Question 4?

2–3 POWER SUPPLY FILTERS AND REGULATORS

A power supply filter ideally eliminates the fluctuations in the output voltage of a half-wave or full-wave rectifier and produces a constant-level dc voltage. Filtering is necessary because electronic circuits require a constant source of dc voltage and current to provide power and biasing for proper operation. Filters are implemented with capacitors, as you will see in this section. Voltage regulation in power supplies is usually done with integrated circuit voltage regulators. A voltage regulator prevents changes in the filtered dc voltage due to variations in input voltage or load.

After completing this section, you should be able to

- **Explain and analyze the operation and characteristics of power supply filters and regulators**
  - Explain the purpose of a filter
  - Describe the capacitor-input filter
  - Define *ripple voltage* and calculate the ripple factor
  - Discuss surge current in a capacitor-input filter
  - Discuss voltage regulation

In most power supply applications, the standard 60 Hz ac power line voltage must be converted to an approximately constant dc voltage. The 60 Hz pulsating dc output of a half-wave rectifier or the 120 Hz pulsating output of a full-wave rectifier must be filtered to reduce the large voltage variations. Figure 2–24 illustrates the filtering concept showing a

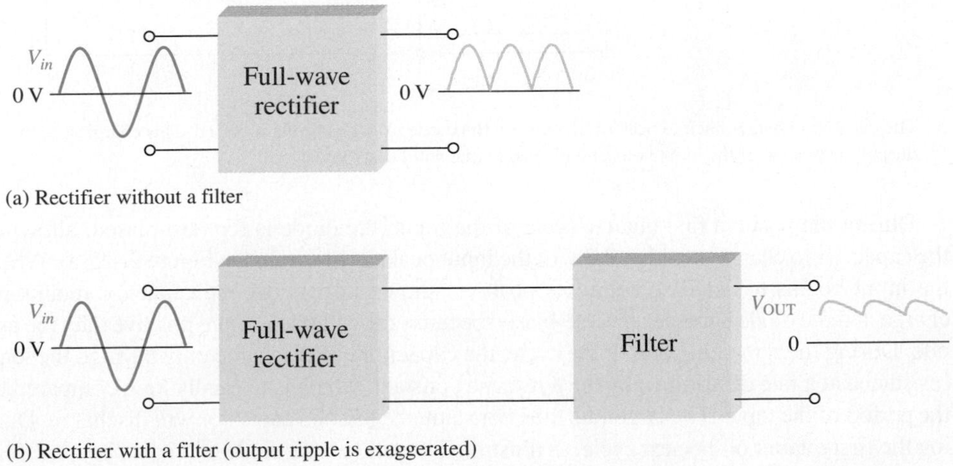

FIGURE 2–24
Power supply filtering.

SAFETY NOTE

When installing polarized capacitors in a circuit, be sure to observe the proper polarity. The positive lead always connects to the more positive side of the circuit. An incorrectly connected polarized capacitor can explode.

nearly smooth dc output voltage from the filter. The small amount of fluctuation in the filter output voltage is called *ripple*.

Capacitor-Input Filter

A half-wave rectifier with a capacitor-input filter is shown in Figure 2–25. The filter is simply a capacitor connected from the rectifier output to ground. R_L represents the equivalent resistance of a load. We will use the half-wave rectifier to illustrate the basic principle and then expand the concept to full-wave rectification.

FIGURE 2–25

Operation of a half-wave rectifier with a capacitor-input filter. The current indicates charging or discharging of the capacitor.

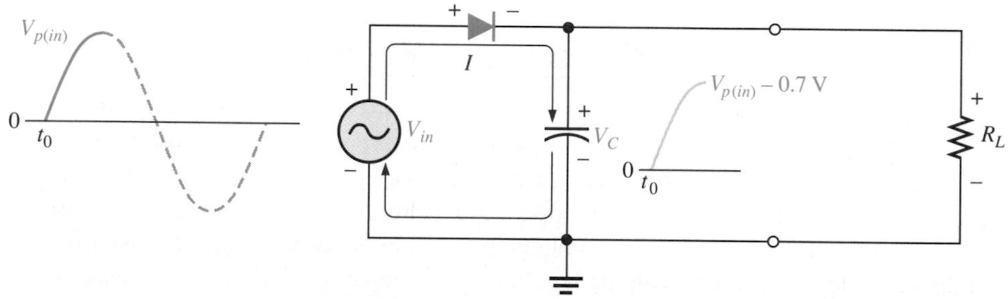

(a) Initial charging of the capacitor (diode is forward-biased) happens only once when power is turned on.

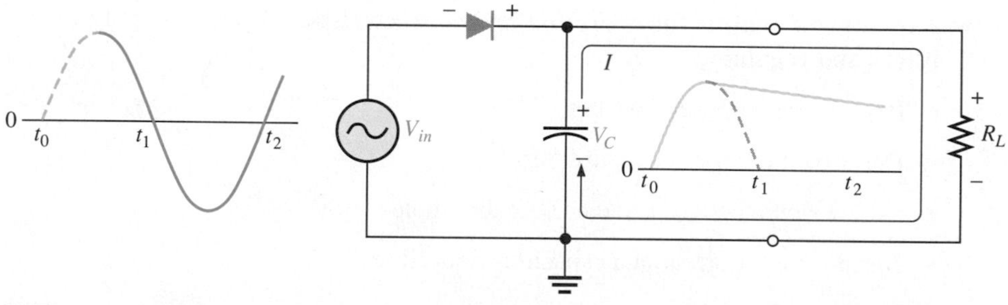

(b) The capacitor discharges through R_L after peak of positive alternation when the diode is reverse-biased. This discharging occurs during the portion of the input voltage indicated by the solid dark blue curve.

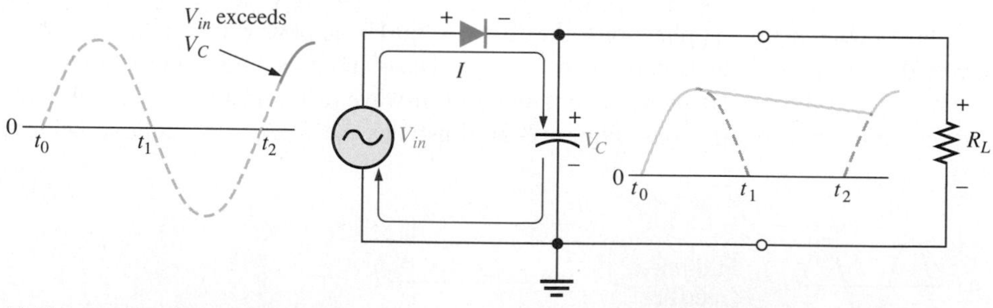

(c) The capacitor charges back to peak of input when the diode becomes forward-biased. This charging occurs during the portion of the input voltage indicated by the solid dark blue curve.

During the positive first quarter-cycle of the input, the diode is forward-biased, allowing the capacitor to charge to within 0.7 V of the input peak, as illustrated in Figure 2–25(a). When the input begins to decrease below its peak, as shown in part (b), the capacitor retains its charge and the diode becomes reverse-biased because the cathode is more positive than the anode. During the remaining part of the cycle, the capacitor can discharge only through the load resistance at a rate determined by the R_LC time constant, which is normally long compared to the period of the input. The larger the time constant, the less the capacitor will discharge. During the first quarter of the next cycle, as illustrated in part (c), the diode will again become forward-biased when the input voltage exceeds the capacitor voltage by approximately 0.7 V.

Ripple Voltage As you have seen, the capacitor quickly charges at the beginning of a cycle and slowly discharges through R_L after the positive peak of the input voltage (when the diode is reverse-biased). The variation in the capacitor voltage due to the charging and discharging is called the **ripple voltage**. Generally, ripple is undesirable; thus, the smaller the ripple, the better the filtering action, as illustrated in Figure 2–26.

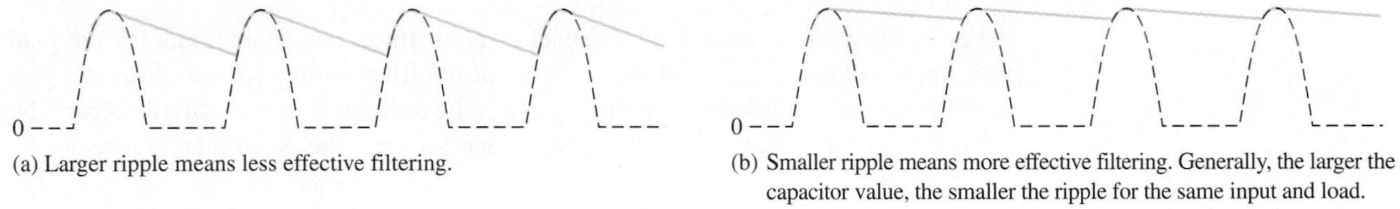

▲ **FIGURE 2–26**

Half-wave ripple voltage (blue line).

For a given input frequency, the output frequency of a full-wave rectifier is twice that of a half-wave rectifier, as illustrated in Figure 2–27. This makes a full-wave rectifier easier to filter because of the shorter time between peaks. When filtered, the full-wave rectified voltage has a smaller ripple than does a half-wave voltage for the same load resistance and capacitor values. The capacitor discharges less during the shorter interval between full-wave pulses, as shown in Figure 2–28.

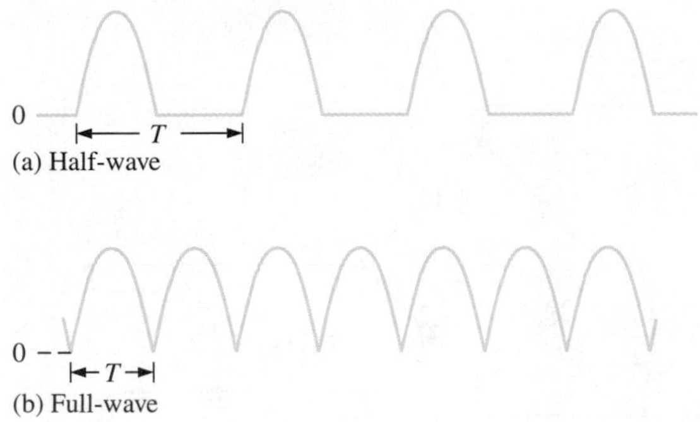

◀ **FIGURE 2–27**

The period of a full-wave rectified voltage is half that of a half-wave rectified voltage. The output frequency of a full-wave rectifier is twice that of a half-wave rectifier.

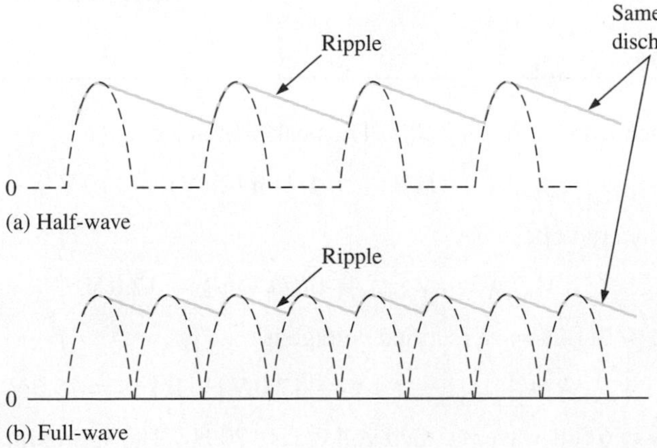

◀ **FIGURE 2–28**

Comparison of ripple voltages for half-wave and full-wave rectified voltages with the same filter capacitor and load and derived from the same sinusoidal input voltage.

Ripple Factor The **ripple factor** (*r*) is an indication of the effectiveness of the filter and is defined as

$$r = \frac{V_{r(pp)}}{V_{DC}}$$

Equation 2–9

where $V_{r(pp)}$ is the peak-to-peak ripple voltage and V_{DC} is the dc (average) value of the filter's output voltage, as illustrated in Figure 2–29. The lower the ripple factor, the better the filter. The ripple factor can be lowered by increasing the value of the filter capacitor or increasing the load resistance.

▶ FIGURE 2–29

V_r and V_{DC} determine the ripple factor.

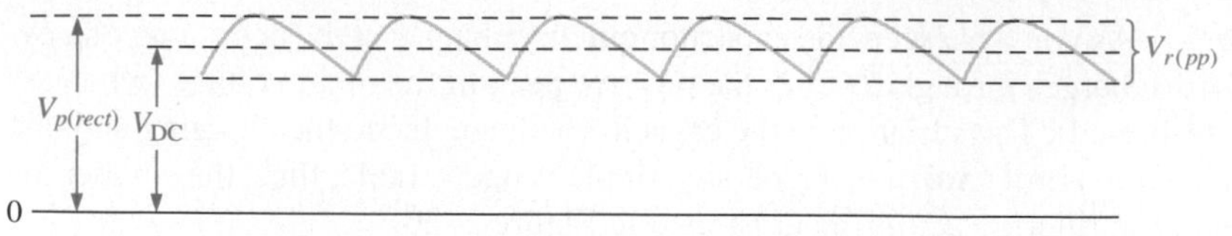

For a full-wave rectifier with a capacitor-input filter, approximations for the peak-to-peak ripple voltage, $V_{r(pp)}$, and the dc value of the filter output voltage, V_{DC}, are given in the following equations. The variable $V_{p(rect)}$ is the unfiltered peak rectified voltage. Notice that if R_L or C increases, the ripple voltage decreases and the dc voltage increases.

Equation 2–10

$$V_{r(pp)} \cong \left(\frac{1}{fR_LC}\right)V_{p(rect)}$$

Equation 2–11

$$V_{DC} \cong \left(1 - \frac{1}{2fR_LC}\right)V_{p(rect)}$$

These equations are derived in Appendix B.

EXAMPLE 2–7

Determine the ripple factor for the filtered bridge rectifier with a load as indicated in Figure 2–30.

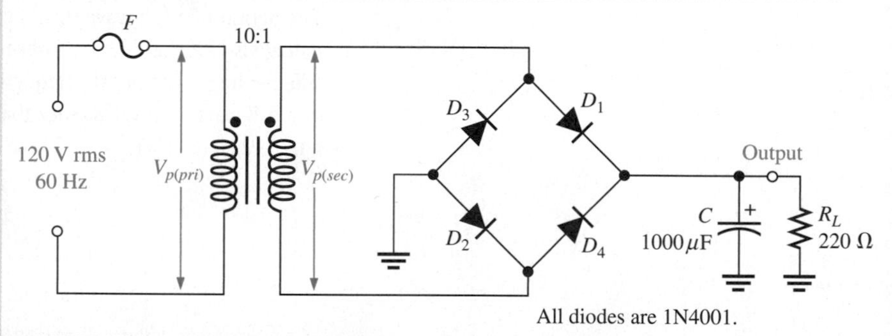

▲ FIGURE 2–30

Solution The transformer turns ratio is $n = 0.1$. The peak primary voltage is

$$V_{p(pri)} = 1.414V_{rms} = 1.414(120\text{ V}) = 170\text{ V}$$

The peak secondary voltage is

$$V_{p(sec)} = nV_{p(pri)} = 0.1(170\text{ V}) = 17.0\text{ V}$$

The unfiltered peak full-wave rectified voltage is

$$V_{p(rect)} = V_{p(sec)} - 1.4\text{ V} = 17.0\text{ V} - 1.4\text{ V} = 15.6\text{ V}$$

The frequency of a full-wave rectified voltage is 120 Hz. The approximate peak-to-peak ripple voltage at the output is

$$V_{r(pp)} \cong \left(\frac{1}{fR_LC}\right)V_{p(rect)} = \left(\frac{1}{(120\text{ Hz})(220\ \Omega)(1000\ \mu\text{F})}\right)15.6\text{ V} = 0.591\text{ V}$$

The approximate dc value of the output voltage is determined as follows:

$$V_{DC} = \left(1 - \frac{1}{2fR_LC}\right)V_{p(rect)} = \left(1 - \frac{1}{(240\text{ Hz})(220\ \Omega)(1000\ \mu\text{F})}\right)15.6\text{ V} = 15.3\text{ V}$$

The resulting ripple factor is

$$r = \frac{V_{r(pp)}}{V_{DC}} = \frac{0.591\ \text{V}}{15.3\ \text{V}} = \mathbf{0.039}$$

The percent ripple is 3.9%.

Related Problem Determine the peak-to-peak ripple voltage if the filter capacitor in Figure 2–30 is increased to 2200 μF and the load resistance changes to 2.2 kΩ.

Open the Multisim file E02-07 in the Examples folder on your CD-ROM. For the specified input voltage, measure the peak-to-peak ripple voltage and the dc value at the output. Do the results agree closely with the calculated values? If not, can you explain why?

Surge Current in the Capacitor-Input Filter Before the switch in Figure 2–31 is closed, the filter capacitor is uncharged. At the instant the switch is closed, voltage is connected to the bridge and the uncharged capacitor appears as a short, as shown. This produces an initial surge of current, I_{surge}, through the two forward-biased diodes D_1 and D_2. The worst-case situation occurs when the switch is closed at a peak of the secondary voltage and a maximum surge current, $I_{surge(max)}$, is produced, as illustrated in the figure.

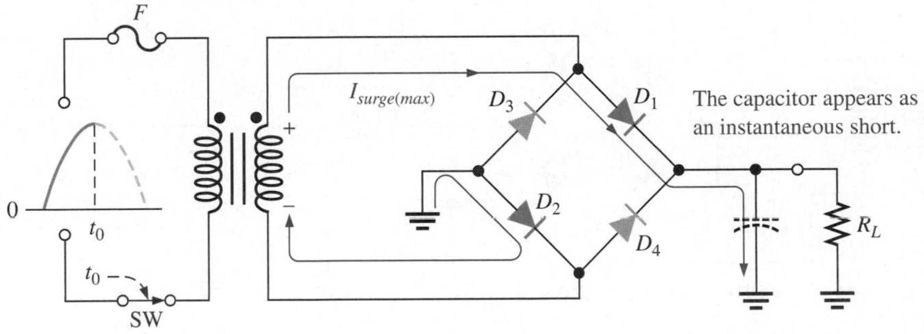

FIGURE 2–31

Surge current in a capacitor-input filter.

In dc power supplies, a **fuse** is always placed in the primary circuit of the transformer, as shown in Figure 2–31. A slow-blow type fuse is generally used because of the surge current that initially occurs when power is first turned on. The fuse rating is determined by calculating the power in the power supply load, which is the output power. Since $P_{in} = P_{out}$ in an ideal transformer, the primary current can be calculated as

$$I_{pri} = \frac{P_{in}}{120\ \text{V}}$$

The fuse rating should be at least 20% larger than the calculated value of I_{pri}.

Voltage Regulators

While filters can reduce the ripple from power supplies to a low value, the most effective approach is a combination of a capacitor-input filter used with a voltage regulator. A voltage regulator is connected to the output of a filtered rectifier and maintains a constant output voltage (or current) despite changes in the input, the load current, or the temperature. The capacitor-input filter reduces the input ripple to the regulator to an acceptable level. The combination of a large capacitor and a voltage regulator helps produce an excellent power supply.

Most regulators are integrated circuits and have three terminals—an input terminal, an output terminal, and a reference (or adjust) terminal. The input to the regulator is first filtered with a capacitor to reduce the ripple to <10%. The regulator reduces the ripple to a negligible amount. In addition, most regulators have an internal voltage reference, short-circuit protection, and thermal shutdown circuitry. They are available in a variety of voltages, including positive and negative outputs, and can be designed for variable outputs with a minimum of external components. Typically, voltage regulators can furnish a constant output of one or more amps of current with high ripple rejection.

Three-terminal regulators designed for fixed output voltages require only external capacitors to complete the regulation portion of the power supply, as shown in Figure 2–32. Filtering is accomplished by a large-value capacitor between the input voltage and ground. An output capacitor (typically 0.1 μF to 1.0 μF) is connected from the output to ground to improve the transient response.

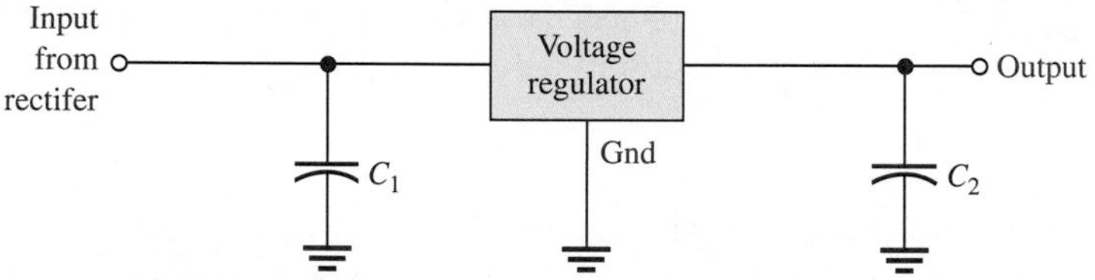

FIGURE 2–32

A voltage regulator with input and output capacitors.

A basic fixed power supply with a +5 V voltage regulator is shown in Figure 2–33. Specific integrated circuit three-terminal regulators with fixed output voltages are covered in Chapter 17.

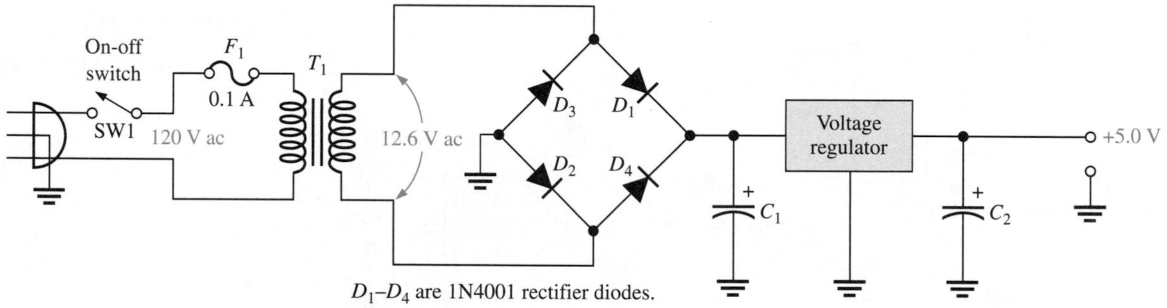

FIGURE 2–33

A basic +5.0 V regulated power supply.

Percent Regulation

The regulation expressed as a percentage is a figure of merit used to specify the performance of a voltage regulator. It can be in terms of input (line) regulation or load regulation. **Line regulation** specifies how much change occurs in the output voltage for a given change in the input voltage. It is typically defined as a ratio of a change in output voltage for a corresponding change in the input voltage expressed as a percentage.

Equation 2–12

$$\textbf{Line regulation} = \left(\frac{\Delta V_{OUT}}{\Delta V_{IN}}\right)100\%$$

Load regulation specifies how much change occurs in the output voltage over a certain range of load current values, usually from minimum current (no load, NL) to maximum current (full load, FL). It is normally expressed as a percentage and can be calculated with the following formula:

Equation 2–13

$$\textbf{Load regulation} = \left(\frac{V_{NL} - V_{FL}}{V_{FL}}\right)100\%$$

where V_{NL} is the output voltage with no load and V_{FL} is the output voltage with full (maximum) load.

EXAMPLE 2–8

A certain 7805 regulator has a measured no-load output voltage of 5.18 V and a full-load output of 5.15 V. What is the load regulation expressed as a percentage?

Solution

$$\text{Load regulation} = \left(\frac{V_{NL} - V_{FL}}{V_{FL}}\right)100\% = \left(\frac{5.18\text{ V} - 5.15\text{ V}}{5.15\text{ V}}\right)100\% = \mathbf{0.58\%}$$

Related Problem

If the no-load output voltage of a regulator is 24.8 V and the full-load output is 23.9 V, what is the load regulation expressed as a percentage?

SECTION 2–3 CHECKUP

1. When a 60 Hz sinusoidal voltage is applied to the input of a half-wave rectifier, what is the output frequency?
2. When a 60 Hz sinusoidal voltage is applied to the input of a full-wave rectifier, what is the output frequency?
3. What causes the ripple voltage on the output of a capacitor-input filter?
4. If the load resistance connected to a filtered power supply is decreased, what happens to the ripple voltage?
5. Define *ripple factor.*
6. What is the difference between input (line) regulation and load regulation?

2–4 DIODE LIMITING AND CLAMPING CIRCUITS

Diode circuits, called limiters or clippers, are sometimes used to clip off portions of signal voltages above or below certain levels. Another type of diode circuit, called a clamper, is used to add or restore a dc level to an electrical signal. Both limiter and clamper diode circuits will be examined in this section.

After completing this section, you should be able to

- **Explain and analyze the operation of diode limiting and clamping circuits**
  - Explain the operation of diode limiters
  - Determine the output voltage of a biased limiter
  - Use voltage-divider bias to set the limiting level
  - Explain the operation of diode clampers

Diode Limiters

Figure 2–34(a) shows a diode positive **limiter** (also called **clipper**) that limits or clips the positive part of the input voltage. As the input voltage goes positive, the diode becomes forward-biased and conducts current. Point A is limited to $+0.7$ V when the input voltage exceeds this value. When the input voltage goes back below 0.7 V, the diode is reverse-biased and appears as an open. The output voltage looks like the negative part of the input voltage, but with a magnitude determined by the voltage divider formed by R_1 and the load resistor, R_L, as follows:

$$V_{out} = \left(\frac{R_L}{R_1 + R_L}\right)V_{in}$$

If R_1 is small compared to R_L, then $V_{out} \cong V_{in}$.

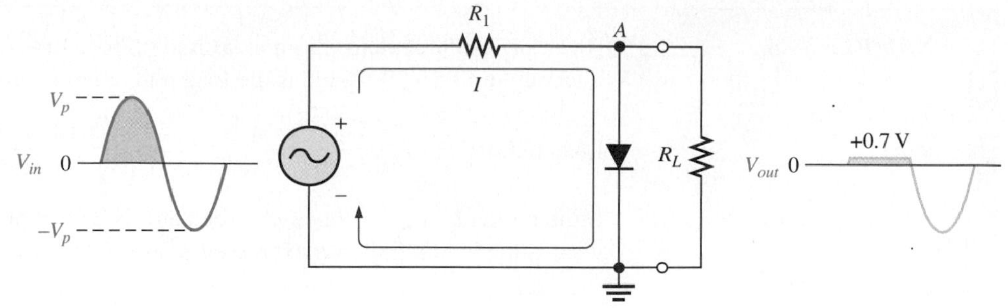

(a) Limiting of the positive alternation. The diode is forward-biased during the positive alternation (above 0.7 V) and reverse-biased during the negative alternation.

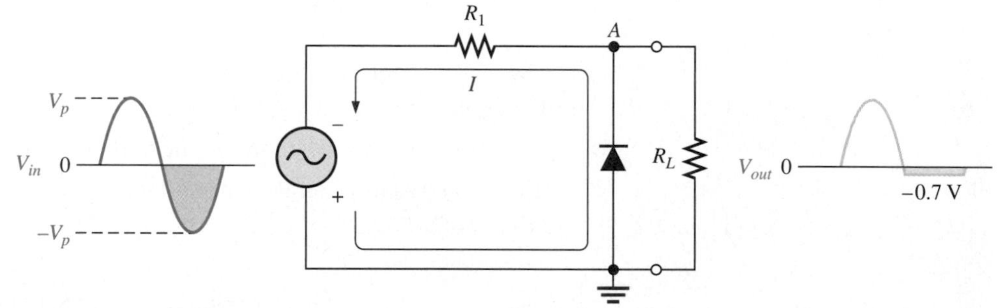

(b) Limiting of the negative alternation. The diode is forward-biased during the negative alternation (below −0.7 V) and reverse-biased during the positive alternation.

▲ FIGURE 2–34

Examples of diode limiters (clippers).

If the diode is turned around, as in Figure 2–34(b), the negative part of the input voltage is clipped off. When the diode is forward-biased during the negative part of the input voltage, point *A* is held at −0.7 V by the diode drop. When the input voltage goes above −0.7 V, the diode is no longer forward-biased; and a voltage appears across R_L proportional to the input voltage.

EXAMPLE 2–9

What would you expect to see displayed on an oscilloscope connected across R_L in the limiter shown in Figure 2–35?

FIGURE 2–35

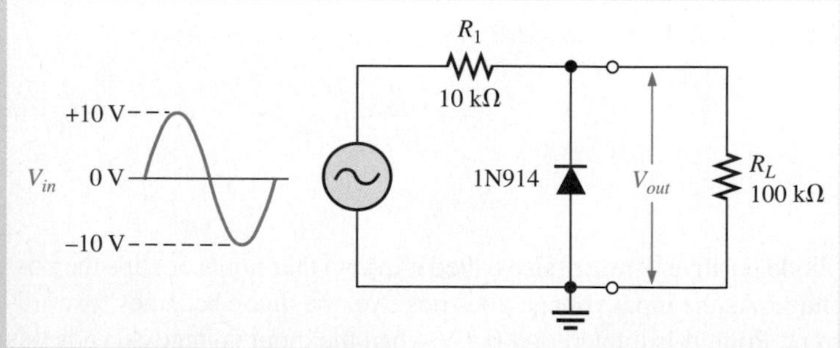

Solution The diode is forward-biased and conducts when the input voltage goes below −0.7 V. So, for the negative limiter, determine the peak output voltage across R_L by the following equation:

$$V_{p(out)} = \left(\frac{R_L}{R_1 + R_L}\right)V_{p(in)} = \left(\frac{100\text{ k}\Omega}{110\text{ k}\Omega}\right)10\text{ V} = 9.09\text{ V}$$

The scope will display an output waveform as shown in Figure 2–36.

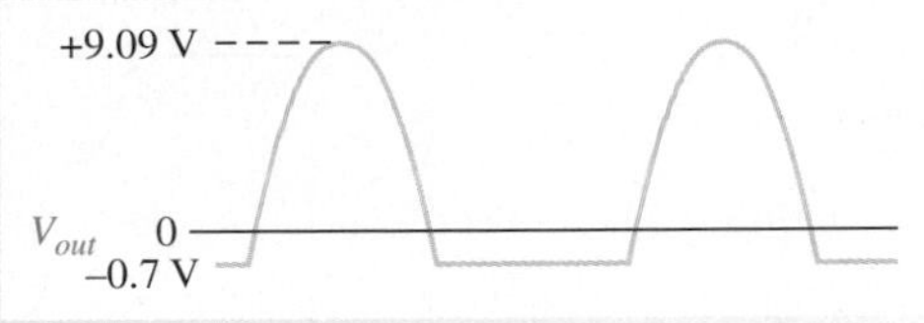

FIGURE 2–36

Output voltage waveform for Figure 2–35.

Related Problem Describe the output waveform for Figure 2–35 if R_1 is changed to 1 kΩ.

Open the Multisim file E02-09 in the Examples folder on your CD-ROM. For the specified input, measure the resulting output waveform. Compare with the waveform shown in the example.

Biased Limiters The level to which an ac voltage is limited can be adjusted by adding a bias voltage, V_{BIAS}, in series with the diode, as shown in Figure 2–37. The voltage at point *A* must equal $V_{BIAS} + 0.7$ V before the diode will become forward-biased and conduct. Once the diode begins to conduct, the voltage at point *A* is limited to $V_{BIAS} + 0.7$ V so that all input voltage above this level is clipped off.

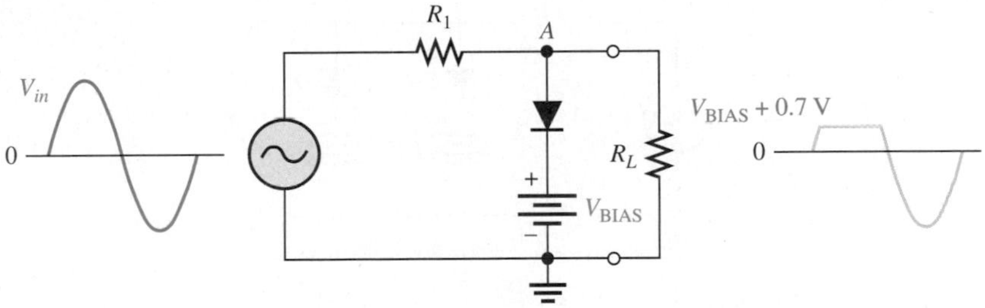

FIGURE 2–37

A positive limiter.

To limit a voltage to a specified negative level, the diode and bias voltage must be connected as in Figure 2–38. In this case, the voltage at point *A* must go below $-V_{BIAS} - 0.7$ V to forward-bias the diode and initiate limiting action as shown.

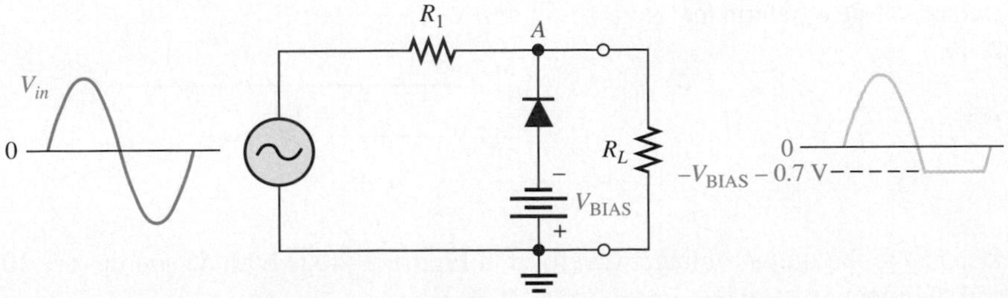

FIGURE 2–38

A negative limiter.

By turning the diode around, the positive limiter can be modified to limit the output voltage to the portion of the input voltage waveform above $V_{BIAS} - 0.7$ V, as shown by the output waveform in Figure 2–39(a). Similarly, the negative limiter can be modified to limit the output voltage to the portion of the input voltage waveform below $-V_{BIAS} + 0.7$ V, as shown by the output waveform in part (b).

FIGURE 2–39

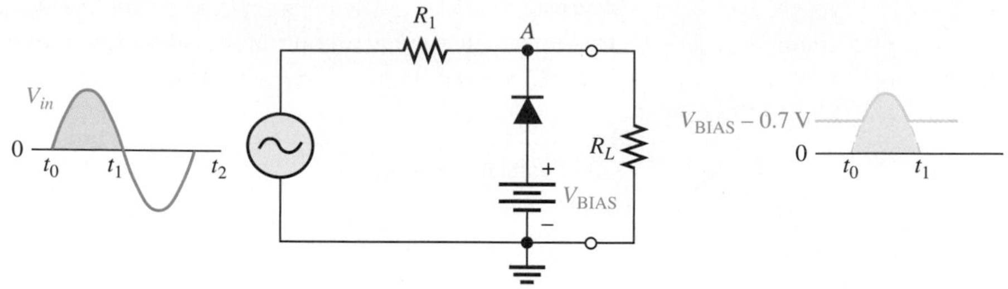

(a)

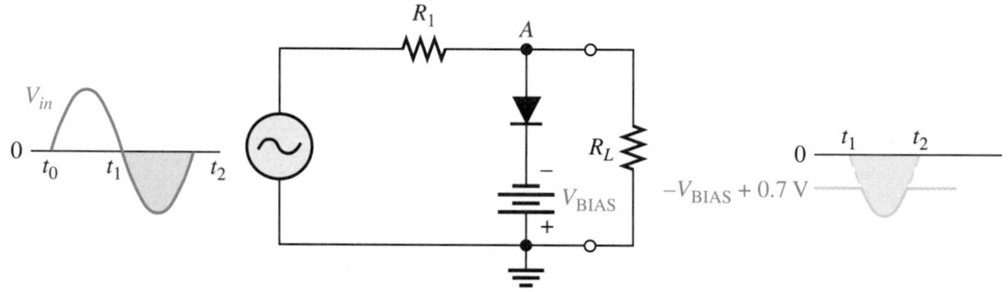

(b)

EXAMPLE 2–10

Figure 2–40 shows a circuit combining a positive limiter with a negative limiter. Determine the output voltage waveform.

FIGURE 2–40

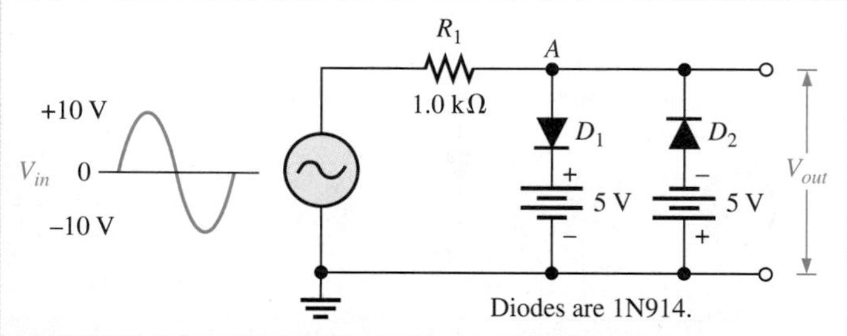

Solution When the voltage at point A reaches $+\ 5.7$ V, diode D_1 conducts and limits the waveform to $+5.7$ V. Diode D_2 does not conduct until the voltage reaches -5.7 V. Therefore, positive voltages above $+5.7$ V and negative voltages below -5.7 V are clipped off. The resulting output voltage waveform is shown in Figure 2–41.

FIGURE 2–41

Output voltage waveform for Figure 2–40.

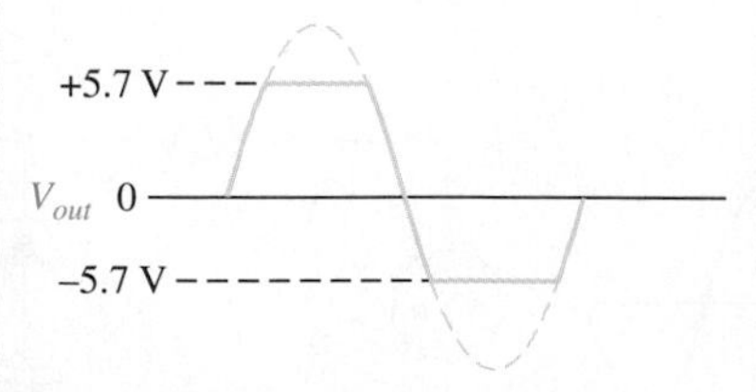

Related Problem Determine the output voltage waveform in Figure 2–40 if both dc sources are 10 V and the input voltage has a peak value of 20 V.

Open the Multisim file E02-10 in the Examples folder on your CD-ROM. For the specified input, measure the resulting output waveform. Compare with the waveform shown in the example.

Voltage-Divider Bias The bias voltage sources that have been used to illustrate the basic operation of diode limiters can be replaced by a resistive voltage divider that derives the desired bias voltage from the dc supply voltage, as shown in Figure 2–42. The bias voltage is set by the resistor values according to the voltage-divider formula.

$$V_{BIAS} = \left(\frac{R_3}{R_2 + R_3}\right)V_{SUPPLY}$$

A positively biased limiter is shown in Figure 2–42(a), a negatively biased limiter is shown in part (b), and a variable positive bias circuit using a potentiometer voltage divider is shown in part (c). The bias resistors must be small compared to R_1 so that the forward current through the diode will not affect the bias voltage.

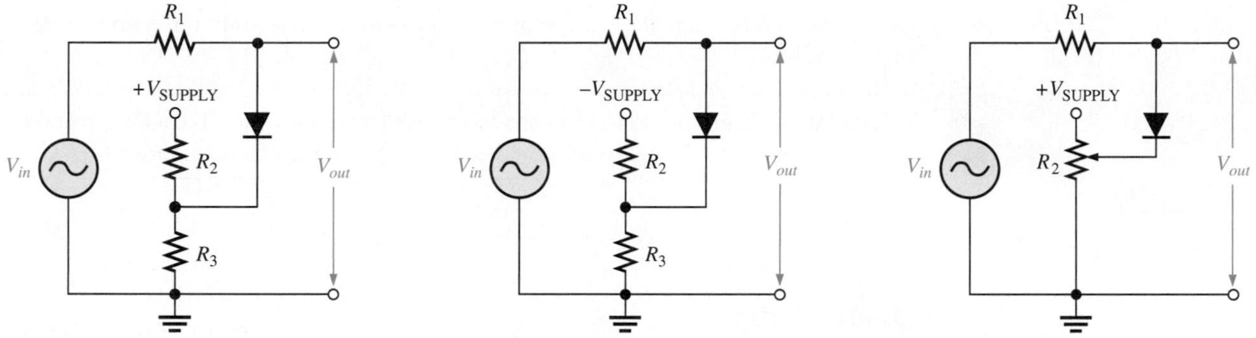

(a) Positive limiter (b) Negative limiter (c) Variable positive limiter

FIGURE 2–42

Diode limiters implemented with voltage-divider bias.

A Limiter Application Many circuits have certain restrictions on the input level to avoid damaging the circuit. For example, almost all digital circuits should not have an input level that exceeds the power supply voltage. An input of a few volts more than this could damage the circuit. To prevent the input from exceeding a specific level, you may see a diode limiter across the input signal path in many digital circuits.

EXAMPLE 2–11 Describe the output voltage waveform for the diode limiter in Figure 2–43.

FIGURE 2–43

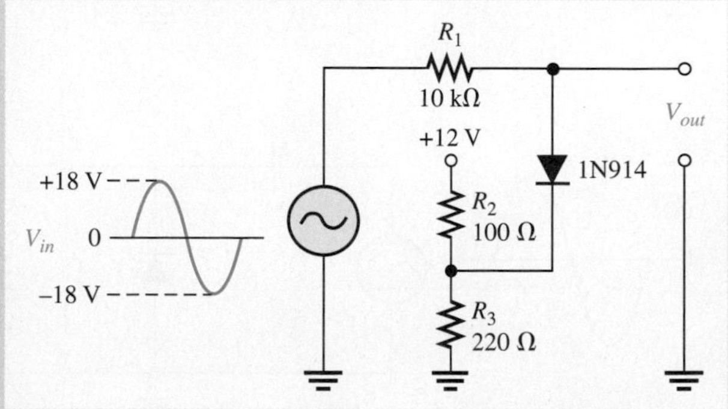

Solution The circuit is a positive limiter. Use the voltage-divider formula to determine the bias voltage.

$$V_{BIAS} = \left(\frac{R_3}{R_2 + R_3}\right)V_{SUPPLY} = \left(\frac{220\ \Omega}{100\ \Omega + 220\ \Omega}\right)12\text{ V} = 8.25\text{ V}$$

The output voltage waveform is shown in Figure 2–44. The positive part of the output voltage waveform is limited to $V_{BIAS} + 0.7$ V.

FIGURE 2–44

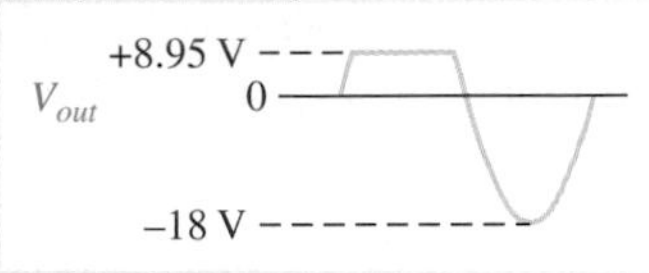

Related Problem How would you change the voltage divider in Figure 2–43 to limit the output voltage to +6.7 V?

Open the Multisim file E02-11 in the Examples folder on your CD-ROM. Observe the output voltage on the oscilloscope and compare to the calculated result.

Diode Clampers

A clamper adds a dc level to an ac voltage. **Clampers** are sometimes known as *dc restorers.* Figure 2–45 shows a diode clamper that inserts a positive dc level in the output waveform. The operation of this circuit can be seen by considering the first negative half-cycle of the input voltage. When the input voltage initially goes negative, the diode is forward-biased, allowing the capacitor to charge to near the peak of the input ($V_{p(in)} - 0.7$ V), as shown in Figure 2–45(a). Just after the negative peak, the diode is reverse-biased. This is because the cathode is held near $V_{p(in)} - 0.7$ V by the charge on the capacitor. The capacitor can only discharge through the high resistance of R_L. So, from the peak of one negative

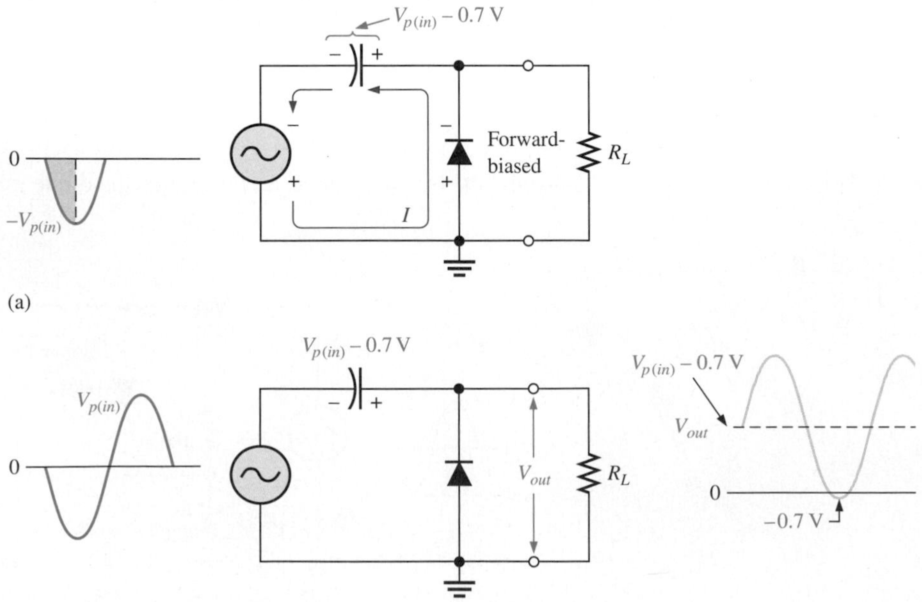

FIGURE 2–45

Positive clamper operation.

half-cycle to the next, the capacitor discharges very little. The amount that is discharged, of course, depends on the value of R_L.

If the capacitor discharges during the period of the input wave, clamping action is affected. If the *RC* time constant is 100 times the period, the clamping action is excellent. An *RC* time constant of ten times the period will have a small amount of distortion at the ground level due to the charging current.

The net effect of the clamping action is that the capacitor retains a charge approximately equal to the peak value of the input less the diode drop. The capacitor voltage acts essentially as a battery in series with the input voltage. The dc voltage of the capacitor adds to the input voltage by superposition, as in Figure 2–45(b).

If the diode is turned around, a negative dc voltage is added to the input voltage to produce the output voltage as shown in Figure 2–46.

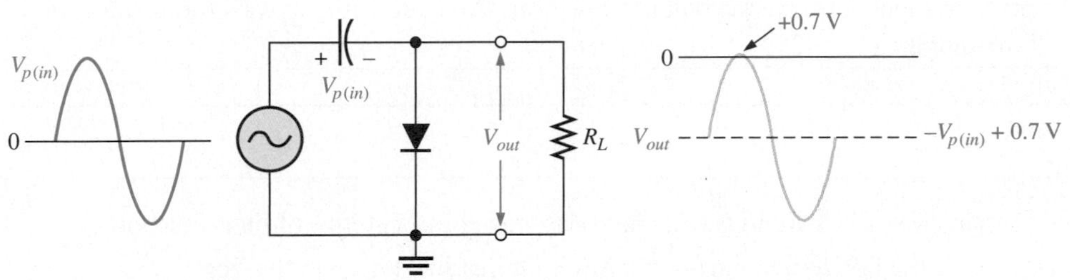

FIGURE 2–46

Negative clamper.

Clamper Applications A clamping circuit is often used in analog television receivers as a dc restorer. The incoming composite video signal is normally processed through capacitively coupled amplifiers that eliminate the dc component, thus losing the black and white reference levels and the blanking level. Before being applied to the picture tube, these reference levels must be restored. Another application is to prevent a signal from going negative and damaging a sensitive input circuit.

EXAMPLE 2–12

What is the output voltage that you would expect to observe across R_L in the clamping circuit of Figure 2–47? Assume that *RC* is large enough to prevent significant capacitor discharge.

FIGURE 2–47

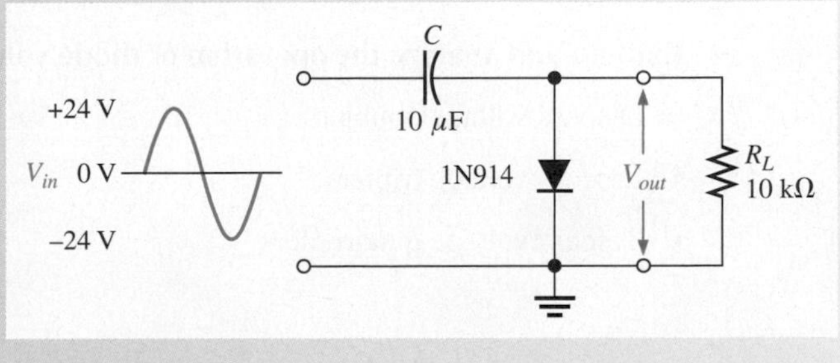

Solution Ideally, a negative dc value equal to the input peak less the diode drop is inserted by the clamping circuit.

$$V_{DC} \cong -(V_{p(in)} - 0.7\text{ V}) = -(24\text{ V} - 0.7\text{ V}) = \mathbf{-23.3\text{ V}}$$

Actually, the capacitor will discharge slightly between peaks, and, as a result, the output voltage will have an average value of slightly less than that calculated above. The output waveform goes to approximately +0.7 V, as shown in Figure 2–48.

FIGURE 2–48

Output waveform across R_L for Figure 2–47.

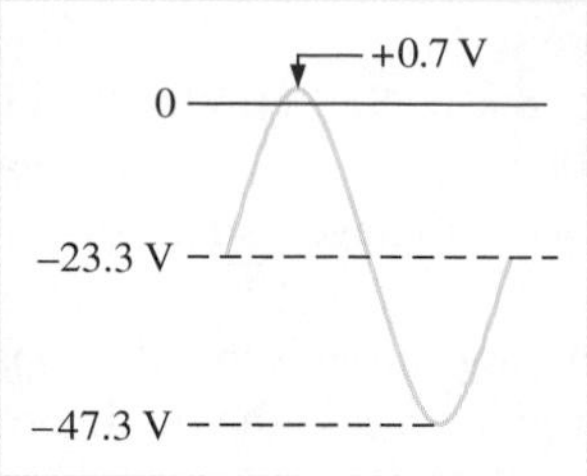

Related Problem What is the output voltage that you would observe across R_L in Figure 2–47 for $C = 22\ \mu F$ and $R_L = 18\ k\Omega$?

Open the Multisim file E02-12 in the Examples folder on your CD-ROM. For the specified input, measure the output waveform. Compare with the waveform shown in the example.

SECTION 2–4 CHECKUP

1. Discuss how diode limiters and diode clampers differ in terms of their function.
2. What is the difference between a positive limiter and a negative limiter?
3. What is the maximum voltage across an unbiased positive silicon diode limiter during the positive alternation of the input voltage?
4. To limit the output voltage of a positive limiter to 5 V when a 10 V peak input is applied, what value must the bias voltage be?
5. What component in a clamping circuit effectively acts as a battery?

2–5 VOLTAGE MULTIPLIERS

Voltage multipliers use clamping action to increase peak rectified voltages without the necessity of increasing the transformer's voltage rating. Multiplication factors of two, three, and four are common. Voltage multipliers are used in high-voltage, low-current applications such as cathode-ray tubes (CRTs) and particle accelerators.

After completing this section, you should be able to

- **Explain and analyze the operation of diode voltage multipliers**
  - Discuss voltage doublers
  - Discuss voltage triplers
  - Discuss voltage quadruplers

Voltage Doubler

Half-Wave Voltage Doubler A voltage doubler is a **voltage multiplier** with a multiplication factor of two. A half-wave voltage doubler is shown in Figure 2–49. During the positive half-cycle of the secondary voltage, diode D_1 is forward-biased and D_2 is reverse-biased. Capacitor C_1 is charged to the peak of the secondary voltage (V_p) less the diode drop with the polarity shown in part (a). During the negative half-cycle, diode D_2 is forward-biased and D_1 is reverse-biased, as shown in part (b). Since C_1 can't discharge, the peak voltage on C_1 adds

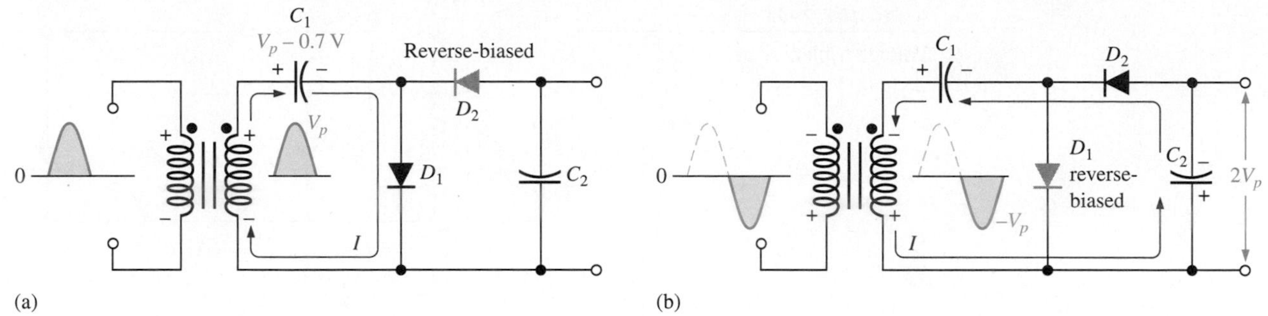

▲ FIGURE 2–49

Half-wave voltage doubler operation. V_p is the peak secondary voltage.

to the secondary voltage to charge C_2 to approximately $2V_p$. Applying Kirchhoff's law around the loop as shown in part (b), the voltage across C_2 is

$$V_{C1} - V_{C2} + V_p = 0$$
$$V_{C2} = V_p + V_{C1}$$

Neglecting the diode drop of D_2, $V_{C1} = V_p$. Therefore,

$$V_{C2} = V_p + V_p = 2V_p$$

Under a no-load condition, C_2 remains charged to approximately $2V_p$. If a load resistance is connected across the output, C_2 discharges slightly through the load on the next positive half-cycle and is again recharged to $2V_p$ on the following negative half-cycle. The resulting output is a half-wave, capacitor-filtered voltage. The peak inverse voltage across each diode is $2V_p$.

Full-Wave Voltage Doubler A full-wave doubler is shown in Figure 2–50. When the secondary voltage is positive, D_1 is forward-biased and C_1 charges to approximately V_p, as shown in part (a). During the negative half-cycle, D_2 is forward-biased and C_2 charges to approximately V_p, as shown in part (b). The output voltage, $2V_p$, is taken across the two capacitors in series.

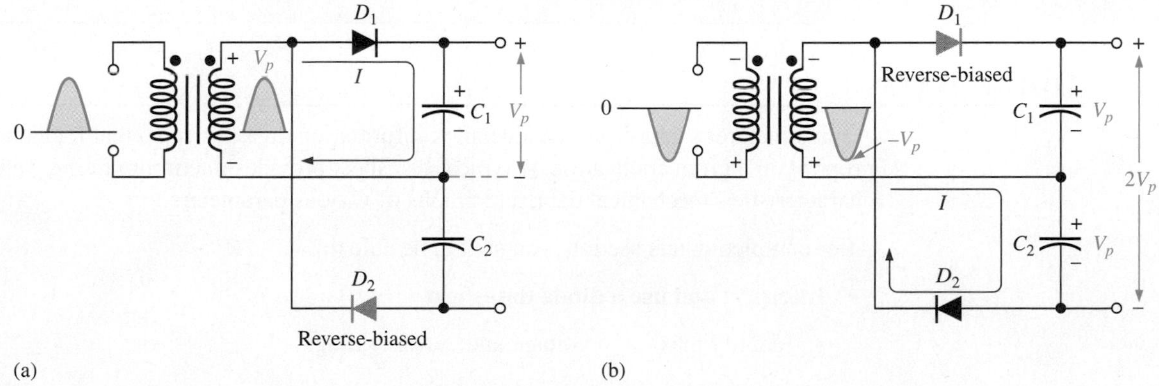

▲ FIGURE 2–50

Full-wave voltage doubler operation.

Voltage Tripler

The addition of another diode-capacitor section to the half-wave voltage doubler creates a voltage tripler, as shown in Figure 2–51. The operation is as follows: On the positive half-cycle of the secondary voltage, C_1 charges to V_p through D_1. During the negative half-cycle, C_2

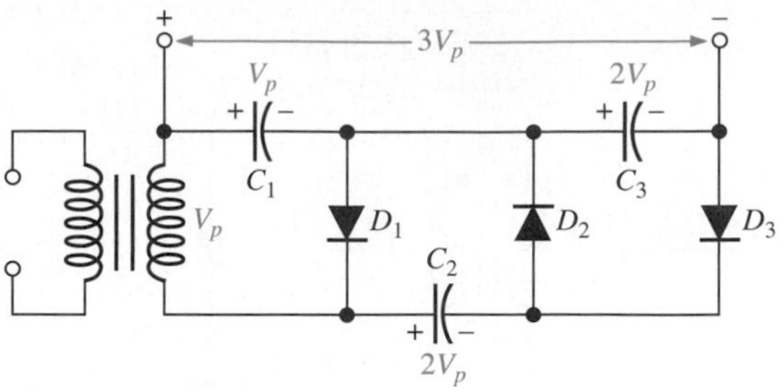

FIGURE 2–51
Voltage tripler.

charges to $2V_p$ through D_2, as described for the doubler. During the next positive half-cycle, C_3 charges to $2V_p$ through D_3. The tripler output is taken across C_1 and C_3, as shown in the figure.

Voltage Quadrupler

The addition of still another diode-capacitor section, as shown in Figure 2–52, produces an output four times the peak secondary voltage. C_4 charges to $2V_p$ through D_4 on a negative half-cycle. The $4V_p$ output is taken across C_2 and C_4, as shown. In both the tripler and quadrupler circuits, the PIV of each diode is $2V_p$.

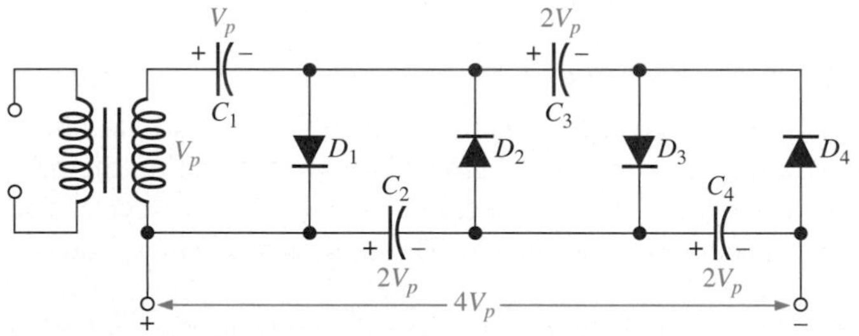

FIGURE 2–52
Voltage quadrupler.

SECTION 2–5 CHECKUP

1. What must be the peak voltage rating of the transformer secondary for a voltage doubler that produces an output of 200 V?
2. The output voltage of a quadrupler is 620 V. What minimum PIV rating must each diode have?

2–6 THE DIODE DATASHEET

A manufacturer's datasheet gives detailed information on a device so that it can be used properly in a given application. A typical datasheet provides maximum ratings, electrical characteristics, mechanical data, and graphs of various parameters.

After completing this section, you should be able to

- **Interpret and use a diode datasheet**
  - Identify maximum voltage and current ratings
  - Determine the electrical characteristics of a diode
  - Analyze graphical data
  - Select an appropriate diode for a given set of specifications

Figure 2–53 shows a typical rectifier diode datasheet. The presentation of information on datasheets may vary from one manufacturer to another, but they basically all convey the

FAIRCHILD
SEMICONDUCTOR®

1N4001 - 1N4007

Features

- Low forward voltage drop.
- High surge current capability.

DO-41
COLOR BAND DENOTES CATHODE

General Purpose Rectifiers

Absolute Maximum Ratings*

$T_A = 25°C$ unless otherwise noted

Symbol	Parameter	Value							Units
		4001	4002	4003	4004	4005	4006	4007	
V_{RRM}	Peak Repetitive Reverse Voltage	50	100	200	400	600	800	1000	V
$I_{F(AV)}$	Average Rectified Forward Current, .375 " lead length @ $T_A = 75°C$	1.0							A
I_{FSM}	Non-repetitive Peak Forward Surge Current 8.3 ms Single Half-Sine-Wave	30							A
T_{stg}	Storage Temperature Range	-55 to +175							°C
T_J	Operating Junction Temperature	-55 to +175							°C

*These ratings are limiting values above which the serviceability of any semiconductor device may be impaired.

Thermal Characteristics

Symbol	Parameter	Value	Units
P_D	Power Dissipation	3.0	W
$R_{\theta JA}$	Thermal Resistance, Junction to Ambient	50	°C/W

Electrical Characteristics

$T_A = 25°C$ unless otherwise noted

Symbol	Parameter	Device							Units
		4001	4002	4003	4004	4005	4006	4007	
V_F	Forward Voltage @ 1.0 A	1.1							V
I_{rr}	Maximum Full Load Reverse Current, Full Cycle $T_A = 75°C$	30							μA
I_R	Reverse Current @ rated V_R $T_A = 25°C$ / $T_A = 100°C$	5.0 / 500							μA / μA
C_T	Total Capacitance $V_R = 4.0$ V, f = 1.0 MHz	15							pF

Typical Characteristics

Forward Current Derating Curve

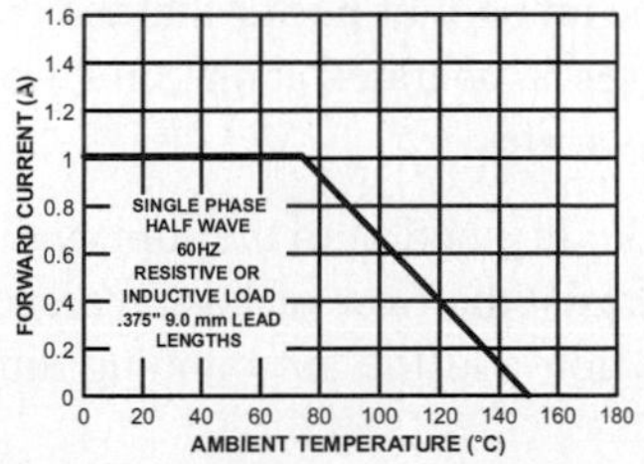

Forward Characteristics

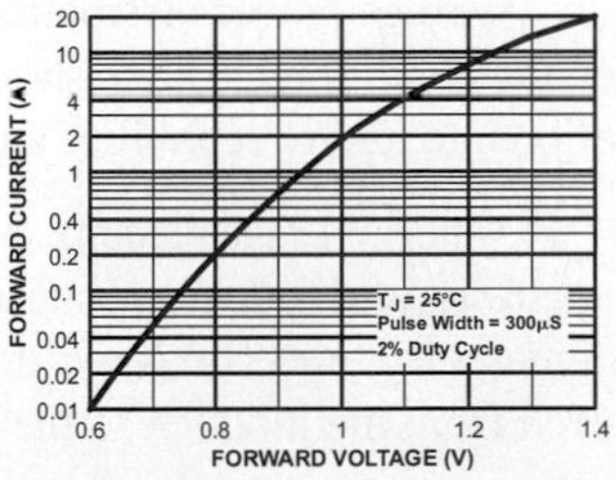

Non-Repetitive Surge Current

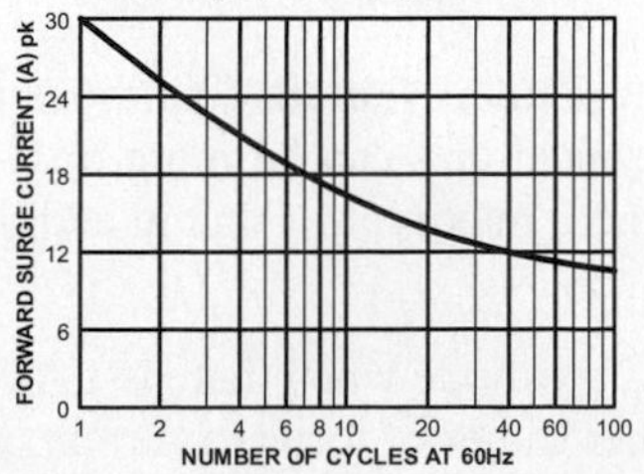

Reverse Characteristics

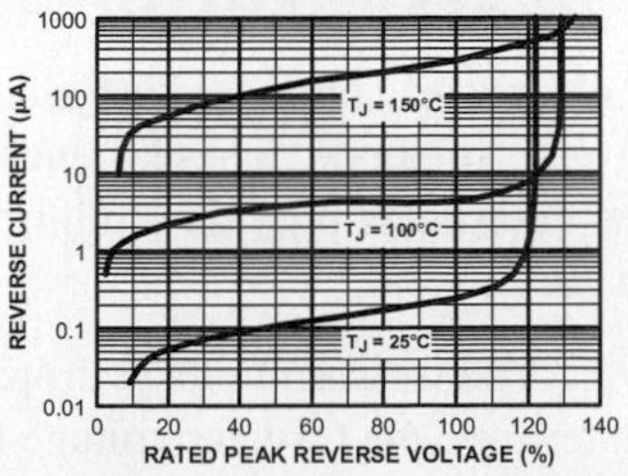

FIGURE 2–53

©2003 Fairchild Semiconductor Corporation. Used by permission.

same information; some datasheets may have more or less data than others. The mechanical information, such as package dimensions, are not shown on this particular datasheet but are generally available from the manufacturer. Notice on this datasheet that there are three categories of data given in table form and four types of characteristics shown in graphical form.

Absolute Maximum Ratings

The absolute maximum ratings indicate the maximum values of the several parameters under which the diode can be operated without damage or degradation. For greatest reliability and longer life, the diode should be operated well under these maximums. Generally, the maximum ratings are specified for an operating ambient temperature (T_A) of 25°C unless otherwise stated. Ambient temperature is the temperature of the air surrounding the device. The parameters given in Figure 2–53 are as follows:

V_{RRM} The peak reverse voltage that can be applied repetitively across the diode. Notice that it is 50 V for the 1N4001 and 1000 V for the 1N4007. This rating is the same as the PIV.

$I_{F(AV)}$ The maximum average value of a 60 Hz half-wave rectified forward current. This current parameter is 1.0 A for all of the diode types and is specified for an ambient temperature of 75°C.

I_{FSM} The maximum peak value of nonrepetitive single half-sine-wave forward surge current with a duration of 8.3 ms. This current parameter is 30 A for all of the diode types.

T_{stg} The allowable range of temperatures at which the device can be kept when not operating or connected to a circuit.

T_J The allowable range of temperatures for the *pn* junction when the diode is operated in a circuit.

Thermal Characteristics

All devices have a limit on the amount of heat that they can tolerate without failing in some way.

P_D Average power dissipation is the amount of power that the diode can dissipate under any condition. A diode should never be operated at maximum power, except for brief periods, to assure reliability and longer life.

$R_{\theta JA}$ Thermal resistance from the diode junction to the surrounding air. This indicates the ability of the device material to resist the flow of heat and specifies the number of degrees difference between the junction and the surrounding air for each watt transferred from the junction to the air.

Electrical Characteristics

The electrical characteristics are specified under certain conditions and are the same for each type of diode. These values are typical and can be more or less for a given diode. Some datasheets provide a minimum and a maximum value in addition to a typical value for a parameter.

V_F The forward voltage drop across the diode when there is 1 A of forward current. To determine the forward voltage for other values of forward current, you must examine the forward characteristics graph.

I_{rr} Maximum full load reverse current averaged over a full ac cycle at 75°C.

I_R The reverse current at the rated reverse voltage (V_{RRM}). Values are specified at two different ambient temperatures.

C_T This is the total diode capacitance including the junction capacitance in reverse bias at a frequency of 1 MHz. Most of the time this parameter is not important in low-frequency applications, such as power supply rectifiers.

The Forward Current Derating Curve

This curve on the datasheet in Figure 2–53 shows maximum forward diode current $I_{F(AV)}$ in amps versus the ambient temperature. Up to about 75°C, the diode can handle a maximum of 1 A. Above 75°C, the diode cannot handle 1 A, so the maximum current must be derated as shown by the curve. For example, if a diode is operating in an ambient temperature of 120°C, it can handle only a maximum of 0.4 A, as shown in Figure 2–54.

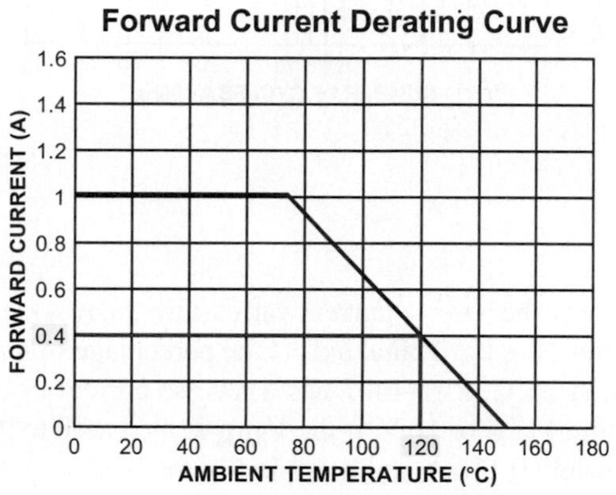

FIGURE 2–54

Forward Characteristics Curve

Another graph from the datasheet shows instantaneous forward current as a function of instantaneous forward voltage. As indicated, data for this curve is derived by applying 300 μs pulses with a duty cycle of 2%. Notice that this graph is for $T_J = 25°C$. For example, a forward current of 1 A corresponds to a forward voltage of about 0.93 V, as shown in Figure 2–55.

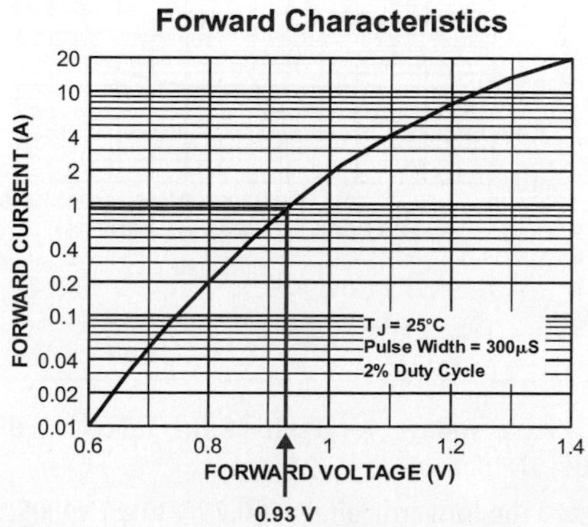

FIGURE 2–55

Nonrepetitive Surge Current

This graph from the datasheet shows I_{FSM} as a function of the number of cycles at 60 Hz. For a one-time surge, the diode can withstand 30 A. However, if the surges are repeated at a frequency of 60 Hz, the maximum surge current decreases. For example, if the surge is repeated 7 times, the maximum current is 18 A, as shown in Figure 2–56.

FIGURE 2–56

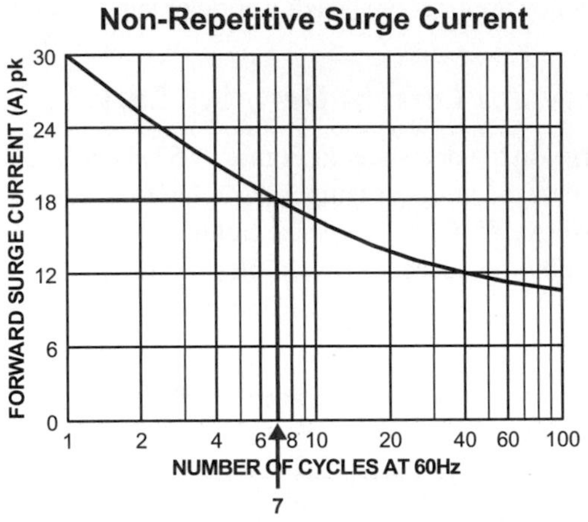

Reverse Characteristics

This graph from the datasheet shows how the reverse current varies with the reverse voltage for three different junction temperatures. The horizontal axis is the percentage of maximum reverse voltage, V_{RRM}. For example, at 25°C, a 1N4001 has a reverse current of approximately 0.04 μA at 20% of its maximum V_{RRM} or 10 V. If the V_{RRM} is increased to 90%, the reverse current increases to approximately 0.11μA, as shown in Figure 2–57.

FIGURE 2–57

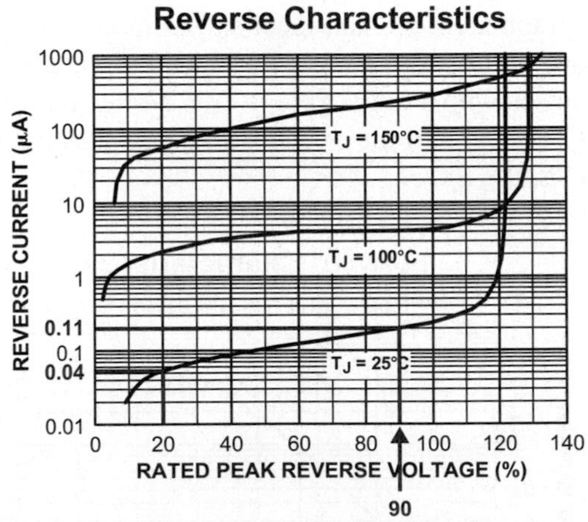

SECTION 2–6 CHECKUP

1. Determine the peak repetitive reverse voltage for each of the following diodes: 1N4002, 1N4003, 1N4004, 1N4005, 1N4006.
2. If the forward current is 800 mA and the forward voltage is 0.75 V in a 1N4005, is the power rating exceeded?
3. What is $I_{F(AV)}$ for a 1N4001 at an ambient temperature of 100°C?
4. What is I_{FSM} for a 1N4003 if the surge is repeated 40 times at 60 Hz?

2–7 TROUBLESHOOTING

This section provides a general overview and application of an approach to troubleshooting. Specific troubleshooting examples of the power supply and diode circuits are covered.

After completing this section, you should be able to

- **Troubleshoot power supplies and diode circuits**
  - Use analysis to evaluate a problem based on symptoms
  - Eliminate basic problems that can be detected by observation
  - Plan an approach to determining what the fault is in a circuit or system
  - Make appropriate measurements to isolate a fault
  - Recognize symptoms caused by certain types of component failures

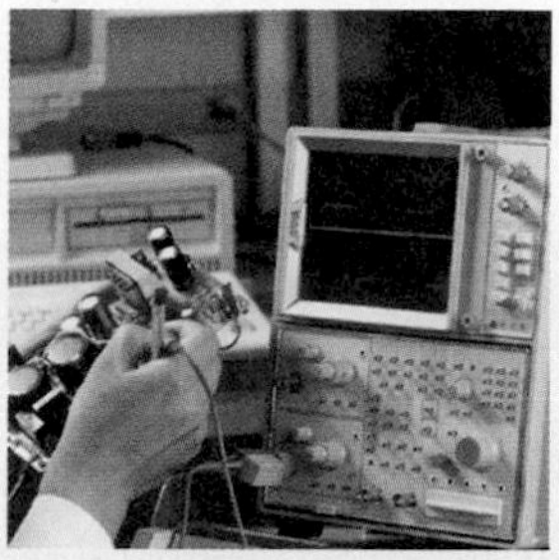

Troubleshooting is the application of logical thinking combined with a thorough knowledge of circuit or system operation to identify and correct a malfunction. A systematic approach to troubleshooting consists of three steps: *analysis, planning,* and *measuring.*

A defective circuit or system is one with a known good input but with no output or an incorrect output. For example, in Figure 2–58(a), a properly functioning dc power supply is represented by a single block with a known input voltage and a correct output voltage. A defective dc power supply is represented in part (b) as a block with an input voltage and an incorrect output voltage.

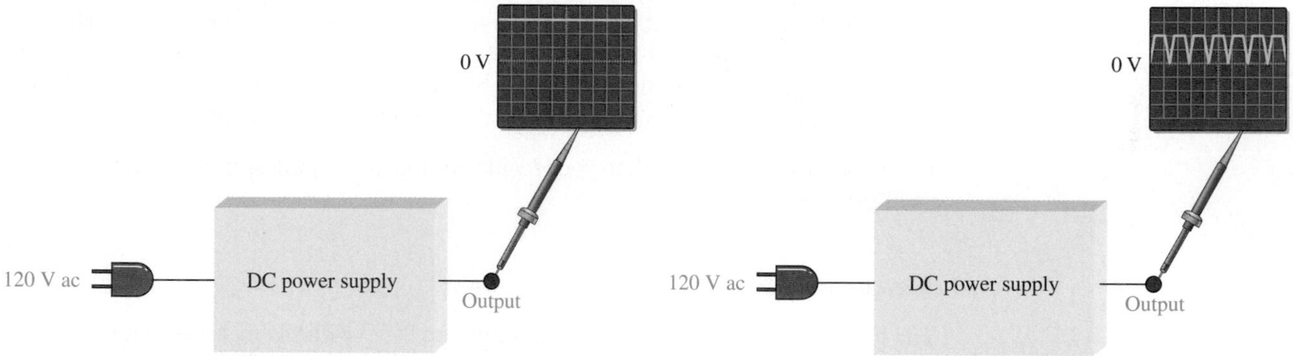

(a) The correct dc output voltage is measured.

(b) An incorrect voltage is measured at the output.

▲ FIGURE 2–58

Block representations of functioning and nonfunctioning power supplies.

Analysis

The first step in troubleshooting a defective circuit or system is to analyze the problem, which includes identifying the symptom and eliminating as many causes as possible. In the case of the power supply example illustrated in Figure 2–58(b), the symptom is that the output voltage is not a constant regulated dc voltage. This symptom does not tell you much about what the specific cause may be. In other situations, however, a particular symptom may point to a given area where a fault is most likely.

The first thing you should do in analyzing the problem is to try to eliminate any obvious causes. In general, you should start by making sure the power cord is plugged into an active outlet and that the fuse is not blown. In the case of a battery-powered system, make sure the battery is good. Something as simple as this is sometimes the cause of a problem. However, in this case, there must be power because there is an output voltage.

SAFETY NOTE

When working with low-voltage power supplies, be careful not to come in contact with the 120 V ac line. Severe shock or worse could result. To verify input voltage to a rectifier, it is always better to check at the transformer secondary instead of trying to measure the line voltage directly. If it becomes necessary to measure the line voltage, use a multimeter and be careful.

Beyond the power check, use your senses to detect obvious defects, such as a burned resistor, broken wire, loose connection, or an open fuse. Since some failures are temperature dependent, you can sometimes find an overheated component by touch. However, be very cautious in a live circuit to avoid possible burn or shock. For intermittent failures, the circuit may work properly for awhile and then fail due to heat buildup. As a rule, you should always do a sensory check as part of the analysis phase before proceeding.

Planning

In this phase, you must consider how you will attack the problem. There are three possible approaches to troubleshooting most circuits or systems.

1. Start at the input (the transformer secondary in the case of a dc power supply) where there is a known input voltage and work toward the output until you get an incorrect measurement. When you find no voltage or an incorrect voltage, you have narrowed the problem to the part of the circuit between the last test point where the voltage was good and the present test point. In all troubleshooting approaches, you must know what the voltage is supposed to be at each point in order to recognize an incorrect measurement when you see it.
2. Start at the output of a circuit and work toward the input. Check for voltage at each test point until you get a correct measurement. At this point, you have isolated the problem to the part of the circuit between the last test point and the current test point where the voltage is correct.
3. Use the half-splitting method and start in the middle of the circuit. If this measurement shows a correct voltage, you know that the circuit is working properly from the input to that test point. This means that the fault is between the current test point and the output point, so begin tracing the voltage from that point toward the output. If the measurement in the middle of the circuit shows no voltage or an incorrect voltage, you know that the fault is between the input and that test point. Therefore, begin tracing the voltage from the test point toward the input.

For illustration, let's say that you decide to apply the half-splitting method using an oscilloscope.

Measurement

The half-splitting method is illustrated in Figure 2–59 with the measurements indicating a particular fault (open filter capacitor in this case). At test point 2 (TP2) you observe a full-wave rectified voltage that indicates that the transformer and rectifier are working properly.

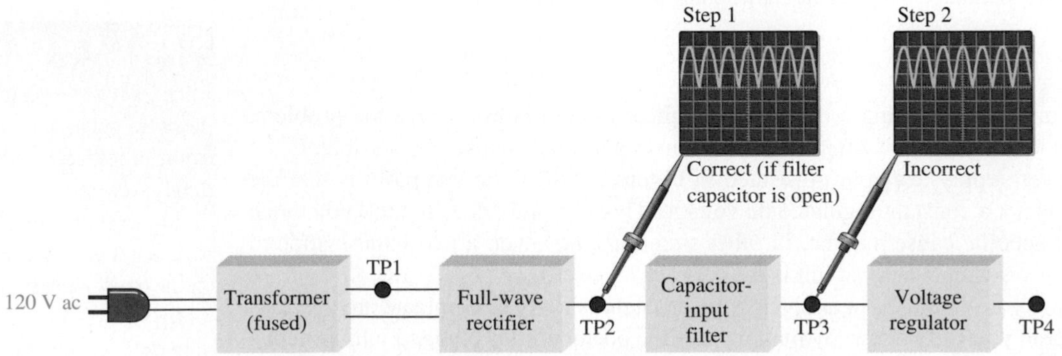

▲ **FIGURE 2–59**

Example of the half-splitting approach. An open filter capacitor is indicated.

This measurement also indicates that the filter capacitor is open, which is verified by the full-wave voltage at TP3. If the filter were working properly, you would measure a dc voltage at both TP2 and TP3. If the filter capacitor were shorted, you would observe no voltage at all of the test points because the fuse would most likely be blown. A short anywhere in the system is very difficult to isolate because, if the system is properly fused, the fuse will blow immediately when a short to ground develops.

For the case illustrated in Figure 2–59, the half-splitting method took two measurements to isolate the fault to the open filter capacitor. If you had started from the transformer output, it would have taken three measurements; and if you had started at the final output, it would have also taken three measurements, as illustrated in Figure 2–60.

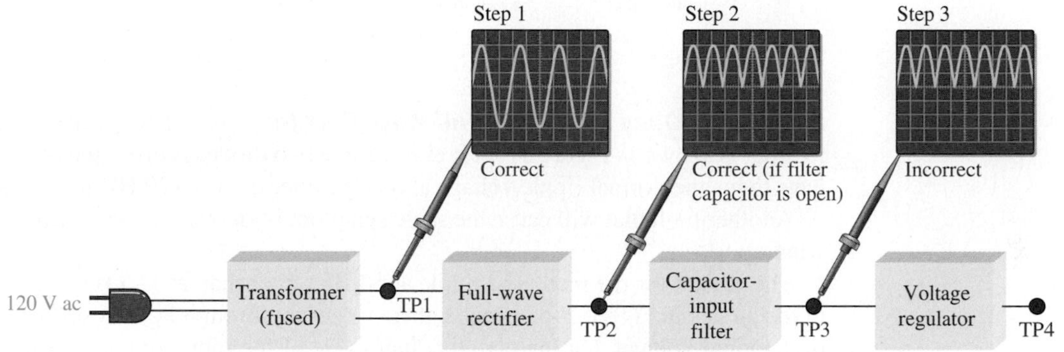

(a) Measurements starting at the transformer output.

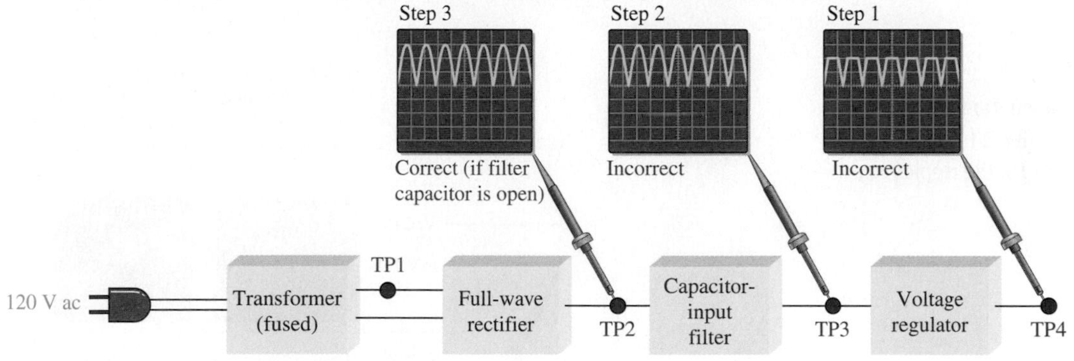

(b) Measurements starting at the regulator output.

▲ FIGURE 2–60

In this particular case, the two other approaches require more measurements than the half-splitting approach in Figure 2–59.

Fault Analysis

In some cases, after isolating a fault to a particular circuit, it may be necessary to isolate the problem to a single component in the circuit. In this event, you have to apply logical thinking and your knowledge of the symptoms caused by certain component failures. Some typical component failures and the symptoms they produce are now discussed.

Effect of an Open Diode in a Half-Wave Rectifier A half-wave filtered rectifier with an open diode is shown in Figure 2–61. The resulting symptom is zero output voltage as indicated. This is obvious because the open diode breaks the current path from the transformer secondary winding to the filter and load resistor and there is no load current.

Other faults that will cause the same symptom in this circuit are an open transformer winding, an open fuse, or no input voltage.

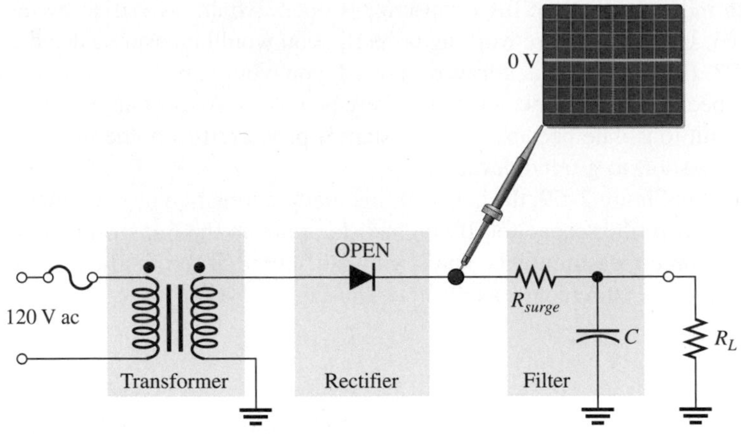

▶ **FIGURE 2–61**

The effect of an open diode in a half-wave rectifier is an output of 0 V.

Effect of an Open Diode in a Full-Wave Rectifier A full-wave center-tapped filtered rectifier is shown in Figure 2–62. If either of the two diodes is open, the output voltage will have twice the normal ripple voltage at 60 Hz rather than at 120 Hz, as indicated.

Another fault that will cause the same symptom is an open in the transformer secondary winding.

The reason for the increased ripple at 60 Hz rather than at 120 Hz is as follows. If one of the diodes in Figure 2–62 is open, there is current through R_L only during one half-cycle of the input voltage. During the other half-cycle of the input, the open path caused by the open diode prevents current through R_L. The result is half-wave rectification, as shown in Figure 2–62, which produces the larger ripple voltage with a frequency of 60 Hz.

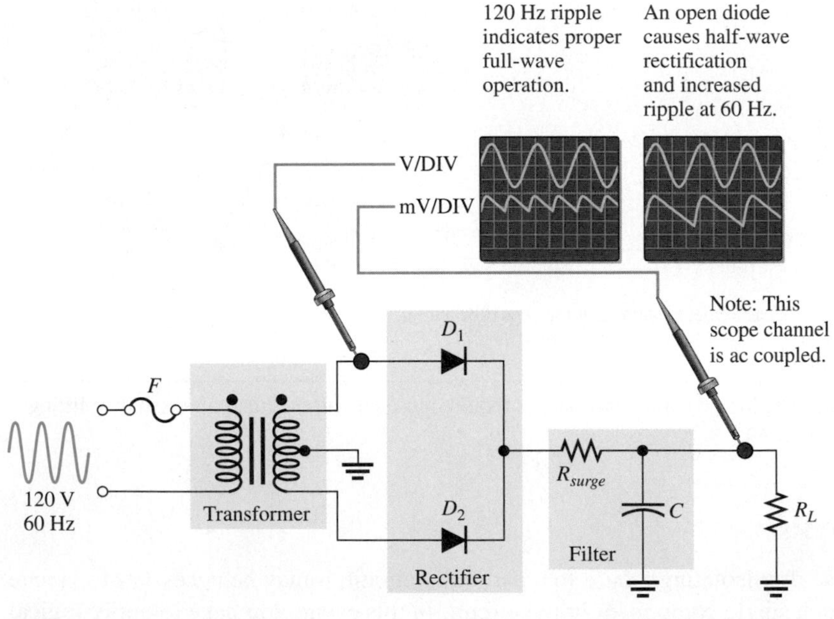

▶ **FIGURE 2–62**

The effect of an open diode in a center-tapped rectifier is half-wave rectification and twice the ripple voltage at 60 Hz.

An open diode in a full-wave bridge rectifier will produce the same symptom as in the center-tapped circuit, as shown in Figure 2–63. The open diode prevents current through R_L during half of the input voltage cycle. The result is half-wave rectification, which produces double the ripple voltage at 60 Hz.

Effects of a Faulty Filter Capacitor Three types of defects of a filter capacitor are illustrated in Figure 2–64.

- *Open* If the filter capacitor for a full-wave rectifier opens, the output is a full-wave rectified voltage.

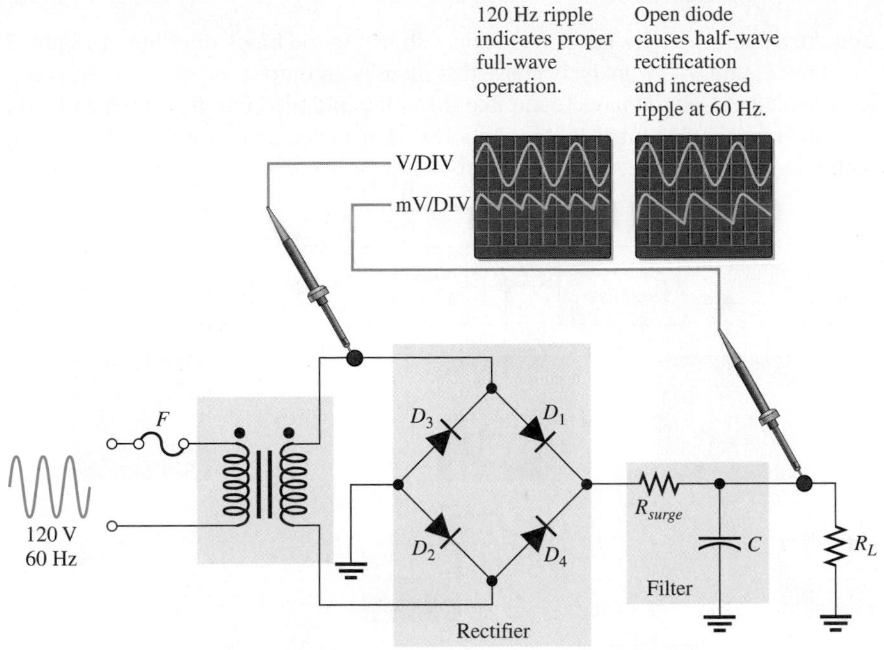

FIGURE 2–63

Effect of an open diode in a bridge rectifier.

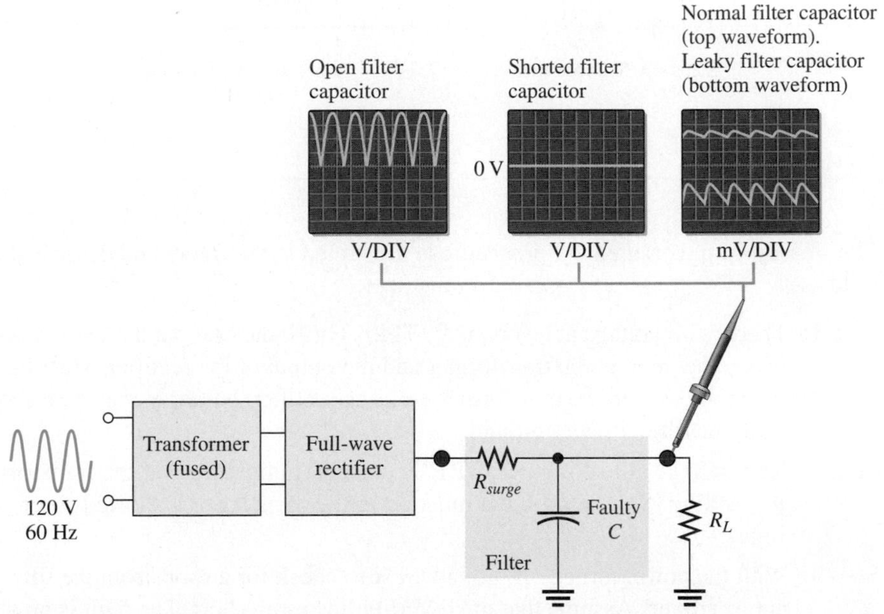

FIGURE 2–64

Effects of a faulty filter capacitor.

- *Shorted* If the filter capacitor shorts, the output is 0 V. A shorted capacitor should cause the fuse to blow open. If not properly fused, a shorted capacitor may cause some or all of the diodes in the rectifier to burn open due to excessive current. In any event, the output is 0 V.
- *Leaky* A leaky filter capacitor is equivalent to a capacitor with a parallel leakage resistance. The effect of the leakage resistance is to reduce the time constant and allow the capacitor to discharge more rapidly than normal. This results in an increase in the ripple voltage on the output. This fault is rare.

Effects of a Faulty Transformer An open primary or secondary winding of a power supply transformer results in an output of 0 V, as mentioned before.

EXAMPLE 2–13

You are troubleshooting the power supply shown in the block diagram of Figure 2–65. You have found in the analysis phase that there is no output voltage from the regulator, as indicated. Also, you have found that the unit is plugged into the outlet and have verified the input to the transformer with a DMM. You decide to use the half-splitting method using the scope. What is the problem?

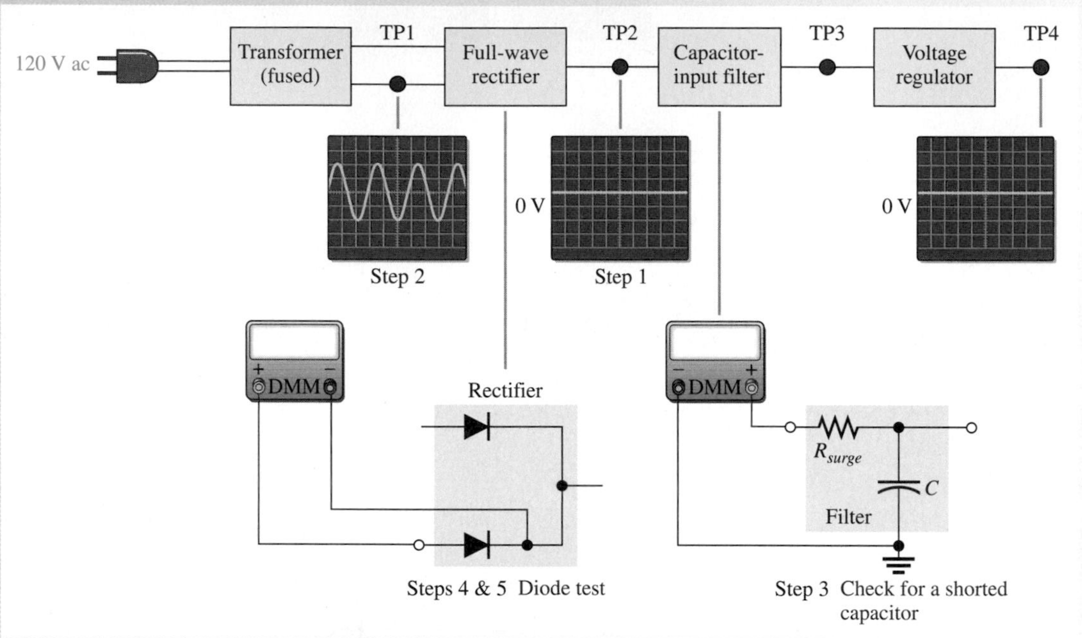

FIGURE 2–65

Solution The step-by-step measurement procedure is illustrated in the figure and described as follows.

Step 1: There is no voltage at test point 2 (TP2). This indicates that the fault is between the input to the transformer and the output of the rectifier. Most likely, the problem is in the transformer or in the rectifier, but there may be a short from the filter input to ground.

Step 2: The voltage at test point 1 (TP1) is correct, indicating that the transformer is working. So, the problem must be in the rectifier or a shorted filter input.

Step 3: With the power turned off, use a DMM to check for a short from the filter input to ground. Assume that the DMM indicates no short. The fault is now isolated to the rectifier.

Step 4: Apply fault analysis to the rectifier circuit. Determine the component failure in the rectifier that will produce a 0 V input. If only one of the diodes in the rectifier is open, there should be a half-wave rectified output voltage, so this is not the problem. In order to have a 0 V output, there must be an open in the rectifier circuit.

Step 5: With the power off, use the DMM in the diode test mode to check each diode. Replace the defective diodes, turn the power on, and check for proper operation. Assume this corrects the problem.

Related Problem Suppose you had found a short in Step 3, what would have been the logical next step?

Multisim Troubleshooting Exercises

These file circuits are in the Troubleshooting Exercises folder on your CD-ROM.

1. Open file TSE02-01. Determine if the circuit is working properly and, if not, determine the fault.
2. Open file TSE02-02. Determine if the circuit is working properly and, if not, determine the fault.
3. Open file TSE02-03. Determine if the circuit is working properly and, if not, determine the fault.

SECTION 2–7 CHECKUP

1. What effect does an open diode have on the output voltage of a half-wave rectifier?
2. What effect does an open diode have on the output voltage of a full-wave rectifier?
3. If one of the diodes in a bridge rectifier shorts, what are some possible consequences?
4. What happens to the output voltage of a rectifier if the filter capacitor becomes very leaky?
5. The primary winding of the transformer in a power supply opens. What will you observe on the rectifier output?
6. The dc output voltage of a filtered rectifier is less than it should be. What may be the problem?

On the Job

Application Activity: *DC Power Supply*

Assume that you are working for a company that designs, tests, manufactures, and markets various electronic instruments including dc power supplies. Your first assignment is to develop and test a basic unregulated power supply using the knowledge that you have acquired so far. Later modifications will include the addition of a regulator. The power supply must meet or exceed the following specifications:

- Input voltage: 120 V rms @60 Hz
- Output voltage: 16 V dc ±10%
- Ripple factor (max): 3.00%
- Load current (max): 250 mA

Design of the Power Supply

The Rectifier Circuit A full-wave rectifier has less ripple for a given filter capacitor than a half-wave rectifier. A full-wave bridge rectifier is probably the best choice because it provides the most output voltage for a given input voltage and the PIV is less than for a center-tapped rectifier. Also, the full-wave bridge does not require a center-tapped transformer.

1. Compare Equations 2–5 and 2–7 for output voltages.
2. Compare Equations 2–6 and 2–8 for PIV.

The full-wave bridge rectifier circuit is shown in Figure 2–66.

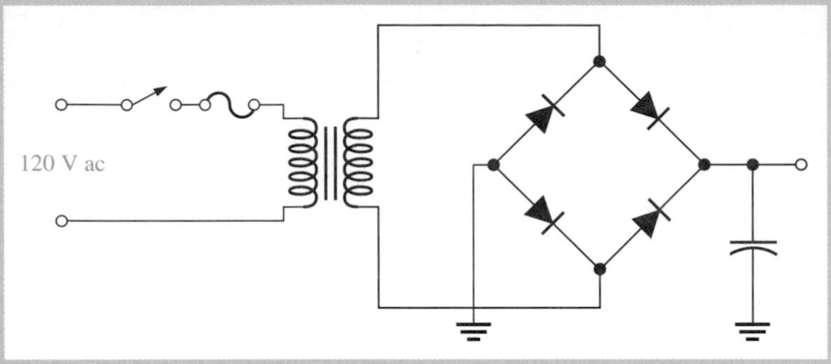

FIGURE 2–66

Power supply with full-wave bridge rectifier and capacitor filter.

The Rectifier Diodes There are two approaches for implementing the full-wave bridge: Four individual diodes, as shown in Figure 2–67(a) or a single IC package containing four diodes connected as a bridge rectifier, as shown in part (b).

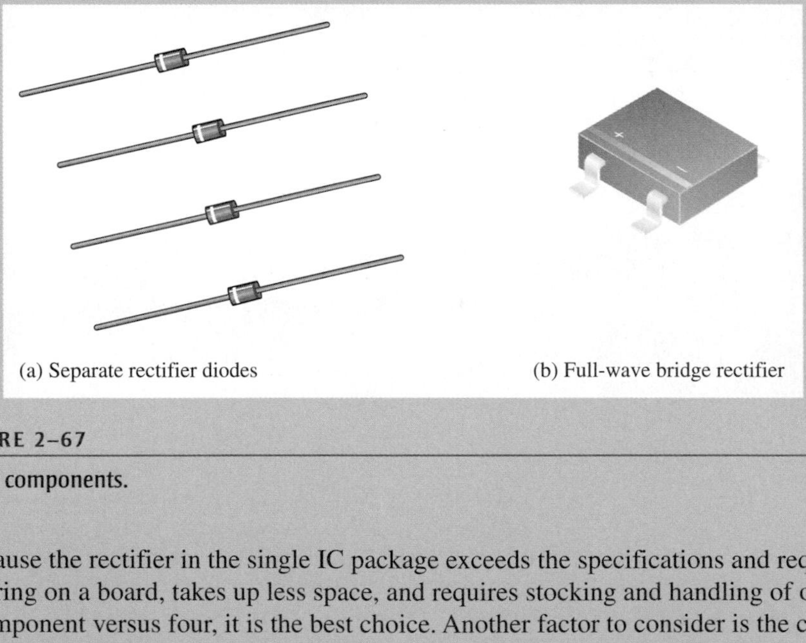

(a) Separate rectifier diodes (b) Full-wave bridge rectifier

FIGURE 2–67

Rectifier components.

Because the rectifier in the single IC package exceeds the specifications and requires less wiring on a board, takes up less space, and requires stocking and handling of only one component versus four, it is the best choice. Another factor to consider is the cost. Requirements for the diodes in the bridge are

- Forward current rating must be equal or greater than 250 mA (maximum load current).
- PIV must be greater than the minimum calculated value of 16.7 V ($PIV = V_{p(out)} + 0.7\text{ V}$).

By reviewing manufacturer's datasheets on-line, a specific device can be chosen. Figure 2–68 shows a partial datasheet for the rectifier to be used for this power supply. Notice that it exceeds the specified requirements. Four possible websites for rectifiers and diodes are *fairchildsemiconductor.com; onsemi.com*; *semiconductor.phillips.com*; and *rectron.com.*

FAIRCHILD
SEMICONDUCTOR®

MB1S - MB8S

Features

- Low leakage
- Surge overload rating: 35 amperes peak.
- Ideal for printed circuit board.
- UL certified, UL #E111753.

SOIC-4
Polarity symbols molded or marking on body

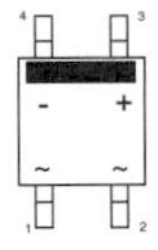

Bridge Rectifiers

Absolute Maximum Ratings* T_A = 25°C unless otherwise noted

Symbol	Parameter	Value 1S	2S	4S	6S	8S	Units
V_{RRM}	Maximum Repetitive Reverse Voltage	100	200	400	600	800	V
V_{RMS}	Maximum RMS Bridge Input Voltage	70	140	280	420	560	V
V_R	DC Reverse Voltage (Rated V_R)	100	200	400	600	800	V
$I_{F(AV)}$	Average Rectified Forward Current, @ T_A = 50°C	0.5					A
I_{FSM}	Non-repetitive Peak Forward Surge Current 8.3 ms Single Half-Sine-Wave	35					A
T_{stg}	Storage Temperature Range	-55 to +150					°C
T_J	Operating Junction Temperature	-55 to +150					°C

*These ratings are limiting values above which the serviceability of any semiconductor device may be impaired.

Thermal Characteristics

Symbol	Parameter	Value	Units
P_D	Power Dissipation	1.4	W
$R_{\theta JA}$	Thermal Resistance, Junction to Ambient,* per leg	85	°C/W
$R_{\theta JL}$	Thermal Resistance, Junction to Lead,* per leg	20	°C/W

*Device mounted on PCB with 0.5-0.5" (13x13 mm) lead length.

Electrical Characteristics T_A = 25°C unless otherwise noted

Symbol	Parameter	Device	Units
V_F	Forward Voltage, per bridge @ 0.5 A	1.0	V
I_R	Reverse Current, per leg @ rated V_R T_A = 25°C / T_A = 125°C	5.0 / 0.5	A / mA
	I^2t rating for fusing t < 8.3 ms	5.0	A^2s
C_T	Total Capacitance, per leg V_R = 4.0 V, f = 1.0 MHz	13	pF

FIGURE 2–68

Rectifier datasheet. You can view the entire datasheet at www.fairchildsemiconductor.com. © 2003 Fairchild Semiconductor Corporation. Used by permission.

The Transformer The transformer must convert the 120 V line voltage to an ac voltage that will result in a rectified voltage that will produce 16 V±10% when filtered. A typical power transformer for mounting on a printed circuit board and a portion of a datasheet for the series are shown in Figure 2–69. Notice that transformer power is measured in VA (volt-amps), not watts.

3. Use Equation 2–7 to calculate the required transformer secondary rms voltage.
4. From the partial datasheet in Figure 2–69, select an appropriate transformer based on its secondary voltage (series) and a VA specification that meets the requirement.
5. Determine the fuse rating.

The Filter Capacitor The capacitance of the filter capacitor must be sufficiently large to provide the specified ripple.

6. Use Equation 2–9 to calculate the peak-to-peak ripple voltage, assuming V_{DC} = 16 V.
7. Use Equation 2–10 to calculate the minimum capacitance value. Use R_L = 64 Ω, calculated on next page.

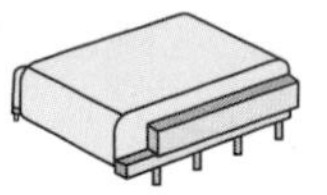

	Secondary		Dimensions					Wt.
VA	Series	Parallel	H	W	L	A	B	Oz.
2.5	10.0V CT @ 0.25A	5.0V @ 0.5A	0.650	1.562	1.875	1.600	0.375	5
2.5	12.6V CT @ 0.2A	6.3V @ 0.4A	0.650	1.562	1.875	1.600	0.375	5
2.5	16.0V CT @ 0.15A	8.0V @ 0.3A	0.650	1.562	1.875	1.600	0.375	5
2.5	20.0V CT @ 0.125A	10.0V @ 0.25A	0.650	1.562	1.875	1.600	0.375	5
2.5	24.0V CT @ 0.1A	12.0V @ 0.2A	0.650	1.562	1.875	1.600	0.375	5
2.5	30.0V CT @ 0.08A	15.0V @ 0.16A	0.650	1.562	1.875	1.600	0.375	5
2.5	34.0V CT @ 0.076A	17.0V @ 0.15A	0.650	1.562	1.875	1.600	0.375	5
2.5	40.0V CT @ 0.06A	20.0V @ 0.12A	0.650	1.562	1.875	1.600	0.375	5
2.5	56.0V CT @ 0.045A	28.0V @ 0.09A	0.650	1.562	1.875	1.600	0.375	5
2.5	88.0V CT @ 0.028A	44.0V @ 0.056A	0.650	1.562	1.875	1.600	0.375	5
2.5	120.0V CT @ 0.02A	60.0V @ 0.04A	0.650	1.562	1.875	1.600	0.375	5
2.5	230.0V CT @ 0.01A	115.0V @ 0.02A	0.650	1.562	1.875	1.600	0.375	5
6.0	10.0V CT @ 0.6A	5.0V @ 1.2A	0.875	1.562	1.875	1.600	0.375	7
6.0	12.0V CT @ 0.475A	6.3V @ 0.95A	0.875	1.562	1.875	1.600	0.375	7
6.0	16.0V CT @ 0.375A	8.0V @ 0.75A	0.875	1.562	1.875	1.600	0.375	7
6.0	20.0V CT @ 0.3A	10.0V @ 0.6A	0.875	1.562	1.875	1.600	0.375	7
6.0	24.0V CT @ 0.25A	12.0V @ 0.5A	0.875	1.562	1.875	1.600	0.375	7

FIGURE 2–69

Typical pc-mounted power transformer and data. Volts are rms.

Simulation

In the development of a new circuit, it is sometimes helpful to simulate the circuit using a software program before actually building it and committing it to hardware. We will use Multisim to simulate this power supply circuit. Figure 2–70 shows the simulated power supply circuit with a load connected and scope displays of the output voltage with and without the filter capacitor connected. The filter capacitor value of 6800 μF is the next highest standard value closest to the minimum calculated value required. A load resistor value was chosen to draw a current equal to or greater than the specified maximum load current.

$$R_L = \frac{16\text{ V}}{250\text{ mA}} = 64\ \Omega$$

The closest standard value is 62 Ω, which draws 258 mA at 16 V and which meets and exceeds the load current specification.

8. Determine the power rating for the load resistor.

To produce a dc output of 16 V, a peak secondary voltage of 16 V + 1.4 V = 17.4 V is required. The rms secondary voltage must be

$$V_{rms(sec)} = 0.707V_{p(sec)} = 0.707(16\text{ V} + 1.4\text{ V}) = 12.3\text{ V}$$

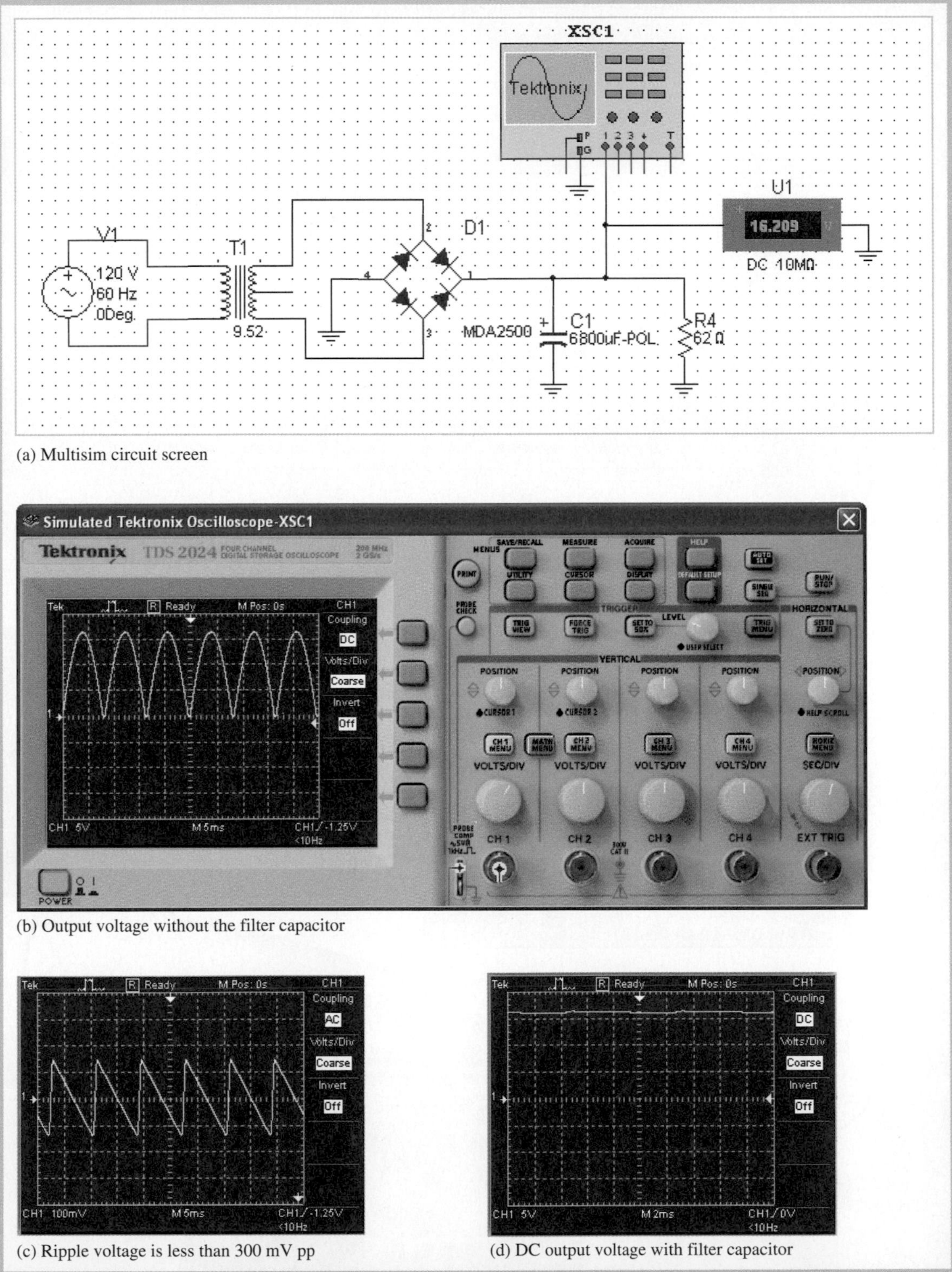

(a) Multisim circuit screen

(b) Output voltage without the filter capacitor

(c) Ripple voltage is less than 300 mV pp

(d) DC output voltage with filter capacitor

FIGURE 2–70

Power supply simulation.

A standard transformer rms output voltage is 12.6 V. The transformer specification required by Multisim is

$$120\text{ V}:12.6\text{ V} = 9.52:1$$

The dc voltmeter in Figure 2–70(a) indicates an output voltage of 16.209 V, which is well within the 16 V±10% requirement. In part (c), the scope is AC coupled and set at 100 mV/division. You can see that the peak-to-peak ripple voltage is less than 300 mV, which is less than 480 mV, corresponding to the specified maximum ripple factor of 3%.

Simulate the circuit using your Multisim software. Observe the operation with the virtual oscilloscope and voltmeter.

Prototyping and Testing

Now that all the components have been selected, the prototype circuit is constructed and tested. After the circuit is successfully tested, it is ready to be finalized on a printed circuit board.

Lab Experiment

To build and test a similar circuit, go to Experiment 2 in your lab manual (*Laboratory Exercises for Electronic Devices* by David Buchla and Steven Wetterling).

The Printed Circuit Board

The circuit board is shown in Figure 2–71. There are additional traces and connection points on the board for expansion to a regulated power supply, which will be done in Chapter 3. The

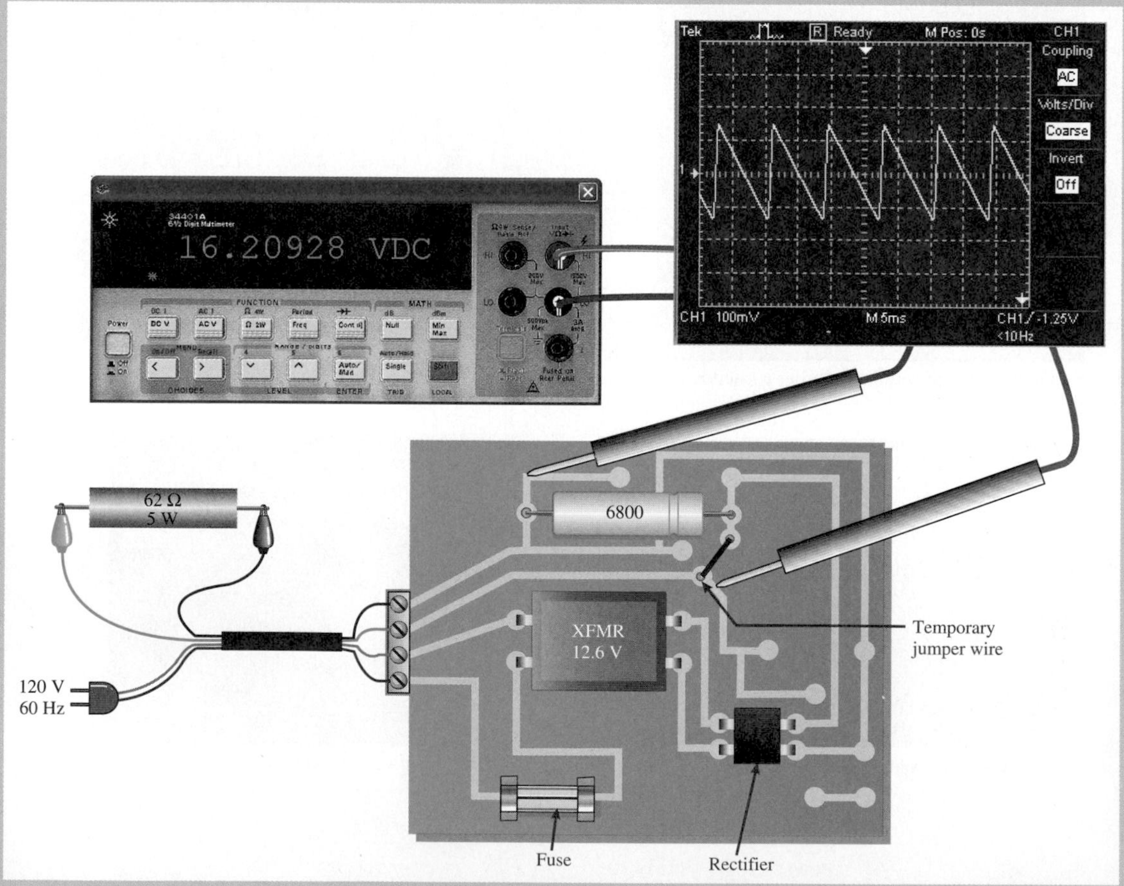

FIGURE 2–71

Testing the power supply printed circuit board. The 62 Ω load is a temporary test load to check ripple when the power supply is used at its maximum rated current.

SAFETY NOTE

Be very careful to not touch the line voltage connections to the transformer primary. In normal practice, the board is housed in a protective box to prevent the possibility of contact with the 120 V ac line.

circuit board is connected to the ac voltage and to a power load resistor via a cable. The power switch shown in the original schematic will be on the PC board housing and is not shown for the test setup. A DMM measurement of the output voltage indicates a correct value. Oscilloscope measurement of the ripple shows that it is within specifications.

Troubleshooting

For each of the scope output voltage measurements in Figure 2–72, determine the likely fault or faults, if any.

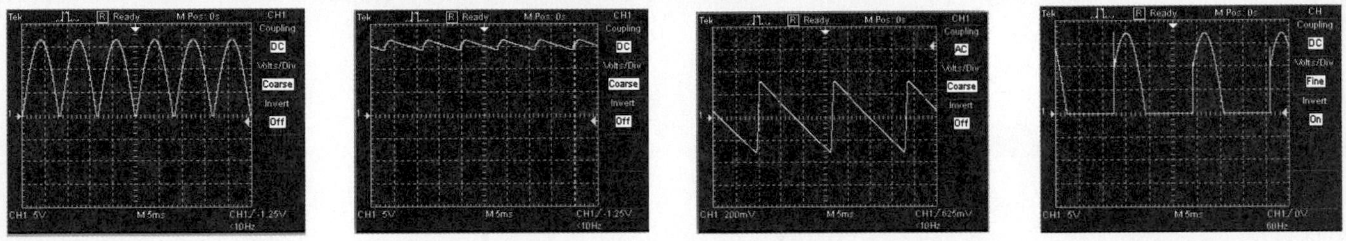

(a) (b) (c) (d)

FIGURE 2–72

Output voltage measurements on the power supply circuit.

SUMMARY OF POWER SUPPLY RECTIFIERS

HALF-WAVE RECTIFIER

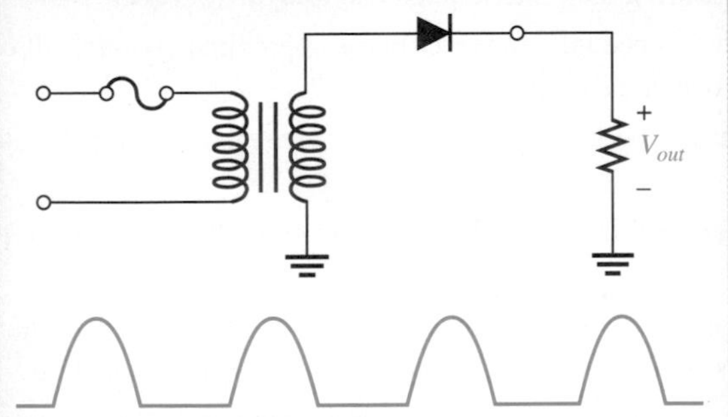

Output voltage waveform

- Peak value of output:

$$V_{p(out)} = V_{p(sec)} - 0.7\text{ V}$$

- Average value of output:

$$V_{AVG} = \frac{V_{p(out)}}{\pi}$$

- Diode peak inverse voltage:

$$PIV = V_{p(sec)}$$

CENTER-TAPPED FULL-WAVE RECTIFIER

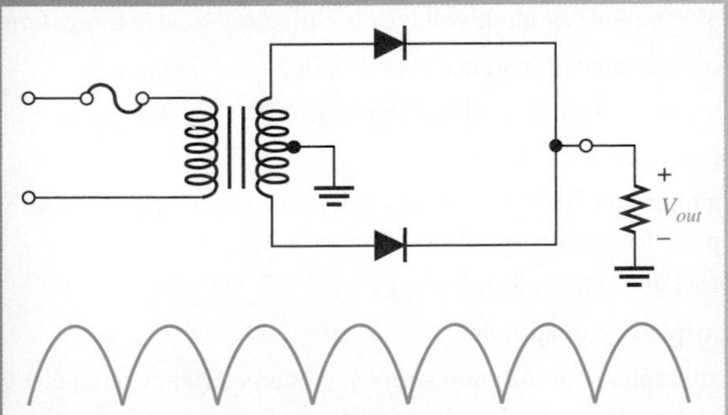

Output voltage waveform

- Peak value of output:

$$V_{p(out)} = \frac{V_{p(sec)}}{2} - 0.7\text{ V}$$

- Average value of output:

$$V_{AVG} = \frac{2V_{p(out)}}{\pi}$$

- Diode peak inverse voltage:

$$PIV = 2V_{p(out)} + 0.7\text{ V}$$

BRIDGE FULL-WAVE RECTIFIER

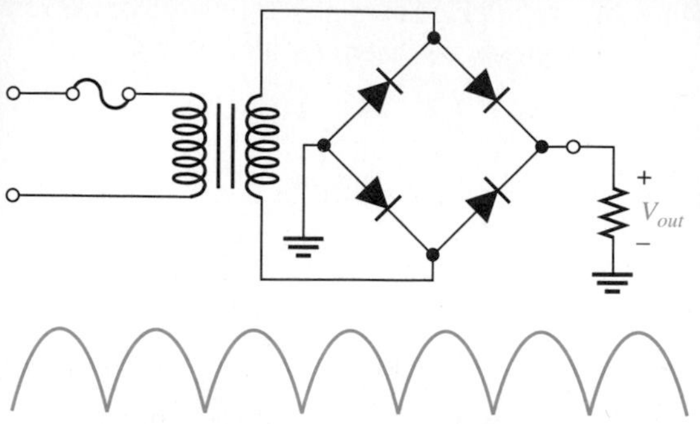

Output voltage waveform

- Peak value of output:

$$V_{p(out)} = V_{p(sec)} - 1.4\text{ V}$$

- Average value of output:

$$V_{AVG} = \frac{2V_{p(out)}}{\pi}$$

- Diode peak inverse voltage:

$$\text{PIV} = V_{p(out)} + 0.7\text{ V}$$

SUMMARY

Section 2–1
- A dc power supply typically consists of a transformer, a diode rectifier, a filter, and a regulator.
- The single diode in a half-wave rectifier is forward-biased and conducts for 180° of the input cycle.
- The output frequency of a half-wave rectifier equals the input frequency.
- PIV (peak inverse voltage) is the maximum voltage appearing across the diode in reverse bias.

Section 2–2
- Each diode in a full-wave rectifier is forward-biased and conducts for 180° of the input cycle.
- The output frequency of a full-wave rectifier is twice the input frequency.
- The two basic types of full-wave rectifier are center-tapped and bridge.
- The peak output voltage of a center-tapped full-wave rectifier is approximately one-half of the total peak secondary voltage less one diode drop.
- The PIV for each diode in a center-tapped full-wave rectifier is twice the peak output voltage plus one diode drop.
- The peak output voltage of a bridge rectifier equals the total peak secondary voltage less two diode drops.
- The PIV for each diode in a bridge rectifier is approximately half that required for an equivalent center-tapped configuration and is equal to the peak output voltage plus one diode drop.

Section 2–3
- A capacitor-input filter provides a dc output approximately equal to the peak of its rectified input voltage.
- Ripple voltage is caused by the charging and discharging of the filter capacitor.
- The smaller the ripple voltage, the better the filter.
- Regulation of output voltage over a range of input voltages is called *input* or *line regulation.*
- Regulation of output voltage over a range of load currents is called *load regulation.*

Section 2–4
- Diode limiters cut off voltage above or below specified levels. Limiters are also called *clippers.*
- Diode clampers add a dc level to an ac voltage.

Section 2–5
- Voltage multipliers are used in high-voltage, low-current applications such as for electron beam acceleration in CRTs and for particle accelerators.
- A voltage multiplier uses a series of diode-capacitor stages.
- Input voltage can be doubled, tripled, or quadrupled.

Section 2–6
- A datasheet provides key information about the parameters and characteristics of an electronic device.
- A diode should always be operated below the absolute maximum ratings specified on the datasheet.

Section 2–7

- Troubleshooting is the application of logical thought combined with a thorough knowledge of the circuit or system to identify and correct a malfunction.
- Troubleshooting is a three-step process of analysis, planning, and measurement.
- Fault analysis is the isolation of a fault to a particular circuit or portion of a circuit.

KEY TERMS

Key terms and other bold terms in the chapter are defined in the end-of-book glossary.

Clamper A circuit that adds a dc level to an ac voltage using a diode and a capacitor.

DC power supply A circuit that converts ac line voltage to dc voltage and supplies constant power to operate a circuit or system.

Filter In a power supply, the capacitor used to reduce the variation of the output voltage from a rectifier.

Full-wave rectifier A circuit that converts an ac sinusoidal input voltage into a pulsating dc voltage with two output pulses occurring for each input cycle.

Half-wave rectifier A circuit that converts an ac sinusoidal input voltage into a pulsating dc voltage with one output pulse occurring for each input cycle.

Limiter A diode circuit that clips off or removes part of a waveform above and/or below a specified level.

Line regulation The change in output voltage of a regulator for a given change in input voltage, normally expressed as a percentage.

Load regulation The change in output voltage of a regulator for a given range of load currents, normally expressed as a percentage.

Peak inverse voltage (PIV) The maximum value of reverse voltage across a diode that occurs at the peak of the input cycle when the diode is reverse-biased.

Rectifier An electronic circuit that converts ac into pulsating dc; one part of a power supply.

Regulator An electronic device or circuit that maintains an essentially constant output voltage for a range of input voltage or load values; one part of a power supply.

Ripple voltage The small variation in the dc output voltage of a filtered rectifier caused by the charging and discharging of the filter capacitor.

Troubleshooting A systematic process of isolating, identifying, and correcting a fault in a circuit or system.

KEY FORMULAS

2–1	$V_{AVG} = \dfrac{V_p}{\pi}$	Half-wave average value
2–2	$V_{p(out)} = V_{p(in)} - 0.7\text{ V}$	Peak half-wave rectifier output (silicon)
2–3	$\text{PIV} = V_{p(in)}$	Peak inverse voltage, half-wave rectifier
2–4	$V_{AVG} = \dfrac{2V_p}{\pi}$	Full-wave average value
2–5	$V_{out} = \dfrac{V_{sec}}{2} - 0.7\text{ V}$	Center-tapped full-wave output
2–6	$\text{PIV} = 2V_{p(out)} + 0.7\text{ V}$	Peak inverse voltage, center-tapped rectifier
2–7	$V_{p(out)} = V_{p(sec)} - 1.4\text{ V}$	Bridge full-wave output
2–8	$\text{PIV} = V_{p(out)} + 0.7\text{ V}$	Peak inverse voltage, bridge rectifier
2–9	$r = \dfrac{V_{r(pp)}}{V_{DC}}$	Ripple factor
2–10	$V_{r(pp)} \cong \left(\dfrac{1}{fR_LC}\right)V_{p(rect)}$	Peak-to-peak ripple voltage, capacitor-input filter

2–11 $$V_{DC} = \left(1 - \frac{1}{2fR_LC}\right)V_{p(rect)}$$ DC output voltage, capacitor-input filter

2–12 $$\text{Line regulation} = \left(\frac{\Delta V_{OUT}}{\Delta V_{IN}}\right)100\%$$

2–13 $$\text{Load regulation} = \left(\frac{V_{NL} - V_{FL}}{V_{FL}}\right)100\%$$

TRUE/FALSE QUIZ

Answers are at the end of the chapter.

1. A basic half-wave rectifier consists of one diode.
2. The output frequency of a half-wave rectifier is twice the input frequency.
3. The diode in a half-wave rectifier conducts for half the input cycle.
4. PIV stands for positive inverse voltage.
5. Each diode in a full-wave rectifier conducts for the entire input cycle.
6. The output frequency of a full-wave rectifier is twice the input frequency.
7. A bridge rectifier uses four diodes.
8. In a bridge rectifier, two diodes conduct during each half cycle of the input.
9. The purpose of the capacitor filter in a rectifier is to convert ac to dc.
10. The output voltage of a filtered rectifier always has some ripple voltage.
11. A smaller filter capacitor reduces the ripple.
12. Line and load regulation are the same.
13. A diode limiter is also known as a clipper.
14. The purpose of a clamper is to remove a dc level from a waveform.
15. Voltage multipliers use diodes and capacitors.

CIRCUIT-ACTION QUIZ

Answers are at the end of the chapter.

1. If the input voltage in Figure 2–10 is increased, the peak inverse voltage across the diode will
   (a) increase **(b)** decrease **(c)** not change
2. If the turns ratio of the transformer in Figure 2–10 is decreased, the forward current through the diode will
   (a) increase **(b)** decrease **(c)** not change
3. If the frequency of the input voltage in Figure 2–18 is increased, the output voltage will
   (a) increase **(b)** decrease **(c)** not change
4. If the PIV rating of the diodes in Figure 2–18 is increased, the current through R_L will
   (a) increase **(b)** decrease **(c)** not change
5. If one of the diodes in Figure 2–23 opens, the average voltage to the load will
   (a) increase **(b)** decrease **(c)** not change
6. If the value of R_L in Figure 2–23 is decreased, the current through each diode will
   (a) increase **(b)** decrease **(c)** not change
7. If the capacitor value in Figure 2–30 is decreased, the output ripple voltage will
   (a) increase **(b)** decrease **(c)** not change
8. If the line voltage in Figure 2–33 is increased, ideally the +5 V output will
   (a) increase **(b)** decrease **(c)** not change
9. If the bias voltage in Figure 2–37 is decreased, the positive portion of the output voltage will
   (a) increase **(b)** decrease **(c)** not change
10. If the bias voltage in Figure 2–37 is increased, the negative portion of the output voltage will
    (a) increase **(b)** decrease **(c)** not change

11. If the value of R_3 in Figure 2–43 is decreased, the positive output voltage will
(a) increase (b) decrease (c) not change

12. If the input voltage in Figure 2–47 is increased, the peak negative value of the output voltage will
(a) increase (b) decrease (c) not change

SELF-TEST

Answers are at the end of the chapter.

Section 2–1

1. The average value of a half-wave rectified voltage with a peak value of 200 V is
(a) 63.7 V (b) 127.3 V (c) 141 V (d) 0 V

2. When a 60 Hz sinusoidal voltage is applied to the input of a half-wave rectifier, the output frequency is
(a) 120 Hz (b) 30 Hz (c) 60 Hz (d) 0 Hz

3. The peak value of the input to a half-wave rectifier is 10 V. The approximate peak value of the output is
(a) 10 V (b) 3.18 V (c) 10.7 V (d) 9.3 V

4. For the circuit in Question 3, the diode must be able to withstand a reverse voltage of
(a) 10 V (b) 5 V (c) 20 V (d) 3.18 V

Section 2–2

5. The average value of a full-wave rectified voltage with a peak value of 75 V is
(a) 53 V (b) 47.8 V (c) 37.5 V (d) 23.9 V

6. When a 60 Hz sinusoidal voltage is applied to the input of a full-wave rectifier, the output frequency is
(a) 120 Hz (b) 60 Hz (c) 240 Hz (d) 0 Hz

7. The total secondary voltage in a center-tapped full-wave rectifier is 125 V rms. Neglecting the diode drop, the rms output voltage is
(a) 125 V (b) 177 V (c) 100 V (d) 62.5 V

8. When the peak output voltage is 100 V, the PIV for each diode in a center-tapped full-wave rectifier is (neglecting the diode drop)
(a) 100 V (b) 200 V (c) 141 V (d) 50 V

9. When the rms output voltage of a bridge full-wave rectifier is 20 V, the peak inverse voltage across the diodes is (neglecting the diode drop)
(a) 20 V (b) 40 V (c) 28.3 V (d) 56.6 V

Section 2–3

10. The ideal dc output voltage of a capacitor-input filter is equal to
(a) the peak value of the rectified voltage
(b) the average value of the rectified voltage
(c) the rms value of the rectified voltage

11. A certain power-supply filter produces an output with a ripple of 100 mV peak-to-peak and a dc value of 20 V. The ripple factor is
(a) 0.05 (b) 0.005 (c) 0.00005 (d) 0.02

12. A 60 V peak full-wave rectified voltage is applied to a capacitor-input filter. If $f = 120$ Hz, $R_L = 10\ \text{k}\Omega$, and $C = 10\ \mu\text{F}$, the ripple voltage is
(a) 0.6 V (b) 6 mV (c) 5.0 V (d) 2.88 V

13. If the load resistance of a capacitor-filtered full-wave rectifier is reduced, the ripple voltage
(a) increases (b) decreases (c) is not affected (d) has a different frequency

14. Line regulation is determined by
(a) load current
(b) zener current and load current
(c) changes in load resistance and output voltage
(d) changes in output voltage and input voltage

15. Load regulation is determined by
   - (a) changes in load current and input voltage
   - (b) changes in load current and output voltage
   - (c) changes in load resistance and input voltage
   - (d) changes in zener current and load current

Section 2–4

16. A 10 V peak-to-peak sinusoidal voltage is applied across a silicon diode and series resistor. The maximum voltage across the diode is

   (a) 9.3 V **(b)** 5 V **(c)** 0.7 V **(d)** 10 V **(e)** 4.3 V

17. In a certain biased limiter, the bias voltage is 5 V and the input is a 10 V peak sine wave. If the positive terminal of the bias voltage is connected to the cathode of the diode, the maximum voltage at the anode is

   (a) 10 V **(b)** 5 V **(c)** 5.7 V **(d)** 0.7 V

18. In a certain positive clamper circuit, a 120 V rms sine wave is applied to the input. The dc value of the output is

   (a) 119.3 V **(b)** 169 V **(c)** 60 V **(d)** 75.6 V

Section 2–5

19. If the input voltage to a voltage tripler has an rms value of 12 V, the dc output voltage is approximately

   (a) 36 V **(b)** 50.9 V **(c)** 33.9 V **(d)** 32.4 V

20. In a rectifier circuit, if the secondary winding in the transformer opens, the output is

   (a) 0 V **(b)** 120 V **(c)** less than it should be **(d)** unaffected

Section 2–7

21. If one of the diodes in a bridge full-wave rectifier opens, the output is
   - (a) 0 V
   - (b) one-fourth the amplitude of the input voltage
   - (c) a half-wave rectified voltage
   - (d) a 120 Hz voltage

22. If you are checking a 60 Hz full-wave bridge rectifier and observe that the output has a 60 Hz ripple,
   - (a) the circuit is working properly
   - (b) there is an open diode
   - (c) the transformer secondary is shorted
   - (d) the filter capacitor is leaky

PROBLEMS

Answers to all odd-numbered problems are at the end of the book.

BASIC PROBLEMS

Section 2–1

Half-Wave Rectifiers

1. Draw the output voltage waveform for each circuit in Figure 2–73 and include the voltage values.

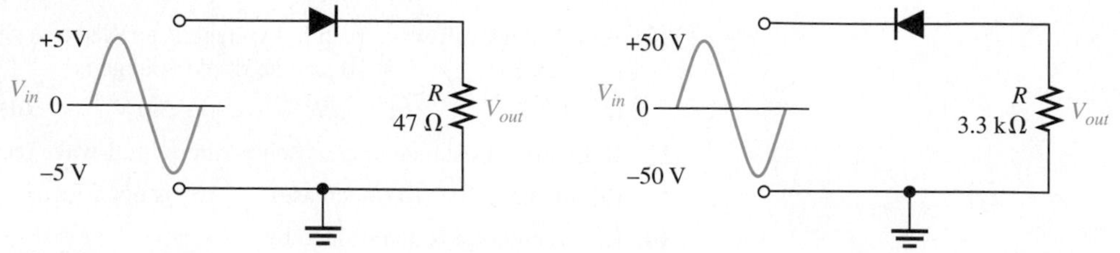

(a) (b)

▲ **FIGURE 2–73**

Multisim file circuits are identified with a CD logo and are in the Problems folder on your CD-ROM. Filenames correspond to figure numbers (e.g., F02-73).

2. What is the peak inverse voltage across each diode in Figure 2–73?
3. Calculate the average value of a half-wave rectified voltage with a peak value of 200 V.
4. What is the peak forward current through each diode in Figure 2–73?
5. A power-supply transformer has a turns ratio of 5:1. What is the secondary voltage if the primary is connected to a 120 V rms source?
6. Determine the peak and average power delivered to R_L in Figure 2–74.

FIGURE 2–74

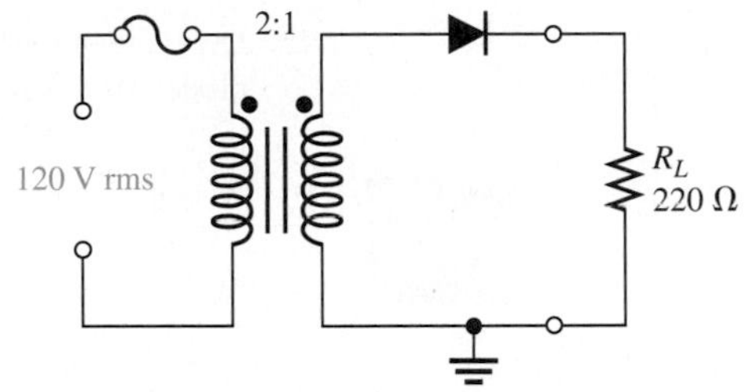

Section 2–2 Full-Wave Rectifiers

7. Find the average value of each voltage in Figure 2–75.

FIGURE 2–75

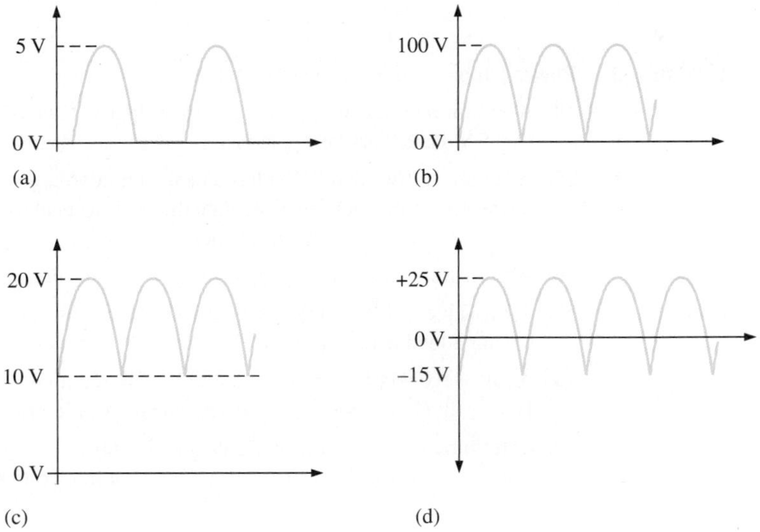

8. Consider the circuit in Figure 2–76.
   (a) What type of circuit is this?
   (b) What is the total peak secondary voltage?
   (c) Find the peak voltage across each half of the secondary.
   (d) Sketch the voltage waveform across R_L.
   (e) What is the peak current through each diode?
   (f) What is the PIV for each diode?

FIGURE 2–76

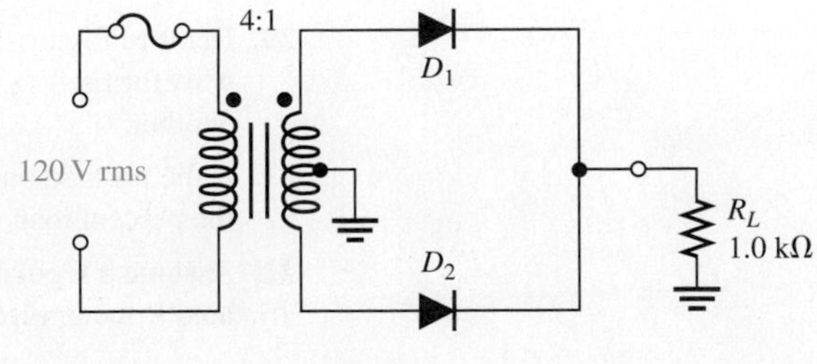

9. Calculate the peak voltage across each half of a center-tapped transformer used in a full-wave rectifier that has an average output voltage of 120 V.

10. Show how to connect the diodes in a center-tapped rectifier in order to produce a negative-going full-wave voltage across the load resistor.

11. What PIV rating is required for the diodes in a bridge rectifier that produces an average output voltage of 50 V?

12. The rms output voltage of a bridge rectifier is 20 V. What is the peak inverse voltage across the diodes?

13. Draw the output voltage waveform for the bridge rectifier in Figure 2–77. Notice that all the diodes are reversed from circuits shown earlier in the chapter.

FIGURE 2–77

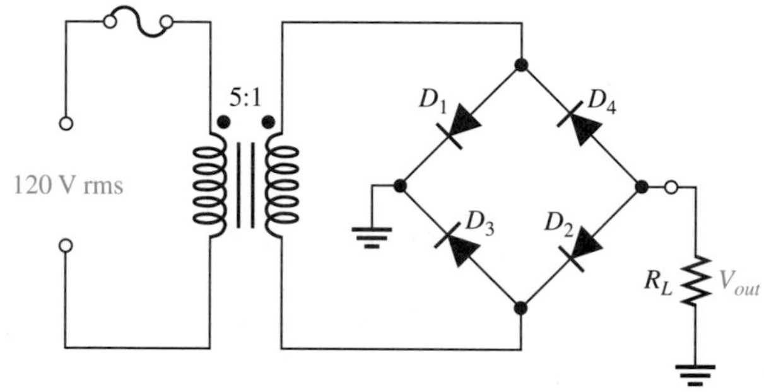

Section 2–3 **Power Supply Filters and Regulators**

14. A certain rectifier filter produces a dc output voltage of 75 V with a peak-to-peak ripple voltage of 0.5 V. Calculate the ripple factor.

15. A certain full-wave rectifier has a peak output voltage of 30 V. A 50 μF capacitor-input filter is connected to the rectifier. Calculate the peak-to-peak ripple and the dc output voltage developed across a 600 Ω load resistance.

16. What is the percentage of ripple for the rectifier filter in Problem 15?

17. What value of filter capacitor is required to produce a 1% ripple factor for a full-wave rectifier having a load resistance of 1.5 kΩ? Assume the rectifier produces a peak output of 18 V.

18. A full-wave rectifier produces an 80 V peak rectified voltage from a 60 Hz ac source. If a 10 μF filter capacitor is used, determine the ripple factor for a load resistance of 10 kΩ.

19. Determine the peak-to-peak ripple and dc output voltages in Figure 2–78. The transformer has a 36 V rms secondary voltage rating, and the line voltage has a frequency of 60 Hz.

FIGURE 2–78

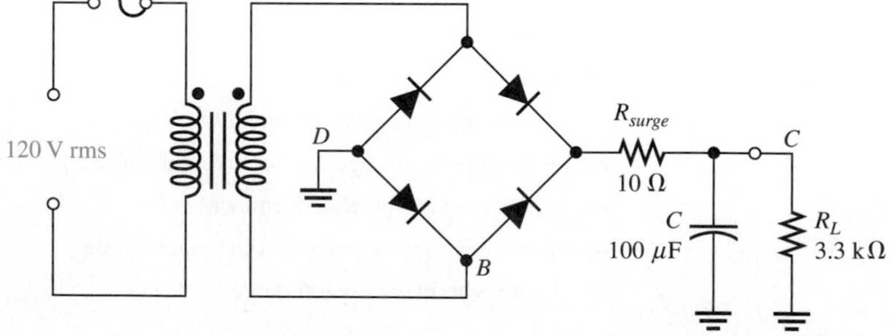

20. Refer to Figure 2–78 and draw the following voltage waveforms in relationship to the input waveforms: V_{AB}, V_{AD}, and V_{CD}. A double letter subscript indicates a voltage from one point to another.

21. If the no-load output voltage of a regulator is 15.5 V and the full-load output is 14.9 V, what is the percent load regulation?

22. Assume a regulator has a percent load regulation of 0.5%. What is the output voltage at full-load if the unloaded output is 12.0 V?

Section 2–4 **Dioде Limiting and Clamping Circuits**

23. Determine the output waveform for the circuit of Figure 2–79.

▶ FIGURE 2–79

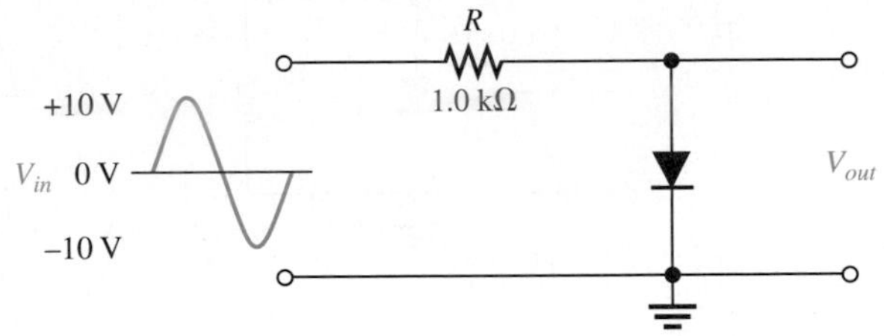

24. Determine the output voltage for the circuit in Figure 2–80(a) for each input voltage in (b), (c), and (d).

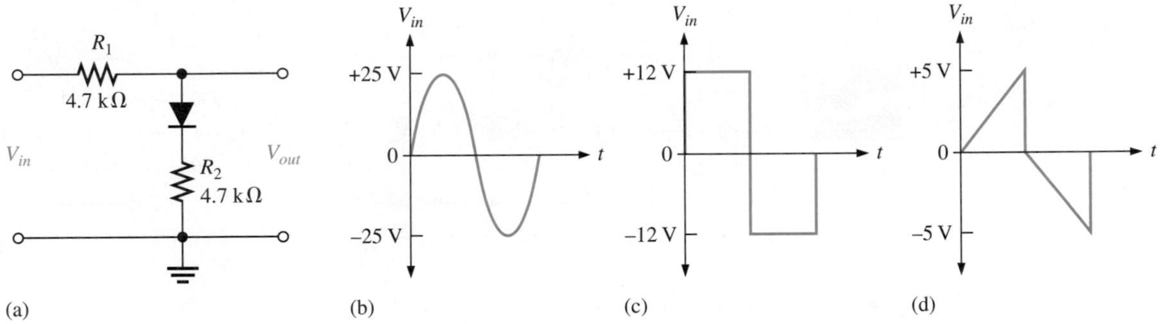

▲ FIGURE 2–80

25. Determine the output voltage waveform for each circuit in Figure 2–81.

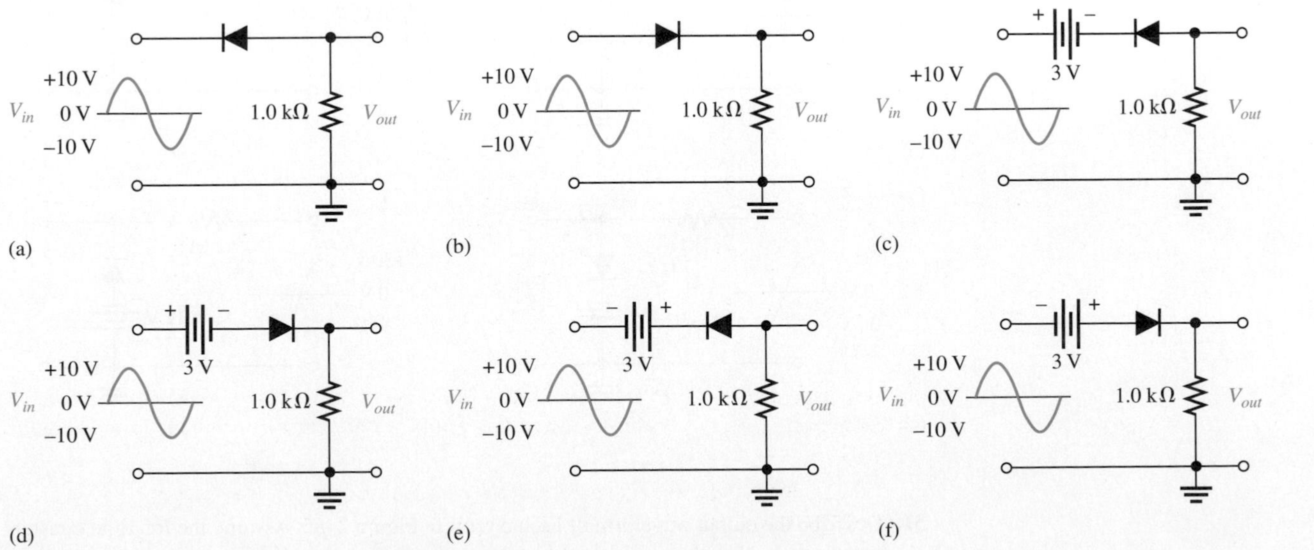

▲ FIGURE 2–81

26. Determine the R_L voltage waveform for each circuit in Figure 2–82.

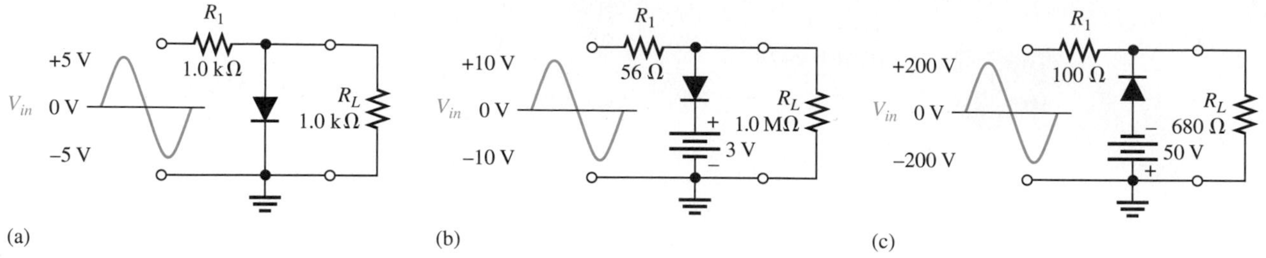

▲ FIGURE 2–82

27. Draw the output voltage waveform for each circuit in Figure 2–83.

28. Determine the peak forward current through each diode in Figure 2–83.

► FIGURE 2–83

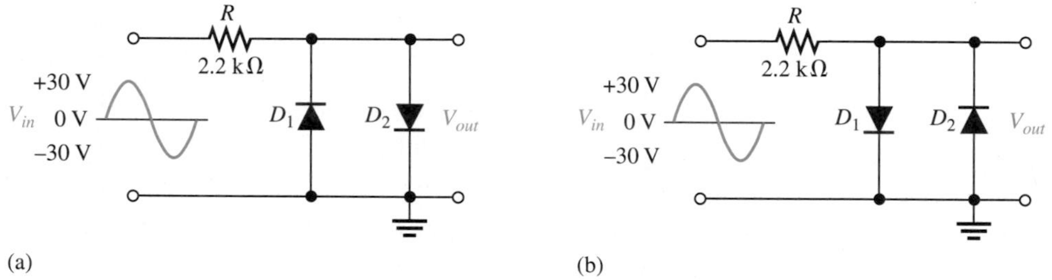

29. Determine the peak forward current through each diode in Figure 2–84.

30. Determine the output voltage waveform for each circuit in Figure 2–84.

► FIGURE 2–84

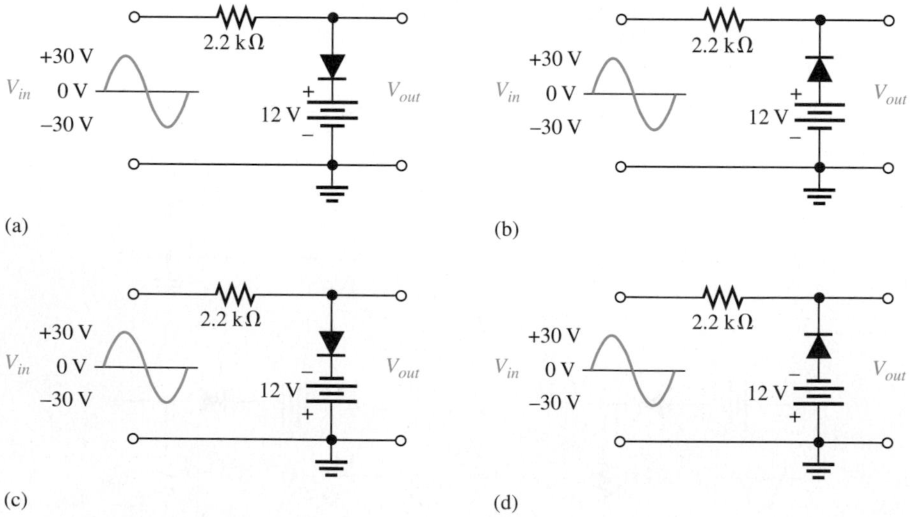

31. Describe the output waveform of each circuit in Figure 2–85. Assume the *RC* time constant is much greater than the period of the input.

32. Repeat Problem 31 with the diodes turned around.

FIGURE 2–85

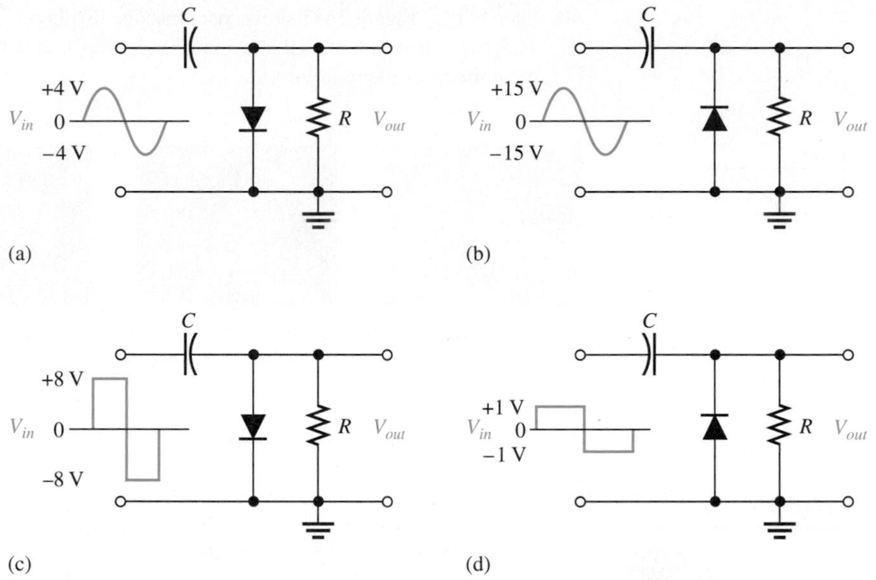

Section 2–5 **Voltage Multipliers**

33. A certain voltage doubler has 20 V rms on its input. What is the output voltage? Draw the circuit, indicating the output terminals and PIV rating for the diode.

34. Repeat Problem 33 for a voltage tripler and quadrupler.

Section 2–6 **The Diode Datasheet**

35. From the datasheet in Figure 2–53, determine how much peak inverse voltage that a 1N4002 diode can withstand.

36. Repeat Problem 35 for a 1N4007.

37. If the peak output voltage of a bridge full-wave rectifier is 50 V, determine the minimum value of the load resistance that can be used when 1N4002 diodes are used.

Section 2–7 **Troubleshooting**

38. If one of the diodes in a bridge rectifier opens, what happens to the output?

39. From the meter readings in Figure 2–86, determine if the rectifier is functioning properly. If it is not, determine the most likely failure(s).

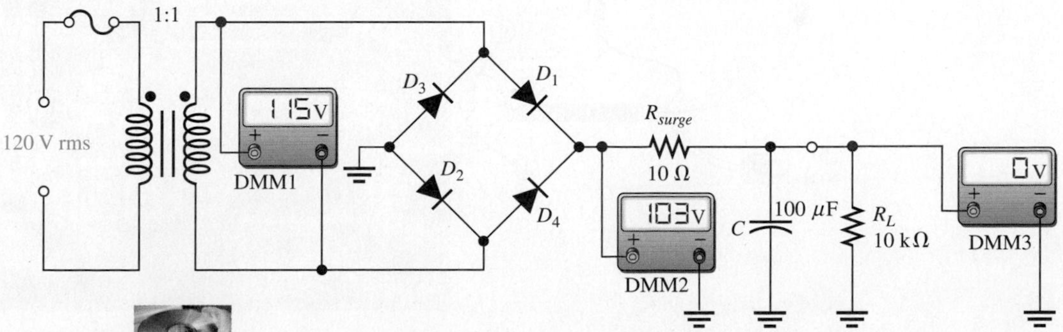

FIGURE 2–86

40. Each part of Figure 2–87 shows oscilloscope displays of various rectifier output voltages. In each case, determine whether or not the rectifier is functioning properly and if it is not, determine the most likely failure(s).

FIGURE 2–87

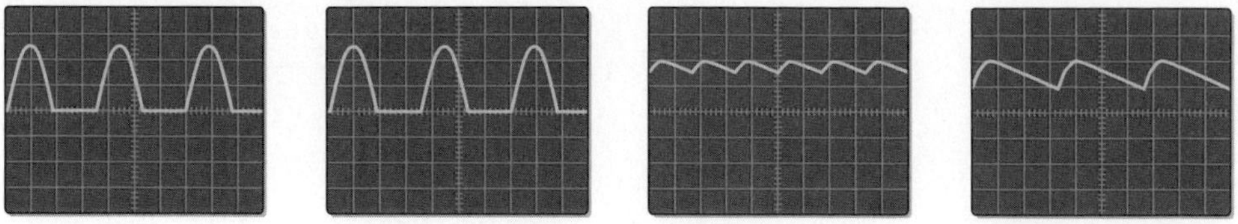

(a) Output of a half-wave unfiltered rectifier (b) Output of a full-wave unfiltered rectifier (c) Output of a full-wave filter (d) Output of same full-wave filter as part (c)

41. Based on the values given, would you expect the circuit in Figure 2–88 to fail? If so, why?

FIGURE 2–88

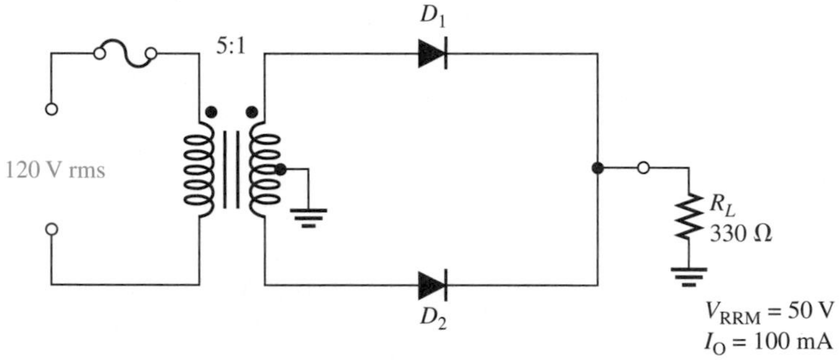

APPLICATION ACTIVITY PROBLEMS

42. Determine the most likely failure(s) in the circuit of Figure 2–89 for each of the following symptoms. State the corrective action you would take in each case. The transformer has a rated output of 10 V rms.

(a) No voltage from test point 1 to test point 2

(b) No voltage from test point 3 to test point 4

(c) 8 V rms from test point 3 to test point 4

(d) Excessive 120 Hz ripple voltage at test point 6

(e) There is a 60 Hz ripple voltage at test point 6

(f) No voltage at test point 6

FIGURE 2–89

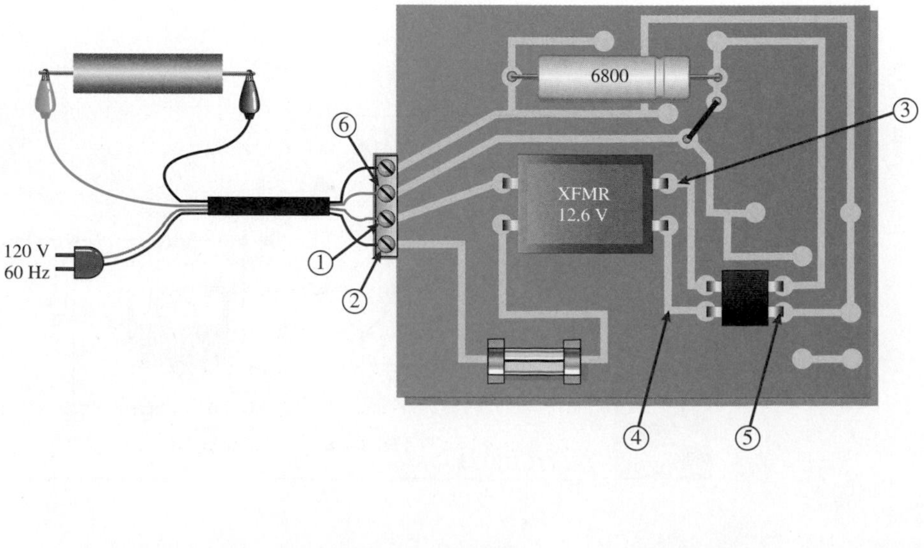

43. In testing the power supply circuit in Figure 2–89 with a 10 kΩ load resistor connected, you find the voltage at the positive side of the filter capacitor to have a 60 Hz ripple voltage. You replace the bridge rectifier and check the point again but it still has the 60 Hz ripple. What now?

44. Suppose the bridge rectifier in Figure 2–89 is connected backwards such that the transformer secondary is now connected to the output pins instead of the input pins. What will be observed at test point 6?

ADVANCED PROBLEMS

45. A full-wave rectifier with a capacitor-input filter provides a dc output voltage of 35 V to a 3.3 kΩ load. Determine the minimum value of filter capacitor if the maximum peak-to-peak ripple voltage is to be 0.5 V.

46. A certain unfiltered full-wave rectifier with 120 V, 60 Hz input produces an output with a peak of 15 V. When a capacitor-input filter and a 1.0 kΩ load are connected, the dc output voltage is 14 V. What is the peak-to-peak ripple voltage?

47. For a certain full-wave rectifier, the measured surge current in the capacitor filter is 50 A. The transformer is rated for a secondary voltage of 24 V with a 120 V, 60 Hz input. Determine the value of the surge resistor in this circuit.

48. Design a full-wave rectifier using an 18 V center-tapped transformer. The output ripple is not to exceed 5% of the output voltage with a load resistance of 680 Ω. Specify the $I_{F(AV)}$ and PIV ratings of the diodes and select an appropriate diode from Figure 2–53.

49. Design a filtered power supply that can produce dc output voltages of +9 V ± 10% and −9 V ± 10% with a maximum load current of 100 mA. The voltages are to be switch selectable across one set of output terminals. The ripple voltage must not exceed 0.25 V rms.

50. Design a circuit to limit a 20 V rms sinusoidal voltage to a maximum positive amplitude of 10 V and a maximum negative amplitude of −5 V using a single 14 V dc voltage source.

51. Determine the voltage across each capacitor in the circuit of Figure 2–90.

▶ **FIGURE 2–90**

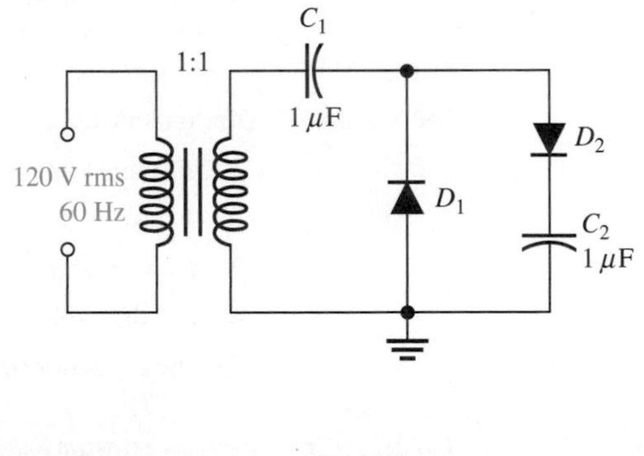

MULTISIM TROUBLESHOOTING PROBLEMS

These file circuits are in the Troubleshooting Problems folder on your CD-ROM.

52. Open file TSP02-52 and determine the fault.

53. Open file TSP02-53 and determine the fault.

54. Open file TSP02-54 and determine the fault.

55. Open file TSP02-55 and determine the fault.

56. Open file TSP02-56 and determine the fault.

57. Open file TSP02-57 and determine the fault.

58. Open file TSP02-58 and determine the fault.

59. Open file TSP02-59 and determine the fault.

60. Open file TSP02-60 and determine the fault.

ANSWERS

SECTION CHECKUPS

Section 2–1 Half-Wave Rectifiers

1. PIV across the diode occurs at the peak of the input when the diode is reversed biased.
2. There is current through the load for approximately half (50%) of the input cycle.
3. The average value is $10\text{ V}/\pi = 3.18\text{ V}$.
4. The peak output voltage is $25\text{ V} - 0.7\text{ V} = 24.3\text{ V}$.
5. The PIV sating must be at least 60 V.

Section 2–2 Full-Wave Rectifiers

1. A full-wave voltage occurs on each half of the input cycle and has a frequency of twice the input frequency. A half-wave voltage occurs once each input cycle and has a frequency equal to the input frequency.
2. The average value of $2(60\text{ V})/\pi = 38.12\text{ V}$
3. The bridge rectifier has the greater output voltage.
4. The 50 V diodes must be used in the bridge rectifier.
5. In the center-tapped rectifier, diodes with a PIV rating of at least 90 V would be required.

Section 2–3 Power Supply Filters and Regulators

1. The output frequency is 60 Hz.
2. The output frequency is 120 Hz.
3. The ripple voltage is caused by the slight charging and discharging of the capacitor through the load resistor.
4. The ripple voltage amplitude increases when the load resistance decreases.
5. Ripple factor is the ratio of the ripple voltage to the average or dc voltage.
6. Input regulation is a measure of the variation in output voltage over a range of input voltages. Load regulation is a measure of the variation in output voltage over a range of load current values.

Section 2–4 Diode Limiting and Clamping Circuits

1. Limiters clip off or remove portions of a waveform. Clampers insert a dc level.
2. A positive limiter clips off positive voltages. A negative limiter clips off negative voltages.
3 0.7 V appears across the diode.
4. The bias voltage must be $5\text{ V} - 0.7\text{ V} = 4.3\text{ V}$.
5. The capacitor acts as a battery.

Section 2–5 Voltage Multipliers

1. The peak voltage rating must be 100 V.
2. The PIV rating must be at least 310 V.

Section 2–6 The Diode Datasheet

1. 1N4002: 100 V; 1N4003: 200 V; 1N4004: 400 V; 1N4005: 600 V; 1N4006: 800 V
2. No
3. Approximately 0.65 A
4. 12 A

Section 2–7 Troubleshooting

1. An open diode results in no output voltage.
2. An open diode produces a half-wave output voltage.

3. The shorted diode may burn open. Transformer will be damaged. Fuse will blow.

4. The amplitude of the ripple voltage increases with a leaky filter capacitor.

5. There will be no output voltage when the primary opens.

6. The problem may be a partially shorted secondary winding.

RELATED PROBLEMS FOR EXAMPLES

2–1 3.82 V

2–2 **(a)** 2.3 V **(b)** 49.3 V

2–3 **(a)** 623.3 V **(b)** 624 V **(c)** negative half-cycles rather than positive half cycles

2–4 98.7 V

2–5 79.3 V including diode drop

2–6 41.0 V; 41.7 V

2–7 26.9 mV

2–8 3.7%

2–9 A positive peak of 9.9 V and clipped at −0.7 V

2–10 Limited at +10.7 V and −10.7 V

2–11 Change R_3 to 100 Ω or R_2 to 220 Ω.

2–12 Same voltage waveform as Figure 2–48

2–13 Verify *C* is shorted and replace it.

TRUE/FALSE QUIZ

1. T **2.** F **3.** T **4.** F **5.** F **6.** T **7.** T **8.** T

9. F **10.** T **11.** F **12.** F **13.** T **14.** F **15.** T

CIRCUIT-ACTION QUIZ

1. (a) **2.** (b) **3.** (c) **4.** (c) **5.** (b) **6.** (a)

7. (a) **8.** (c) **9.** (b) **10.** (c) **11.** (b) **12.** (a)

SELF-TEST

1. (a) **2.** (c) **3.** (d) **4.** (a) **5.** (b) **6.** (a) **7.** (d) **8.** (b)

9. (c) **10.** (a) **11.** (b) **12.** (c) **13.** (a) **14.** (d) **15.** (b) **16.** (d)

17. (c) **18.** (b) **19.** (b) **20.** (a) **21.** (c) **22.** (b)

3 SPECIAL-PURPOSE DIODES

CHAPTER OUTLINE

CHAPTER OBJECTIVES

- Describe the characteristics of a zener diode and analyze its operation
- Explain how a zener can be used in voltage reference, regulation, and limiting applications
- Describe the variable-capacitance characteristics of a varactor diode and analyze its operation in a typical circuit
- Discuss the operation and characteristics of LEDs and photodiodes
- Discuss the basic characteristics of laser, Schottky, *pin*, step-recovery, tunnel, and current regulator diodes
- Troubleshoot zener diode regulators

KEY TERMS

- Zener diode
- Zener breakdown
- Varactor
- Light-emitting diode (LED)
- Electroluminescence
- Pixel
- Photodiode
- Laser

VISIT THE COMPANION WEBSITE

Study aids for this chapter are available at http://www.prenhall.com/floyd

INTRODUCTION

Chapter 2 was devoted to general-purpose and rectifier diodes, which are the most widely used types. In this chapter, we will cover several other types of diodes that are designed for specific applications, including the zener, varactor (variable-capacitance), light-emitting, photo, laser, Schottky, tunnel, *pin*, step-recovery, and current regulator diodes.

APPLICATION ACTIVITY PREVIEW

The Application Activity in this chapter is the expansion of the 16 V power supply developed in Chapter 2 into a 12 V regulated power supply with an LED power-on indicator. The new circuit will incorporate a voltage regulator IC, which is introduced in this chapter.

3–1 THE ZENER DIODE

A major application for zener diodes is as a type of voltage regulator for providing stable reference voltages for use in power supplies, voltmeters, and other instruments. In this section, you will see how the zener diode maintains a nearly constant dc voltage under the proper operating conditions. You will learn the conditions and limitations for properly using the zener diode and the factors that affect its performance.

After completing this section, you should be able to

- **Describe the characteristics of a zener diode and analyze its operation**
  - Identify a zener diode by its symbol
  - Discuss avalanche and zener breakdown
  - Analyze the *V-I* characteristic curve of a zener diode
  - Discuss the zener equivalent circuits
  - Define temperature coefficient and apply it to zener analysis
  - Discuss power dissipation in a zener and apply derating
  - Interpret a zener diode datasheet

The symbol for a zener diode is shown in Figure 3–1. Instead of a straight line representing the cathode, the zener diode has a bent line that reminds you of the letter Z (for zener). A **zener diode** is a silicon *pn* junction device that is designed for operation in the reverse-breakdown region. The breakdown voltage of a zener diode is set by carefully controlling the doping level during manufacture. Recall, from the discussion of the diode characteristic curve in Chapter 1, that when a diode reaches reverse breakdown, its voltage remains almost constant even though the current changes drastically, and this is the key to zener diode operation. This volt-ampere characteristic is shown again in Figure 3–2 with the normal operating region for zener diodes shown as a shaded area.

▲ **FIGURE 3–1**

Zener diode symbol.

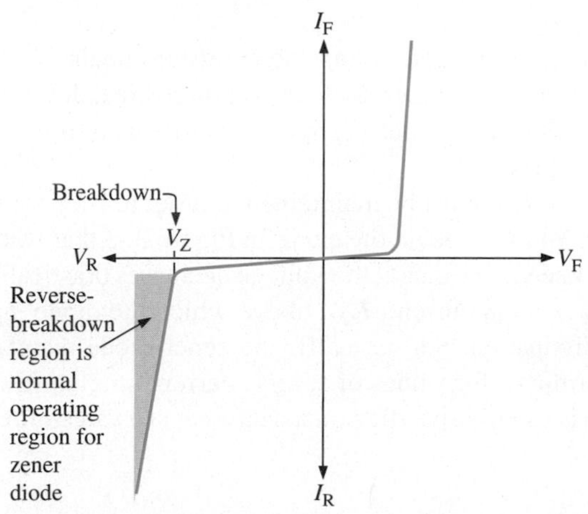

◄ **FIGURE 3–2**

General zener diode *V-I* characteristic.

Zener Breakdown

Zener diodes are designed to operate in reverse breakdown. Two types of reverse breakdown in a zener diode are *avalanche* and *zener*. The avalanche effect, discussed in Chapter 1, occurs in both rectifier and zener diodes at a sufficiently high reverse voltage. **Zener breakdown**

HISTORY NOTE

Clarence Melvin Zener, an American physicist, was born in Indianapolis and earned his PhD from Harvard in 1930. He was first to describe the properties of reverse breakdown that are exploited by the zener diode. As a result, Bell Labs, where the device was developed, named the diode after him. He was also involved in areas of superconductivity, metallurgy, and geometric programming.

occurs in a zener diode at low reverse voltages. A zener diode is heavily doped to reduce the breakdown voltage. This causes a very thin depletion region. As a result, an intense electric field exists within the depletion region. Near the zener breakdown voltage (V_Z), the field is intense enough to pull electrons from their valence bands and create current.

Zener diodes with breakdown voltages of less than approximately 5 V operate predominately in zener breakdown. Those with breakdown voltages greater than approximately 5 V operate predominately in **avalanche breakdown.** Both types, however, are called *zener diodes.* Zeners are commercially available with breakdown voltages from less than 1 V to more than 250 V with specified tolerances from 1% to 20%.

Breakdown Characteristics

Figure 3–3 shows the reverse portion of a zener diode's characteristic curve. Notice that as the reverse voltage (V_R) is increased, the reverse current (I_R) remains extremely small up to the "knee" of the curve. The reverse current is also called the zener current, I_Z. At this point, the breakdown effect begins; the internal zener resistance, also called zener impedance (Z_Z), begins to decrease as the reverse current increases rapidly. From the bottom of the knee, the zener breakdown voltage (V_Z) remains essentially constant although it increases slightly as the zener current, I_Z, increases.

▶ FIGURE 3–3

Reverse characteristic of a zener diode. V_Z is usually specified at a value of the zener current known as the test current.

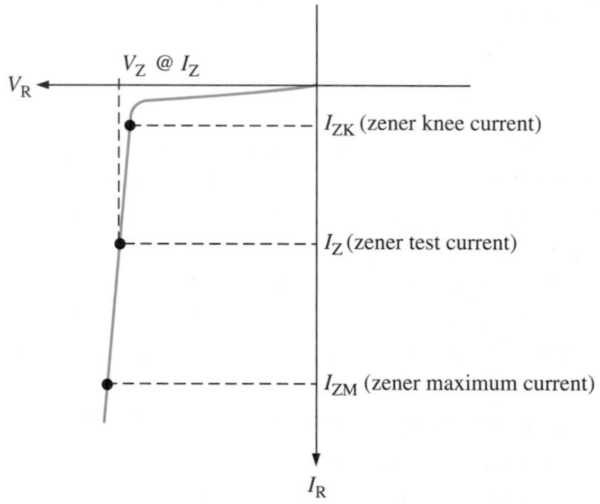

Zener Regulation The ability to keep the reverse voltage across its terminals essentially constant is the key feature of the zener diode. A zener diode operating in breakdown acts as a voltage regulator because it maintains a nearly constant voltage across its terminals over a specified range of reverse-current values.

A minimum value of reverse current, I_{ZK}, must be maintained in order to keep the diode in breakdown for voltage regulation. You can see on the curve in Figure 3–3 that when the reverse current is reduced below the knee of the curve, the voltage decreases drastically and regulation is lost. Also, there is a maximum current, I_{ZM}, above which the diode may be damaged due to excessive power dissipation. So, basically, the zener diode maintains a nearly constant voltage across its terminals for values of reverse current ranging from I_{ZK} to I_{ZM}. A nominal zener voltage, V_Z, is usually specified on a datasheet at a value of reverse current called the *zener test current.*

Ideal Zener Equivalent Circuit

Figure 3–4 shows the ideal model (first approximation) of a zener diode in reverse breakdown and its ideal characteristic curve. It has a constant voltage drop equal to the nominal zener voltage. This constant voltage drop across the zener diode produced by reverse breakdown is represented by a dc voltage symbol even though the zener diode does not produce a voltage.

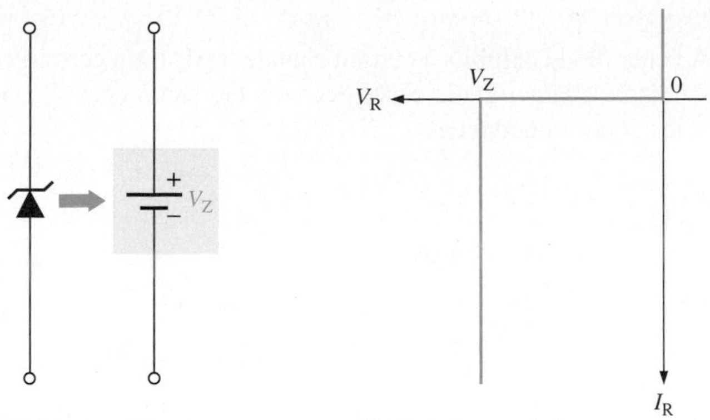

(a) Ideal model

(b) Ideal characteristic curve

FIGURE 3–4

Ideal zener diode equivalent circuit model and the characteristic curve.

Practical Zener Equivalent Circuit

Figure 3–5(a) represents the practical model (second approximation) of a zener diode, where the zener impedance (resistance), Z_Z, is included. Since the actual voltage curve is not ideally vertical, a change in zener current (ΔI_Z) produces a small change in zener voltage (ΔV_Z), as illustrated in Figure 3–5(b). By Ohm's law, the ratio of ΔV_Z to ΔI_Z is the impedance, as expressed in the following equation:

$$Z_Z = \frac{\Delta V_Z}{\Delta I_Z}$$

Equation 3–1

Normally, Z_Z is specified at the zener test current. In most cases, you can assume that Z_Z is a small constant over the full range of zener current values and is purely resistive. It is best to avoid operating a zener diode near the knee of the curve because the impedance changes dramatically in that area.

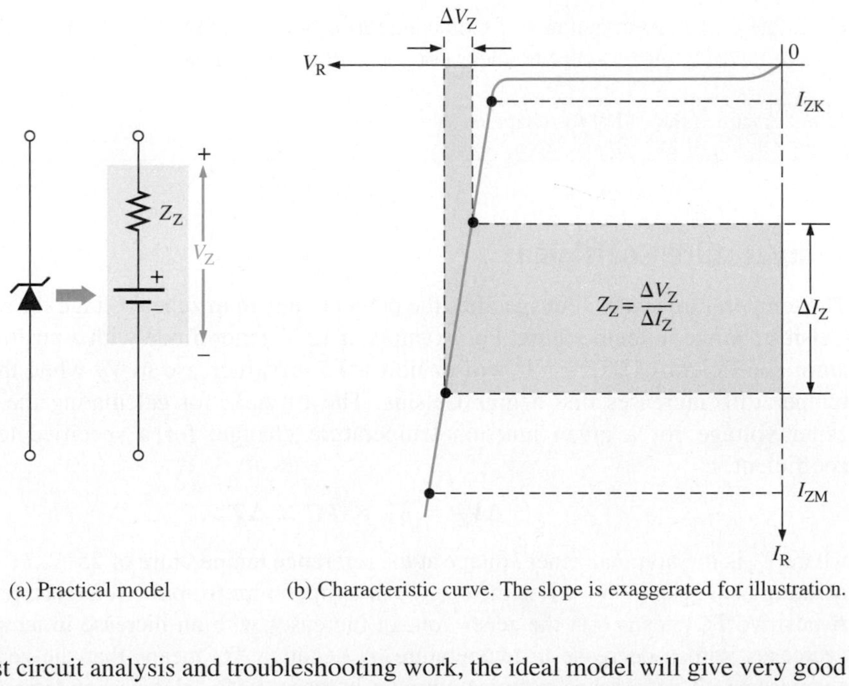

(a) Practical model

(b) Characteristic curve. The slope is exaggerated for illustration.

FIGURE 3–5

Practical zener diode equivalent circuit and the characteristic curve illustrating Z_Z.

For most circuit analysis and troubleshooting work, the ideal model will give very good results and is much easier to use than more complicated models. When a zener diode is operating normally, it will be in reverse breakdown and you should observe the nominal breakdown voltage across it. Most **schematics** will indicate on the drawing what this voltage should be.

EXAMPLE 3–1

A zener diode exhibits a certain change in V_Z for a certain change in I_Z on a portion of the linear characteristic curve between I_{ZK} and I_{ZM} as illustrated in Figure 3–6. What is the zener impedance?

FIGURE 3–6

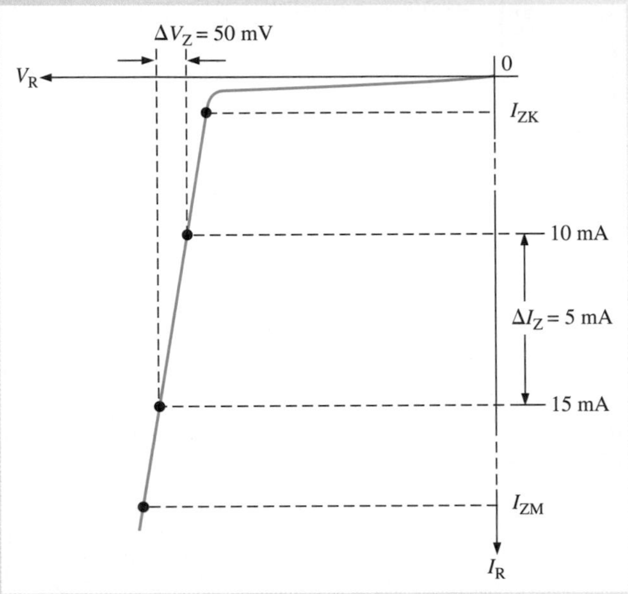

Solution

$$Z_Z = \frac{\Delta V_Z}{\Delta I_Z} = \frac{50 \text{ mV}}{5 \text{ mA}} = \mathbf{10\ \Omega}$$

*Related Problem** Calculate the zener impedance if the change in zener voltage is 100 mV for a 20 mA change in zener current on the linear portion of the characteristic curve.

*Answers are at the end of the chapter.

Temperature Coefficient

The temperature coefficient specifies the percent change in zener voltage for each degree Celsius change in temperature. For example, a 12 V zener diode with a positive temperature coefficient of 0.01%/°C will exhibit a 1.2 mV increase in V_Z when the junction temperature increases one degree Celsius. The formula for calculating the change in zener voltage for a given junction temperature change, for a specified temperature coefficient, is

Equation 3–2

$$\Delta V_Z = V_Z \times TC \times \Delta T$$

where V_Z is the nominal zener voltage at the reference temperature of 25°C, TC is the temperature coefficient, and ΔT is the change in temperature from the reference temperature. A positive TC means that the zener voltage increases with an increase in temperature or decreases with a decrease in temperature. A negative TC means that the zener voltage decreases with an increase in temperature or increases with a decrease in temperature.

In some cases, the temperature coefficient is expressed in mV/°C rather than as %/°C. For these cases, ΔV_Z is calculated as

Equation 3–3

$$\Delta V_Z = TC \times \Delta T$$

EXAMPLE 3–2 An 8.2 V zener diode (8.2 V at 25°C) has a positive temperature coefficient of 0.05%/°C. What is the zener voltage at 60°C?

Solution The change in zener voltage is

$$\Delta V_Z = V_Z \times TC \times \Delta T = (8.2\text{ V})(0.05\%/^\circ\text{C})(60^\circ\text{C} - 25^\circ\text{C})$$
$$= (8.2\text{ V})(0.0005/^\circ\text{C})(35^\circ\text{C}) = 144\text{ mV}$$

Notice that 0.05%/°C was converted to 0.0005/°C. The zener voltage at 60°C is

$$V_Z + \Delta V_Z = 8.2\text{ V} + 144\text{ mV} = \mathbf{8.34\text{ V}}$$

Related Problem A 12 V zener has a positive temperature coefficient of 0.075%/°C. How much will the zener voltage change when the junction temperature decreases 50 degrees Celsius?

Zener Power Dissipation and Derating

Zener diodes are specified to operate at a maximum power called the maximum dc power dissipation, $P_{D(max)}$. For example, the 1N746 zener is rated at a $P_{D(max)}$ of 500 mW and the 1N3305A is rated at a $P_{D(max)}$ of 50 W. The dc power dissipation is determined by the formula,

$$P_D = V_Z I_Z$$

Power Derating The maximum power dissipation of a zener diode is typically specified for temperatures at or below a certain value (50°C, for example). Above the specified temperature, the maximum power dissipation is reduced according to a derating factor. The derating factor is expressed in mW/°C. The maximum derated power can be determined with the following formula:

$$P_{D(derated)} = P_{D(max)} - (\text{mW}/^\circ\text{C})\Delta T$$

EXAMPLE 3–3 A certain zener diode has a maximum power rating of 400 mW at 50°C and a derating factor of 3.2 mW/°C. Determine the maximum power the zener can dissipate at a temperature of 90°C.

Solution

$$P_{D(derated)} = P_{D(max)} - (\text{mW}/^\circ\text{C})\Delta T$$
$$= 400\text{ mW} - (3.2\text{ mW}/^\circ\text{C})(90^\circ\text{C} - 50^\circ\text{C})$$
$$= 400\text{ mW} - 128\text{ mW} = \mathbf{272\text{ mW}}$$

Related Problem A certain 50 W zener diode must be derated with a derating factor of 0.5 W/°C above 75°C. Determine the maximum power it can dissipate at 160°C.

Zener Diode Datasheet Information

The amount and type of information found on datasheets for zener diodes (or any category of electronic device) varies from one type of diode to the next. The datasheet for some zeners contains more information than for others. Figure 3–7 gives an example of the type of information you have studied that can be found on a typical datasheet. This particular information is for a popular zener series, the 1N4728A–1N4764A.

▶ FIGURE 3–7

Partial datasheet for the 1N4728A–1N4764A series 1 W zener diodes. © 2005 Fairchild Semiconductor Corporation. Used by permission.

FAIRCHILD
SEMICONDUCTOR®

January 2005

1N4728A - 1N4764A

Zeners

DO-41 Glass case
COLOR BAND DENOTES CATHODE

Absolute Maximum Ratings * T_a = 25°C unless otherwise noted

Symbol	Parameter	Value	Units
P_D	Power Dissipation @ TL ≤ 50°C, Lead Length = 3/8"	1.0	W
	Derate above 50°C	6.67	mW/°C
T_J, T_{STG}	Operating and Storage Temperature Range	-65 to +200	°C

* These ratings are limiting values above which the serviceability of the diode may be impaired.

Electrical Characteristics T_a = 25°C unless otherwise noted

Device	V_Z (V) @ I_Z (Note 1)			Test Current I_Z (mA)	Max. Zener Impedance			Leakage Current	
	Min.	Typ.	Max.		Z_Z @ I_Z (Ω)	Z_{ZK} @ I_{ZK} (Ω)	I_{ZK} (mA)	I_R (μA)	V_R (V)
1N4728A	3.315	3.3	3.465	76	10	400	1	100	1
1N4729A	3.42	3.6	3.78	69	10	400	1	100	1
1N4730A	3.705	3.9	4.095	64	9	400	1	50	1
1N4731A	4.085	4.3	4.515	58	9	400	1	10	1
1N4732A	4.465	4.7	4.935	53	8	500	1	10	1
1N4733A	4.845	5.1	5.355	49	7	550	1	10	1
1N4734A	5.32	5.6	5.88	45	5	600	1	10	2
1N4735A	5.89	6.2	6.51	41	2	700	1	10	3
1N4736A	6.46	6.8	7.14	37	3.5	700	1	10	4
1N4737A	7.125	7.5	7.875	34	4	700	0.5	10	5
1N4738A	7.79	8.2	8.61	31	4.5	700	0.5	10	6
1N4739A	8.645	9.1	9.555	28	5	700	0.5	10	7
1N4740A	9.5	10	10.5	25	7	700	0.25	10	7.6
1N4741A	10.45	11	11.55	23	8	700	0.25	5	8.4
1N4742A	11.4	12	12.6	21	9	700	0.25	5	9.1
1N4743A	12.35	13	13.65	19	10	700	0.25	5	9.9
1N4744A	14.25	15	15.75	17	14	700	0.25	5	11.4
1N4745A	15.2	16	16.8	15.5	16	700	0.25	5	12.2
1N4746A	17.1	18	18.9	14	20	750	0.25	5	13.7
1N4747A	19	20	21	12.5	22	750	0.25	5	15.2

Electrical Characteristics T_C = 25°C unless otherwise noted

Device	V_Z (V) @ I_Z (Note 1)			Test Current I_Z (mA)	Max. Zener Impedance			Leakage Current	
	Min.	Typ.	Max.		Z_Z @ I_Z (Ω)	Z_{ZK} @ I_{ZK} (Ω)	I_{ZK} (mA)	I_R (μA)	V_R (V)
1N4748A	20.9	22	23.1	11.5	23	750	0.25	5	16.7
1N4749A	22.8	24	25.2	10.5	25	750	0.25	5	18.2
1N4750A	25.65	27	28.35	9.5	35	750	0.25	5	20.6
1N4751A	28.5	30	31.5	8.5	40	1000	0.25	5	22.8
1N4752A	31.35	33	34.65	7.5	45	1000	0.25	5	25.1
1N4753A	34.2	36	37.8	7	50	1000	0.25	5	27.4
1N4754A	37.05	39	40.95	6.5	60	1000	0.25	5	29.7
1N4755A	40.85	43	45.15	6	70	1500	0.25	5	32.7
1N4756A	44.65	47	49.35	5.5	80	1500	0.25	5	35.8
1N4757A	48.45	51	53.55	5	95	1500	0.25	5	38.8
1N4758A	53.2	56	58.8	4.5	110	2000	0.25	5	42.6
1N4759A	58.9	62	65.1	4	125	2000	0.25	5	47.1
1N4760A	64.6	68	71.4	3.7	150	2000	0.25	5	51.7
1N4761A	71.25	75	78.75	3.3	175	2000	0.25	5	56
1N4762A	77.9	82	86.1	3	200	3000	0.25	5	62.2
1N4763A	86.45	91	95.55	2.8	250	3000	0.25	5	69.2
1N4764A	95	100	105	2.5	350	3000	0.25	5	76

Notes:
1. Zener Voltage (V_Z)
The zener voltage is measured with the device junction in the thermal equilibrium at the lead temperature (T_L) at 30°C ± 1°C and 3/8" lead length.

Absolute Maximum Ratings The maximum power dissipation, P_D, is specified as 1.0 W up to 50°C. Generally, the zener diode should be operated at least 20% below this maximum to assure reliability and longer life. The power dissipation is derated as shown on the datasheet at 6.67 mW for each degree above 50°C. For example, using the procedure illustrated in Example 3–3, the maximum power dissipation at 60°C is

$$P_D = 1\text{ W} - 10°\text{C}(6.67\text{ mW}/°\text{C}) = 1\text{ W} - 66.7\text{ mW} = 0.9933\text{ W}$$

At 125°C, the maximum power dissipation is

$$P_D = 1\text{ W} - 75°\text{C}(6.67\text{ mW}/°\text{C}) = 1\text{ W} - 500.25\text{ mW} = 0.4998\text{ W}$$

Notice that a maximum reverse current is not specified but can be determined from the maximum power dissipation for a given value of V_Z. For example, at 50°C, the maximum zener current for a zener voltage of 3.3 V is

$$I_{ZM} = \frac{P_D}{V_Z} = \frac{1\text{ W}}{3.3\text{ V}} = 303\text{ mA}$$

The operating junction temperature, T_J, and the storage temperature, T_{STG}, have a range of from −65°C to 200°C.

Electrical Characteristics The first column in the datasheet lists the zener type numbers, 1N4728A through 1N4764A.

Zener voltage, V_Z, and zener test current, I_Z For each device type, the minimum, typical, and maximum zener voltages are listed. V_Z is measured at the specified zener test current, I_Z. For example, the zener voltage for a 1N4728A can range from 3.315 V to 3.465 V with a typical value of 3.3 V at a test current of 76 mA.

Maximum zener impedance Z_Z is the maximum zener impedance at the specified test current, I_Z. For example, for a 1N4728A, Z_Z is 10 Ω at 76 mA. The maximum zener impedance, Z_{ZK}, at the knee of the characteristic curve is specified at I_{ZK}, which is the current at the knee of the curve. For example, Z_{ZK} is 400 Ω at 1 mA for a 1N4728A.

Leakage current Reverse leakage current is specified for a reverse voltage that is less than the knee voltage. This means that the zener is not in reverse breakdown for these measurements. For example I_R is 100 μA for a reverse voltage of 1 V in a 1N4728A.

EXAMPLE 3–4

From the datasheet in Figure 3–7, a 1N4736A zener diode has a Z_Z of 3.5 Ω. The datasheet gives V_Z = 6.8 V at a test current, I_Z, of 37 mA. What is the voltage across the zener terminals when the current is 50 mA? When the current is 25 mA? Figure 3–8 represents the zener diode.

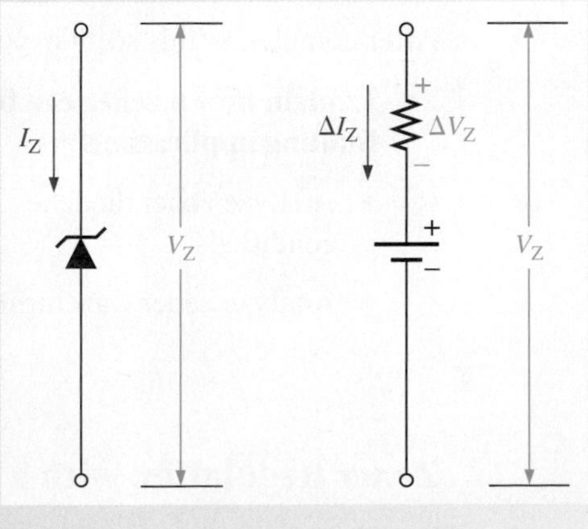

FIGURE 3–8

Solution For $I_Z = 50$ mA: The 50 mA current is a 13 mA increase above the test current, I_Z, of 37 mA.

$$\Delta I_Z = I_Z - 37\text{ mA} = 50\text{ mA} - 37\text{ mA} = +13\text{ mA}$$

$$\Delta V_Z = \Delta I_Z Z_Z = (13\text{ mA})(3.5\ \Omega) = +45.5\text{ mV}$$

The change in voltage due to the increase in current above the I_Z value causes the zener terminal voltage to increase. The zener voltage for $I_Z = 50$ mA is

$$V_Z = 6.8\text{ V} + \Delta V_Z = 6.8\text{ V} + 45.5\text{ mV} = \mathbf{6.85\ V}$$

For $I_Z = 25$ mA: The 25 mA current is a 12 mA decrease below the test current, I_Z, of 37 mA.

$$\Delta I_Z = -12\text{ mA}$$

$$\Delta V_Z = \Delta I_Z Z_Z = (-12\text{ mA})(3.5\ \Omega) = -42\text{ mV}$$

The change in voltage due to the decrease in current below the test current causes the zener terminal voltage to decrease. The zener voltage for $I_Z = 25$ mA is

$$V_Z = 6.8\text{ V} - \Delta V_Z = 6.8\text{ V} - 42\text{ mV} = \mathbf{6.76\ V}$$

Related Problem Repeat the analysis for $I_Z = 10$ mA and for $I_Z = 30$ mA using a 1N4742A zener with $V_Z = 12$ V at $I_Z = 21$ mA and $Z_Z = 9\ \Omega$.

SECTION 3–1 CHECKUP
Answers are at the end of the chapter.

1. In what region of their characteristic curve are zener diodes operated?
2. At what value of zener current is the zener voltage normally specified?
3. How does the zener impedance affect the voltage across the terminals of the device?
4. What does a positive temperature coefficient of 0.05%/°C mean?
5. Explain power derating.

3–2 ZENER DIODE APPLICATIONS

The zener diode can be used as a type of voltage regulator for providing stable reference voltages. In this section, you will see how zeners can be used as voltage references, regulators, and as simple limiters or clippers.

After completing this section, you should be able to

- **Explain how a zener can be used in voltage reference, regulation, and limiting applications**
  - Analyze zener diode regulators under varying input and varying load conditions
  - Analyze zener waveform-limiting circuits

Zener Regulation with a Varying Input Voltage

Zener diode regulators can provide a reasonably constant dc level at the output, but they are not particularly efficient. For this reason, they are limited to applications that require only low current to the load. Figure 3–9 illustrates how a zener diode can be used to regulate a dc

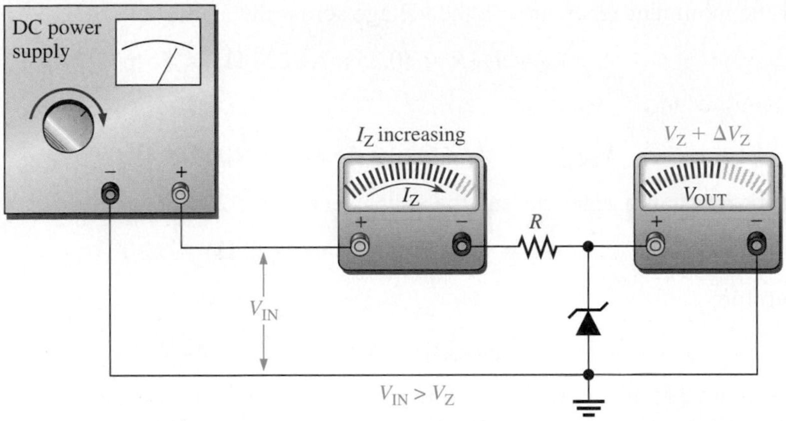

(a) As the input voltage increases, the output voltage remains nearly constant ($I_{ZK} < I_Z < I_{ZM}$).

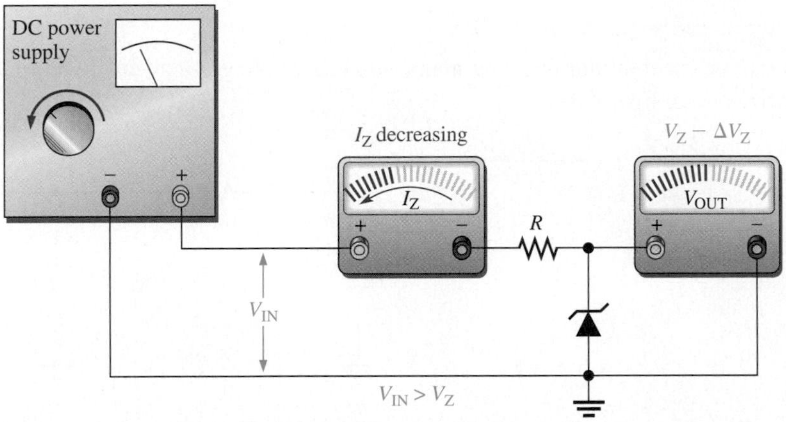

(b) As the input voltage decreases, the output voltage remains nearly constant ($I_{ZK} < I_Z < I_{ZM}$).

FIGURE 3–9

Zener regulation of a varying input voltage.

voltage. As the input voltage varies (within limits), the zener diode maintains a nearly constant output voltage across its terminals. However, as V_{IN} changes, I_Z will change proportionally so that the limitations on the input voltage variation are set by the minimum and maximum current values (I_{ZK} and I_{ZM}) with which the zener can operate. Resistor R is the series current-limiting resistor. The meters indicate the relative values and trends.

To illustrate regulation, let's use the ideal model of the 1N4740A zener diode (ignoring the zener resistance) in the circuit of Figure 3–10. The absolute lowest current that will maintain regulation is specified at I_{ZK}, which for the 1N4740A is 0.25 mA and represents the no-load current. The maximum current is not given on the datasheet but can be calculated from the power specification of 1 W, which is given on the datasheet. Keep in mind that both the minimum and maximum values are at the operating extremes and represent worst-case operation.

$$I_{ZM} = \frac{P_{D(max)}}{V_Z} = \frac{1\text{ W}}{10\text{ V}} = 100\text{ mA}$$

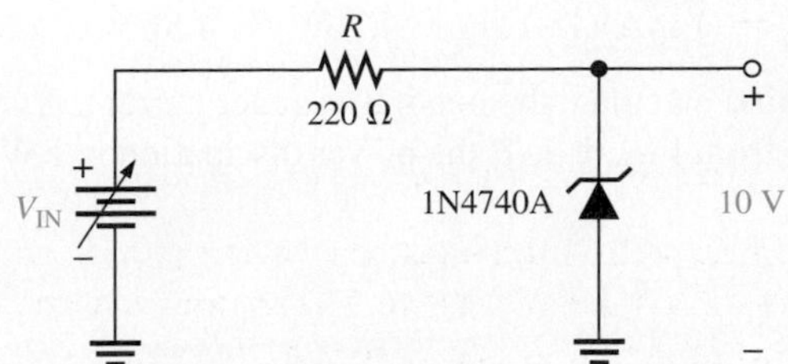

FIGURE 3–10

For the minimum zener current, the voltage across the 220 Ω resistor is

$$V_R = I_{ZK}R = (0.25 \text{ mA})(220 \ \Omega) = 55 \text{ mV}$$

Since $V_R = V_{IN} - V_Z$,

$$V_{IN(min)} = V_R + V_Z = 55 \text{ mV} + 10 \text{ V} = 10.055 \text{ V}$$

For the maximum zener current, the voltage across the 220 Ω resistor is

$$V_R = I_{ZM}R = (100 \text{ mA})(220 \ \Omega) = 22 \text{ V}$$

Therefore,

$$V_{IN(max)} = 22 \text{ V} + 10 \text{ V} = 32 \text{ V}$$

This shows that this zener diode can ideally regulate an input voltage from 10.055 V to 32 V and maintain an approximate 10 V output. The output will vary slightly because of the zener impedance, which has been neglected in these calculations.

EXAMPLE 3–5

Determine the minimum and the maximum input voltages that can be regulated by the zener diode in Figure 3–11.

FIGURE 3–11

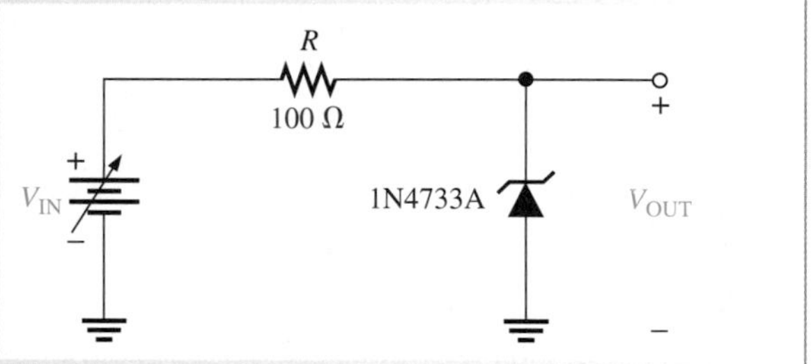

Solution From the datasheet in Figure 3–7 for the 1N4733A: $V_Z = 5.1$ V at $I_Z = 49$ mA, $I_{ZK} =$ 1 mA, and $Z_Z = 7 \ \Omega$ at I_Z. For simplicity, assume this value of Z_Z over the range of current values. The equivalent circuit is shown in Figure 3–12.

FIGURE 3–12

Equivalent of circuit in Figure 3–11.

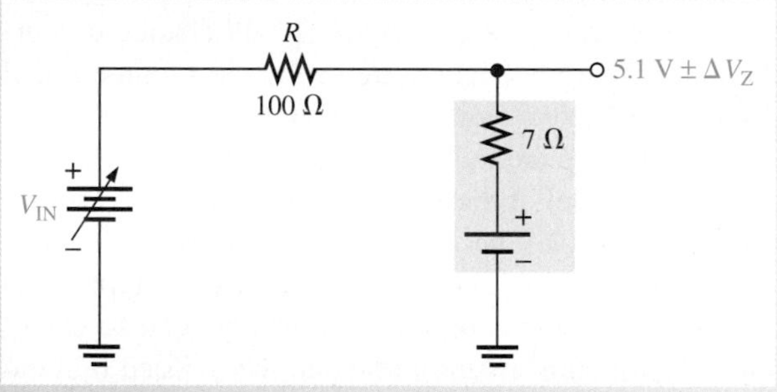

At $I_{ZK} = 1$ mA, the output voltage is

$$V_{OUT} \cong 5.1 \text{ V} - \Delta V_Z = 5.1 \text{ V} - (I_Z - I_{ZK})Z_Z = 5.1 \text{ V} - (49 \text{ mA} - 1 \text{ mA})(7 \ \Omega)$$
$$= 5.1 \text{ V} - (48 \text{ mA})(7 \ \Omega) = 5.1 \text{ V} - 0.336 \text{ V} = 4.76 \text{ V}$$

Therefore,

$$V_{IN(min)} = I_{ZK}R + V_{OUT} = (1 \text{ mA})(100 \ \Omega) + 4.76 \text{ V} = \mathbf{4.86 \ V}$$

To find the maximum input voltage, first calculate the maximum zener current. Assume the temperature is 50°C or below; so from Figure 3–7, the power dissipation is 1 W.

$$I_{ZM} = \frac{P_{D(max)}}{V_Z} = \frac{1 \text{ W}}{5.1 \text{ V}} = 196 \text{ mA}$$

At I_{ZM}, the output voltage is

$$V_{OUT} \cong 5.1\text{ V} + \Delta V_Z = 5.1\text{ V} + (I_{ZM} - I_Z)Z_Z$$
$$= 5.1\text{ V} + (147\text{ mA})(7\ \Omega) = 5.1\text{ V} + 1.03\text{ V} = 6.13\text{ V}$$

Therefore,

$$V_{IN(max)} = I_{ZM}R + V_{OUT} = (196\text{ mA})(100\ \Omega) + 6.13\text{ V} = \mathbf{25.7\text{ V}}$$

Related Problem Determine the minimum and maximum input voltages that can be regulated if a 1N4736A zener diode is used in Figure 3–11.

Open the Multisim file E03-05 in the Examples folder on your CD-ROM. For the calculated minimum and maximum dc input voltages, measure the resulting output voltages. Compare with the calculated values.

Zener Regulation with a Variable Load

Figure 3–13 shows a zener voltage regulator with a variable load resistor across the terminals. The zener diode maintains a nearly constant voltage across R_L as long as the zener current is greater than I_{ZK} and less than I_{ZM}.

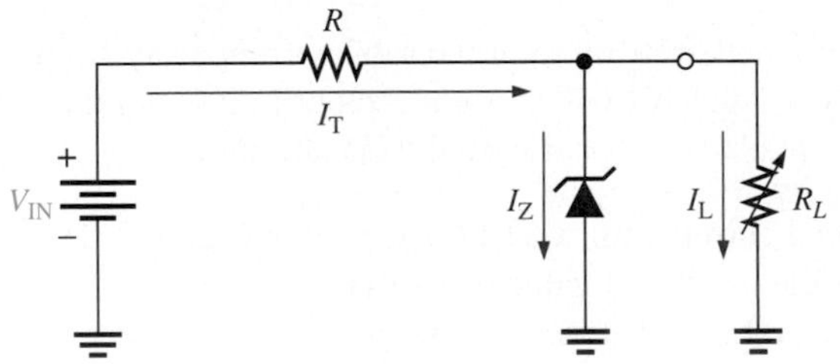

FIGURE 3–13

Zener regulation with a variable load.

From No Load to Full Load

When the output terminals of the zener regulator are open ($R_L = \infty$), the load current is zero and *all* of the current is through the zener; this is a no-load condition. When a load resistor (R_L) is connected, part of the total current is through the zener and part through R_L. The total current through R remains essentially constant as long as the zener is regulating. As R_L is decreased, the load current, I_L, increases and I_Z decreases. The zener diode continues to regulate the voltage until I_Z reaches its minimum value, I_{ZK}. At this point the load current is maximum, and a full-load condition exists. The following example will illustrate this.

EXAMPLE 3–6

Determine the minimum and the maximum load currents for which the zener diode in Figure 3–14 will maintain regulation. What is the minimum value of R_L that can be used? $V_Z = 12$ V, $I_{ZK} = 1$ mA, and $I_{ZM} = 50$ mA. Assume an ideal zener diode where $Z_Z = 0\ \Omega$ and V_Z remains a constant 12 V over the range of current values, for simplicity.

FIGURE 3–14

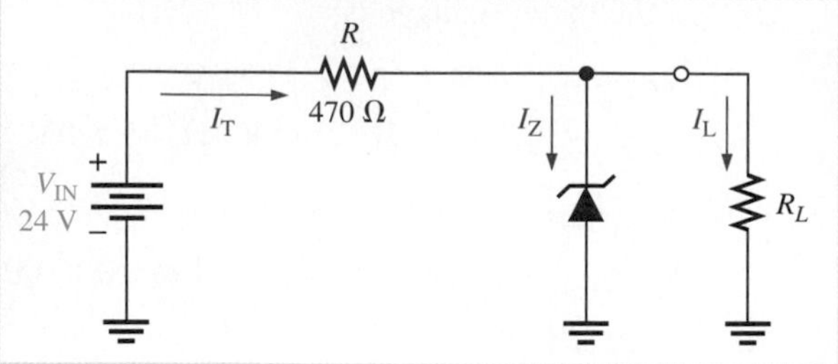

Solution When $I_L = 0$ A ($R_L = \infty$), I_Z is maximum and equal to the total circuit current I_T.

$$I_{Z(max)} = I_T = \frac{V_{IN} - V_Z}{R} = \frac{24\text{ V} - 12\text{ V}}{470\ \Omega} = 25.5\text{ mA}$$

If R_L is removed from the circuit, the load current is 0 A. Since $I_{Z(max)}$ is less than I_{ZM}, 0 A is an acceptable minimum value for I_L because the zener can handle all of the 25.5 mA.

$$I_{L(min)} = \mathbf{0\ A}$$

The maximum value of I_L occurs when I_Z is minimum ($I_Z = I_{ZK}$), so

$$I_{L(max)} = I_T - I_{ZK} = 25.5\text{ mA} - 1\text{ mA} = \mathbf{24.5\ mA}$$

The minimum value of R_L is

$$R_{L(min)} = \frac{V_Z}{I_{L(max)}} = \frac{12\text{ V}}{24.5\text{ mA}} = \mathbf{490\ \Omega}$$

Therefore, if R_L is less than 490 Ω, R_L will draw more of the total current away from the zener and I_Z will be reduced below I_{ZK}. This will cause the zener to lose regulation. Regulation is maintained for any value of R_L between 490 Ω and infinity.

Related Problem Find the minimum and maximum load currents for which the circuit in Figure 3–14 will maintain regulation. Determine the minimum value of R_L that can be used. $V_Z = 3.3$ V (constant), $I_{ZK} = 1$ mA, and $I_{ZM} = 150$ mA. Assume an ideal zener.

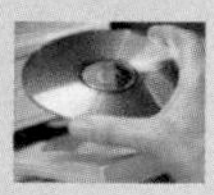

Open the Multisim file E03-06 in the Examples folder on your CD-ROM. For the calculated minimum value of load resistance, verify that regulation occurs.

In the last example, we assumed that Z_Z was zero and, therefore, the zener voltage remained constant over the range of currents. We made this assumption to demonstrate the concept of how the regulator works with a varying load. Such an assumption is often acceptable and in many cases produces results that are reasonably accurate. In Example 3–7, we will take the zener impedance into account.

EXAMPLE 3–7

For the circuit in Figure 3–15:

(a) Determine V_{OUT} at I_{ZK} and at I_{ZM}.

(b) Calculate the value of R that should be used.

(c) Determine the minimum value of R_L that can be used.

FIGURE 3–15

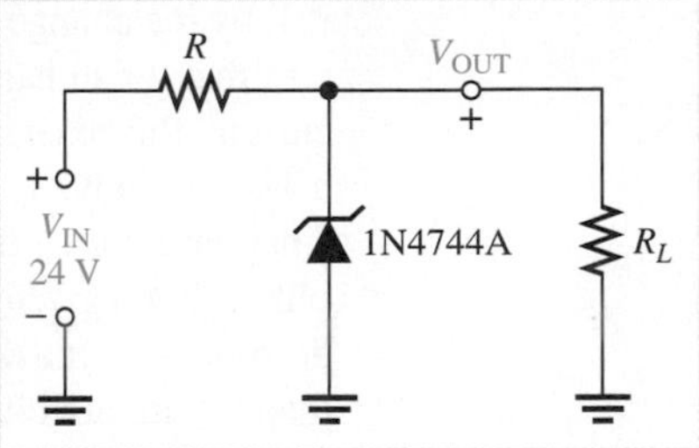

Solution First, review Example 3–6. The 1N4744A zener used in the regulator circuit of Figure 3–15 is a 15 V diode. The datasheet in Figure 3–7 gives the following information: $V_Z = 15\text{ V} @ I_Z = 17\text{ mA}$, $I_{ZK} = 0.25\text{ mA}$, and $Z_Z = 14\ \Omega$.

(a) For I_{ZK}:

$$V_{OUT} = V_Z = 15\text{ V} - \Delta I_Z Z_Z = 15\text{ V} - (I_Z - I_{ZK})Z_Z$$
$$= 15\text{ V} - (16.75\text{ mA})(14\ \Omega) = 15\text{ V} - 0.235\text{ V} = \mathbf{14.76\text{ V}}$$

Calculate the zener maximum current. The power dissipation is 1 W.

$$I_{ZM} = \frac{P_{D(max)}}{V_Z} = \frac{1\text{ W}}{15\text{ V}} = 66.7\text{ mA}$$

For I_{ZM}:

$$V_{OUT} = V_Z = 15\text{ V} + \Delta I_Z Z_Z$$
$$= 15\text{ V} + (I_{ZM} - I_Z)Z_Z = 15\text{ V} + (49.7\text{ mA})(14\ \Omega) = \mathbf{15.7\text{ V}}$$

(b) Calculate the value of R for the maximum zener current that occurs when there is no load as shown in Figure 3–16(a).

$$R = \frac{V_{IN} - V_Z}{I_{ZM}} = \frac{24\text{ V} - 15.7\text{ V}}{66.7\text{ mA}} = 124\ \Omega$$

$R = \mathbf{130\ \Omega}$ (nearest larger standard value).

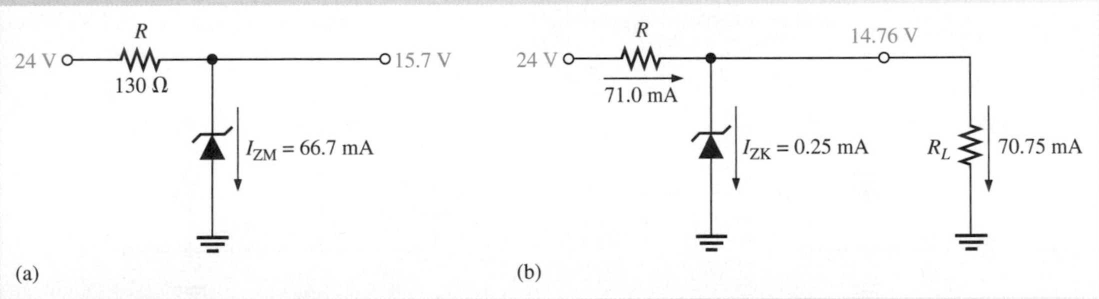

FIGURE 3–16

(c) For the minimum load resistance (maximum load current), the zener current is minimum ($I_{ZK} = 0.25\text{ mA}$) as shown in Figure 3–16(b).

$$I_T = \frac{V_{IN} - V_{OUT}}{R} = \frac{24\text{ V} - 14.76\text{ V}}{130\ \Omega} = 71.0\text{ mA}$$
$$I_L = I_T - I_{ZK} = 71.0\text{ mA} - 0.25\text{ mA} = 70.75\text{ mA}$$
$$R_{L(min)} = \frac{V_{OUT}}{I_L} = \frac{14.76\text{ V}}{70.75\text{ mA}} = \mathbf{209\ \Omega}$$

Related Problem Repeat each part of the preceding analysis if the zener is changed to a 1N4742A 12 V device.

You have seen how the zener diode regulates voltage. Its regulating ability is somewhat limited by the change in zener voltage over a range of current values, which restricts the load current that it can handle. To achieve better regulation and provide for greater variations in load current, the zener diode is combined as a key element with other circuit components to create a 3-terminal linear voltage regulator. Three-terminal voltage regulators that were introduced in Chapter 2 are IC devices that use the zener to provide a reference voltage for an internal amplifier. For a given dc input voltage, the 3-terminal regulator maintains an essentially constant dc voltage over a range of input voltages and load currents. The dc output voltage is always less than the input voltage. The details of this type of regulator are covered in Chapter 17. Figure 3–17 illustrates a basic 3-terminal regulator showing where the zener diode is used.

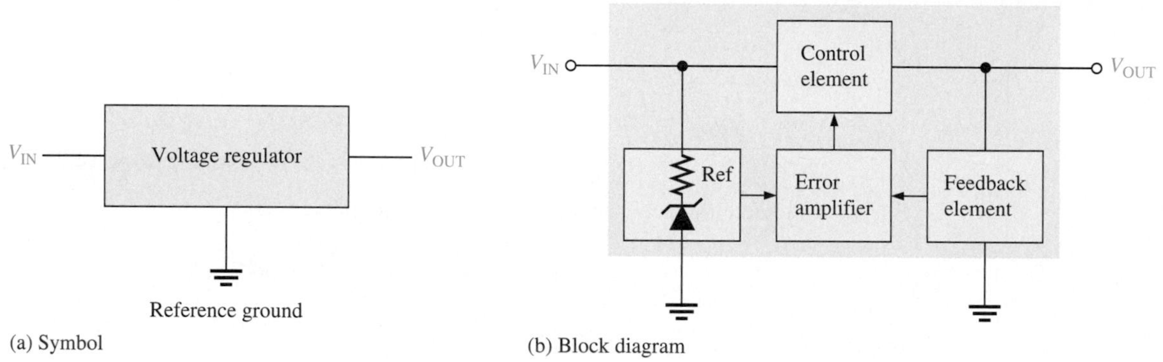

▲ FIGURE 3–17

Three-terminal voltage regulators.

Zener Limiting

In addition to voltage regulation applications, zener diodes can be used in ac applications to limit voltage swings to desired levels. Figure 3–18 shows three basic ways the limiting action of a zener diode can be used. Part (a) shows a zener used to limit the positive peak of a signal voltage to the selected zener voltage. During the negative alternation, the zener acts as a forward-biased diode and limits the negative voltage to −0.7 V. When the zener is

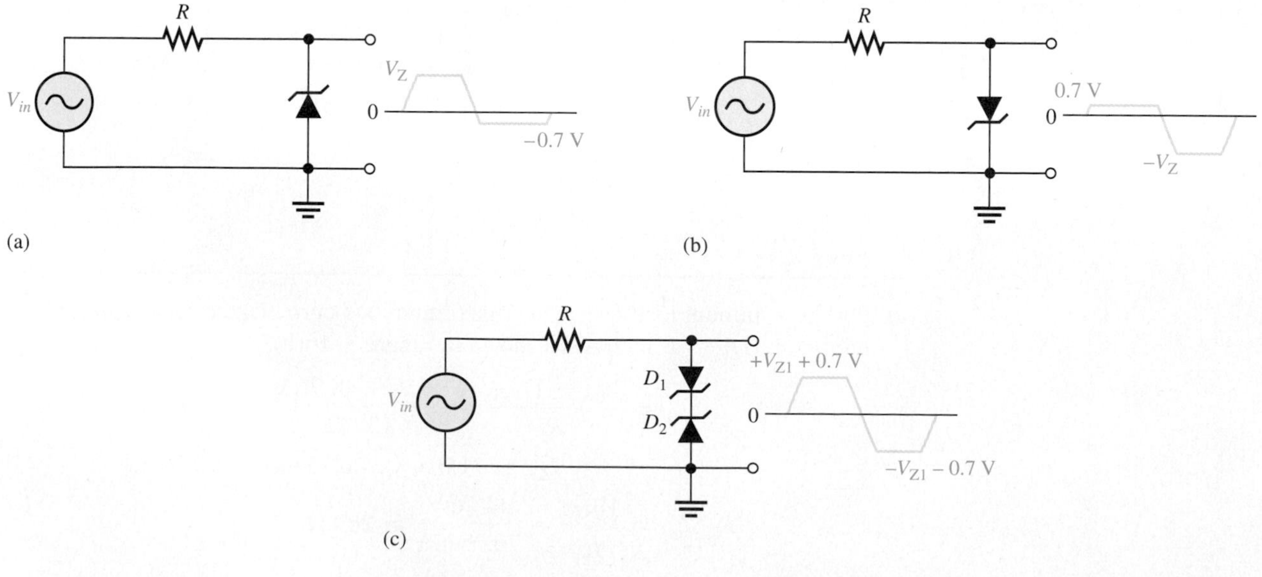

▲ FIGURE 3–18

Basic zener limiting action with a sinusoidal input voltage.

turned around, as in part (b), the negative peak is limited by zener action and the positive voltage is limited to +0.7 V. Two back-to-back zeners limit both peaks to the zener voltage ±0.7 V, as shown in part (c). During the positive alternation, D_2 is functioning as the zener limiter and D_1 is functioning as a forward-biased diode. During the negative alternation, the roles are reversed.

EXAMPLE 3–8

Determine the output voltage for each zener limiting circuit in Figure 3–19.

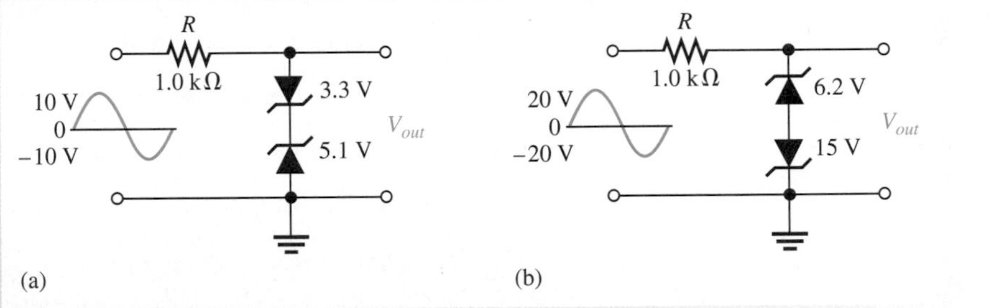

FIGURE 3–19

Solution See Figure 3–20 for the resulting output voltages. Remember, when one zener is operating in breakdown, the other one is forward-biased with approximately 0.7 V across it.

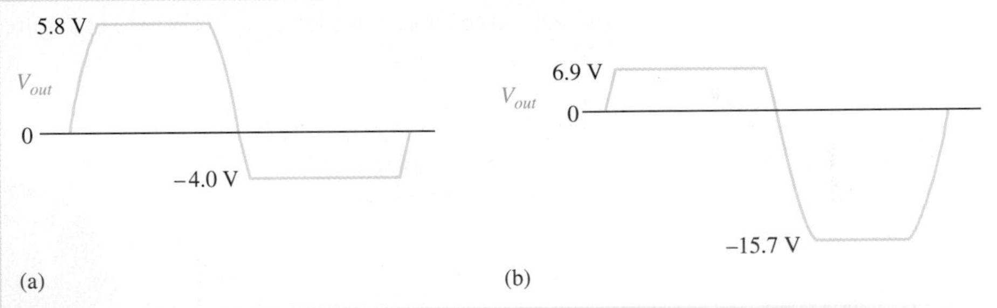

FIGURE 3–20

Related Problem

(a) What is the output in Figure 3–19(a) if the input voltage is increased to a peak value of 20 V?

(b) What is the output in Figure 3–19(b) if the input voltage is decreased to a peak value of 5 V?

Open the Multisim file E03-08 in the Examples folder on your CD-ROM. For the specified input voltages, measure the resulting output waveforms. Compare with the waveforms shown in the example.

SECTION 3–2 CHECKUP

1. In a zener diode regulator, what value of load resistance results in the maximum zener current?
2. Explain the terms *no-load* and *full-load.*
3. How much voltage appears across a zener diode when it is forward-biased?

3–3 THE VARACTOR DIODE

The junction capacitance of diodes varies with the amount of reverse bias. Varactor diodes are specially designed to take advantage of this characteristic and are used as voltage-controlled capacitors rather than traditional diodes. These devices are commonly used in communication systems. Varactor diodes are also referred to as *varicaps* or *tuning diodes.*

After completing this section, you should be able to

- **Describe the variable-capacitance characteristics of a varactor diode and analyze its operation in a typical circuit**
  - Identify a varactor diode symbol
  - Explain why a reverse-biased varactor exhibits capacitance
  - Discuss how the capacitance varies with reverse-bias voltage
  - Interpret a varactor datasheet
  - Define *capacitance ratio*
  - Analyze a varactor-tuned band-pass filter

A **varactor** is a diode that always operates in reverse-bias and is doped to maximize the inherent capacitance of the depletion region. The depletion region acts as a capacitor dielectric because of its nonconductive characteristic. The *p* and *n* regions are conductive and act as the capacitor plates, as illustrated in Figure 3–21.

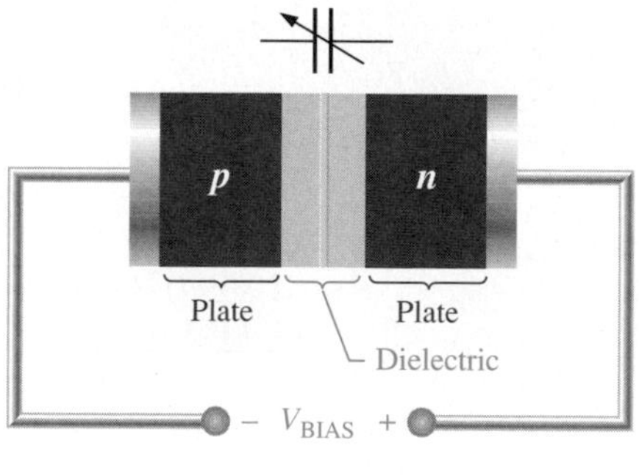

FIGURE 3–21

The reverse-biased varactor diode acts as a variable capacitor.

Basic Operation

Recall that capacitance is determined by the parameters of plate area (A), dielectric constant (ϵ), and plate separation (d), as expressed in the following formula:

$$C = \frac{A\epsilon}{d}$$

As the reverse-bias voltage increases, the depletion region widens, effectively increasing the plate separation, thus decreasing the capacitance. When the reverse-bias voltage decreases, the depletion region narrows, thus increasing the capacitance. This action is shown in Figure 3–22(a) and (b). A graph of diode capacitance (C_T) versus reverse voltage for a certain varactor is shown in Figure 3–22(c). For this particular device, C_T varies from 30 pF to slightly less than 4 pF as V_R varies from 1 V to 30 V.

In a varactor diode, these capacitance parameters are controlled by the method of doping near the *pn* junction and the size and geometry of the diode's construction. Nominal varactor

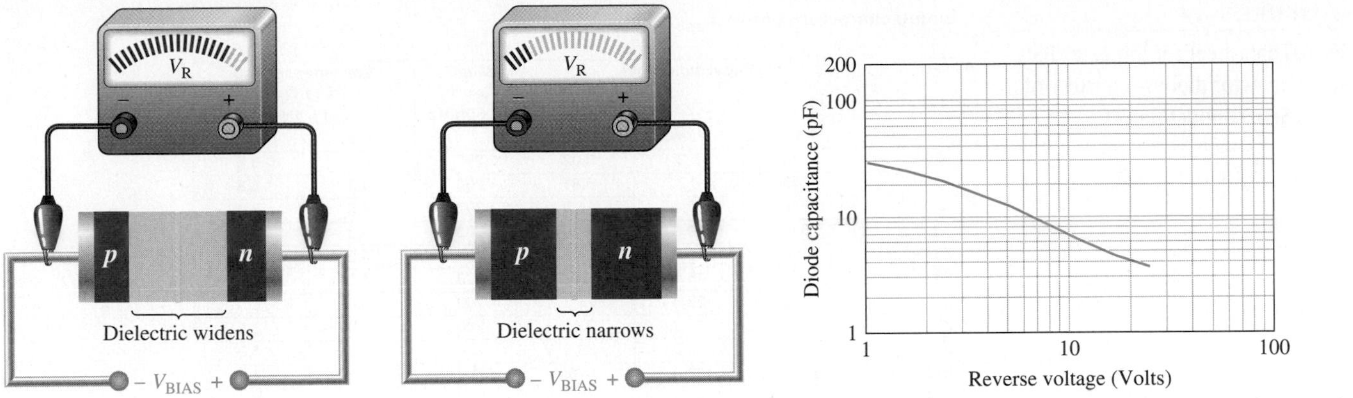

(a) Greater reverse bias, less capacitance (b) Less reverse bias, greater capacitance (c) Example of a diode capacitance versus reverse voltage graph

FIGURE 3–22

Varactor diode capacitance varies with reverse voltage.

FIGURE 3–23

Varactor diode symbol.

capacitances are typically available from a few picofarads to several hundred picofarads. Figure 3–23 shows a common symbol for a varactor.

Varactor Datasheet Information

A partial datasheet for a specific series of varactor diode (Zetex 830 series) is shown in Figure 3–24.

Capacitance Tolerance Range The minimum, nominal, and maximum values of capacitance are shown on the datasheet. For example, when reverse-biased at 3 V, the 832A can exhibit a capacitance anywhere between 19.8 pF and 24.2 pF. This tolerance range should not be confused with the range of capacitance values that result from varying the reverse bias as determined by the capacitance ratio.

Capacitance Ratio The varactor **capacitance ratio** is also known as the *tuning ratio.* It is the ratio of the diode capacitance at a minimum reverse voltage to the diode capacitance at a maximum reverse voltage. For the varactor diodes represented in Figure 3–24, the capacitance ratio is the ratio of C measured at a V_R of 2 V divided by C measured at a V_R of 20 V. The capacitance ratio is designated as C_2/C_{20} in this case.

For the 832A, the minimum capacitance ratio is 5.0. This means that the capacitance value decreases by a factor of 5.0 as V_R is increased from 2 V to 20 V. The following calculation illustrates how to use the capacitance ratio (CR) to find the capacitance range for the 832A. If $C_2 = 22$ pF and the minimum $CR = C_2/C_{20} = 5.0$,

$$C_{20} = \frac{C_2}{CR} = \frac{22\text{ pF}}{5} = 4.4\text{ pF}$$

The diode capacitance varies from 22 pF to 4.4 pF when V_R is increased from 2 V to 20 V.

The Zetex 830 series of varactor diodes are hyper-abrupt junction devices. The doping in the n and p regions is made uniform so that at the pn junction there is a very abrupt change from n to p instead of the more gradual change found in the rectifier diodes. The abruptness of the pn junction determines the capacitance ratio.

Back-to-Back Configuration One of the drawbacks of using just a single varactor diode in certain applications, such as rf tuning, is that if the diode is forward-biased by the rf signal during part of the ac cycle, its reverse leakage will increase momentarily. Also, a type of distortion called *harmonic distortion* is produced if the varactor is alternately biased positively and negatively. To avoid harmonic distortion, you will often see two varactor

▶ **FIGURE 3–24**

Partial datasheet for the Zetex 830 series varactor diodes. Courtesy of Zetex Semiconductors PLC.

Tuning characteristics at T_{amb} = 25°C

Part	Capacitance (pF)			Min Q V_R = 3V f = 50MHz	Capacitance ratio C_2 / C_{20} @ f = 1MHz	
	Min.	Nom.	Max.		Min.	Max.
829A	7.38	8.2	9.02	250	4.3	5.8
829B	7.79	8.2	8.61	250	4.3	5.8
830A	9.0	10.0	11.0	300	4.5	6.0
830B	9.5	10.0	10.5	300	4.5	6.0
831A	13.5	15.0	16.5	300	4.5	6.0
831B	14.25	15.0	15.75	300	4.5	6.0
832A	19.8	22.0	24.2	200	5.0	6.5
832B	20.9	22.0	23.1	200	5.0	6.5
833A	29.7	33.0	36.3	200	5.0	6.5
833B	31.35	33.0	34.65	200	5.0	6.5
834A	42.3	47.0	51.7	200	5.0	6.5
834B	44.65	47.0	49.35	200	5.0	6.5
835A	61.2	68.0	74.8	100	5.0	6.5
835B	64.6	68.0	71.4	100	5.0	6.5
836A	90.0	100.0	110.0	100	5.0	6.5
836B	95.0	100.0	105.0	100	5.0	6.5

Absolute maximum ratings

Parameter	Symbol	Max.	Unit
Forward current	I_F	200	mA
Power dissipation at T_{amb} = 25°C SOT23	P_{tot}	330	mW
Power dissipation at T_{amb} = 25°C SOD323	P_{tot}	330	mW
Power dissipation at T_{amb} = 25°C SOD523	P_{tot}	250	mW
Operating and storage temperature range		-55 to +150	°C

Electrical characteristics at T_{amb} = 25°C

Paramater	Conditions	Min.	Typ.	Max.	Unit
Reverse breakdown voltage	I_R = 10 A	25			V
Reverse voltage leakage	V_R = 20V		0.2	20	nA
Temperature coefficient of capacitance	V_R = 3V, f = 1MHz		300	400	ppCm/°C

diodes back to back, as shown in Figure 3–25(a) with the reverse dc voltage applied to both devices simultaneously. The two tuning diodes will be driven alternately into high and low capacitance, and the net capacitance will remain constant and is unaffected by the rf signal amplitude. The Zetex 832A varactor diode is available in a back-to-back configuration in an SOT-23 package or as a single diode in an SOD523 package, as shown in Figure 3–25(b). Although the cathodes in the back-to-back configuration are connected to a common pin, each diode can also be used individually.

▶ **FIGURE 3–25**

Varactor diodes and typical packages.

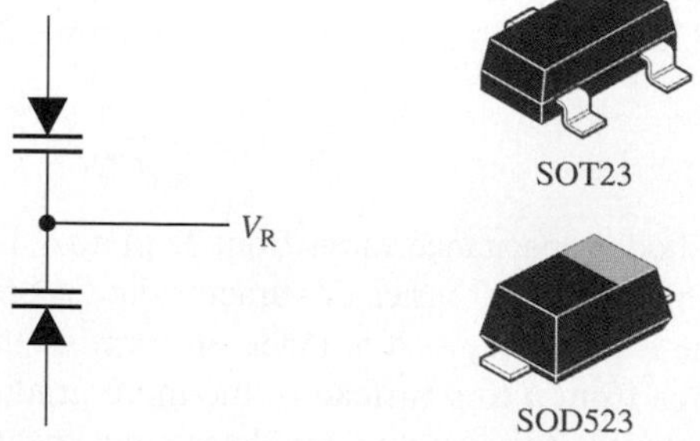

(a) Back-to-back configuration (b)

An Application

A major application of varactors is in tuning circuits. For example, VHF, UHF, and satellite receivers utilize varactors. Varactors are also used in cellular communications. When used in a parallel resonant circuit, as illustrated in Figure 3–26, the varactor acts as a vari-

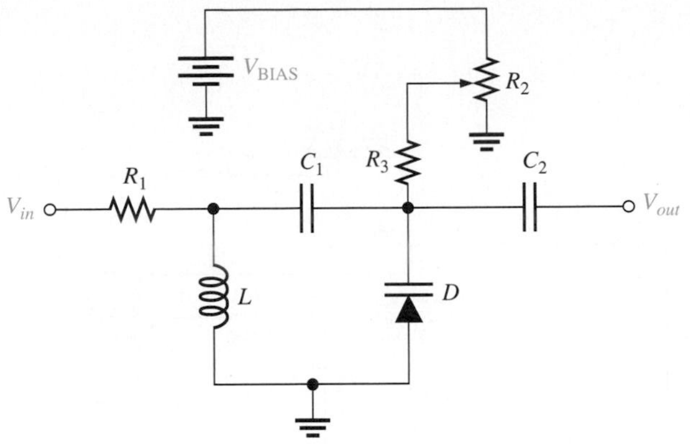

FIGURE 3–26

A resonant band-pass filter using a varactor diode for adjusting the resonant frequency over a specified range.

able capacitor, thus allowing the resonant frequency to be adjusted by a variable voltage level. The varactor diode provides the total variable capacitance in the parallel resonant band-pass filter. The varactor diode and the inductor form a parallel resonant circuit from the output to ac ground. The capacitors C_1 and C_2 have no effect on the filter's frequency response because their reactances are negligible at the resonant frequencies. C_1 prevents a dc path from the potentiometer wiper back to the ac source through the inductor and R_1. C_2 prevents a dc path from the wiper of the potentiometer to a load on the output. The potentiometer R_2 forms a variable dc voltage for biasing the varactor. The reverse-bias voltage across the varactor can be varied with the potentiometer.

Recall that the parallel resonant frequency is

$$f_r \cong \frac{1}{2\pi\sqrt{LC}}$$

EXAMPLE 3–9

(a) Given that the capacitance of a Zetex 832A varactor is approximately 40 pF at 0 V bias and that the capacitance at a 2 V reverse bias is 22 pF, determine the capacitance at a reverse bias of 20 V using the specified minimum capacitance ratio.

(b) Using the capacitances at bias voltages of 0 V and 20 V, calculate the resonant frequencies at the bias extremes for the circuit in Figure 3–26 if $L = 2$ mH.

(c) Verify the frequency calculations by simulating the circuit in Figure 3–26 for the following component values: $R_1 = 47\text{ k}\Omega$, $R_2 = 10\text{ k}\Omega$, $R_3 = 5.1\text{ M}\Omega$, $C_1 = 10\text{ nF}$, $C_2 = 10\text{ nF}$, $L = 2\text{ mH}$, and $V_{BIAS} = 20\text{ V}$.

Solution

(a) $$C_{20} = \frac{C_2}{CR} = \frac{22\text{ pF}}{5.0} = \mathbf{4.4\ pF}$$

(b) $$f_0 = \frac{1}{2\pi\sqrt{LC}} = \frac{1}{2\pi\sqrt{(2\text{ mH})(40\text{ pF})}} = \mathbf{563\ kHz}$$

$$f_{20} = \frac{1}{2\pi\sqrt{LC}} = \frac{1}{2\pi\sqrt{(2\text{ mH})(4.4\text{ pF})}} = \mathbf{1.7\ MHz}$$

(c) The Multisim simulation of the circuit is shown in Figure 3–27. The Bode plotters show the frequency responses at 0 V and 20 V reverse bias. The center of the 0 V bias response curve is at 553.64 kHz and the center of the 20 V bias response curve is at 1.548 MHz. These results agree reasonably well with the calculated values.

These results show that this circuit can be tuned over most of the AM broadcast band.

Related Problem How could you increase the tuning range of the circuit?

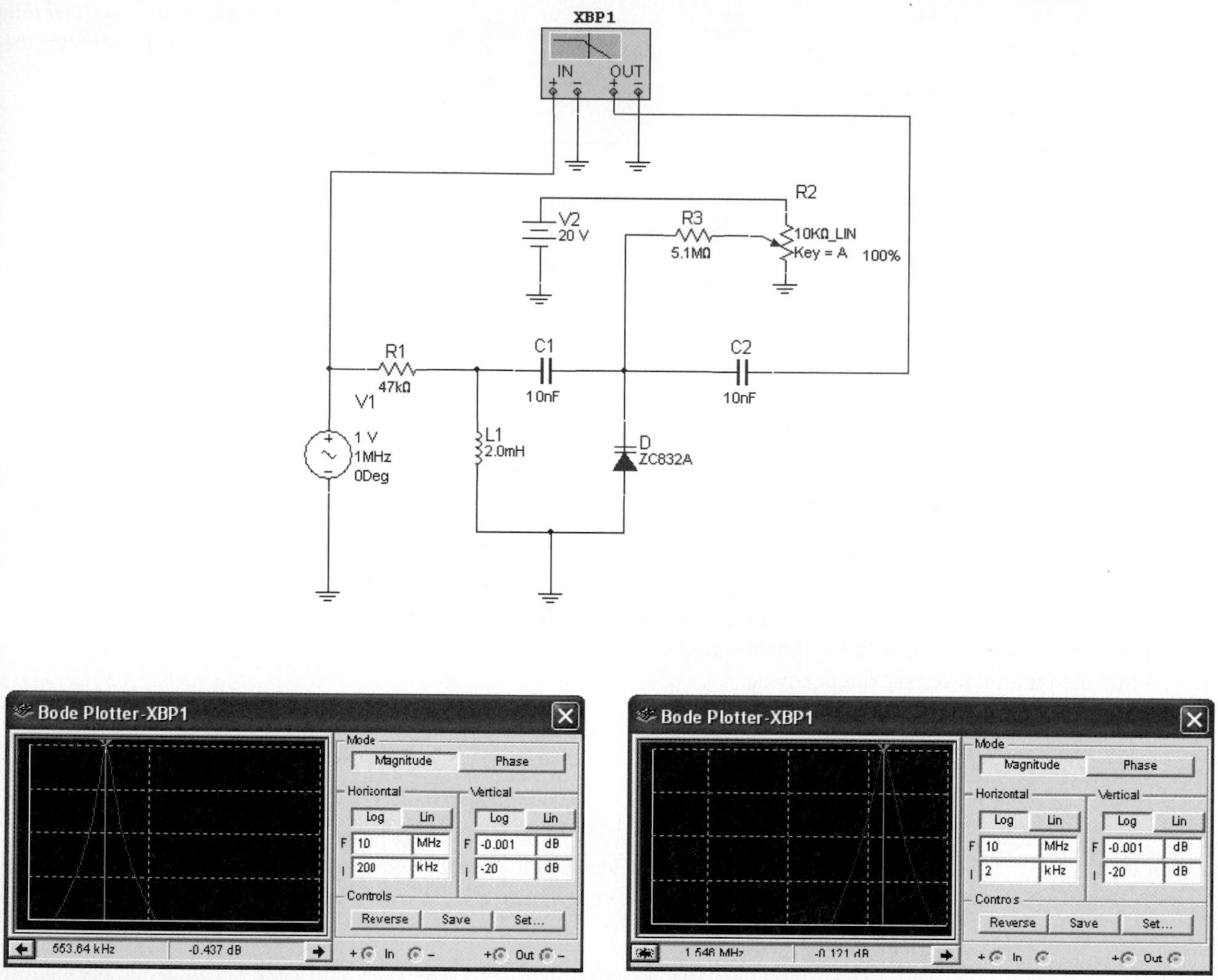

Frequency response for 0 V varactor bias

Frequency response for 20 V reverse varactor bias

FIGURE 3–27

Multisim simulation.

SECTION 3–3 CHECKUP

1. What is the key feature of a varactor diode?
2. Under what bias condition is a varactor operated?
3. What part of the varactor produces the capacitance?
4. Based on the graph in Figure 3–22(c), what happens to the diode capacitance when the reverse voltage is increased?
5. Define *capacitance ratio*.

3–4 OPTICAL DIODES

In this section, two types of optoelectronic devices—the light-emitting diode (LED) and the photodiode—are introduced. As the name implies, the LED is a light emitter. The photodiode, on the other hand, is a light detector.

After completing this section, you should be able to

- **Discuss the operation and characteristics of LEDs and photodiodes**
  - Identify LED and photodiode symbols
  - Explain basically how an LED emits light
  - Analyze the spectral output curves and radiation patterns of LEDs
  - Interpret an LED datasheet
  - Define *radiant intensity* and *irradiance*
  - Use an LED seven-segment display
  - Explain how a photodiode detects light
  - Analyze the response curve of a photodiode
  - Interpret a photodiode datasheet
  - Discuss photodiode sensitivity

The Light-Emitting Diode (LED)

The symbol for an LED is shown in Figure 3–28.

The basic operation of the **light-emitting diode (LED)** is as follows. When the device is forward-biased, electrons cross the *pn* junction from the *n*-type material and recombine with holes in the *p*-type material. Recall from Chapter 1 that these free electrons are in the conduction band and at a higher energy than the holes in the valence band. When recombination takes place, the recombining electrons release energy in the form **photons.** A large exposed surface area on one layer of the semiconductive material permits the photons to be emitted as visible light. This process, called **electroluminescence**, is illustrated in Figure 3–29. Various impurities are added during the doping process to establish the **wavelength** of the emitted light. The wavelength determines the color of visible light. Some LEDs emit photons that are not part of the visible spectrum but have longer wavelengths and are in the **infrared** (IR) portion of the spectrum.

▲ FIGURE 3–28

Symbol for an LED. When forward-biased, it emits light.

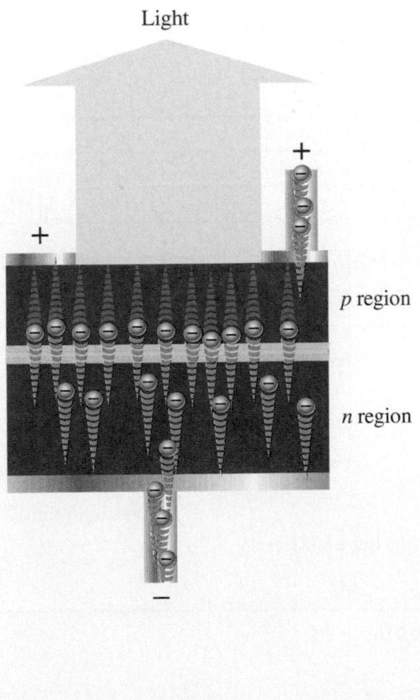

◄ FIGURE 3–29

Electroluminescence in a forward-biased LED.

LED Semiconductor Materials The semiconductor gallium arsenide (GaAs) was used in early LEDs and emits IR radiation, which is invisible. The first visible red LEDs were produced using gallium arsenide phosphide (GaAsP) on a GaAs substrate. The efficiency was increased using a gallium phosphide (GaP) substrate, resulting in brighter red LEDs and also allowing orange LEDs.

Later, GaP was used as the light-emitter to achieve pale green light. By using a red and a green chip, LEDs were able to produce yellow light. The first super-bright red, yellow, and green LEDs were produced using gallium aluminum arsenide phosphide (GaAlAsP). By the early 1990s ultrabright LEDs using indium gallium aluminum phosphide (InGaAlP) were available in red, orange, yellow, and green.

Blue LEDs using silicon carbide (SiC) and ultrabright blue LEDs made of gallium nitride (GaN) became available. High intensity LEDs that produce green and blue are also made using indium gallium nitride (InGaN). High-intensity white LEDs are formed using ultrabright blue GaN coated with fluorescent phosphors that absorb the blue light and reemit it as white light.

LED Biasing The forward voltage across an LED is considerably greater than for a silicon diode. Typically, the maximum V_F for LEDs is between 1.2 V and 3.2 V, depending on the material. Reverse breakdown for an LED is much less than for a silicon rectifier diode (3 V to 10 V is typical).

The LED emits light in response to a sufficient forward current, as shown in Figure 3–30(a). The amount of power output translated into light is directly proportional to the forward current, as indicated in Figure 3–30(b). An increase in I_F corresponds proportionally to an increase in light output.

Light Emission An LED emits light over a specified range of wavelengths as indicated by the **spectral** output curves in Figure 3–31. The curves in part (a) represent the light

FIGURE 3–30

Basic operation of an LED.

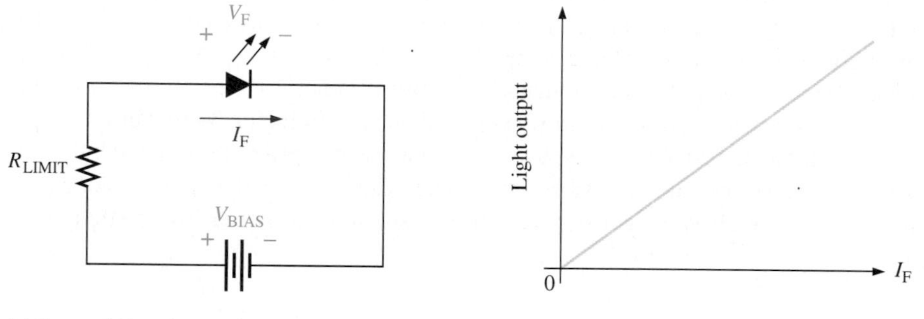

(a) Forward-biased operation

(b) General light output versus forward current

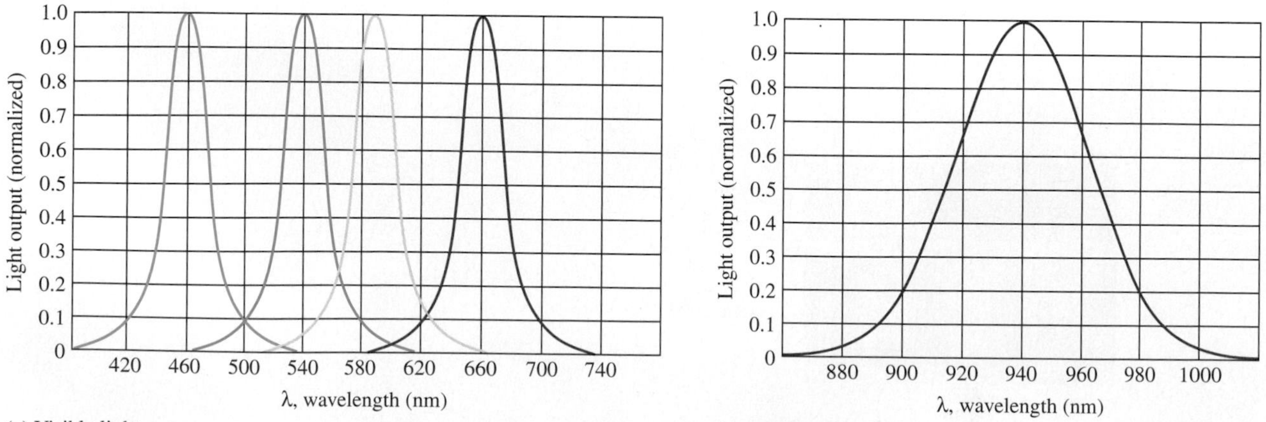

(a) Visible light

(b) Infrared (IR)

FIGURE 3–31

Examples of typical spectral output curves for LEDs.

output versus wavelength for typical visible LEDs, and the curve in part (b) is for a typical infrared LED. The wavelength (λ) is expressed in nanometers (nm). The normalized output of the visible red LED peaks at 660 nm, the yellow at 590 nm, green at 540 nm, and blue at 460 nm. The output for the infrared LED peaks at 940 nm.

The graph in Figure 3–32 is the **radiation** pattern for a typical LED. It shows how directional the emitted light is. The radiation pattern depends on the type of lens structure of the LED. The narrower the radiation pattern, the more the light is concentrated in a particular direction. Also, colored lenses are used to enhance the color.

Typical LEDs are shown in Figure 3–33.

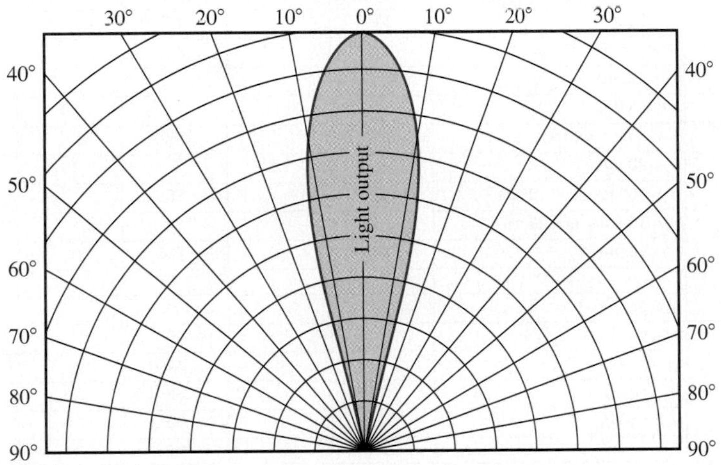

FIGURE 3–32

General radiation pattern of a typical LED.

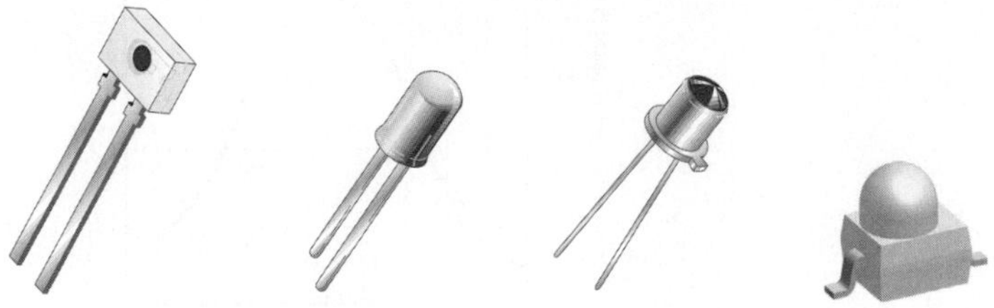

FIGURE 3–33

Typical LEDs.

LED Datasheet Information

A partial datasheet for an TSMF1000 infrared (IR) light-emitting diode is shown in Figure 3–34. Notice that the maximum reverse voltage is only 5 V, the maximum forward current is 100 mA, and the forward voltage drop is approximately 1.3 V for I_F = 20 mA.

From the graph in part (c), you can see that the peak power output for this device occurs at a wavelength of 870 nm; its radiation pattern is shown in part (d).

Radiant Intensity and Irradiance In Figure 3–34(a), the **radiant intensity,** I_e (symbol not to be confused with current), is the output power per steradian and is specified as 5 mW/sr at I_F = 20 mA. The steradian (sr) is the unit of solid angular measurement. **Irradiance,** E, is the power per unit area at a given distance from an LED source expressed in mW/cm^2. Irradiance is important because the response of a detector (photodiode) used in conjunction with an LED depends on the irradiance of the light it receives.

Absolute Maximum Ratings

T_{amb} = 25°C, unless otherwise specified

Parameter	Test condition	Symbol	Value	Unit
Reverse Voltage		V_R	5	V
Forward current		I_F	100	mA
Peak Forward Current	t_p/T = 0.5, t_p = 100 μs	I_{FM}	200	mA
Surge Forward Current	t_p = 100 μs	I_{FSM}	0.8	A
Power Dissipation		P_V	190	mW
Junction Temperature		T_j	100	°C
Operating Temperature Range		T_{amb}	- 40 to + 85	°C

Basic Characteristics

T_{amb} = 25°C, unless otherwise specified
T_{amb} = 25°C, unless otherwise specified

Parameter	Test condition	Symbol	Min	Typ.	Max	Unit
Forward Voltage	I_F = 20 mA	V_F		1.3	1.5	V
	I_F = 1 A, t_p = 100 μs	V_F		2.4		V
Temp. Coefficient of V_F	I_F = 1.0 mA	TK_{VF}		- 1.7		mV/K
Reverse Current	V_R = 5 V	I_R			10	μA
Junction capacitance	V_R = 0 V, f = 1 MHz, E = 0	C_j		160		pF
Radiant Intensity	I_F = 20 mA	I_e	2.5	5	13	mW/sr
	I_F = 100 mA, t_p = 100 μs	I_e		25		mW/sr
Radiant Power	I_F = 100 mA, t_p = 20 ms	ϕ_e		35		mW
Temp. Coefficient of ϕ_e	I_F = 20 mA	$TK\phi_e$		- 0.6		%/K
Angle of Half Intensity		ϕ		± 17		deg
Peak Wavelength	I_F = 20 mA	λ_p		870		nm
Spectral Bandwidth	I_F = 20 mA	$\Delta\lambda$		40		nm
Temp. Coefficient of λ_p	I_F = 20 mA	$TK\lambda_p$		0.2		nm/K
Rise Time	I_F = 20 mA	t_r		30		ns
Fall Time	I_F = 20 mA	t_f		30		ns
Virtual Source Diameter		∅		1.2		mm

(a)

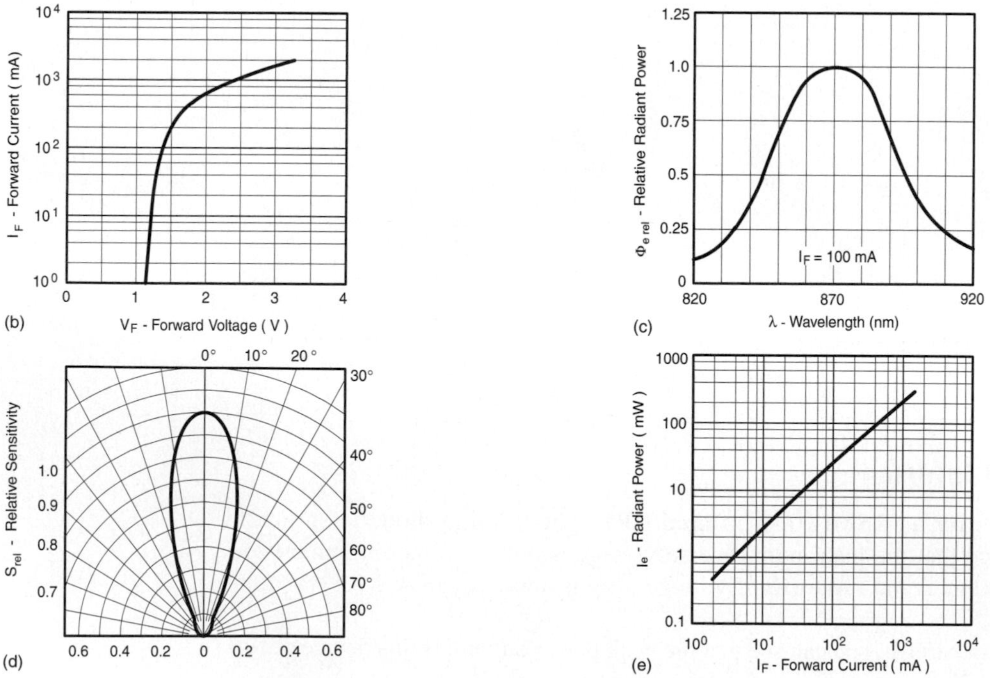

▲ **FIGURE 3–34**

Partial datasheet for an TSMF1000 IR light-emitting diode. Datasheet courtesy of Vishay Intertechnology, Inc.

EXAMPLE 3–10

From the LED datasheet in Figure 3–34 determine the following:

(a) The radiant power at 910 nm if the maximum output is 35 mW.

(b) The forward voltage drop for $I_F = 20$ mA.

(c) The radiant intensity for $I_F = 40$ mA.

Solution

(a) From the graph in Figure 3–34(c), the relative radiant power at 910 nm is approximately 0.25 and the peak radiant power is 35 mW. Therefore, the radiant power at 910 nm is

$$\phi_e = 0.25(35 \text{ mW}) = \mathbf{8.75\ mW}$$

(b) From the graph in part (b), $V_F \cong$ **1.25 V** for $I_F = 20$ mA.

(c) From the graph in part (e), $I_e \cong$ **10 mW/sr** for $I_F = 40$ mA.

Related Problem Determine the relative radiant power at 850 nm.

Applications Standard LEDs are used for indicator lamps and readout displays on a wide variety of instruments, ranging from consumer appliances to scientific apparatus. A common type of display device using LEDs is the seven-segment display. Combinations of the segments form the ten decimal digits as illustrated in Figure 3–35. Each segment in the display is an LED. By forward-biasing selected combinations of segments, any decimal digit and a decimal point can be formed. Two types of LED circuit arrangements are the common anode and common cathode as shown.

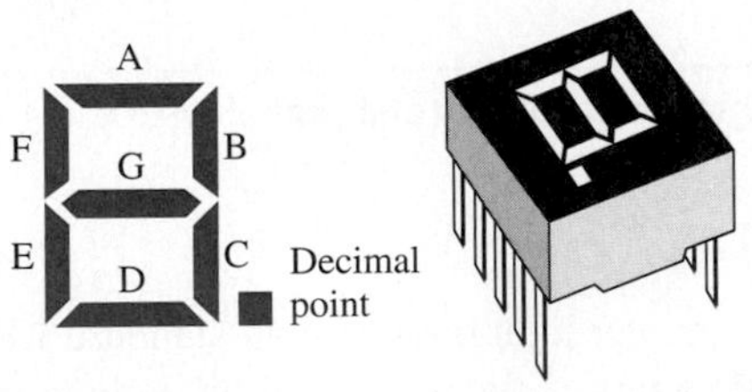

(a) LED segment arrangement and typical device

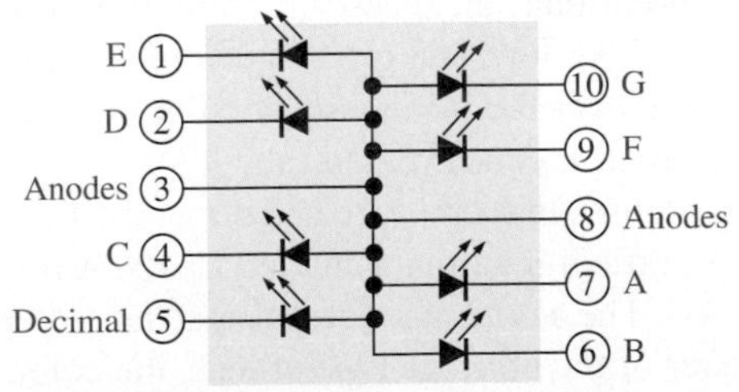

(b) Common anode

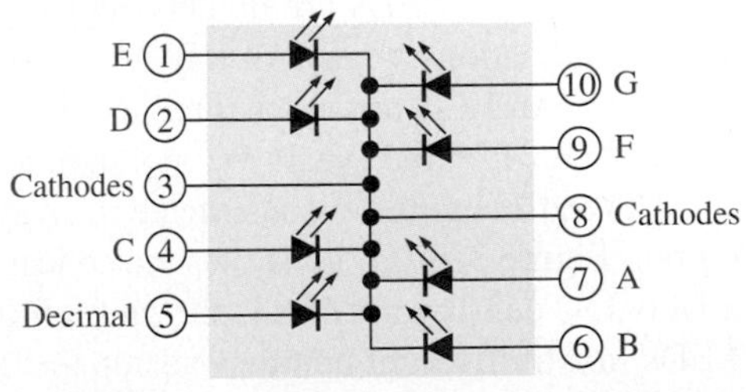

(c) Common cathode

FIGURE 3–35

The 7-segment LED display.

One common application of an infrared LED is in remote control units for TV, DVD, gate openers, etc. The IR LED sends out a beam of invisible light that is sensed by the receiver in your TV, for example. For each button on the remote control unit, there is a unique code. When a specific button is pressed, a coded electrical signal is generated that goes to the LED, which converts the electrical signal to a coded infrared light signal. The TV receiver recognizes the code and takes appropriate action, such as changing the channel or increasing the volume.

Also, IR light-emitting diodes are used in optical coupling applications, often in conjunction with fiber optics. Areas of application include industrial processing and control, position encoders, bar graph readers, and optical switching.

An example of how an IR LED could be used in an industrial application is illustrated in Figure 3–36. This particular system is used to count baseballs as they are fed down a chute into a box for shipping. As each ball passes through the chute, the IR beam emitted by the LED is interrupted. This is detected by the photodiode (discussed later) and the resulting change in current is sensed by a detector circuit. An electronic circuit counts each time that the beam is interrupted; and when a preset number of balls pass through the chute, the "stop" mechanism is activated to stop the flow of balls until the next empty box is automatically moved into place on the conveyor. When the next box is in place, the "stop" mechanism is deactivated and the balls begin to roll again. This idea can also be applied to inventory and packing control for many other types of products.

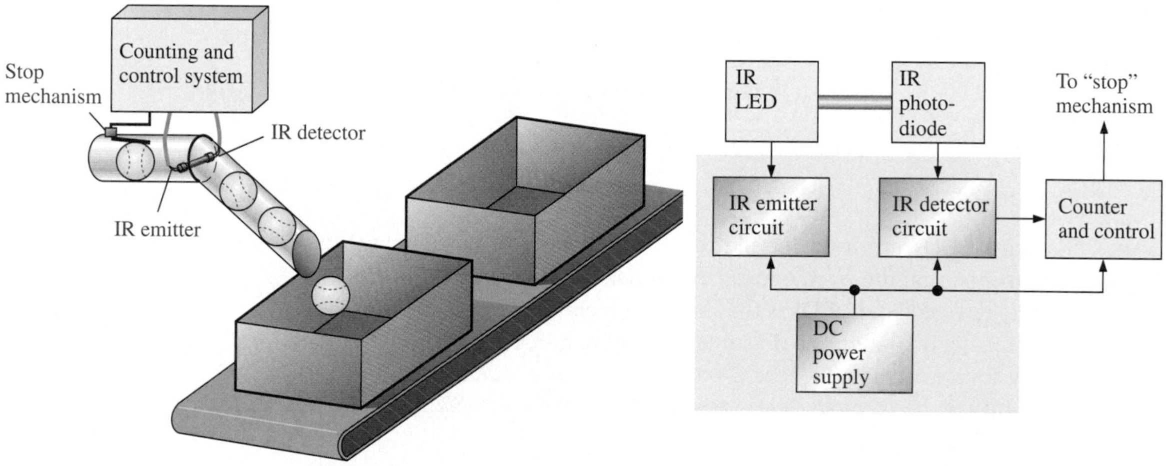

FIGURE 3–36

Basic concept and block diagram of a counting and control system.

High-Intensity LEDs

LEDs that produce much greater light outputs than standard LEDs are found in many applications including traffic lights, automotive lighting, indoor and outdoor advertising and informational signs, and home lighting.

Traffic Lights LEDs are quickly replacing the traditional incandescent bulbs in traffic signal applications. Arrays of tiny LEDs form the red, yellow, and green lights in a traffic light unit. An LED array has three major advantages over the incandescent bulb: brighter light, longer lifetime (years vs. months), and less energy consumption (about 90% less).

LED traffic lights are constructed in arrays with lenses that optimize and direct the light output. Figure 3–37(a) illustrates the concept of a traffic light array using red LEDs. A relatively low density of LEDs is shown for illustration. The actual number and spacing of the LEDs in a traffic light unit depends on the diameter of the unit, the type of lens, the color, and the required light intensity. With an appropriate LED density and a lens, an 8- or 12-inch traffic light will appear essentially as a solid-color circle.

LEDs in an array are usually connected either in a series-parallel or a parallel arrangement. A series connection is not practical because if one LED fails open, then all the LEDs are disabled. For a parallel connection, each LED requires a limiting resistor. To reduce the number of limiting resistors, a series-parallel connection can be used, as shown in Figure 3–37(b).

Some LED traffic arrays use small reflectors for each LED to help maximize the effect of the light output. Also, an optical lens covers the front of the array to direct the light from each individual diode to prevent improper dispersion of light and to optimize the visibility. Figure 3–38 illustrates how a lens is used to direct the light toward the viewer.

The particular LED circuit configuration depends on the voltage and the color of the LED. Different color LEDs require different forward voltages to operate. Red LEDs take the least;

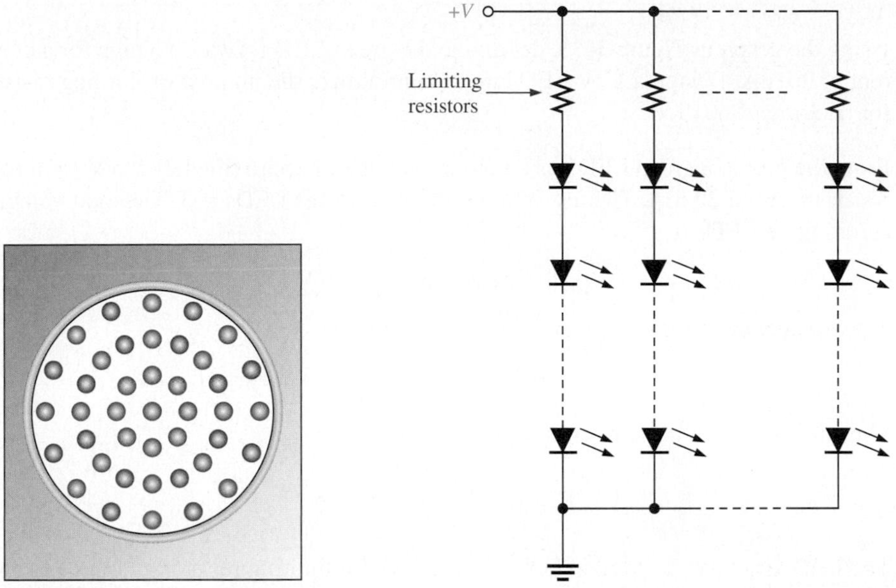

(a) LED array (b) Circuit

◀ **FIGURE 3–37**

LED traffic light.

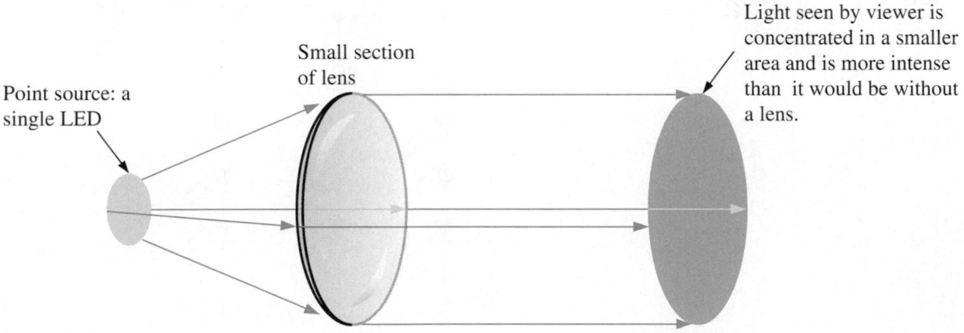

◀ **FIGURE 3–38**

The lens directs the light emitted from the LED to optimize visibility.

and as the color moves up the color spectrum toward blue, the voltage requirement increases. Typically, a red LED requires about 2 V, while blue LEDs require between 3 V and 4 V. Generally, LEDs, however, need 20 mA to 30 mA of current, regardless of their voltage requirements. Typical *V-I* curves for red, yellow, green, and blue LEDs are shown in Figure 3–39.

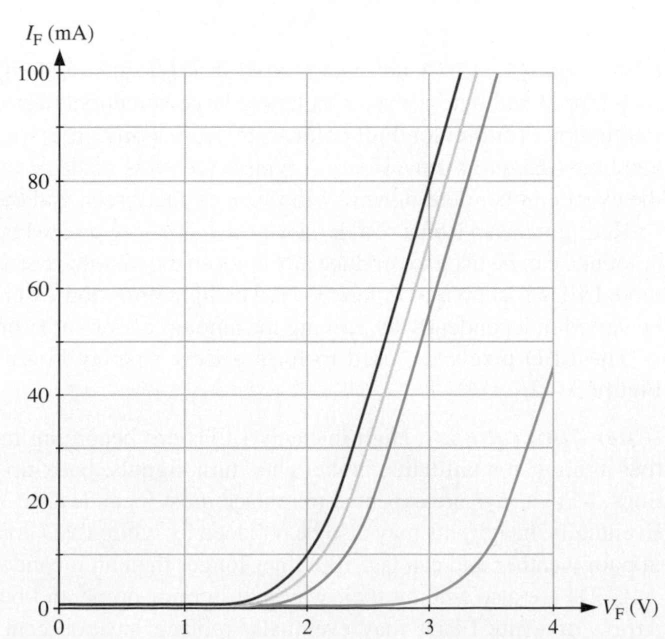

◀ **FIGURE 3–39**

V-I characteristic curves for visible-light LEDs.

EXAMPLE 3–11

Using the graph in Figure 3–39, determine the green LED forward voltage for a current of 20 mA. Design a 12 V LED circuit to minimize the number of limiting resistors for an array of 60 diodes.

Solution From the graph, a green LED has a forward voltage of approximately 2.5 V for a forward current of 20 mA. The maximum number of series LEDs is 3. The total voltage across three LEDs is

$$V = 3 \times 2.5\text{ V} = 7.5\text{ V}$$

The voltage drop across the series-limiting resistor is

$$V = 12\text{ V} - 7.5\text{ V} = 4.5\text{ V}$$

The value of the limiting resistor is

$$R_{\text{LIMIT}} = \frac{4.5\text{ V}}{20\text{ mA}} = 225\ \Omega$$

The LED array has 20 parallel branches each with a limiting resistor and three LEDs, as shown in Figure 3–40.

FIGURE 3–40

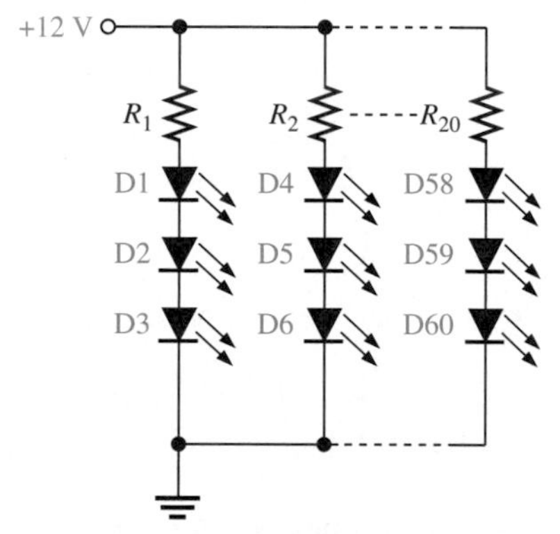

Related Problem Design a 12 V red LED array with minimum limiting resistors, a forward current of 30 mA, and containing 64 diodes.

LED Displays LEDs are widely used in large and small signs and message boards for both indoor and outdoor uses, including large-screen television. Signs can be single-color, multicolor, or full-color. Full-color screens use a tiny grouping of high-intensity red, green, and blue LEDs to form a **pixel**. A typical screen is made of thousands of RGB pixels with the exact number determined by the sizes of the screen and the pixel.

Red, green, and blue (RGB) are primary colors and when mixed together in varying amounts, can be used to produce any color in the visible spectrum. A basic pixel formed by three LEDs is shown in Figure 3–41. The light emission from each of the three diodes can be varied independently by varying the amount of forward current.

The LED pixels are used to form a large display board or screen, as illustrated in Figure 3–42.

Other Applications High-intensity LEDs are becoming more widely used in automotive lighting for taillights, brakelights, turn signals, back-up lights, and interior applications. LED arrays are expected to replace most incandescent bulbs in automotive lighting. Eventually, headlights may also be replaced by white LED arrays. LEDs can be seen better in poor weather and can last 100 times longer than an incandescent bulb.

LEDs are also finding their way into interior home and business lighting applications. Arrays of white LEDs may eventually replace incandescent light bulbs and flourescent

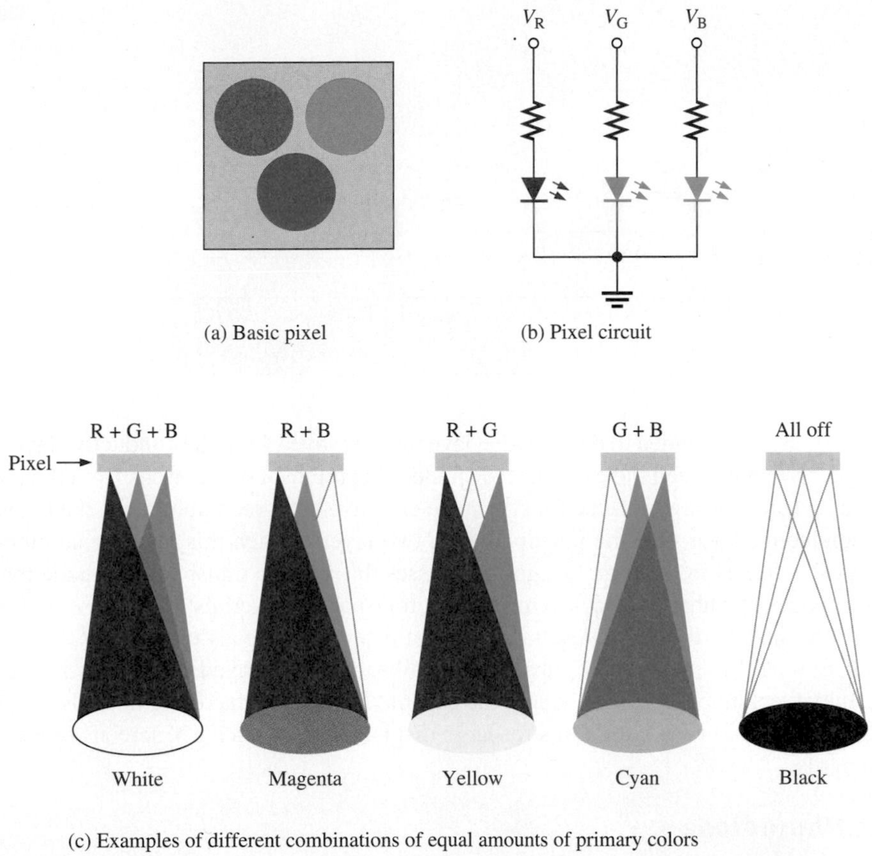

FIGURE 3–41

The concept of an RGB pixel used in LED display screens.

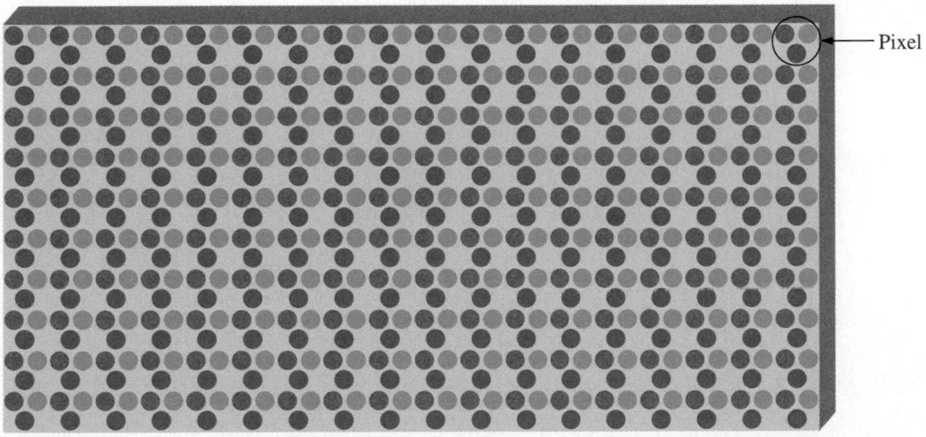

FIGURE 3–42

Large display screen formed by many RGB pixels. Pixel size is greatly exaggerated for illustration.

lighting in interior living and work areas. As previously mentioned, most white LEDs use a blue GaN (gallium nitride) LED covered by a yellowish phosphor coating made of a certain type of crystals that have been powdered and bound in a type of viscous adhesive. Since yellow light stimulates the red and green receptors of the eye, the resulting mix of blue and yellow light gives the appearance of white.

The Organic LED (OLED)

An **OLED** is a device that consists of two or three layers of materials composed of organic molecules or polymers that emit light with the application of voltage. OLEDs produce light through the process of electrophosphorescence. The color of the light depends on the type of organic molecule in the emissive layer. The basic structure of a 2-layer OLED is shown in Figure 3–43.

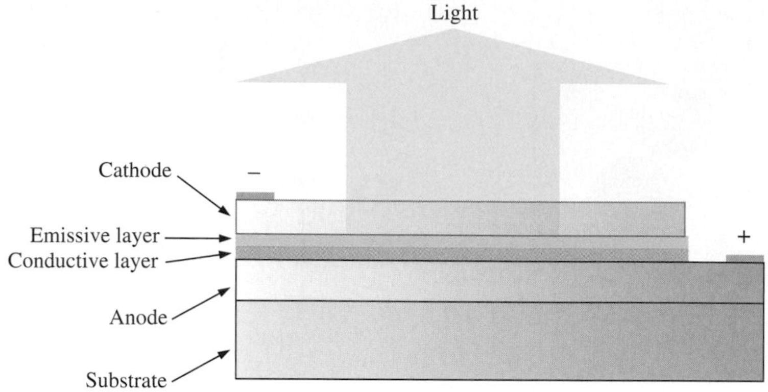

FIGURE 3–43

Basic structure of a top-emitting 2-layer OLED.

TECH NOTE

OLED technology was developed by Eastman Kodak. It is beginning to replace LCD (liquid crystal display) technology in handheld devices such as PDAs and cellular phones. OLEDs are brighter, thinner, faster, and lighter than conventional LEDs or LCDs. They also use less power and are cheaper to manufacture.

Electrons are provided to the emissive layer and removed from the conductive layer when there is current between the cathode and anode. This removal of electrons from the conductive layer leaves holes. The electrons from the emissive layer recombine with the holes from the conductive layer near the junction of the two layers. When this recombination occurs, energy is released in the form of light that passes through the transparent cathode material. If the anode and substrate are also made from transparent materials, light is emitted in both directions, making the OLED useful in applications such as heads-up displays.

OLEDs can be sprayed onto substrates just like inks are sprayed onto paper during printing. Inkjet technology greatly reduces the cost of OLED manufacturing and allows OLEDs to be printed onto very large films for large displays like 80-inch TV screens or electronic billboards.

The Photodiode

The **photodiode** is a device that operates in reverse bias, as shown in Figure 3–44(a), where I_λ is the reverse light current. The photodiode has a small transparent window that allows light to strike the *pn* junction. Some typical photodiodes are shown in Figure 3–44(b). An alternate photodiode symbol is shown in Figure 3–44(c).

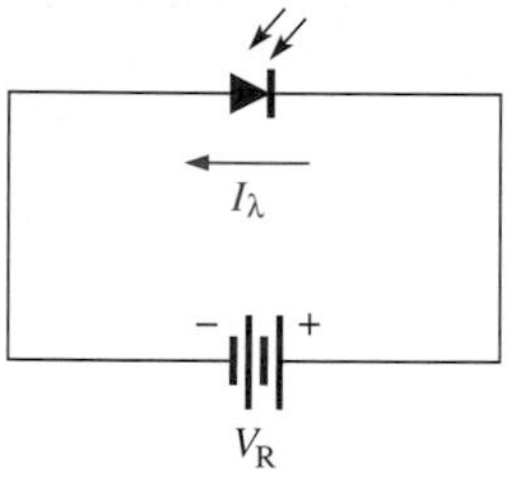

(a) Reverse-bias operation using standard symbol

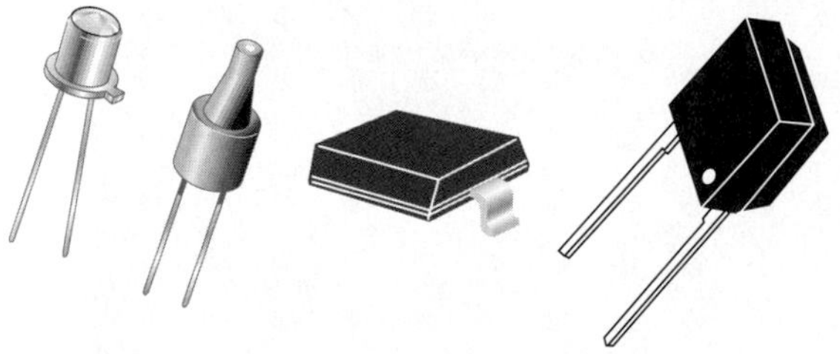

(b) Typical devices

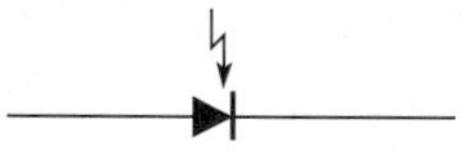

(c) Alternate symbol

FIGURE 3–44

Photodiode.

Recall that when reverse-biased, a rectifier diode has a very small reverse leakage current. The same is true for a photodiode. The reverse-biased current is produced by thermally generated electron-hole pairs in the depletion region, which are swept across the *pn* junction by the electric field created by the reverse voltage. In a rectifier diode, the reverse leakage current increases with temperature due to an increase in the number of electron-hole pairs.

A photodiode differs from a rectifier diode in that when its *pn* junction is exposed to light, the reverse current increases with the light intensity. When there is no incident light, the reverse current, I_λ, is almost negligible and is called the **dark current.** An increase in the amount of light intensity, expressed as irradiance (mW/cm^2), produces an increase in the reverse current, as shown by the graph in Figure 3–45(a).

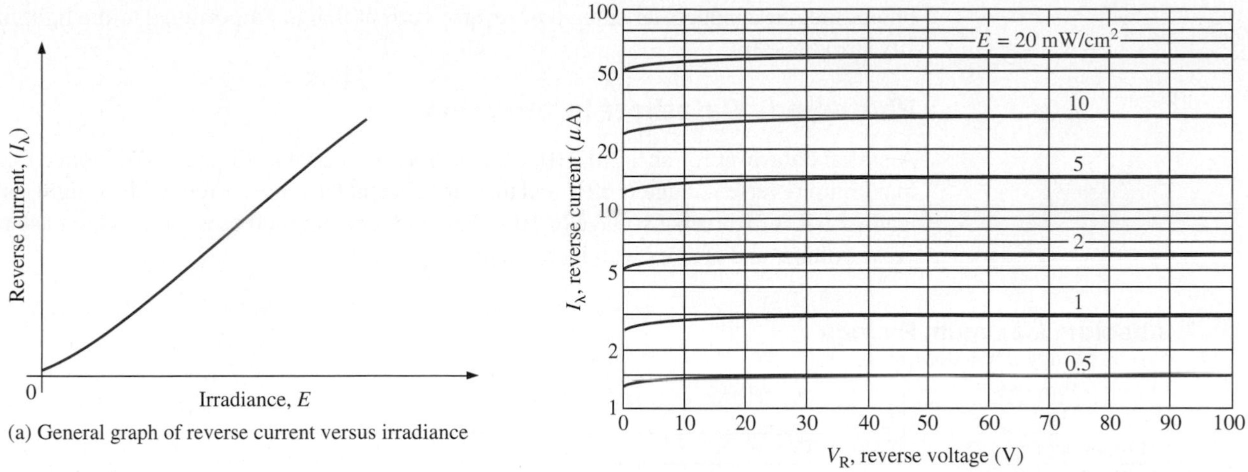

(a) General graph of reverse current versus irradiance

(b) Example of a graph of reverse current versus reverse voltage for several values of irradiance

FIGURE 3–45

Typical photodiode characteristics.

From the graph in Figure 3–45(b), you can see that the reverse current for this particular device is approximately 1.4 μA at a reverse-bias voltage of 10 V with an irradiance of 0.5 mW/cm^2. Therefore, the resistance of the device is

$$R_R = \frac{V_R}{I_\lambda} = \frac{10\text{ V}}{1.4\ \mu\text{A}} = 7.14\text{ M}\Omega$$

At 20 mW/cm^2, the current is approximately 55 μA at $V_R = 10$ V. The resistance under this condition is

$$R_R = \frac{V_R}{I_\lambda} = \frac{10\text{ V}}{55\ \mu\text{A}} = 182\text{ k}\Omega$$

These calculations show that the photodiode can be used as a variable-resistance device controlled by light intensity.

Figure 3–46 illustrates that the photodiode allows essentially no reverse current (except for a very small dark current) when there is no incident light. When a light beam strikes the

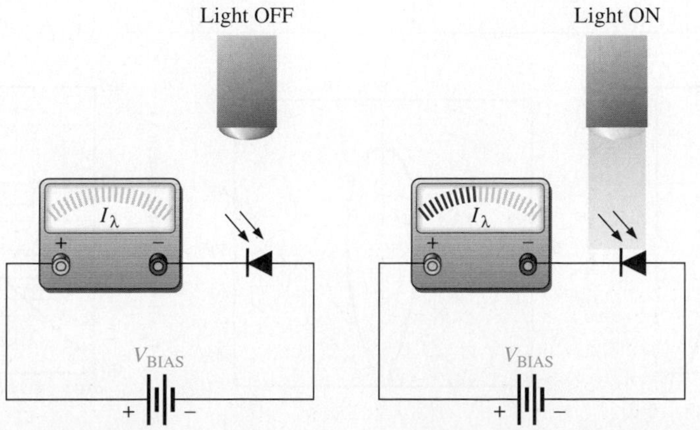

(a) No light, no current except negligible dark current

(b) Where there is incident light, resistance decreases and there is reverse current.

FIGURE 3–46

Operation of a photodiode.

photodiode, it conducts an amount of reverse current that is proportional to the light intensity (irradiance).

Photodiode Datasheet Information

A partial datasheet for an TEMD1000 photodiode is shown in Figure 3–47. Notice that the maximum reverse voltage is 60 V and the dark current (reverse current with no light) is typically 1 nA for a reverse voltage of 10 V. The dark current increases with an increase in reverse voltage and also with an increase in temperature.

Absolute Maximum Ratings

T_{amb} = 25°C, unless otherwise specified

Parameter	Test condition	Symbol	Value	Unit
Reverse Voltage		V_R	60	V
Power Dissipation	$T_{amb} \leq 25°C$	P_V	75	mW
Junction Temperature		T_j	100	°C
Storage Temperature Range		T_{stg}	- 40 to + 100	°C
Operating Temperature Range		T_{stg}	- 40 to + 85	°C
Soldering Temperature	$t \leq 5$ s	T_{sd}	< 260	°C

Basic Characteristics

T_{amb} = 25 °C, unless otherwise specified
T_{amb} = 25 °C, unless otherwise specified

Parameter	Test condition	Symbol	Min	Typ.	Max	Unit
Forward Voltage	I_F = 50 mA	V_F		1.0	1.3	V
Breakdown Voltage	I_R = 100 μA, E = 0	$V_{(BR)}$	60			V
Reverse Dark Current	V_R = 10 V, E = 0	I_{ro}		1	10	nA
Diode capacitance	V_R = 5 V, f = 1 MHz, E = 0	C_D		1.8		pF
Reverse Light Current	E_e = 1 mW/cm^2, λ = 870 nm, V_R = 5 V	I_{ra}		10		μA
	E_e = 1 mW/cm^2, λ = 950 nm, V_R = 5 V	I_{ra}	5	12		μA

Parameter	Test condition	Symbol	Min	Typ.	Max	Unit
Temp. Coefficient of I_{ra}	V_R = 5 V, = 870 nm	TK_{Ira}		0.2		%/K
Absolute Spectral Sensitivity	V_R = 5 V, = 870 nm	s(λ)		0.60		A/W
	V_R = 5 V, = 950 nm	s(λ)		0.55		A/W
Angle of Half Sensitivity		ϕ		±15		deg
Wavelength of Peak Sensitivity		λ_p		900		nm
Range of Spectral Bandwidth		$\lambda_{0.5}$		840 to 1050		nm
Rise Time	V_R = 10 V, R_L = 50, Ω λ = 820 nm	t_r		4		ns
Fall Time	V_R = 10 V, R_L = 50, Ω λ = 820 nm	t_f		4		ns

(a)

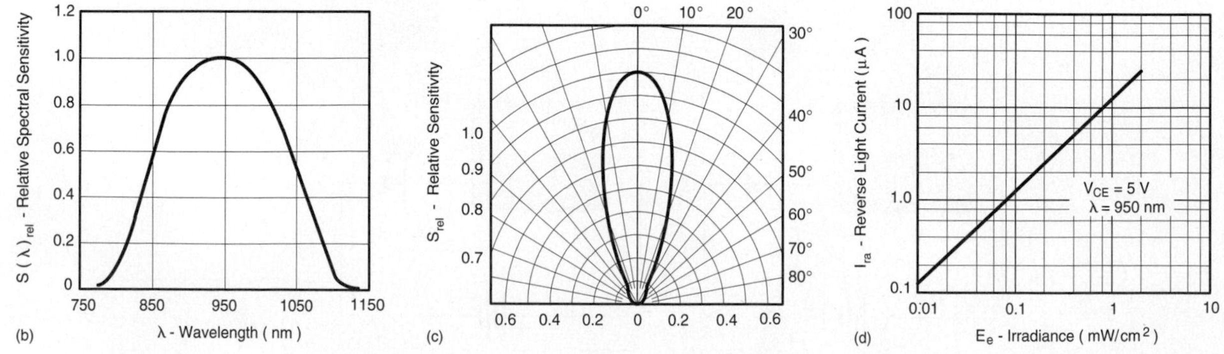

▲ FIGURE 3–47

Partial datasheet for the TEMD1000 photodiode. Datasheet courtesy of Vishay Intertechnology, Inc.

Sensitivity From the graph in part (b), you can see that the maximum sensitivity for this device occurs at a wavelength of 950 nm. The angular response graph in part (c) shows an area of response measured as relative sensitivity. At 10° on either side of the maximum orientation, the sensitivity drops to approximately 82% of maximum.

EXAMPLE 3–12

For a TEMD1000 photodiode,

(a) Determine the maximum dark current for $V_R = 10$ V.

(b) Determine the reverse light current for an irradiance of 1 mW/cm^2 at a wavelength of 850 nm if the device angle is oriented at 10° with respect to the maximum irradiance and the reverse voltage is 5 V.

Solution

(a) From Figure 3–47(a), the maximum dark current I_{ro} = **10 nA.**

(b) From the graph in Figure 3–47(d), the reverse light current is 12 μA at 950 nm. From Figure 3–47(b), the relative sensitivity is 0.6 at 850 nm. Therefore, the reverse light current is

$$I_\lambda = I_{ra} = 0.6(12\ \mu\text{A}) = 72\ \mu\text{A}$$

For an angle of 10°, the relative sensitivity is reduced to 0.92 of its value at 0°.

$$I_\lambda = I_{ra} = 0.92\ (7.2\ \mu\text{A}) = \mathbf{6.62\ \mu A}$$

Related Problem What is the reverse current if the wavelength is 1050 nm and the angle is 0°?

SECTION 3–4 CHECKUP

1. Name two types of LEDs in terms of their light-emission spectrum.
2. Which has the greater wavelength, visible light or infrared?
3. In what bias condition is an LED normally operated?
4. What happens to the light emission of an LED as the forward current increases?
5. The forward voltage drop of an LED is 0.7 V. (true or false)
6. What is a pixel?
7. In what bias condition is a photodiode normally operated?
8. When the intensity of the incident light (irradiance) on a photodiode increases, what happens to its internal reverse resistance?
9. What is *dark current*?

3–5 OTHER TYPES OF DIODES

In this section, several types of diodes that you are less likely to encounter as a technician but are nevertheless important are introduced. Among these are the laser diode, the Schottky diode, the *pin* diode, the step-recovery diode, the tunnel diode, and the current regulator diode.

After completing this section, you should be able to

- **Discuss the basic characteristics of laser, Schottky, *pin*, step-recovery, tunnel, and current regulator diodes**
  - Identify the various diode symbols
  - Describe the laser diode and how it differs from an LED

- Describe the characteristics of the Schottky diode
- Describe the characteristics of the *pin* diode
- Describe the characteristics of the step-recovery diode
- Describe the characteristics of the tunnel diode and explain its negative resistance
- Discuss how the current regulator diode maintains a constant forward current

The Laser Diode

The term **laser** stands for *l*ight *a*mplification by *s*timulated *e*mission of *r*adiation. Laser light is **monochromatic,** which means that it consists of a single color and not a mixture of colors. Laser light is also called **coherent light,** a single wavelength, as compared to incoherent light, which consists of a wide band of wavelengths. The laser diode normally emits coherent light, whereas the LED emits incoherent light. The symbols are the same as shown in Figure 3–48(a).

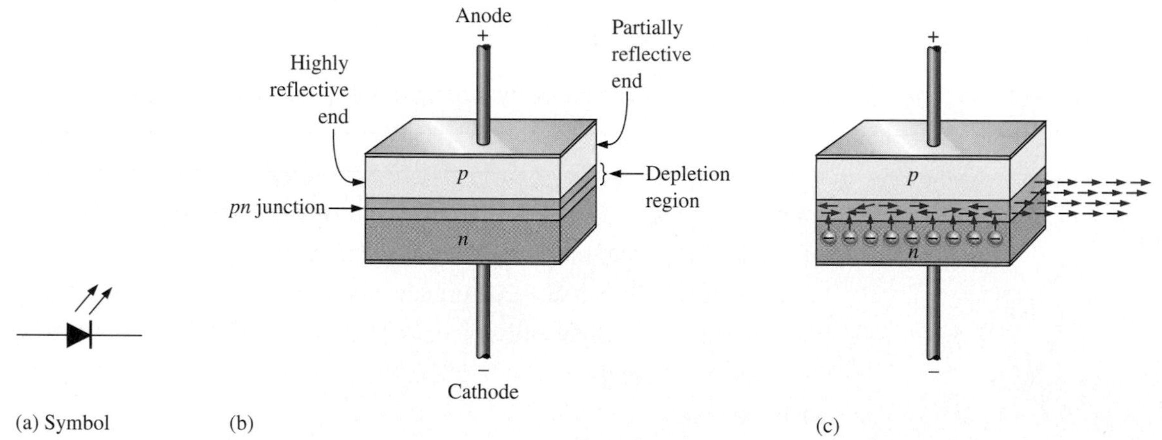

FIGURE 3–48

Basic laser diode construction and operation.

The basic construction of a laser diode is shown in Figure 3–48(b). A *pn* junction is formed by two layers of doped gallium arsenide. The length of the *pn* junction bears a precise relationship with the wavelength of the light to be emitted. There is a highly reflective surface at one end of the *pn* junction and a partially reflective surface at the other end, forming a resonant cavity for the photons. External leads provide the anode and cathode connections.

The basic operation is as follows. The laser diode is forward-biased by an external voltage source. As electrons move through the junction, recombination occurs just as in an ordinary diode. As electrons fall into holes to recombine, photons are released. A released photon can strike an atom, causing another photon to be released. As the forward current is increased, more electrons enter the depletion region and cause more photons to be emitted. Eventually some of the photons that are randomly drifting within the depletion region strike the reflected surfaces perpendicularly. These reflected photons move along the depletion region, striking atoms and releasing additional photons due to the avalanche effect. This back-and-forth movement of photons increases as the generation of photons "snowballs" until a very intense beam of laser light is formed by the photons that pass through the partially reflective end of the *pn* junction.

Each photon produced in this process is identical to the other photons in energy level, phase relationship, and frequency. So a single wavelength of intense light emerges from the laser diode, as indicated in Figure 3–48(c). Laser diodes have a threshold level of current above which the laser action occurs and below which the diode behaves essentially as an LED, emitting incoherent light.

An Application Laser diodes and photodiodes are used in the pick-up system of compact disk (CD) players. Audio information (sound) is digitally recorded in stereo on the surface of a compact disk in the form of microscopic "pits" and "flats." A lens arrangement focuses the laser beam from the diode onto the CD surface. As the CD rotates, the lens and beam follow the track under control of a servomotor. The laser light, which is altered by the pits and flats along the recorded track, is reflected back from the track through a lens and optical system to infrared photodiodes. The signal from the photodiodes is then used to reproduce the digitally recorded sound. Laser diodes are also used in laser printers and fiber-optic systems.

The Schottky Diode

Schottky diodes are high-current diodes used primarily in high-frequency and fast-switching applications. They are also known as *hot-carrier diodes*. The term *hot-carrier* is derived from the higher energy level of electrons in the *n* region compared to those in the metal region. A Schottky diode symbol is shown in Figure 3–49. A Schottky diode is formed by joining a doped semiconductor region (usually *n*-type) with a metal such as gold, silver, or platinum. Rather than a *pn* junction, there is a metal-to-semiconductor junction, as shown in Figure 3–50. The forward voltage drop is typically around 0.3 V because there is no depletion region as in a *pn* junction diode.

▲ **FIGURE 3–49**

Schottky diode symbol.

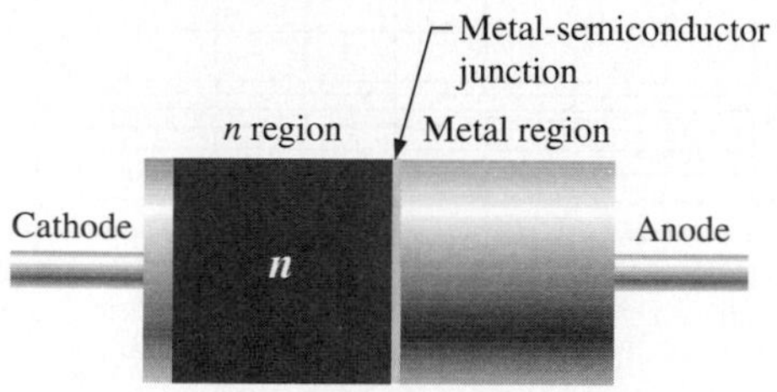

◀ **FIGURE 3–50**

Basic internal construction of a Schottky diode.

The Schottky diode operates only with majority carriers. There are no minority carriers and thus no reverse leakage current as in other types of diodes. The metal region is heavily occupied with conduction-band electrons, and the *n*-type semiconductor region is lightly doped. When forward-biased, the higher energy electrons in the *n* region are injected into the metal region where they give up their excess energy very rapidly. Since there are no minority carriers, as in a conventional rectifier diode, there is a very rapid response to a change in bias. The Schottky is a fast-switching diode, and most of its applications make use of this property. It can be used in high-frequency applications and in many digital circuits to decrease switching times. The LS family of TTL logic (LS stands for low-power Schottky) is one type of digital integrated circuit that uses the Schottky diode.

The *PIN* Diode

The *pin* diode consists of heavily doped *p* and *n* regions separated by an intrinsic (*i*) region, as shown in Figure 3–51(a). When reverse-biased, the *pin* diode acts like a nearly constant capacitance. When forward-biased, it acts like a current-controlled variable resistance. This is shown in Figure 3–51(b) and (c). The low forward resistance of the intrinsic region decreases with increasing current.

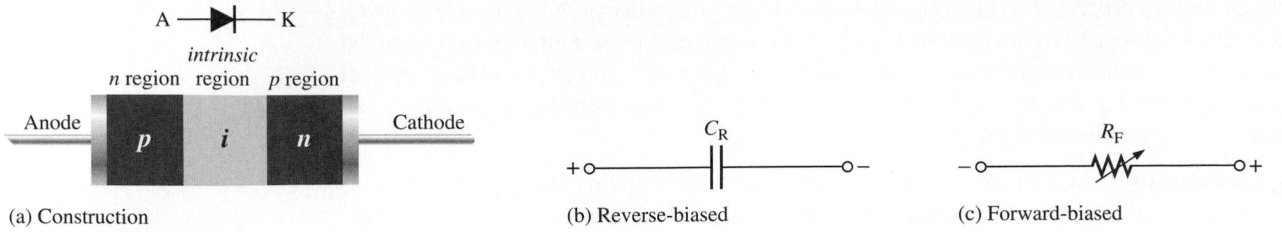

FIGURE 3–51

PIN diode.

The forward series resistance characteristic and the reverse capacitance characteristic are shown graphically in Figure 3–52 for a typical *pin* diode.

The *pin* diode is used as a dc-controlled microwave switch operated by rapid changes in bias or as a modulating device that takes advantage of the variable forward-resistance characteristic. Since no rectification occurs at the *pn* junction, a high-frequency signal can be modulated (varied) by a lower-frequency bias variation. A *pin* diode can also be used in attenuator applications because its resistance can be controlled by the amount of current. Certain types of *pin* diodes are used as photodetectors in fiber-optic systems.

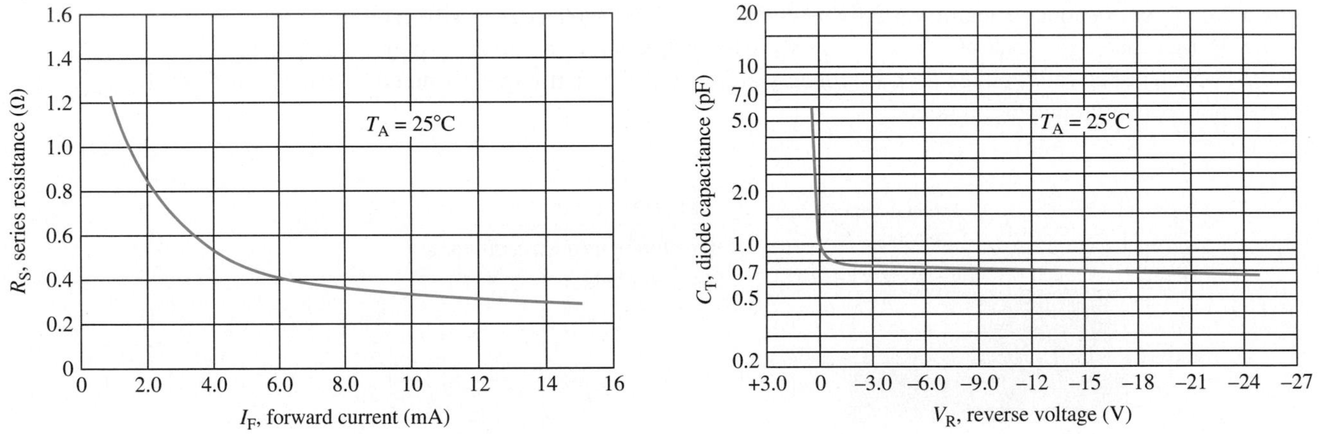

FIGURE 3–52

PIN diode characteristics.

The Step-Recovery Diode

The step-recovery diode uses graded doping where the doping level of the semiconductive materials is reduced as the *pn* junction is approached. This produces an abrupt turn-off time by allowing a fast release of stored charge when switching from forward to reverse bias. It also allows a rapid re-establishment of forward current when switching from reverse to forward bias. This diode is used in very high frequency (VHF) and fast-switching applications.

The Tunnel Diode

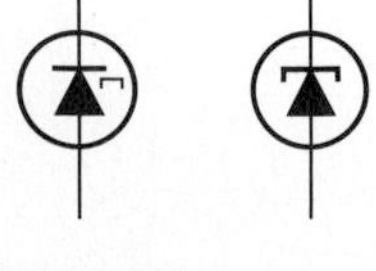

FIGURE 3–53

Tunnel diode symbols.

The tunnel diode exhibits a special characteristic known as *negative resistance.* This feature makes it useful in oscillator and microwave amplifier applications. Two alternate symbols are shown in Figure 3–53. Tunnel diodes are constructed with germanium or gallium arsenide by doping the *p* and *n* regions much more heavily than in a conventional rectifier diode. This heavy doping results in an extremely narrow depletion region. The heavy doping allows conduction for all reverse voltages so that there is no breakdown effect as with the conventional rectifier diode. This is shown in Figure 3–54.

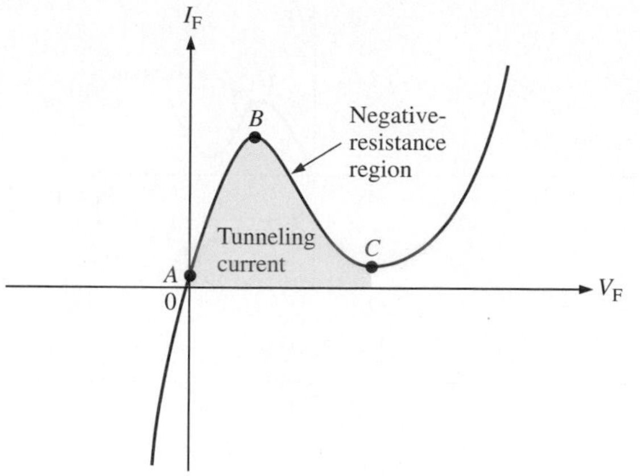

◀ **FIGURE 3–54**

Tunnel diode characteristic curve.

Also, the extremely narrow depletion region permits electrons to "tunnel" through the *pn* junction at very low forward-bias voltages, and the diode acts as a conductor. This is shown in Figure 3–54 between points *A* and *B*. At point *B*, the forward voltage begins to develop a barrier, and the current begins to decrease as the forward voltage continues to increase. This is the *negative-resistance region.*

$$R_F = \frac{\Delta V_F}{\Delta I_F}$$

This effect is opposite to that described in Ohm's law, where an increase in voltage results in an increase in current. At point *C*, the diode begins to act as a conventional forward-biased diode.

HISTORY NOTE

Leo Esaki won the Nobel Prize in physics in 1973 for the invention of the tunnel diode in the late 1950s. Surprisingly, in 1976 Robert Noyce, cofounder of Intel Corp., revealed in a talk before the MIT Club of New York that he had in his notebooks from 1956 a complete description of the tunnel diode. However, credit for the invention is given to Esaki and the tunnel diode is also known as the Esaki diode in his honor.

An Application A parallel resonant circuit can be represented by a capacitance, inductance, and resistance in parallel, as in Figure 3–55(a). R_P is the parallel equivalent of the series winding resistance of the coil. When the tank circuit is "shocked" into oscillation by an application of voltage as in Figure 3–55(b), a damped sinusoidal output results. The damping is due to the resistance of the tank, which prevents a sustained oscillation because energy is lost when there is current through the resistance.

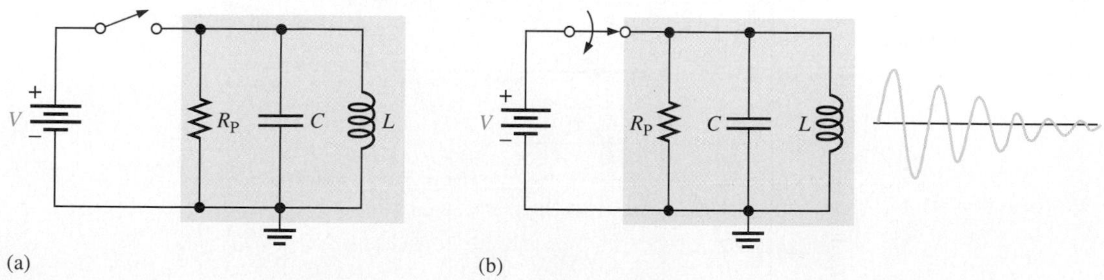

▲ **FIGURE 3–55**

Parallel resonant circuit.

If a tunnel diode is placed in series with the tank circuit and biased at the center of the negative-resistance portion of its characteristic curve, as shown in Figure 3–56, a sustained oscillation (constant sinusoidal voltage) will result on the output. This is because the negative-resistance characteristic of the tunnel diode counteracts the positive-resistance characteristic of the tank resistance. The tunnel diode is only used at very high frequencies.

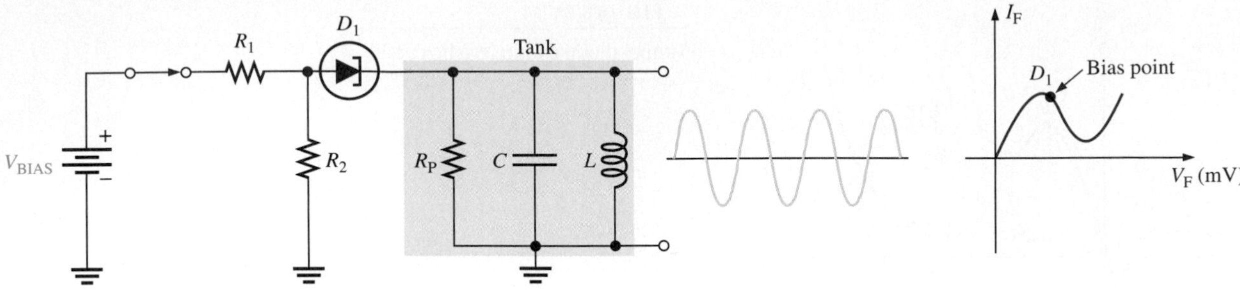

▲ **FIGURE 3–56**

Basic tunnel diode oscillator.

Current Regulator Diode

The current regulator diode is often referred to as a constant-current diode. Rather than maintaining a constant voltage, as the zener diode does, this diode maintains a constant current. The symbol is shown in Figure 3–57.

► **FIGURE 3–57**

Symbol for a current regulator diode.

Figure 3–58 shows a typical characteristic curve. The current regulator diode operates in forward bias (shaded region), and the forward current becomes a specified constant value at forward voltages ranging from about 1.5 V to about 6 V, depending on the diode type. The constant forward current is called the *regulator current* and is designated I_P. For example, the 1N5283–1N5314 series of diodes have nominal regulator currents ranging from 220 μA to 4.7 mA. These diodes may be used in parallel to obtain higher currents. This diode does not have a sharply defined reverse breakdown, so the reverse current begins to increase for V_{AK} values of less than 0 V (unshaded region of the figure). This device should never be operated in reverse bias.

► **FIGURE 3–58**

Typical characteristic curve for a current regulator diode.

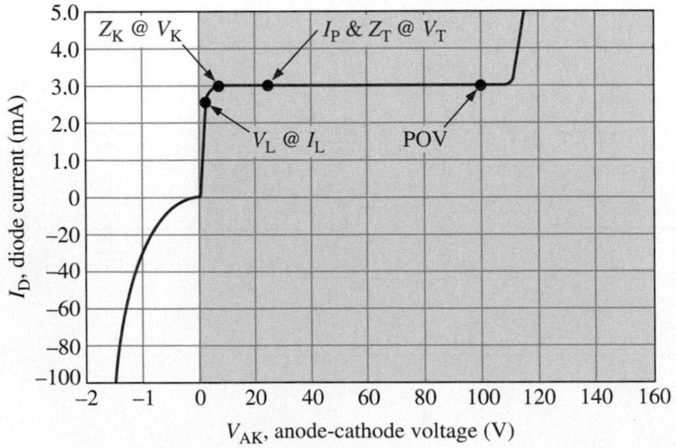

In forward bias, the diode regulation begins at the limiting voltage, V_L, and extends up to the POV (peak operating voltage). Notice that between V_K and POV, the current is essentially constant. V_T is the test voltage at which I_P and the diode impedance, Z_T, are specified on a datasheet. The impedance Z_T has very high values ranging from 235 kΩ to 25 MΩ for the diode series mentioned before.

SECTION 3–5 CHECKUP

1. What does *laser* mean?
2. What is the difference between incoherent and coherent light and which is produced by a laser diode?
3. What are the primary application areas for Schottky diodes?
4. What is a hot-carrier diode?
5. What is the key characteristic of a tunnel diode?
6. What is one application for a tunnel diode?
7. Name the three regions of a *pin* diode.
8. Between what two voltages does a current regulator diode operate?

3–6 TROUBLESHOOTING

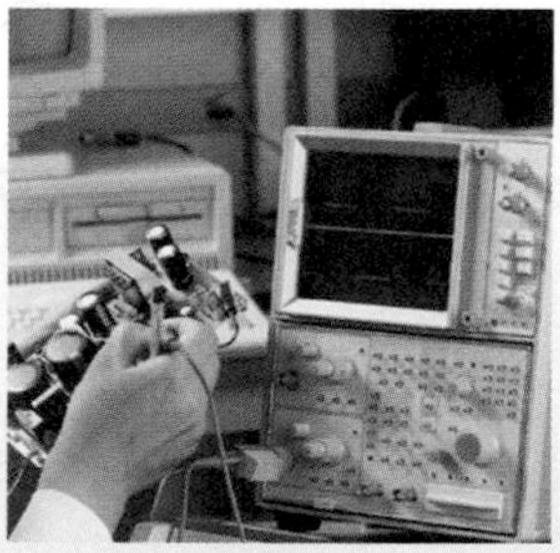

In this section, you will see how a faulty zener diode can affect the output of a regulated dc power supply. Although IC regulators are generally used for power supply outputs, the zener is occasionally used when less precise regulation and low current is acceptable. Like other diodes, the zener can fail open, it can exhibit degraded performance, or it can short out.

After completing this section, you should be able to

- **Troubleshoot zener diode regulators**
  - Recognize the effects of an open zener
  - Recognize the effects of a zener with degraded performance or shorted

A Zener-Regulated DC Power Supply

Figure 3–59 shows a filtered dc power supply that produces a constant 24 V before it is regulated down to 15 V by the zener regulator. The 1N4744A zener diode is the same as the one in Example 3–7. A no-load check of the regulated output voltage shows 15.5 V as indicated in part (a). The typical voltage expected at the zener test current for this particular diode is 15 V. In part (b), a potentiometer is connected to provide a variable load resistance. It is adjusted to a minimum value for a full-load test as determined by the following calculations. The full-load test is at minimum zener current (I_{ZK}). The meter reading of 14.8 V indicates approximately the expected output voltage of 15.0 V.

$$I_T = \frac{24\text{ V} - 14.8\text{ V}}{180\ \Omega} = 51.1\text{ mA}$$

$$I_L = I_T - I_Z = 51.1\text{ mA} - 0.25\text{ mA} = 50.9\text{ mA}$$

$$R_{L(min)} = \frac{14.8\text{ V}}{50.9\text{ mA}} = 291\ \Omega$$

Case 1: Zener Diode Open If the zener diode fails open, the power supply test gives the approximate results indicated in Figure 3–60. In the no-load check shown in part (a), the output voltage is 24 V because there is no voltage dropped between the filtered output of the power supply and the output terminal. This definitely indicates an open between the output terminal and ground. In the full-load check, the voltage of 14.8 V results from the

39. Refer to the photodiode datasheet in Figure 3–47.

(a) An TEMD1000 is connected in series with a 1 kΩ resistor and a reverse-bias voltage source. There is no incident light on the diode. What is the maximum voltage drop across the resistor?

(b) At what wavelength will the reverse current be the greatest for a given irradiance?

(c) At what wavelength is relative spectral sensitivity of the TEMD1000 equal to 0.4?

ADVANCED PROBLEMS

40. Develop the schematic for the circuit board in Figure 3–79 and determine what type of circuit it is.

FIGURE 3–79

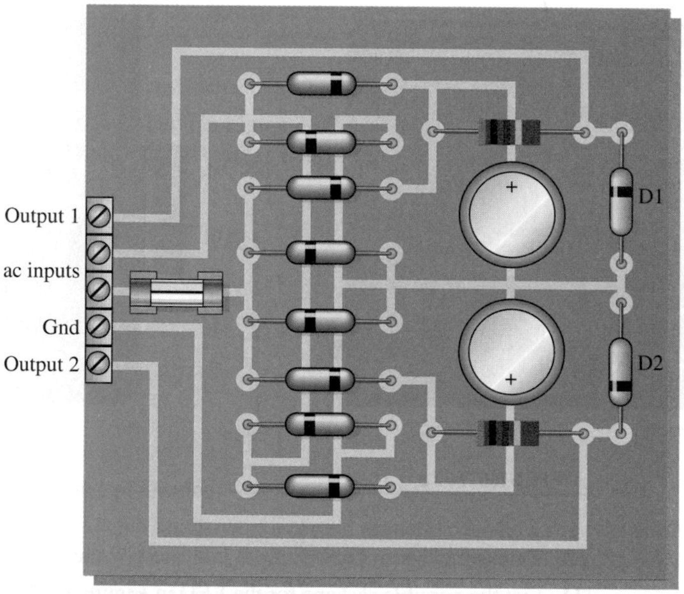

Rectifier diodes: 1N4001A
Zener diodes: D1-1N4736A, D2-1N4749A
Filter capacitors: 100 μF

41. If a 30 V rms, 60 Hz input voltage is connected to the ac inputs, determine the output voltages on the circuit board in Figure 3–79.

42. If each output of the board in Figure 3–79 is loaded with 10 kΩ, what fuse rating should be used?

43. Design a zener voltage regulator to meet the following specifications: The input voltage is 24 V dc, the load current is 35 mA, and the load voltage is 8.2 V.

44. The varactor-tuned band-pass filter in Figure 3–27 is to be redesigned to produce a bandwidth of from 350 kHz to 850 kHz within a 10% tolerance. Specify what change you would have to make using the graph in Figure 3–80.

FIGURE 3–80

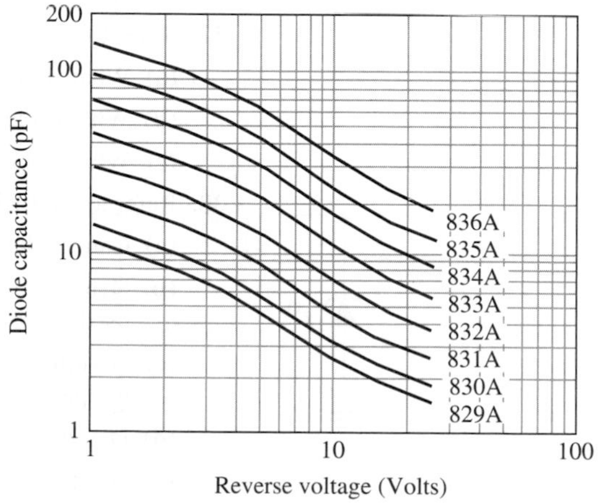

45. Design a seven-segment red LED display circuit in which any of the ten digits can be displayed using a set of switches. Each LED segment is to have a current of 20 mA ± 10% from a 12 V source and the circuit must be designed with a minimum number of switches.
46. If you used a common-anode seven-segment display in Problem 45, redesign it for a common-cathode display or vice versa.

MULTISIM TROUBLESHOOTING PROBLEMS

These file circuits are in the Troubleshooting Problems folder on your CD-ROM.

47. Open file TSP03-47 and determine the fault.
48. Open file TSP03-48 and determine the fault.
49. Open file TSP03-49 and determine the fault.
50. Open file TSP03-50 and determine the fault.

ANSWERS

SECTION CHECKUPS

Section 3–1 The Zener Diode

1. Zener diodes are operated in the reverse-breakdown region.
2. The test current, I_Z
3. The zener impedance causes the voltage to vary slightly with current.
4. The zener voltage increases (or decreases) 0.05% for each degree centigrade increase (or decrease).
5. Power derating is the reduction in the power rating of a device as a result of an increase in temperature.

Section 3–2 Zener Diode Applications

1. An infinite resistance (open)
2. With no load, there is no current to a load. With full load, there is maximum current to the load.
3. Approximately 0.7 V, just like a rectifier diode

Section 3–3 The Varactor Diode

1. A varactor exhibits variable capacitance.
2. A varactor is operated in reverse bias.
3. The depletion region
4. Capacitance decreases with more reverse bias.
5. The capacitance ratio is the ratio of a varactor's capacitance at a specified minimum voltage to the capacitance at a specified maximum voltage.

Section 3–4 Optical Diodes

1. Infrared and visible light
2. Infrared has the greater wavelength.
3. An LED operates in forward bias.
4. Light emission increases with forward current.
5. False, V_F of an LED is usually greater than 1.2 V.
6. A tiny grouping of red, green, and blue LEDs.
7. A photodiode operates in reverse bias.
8. The internal resistance decreases.
9. Dark current is the reverse photodiode current when there is no light.

Section 3–5 Other Types of Diodes

1. *l*ight *a*mplification by *s*timulated *e*mission of *r*adiation
2. Coherent light has only a single wavelength, but incoherent light has a wide band of wavelengths. A laser diode produces coherent light.
3. High-frequency and fast-switching circuits
4. *Hot carrier* is another name for Schottky diodes.
5. Tunnel diodes have negative resistance.
6. Oscillators
7. *p* region, *n* region, and intrinsic (*i*) region
8. A current regulator operates between V_L (limiting voltage) and POV (peak operating voltage).

Section 3–6 Troubleshooting

1. The output voltage is too high and equal to the rectifier output.
2. More
3. Series limiting resistor open, fuse blown. A shorted zener could have caused this.
4. The output voltage changes as the load resistance changes.

RELATED PROBLEMS FOR EXAMPLES

3–1 5 Ω

3–2 The voltage will decrease by 0.45 V.

3–3 7.5 W

3–4 $V_Z = -11.9$ V at 10 mA; $V_Z = 12.08$ V at 30 mA

3–5 $V_{IN(min)} = 6.77$ V; $V_{IN(max)} = 21.9$ V

3–6 $I_{L(min)} = 0$ A; $I_{L(max)} = 43$ mA; $R_{L(min)} = 76.7\ \Omega$

3–7 **(a)** 11.8 V at I_{ZK}; 12.9 V at I_{ZM}

(b) 133 Ω

(c) 151 Ω

3–8 **(a)** A waveform identical to Figure 3–20(a)

(b) A sine wave with a peak value of 5 V

3–9 Increase V_2.

3–10 approximately 0.7

3–11 Sixteen parallel branches with four LEDs in each branch.
$R_{LIMIT} = (12\text{ V} - 9.2\text{ V})/30\text{ mA} = 93\ \Omega$.

3–12 approximately 6.6 μA

TRUE/FALSE QUIZ

1. T **2.** T **3.** F **4.** F **5.** T **6.** F
7. T **8.** T **9.** F **10.** T **11.** T **12.** T

CIRCUIT-ACTION QUIZ

1. (c) **2.** (b) **3.** (a) **4.** (a) **5.** (b) **6.** (c) **7.** (c) **8.** (b)
9. (a) **10.** (a) **11.** (b)

SELF-TEST

1. (a) **2.** (b) **3.** (c) **4.** (b) **5.** (d) **6.** (a) **7.** (c) **8.** (d)
9. (d) **10.** (d) **11.** (b) **12.** (b) **13.** (d) **14.** (b) **15.** (d)

BIPOLAR JUNCTION TRANSISTORS

4

CHAPTER OUTLINE

CHAPTER OBJECTIVES

- Describe the basic structure of the BJT (bipolar junction transistor)
- Explain how a BJT is biased and discuss the transistor currents and their relationships
- Discuss BJT parameters and characteristics and use these to analyze a transistor circuit
- Discuss how a BJT is used as a voltage amplifier
- Discuss how a BJT is used as an electronic switch
- Describe a phototransistor and its operation
- Identify various types of transistor package configurations
- Troubleshoot various faults in transistor circuits

KEY TERMS

- BJT (bipolar junction transistor)
- Emitter
- Base
- Collector
- Gain
- Beta
- Saturation
- Linear region
- Cutoff
- Amplification
- Phototransistor

VISIT THE COMPANION WEBSITE

Study aids for this chapter are available at
http://www.prenhall.com/floyd

INTRODUCTION

The invention of the transistor was the beginning of a technological revolution that is still continuing. All of the complex electronic devices and systems today are an outgrowth of early developments in semiconductor transistors.

Two basic types of transistors are the bipolar junction transistor (BJT), which we will begin to study in this chapter, and the field-effect transistor (FET), which we will cover in later chapters. The BJT is used in two broad areas—as a linear amplifier to boost or amplify an electrical signal and as an electronic switch. Both of these applications are introduced in this chapter.

APPLICATION ACTIVITY PREVIEW

Suppose you work for a company that makes a security alarm system for protecting homes and businesses against illegal entry. You are given the responsibility for final development and for testing each system before it is shipped out. The first step is to learn all you can about transistor operation. You will then apply your knowledge to the Application Activity at the end of the chapter.

4–1 BJT STRUCTURE

The basic structure of the bipolar junction transistor (BJT) determines its operating characteristics. In this section, you will see how semiconductive materials are used to form a BJT, and you will learn the standard BJT symbols.

After completing this section, you should be able to

- **Describe the basic structure of the BJT (bipolar junction transistor)**
  - Explain the difference between the structure of an *npn* and a *pnp* transistor
  - Identify the symbols for *npn* and *pnp* transistors
  - Name the three regions of a BJT and their labels

The **BJT (bipolar junction transistor)** is constructed with three doped semiconductor regions separated by two *pn* junctions, as shown in the epitaxial planar structure in Figure 4–1(a). The three regions are called **emitter**, **base**, and **collector**. Physical representations of the two types of BJTs are shown in Figure 4–1(b) and (c). One type consists of two *n* regions separated by a *p* region (*npn*), and the other type consists of two *p* regions separated by an *n* region (*pnp*). The term **bipolar** refers to the use of both holes and electrons as current carriers in the transistor structure.

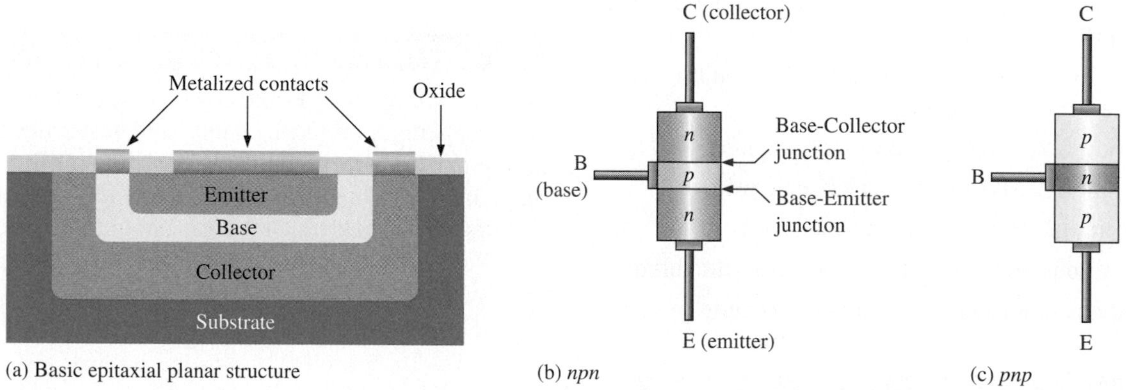

▲ **FIGURE 4–1**

Basic BJT construction.

HISTORY NOTE

The transistor was invented in 1947 by a team of scientists from Bell Laboratories. William Shockley, Walter Brattain, and John Bardeen developed the solid-state device that replaced the vacuum tube. Each received the Nobel prize in 1956. The transistor is arguably the most significant invention of the twentieth century.

The *pn* junction joining the base region and the emitter region is called the *base-emitter junction*. The *pn* junction joining the base region and the collector region is called the *base-collector junction*, as indicated in Figure 4–1(b). A wire lead connects to each of the three regions, as shown. These leads are labeled E, B, and C for emitter, base, and collector, respectively. The base region is lightly doped and very thin compared to the heavily doped emitter and the moderately doped collector regions. (The reason for this is discussed in the next section.) Figure 4–2 shows the schematic symbols for the *npn* and *pnp* bipolar junction transistors.

► **FIGURE 4–2**

Standard BJT (bipolar junction transistor) symbols.

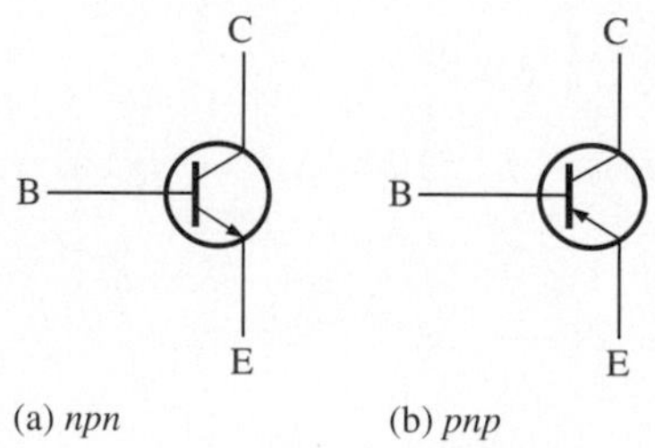

SECTION 4–1 CHECKUP
Answers are at the end of the chapter.

1. Name the two types of BJTs according to their structure.
2. The BJT is a three-terminal device. Name the three terminals.
3. What separates the three regions in a BJT?

4–2 BASIC BJT OPERATION

In order for a BJT to operate properly as an amplifier, the two *pn* junctions must be correctly biased with external dc voltages. In this section, we mainly use the *npn* transistor for illustration. The operation of the *pnp* is the same as for the *npn* except that the roles of the electrons and holes, the bias voltage polarities, and the current directions are all reversed.

After completing this section, you should be able to

- **Explain how a BJT is biased and discuss the transistor currents and their relationships**
  - Describe forward-reverse bias
  - Show how to connect a BJT to the bias-voltage sources
  - Describe the basic internal operation of a BJT
  - State the formula relating the collector, emitter, and base currents in a BJT

Figure 4–3 shows a bias arrangement for both *npn* and *pnp* BJTs for operation as an **amplifier**. Notice that in both cases the base-emitter (BE) junction is forward-biased and the base-collector (BC) junction is reverse-biased. This condition is called *forward-reverse bias.*

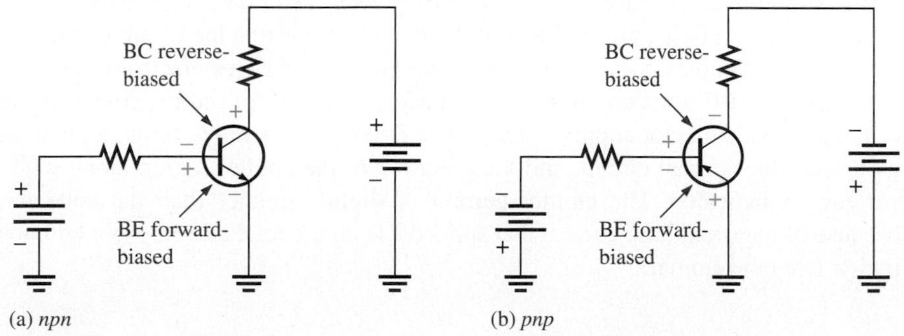

◄ **FIGURE 4–3**
Forward-reverse bias of a BJT.

To understand how a transistor operates, let's examine what happens inside the *npn* structure. The heavily doped *n*-type emitter region has a very high density of conduction-band (free) electrons, as indicated in Figure 4–4. These free electrons easily diffuse through the forward-based BE junction into the lightly doped and very thin *p*-type base region, as indicated by the wide arrow. The base has a low density of holes, which are the majority carriers, as represented by the white circles. A small percentage of the total number of free electrons injected into the base region recombine with holes and move as valence electrons through the base region and into the emitter region as hole current, indicated by the red arrows.

FIGURE 4–4

BJT operation showing electron flow.

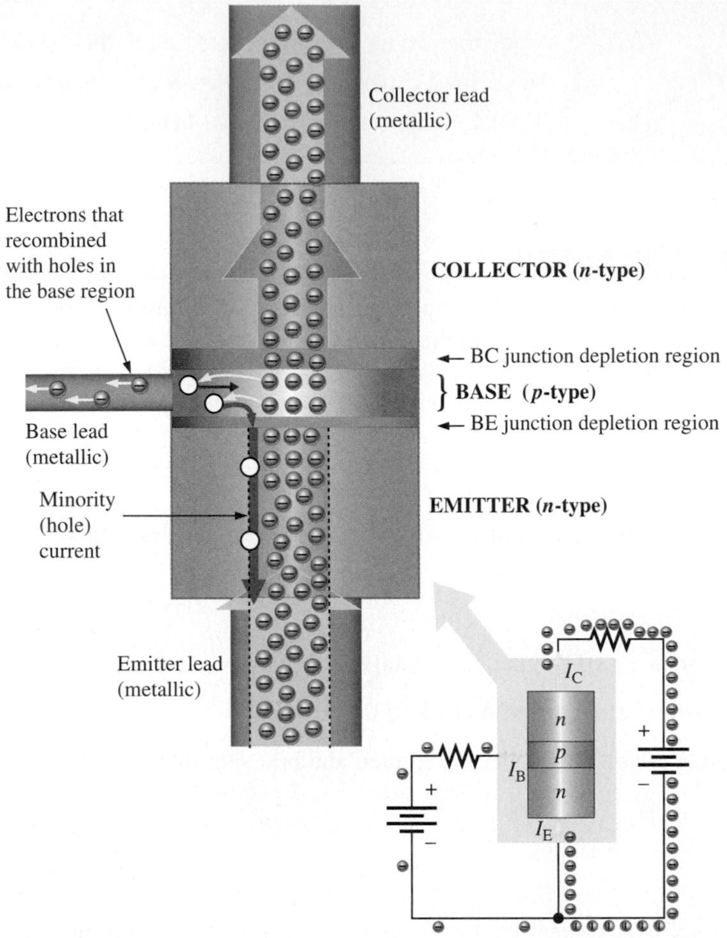

When the electrons that have recombined with holes as valence electrons leave the crystalline structure of the base, they become free electrons in the metallic base lead and produce the external base current. Most of the free electrons that have entered the base do not recombine with holes because the base is very thin. As the free electrons move toward the reverse-biased BC junction, they are swept across into the collector region by the attraction of the positive collector supply voltage. The free electrons move through the collector region, into the external circuit, and then return into the emitter region along with the base current, as indicated. The emitter current is slightly greater than the collector current because of the small base current that splits off from the total current injected into the base region from the emitter.

Transistor Currents

The directions of the currents in an *npn* transistor and its schematic symbol are as shown in Figure 4–5(a); those for a *pnp* transistor are shown in Figure 4–5(b). Notice that the arrow on the emitter inside the transistor symbols points in the direction of conventional current. These diagrams show that the emitter current (I_E) is the sum of the collector current (I_C) and the base current (I_B), expressed as follows:

Equation 4–1

$$I_E = I_C + I_B$$

As mentioned before, I_B is very small compared to I_E or I_C. The capital-letter subscripts indicate dc values.

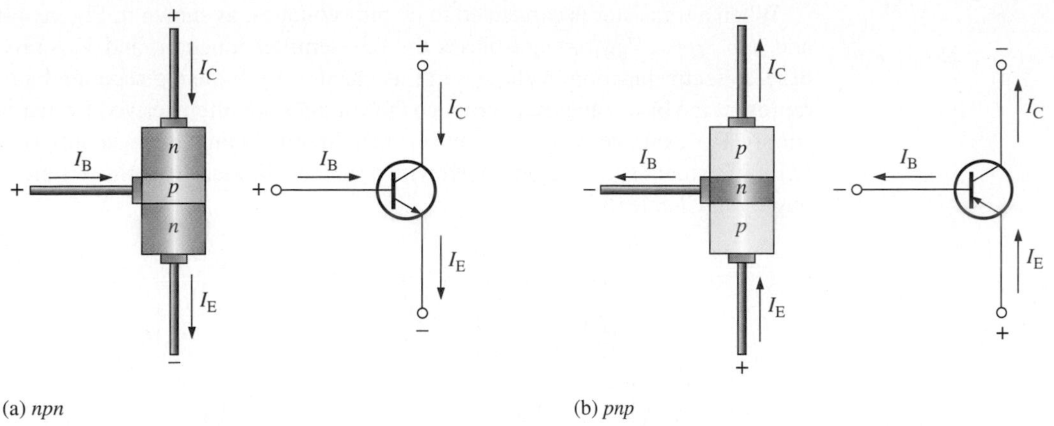

▲ FIGURE 4–5

Transistor currents.

SECTION 4–2 CHECKUP

1. What are the bias conditions of the base-emitter and base-collector junctions for a transistor to operate as an amplifier?
2. Which is the largest of the three transistor currents?
3. Is the base current smaller or larger than the emitter current?
4. Is the base region much thinner or much wider than the collector and emitter regions?
5. If the collector current is 1 mA and the base current is 10 μA, what is the emitter current?

4–3 BJT CHARACTERISTICS AND PARAMETERS

Two important parameters, β_{DC} (dc current gain) and α_{DC} are introduced and used to analyze a BJT circuit. Also, transistor characteristic curves are covered, and you will learn how a BJT's operation can be determined from these curves. Finally, maximum ratings of a BJT are discussed.

After completing this section, you should be able to

- **Discuss BJT parameters and characteristics and use these to analyze a transistor circuit**
  - Define dc *beta* (β_{DC})
  - Define dc *alpha* (α_{DC})
  - Identify all currents and voltages in a BJT circuit
  - Analyze a basic BJT dc circuit
  - Interpret collector characteristic curves and use a dc load line
  - Describe how β_{DC} varies with temperature and collector current
  - Discuss and apply maximum transistor ratings
  - Derate a transistor for power dissipation
  - Interpret a transistor datasheet

When a transistor is connected to dc bias voltages, as shown in Figure 4–6 for both *npn* and *pnp* types, V_{BB} forward-biases the base-emitter junction, and V_{CC} reverse-biases the base-collector junction. Although in this chapter we are using separate battery symbols to represent the bias voltages, in practice the voltages are often derived from a single dc power supply. For example, V_{CC} is normally taken directly from the power supply output and V_{BB} (which is smaller) can be produced with a voltage divider. Bias circuits are examined thoroughly in Chapter 5.

FIGURE 4–6

Transistor dc bias circuits.

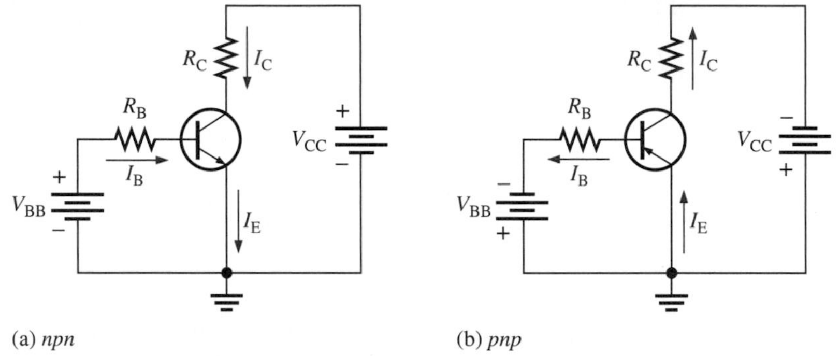

DC Beta (β_{DC}) and DC Alpha (α_{DC})

The dc current **gain** of a transistor is the ratio of the dc collector current (I_C) to the dc base current (I_B) and is designated dc **beta** (β_{DC}).

Equation 4–2

$$\boldsymbol{\beta_{DC} = \frac{I_C}{I_B}}$$

Typical values of β_{DC} range from less than 20 to 200 or higher. β_{DC} is usually designated as an equivalent hybrid (h) parameter, h_{FE}, on transistor datasheets. h-parameters are covered in Chapter 6. All you need to know now is that

$$h_{FE} = \beta_{DC}$$

The ratio of the dc collector current (I_C) to the dc emitter current (I_E) is the dc **alpha** (α_{DC}). The alpha is a less-used parameter than beta in transistor circuits.

$$\alpha_{DC} = \frac{I_C}{I_E}$$

Typically, values of α_{DC} range from 0.95 to 0.99 or greater, but α_{DC} is always less than 1. The reason is that I_C is always slightly less than I_E by the amount of I_B. For example, if I_E = 100 mA and I_B = 1 mA, then I_C = 99 mA and α_{DC} = 0.99.

EXAMPLE 4–1

Determine the dc current gain β_{DC} and the emitter current I_E for a transistor where $I_B = 50\ \mu A$ and $I_C = 3.65$ mA.

Solution

$$\beta_{DC} = \frac{I_C}{I_B} = \frac{3.65\text{ mA}}{50\ \mu\text{A}} = \mathbf{73}$$

$$I_E = I_C + I_B = 3.65\text{ mA} + 50\ \mu\text{A} = \mathbf{3.70\ mA}$$

*Related Problem** A certain transistor has a β_{DC} of 200. When the base current is 50 μA, determine the collector current.

*Answers are at the end of the chapter.

Transistor DC Model

You can view the unsaturated BJT as a device with a current input and a dependent current source in the output circuit, as shown in Figure 4–7 for an *npn*. The input circuit is a forward-biased diode through which there is base current. The output circuit is a dependent current source (diamond-shaped element) with a value that is dependent on the base current, I_B, and equal to $\beta_{DC}I_B$. Recall that independent current source symbols have a circular shape.

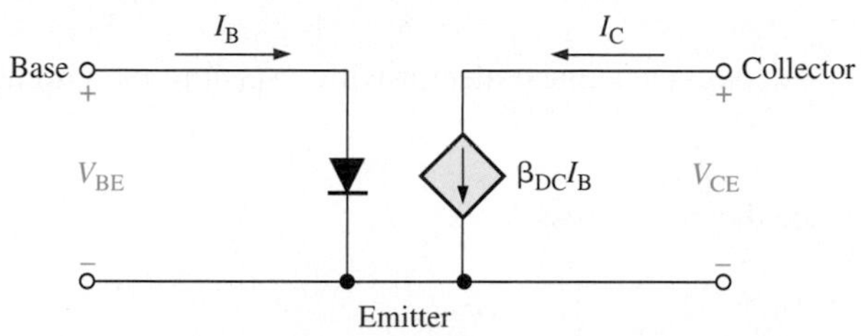

FIGURE 4–7

Ideal dc model of an *npn* transistor.

BJT Circuit Analysis

Consider the basic transistor bias circuit configuration in Figure 4–8. Three transistor dc currents and three dc voltages can be identified.

I_B: dc base current

I_E: dc emitter current

I_C: dc collector current

V_{BE}: dc voltage at base with respect to emitter

V_{CB}: dc voltage at collector with respect to base

V_{CE}: dc voltage at collector with respect to emitter

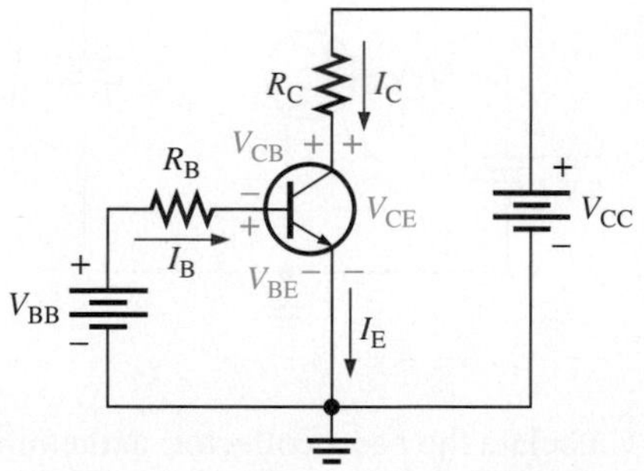

FIGURE 4–8

Transistor currents and voltages.

The base-bias voltage source, V_{BB}, forward-biases the base-emitter junction, and the collector-bias voltage source, V_{CC}, reverse-biases the base-collector junction. When the base-emitter junction is forward-biased, it is like a forward-biased diode and has a nominal forward voltage drop of

$$V_{BE} \cong 0.7\text{ V} \quad \text{Equation 4–3}$$

Although in an actual transistor V_{BE} can be as high as 0.9 V and is dependent on current, we will use 0.7 V throughout this text in order to simplify the analysis of the basic concepts. Keep in mind that the characteristic of the base-emitter junction is the same as a normal diode curve like the one in Figure 1–28.

Since the emitter is at ground (0 V), by Kirchhoff's voltage law, the voltage across R_B is

$$V_{R_B} = V_{BB} - V_{BE}$$

Also, by Ohm's law,

$$V_{R_B} = I_B R_B$$

Substituting for V_{R_B} yields

$$I_B R_B = V_{BB} - V_{BE}$$

Solving for I_B,

Equation 4–4

$$\mathbf{I_B = \frac{V_{BB} - V_{BE}}{R_B}}$$

The voltage at the collector with respect to the grounded emitter is

$$V_{CE} = V_{CC} - V_{R_C}$$

Since the drop across R_C is

$$V_{R_C} = I_C R_C$$

the voltage at the collector with respect to the emitter can be written as

Equation 4–5

$$\mathbf{V_{CE} = V_{CC} - I_C R_C}$$

where $I_C = \beta_{DC} I_B$.

The voltage across the reverse-biased collector-base junction is

Equation 4–6

$$\mathbf{V_{CB} = V_{CE} - V_{BE}}$$

EXAMPLE 4–2

Determine I_B, I_C, I_E, V_{BE}, V_{CE}, and V_{CB} in the circuit of Figure 4–9. The transistor has a $\beta_{DC} = 150$.

FIGURE 4–9

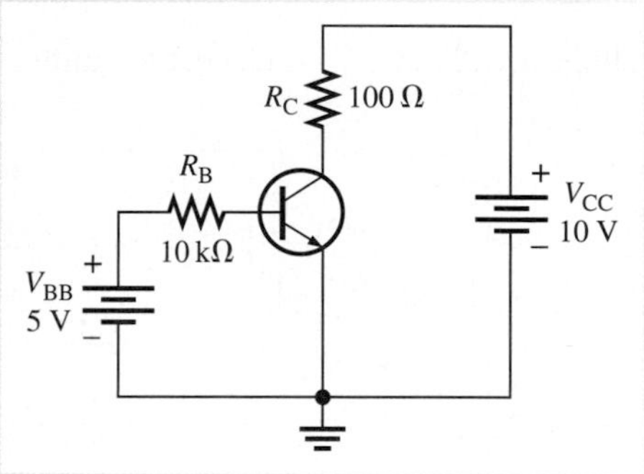

Solution From Equation 4–3, $V_{BE} \cong$ **0.7 V**. Calculate the base, collector, and emitter currents as follows:

$$I_B = \frac{V_{BB} - V_{BE}}{R_B} = \frac{5\text{ V} - 0.7\text{ V}}{10\text{ k}\Omega} = \mathbf{430\ \mu A}$$

$$I_C = \beta_{DC} I_B = (150)(430\ \mu\text{A}) = \mathbf{64.5\text{ mA}}$$

$$I_E = I_C + I_B = 64.5\text{ mA} + 430\ \mu\text{A} = \mathbf{64.9\text{ mA}}$$

Solve for V_{CE} and V_{CB}.

$$V_{CE} = V_{CC} - I_C R_C = 10\text{ V} - (64.5\text{ mA})(100\ \Omega) = 10\text{ V} - 6.45\text{ V} = \mathbf{3.55\text{ V}}$$

$$V_{CB} = V_{CE} - V_{BE} = 3.55\text{ V} - 0.7\text{ V} = \mathbf{2.85\text{ V}}$$

Since the collector is at a higher voltage than the base, the collector-base junction is reverse-biased.

Related Problem Determine I_B, I_C, I_E, V_{CE}, and V_{CB} in Figure 4–9 for the following values: $R_B = 22\text{ k}\Omega$, $R_C = 220\ \Omega$, $V_{BB} = 6\text{ V}$, $V_{CC} = 9\text{ V}$, and $\beta_{DC} = 90$.

Open the Multisim file E04-02 in the Examples folder on your CD-ROM. Measure each current and voltage and compare with the calculated values.

Collector Characteristic Curves

Using a circuit like that shown in Figure 4–10(a), a set of *collector characteristic curves* can be generated that show how the collector current, I_C, varies with the collector-to-emitter voltage, V_{CE}, for specified values of base current, I_B. Notice in the circuit diagram that both V_{BB} and V_{CC} are variable sources of voltage.

Assume that V_{BB} is set to produce a certain value of I_B and V_{CC} is zero. For this condition, both the base-emitter junction and the base-collector junction are forward-biased because the base is at approximately 0.7 V while the emitter and the collector are at 0 V. The base current is through the base-emitter junction because of the low impedance path to

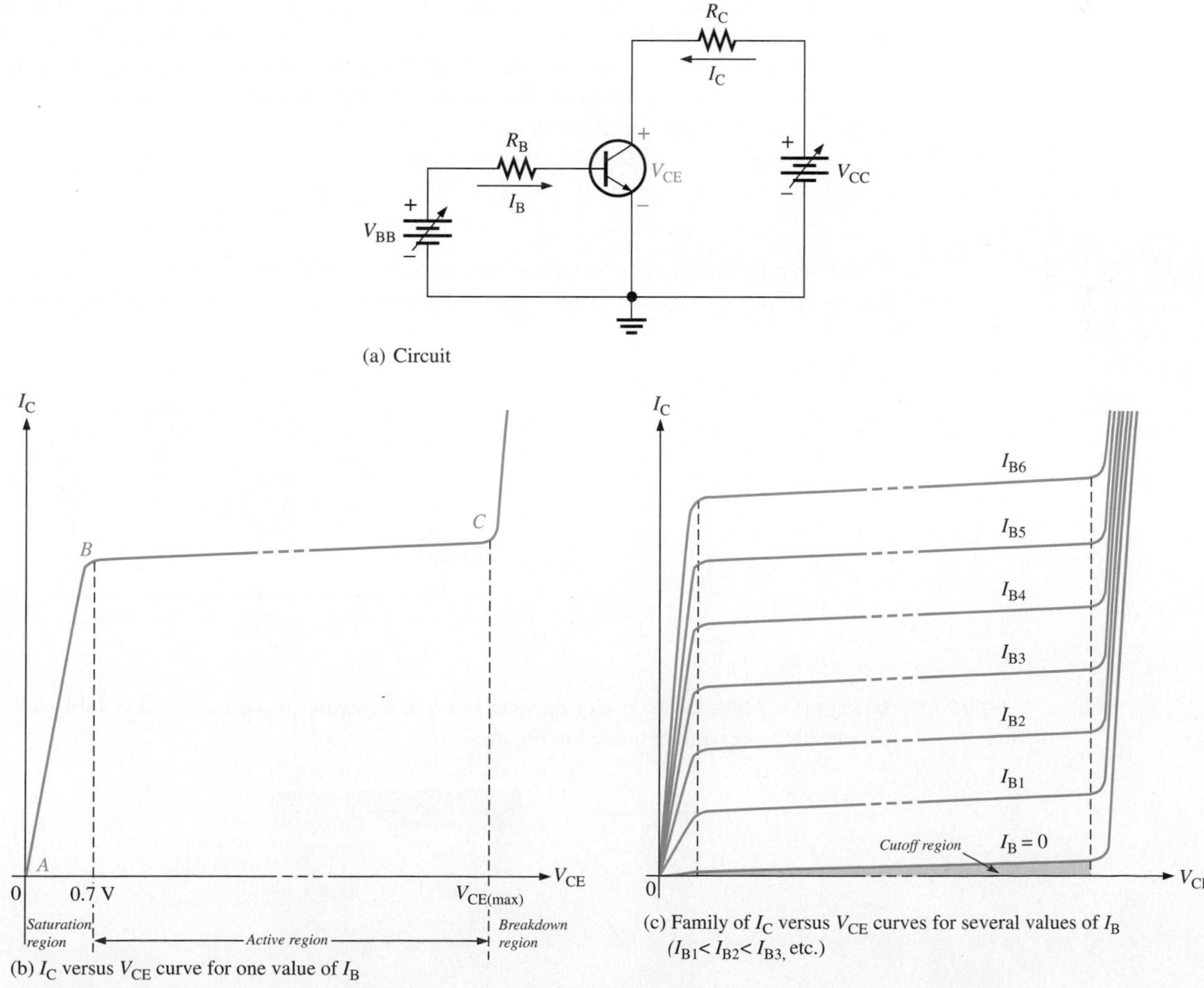

▲ **FIGURE 4–10**

Collector characteristic curves.

ground and, therefore, I_C is zero. When both junctions are forward-biased, the transistor is in the saturation region of its operation. **Saturation** is the state of a BJT in which the collector current has reached a maximum and is independent of the base current.

As V_{CC} is increased, V_{CE} increases as the collector current increases. This is indicated by the portion of the characteristic curve between points *A* and *B* in Figure 4–10(b). I_C increases as V_{CC} is increased because V_{CE} remains less than 0.7 V due to the forward-biased base-collector junction.

Ideally, when V_{CE} exceeds 0.7 V, the base-collector junction becomes reverse-biased and the transistor goes into the *active* or **linear region** of its operation. Once the base-collector junction is reverse-biased, I_C levels off and remains essentially constant for a given value of I_B as V_{CE} continues to increase. Actually, I_C increases very slightly as V_{CE} increases due to widening of the base-collector depletion region. This results in fewer holes for recombination in the base region which effectively causes a slight increase in β_{DC}. This is shown by the portion of the characteristic curve between points *B* and *C* in Figure 4–10(b). For this portion of the characteristic curve, the value of I_C is determined only by the relationship expressed as $I_C = \beta_{DC} I_B$.

When V_{CE} reaches a sufficiently high voltage, the reverse-biased base-collector junction goes into breakdown; and the collector current increases rapidly as indicated by the part of the curve to the right of point *C* in Figure 4–10(b). A transistor should never be operated in this breakdown region.

A family of collector characteristic curves is produced when I_C versus V_{CE} is plotted for several values of I_B, as illustrated in Figure 4–10(c). When $I_B = 0$, the transistor is in the cutoff region although there is a very small collector leakage current as indicated. **Cutoff** is the nonconducting state of a transistor. The amount of collector leakage current for $I_B = 0$ is exaggerated on the graph for illustration.

EXAMPLE 4–3

Sketch an ideal family of collector curves for the circuit in Figure 4–11 for $I_B = 5\ \mu A$ to $25\ \mu A$ in $5\ \mu A$ increments. Assume $\beta_{DC} = 100$ and that V_{CE} does not exceed breakdown.

FIGURE 4–11

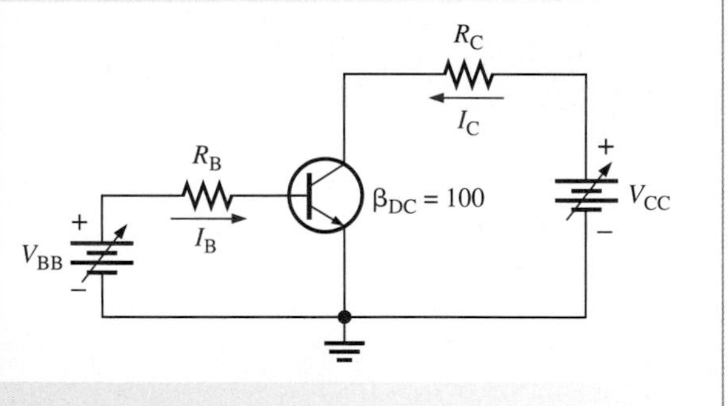

Solution Using the relationship $I_C = \beta_{DC} I_B$, values of I_C are calculated and tabulated in Table 4–1. The resulting curves are plotted in Figure 4–12.

TABLE 4–1

I_B	I_C
5 μA	0.5 mA
10 μA	1.0 mA
15 μA	1.5 mA
20 μA	2.0 mA
25 μA	2.5 mA

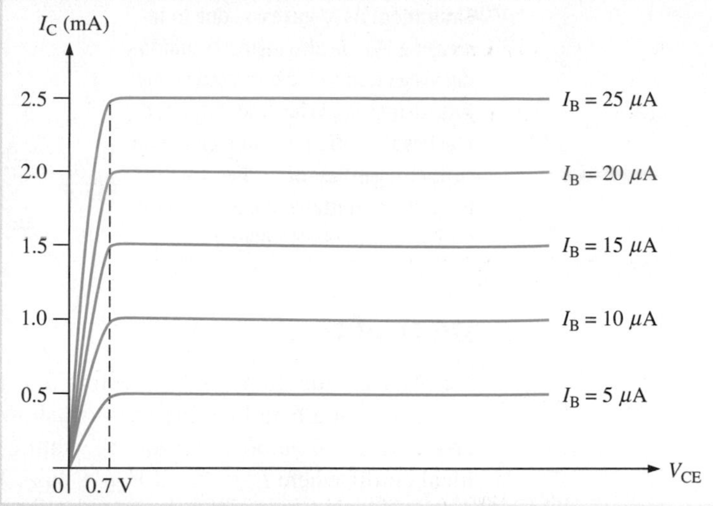

FIGURE 4–12

Related Problem Where would the curve for $I_B = 0$ appear on the graph in Figure 4–12, neglecting collector leakage current?

Cutoff

As previously mentioned, when $I_B = 0$, the transistor is in the cutoff region of its operation. This is shown in Figure 4–13 with the base lead open, resulting in a base current of zero. Under this condition, there is a very small amount of collector leakage current, I_{CEO}, due mainly to thermally produced carriers. Because I_{CEO} is extremely small, it will usually be neglected in circuit analysis so that $V_{CE} = V_{CC}$. In cutoff, neither the base-emitter nor the base-collector junctions are forward-biased. The subscript CEO represents collector-to-emitter with the base open.

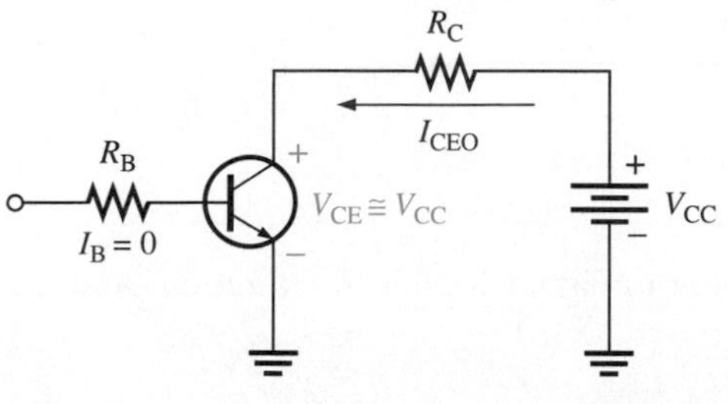

FIGURE 4–13

Cutoff: Collector leakage current (I_{CEO}) is extremely small and is usually neglected. Base-emitter and base-collector junctions are reverse-biased.

Saturation

When the base-emitter junction becomes forward-biased and the base current is increased, the collector current also increases ($I_C = \beta_{DC}I_B$) and V_{CE} decreases as a result of more drop across the collector resistor ($V_{CE} = V_{CC} - I_CR_C$). This is illustrated in Figure 4–14. When V_{CE} reaches its saturation value, $V_{CE(sat)}$, the base-collector junction becomes forward-biased and I_C can increase no further even with a continued increase in I_B. At the point of saturation, the relation $I_C = \beta_{DC}I_B$ is no longer valid. $V_{CE(sat)}$ for a transistor occurs somewhere below the knee of the collector curves, and it is usually only a few tenths of a volt.

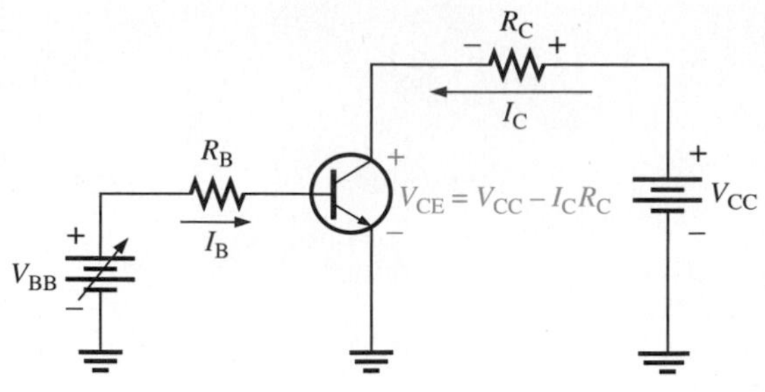

FIGURE 4–14

Saturation: As I_B increases due to increasing V_{BB}, I_C also increases and V_{CE} decreases due to the increased voltage drop across R_C. When the transistor reaches saturation, I_C can increase no further regardless of further increase in I_B. Base-emitter and base-collector junctions are forward-biased.

DC Load Line

Cutoff and saturation can be illustrated in relation to the collector characteristic curves by the use of a load line. Figure 4–15 shows a dc load line drawn on a family of curves connecting the cutoff point and the saturation point. The bottom of the load line is at ideal cutoff where $I_C = 0$ and $V_{CE} = V_{CC}$. The top of the load line is at saturation where $I_C = I_{C(sat)}$ and $V_{CE} = V_{CE(sat)}$. In between cutoff and saturation along the load line is the *active region* of the transistor's operation. Load line operation is discussed more in Chapter 5.

FIGURE 4–15

DC load line on a family of collector characteristic curves illustrating the cutoff and saturation conditions.

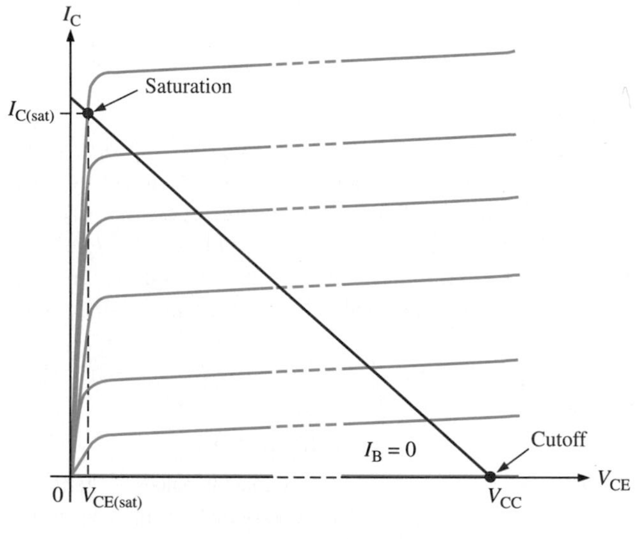

EXAMPLE 4–4

Determine whether or not the transistor in Figure 4–16 is in saturation. Assume $V_{CE(sat)} = 0.2$ V.

FIGURE 4–16

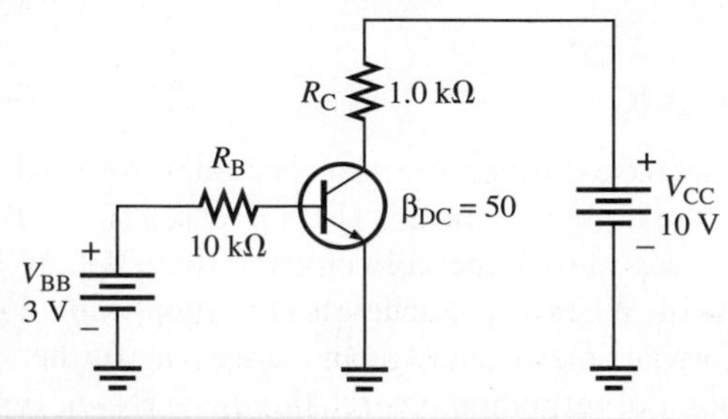

Solution First, determine $I_{C(sat)}$.

$$I_{C(sat)} = \frac{V_{CC} - V_{CE(sat)}}{R_C} = \frac{10\text{ V} - 0.2\text{ V}}{1.0\text{ k}\Omega} = \frac{9.8\text{ V}}{1.0\text{ k}\Omega} = 9.8\text{ mA}$$

Now, see if I_B is large enough to produce $I_{C(sat)}$.

$$I_B = \frac{V_{BB} - V_{BE}}{R_B} = \frac{3\text{ V} - 0.7\text{ V}}{10\text{ k}\Omega} = \frac{2.3\text{ V}}{10\text{ k}\Omega} = 0.23\text{ mA}$$

$$I_C = \beta_{DC}I_B = (50)(0.23\text{ mA}) = 11.5\text{ mA}$$

This shows that with the specified β_{DC}, this base current is capable of producing an I_C greater than $I_{C(sat)}$. Therefore, the **transistor is saturated,** and the collector current value of 11.5 mA is never reached. If you further increase I_B, the collector current remains at its saturation value of 9.8 mA.

Related Problem Determine whether or not the transistor in Figure 4–16 is saturated for the following values: $\beta_{DC} = 125$, $V_{BB} = 1.5$ V, $R_B = 6.8$ kΩ, $R_C = 180$ Ω, and $V_{CC} = 12$ V.

Open the Multisim file E04-04 in the Examples folder on your CD-ROM. Determine if the transistor is in saturation and explain how you did this.

More About β_{DC}

The β_{DC} or h_{FE} is an important BJT parameter that we need to examine further. β_{DC} is not truly constant but varies with both collector current and with temperature. Keeping the junction temperature constant and increasing I_C causes β_{DC} to increase to a maximum. A further increase in I_C beyond this maximum point causes β_{DC} to decrease. If I_C is held constant and the temperature is varied, β_{DC} changes directly with the temperature. If the temperature goes up, β_{DC} goes up and vice versa. Figure 4–17 shows the variation of β_{DC} with I_C and junction temperature (T_J) for a typical BJT.

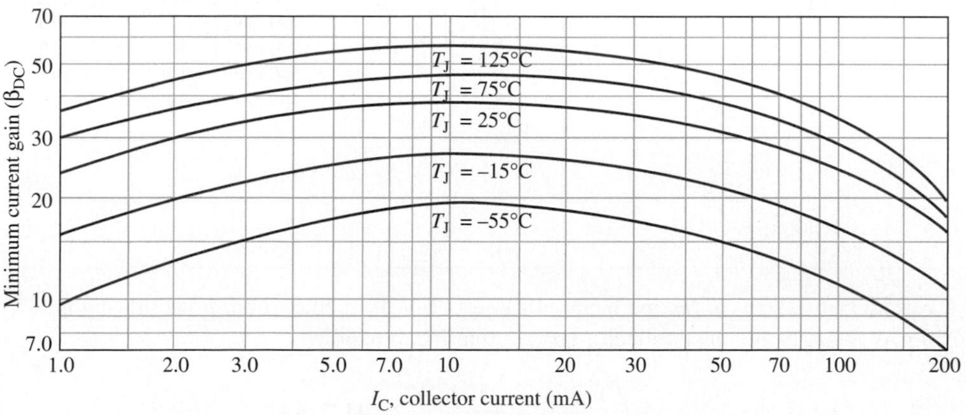

FIGURE 4–17

Variation of β_{DC} with I_C for several temperatures.

A transistor datasheet usually specifies β_{DC} (h_{FE}) at specific I_C values. Even at fixed values of I_C and temperature, β_{DC} varies from one device to another for a given type of transistor due to inconsistencies in the manufacturing process that are unavoidable. The β_{DC} specified at a certain value of I_C is usually the minimum value, $\beta_{DC(min)}$, although the maximum and typical values are also sometimes specified.

Maximum Transistor Ratings

A BJT, like any other electronic device, has limitations on its operation. These limitations are stated in the form of maximum ratings and are normally specified on the manufacturer's datasheet. Typically, maximum ratings are given for collector-to-base voltage, collector-to-emitter voltage, emitter-to-base voltage, collector current, and power dissipation.

The product of V_{CE} and I_C must not exceed the maximum power dissipation. Both V_{CE} and I_C cannot be maximum at the same time. If V_{CE} is maximum, I_C can be calculated as

$$I_C = \frac{P_{D(max)}}{V_{CE}}$$

If I_C is maximum, V_{CE} can be calculated by rearranging the previous equation as follows:

$$V_{CE} = \frac{P_{D(max)}}{I_C}$$

For any given transistor, a maximum power dissipation curve can be plotted on the collector characteristic curves, as shown in Figure 4–18(a). These values are tabulated in Figure 4–18(b). Assume $P_{D(max)}$ is 500 mW, $V_{CE(max)}$ is 20 V, and $I_{C(max)}$ is 50 mA. The curve shows that this particular transistor cannot be operated in the shaded portion of the graph. $I_{C(max)}$ is the limiting rating between points A and B, $P_{D(max)}$ is the limiting rating between points B and C, and $V_{CE(max)}$ is the limiting rating between points C and D.

FIGURE 4–18

Maximum power dissipation curve and tabulated values.

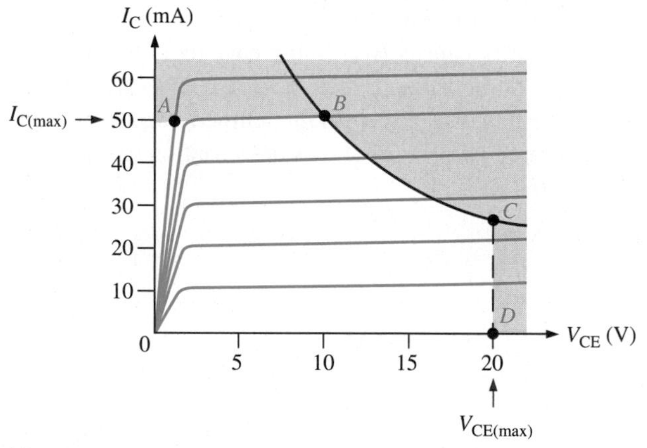

(a)

$P_{D(max)}$	V_{CE}	I_C
500 mW	5 V	100 mA
500 mW	10 V	50 mA
500 mW	15 V	33 mA
500 mW	20 V	25 mA

(b)

EXAMPLE 4–5

A certain transistor is to be operated with $V_{CE} = 6$ V. If its maximum power rating is 250 mW, what is the most collector current that it can handle?

Solution

$$I_C = \frac{P_{D(max)}}{V_{CE}} = \frac{250\text{ mW}}{6\text{ V}} = \mathbf{41.7\ mA}$$

This is the maximum current for this particular value of V_{CE}. The transistor can handle more collector current if V_{CE} is reduced, as long as $P_{D(max)}$ and $I_{C(max)}$ are not exceeded.

Related Problem If $P_{D(max)} = 1$ W, how much voltage is allowed from collector to emitter if the transistor is operating with $I_C = 100$ mA?

EXAMPLE 4–6

The transistor in Figure 4–19 has the following maximum ratings: $P_{D(max)} = 800$ mW, $V_{CE(max)} = 15$ V, and $I_{C(max)} = 100$ mA. Determine the maximum value to which V_{CC} can be adjusted without exceeding a rating. Which rating would be exceeded first?

FIGURE 4–19

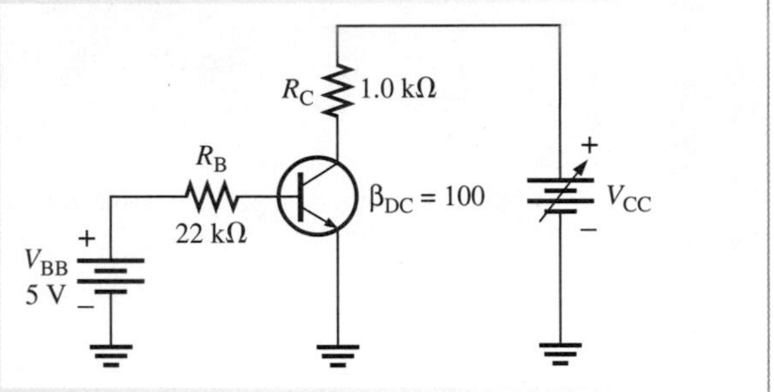

Solution First, find I_B so that you can determine I_C.

$$I_B = \frac{V_{BB} - V_{BE}}{R_B} = \frac{5\text{ V} - 0.7\text{ V}}{22\text{ k}\Omega} = 195\ \mu\text{A}$$

$$I_C = \beta_{DC} I_B = (100)(195\ \mu\text{A}) = 19.5\text{ mA}$$

I_C is much less than $I_{C(max)}$ and ideally will not change with V_{CC}. It is determined only by I_B and β_{DC}.

The voltage drop across R_C is

$$V_{R_C} = I_C R_C = (19.5\text{ mA})(1.0\text{ k}\Omega) = 19.5\text{ V}$$

Now you can determine the value of V_{CC} when $V_{CE} = V_{CE(max)} = 15$ V.

$$V_{R_C} = V_{CC} - V_{CE}$$

So,

$$V_{CC(max)} = V_{CE(max)} + (V_{R_C} = 15\text{ V} + 19.5\text{ V} = \mathbf{34.5\ V}$$

V_{CC} can be increased to 34.5 V, under the existing conditions, before $V_{CE(max)}$ is exceeded. However, at this point it is not known whether or not $P_{D(max)}$ has been exceeded.

$$P_D = V_{CE(max)} I_C = (15\text{ V})(19.5\text{ mA}) = 293\text{ mW}$$

Since $P_{D(max)}$ is 800 mW, it is *not* exceeded when $V_{CC} = 34.5$ V. So, $V_{CE(max)} = 15$ V is the limiting rating in this case. If the base current is removed causing the transistor to turn off, $\boldsymbol{V_{CE(max)}}$ **will be exceeded first** because the entire supply voltage, V_{CC}, will be dropped across the transistor.

Related Problem The transistor in Figure 4–19 has the following maximum ratings: $P_{D(max)} = 500$ mW, $V_{CE(max)} = 25$ V, and $I_{C(max)} = 200$ mA. Determine the maximum value to which V_{CC} can be adjusted without exceeding a rating. Which rating would be exceeded first?

Derating $P_{D(max)}$

$P_{D(max)}$ is usually specified at 25°C. For higher temperatures, $P_{D(max)}$ is less. Datasheets often give derating factors for determining $P_{D(max)}$ at any temperature above 25°C. For example, a derating factor of 2 mW/°C indicates that the maximum power dissipation is reduced 2 mW for each degree Celsius increase in temperature.

EXAMPLE 4–7

A certain transistor has a $P_{D(max)}$ of 1 W at 25°C. The derating factor is 5 mW/°C. What is the $P_{D(max)}$ at a temperature of 70°C?

Solution The change (reduction) in $P_{D(max)}$ is

$$\Delta P_{D(max)} = (5\text{ mW/°C})(70°\text{C} - 25°\text{C}) = (5\text{ mW/°C})(45°\text{C}) = 225\text{ mW}$$

Therefore, the $P_{D(max)}$ at 70°C is

$$1\text{ W} - 225\text{ mW} = \mathbf{775\ mW}$$

Related Problem A transistor has a $P_{D(max)} = 5$ W at 25°C. The derating factor is 10 mW/°C. What is the $P_{D(max)}$ at 70°C?

Transistor Datasheet

A partial datasheet for the 2N3904 *npn* transistor is shown in Figure 4–20. Notice that the maximum collector-emitter voltage (V_{CEO}) is 40 V. The CEO subscript indicates that the voltage is measured from collector (C) to emitter (E) with the base open (O). In the text, we use $V_{CE(max)}$ for this parameter. Also notice that the maximum collector current is 200 mA.

The β_{DC} (h_{FE}) is specified for several values of I_C. As you can see, h_{FE} varies with I_C as we previously discussed.

The collector-emitter saturation voltage, $V_{CE(sat)}$ is 0.2 V maximum for $I_{C(sat)} = 10$ mA and increases with the current.

EXAMPLE 4–8

A 2N3904 transistor is used in the circuit of Figure 4–19 (Example 4–6). Determine the maximum value to which V_{CC} can be adjusted without exceeding a rating. Which rating would be exceeded first? Refer to the datasheet in Figure 4–20.

Solution From the datasheet,

$$P_{D(max)} = P_D = 625\text{ mW}$$
$$V_{CE(max)} = V_{CEO} = 40\text{ V}$$
$$I_{C(max)} = I_C = 200\text{ mA}$$

Assume $\beta_{DC} = 100$. This is a reasonably valid assumption based on the datasheet $h_{FE} = 100$ minimum for specified conditions (β_{DC} and h_{FE} are the same parameter). As you have learned, the β_{DC} has considerable variations for a given transistor, depending on circuit conditions. Under this assumption, $I_C = 19.5$ mA and $V_{R_C} = 19.5$ V from Example 4–6.

Since I_C is much less than $I_{C(max)}$ and, ideally, will not change with V_{CC}, the maximum value to which V_{CC} can be increased before $V_{CE(max)}$ is exceeded is

$$V_{CC(max)} = V_{CE(max)} + V_{R_C} = 40\text{ V} + 19.5\text{ V} = 59.5\text{ V}$$

However, at the maximum value of V_{CE}, the power dissipation is

$$P_D = V_{CE(max)}I_C = (40\text{ V})(19.5\text{ mA}) = 780\text{ mW}$$

Power dissipation exceeds the maximum of 625 mW specified on the datasheet.

Related Problem Use the datasheet in Figure 4–20 to find the maximum P_D at 50°C.

FAIRCHILD SEMICONDUCTOR™

2N3904 **MMBT3904** **PZT3904**

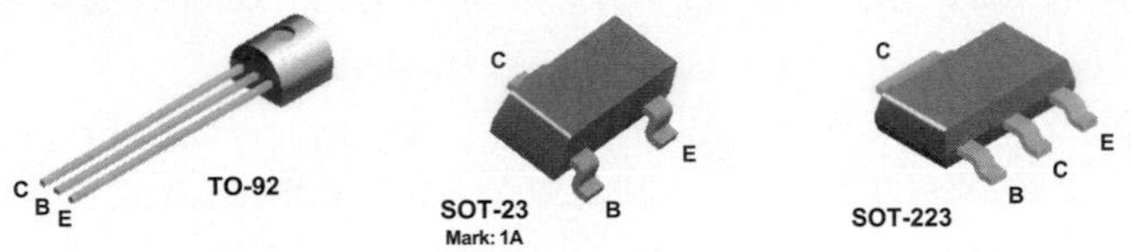

NPN General Purpose Amplifier

This device is designed as a general purpose amplifier and switch. The useful dynamic range extends to 100 mA as a switch and to 100 MHz as an amplifier.

Absolute Maximum Ratings* $T_A = 25°C$ unless otherwise noted

Symbol	Parameter	Value	Units
V_{CEO}	Collector-Emitter Voltage	40	V
V_{CBO}	Collector-Base Voltage	60	V
V_{EBO}	Emitter-Base Voltage	6.0	V
I_C	Collector Current - Continuous	200	mA
T_J, T_{stg}	Operating and Storage Junction Temperature Range	-55 to +150	°C

*These ratings are limiting values above which the serviceability of any semiconductor device may be impaired.

NOTES:
1) These ratings are based on a maximum junction temperature of 150 degrees C.
2) These are steady state limits. The factory should be consulted on applications involving pulsed or low duty cycle operations.

Thermal Characteristics $T_A = 25°C$ unless otherwise noted

Symbol	Characteristic	Max			Units
		2N3904	*MMBT3904	**PZT3904	
P_D	Total Device Dissipation Derate above 25°C	625 5.0	350 2.8	1,000 8.0	mW mW/°C
$R_{\theta JC}$	Thermal Resistance, Junction to Case	83.3			°C/W
$R_{\theta JA}$	Thermal Resistance, Junction to Ambient	200	357	125	°C/W

*Device mounted on FR-4 PCB 1.6" X 1.6" X 0.06."
**Device mounted on FR-4 PCB 36 mm X 18 mm X 1.5 mm; mounting pad for the collector lead min. 6 cm².

Electrical Characteristics $T_A = 25°C$ unless otherwise noted

Symbol	Parameter	Test Conditions	Min	Max	Units
OFF CHARACTERISTICS					
$V_{(BR)CEO}$	Collector-Emitter Breakdown Voltage	$I_C = 1.0$ mA, $I_B = 0$	40		V
$V_{(BR)CBO}$	Collector-Base Breakdown Voltage	$I_C = 10\ \mu A$, $I_E = 0$	60		V
$V_{(BR)EBO}$	Emitter-Base Breakdown Voltage	$I_E = 10\ \mu A$, $I_C = 0$	6.0		V
I_{BL}	Base Cutoff Current	$V_{CE} = 30$ V, $V_{EB} = 3$V		50	nA
I_{CEX}	Collector Cutoff Current	$V_{CE} = 30$ V, $V_{EB} = 3$V		50	nA
ON CHARACTERISTICS*					
h_{FE}	DC Current Gain	$I_C = 0.1$ mA, $V_{CE} = 1.0$ V $I_C = 1.0$ mA, $V_{CE} = 1.0$ V $I_C = 10$ mA, $V_{CE} = 1.0$ V $I_C = 50$ mA, $V_{CE} = 1.0$ V $I_C = 100$ mA, $V_{CE} = 1.0$ V	40 70 100 60 30	 300 	
$V_{CE(sat)}$	Collector-Emitter Saturation Voltage	$I_C = 10$ mA, $I_B = 1.0$ mA $I_C = 50$ mA, $I_B = 5.0$ mA		0.2 0.3	V V
$V_{BE(sat)}$	Base-Emitter Saturation Voltage	$I_C = 10$ mA, $I_B = 1.0$ mA $I_C = 50$ mA, $I_B = 5.0$ mA	0.65	0.85 0.95	V V
SMALL SIGNAL CHARACTERISTICS					
f_T	Current Gain - Bandwidth Product	$I_C = 10$ mA, $V_{CE} = 20$ V, f = 100 MHz	300		MHz
C_{obo}	Output Capacitance	$V_{CB} = 5.0$ V, $I_E = 0$, f = 1.0 MHz		4.0	pF
C_{ibo}	Input Capacitance	$V_{EB} = 0.5$ V, $I_C = 0$, f = 1.0 MHz		8.0	pF
NF	Noise Figure	$I_C = 100\ \mu A$, $V_{CE} = 5.0$ V, $R_S = 1.0\ k\Omega$, f=10 Hz to 15.7kHz		5.0	dB
SWITCHING CHARACTERISTICS					
t_d	Delay Time	$V_{CC} = 3.0$ V, $V_{BE} = 0.5$ V,		35	ns
t_r	Rise Time	$I_C = 10$ mA, $I_{B1} = 1.0$ mA		35	ns
t_s	Storage Time	$V_{CC} = 3.0$ V, $I_C = 10$mA		200	ns
t_f	Fall Time	$I_{B1} = I_{B2} = 1.0$ mA		50	ns

*Pulse Test: Pulse Width ≤ 300 μs, Duty Cycle ≤ 2.0%

◄ **FIGURE 4–20**

Partial datasheet. For a complete 2N3904 datasheet, go to *http://www.fairchildsemi.com/ds/2N%2F2N3904.pdf*. © 2003 Fairchild Semiconductor Corporation. Used by permission.

SECTION 4–3 CHECKUP

1. Define β_{DC} and α_{DC}. What is h_{FE}?
2. If the dc current gain of a transistor is 100, determine β_{DC} and α_{DC}.
3. What two variables are plotted on a collector characteristic curve?
4. What bias conditions must exist for a transistor to operate as an amplifier?
5. Does β_{DC} increase or decrease with temperature?
6. For a given type of transistor, can β_{DC} be considered to be a constant?

4–4 THE BJT AS AN AMPLIFIER

Amplification is the process of linearly increasing the amplitude of an electrical signal and is one of the major properties of a transistor. As you learned, a BJT exhibits current gain (called β). When a BJT is biased in the active (or linear) region, as previously described, the BE junction has a low resistance due to forward bias and the BC junction has a high resistance due to reverse bias.

After completing this section, you should be able to

- **Discuss how a BJT is used as a voltage amplifier**
    - Describe amplification
    - Develop the ac equivalent circuit for a basic BJT amplifier
    - Determine the voltage gain of a basic BJT amplifier

DC and AC Quantities

Before discussing the concept of transistor amplification, the designations that we will use for the circuit quantities of current, voltage, and resistance must be explained because amplifier circuits have both dc and ac quantities.

In this text, italic capital letters are used for both dc and ac currents (I) and voltages (V). This rule applies to rms, average, peak, and peak-to-peak ac values. AC current and voltage values are always rms unless stated otherwise. Although some texts use lowercase i and v for ac current and voltage, we reserve the use of lowercase i and v only for instantaneous values. In this text, the distinction between a dc current or voltage and an ac current or voltage is in the subscript.

DC quantities always carry an uppercase roman (nonitalic) subscript. For example, I_{B}, I_{C}, and I_{E} are the dc transistor currents. V_{BE}, V_{CB}, and V_{CE} are the dc voltages from one transistor terminal to another. Single subscripted voltages such as V_{B}, V_{C}, and V_{E} are dc voltages from the transistor terminals to ground.

AC and all time-varying quantities always carry a lowercase italic subscript. For example, I_b, I_c, and I_e are the ac transistor currents. V_{be}, V_{cb}, and V_{ce} are the ac voltages from one transistor terminal to another. Single subscripted voltages such as V_b, V_c, and V_e are ac voltages from the transistor terminals to ground.

The rule is different for *internal* transistor resistances. As you will see later, transistors have internal ac resistances that are designated by lowercase r' with an appropriate subscript. For example, the internal ac emitter resistance is designated as r'_e.

Circuit resistances external to the transistor itself use the standard italic capital R with a subscript that identifies the resistance as dc or ac (when applicable), just as for current and voltage. For example R_{E} is an external dc emitter resistance and R_e is an external ac emitter resistance.

Transistor Voltage Amplification

As you have learned, a transistor amplifies current because the collector current is equal to the base current multiplied by the current gain, β. The base current in a transistor is very small compared to the collector and emitter currents. Because of this, the collector current is approximately equal to the emitter current.

With this in mind, let's look at the circuit in Figure 4–21. An ac voltage, V_s, is superimposed on the dc bias voltage V_{BB} by capacitive coupling as shown. The dc bias voltage V_{CC} is connected to the collector through the collector resistor, R_C.

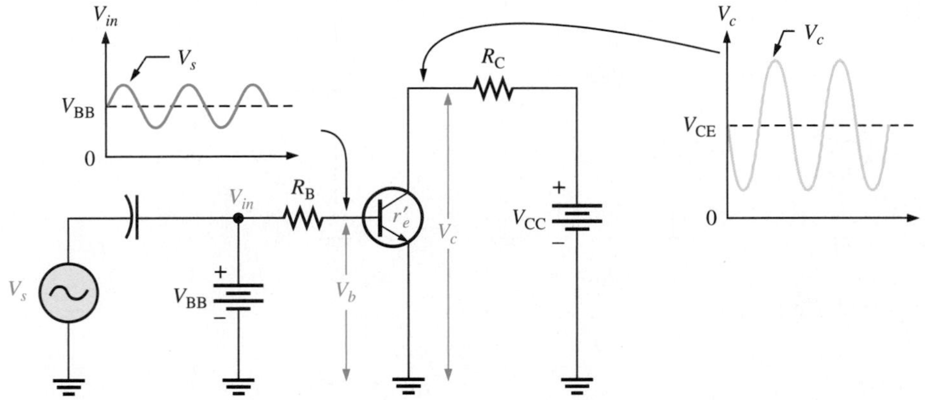

FIGURE 4–21

Basic transistor amplifier circuit with ac source voltage V_s and dc bias voltage V_{BB} superimposed.

The ac input voltage produces an ac base current, which results in a much larger ac collector current. The ac collector current produces an ac voltage across R_C, thus producing an amplified, but inverted, reproduction of the ac input voltage in the active region of operation, as illustrated in Figure 4–21.

The forward-biased base-emitter junction presents a very low resistance to the ac signal. This internal ac emitter resistance is designated r'_e in Figure 4–21 and appears in series with R_B. The ac base voltage is

$$V_b = I_e r'_e$$

The ac collector voltage, V_c, equals the ac voltage drop across R_C.

$$V_c = I_c R_C$$

Since $I_c \cong I_e$, the ac collector voltage is

$$V_c \cong I_e R_C$$

V_b can be considered the transistor ac input voltage where $V_b = V_s - I_b R_B$. V_c can be considered the transistor ac output voltage. Since *voltage gain* is defined as the ratio of the output voltage to the input voltage, the ratio of V_c to V_b is the ac voltage gain, A_v, of the transistor.

$$A_v = \frac{V_c}{V_b}$$

Substituting $I_e R_C$ for V_c and $I_e r'_e$ for V_b yields

$$A_v = \frac{V_c}{V_b} \cong \frac{I_e R_C}{I_e r'_e}$$

The I_e terms cancel; therefore,

$$\mathbf{A_v \cong \frac{R_C}{r'_e}}$$ **Equation 4–7**

Equation 4–7 shows that the transistor in Figure 4–21 provides amplification in the form of voltage gain, which is dependent on the values of R_C and r'_e.

Since R_C is always considerably larger in value than r'_e, the output voltage for this configuration is greater than the input voltage. Various types of amplifiers are covered in detail in later chapters.

EXAMPLE 4–9

Determine the voltage gain and the ac output voltage in Figure 4–22 if $r'_e = 50\ \Omega$.

FIGURE 4–22

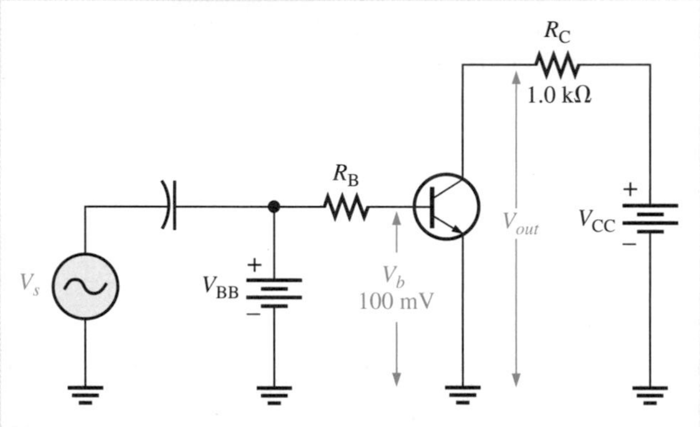

Solution The voltage gain is

$$A_v \cong \frac{R_C}{r'_e} = \frac{1.0\ \text{k}\Omega}{50\ \Omega} = \mathbf{20}$$

Therefore, the ac output voltage is

$$V_{out} = A_v V_b = (20)(100\ \text{mV}) = \mathbf{2\ V\ rms}$$

Related Problem What value of R_C in Figure 4–22 will it take to have a voltage gain of 50?

SECTION 4–4 CHECKUP

1. What is amplification?
2. How is voltage gain defined?
3. Name two factors that determine the voltage gain of an amplifier.
4. What is the voltage gain of a transistor amplifier that has an output of 5 V rms and an input of 250 mV rms?
5. A transistor connected as in Figure 4–22 has an $r'_e = 20\ \Omega$. If R_C is 1200 Ω, what is the voltage gain?

4–5 The BJT as a Switch

In the previous section, you saw how a BJT can be used as a linear amplifier. The second major application area is switching applications. When used as an electronic switch, a BJT is normally operated alternately in cutoff and saturation. Many digital circuits use the BJT as a switch.

After completing this section, you should be able to

- **Discuss how a BJT is used as an electronic switch**
  - Analyze a BJT switching circuit for cutoff and saturation
  - Describe the conditions that produce cutoff
  - Describe the conditions that produce saturation
  - Discuss a basic application of a transistor switching circuit

Figure 4–23 illustrates the basic operation of a BJT as a switching device. In part (a), the transistor is in the cutoff region because the base-emitter junction is not forward-biased. In this condition, there is, ideally, an *open* between collector and emitter, as indicated by the switch equivalent. In part (b), the transistor is in the saturation region because the base-emitter junction and the base-collector junction are forward-biased and the base current is made large enough to cause the collector current to reach its saturation value. In this condition, there is, ideally, a *short* between collector and emitter, as indicated by the switch equivalent. Actually, a small voltage drop across the transister of up to a few tenths of a volt normally occurs, which is the saturation voltage, $V_{CE(sat)}$.

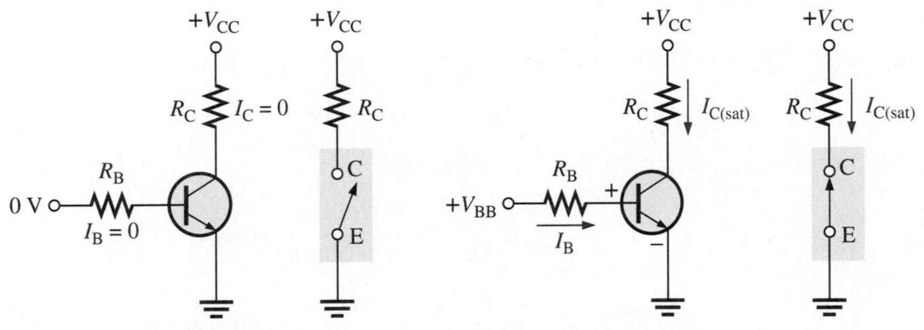

FIGURE 4–23

Switching action of an ideal transistor.

Conditions in Cutoff

As mentioned before, a transistor is in the cutoff region when the base-emitter junction is not forward-biased. Neglecting leakage current, all of the currents are zero, and V_{CE} is equal to V_{CC}.

$$\mathbf{V_{CE(cutoff)} = V_{CC}}$$

Equation 4–8

Conditions in Saturation

As you have learned, when the base-emitter junction is forward-biased and there is enough base current to produce a maximum collector current, the transistor is saturated. The formula for collector saturation current is

$$\mathbf{I_{C(sat)} = \frac{V_{CC} - V_{CE(sat)}}{R_C}}$$

Equation 4–9

Since $V_{CE(sat)}$ is very small compared to V_{CC}, it can usually be neglected.

The minimum value of base current needed to produce saturation is

$$\mathbf{I_{B(min)} = \frac{I_{C(sat)}}{\beta_{DC}}}$$

Equation 4–10

Normally, I_B should be significantly greater than $I_{B(min)}$ to ensure that the transistor is saturated.

EXAMPLE 4–10

(a) For the transistor circuit in Figure 4–24, what is V_{CE} when $V_{IN} = 0$ V?

(b) What minimum value of I_B is required to saturate this transistor if β_{DC} is 200? Neglect $V_{CE(sat)}$.

(c) Calculate the maximum value of R_B when $V_{IN} = 5$ V.

FIGURE 4–24

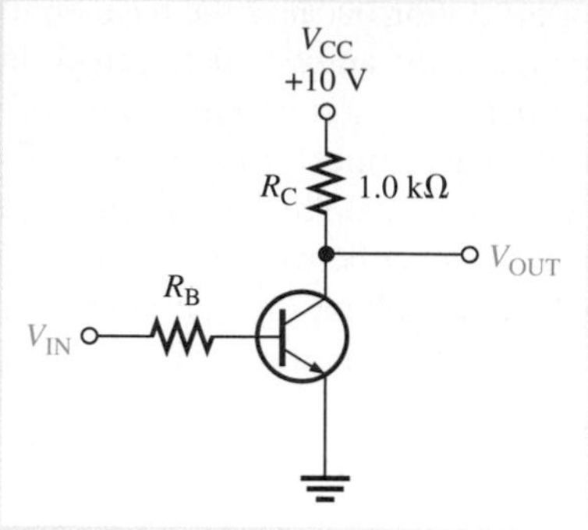

Solution **(a)** When $V_{IN} = 0$ V, the transistor is in cutoff (acts like an open switch) and

$$V_{CE} = V_{CC} = \mathbf{10\ V}$$

(b) Since $V_{CE(sat)}$ is neglected (assumed to be 0 V),

$$I_{C(sat)} = \frac{V_{CC}}{R_C} = \frac{10\text{ V}}{1.0\text{ k}\Omega} = 10\text{ mA}$$

$$I_{B(min)} = \frac{I_{C(sat)}}{\beta_{DC}} = \frac{10\text{ mA}}{200} = \mathbf{50\ \mu A}$$

This is the value of I_B necessary to drive the transistor to the point of saturation. Any further increase in I_B will ensure the transistor remains in saturation but there cannot be any further increase in I_C.

(c) When the transistor is on, $V_{BE} \cong 0.7$ V. The voltage across R_B is

$$V_{R_B} = V_{IN} - V_{BE} \cong 5\text{ V} - 0.7\text{ V} = 4.3\text{ V}$$

Calculate the maximum value of R_B needed to allow a minimum I_B of 50 μA using Ohm's law as follows:

$$R_{B(max)} = \frac{V_{R_B}}{I_{B(min)}} = \frac{4.3\text{ V}}{50\ \mu\text{A}} = \mathbf{86\ k\Omega}$$

Related Problem Determine the minimum value of I_B required to saturate the transistor in Figure 4–24 if β_{DC} is 125 and $V_{CE(sat)}$ is 0.2 V.

A Simple Application of a Transistor Switch

The transistor in Figure 4–25 is used as a switch to turn the LED on and off. For example, a square wave input voltage with a period of 2 s is applied to the input as indicated. When

FIGURE 4–25

A transistor used to switch an LED on and off.

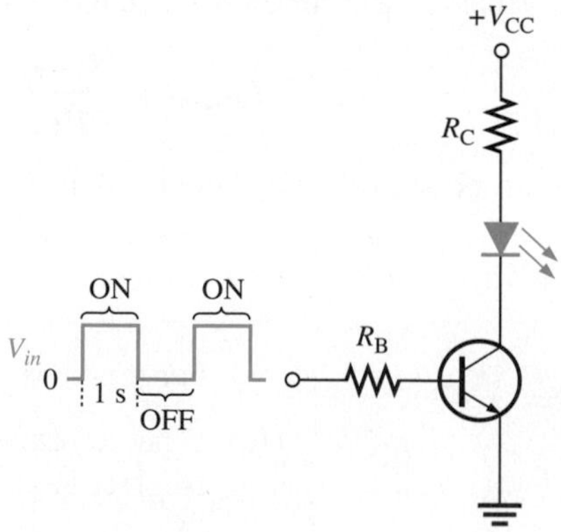

the square wave is at 0 V, the transistor is in cutoff; and since there is no collector current, the LED does not emit light. When the square wave goes to its high level, the transistor saturates. This forward-biases the LED, and the resulting collector current through the LED causes it to emit light. Thus, the LED is on for 1 second and off for 1 second.

EXAMPLE 4–11

The LED in Figure 4–25 requires 30 mA to emit a sufficient level of light. Therefore, the collector current should be approximately 30 mA. For the following circuit values, determine the amplitude of the square wave input voltage necessary to make sure that the transistor saturates. Use double the minimum value of base current as a safety margin to ensure saturation. $V_{CC} = 9$ V, $V_{CE(sat)} = 0.3$ V, $R_C = 220\ \Omega$, $R_B = 3.3\ \text{k}\Omega$, $\beta_{DC} = 50$, and $V_{LED} = 1.6$ V.

Solution

$$I_{C(sat)} = \frac{V_{CC} - V_{LED} - V_{CE(sat)}}{R_C} = \frac{9\text{ V} - 1.6\text{ V} - 0.3\text{ V}}{220\ \Omega} = 32.3\text{ mA}$$

$$I_{B(min)} = \frac{I_{C(sat)}}{\beta_{DC}} = \frac{32.3\text{ mA}}{50} = 646\ \mu\text{A}$$

To ensure saturation, use twice the value of $I_{B(min)}$, which is 1.29 mA. Use Ohm's law to solve for V_{in}.

$$I_B = \frac{V_{R_B}}{R_B} = \frac{V_{in} - V_{BE}}{R_B} = \frac{V_{in} - 0.7\text{ V}}{3.3\text{ k}\Omega}$$

$$V_{in} - 0.7\text{ V} = 2I_{B(min)}R_B = (1.29\text{ mA})(3.3\text{ k}\Omega)$$

$$V_{in} = (1.29\text{ mA})(3.3\text{ k}\Omega) + 0.7\text{ V} = \mathbf{4.96\ V}$$

Related Problem If you change the LED in Figure 4–25 to one that requires 50 mA for a specified light emission and you can't increase the input amplitude above 5 V or V_{CC} above 9 V, how would you modify the circuit? Specify the component(s) to be changed and the value(s).

Open the Multisim file E04-11 in the Examples folder on your CD-ROM. Using a 0.5 Hz square wave input with the calculated amplitude, verify that the transistor is switching between cutoff and saturation and that the LED is alternately turning on and off.

SECTION 4–5 CHECKUP

1. When a transistor is used as a switch, in what two states is it operated?
2. When is the collector current maximum?
3. When is the collector current approximately zero?
4. Under what condition is $V_{CE} = V_{CC}$?
5. When is V_{CE} minimum?

4–6 THE PHOTOTRANSISTOR

A phototransistor is similar to a regular BJT except that the base current is produced and controlled by light instead of a voltage source. The phototransistor effectively converts light energy to an electrical signal.

After completing this section, you should be able to

- **Describe a phototransistor and its operation**
  - Explain how the base current is produced
  - Discuss how phototransistors are used
  - Describe an optocoupler

In a **phototransistor** the base current is produced when light strikes the photosensitive semiconductor base region. The collector-base *pn* junction is exposed to incident light through a lens opening in the transistor package. When there is no incident light, there is only a small thermally generated collector-to-emitter leakage current, I_{CEO}; this dark current is typically in the nA range. When light strikes the collector-base *pn* junction, a base current, I_λ, is produced that is directly proportional to the light intensity. This action produces a collector current that increases with I_λ. Except for the way base current is generated, the phototransistor behaves as a conventional BJT. In many cases, there is no electrical connection to the base.

The relationship between the collector current and the light-generated base current in a phototransistor is

Equation 4–11

$$I_C = \beta_{DC} I_\lambda$$

The schematic symbol and some typical phototransistors are shown in Figure 4–26. Since the actual photogeneration of base current occurs in the collector-base region, the larger the physical area of this region, the more base current is generated. Thus, a typical phototransistor is designed to offer a large area to the incident light, as the simplified structure diagram in Figure 4–27 illustrates.

FIGURE 4–26

Phototransistor.

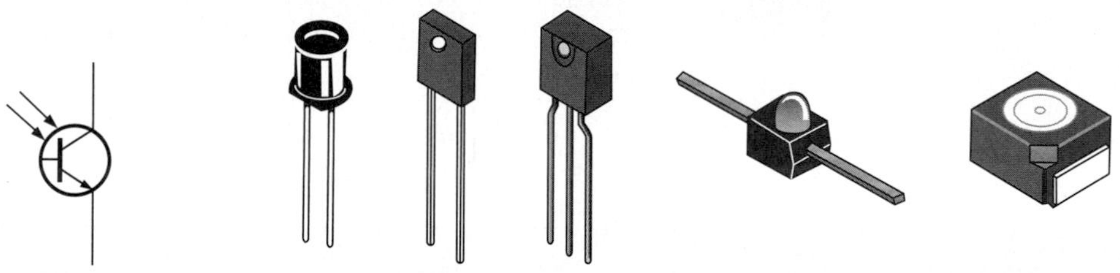

(a) Schematic symbol (b) Typical packages

FIGURE 4–27

Typical phototransistor structure.

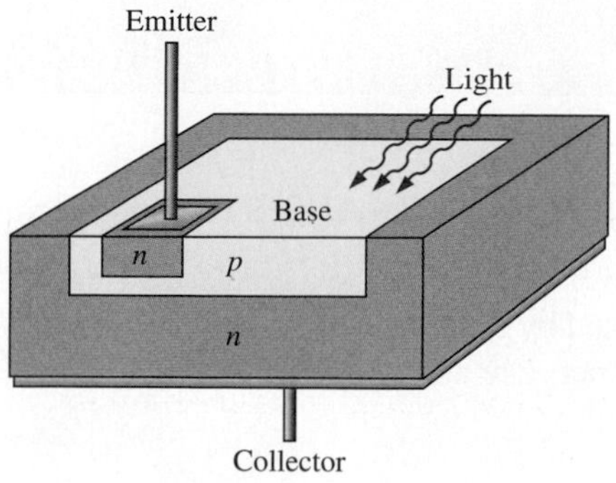

A phototransistor can be either a two-lead or a three-lead device. In the three-lead configuration, the base lead is brought out so that the device can be used as a conventional BJT with or without the additional light-sensitivity feature. In the two-lead configuration, the base is not electrically available, and the device can be used only with light as the input. In many applications, the phototransistor is used in the two-lead version.

Figure 4–28 shows a phototransistor with a biasing circuit and typical collector characteristic curves. Notice that each individual curve on the graph corresponds to a certain value of light intensity (in this case, the units are mW/cm^2) and that the collector current in creases with light intensity.

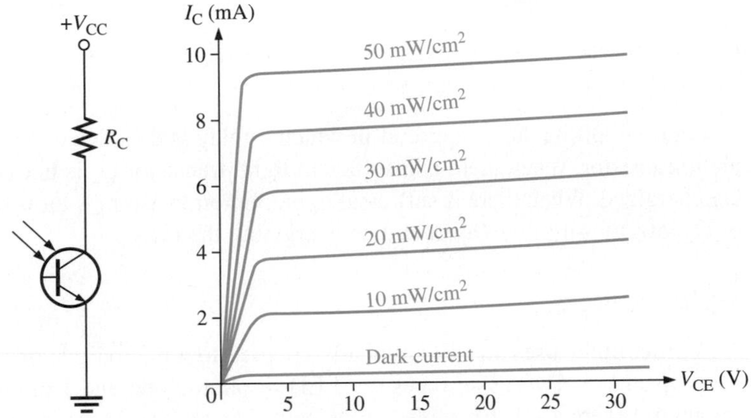

FIGURE 4–28

Phototransistor circuit and typical collector characteristic curves.

Phototransistors are not sensitive to all light but only to light within a certain range of wavelengths. They are most sensitive to particular wavelengths in the red and infrared part of the spectrum, as shown by the peak of the infrared spectral response curve in Figure 4–29.

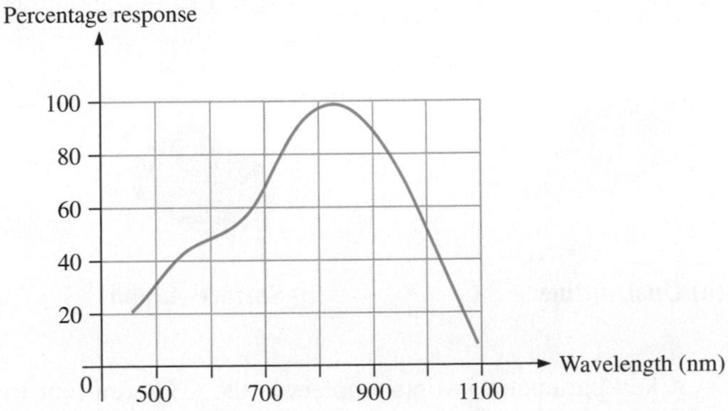

FIGURE 4–29

Typical phototransistor spectral response.

Applications

Phototransistors are used in a variety of applications. A light-operated relay circuit is shown in Figure 4–30(a). The phototransistor Q_1 drives the BJT Q_2. When there is sufficient incident light on Q_1, transistor Q_2 is driven into saturation, and collector current through the relay coil energizes the relay. The diode across the relay coil prevents, by its limiting action, a large voltage transient from occurring at the collector of Q_2 when the transistor turns off.

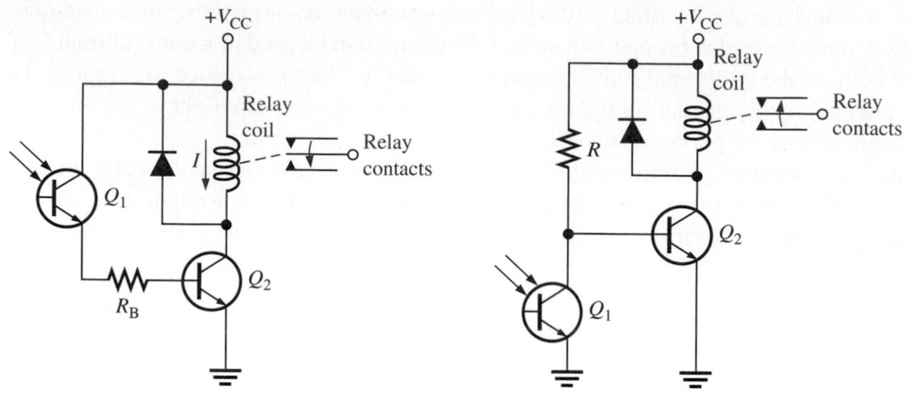

(a) Light activated (b) Light deactivated

FIGURE 4–30

Relay circuits driven by a phototransistor.

Figure 4–30(b) shows a circuit in which a relay is deactivated by incident light on the phototransistor. When there is insufficient light, transistor Q_2 is biased on, keeping the relay energized. When there is sufficient light, phototransistor Q_1 turns on; this pulls the base of Q_2 low, thus turning Q_2 off and de-energizing the relay.

Optocouplers

An **optocoupler** uses an LED optically coupled to a photodiode or a phototransistor in a single package. Two basic types are LED-to-photodiode and LED-to-phototransistor, as shown in Figure 4–31. Examples of typical packages are shown in Figure 4–32.

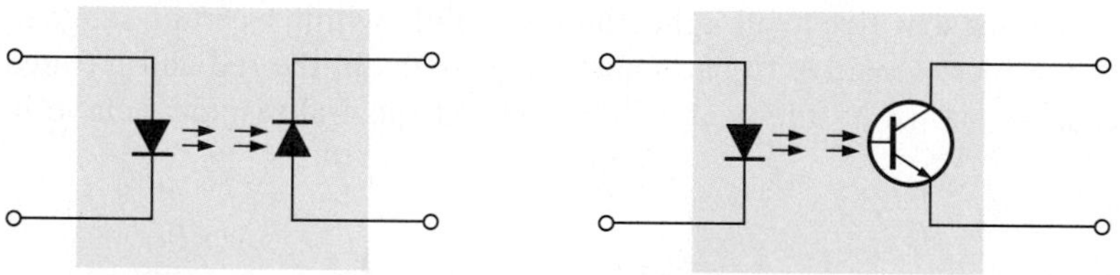

(a) LED-to-photodiode (b) LED-to-phototransistor

FIGURE 4–31

Basic optocouplers.

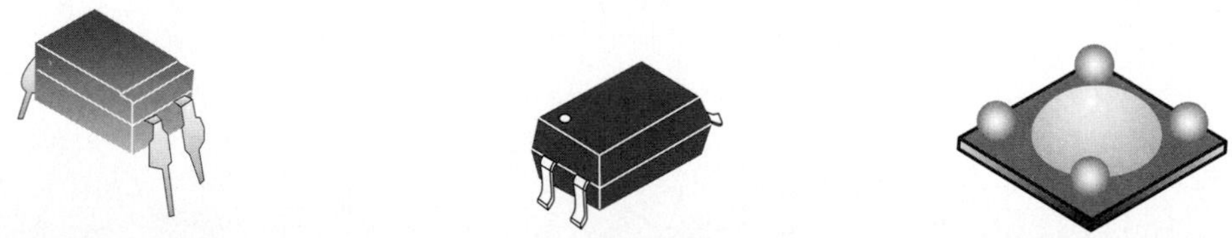

(a) Dual-in-line (b) Surface-mount (c) Ball-grid

FIGURE 4–32

Examples of optocoupler packages.

A key parameter in optocouplers is the CTR (current transfer ratio). The CTR is an indication of how efficiently a signal is coupled from input to output and is expressed as the ratio of a change in the LED current to the corresponding change in the photodiode or phototransistor current. It is usually expressed as a percentage. Figure 4–33 shows a typical graph of CTR versus forward LED current. For this case, it varies from about 50% to about 110%.

Optocouplers are used to isolate sections of a circuit that are incompatible in terms of the voltage levels or currents required. For example, they are used to protect hospital patients from shock when they are connected to monitoring instruments or other devices. They are also used to isolate low-current control or signal circuits from noisy power supply circuits or higher-current motor and machine circuits.

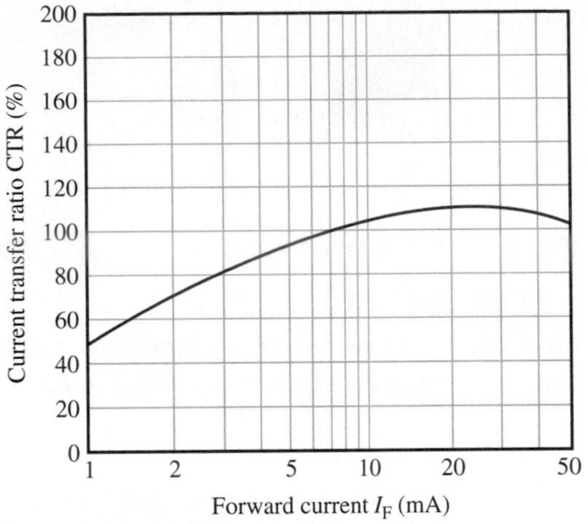

FIGURE 4–33

CTR versus I_F for a typical optocoupler.

SECTION 4–6 CHECKUP

1. How does a phototransistor differ from a conventional BJT?
2. A three-lead phototransistor has an external (emitter, base, collector) lead.
3. The collector current in a phototransistor circuit depends on what two factors?
4. What is the optocoupler parameter, OTR?

4–7 TRANSISTOR CATEGORIES AND PACKAGING

BJTs are available in a wide range of package types for various applications. Those with mounting studs or heat sinks are usually power transistors. Low-power and medium-power transistors are usually found in smaller metal or plastic cases. Still another package classification is for high-frequency devices. You should be familiar with common transistor packages and be able to identify the emitter, base, and collector terminals.

After completing this section, you should be able to

- **Identify various types of transistor package configurations**
  - List three broad categories of transistors
  - Recognize various types of cases and identify the pin configurations

Transistor Categories

Manufacturers generally classify bipolar junction transistors into three broad categories: general-purpose/small-signal devices, power devices, and RF (radio frequency/microwave) devices. Although each of these categories, to a large degree, has its own unique package types, you will find certain types of packages used in more than one device category. Let's look at transistor packages for each of the three categories so that you will be able to recognize a transistor when you see one on a circuit board and have a good idea of what general category it is in.

General-Purpose/Small-Signal Transistors General-purpose/small-signal transistors are generally used for low- or medium-power amplifiers or switching circuits. The packages are either plastic or metal cases. Certain types of packages contain multiple transistors. Figure 4–34 illustrates two common plastic cases and a metal can package.

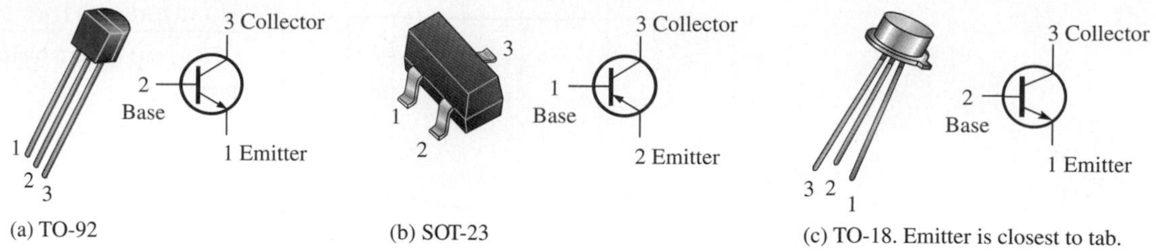

▲ **FIGURE 4–34**

Plastic and metal cases for general-purpose/small-signal transistors. Pin configurations may vary. Always check the datasheet (*http://fairchildsemiconductor.com/*).

Figure 4–35 shows multiple-transistor packages. Some of the multiple-transistor packages such as the dual in-line (DIP) and the small-outline (SO) are the same as those used for many integrated circuits. Typical pin connections are shown so you can identify the emitter, base, and collector.

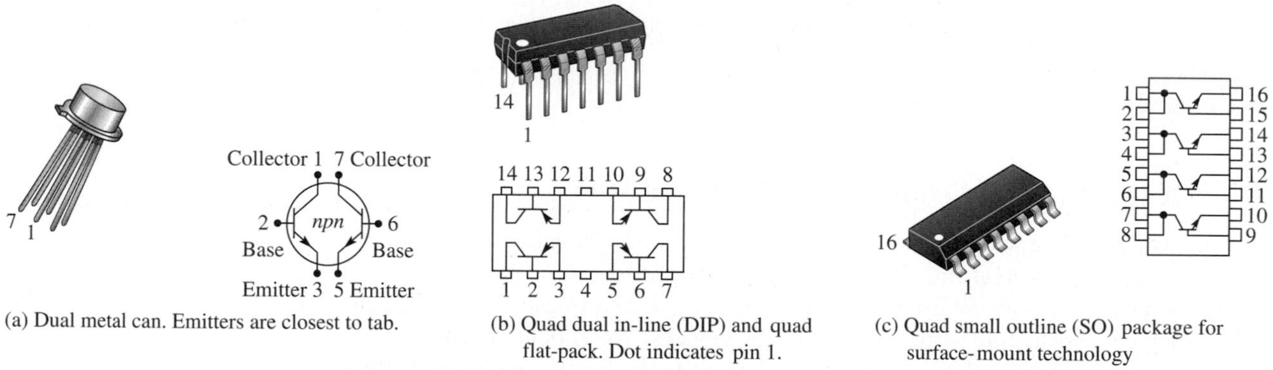

▲ **FIGURE 4–35**

Examples of multiple-transistor packages.

Power Transistors Power transistors are used to handle large currents (typically more than 1 A) and/or large voltages. For example, the final audio stage in a stereo system uses a power transistor amplifier to drive the speakers. Figure 4–36 shows some common package configurations. The metal tab or the metal case is common to the collector and is thermally connected to a heat sink for heat dissipation. Notice in part (e) how the small transistor chip is mounted inside the much larger package.

RF Transistors RF transistors are designed to operate at extremely high frequencies and are commonly used for various purposes in communications systems and other high-frequency applications. Their unusual shapes and lead configurations are designed to optimize certain high-frequency parameters. Figure 4–37 shows some examples.

SECTION 4–7 CHECKUP

1. **List the three broad categories of bipolar junction transistors.**
2. **In a metal can package of a general-purpose BJT, how is the emitter identified?**
3. **In power transistors, the metal mounting tab or case is connected to which transistor region?**

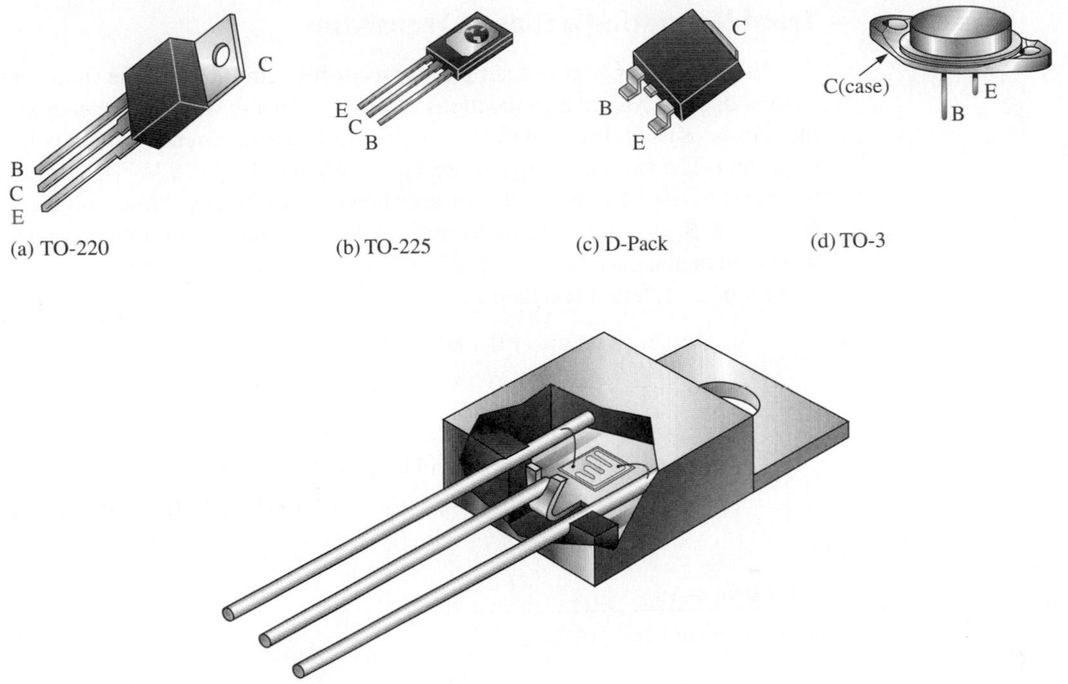

(e) Greatly enlarged cutaway view of tiny transistor chip mounted in the encapsulated package

▲ **FIGURE 4–36**

Examples of power transistors and packages.

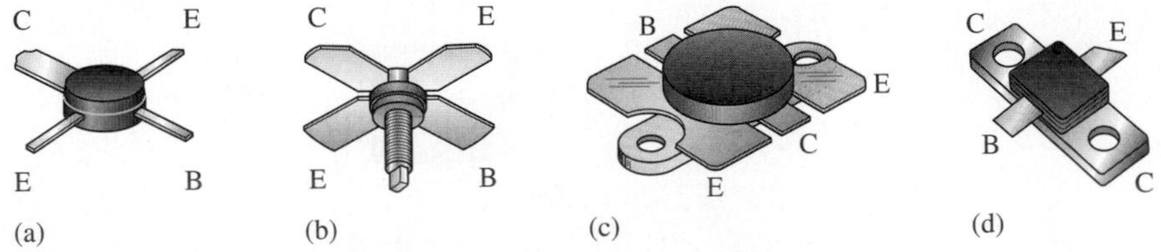

◀ **FIGURE 4–37**

Examples of RF transistor packages.

4–8 TROUBLESHOOTING

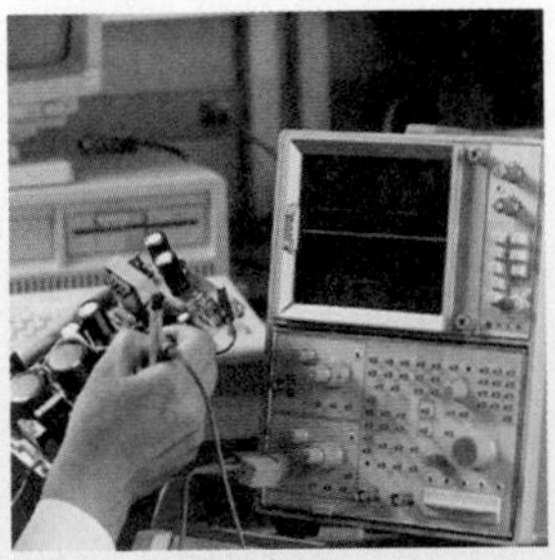

As you already know, a critical skill in electronics work is the ability to identify a circuit malfunction and to isolate the failure to a single component if necessary. In this section, the basics of troubleshooting transistor bias circuits and testing individual transistors are covered.

After completing this section, you should be able to

- **Troubleshoot various faults in transistor circuits**
  - Explain floating point measurement
  - Use voltage measurements to identify a fault in a transistor circuit
  - Use a DMM to test a transistor
  - Explain how a transistor can be viewed in terms of a diode equivalent
  - Discuss in-circuit and out-of-circuit testing
  - Discuss point-of-measurement in troubleshooting
  - Discuss leakage and gain measurements

Troubleshooting a Biased Transistor

Several faults can occur in a simple transistor bias circuit. Possible faults are open bias resistors, open or resistive connections, shorted connections, and opens or shorts internal to the transistor itself. Figure 4–38 is a basic transistor bias circuit with all voltages referenced to ground. The two bias voltages are $V_{BB} = 3\text{ V}$ and $V_{CC} = 9\text{ V}$. The correct voltage measurements at the base and collector are shown. Analytically, these voltages are verified as follows. A $\beta_{DC} = 200$ is taken as midway between the minimum and maximum values of h_{FE} given on the datasheet for the 2N3904 in Figure 4–20. A different h_{FE} (β_{DC}), of course, will produce different results for the given circuit.

$$V_B = V_{BE} = 0.7\text{ V}$$

$$I_B = \frac{V_{BB} - 0.7\text{ V}}{R_B} = \frac{3\text{ V} - 0.7\text{ V}}{56\text{ k}\Omega} = \frac{2.3\text{ V}}{56\text{ k}\Omega} = 41.1\ \mu\text{A}$$

$$I_C = \beta_{DC}I_B = 200(41.1\ \mu\text{A}) = 8.2\text{ mA}$$

$$V_C = 9\text{ V} - I_CR_C = 9\text{ V} - (8.2\text{ mA})(560\ \Omega) = 4.4\text{ V}$$

FIGURE 4–38

A basic transistor bias circuit.

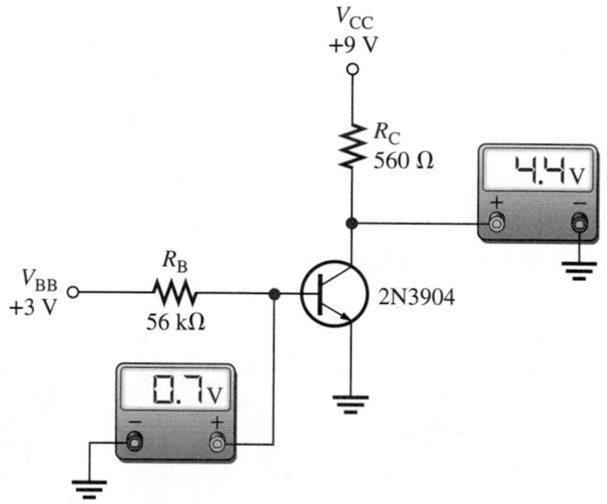

Several faults that can occur in the circuit and the accompanying symptoms are illustrated in Figure 4–39. Symptoms are shown in terms of measured voltages that are incorrect. If a transistor circuit is not operating correctly, it is a good idea to verify that V_{CC} and ground are connected and operating. A simple check at the top of the collector resistor and at the collector itself will quickly ascertain if V_{CC} is present and if the transistor is conducting normally or is in cutoff or saturation. If it is in cutoff, the collector voltage will equal V_{CC}; if it is in saturation, the collector voltage will be near zero. Another faulty measurement can be seen if the there is an open in the collector path. The term **floating point** refers to a point in the circuit that is not electrically connected to ground or a "solid" voltage. Normally, very small and sometimes fluctuating voltages in the μV to low mV range are generally measured at floating points. The faults in Figure 4–39 are typical but do not represent all possible faults that may occur.

Testing a Transistor with a DMM

A digital multimeter can be used as a fast and simple way to check a transistor for open or shorted junctions. For this test, you can view the transistor as two diodes connected as shown in Figure 4–40 for both *npn* and *pnp* transistors. The base-collector junction is one diode and the base-emitter junction is the other.

Recall that a good diode will show an extremely high resistance (or open) with reverse bias and a very low resistance with forward bias. A defective open diode will show an extremely high resistance (or open) for both forward and reverse bias. A defective shorted or

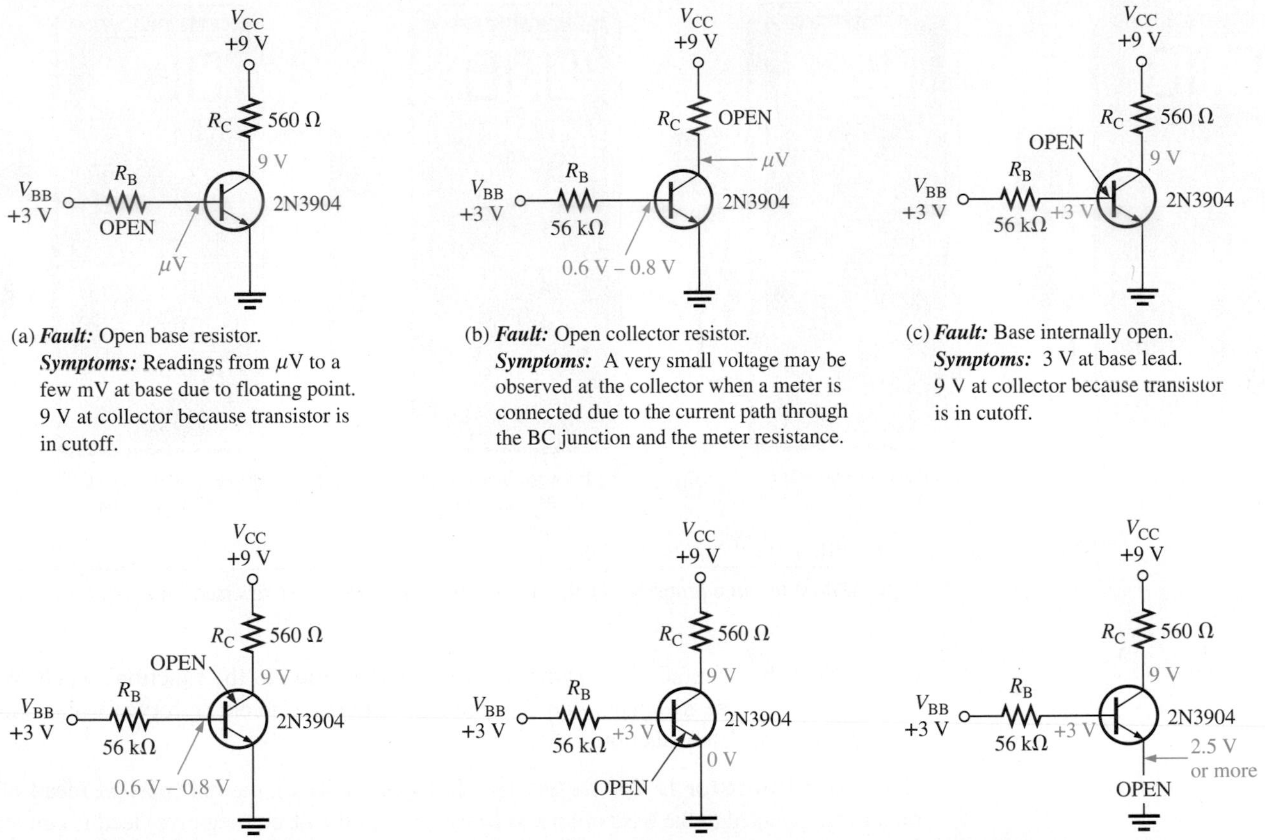

(a) ***Fault:*** Open base resistor. ***Symptoms:*** Readings from μV to a few mV at base due to floating point. 9 V at collector because transistor is in cutoff.

(b) ***Fault:*** Open collector resistor. ***Symptoms:*** A very small voltage may be observed at the collector when a meter is connected due to the current path through the BC junction and the meter resistance.

(c) ***Fault:*** Base internally open. ***Symptoms:*** 3 V at base lead. 9 V at collector because transistor is in cutoff.

(d) ***Fault:*** Collector internally open. ***Symptoms:*** 0.6 V – 0.8 V at base lead due to forward voltage drop across base-emitter junction. 9 V at collector because the open prevents collector current.

(e) ***Fault:*** Emitter internally open. ***Symptoms:*** 3 V at base lead. 9 V at collector because there is no collector current. 0 V at the emitter as normal.

(f) ***Fault:*** Open ground connection. ***Symptoms:*** 3 V at base lead. 9 V at collector because there is no collector current. 2.5 V or more at the emitter due to the forward voltage drop across the base-emitter junction. The measuring voltmeter provides a forward current path through its internal resistance.

▲ **FIGURE 4–39**

Examples of faults and symptoms in the basic transistor bias circuit.

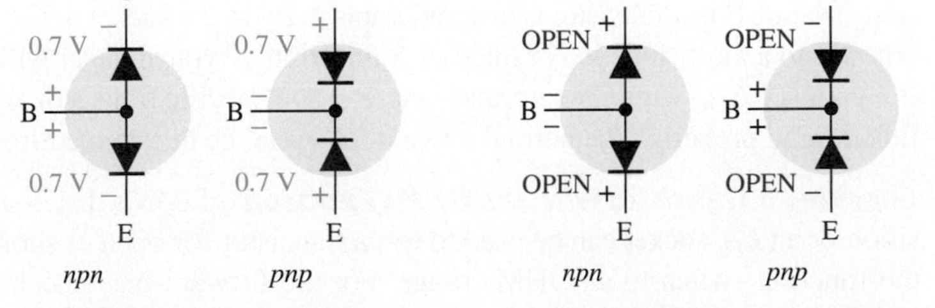

(a) Both junctions should typically read 0.7 V when forward-biased.

(b) Both junctions should ideally read OPEN when reverse-biased.

◀ **FIGURE 4–40**

A transistor viewed as two diodes.

resistive diode will show zero or a very low resistance for both forward and reverse bias. An open diode is the most common type of failure. Since the transistor *pn* junctions are, in effect diodes, the same basic characteristics apply.

The DMM Diode Test Position Many digital multimeters (DMMs) have a *diode test* position that provides a convenient way to test a transistor. A typical DMM, as shown in

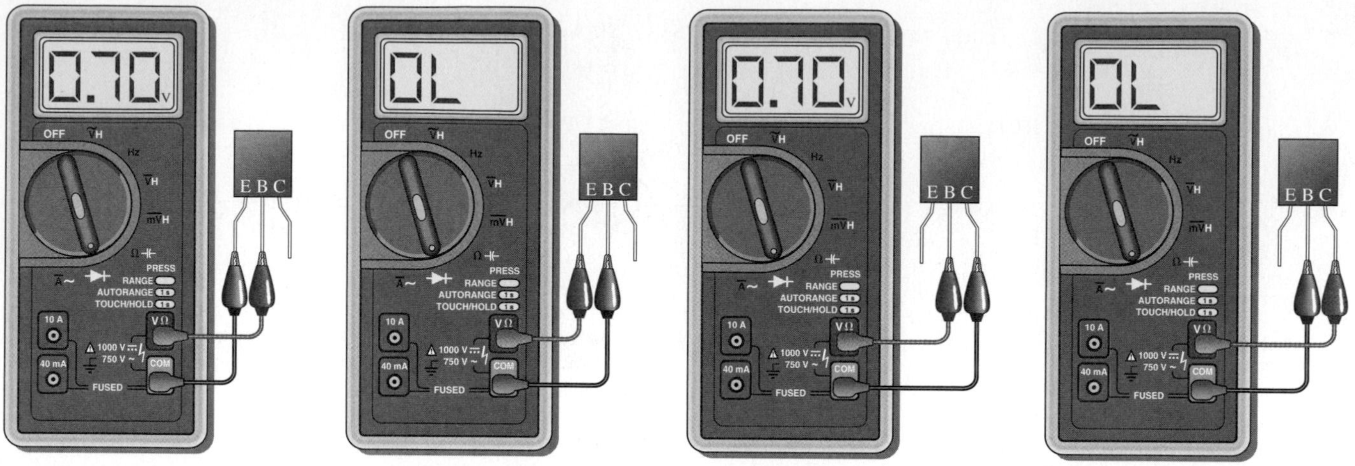

(a) Forward-bias test of the BE junction

(b) Reverse-bias test of the BE junction

(c) Forward-bias test of the BC junction

(d) Reverse-bias test of the BC junction

▲ **FIGURE 4–41**

Typical DMM test of a properly functioning *npn* transistor. Leads are reversed for a *pnp* transistor.

Figure 4–41, has a small diode symbol to mark the position of the function switch. When set to diode test, the meter provides an internal voltage sufficient to forward-bias and reverse-bias a transistor junction.

When the Transistor Is Not Defective In Figure 4–41(a), the red (positive) lead of the meter is connected to the base of an *npn* transistor and the black (negative) lead is connected to the emitter to forward-bias the base-emitter junction. If the junction is good, you will get a reading of between approximately 0.6 V and 0.8 V, with 0.7 V being typical for forward bias.

In Figure 4–41(b), the leads are switched to reverse-bias the base-emitter junction, as shown. If the transistor is working properly, you will typically get an OL indication.

The process just described is repeated for the base-collector junction as shown in Figure 4–41(c) and (d). For a *pnp* transistor, the polarity of the meter leads are reversed for each test.

When the Transistor Is Defective When a transistor has failed with an open junction or internal connection, you get an open circuit voltage reading (OL) for both the forward-bias and the reverse-bias conditions for that junction, as illustrated in Figure 4–42(a). If a junction is shorted, the meter reads 0 V in both forward- and reverse-bias tests, as indicated in part (b).

Some DMMs provide a test socket on their front panel for testing a transistor for the h_{FE} (β_{DC}) value. If the transistor is inserted improperly in the socket or if it is not functioning properly due to a faulty junction or internal connection, a typical meter will flash a 1 or display a 0. If a value of β_{DC} within the normal range for the specific transistor is displayed, the device is functioning properly. The normal range of β_{DC} can be determined from the datasheet.

Checking a Transistor with the OHMs Function DMMs that do not have a diode test position or an h_{FE} socket can be used to test a transistor for open or shorted junctions by setting the function switch to an OHMs range. For the forward-bias check of a good transistor *pn* junction, you will get a resistance reading that can vary depending on the meter's internal battery. Many DMMs do not have sufficient voltage on the OHMs range to fully forward-bias a junction, and you may get a reading of from several hundred to several thousand ohms.

For the reverse-bias check of a good transistor, you will get an out-of-range indication on most DMMs because the reverse resistance is too high to measure. An out-of-range indication may be a flashing 1 or a display of dashes, depending on the particular DMM.

Even though you may not get accurate forward and reverse resistance readings on a DMM, the relative readings are sufficient to indicate a properly functioning transistor *pn* junction. The out-of-range indication shows that the reverse resistance is very high, as you

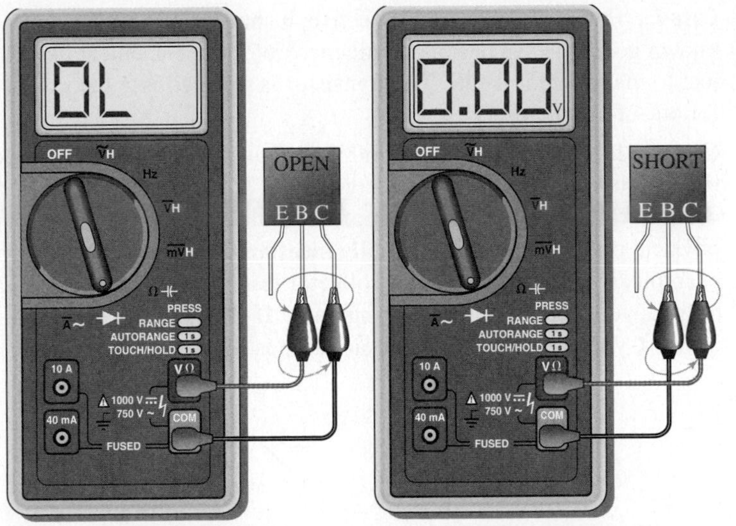

(a) Forward-bias test and reverse-bias test give the same reading (OL is typical) for an open BC junction.

(b) Forward- and reverse-bias tests for a shorted junction give the same 0 V reading.

FIGURE 4–42

Testing a defective *npn* transistor. Leads are reversed for a *pnp* transistor.

expect. The reading of a few hundred to a few thousand ohms for forward bias indicates that the forward resistance is small compared to the reverse resistance, as you expect.

Transistor Testers

An individual transistor can be tested either in-circuit or out-of-circuit with a transistor tester. For example, let's say that an amplifier on a particular printed circuit (PC) board has malfunctioned. Good troubleshooting practice dictates that you do not unsolder a component from a circuit board unless you are reasonably sure that it is bad or you simply cannot isolate the problem down to a single component. When components are removed, there is a risk of damage to the PC board contacts and traces.

You can perform an in-circuit check of the transistor using a transistor tester similar to the one shown in Figure 4–43. The three clip-leads are connected to the transistor terminals and the tester gives a positive indication if the transistor is good.

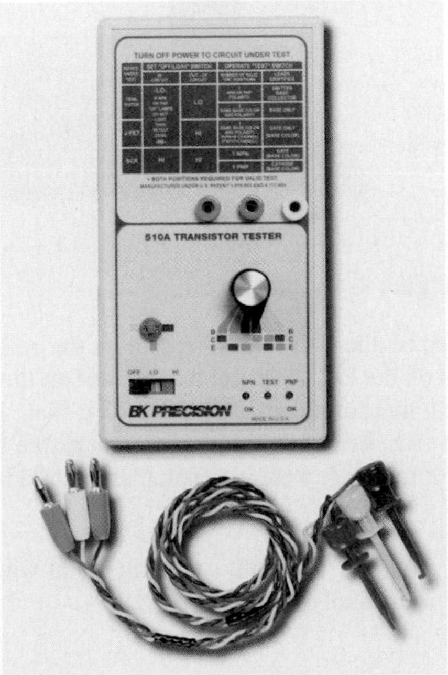

FIGURE 4–43

Transistor tester (courtesy of B+K Precision).

Case 1 If the transistor tests defective, it should be carefully removed and replaced with a known good one. An out-of-circuit check of the replacement device is usually a good idea, just to make sure it is OK. The transistor is plugged into the socket on the transistor tester for out-of-circuit tests.

Case 2 If the transistor tests good in-circuit but the circuit is not working properly, examine the circuit board for a poor connection at the collector pad or for a break in the connecting trace. A poor solder joint often results in an open or a highly resistive contact. The physical point at which you actually measure the voltage is very important in this case. For example, if you measure on the collector lead when there is an external open at the collector pad, you will measure a floating point. If you measure on the connecting trace or on the R_C lead, you will read V_{CC}. This situation is illustrated in Figure 4–44.

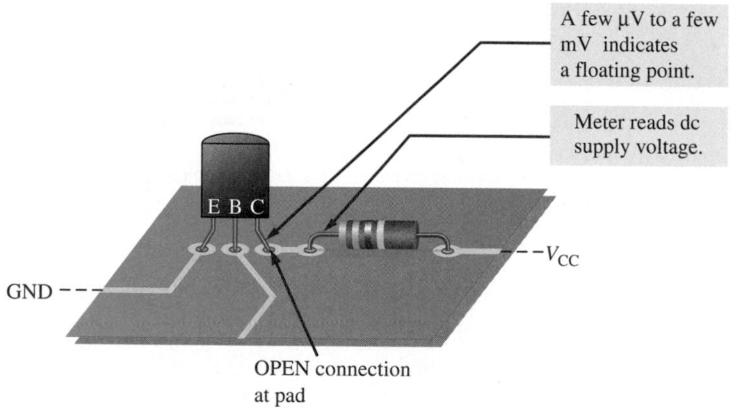

FIGURE 4–44

The indication of an open, when it is in the circuit external to the transistor, depends on where you measure.

Importance of Point-of-Measurement in Troubleshooting In case 2, if you had taken the initial measurement on the transistor lead itself and the open were *internal* to the transistor as shown in Figure 4–45, you would have measured V_{CC}. This indicates a defective transistor even before the tester was used, assuming the base-to-emitter voltage is normal. This simple concept emphasizes the importance of point-of-measurement in certain troubleshooting situations.

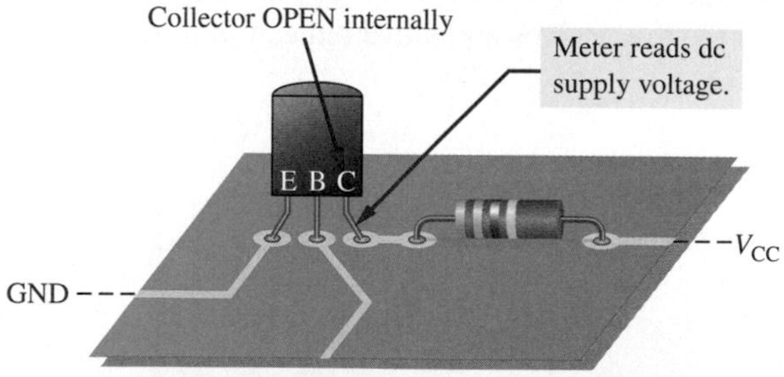

FIGURE 4–45

Illustration of an internal open. Compare with Figure 4–44.

EXAMPLE 4–12 What fault do the measurements in Figure 4–46 indicate?

Solution The transistor is in cutoff, as indicated by the 10 V measurement on the collector lead. The base bias voltage of 3 V appears on the PC board contact but not on the transistor lead, as indicated by the floating point measurement. This shows that there is an open external to the transistor between the two measured base points. Check the solder joint at the base contact on the PC board. If the open were internal, there would be 3 V on the base lead.

Related Problem If the meter in Figure 4–46 that now reads 3 V indicates a floating point when touching the circuit board pad, what is the most likely fault?

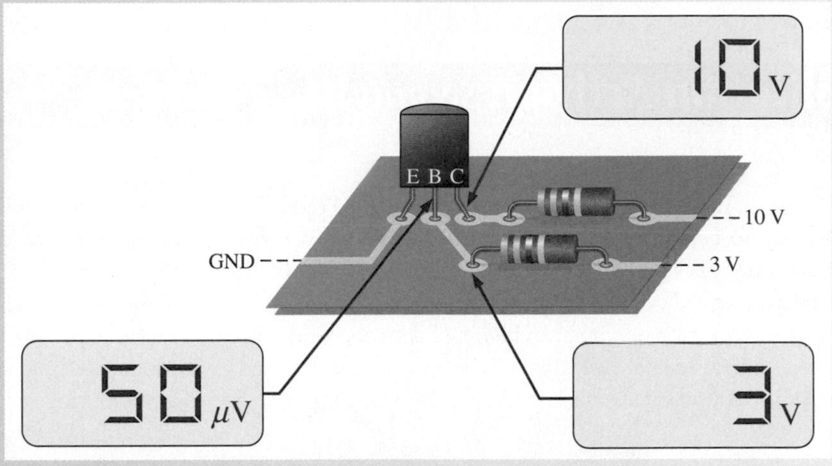

FIGURE 4–46

Leakage Measurement Very small leakage currents exist in all transistors and in most cases are small enough to neglect (usually nA). When a transistor is connected with the base open ($I_B = 0$), it is in cutoff. Ideally $I_C = 0$; but actually there is a small current from collector to emitter, as mentioned earlier, called I_{CEO} (collector-to-emitter current with base open). This leakage current is usually in the nA range. A faulty transistor will often have excessive leakage current and can be checked in a transistor tester. Another leakage current in transistors is the reverse collector-to-base current, I_{CBO}. This is measured with the emitter open. If it is excessive, a shorted collector-base junction is likely.

Gain Measurement In addition to leakage tests, the typical transistor tester also checks the β_{DC}. A known value of I_B is applied, and the resulting I_C is measured. The reading will indicate the value of the I_C/I_B ratio, although in some units only a relative indication is given. Most testers provide for an in-circuit β_{DC} check, so that a suspected device does not have to be removed from the circuit for testing.

Curve Tracers

A *curve tracer* is an oscilloscope type of instrument that can display transistor characteristics such as a family of collector curves. In addition to the measurement and display of various transistor characteristics, diode curves can also be displayed.

Multisim Troubleshooting Exercises

These file circuits are in the Troubleshooting Exercises folder on your CD-ROM.

1. Open file TSE04-01. Determine if the circuit is working properly and, if not, determine the fault.
2. Open file TSE04-02. Determine if the circuit is working properly and, if not, determine the fault.
3. Open file TSE04-03. Determine if the circuit is working properly and, if not, determine the fault.

SECTION 4–8 CHECKUP

1. If a transistor on a circuit board is suspected of being faulty, what should you do?
2. In a transistor bias circuit, such as the one in Figure 4–38; what happens if R_B opens?
3. In a circuit such as the one in Figure 4–38, what are the base and collector voltages if there is an external open between the emitter and ground?

On the Job

Application Activity: *Security Alarm System*

A circuit using transistor switches will be developed for use in an alarm system for detecting forced entry into a building. In its simplest form, the alarm system will accommodate four zones with any number of openings. It can be expanded to cover additional zones. For the purposes of this application, a zone is one room in a house or other building. The sensor used for each opening can be either a mechanical switch, a magnetically operated switch, or an optical sensor. Detection of an intrusion can be used to initiate an audible alarm signal and/or to initiate transmission of a signal over the phone line to a monitoring service.

Designing the Circuit

A basic block diagram of the system is shown in Figure 4–47. The sensors for each zone are connected to the switching circuits, and the output of the switching circuit goes to an audible alarm circuit and/or to a telephone dialing circuit. The focus of this application is the transistor switching circuits.

FIGURE 4–47

Block diagram of security alarm system.

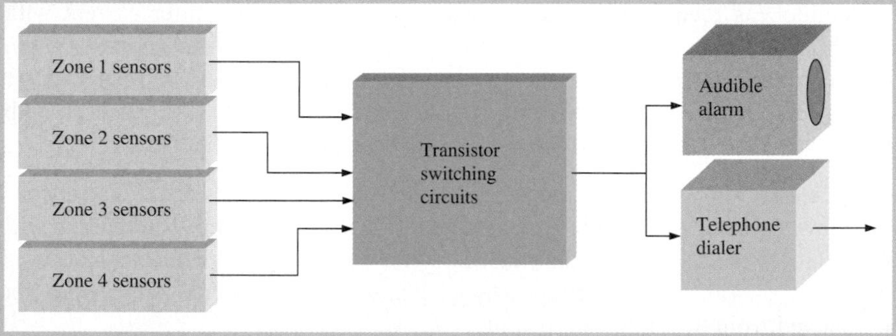

A zone sensor detects when a window or door is opened. They are normally in a closed position and are connected in series to a dc voltage source, as shown in Figure 4–48(a). When a window or door is opened, the corresponding sensor creates an open circuit, as shown in part (b). The sensors are represented by switch symbols.

FIGURE 4–48

Zone sensor configuration.

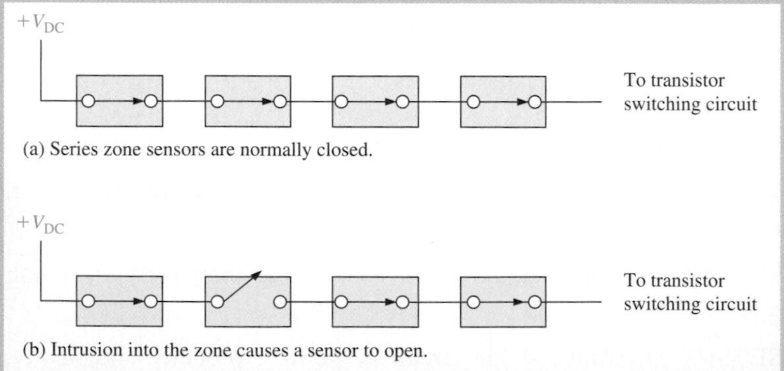

A circuit for one zone is shown in Figure 4–49. It consists of two BJTs, Q_1 and Q_2. As long as the zone sensors are closed, Q_1 is in the *on* state (saturated). The very low saturation voltage at the Q_1 collector keeps Q_2 *off.* Notice that the collector of Q_2 is left open with no load connected. This allows for all four of the zone circuit outputs to be tied together and a common load connected externally to drive the alarm and/or dialing circuits. If one of the zone sensors opens, indicating a break-in, Q_1 turns *off* and its collector voltage goes to V_{CC}. This turns on Q_2, causing it to saturate. The *on* state of Q_2 will then activate the audible alarm and the telephone dialing sequence.

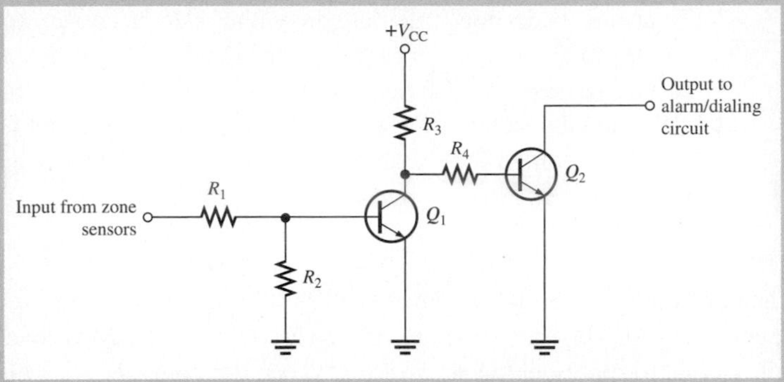

FIGURE 4–49

One of the four identical transistor switching circuits.

1. Refer to the partial datasheet for the 2N2222A in Figure 4–50 and determine the value of the collector resistor R_3 to limit the current to 10 mA with a +12 V dc supply voltage.
2. Using the minimum β_{DC} or h_{FE} from the datasheet, determine the base current required to saturate Q_1 at $I_C = 10$ mA.

Absolute Maximum Ratings * T_a=25°C unless otherwise noted

Symbol	Parameter	Value	Units
V_{CEO}	Collector-Emitter Voltage	40	V
V_{CBO}	Collector-Base Voltage	75	V
V_{EBO}	Emitter-Base Voltage	6.0	V
I_C	Collector Current	1.0	A
T_{STG}	Operating and Storage Junction Temperature Range	- 55 ~ 150	°C

* These ratings are limiting values above which the serviceability of any semiconductor device may be impaired

NOTES:
1) These ratings are based on a maximum junction temperature of 150 degrees C.
2) These are steady state limits. The factory should be consulted on applications involving pulsed or low duty cycle operations

Electrical Characteristics T_a=25°C unless otherwise noted

Symbol	Parameter	Test Condition	Min.	Max.	Units
Off Characteristics					
$BV_{(BR)CEO}$	Collector-Emitter Breakdown Voltage *	I_C = 10mA, I_B = 0	40		V
$BV_{(BR)CBO}$	Collector-Base Breakdown Voltage	I_C = 10μA, I_E = 0	75		V
$BV_{(BR)EBO}$	Emitter-Base Breakdown Voltage	I_E = 10μA, I_C = 0	6.0		V
I_{CEX}	Collector Cutoff Current	V_{CE} = 60V, $V_{EB(off)}$ = 3.0V		10	nA
I_{CBO}	Collector Cutoff Current	V_{CB} = 60V, I_E = 0 V_{CB} = 60V, I_E = 0, T_a = 125°C		0.01 10	μA μA
I_{EBO}	Emitter Cutoff Current	V_{EB} = 3.0V, I_C = 0		10	μA
I_{BL}	Base Cutoff Current	V_{CE} = 60V, $V_{EB(off)}$ = 3.0V		20	μA
On Characteristics					
h_{FE}	DC Current Gain	I_C = 0.1mA, V_{CE} = 10V I_C = 1.0mA, V_{CE} = 10V I_C = 10mA, V_{CE} = 10V I_C = 10mA, V_{CE} = 10V, T_a = -55°C I_C = 150mA, V_{CE} = 10V * I_C = 150mA, V_{CE} = 10V * I_C = 500mA, V_{CE} = 10V *	35 50 75 35 100 50 40	 300 	
$V_{CE(sat)}$	Collector-Emitter Saturation Voltage *	I_C = 150mA, V_{CE} = 10V I_C = 500mA, V_{CE} = 10V		0.3 1.0	V V
$V_{BE(sat)}$	Base-Emitter Saturation Voltage *	I_C = 150mA, V_{CE} = 10V I_C = 500mA, V_{CE} = 10V	0.6	1.2 2.0	V V

* Pulse Test: Pulse Width ≤ 300μs, Duty Cycle ≤ 2.0%

FIGURE 4–50

Partial datasheet for the 2N2222A transistor. © 2003 Fairchild Semiconductor Corporation. Used by permission.

3. To ensure saturation, calculate the value of R_1 necessary to provide sufficient base current to Q_1 from the +12 V sensor input. R_2 can be any arbitrarily high value to assure the base of Q_1 is near ground when there is no input voltage.
4. Calculate the value of R_4 so that a sufficient base current is supplied to Q_2 to ensure saturation for a load of 620 Ω. This simulates the actual load of the alarm and dialing circuits.

Simulation

The switching circuit is simulated with Multisim, as shown in Figure 4–51. A switch connected to a 12 V source simulates the zone input and a 620 Ω load resistor is connected to the output to represent the actual load. When the zone switch is open, Q_2 is saturated as indicated by 0.126 V at its collector. When the zone switch is closed, Q_2 is *off* as indicated by the 11.999 V at its collector.

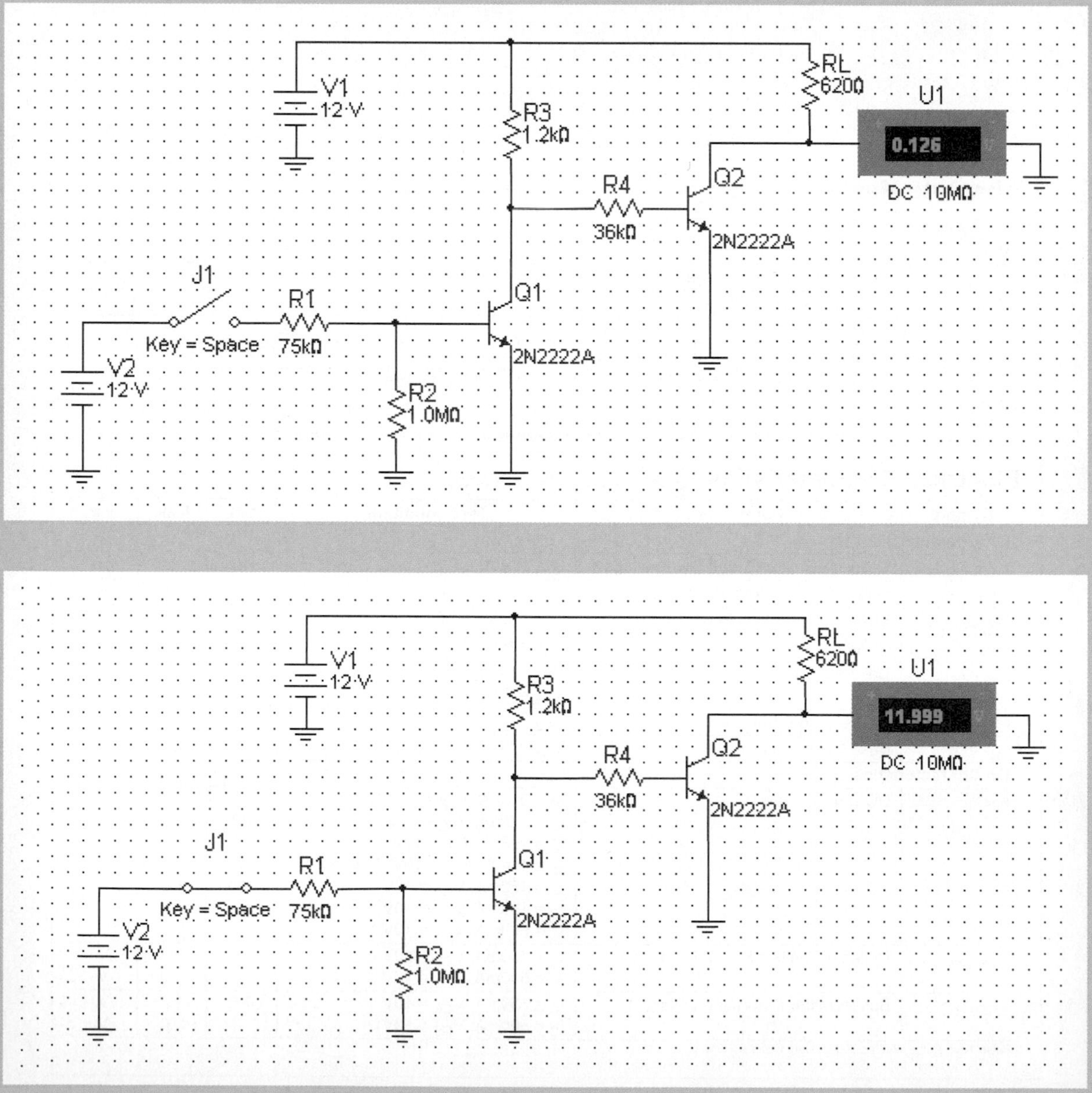

FIGURE 4–51

Simulation of the switching circuit.

5. How does the Q_2 saturation voltage compare to the value specified on the datasheet?

Simulate the circuit using your Multisim software. Observe the operation with the virtual multimeter.

Prototyping and Testing

Now that the circuit has been simulated, it is connected on a protoboard and tested for proper operation.

To build and test a similar circuit, go to Experiment 4 in your lab manual (*Laboratory Exercises for Electronic Devices* by David Buchla and Steven Wetterling).

Lab Experiment

Printed Circuit Board

The transistor switching circuit prototype has been built and tested. It is now committed to a printed circuit layout, as shown in Figure 4–52. Notice that there are four identical circuits on the board, one for each zone to be monitored. The outputs are externally connected to form a single input.

6. Compare the printed circuit board to the schematic in Figure 4–49 and verify that they agree. Identify each component.
7. Compare the resistor values on the printed circuit board to those that you calculated previously. They should closely agree.
8. Label the input and output pins on the printed circuit board according to their function.
9. Describe how you would test the circuit board.
10. Explain how the system can be expanded to monitor six zones instead of four.

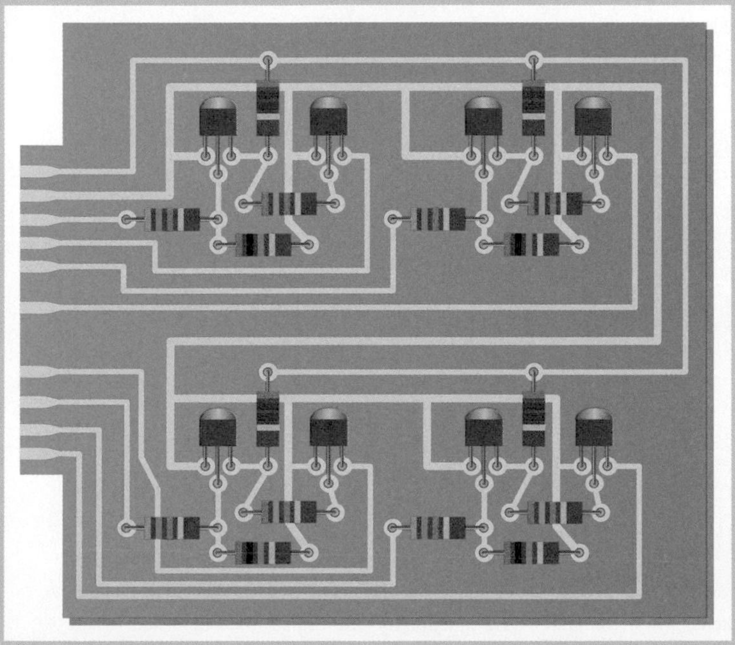

FIGURE 4–52

The 4-zone transistor switching circuit board.

SUMMARY OF BIPOLAR JUNCTION TRANSISTORS

SYMBOLS

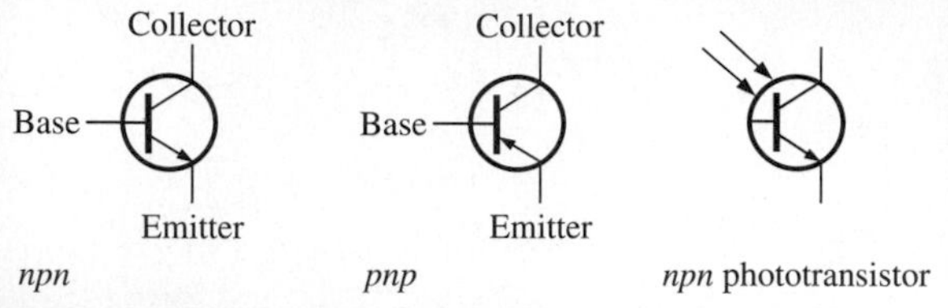

CURRENTS AND VOLTAGES

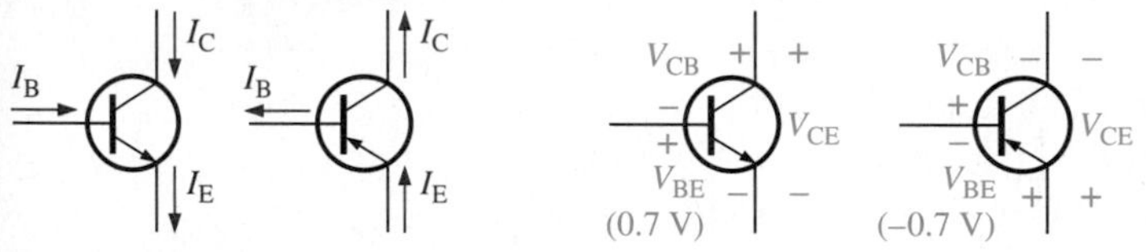

$I_E = I_C + I_B$

AMPLIFICATION

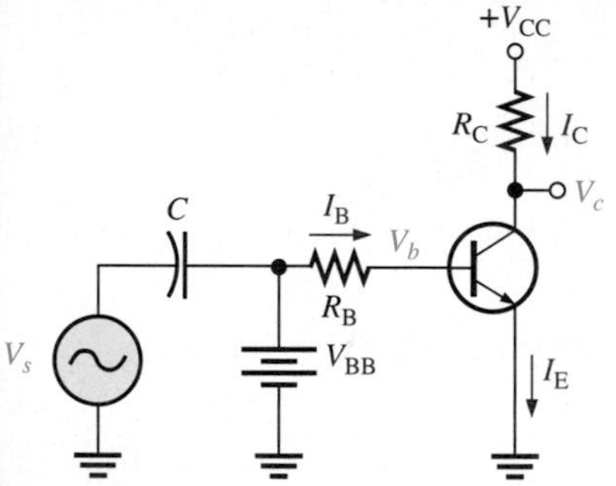

- DC current gain

$$I_C = \beta_{DC} I_B$$

- AC voltage gain

$$A_v = \frac{V_c}{V_b} = \frac{R_C}{r'_e}$$

BE junction forward-biased
BC junction reverse-biased

SWITCHING

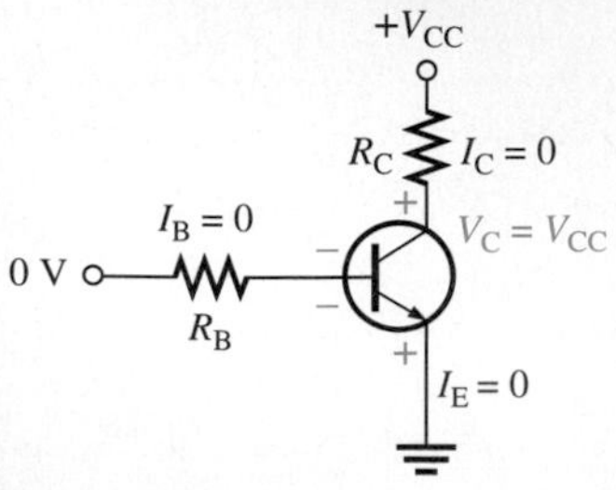

Cutoff: BE junction reverse-biased
BC junction reverse-biased

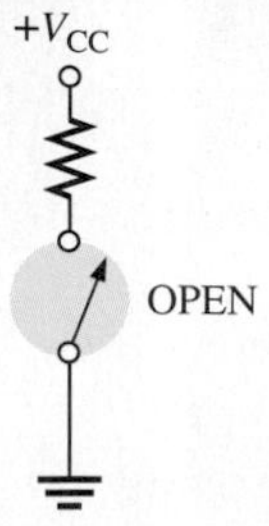

Ideal switch equivalent for cutoff

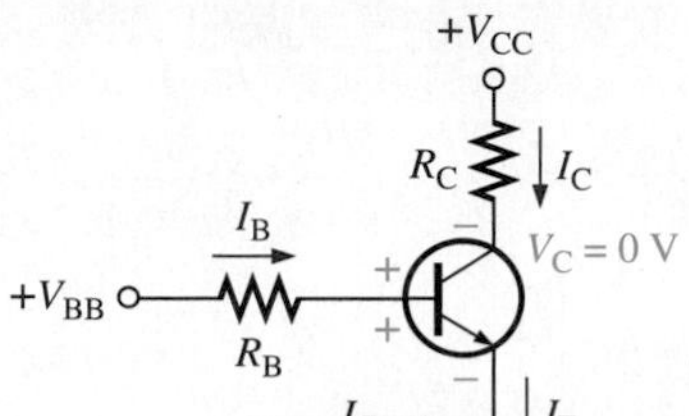

$$I_B \geq \frac{I_{C(sat)}}{\beta_{DC}}$$

Saturation: BE junction forward-biased
BC junction forward-biased

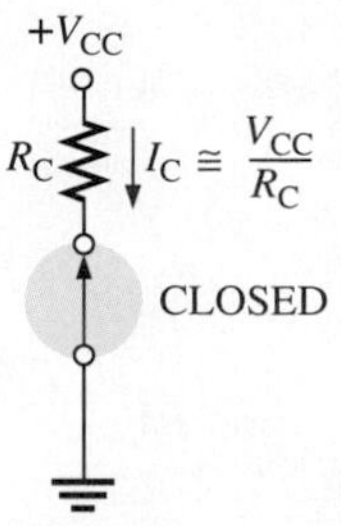

Ideal switch equivalent for saturation

SUMMARY

Section 4–1
- The BJT (bipolar junction transistor) is constructed with three regions: base, collector, and emitter.
- The BJT has two *pn* junctions, the base-emitter junction and the base-collector junction.
- Current in a BJT consists of both free electrons and holes, thus the term *bipolar*.
- The base region is very thin and lightly doped compared to the collector and emitter regions.
- The two types of bipolar junction transistor are the *npn* and the *pnp*.

Section 4–2
- To operate as an amplifier, the base-emitter junction must be forward-biased and the base-collector junction must be reverse-biased. This is called *forward-reverse bias*.
- The three currents in the transistor are the base current (I_B), emitter current (I_E), and collector current (I_C).
- I_B is very small compared to I_C and I_E.

Section 4–3
- The dc current gain of a transistor is the ratio of I_C to I_B and is designated β_{DC}. Values typically range from less than 20 to several hundred.
- β_{DC} is usually referred to as h_{FE} on transistor datasheets.
- The ratio of I_C to I_E is called α_{DC}. Values typically range from 0.95 to 0.99.
- There is a variation in β_{DC} over temperature and also from one transistor to another of the same type.

Section 4–4
- When a transistor is forward-reverse biased, the voltage gain depends on the internal emitter resistance and the external collector resistance.
- Voltage gain is the ratio of output voltage to input voltage.
- Internal transistor resistances are represented by a lowercase *r*.

Section 4–5
- A transistor can be operated as an electronic switch in cutoff and saturation.
- In cutoff, both *pn* junctions are reverse-biased and there is essentially no collector current. The transistor ideally behaves like an open switch between collector and emitter.
- In saturation, both *pn* junctions are forward-biased and the collector current is maximum. The transistor ideally behaves like a closed switch between collector and emitter.

Section 4–6
- In a phototransistor, base current is produced by incident light.
- A phototransistor can be either a two-lead or a three-lead device.
- An optocoupler consists of an LED and a photodiode or phototransistor.
- Optocouplers are used to electrically isolate circuits.

Section 4–7
- There are many types of transistor packages using plastic, metal, or ceramic.
- Two basic package types are through-hole and surface mount.

Section 4–8
- It is best to check a transistor in-circuit before removing it.
- Common faults in transistor circuits are open junctions, low β_{DC}, excessive leakage currents, and external opens and shorts on the circuit board.

KEY TERMS

Key terms and other bold terms in the chapter are defined in the end-of-book glossary.

Amplification The process of increasing the power, voltage, or current by electronic means.

Base One of the semiconductor regions in a BJT. The base is very thin and lightly doped compared to the other regions.

Beta (β) The ratio of dc collector current to dc base current in a BJT; current gain from base to collector.

BJT (bipolar junction transistor) A transistor constructed with three doped semiconductor regions separated by two *pn* junctions.

Collector The largest of the three semiconductor regions of a BJT.

Cutoff The nonconducting state of a transistor.

Emitter The most heavily doped of the three semiconductor regions of a BJT.

Gain The amount by which an electrical signal is increased or amplified.

Linear region Characterized by a straight-line relationship of the transistor currents.

Phototransistor A transistor in which base current is produced when light strikes the photosensitive semiconductor base region.

Saturation The state of a BJT in which the collector current has reached a maximum and is independent of the base current.

KEY FORMULAS

4–1	$I_E = I_C + I_B$	Transistor currents
4–2	$\beta_{DC} = \dfrac{I_C}{I_B}$	DC current gain
4–3	$V_{BE} \cong 0.7\text{ V}$	Base-to-emitter voltage (silicon)
4–4	$I_B = \dfrac{V_{BB} - V_{BE}}{R_B}$	Base current
4–5	$V_{CE} = V_{CC} - I_C R_C$	Collector-to-emitter voltage (common-emitter)
4–6	$V_{CB} = V_{CE} - V_{BE}$	Collector-to-base voltage
4–7	$A_v \cong \dfrac{R_C}{r'_e}$	Approximate ac voltage gain
4–8	$V_{CE(cutoff)} = V_{CC}$	Cutoff condition
4–9	$I_{C(sat)} = \dfrac{V_{CC} - V_{CE(sat)}}{R_C}$	Collector saturation current
4–10	$I_{B(min)} = \dfrac{I_{C(sat)}}{\beta_{DC}}$	Minimum base current for saturation
4–11	$I_C = \beta_{DC} I_\lambda$	Phototransistor collector current

TRUE/FALSE QUIZ

Answers are at the end of the chapter.

1. A bipolar junction transistor has three terminals.
2. The three regions of a BJT are base, emitter, and cathode.
3. For operation in the linear or active region, the base-emitter junction of a transistor is forward-biased.
4. Two types of BJT are *npn* and *pnp*.
5. The base current and collector current are approximately equal.
6. The dc voltage gain of a transistor is designated β_{DC}.
7. Cutoff and saturation are the two normal states of a linear transistor amplifier.
8. When a transistor is saturated, the collector current is maximum.
9. β_{DC} and h_{FE} are two different transistor parameters.
10. Voltage gain of a transistor amplifier depends on the collector resistor and the internal ac resistance.
11. Amplification is the output voltage divided by the input current.
12. A transistor in cutoff acts as an open switch.

CIRCUIT-ACTION QUIZ

Answers are at the end of the chapter.

1. If a transistor with a higher β_{DC} is used in Figure 4–9, the collector current will
   (a) increase **(b)** decrease **(c)** not change
2. If a transistor with a higher β_{DC} is used in Figure 4–9, the emitter current will
   (a) increase **(b)** decrease **(c)** not change

3. If a transistor with a higher β_{DC} is used in Figure 4–9, the base current will
   (a) increase (b) decrease (c) not change
4. If V_{BB} is reduced in Figure 4–16, the collector current will
   (a) increase (b) decrease (c) not change
5. If V_{CC} in Figure 4–16 is increased, the base current will
   (a) increase (b) decrease (c) not change
6. If the amplitude of V_{in} in Figure 4–22 is decreased, the ac output voltage amplitude will
   (a) increase (b) decrease (c) not change
7. If the transistor in Figure 4–24 is saturated and the base current is increased, the collector current will
   (a) increase (b) decrease (c) not change
8. If R_C in Figure 4–24 is reduced in value, the value of $I_{C(sat)}$ will
   (a) increase (b) decrease (c) not change
9. If the transistor in Figure 4–38 is open from collector to emitter, the voltage across R_C will
   (a) increase (b) decrease (c) not change
10. If the transistor in Figure 4–38 is open from collector to emitter, the collector voltage will
    (a) increase (b) decrease (c) not change
11. If the base resistor in Figure 4–38 is open, the transistor collector voltage will
    (a) increase (b) decrease (c) not change
12. If the emitter in Figure 4–38 becomes disconnected from ground, the collector voltage will
    (a) increase (b) decrease (c) not change

SELF-TEST

Answers are at the end of the chapter.

Section 4–1

1. The three terminals of a bipolar junction transistor are called
   (a) *p, n, p* (b) *n, p, n* (c) input, output, ground (d) base, emitter, collector
2. In a *pnp* transistor, the *p* regions are
   (a) base and emitter (b) base and collector (c) emitter and collector

Section 4–2

3. For operation as an amplifier, the base of an *npn* transistor must be
   (a) positive with respect to the emitter (b) negative with respect to the emitter
   (c) positive with respect to the collector (d) 0 V
4. The emitter current is always
   (a) greater than the base current (b) less than the collector current
   (c) greater than the collector current (d) answers (a) and (c)

Section 4–3

5. The β_{DC} of a transistor is its
   (a) current gain (b) voltage gain (c) power gain (d) internal resistance
6. If I_C is 50 times larger than I_B, then β_{DC} is
   (a) 0.02 (b) 100 (c) 50 (d) 500
7. The approximate voltage across the forward-biased base-emitter junction of a silicon BJT is
   (a) 0 V (b) 0.7 V (c) 0.3 V (d) V_{BB}
8. The bias condition for a transistor to be used as a linear amplifier is called
   (a) forward-reverse (b) forward-forward
   (c) reverse-reverse (d) collector bias

Section 4–4

9. If the output of a transistor amplifier is 5 V rms and the input is 100 mV rms, the voltage gain is
   (a) 5 (b) 500 (c) 50 (d) 100
10. When a lowercase r' is used in relation to a transistor, it refers to
    (a) a low resistance (b) a wire resistance
    (c) an internal ac resistance (d) a source resistance

11. In a given transistor amplifier, $R_C = 2.2\ \text{k}\Omega$ and $r'_e = 20\ \Omega$, the voltage gain is
    (a) 2.2 (b) 110 (c) 20 (d) 44

Section 4–5

12. When operated in cutoff and saturation, the transistor acts like a
    (a) linear amplifier (b) switch (c) variable capacitor (d) variable resistor
13. In cutoff, V_{CE} is
    (a) 0 V (b) minimum (c) maximum
    (d) equal to V_{CC} (e) answers (a) and (b) (f) answers (c) and (d)
14. In saturation, V_{CE} is
    (a) 0.7 V (b) equal to V_{CC} (c) minimum (d) maximum
15. To saturate a BJT,
    (a) $I_B = I_{C(sat)}$ (b) $I_B > I_{C(sat)}/\beta_{DC}$
    (c) V_{CC} must be at least 10 V (d) the emitter must be grounded
16. Once in saturation, a further increase in base current will
    (a) cause the collector current to increase (b) not affect the collector current
    (c) cause the collector current to decrease (d) turn the transistor off

Section 4–6

17. In a phototransistor, base current is
    (a) set by a bias voltage (b) directly proportional to light intensity
    (c) inversely proportional to light intensity (d) not a factor
18. The relationship between the collector current and a light-generated base current is
    (a) $I_C = \beta_{DC}I_\lambda$ (b) $I_C = \alpha_{DC}I_\lambda$ (c) $I_C = \lambda I_\lambda$ (d) $I_C = \beta^2_{DC}I_\lambda$
19. An optocoupler usually consists of
    (a) two LEDs (b) an LED and a photodiode
    (c) an LED and a phototransistor (d) both (b) and (c)

Section 4–8

20. In a transistor amplifier, if the base-emitter junction is open, the collector voltage is
    (a) V_{CC} (b) 0 V (c) floating (d) 0.2 V
21. A DMM measuring on open transistor junction shows
    (a) 0 V (b) 0.7 V (c) OL (d) V_{CC}

PROBLEMS

Answers to all odd-numbered problems are at the end of the book.

BASIC PROBLEMS

Section 4–1

BJT Structure

1. What are the majority carriers in the base region of an *npn* transistor called?
2. Explain the purpose of a thin, lightly doped base region.

Section 4–2

Basic BJT Operation

3. Why is the base current in a transistor so much less than the collector current?
4. In a certain transistor circuit, the base current is 2 percent of the 30 mA emitter current. Determine the collector current.
5. For normal operation of a *pnp* transistor, the base must be (+ or −) with respect to the emitter, and (+ or −) with respect to the collector.
6. What is the value of I_C for $I_E = 5.34$ mA and $I_B = 475\ \mu$A?

Section 4–3

BJT Characteristics and Parameters

7. What is the α_{DC} when $I_C = 8.23$ mA and $I_E = 8.69$ mA?
8. A certain transistor has an $I_C = 25$ mA and an $I_B = 200\ \mu$A. Determine the β_{DC}.
9. What is the β_{DC} of a transistor if $I_C = 20.3$ mA and $I_E = 20.5$ mA?
10. What is the α_{DC} if $I_C = 5.35$ mA and $I_B = 50\ \mu$A?
11. A certain transistor exhibits an α_{DC} of 0.96. Determine I_C when $I_E = 9.35$ mA.

12. A base current of 50 μA is applied to the transistor in Figure 4–53, and a voltage of 5 V is dropped across R_C. Determine the β_{DC} of the transistor.

▶ FIGURE 4–53

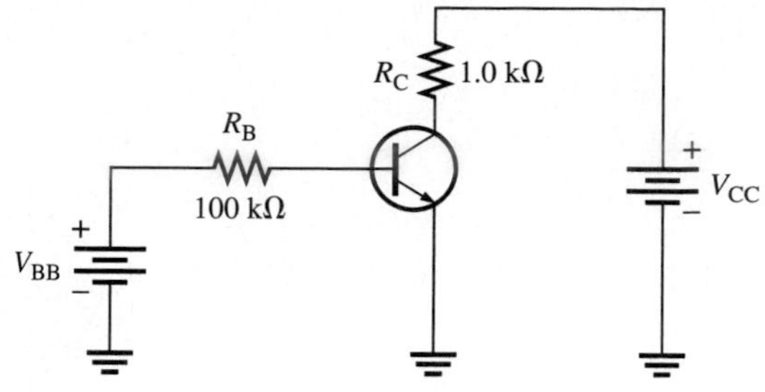

13. Calculate α_{DC} for the transistor in Problem 12.

14. Assume that the transistor in the circuit of Figure 4–53 is replaced with one having a β_{DC} of 200. Determine I_B, I_C, I_E, and V_{CE} given that $V_{CC} = 10$ V and $V_{BB} = 3$ V.

15. If V_{CC} is increased to 15 V in Figure 4–53, how much do the currents and V_{CE} change?

16. Determine each current in Figure 4–54. What is the β_{DC}?

▶ FIGURE 4–54

Multisim file circuits are identified with a CD logo and are in the Problems folder on your CD-ROM. Filenames correspond to figure numbers (e.g., F04-54).

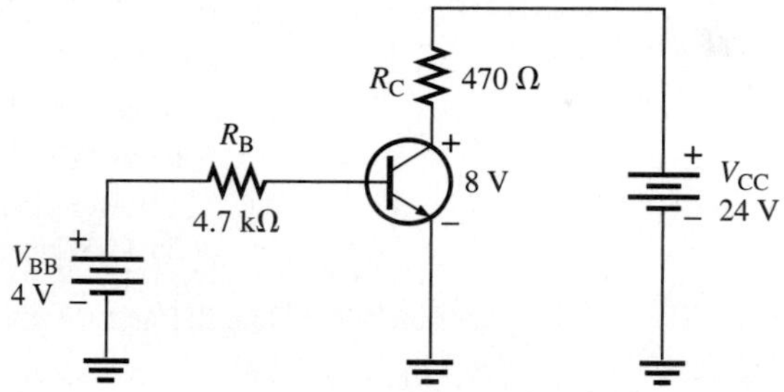

17. Find V_{CE}, V_{BE}, and V_{CB} in both circuits of Figure 4–55.

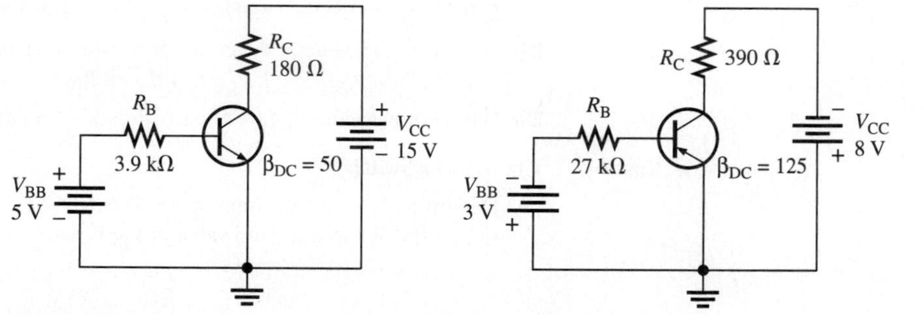

(a) (b)

▲ FIGURE 4–55

18. Determine whether or not the transistors in Figure 4–55 are saturated.

19. Find I_B, I_E, and I_C in Figure 4–56. $\alpha_{DC} = 0.98$.

▶ FIGURE 4–56

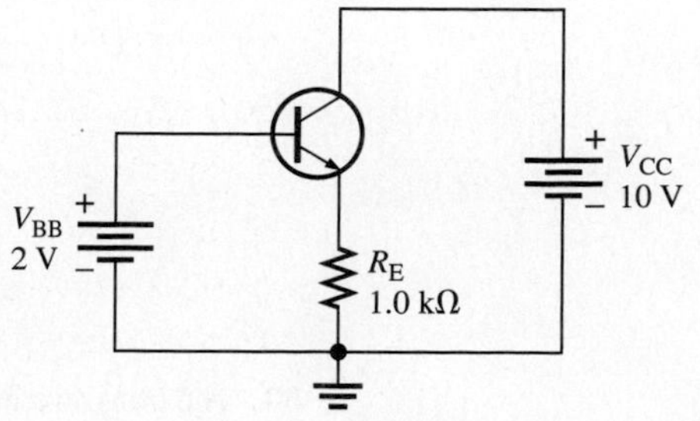

20. Determine the terminal voltages of each transistor with respect to ground for each circuit in Figure 4–57. Also determine V_{CE}, V_{BE}, and V_{CB}.

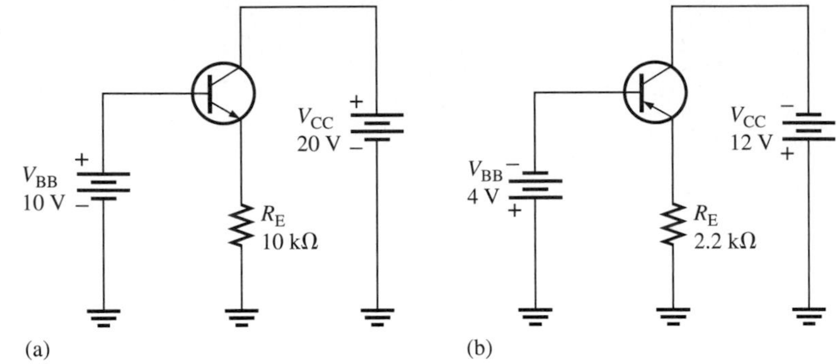

▲ FIGURE 4–57

21. If the β_{DC} in Figure 4–57(a) changes from 100 to 150 due to a temperature increase, what is the change in collector current?

22. A certain transistor is to be operated at a collector current of 50 mA. How high can V_{CE} go without exceeding a $P_{D(max)}$ of 1.2 W?

23. The power dissipation derating factor for a certain transistor is 1 mW/°C. The $P_{D(max)}$ is 0.5 W at 25°C. What is $P_{D(max)}$ at 100°C?

Section 4–4 The BJT as an Amplifier

24. A transistor amplifier has a voltage gain of 50. What is the output voltage when the input voltage is 100 mV?

25. To achieve an output of 10 V with an input of 300 mV, what voltage gain is required?

26. A 50 mV signal is applied to the base of a properly biased transistor with $r'_e = 10\ \Omega$ and $R_C = 560\ \Omega$. Determine the signal voltage at the collector.

27. Determine the value of the collector resistor in an *npn* transistor amplifier with $\beta_{DC} = 250$, $V_{BB} = 2.5$ V, $V_{CC} = 9$ V, $V_{CE} = 4$ V, and $R_B = 100$ kΩ.

28. What is the dc current gain of each circuit in Figure 4–55?

Section 4–5 The BJT as a Switch

29. Determine $I_{C(sat)}$ for the transistor in Figure 4–58. What is the value of I_B necessary to produce saturation? What minimum value of V_{IN} is necessary for saturation? Assume $V_{CE(sat)} = 0$ V.

▶ FIGURE 4–58

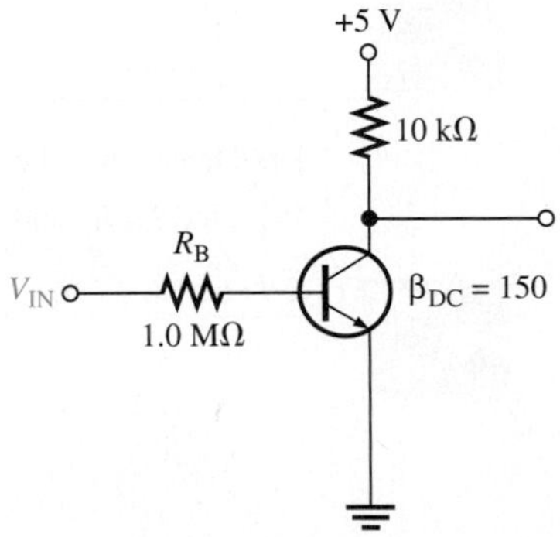

30. The transistor in Figure 4–59 has a β_{DC} of 50. Determine the value of R_B required to ensure saturation when V_{IN} is 5 V. What must V_{IN} be to cut off the transistor? Assume $V_{CE(sat)} = 0$ V.

FIGURE 4–59

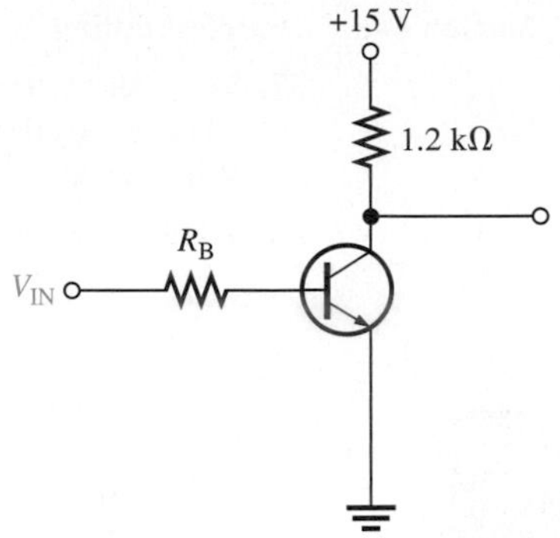

Section 4–6 The Phototransistor

31. A certain phototransistor in a circuit has a $\beta_{DC} = 200$. If $I_\lambda = 100\ \mu A$, what is the collector current?

32. Determine the emitter current in the phototransistor circuit in Figure 4–60 if, for each lm/m² of light intensity, 1 μA of base current is produced in the phototransistor.

FIGURE 4–60

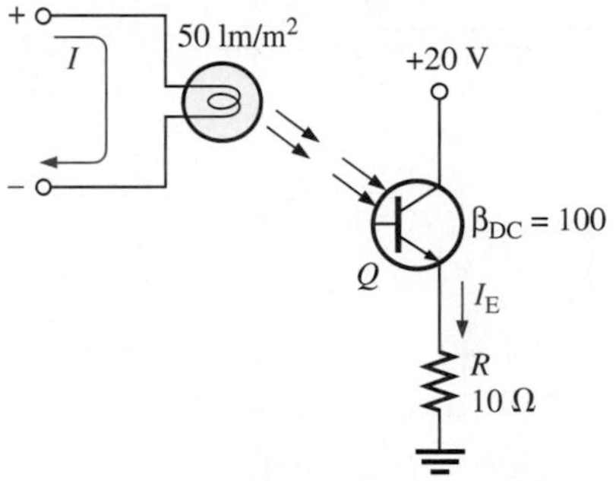

33. A particular optical coupler has a current transfer ratio of 30 percent. If the input current is 100 mA, what is the output current?

34. The optical coupler shown in Figure 4–61 is required to deliver at least 10 mA to the external load. If the current transfer ratio is 60 percent, how much current must be supplied to the input?

FIGURE 4–61

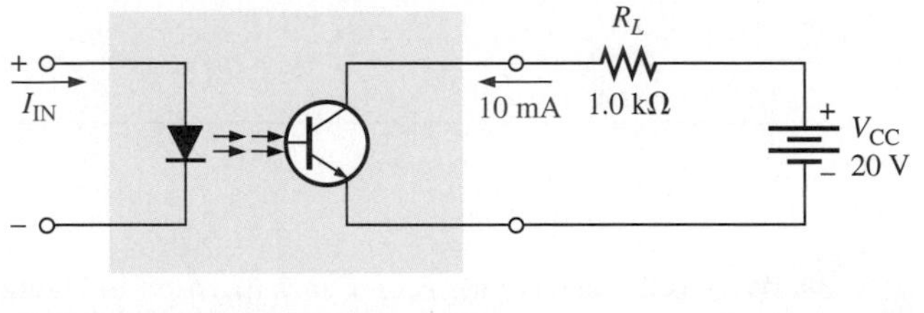

Section 4–7 Transistor Categories and Packaging

35. Identify the leads on the transistors in Figure 4–62. Bottom views are shown.

FIGURE 4–62

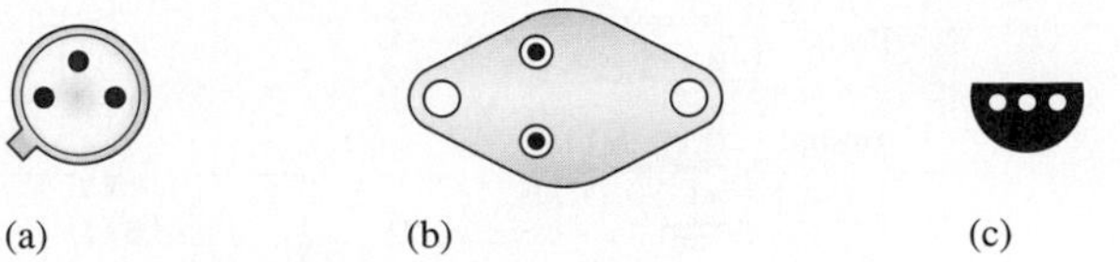

36. What is the most probable category of each transistor in Figure 4–63?

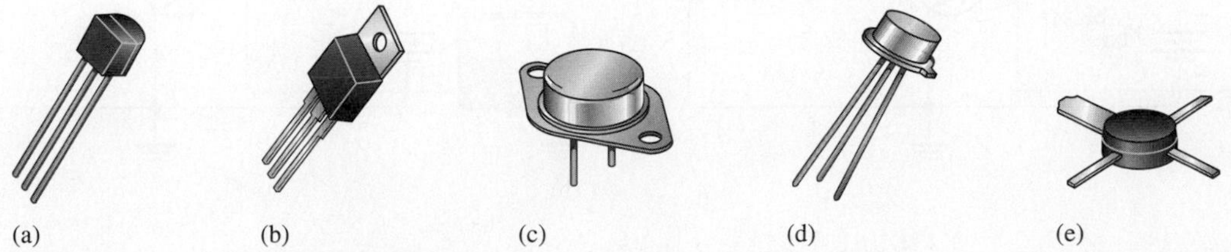

FIGURE 4–63

Section 4–8 Troubleshooting

37. In an out-of-circuit test of a good *npn* transistor, what should an analog ohmmeter indicate when its positive probe is touching the emitter and the negative probe is touching the base? When its positive probe is touching the base and the negative probe is touching the collector?

38. What is the most likely problem, if any, in each circuit of Figure 4–64? Assume a β_{DC} of 75.

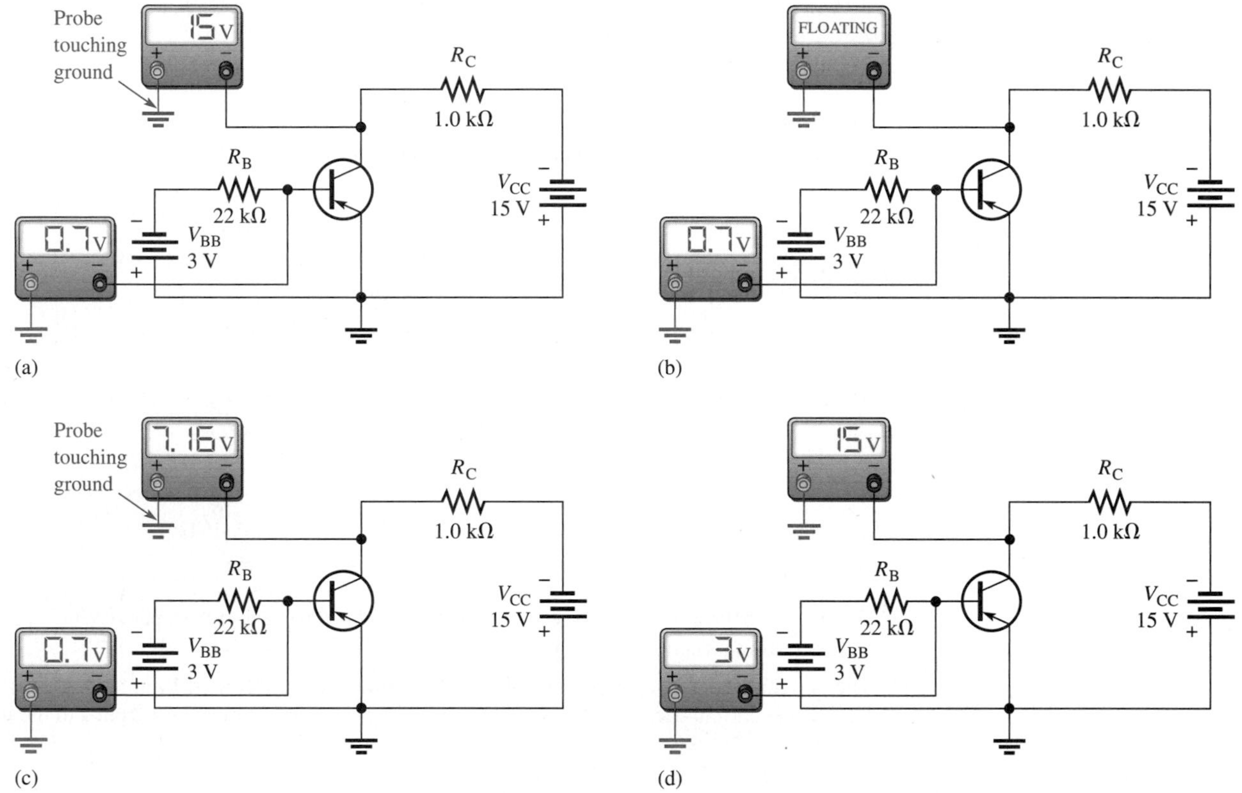

FIGURE 4–64

39. What is the value of the β_{DC} of each transistor in Figure 4–65?

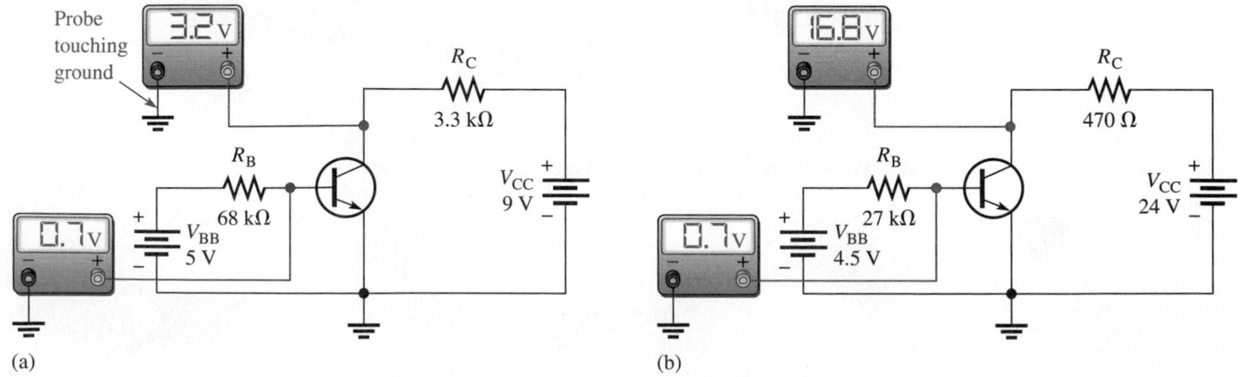

FIGURE 4–65

APPLICATION ACTIVITY PROBLEMS

40. Calculate the power dissipation in each resistor in Figure 4–51 for both states of the circuit.

41. Determine the minimum value of load resistance that Q_2 can drive without exceeding the maximum collector current specified on the datasheet.

42. Develop a wiring diagram for the printed circuit board in Figure 4–52 for connecting it in the security alarm system. The input/output pins are numbered from 1 to 10 starting at the top.

DATASHEET PROBLEMS

43. Refer to the partial transistor datasheet in Figure 4–20.

(a) What is the maximum collector-to-emitter voltage for a 2N3904?

(b) How much continuous collector current can the 2N3904 handle?

(c) How much power can a 2N3904 dissipate if the ambient temperature is 25°C?

(d) How much power can a 2N3904 dissipate if the ambient temperature is 50°C?

(e) What is the minimum h_{FE} of a 2N3904 if the collector current is 1 mA?

44. Refer to the transistor datasheet in Figure 4–20. A MMBT3904 is operating in an environment where the ambient temperature is 65°C. What is the most power that it can dissipate?

45. Refer to the transistor datasheet in Figure 4–20. A PZT3904 is operating with an ambient temperature of 45°C. What is the most power that it can dissipate?

46. Refer to the transistor datasheet in Figure 4–20. Determine if any rating is exceeded in each circuit of Figure 4–66 based on minimum specified values.

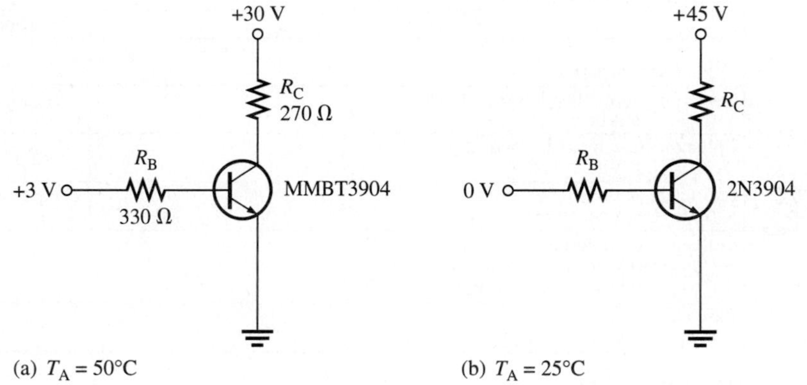

▲ FIGURE 4–66

47. Refer to the transistor datasheet in Figure 4–20. Determine whether or not the transistor is saturated in each circuit of Figure 4–67 based on the maximum specified value of h_{FE}.

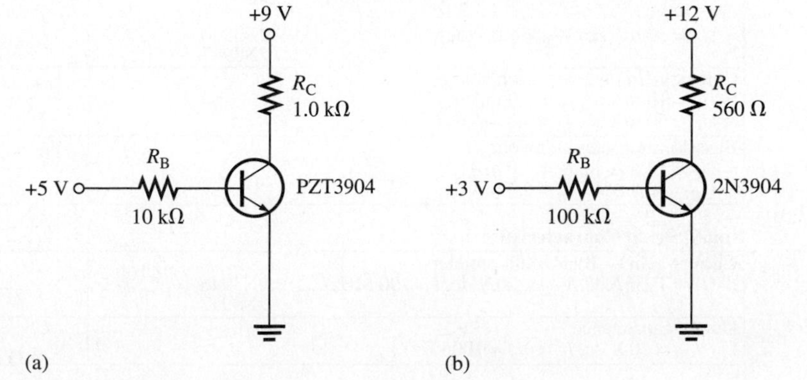

▲ FIGURE 4–67

48. Refer to the partial transistor datasheet in Figure 4–68. Determine the minimum and maximum base currents required to produce a collector current of 10 mA in a 2N3946. Assume that the transistor is not in saturation and $V_{CE} = 1$ V.

2N3946
2N3947

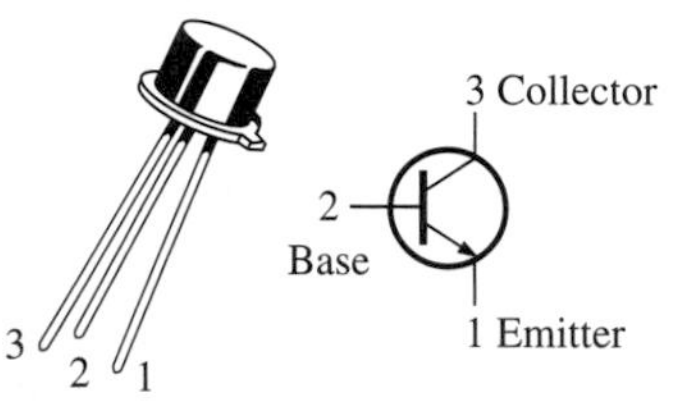

General-Purpose Transistors

NPN Silicon

Maximum Ratings

Rating	Symbol	Value	Unit
Collector-Emitter voltage	V_{CEO}	40	V dc
Collector-Base voltage	V_{CBO}	60	V dc
Emitter-Base voltage	V_{EBO}	6.0	V dc
Collector current — continuous	I_C	200	mA dc
Total device dissipation @ $T_A = 25°C$ Derate above 25°C	P_D	0.36 2.06	Watts mW/°C
Total device dissipation @ $T_C = 25°C$ Derate above 25°C	P_D	1.2 6.9	Watts mW/°C
Operating and storage junction Temperature range	T_J, T_{stg}	−65 to +200	°C

Thermal Characteristics

Characteristic	Symbol	Max	Unit
Thermal resistance, junction to case	$R_{\theta JC}$	0.15	°C/mW
Thermal resistance, junction to ambient	$R_{\theta JA}$	0.49	°C/mW

Electrical Characteristics ($T_A = 25°C$ unless otherwise noted.)

Characteristic		Symbol	Min	Max	Unit
OFF Characteristics					
Collector-Emitter breakdown voltage ($I_C = 10$ mA dc)		$V_{(BR)CEO}$	40	–	V dc
Collector-Base breakdown voltage ($I_C = 10\ \mu$A dc, $I_E = 0$)		$V_{(BR)CBO}$	60	–	V dc
Emitter-Base breakdown voltage ($I_E = 10\ \mu$A dc, $I_C = 0$)		$V_{(BR)EBO}$	6.0	–	V dc
Collector cutoff current ($V_{CE} = 40$ V dc, $V_{OB} = 3.0$ V dc) ($V_{CE} = 40$ V dc, $V_{OB} = 3.0$ V dc, $T_A = 150°C$)		I_{CEX}	– –	0.010 15	μA dc
Base cutoff current ($V_{CE} = 40$ V dc, $V_{OB} = 3.0$ V dc)		I_{BL}	–	.025	μA dc
ON Characteristics					
DC current gain ($I_C = 0.1$ mA dc, $V_{CE} = 1.0$ V dc)	2N3946 2N3947	h_{FE}	30 60	– –	–
($I_C = 1.0$ mA dc, $V_{CE} = 1.0$ V dc)	2N3946 2N3947		45 90	– –	
($I_C = 10$ mA dc, $V_{CE} = 1.0$ V dc)	2N3946 2N3947		50 100	150 300	
($I_C = 50$ mA dc, $V_{CE} = 1.0$ V dc)	2N3946 2N3947		20 40	– –	
Collector-Emitter saturation voltage ($I_C = 10$ mA dc, $I_B = 1.0$ mA dc) ($I_C = 50$ mA dc, $I_B = 5.0$ mA dc)		$V_{CE(sat)}$	– –	0.2 0.3	V dc
Base-Emitter saturation voltage ($I_C = 10$ mA dc, $I_B = 1.0$ mA dc) ($I_C = 50$ mA dc, $I_B = 5.0$ mA dc)		$V_{BE(sat)}$	0.6 –	0.9 1.0	V dc
Small-Signal Characteristics					
Current gain — Bandwidth product ($I_C = 10$ mA dc, $V_{CE} = 20$ V dc, $f = 100$ MHz)	2N3946 2N3947	f_T	250 300	– –	MHz
Output capacitance ($V_{CB} = 10$ V dc, $I_E = 0$, $f = 100$ kHz)		C_{obo}	–	4.0	pF

▲ FIGURE 4–68

49. For each of the circuits in Figure 4–69, determine if there is a problem based on the datasheet information in Figure 4–68. Use the maximum specified h_{FE}.

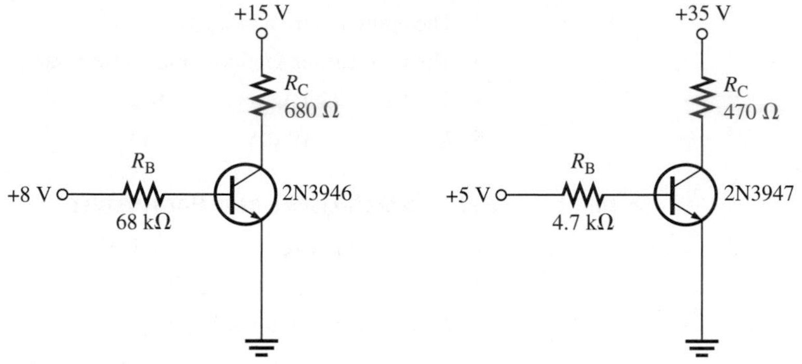

(a) $T_A = 40°C$ (b) $T_A = 25°C$

▲ **FIGURE 4–69**

ADVANCED PROBLEMS

50. Derive a formula for α_{DC} in terms of β_{DC}.

51. A certain 2N3904 dc bias circuit with the following values is in saturation. $I_B = 500\ \mu A$, $V_{CC} = 10\ V$, and $R_C = 180\ \Omega$, $h_{FE} = 150$. If you increase V_{CC} to 15 V, does the transistor come out of saturation? If so, what is the collector-to-emitter voltage and the collector current?

52. Design a dc bias circuit for a 2N3904 operating from a collector supply voltage of 9 V and a base-bias voltage of 3 V that will supply 150 mA to a resistive load that acts as the collector resistor. The circuit must not be in saturation. Assume the minimum specified β_{DC} from the datasheet.

53. Modify the design in Problem 52 to use a single 9 V dc source rather than two different sources. Other requirements remain the same.

54. Design a dc bias circuit for an amplifier in which the voltage gain is to be a minimum of 50 and the output signal voltage is to be "riding" on a dc level of 5 V. The maximum input signal voltage at the base is 10 mV rms. $V_{CC} = 12\ V$, and $V_{BB} = 4\ V$. Assume $r'_e = 8\ \Omega$.

MULTISIM TROUBLESHOOTING PROBLEMS

These file circuits are in the Troubleshooting Problems folder on your CD-ROM.

55. Open file TSP04-55 and determine the fault.

56. Open file TSP04-56 and determine the fault.

57. Open file TSP04-57 and determine the fault.

58. Open file TSP04-58 and determine the fault.

59. Open file TSP04-59 and determine the fault.

60. Open file TSP04-60 and determine the fault.

61. Open file TSP04-61 and determine the fault.

62. Open file TSP04-62 and determine the fault.

ANSWERS

SECTION CHECKUPS

Section 4–1 **BJT Structure**

1. The two types of BJTs are *npn* and *pnp*.

2. The terminals of a BJT are base, collector, and emitter.

3. The three regions of a BJT are separated by two *pn* junctions.

Section 4–2 Basic BJT Operation

1. To operate as an amplifier, the base-emitter is forward-biased and the base-collector is reverse-biased.
2. The emitter current is the largest.
3. The base current is much smaller than the emitter current.
4. The base region is very narrow compared to the other two regions.
5. $I_E = 1\text{ mA} + 10\ \mu\text{A} = 1.01\text{ mA}$

Section 4–3 BJT Characteristics and Parameters

1. $\beta_{DC} = I_C/I_B$; $\alpha_{DC} = I_C/I_E$; h_{FE} is β_{DC}.
2. $\beta_{DC} = 100$; $\alpha_{DC} = 100/(100 + 1) = 0.99$
3. I_C is plotted versus V_{CE}.
4. Forward-reverse bias is required for amplifier operation.
5. β_{DC} increases with temperature.
6. No. β_{DC} generally varies some from one device to the next for a given type.

Section 4–4 The BJT as an Amplifier

1. Amplification is the process where a smaller signal is used to produce a larger identical signal.
2. Voltage gain is the ratio of output voltage to input voltage.
3. R_C and r'_e determine the voltage gain.
4. $A_v = 5\text{ V}/250\text{ mV} = 20$
5. $A_v = 1200\ \Omega/20\ \Omega = 60$

Section 4–5 The BJT as a Switch

1. A transistor switch operates in cutoff and saturation.
2. The collector current is maximum in saturation.
3. The collector current is approximately zero in cutoff.
4. $V_{CE} = V_{CC}$ in cutoff.
5. V_{CE} is minimum in saturation.

Section 4–6 The Phototransistor

1. The base current of a phototransistor is light induced.
2. Base
3. The collector current depends on β_{DC} and I_λ.
4. Current transfer ratio

Section 4–7 Transistor Categories and Packaging

1. Three categories of BJTs are small signal/general purpose, power, and RF.
2. Emitter is the lead closest to the tab.
3. The metal mounting tab or case in power transistors is the collector.

Section 4–8 Troubleshooting

1. First, test it in-circuit.
2. If R_B opens, the transistor is in cutoff.
3. The base voltage is +3 V and the collector voltage is +9 V.

RELATED PROBLEMS FOR EXAMPLES

4–1 10 mA

4–2 $I_B = 241\ \mu\text{A}$; $I_C = 21.7\text{ mA}$; $I_E = 21.94\text{ mA}$; $V_{CE} = 4.23\text{ V}$; $V_{CB} = 3.53\text{ V}$

4–3 Along the horizontal axis

4–4 Not saturated

4–5 10 V

4–6. $V_{CC(max)} = 44.5$ V; $V_{CE(max)}$ is exceeded first.

4–7 4.55 W

4–8 $P_{D(max)} = 500$ mW @ 50°C

4–9 2.5 kΩ

4–10 78.4 μA

4–11 Reduce R_C to 140 Ω and R_B to 2.2 kΩ.

4–12 R_B open

TRUE/FALSE QUIZ

1. T	**2.** F	**3.** T	**4.** T	**5.** F	**6.** F
7. F	**8.** T	**9.** F	**10.** T	**11.** F	**12.** T

CIRCUIT-ACTION QUIZ

1. (a)	**2.** (a)	**3.** (c)	**4.** (b)	**5.** (c)	**6.** (b)
7. (c)	**8.** (a)	**9.** (b)	**10.** (a)	**11.** (a)	**12.** (a)

SELF-TEST

1. (d)	**2.** (c)	**3.** (a)	**4.** (d)	**5.** (a)	**6.** (c)	**7.** (b)
8. (a)	**9.** (c)	**10.** (c)	**11.** (b)	**12.** (b)	**13.** (f)	**14.** (c)
15. (b)	**16.** (b)	**17.** (b)	**18.** (a)	**19.** (d)	**20.** (a)	**21.** (c)

5 Transistor Bias Circuits

CHAPTER OUTLINE

5–1 The DC Operating Point
5–2 Voltage-Divider Bias
5–3 Other Bias Methods
5–4 Troubleshooting
Application Activity

CHAPTER OBJECTIVES

- Discuss the concept of dc bias in a linear amplifier
- Analyze a voltage-divider bias circuit
- Analyze an emitter bias circuit, a base bias circuit, an emitter-feedback bias circuit, and a collector-feedback bias circuit
- Troubleshoot various faults in transistor bias circuits

KEY TERMS

- Q-point
- DC load line
- Linear region
- Stiff voltage divider
- Feedback

VISIT THE COMPANION WEBSITE

Study aids for this chapter are available at
http://www.prenhall.com/floyd

INTRODUCTION

As you learned in Chapter 4, a transistor must be properly biased in order to operate as an amplifier. DC biasing is used to establish fixed dc values for the transistor currents and voltages called the *dc operating point* or *quiescent point* (*Q-point*). In this chapter, several types of bias circuits are discussed. This material lays the groundwork for the study of amplifiers, and other circuits that require proper biasing.

APPLICATION ACTIVITY PREVIEW

The Application Activity focuses on a system for controlling temperature in an industrial chemical process. You will be dealing with a circuit that converts a temperature measurement to a proportional voltage that is used to adjust the temperature of a liquid in a storage tank. The first step is to learn all you can about transistor operation. You will then apply your knowledge to the Application Activity at the end of the chapter.

5–1 THE DC OPERATING POINT

A transistor must be properly biased with a dc voltage in order to operate as a linear amplifier. A dc operating point must be set so that signal variations at the input terminal are amplified and accurately reproduced at the output terminal. As you learned in Chapter 4, when you bias a transistor, you establish the dc voltage and current values. This means, for example, that at the dc operating point, I_C and V_{CE} have specified values. The dc operating point is often referred to as the Q-point (quiescent point).

After completing this section, you should be able to

- **Discuss the concept of dc bias in a linear amplifier**
  - Describe how to generate collector characteristic curves for a biased transistor
  - Draw a dc load line for a given biased transistor circuit
  - Explain Q-point
  - Explain the conditions for linear operation
  - Explain the conditions for saturation and cutoff
  - Discuss the reasons for output waveform distortion

DC Bias

Bias establishes the dc operating point (**Q-point**) for proper linear operation of an amplifier. If an amplifier is not biased with correct dc voltages on the input and output, it can go into saturation or cutoff when an input signal is applied. Figure 5–1 shows the effects of proper and improper dc biasing of an inverting amplifier. In part (a), the output signal is an amplified replica of the input signal except that it is inverted, which means that it is 180° out of phase with the input. The output signal swings equally above and below the dc bias level of the output, $V_{DC(out)}$. Improper biasing can cause distortion in the output signal, as illustrated in parts (b) and (c). Part (b) illustrates limiting of the positive portion of the output voltage as a result of a Q-point (dc operating point) being too close to cutoff. Part (c) shows limiting of the negative portion of the output voltage as a result of a dc operating point being too close to saturation.

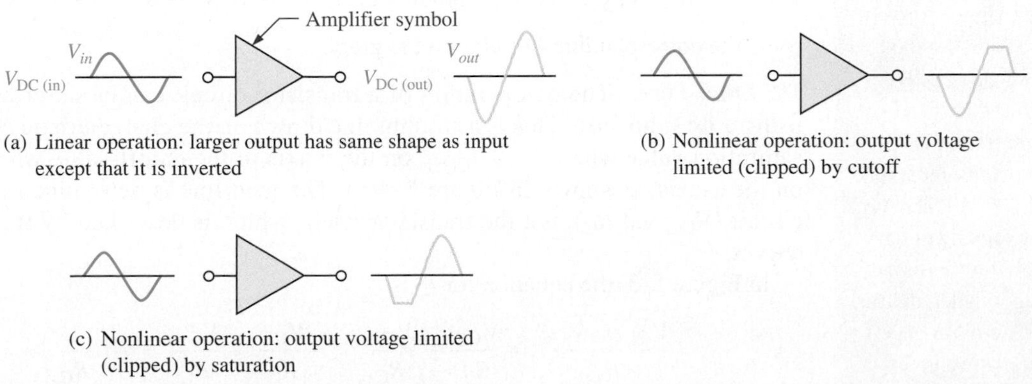

(a) Linear operation: larger output has same shape as input except that it is inverted

(b) Nonlinear operation: output voltage limited (clipped) by cutoff

(c) Nonlinear operation: output voltage limited (clipped) by saturation

▲ FIGURE 5–1

Examples of linear and nonlinear operation of an inverting amplifier (the triangle symbol).

Graphical Analysis The transistor in Figure 5–2(a) is biased with V_{CC} and V_{BB} to obtain certain values of I_B, I_C, I_E, and V_{CE}. The collector characteristic curves for this particular transistor are shown in Figure 5–2(b); we will use these curves to graphically illustrate the effects of dc bias.

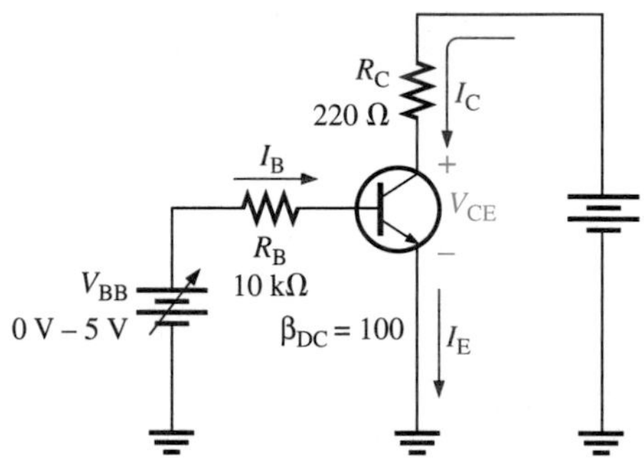

(a) DC biased circuit

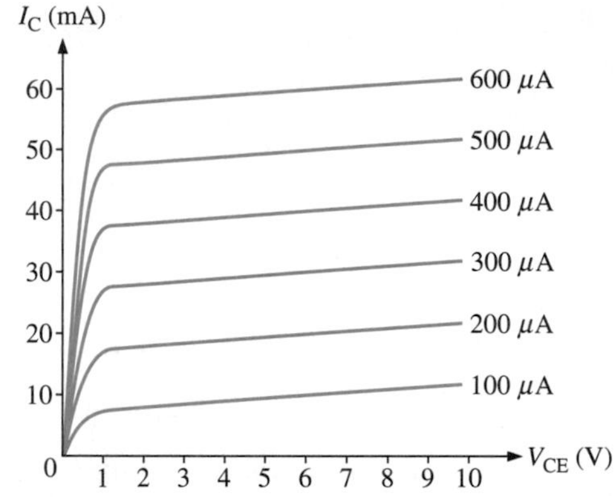

(b) Collector characteristic curves

▲ **FIGURE 5–2**

A dc-biased transistor circuit with variable bias voltage (V_{BB}) for generating the collector characteristic curves shown in part (b).

In Figure 5–3, we assign three values to I_B and observe what happens to I_C and V_{CE}. First, V_{BB} is adjusted to produce an I_B of 200 μA, as shown in Figure 5–3(a). Since $I_C = \beta_{DC}I_B$, the collector current is 20 mA, as indicated, and

$$V_{CE} = V_{CC} - I_CR_C = 10\text{ V} - (20\text{ mA})(220\ \Omega) = 10\text{ V} - 4.4\text{ V} = 5.6\text{ V}$$

This Q-point is shown on the graph of Figure 5–3(a) as Q_1.

Next, as shown in Figure 5–3(b), V_{BB} is increased to produce an I_B of 300 μA and an I_C of 30 mA.

$$V_{CE} = 10\text{ V} - (30\text{ mA})(220\ \Omega) = 10\text{ V} - 6.6\text{ V} = 3.4\text{ V}$$

The Q-point for this condition is indicated by Q_2 on the graph.

Finally, as in Figure 5–3(c), V_{BB} is increased to give an I_B of 400 μA and an I_C of 40 mA.

$$V_{CE} = 10\text{ V} - (40\text{ mA})(220\ \Omega) = 10\text{ V} - 8.8\text{ V} = 1.2\text{ V}$$

Q_3 is the corresponding Q-point on the graph.

DC Load Line The dc operation of a transistor circuit can be described graphically using a **dc load line**. This is a straight line drawn on the characteristic curves from the saturation value where $I_C = I_{C(sat)}$ on the y-axis to the cutoff value where $V_{CE} = V_{CC}$ on the x-axis, as shown in Figure 5–4(a). The load line is determined by the external circuit (V_{CC} and R_C), not the transistor itself, which is described by the characteristic curves.

In Figure 5–3, the equation for I_C is

$$I_C = \frac{V_{CC} - V_{CE}}{R_C} = \frac{V_{CC}}{R_C} - \frac{V_{CE}}{R_C} = -\frac{V_{CE}}{R_C} + \frac{V_{CC}}{R_C} = -\left(\frac{1}{R_C}\right)V_{CE} + \frac{V_{CC}}{R_C}$$

This is the equation of a straight line with a slope of $-1/R_C$, an x intercept of $V_{CE} = V_{CC}$, and a y intercept of V_{CC}/R_C, which is $I_{C(sat)}$.

TECH NOTE

In 1965, a single transistor cost more than a dollar. By 1975, the cost of a transistor had dropped to less than a penny, while transistor size allowed for almost 100,000 transistors on a single chip. From 1979 to 1999, processor performance went from about 1.5 million instructions per second (MIPS) to over 1,000 MIPS. Today's processors, some topping out at well above one billion transistors, run at 3.2 GHz and higher, deliver over 10,000 MIPS, and can be manufactured in high volumes with transistors that cost less than 1/10,000th of a cent.

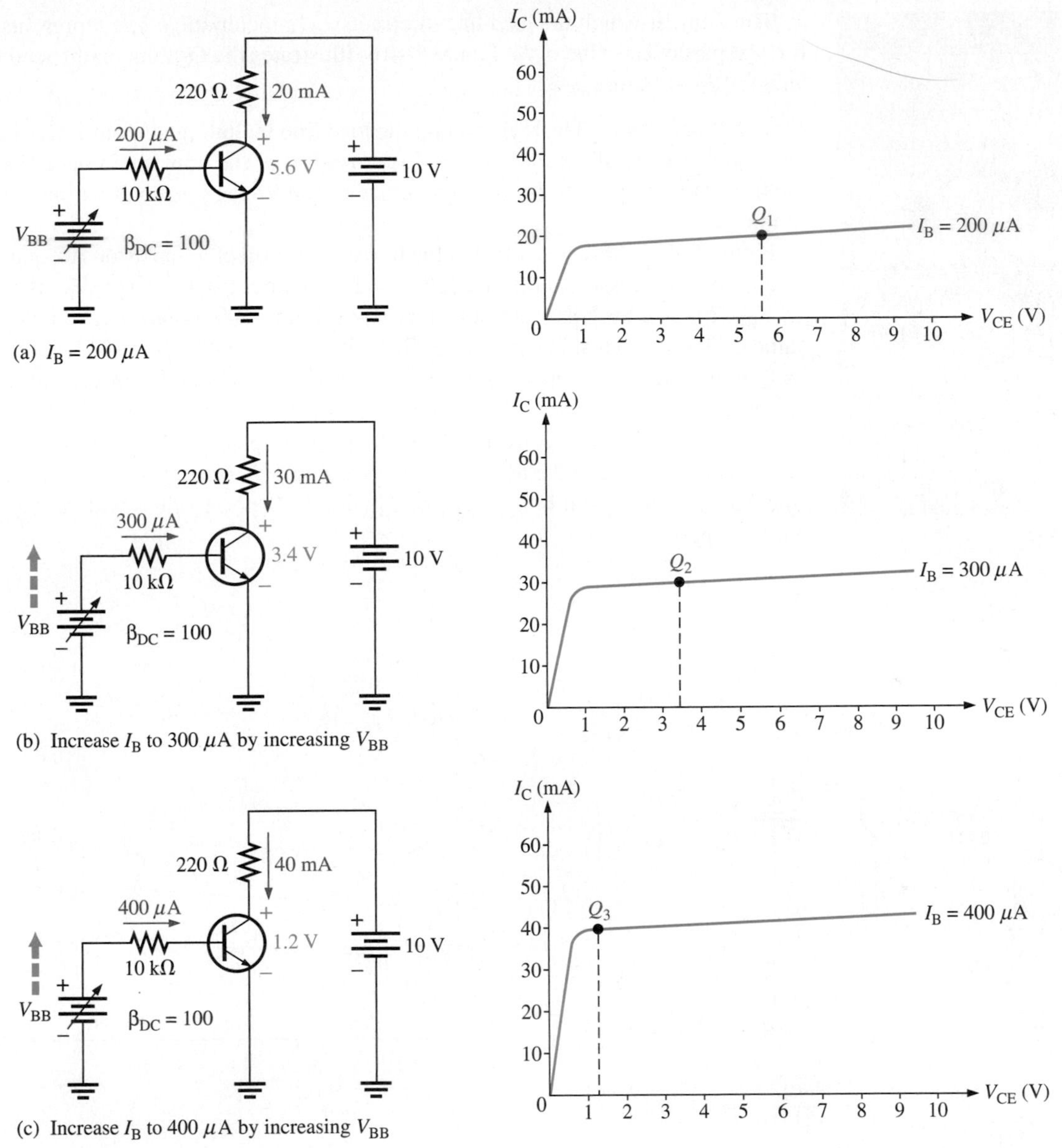

▲ **FIGURE 5–3**

Illustration of Q-point adjustment.

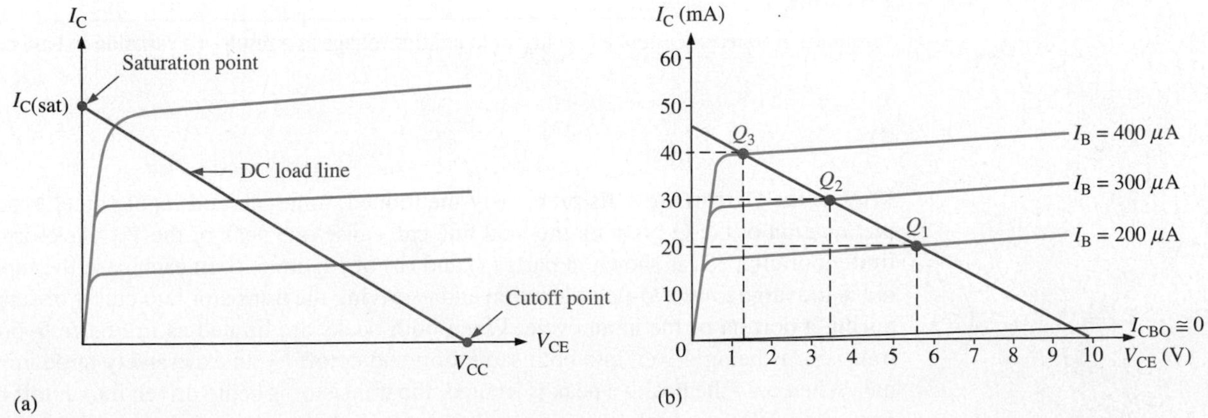

▲ **FIGURE 5–4**

The dc load line.

> **TECH NOTE**
>
> Gordon Moore, one of the founders of Intel, observed in an article in the April, 1965, issue of *Electronics* magazine that innovations in technology would allow a doubling of the number of transistors in a given space every year (in an update article in 1975, Moore adjusted the rate to every two years to account for the growing complexity of chips), and that the speed of those transistors would increase. This prediction has become widely known as Moore's law.

The point at which the load line intersects a characteristic curve represents the Q-point for that particular value of I_B. Figure 5–4(b) illustrates the Q-point on the load line for each value of I_B in Figure 5–3.

Linear Operation The region along the load line including all points between saturation and cutoff is generally known as the **linear region** of the transistor's operation. As long as the transistor is operated in this region, the output voltage is ideally a linear reproduction of the input.

Figure 5–5 shows an example of the linear operation of a transistor. AC quantities are indicated by lowercase italic subscripts. Assume a sinusoidal voltage, V_{in}, is superimposed on V_{BB}, causing the base current to vary sinusoidally 100 μA above and below its Q-point value of 300 μA. This, in turn, causes the collector current to vary 10 mA above and below its Q-point value of 30 mA. As a result of the variation in collector current, the collector-to-emitter voltage varies 2.2 V above and below its Q-point value of 3.4 V. Point *A* on the load line in Figure 5–5 corresponds to the positive peak of the sinusoidal input voltage. Point *B* corresponds to the negative peak, and point *Q* corresponds to the zero value of the sine wave, as indicated. V_{CEQ}, I_{CQ}, and I_{BQ} are dc Q-point values with no input sinusoidal voltage applied.

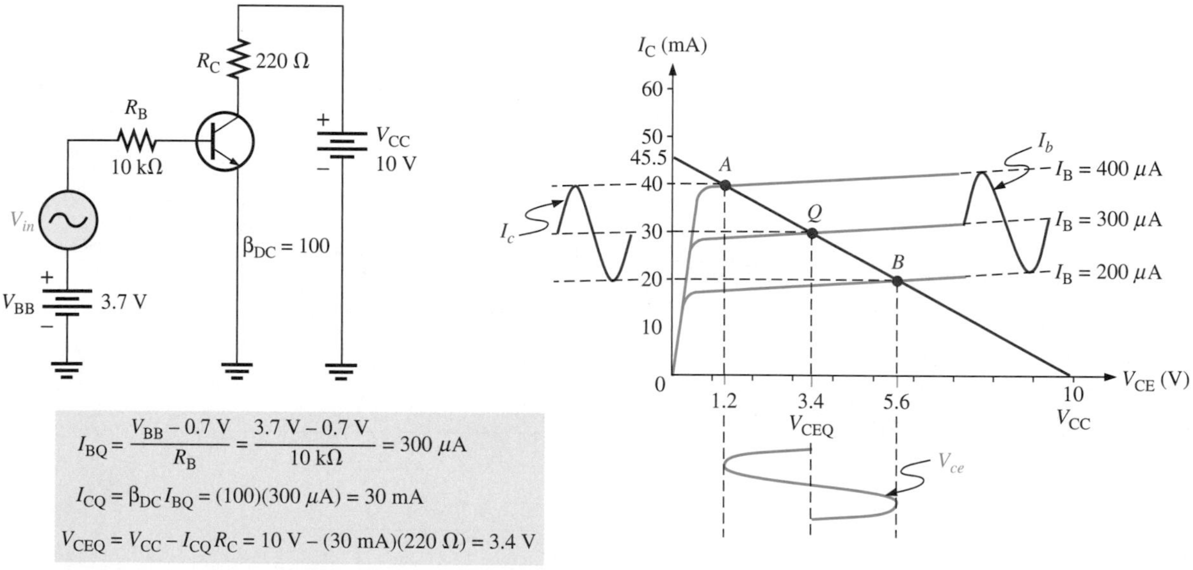

$$I_{BQ} = \frac{V_{BB} - 0.7\text{ V}}{R_B} = \frac{3.7\text{ V} - 0.7\text{ V}}{10\text{ k}\Omega} = 300\ \mu\text{A}$$

$$I_{CQ} = \beta_{DC} I_{BQ} = (100)(300\ \mu\text{A}) = 30\text{ mA}$$

$$V_{CEQ} = V_{CC} - I_{CQ}R_C = 10\text{ V} - (30\text{ mA})(220\ \Omega) = 3.4\text{ V}$$

▲ **FIGURE 5–5**

Variations in collector current and collector-to-emitter voltage as a result of a variation in base current.

Waveform Distortion As previously mentioned, under certain input signal conditions the location of the Q-point on the load line can cause one peak of the V_{ce} waveform to be limited or clipped, as shown in parts (a) and (b) of Figure 5–6. In each case the input signal is too large for the Q-point location and is driving the transistor into cutoff or saturation during a portion of the input cycle. When both peaks are limited as in Figure 5–6(c), the transistor is being driven into both saturation and cutoff by an excessively large input signal. When only the positive peak is limited, the transistor is being driven into cutoff but not saturation. When only the negative peak is limited, the transistor is being driven into saturation but not cutoff.

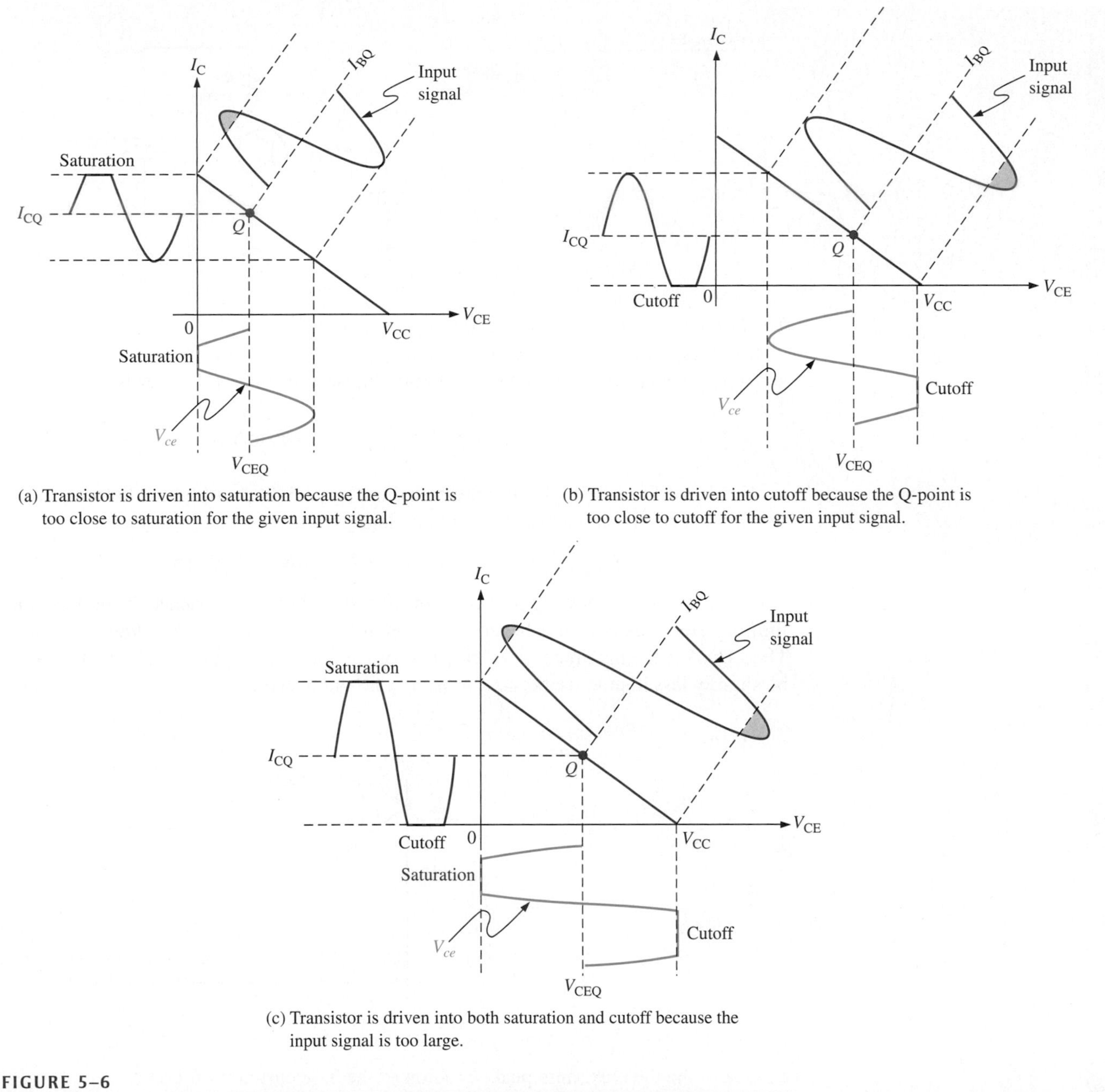

▲ **FIGURE 5–6**

Graphical load line illustration of a transistor being driven into saturation and/or cutoff.

EXAMPLE 5–1

Determine the Q-point for the circuit in Figure 5–7 and draw the dc load line. Find the maximum peak value of base current for linear operation. Assume $\beta_{DC} = 200$.

Solution The Q-point is defined by the values of I_C and V_{CE}.

$$I_B = \frac{V_{BB} - V_{BE}}{R_B} = \frac{10\text{ V} - 0.7\text{ V}}{47\text{ k}\Omega} = 198\ \mu\text{A}$$

$$I_C = \beta_{DC} I_B = (200)(198\ \mu\text{A}) = \mathbf{39.6\ mA}$$

$$V_{CE} = V_{CC} - I_C R_C = 20\text{ V} - 13.07\text{ V} = \mathbf{6.93\ V}$$

FIGURE 5–7

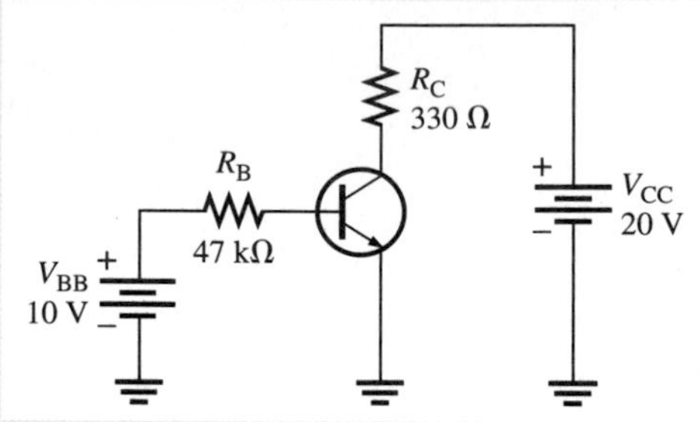

The Q-point is at $I_C = 39.6$ mA and at $V_{CE} = 6.93$ V.

Since $I_{C(cutoff)} = 0$, you need to know $I_{C(sat)}$ to determine how much variation in collector current can occur and still maintain linear operation of the transistor.

$$I_{C(sat)} = \frac{V_{CC}}{R_C} = \frac{20\text{ V}}{330\ \Omega} = 60.6\text{ mA}$$

The dc load line is graphically illustrated in Figure 5–8, showing that before saturation is reached, I_C can increase an amount ideally equal to

$$I_{C(sat)} - I_{CQ} = 60.6\text{ mA} - 39.6\text{ mA} = 21.0\text{ mA}$$

However, I_C can decrease by 39.6 mA before cutoff ($I_C = 0$) is reached. Therefore, the limiting excursion is 21 mA because the *Q-point is closer to saturation than to cutoff.* The 21 mA is the maximum peak variation of the collector current. Actually, it would be slightly less in practice because $V_{CE(sat)}$ is not quite zero.

FIGURE 5–8

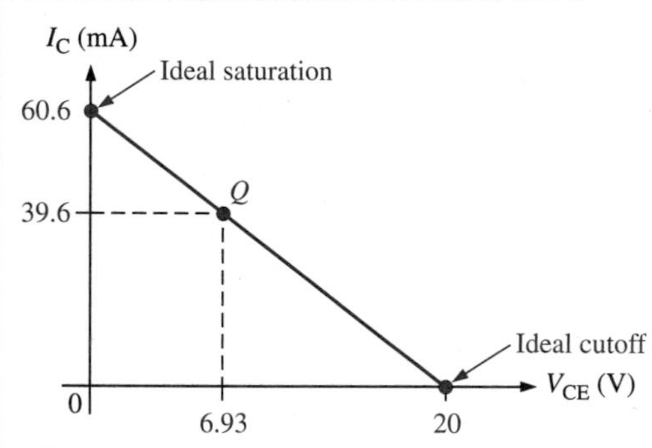

Determine the maximum peak variation of the base current as follows:

$$I_{b(peak)} = \frac{I_{c(peak)}}{\beta_{DC}} = \frac{21\text{ mA}}{200} = \mathbf{105\ \mu A}$$

*Related Problem** Find the Q-point for the circuit in Figure 5–7, and determine the maximum peak value of base current for linear operation for the following circuit values: $\beta_{DC} = 100$, $R_C = 1.0\text{ k}\Omega$, and $V_{CC} = 24$ V.

*Answers are at the end of the chapter.

Open the Multisim file E05-01 in the Examples folder on your CD-ROM. Measure I_C and V_{CE} and compare with the calculated values.

SECTION 5–1 CHECKUP
Answers are at the end of the chapter.

1. What are the upper and lower limits on a dc load line in terms of V_{CE} and I_C?
2. Define *Q-point.*
3. At what point on the load line does saturation occur? At what point does cutoff occur?
4. For maximum V_{ce}, where should the Q-point be placed?

5–2 VOLTAGE-DIVIDER BIAS

You will now study a method of biasing a transistor for linear operation using a single-source resistive voltage divider. This is the most widely used biasing method. Four other methods are covered in Section 5–3.

After completing this section, you should be able to

- **Analyze a voltage-divider bias circuit**
  - Discuss the effect of the input resistance on the bias circuit
  - Discuss the loading effects in voltage-divider bias
  - Explain how to minimize or essentially eliminate the effects of β_{DC} and V_{BE} on the stability of the Q-point
  - Discuss voltage-divider bias for a *pnp* transistor

Up to this point a separate dc source, V_{BB}, was used to bias the base-emitter junction because it could be varied independently of V_{CC} and it helped to illustrate transistor operation. A more practical bias method is to use V_{CC} as the single bias source, as shown in Figure 5–9. To simplify the schematic, the battery symbol is omitted and replaced by a line termination circle with a voltage indicator (V_{CC}) as shown.

A dc bias voltage at the base of the transistor can be developed by a resistive voltage-divider that consists of R_1 and R_2, as shown in Figure 5–9. V_{CC} is the dc collector supply voltage. Two current paths are between point *A* and ground: one through R_2 and the other through the base-emitter junction of the transistor and R_E.

Generally, voltage-divider bias circuits are designed so that the base current is much smaller than the current (I_2) through R_2 in Figure 5–9. In this case, the voltage-divider circuit is very straightforward to analyze because the loading effect of the base current can be ignored. A voltage divider in which the base current is small compared to the current in R_2 is said to be a **stiff voltage divider** because the base voltage is relatively independent of different transistors and temperature effects.

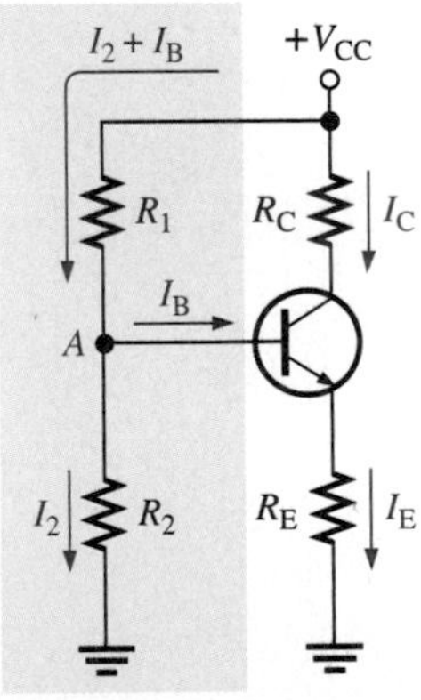

▲ FIGURE 5–9

Voltage-divider bias.

To analyze a voltage-divider circuit in which I_B is small compared to I_2, first calculate the voltage on the base using the unloaded voltage-divider rule:

$$V_B \cong \left(\frac{R_2}{R_1 + R_2}\right)V_{CC} \qquad \text{Equation 5–1}$$

Once you know the base voltage, you can find the voltages and currents in the circuit, as follows:

$$V_E = V_B - V_{BE} \qquad \text{Equation 5–2}$$

and

$$I_C \cong I_E = \frac{V_E}{R_E} \qquad \text{Equation 5–3}$$

Then,

Equation 5–4

$$V_C = V_{CC} - I_C R_C$$

Once you know V_C and V_E, you can determine V_{CE}.

$$V_{CE} = V_C - V_E$$

EXAMPLE 5–2

Determine V_{CE} and I_C in the stiff voltage-divider biased transistor circuit of Figure 5–10 if $\beta_{DC} = 100$.

FIGURE 5–10

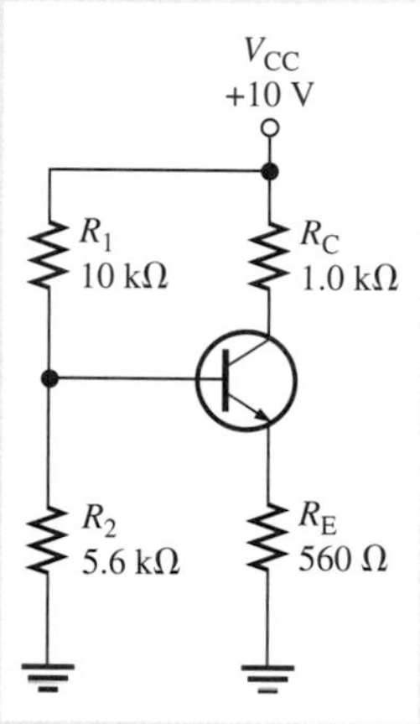

Solution The base voltage is

$$V_B \cong \left(\frac{R_2}{R_1 + R_2}\right)V_{CC} = \left(\frac{5.6\text{ k}\Omega}{15.6\text{ k}\Omega}\right)10\text{ V} = 3.59\text{ V}$$

So,

$$V_E = V_B - V_{BE} = 3.59\text{ V} - 0.7\text{ V} = 2.89\text{ V}$$

and

$$I_E = \frac{V_E}{R_E} = \frac{2.89\text{ V}}{560\ \Omega} = 5.16\text{ mA}$$

Therefore,

$$I_C \cong I_E = \mathbf{5.16\ mA}$$

and

$$V_C = V_{CC} - I_C R_C = 10\text{ V} - (5.16\text{ mA})(1.0\text{ k}\Omega) = 4.84\text{ V}$$

$$V_{CE} = V_C - V_E = 4.84\text{ V} - 2.89\text{ V} = \mathbf{1.95\ V}$$

Related Problem If the voltage divider in Figure 5–10 was not stiff, how would V_B be affected?

Open the Multisim file E05-02 in the Examples folder on your CD-ROM. Measure I_C and V_{CE}. Your results should agree more closely with those in the Related Problem than with those calculated in the example. Can you explain this?

The basic analysis developed in Example 5–2 is all that is needed for most voltage-divider circuits, but there may be cases where you need to analyze the circuit with more accuracy. Ideally, a voltage-divider circuit is stiff, which means that the transistor does not appear as a significant load. All circuit design involves trade-offs; and one trade-off is that stiff voltage dividers require smaller resistors, which are not always desirable because of potential loading effects on other circuits and added power requirements. If the circuit designer wanted to raise the input resistance, the divider string may not be stiff; and more detailed analysis is required to calculate circuit parameters. To determine if the divider is stiff, you need to examine the dc input resistance looking in at the base as shown in Figure 5–11.

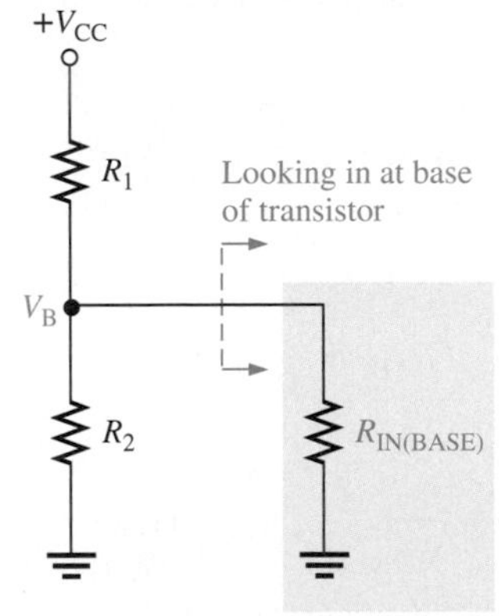

Stiff:

$R_{IN(BASE)} => 10R_2$

$V_B \cong \left(\frac{R_2}{R_1 + R_2}\right) V_{CC}$

Not stiff:

$R_{IN(BASE)} < 10R_2$

$V_B = \left(\frac{R_2 \parallel R_{IN(BASE)}}{R_1 + R_2 \parallel R_{IN(BASE)}}\right) V_{CC}$

FIGURE 5–11

Voltage divider with load.

Loading Effects of Voltage-Divider Bias

DC Input Resistance at the Transistor Base The dc input resistance of the transistor is proportional to β_{DC}, so it will change for different transistors. When a transistor is operating in its linear region, the emitter current is $\beta_{DC}I_B$. When the emitter resistor is viewed from the base circuit, the resistor appears to be larger than its actual value by a factor of β_{DC} because of the current gain in the transistor. That is,

$$\mathbf{R_{IN(BASE)} = \beta_{DC}R_E}$$ Equation 5–5

This is the effective load on the voltage divider illustrated in Figure 5–11. A derivation of this formula is given in Appendix B.

You can quickly estimate the loading effect by comparing $\beta_{DC}R_E$ to the resistor R_2 in the voltage divider. As long as $\beta_{DC}R_E$ is at least ten times larger than R_2, the loading effect will be 10% or less and the voltage divider is stiff. If $\beta_{DC}R_E$ is less than ten times R_2, it should be combined in parallel with R_2.

EXAMPLE 5–3

Determine the dc input resistance looking in at the base of the transistor in Figure 5–12. $\beta_{DC} = 125$.

FIGURE 5–12

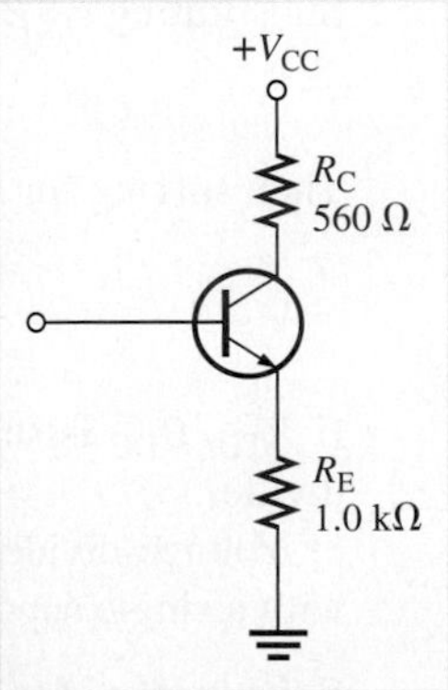

Solution $R_{IN(BASE)} = \beta_{DC}R_E = (125)(1.0\text{ k}\Omega) = \mathbf{125\text{ k}\Omega}$

Related Problem What is $R_{IN(BASE)}$ in Figure 5–12 if $\beta_{DC} = 60$ and $R_E = 910\ \Omega$?

Stability of Voltage-Divider Bias To analyze a voltage-divider biased transistor circuit for base current loading effects, apply Thevenin's theorem. We will use this method to evaluate the circuit. First, let's get an equivalent base-emitter circuit for the circuit in Figure 5–13(a) using Thevenin's theorem. Looking out from the base terminal, the bias circuit can be redrawn as shown in Figure 5–13(b). Apply Thevenin's theorem to the circuit left of point A, with V_{CC} replaced by a short to ground and the transistor disconnected from the circuit. The voltage at point A with respect to ground is

$$V_{TH} = \left(\frac{R_2}{R_1 + R_2}\right)V_{CC}$$

and the resistance is

$$R_{TH} = \frac{R_1R_2}{R_1 + R_2}$$

▶ FIGURE 5–13

Thevenizing the bias circuit.

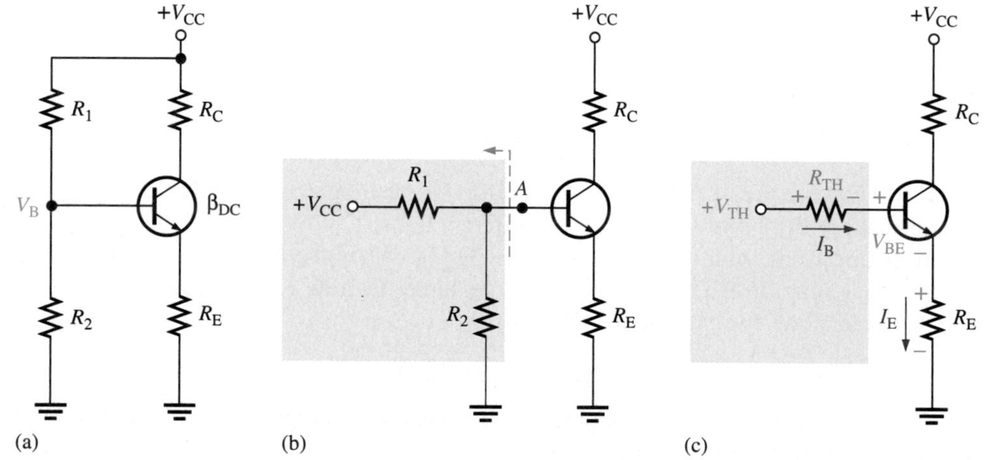

The Thevenin equivalent of the bias circuit, connected to the transistor base, is shown in the beige box in Figure 5–13(c). Applying Kirchhoff's voltage law around the equivalent base-emitter loop gives

$$V_{TH} - V_{R_{TH}} - V_{BE} - V_{R_E} = 0$$

Substituting, using Ohm's law, and solving for V_{TH},

$$V_{TH} = I_BR_{TH} + V_{BE} + I_ER_E$$

Substituting I_E/β_{DC} for I_B,

$$V_{TH} = I_E(R_E + R_{TH}/\beta_{DC}) + V_{BE}$$

Then solving for I_E,

Equation 5–6

$$\mathbf{I_E = \frac{V_{TH} - V_{BE}}{R_E + R_{TH}/\beta_{DC}}}$$

If R_{TH}/β_{DC} is small compared to R_E, the result is the same as for an unloaded voltage divider.

Voltage-divider bias is widely used because reasonably good bias stability is achieved with a single supply voltage.

Voltage-Divider Biased PNP Transistor As you know, a *pnp* transistor requires bias polarities opposite to the *npn*. This can be accomplished with a negative collector supply voltage, as in Figure 5–14(a), or with a positive emitter supply voltage, as in Figure 5–14(b). In a schematic, the *pnp* is often drawn upside down so that the supply voltage line can be drawn

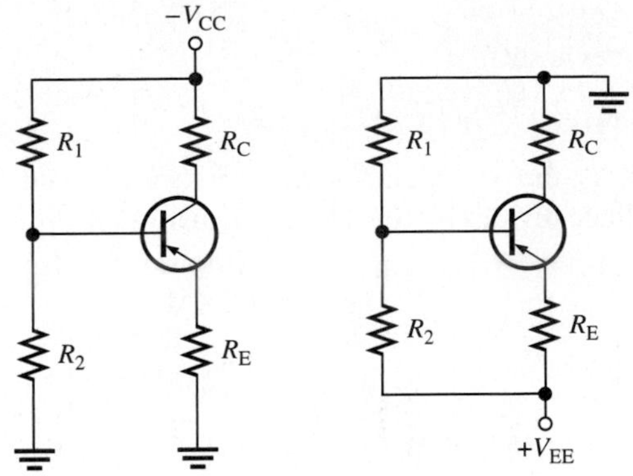

(a) Negative collector supply voltage, V_{CC} (b) Positive emitter supply voltage, V_{EE}

◀ FIGURE 5–14

Voltage-divider biased *pnp* transistor.

across the top of the schematic and ground at the bottom, as in Figure 5–15. The analysis procedure is basically the same as for an *npn* transistor circuit, as demonstrated in the following steps with reference to Figure 5–15. For a stiff voltage divider (ignoring loading effects), the base voltage is

$$V_B \cong \left(\frac{R_1}{R_1 + R_2}\right)V_{EE}$$

and

$$V_E = V_B + V_{BE}$$

By Ohm's law,

$$I_E = \frac{V_{EE} - V_E}{R_E}$$

and

$$V_C = I_C R_C$$

Therefore,

$$V_{EC} = V_E - V_C$$

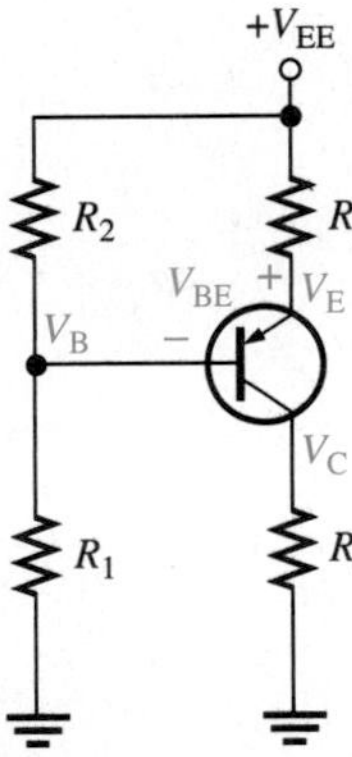

▲ FIGURE 5–15

EXAMPLE 5–4

Find I_C and V_{EC} for the *pnp* transistor circuit in Figure 5–16.

▶ FIGURE 5–16

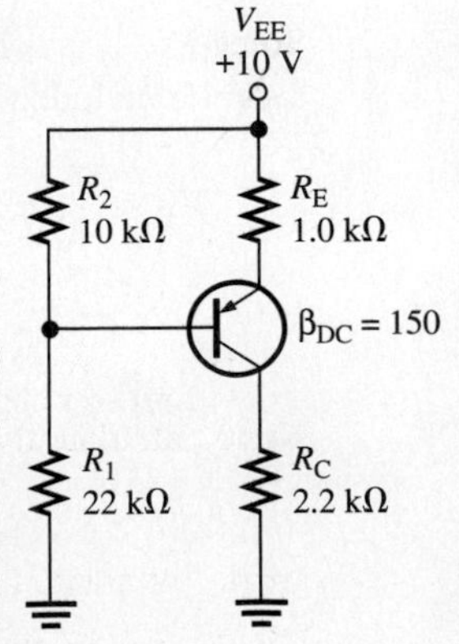

Solution First, check to see if the voltage divider is stiff.

$$R_{IN(BASE)} = \beta_{DC}R_E = (150)(1.0\text{ k}\Omega) = 150\text{ k}\Omega$$

Since 150 kΩ is more than ten times R_2, the condition $\beta_{DC}R_E \gg R_2$ is met and $R_{IN(base)}$ can be neglected; so the voltage divider is stiff. Now, calculate V_B.

$$V_B \cong \left(\frac{R_1}{R_1 + R_2}\right)V_{EE} = \left(\frac{22\text{ k}\Omega}{32\text{ k}\Omega}\right)10\text{ V} = 6.88\text{ V}$$

Then

$$V_E = V_B + V_{BE} = 6.88\text{ V} + 0.7\text{ V} = 7.58\text{ V}$$

and

$$I_E = \frac{V_{EE} - V_E}{R_E} = \frac{10\text{ V} - 7.58\text{ V}}{1.0\text{ k}\Omega} = 2.42\text{ mA}$$

From I_E, you can determine I_C and V_{CE} as follows:

$$I_C \cong I_E = \mathbf{2.42\text{ mA}}$$

and

$$V_C = I_C R_C = (2.42\text{ mA})(2.2\text{ k}\Omega) = 5.32\text{ V}$$

Therefore,

$$V_{EC} = V_E - V_C = 7.58\text{ V} - 5.32\text{ V} = \mathbf{2.26\text{ V}}$$

Related Problem Determine I_C and V_{EC} in Figure 5–16 with $R_{IN(BASE)}$ taken into account.

Open the Multisim file E05-04 in the Examples folder on your CD-ROM. Measure I_C and V_{EC}. Your results should agree more closely with those in the Related Exercise than with those calculated in the example. Can you explain?

EXAMPLE 5–5

Find I_C and V_{CE} for a *pnp* transistor circuit with these values: $R_1 = 68\text{ k}\Omega$, $R_2 = 47\text{ k}\Omega$, $R_C = 1.8\text{ k}\Omega$, $R_E = 2.2\text{ k}\Omega$, $V_{CC} = -6\text{ V}$, and $\beta_{DC} = 75$. Refer to Figure 5–14(a), which shows the schematic with a negative supply voltage.

Solution

$$R_{IN(BASE)} = \beta_{DC}R_E = 75(2.2\text{ k}\Omega) = 165\text{ k}\Omega$$

Since $R_{IN(base)}$ is less than ten times R_2, the voltage divider is not stiff and $R_{IN(BASE)}$ must be included in parallel with R_2. Calculate V_B.

$$V_B = \left(\frac{R_2 \| R_{IN(BASE)}}{R_1 + R_2 \| R_{IN(BASE)}}\right)V_{CC} = \left(\frac{47\text{ k}\Omega \| 165\text{ k}\Omega}{68\text{ k}\Omega + 47\text{ k}\Omega \| 165\text{ k}\Omega}\right)(-6\text{ V})$$

$$= \left(\frac{36.6\text{ k}\Omega}{68\text{ k}\Omega + 36.6\text{ k}\Omega}\right)(-6\text{ V}) = -2.1\text{ V}$$

Next, calculate the emitter voltage and current.

$$V_E = V_B + V_{BE} = -2.1\text{ V} + 0.7\text{ V} = -1.4\text{ V}$$

$$I_E = \frac{V_E}{R_E} = \frac{-1.4\text{ V}}{2.2\text{ k}\Omega} = -636\ \mu\text{A}$$

From I_E, you can determine I_C and V_{CE} as follows:

$$I_C \cong I_E = \mathbf{-636\ \mu A}$$
$$V_C = V_{CC} - I_C R_C = -6\text{ V} - (-636\ \mu\text{A})(1.8\text{ k}\Omega) = -4.86\text{ V}$$
$$V_{CE} = V_C - V_E = -4.86\text{ V} - (-1.4\text{ V}) = \mathbf{-3.46\ V}$$

Related Problem What value of β_{DC} is required in this example in order to neglect $R_{IN(BASE)}$ in keeping with the basic ten-times rule for a stiff voltage divider?

SECTION 5–2 CHECKUP

1. If the voltage at the base of a transistor is 5 V and the base current is 5 μA, what is the dc input resistance at the base?
2. If a transistor has a dc beta of 190 and its emitter resistor is 1.0 kΩ, what is the dc input resistance at the base?
3. What bias voltage is developed at the base of a transistor if both resistors in a stiff voltage divider are equal and V_{CC} = +10 V?
4. What are two advantages of voltage-divider bias?

5–3 OTHER BIAS METHODS

In this section, four additional methods for dc biasing a transistor circuit are discussed. Although these methods are not as common as voltage-divider bias, you should be able to recognize them when you see them and understand the basic differences.

After completing this section, you should be able to

- **Analyze four additional types of bias circuits**
  - Recognize emitter bias
  - Recognize base bias
  - Recognize emitter-feedback bias
  - Recognize collector-feedback bias
  - Discuss the stability of each bias circuit and compare with the voltage-divider bias

Emitter Bias

Emitter bias provides excellent bias stability in spite of changes in β or temperature. It uses both a positive and a negative supply voltage. To obtain a reasonable estimate of the key dc values in an emitter-biased circuit, analysis is quite easy. In an *npn* circuit, such as shown in Figure 5–18, the small base current causes the base voltage to be slightly below ground. The emitter voltage is one diode drop less than this. The combination of this small drop across R_B and V_{BE} forces the emitter to be at approximately −1 V. Using this approximation, you can obtain the emitter current as

$$I_E = \frac{-V_{EE} - 1\text{ V}}{R_E}$$

V_{EE} is entered as a negative value in this equation.

You can apply the approximation that $I_C \cong I_E$ to calculate the collector voltage.

$$V_C = V_{CC} - I_C R_C$$

The approximation that $V_E \cong -1$ V is useful for troubleshooting because you won't need to perform any detailed calculations. As in the case of voltage-divider bias, there is a more rigorous calculation for cases where you need a more exact result.

EXAMPLE 5–6

Calculate I_E and V_{CE} for the circuit in Figure 5–17 using the approximations $V_E \cong -1$ V and $I_C \cong I_E$.

FIGURE 5–17

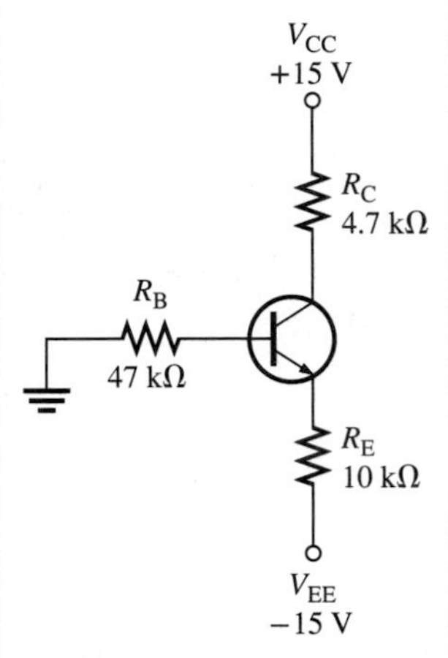

Solution

$$V_E \cong -1\text{ V}$$

$$I_E = \frac{-V_{EE} - 1\text{ V}}{R_E} = \frac{-(-15\text{ V}) - 1\text{ V}}{10\text{ k}\Omega} = \frac{14\text{ V}}{10\text{ k}\Omega} = \mathbf{1.4\text{ mA}}$$

$$V_C = V_{CC} - I_C R_C = +15\text{ V} - (1.4\text{ mA})(4.7\text{ k}\Omega) = 8.4\text{ V}$$

$$V_{CE} = 8.4\text{ V} - (-1) = \mathbf{9.4\text{ V}}$$

Related Problem If V_{EE} is changed to −12 V, what is the new value of V_{CE}?

The approximation that $V_E \cong -1$ V and the neglect of β_{DC} may not be accurate enough for design work or detailed analysis. In this case, Kirchhoff's voltage law can be applied as follows to develop a more detailed formula for I_E. Kirchhoff's voltage law applied around the base-emitter circuit in Figure 5–18(a), which has been redrawn in part (b) for analysis, gives the following equation:

$$V_{EE} + V_{R_B} + V_{BE} + V_{R_E} = 0$$

Substituting, using Ohm's law,

$$V_{EE} + I_B R_B + V_{BE} + I_E R_E = 0$$

Substituting for $I_B \cong I_E/\beta_{DC}$ and transposing V_{EE},

$$\left(\frac{I_E}{\beta_{DC}}\right)R_B + I_E R_E + V_{BE} = -V_{EE}$$

Factoring out I_E and solving for I_E,

Equation 5–7

$$\mathbf{I_E = \frac{-V_{EE} - V_{BE}}{R_E + R_B/\beta_{DC}}}$$

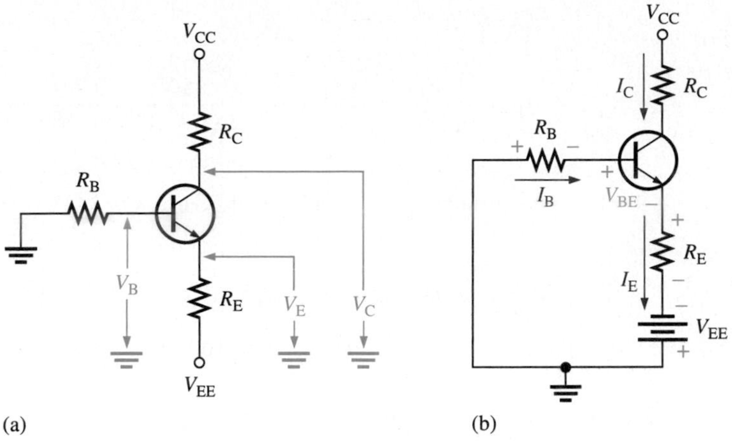

FIGURE 5–18

An *npn* transistor with emitter bias. Polarities are reversed for a *pnp* transistor. Single subscripts indicate voltages with respect to ground.

Voltages with respect to ground are indicated by a single subscript. The emitter voltage with respect to ground is

$$V_E = V_{EE} + I_E R_E$$

The base voltage with respect to ground is

$$V_B = V_E + V_{BE}$$

The collector voltage with respect to ground is

$$V_C = V_{CC} - I_C R_C$$

EXAMPLE 5–7

Determine how much the Q-point (I_C, V_{CE}) for the circuit in Figure 5–19 will change if β_{DC} increases from 100 to 200 when one transistor is replaced by another.

FIGURE 5–19

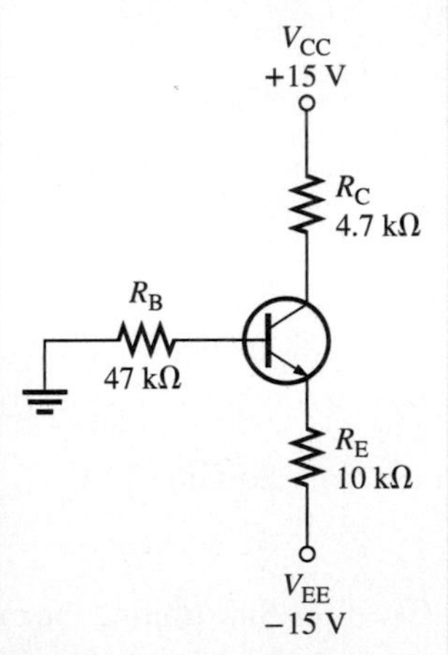

Solution For $\beta_{DC} = 100$,

$$I_{C(1)} \cong I_E = \frac{-V_{EE} - V_{BE}}{R_E + R_B/\beta_{DC}} = \frac{-(-15\text{ V}) - 0.7\text{ V}}{10\text{ k}\Omega + 47\text{ k}\Omega/100} = 1.37\text{ mA}$$

$$V_C = V_{CC} - I_{C(1)}R_C = 15\text{ V} - (1.37\text{ mA})(4.7\text{ k}\Omega) = 8.56\text{ V}$$

$$V_E = V_{EE} + I_E R_E = -15\text{ V} + (1.37\text{ mA})(10\text{ k}\Omega) = -1.3\text{ V}$$

Therefore,

$$V_{CE(1)} = V_C - V_E = 8.56\text{ V} - (-1.3\text{ V}) = 9.83\text{ V}$$

For $\beta_{DC} = 200$,

$$I_{C(2)} \cong I_E = \frac{-V_{EE} - V_{BE}}{R_E + R_B/\beta_{DC}} = \frac{-(-15\text{ V}) - 0.7\text{ V}}{10\text{ k}\Omega + 47\text{ k}\Omega/200} = 1.38\text{ mA}$$

$$V_C = V_{CC} - I_{C(2)}R_C = 15\text{ V} - (1.38\text{ mA})(4.7\text{ k}\Omega) = 8.51\text{ V}$$

$$V_E = V_{EE} + I_E R_E = -15\text{ V} + (1.38\text{ mA})(10\text{ k}\Omega) = -1.2\text{ V}$$

Therefore,

$$V_{CE(2)} = V_C - V_E = 8.51\text{ V} - (-1.2\text{ V}) = 9.71\text{ V}$$

The percent change in I_C as β_{DC} changes from 100 to 200 is

$$\%\Delta I_C = \left(\frac{I_{C(2)} - I_{C(1)}}{I_{C(1)}}\right)100\% = \left(\frac{1.38\text{ mA} - 1.37\text{ mA}}{1.37\text{ mA}}\right)100\% = 0.730\%$$

The percent change in V_{CE} is

$$\%\Delta V_{CE} = \left(\frac{V_{CE(2)} - V_{CE(1)}}{V_{CE(1)}}\right)100\% = \left(\frac{9.71\text{ V} - 9.83\text{ V}}{9.83\text{ V}}\right)100\% = \mathbf{-1.22\%}$$

Related Problem Determine the Q-point in Figure 5–19 if β_{DC} increases to 300.

Base Bias

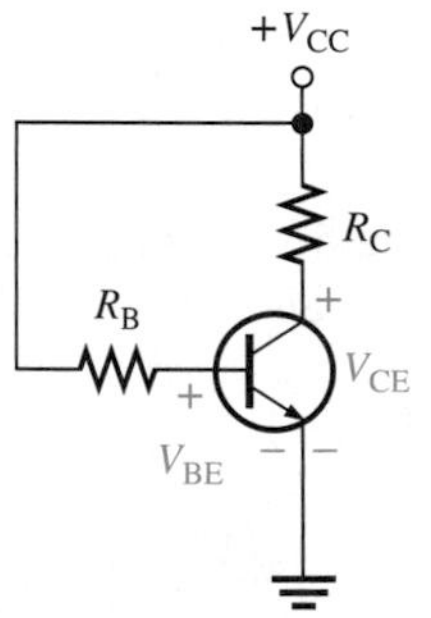

▲ FIGURE 5–20

Base bias.

This method of biasing is common in switching circuits. Figure 5–20 shows a base-biased transistor. The analysis of this circuit for the linear region shows that it is directly dependent on β_{DC}. Starting with Kirchhoff's voltage law around the base circuit,

$$V_{CC} - V_{R_B} - V_{BE} = 0$$

Substituting $I_B R_B$ for V_{R_B}, you get

$$V_{CC} - I_B R_B - V_{BE} = 0$$

Then solving for I_B,

$$I_B = \frac{V_{CC} - V_{BE}}{R_B}$$

Kirchhoff's voltage law applied around the collector circuit in Figure 5–20 gives the following equation:

$$V_{CC} - I_C R_C - V_{CE} = 0$$

Solving for V_{CE},

Equation 5–8

$$\mathbf{V_{CE} = V_{CC} - I_C R_C}$$

Substituting the expression for I_B into the formula $I_C = \beta_{DC} I_B$ yields

Equation 5–9

$$\mathbf{I_C = \beta_{DC}\left(\frac{V_{CC} - V_{BE}}{R_B}\right)}$$

Q-Point Stability of Base Bias Notice that Equation 5–9 shows that I_C is dependent on β_{DC}. The disadvantage of this is that a variation in β_{DC} causes I_C and, as a result, V_{CE} to change, thus changing the Q-point of the transistor. This makes the base bias circuit extremely beta-dependent and unpredictable.

Recall that β_{DC} varies with temperature and collector current. In addition, there is a large spread of β_{DC} values from one transistor to another of the same type due to manufacturing variations. For these reasons, base bias is rarely used in linear circuits but is discussed here so you will be familiar with it.

EXAMPLE 5–8

Determine how much the Q-point (I_C, V_{CE}) for the circuit in Figure 5–21 will change over a temperature range where β_{DC} increases from 100 to 200.

FIGURE 5–21

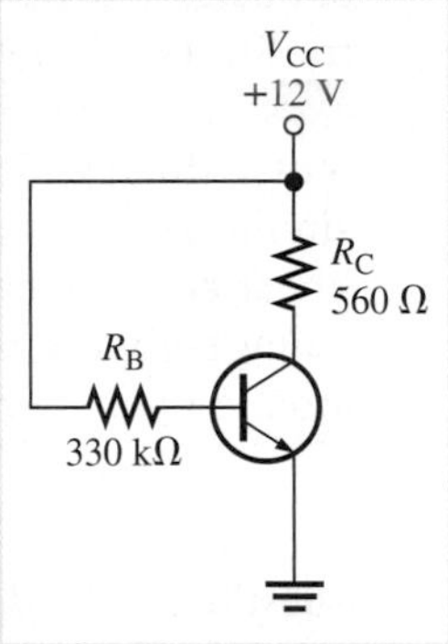

Solution For $\beta_{DC} = 100$,

$$I_{C(1)} = \beta_{DC}\left(\frac{V_{CC} - V_{BE}}{R_B}\right) = 100\left(\frac{12\text{ V} - 0.7\text{ V}}{330\text{ k}\Omega}\right) = 3.42\text{ mA}$$

$$V_{CE(1)} = V_{CC} - I_{C(1)}R_C = 12\text{ V} - (3.42\text{ mA})(560\ \Omega) = 10.1\text{ V}$$

For $\beta_{DC} = 200$,

$$I_{C(2)} = \beta_{DC}\left(\frac{V_{CC} - V_{BE}}{R_B}\right) = 200\left(\frac{12\text{ V} - 0.7\text{ V}}{330\text{ k}\Omega}\right) = 6.84\text{ mA}$$

$$V_{CE(2)} = V_{CC} - I_{C(2)}R_C = 12\text{ V} - (6.84\text{ mA})(560\ \Omega) = 8.17\text{ V}$$

The percent change in I_C as β_{DC} changes from 100 to 200 is

$$\%\Delta I_C = \left(\frac{I_{C(2)} - I_{C(1)}}{I_{C(1)}}\right)100\%$$

$$= \left(\frac{6.84\text{ mA} - 3.42\text{ mA}}{3.42\text{ mA}}\right)100\% = \mathbf{100\%}\text{ (an increase)}$$

The percent change in V_{CE} is

$$\%\Delta V_{CE} = \left(\frac{V_{CE(2)} - V_{CE(1)}}{V_{CE(1)}}\right)100\%$$

$$= \left(\frac{8.17\text{ V} - 10.1\text{ V}}{10.1\text{ V}}\right)100\% = \mathbf{-19.1\%}\text{ (a decrease)}$$

As you can see, the Q-point is very dependent on β_{DC} in this circuit and therefore makes the base bias arrangement very unreliable. Consequently, base bias is not normally used if linear operation is required. However, it can be used in switching applications.

Related Problem Determine I_C if β_{DC} increases to 300.

Open the Multisim file E05-08 in the Examples folder on your CD-ROM. Set $\beta_{DC} = 100$ and measure I_C and V_{CE}. Next, set $\beta_{DC} = 200$ and measure I_C and V_{CE}. Compare results with the calculated values.

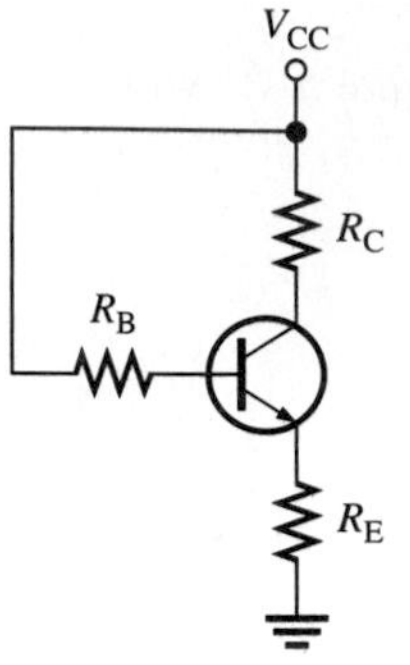

▲ FIGURE 5–22

Emitter-feedback bias.

Emitter-Feedback Bias

If an emitter resistor is added to the base-bias circuit in Figure 5–21, the result is emitter-feedback bias, as shown in Figure 5–22. The idea is to help make base bias more predictable with negative **feedback**, which negates any attempted change in collector current with an opposing change in base voltage. If the collector current tries to increase, the emitter voltage increases, causing an increase in base voltage because $V_B = V_E + V_{BE}$. This increase in base voltage reduces the voltage across R_B, thus reducing the base current and keeping the collector current from increasing. A similar action occurs if the collector current tries to decrease. While this is better for linear circuits than base bias, it is still dependent on β_{DC} and is not as predictable as voltage-divider bias. To calculate I_E, you can write Kirchhoff's voltage law (KVL) around the base circuit.

$$-V_{CC} + I_B R_B + V_{BE} + I_E R_E = 0$$

Substituting I_E/β_{DC} for I_B, you can see that I_E is still dependent on β_{DC}.

Equation 5–10

$$I_E = \frac{V_{CC} - V_{BE}}{R_E + R_B/\beta_{DC}}$$

EXAMPLE 5–9

The base-bias circuit from Example 5–8 is converted to emitter-feedback bias by the addition of a 1 kΩ emitter resistor. All other values are the same, and a transistor with a $\beta_{DC} = 100$ is used. Determine how much the Q-point will change if the first transistor is replaced with one having a $\beta_{DC} = 200$. Compare the results to those of the base-bias circuit.

Solution For $\beta_{DC} = 100$,

$$I_{C(1)} = I_E = \frac{V_{CC} - V_{BE}}{R_E + R_B/\beta_{DC}} = \frac{12\text{ V} - 0.7\text{ V}}{1\text{ k}\Omega + 330\text{ k}\Omega/100} = 2.63\text{ mA}$$

$$V_{CE(1)} = V_{CC} - I_{C(1)}(R_C + R_E) = 12\text{ V} - (2.63\text{ mA})(560\ \Omega + 1\text{ k}\Omega) = 7.90\text{ V}$$

For $\beta_{DC} = 200$,

$$I_{C(2)} = I_E = \frac{V_{CC} - V_{BE}}{R_E + R_B/\beta_{DC}} = \frac{12\text{ V} - 0.7\text{ V}}{1\text{ k}\Omega + 330\text{ k}\Omega/200} = 4.26\text{ mA}$$

$$V_{CE(2)} = V_{CC} - I_{C(2)}(R_C + R_E) = 12\text{ V} - (4.26\text{ mA})(560\ \Omega + 1\text{ k}\Omega) = 5.35\text{ V}$$

The percent change in I_C is

$$\%\Delta I_C = \left(\frac{I_{C(2)} - I_{C(1)}}{I_{C(1)}}\right)100\% = \left(\frac{4.26\text{ mA} - 2.63\text{ mA}}{2.63\text{ mA}}\right)100\% = \mathbf{62.0\%}$$

$$\%\Delta V_{CE} = \left(\frac{V_{CE(2)} - V_{CE(1)}}{V_{CE(1)}}\right)100\% = \left(\frac{7.90\text{ V} - 5.35\text{ V}}{7.90\text{ V}}\right)100\% = \mathbf{-32.3\%}$$

Although the emitter-feedback bias significantly improved the stability of the bias for a change in β_{DC} compared to base bias, it still does not provide a reliable Q-point.

Related Problem Determine I_C if a transistor with $\beta_{DC} = 300$ is used in the circuit.

Collector-Feedback Bias

In Figure 5–23, the base resistor R_B is connected to the collector rather than to V_{CC}, as it was in the base bias arrangement discussed earlier. The collector voltage provides the bias for the base-emitter junction. The negative feedback creates an "offsetting" effect that tends to keep the Q-point stable. If I_C tries to increase, it drops more voltage across R_C, thereby causing V_C to decrease. When V_C decreases, there is a decrease in voltage across R_B, which decreases I_B. The decrease in I_B produces less I_C which, in turn, drops less voltage across R_C and thus offsets the decrease in V_C.

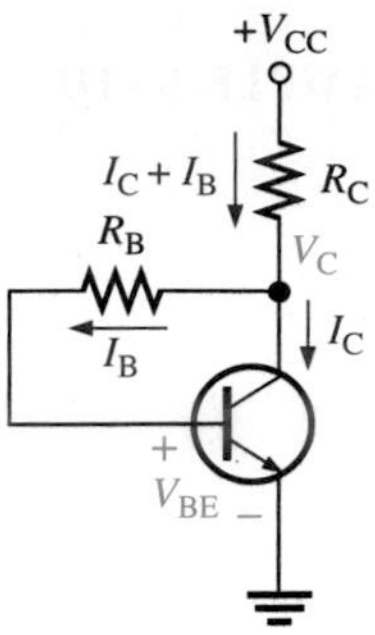

▲ **FIGURE 5–23**

Collector-feedback bias.

Analysis of a Collector-Feedback Bias Circuit By Ohm's law, the base current can be expressed as

$$I_B = \frac{V_C - V_{BE}}{R_B}$$

Let's assume that $I_C \gg I_B$. The collector voltage is

$$V_C \cong V_{CC} - I_C R_C$$

Also,

$$I_B = \frac{I_C}{\beta_{DC}}$$

Substituting for V_C in the equation $I_B = (V_C - V_{BE})/R_B$,

$$\frac{I_C}{\beta_{DC}} = \frac{V_{CC} - I_C R_C - V_{BE}}{R_B}$$

The terms can be arranged so that

$$\frac{I_C R_B}{\beta_{DC}} + I_C R_C = V_{CC} - V_{BE}$$

Then you can solve for I_C as follows:

$$I_C\left(R_C + \frac{R_B}{\beta_{DC}}\right) = V_{CC} - V_{BE}$$

$$\mathbf{I_C = \frac{V_{CC} - V_{BE}}{R_C + R_B/\beta_{DC}}} \qquad \text{Equation 5–11}$$

Since the emitter is ground, $V_{CE} = V_C$.

$$\mathbf{V_{CE} = V_{CC} - I_C R_C} \qquad \text{Equation 5–12}$$

Q-Point Stability Over Temperature Equation 5–11 shows that the collector current is dependent to some extent on β_{DC} and V_{BE}. This dependency, of course, can be minimized by making $R_C \gg R_B/\beta_{DC}$ and $V_{CC} \gg V_{BE}$. An important feature of collector-feedback bias is that it essentially eliminates the β_{DC} and V_{BE} dependency even if the stated conditions are met.

As you have learned, β_{DC} varies directly with temperature, and V_{BE} varies inversely with temperature. As the temperature goes up in a collector-feedback circuit, β_{DC} goes up and V_{BE} goes down. The increase in β_{DC} acts to increase I_C. The decrease in V_{BE} acts to increase I_B which, in turn also acts to increase I_C. As I_C tries to increase, the voltage drop across R_C also tries to increase. This tends to reduce the collector voltage and therefore the voltage across R_B, thus reducing I_B and offsetting the attempted increase in I_C and the attempted decrease in V_C. The result is that the collector-feedback circuit maintains a relatively stable Q-point. The reverse action occurs when the temperature decreases.

3. Open file TSE05-03. Determine if the circuit is working properly and, if not, determine the fault.

4. Open file TSE05-04. Determine if the circuit is working properly and, if not, determine the fault.

SECTION 5–4 CHECKUP

1. How do you determine when a transistor is saturated? When a transistor is in cutoff?
2. In a voltage-divider biased *npn* transistor circuit, you measure V_{CC} at the collector and an emitter voltage 0.7 V less than the base voltage. Is the transistor functioning in cutoff, or is R_E open?
3. What symptoms does an open R_C produce?

On the Job

Application Activity: *Temperature to Voltage Conversion*

The focus of this Application Activity is a temperature-sensing circuit that converts the temperature of a liquid to a proportional voltage for the purpose of maintaining the temperature of the liquid within a specified range. Figure 5–29 illustrates the temperature-control system. The temperature sensor is a **thermistor**, which is a device whose resistance changes with temperature. The thermistor is connected to a transistor circuit that is biased for linear operation. The output voltage of the circuit is proportional to the thermistor resistance and thus to the temperature of the liquid in the tank. The output voltage goes to an interface circuit that controls the valve to control the flow of fuel to the burner based on the voltage. If the temperature of the liquid is below a set value, the fuel is increased and if it is above that value, the fuel is decreased. The temperature is to be maintained at 70°C ± 5°C.

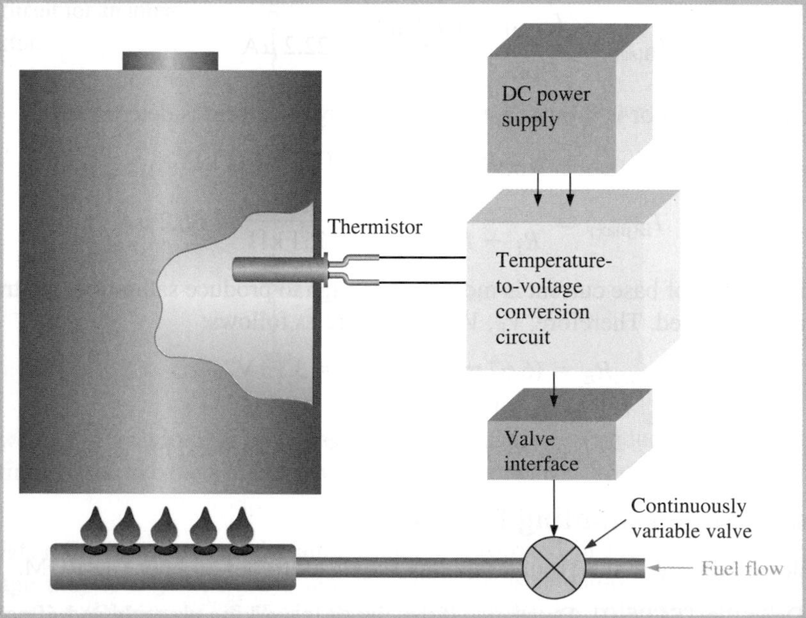

FIGURE 5–29

Temperature-control system.

Designing the Circuit

Circuit Configuration A voltage-divider biased linear amplifier is used for the temperature-to-voltage conversion. The thermistor is used as one of the resistors in the voltage-divider bias. This thermistor has a positive temperature coefficient so, if the temperature increases, the resistance of the thermistor increases and if the temperature decreases, the resistance decreases. The base voltage changes proportionally to the change in thermistor resistance. The output voltage is inversely proportional to the base voltage, so as the temperature goes up, the output voltage decreases and reduces the fuel flow to the burner. As the temperature goes down, the output voltage increases and allows more fuel to the burner.

Components As shown in Figure 5–30(a), the circuit is implemented with a 2N3904 transistor, four resistors with the values shown, and a +9 V dc source. The thermistor has the temperature characteristic shown in part (b).

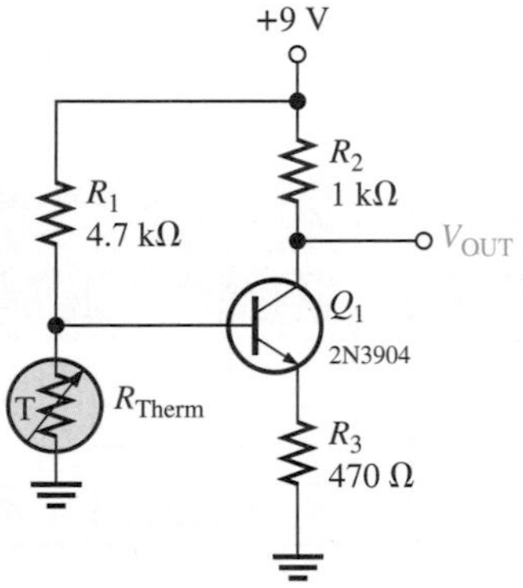

(a) Circuit

TEMPERATURE, °C	THERMISTOR RESISTANCE, kΩ
60	1.256
65	1.481
70	1.753
75	2.084
80	2.490

(b) Temperature characteristic of the thermistor for the specified range

FIGURE 5–30

Temperature-to-voltage conversion circuit.

1. Plot a graph of the thermistor temperature characteristic.
2. Refer to Figure 5–30 and calculate the emitter and collector currents for each temperature shown.
3. Calculate the output voltage for each temperature shown in Figure 5–30.

Simulation

The temperature-to-voltage conversion circuit is simulated to determine how the output voltage changes with temperature, as shown in Figure 5–31. The thermistor is represented by a resistor with values corresponding to each specified temperature.

4. Compare your calculations for the output voltage with the simulated values.

Simulate the circuit using your Multisim software. Observe the operation with the virtual multimeter.

Prototyping and Testing

Now that all the components have been selected, the prototype circuit is constructed and tested. After the circuit is successfully tested, it is ready to be finalized on a printed circuit board.

DATASHEET PROBLEMS

42. Analyze the temperature-to-voltage conversion circuit in Figure 5–48 at the temperature extremes indicated on the graph in Figure 5–47 for both minimum and maximum specified datasheet values of h_{FE}. Refer to the partial datasheet in Figure 5–49.

43. Verify that no maximum ratings are exceeded in the temperature-to-voltage conversion circuit in Figure 5–48. Refer to the partial datasheet in Figure 5–49.

FIGURE 5–48

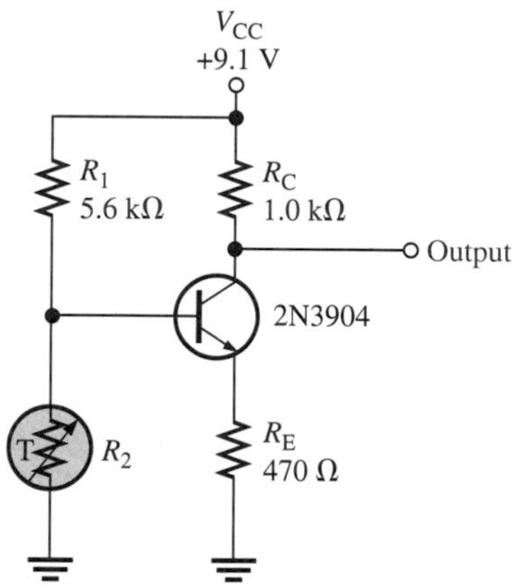

FIGURE 5–49

Partial datasheet for the 2N3904 transistor. © 2003 Fairchild Semiconductor Corporation. Used by permission.

Absolute Maximum Ratings* T_A = 25°C unless otherwise noted

Symbol	Parameter	Value	Units
V_{CEO}	Collector-Emitter Voltage	40	V
V_{CBO}	Collector-Base Voltage	60	V
V_{EBO}	Emitter-Base Voltage	6.0	V
I_C	Collector Current - Continuous	200	mA
T_J, T_{stg}	Operating and Storage Junction Temperature Range	-55 to +150	°C

*These ratings are limiting values above which the serviceability of any semiconductor device may be impaired.

NOTES:
1) These ratings are based on a maximum junction temperature of 150 degrees C.
2) These are steady state limits. The factory should be consulted on applications involving pulsed or low duty cycle operations.

ON CHARACTERISTICS*

h_{FE}	DC Current Gain	I_C = 0.1 mA, V_{CE} = 1.0 V	40		
		I_C = 1.0 mA, V_{CE} = 1.0 V	70		
		I_C = 10 mA, V_{CE} = 1.0 V	100	300	
		I_C = 50 mA, V_{CE} = 1.0 V	60		
		I_C = 100 mA, V_{CE} = 1.0 V	30		
$V_{CE(sat)}$	Collector-Emitter Saturation Voltage	I_C = 10 mA, I_B = 1.0 mA		0.2	V
		I_C = 50 mA, I_B = 5.0 mA		0.3	V
$V_{BE(sat)}$	Base-Emitter Saturation Voltage	I_C = 10 mA, I_B = 1.0 mA	0.65	0.85	V
		I_C = 50 mA, I_B = 5.0 mA		0.95	V

44. Refer to the partial datasheet in Figure 5–50.

(a) What is the maximum collector current for a 2N2222A?

(b) What is the maximum reverse base-emitter voltage for a 2N2218A?

45. Determine the maximum power dissipation for a 2N2222A at 100°C.

46. When you increase the collector current in a 2N2219A from 1 mA to 500 mA, how much does the minimum β_{DC} (h_{FE}) change?

ADVANCED PROBLEMS

47. Design a circuit using base bias that operates from a 15 V dc voltage and draws a maximum current from the dc source ($I_{CC(max)}$) of 10 mA. The Q-point values are to be I_C = 5 mA and V_{CE} = 5 V. The transistor is a 2N3904. Assume a midpoint value for β_{DC}.

Maximum Ratings

Rating	Symbol	2N2218 2N2219 2N2221 2N2222	2N2218A 2N2219A 2N2221A 2N2222A	2N5581 2N5582	Unit
Collector-Emitter voltage	V_{CEO}	30	40	40	V dc
Collector-Base voltage	V_{CBO}	60	75	75	V dc
Emitter-Base voltage	V_{EBO}	5.0	6.0	6.0	V dc
Collector current — continuous	I_C	800	800	800	mA dc
		2N2218,A 2N2219,A	**2N2221,A 2N2222,A**	**2N5581 2N5582**	
Total device dissipation @ $T_A = 25°C$ Derate above 25°C	P_D	 0.8 4.57	 0.5 2.28	 0.6 3.33	 Watt mW/°C
Total device dissipation @ $T_C = 25°C$ Derate above 25°C	P_D	 3.0 17.1	 1.2 6.85	 2.0 11.43	 Watt mW/°C
Operating and storage junction Temperature range	T_J, T_{stg}		−65 to +200		°C

Electrical Characteristics ($T_A = 25°C$ unless otherwise noted.)

Characteristic		Symbol	Min	Max	Unit
Off Characteristics					
Collector-Emitter breakdown voltage ($I_C = 10$ mA dc, $I_B = 0$)	 Non-A Suffix A-Suffix, 2N5581, 2N5582	$V_{(BR)CEO}$	 30 40	 — —	V dc
Collector-Base breakdown voltage ($I_C = 10$ μA dc, $I_E = 0$)	 Non-A Suffix A-Suffix, 2N5581, 2N5582	$V_{(BR)CBO}$	 60 75	 — —	V dc
Emitter-Base breakdown voltage ($I_E = 10$ μA dc, $I_C = 0$)	 Non-A Suffix A-Suffix, 2N5581, 2N5582	$V_{(BR)EBO}$	 5.0 6.0	 — —	V dc
Collector cutoff current ($V_{CE} = 60$ V dc, $V_{EB(off)} = 3.0$ V dc	A-Suffix, 2N5581, 2N5582	I_{CEX}	—	10	nA dc
Collector cutoff current ($V_{CB} = 50$ V dc, $I_E = 0$) ($V_{CB} = 60$ V dc, $I_E = 0$) ($V_{CB} = 50$ V dc, $I_E = 0$, $T_A = 150°C$) ($V_{CB} = 60$ V dc, $I_E = 0$, $T_A = 150°C$)	 Non-A Suffix A-Suffix, 2N5581, 2N5582 Non-A Suffix A-Suffix, 2N5581, 2N5582	I_{CBO}	 — — — —	 0.01 0.01 10 10	μA dc
Emitter cutoff current ($V_{EB} = 3.0$ V dc, $I_C = 0$)	A-Suffix, 2N5581, 2N5582	I_{EBO}	—	10	nA dc
Base cutoff current ($V_{CE} = 60$ V dc, $V_{EB(off)} = 3.0$ V dc)	A-Suffix	I_{BL}	—	20	nA dc
On Characteristics					
DC current gain ($I_C = 0.1$ mA dc, $V_{CE} = 10$ V dc)	 2N2218,A, 2N2221,A, 2N5581(1) 2N2219,A, 2N2222,A, 2N5582(1)	h_{FE}	 20 35	 — —	—
($I_C = 1.0$ mA dc, $V_{CE} = 10$ V dc)	2N2218,A, 2N2221,A, 2N5581 2N2219,A, 2N2222,A, 2N5582		25 50	— —	
($I_C = 10$ mA dc, $V_{CE} = 10$ V dc)	2N2218,A, 2N2221,A, 2N5581(1) 2N2219,A, 2N2222,A, 2N5582(1)		35 75	— —	
($I_C = 10$ mA dc, $V_{CE} = 10$ V dc, $T_A = -55°C$)	2N2218,A, 2N2221,A, 2N5581 2N2219,A, 2N2222,A, 2N5582		15 35	— —	
($I_C = 150$ mA dc, $V_{CE} = 10$ V dc)	2N2218,A, 2N2221,A, 2N5581 2N2219,A, 2N2222,A, 2N5582		40 100	120 300	
($I_C = 150$ mA dc, $V_{CE} = 1.0$ V dc)	2N2218,A, 2N2221,A, 2N5581 2N2219,A, 2N2222,A, 2N5582		20 50	— —	
($I_C = 500$ mA dc, $V_{CE} = 10$ V dc)	2N2218, 2N2221 2N2219, 2N2222 2N2218A, 2N2221A, 2N5581 2N2219A, 2N2222A, 2N5582		20 30 25 40	— — — —	
Collector-Emitter saturation voltage ($I_C = 150$ mA dc, $I_B = 15$ mA dc)	 Non-A Suffix A-Suffix, 2N5581, 2N5582	$V_{CE(sat)}$	 — —	 0.4 0.3	V dc
($I_C = 500$ mA dc, $I_B = 50$ mA dc)	Non-A Suffix A-Suffix, 2N5581, 2N5582		— —	1.6 1.0	
Base-Emitter saturation voltage ($I_C = 150$ mA dc, $I_B = 15$ mA dc)	 Non-A Suffix A-Suffix, 2N5581, 2N5582	$V_{BE(sat)}$	 0.6 0.6	 1.3 1.2	V dc
($I_C = 500$ mA dc, $I_B = 50$ mA dc)	Non-A Suffix A-Suffix, 2N5581, 2N5582		— —	2.6 2.0	

▲ **FIGURE 5–50**

Partial datasheet for 2N2218A–2N2222A.

48. Design a circuit using emitter bias that operates from dc voltages of $+12$ V and -12 V. The maximum I_{CC} is to be 20 mA and the Q-point is at 10 mA and 4 V. The transistor is a 2N3904.

49. Design a circuit using voltage-divider bias for the following specifications: $V_{CC} = 9$ V, $I_{CC(max)} = 5$ mA, $I_C = 1.5$ mA, and $V_{CE} = 3$ V. The transistor is a 2N3904.

50. Design a collector-feedback circuit using a 2N2222A with $V_{CC} = 5$ V, $I_C = 10$ mA, and $V_{CE} = 1.5$ V.

51. Can you replace the 2N3904 in Figure 5–48 with a 2N2222A and maintain the same range of output voltage over a temperature range from 45°C to 55°C?

52. Refer to the datasheet graph in Figure 5–51 and the partial datasheet in Figure 5–50. Determine the minimum dc current gain for a 2N2222A at −55°C, 25°C, and 175°C for $V_{CE} = 1$ V.

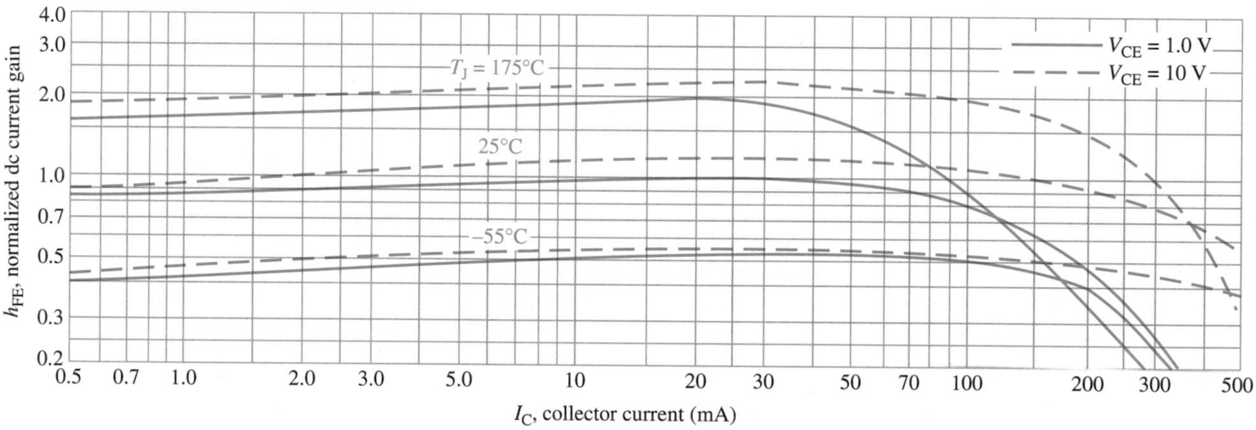

▲ FIGURE 5–51

53. A design change is required in the valve interface circuit of the temperature-control system shown in Figure 5–29. The new design will have a valve interface input resistance of 10 kΩ. Determine the effect this change has on the temperature-to-voltage conversion circuit.

54. Investigate the feasibility of redesigning the temperature-to-voltage conversion circuit in Figure 5–30 to operate from a dc supply voltage of 5.1 V and produce the same range of output voltages determined in the Application Activity over the required thermistor temperature range from 60°C to 80°C.

MULTISIM TROUBLESHOOTING PROBLEMS

These file circuits are in the Troubleshooting Problems folder on your CD-ROM.

55. Open file TSP05-55 and determine the fault.

56. Open file TSP05-56 and determine the fault.

57. Open file TSP05-57 and determine the fault.

58. Open file TSP05-58 and determine the fault.

59. Open file TSP05-59 and determine the fault.

60. Open file TSP05-60 and determine the fault.

ANSWERS

SECTION CHECKUPS

Section 5–1 The DC Operating Point

1. The upper load line limit is $I_{C(sat)}$ and $V_{CE(sat)}$. The lower limit is $I_C = 0$ and $V_{CE(cutoff)}$.

2. The Q-point is the dc point at which a transistor is biased and is specified by V_{CE} and I_C.

3. Ideally, saturation occurs at the intersection of the load line and the y-axis ($V_{CE} = 0$ V). Cutoff occurs at the intersection of the load line and the $I_B = 0$ curve.
4. The Q-point must be centered on the load line for maximum V_{ce}.

Section 5–2 Voltage-Divider Bias

1. $R_{IN(BASE)} = V_{IN}/I_{IN} = 5\text{ V}/5\ \mu\text{A} = 1\text{ M}\Omega$
2. $R_{IN(BASE)} = \beta_{DC}R_E = 190(1.0\text{ k}\Omega) = 190\text{ k}\Omega$
3. $V_B = 5$ V
4. Voltage-divider bias is stable and requires only one supply voltage.

Section 5–3 Other Bias Methods

1. Emitter bias is much less dependent on the value of beta than is base bias.
2. Emitter bias requires two separate supply voltages.
3. I_C increases with β_{DC}, causing a reduction in V_C and, therefore, less voltage across R_B, thus less I_B.
4. Base bias is beta-dependent.
5. The Q-point changes due to changes in β_{DC} and V_{CE} over temperature.
6. Emitter-feedback improves stability.

Section 5–4 Troubleshooting

1. A transistor is saturated when $V_{CE} = 0$ V. A transistor is in cutoff when $V_{CE} = V_{CC}$.
2. R_E is open because the BE junction of the transistor is still forward-biased.
3. If R_C is open, V_C is about 0.7 V less than V_B.

RELATED PROBLEMS FOR EXAMPLES

5–1 $I_{CQ} = 19.8$ mA; $V_{CEQ} = 4.2$ V; $I_{b(peak)} = 42\ \mu$A
5–2 The voltage divider would be loaded, so V_B would decrease.
5–3 54.6 kΩ
5–4 $I_C \cong 2.29$ mA, $V_{EC} = 2.67$ V
5–5 214
5–6 7.83 V
5–7 $I_C = 1.41$ mA; $V_{CE} = 9.27$ V
5–8 10.3 mA
5–9 5.38 mA
5–10 $I_C = 853\ \mu$A; $V_{CE} = 1.47$ V
$\%\Delta I_C = 8.2\%$; $\%\Delta V_{CE} = -30.7\%$

TRUE/FALSE QUIZ

1. T	**2.** F	**3.** T	**4.** F	**5.** T	**6.** F
7. T	**8.** T	**9.** F	**10.** T	**11.** T	**12.** T

CIRCUIT-ACTION QUIZ

1. (a)	**2.** (b)	**3.** (b)	**4.** (b)	**5.** (a)
6. (c)	**7.** (c)	**8.** (a)	**9.** (b)	**10.** (a)

SELF-TEST

1. (b)	**2.** (c)	**3.** (d)	**4.** (d)	**5.** (c)	**6.** (a)	**7.** (b)	**8.** (d)
9. (a)	**10.** (c)	**11.** (c)	**12.** (a)	**13.** (c)	**14.** (f)	**15.** (b)	**16.** (d)

6 BJT Amplifiers

CHAPTER OUTLINE

CHAPTER OBJECTIVES

- Describe the amplifier concept
- Identify and apply internal transistor parameters
- Analyze the operation of common-emitter amplifiers
- Describe and analyze the operation of common-collector amplifiers
- Analyze the operation of common-base amplifiers
- Discuss multistage amplifiers and analyze their operation
- Discuss the differential amplifier and its operation
- Troubleshoot amplifier circuits

KEY TERMS

- *r* parameter
- Common-emitter
- ac ground
- Input resistance
- Output resistance
- Attenuation
- Bypass capacitor
- Common-collector
- Emitter-follower
- Common-base
- Decibel
- Differential amplifier
- Common mode
- CMRR (Common-mode rejection ratio)

APPLICATION ACTIVITY PREVIEW

The Application Activity in this chapter involves a preamplifier circuit for a public address system. The complete system includes the preamplifier, a power amplifier, and a dc power supply. You will focus on the preamplifier in this chapter and then on the power amplifier in Chapter 7.

VISIT THE COMPANION WEBSITE

Study aids for this chapter are available at http://www.prenhall.com/floyd

INTRODUCTION

The things you learned about biasing a transistor in Chapter 5 are now applied in this chapter where bipolar junction transistor (BJT) circuits are used as small-signal amplifiers. The term *small-signal* refers to the use of signals that take up a relatively small percentage of an amplifier's operational range. Additionally, you will learn how to reduce an amplifier to an equivalent dc and ac circuit for easier analysis, and you will learn about multistage amplifiers. The differential amplifier is also covered.

6–1 AMPLIFIER OPERATION

The biasing of a transistor is purely a dc operation. The purpose of biasing is to establish a Q-point about which variations in current and voltage can occur in response to an ac input signal. In applications where small signal voltages must be amplified—such as from an antenna or a microphone—variations about the Q-point are relatively small. Amplifiers designed to handle these small ac signals are often referred to as *small-signal amplifiers.*

After completing this section, you should be able to

- **Describe the amplifier concept**
  - Interpret labels used for dc and ac voltages and currents
  - Discuss the general operation of a small-signal amplifier
  - Analyze ac load line operation
  - Describe phase inversion

AC Quantities

In the previous chapters, dc quantities were identified by nonitalic uppercase (capital) subscripts such as I_{C}, I_{E}, V_{C}, and V_{CE}. Lowercase italic subscripts are used to indicate ac quantities of rms, peak, and peak-to-peak currents and voltages: for example, I_c, I_e, I_b, V_c, and V_{ce} (rms values are assumed unless otherwise stated). Instantaneous quantities are represented by both lowercase letters and subscripts such as i_c, i_e, i_b, and v_{ce}. Figure 6–1 illustrates these quantities for a specific voltage waveform.

HISTORY NOTE

The American inventor Lee De Forest (1873–1961) is one of several pioneers of radio development. De Forest experimented with receiving long-distance radio signals and in 1907 patented an electronic device named the *audion,* which was the first amplifier. De Forest's new three-electrode (triode) vacuum tube boosted radio waves as they were received and made possible what was then called "wireless telephony," which allowed the human voice, music, or any broadcast signal to be heard.

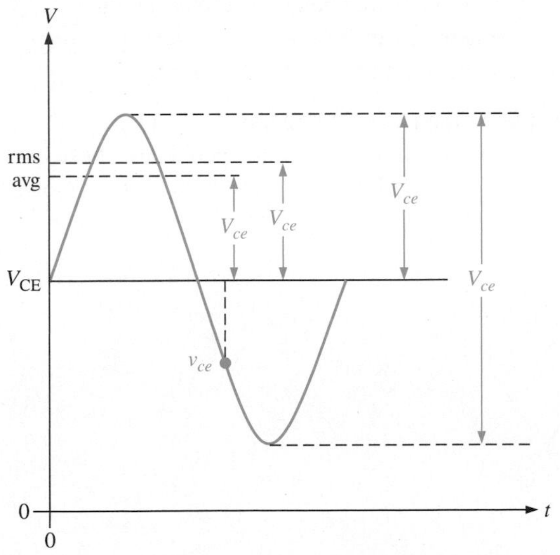

FIGURE 6–1

V_{ce} can represent rms, average, peak, or peak-to-peak, but rms will be assumed unless stated otherwise. v_{ce} can be any instantaneous value on the curve.

In addition to currents and voltages, resistances often have different values when a circuit is analyzed from an ac viewpoint as opposed to a dc viewpoint. Lowercase subscripts are used to identify ac resistance values. For example, R_c is the ac collector resistance, and R_{C} is the dc collector resistance. You will see the need for this distinction later. Resistance values *internal* to the transistor use a lowercase r' to show it is an ac resistance. An example is the internal ac emitter resistance, r'_e.

The Linear Amplifier

A linear amplifier provides amplification of a signal without any distortion so that the output signal is an exact amplified replica of the input signal. A voltage-divider biased transistor with a sinusoidal ac source capacitively coupled to the base through C_1 and a load capacitively coupled to the collector through C_2 is shown in Figure 6–2. The coupling capacitors block dc and thus prevent the internal source resistance, R_s, and the load resistance, R_L, from changing the dc bias voltages at the base and collector. The capacitors ideally appear as shorts to the signal voltage. The sinusoidal source voltage causes the base voltage to vary sinusoidally above and below its dc bias level, V_{BQ}. The resulting variation in base current produces a larger variation in collector current because of the current gain of the transistor.

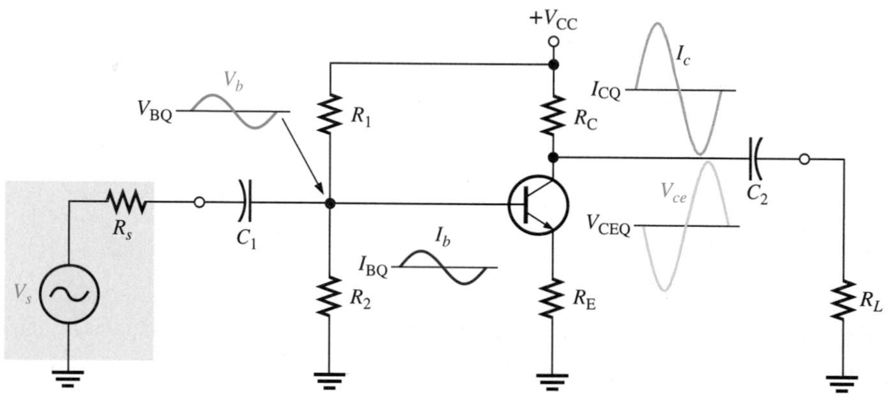

FIGURE 6–2
An amplifier with voltage-divider bias driven by an ac voltage source with an internal resistance, R_s.

As the sinusoidal collector current increases, the collector voltage decreases. The collector current varies above and below its Q-point value, I_{CQ}, in phase with the base current. The sinusoidal collector-to-emitter voltage varies above and below its Q-point value, V_{CEQ}, 180° out of phase with the base voltage, as illustrated in Figure 6–2. A transistor always produces a phase inversion between the base voltage and the collector voltage.

A Graphical Picture The operation just described can be illustrated graphically on the ac load line, as shown in Figure 6–3. The sinusoidal voltage at the base produces a base current that varies above and below the Q-point on the ac load line, as shown by the arrows.

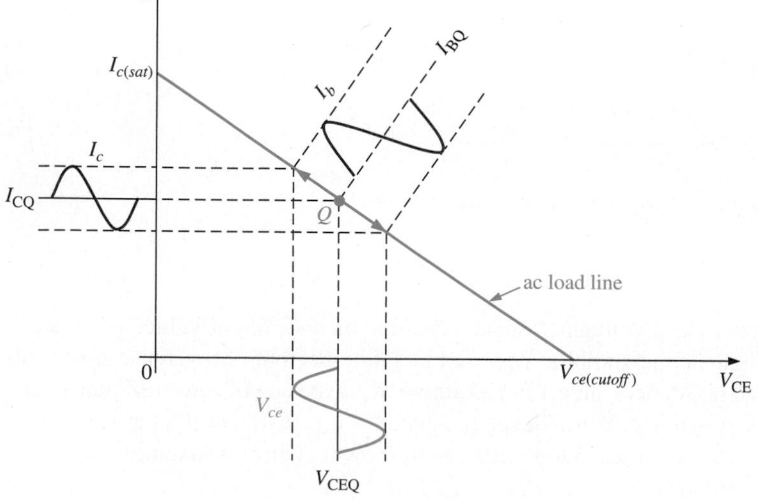

FIGURE 6–3
Graphical ac load line operation of the amplifier showing the variation of the base current, collector current, and collector-to-emitter voltage about their dc Q-point values. I_b and I_c are on different scales.

Lines projected from the peaks of the base current, across to the I_C axis, and down to the V_{CE} axis, indicate the peak-to-peak variations of the collector current and collector-to-emitter voltage, as shown. The ac load line differs from the dc load line because the effective ac collector resistance is R_L in parallel with R_C and is less than the dc collector resistance R_C alone. This difference between the dc and the ac load lines is covered in Chapter 7 in relation to power amplifiers.

EXAMPLE 6–1

The ac load line operation of a certain amplifier extends 10 μA above and below the Q-point base current value of 50 μA, as shown in Figure 6–4. Determine the resulting peak-to-peak values of collector current and collector-to-emitter voltage from the graph.

FIGURE 6–4

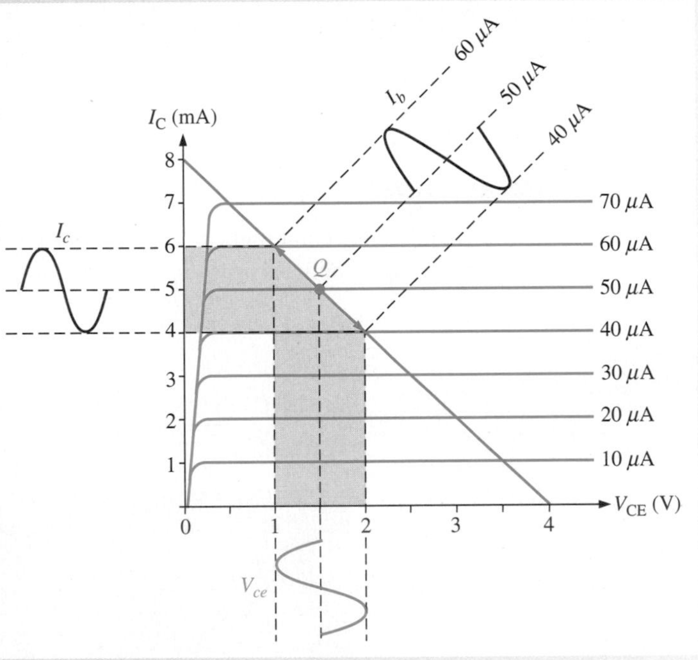

Solution Projections on the graph of Figure 6–4 show the collector current varying from 6 mA to 4 mA for a peak-to-peak value of **2 mA** and the collector-to-emitter voltage varying from 1 V to 2 V for a peak-to-peak value of **1 V**.

*Related Problem** What are the Q-point values of I_C and V_{CE} in Figure 6–4?

*Answers are at the end of the chapter.

SECTION 6–1 CHECKUP
Answers are at the end of the chapter.

1. When I_b is at its positive peak, I_c is at its ______ peak, and V_{ce} is at its ______ peak.
2. What is the difference between V_{CE} and V_{ce}?
3. What is the difference between R_e and r_e'?

6–2 TRANSISTOR AC MODELS

To visualize the operation of a transistor in an amplifier circuit, it is often useful to represent the device by a model circuit. A transistor model circuit uses various internal transistor parameters to represent its operation. Transistor models are described in this section based on resistance or *r* parameters. Another system of parameters, called *h* parameters, is briefly described.

After completing this section, you should be able to

- **Identify and apply internal transistor parameters**
  - Define the *r* parameters
  - Represent a transistor by an *r*-parameter model
  - Distinguish between the dc beta and the ac beta
  - Define the *h* parameters

r Parameters

The five *r* parameters commonly used for BJTs are given in Table 6–1. The italic lowercase letter *r* with a prime denotes resistances internal to the transistor.

TABLE 6–1

r parameters.

r PARAMETER	DESCRIPTION
α_{ac}	ac alpha (I_c/I_e)
β_{ac}	ac beta (I_c/I_b)
r'_e	ac emitter resistance
r'_b	ac base resistance
r'_c	ac collector resistance

r-Parameter Transistor Model

An ***r*-parameter** model for a BJT is shown in Figure 6–5(a). For most general analysis work, it can be simplified as follows: The effect of the ac base resistance (r'_b) is usually small enough to neglect, so it can be replaced by a short. The ac collector resistance (r'_c) is usually several hundred kilohms and can be replaced by an open. The resulting simplified *r*-parameter equivalent circuit is shown in Figure 6–5(b).

The interpretation of this model circuit in terms of a transistor's ac operation is as follows: A resistance (r'_e) appears between the emitter and base terminals. This is the resistance "seen" looking into the emitter of a forward-biased transistor. The collector effectively acts as a dependent current source of $\alpha_{ac}I_e$ or, equivalently, $\beta_{ac}I_b$, represented by the diamond-shaped symbol. These factors are shown with a transistor symbol in Figure 6–6.

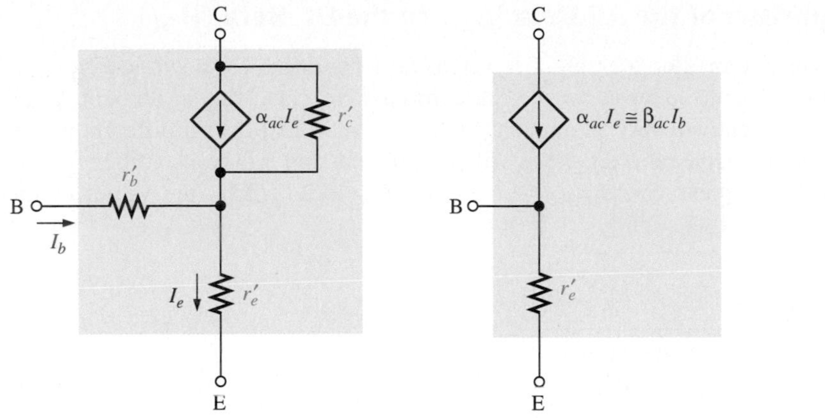

(a) Generalized r-parameter model for a BJT (b) Simplified r-parameter model for a BJT

▲ **FIGURE 6–5**

r-parameter transistor model.

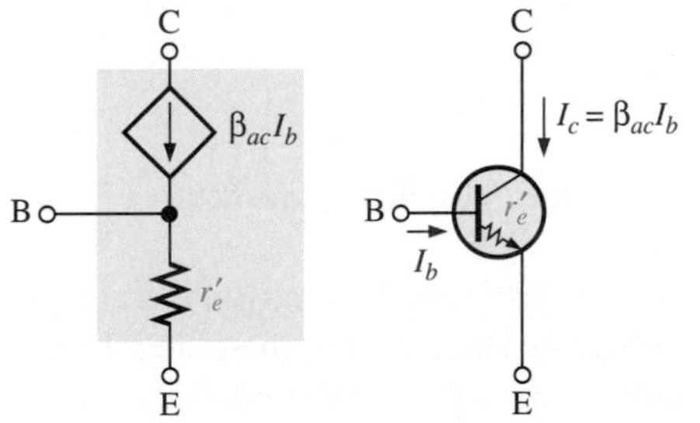

◄ **FIGURE 6–6**

Relation of transistor symbol to r-parameter model.

Determining r'_e by a Formula

For amplifier analysis, the ac emitter resistance, r'_e, is the most important of the r parameters. To calculate the approximate value of r'_e, you can use Equation 6–1, which is derived assuming an abrupt junction between the n and p regions. It is also temperature dependent and is based on an ambient temperature of 20°C.

$$r'_e \cong \frac{25\text{ mV}}{I_E}$$ **Equation 6–1**

The numerator will be slightly larger for higher temperatures or transistors with a gradual (instead of an abrupt) junction. Although these cases will yield slightly different results, most designs are not critically dependent on the value of r'_e, and you will generally obtain excellent agreement with actual circuits using the equation as given. Although Equation 6–1 is simple, its derivation is not and is therefore reserved for Appendix B.

EXAMPLE 6–2

Determine the r'_e of a transistor that is operating with a dc emitter current of 2 mA.

Solution

$$r'_e \cong \frac{25\text{ mV}}{I_E} = \frac{25\text{ mV}}{2\text{ mA}} = \mathbf{12.5\ \Omega}$$

Related Problem What is I_E if $r'_e = 8\ \Omega$?

Comparison of the AC Beta (β_{ac}) to the DC Beta (β_{DC})

For a typical transistor, a graph of I_C versus I_B is nonlinear, as shown in Figure 6–7(a). If you pick a Q-point on the curve and cause the base current to vary an amount ΔI_B, then the collector current will vary an amount ΔI_C as shown in part (b). At different points on the nonlinear curve, the ratio $\Delta I_C/\Delta I_B$ will be different, and it may also differ from the I_C/I_B ratio at the Q-point. Since $\beta_{DC} = I_C/I_B$ and $\beta_{ac} = \Delta I_C/\Delta I_B$, the values of these two quantities can differ slightly.

FIGURE 6–7

I_C-versus-I_B curve illustrates the difference between $\beta_{DC} = I_C/I_B$ and $\beta_{ac} = \Delta I_C/\Delta I_B$.

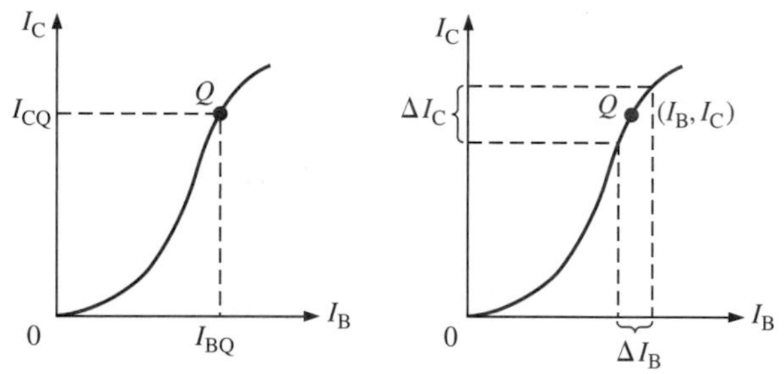

(a) $\beta_{DC} = I_C/I_B$ at Q-point (b) $\beta_{ac} = \Delta I_C/\Delta I_B$

h Parameters

A manufacturer's datasheet typically specifies *h* (hybrid) parameters (h_i, h_r, h_f, and h_o) because they are relatively easy to measure.

The four basic ac *h* parameters and their descriptions are given in Table 6–2. Each of the four *h* parameters carries a second subscript letter to designate the common-emitter (*e*), common-base (*b*), or common-collector (*c*) amplifier configuration, as listed in Table 6–3. The term *common* refers to one of the three terminals (E, B, or C) that is referenced to ac ground for both input and output signals. The characteristics of each of these three BJT amplifier configurations are covered later in this chapter.

TABLE 6–2

Basic ac *h* parameters.

h PARAMETER	DESCRIPTION	CONDITION
h_i	Input impedance (resistance)	Output shorted
h_r	Voltage feedback ratio	Input open
h_f	Forward current gain	Output shorted
h_o	Output admittance (conductance)	Input open

TABLE 6–3

Subscripts of *h* parameters for each of the three amplifier configurations.

CONFIGURATION	*h* PARAMETERS
Common-Emitter	h_{ie}, h_{re}, h_{fe}, h_{oe}
Common-Base	h_{ib}, h_{rb}, h_{fb}, h_{ob}
Common-Collector	h_{ic}, h_{rc}, h_{fc}, h_{oc}

Relationships of *h* Parameters and *r* Parameters

The ac current ratios, α_{ac} and β_{ac}, convert directly from *h* parameters as follows:

$$\alpha_{ac} = h_{fb}$$
$$\beta_{ac} = h_{fe}$$

Because datasheets often provide only common-emitter h parameters, the following formulas show how to convert them to r parameters. We will use r parameters throughout the text because they are easier to apply and more practical.

$$r'_e = \frac{h_{re}}{h_{oe}}$$

$$r'_c = \frac{h_{re} + 1}{h_{oe}}$$

$$r'_b = h_{ie} - \frac{h_{re}}{h_{oe}}(1 + h_{fe})$$

SECTION 6–2 CHECKUP

1. Define each of the parameters: α_{ac}, β_{ac}, r'_e, r'_b, and r'_c.
2. Which h parameter is equivalent to β_{ac}?
3. If $I_E = 15$ mA, what is the approximate value of r'_e?

6–3 THE COMMON-EMITTER AMPLIFIER

As you have learned, a BJT can be represented in an ac model circuit. Three amplifier configurations are the common-emitter, the common-base, and the common-collector. The common-emitter (CE) configuration has the emitter as the common terminal, or ground, to an ac signal. CE amplifiers exhibit high voltage gain and high current gain. The common-collector and common-base configurations are covered in the Sections 6–4 and 6–5.

After completing this section, you should be able to

- **Analyze the operation of common-emitter amplifiers**
  - Represent a CE amplifier by its dc equivalent circuit
  - Discuss phase inversion in a CE amplifier
  - Analyze the dc operation of a CE amplifier
  - Represent a CE amplifier by its ac equivalent circuit
  - Analyze the ac operation of a CE amplifier
  - Determine the input resistance and the output resistance
  - Determine the voltage gain
  - Explain the effects of an emitter bypass capacitor
  - Describe swamping and discuss its purpose and effects
  - Describe the effect of a load resistor on the voltage gain
  - Determine current gain and power gain

Figure 6–8 shows a **common-emitter** amplifier with voltage-divider bias and coupling capacitors C_1 and C_3 on the input and output and a bypass capacitor, C_2, from emitter to ground. The input signal, V_{in}, is capacitively coupled to the base terminal, the output signal, V_{out}, is capacitively coupled from the collector to the load. The amplified output is 180° out of phase with the input. Because the ac signal is applied to the base terminal as the input and taken from the collector terminal as the output, the emitter is common to both the

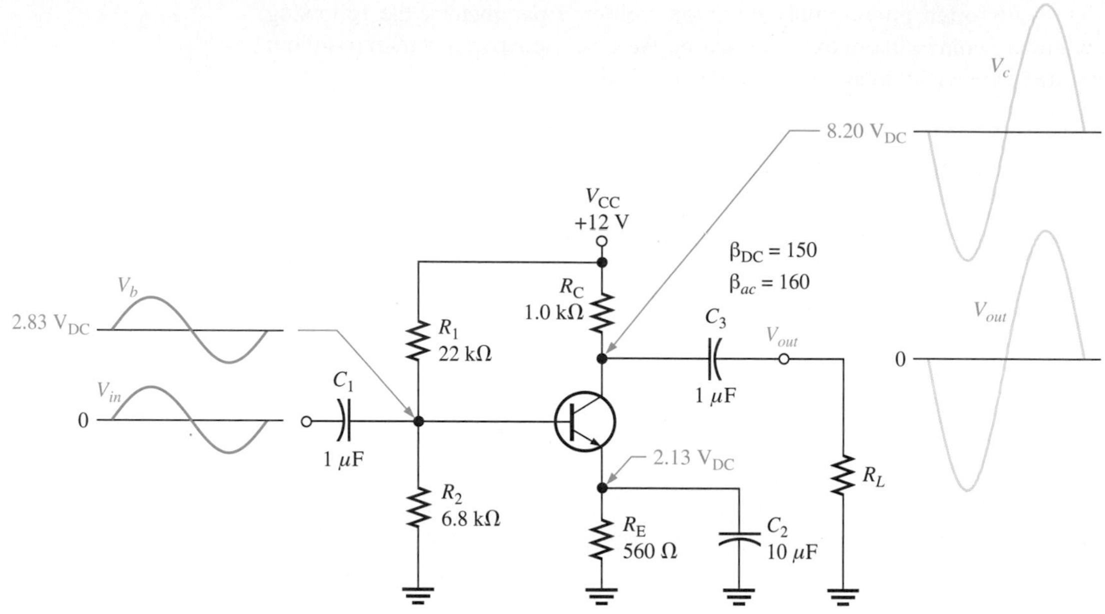

▲ FIGURE 6–8

A common-emitter amplifier.

input and output signals. There is no signal at the emitter because the bypass capacitor effectively shorts the emitter to ground at the signal frequency. All amplifiers have a combination of both ac and dc operation, which must be considered, but keep in mind that the common-emitter designation refers to the ac operation.

Phase Inversion The output signal is 180° out of phase with the input signal. As the input signal voltage changes, it causes the ac base current to change, resulting in a change in the collector current from its Q-point value. If the base current increases, the collector current increases above its Q-point value, causing an increase in the voltage drop across R_C. This increase in the voltage across R_C means that the voltage at the collector decreases from its Q-point. So, any change in input signal voltage results in an opposite change in collector signal voltage, which is a phase inversion.

DC Analysis

To analyze the amplifier in Figure 6–8, the dc bias values must first be determined. To do this, a dc equivalent circuit is developed by removing the coupling and bypass capacitors because they appear open as far as the dc bias is concerned. This also removes the load resistor and signal source. The dc equivalent circuit is shown in Figure 6–9.

The dc input resistance at the base is determined as follows:

$$R_{IN(BASE)} = \beta_{DC}R_E = (150)(560\ \Omega) = 84\ \text{k}\Omega$$

Since $R_{IN(BASE)}$ is more than ten times R_2 in this case, the voltage divider is stiff. The dc base voltage is calculated as

$$V_B \cong \left(\frac{R_2}{R_1 + R_2}\right)V_{CC} = \left(\frac{6.8\ \text{k}\Omega}{28.8\ \text{k}\Omega}\right)12\ \text{V} = 2.83\ \text{V}$$

and

$$V_E = V_B - V_{BE} = 2.83\ \text{V} - 0.7\ \text{V} = 2.13\ \text{V}$$

Therefore,

$$I_E = \frac{V_E}{R_E} = \frac{2.13\ \text{V}}{560\ \Omega} = 3.80\ \text{mA}$$

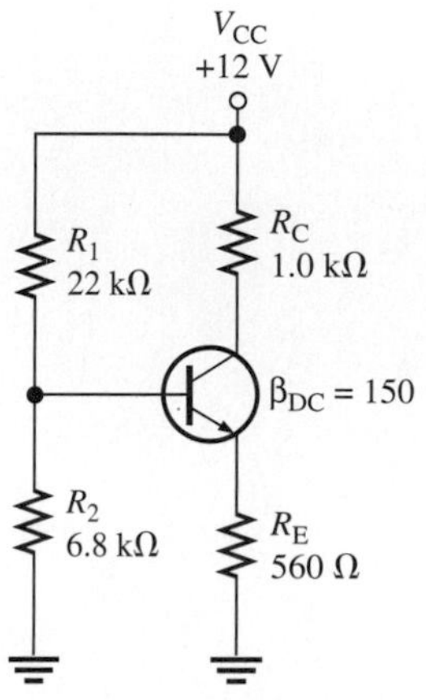

▲ FIGURE 6–9

DC equivalent circuit for the amplifier in Figure 6–8.

Since $I_C \cong I_E$, then

$$V_C = V_{CC} - I_C R_C = 12\text{ V} - (3.80\text{ mA})(1.0\text{ k}\Omega) = 12\text{ V} - 3.80\text{ V} = 8.20\text{ V}$$

Finally,

$$V_{CE} = V_C - V_E = 8.20\text{ V} - 2.13\text{ V} = 6.07\text{ V}$$

AC Analysis

To analyze the ac signal operation of an amplifier, an ac equivalent circuit is developed as follows:

1. The capacitors C_1, C_2, and C_3 are replaced by effective shorts because their values are selected so that X_C is negligible at the signal frequency and can be considered to be 0 Ω.

2. The dc source is replaced by ground.

A dc voltage source has an internal resistance of near 0 Ω because it holds a constant voltage independent of the load (within limits); no ac voltage can be developed across it so it appears as an ac short. This is why a dc source is called an **ac ground**.

The ac equivalent circuit for the common-emitter amplifier in Figure 6–8 is shown in Figure 6–10(a). Notice that both R_C and R_1 have one end connected to ac ground (red) because, in the actual circuit, they are connected to V_{CC} which is, in effect, ac ground.

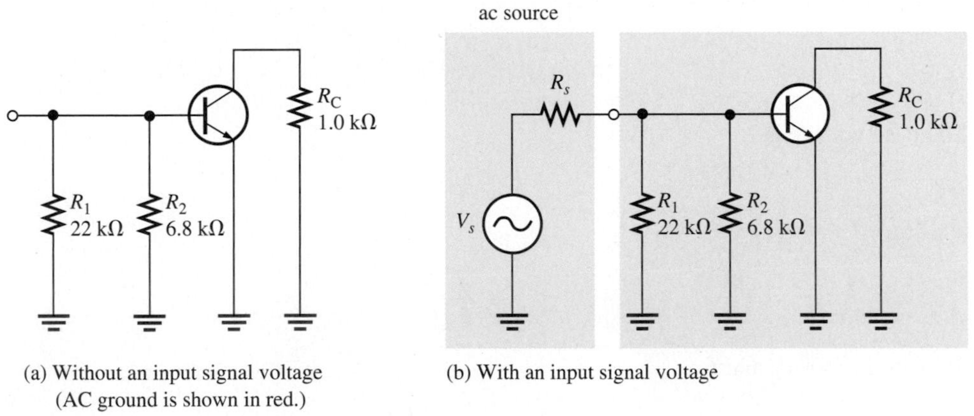

FIGURE 6–10

AC equivalent circuit for the amplifier in Figure 6–8.

In ac analysis, the ac ground and the actual ground are treated as the same point electrically. The amplifier in Figure 6–8 is called a common-emitter amplifier because the bypass capacitor C_2 keeps the emitter at ac ground. Ground is the common point in the circuit.

Signal (AC) Voltage at the Base An ac voltage source, V_s, is shown connected to the input in Figure 6–10(b). If the internal resistance of the ac source is 0 Ω, then all of the source voltage appears at the base terminal. If, however, the ac source has a nonzero internal resistance, then three factors must be taken into account in determining the actual signal voltage at the base. These are the *source resistance* (R_s), the *bias resistance* ($R_1 \| R_2$), and the *ac input resistance* at the base of the transistor ($R_{in(base)}$). This is illustrated in Figure 6–11(a) and is simplified by combining R_1, R_2, and $R_{in(base)}$ in parallel to get the total **input resistance**, $R_{in(tot)}$, which is the resistance "seen" by an ac source connected to the input, as shown in Figure 6–11(b). A high value of input resistance is desirable so that the amplifier will not excessively load the signal source. This is opposite to the requirement for a stable Q-point, which requires smaller resistors. The conflicting requirement for high input resistance and stable biasing is but one of the many trade-offs that must be considered when choosing components for a circuit. The total input resistance is expressed by the following formula:

$$\boldsymbol{R_{in(tot)} = R_1 \| R_2 \| R_{in(base)}} \qquad \text{Equation 6–2}$$

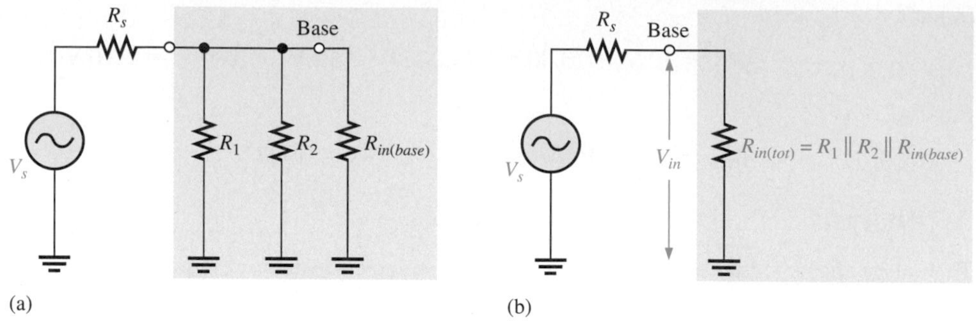

FIGURE 6–11

AC equivalent of the base circuit.

As you can see in the figure, the source voltage, V_s, is divided down by R_s (source resistance) and $R_{in(tot)}$ so that the signal voltage at the base of the transistor is found by the voltage-divider formula as follows:

$$V_b = \left(\frac{R_{in(tot)}}{R_s + R_{in(tot)}}\right)V_s$$

If $R_s \ll R_{in(tot)}$, then $V_b \cong V_s$ where V_b is the input voltage, V_{in}, to the amplifier.

Input Resistance at the Base To develop an expression for the ac input resistance looking in at the base, use the simplified r-parameter model of the transistor. Figure 6–12 shows the transistor model connected to the external collector resistor, R_C. The input resistance looking in at the base is

$$R_{in(base)} = \frac{V_{in}}{I_{in}} = \frac{V_b}{I_b}$$

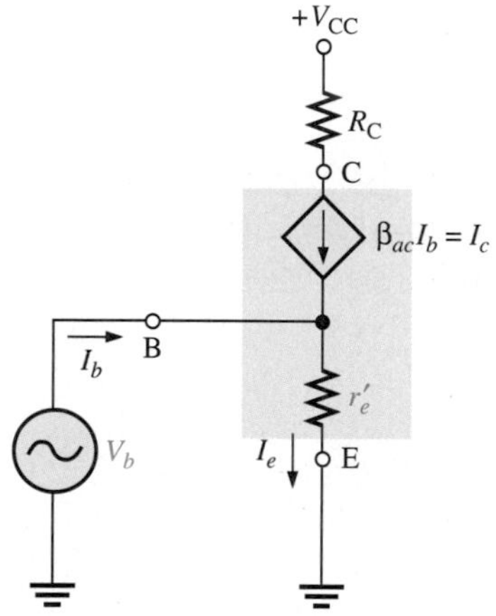

FIGURE 6–12

r-parameter transistor model (inside shaded block) connected to external circuit.

The base voltage is

$$V_b = I_e r'_e$$

and since $I_e \cong I_c$,

$$I_b \cong \frac{I_e}{\beta_{ac}}$$

Substituting for V_b and I_b,

$$R_{in(base)} = \frac{V_b}{I_b} = \frac{I_e r'_e}{I_e/\beta_{ac}}$$

Cancelling I_e,

Equation 6–3

$$\mathbf{R_{in(base)} = \beta_{ac} r'_e}$$

Output Resistance The **output resistance** of the common-emitter amplifier is the resistance looking in at the collector and is approximately equal to the collector resistor.

Equation 6–4

$$\mathbf{R_{out} \cong R_C}$$

Actually, $R_{out} = R_C \| r'_c$, but since the internal ac collector resistance of the transistor, r'_c, is typically much larger than R_C, the approximation is usually valid.

EXAMPLE 6–3

Determine the signal voltage at the base of the transistor in Figure 6–13. This circuit is the ac equivalent of the amplifier in Figure 6–8 with a 10 mV rms, 300 Ω signal source. I_E was previously found to be 3.80 mA.

FIGURE 6–13

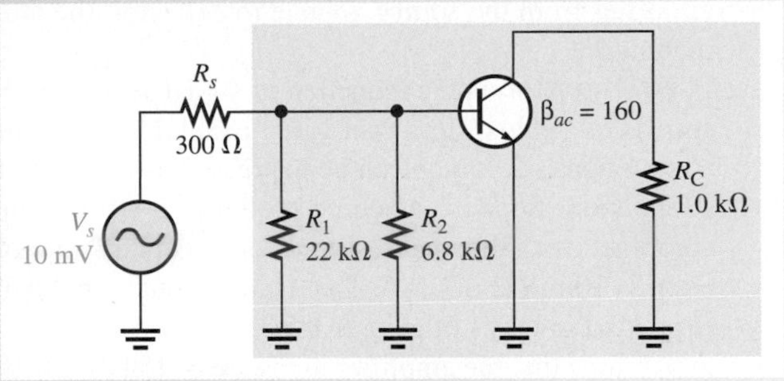

Solution First, determine the ac emitter resistance.

$$r'_e \cong \frac{25\text{ mV}}{I_E} = \frac{25\text{ mV}}{3.80\text{ mA}} = 6.58\ \Omega$$

Then,

$$R_{in(base)} = \beta_{ac}r'_e = 160(6.58\ \Omega) = 1.05\text{ k}\Omega$$

Next, determine the total input resistance viewed from the source.

$$R_{in(tot)} = R_1 \parallel R_2 \parallel R_{in(base)} = \frac{1}{\dfrac{1}{22\text{ k}\Omega} + \dfrac{1}{6.8\text{ k}\Omega} + \dfrac{1}{1.05\text{ k}\Omega}} = 873\ \Omega$$

The source voltage is divided down by R_s and $R_{in(tot)}$, so the signal voltage at the base is the voltage across $R_{in(tot)}$.

$$V_b = \left(\frac{R_{in(tot)}}{R_s + R_{in(tot)}}\right)V_s = \left(\frac{873\ \Omega}{1173\ \Omega}\right)10\text{ mV} = \mathbf{7.44\text{ mV}}$$

As you can see, there is significant attenuation (reduction) of the source voltage due to the source resistance and amplifier's input resistance combining to act as a voltage divider.

Related Problem Determine the signal voltage at the base of Figure 6–13 if the source resistance is 75 Ω and another transistor with an ac beta of 200 is used.

Voltage Gain

The ac voltage gain expression for the common-emitter amplifier is developed using the model circuit in Figure 6–14. The gain is the ratio of ac output voltage at the collector (V_c) to ac input voltage at the base (V_b).

$$A_v = \frac{V_{out}}{V_{in}} = \frac{V_c}{V_b}$$

Notice in the figure that $V_c = \alpha_{ac}I_eR_C \cong I_eR_C$ and $V_b = I_er'_e$. Therefore,

$$A_v = \frac{I_eR_C}{I_er'_e}$$

The I_e terms cancel, so

Equation 6–5

$$\mathbf{A_v = \frac{R_C}{r'_e}}$$

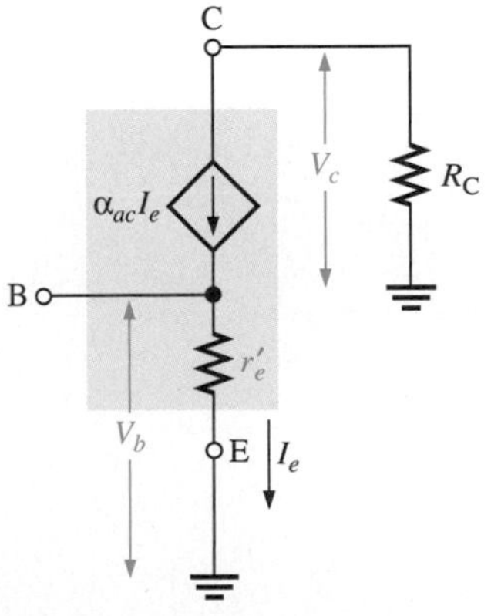

▲ FIGURE 6–14

Model circuit for obtaining ac voltage gain.

Equation 6–5 is the voltage gain from base to collector. To get the overall gain of the amplifier from the source voltage to collector, the attenuation of the input circuit must be included.

Attenuation is the reduction in signal voltage as it passes through a circuit and corresponds to a gain of less than 1. For example, if the signal amplitude is reduced by half, the attenuation is 2, which can be expressed as a gain of 0.5 because gain is the reciprocal of attenuation. Suppose a source produces a 10 mV input signal and the source resistance combined with the load resistance results in a 2 mV output signal. In this case, the attenuation is 10 mV/2 mV = 5. That is, the input signal is reduced by a factor of 5. This can be expressed in terms of gain as 1/5 = 0.2.

Assume that the amplifier in Figure 6–15 has a voltage gain from base to collector of A_v and the attenuation from the source to the base is V_s/V_b. This attenuation is produced by the source resistance and total input resistance of the amplifier acting as a voltage divider and can be expressed as

$$\text{Attenuation} = \frac{V_s}{V_b} = \frac{R_s + R_{in(tot)}}{R_{in(tot)}}$$

The overall voltage gain of the amplifier, A'_v, is the voltage gain from base to collector, A_v, times the reciprocal of the attenuation.

$$A'_v = \left(\frac{V_b}{V_s}\right)\left(\frac{V_c}{V_b}\right) = \frac{V_c}{V_s}$$

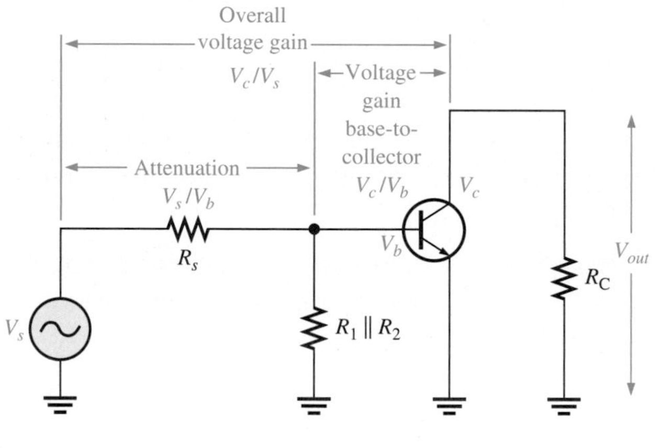

▲ **FIGURE 6–15**

Base circuit attenuation and overall voltage gain.

Effect of the Emitter Bypass Capacitor on Voltage Gain The emitter **bypass capacitor**, which is C_2 in Figure 6–8, provides an effective short to the ac signal around the emitter resistor, thus keeping the emitter at ac ground, as you have seen. With the bypass capacitor, the gain of a given amplifier is maximum and equal to R_C/r'_e.

The value of the bypass capacitor must be large enough so that its reactance over the frequency range of the amplifier is very small (ideally 0 Ω) compared to R_E. A good rule-of-thumb is that X_C of the bypass capacitor should be at least 10 times smaller than R_E at the minimum frequency for which the amplifier must operate.

$$10X_C \leq R_E$$

EXAMPLE 6–4

Select a minimum value for the emitter bypass capacitor, C_2, in Figure 6–16 if the amplifier must operate over a frequency range from 200 Hz to 10 kHz.

FIGURE 6–16

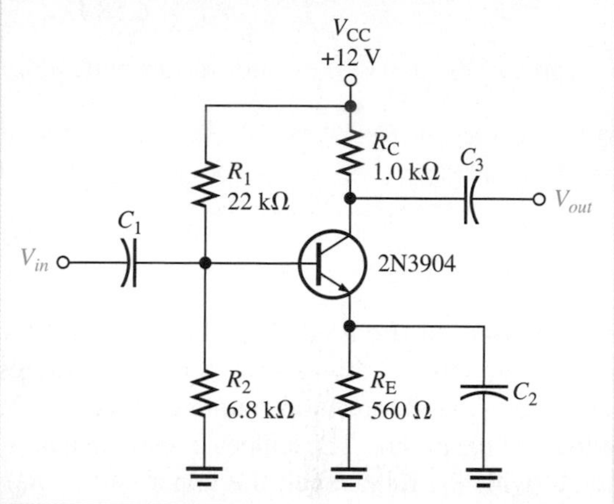

Solution The X_C of the bypass capacitor, C_2, should be at least ten times less than R_E.

$$X_{C2} = \frac{R_E}{10} = \frac{560\ \Omega}{10} = 56\ \Omega$$

Determine the capacitance value at the minimum frequency of 200 Hz as follows:

$$C_2 = \frac{1}{2\pi f X_{C2}} = \frac{1}{2\pi(200\text{ Hz})(56\ \Omega)} = \mathbf{14.2\ \mu F}$$

This is the minimum value for the bypass capacitor for this circuit. You can always use a larger value, although cost and physical size may impose limitations.

Related Problem If the minimum frequency is reduced to 100 Hz, what value of bypass capacitor must you use?

Voltage Gain Without the Bypass Capacitor To see how the bypass capacitor affects ac voltage gain, let's remove it from the circuit in Figure 6–16 and compare voltage gains.

Without the bypass capacitor, the emitter is no longer at ac ground. Instead, R_E is seen by the ac signal between the emitter and ground and effectively adds to r'_e in the voltage gain formula.

$$\mathbf{A_v = \frac{R_C}{r'_e + R_E}}$$ Equation 6–6

The effect of R_E is to decrease the ac voltage gain.

EXAMPLE 6–5

Calculate the base-to-collector voltage gain of the amplifier in Figure 6–16 both without and with an emitter bypass capacitor if there is no load resistor.

Solution From Example 6–3, $r'_e = 6.58\ \Omega$ for this same amplifier. Without C_2, the gain is

$$A_v = \frac{R_C}{r'_e + R_E} = \frac{1.0\text{ k}\Omega}{567\ \Omega} = \mathbf{1.76}$$

With C_2, the gain is

$$A_v = \frac{R_C}{r'_e} = \frac{1.0\text{ k}\Omega}{6.58\ \Omega} = \mathbf{152}$$

As you can see, the bypass capacitor makes quite a difference.

Related Problem Determine the base-to-collector voltage gain in Figure 6–16 with R_E bypassed, for the following circuit values: $R_C = 1.8\text{ k}\Omega$, $R_E = 1.0\text{ k}\Omega$, $R_1 = 33\text{ k}\Omega$, and $R_2 = 6.8\text{ k}\Omega$.

Effect of a Load on the Voltage Gain A **load** is the amount of current drawn from the output of an amplifier or other circuit through a load resistance. When a resistor, R_L, is connected to the output through the coupling capacitor C_3, as shown in Figure 6–17(a), it creates a load on the circuit. The collector resistance at the signal frequency is effectively R_C in parallel with R_L. Remember, the upper end of R_C is effectively at ac ground. The ac equivalent circuit is shown in Figure 6–17(b). The total ac collector resistance is

$$R_c = \frac{R_C R_L}{R_C + R_L}$$

Replacing R_C with R_c in the voltage gain expression gives

Equation 6–7

$$\mathbf{A_v = \frac{R_c}{r'_e}}$$

When $R_c < R_C$ because of R_L, the voltage gain is reduced. However, if $R_L \gg R_C$, then $R_c \cong R_C$ and the load has very little effect on the gain.

FIGURE 6–17

A common-emitter amplifier with an ac (capacitively) coupled load.

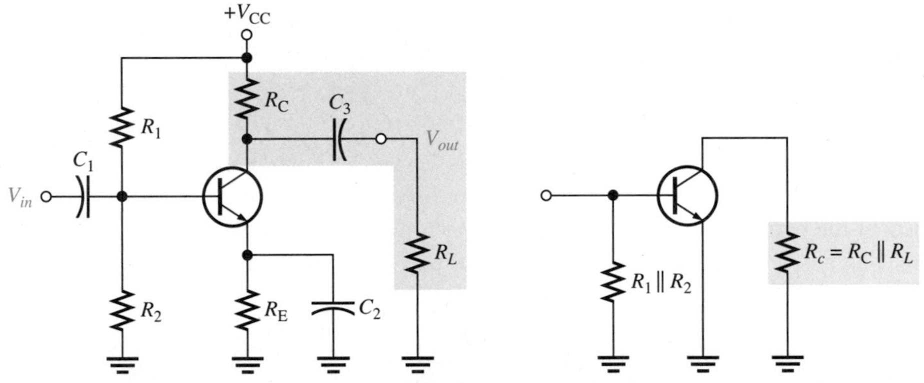

(a) Complete amplifier

(b) AC equivalent ($X_{C1} = X_{C2} = X_{C3} = 0$)

EXAMPLE 6–6

Calculate the base-to-collector voltage gain of the amplifier in Figure 6–16 when a load resistance of 5 kΩ is connected to the output. The emitter is effectively bypassed and $r'_e = 6.58\ \Omega$.

Solution The ac collector resistance is

$$R_c = \frac{R_C R_L}{R_C + R_L} = \frac{(1.0\text{ k}\Omega)(5\text{ k}\Omega)}{6\text{ k}\Omega} = 833\ \Omega$$

Therefore,

$$A_v = \frac{R_c}{r'_e} = \frac{833\ \Omega}{6.58\ \Omega} = \mathbf{127}$$

The unloaded gain was found to be 152 in Example 6–5.

Related Problem Determine the base-to-collector voltage gain in Figure 6–16 when a 10 kΩ load resistance is connected from collector to ground. Change the resistance values as follows: $R_C = 1.8\ k\Omega$, $R_E = 1.0\ k\Omega$, $R_1 = 33\ k\Omega$, and $R_2 = 6.8\ k\Omega$. The emitter resistor is effectively bypassed and $r'_e = 18.5\ \Omega$.

Stability of the Voltage Gain

Stability is a measure of how well an amplifier maintains its design values over changes in temperature or for a transistor with a different β. Although bypassing R_E does produce the maximum voltage gain, there is a stability problem because the ac voltage gain is dependent on r'_e since $A_v = R_C/r'_e$. Also, r'_e depends on I_E and on temperature. This causes the gain to be unstable over changes in temperature because when r'_e increases, the gain decreases and vice versa.

With no bypass capacitor, the gain is decreased because R_E is now in the ac circuit $(A_v = R_C/(r'_e + R_E))$. However, with R_E unbypassed, the gain is much less dependent on r'_e. If $R_E \gg r'_e$, the gain is essentially independent of r'_e because

$$A_v \cong \frac{R_C}{R_E}$$

Swamping r'_e to Stabilize the Voltage Gain *Swamping* is a method used to minimize the effect of r'_e without reducing the voltage gain to its minimum value. This method "swamps" out the effect of r'_e on the voltage gain. Swamping is, in effect, a compromise between having a bypass capacitor across R_E and having no bypass capacitor at all.

In a swamped amplifier, R_E is partially bypassed so that a reasonable gain can be achieved, and the effect of r'_e on the gain is greatly reduced or eliminated. The total external emitter resistance, R_E, is formed with two separate emitter resistors, R_{E1} and R_{E2}, as indicated in Figure 6–18. One of the resistors, R_{E2}, is bypassed and the other is not.

Both resistors $(R_{E1} + R_{E2})$ affect the dc bias while only R_{E1} affects the ac voltage gain.

$$A_v = \frac{R_C}{r'_e + R_{E1}}$$

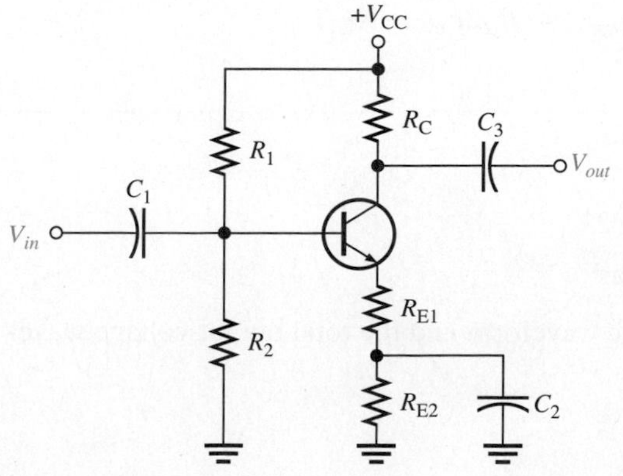

FIGURE 6–18

A swamped amplifier uses a partially bypassed emitter resistance to minimize the effect of r'_e on the gain in order to achieve gain stability.

If R_{E1} is at least ten times larger than r'_e, then the effect of r'_e is minimized and the approximate voltage gain for the swamped amplifier is

Equation 6–8

$$A_v \cong \frac{R_C}{R_{E1}}$$

EXAMPLE 6–7

Determine the voltage gain of the swamped amplifier in Figure 6–19. Assume that the bypass capacitor has a negligible reactance for the frequency at which the amplifier is operated. Assume $r'_e = 20\ \Omega$.

FIGURE 6–19

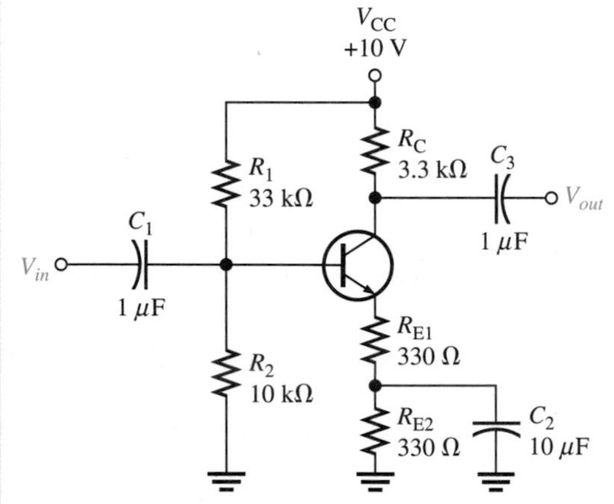

Solution R_{E2} is bypassed by C_2. R_{E1} is more than ten times r'_e so the approximate voltage gain is

$$A_v \cong \frac{R_C}{R_{E1}} = \frac{3.3\ \text{k}\Omega}{330\ \Omega} = \mathbf{10}$$

Related Problem What would be the voltage gain without C_2? What would be the voltage gain with C_2 bypassing both R_{E1} and R_{E2}?

The Effect of Swamping on the Amplifier's Input Resistance The ac input resistance, looking in at the base of a common-emitter amplifier with R_E completely bypassed, is $R_{in} = \beta_{ac}r'_e$. When the emitter resistance is partially bypassed, the portion of the resistance that is unbypassed is seen by the ac signal and results in an increase in the ac input resistance by appearing in series with r'_e. The formula is

Equation 6–9

$$R_{in(base)} = \beta_{ac}(r'_e + R_{E1})$$

EXAMPLE 6–8

For the amplifier in Figure 6–20,

(a) Determine the dc collector voltage.

(b) Determine the ac collector voltage.

(c) Draw the total collector voltage waveform and the total output voltage waveform.

FIGURE 6–20

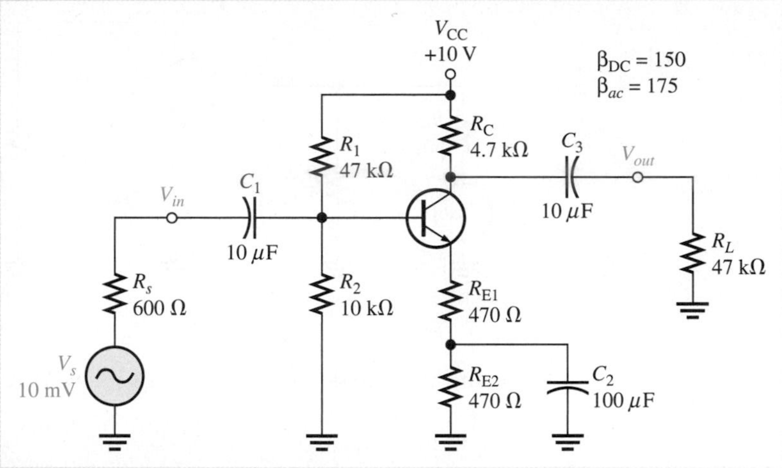

Solution **(a)** Determine the dc bias values using the dc equivalent circuit in Figure 6–21.

FIGURE 6–21

DC equivalent for the circuit in Figure 6–20.

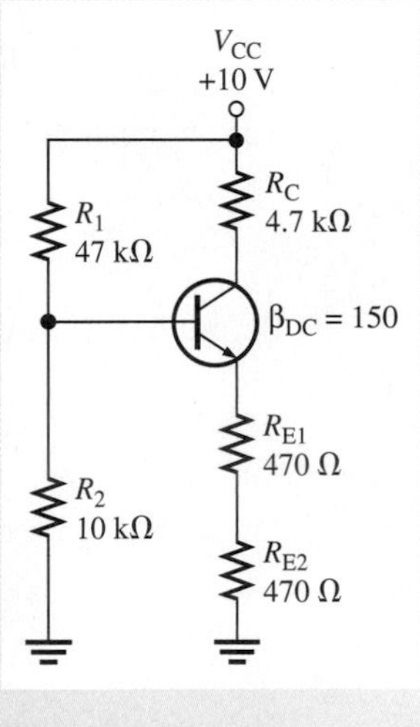

$$R_{IN(BASE)} = \beta_{DC}(R_{E1} + R_{E2}) = 150(940\ \Omega) = 141\ k\Omega$$

Since $R_{IN(BASE)}$ is more than ten times larger than R_2, you have a stiff voltage divider and $R_{IN(BASE)}$ can be neglected in the dc base voltage calculation.

$$V_B \cong \left(\frac{R_2}{R_1 + R_2}\right)V_{CC} = \left(\frac{10\ k\Omega}{47\ k\Omega + 10\ k\Omega}\right)10\ V = 1.75\ V$$

$$V_E = V_B - 0.7\ V = 1.75\ V - 0.7\ V = 1.05\ V$$

$$I_E = \frac{V_E}{R_{E1} + R_{E2}} = \frac{1.05\ V}{940\ \Omega} = 1.12\ mA$$

$$V_C = V_{CC} - I_C R_C = 10\ V - (1.12\ mA)(4.7\ k\Omega) = \mathbf{4.74\ V}$$

(b) The ac analysis is based on the ac equivalent circuit in Figure 6–22.

FIGURE 6–22

AC equivalent for the circuit in Figure 6–20.

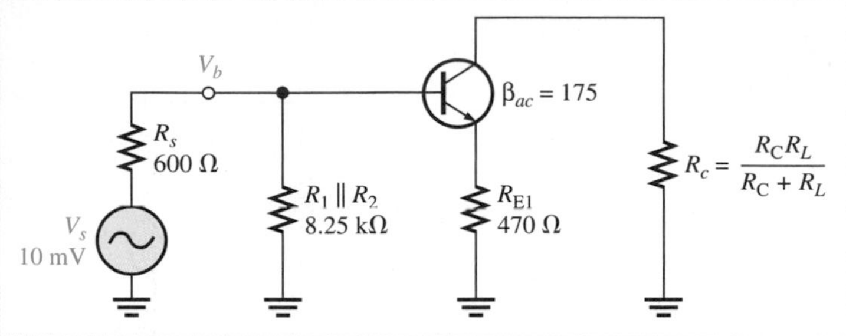

The first thing to do in the ac analysis is calculate r'_e.

$$r'_e \cong \frac{25\text{ mV}}{I_E} = \frac{25\text{ mV}}{1.12\text{ mA}} = 22\ \Omega$$

Next, determine the attenuation in the base circuit. Looking from the 600 Ω source, the total R_{in} is

$$R_{in(tot)} = R_1 \parallel R_2 \parallel R_{in(base)}$$
$$R_{in(base)} = \beta_{ac}(r'_e + R_{E1}) = 175(492\ \Omega) = 86.1\text{ k}\Omega$$

Therefore,

$$R_{in(tot)} = 47\text{ k}\Omega \parallel 10\text{ k}\Omega \parallel 86.1\text{ k}\Omega = 7.53\text{ k}\Omega$$

The attenuation from source to base is

$$\text{Attenuation} = \frac{V_s}{V_b} = \frac{R_s + R_{in(tot)}}{R_{in(tot)}} = \frac{600\ \Omega + 7.53\text{ k}\Omega}{7.53\text{ k}\Omega} = 1.08$$

Before A_v can be determined, you must know the ac collector resistance R_c.

$$R_c = \frac{R_C R_L}{R_C + R_L} = \frac{(4.7\text{ k}\Omega)(47\text{ k}\Omega)}{4.7\text{ k}\Omega + 47\text{ k}\Omega} = 4.27\text{ k}\Omega$$

The voltage gain from base to collector is

$$A_v \cong \frac{R_c}{R_{E1}} = \frac{4.27\text{ k}\Omega}{470\ \Omega} = 9.09$$

The overall voltage gain is the reciprocal of the attenuation times the amplifier voltage gain.

$$A'_v = \left(\frac{V_b}{V_s}\right)A_v = (0.93)(9.09) = 8.45$$

The source produces 10 mV rms, so the rms voltage at the collector is

$$V_c = A'_v V_s = (8.45)(10\text{ mV}) = \mathbf{84.5\ mV}$$

(c) The total collector voltage is the signal voltage of 84.5 mV rms riding on a dc level of 4.74 V, as shown in Figure 6–23(a), where approximate peak values are determined as follows:

$$\text{Max } V_{c(p)} = V_C + 1.414\,V_c = 4.74\text{ V} + (84.5\text{ mV})(1.414) = 4.86\text{ V}$$
$$\text{Min } V_{c(p)} = V_C - 1.414\,V_c = 4.74\text{ V} - (84.5\text{ mV})(1.414) = 4.62\text{ V}$$

The coupling capacitor, C_3, keeps the dc level from getting to the output. So, V_{out} is equal to the ac component of the collector voltage ($V_{out(p)} = (84.5\text{ mV})(1.414) = 119\text{ mV}$),

FIGURE 6–23

Voltages for Figure 6–20.

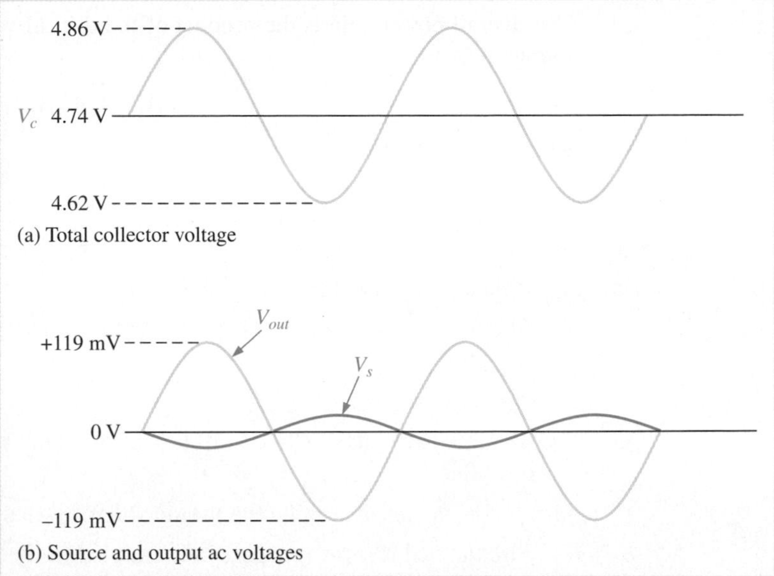

as indicated in Figure 6–23(b). The source voltage is shown to emphasize the phase inversion.

Related Problem What is A_v in Figure 6–20 with R_L removed?

Open the Multisim file E06-08 in the Examples folder on your CD-ROM. Measure the dc and the ac values of the collector voltage and compare with the calculated values.

Current Gain

The current gain from base to collector is I_c/I_b or β_{ac}. However, the overall current gain of the common-emitter amplifier is

$$A_i = \frac{I_c}{I_s}$$

Equation 6–10

I_s is the total signal input current produced by the source, part of which (I_b) is base current and part of which (I_{bias}) goes through the bias circuit ($R_1 \| R_2$), as shown in Figure 6–24. The source "sees" a total resistance of $R_s + R_{in(tot)}$. The total current produced by the source is

$$I_s = \frac{V_s}{R_s + R_{in(tot)}}$$

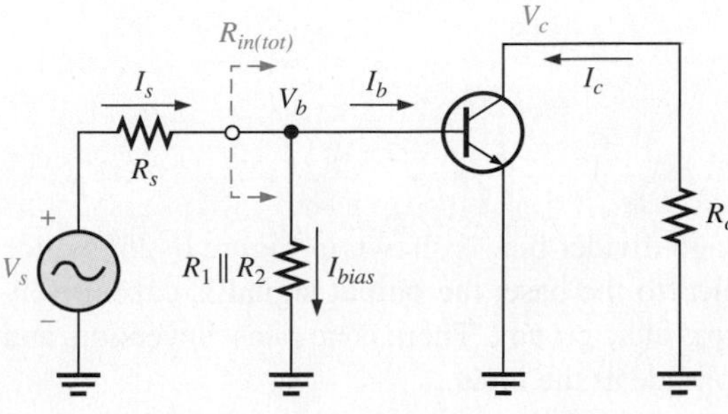

FIGURE 6–24

Signal currents (directions shown are for the positive half-cycle of V_s).

Power Gain

The overall power gain is the product of the overall voltage gain (A'_v) and the overall current gain (A_i).

Equation 6–11

$$A_p = A'_v A_i$$

where $A'_v = V_c/V_s$.

SECTION 6–3 CHECKUP

1. In the dc equivalent circuit of an amplifier, how are the capacitors treated?
2. When the emitter resistor is bypassed with a capacitor, how is the gain of the amplifier affected?
3. Explain swamping.
4. List the elements included in the total input resistance of a common-emitter amplifier.
5. What elements determine the overall voltage gain of a common-emitter amplifier?
6. When a load resistor is capacitively coupled to the collector of a CE amplifier, is the voltage gain increased or decreased?
7. What is the phase relationship of the input and output voltages of a CE amplifier?

6–4 THE COMMON-COLLECTOR AMPLIFIER

The **common-collector** (CC) amplifier is usually referred to as an emitter-follower (EF). The input is applied to the base through a coupling capacitor, and the output is at the emitter. The voltage gain of a CC amplifier is approximately 1, and its main advantages are its high input resistance and current gain.

After completing this section, you should be able to

- **Describe and analyze the operation of common-collector amplifiers**
  - Represent a CC amplifier by its dc and ac equivalent circuits
  - Analyze the dc and ac operation of a CC amplifier
  - Determine the voltage gain
  - Determine the input resistance
  - Determine the output resistance
  - Determine the current gain
  - Determine the power gain
  - Discuss the Darlington pair and the complementary Darlington pair (Sziklai pair)

An **emitter-follower** circuit with voltage-divider bias is shown in Figure 6–25. Notice that the input signal is capacitively coupled to the base, the output signal is capacitively coupled from the emitter, and the collector is at ac ground. There is no phase inversion, and the output is approximately the same amplitude as the input.

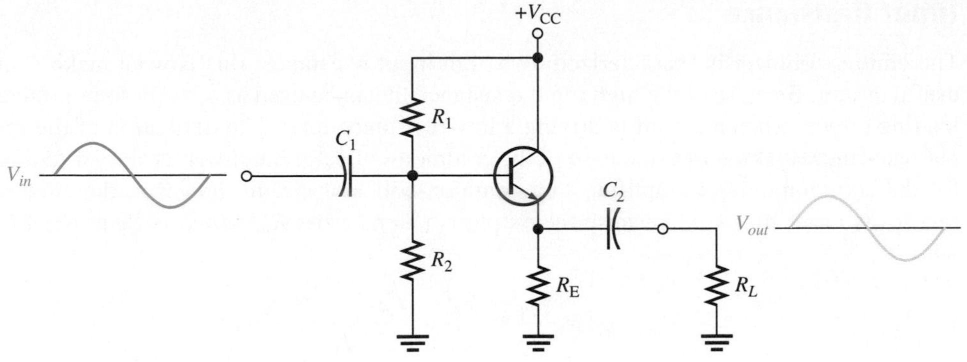

FIGURE 6–25
Emitter-follower with voltage-divider bias.

Voltage Gain

As in all amplifiers, the voltage gain is $A_v = V_{out}/V_{in}$. The capacitive reactances are assumed to be negligible at the frequency of operation. For the emitter-follower, as shown in the ac model in Figure 6–26,

$$V_{out} = I_e R_e$$

and

$$V_{in} = I_e(r'_e + R_e)$$

Therefore, the voltage gain is

$$A_v = \frac{I_e R_e}{I_e(r'_e + R_e)}$$

The I_e current terms cancel, and the base-to-emitter voltage gain expression simplifies to

$$A_v = \frac{R_e}{r'_e + R_e}$$

where R_e is the parallel combination of R_E and R_L. If there is no load, then $R_e = R_E$. Notice that the gain is always less than 1. If $R_e \gg r'_e$, then a good approximation is

$$A_v \cong 1$$ Equation 6–12

Since the output voltage is at the emitter, it is in phase with the base voltage, so there is no inversion from input to output. Because there is no inversion and because the voltage gain is approximately 1, the output voltage closely follows the input voltage in both phase and amplitude; thus the term *emitter-follower*.

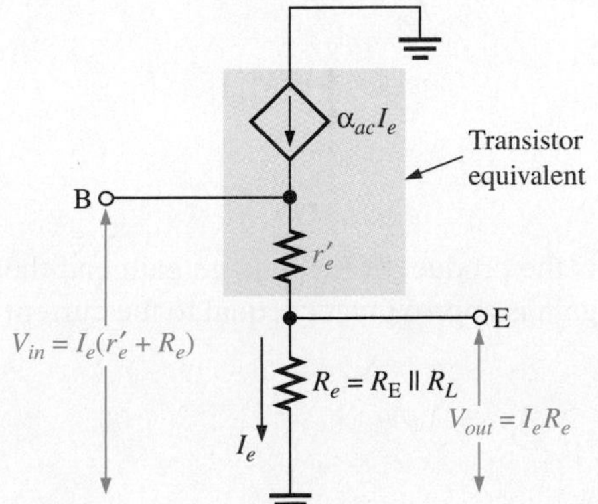

FIGURE 6–26
Emitter-follower model for voltage gain derivation.

Input Resistance

The emitter-follower is characterized by a high input resistance; this is what makes it a useful circuit. Because of the high input resistance, it can be used as a buffer to minimize loading effects when a circuit is driving a low-resistance load. The derivation of the input resistance, looking in at the base of the common-collector amplifier, is similar to that for the common-emitter amplifier. In a common-collector circuit, however, the emitter resistor is *never* bypassed because the output is taken across R_e, which is R_E in parallel with R_L.

$$R_{in(base)} = \frac{V_{in}}{I_{in}} = \frac{V_b}{I_b} = \frac{I_e(r'_e + R_e)}{I_b}$$

Since $I_e \cong I_c = \beta_{ac}I_b$,

$$R_{in(base)} \cong \frac{\beta_{ac}I_b(r'_e + R_e)}{I_b}$$

The I_b terms cancel; therefore,

$$R_{in(base)} \cong \beta_{ac}(r'_e + R_e)$$

If $R_e \gg r'_e$, then the input resistance at the base is simplified to

Equation 6–13

$$\boldsymbol{R_{in(base)} \cong \beta_{ac}R_e}$$

The bias resistors in Figure 6–25 appear in parallel with $R_{in(base)}$, looking from the input source; and just as in the common-emitter circuit, the total input resistance is

$$R_{in(tot)} = R_1 \,\|\, R_2 \,\|\, R_{in(base)}$$

Output Resistance

With the load removed, the output resistance, looking into the emitter of the emitter-follower, is approximated as follows:

Equation 6–14

$$\boldsymbol{R_{out} \cong \left(\frac{R_s}{\beta_{ac}}\right) \| R_E}$$

R_s is the resistance of the input source. The derivation of this expression is relatively involved and several simplifying assumptions have been made, as shown in Appendix B. The output resistance is very low, making the emitter-follower useful for driving low-resistance loads.

Current Gain

The current gain for the emitter-follower in Figure 6–25 is

Equation 6–15

$$\boldsymbol{A_i = \frac{I_e}{I_{in}}}$$

where $I_{in} = V_{in}/R_{in(tot)}$.

Power Gain

The common-collector power gain is the product of the voltage gain and the current gain. For the emitter-follower, the power gain is approximately equal to the current gain because the voltage gain is approximately 1.

$$A_p = A_vA_i$$

Since $A_v \cong 1$, the power gain is

Equation 6–16

$$\boldsymbol{A_p \cong A_i}$$

EXAMPLE 6–9

Determine the total input resistance of the emitter-follower in Figure 6–27. Also find the voltage gain, current gain, and power gain in terms of power delivered to the load, R_L. Assume $\beta_{ac} = 175$ and that the capacitive reactances are negligible at the frequency of operation.

FIGURE 6–27

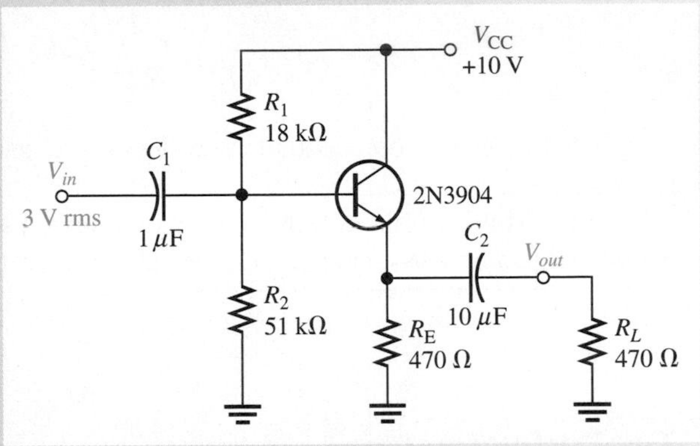

Solution The ac emitter resistance external to the transistor is

$$R_e = R_E \| R_L = 470\ \Omega \| 470\ \Omega = 235\ \Omega$$

The approximate resistance, looking in at the base, is

$$R_{in(base)} \cong \beta_{ac} R_e = (175)(235\ \Omega) = 41.1\ \text{k}\Omega$$

The total input resistance is

$$R_{in(tot)} = R_1 \| R_2 \| R_{in(base)} = 18\ \text{k}\Omega \| 51\ \text{k}\Omega \| 41.1\ \text{k}\Omega = \mathbf{10.1\ k\Omega}$$

The voltage gain is $A_v \cong 1$. By using r'_e, you can determine a more precise value of A_v if necessary.

$$V_E = \left(\frac{R_2}{R_1 + R_2}\right)V_{CC} - V_{BE} = \left(\frac{51\ \text{k}\Omega}{18\ \text{k}\Omega + 51\ \text{k}\Omega}\right)10\ \text{V} - 0.7\ \text{V}$$

$$= (0.739)(10\ \text{V}) - 0.7\ \text{V} = 6.69\ \text{V}$$

Therefore,

$$I_E = \frac{V_E}{R_E} = \frac{6.69\ \text{V}}{470\ \Omega} = 14.2\ \text{mA}$$

and

$$r'_e \cong \frac{25\ \text{mV}}{I_E} = \frac{25\ \text{mV}}{14.2\ \text{mA}} = 1.76\ \Omega$$

So,

$$A_v = \frac{R_e}{r'_e + R_e} = \frac{235\ \Omega}{237\ \Omega} = \mathbf{0.992}$$

The small difference in A_v as a result of considering r'_e is insignificant in most cases.

The current gain is $A_i = I_e/I_{in}$. The calculations are as follows:

$$I_e = \frac{V_e}{R_e} = \frac{A_v V_b}{R_e} \cong \frac{(0.992)(3\ \text{V})}{235\ \Omega} = \frac{2.98\ \text{V}}{235\ \Omega} = 12.7\ \text{mA}$$

$$I_{in} = \frac{V_{in}}{R_{in(tot)}} = \frac{3\ \text{V}}{10.1\ \text{k}\Omega} = 297\ \mu\text{A}$$

$$A_i = \frac{I_e}{I_{in}} = \frac{12.7\ \text{mA}}{297\ \mu\text{A}} = \mathbf{42.8}$$

The power gain is

$$A_p \cong A_i = 42.8$$

Since $R_L = R_E$, one-half of the power is dissipated in R_E and one-half in R_L. Therefore, in terms of power to the load, the power gain is

$$A_{p(load)} = \frac{A_p}{2} = \frac{42.8}{2} = \mathbf{21.4}$$

Related Problem If R_L in Figure 6–27 is decreased in value, does power gain to the load increase or decrease?

Open the Multisim file E06-09 in the Examples folder on your CD-ROM. Measure the voltage gain and compare with the calculated value.

HISTORY NOTE

Sidney Darlington (1906–1997) was a renowned electrical engineer, whose name lives on through the transistor configuration he patented in 1953. He also had inventions in chirp radar, bombsights, and gun and rocket guidance. In 1945, he was awarded the Presidential Medal of Freedom and in 1975, he received IEEE's Edison Medal "for basic contributions to network theory and for important inventions in radar systems and electronic circuits" and the IEEE Medal of Honor in 1981 "for fundamental contributions to filtering and signal processing leading to chirp radar."

The Darlington Pair

As you have seen, β_{ac} is a major factor in determining the input resistance of an amplifier. The β_{ac} of the transistor limits the maximum achievable input resistance you can get from a given emitter-follower circuit.

One way to boost input resistance is to use a **Darlington pair,** as shown in Figure 6–28. The collectors of two transistors are connected, and the emitter of the first drives the base of the second. This configuration achieves β_{ac} multiplication as shown in the following steps. The emitter current of the first transistor is

$$I_{e1} \cong \beta_{ac1} I_{b1}$$

This emitter current becomes the base current for the second transistor, producing a second emitter current of

$$I_{e2} \cong \beta_{ac2} I_{e1} = \beta_{ac1} \beta_{ac2} I_{b1}$$

Therefore, the effective current gain of the Darlington pair is

$$\beta_{ac} = \beta_{ac1} \beta_{ac2}$$

Neglecting r'_e by assuming that it is much smaller than R_E, the input resistance is

Equation 6–17

$$\boldsymbol{R_{in} = \beta_{ac1} \beta_{ac2} R_E}$$

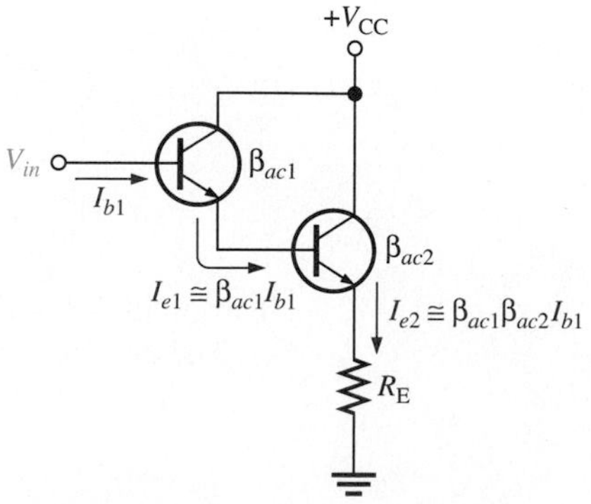

FIGURE 6–28

A Darlington pair multiplies β_{ac}.

An Application

The emitter-follower is often used as an interface between a circuit with a high output resistance and a low-resistance load. In such an application, the emitter-follower is called a *buffer.*

Suppose a common-emitter amplifier with a 1.0 kΩ collector resistance must drive a low-resistance load such as an 8 Ω low-power speaker. If the speaker is capacitively coupled to the output of the amplifier, the 8 Ω load appears—to the ac signal—in parallel with the 1.0 kΩ collector resistor. This results in an ac collector resistance of

$$R_c = R_C \| R_L = 1.0\,\text{k}\Omega \| 8\,\Omega = 7.94\,\Omega$$

Obviously, this is not acceptable because most of the voltage gain is lost ($A_v = R_c/r'_e$). For example, if $r'_e = 5\,\Omega$, the voltage gain is reduced from

$$A_v = \frac{R_C}{r'_e} = \frac{1.0\,\text{k}\Omega}{5\,\Omega} = 200$$

with no load to

$$A_v = \frac{R_c}{r'_e} = \frac{7.94\,\Omega}{5\,\Omega} = 1.59$$

with an 8 Ω speaker load.

An emitter-follower using a Darlington pair can be used to interface the amplifier and the speaker, as shown in Figure 6–29.

TECH NOTE

The circuit arrangement in Figure 6–29 is useful for low-power applications (<1 W load power) but is inefficient and wasteful for higher power requirements.

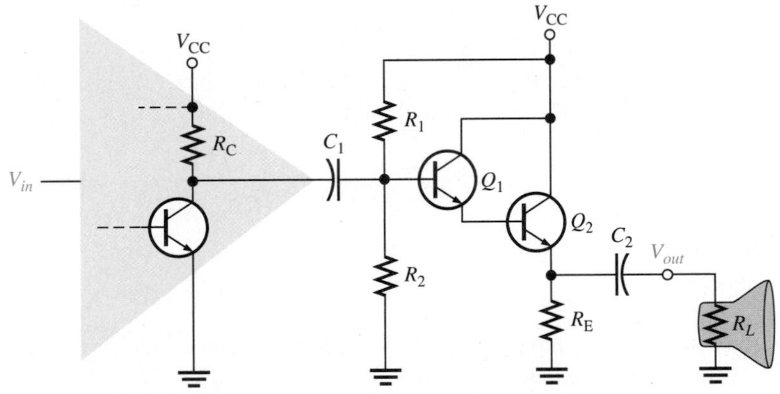

FIGURE 6–29

A Darlington emitter-follower used as a buffer between a common-emitter amplifier and a low-resistance load such as a speaker.

EXAMPLE 6–10

In Figure 6–29 for the common-emitter amplifier, $V_{CC} = 12\text{ V}$, $R_C = 1.0\,\text{k}\Omega$ and $r'_e = 5\,\Omega$. For the Darlington emitter-follower, $R_1 = 10\,\text{k}\Omega$, $R_2 = 22\,\text{k}\Omega$, $R_E = 22\,\Omega$, $R_L = 8\,\Omega$, $V_{CC} = 12\text{ V}$, and $\beta_{DC} = \beta_{ac} = 100$ for each transistor.

(a) Determine the voltage gain of the common-emitter amplifier.

(b) Determine the voltage gain of the Darlington emitter-follower.

(c) Determine the overall voltage gain and compare to the gain of the common-emitter amplifier driving the speaker directly without the Darlington emitter-follower.

Solution **(a)** To determine A_v for the common-emitter amplifier, first find r'_e for the Darlington emitter-follower.

$$V_B = \left(\frac{R_2 \| \beta_{DC}^2 R_E}{R_1 + R_2 \| \beta_{DC}^2 R_E}\right)V_{CC} = \left(\frac{20\,\text{k}\Omega}{30\,\text{k}\Omega}\right)12\text{ V} = 8.0\text{ V}$$

$$I_E = \frac{V_E}{R_E} = \frac{V_B - 2V_{BE}}{R_E} = \frac{8.0\text{ V} - 1.4\text{ V}}{22\,\Omega} = \frac{6.6\text{ V}}{22\,\Omega} = 300\text{ mA}$$

$$r'_e = \frac{25\text{ mV}}{I_E} = \frac{25\text{ mV}}{300\text{ mA}} = 83\text{ m}\Omega$$

Note that R_E must dissipate a power of

$$P_{R_E} = I_E^2 R_E = (300\text{ mA})^2(22\ \Omega) = 1.98\text{ W}$$

and transistor Q_2 must dissipate

$$P_{Q2} = (V_{CC} - V_E)I_E = (5.4\text{ V})(300\text{ mA}) = 1.62\text{ W}$$

Next, the ac emitter resistance of the Darlington emitter-follower is

$$R_e = R_E \| R_L = 22\ \Omega \| 8\ \Omega = 5.87\ \Omega$$

The total input resistance of the Darlington emitter-follower is

$$R_{in(tot)} = R_1 \| R_2 \| \beta_{ac}^2(r'_e + R_e)$$
$$= 10\text{ k}\Omega \| 22\text{ k}\Omega \| 100^2(83\text{ m}\Omega + 5.87\ \Omega) = 6.16\text{ k}\Omega$$

The effective ac collector resistance of the common-emitter amplifier is

$$R_c = R_C \| R_{in(tot)} = 1.0\text{ k}\Omega \| 6.16\text{ k}\Omega = 860\ \Omega$$

The voltage gain of the common-emitter amplifier is

$$A_v = \frac{R_c}{r'_e} = \frac{860\ \Omega}{5\ \Omega} = \mathbf{172}$$

(b) The effective ac emitter resistance was found in part (a) to be 5.87 Ω. The voltage gain for the Darlington emitter-follower is

$$A_v = \frac{R_e}{r'_e + R_e} = \frac{5.87\ \Omega}{83\text{ m}\Omega + 5.87\ \Omega} = \mathbf{0.99}$$

(c) The overall voltage gain is

$$A'_v = A_{v(EF)}A_{v(CE)} = (0.99)(172) = \mathbf{170}$$

If the common-emitter amplifier drives the speaker directly, the gain is 1.59 as we previously calculated.

Related Problem Using the same circuit values, determine the voltage gain of the common-emitter amplifier in Figure 6–29 if a single transistor is used in the emitter-follower in place of the Darlington pair. Assume $\beta_{DC} = \beta_{ac} = 100$. Explain the difference in the voltage gain without the Darlington pair.

The Sziklai Pair

The **Sziklai pair,** shown in Figure 6–30, is similar to the Darlington pair except that it consists of two types of transistors, an *npn* and a *pnp*. This configuration is sometimes known as a *complementary Darlington* or a *compound transistor.* The current gain is

HISTORY NOTE

George Clifford Sziklai, born in Hungary in 1909, was an electronics engineer, who emigrated to New York in 1930. Among many other contributions to radio and TV, he invented the transistor configuration named after him, the *Sziklai pair,* also known as the complementary Darlington. Sziklai is also credited with constructing the first image orthicon television camera and inventing a high-speed elevator in addition to some 200 other patents.

▶ **FIGURE 6–30**

The Sziklai pair.

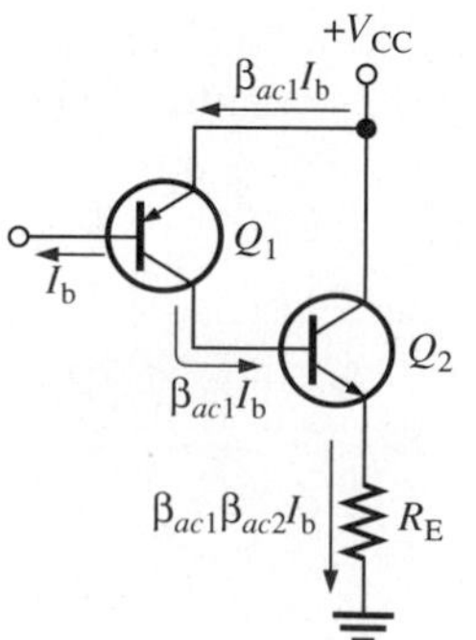

about the same as in the Darlington pair, as illustrated. The difference is that the Q_2 base current is the Q_1 collector current instead of emitter current, as in the Darlington arrangement.

An advantage of the Sziklai pair, compared to the Darlington, is that it takes less voltage to turn it on because only one barrier potential has to be overcome. A Sziklai pair is sometimes used in conjunction with a Darlington pair as the output stage of power amplifiers. In this case, the output power transistors are both the same type (two *npn* or two *pnp* transistors). This makes it easier to obtain exact matches of the output transistors, resulting in improved thermal stability and better sound quality in audio applications.

SECTION 6–4 CHECKUP

1. What is a common-collector amplifier called?
2. What is the ideal maximum voltage gain of a common-collector amplifier?
3. What characteristic of the common-collector amplifier makes it a useful circuit?

6–5 THE COMMON-BASE AMPLIFIER

The common-base (CB) amplifier provides high voltage gain with a maximum current gain of 1. Since it has a low input resistance, the CB amplifier is the most appropriate type for certain applications where sources tend to have very low-resistance outputs.

After completing this section, you should be able to

- **Analyze the operation of common-base amplifiers**
  - Represent a CB amplifier by its dc and ac equivalent circuits
  - Analyze the dc and ac operation of a CB amplifier
  - Determine the voltage gain
  - Determine the input resistance
  - Determine the output resistance
  - Determine the current gain
  - Determine the power gain

A typical **common-base** amplifier is shown in Figure 6–31. The base is the common terminal and is at ac ground because of capacitor C_2. The input signal is capacitively coupled to the emitter. The output is capacitively coupled from the collector to a load resistor.

TECHNOTE

The CB amplifier is useful at high frequencies when impedance matching is required because input impedance can be controlled and because noninverting amps have better frequency response.

Voltage Gain

The voltage gain from emitter to collector is developed as follows ($V_{in} = V_e$, $V_{out} = V_c$).

$$A_v = \frac{V_{out}}{V_{in}} = \frac{V_c}{V_e} = \frac{I_cR_c}{I_e(r'_e \| R_E)} \cong \frac{I_eR_c}{I_e(r'_e \| R_E)}$$

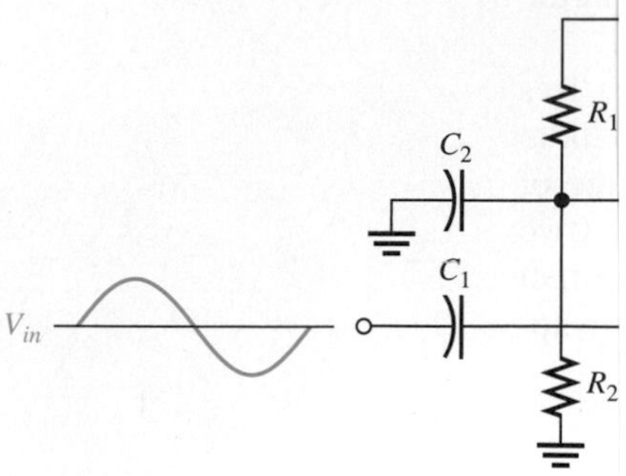

(a) Complete circuit with load

FIG
Comm

If R_E

Equation 6–18

where
ter am

Inpu
The re

If R_E

Equation 6–19

R_E is t
The in

Outp
Lookin
you ha
than R

Equation 6–20

Curr
The cu
rent, a

Equation 6–21

Powe
Since t
the pov

Equation 6–22

FIGURE 6–39

Single-ended differential input operation.

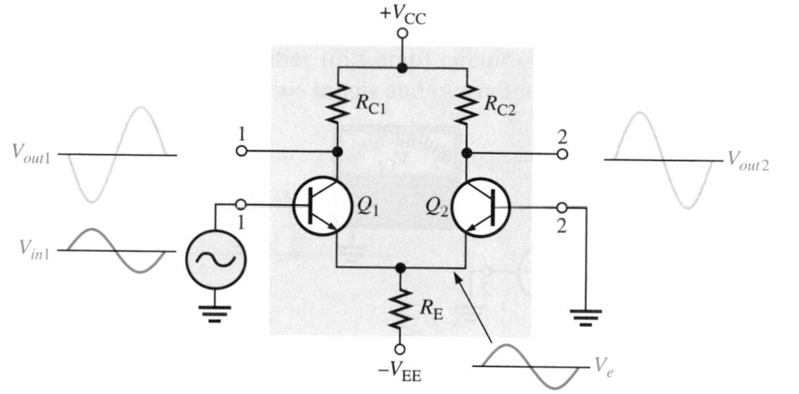

(a)

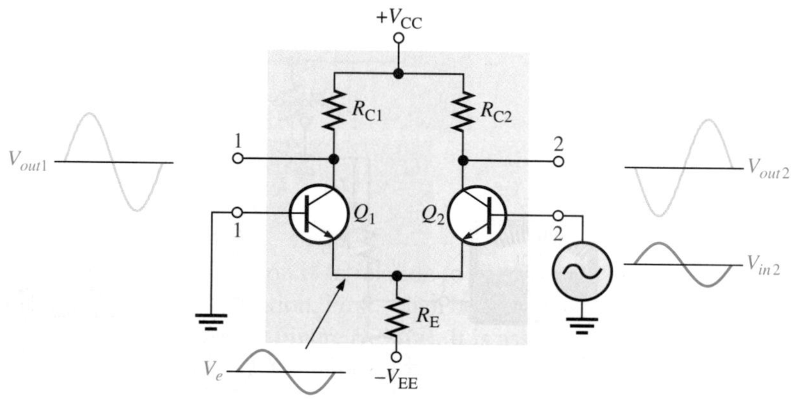

(b)

Double-Ended Differential Inputs In this input configuration, two opposite-polarity (out-of-phase) signals are applied to the inputs, as shown in Figure 6–40(a). Each input affects the outputs, as you will see in the following discussion.

Figure 6–40(b) shows the output signals due to the signal on input 1 acting alone as a single-ended input. Figure 6–40(c) shows the output signals due to the signal on input 2 acting alone as a single-ended input. Notice in parts (b) and (c) that the signals on output 1 are of the same polarity. The same is also true for output 2. By superimposing both output 1 signals and both output 2 signals, you get the total output signals, as shown in Figure 6–40(d).

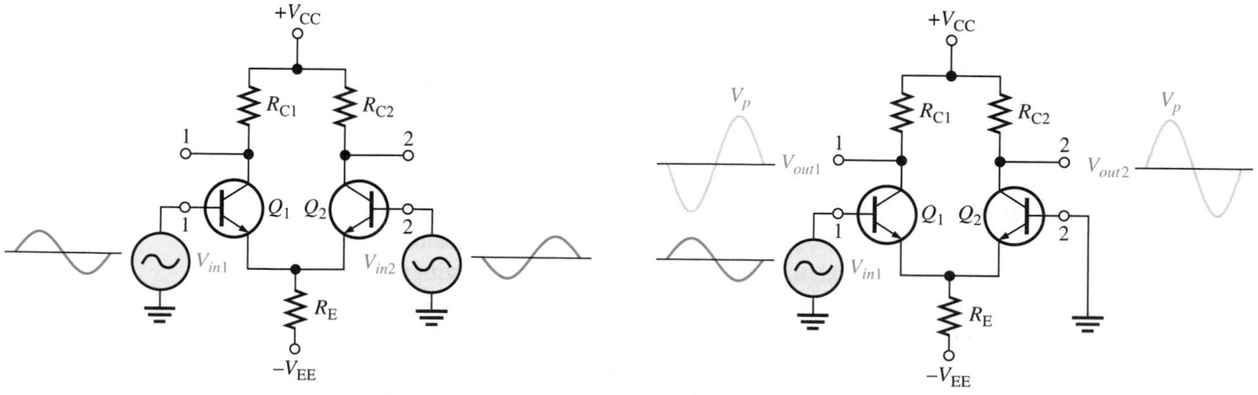

(a) Differential inputs (180° out of phase)

(b) Outputs due to V_{in1}

FIGURE 6–40

Double-ended differential operation.

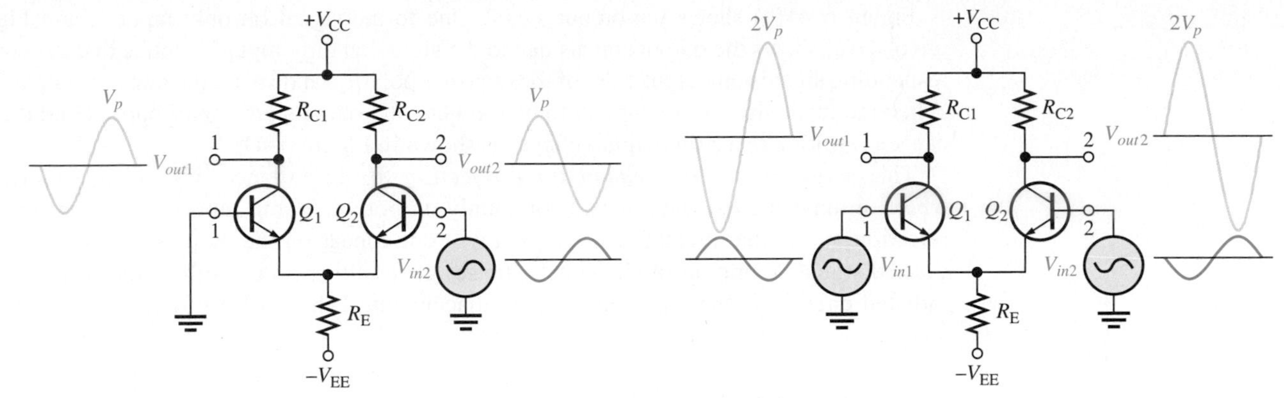

(c) Outputs due to V_{in2}

(d) Total outputs

▲ **FIGURE 6–40**

(continued)

Common-Mode Inputs One of the most important aspects of the operation of a diff-amp can be seen by considering the **common-mode** condition where two signal voltages of the same phase, frequency, and amplitude are applied to the two inputs, as shown in Figure 6–41(a). Again, by considering each input signal as acting alone, the basic operation can be understood.

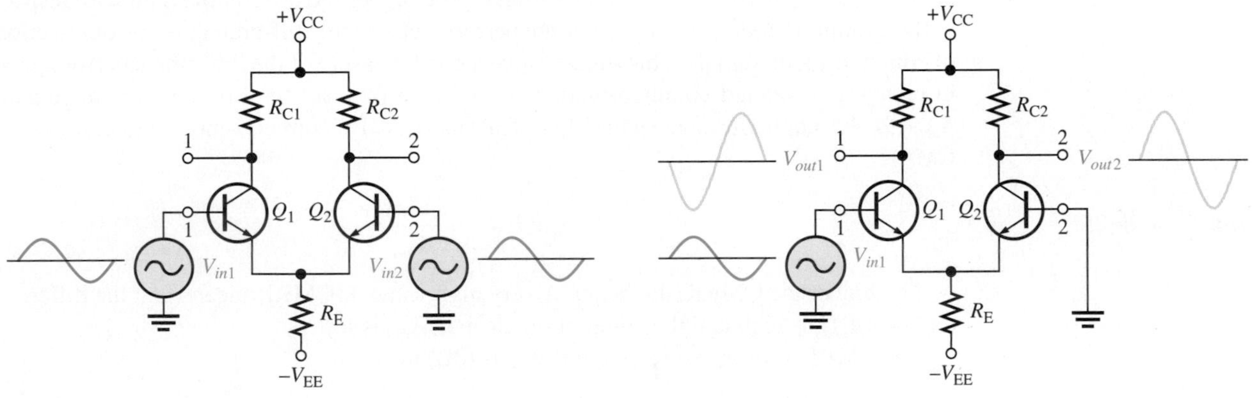

(a) Common-mode inputs (in phase)

(b) Outputs due to V_{in1}

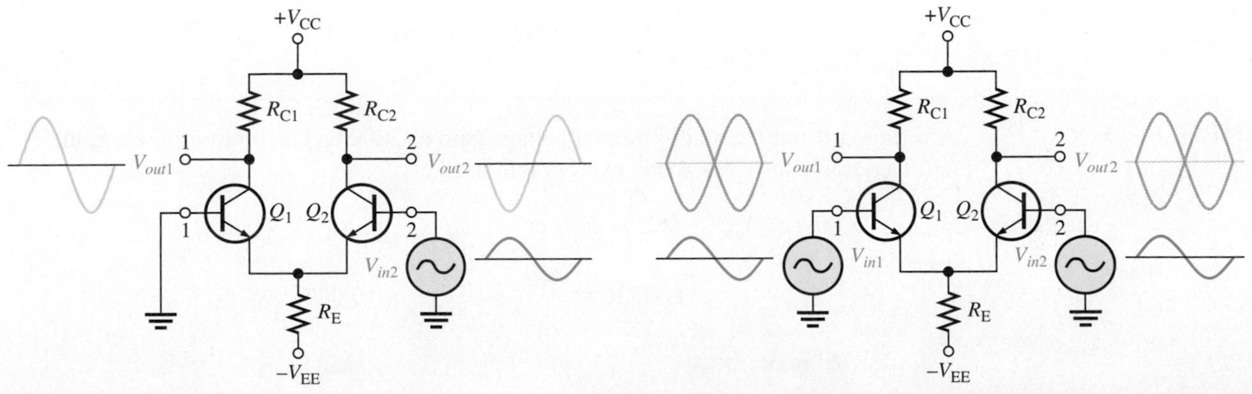

(c) Outputs due to V_{in2}

(d) Outputs due to V_{in1} and V_{in2} cancel because they are equal in amplitude but opposite in phase. The resulting outputs are 0 V ac.

▲ **FIGURE 6–41**

Common-mode operation of a differential amplifier.

Figure 6–41(b) shows the output signals due to the signal on only input 1, and Figure 6–41(c) shows the output signals due to the signal on only input 2. Notice that the corresponding signals on output 1 are of the opposite polarity, and so are the ones on output 2. When the input signals are applied to both inputs, the outputs are superimposed and they cancel, resulting in a zero output voltage, as shown in Figure 6–41(d).

This action is called *common-mode rejection.* Its importance lies in the situation where an unwanted signal appears commonly on both diff-amp inputs. Common-mode rejection means that this unwanted signal will not appear on the outputs and distort the desired signal. Common-mode signals (noise) generally are the result of the pick-up of radiated energy on the input lines from adjacent lines, the 60 Hz power line, or other sources.

Common-Mode Rejection Ratio

Desired signals appear on only one input or with opposite polarities on both input lines. These desired signals are amplified and appear on the outputs as previously discussed. Unwanted signals (noise) appearing with the same polarity on both input lines are essentially cancelled by the diff-amp and do not appear on the outputs. The measure of an amplifier's ability to reject common-mode signals is a parameter called the **CMRR (common-mode rejection ratio)**.

Ideally, a diff-amp provides a very high gain for desired signals (single-ended or differential) and zero gain for common-mode signals. Practical diff-amps, however, do exhibit a very small common-mode gain (usually much less than 1), while providing a high differential voltage gain (usually several thousand). The higher the differential gain with respect to the common-mode gain, the better the performance of the diff-amp in terms of rejection of common-mode signals. This suggests that a good measure of the diff-amp's performance in rejecting unwanted common-mode signals is the ratio of the differential voltage gain $A_{v(d)}$ to the common-mode gain, A_{cm}. This ratio is the common-mode rejection ratio, CMRR.

Equation 6–25

$$\mathbf{CMRR} = \frac{A_{v(d)}}{A_{cm}}$$

The higher the CMRR, the better. A very high value of CMRR means that the differential gain $A_{v(d)}$ is high and the common-mode gain A_{cm} is low.

The CMRR is often expressed in decibels (dB) as

Equation 6–26

$$\mathbf{CMRR} = 20 \log \left(\frac{A_{v(d)}}{A_{cm}} \right)$$

EXAMPLE 6–13

A certain diff-amp has a differential voltage gain of 2000 and a common-mode gain of 0.2. Determine the CMRR and express it in decibels.

Solution $A_{v(d)} = 2000$, and $A_{cm} = 0.2$. Therefore,

$$\text{CMRR} = \frac{A_{v(d)}}{A_{cm}} = \frac{2000}{0.2} = \mathbf{10{,}000}$$

Expressed in decibels,

$$\text{CMRR} = 20 \log (10{,}000) = \mathbf{80\ dB}$$

Related Problem Determine the CMRR and express it in dB for an amplifier with a differential voltage gain of 8500 and a common-mode gain of 0.25.

A CMRR of 10,000 means that the desired input signal (differential) is amplified 10,000 times more than the unwanted noise (common-mode). For example, if the amplitudes of the differential input signal and the common-mode noise are equal, the desired signal will appear on the output 10,000 times greater in amplitude than the noise. Thus, the noise or interference has been essentially eliminated.

SECTION 6–7 CHECKUP

1. Distinguish between double-ended and single-ended differential inputs.
2. Define *common-mode rejection.*
3. For a given value of differential gain, does a higher CMRR result in a higher or lower common-mode gain?

6–8 TROUBLESHOOTING

In working with any circuit, you must first know how it is supposed to work before you can troubleshoot it for a failure. The two-stage capacitively coupled amplifier discussed in Section 6–6 is used to illustrate a typical troubleshooting procedure.

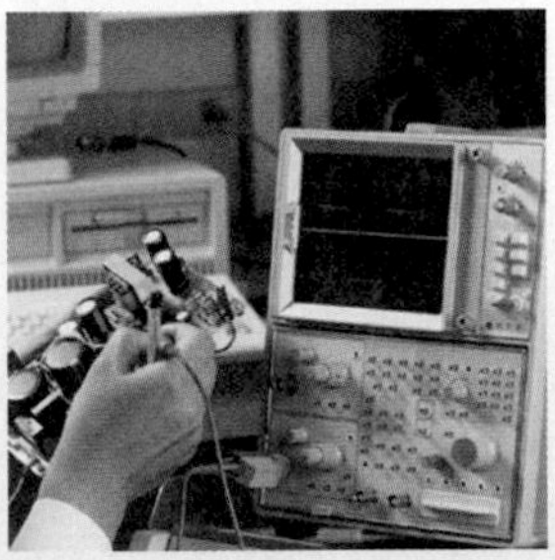

After completing this section, you should be able to

- **Troubleshoot amplifier circuits**
    - Discuss the complete troubleshooting process
    - Apply the troubleshooting process to a two-stage amplifier
    - Use the signal-tracing method
    - Apply fault analysis

At this point, you should review the general troubleshooting techniques discussed in Chapter 2. The two-stage common-emitter amplifier that was discussed in Section 6–6 is used to illustrate basic multistage amplifier troubleshooting.

When you are faced with having to troubleshoot a circuit, the first thing you need is a schematic with the proper dc and signal voltages labeled. You must know what the correct voltages in the circuit should be before you can identify an incorrect voltage. Schematics of some circuits are available with voltages indicated at certain points. If this is not the case, you must use your knowledge of the circuit operation to determine the correct voltages. Figure 6–42 is the schematic for the two-stage amplifier that was analyzed in Section 6–6. The correct voltages are indicated at each point.

Troubleshooting Procedure

The analysis, planning, and measurement approach to troubleshooting will be used.

Analysis It has been found that there is no output voltage, V_{out}. You have also determined that the circuit did work properly and then failed. A visual check of the circuit board or assembly for obvious problems such as broken or poor connections, solder splashes, wire

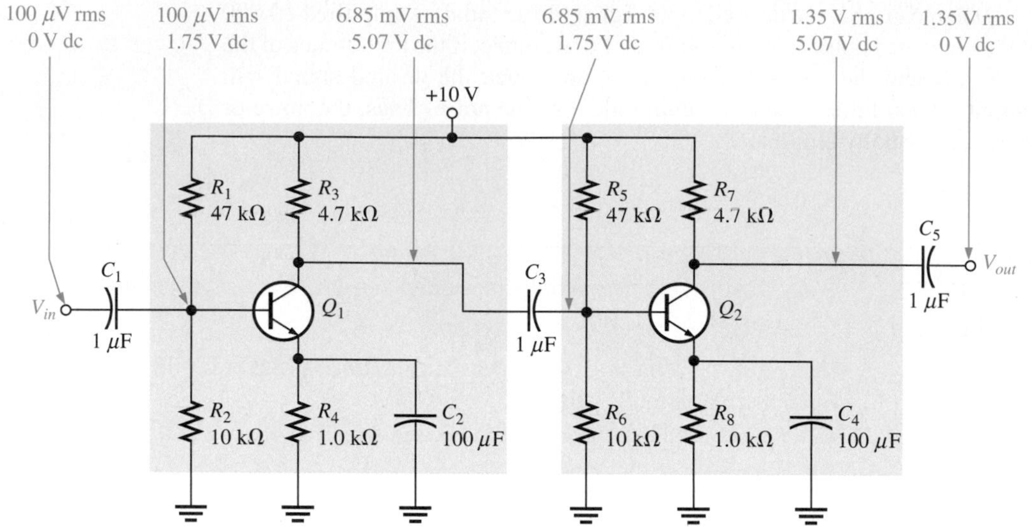

▲ **FIGURE 6–42**

A two-stage common-emitter amplifier with correct voltages indicated. Both transistors have dc and ac betas of 150. Different values of β will produce slightly different results.

clippings, or burned components turns up nothing. You conclude that the problem is most likely a faulty component in the amplifier circuit or an open connection. Also, the dc supply voltage may not be correct or may be missing.

Planning You decide to use an oscilloscope to check the dc levels and the ac signals (you prefer to use a DMM to measure the dc voltages) at certain test points. Also, you decide to apply the half-splitting method to trace the voltages in the circuit and use an in-circuit transistor tester if a transistor is suspected of being faulty.

Measurement To determine the faulty component in a multistage amplifier, use the general five-step troubleshooting procedure which is illustrated as follows.

Step 1: *Perform a power check.* Assume the dc supply voltage is correct as indicated in Figure 6–43.

Step 2: *Check the input and output voltages.* Assume the measurements indicate that the input signal voltage is correct. However, there is no output signal voltage or the output signal voltage is much less than it should be, as shown by the diagram in Figure 6–43.

► **FIGURE 6–43**

Initial check of a faulty two-stage amplifier.

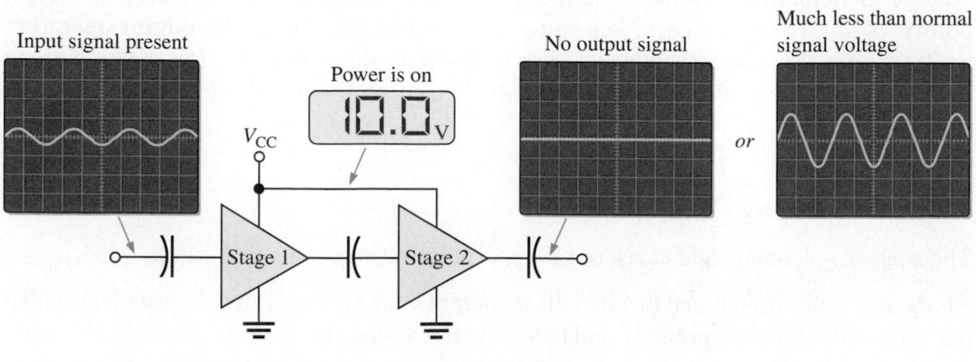

Step 3: *Apply the half-splitting method of signal tracing.* Check the voltages at the output of the first stage. No signal voltage or a much less than normal signal voltage indicates that the problem is in the first stage. An incorrect dc voltage also indicates a first-stage problem. If the signal voltage and the dc voltage are correct at the output of the first stage, the problem is in the second stage. After this check, you have narrowed the problem to one of the two stages. This step is illustrated in Figure 6–44.

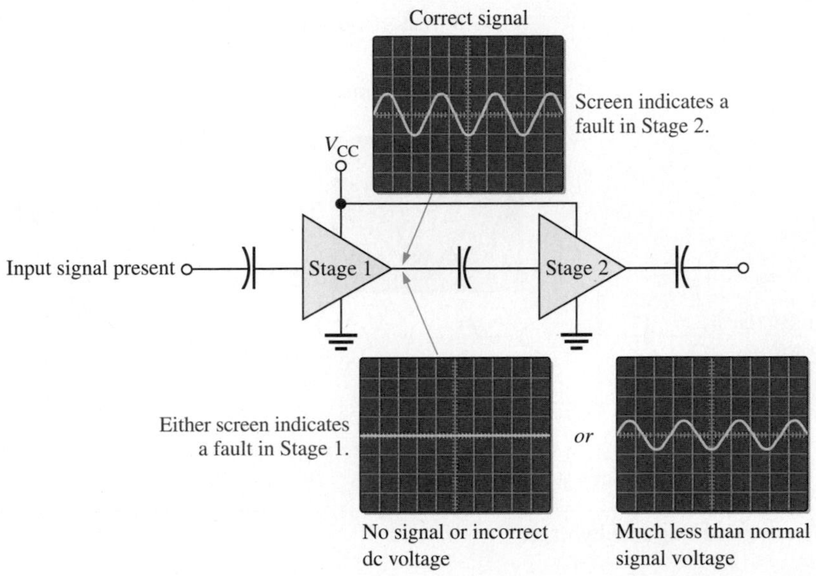

FIGURE 6–44

Half-splitting signal tracing isolates the faulty stage.

Step 4: *Apply fault analysis.* Focus on the faulty stage and determine the component failure that can produce the incorrect output.

Symptom: DC voltages incorrect.

Likely faults: A failure of any resistor or the transistor will produce an incorrect dc bias voltage. A leaky bypass or coupling capacitor will also affect the dc bias voltages. Further measurements in the stage are necessary to isolate the faulty component.

Incorrect ac voltages and the most likely fault(s) are illustrated in Figure 6–45 as follows:

(a) *Symptom 1:* Signal voltage at output missing; dc voltage correct.
Symptom 2: Signal voltage at base missing; dc voltage correct.
Likely fault: Input coupling capacitor open. This prevents the signal from getting to the base.

(b) *Symptom:* Correct signal at base but no output signal.
Likely fault: Transistor base open.

(c) *Symptom:* Signal voltage at output much less than normal; dc voltage correct.
Likely fault: Bypass capacitor open.

Step 5: *Replace or repair.* With the power turned off, replace the defective component or repair the defective connection. Turn on the power, and check for proper operation.

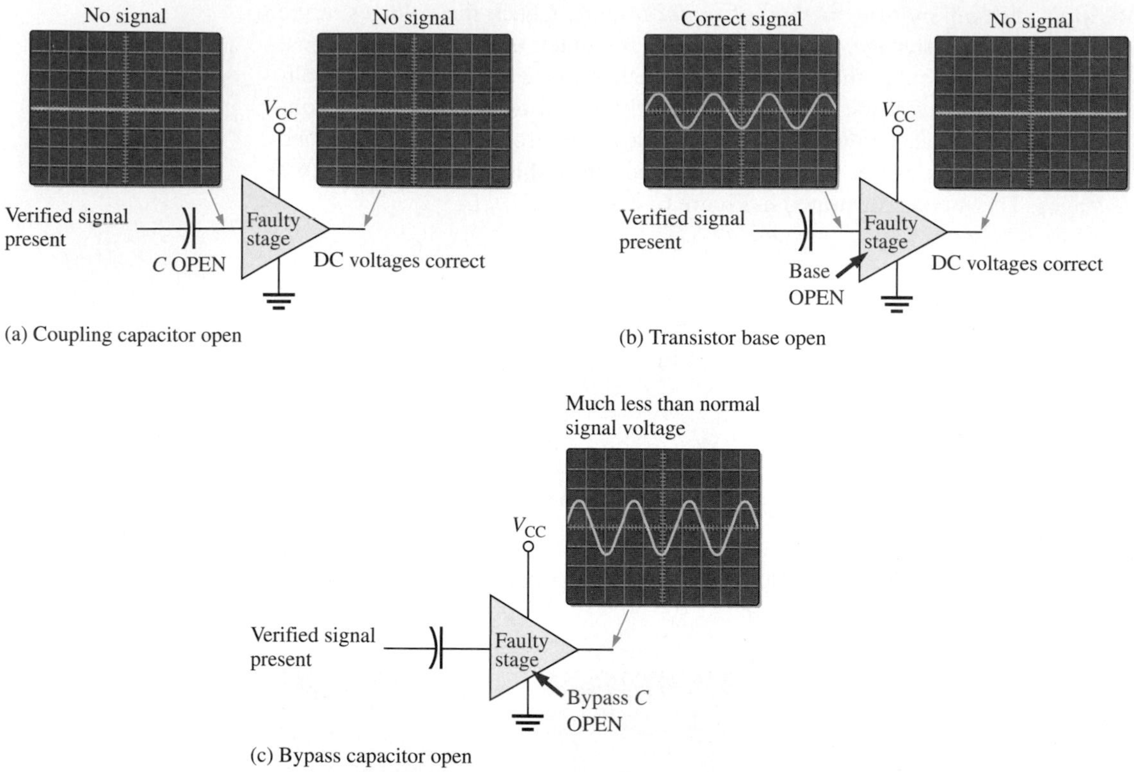

▲ **FIGURE 6–45**

Troubleshooting a faulty stage.

EXAMPLE 6–14

The two-stage amplifier in Figure 6–42 has malfunctioned. Specify the step-by-step troubleshooting procedure for an assumed fault.

Solution Assume there are no visual or other indications of a problem such as a charred resistor, solder splash, wire clipping, broken connection, or extremely hot component. The troubleshooting procedure for a certain fault scenario is as follows:

Step 1: There is power to the circuit as indicated by a correct V_{CC} measurement.

Step 2: There is a verified input signal voltage, but no output signal voltage is measured.

Step 3: The signal voltage and the dc voltage at the collector of Q_1 are correct. This means that the problem is in the second stage or the coupling capacitor C_3 between the stages.

Step 4: The correct signal voltage and dc bias voltage are measured at the base of Q_2. This eliminates the possibility of a fault in C_3 or the second stage bias circuit.

The collector of Q_2 is at 10 V and there is no signal voltage. This measurement, made directly on the transistor collector, indicates that either the collector is shorted to V_{CC} or the transistor is internally open. It is unlikely that the collector resistor R_7 is shorted but to verify, turn off the power and use an ohmmeter to check.

The possibility of a short is eliminated by the ohmmeter check. The other possible faults are (a) transistor Q_2 internally open or (b) emitter resistor or connection open. Use a transistor tester and/or ohmmeter to check each of these possible faults with power off.

Step 5: Replace the faulty component or repair open connection and retest the circuit for proper operation.

Related Problem Determine the possible fault(s) if, in Step 4, you find no signal voltage at the base of Q_2 but the dc voltage is correct.

Multisim Troubleshooting Exercises

These file circuits are in the Troubleshooting Exercises folder on your CD-ROM.

1. Open file TSE06-01. Determine if the circuit is working properly and, if not, determine the fault.
2. Open file TSE06-02. Determine if the circuit is working properly and, if not, determine the fault.
3. Open file TSE06-03. Determine if the circuit is working properly and, if not, determine the fault.
4. Open file TSE06-04. Determine if the circuit is working properly and, if not, determine the fault.

SECTION 6–8 CHECKUP

1. If C_4 in Figure 6–42 were open, how would the output signal be affected? How would the dc level at the collector of Q_2 be affected?
2. If R_5 in Figure 6–42 were open, how would the output signal be affected?
3. If the coupling capacitor C_3 in Figure 6–42 shorted out, would any of the dc voltages in the amplifier be changed? If so, which ones?

On the Job

Application Activity: *Audio Preamplifier for PA System*

An audio preamplifier is to be developed for use in a small portable public address (PA) system. The preamplifier will have a microphone input, and its output will drive a power amplifier to be developed in Chapter 7. A block diagram of the complete PA system is shown in Figure 6–46(a), and its physical configuration is shown in part (b). The dc supply voltages are provided by a battery pack or by an electronic power supply.

The Circuit

A 2-stage audio voltage preamplifier is shown in Figure 6–47. The first stage is a common-emitter *pnp* with voltage-divider bias, and the second stage is a common-emitter *npn* with voltage-divider bias. It has been decided that the amplifier should operate from 30 V dc to

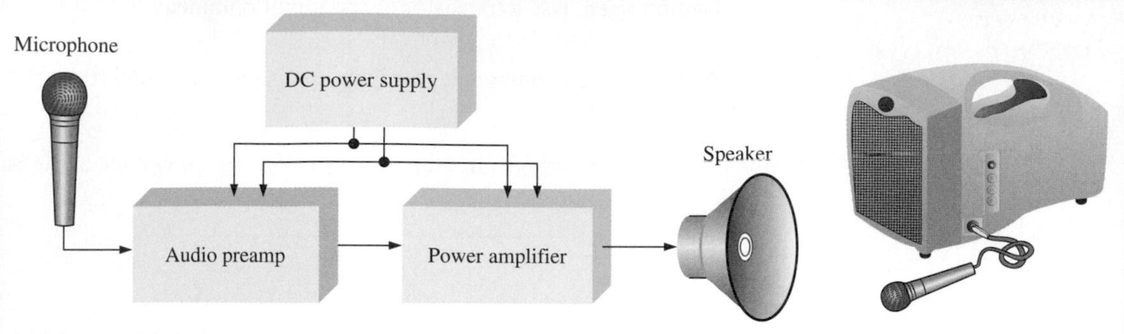

(a) PA system block diagram

(b) Physical configuration

FIGURE 6–46

The public address system.

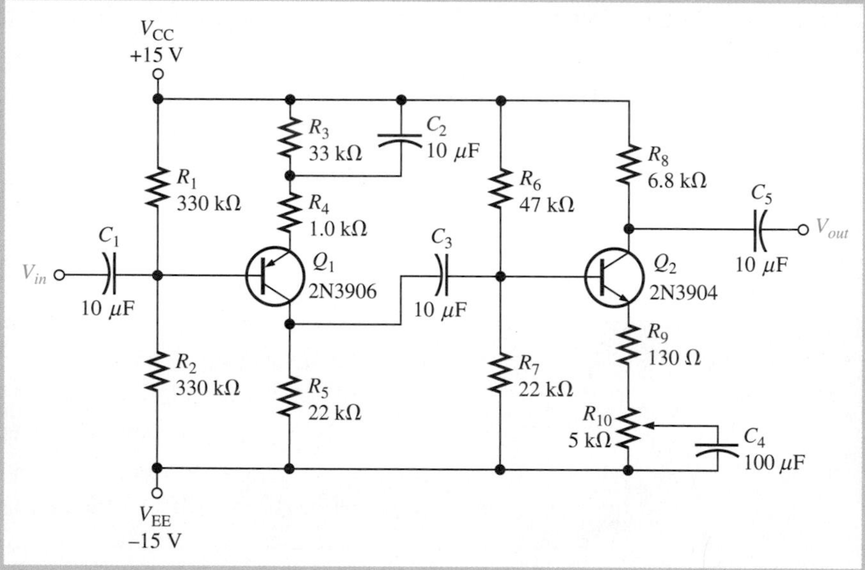

FIGURE 6–47

Two-stage voltage preamplifier.

get a large enough signal voltage swing to provide a maximum of 6 W to the speaker. Because small IC regulators such as the 78xx and 79xx series are not available above 24 V, dual ±15 V dc supplies are used in this particular system instead of a single supply. The operation is essentially the same as if a single +30 V dc source had been used. The potentiometer at the output provides gain adjustment for volume control. The input to the first stage is from the microphone, and the output of the second stage will drive a power amplifier to be developed in Chapter 7. The power amplifier will drive the speaker. The preamp is to operate with a peak input signal range of from 25 mV to 50 mV. The minimum range of voltage gain adjustment is from 90 to 170.

1. Calculate the theoretical voltage gain of the first stage when the second stage is set for maximum gain.
2. Calculate the theoretical maximum voltage gain of the second stage.
3. Determine the overall theoretical voltage gain.
4. Calculate the circuit power dissipation with no signal (quiescent).

Simulation

The preamp is simulated with a peak input signal of 45 mV using Multisim. The results are shown in Figure 6–48.

5. Determine the voltage gain of the simulated circuit based on the voltage measurements.
6. Compare the measured voltage gain with the calculated voltage gain.

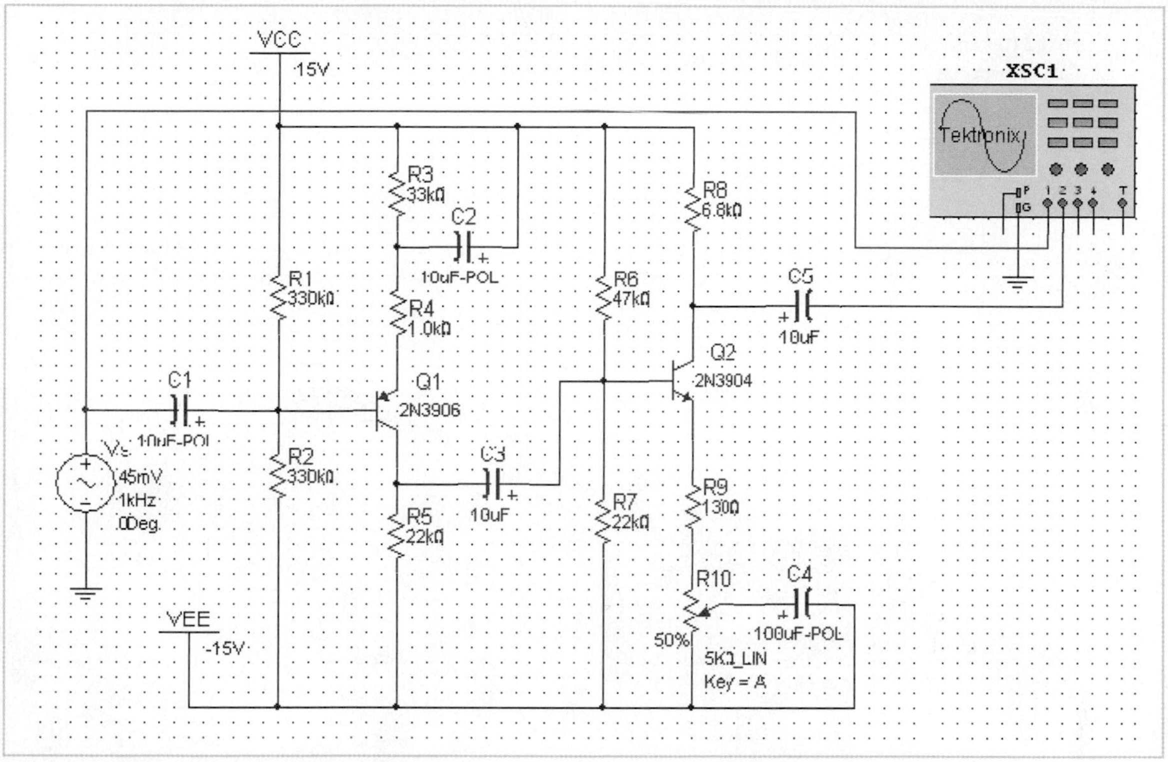

(a) Circuit screen

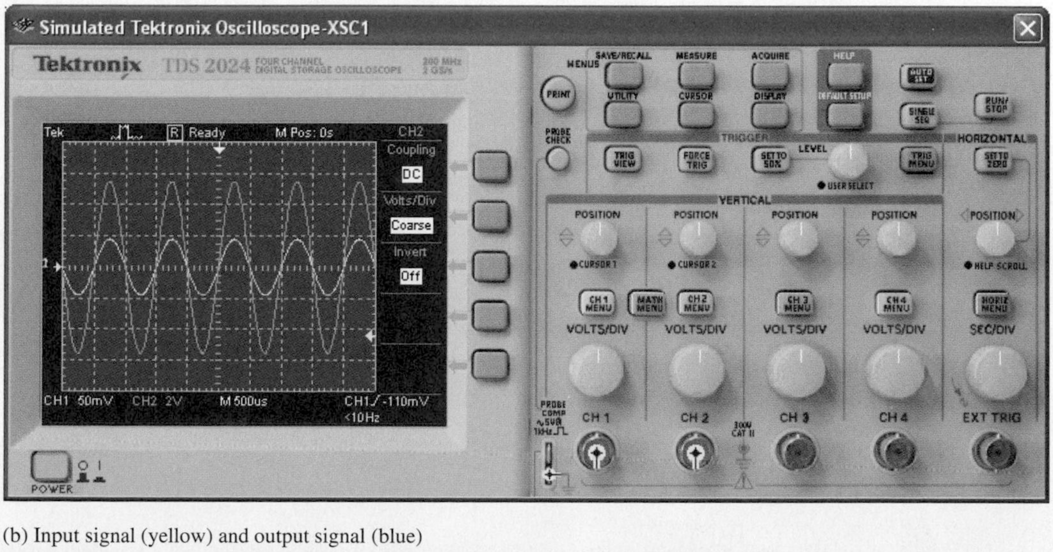

(b) Input signal (yellow) and output signal (blue)

FIGURE 6–48

Preamp input and output signals.

Simulate the preamp circuit using your Multisim software. Observe the operation with the virtual oscilloscope.

Prototyping and Testing*

Now that the circuit has been simulated, the prototype circuit is constructed and tested. After the circuit is successfully tested on a protoboard, it is ready to be finalized on a printed circuit board.

Lab Experiment

To build and test a similar circuit, go to Experiment 6 in your lab manual (*Laboratory Exercises for Electronic Devices* by David Buchla and Steven Wetterling).

Circuit Board

The preamp is implemented on a printed circuit board as shown in Figure 6–49.

7. Check the printed circuit board and verify that it agrees with the schematic in Figure 6–47. The volume control potentiometer is mounted off the PC board for easy access.
8. Label each input and output pin according to function.

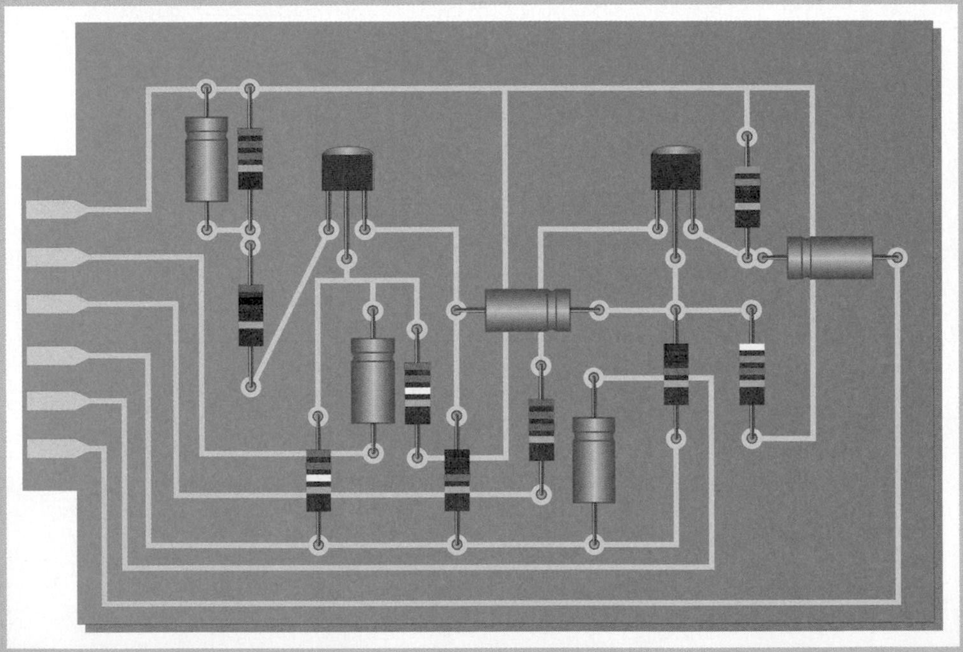

FIGURE 6–49

Preamp circuit board.

Troubleshooting

Two preamp circuit boards have failed the production test. You will troubleshoot the boards based on the scope measurements shown in Figure 6–50.

9. List possible faults for board 1.
10. List possible faults for board 2.

---

*An example of a combined software/hardware approach to simulating and prototyping a circuit is NI ELVIS (National Instrument Educational Laboratory Virtual Instrumentation Suite), which combines Multisim software with actual prototyping hardware. See Appendix C for a brief introductory tutorial using the amplifier from this Application Activity.

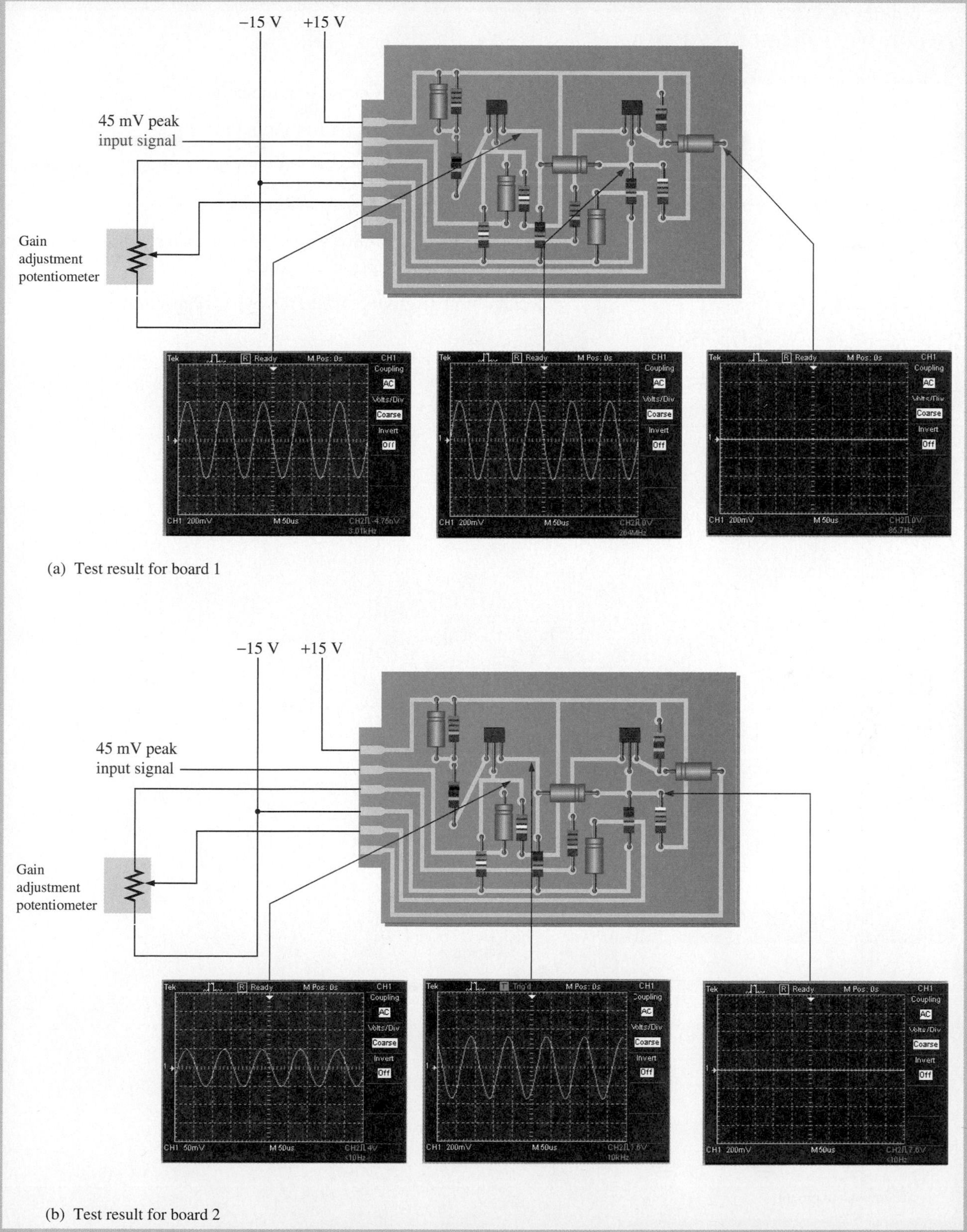

(a) Test result for board 1

(b) Test result for board 2

FIGURE 6–50

Test of two faulty preamp boards.

SUMMARY OF THE COMMON-EMITTER AMPLIFIER

CIRCUIT WITH VOLTAGE-DIVIDER BIAS

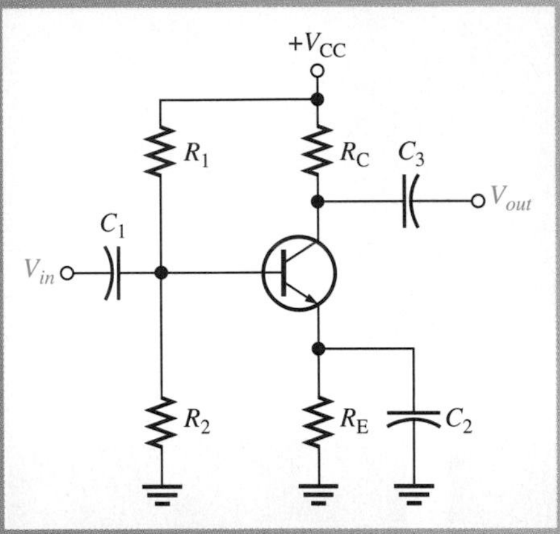

- Input is at the base. Output is at the collector.
- There is a phase inversion from input to output.
- C_1 and C_3 are coupling capacitors for the input and output signals.
- C_2 is the emitter-bypass capacitor.
- All capacitors must have a negligible reactance at the frequency of operation, so they appear as shorts.
- Emitter is at ac ground due to the bypass capacitor.

EQUIVALENT CIRCUITS AND FORMULAS

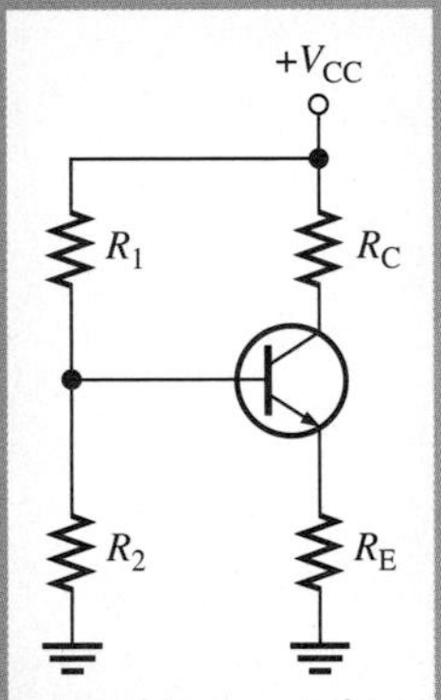

DC equivalent circuit

- DC formulas:

$$V_B \cong \left(\frac{R_2}{R_1 + R_2}\right)V_{CC} \quad \text{when } R_{IN(BASE)} \geqq 10R_2$$

$$V_B = \left(\frac{R_2 \| \beta_{DC}R_E}{R_1 + R_2 \| \beta_{DC}R_E}\right)V_{CC} \quad \text{when } R_{IN(BASE)} < 10R_2$$

$$V_E = V_B - V_{BE}$$

$$I_E = \frac{V_E}{R_E}$$

$$V_C = V_{CC} - I_C R_C$$

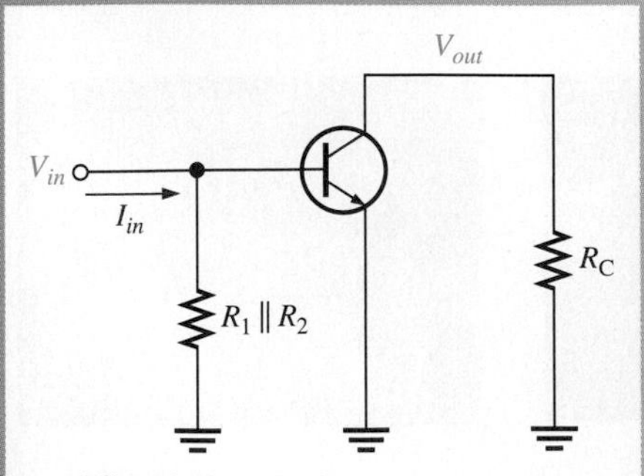

AC equivalent circuit

- AC formulas:

$$r'_e = \frac{25\text{ mV}}{I_E}$$

$$R_{in(base)} = \beta_{ac} r'_e$$

$$R_{out} \cong R_C$$

$$A_v = \frac{R_C}{r'_e}$$

$$A_i = \frac{I_c}{I_{in}}$$

$$A_p = A'_v A_i$$

SWAMPED AMPLIFIER WITH RESISTIVE LOAD

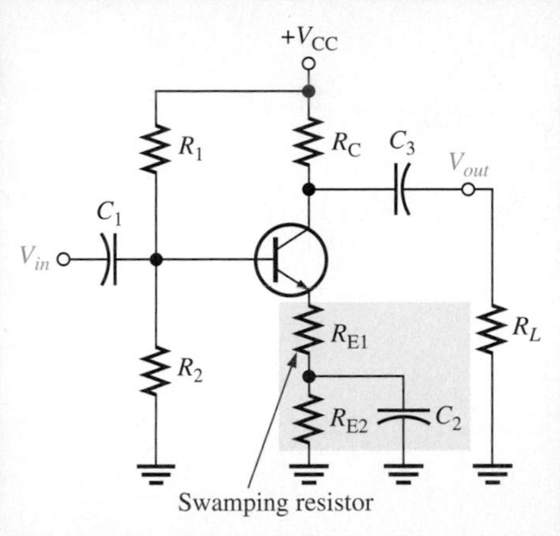

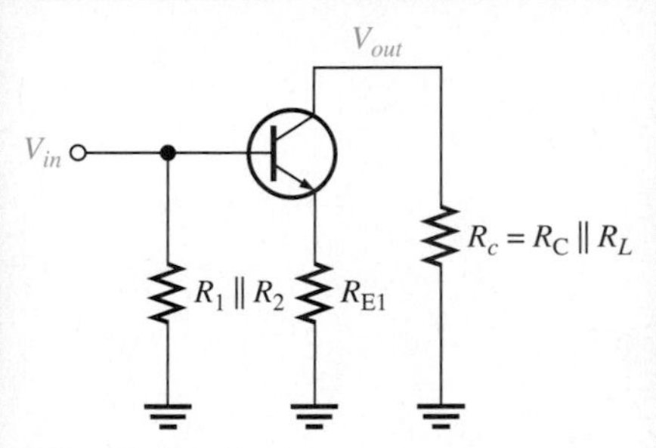

AC equivalent circuit

- AC formulas:

$$A_v \cong \frac{R_c}{R_{E1}}$$

where $R_c = R_C \| R_L$

$$R_{in(base)} = \beta_{ac}(r'_e + R_{E1})$$

- Swamping stabilizes gain by minimizing the effect of r'_e.
- Swamping reduces the voltage gain from its unswamped value.
- Swamping increases input resistance.
- The load resistance reduces the voltage gain. The smaller the load resistance, the less the gain.

SUMMARY OF THE COMMON-COLLECTOR AMPLIFIER

CIRCUIT WITH VOLTAGE-DIVIDER BIAS

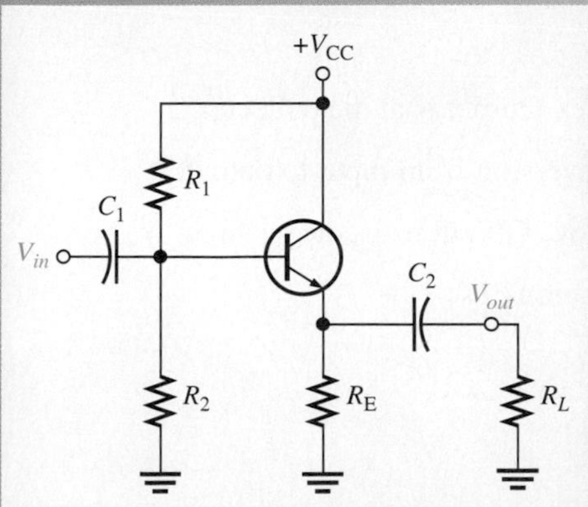

- Input is at the base. Output is at the emitter.
- There is no phase inversion from input to output.
- Input resistance is high. Output resistance is low.
- Maximum voltage gain is 1.
- Collector is at ac ground.
- Coupling capacitors must have a negligible reactance at the frequency of operation.

EQUIVALENT CIRCUITS AND FORMULAS

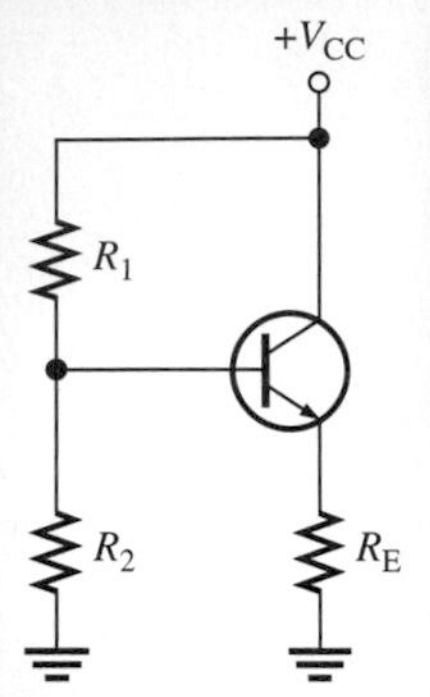

DC equivalent circuit

- DC formulas:

$$V_B \cong \left(\frac{R_2}{R_1 + R_2}\right)V_{CC} \quad \text{when } R_{IN(BASE)} \geqq 10R_2$$

$$V_B = \left(\frac{R_2 \,\|\, \beta_{DC}R_E}{R_1 + R_2 \,\|\, \beta_{DC}R_E}\right)V_{CC} \quad \text{when } R_{IN(BASE)} < 10R_2$$

$$V_E = V_B - V_{BE}$$

$$I_E = \frac{V_E}{R_E}$$

$$V_C = V_{CC}$$

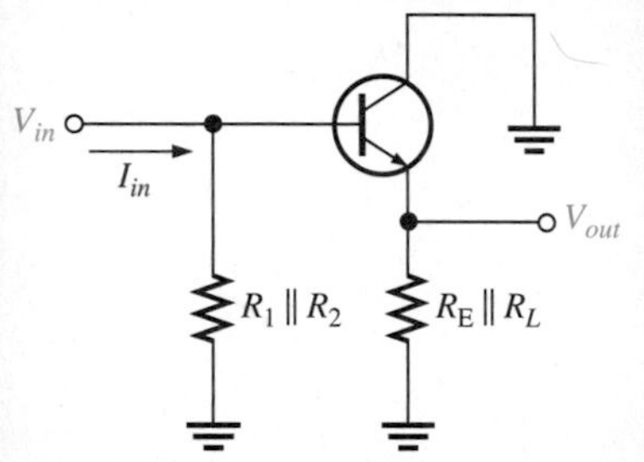

AC equivalent circuit

- AC formulas:

$$r'_e = \frac{25\text{ mV}}{I_E}$$

$$R_{in(base)} = \beta_{ac}(r'_e + R_e) \cong \beta_{ac}R_e$$

$$R_{out} = \left(\frac{R_s}{\beta_{ac}}\right) \| R_E$$

$$A_v = \frac{R_e}{r'_e + R_e} \cong 1$$

$$A_i = \frac{I_e}{I_{in}}$$

$$A_p \cong A_i$$

SUMMARY OF COMMON-BASE AMPLIFIER

CIRCUIT WITH VOLTAGE-DIVIDER BIAS

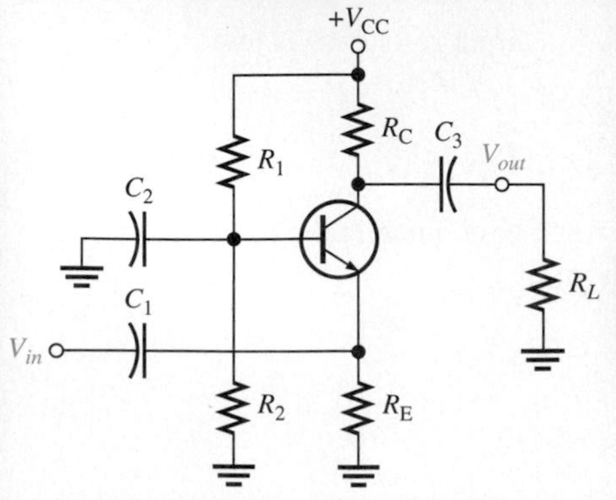

- Input is at the emitter. Output is at the collector.
- There is no phase inversion from input to output.
- Input resistance is low. Output resistance is high.
- Maximum current gain is 1.
- Base is at ac ground.

EQUIVALENT CIRCUITS AND FORMULAS

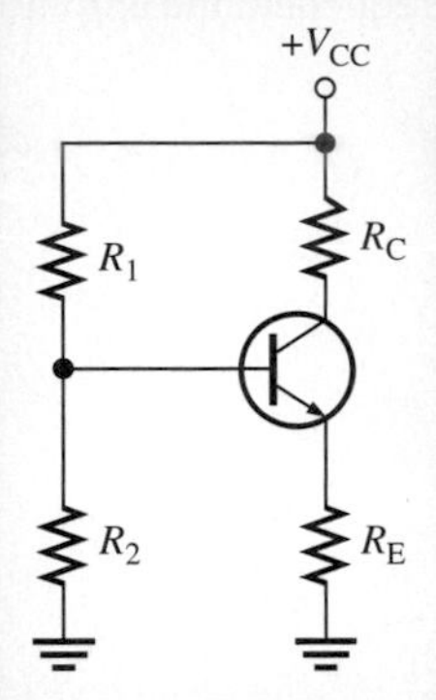

DC equivalent circuit

- DC formulas:

$$V_B \cong \left(\frac{R_2}{R_1 + R_2}\right)V_{CC} \quad \text{when } R_{IN(BASE)} \geqq 10R_2$$

$$V_B = \left(\frac{R_2 \| \beta_{DC}R_E}{R_1 + R_2 \| \beta_{DC}R_E}\right)V_{CC} \quad \text{when } R_{IN(BASE)} < 10R_2$$

$$V_E = V_B - V_{BE}$$

$$I_E = \frac{V_E}{R_E}$$

$$V_C = V_{CC} - I_C R_C$$

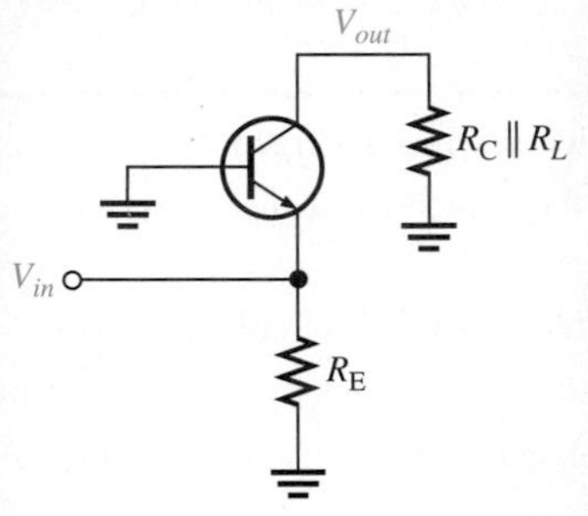

AC equivalent circuit

- AC formulas:

$$r'_e = \frac{25\text{ mV}}{I_E}$$

$$R_{in(emitter)} \cong r'_e$$

$$R_{out} \cong R_C$$

$$A_v \cong \frac{R_c}{r'_e}$$

$$A_i \cong 1$$

$$A_p \cong A_v$$

SUMMARY OF DIFFERENTIAL AMPLIFIER

CIRCUIT WITH DIFFERENTIAL INPUTS

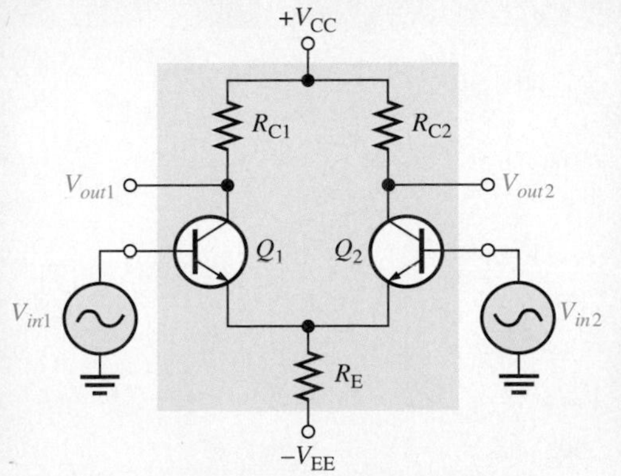

- Double-ended differential inputs (shown)

  Signal on both inputs

  Input signals are out of phase

- Single-ended differential inputs (not shown)

  Signal on one input only

  One input connected to ground

CIRCUIT WITH COMMON-MODE INPUTS

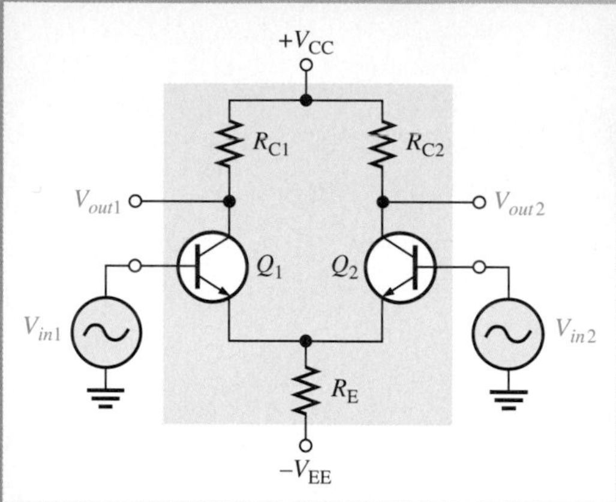

- Both input signals are the same phase, frequency, and amplitude.
- Common-mode rejection ratio:

$$\text{CMRR} = \frac{A_{v(d)}}{A_{cm}}$$

$$\text{CMRR} = 20 \log\left(\frac{A_{v(d)}}{A_{cm}}\right)$$

SUMMARY

Section 6–1
- ◆ A small-signal amplifier uses only a small portion of its load line under signal conditions.
- ◆ The ac load line differs from the dc load line because the effective ac collector resistance is less than the dc collector resistance.

Section 6–2
- ◆ *r* parameters are easily identifiable and applicable with a transistor's circuit operation.
- ◆ *h* parameters are important because manufacturers' datasheets specify transistors using *h* parameters.

Section 6–3
- ◆ A common-emitter amplifier has good voltage, current, and power gains, but a relatively low input resistance.
- ◆ Swamping is a method of stabilizing the voltage gain.

Section 6–4
- ◆ A common-collector amplifier has high input resistance and good current gain, but its voltage gain is approximately 1.
- ◆ A Darlington pair provides beta multiplication for increased input resistance.
- ◆ A common-collector amplifier is known as an emitter-follower.

Section 6–5
- ◆ The common-base amplifier has a good voltage gain, but it has a very low input resistance and its current gain is approximately 1.
- ◆ Common-emitter, common-collector, and common-base amplifier configurations are summarized in Table 6–4.

TABLE 6–4

Relative comparison of amplifier configurations. The current gains and the input and output resistances are the approximate maximum achievable values, with the bias resistors neglected.

	CE	CC	CB
Voltage gain, A_v	High R_C/r'_e	Low $\cong 1$	High R_C/r'_e
Current gain, $A_{i(max)}$	High β_{ac}	High β_{ac}	Low $\cong 1$
Power gain, A_p	Very high A_iA_v	High $\cong A_i$	High $\cong A_v$
Input resistance, $R_{in(max)}$	Low $\beta_{ac}r'_e$	High $\beta_{ac}R_E$	Very low r'_e
Output resistance, R_{out}	High R_C	Very low $(R_s/\beta_{ac}) \parallel R_E$	High R_C

Section 6–6

- The total gain of a multistage amplifier is the product of the individual gains (sum of dB gains).
- Single-stage amplifiers can be connected in sequence with capacitively-coupling and direct coupling methods to form multistage amplifiers.

Section 6–7

- A differential input voltage appears between the inverting and noninverting inputs of a differential amplifier.
- In the differential mode, a diff-amp can be operated with single-ended or double-ended inputs.
- In single-ended operation, there is a signal on one input and the other input is grounded.
- In double-ended operation, two signals that are 180° out of phase are on the inputs.
- Common-mode occurs when equal in-phase voltages are applied to both input terminals.

KEY TERMS

Key terms and other bold terms in the chapter are defined in the end-of-book glossary.

ac ground A point in a circuit that appears as ground to ac signals only.

Attenuation The reduction in the level of power, current, or voltage.

Bypass capacitor A capacitor placed across the emitter resistor of an amplifier.

CMRR (common-mode rejection ratio) A measure of a differential amplifier's ability to reject common-mode signals.

Common-base (CB) A BJT amplifier configuration in which the base is the common terminal to an ac signal or ground.

Common-collector (CC) A BJT amplifier configuration in which the collector is the common terminal to an ac signal or ground.

Common-emitter (CE) A BJT amplifier configuration in which the emitter is the common terminal to an ac signal or ground.

Common mode A condition where two signals applied to differential inputs are of the same phase, frequency, and amplitude.

Decibel A logarithmic measure of the ratio of one voltage to another or one power to another.

Differential amplifier An amplifier in which the output is a function of the difference between two input voltages.

Emitter-follower A popular term for a common-collector amplifier.

Input resistance The resistance seen by an ac source connected to the input of an amplifier.

Output resistance The ac resistance looking in at the output of an amplifier.

***r* parameter** One of a set of BJT characteristic parameters that include α_{ac}, β_{ac}, r'_e, r'_b, and r'_c.

KEY FORMULAS

6–1	$r'_e \cong \dfrac{25\text{ mV}}{I_E}$	Internal ac emitter resistance

Common-Emitter

6–2	$R_{in(base)} = R_1 \| R_2 \| R_{in(base)}$	Total amplifier input resistance, voltage-divider bias
6–3	$R_{in(base)} = \beta_{ac} r'_e$	Input resistance at base
6–4	$R_{out} \cong R_C$	Output resistance
6–5	$A_v = \dfrac{R_C}{r'_e}$	Voltage gain, base-to-collector, unloaded
6–6	$A_v = \dfrac{R_C}{r'_e + R_E}$	Voltage gain without bypass capacitor
6–7	$A_v = \dfrac{R_c}{r'_e}$	Voltage gain, base-to-collector, loaded, bypassed R_E

6–8	$A_v \cong \dfrac{R_C}{R_{E1}}$	Voltage gain, swamped amplifier
6–9	$R_{in(base)} = \beta_{ac}(r'_e + R_{E1})$	Input resistance at base, swamped amplifier
6–10	$A_i = \dfrac{I_c}{I_s}$	Current gain, input source to collector
6–11	$A_p = A'_v A_i$	Power gain

Common-Collector (Emitter-Follower)

6–12	$A_v \cong 1$	Voltage gain, base-to-emitter
6–13	$R_{in(base)} \cong \beta_{ac}R_e$	Input resistance at base, loaded
6–14	$R_{out} \cong \left(\dfrac{R_s}{\beta_{ac}}\right) \parallel R_E$	Output resistance
6–15	$A_i = \dfrac{I_e}{I_{in}}$	Current gain
6–16	$A_p \cong A_i$	Power gain
6–17	$R_{in} = \beta_{ac1}\beta_{ac2}R_E$	Input resistance, Darlington pair

Common-Base

6–18	$A_v \cong \dfrac{R_c}{r'_e}$	Voltage gain, emitter-to-collector
6–19	$R_{in(emitter)} \cong r'_e$	Input resistance at emitter
6–20	$R_{out} \cong R_C$	Output resistance
6–21	$A_i \cong 1$	Current gain
6–22	$A_p \cong A_v$	Power gain

Multistage Amplifier

6–23	$A'_v = A_{v1}A_{v2}A_{v3} \dots A_{vn}$	Overall voltage gain
6–24	$A_{v(dB)} = 20 \log A_v$	Voltage gain expressed in dB

Differential Amplifier

6–25	$\text{CMRR} = \dfrac{A_{v(d)}}{A_{cm}}$	Common-mode rejection ratio
6–26	$\text{CMRR} = 20 \log \left(\dfrac{A_{v(d)}}{A_{cm}}\right)$	Common mode rejection ratio in dB

TRUE/FALSE QUIZ

Answers are at the end of the chapter.

1. In an amplifier, a coupling capacitor should appear ideally as a short to the signal.
2. r parameters include β_{ac} and r'_e.
3. h parameters are never specified on a datasheet.
4. The r parameter β_{ac} is the same as the h parameter h_{fe}.
5. A bypass capacitor in a CE amplifier decreases the voltage gain.
6. If R_C in a CE amplifier is increased, the voltage gain is reduced.
7. The load is the amount of current drawn from the output of an amplifier.
8. In a CE amplifier, the gain can be stabilized by using a swamping resistor.

9. An emitter-follower is a CC amplifier.
10. A CC amplifier has high voltage gain.
11. A Darlington pair consists essentially of two CC amplifiers.
12. A CB amplifier has high current gain.
13. The overall voltage gain of a multistage amplifier is the product of the gains of each stage.
14. A differential amplifier amplifies the difference of two input signals.
15. CMRR is the common-mode resistance ratio.

CIRCUIT-ACTION QUIZ

Answers are at the end of the chapter.

1. If the transistor in Figure 6–8 is exchanged for one with higher betas, V_{out} will
(a) increase **(b)** decrease **(c)** not change

2. If C_2 is removed from the circuit in Figure 6–8, V_{out} will
(a) increase **(b)** decrease **(c)** not change

3. If the value of R_C in Figure 6–8 is increased, V_{out} will
(a) increase **(b)** decrease **(c)** not change

4. If the amplitude of V_{in} in Figure 6–8 is decreased, V_{out} will
(a) increase **(b)** decrease **(c)** not change

5. If C_2 in Figure 6–27 is shorted, the average value of the output voltage will
(a) increase **(b)** decrease **(c)** not change

6. If the value of R_E in Figure 6–27 is increased, the voltage gain will
(a) increase **(b)** decrease **(c)** not change

7. If the value of C_1 in Figure 6–27 is increased, V_{out} will
(a) increase **(b)** decrease **(c)** not change

8. If the value of R_C in Figure 6–32 is increased, the current gain will
(a) increase **(b)** decrease **(c)** not change

9. If C_2 and C_4 in Figure 6–34 are increased in value, V_{out} will
(a) increase **(b)** decrease **(c)** not change

10. If the value of R_4 in Figure 6–34 is reduced, the overall voltage gain will
(a) increase **(b)** decrease **(c)** not change

SELF-TEST

Answers are at the end of the chapter.

Section 6–1

1. A small-signal amplifier
(a) uses only a small portion of its load line
(b) always has an output signal in the mV range
(c) goes into saturation once on each input cycle
(d) is always a common-emitter amplifier

Section 6–2

2. The parameter h_{fe} corresponds to
(a) β_{DC} **(b)** β_{ac} **(c)** r'_e **(d)** r'_c

3. If the dc emitter current in a certain transistor amplifier is 3 mA, the approximate value of r'_e is
(a) 3 kΩ **(b)** 3 Ω **(c)** 8.33 Ω **(d)** 0.33 kΩ

Section 6–3

4. A certain common-emitter amplifier has a voltage gain of 100. If the emitter bypass capacitor is removed,
(a) the circuit will become unstable **(b)** the voltage gain will decrease
(c) the voltage gain will increase **(d)** the Q-point will shift

5. For a common-emitter amplifier, $R_C = 1.0\text{ k}\Omega$, $R_E = 390\ \Omega$, $r'_e = 15\ \Omega$, and $\beta_{ac} = 75$. Assuming that R_E is completely bypassed at the operating frequency, the voltage gain is

   (a) 66.7 **(b)** 2.56 **(c)** 2.47 **(d)** 75

6. In the circuit of Question 5, if the frequency is reduced to the point where $X_{C(bypass)} = R_E$, the voltage gain

   (a) remains the same **(b)** is less **(c)** is greater

7. In a common-emitter amplifier with voltage-divider bias, $R_{in(base)} = 68\text{ k}\Omega$, $R_1 = 33\text{ k}\Omega$, and $R_2 = 15\text{ k}\Omega$. The total input resistance is

   (a) 68 kΩ **(b)** 8.95 kΩ **(c)** 22.2 kΩ **(d)** 12.3 kΩ

8. A CE amplifier is driving a 10 kΩ load. If $R_C = 2.2\text{ k}\Omega$ and $r'_e = 10\ \Omega$, the voltage gain is approximately

   (a) 220 **(b)** 1000 **(c)** 10 **(d)** 180

Section 6–4

9. For a common-collector amplifier, $R_E = 100\ \Omega$, $r'_e = 10\ \Omega$, and $\beta_{ac} = 150$. The ac input resistance at the base is

   (a) 1500 Ω **(b)** 15 kΩ **(c)** 110 Ω **(d)** 16.5 kΩ

10. If a 10 mV signal is applied to the base of the emitter-follower circuit in Question 9, the output signal is approximately

    (a) 100 mV **(b)** 150 mV **(c)** 1.5 V **(d)** 10 mV

11. In a certain emitter-follower circuit, the current gain is 50. The power gain is approximately

    (a) $50A_v$ **(b)** 50 **(c)** 1 **(d)** answers (a) and (b)

12. In a Darlington pair configuration, each transistor has an ac beta of 125. If R_E is 560 Ω, the input resistance is

    (a) 560 Ω **(b)** 70 kΩ **(c)** 8.75 MΩ **(d)** 140 kΩ

Section 6–5

13. The input resistance of a common-base amplifier is

    (a) very low **(b)** very high
    (c) the same as a CE **(d)** the same as a CC

Section 6–6

14. Each stage of a four-stage amplifier has a voltage gain of 15. The overall voltage gain is

    (a) 60 **(b)** 15 **(c)** 50,625 **(d)** 3078

15. The overall gain found in Question 14 can be expressed in decibels as

    (a) 94.1 dB **(b)** 47.0 dB **(c)** 35.6 dB **(d)** 69.8 dB

Section 6–7

16. A differential amplifier

    (a) is used in op-amps **(b)** has one input and one output
    (c) has two outputs **(d)** answers (a) and (c)

17. When a differential amplifier is operated single-ended,

    (a) the output is grounded
    (b) one input is grounded and a signal is applied to the other
    (c) both inputs are connected together
    (d) the output is not inverted

18. In the double-ended differential mode,

    (a) opposite polarity signals are applied to the inputs
    (b) the gain is 1
    (c) the outputs are different amplitudes
    (d) only one supply voltage is used

19. In the common mode,

    (a) both inputs are grounded
    (b) the outputs are connected together
    (c) an identical signal appears on both inputs
    (d) the output signals are in-phase

PROBLEMS

Answers to all odd-numbered problems are at the end of the book.

BASIC PROBLEMS

Section 6–1 Amplifier Operation

1. What is the lowest value of dc collector current to which a transistor having the characteristic curves in Figure 6–4 can be biased and still retain linear operation with a peak-to-peak base current swing of 20 μA?

2. What is the highest value of I_C under the conditions described in Problem 1?

Section 6–2 Transistor AC Models

3. If the dc emitter current in a transistor is 3 mA, what is the value of r'_e?

4. If the h_{fe} of a transistor is specified as 200, determine β_{ac}.

5. A certain transistor has a dc beta (h_{FE}) of 130. If the dc base current is 10 μA, determine r'_e. $\alpha_{DC} = 0.99$.

6. At the dc bias point of a certain transistor circuit, $I_B = 15\ \mu$A and $I_C = 2$ mA. Also, a variation in I_B of 3 μA about the Q-point produces a variation in I_C of 0.35 mA about the Q-point. Determine β_{DC} and β_{ac}.

Section 6–3 The Common-Emitter Amplifier

7. Draw the dc equivalent circuit and the ac equivalent circuit for the unloaded amplifier in Figure 6–51.

8. Determine the following dc values for the amplifier in Figure 6–51.

(a) V_B **(b)** V_E **(c)** I_E **(d)** I_C **(e)** V_C

9. Calculate the quiescent power dissipation in Figure 6–51.

10. Determine the following values for the amplifier in Figure 6–51.

(a) $R_{in(base)}$ **(b)** $R_{in(tot)}$ **(c)** A_v

11. Connect a bypass capacitor across R_E in Figure 6–51, and repeat Problem 10.

12. Connect a 10 kΩ load resistor to the output in Figure 6–51, and repeat Problem 11.

13. Determine the following dc values for the amplifier in Figure 6–52.

(a) V_B **(b)** V_E **(c)** I_E **(d)** I_C **(e)** V_C **(f)** V_{CE}

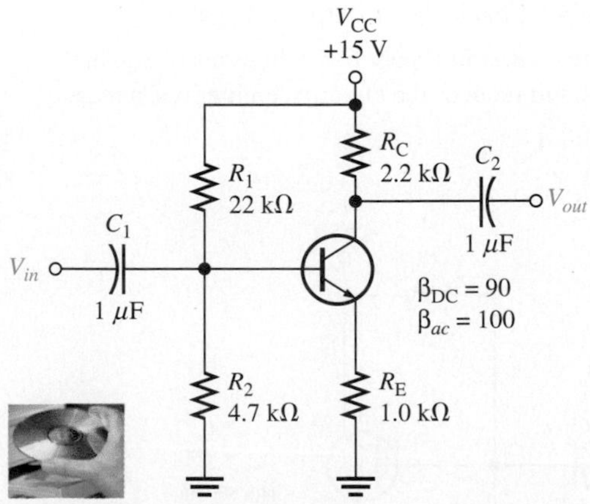

▲ FIGURE 6–51

Multisim file circuits are identified with a CD logo and are in the Problems folder on your CD-ROM. Filenames correspond to figure numbers (e.g., F06-51).

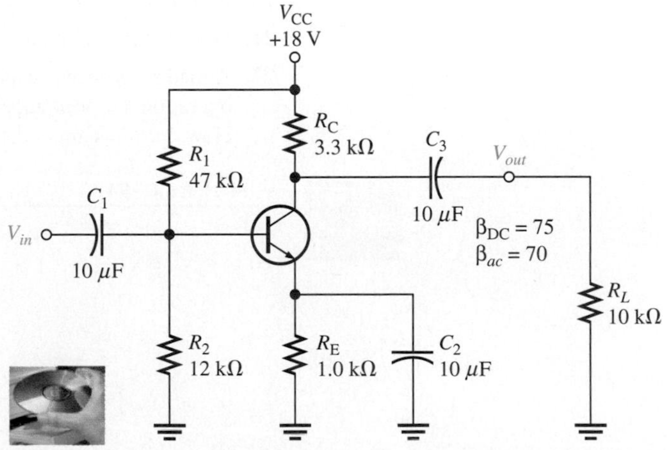

▲ FIGURE 6–52

14. Determine the following ac values for the amplifier in Figure 6–52.

(a) $R_{in(base)}$ **(b)** R_{in} **(c)** A_v **(d)** A_i **(e)** A_p

15. Assume that a 600 Ω, 12 μV rms voltage source is driving the amplifier in Figure 6–52. Determine the overall voltage gain by taking into account the attenuation in the base circuit, and find the *total* output voltage (ac and dc). What is the phase relationship of the collector signal voltage to the base signal voltage?

16. The amplifier in Figure 6–53 has a variable gain control, using a 100 Ω potentiometer for R_E with the wiper ac-grounded. As the potentiometer is adjusted, more or less of R_E is bypassed to ground, thus varying the gain. The total R_E remains constant to dc, keeping the bias fixed. Determine the maximum and minimum gains for this unloaded amplifier.

17. If a load resistance of 600 Ω is placed on the output of the amplifier in Figure 6–53, what are the maximum and minimum gains?

18. Find the overall maximum voltage gain for the amplifier in Figure 6–53 with a 1.0 kΩ load if it is being driven by a 300 kΩ source.

FIGURE 6–53

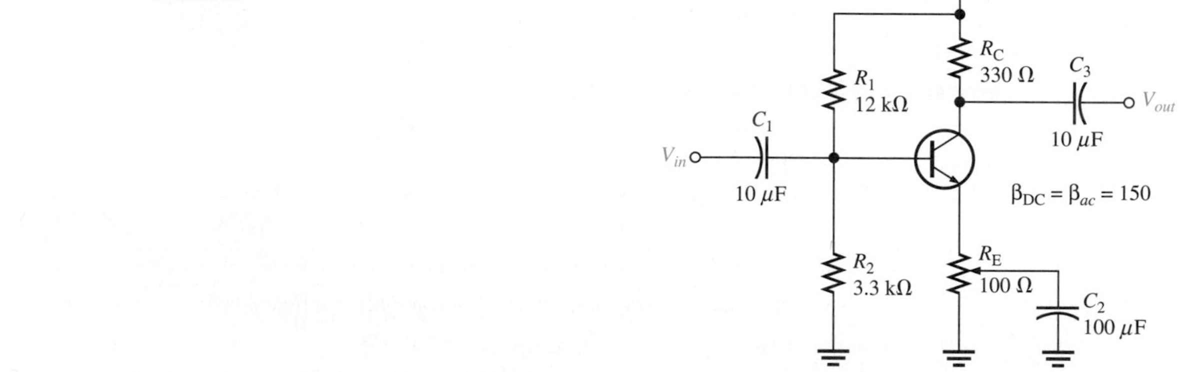

19. Modify the schematic to show how you would "swamp out" the temperature effects of r'_e in Figure 6–52 by making R_e at least ten times larger than r'_e. Keep the same total R_E. How does this affect the voltage gain?

Section 6–4 The Common-Collector Amplifier

20. Determine the *exact* voltage gain for the unloaded emitter-follower in Figure 6–54.

21. What is the total input resistance in Figure 6–54? What is the dc output voltage?

22. A load resistance is capacitively coupled to the emitter in Figure 6–54. In terms of signal operation, the load appears in parallel with R_E and reduces the effective emitter resistance. How does this affect the voltage gain?

FIGURE 6–54

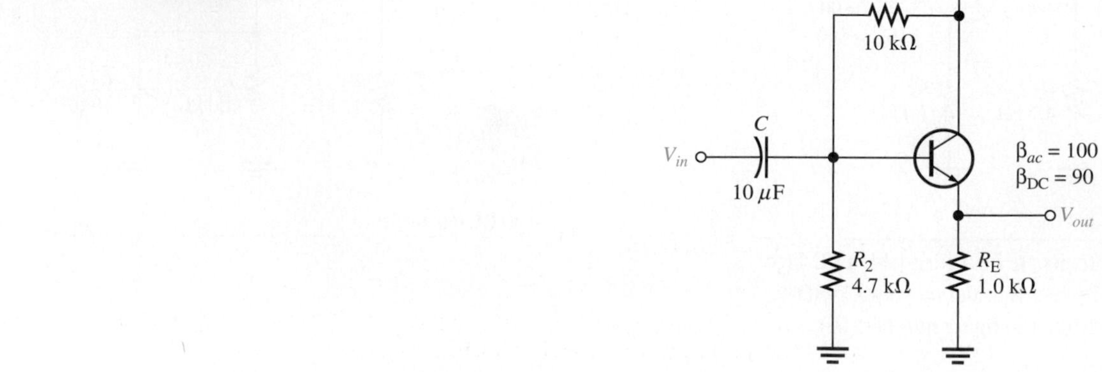

23. In Problem 22, what value of R_L will cause the voltage gain to drop to 0.9?

24. For the circuit in Figure 6–55, determine the following:
   (a) Q_1 and Q_2 dc terminal voltages
   (b) overall β_{ac}
   (c) r'_e for each transistor
   (d) total input resistance

25. Find the overall current gain A_i in Figure 6–55.

FIGURE 6–55

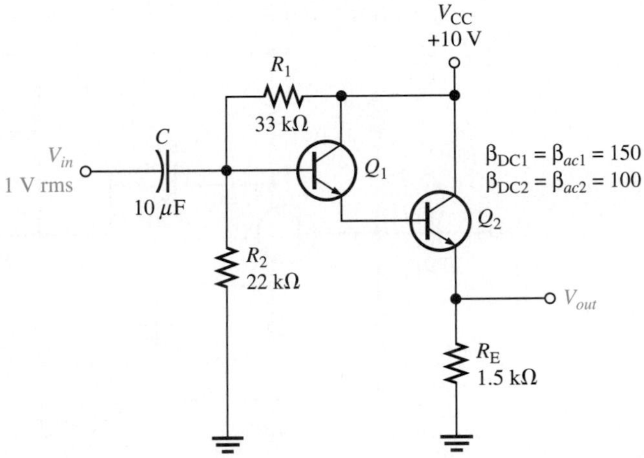

Section 6–5 The Common-Base Amplifier

26. What is the main disadvantage of the common-base amplifier compared to the common-emitter and the emitter-follower amplifiers?

27. Find $R_{in(emitter)}$, A_v, A_i, and A_p for the unloaded amplifier in Figure 6–56.

28. Match the following generalized characteristics with the appropriate amplifier configuration.
   (a) Unity current gain, good voltage gain, very low input resistance
   (b) Good current gain, good voltage gain, low input resistance
   (c) Good current gain, unity voltage gain, high input resistance

FIGURE 6–56

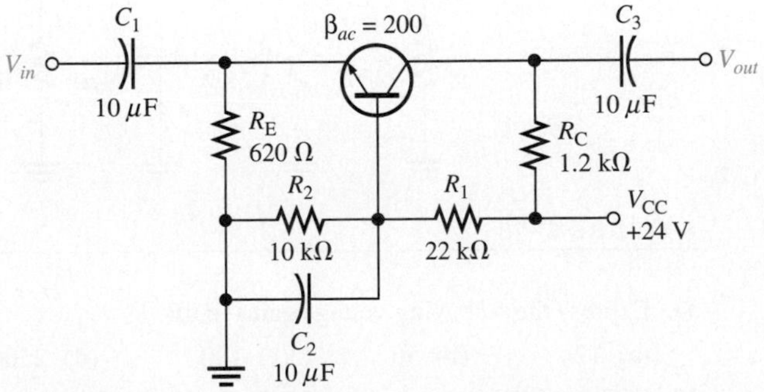

Section 6–6 Multistage Amplifiers

29. Each of two cascaded amplifier stages has an $A_v = 20$. What is the overall gain?

30. Each of three cascaded amplifier stages has a dB voltage gain of 10 dB. What is the overall voltage gain in dB? What is the actual overall voltage gain?

31. For the two-stage, capacitively coupled amplifier in Figure 6–57, find the following values:

(a) voltage gain of each stage

(b) overall voltage gain

(c) Express the gains found in (a) and (b) in dB.

32. If the multistage amplifier in Figure 6–57 is driven by a 75 Ω, 50 μV source and the second stage is loaded with an $R_L = 18\ \text{k}\Omega$, determine

(a) voltage gain of each stage

(b) overall voltage gain

(c) Express the gains found in (a) and (b) in dB.

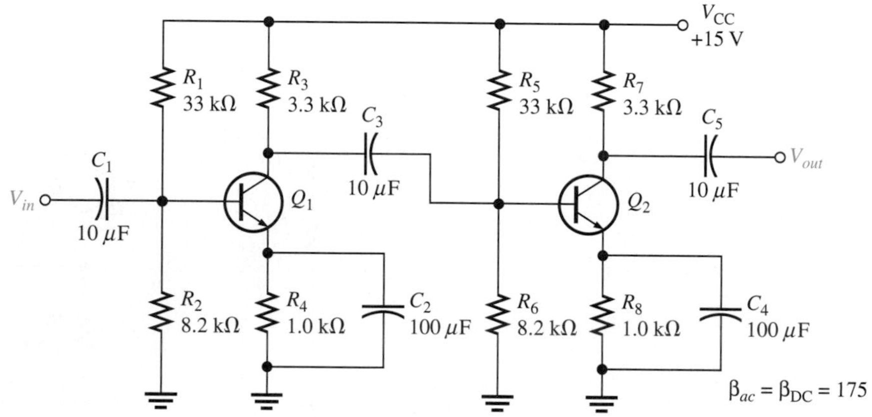

▲ **FIGURE 6–57**

33. Figure 6–58 shows a direct-coupled (that is, with no coupling capacitors between stages) two-stage amplifier. The dc bias of the first stage sets the dc bias of the second. Determine all dc voltages for both stages and the overall ac voltage gain.

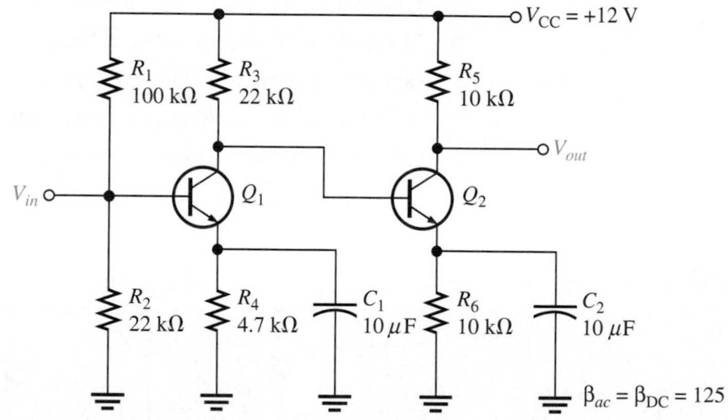

▲ **FIGURE 6–58**

34. Express the following voltage gains in dB:

(a) 12 **(b)** 50 **(c)** 100 **(d)** 2500

35. Express the following voltage gains in dB as standard voltage gains:

(a) 3 dB **(b)** 6 dB **(c)** 10 dB **(d)** 20 dB **(e)** 40 dB

Section 6–7 The Differential Amplifier

36. The dc base voltages in Figure 6–59 are zero. Using your knowledge of transistor analysis, determine the dc differential output voltage. Assume that Q_1 has an $\alpha = 0.980$ and Q_2 has an $\alpha = 0.975$.

FIGURE 6–59

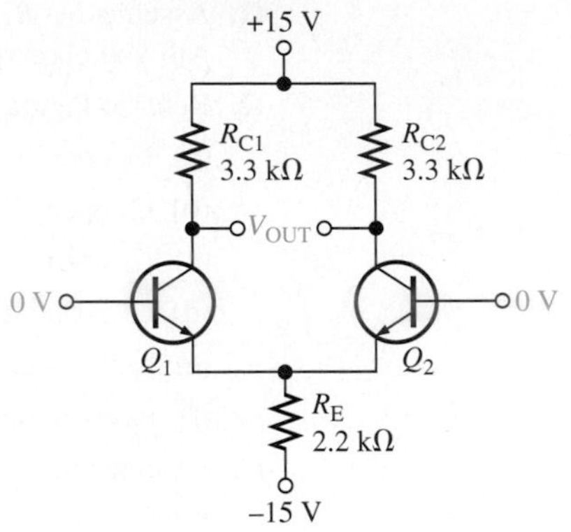

37. Identify the quantity being measured by each meter in Figure 6–60.

FIGURE 6–60

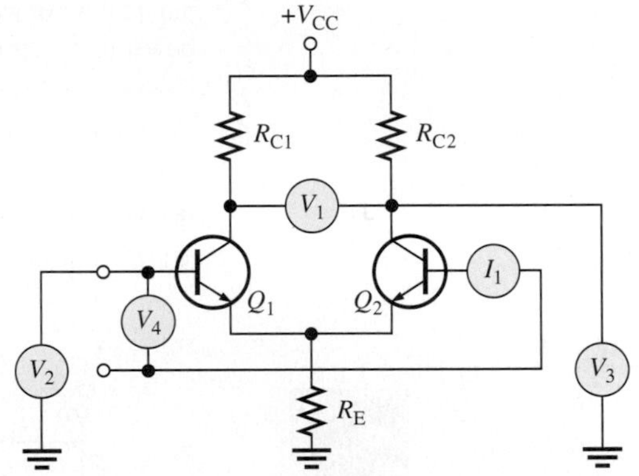

38. A differential amplifier stage has collector resistors of 5.1 kΩ each. If $I_{C1} = 1.35$ mA and $I_{C2} = 1.29$ mA, what is the differential output voltage?

39. Identify the type of input and output configuration for each basic differential amplifier in Figure 6–61.

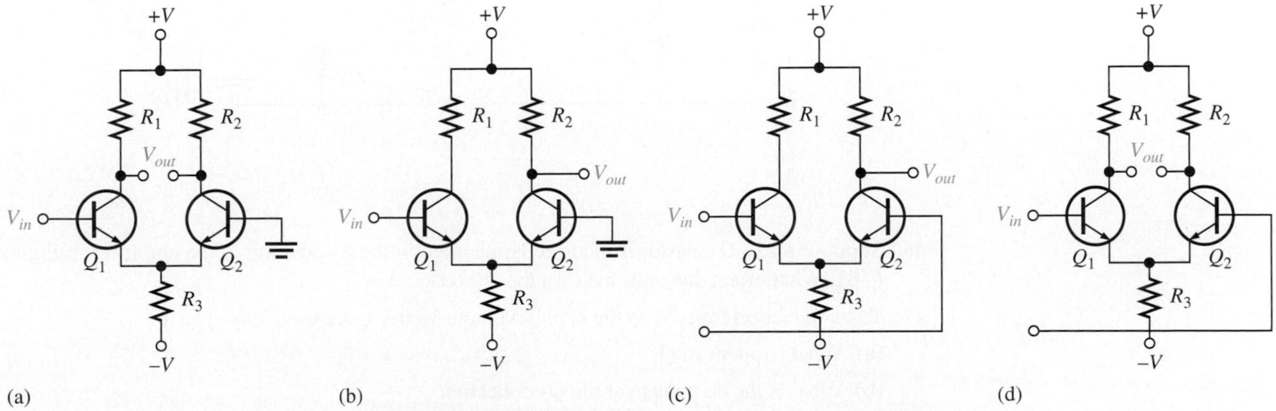

FIGURE 6–61

Section 6–8 Troubleshooting

40. Assume that the coupling capacitor C_3 is shorted in Figure 6–34. What dc voltage will appear at the collector of Q_1?

41. Assume that R_5 opens in Figure 6–34. Will Q_2 be in cutoff or in conduction? What dc voltage will you observe at the Q_2 collector?

42. Refer to Figure 6–57 and determine the general effect of each of the following failures:
   - **(a)** C_2 open
   - **(b)** C_3 open
   - **(c)** C_4 open
   - **(d)** C_2 shorted
   - **(e)** base-collector junction of Q_1 open
   - **(f)** base-emitter junction of Q_2 open

43. Assume that you must troubleshoot the amplifier in Figure 6–57. Set up a table of test point values, input, output, and all transistor terminals that include both dc and rms values that you expect to observe when a 300 Ω test signal source with a 25 μV rms output is used.

APPLICATION ACTIVITY PROBLEMS

44. Refer to the public address system block diagram in Figure 6–47. You are asked to repair a system that is not working. After a preliminary check, you find that there is no output signal from the power amplifier or from the preamplifier. Based on this check and assuming that only one of the blocks is faulty, which block can you eliminate as the faulty one? What would you check next?

45. What effect would each of the following faults in the amplifier of Figure 6–62 have on the output signal?

   (a) Open C_1 **(b)** Open C_2 **(c)** Open C_3 **(d)** Open C_4

   (e) Q_1 collector internally open **(f)** Q_2 emitter shorted to ground

FIGURE 6–62

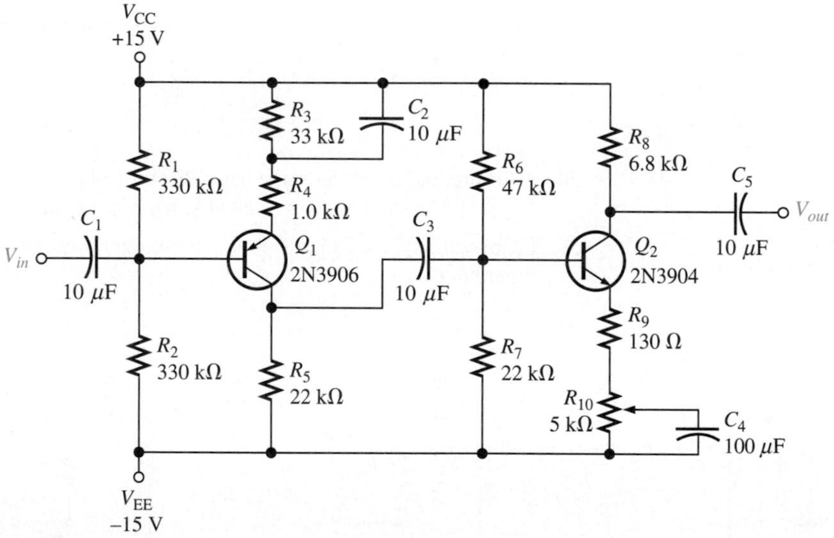

46. Suppose a 220 Ω resistor is incorrectly installed in the R_7 position of the amplifier in Figure 6–62. What effect does this have on the circuit?

47. The connection from R_1 to the supply voltage V_1 in Figure 6–62 has opened.
   - **(a)** What happens to Q_1?
   - **(b)** What is the dc voltage at the Q_1 collector?
   - **(c)** What is the dc voltage at the Q_2 collector?

DATASHEET PROBLEMS

48. Refer to the 2N3946/2N3947 partial datasheet in Figure 6–63. Determine the minimum value for each of the following *r* parameters:

   (a) β_{ac} **(b)** r'_e **(c)** r'_c

Electrical Characteristics (T_A = 25°C unless otherwise noted.)

Characteristic		Symbol	Min	Max	Unit
Input capacitance (V_{EB} = 1.0 V dc, I_C = 0, f = 1.0 MHz)		C_{ibo}	–	8.0	pF
Input impedance (I_C = 1.0 mA, V_{CE} = 10 V, f = 1.0 kHz)	2N3946 2N3947	h_{ie}	0.5 2.0	6.0 12	kohms
Voltage feedback ratio (I_C = 1.0 mA, V_{CE} = 10 V, f = 1.0 kHz)	2N3946 2N3947	h_{re}	– –	10 20	$\times 10^{-4}$
Small-signal current gain (I_C = 1.0 mA, V_{CE} = 10 V, f = 1.0 kHz)	2N3946 2N3947	h_{fe}	50 100	250 700	–
Output admittance (I_C = 1.0 mA, V_{CE} = 10 V, f = 1.0 kHz)	2N3946 2N3947	h_{oe}	1.0 5.0	30 50	μmhos
Collector base time constant (I_C = 10 mA, V_{CE} = 20 V, f = 31.8 MHz)		$rb'C_c$	–	200	ps
Noise figure (I_C = 100 μA, V_{CE} = 5.0 V, R_G = 1.0 kΩ, f = 1.0 kHz)		NF	–	5.0	dB

Switching Characteristics

Characteristic	Conditions		Symbol	Min	Max	Unit
Delay time	V_{CC} = 3.0 V dc, V_{OB} = 0.5 V dc,		t_d	–	35	ns
Rise time	I_C = 10 mA dc, I_{B1} = 1.0 mA		t_r	–	35	ns
Storage time	V_{CC} = 3.0 V, I_C = 10 mA,	2N3946 2N3947	t_s	– –	300 375	ns
Fall time	$I_{B1} = I_{B2}$ = 1.0 mA dc		t_f	–	75	ns

(1) Pulse test: $PW \leq 300$ μs, Duty Cycle ≤ 2%.

▲ **FIGURE 6–63**

Partial datasheet for the 2N3946/2N3947.

49. Repeat Problem 48 for maximum values.

50. Should you use a 2N3946 or a 2N3947 transistor in a certain application if the criteria is maximum current gain?

ADVANCED PROBLEMS

51. In an amplifier such as the one in Figure 6–62, explain the general effect that a leaky coupling capacitor would have on circuit performance.

52. Draw the dc and ac equivalent circuits for the amplifier in Figure 6–62.

53. Modify the 2-stage amplifier in Figure 6–62 to drive a load of 10 kΩ and maintain the same voltage gain.

54. Design a single-stage common-emitter amplifier with a voltage gain of 40 dB that operates from a dc supply voltage of +12 V. Use a 2N2222 transistor, voltage-divider bias, and a 330 Ω swamping resistor. The maximum input signal is 25 mV rms.

55. Design an emitter-follower with a minimum input resistance of 50 kΩ using a 2N3904 *npn* transistor with a $\beta_{ac} = 100$.

56. Repeat Problem 55 using a 2N3906 with a $\beta_{ac} = 100$.

57. Design a single-stage common-base amplifier for a voltage gain of 75. Use a 2N3904 with emitter bias. The dc supply voltages are to be ±6 V.

58. Refer to the amplifier in Figure 6–62 and determine the minimum value of coupling capacitors necessary for the amplifier to produce the same output voltage at 100 Hz that it does at 5000 Hz.

59. Prove that for any unloaded common-emitter amplifier with a collector resistor R_C and R_E bypassed, the voltage gain is $A_v \cong 40\ V_{R_C}$.

MULTISIM TROUBLESHOOTING PROBLEMS

These file circuits are in the Troubleshooting Problems folder on your CD-ROM.

60. Open file TSP06-60 and determine the fault.

61. Open file TSP06-61 and determine the fault.

62. Open file TSP06-62 and determine the fault.
63. Open file TSP06-63 and determine the fault.
64. Open file TSP06-64 and determine the fault.
65. Open file TSP06-65 and determine the fault.

ANSWERS

SECTION CHECKUPS

Section 6–1 **Amplifier Operation**

1. Positive, negative
2. V_{CE} is a dc quantity and V_{ce} is an ac quantity.
3. R_e is the external emitter ac resistance, r'_e is the internal emitter ac resistance.

Section 6–2 **Transistor AC Models**

1. α_{ac}—ac alpha, I_c/I_e; β_{ac}—ac beta, I_c/I_b; r'_e—ac emitter resistance; r'_b—ac base resistance; r'_e—ac collector resistor.
2. h_{fe} is equivalent to β_{ac}.
3. $r'_e = 25\text{ mV}/15\text{ mA} = 1.67\ \Omega$

Section 6–3 **The Common-Emitter Amplifier**

1. The capacitors are treated as opens.
2. The gain increases with a bypass capacitor.
3. Swamping eliminates the effects of r'_e by partially bypassing R_E.
4. Total input resistance includes the bias resistors, r'_e, and any unbypassed R_E.
5. The gain is determined by R_c, r'_e, and any unbypassed R_E.
6. The voltage gain decreases with a load.
7. The input and output voltages are 180° out of phase.

Section 6–4 **The Common-Collector Amplifier**

1. A common-collector amplifier is an emitter-follower.
2. The maximum voltage gain of a common-collector amplifier is 1.
3. A common-collector amplifier has a high input resistance.

Section 6–5 **The Common-Base Amplifier**

1. Yes
2. The common-base amplifier has a low input resistance.
3. The maximum current gain is 1 in a CB amplifier.

Section 6–6 **Multistage Amplifiers**

1. A stage is one amplifier in a cascaded arrangement.
2. The overall voltage gain is the product of the individual gains.
3. $20 \log (500) = 54.0\text{ dB}$
4. At lower frequencies, X_C becomes large enough to affect the gain.

Section 6–7 **The Differential Amplifier**

1. Double-ended differential input is between two input terminals. Single-ended differential input is from one input terminal to ground (with other input grounded).

2. Common-mode rejection is the ability of an op-amp to produce very little output when the same signal is applied to both inputs.
3. A higher CMRR results in a lower common-mode gain.

Section 6–8 Troubleshooting

1. If C_4 opens, the gain drops. The dc level would not be affected.
2. Q_2 would be biased in cutoff.
3. The collector voltage of Q_1 and the base, emitter, and collector voltages of Q_2 would change.

RELATED PROBLEMS FOR EXAMPLES

6–1 $I_C = 5$ mA; $V_{CE} = 1.5$ V

6–2 3.13 mA

6–3 9.3 mV

6–4 $C_2 = 28.4\ \mu$F

6–5 97.3

6–6 83

6–7 5; 165

6–8 9.56

6–9 Increases

6–10 71. A single transistor loads the CE amplifier much more than the Darlington pair.

6–11 55.9

6–12 $A'_v = 1500$; $A_{v1(dB)} = 27.96$ dB; $A_{v2(dB)} = 13.98$ dB; $A_{v3(dB)} = 21.58$ dB; $A'_{v(db)} = 63.52$ dB

6–13 34,000; 90.6 dB

6–14 C_3 open

TRUE/FALSE QUIZ

1. T **2.** T **3.** F **4.** T **5.** F **6.** F **7.** T **8.** T
9. T **10.** F **11.** T **12.** F **13.** T **14.** T **15.** F

CIRCUIT-ACTION QUIZ

1. (a) **2.** (b) **3.** (a) **4.** (b) **5.** (a) **6.** (c) **7.** (c) **8.** (c)
9. (c) **10.** (c)

SELF-TEST

1. (a) **2.** (b) **3.** (c) **4.** (b) **5.** (a) **6.** (b) **7.** (b) **8.** (d)
9. (d) **10.** (d) **11.** (d) **12.** (c) **13.** (a) **14.** (c) **15.** (a) **16.** (d)
17. (b) **18.** (a) **19.** (c)

7 POWER AMPLIFIERS

CHAPTER OUTLINE

7–1 The Class A Power Amplifier
7–2 The Class B and Class AB Push-Pull Amplifiers
7–3 The Class C Amplifier
7–4 Troubleshooting
Application Activity

CHAPTER OBJECTIVES

- Explain and analyze the operation of class A power amplifiers
- Explain and analyze the operation of class B and class AB amplifiers
- Discuss and analyze the operation of class C amplifiers
- Troubleshoot power amplifiers

KEY TERMS

- Class A
- Power gain
- Efficiency
- Class B
- Push-pull
- Class AB
- Class C

APPLICATION ACTIVITY PREVIEW

The Application Activity in this chapter continues with the public address system started in Chapter 6. Recall that the complete system includes the preamplifier, a power amplifier, and a dc power supply. You will focus on the power amplifier in this chapter and complete the total system by combining the three component parts.

VISIT THE COMPANION WEBSITE

Study aids for this chapter are available at http://www.prenhall.com/floyd

INTRODUCTION

Power amplifiers are large-signal amplifiers. This generally means that a much larger portion of the load line is used during signal operation than in a small-signal amplifier. In this chapter, we will cover four classes of power amplifiers: class A, class B, class AB, and class C. These amplifier classifications are based on the percentage of the input cycle for which the amplifier operates in its linear region. Each class has a unique circuit configuration because of the way it must be operated. The emphasis is on power amplification.

Power amplifiers are normally used as the final stage of a communications receiver or transmitter to provide signal power to speakers or to a transmitting antenna.

7–1 THE CLASS A POWER AMPLIFIER

When an amplifier is biased such that it always operates in the linear region where the output signal is an amplified replica of the input signal, it is a **class A** amplifier. The discussion of amplifiers in the previous chapters apply to class A operation. Power amplifiers are those amplifiers that have the objective of delivering power to a load. This means that components must be considered in terms of their ability to dissipate heat.

After completing this section, you should be able to

- **Explain and analyze the operation of class A power amplifiers**
  - Explain why a centered Q-point is important for a class A power amplifier
  - Determine the voltage gain and power gain for a multistage amplifier
  - Determine the efficiency of a class A power amplifier

In a small-signal amplifier, the ac signal moves over a small percentage of the total ac load line. When the output signal is larger and approaches the limits of the ac load line, the amplifier is a **large-signal** type. Both large-signal and small-signal amplifiers are considered to be class A if they operate in the linear region at all times, as illustrated in Figure 7–1. Class A power amplifiers are large-signal amplifiers with the objective of providing power (rather than voltage) to a load. As a rule of thumb, an amplifier may be considered to be a power amplifier if it is rated for more than 1 W and it is necessary to consider the problem of heat dissipation in components.

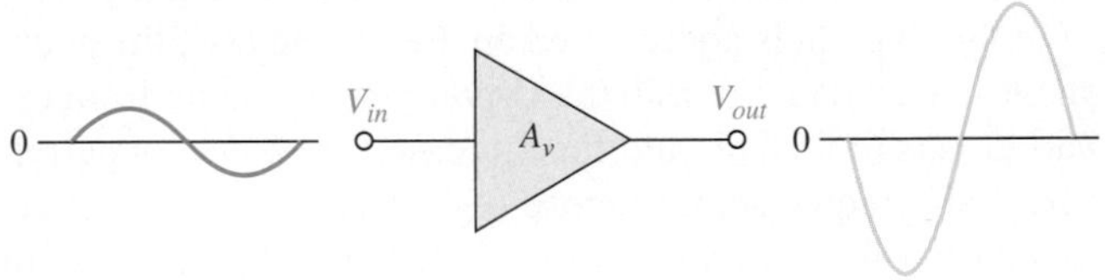

FIGURE 7–1

Basic class A amplifier operation. Output is shown 180° out of phase with the input (inverted).

Heat Dissipation

Power transistors (and other power devices) must dissipate a large amount of internally generated heat. For BJT power transistors, the collector terminal is the critical junction; for this reason, the transistor's case is always connected to the collector terminal. The case of all power transistors is designed to provide a large contact area between it and an external heat sink. Heat from the transistor flows through the case to the heat sink and then dissipates in the surrounding air. Heat sinks vary in size, number of fins, and type of material. Their size depends on the heat dissipation requirement and the maximum ambient temperature in which the transistor is to operate. In high-power applications (a few hundred watts), a cooling fan may be necessary.

Centered Q-Point

Recall that the dc and ac load lines intersect at the Q-point. When the Q-point is at the center of the ac load line, a maximum class A signal can be obtained. You can see this concept by examining the graph of the load line for a given amplifier in Figure 7–2(a). This graph shows the ac load line with the Q-point at its center. The collector current can vary from its Q-point value, I_{CQ}, up to its saturation value, $I_{c(sat)}$, and down to its cutoff value of zero. Likewise, the collector-to-emitter voltage can swing from its Q-point value, V_{CEQ}, up to its cutoff value, $V_{ce(cutoff)}$, and down to its saturation value of near zero. This operation is indicated in Figure 7–2(b). The peak value of the collector current equals I_{CQ}, and the peak

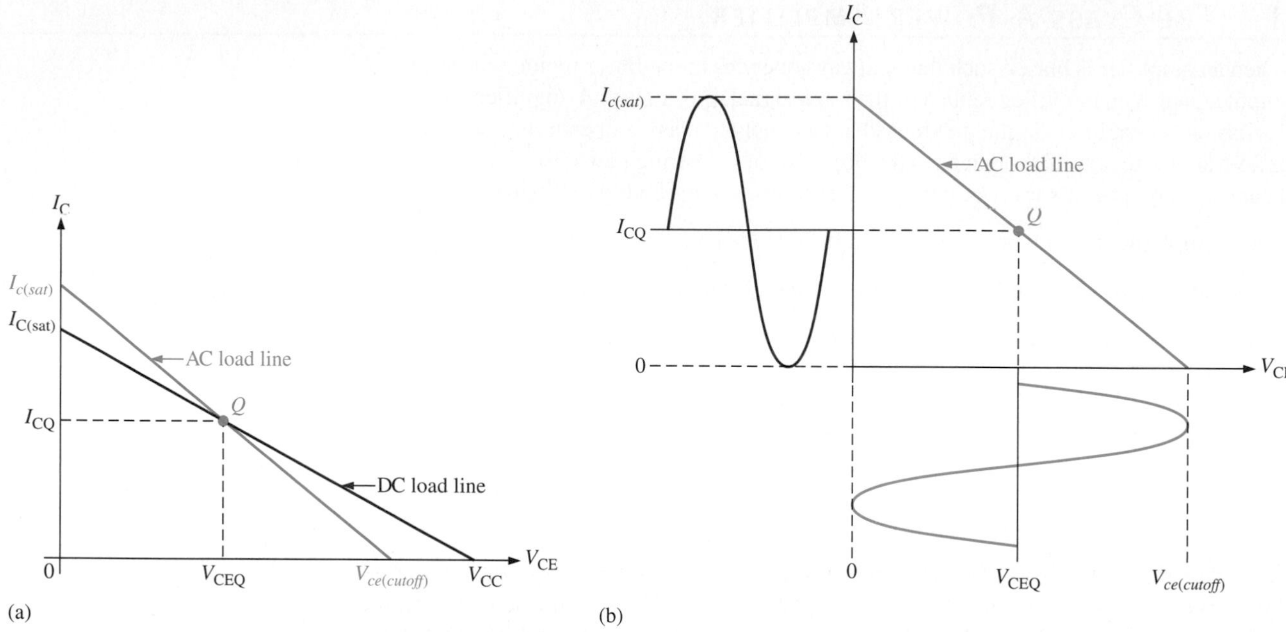

▲ **FIGURE 7–2**

Maximum class A output occurs when the Q-point is centered on the ac load line.

value of the collector-to-emitter voltage equals V_{CEQ} in this case. This signal is the maximum that can be obtained from the class A amplifier. Actually, the output cannot quite reach saturation or cutoff, so the practical maximum is slightly less.

If the Q-point is not centered on the ac load line, the output signal is limited. Figure 7–3 shows an ac load line with the Q-point moved away from center toward cutoff. The output variation is limited by cutoff in this case. The collector current can only swing down to near zero and an equal amount above I_{CQ}. The collector-to-emitter voltage can only swing up to its cutoff value and an equal amount below V_{CEQ}. This situation is illustrated in Figure 7–3(a). If the amplifier is driven any further than this, it will "clip" at cutoff, as shown in Figure 7–3(b).

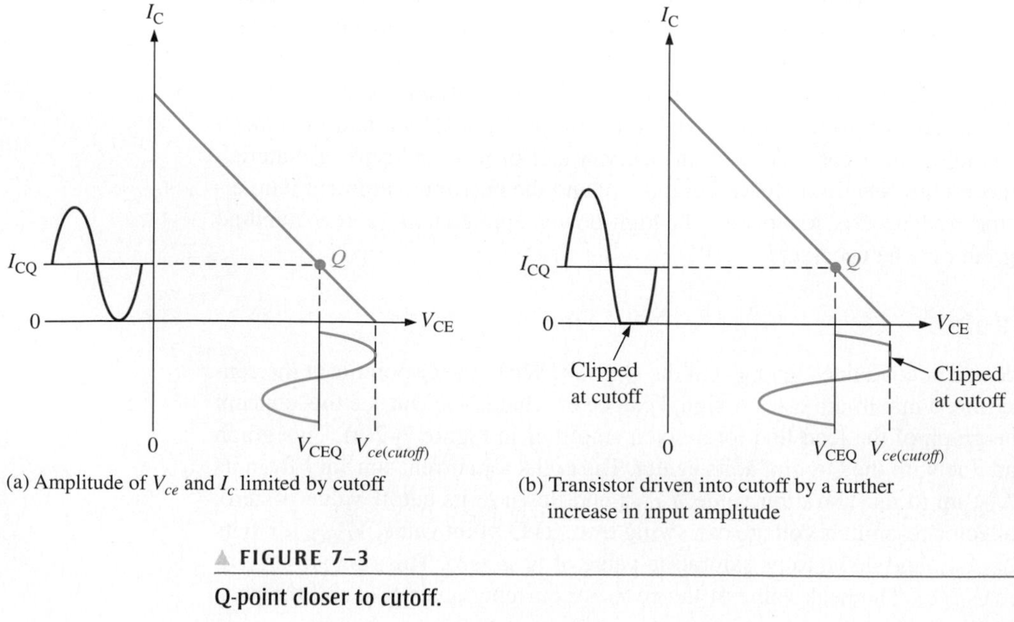

▲ **FIGURE 7–3**

Q-point closer to cutoff.

Figure 7–4 shows an ac load line with the Q-point moved away from center toward saturation. In this case, the output variation is limited by saturation. The collector current can only swing up to near saturation and an equal amount below I_{CQ}. The collector-to-emitter voltage can only swing down to its saturation value and an equal amount above V_{CEQ}. This situation is illustrated in Figure 7–4(a). If the amplifier is driven any further, it will "clip" at saturation, as shown in Figure 7–4(b).

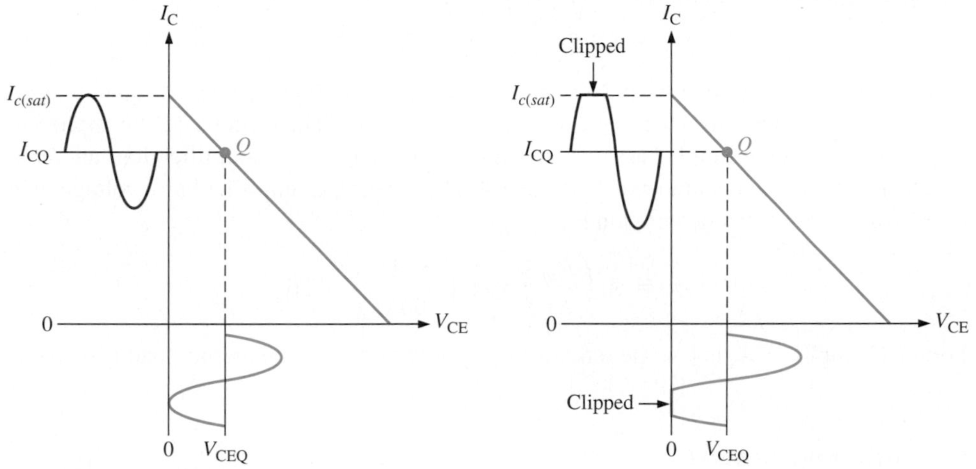

(a) Amplitude of V_{ce} and I_c limited by saturation

(b) Transistor driven into saturation by a further increase in input amplitude

FIGURE 7–4

Q-point closer to saturation.

Power Gain

A power amplifier delivers power to a load. The **power gain** of an amplifier is the ratio of the output power (power delivered to the load) to the input power. In general, power gain is

$$A_p = \frac{P_L}{P_{in}} \qquad \text{Equation 7–1}$$

where A_p is the power gain, P_L is signal power delivered to the load, and P_{in} is signal power delivered to the amplifier.

The power gain can be computed by any of several formulas, depending on what is known. Frequently, the easiest way to obtain power gain is from input resistance, load resistance, and voltage gain. To see how this is done, recall that power can be expressed in terms of voltage and resistance as

$$P = \frac{V^2}{R}$$

For ac power, the voltage is expressed as rms. The output power delivered to the load is

$$P_L = \frac{V_L^2}{R_L}$$

The input power delivered to the amplifier is

$$P_{in} = \frac{V_{in}^2}{R_{in}}$$

By substituting into Equation 7–1, the following useful relationship is produced:

$$A_p = \frac{V_L^2}{V_{in}^2}\left(\frac{R_{in}}{R_L}\right)$$

Since $V_L/V_{in} = A_v$,

Equation 7–2

$$A_p = A_v^2\left(\frac{R_{in}}{R_L}\right)$$

Recall from Chapter 6 that for a voltage-divider biased amplifier,

$$R_{in(tot)} = R_1 \| R_2 \| R_{in(base)}$$

and that for a CE or CC amplifier,

$$R_{in(base)} = \beta_{ac}R_e$$

Equation 7–2 shows that the power gain of an amplifier is the voltage gain squared times the ratio of the input resistance to the output load resistance. The formula can be applied to any amplifier. For example, assume a common-collector (CC) amplifier has an input resistance of 5 kΩ and a load resistance of 100 Ω. Since a CC amplifier has a voltage gain of approximately 1, the power gain is

$$A_p = A_v^2\left(\frac{R_{in}}{R_L}\right) = 1^2\left(\frac{5\text{ k}\Omega}{100\ \Omega}\right) = 50$$

For a CC amplifier, A_p is just the ratio of the input resistance to the output load resistance.

DC Quiescent Power

The power dissipation of a transistor with no signal input is the product of its Q-point current and voltage.

Equation 7–3

$$P_{DQ} = I_{CQ}V_{CEQ}$$

The only way a class A power amplifier can supply power to a load is to maintain a quiescent current that is at least as large as the peak current requirement for the load current. A signal will not increase the power dissipated by the transistor but actually causes less total power to be dissipated. The quiescent power, given in Equation 7–3, is the maximum power that a class A amplifier must handle. The transistor's power rating must exceed this value.

Output Power

In general, the output signal power is the product of the rms load current and the rms load voltage. The maximum unclipped ac signal occurs when the Q-point is centered on the ac load line. For a CE amplifier with a centered Q-point, the maximum peak voltage swing is

$$V_{c(max)} = I_{CQ}R_c$$

The rms value is $0.707V_{c(max)}$.

The maximum peak current swing is

$$I_{c(max)} = \frac{V_{CEQ}}{R_c}$$

The rms value is $0.707I_{c(max)}$.

To find the maximum signal power output, use the rms values of maximum current and voltage. The maximum power out from a class A amplifier is

$$P_{out(max)} = (0.707I_c)(0.707V_c)$$

Equation 7–4

$$P_{out(max)} = 0.5I_{CQ}V_{CEQ}$$

EXAMPLE 7–1

Determine the voltage gain and the power gain of the class A power amplifier in Figure 7–5. Assume $\beta_{ac} = 200$ for all transistors.

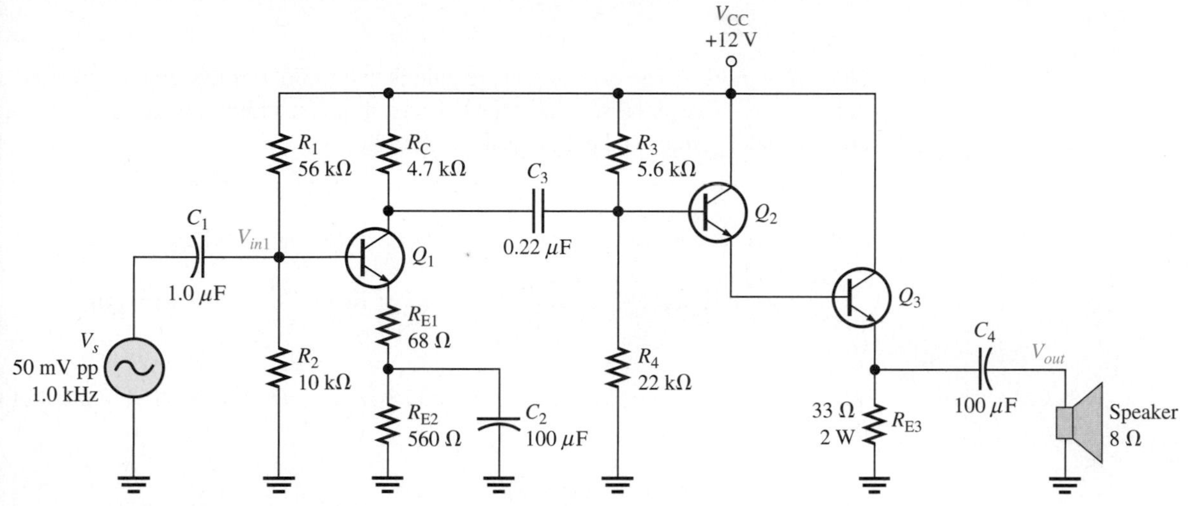

FIGURE 7–5

Solution Notice that the first stage (Q_1) is a voltage-divider biased common-emitter with a swamping resistor (R_{E1}). The second stage (Q_2 and Q_3) is a Darlington voltage-follower configuration. The speaker is the load.

First stage: The ac collector resistance of the first stage is R_C in parallel with the input resistance to the second stage.

$$R_{c1} \cong R_C \| (R_3 \| R_4) = 4.7\text{ k}\Omega \| 5.6\text{ k}\Omega \| 22\text{ k}\Omega = 2.29\text{ k}\Omega$$

The voltage gain of the first stage is the ac collector resistance, R_{c1}, divided by the ac emitter resistance, which is the sum of $R_{E1} + r'_{e(Q1)}$. The approximate value of $r'_{e(Q1)}$ is determined by first finding I_E.

$$V_B \cong \left(\frac{R_2}{R_1 + R_2}\right)V_{CC} = \left(\frac{10\text{ k}\Omega}{66\text{ k}\Omega}\right)12\text{ V} = 1.82\text{ V}$$

$$I_E = \frac{V_B - 0.7\text{ V}}{R_{E1} + R_{E2}} = \frac{1.82\text{ V} - 0.7\text{ V}}{628\ \Omega} = 1.78\text{ mA}$$

$$r'_{e(Q1)} = \frac{25\text{ mV}}{I_E} = \frac{25\text{ mV}}{1.78\text{ mA}} = 14\ \Omega$$

Using the value of r'_e, determine the voltage gain of the first stage with the loading of the second stage taken into account.

$$A_{v1} = -\frac{R_{c1}}{R_{E1} + r'_{e(Q1)}} = -\frac{2.29\text{ k}\Omega}{68\ \Omega + 14\ \Omega} = -27.9$$

The negative sign is for inversion.

The total input resistance of the first stage is equal to the bias resistors in parallel with the ac input resistance at the base of Q_1.

$$R_{in(tot)1} = R_1 \| R_2 \| \beta_{ac(Q1)}(R_{E1} + r'_{e(Q1)})$$
$$= 56\text{ k}\Omega \| 10\text{ k}\Omega \| 200(68\ \Omega + 14\ \Omega) = 8.4\text{ k}\Omega$$

Second stage: The voltage gain of the Darlington emitter-follower is approximately equal to 1.

$$A_{v2} \cong 1$$

Overall amplifier: The overall voltage gain is the product of the first and second stage voltage gains. Since the second stage has a gain of approximately 1, the overall gain is approximately equal to the gain of the first stage.

$$A_{v(tot)} = A_{v1}A_{v2} = (-27.9)(1) = -27.9$$

Power gain: The power gain of the amplifier can be calculated using Equation 7–2.

$$A_p = A^2_{v(tot)}\left(\frac{R_{in(tot)1}}{R_L}\right) = (-27.9)^2\left(\frac{8.4\text{ k}\Omega}{8\ \Omega}\right) = \mathbf{817{,}330}$$

*Related Problem** What happens to the power gain if a second 8 Ω speaker is connected in parallel with the first one?

*Answers are at the end of the chapter.

Efficiency

The **efficiency** of any amplifier is the ratio of the output signal power supplied to a load to the total power from the dc supply. The maximum output signal power that can be obtained is given by Equation 7–4. The average power supply current, I_{CC}, is equal to I_{CQ} and the supply voltage is at least $2V_{CEQ}$. Therefore, the total dc power is

$$P_{DC} = I_{CC}V_{CC} = 2I_{CQ}V_{CEQ}$$

The maximum efficiency, η_{max}, of a capacitively coupled class A amplifier is

$$\eta_{max} = \frac{P_{out}}{P_{DC}} = \frac{0.5I_{CQ}V_{CEQ}}{2I_{CQ}V_{CEQ}} = 0.25$$

The maximum efficiency of a capacitively coupled class A amplifier cannot be higher than 0.25, or 25%, and, in practice, is usually considerably less (about 10%). Although the efficiency can be made higher by transformer coupling the signal to the load, there are drawbacks to transformer coupling. These drawbacks include the size and cost of transformers as well as potential distortion problems when the transformer core begins to saturate. In general, the low efficiency of class A amplifiers limits their usefulness to small power applications that require usually less than 1 W.

EXAMPLE 7–2 Determine the efficiency of the power amplifier in Figure 7–5 (Example 7–1).

Solution The efficiency is the ratio of the signal power in the load to the power supplied by the dc source. The input voltage is 50 mV peak-to-peak which is 35.4 mV rms. The input power is, therefore,

$$P_{in} = \frac{V^2_{in}}{R_{in}} = \frac{(35.4\text{ mV})^2}{8.4\text{ k}\Omega} = 149\text{ nW}$$

The output power is

$$P_{out} = P_{in}A_p = (149\text{ nW})(817{,}330) = 122\text{ mW}$$

Most of the power from the dc source is supplied to the output stage. The current in the output stage can be computed from the dc emitter voltage of Q_3.

$$V_{E(Q3)} \cong \left(\frac{22\text{ k}\Omega}{27.6\text{ k}\Omega}\right)12\text{ V} - 1.4\text{ V} = 8.2\text{ V}$$

$$I_{E(Q3)} = \frac{V_{E(Q3)}}{R_E} = \frac{8.2\text{ V}}{33\ \Omega} = 0.25$$

Neglecting the other transistor and bias currents, which are very small, the total dc supply current is about 0.25 A. The power from the dc source is

$$P_{DC} = I_{CC}V_{CC} = (0.25\text{ A})(12\text{ V}) = 3\text{ W}$$

Therefore, the efficiency of the amplifier for this input is

$$\eta = \frac{P_{out}}{P_{DC}} = \frac{122\text{ mW}}{3\text{ W}} \cong \mathbf{0.04}$$

This represents an efficiency of 4% and illustrates why class A is not a good choice for a power amplifier.

Related Problem Explain what happens to the efficiency if R_{E3} were replaced with the speaker. What problem does this have?

SECTION 7–1 CHECKUP
Answers are at the end of the chapter.

1. What is the purpose of a heat sink?
2. Which lead of a BJT is connected to the case?
3. What are the two types of clipping with a class A power amplifier?
4. What is the maximum efficiency for a class A amplifier?
5. How can the power gain of a CC amplifier be expressed in terms of a ratio of resistances?

7–2 THE CLASS B AND CLASS AB PUSH-PULL AMPLIFIERS

When an amplifier is biased at cutoff so that it operates in the linear region for 180° of the input cycle and is in cutoff for 180°, it is a **class B** amplifier. Class AB amplifiers are biased to conduct for slightly more than 180°. The primary advantage of a class B or class AB amplifier over a class A amplifier is that either one is more efficient than a class A amplifier; you can get more output power for a given amount of input power. A disadvantage of class B or class AB is that it is more difficult to implement the circuit in order to get a linear reproduction of the input waveform. The term *push-pull* refers to a common type of class B or class AB amplifier circuit in which two transistors are used on alternating half-cycles to reproduce the input waveform at the output.

After completing this section, you should be able to

- **Explain and analyze the operation of class B and class AB amplifiers**
  - Explain class B operation
  - Discuss Q-point location for class B amplifiers
  - Analyze class B push-pull operation
  - Explain crossover distortion and its cause
  - Explain class AB operation
  - Analyze class AB push-pull amplifiers
  - Determine class B maximum efficiency
  - Describe a Darlington push-pull amplifier
  - Describe a complementary Darlington amplifier

Class B Operation

The class B operation is illustrated in Figure 7–6, where the output waveform is shown relative to the input in terms of time (t).

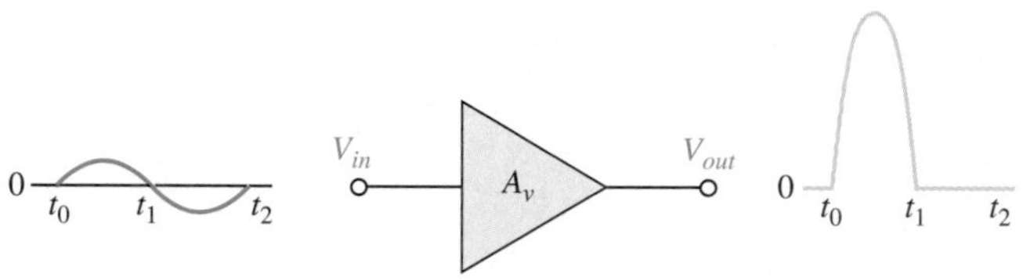

▶ FIGURE 7–6

Basic class B amplifier operation (noninverting).

The Q-Point Is at Cutoff The class B amplifier is biased at the cutoff point so that $I_{CQ} = 0$ and $V_{CEQ} = V_{CE(cutoff)}$. It is brought out of cutoff and operates in its linear region when the input signal drives the transistor into conduction. This is illustrated in Figure 7–7 with an emitter-follower circuit where the output is not a replica of the input.

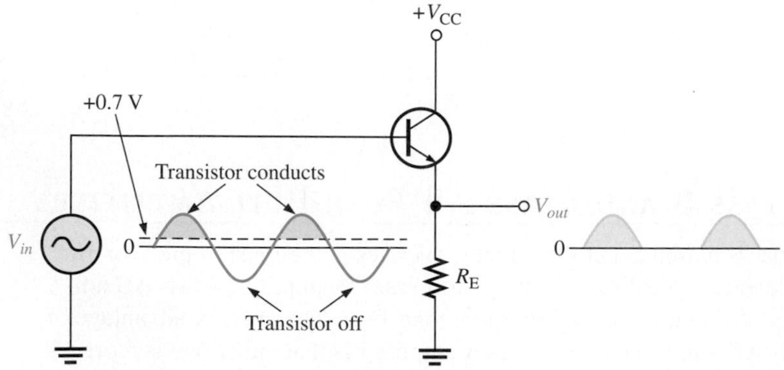

▶ FIGURE 7–7

Common-collector class B amplifier.

Class B Push-Pull Operation

As you can see, the circuit in Figure 7–7 only conducts for the positive half of the cycle. To amplify the entire cycle, it is necessary to add a second class B amplifier that operates on the negative half of the cycle. The combination of two class B amplifiers working together is called **push-pull** operation.

There are two common approaches for using push-pull amplifiers to reproduce the entire waveform. The first approach uses transformer coupling. The second uses two **complementary symmetry transistors;** these are a matching pair of *npn*/*pnp* BJTs.

Transformer Coupling Transformer coupling is illustrated in Figure 7–8. The input transformer has a center-tapped secondary that is connected to ground, producing phase inversion of one side with respect to the other. The input transformer thus converts the input signal to two out-of-phase signals for the transistors. Notice that both transistors are *npn* types. Because of the signal inversion, Q_1 will conduct on the positive part of the cycle and Q_2 will conduct on the negative part. The output transformer combines the signals by permitting current in both directions, even though one transistor is always cut off. The positive power supply signal is connected to the center tap of the output transformer.

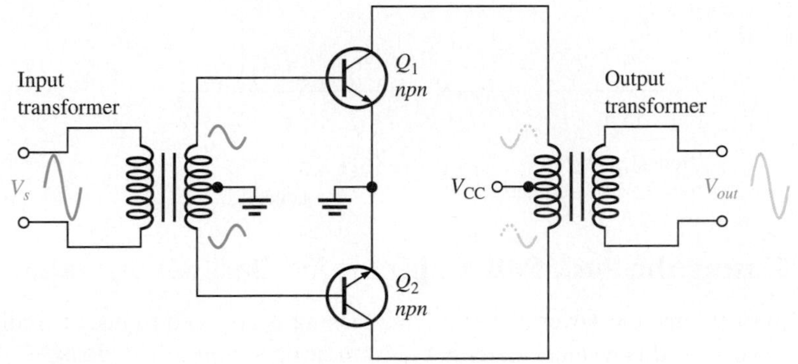

FIGURE 7–8

Transformer coupled push-pull amplifiers. Q_1 conducts during the positive half-cycle; Q_2 conducts during the negative half-cycle. The two halves are combined by the output transformer.

Complementary Symmetry Transistors Figure 7–9 shows one of the most popular types of push-pull class B amplifiers using two emitter-followers and both positive and negative power supplies. This is a complementary amplifier because one emitter-follower uses an *npn* transistor and the other a *pnp,* which conduct on opposite alternations of the input cycle. Notice that there is no dc base bias voltage ($V_B = 0$). Thus, only the signal voltage drives the transistors into conduction. Transistor Q_1 conducts during the positive half of the input cycle, and Q_2 conducts during the negative half.

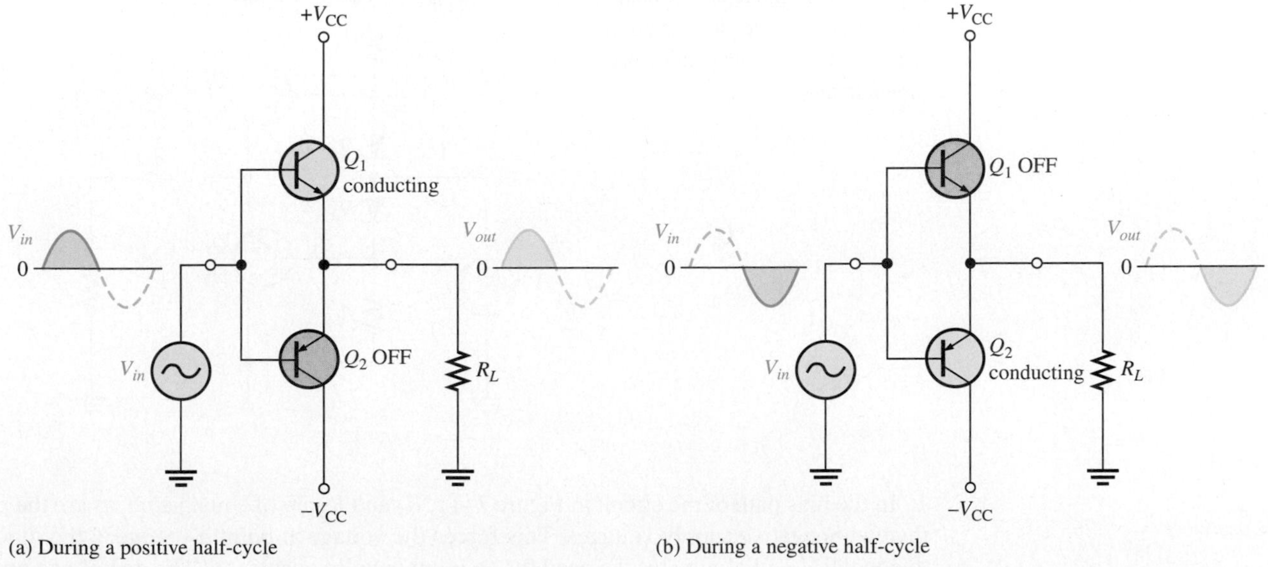

FIGURE 7–9

Class B push-pull ac operation.

Crossover Distortion When the dc base voltage is zero, both transistors are off and the input signal voltage must exceed V_{BE} before a transistor conducts. Because of this, there is a time interval between the positive and negative alternations of the input when neither transistor is conducting, as shown in Figure 7–10. The resulting distortion in the output waveform is called **crossover distortion.**

FIGURE 7–10

Illustration of crossover distortion in a class B push pull amplifier. The transistors conduct only during portions of the input indicated by the shaded areas.

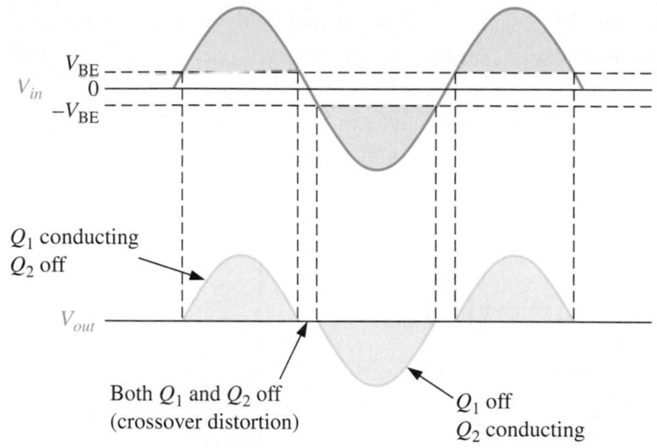

Biasing the Push-Pull Amplifier for Class AB Operation

To overcome crossover distortion, the biasing is adjusted to just overcome the V_{BE} of the transistors; this results in a modified form of operation called **class AB**. In class AB operation, the push-pull stages are biased into slight conduction, even when no input signal is present. This can be done with a voltage-divider and diode arrangement, as shown in Figure 7–11. When the diode characteristics of D_1 and D_2 are closely matched to the characteristics of the transistor base-emitter junctions, the current in the diodes and the current in the transistors are the same; this is called a **current mirror.** This current mirror produces the desired class AB operation and eliminates crossover distortion.

FIGURE 7–11

Biasing the push-pull amplifier with current-mirror diode bias to eliminate crossover distortion. The transistors form a complementary pair (one *npn* and one *pnp*).

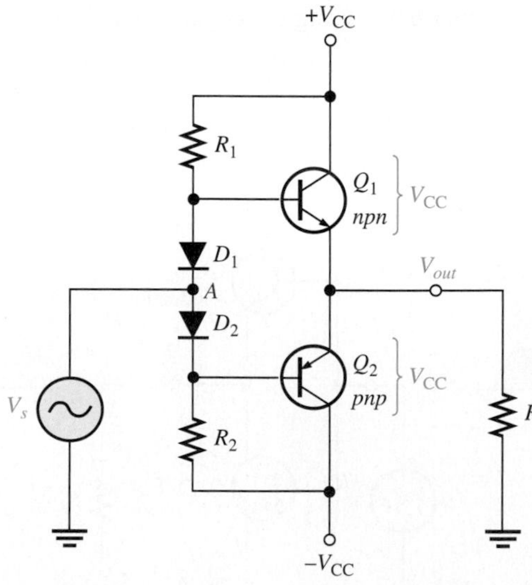

In the bias path of the circuit in Figure 7–11, R_1 and R_2 are of equal value, as are the positive and negative supply voltages. This forces the voltage at point A (between the diodes) to equal 0 V and eliminates the need for an input coupling capacitor. The dc voltage on the output is also 0 V. Assuming that both diodes and both complementary transistors are identical, the drop across D_1 equals the V_{BE} of Q_1, and the drop across D_2 equals the V_{BE} of Q_2.

Since they are matched, the diode current will be the same as I_{CQ}. The diode current and I_{CQ} can be found by applying Ohm's law to either R_1 or R_2 as follows:

$$I_{CQ} = \frac{V_{CC} - 0.7\text{ V}}{R_1}$$

This small current required of class AB operation eliminates the crossover distortion but has the potential for thermal instability if the transistor's V_{BE} drops are not matched to the diode drops or if the diodes are not in thermal equilibrium with the transistors. Heat in the power transistors decreases the base-emitter voltage and tends to increase current. If the diodes are warmed the same amount, the current is stabilized; but if the diodes are in a cooler environment, they cause I_{CQ} to increase even more. More heat is produced in an unrestrained cycle known as *thermal runaway*. To keep this from happening, the diodes should have the same thermal environment as the transistors. In some cases, a small resistor in the emitter of each transistor can alleviate thermal runaway.

Crossover distortion also occurs in transformer-coupled amplifiers like the one shown in Figure 7–8. To eliminate it in this case, 0.7 V is applied to the input transformer's secondary that just biases both transistors into conduction. The bias voltage to produce this drop can be derived from the power supply using a single diode as shown in Figure 7–12.

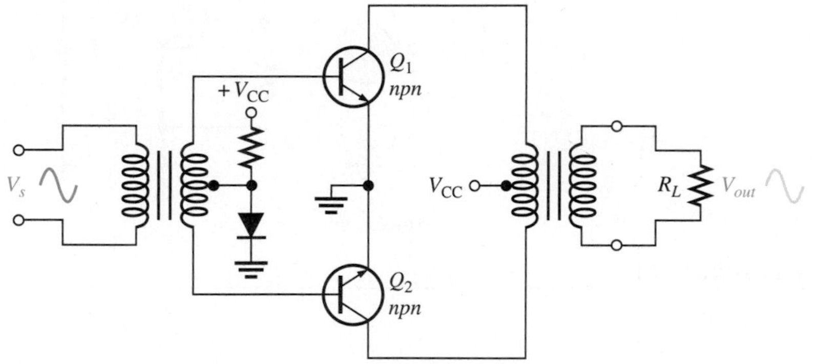

FIGURE 7–12

Eliminating crossover distortion in a transformer-coupled push-pull amplifier. The biased diode compensates for the base-emitter drop of the transistors and produces class AB operation.

AC Operation Consider the ac load line for Q_1 of the class AB amplifier in Figure 7–11. The Q-point is slightly above cutoff. (In a true class B amplifier, the Q-point is at cutoff.) The ac cutoff voltage for a two-supply operation is at V_{CC} with an I_{CQ} as given earlier. The ac saturation current for a two-supply operation with a push-pull amplifier is

$$I_{c(sat)} = \frac{V_{CC}}{R_L}$$

Equation 7–5

The ac load line for the *npn* transistor is as shown in Figure 7–13. The dc load line can be found by drawing a line that passes through V_{CEQ} and the dc saturation current, $I_{C(sat)}$. However, the saturation current for dc is the current if the collector to emitter is shorted on

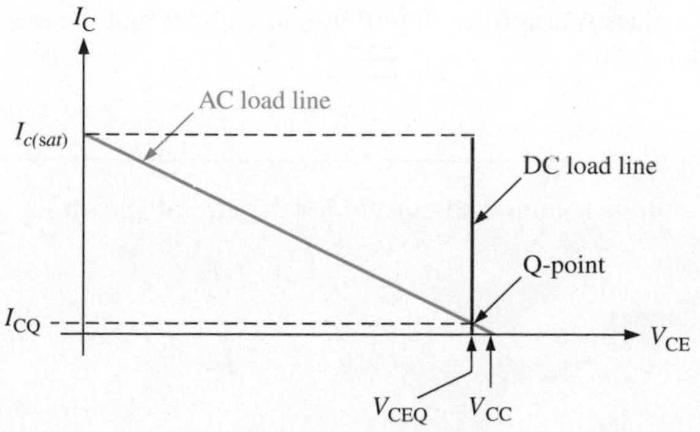

FIGURE 7–13

Load lines for a complementary symmetry push-pull amplifier. Only the load lines for the *npn* transistor are shown.

both transistors! This assumed short across the power supplies obviously would cause maximum current from the supplies and implies the dc load line passes almost vertically through the cutoff as shown. Operation along the dc load line, such as caused by thermal runaway, could produce such a high current that the transistors are destroyed.

Figure 7–14(a) illustrates the ac load line for Q_1 of the class AB amplifier in Figure 7–14(b). In the case illustrated, a signal is applied that swings over the region of the ac load line shown in bold. At the upper end of the ac load line, the voltage across the transistor (V_{ce}) is a minimum, and the output voltage is maximum.

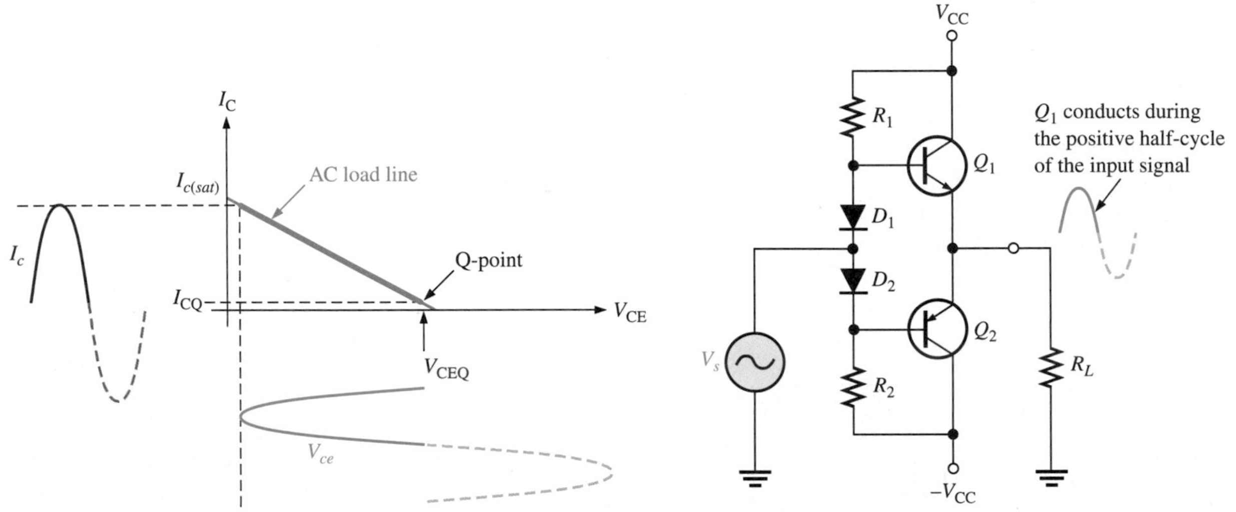

(a) AC load line for Q_1 (b) Circuit

▲ **FIGURE 7–14**

Under maximum conditions, transistors Q_1 and Q_2 are alternately driven from near cutoff to near saturation. During the positive alternation of the input signal, the Q_1 emitter is driven from its Q-point value of 0 to nearly V_{CC}, producing a positive peak voltage a little less than V_{CC}. Likewise, during the negative alternation of the input signal, the Q_2 emitter is driven from its Q-point value of 0 V, to near $-V_{CC}$, producing a negative peak voltage almost equal to $-V_{CC}$. Although it is possible to operate close to the saturation current, this type of operation results in increased distortion of the signal.

The ac saturation current (Equation 7–5) is also the peak output current. Each transistor can essentially operate over its entire load line. Recall that in class A operation, the transistor can also operate over the entire load line but with a significant difference. In class A operation, the Q-point is near the middle and there is significant current in the transistors even with no signal. In class B operation, when there is no signal, the transistors have only a very small current and therefore dissipate very little power. Thus, the efficiency of a class B amplifier can be much higher than a class A amplifier. It will be shown later that the maximum efficiency of a class B amplifier is 79%.

EXAMPLE 7–3

Determine the ideal maximum peak output voltage and current for the circuit shown in Figure 7–15.

Solution The ideal maximum peak output voltage is

$$V_{out(peak)} \cong V_{CEQ} \cong V_{CC} = \mathbf{20\ V}$$

FIGURE 7–15

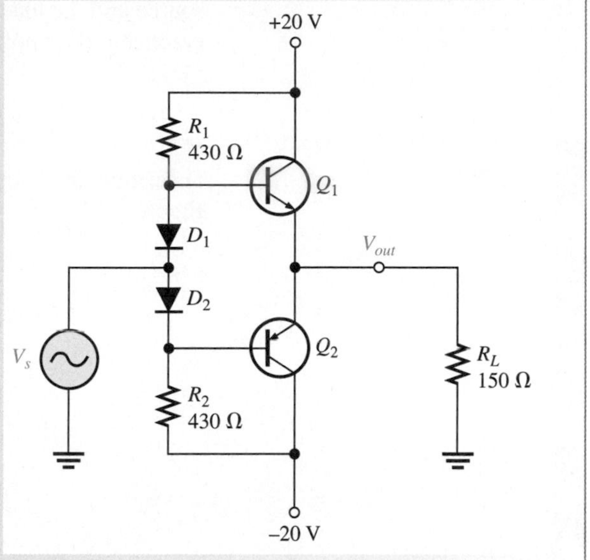

The ideal maximum peak current is

$$I_{out(peak)} \cong I_{c(sat)} \cong \frac{V_{CC}}{R_L} = \frac{20\text{ V}}{150\ \Omega} = \mathbf{133\ mA}$$

The actual maximum values of voltage and current are slightly smaller.

Related Problem What is the maximum peak output voltage and current if the supply voltages are changed to +15 V and −15 V?

Open the Multisim file E07-03 in the Examples folder on your CD-ROM. Measure the maximum peak-to-peak output voltage.

Single-Supply Push-Pull Amplifier

Push-pull amplifiers using complementary symmetry transistors can be operated from a single voltage source as shown in Figure 7–16. The circuit operation is the same as that described previously, except the bias is set to force the output emitter voltage to be $V_{CC}/2$ instead of zero volts used with two supplies. Because the output is not biased at zero volts,

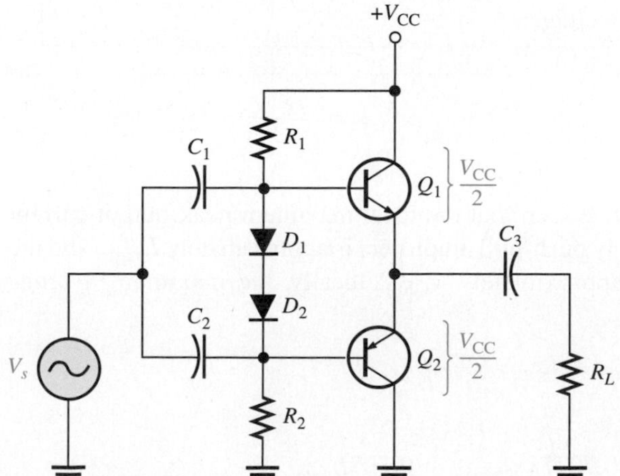

FIGURE 7–16

Single-ended push-pull amplifier.

capacitive coupling for the input and output is necessary to block the bias voltage from the source and the load resistor. Ideally, the output voltage can swing from zero to V_{CC}, but in practice it does not quite reach these ideal values.

EXAMPLE 7–4

Determine the maximum ideal peak values for the output voltage and current in Figure 7–17.

FIGURE 7–17

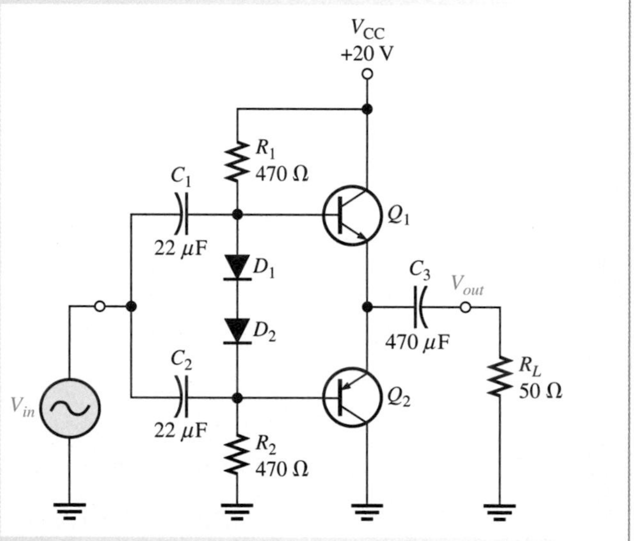

Solution The maximum peak output voltage is

$$V_{out(peak)} \cong V_{CEQ} = \frac{V_{CC}}{2} = \frac{20\text{ V}}{2} = \mathbf{10\ V}$$

The maximum peak output current is

$$I_{out(peak)} \cong I_{c(sat)} = \frac{V_{CEQ}}{R_L} = \frac{10\text{ V}}{50\ \Omega} = \mathbf{200\ mA}$$

Related Problem Find the maximum peak values for the output voltage and current in Figure 7–17 if V_{CC} is lowered to 15 V and the load resistance is changed to 30 Ω.

Open the Multisim file E07-04 in the Examples folder on your CD-ROM. Measure the maximum peak-to-peak output voltage.

Class B/AB Power

Maximum Output Power You have seen that the ideal maximum peak output current for both dual-supply and single-supply push-pull amplifiers is approximately $I_{c(sat)}$, and the maximum peak output voltage is approximately V_{CEQ}. Ideally, the maximum *average* output power is, therefore,

$$P_{out} = I_{out(rms)}V_{out(rms)}$$

Since

$$I_{out(rms)} = 0.707I_{out(peak)} = 0.707I_{c(sat)}$$

and

$$V_{out(rms)} = 0.707V_{out(peak)} = 0.707V_{CEQ}$$

then

$$P_{out} = 0.5I_{c(sat)}V_{CEQ}$$

Substituting $V_{CC}/2$ for V_{CEQ}, the maximum average output power is

$$\mathbf{P_{out} = 0.25I_{c(sat)}V_{CC}}$$

Equation 7–6

DC Input Power The dc input power comes from the V_{CC} supply and is

$$P_{DC} = I_{CC}V_{CC}$$

Since each transistor draws current for a half-cycle, the current is a half-wave signal with an average value of

$$I_{CC} = \frac{I_{c(sat)}}{\pi}$$

So,

$$P_{DC} = \frac{I_{c(sat)}V_{CC}}{\pi}$$

Efficiency An advantage of push-pull class B and class AB amplifiers over class A is a much higher efficiency. This advantage usually overrides the difficulty of biasing the class AB push-pull amplifier to eliminate crossover distortion. Recall that efficiency, η, is defined as the ratio of ac output power to dc input power.

$$\eta = \frac{P_{out}}{P_{DC}}$$

The maximum efficiency, η_{max}, for a class B amplifier (class AB is slightly less) is developed as follows, starting with Equation 7–6.

$$P_{out} = 0.25I_{c(sat)}V_{CC}$$

$$\eta_{max} = \frac{P_{out}}{P_{DC}} = \frac{0.25I_{c(sat)}V_{CC}}{I_{c(sat)}V_{CC}/\pi} = 0.25\pi$$

$$\boldsymbol{\eta_{max} = 0.79}$$

Equation 7–7

or, as a percentage,

$$\eta_{max} = 79\%$$

Recall that the maximum efficiency for class A is 0.25 (25 percent).

EXAMPLE 7–5

Find the maximum ac output power and the dc input power of the amplifier in Figure 7–18.

Solution The ideal maximum peak output voltage is

$$V_{out(peak)} \cong V_{CEQ} = \frac{V_{CC}}{2} = \frac{20\text{ V}}{2} = 10\text{ V}$$

The maximum peak output current is

$$I_{out(peak)} \cong I_{c(sat)} = \frac{V_{CEQ}}{R_L} = \frac{10\text{ V}}{8\ \Omega} = 1.25\text{ A}$$

FIGURE 7–18

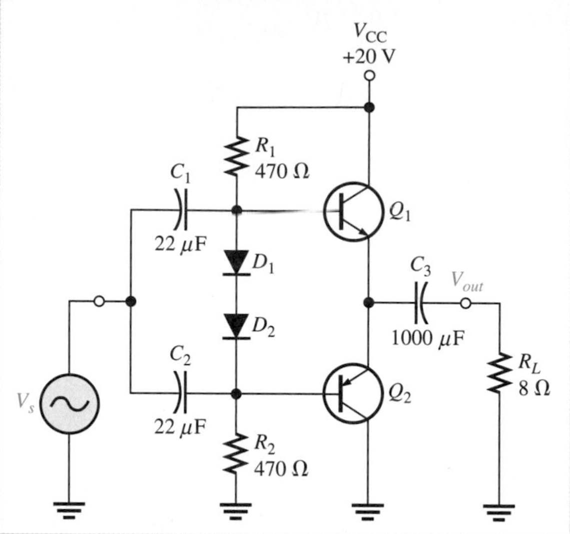

The ac output power and the dc input power are

$$P_{out} = 0.25I_{c(sat)}V_{CC} = 0.25(1.25\text{ A})(20\text{ V}) = \mathbf{6.25\ W}$$

$$P_{DC} = \frac{I_{c(sat)}V_{CC}}{\pi} = \frac{(1.25\text{ A})(20\text{ V})}{\pi} = \mathbf{7.96\ W}$$

Related Problem Determine the maximum ac output power and the dc input power in Figure 7–18 for $V_{CC} = 15$ V and $R_L = 16\ \Omega$.

Input Resistance The complementary push-pull configuration used in class B/class AB amplifiers is, in effect, two emitter-followers. The input resistance for the emitter-follower, where R_1 and R_2 are the bias resistors, is

$$R_{in} = \beta_{ac}(r'_e + R_E)\|R_1\|R_2$$

Since $R_E = R_L$, the formula is

Equation 7–8

$$\mathbf{R_{in} = \beta_{ac}(r'_e + R_L)\|R_1\|R_2}$$

EXAMPLE 7–6

Assume that a preamplifier stage with an output signal voltage of 3 V rms and an output resistance of 50 Ω is driving the push-pull power amplifier in Figure 7–18 (Example 7–5). Q_1 and Q_2 in the power amplifier have a β_{ac} of 100 and an r'_e of 1.6 Ω. Determine the loading effect that the power amplifier has on the preamp stage.

Solution Looking from the input signal source, the bias resistors appear in parallel because both go to ac ground and the ac resistance of the forward-biased diodes is very small and can be ignored. The input resistance at the emitter of either transistor is $\beta_{ac}(r'_e + R_L)$. So, the signal source sees R_1, R_2, and $\beta_{ac}(r'_e + R_L)$ all in parallel.

The ac input resistance of the power amplifier is

$$R_{in} = \beta_{ac}(r'_e + R_L)\|R_1\|R_2 = 100(9.6\ \Omega)\|470\ \Omega\|470\ \Omega = 188\ \Omega$$

Obviously, this will have an effect on the preamp driver stage. The output resistance of the preamp stage and the input resistance of the power amp effectively form a voltage divider that reduces the output signal from the preamp. The actual signal at the power amp is

$$V_{in} = \left(\frac{R_{in}}{R_s + R_{in}}\right)V_s = \left(\frac{188\ \Omega}{238\ \Omega}\right)3\text{ V} = 2.37\text{ V}$$

Related Problem What would be the effect of raising the bias resistors in the circuit?

Darlington Class AB Amplifier

In many applications where the push-pull configuration is used, the load resistance is relatively small. For example, an 8 Ω speaker is a common load for a class AB push-pull amplifier.

As you saw in the previous example, push-pull amplifiers can present a quite low input resistance to the preceding amplifier that drives it. Depending on the output resistance of the preceding amplifier, the low push-pull input resistance can load it severely and significantly reduce the voltage gain. As an example, if each bias resistor is 1 kΩ and if the complementary transistors in a push-pull amplifier exhibit an ac beta of 50 and the load resistance is 8 Ω, the input resistance (assuming $r'_e = 1\ \Omega$) is

$$R_{in} = \beta_{ac}(r'_e + R_L)\|R_1\|R_2 = 50(1\ \Omega + 8\ \Omega)\|1\text{ k}\Omega\|1\text{ k}\Omega = 236\ \Omega$$

If the collector resistance of the driving amplifier is, for example, 1.0 kΩ, the input resistance of the push-pull amplifier reduces the effective collector resistance of the driving amplifier (assuming a common-emitter) to $R_c = R_C\|R_{in} = 1.0\text{ k}\Omega\|236\ \Omega = 190\ \Omega$. This drastically reduces the voltage gain of the driving amplifier because its gain is R_c/r'_e.

In certain applications with low-resistance loads, a push-pull amplifier using Darlington transistors can be used to increase the input resistance presented to the driving amplifier and avoid severely reducing the voltage gain. The overall ac beta of a Darlington pair is generally in excess of a thousand. Also, the bias resistors can be greater because less base current is required.

In the previous case, for example, if $\beta_{ac} = 50$ for each transistor in a Darlington pair, the overall ac beta is $\beta_{ac} = (50)(50) = 2500$. If the bias resistors are 10 kΩ, the input resistance is greatly increased, as the following calculation shows.

$$R_{in} = \beta_{ac}(r'_e + R_L)\|R_1\|R_2 = 2500(1\ \Omega + 8\ \Omega)\|10\text{ k}\Omega\|10\text{ k}\Omega = 4.09\text{ k}\Omega$$

A Darlington class AB push-pull amplifier is shown in Figure 7–19. Four diodes are required in the bias circuit to match the four base-emitter junctions of the two Darlington pairs.

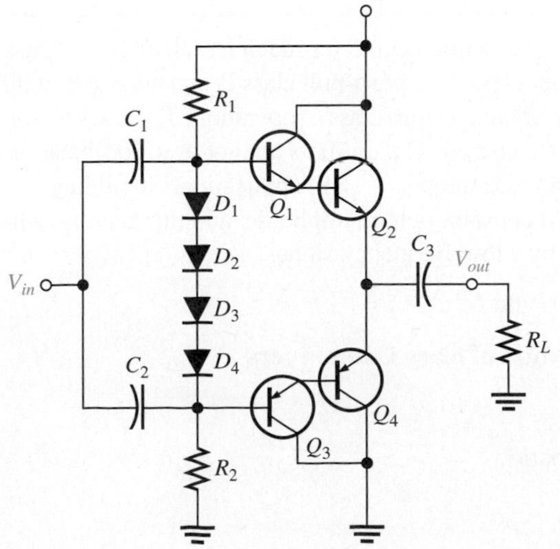

FIGURE 7–19

A Darlington class AB push-pull amplifier.

Darlington/Complementary Darlington Class AB Amplifier

The complementary Darlington, also known as the Sziklai pair, was introduced in Chapter 6. Recall that it is similar to the traditional Darlington pair except it uses complementary transistors (one *npn* and one *pnp*). The complementary Darlington is used when it is determined that output power transistors of the same type should be used (both *npn* or both *pnp*). Figure 7–20 shows a class AB push-pull amplifier with two *npn* output power transistors (Q_2 and Q_4). The upper part of the push-pull configuration is a traditional Darlington, and the lower part is a complementary Darlington.

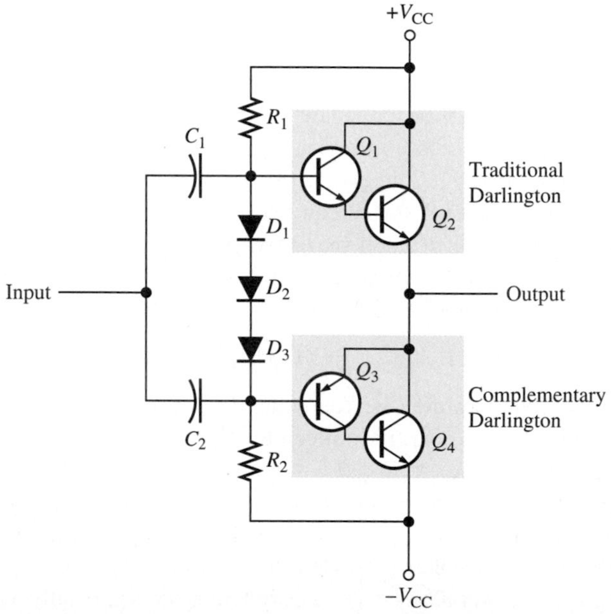

▶ FIGURE 7–20

A Darlington/complementary Darlington class AB push-pull amplifier.

SECTION 7–2 CHECKUP

1. Where is the Q-point for a class B amplifier?
2. What causes crossover distortion?
3. What is the maximum efficiency of a push-pull class B amplifier?
4. Explain the purpose of the push-pull configuration for class B.
5. How does a class AB differ from a class B amplifier?

7–3 THE CLASS C AMPLIFIER

Class C amplifiers are biased so that conduction occurs for much less than 180°. Class C amplifiers are more efficient than either class A or push-pull class B and class AB, which means that more output power can be obtained from class C operation. The output amplitude is a nonlinear function of the input, so class C amplifiers are not used for linear amplification. They are generally used in radio frequency (RF) applications, including circuits, such as oscillators, that have a constant output amplitude, and modulators, where a high-frequency signal is controlled by a low-frequency signal.

After completing this section, you should be able to

- **Discuss and analyze the operation of class C amplifiers**
    - Explain class C operation
    - Discuss class C power dissipation

- Describe tuned operation
- Calculate maximum output power
- Determine efficiency
- Explain clamper bias in a class C amplifier

Basic Class C Operation

The basic concept of class C operation is illustrated in Figure 7–21. A common-emitter class C amplifier with a resistive load is shown in Figure 7–22(a). A class C amplifier is normally operated with a resonant circuit load, so the resistive load is used only for the purpose of illustrating the concept. It is biased below cutoff with the negative V_{BB} supply. The ac source voltage has a peak value that is slightly greater than $|V_{BB}| + V_{BE}$ so that the base voltage exceeds the barrier potential of the base-emitter junction for a short time near the positive peak of each cycle, as illustrated in Figure 7–22(b). During this short interval, the transistor is turned on. When the entire ac load line is used, as shown in Figure 7–22(c), the ideal maximum collector current is $I_{c(sat)}$, and the ideal minimum collector voltage is $V_{ce(sat)}$.

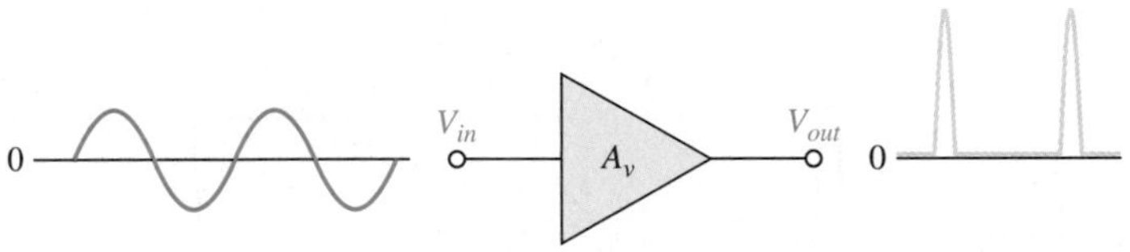

FIGURE 7–21

Basic class C amplifier operation (noninverting).

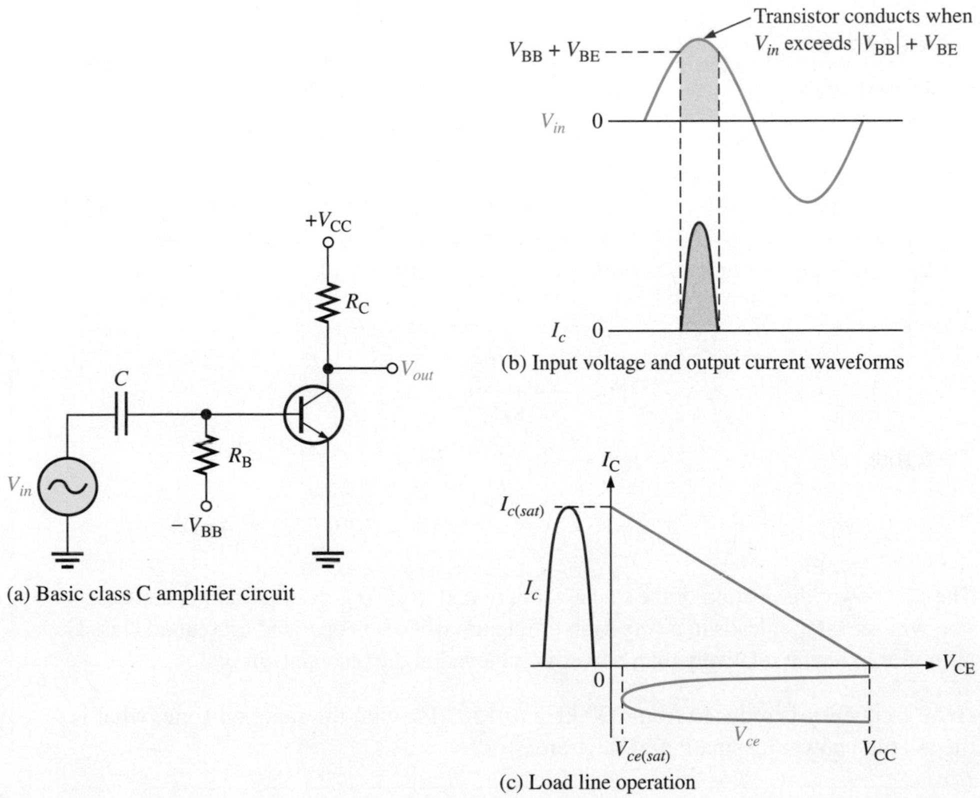

FIGURE 7–22

Basic class C operation.

Power Dissipation

The power dissipation of the transistor in a class C amplifier is low because it is on for only a small percentage of the input cycle. Figure 7–23(a) shows the collector current pulses. The time between the pulses is the period (T) of the ac input voltage. The collector current and the collector voltage during the *on* time of the transistor are shown in Figure 7–23(b). To avoid complex mathematics, we will assume ideal pulse approximations. Using this simplification, if the output swings over the entire load, the maximum current amplitude is $I_{c(sat)}$ and the minimum voltage amplitude is $V_{ce(sat)}$ during the time the transistor is on. The power dissipation during the *on* time is, therefore,

$$P_{\text{D(on)}} = I_{c(sat)}V_{ce(sat)}$$

The transistor is on for a short time, t_{on}, and off for the rest of the input cycle. Therefore, assuming the entire load line is used, the power dissipation averaged over the entire cycle is

$$P_{\text{D(avg)}} = \left(\frac{t_{\text{on}}}{T}\right)P_{\text{D(on)}} = \left(\frac{t_{\text{on}}}{T}\right)I_{c(sat)}V_{ce(sat)}$$

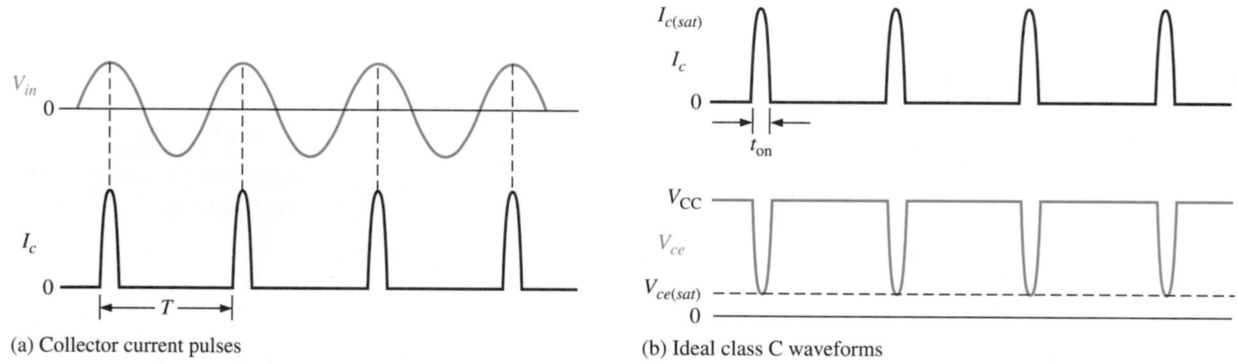

▲ **FIGURE 7–23**

Class C waveforms.

EXAMPLE 7–7

A class C amplifier is driven by a 200 kHz signal. The transistor is on for 1 μs, and the amplifier is operating over 100 percent of its load line. If $I_{c(sat)} = 100$ mA and $V_{ce(sat)} =$ 0.2 V, what is the average power dissipation of the transistor?

Solution The period is

$$T = \frac{1}{200\text{ kHz}} = 5\ \mu\text{s}$$

Therefore,

$$P_{\text{D(avg)}} = \left(\frac{t_{\text{on}}}{T}\right)I_{c(sat)}V_{ce(sat)} = (0.2)(100\text{ mA})(0.2\text{ V}) = \mathbf{4\ mW}$$

The low power dissipation of the transistor operated in class C is important because, as you will see later, it leads to a very high efficiency when it is operated as a tuned class C amplifier in which relatively high power is achieved in the resonant circuit.

Related Problem If the frequency is reduced from 200 kHz to 150 kHz with the same on-time, what is the average power dissipation of the transistor?

Tuned Operation

Because the collector voltage (output) is not a replica of the input, the resistively loaded class C amplifier alone is of no value in linear applications. It is therefore necessary to use a class C amplifier with a parallel resonant circuit (tank), as shown in Figure 7–24(a). The resonant frequency of the tank circuit is determined by the formula $f_r = 1/(2\pi\sqrt{LC})$. The short pulse of collector current on each cycle of the input initiates and sustains the oscillation of the tank circuit so that an output sinusoidal voltage is produced, as illustrated in Figure 7–24(b). The tank circuit has high impedance only near the resonant frequency, so the gain is large only at this frequency.

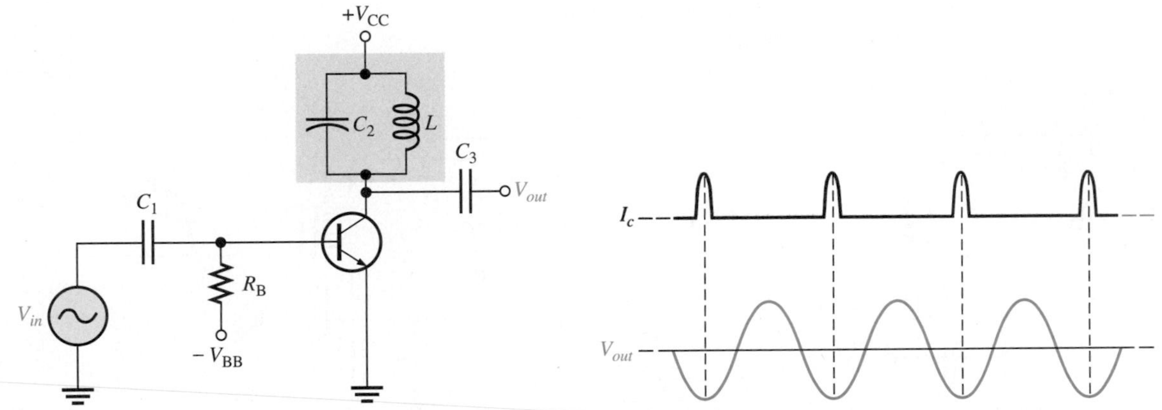

(a) Basic circuit

(b) Output waveforms

▲ **FIGURE 7–24**

Tuned class C amplifier.

The current pulse charges the capacitor to approximately $+V_{CC}$, as shown in Figure 7–25(a). After the pulse, the capacitor quickly discharges, thus charging the inductor. Then, after the capacitor completely discharges, the inductor's magnetic field collapses and then quickly recharges C to near V_{CC} in a direction opposite to the previous charge. This completes one half-cycle of the oscillation, as shown in parts (b) and (c) of Figure 7–25. Next, the capacitor discharges again, increasing the inductor's magnetic field. The inductor then quickly recharges the capacitor back to a positive peak slightly less than the previous one, due to energy loss in the winding resistance. This completes one full cycle, as shown in parts (d) and (e) of Figure 7–25. The peak-to-peak output voltage is therefore approximately equal to $2V_{CC}$.

The amplitude of each successive cycle of the oscillation will be less than that of the previous cycle because of energy loss in the resistance of the tank circuit, as shown in Figure 7–26(a), and the oscillation will eventually die out. However, the regular recurrences of the collector current pulse re-energizes the resonant circuit and sustains the oscillations at a constant amplitude.

When the tank circuit is tuned to the frequency of the input signal (fundamental), re-energizing occurs on each cycle of the tank voltage, V_r, as shown in Figure 7–26(b). When the tank circuit is tuned to the second harmonic of the input signal, re-energizing occurs on alternate cycles as shown in Figure 7–26(c). In this case, a class C amplifier operates as a frequency multiplier ($\times 2$). By tuning the resonant tank circuit to higher harmonics, further frequency multiplication factors are achieved.

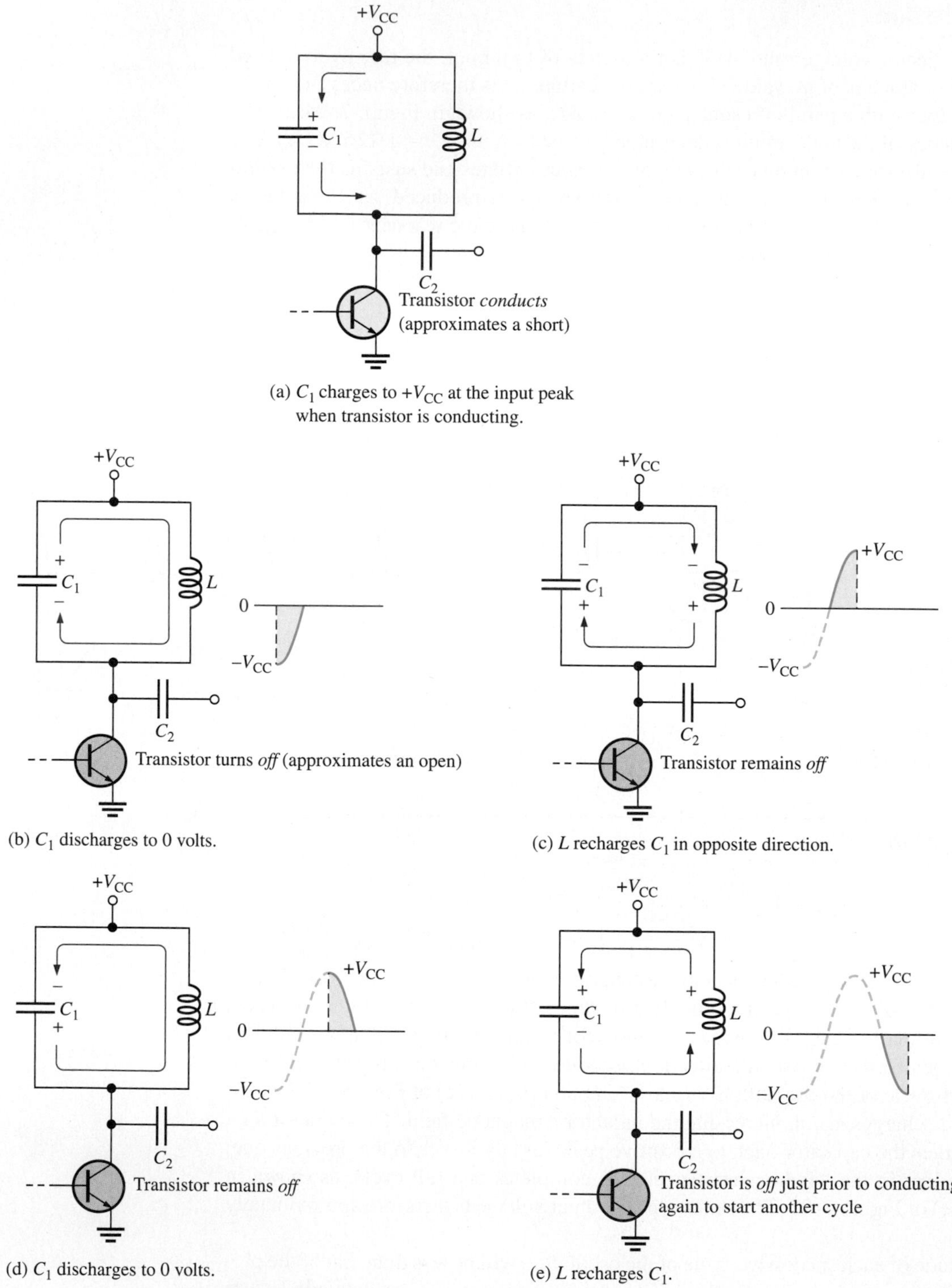

▲ **FIGURE 7–25**

Resonant circuit action.

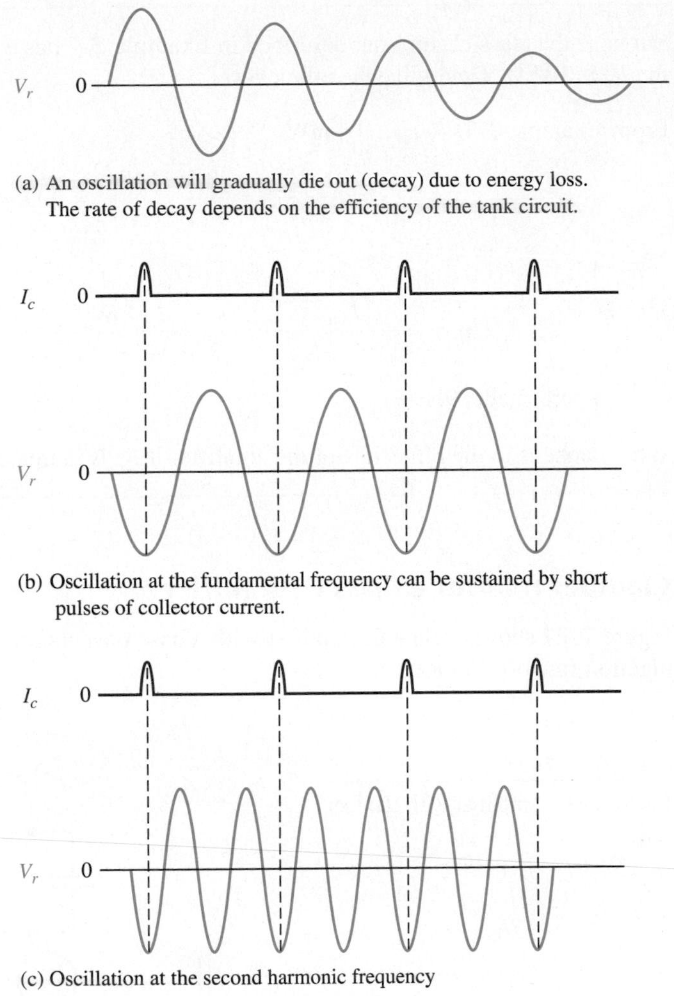

◀ **FIGURE 7–26**

Tank circuit oscillations. V_r is the voltage across the tank circuit.

Maximum Output Power

Since the voltage developed across the tank circuit has a peak-to-peak value of approximately $2V_{CC}$, the maximum output power can be expressed as

$$P_{out} = \frac{V_{rms}^2}{R_c} = \frac{(0.707V_{CC})^2}{R_c}$$

$$\boldsymbol{P_{out} = \frac{0.5V_{CC}^2}{R_c}}$$

Equation 7–9

R_c is the equivalent parallel resistance of the collector tank circuit at resonance and represents the parallel combination of the coil resistance and the load resistance. It usually has a low value. The total power that must be supplied to the amplifier is

$$P_T = P_{out} + P_{D(avg)}$$

Therefore, the efficiency is

$$\boldsymbol{\eta = \frac{P_{out}}{P_{out} + P_{D(avg)}}}$$

Equation 7–10

When $P_{out} >> P_{D(avg)}$, the class C efficiency closely approaches 1 (100 percent).

EXAMPLE 7–8

Suppose the class C amplifier described in Example 7–7 has a V_{CC} equal to 24 V and the R_c is 100 Ω. Determine the efficiency.

Solution From Example 7–7, $P_{D(avg)} = 4$ mW.

$$P_{out} = \frac{0.5V_{CC}^2}{R_c} = \frac{0.5(24\text{ V})^2}{100\ \Omega} = 2.88\text{ W}$$

Therefore,

$$\eta = \frac{P_{out}}{P_{out} + P_{D(avg)}} = \frac{2.88\text{ W}}{2.88\text{ W} + 4\text{ mW}} = \mathbf{0.999}$$

or, as a percentage, 99.9%.

Related Problem What happens to the efficiency of the amplifier if R_c is increased?

Clamper Bias for a Class C Amplifier

Figure 7–27 shows a class C amplifier with a base bias clamping circuit. The base-emitter junction functions as a diode.

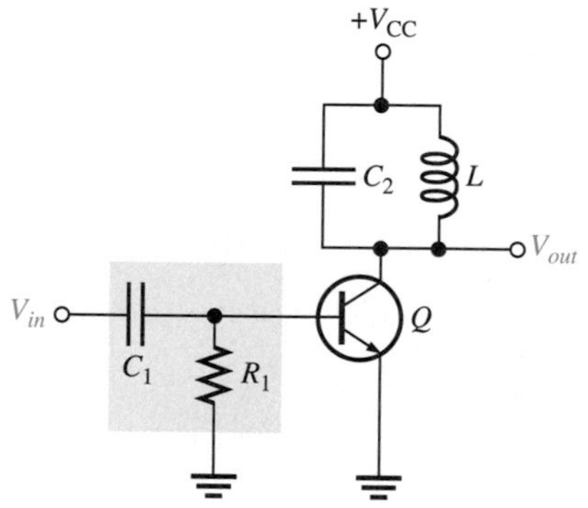

FIGURE 7–27

Tuned class C amplifier with clamper bias.

When the input signal goes positive, capacitor C_1 is charged to the peak value with the polarity shown in Figure 7–28(a). This action produces an average voltage at the base of approximately $-V_p$. This places the transistor in cutoff except at the positive peaks, when the transistor conducts for a short interval. For good clamping action, the R_1C_1 time constant of the clamping circuit must be much greater than the period of the input signal. Parts (b) through (f) of Figure 7–28 illustrate the bias clamping action in more detail. During the time up to the positive peak of the input (t_0 to t_1), the capacitor charges to $V_p - 0.7$ V through the base-emitter diode, as shown in part (b). During the time from t_1 to t_2, as shown in part (c), the capacitor discharges very little because of the large RC time constant. The capacitor, therefore, maintains an average charge slightly less than $V_p - 0.7$ V.

Since the dc value of the input signal is zero (positive side of C_1), the dc voltage at the base (negative side of C_1) is slightly more positive than $-(V_p - 0.7\text{ V})$, as indicated in Figure 7–28(d). As shown in Figure 7–28(e), the capacitor couples the ac input signal through to the base so that the voltage at the transistor's base is the ac signal riding on a dc level slightly more positive than $-(V_p - 0.7\text{ V})$. Near the positive peaks of the input voltage, the base voltage goes slightly above 0.7 V and causes the transistor to conduct for a short time, as shown in Figure 7–28(f).

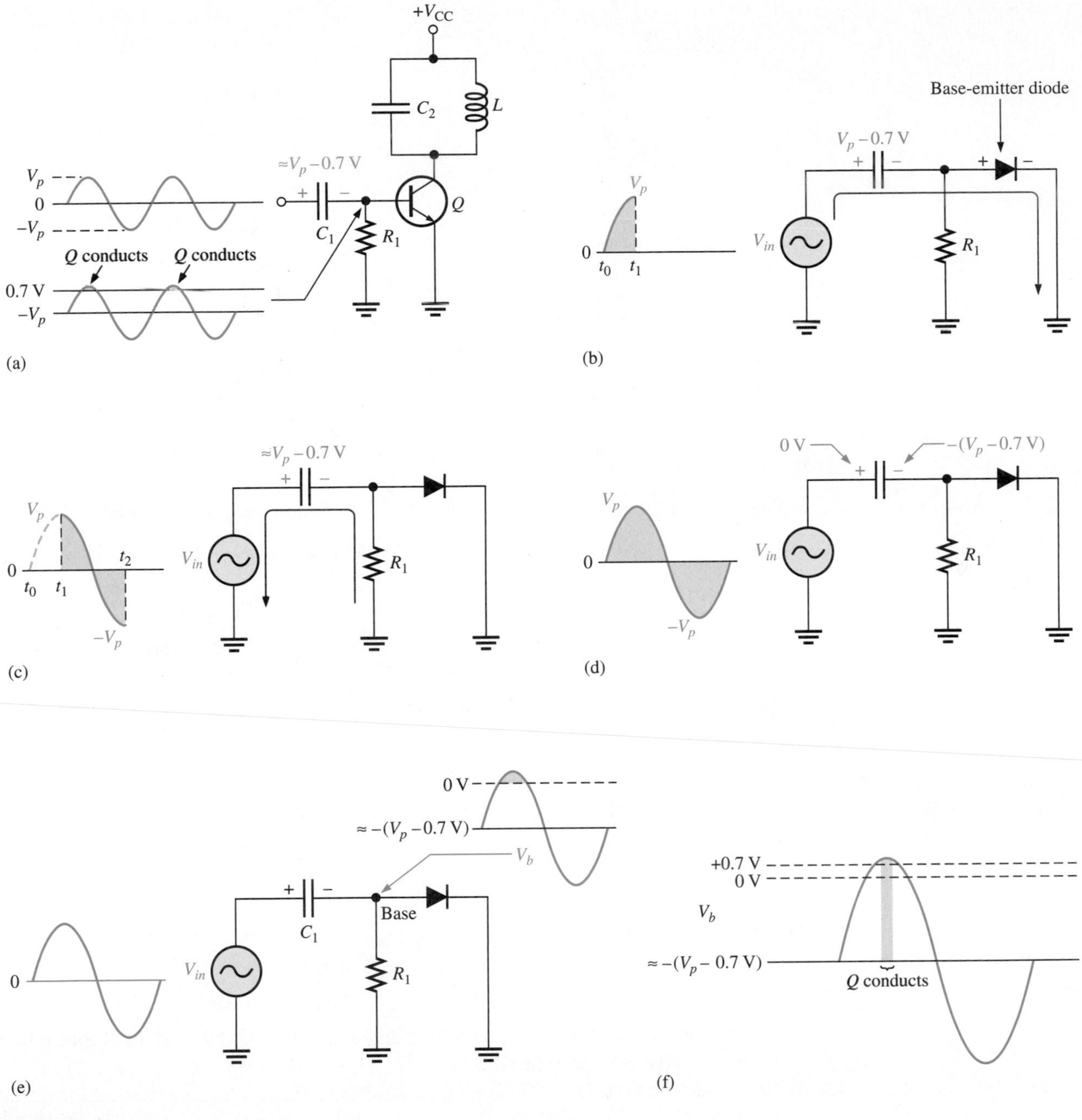

▲ **FIGURE 7–28**

Clamper bias action.

EXAMPLE 7–9

Determine the voltage at the base of the transistor, the resonant frequency, and the peak-to-peak value of the output signal voltage for the class C amplifier in Figure 7–29.

Solution

$$V_{s(p)} = (1.414)(1\text{ V}) \cong 1.4\text{ V}$$

The base is clamped at

$$-(V_{s(p)} - 0.7) = \mathbf{-0.7\text{ V dc}}$$

FIGURE 7–29

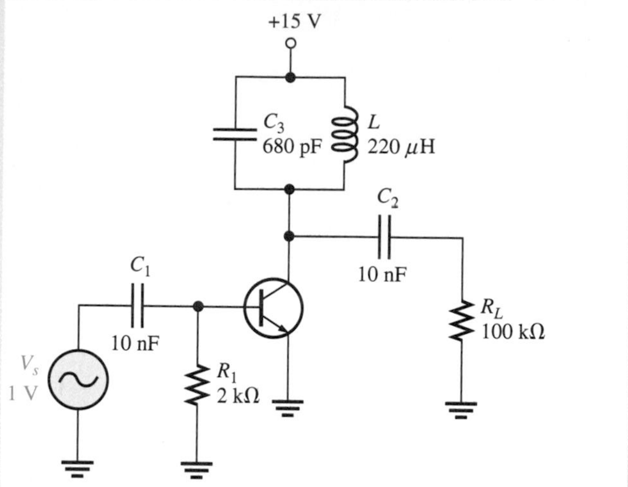

The signal at the base has a positive peak of +0.7 V and a negative peak of

$$-V_{s(p)} + (-0.7\text{ V}) = -1.4\text{ V} - 0.7\text{ V} = \mathbf{-2.1\text{ V}}$$

The resonant frequency is

$$f_r = \frac{1}{2\pi\sqrt{LC}} = \frac{1}{2\pi\sqrt{(220\ \mu\text{H})(680\text{ pF})}} = \mathbf{411\text{ kHz}}$$

The output signal has a peak-to-peak value of

$$V_{pp} = 2V_{CC} = 2(15\text{ V}) = \mathbf{30\text{ V}}$$

Related Problem How could you make the circuit in Figure 7–29 a frequency doubler?

SECTION 7–3 CHECKUP

1. At what point is a class C amplifier normally biased?
2. What is the purpose of the tuned circuit in a class C amplifier?
3. A certain class C amplifier has a power dissipation of 100 mW and an output power of 1 W. What is its percent efficiency?

7–4 TROUBLESHOOTING

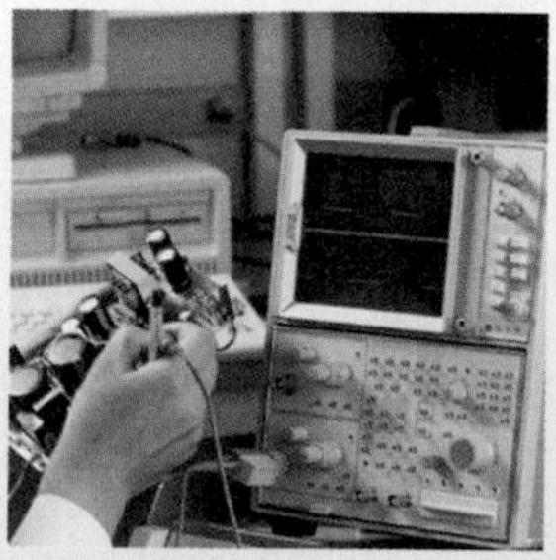

In this section, examples of isolating a component failure in a circuit are presented. We will use a class A amplifier and a class AB amplifier with the output voltage monitored by an oscilloscope. Several incorrect output waveforms will be examined and the most likely faults will be discussed.

After completing this section, you should be able to

- **Troubleshoot power amplifiers**
  - Detect various faults in a class A power amplifier and a class AB amplifier

Case 1: Class A

As shown in Figure 7–30, the class A power amplifier should have a normal sinusoidal output when a sinusoidal input signal is applied.

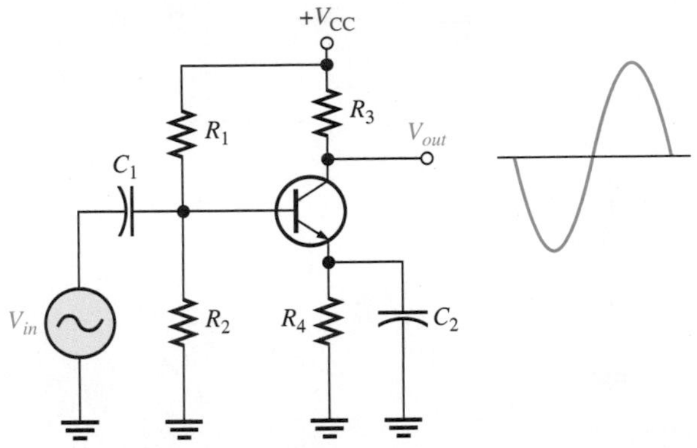

FIGURE 7–30

Class A power amplifier with correct output voltage swing.

Now let's consider four incorrect output waveforms and the most likely causes in each case. In Figure 7–31(a), the scope displays a dc level equal to the dc supply voltage, indicating that the transistor is in cutoff. The two most likely causes of this condition are (1) the transistor has an open *pn* junction, or (2) R_4 is open, preventing collector and emitter current.

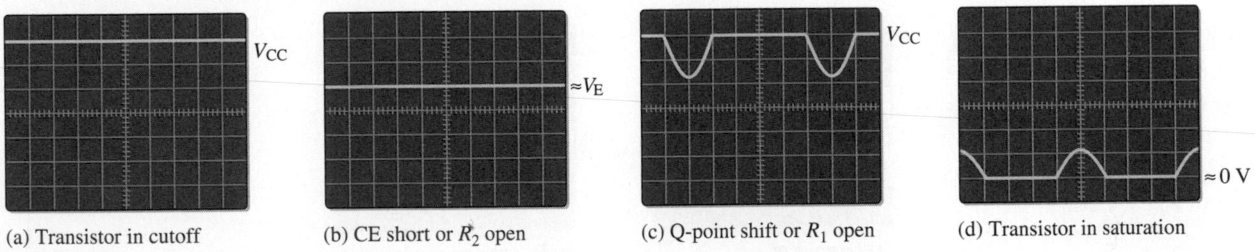

FIGURE 7–31

Oscilloscope displays showing output voltage for the amplifier in Figure 7–30 for several types of failures.

In Figure 7–31(b), the scope displays a dc level at the collector approximately equal to the dc emitter voltage. The two probable causes of this indication are (1) the transistor is shorted from collector to emitter, or (2) R_2 is open, causing the transistor to be biased in saturation. In the second case, a sufficiently large input signal can bring the transistor out of saturation on its negative peaks, resulting in short pulses on the output.

In Figure 7–31(c), the scope displays an output waveform that indicates the transistor is in cutoff except during a small portion of the input cycle. Possible causes of this indication are (1) the Q-point has shifted down due to a drastic out-of-tolerance change in a resistor value, or (2) R_1 is open, biasing the transistor in cutoff. The display shows that the input signal is sufficient to bring it out of cutoff for a small portion of the cycle.

In Figure 7–31(d), the scope displays an output waveform that indicates the transistor is saturated except during a small portion of the input cycle. Again, it is possible that an incorrect resistance value has caused a drastic shift in the Q-point up toward saturation, or R_2 is open, causing the transistor to be biased in saturation, and the input signal is bringing it out of saturation for a small portion of the cycle.

Case 2: Class AB

As shown in Figure 7–32, the class AB push-pull amplifier should have a sinusoidal output when a sinusoidal input signal is applied.

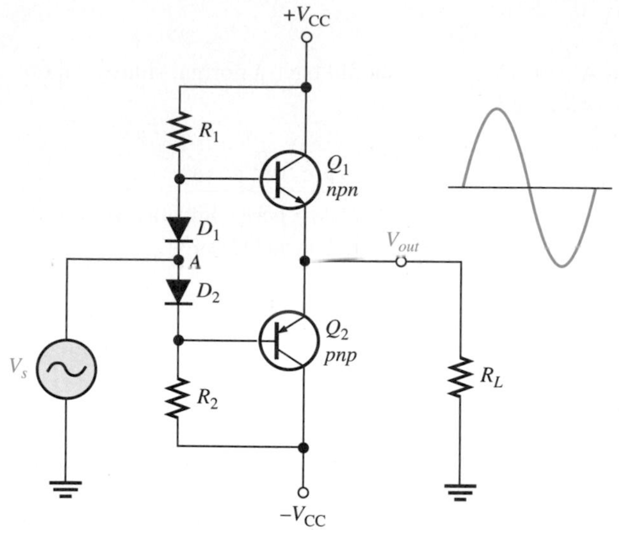

FIGURE 7–32

A class AB push-pull amplifier with correct output voltage.

Two incorrect output waveforms are shown in Figure 7–33. The waveform in part (a) shows that only the positive half of the input signal is present on the output. One possible cause is that diode D_1 is open. If this is the fault, the positive half of the input signal forward-biases D_2 and causes transistor Q_2 to conduct. Another possible cause is that the base-emitter junction of Q_2 is open so only the positive half of the input signal appears on the output because Q_1 is still working.

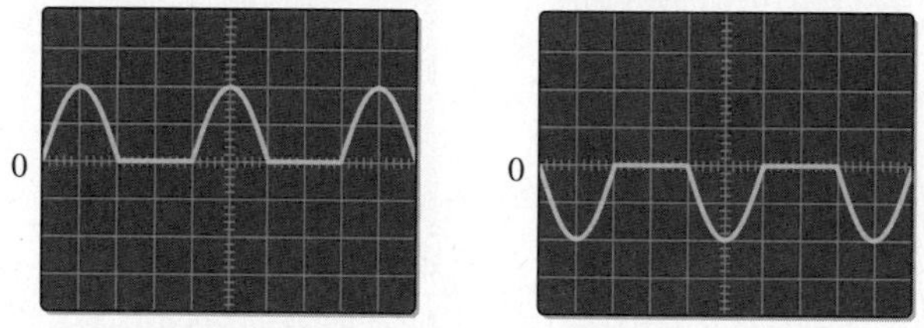

(a) D_1 open or Q_2 base-emitter open

(b) D_2 open or Q_1 base-emitter open

FIGURE 7–33

Incorrect output waveforms for the amplifier in Figure 7–32.

The waveform in Figure 7–33(b) shows that only the negative half of the input signal is present on the output. One possible cause is that diode D_2 is open. If this is the fault, the negative half of the input signal forward-biases D_1 and places the half-wave signal on the base of Q_1. Another possible cause is that the base-emitter junction of Q_1 is open so only the negative half of the input signal appears on the output because Q_2 is still working.

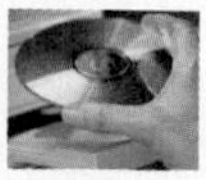

Multisim Troubleshooting Exercises

These file circuits are in the Troubleshooting Exercises folder on your CD-ROM.

1. Open file TSE07-01. Determine if the circuit is working properly and, if not, determine the fault.
2. Open file TSE07-02. Determine if the circuit is working properly and, if not, determine the fault.
3. Open file TSE07-03. Determine if the circuit is working properly and, if not, determine the fault.

SECTION 7–4 CHECKUP

1. What would you check for if you noticed clipping at both peaks of the output waveform?
2. A significant loss of gain in the amplifier of Figure 7–30 would most likely be caused by what type of failure?

On the Job

Application Activity: *The Complete PA System*

The class AB power amplifier follows the audio preamp and drives the speaker as shown in the PA system block diagram in Figure 7–34. In this application, the power amplifier is developed and interfaced with the preamp that was developed in Chapter 6. The maximum signal power to the speaker should be approximately 6 W for a frequency range of 70 Hz to 5 kHz. The dynamic range for the input voltage is up to 40 mV. Finally, the complete PA system is put together.

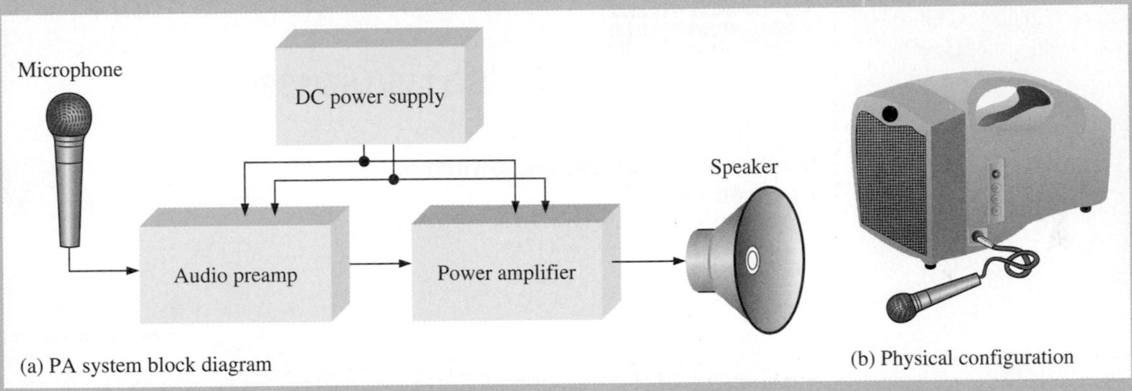

FIGURE 7–34

The Power Amplifier Circuit

The schematic of the push-pull power amplifier is shown in Figure 7–35. The circuit is a class AB amplifier implemented with Darlington configurations and diode current mirror

FIGURE 7–35

Class AB power push-pull amplifier.

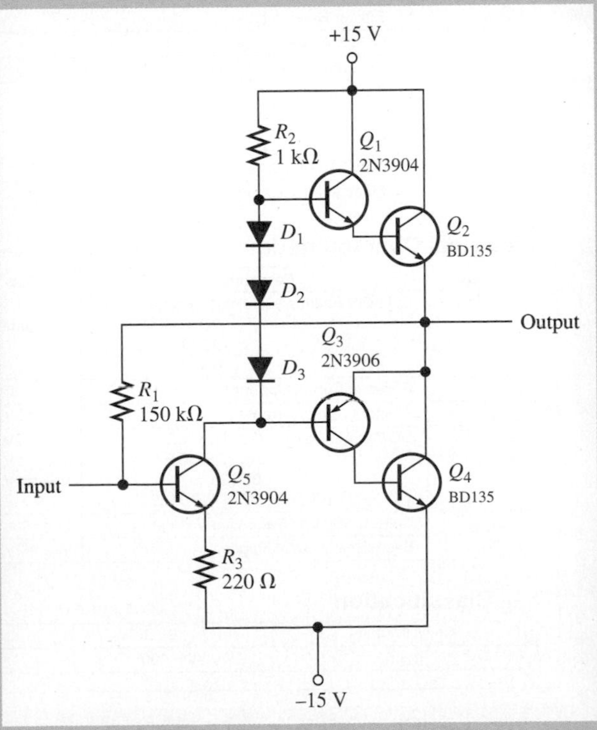

bias. Both a traditional Darlington pair and a complementary Darlington (Sziklai) pair are used to provide sufficient current to an 8 Ω speaker load. The signal from the preamp is capacitively coupled to the driver stage, Q_5, which is used to prevent excessive loading on the preamp and provide additional gain. Notice that Q_5 is biased with the dc output voltage (0 V) fed back through R_1. Also, the signal voltage fed back to the base of Q_5 is out-of-phase with the signal from the preamp and has the effect of stabilizing the gain. This is called *negative feedback*. The amplifier will deliver up to 5 W to an 8 Ω speaker.

A partial datasheet for the BD135 power transistor is shown in Figure 7–36.

FIGURE 7–36

Partial datasheet for the BD135 power transistors. © 2003 Fairchild Semiconductor Corporation. Used by permission.

FAIRCHILD
SEMICONDUCTOR TM

BD135/137/139

Medium Power Linear and Switching Applications

- Complement to BD136, BD138 and BD140 respectively

TO-126
1. Emitter 2.Collector 3.Base

NPN Epitaxial Silicon Transistor

Absolute Maximum Ratings T_C = 25°C unless otherwise noted

Symbol	Parameter		Value	Units
V_{CBO}	Collector-Base Voltage	: BD135	45	V
		: BD137	60	V
		: BD139	80	V
V_{CEO}	Collector-Emitter Voltage	: BD135	45	V
		: BD137	60	V
		: BD139	80	V
V_{EBO}	Emitter-Base Voltage		5	V
I_C	Collector Current (DC)		1.5	A
I_{CP}	Collector Current (Pulse)		3.0	A
I_B	Base Current		0.5	A
P_C	Collector Dissipation (T_C = 25°C)		12.5	W
P_C	Collector Dissipation (T_a = 25°C)		1.25	W
T_J	Junction Temperature		150	°C
T_{STG}	Storage Temperature		- 55 ~ 150	°C

Electrical Characteristics T_C = 25°C unless otherwise noted

Symbol	Parameter	Test Condition	Min.	Typ.	Max.	Units
V_{CEO}(sus)	Collector-Emitter Sustaining Voltage					
	: BD135	I_C = 30mA, I_B = 0	45			V
	: BD137		60			V
	: BD139		80			V
I_{CBO}	Collector Cut-off Current	V_{CB} = 30V, I_E = 0			0.1	μA
I_{EBO}	Emitter Cut-off Current	V_{EB} = 5V, I_C = 0			10	μA
h_{FE1}	DC Current Gain : ALL DEVICE	V_{CE} = 2V, I_C = 5mA	25			
h_{FE2}	: ALL DEVICE	V_{CE} = 2V, I_C = 0.5A	25			
h_{FE3}	: BD135	V_{CE} = 2V, I_C = 150mA	40		250	
	: BD137, BD139		40		160	
V_{CE}(sat)	Collector-Emitter Saturation Voltage	I_C = 500mA, I_B = 50mA			0.5	V
V_{BE}(on)	Base-Emitter ON Voltage	V_{CE} = 2V, I_C = 0.5A			1	V

h_{FE} Classification

Classification	6	10	16
h_{FE3}	40 ~ 100	63 ~ 160	100 ~ 250

1. Estimate the input resistance of the power amplifier in Figure 7–35.
2. Calculate the approximate voltage gain of the power amplifier in Figure 7–35?

Simulation

The power amplifier is simulated using Multisim with a 1 kHz input signal at near its maximum linear operation. The results are shown in Figure 7–37 where an 8.2 Ω resistor is used to closely approximate the 8 Ω speaker.

3. Calculate the power to the load in Figure 7–37.
4. What is the measured voltage gain? The input is a peak value.
5. Compare the measured gain to the calculated gain for the amplifier in Figure 7–35.

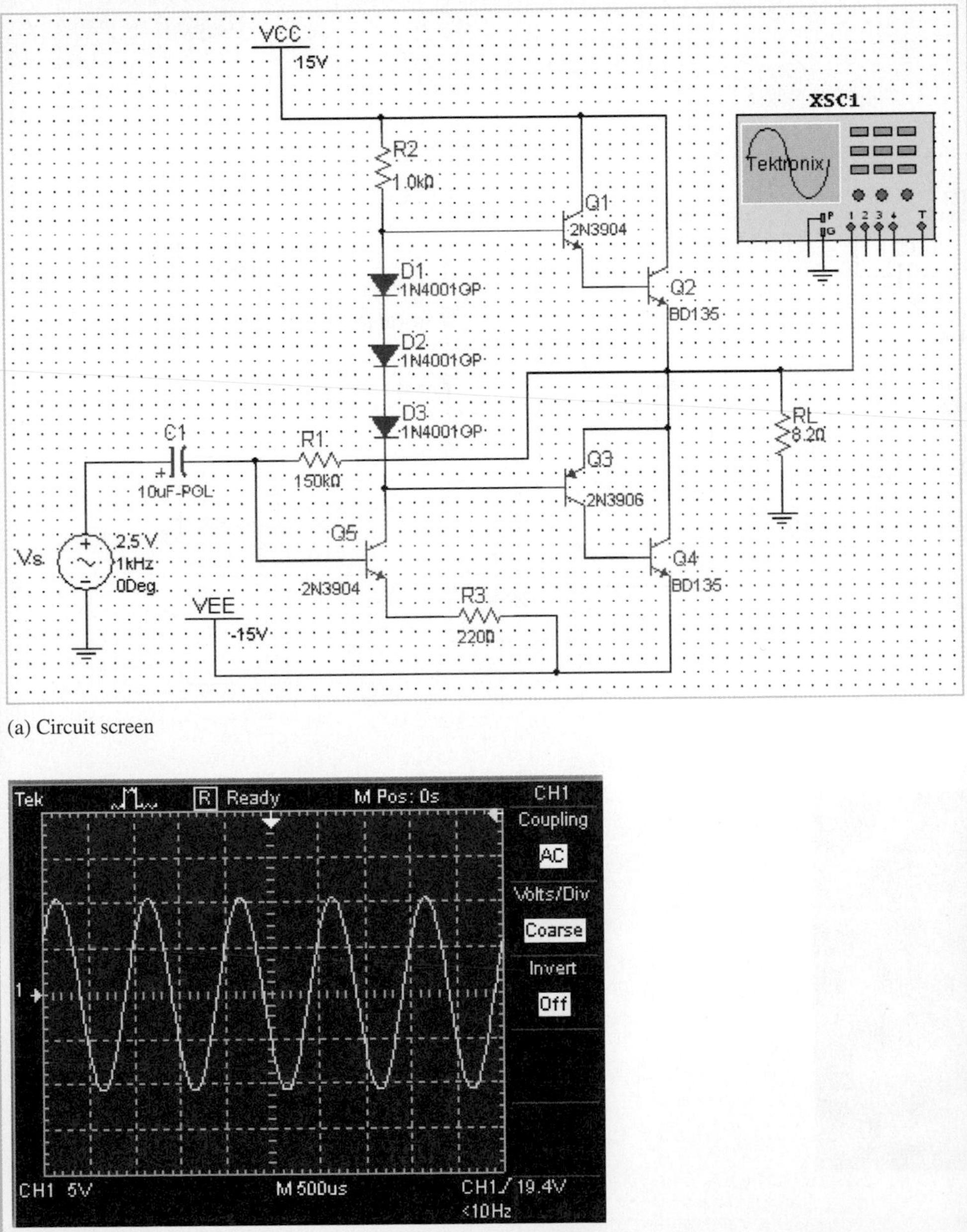

(a) Circuit screen

(b) Output signal

FIGURE 7–37

Simulation of the power amplifier.

The Complete Audio Amplifier

Both the preamp and the power amp have been simulated individually. Now, they must work together to produce the required signal power to the speaker. Figure 7–38 is the simulation of the combined audio preamp and power amp. Components in the power amplifier are now numbered sequentially with the preamp components.

6. Calculate the power to the load in Figure 7–38.
7. What is the measured voltage gain of the power amplifier?
8. What is the measured overall voltage gain?

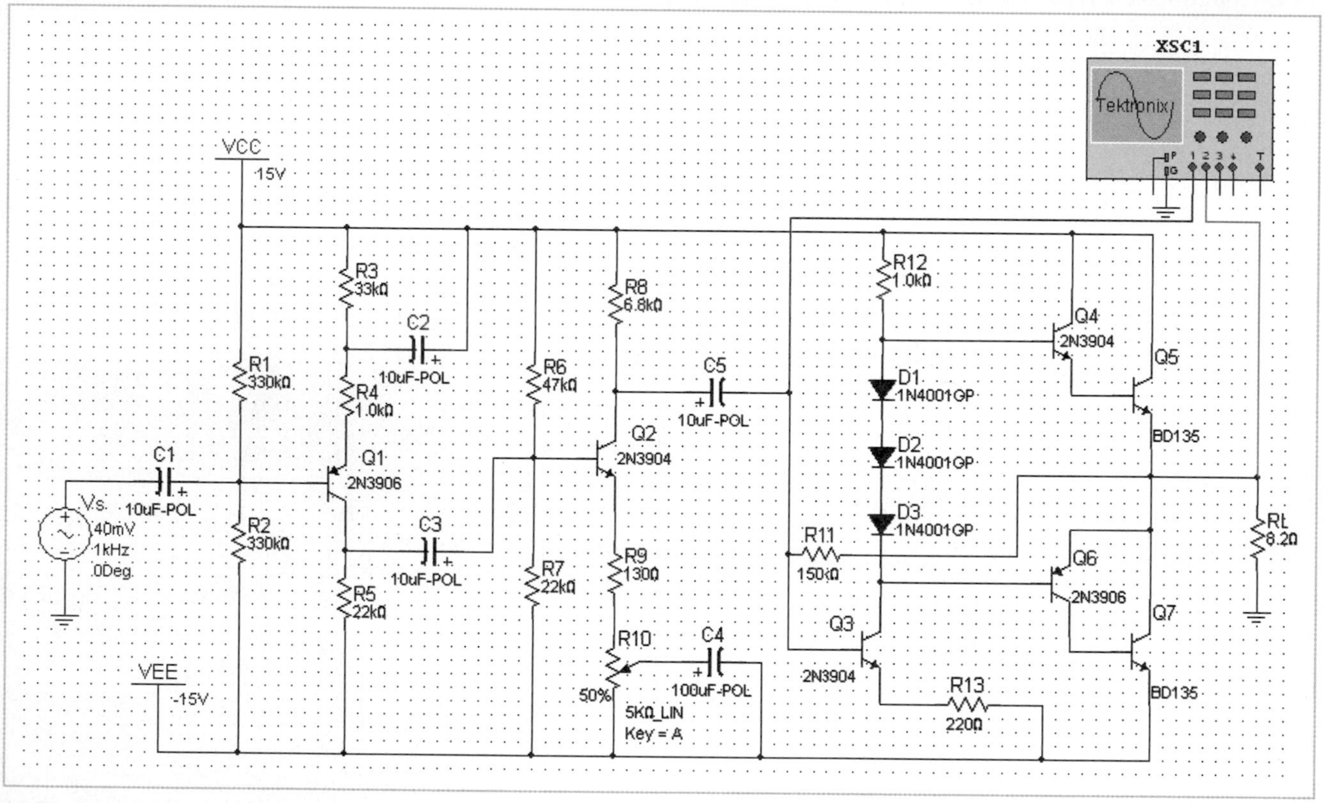

(a) Circuit screen

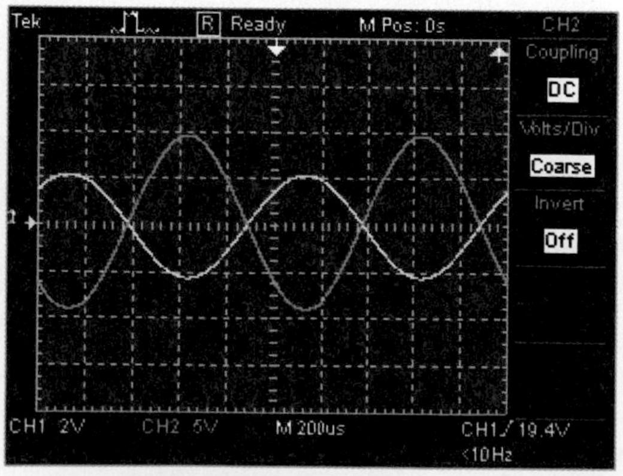

(b) Preamp output and final output

FIGURE 7–38

Simulation of the complete audio amplifier.

Simulate the audio amplifier using your Multisim software. Observe the operation with the virtual oscilloscope.

Prototyping and Testing

Now that the circuit has been simulated, the prototype circuit is constructed and tested. After the circuit is successfully tested on a protoboard, it is ready to be finalized on a printed circuit board.

Lab Experiment

To build and test a similar circuit, go to Experiment 7 in your lab manual (*Laboratory Exercises for Electronic Devices* by David Buchla and Steven Wetterling).

Circuit Board

The power amplifier is implemented on a printed circuit board as shown in Figure 7–39. Heat sinks are used to provide additional heat dissipation from the power transistors.

9. Check the printed circuit board and verify that it agrees with the schematic in Figure 7–35. The volume control potentiometer is mounted off the PC board for easy access.
10. Label each input and output pin according to function. Locate the single back-side trace.

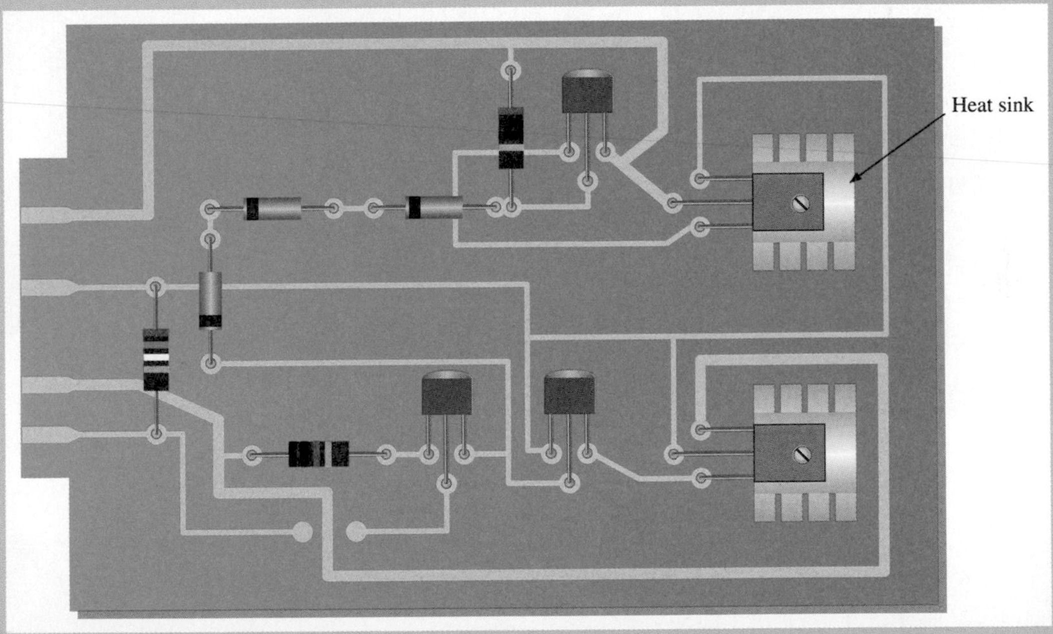

FIGURE 7–39

Power amplifier circuit board.

Troubleshooting the Power Amplifier Board

A power amplifier circuit board has failed the production test. Test results are shown in Figure 7–40.

11. Based on the scope displays, list possible faults for the circuit board.

Putting the System Together

The two circuit boards are interconnected and the dc power supply (battery pack), microphone, speaker, and volume control potentiometer are attached, as shown in Figure 7–41.

12. Verify that the system interconnections are correct.

FIGURE 7–40

Test of faulty power amplifier board.

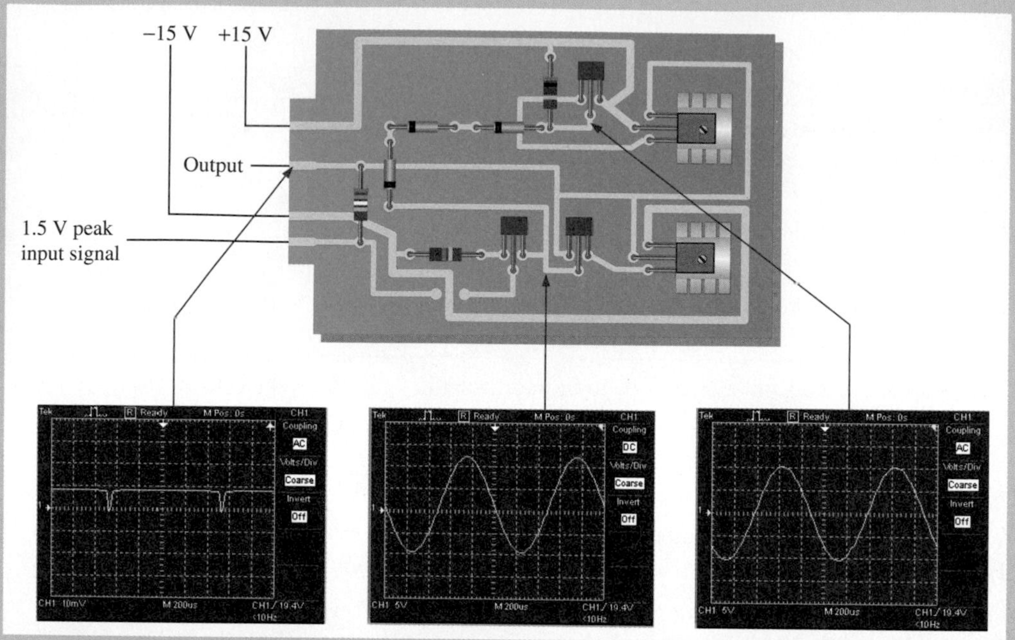

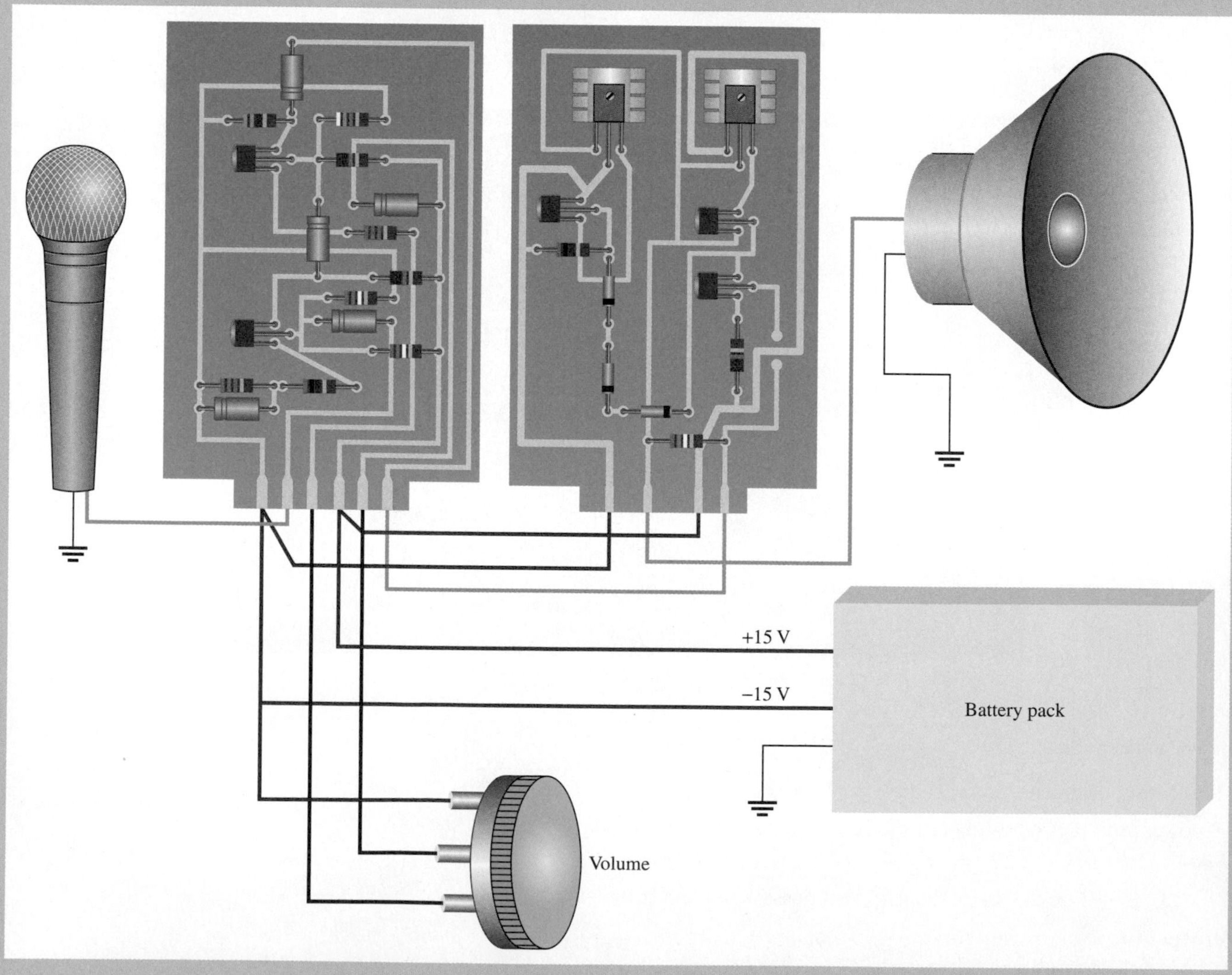

FIGURE 7–41

The complete public address system.

SUMMARY

Section 7–1

- A class A power amplifier operates entirely in the linear region of the transistor's characteristic curves. The transistor conducts during the full 360° of the input cycle.
- The Q-point must be centered on the load line for maximum class A output signal swing.
- The maximum efficiency of a class A power amplifier is 25 percent.

Section 7–2

- A class B amplifier operates in the linear region for half of the input cycle (180°), and it is in cutoff for the other half.
- The Q-point is at cutoff for class B operation.
- Class B amplifiers are normally operated in a push-pull configuration in order to produce an output that is a replica of the input.
- The maximum efficiency of a class B amplifier is 79 percent.
- A class AB amplifier is biased slightly above cutoff and operates in the linear region for slightly more than 180° of the input cycle.
- Class AB eliminates crossover distortion found in pure class B.

Section 7–3

- A class C amplifier operates in the linear region for only a small part of the input cycle.
- The class C amplifier is biased below cutoff.
- Class C amplifiers are normally operated as tuned amplifiers to produce a sinusoidal output.
- The maximum efficiency of a class C amplifier is higher than that of either class A or class B amplifiers. Under conditions of low power dissipation and high output power, the efficiency can approach 100 percent.

KEY TERMS

Key terms and other bold terms in the chapter are defined in the end-of-book glossary.

Class A A type of amplifier that operates entirely in its linear (active) region.

Class AB A type of amplifier that is biased into slight conduction.

Class B A type of amplifier that operates in the linear region for 180° of the input cycle because it is biased at cutoff.

Class C A type of amplifier that operates only for a small portion of the input cycle.

Efficiency The ratio of the signal power delivered to a load to the power from the power supply of an amplifier.

Power gain The ratio of output power to input power of an amplifier.

Push-Pull A type of class B amplifier with two transistors in which one transistor conducts for one half-cycle and the other conducts for the other half-cycle.

KEY FORMULAS

The Class A Power Amplifier

7–1	$A_p = \dfrac{P_L}{P_{in}}$	Power gain
7–2	$A_p = A_v^2\left(\dfrac{R_{in}}{R_L}\right)$	Power gain in terms of voltage gain
7–3	$P_{DQ} = I_{CQ}V_{CEQ}$	DC quiescent power
7–4	$P_{out(max)} = 0.5I_{CQ}V_{CEQ}$	Maximum output power

The Class B/AB Push-Pull Amplifiers

7–5	$I_{c(sat)} = \dfrac{V_{CC}}{R_L}$	AC saturation current
7–6	$P_{out} = 0.25I_{c(sat)}V_{CC}$	Maximum average output power

7–7	$\eta_{max} = 0.79$	Maximum efficiency
7–8	$R_{in} = \beta_{ac}(r'_e + R_L) \| R_1 \| R_2$	Input resistance

The Class C Amplifier

7–9	$P_{out} = \dfrac{0.5V_{CC}^2}{R_c}$	Output power
7–10	$\eta = \dfrac{P_{out}}{P_{out} + P_{D(avg)}}$	Efficiency

TRUE/FALSE QUIZ

Answers are at the end of the chapter.

1. Class A power amplifiers are a type of large-signal amplifier.
2. Ideally, the Q-point should be centered on the load line in a class A amplifier.
3. The quiescent power dissipation occurs when the maximum signal is applied.
4. Efficiency is the ratio of output signal power to total power.
5. Each transistor in a class B amplifier conducts for the entire input cycle.
6. Class AB operation overcomes the problem of crossover distortion.
7. Complementary symmetry transistors must be used in a class AB amplifier.
8. A current mirror is implemented with a laser diode.
9. Darlington transistors can be used to increase the input resistance of a class AB amplifier.
10. The transistor in a class C amplifier conducts for a small portion of the input cycle.
11. The output of a class C amplifier is a replica of the input signal.
12. A class C amplifier usually employs a tuned circuit.

CIRCUIT-ACTION QUIZ

Answers are at the end of the chapter.

1. If the value of R_3 in Figure 7–5 is decreased, the voltage gain of the first stage will
   (a) increase (b) decrease (c) not change
2. If the value of R_{E2} in Figure 7–5 is increased, the voltage gain of the first stage will
   (a) increase (b) decrease (c) not change
3. If C_2 in Figure 7–5 opens, the dc voltage at the emitter of Q_1 will
   (a) increase (b) decrease (c) not change
4. If the value of R_4 in Figure 7–5 is increased, the dc voltage at the base of Q_3 will
   (a) increase (b) decrease (c) not change
5. If V_{CC} in Figure 7–18 is increased, the peak output voltage will
   (a) increase (b) decrease (c) not change
6. If the value of R_L in Figure 7–18 is increased, the ac output power will
   (a) increase (b) decrease (c) not change
7. If the value of R_L in Figure 7–19 is decreased, the voltage gain will
   (a) increase (b) decrease (c) not change
8. If the value of V_{CC} in Figure 7–19 is increased, the ac output power will
   (a) increase (b) decrease (c) not change
9. If the values of R_1 and R_2 in Figure 7–19 are increased, the voltage gain will
   (a) increase (b) decrease (c) not change
10. If the value of C_2 in Figure 7–24 is decreased, the resonant frequency will
    (a) increase (b) decrease (c) not change

SELF-TEST

Answers are at the end of the chapter.

Section 7–1

1. An amplifier that operates in the linear region at all times is
   (a) Class A (b) Class AB (c) Class B (d) Class C
2. A certain class A power amplifier delivers 5 W to a load with an input signal power of 100 mW. The power gain is
   (a) 100 (b) 50 (c) 250 (d) 5
3. The peak current a class A power amplifier can deliver to a load depends on the
   (a) maximum rating of the power supply (b) quiescent current
   (c) current in the bias resistors (d) size of the heat sink
4. For maximum output, a class A power amplifier must maintain a value of quiescent current that is
   (a) one-half the peak load current (b) twice the peak load current
   (c) at least as large as the peak load current (d) just above the cutoff value
5. A certain class A power amplifier has $V_{CEQ} = 12$ V and $I_{CQ} = 1$ A. The maximum signal power output is
   (a) 6 W (b) 12 W (c) 1 W (d) 0.707 W
6. The efficiency of a power amplifier is the ratio of the power delivered to the load to the
   (a) input signal power (b) power dissipated in the last stage
   (c) power from the dc power supply (d) none of these answers
7. The maximum efficiency of a class A power amplifier is
   (a) 25% (b) 50% (c) 79% (d) 98%

Section 7–2

8. The transistors in a class B amplifier are biased
   (a) into cutoff (b) in saturation
   (c) at midpoint of the load line (d) right at cutoff
9. Crossover distortion is a problem for
   (a) class A amplifiers (b) class AB amplifiers
   (c) class B amplifiers (d) all of these amplifiers
10. A BJT class B push-pull amplifier with no transformer coupling uses
    (a) two *npn* transistors (b) two *pnp* transistors
    (c) complementary symmetry transitors (d) none of these
11. A current mirror in a push-pull amplifier should give an I_{CQ} that is
    (a) equal to the current in the bias resistors and diodes
    (b) twice the current in the bias resistors and diodes
    (c) half the current in the bias resistors and diodes
    (d) zero
12. The maximum efficiency of a class B push-pull amplifier is
    (a) 25% (b) 50% (c) 79% (d) 98%
13. The output of a certain two-supply class B push-pull amplifier has a V_{CC} of 20 V. If the load resistance is 50 Ω, the value of $I_{c(sat)}$ is
    (a) 5 mA (b) 0.4 A (c) 4 mA (d) 40 mA
14. The maximum efficiency of a class AB amplifier is
    (a) higher than a class B (b) the same as a class B
    (c) about the same as a class A (d) slightly less than a class B

Section 7–3

15. The power dissipation of a class C amplifier is normally
    (a) very low (b) very high (c) the same as a class B (d) the same as a class A
16. The efficiency of a class C amplifier is
    (a) less than class A (b) less than class B
    (c) less than class AB (d) greater than classes A, B, or AB

17. The transistor in a class C amplifier conducts for
    (a) more than 180° of the input cycle
    (b) one-half of the input cycle
    (c) a very small percentage of the input cycle
    (d) all of the input cycle

PROBLEMS

Answers to all odd-numbered problems are at the end of the book.

BASIC PROBLEMS

Section 7–1 The Class A Power Amplifier

1. Figure 7–42 shows a CE power amplifier in which the collector resistor serves also as the load resistor. Assume $\beta_{DC} = \beta_{ac} = 100$.
   (a) Determine the dc Q-point (I_{CQ} and V_{CEQ}).
   (b) Determine the voltage gain and the power gain.

▶ FIGURE 7–42

Multisim file circuits are identified with a CD logo and are in the Problems folder on your CD-ROM. Filenames correspond to figure numbers (e.g., F07-42).

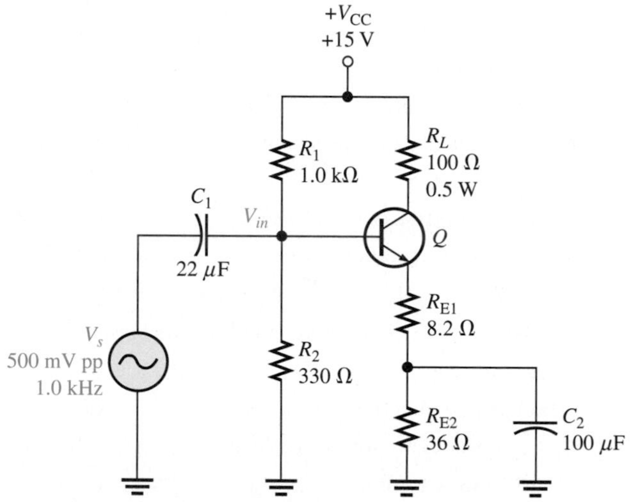

2. For the circuit in Figure 7–42, determine the following:
   (a) the power dissipated in the transistor with no load
   (b) the total power from the power supply with no load
   (c) the signal power in the load with a 500 mV input
3. Refer to the circuit in Figure 7–42. What changes would be necessary to convert the circuit to a *pnp* transistor with a positive supply? What advantage would this have?
4. Assume a CC amplifier has an input resistance of 2.2 kΩ and drives an output load of 50 Ω. What is the power gain?
5. Determine the Q-point for each amplifier in Figure 7–43.

▶ FIGURE 7–43

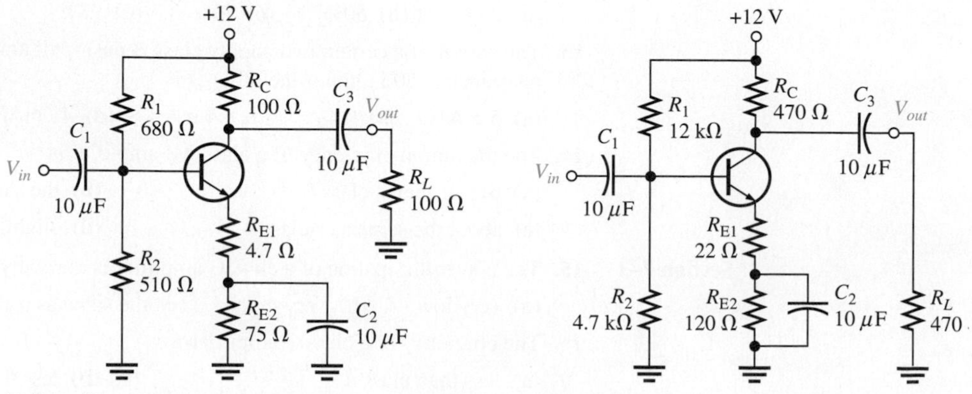

(a) $\beta_{ac} = \beta_{DC} = 125$ (b) $\beta_{ac} = \beta_{DC} = 120$

6. If the load resistor in Figure 7–43(a) is changed to 50 Ω, how much does the Q-point change?
7. What is the maximum peak value of collector current that can be realized in each circuit of Figure 7–43? What is the maximum peak value of output voltage in each circuit?
8. Find the power gain for each circuit in Figure 7–43. Neglect r'_e.
9. Determine the minimum power rating for the transistor in Figure 7–44.
10. Find the maximum output signal power to the load and efficiency for the amplifier in Figure 7–44 with a 500 Ω load resistor.

FIGURE 7–44

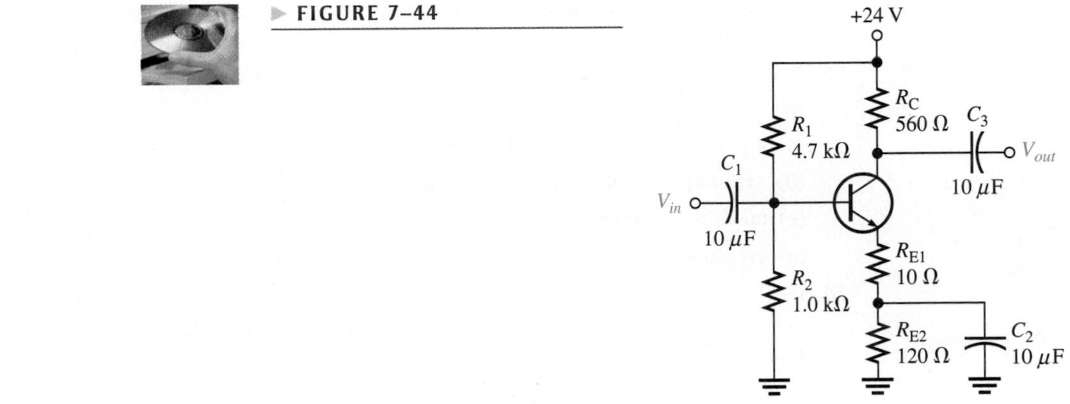

$\beta_{ac} = \beta_{DC} = 90$; $r'_e = 10\ \Omega$

Section 7–2 The Class B and Class AB Push-Pull Amplifiers

11. Refer to the class AB amplifier in Figure 7–45.
    (a) Determine the dc parameters $V_{B(Q1)}$, $V_{B(Q2)}$, V_E, I_{CQ}, $V_{CEQ(Q1)}$, $V_{CEQ(Q2)}$.
    (b) For the 5 V rms input, determine the power delivered to the load resistor.

FIGURE 7–45

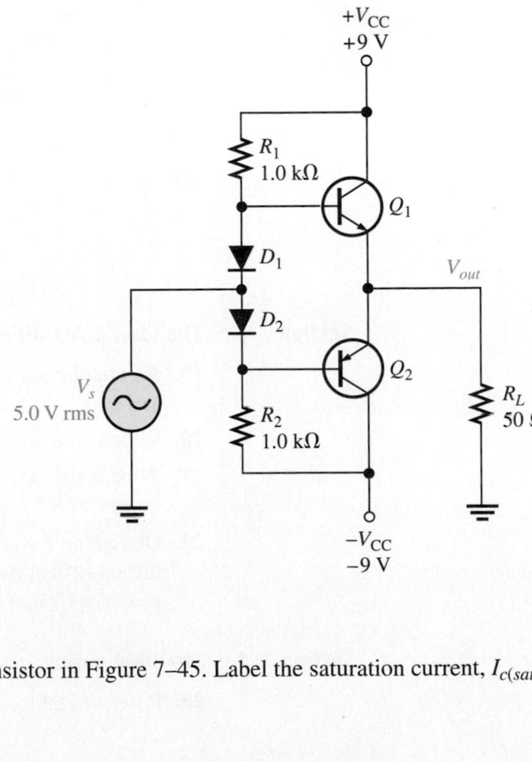

12. Draw the load line for the *npn* transistor in Figure 7–45. Label the saturation current, $I_{c(sat)}$, and show the Q-point.

13. Determine the approximate input resistance seen by the signal source for the amplifier of Figure 7–45 if $\beta_{ac} = 100$.
14. If D_2 has more voltage drop than D_1, what effect does this have on the output?
15. Refer to the class AB amplifier in Figure 7–46 operating with a single power supply.
    - **(a)** Determine the dc parameters $V_{B(Q1)}$, $V_{B(Q2)}$, V_E, I_{CQ}, $V_{CEQ(Q1)}$, $V_{CEQ(Q2)}$.
    - **(b)** Assuming the input voltage is 10 V pp, determine the power delivered to the load resistor.
16. Refer to the class AB amplifier in Figure 7–46.
    - **(a)** What is the maximum power that could be delivered to the load resistor?
    - **(b)** Assume the power supply voltage is raised to 24 V. What is the new maximum power that could be delivered to the load resistor?
17. Refer to the class AB amplifier in Figure 7–46. What fault or faults could account for each of the following troubles?
    - **(a)** a positive half-wave output signal
    - **(b)** zero volts on both bases and the emitters
    - **(c)** no output: emitter voltage = +15 V
    - **(d)** crossover distortion observed on the output waveform
18. If a 1 V rms signal source with an internal resistance of 50 Ω is connected to the amplifier in Figure 7–46, what is the actual rms signal applied to the amplifier input? Assume $\beta_{ac} = 200$.

FIGURE 7–46

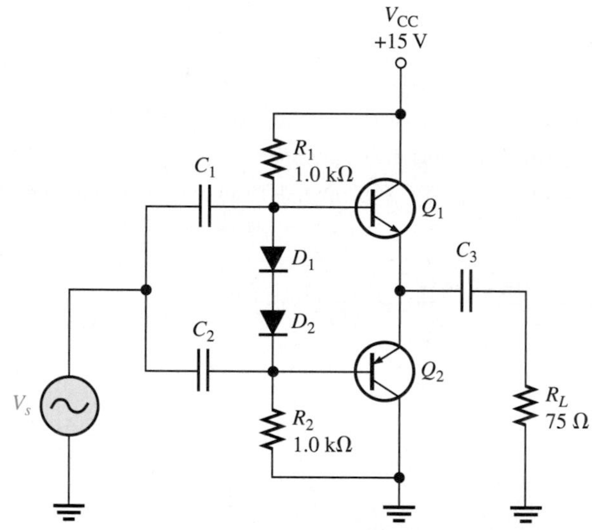

Section 7–3 The Class C Amplifier

19. A certain class C amplifier transistor is on for 10 percent of the input cycle. If $V_{ce(sat)} = 0.18$ V and $I_{c(sat)} = 25$ mA, what is the average power dissipation for maximum output?
20. What is the resonant frequency of a tank circuit with $L = 10$ mH and $C = 0.001\ \mu$F?
21. What is the maximum peak-to-peak output voltage of a tuned class C amplifier with $V_{CC} = 12$ V?
22. Determine the efficiency of the class C amplifier described in Problem 21 if $V_{CC} = 15$ V and the equivalent parallel resistance in the collector tank circuit is 50 Ω. Assume that the transistor is on for 10% of the period.

Section 7–4 Troubleshooting

23. Refer to Figure 7–47. What would you expect to observe across R_L if C_1 opened?

▶ FIGURE 7–47

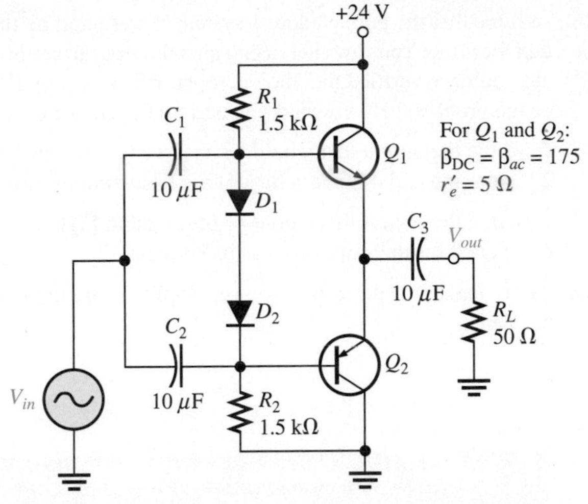

24. Your oscilloscope displays a half-wave output when connected across R_L in Figure 7–47. What is the probable cause?

25. Determine the possible fault or faults, if any, for each circuit in Figure 7–48 based on the indicated dc voltage measurements.

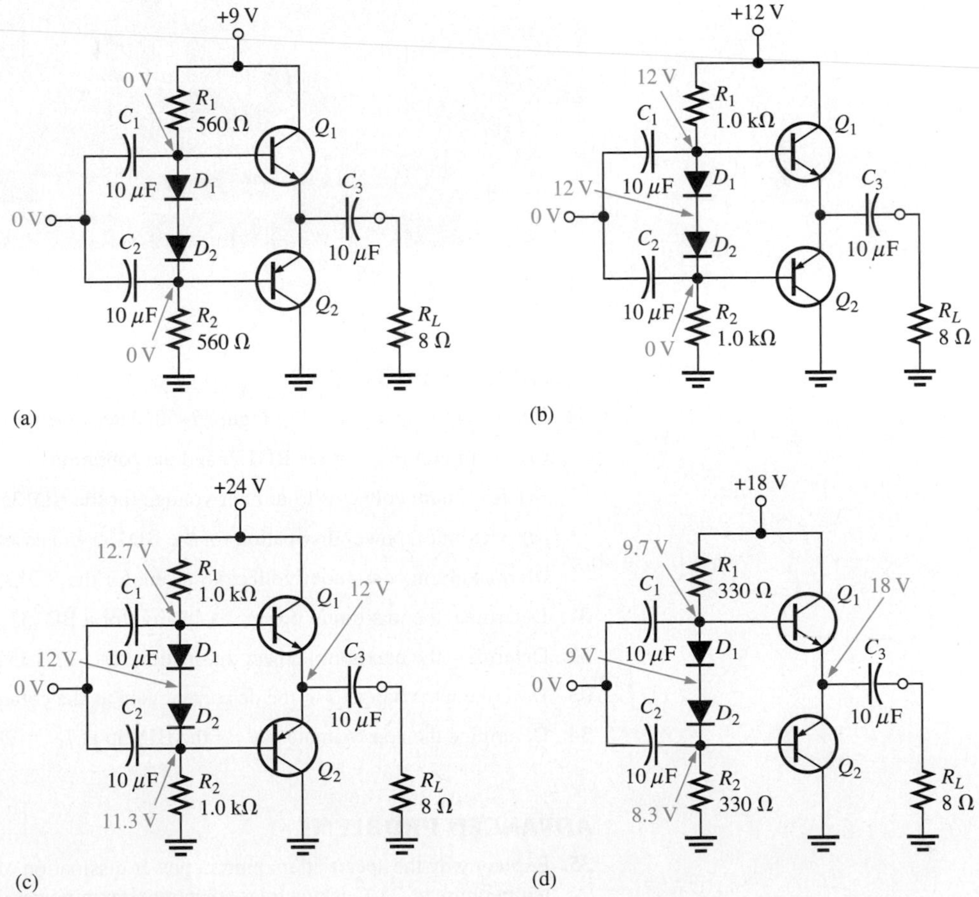

▲ FIGURE 7–48

APPLICATION ACTIVITY PROBLEMS

26. Assume that the public address system represented by the block diagram in Figure 7–34 has quit working. You find there is no signal output from the power amplifier or the preamplifier, but you have verified that the microphone is working. Which two blocks are the most likely to be the problem? How would you narrow the choice down to one block?

27. Describe the output that would be observed in the push-pull amplifier of Figure 7–35 with a 2 V rms sinusoidal input voltage if the base-emitter junction of Q_2 opened.

28. Describe the output that would be observed in Figure 7–35 if the collector-emitter junction of Q_5 opened for the same input as in Problem 27.

29. After visually inspecting the power amplifier circuit board in Figure 7–49, describe any problems.

FIGURE 7–49

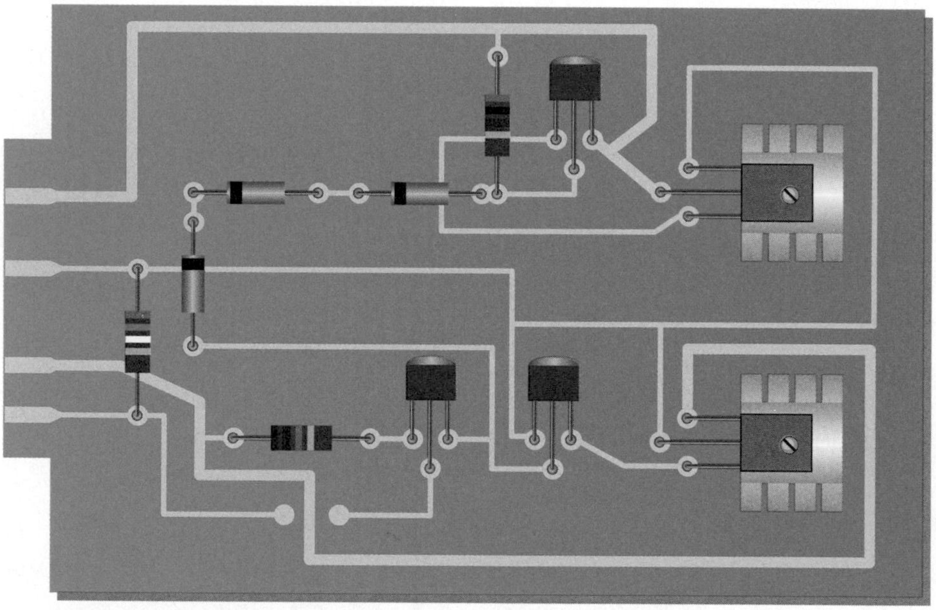

DATASHEET PROBLEMS

30. Referring to the datasheet in Figure 7–50, determine the following:

(a) minimum β_{DC} for the BD135 and the conditions

(b) maximum collector-to-emitter voltage for the BD135

(c) maximum power dissipation for the BD135 at a case temperature of 25°C

(d) maximum continuous collector current for the BD135

31. Determine the maximum power dissipation for a BD135 at a case temperature of 50°C.

32. Determine the maximum power dissipation for a BD135 at an ambient temperature of 50°C.

33. Describe what happens to the dc current gain as the collector current increases.

34. Determine the approximate h_{FE} for the BD135 at $I_C = 20$ mA.

ADVANCED PROBLEMS

35. Explain why the specified maximum power dissipation of a power transistor at an ambient temperature of 25°C is much less than maximum power dissipation at a case temperature of 25°C.

FAIRCHILD
SEMICONDUCTOR™

BD135/137/139

Medium Power Linear and Switching Applications

- Complement to BD136, BD138 and BD140 respectively

1

TO-126

1. Emitter 2.Collector 3.Base

NPN Epitaxial Silicon Transistor

Absolute Maximum Ratings T_C = 25°C unless otherwise noted

Symbol	Parameter	Value	Units
V_{CBO}	Collector-Base Voltage : BD135 : BD137 : BD139	45 60 80	V V V
V_{CEO}	Collector-Emitter Voltage : BD135 : BD137 : BD139	45 60 80	V V V
V_{EBO}	Emitter-Base Voltage	5	V
I_C	Collector Current (DC)	1.5	A
I_{CP}	Collector Current (Pulse)	3.0	A
I_B	Base Current	0.5	A
P_C	Collector Dissipation (T_C = 25°C)	12.5	W
P_C	Collector Dissipation (T_a = 25°C)	1.25	W
T_J	Junction Temperature	150	°C
T_{STG}	Storage Temperature	- 55 ~ 150	°C

Electrical Characteristics T_C = 25°C unless otherwise noted

Symbol	Parameter	Test Condition	Min.	Typ.	Max.	Units
$V_{CEO}(sus)$	Collector-Emitter Sustaining Voltage : BD135 : BD137 : BD139	I_C = 30mA, I_B = 0	 45 60 80			 V V V
I_{CBO}	Collector Cut-off Current	V_{CB} = 30V, I_E = 0			0.1	μA
I_{EBO}	Emitter Cut-off Current	V_{EB} = 5V, I_C = 0			10	μA
h_{FE1} h_{FE2} h_{FE3}	DC Current Gain : ALL DEVICE : ALL DEVICE : BD135 : BD137, BD139	V_{CE} = 2V, I_C = 5mA V_{CE} = 2V, I_C = 0.5A V_{CE} = 2V, I_C = 150mA	25 25 40 40		 250 160	
$V_{CE}(sat)$	Collector-Emitter Saturation Voltage	I_C = 500mA, I_B = 50mA			0.5	V
$V_{BE}(on)$	Base-Emitter ON Voltage	V_{CE} = 2V, I_C = 0.5A			1	V

h_{FE} Classification

Classification	6	10	16
h_{FE3}	40 ~ 100	63 ~ 160	100 ~ 250

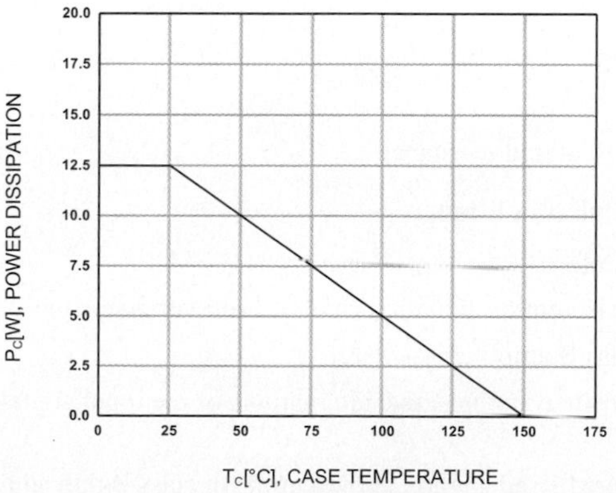

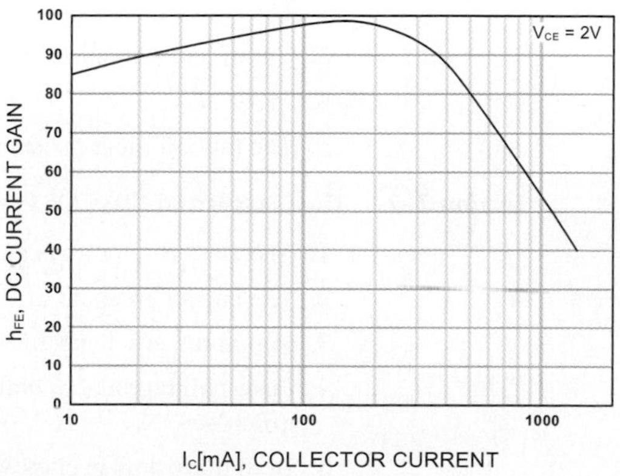

▲ FIGURE 7–50

© 2003 Fairchild Semiconductor Corporation. Used by permission.

36. Draw the dc and the ac load lines for the amplifier in Figure 7–51.
37. Design a swamped class A power amplifier that will operate from a dc supply of +15 V with an approximate voltage gain of 50. The quiescent collector current should be approximately 500 mA, and the total dc current from the supply should not exceed 750 mA. The output power must be at least 1 W.
38. The public address system in Figure 7–34 is a portable unit that is independent of 115 V ac. Determine the ampere-hour rating for the +15 V and the −15 V battery supply necessary for the system to operate for 4 hours on a continuous basis.

FIGURE 7–51

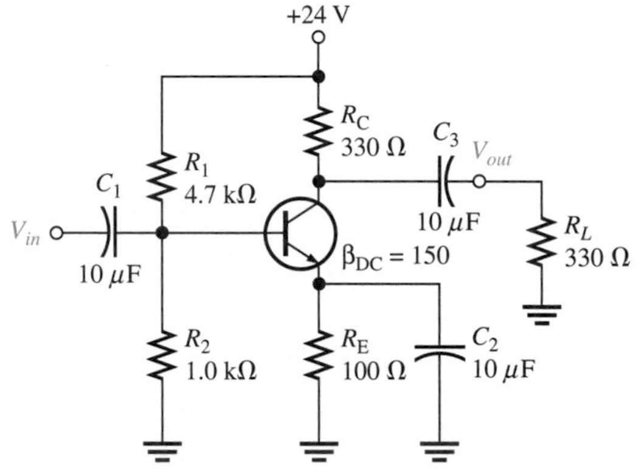

MULTISIM TROUBLESHOOTING PROBLEMS

These file circuits are in the Troubleshooting Problems folder on your CD-ROM.

39. Open file TSP07-39 and determine the fault.
40. Open file TSP07-40 and determine the fault.
41. Open file TSP07-41 and determine the fault.
42. Open file TSP07-42 and determine the fault.
43. Open file TSP07-43 and determine the fault.

ANSWERS

SECTION CHECKUPS

Section 7–1 The Class A Power Amplifier

1. To dissipate excessive heat
2. The collector
3. Cutoff and saturation clipping
4. 25%
5. The ratio of input resistance to output resistance

Section 7–2 The Class B and Class AB Push-Pull Amplifiers

1. The class B Q-point is at cutoff.
2. The barrier potential of the base-emitter junction causes crossover distortion.
3. Maximum efficiency of a class B amplifier is 79%.
4. Push-pull reproduces both positive and negative alternations of the input signal with greater efficiency.
5. Both transistors in class AB are biased slightly above cutoff. In class B they are biased at cutoff.

Section 7–3 The Class C Amplifier

1. Class C is biased well into cutoff.
2. The purpose of the tuned circuit is to produce a sinusoidal voltage output.
3. $\eta = [1\text{ W}/(1\text{ W} + 0.1\text{ W})]100 = 90.9\%$

Section 7–4 Troubleshooting

1. Excess input signal voltage

2. Open bypass capacitor, C_2

RELATED PROBLEMS FOR EXAMPLES

7–1 Power gain increases.

7–2 The efficiency goes up because no power is wasted in R_{E3}. The problem is that the speaker has direct current in the coil and may burn open.

7–3 15 V, 0.1 A

7–4 7.5 V; 0.25 A

7–5 $P_{out} = 1.76$ W; $P_{DC} = 2.24$ W

7–6 R_{in} would increase.

7–7 3 mW

7–8 The efficiency decreases.

7–9 Halve the input frequency.

TRUE/FALSE QUIZ

1. T	**2.** T	**3.** F	**4.** T	**5.** F	**6.** T
7. T	**8.** F	**9.** T	**10.** T	**11.** F	**12.** T

CIRCUIT-ACTION QUIZ

1. (c)	**2.** (c)	**3.** (c)	**4.** (a)	**5.** (a)
6. (b)	**7.** (c)	**8.** (c)	**9.** (c)	**10.** (a)

SELF-TEST

1. (a)	**2.** (b)	**3.** (b)	**4.** (a)	**5.** (b)	**6.** (c)	**7.** (a)	**8.** (e)	**9.** (c)
10. (c)	**11.** (a)	**12.** (c)	**13.** (b)	**14.** (d)	**15.** (a)	**16.** (d)	**17.** (c)	

8 Field-Effect Transistors (FETs)

CHAPTER OUTLINE

CHAPTER OBJECTIVES

- Explain the operation of JFETs
- Define, discuss, and apply important JFET parameters
- Discuss and analyze JFET bias circuits
- Discuss the ohmic region
- Explain the operation of MOSFETs
- Define, discuss, and apply important MOSFET parameters
- Discuss and analyze MOSFET bias circuits
- Explain the operation of IGBTs
- Troubleshoot FET circuits

KEY TERMS

- JFET
- Drain
- Source
- Gate
- Pinch-off voltage
- Transconductance
- Ohmic region
- MOSFET
- Depletion
- Enhancement
- IGBT

APPLICATION ACTIVITY PREVIEW

The Application Activity involves the electronic control circuits for a waste water treatment system. In particular, you will focus on the application of field-effect transistors in the sensing circuits for chemical measurements.

VISIT THE COMPANION WEBSITE

Study aids for this chapter are available at http://www.prenhall.com/floyd

INTRODUCTION

BJTs (bipolar junction transistors) were covered in previous chapters. Now we will discuss the second major type of transistor, the FET (field-effect transistor). **FETs** are unipolar devices because, unlike BJTs that use both electron and hole current, they operate only with one type of charge carrier. The two main types of FETs are the junction field-effect transistor (JFET) and the metal oxide semiconductor field-effect transistor (MOSFET). The term field-effect relates to the depletion region formed in the channel of a FET as a result of a voltage applied on one of its terminals (gate).

Recall that a BJT is a current-controlled device; that is, the base current controls the amount of collector current. A FET is different. It is a voltage-controlled device, where the voltage between two of the terminals (gate and source) controls the current through the device. A major advantage of FETs is their very high input resistance. Because of their nonlinear characteristics, they are generally not as widely used in amplifiers as BJTs except where very high input impedances are required. However, FETs are the preferred device in low-voltage switching applications because they are generally faster than BJTs when turned on and off. The IGBT is generally used in high-voltage switching applications.

8–1 THE JFET

The **JFET** (junction field-effect transistor) is a type of FET that operates with a reverse-biased *pn* junction to control current in a channel. Depending on their structure, JFETs fall into either of two categories, *n* channel or *p* channel.

After completing this section, you should be able to

- **Explain the operation of JFETs**
  - Identify the three terminals of a JFET
  - Explain what a channel is
  - Describe the structural difference between an *n*-channel JFET and a *p*-channel JFET
  - Discuss how voltage controls the current in a JFET
  - Identify the symbols for *n*-channel and *p*-channel JFETs

Figure 8–1(a) shows the basic structure of an *n*-channel JFET (junction field-effect transistor). Wire leads are connected to each end of the *n*-channel; the **drain** is at the upper end, and the **source** is at the lower end. Two *p*-type regions are diffused in the *n*-type material to form a **channel,** and both *p*-type regions are connected to the **gate** lead. For simplicity, the gate lead is shown connected to only one of the *p* regions. A *p*-channel JFET is shown in Figure 8–1(b).

HISTORY NOTE

In 1952, Ian Ross and George Dacey succeeded in making a unipolar device with a structure similar to today's JFET.

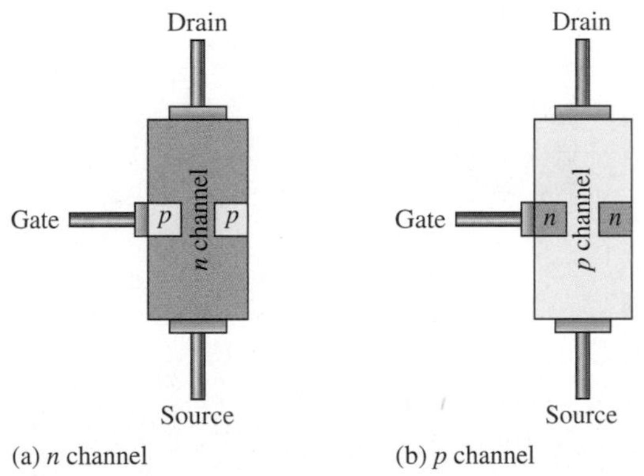

FIGURE 8–1

A representation of the basic structure of the two types of JFET.

Basic Operation

To illustrate the operation of a JFET, Figure 8–2 shows dc bias voltages applied to an *n*-channel device. V_{DD} provides a drain-to-source voltage and supplies current from drain to source. V_{GG} sets the reverse-bias voltage between the gate and the source, as shown.

The JFET is always operated with the gate-source pn junction reverse-biased. Reverse-biasing of the gate-source junction with a negative gate voltage produces a depletion region along the *pn* junction, which extends into the *n* channel and thus increases its resistance by restricting the channel width.

The channel width and thus the channel resistance can be controlled by varying the gate voltage, thereby controlling the amount of drain current, I_D. Figure 8–3 illustrates this concept. The white areas represent the depletion region created by the reverse bias. It is wider toward the drain end of the channel because the reverse-bias voltage between the gate and

FIGURE 8–2

A biased *n*-channel JFET.

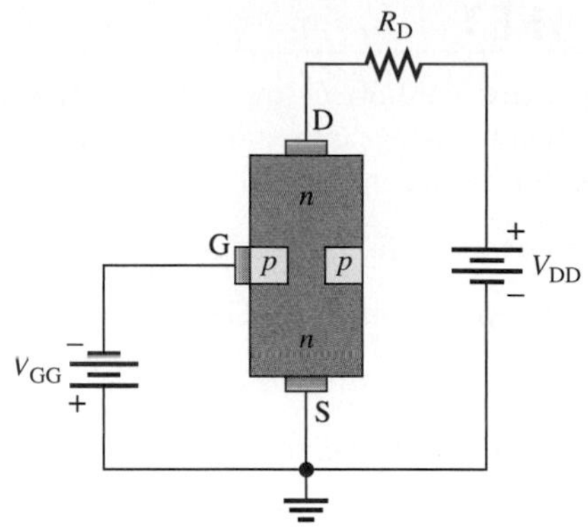

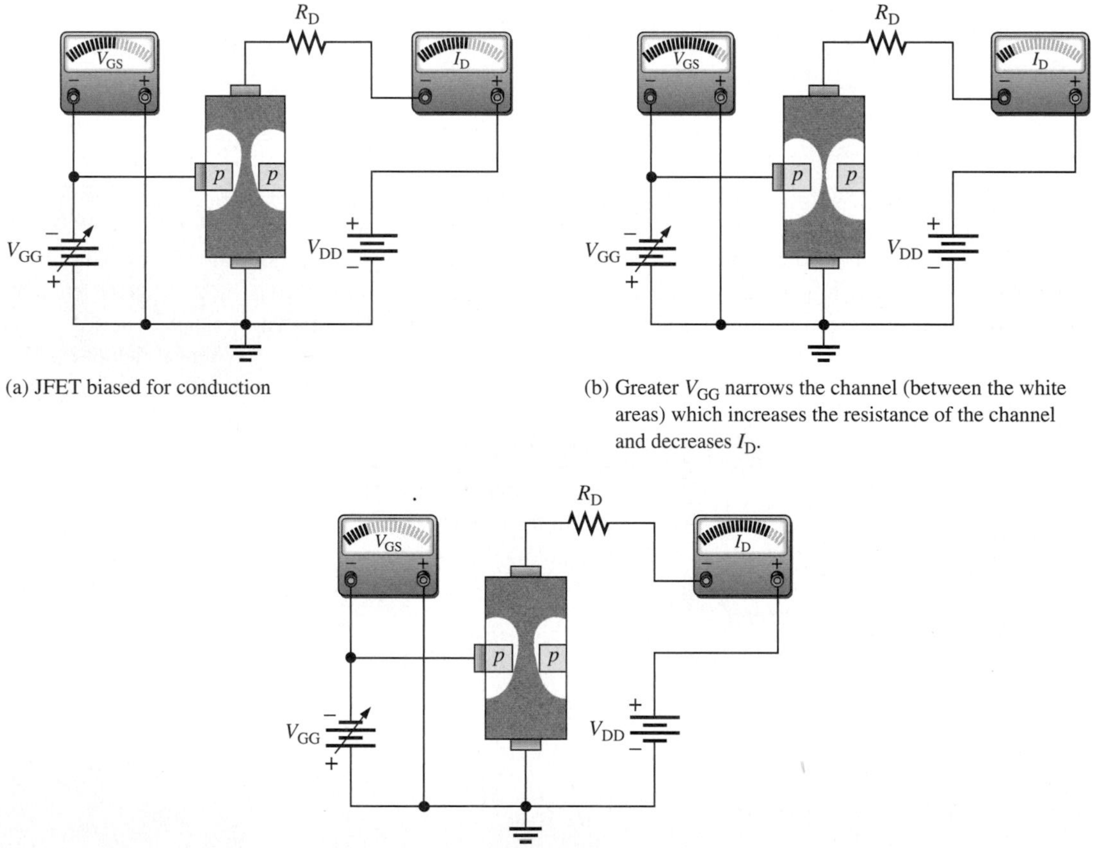

(a) JFET biased for conduction

(b) Greater V_{GG} narrows the channel (between the white areas) which increases the resistance of the channel and decreases I_D.

(c) Less V_{GG} widens the channel (between the white areas) which decreases the resistance of the channel and increases I_D.

FIGURE 8–3

Effects of V_{GS} on channel width, resistance, and drain current ($V_{GG} = V_{GS}$).

the drain is greater than that between the gate and the source. We will discuss JFET characteristic curves and some important parameters in Section 8–2.

JFET Symbols

The schematic symbols for both *n*-channel and *p*-channel JFETs are shown in Figure 8–4. Notice that the arrow on the gate points "in" for *n* channel and "out" for *p* channel.

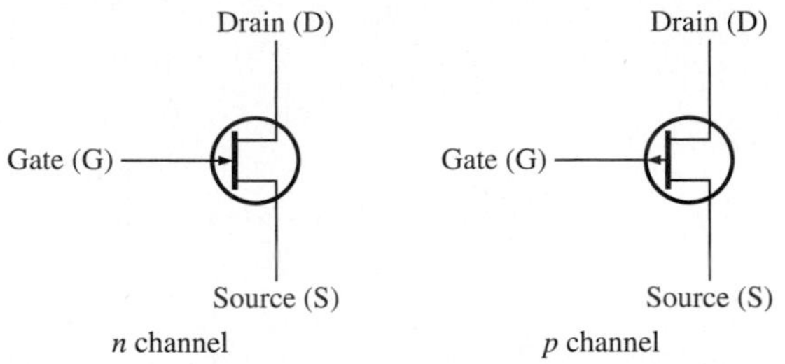

FIGURE 8–4

JFET schematic symbols.

SECTION 8–1 CHECKUP
Answers are at the end of the chapter.

1. Name the three terminals of a JFET.
2. Does an *n*-channel JFET require a positive or negative value for V_{GS}?
3. How is the drain current controlled in a JFET?

8–2 JFET CHARACTERISTICS AND PARAMETERS

The JFET operates as a voltage-controlled, constant-current device. Cutoff and pinch-off as well as JFET transfer characteristics are covered in this section.

After completing this section, you should be able to

- **Define, discuss, and apply important JFET parameters**
  - Explain ohmic region, constant-current region, and breakdown
  - Define *pinch-off voltage*
  - Describe how gate-to-source voltage controls the drain current
  - Define *cutoff voltage*
  - Compare pinch-off and cutoff
  - Analyze a JFET transfer characteristic curve
  - Use the equation for the transfer characteristic to calculate I_D
  - Use a JFET datasheet
  - Define *transconductance*
  - Explain and determine input resistance and capacitance
  - Determine drain-to-source resistance

Consider the case when the gate-to-source voltage is zero ($V_{GS} = 0$ V). This is produced by shorting the gate to the source, as in Figure 8–5(a) where both are grounded. As V_{DD} (and thus V_{DS}) is increased from 0 V, I_D will increase proportionally, as shown in the graph of Figure 8–5(b) between points *A* and *B*. In this area, the channel resistance is essentially constant because the depletion region is not large enough to have significant effect. This is called the ohmic region because V_{DS} and I_D are related by Ohm's law. (Ohmic region is discussed further in Section 8–4.)

At point *B* in Figure 8–5(b), the curve levels off and enters the active region where I_D becomes essentially constant. As V_{DS} increases from point *B* to point *C*, the reverse-bias

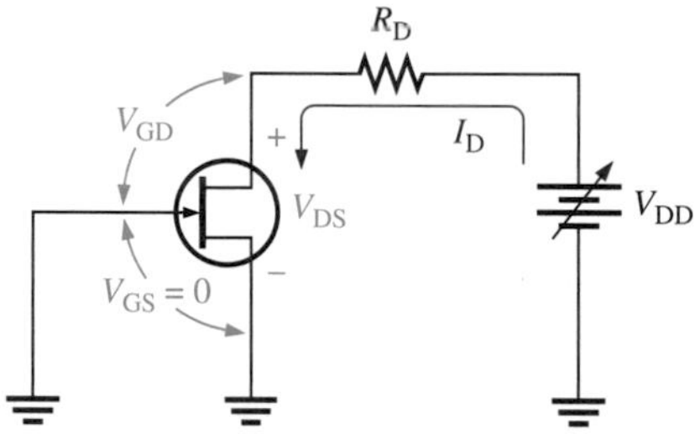

(a) JFET with $V_{GS} = 0$ V and a variable V_{DS} (V_{DD})

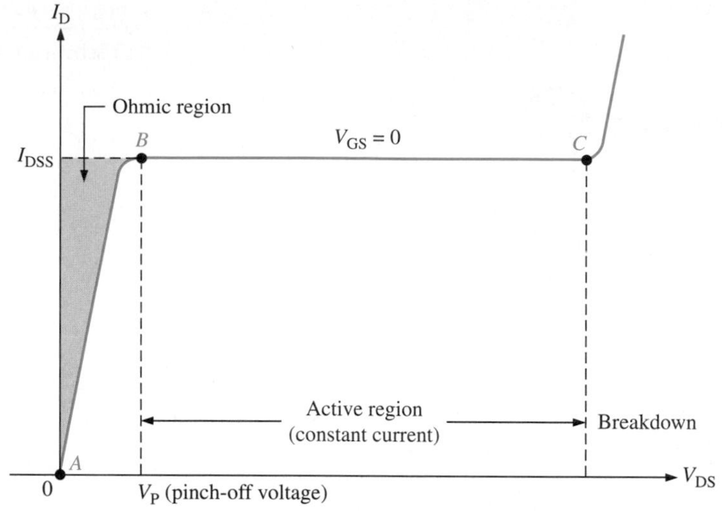

(b) Drain characteristic

▲ FIGURE 8–5

The drain characteristic curve of a JFET for $V_{GS} = 0$ showing pinch-off voltage.

voltage from gate to drain (V_{GD}) produces a depletion region large enough to offset the increase in V_{DS}, thus keeping I_D relatively constant.

Pinch-Off Voltage

For $V_{GS} = 0$ V, the value of V_{DS} at which I_D becomes essentially constant (point *B* on the curve in Figure 8–5(b)) is the **pinch-off voltage**, V_P. For a given JFET, V_P has a fixed value. As you can see, a continued increase in V_{DS} above the pinch-off voltage produces an almost constant drain current. This value of drain current is I_{DSS} (*D*rain to *S*ource current with gate *S*horted) and is always specified on JFET datasheets. I_{DSS} is the *maximum* drain current that a specific JFET can produce regardless of the external circuit, and it is always specified for the condition, $V_{GS} = 0$ V.

Breakdown

As shown in the graph in Figure 8–5(b), **breakdown** occurs at point *C* when I_D begins to increase very rapidly with any further increase in V_{DS}. Breakdown can result in irreversible damage to the device, so JFETs are always operated below breakdown and within the active region (constant current) (between points *B* and *C* on the graph). The JFET action that produces the drain characteristic curve to the point of breakdown for $V_{GS} = 0$ V is illustrated in Figure 8–6.

V_{GS} Controls I_D

Let's connect a bias voltage, V_{GG}, from gate to source as shown in Figure 8–7(a). As V_{GS} is set to increasingly more negative values by adjusting V_{GG}, a family of drain characteristic curves is produced, as shown in Figure 8–7(b). Notice that I_D decreases as the magnitude of V_{GS} is increased to larger negative values because of the narrowing of the channel. Also notice that, for each increase in V_{GS}, the JFET reaches pinch-off (where constant current begins) at values of V_{DS} less than V_P. The term *pinch-off* is not the same as pinch-off voltage, V_P. Therefore, the amount of drain current is controlled by V_{GS}, as illustrated in Figure 8–8.

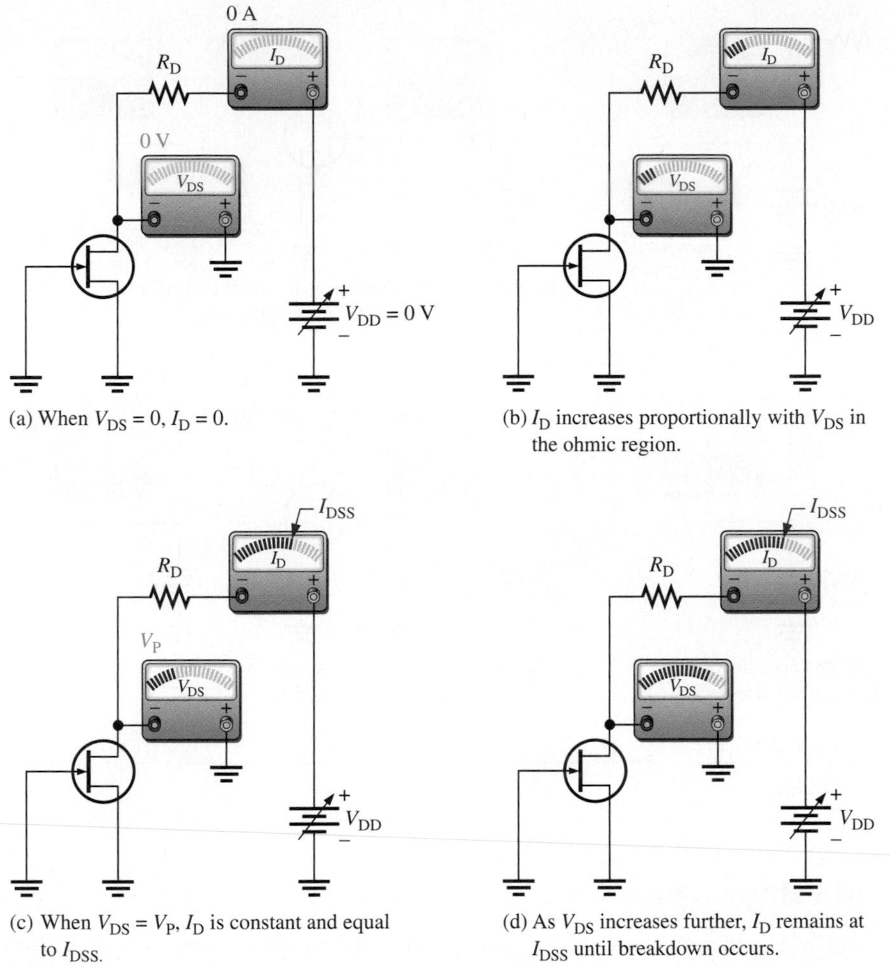

▲ **FIGURE 8–6**

JFET action that producdes the characteristic curve for $V_{GS} = 0$ V.

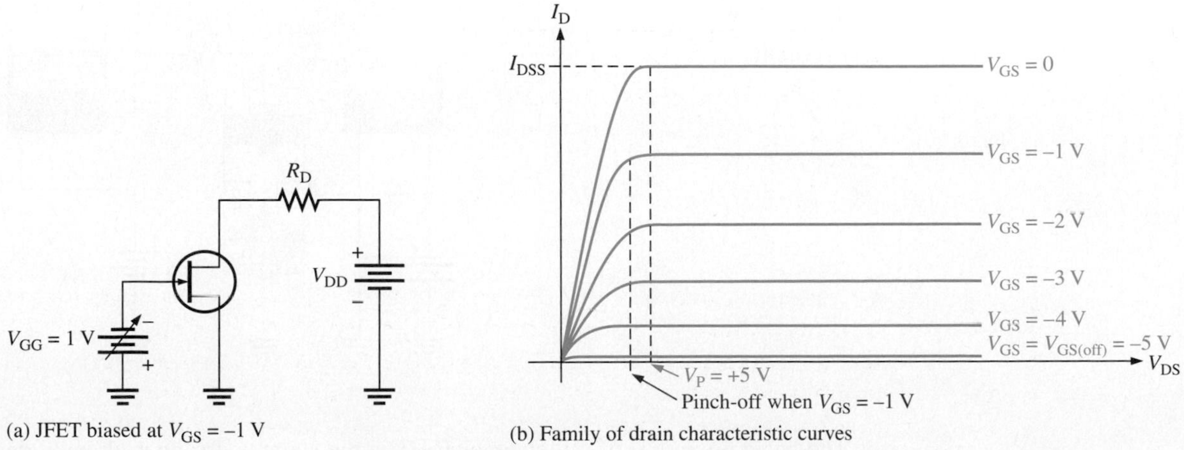

▲ **FIGURE 8–7**

Pinch-off occurs at a lower V_{DS} as V_{GS} is increased to more negative values.

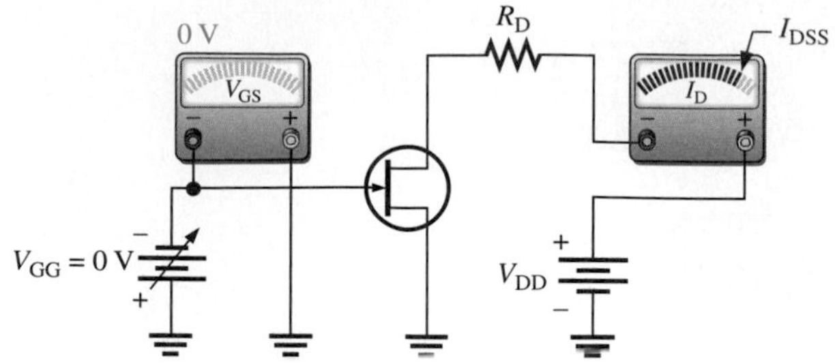

(a) $V_{GS} = 0$ V, $V_{DS} \geq V_P$, $I_D = I_{DSS}$

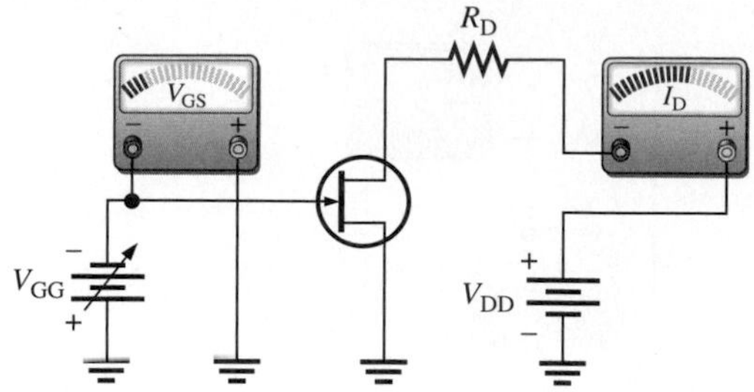

(b) When V_{GS} is negative, I_D decreases and is constant above pinch-off, which is less than V_P.

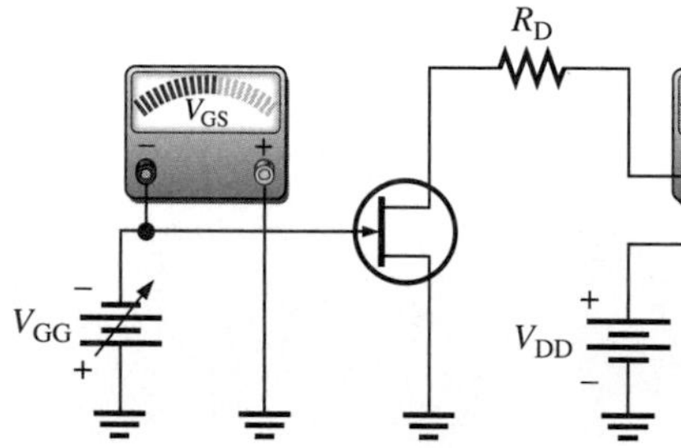

(c) As V_{GS} is made more negative, I_D continues to decrease but is constant above pinch-off, which has also decreased.

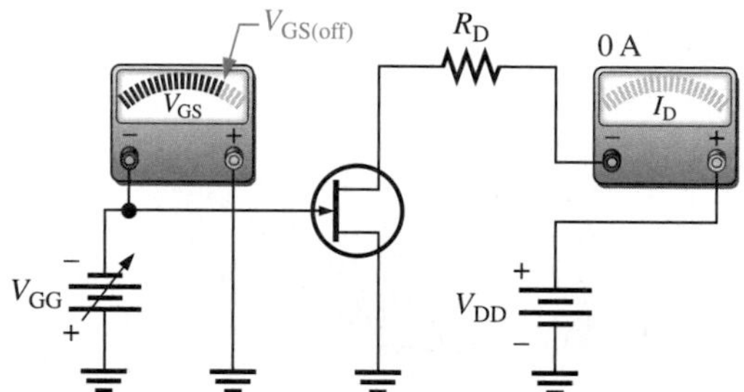

(d) Until $V_{GS} = -V_{GS(off)}$, I_D continues to decrease. When $V_{GS} \geq -V_{GS(off)}$, $I_D \cong 0$.

▲ FIGURE 8–8

V_{GS} controls I_D.

Cutoff Voltage

The value of V_{GS} that makes I_D approximately zero is the **cutoff voltage,** $V_{GS(off)}$, as shown in Figure 8–8(d). The JFET must be operated between $V_{GS} = 0$ V and $V_{GS(off)}$. For this range of gate-to-source voltages, I_D will vary from a maximum of I_{DSS} to a minimum of almost zero.

As you have seen, for an n-channel JFET, the more negative V_{GS} is, the smaller I_D becomes in the active region. When V_{GS} has a sufficiently large negative value, I_D is reduced to zero. This cutoff effect is caused by the widening of the depletion region to a point where it completely closes the channel, as shown in Figure 8–9.

▶ FIGURE 8–9

JFET at cutoff.

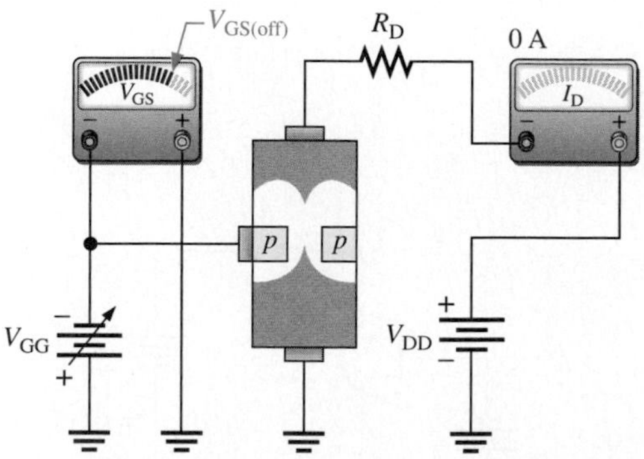

The basic operation of a p-channel JFET is the same as for an n-channel device except that a p-channel JFET requires a negative V_{DD} and a positive V_{GS}, as illustrated in Figure 8–10.

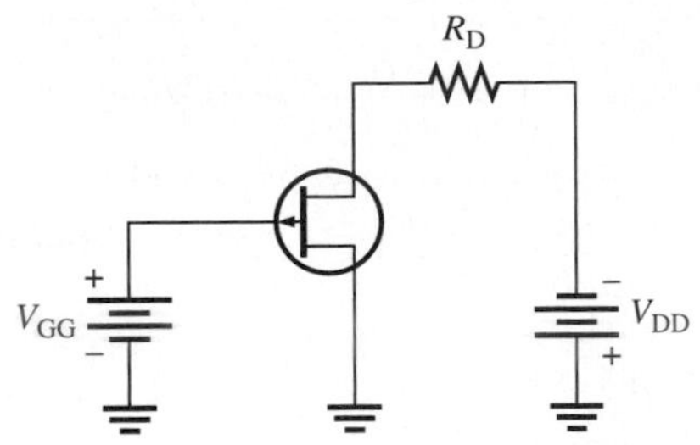

FIGURE 8–10

A biased *p*-channel JFET.

Comparison of Pinch-Off Voltage and Cutoff Voltage

As you have seen, there is a difference between pinch-off and cutoff voltages. There is also a connection. The pinch-off voltage V_P is the value of V_{DS} at which the drain current becomes constant and equal to I_{DSS} and is always measured at $V_{GS} = 0$ V. However, pinch-off occurs for V_{DS} values less than V_P when V_{GS} is nonzero. So, although V_P is a constant, the minimum value of V_{DS} at which I_D becomes constant varies with V_{GS}.

$V_{GS(off)}$ and V_P are always equal in magnitude but opposite in sign. A datasheet usually will give either $V_{GS(off)}$ or V_P, but not both. However, when you know one, you have the other. For example, if $V_{GS(off)} = -5$ V, then $V_P = +5$ V, as shown in Figure 8–7(b).

EXAMPLE 8–1

For the JFET in Figure 8–11, $V_{GS(off)} = -4$ V and $I_{DSS} = 12$ mA. Determine the *minimum* value of V_{DD} required to put the device in the constant-current region of operation when $V_{GS} = 0$ V.

FIGURE 8–11

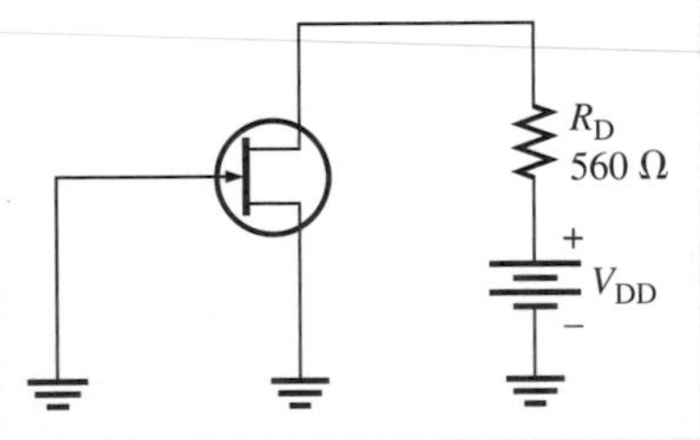

Solution Since $V_{GS(off)} = -4$ V, $V_P = 4$ V. The minimum value of V_{DS} for the JFET to be in its constant-current region is

$$V_{DS} = V_P = 4\text{ V}$$

In the constant-current region with $V_{GS} = 0$ V,

$$I_D = I_{DSS} = 12\text{ mA}$$

The drop across the drain resistor is

$$V_{R_D} = I_D R_D = (12\text{ mA})(560\ \Omega) = 6.72\text{ V}$$

Apply Kirchhoff's law around the drain circuit.

$$V_{DD} = V_{DS} + V_{R_D} = 4\text{ V} + 6.72\text{ V} = \mathbf{10.7\ V}$$

This is the value of V_{DD} to make $V_{DS} = V_P$ and put the device in the constant-current region.

*Related Problem** If V_{DD} is increased to 15 V, what is the drain current?

*Answers are at the end of the chapter.

EXAMPLE 8–2

A particular *p*-channel JFET has a $V_{GS(off)} = +4$ V. What is I_D when $V_{GS} = +6$ V?

Solution The *p*-channel JFET requires a positive gate-to-source voltage. The more positive the voltage, the less the drain current. When $V_{GS} = 4$ V, $I_D = 0$. Any further increase in V_{GS} keeps the JFET cut off, so I_D remains **0**.

Related Problem What is V_P for the JFET described in this example?

JFET Universal Transfer Characteristic

You have learned that a range of V_{GS} values from zero to $V_{GS(off)}$ controls the amount of drain current. For an *n*-channel JFET, $V_{GS(off)}$ is negative, and for a *p*-channel JFET, $V_{GS(off)}$ is positive. Because V_{GS} does control I_D, the relationship between these two quantities is very important. Figure 8–12 is a general transfer characteristic curve that illustrates graphically the relationship between V_{GS} and I_D. This curve is also known as a transconductance curve.

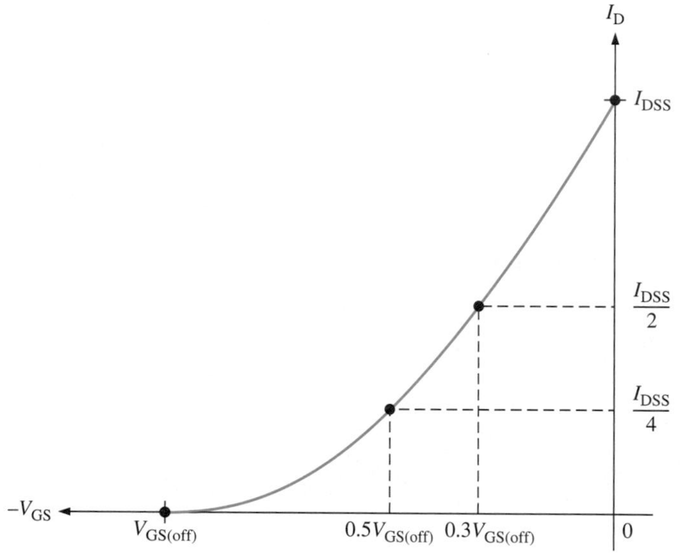

▶ FIGURE 8–12

JFET universal transfer characteristic curve (*n*-channel).

Notice that the bottom end of the curve is at a point on the V_{GS} axis equal to $V_{GS(off)}$, and the top end of the curve is at a point on the I_D axis equal to I_{DSS}. This curve shows that

$$I_D = 0 \quad \text{when } V_{GS} = V_{GS(off)}$$

$$I_D = \frac{I_{DSS}}{4} \quad \text{when } V_{GS} = 0.5V_{GS(off)}$$

$$I_D = \frac{I_{DSS}}{2} \quad \text{when } V_{GS} = 0.3V_{GS(off)}$$

and

$$I_D = I_{DSS} \quad \text{when } V_{GS} = 0$$

The transfer characteristic curve can also be developed from the drain characteristic curves by plotting values of I_D for the values of V_{GS} taken from the family of drain curves at pinch-off, as illustrated in Figure 8–13 for a specific set of curves. Each point on the transfer characteristic curve corresponds to specific values of V_{GS} and I_D on the drain curves. For example, when $V_{GS} = -2$ V, $I_D = 4.32$ mA. Also, for this specific JFET, $V_{GS(off)} = -5$ V and $I_{DSS} = 12$ mA.

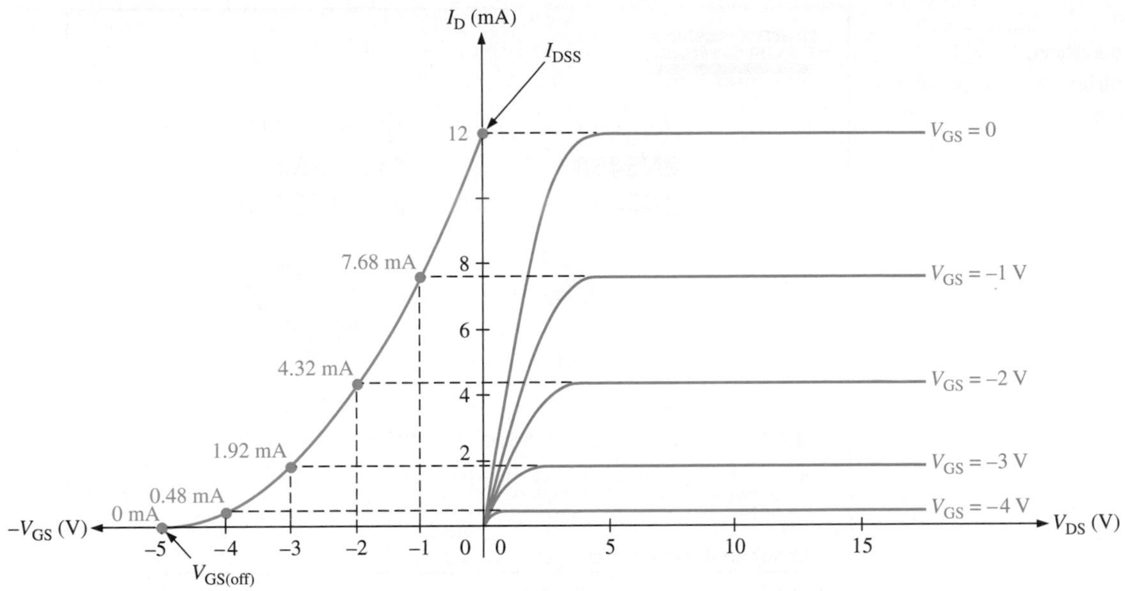

▲ **FIGURE 8–13**

Example of the development of an *n*-channel JFET transfer characteristic curve (blue) from the JFET drain characteristic curves (green).

A JFET transfer characteristic curve is expressed approximately as

$$\mathbf{I_D \cong I_{DSS}\left(1 - \frac{V_{GS}}{V_{GS(off)}}\right)^2} \qquad \text{Equation 8–1}$$

With Equation 8–1, I_D can be determined for any V_{GS} if $V_{GS(off)}$ and I_{DSS} are known. These quantities are usually available from the datasheet for a given JFET. Notice the squared term in the equation. Because of its form, a parabolic relationship is known as a *square law,* and therefore, JFETs and MOSFETs are often referred to as *square-law devices.*

The datasheet for a typical JFET series is shown in Figure 8–14.

EXAMPLE 8–3

The partial datasheet in Figure 8–14 for a 2N5459 JFET indicates that typically $I_{DSS} = 9$ mA and $V_{GS(off)} = -8$ V (maximum). Using these values, determine the drain current for $V_{GS} = 0$ V, -1 V, and -4 V.

Solution For $V_{GS} = 0$ V,

$$I_D = I_{DSS} = \mathbf{9\ mA}$$

For $V_{GS} = -1$ V, use Equation 8–1.

$$I_D \cong I_{DSS}\left(1 - \frac{V_{GS}}{V_{GS(off)}}\right)^2 = (9\text{ mA})\left(1 - \frac{-1\text{ V}}{-8\text{ V}}\right)^2$$

$$= (9\text{ mA})(1 - 0.125)^2 = (9\text{ mA})(0.766) = \mathbf{6.89\ mA}$$

For $V_{GS} = -4$ V,

$$I_D \cong (9\text{ mA})\left(1 - \frac{-4\text{ V}}{-8\text{ V}}\right)^2 = (9\text{ mA})(1 - 0.5)^2 = (9\text{ mA})(0.25) = \mathbf{2.25\ mA}$$

Related Problem Determine I_D for $V_{GS} = -3$ V for the 2N5459 JFET.

FIGURE 8–14

JFET partial datasheet. © 2003 Fairchild Semiconductor Corporation. Used by permission.

FAIRCHILD SEMICONDUCTOR™

2N5457 2N5458 2N5459 MMBF5457 MMBF5458 MMBF5459

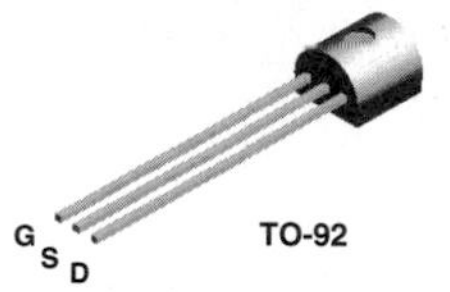

TO-92

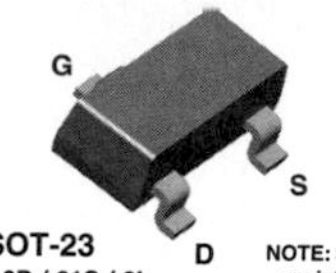

SOT-23
Mark: 6D / 61S / 6L

NOTE: Source & Drain are interchangeable

N-Channel General Purpose Amplifier

This device is a low level audio amplifier and switching transistors, and can be used for analog switching applications. Sourced from Process 55.

Absolute Maximum Ratings* TA = 25°C unless otherwise noted

Symbol	Parameter	Value	Units
V_{DG}	Drain-Gate Voltage	25	V
V_{GS}	Gate-Source Voltage	- 25	V
I_{GF}	Forward Gate Current	10	mA
T_J, T_{stg}	Operating and Storage Junction Temperature Range	-55 to +150	°C

*These ratings are limiting values above which the serviceability of any semiconductor device may be impaired.

NOTES:
1) These ratings are based on a maximum junction temperature of 150 degrees C.
2) These are steady state limits. The factory should be consulted on applications involving pulsed or low duty cycle operations.

Thermal Characteristics TA = 25°C unless otherwise noted

Symbol	Characteristic	Max		Units
		2N5457-5459	*MMBF5457-5459	
P_D	Total Device Dissipation Derate above 25°C	625 5.0	350 2.8	mW mW/°C
$R_{\theta JC}$	Thermal Resistance, Junction to Case	125		°C/W
$R_{\theta JA}$	Thermal Resistance, Junction to Ambient	357	556	°C/W

*Device mounted on FR-4 PCB 1.6" X 1.6" X 0.06."

Electrical Characteristics TA = 25°C unless otherwise noted

Symbol	Parameter	Test Conditions	Min	Typ	Max	Units
OFF CHARACTERISTICS						
$V_{(BR)GSS}$	Gate-Source Breakdown Voltage	I_G = 10 μA, V_{DS} = 0	- 25			V
I_{GSS}	Gate Reverse Current	V_{GS} = -15 V, V_{DS} = 0			- 1.0	nA
		V_{GS} = -15 V, V_{DS} = 0, T_A = 100°C			- 200	nA
$V_{GS(off)}$	Gate-Source Cutoff Voltage	V_{DS} = 15 V, I_D = 10 nA **5457**	- 0.5		- 6.0	V
		5458	- 1.0		- 7.0	V
		5459	- 2.0		- 8.0	V
V_{GS}	Gate-Source Voltage	V_{DS} = 15 V, I_D = 100 μA **5457**		- 2.5		V
		V_{DS} = 15 V, I_D = 200 μA **5458**		- 3.5		V
		V_{DS} = 15 V, I_D = 400 μA **5459**		- 4.5		V
ON CHARACTERISTICS						
I_{DSS}	Zero-Gate Voltage Drain Current*	V_{DS} = 15 V, V_{GS} = 0 **5457**	1.0	3.0	5.0	mA
		5458	2.0	6.0	9.0	mA
		5459	4.0	9.0	16	mA
SMALL SIGNAL CHARACTERISTICS						
g_{fs}	Forward Transfer Conductance*	V_{DS} = 15 V, V_{GS} = 0, f = 1.0 kHz				
		5457	1000		5000	μmhos
		5458	1500		5500	μmhos
		5459	2000		6000	μmhos
g_{os}	Output Conductance*	V_{DS} = 15 V, V_{GS} = 0, f = 1.0 kHz		10	50	μmhos
C_{iss}	Input Capacitance	V_{DS} = 15 V, V_{GS} = 0, f = 1.0 MHz		4.5	7.0	pF
C_{rss}	Reverse Transfer Capacitance	V_{DS} = 15 V, V_{GS} = 0, f = 1.0 MHz		1.5	3.0	pF
NF	Noise Figure	V_{DS} = 15 V, V_{GS} = 0, f = 1.0 kHz, R_G = 1.0 megohm, BW = 1.0 Hz			3.0	dB

*Pulse Test: Pulse Width ≤ 300 ms, Duty Cycle ≤ 2%

JFET Forward Transconductance

The forward **transconductance** (transfer conductance), g_m, is the change in drain current (ΔI_D) for a given change in gate-to-source voltage (ΔV_{GS}) with the drain-to-source voltage constant. It is expressed as a ratio and has the unit of siemens (S).

$$g_m = \frac{\Delta I_D}{\Delta V_{GS}}$$

Other common designations for this parameter are g_{fs} and y_{fs} (forward transfer admittance). As you will see in Chapter 9, g_m is important in FET amplifiers as a major factor in determining the voltage gain.

Because the transfer characteristic curve for a JFET is nonlinear, g_m varies in value depending on the location on the curve as set by V_{GS}. The value for g_m is greater near the top of the curve (near $V_{GS} = 0$) than it is near the bottom (near $V_{GS(off)}$), as illustrated in Figure 8–15.

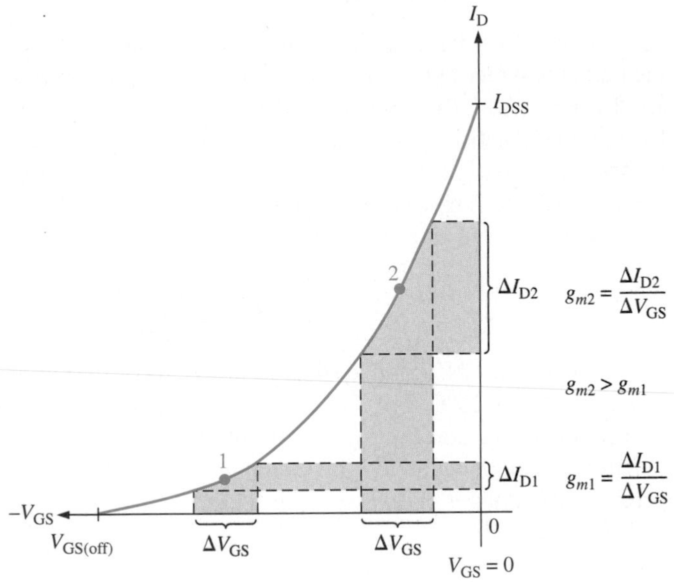

FIGURE 8–15

g_m varies depending on the bias point (V_{GS}).

A datasheet normally gives the value of g_m measured at $V_{GS} = 0$ V (g_{m0}). For example, the datasheet for the 2N5457 JFET specifies a minimum g_{m0} (g_{fs}) of 1000 μmhos (the mho is the same unit as the siemens (S)) with $V_{DS} = 15$ V.

Given g_{m0}, you can calculate an approximate value for g_m at any point on the transfer characteristic curve using the following formula:

$$g_m = g_{m0}\left(1 - \frac{V_{GS}}{V_{GS(off)}}\right)$$

Equation 8–2

When a value of g_{m0} is not available, you can calculate it using values of I_{DSS} and $V_{GS(off)}$. The vertical lines indicate an absolute value (no sign).

$$g_{m0} = \frac{2I_{DSS}}{|V_{GS(off)}|}$$

Equation 8–3

EXAMPLE 8–4

The following information is included on the datasheet in Figure 8–14 for a 2N5457 JFET: typically, $I_{DSS} = 3.0$ mA, $V_{GS(off)} = -6$ V maximum, and $g_{fs(max)} = 5000\ \mu$S. Using these values, determine the forward transconductance for $V_{GS} = -4$ V, and find I_D at this point.

Solution $g_{m0} = g_{fs} = 5000\ \mu S$. Use Equation 8–2 to calculate g_m.

$$g_m = g_{m0}\left(1 - \frac{V_{GS}}{V_{GS(off)}}\right) = (5000\ \mu S)\left(1 - \frac{-4\ V}{-6\ V}\right) = \mathbf{1667\ \mu S}$$

Next, use Equation 8–1 to calculate I_D at $V_{GS} = -4$ V.

$$I_D \cong I_{DSS}\left(1 - \frac{V_{GS}}{V_{GS(off)}}\right)^2 = (3.0\ mA)\left(1 - \frac{-4\ V}{-6\ V}\right)^2 = \mathbf{333\ \mu A}$$

Related Problem A given JFET has the following characteristics: $I_{DSS} = 12$ mA, $V_{GS(off)} = -5$ V, and $g_{m0} = g_{fs} = 3000\ \mu S$. Find g_m and I_D when $V_{GS} = -2$ V.

Input Resistance and Capacitance

As you know, a JFET operates with its gate-source junction reverse-biased, which makes the input resistance at the gate very high. This high input resistance is one advantage of the JFET over the BJT. (Recall that a bipolar junction transistor operates with a forward-biased base-emitter junction.) JFET datasheets often specify the input resistance by giving a value for the gate reverse current, I_{GSS}, at a certain gate-to-source voltage. The input resistance can then be determined using the following equation, where the vertical lines indicate an absolute value (no sign):

$$R_{IN} = \left|\frac{V_{GS}}{I_{GSS}}\right|$$

For example, the 2N5457 datasheet in Figure 8–14 lists a maximum I_{GSS} of -1.0 nA for $V_{GS} = -15$ V at 25°C. I_{GSS} increases with temperature, so the input resistance decreases.

The input capacitance, C_{iss}, is a result of the JFET operating with a reverse-biased *pn* junction. Recall that a reverse-biased *pn* junction acts as a capacitor whose capacitance depends on the amount of reverse voltage. For example, the 2N5457 has a maximum C_{iss} of 7 pF for $V_{GS} = 0$.

EXAMPLE 8–5

A certain JFET has an I_{GSS} of -2 nA for $V_{GS} = -20$ V. Determine the input resistance.

Solution

$$R_{IN} = \left|\frac{V_{GS}}{I_{GSS}}\right| = \frac{20\ V}{2\ nA} = \mathbf{10{,}000\ M\Omega}$$

Related Problem Determine the input resistance for the 2N5458 from the datasheet in Figure 8–14.

AC Drain-to-Source Resistance

You learned from the drain characteristic curve that, above pinch-off, the drain current is relatively constant over a range of drain-to-source voltages. Therefore, a large change in V_{DS} produces only a very small change in I_D. The ratio of these changes is the ac drain-to-source resistance of the device, r'_{ds}.

$$r'_{ds} = \frac{\Delta V_{DS}}{\Delta I_D}$$

Datasheets often specify this parameter in terms of the output conductance, g_{os}, or output admittance, y_{os}, for $V_{GS} = 0$ V.

SECTION 8–2 CHECKUP

1. The drain-to-source voltage at the pinch-off point of a particular JFET is 7 V. If the gate-to-source voltage is zero, what is V_P?
2. The V_{GS} of a certain n-channel JFET is increased negatively. Does the drain current increase or decrease?
3. What value must V_{GS} have to produce cutoff in a p-channel JFET with a $V_p = -3$ V?

8–3 JFET BIASING

Using some of the JFET parameters discussed previously, you will now see how to dc-bias JFETs. Just as with the BJT, the purpose of biasing is to select the proper dc gate-to-source voltage to establish a desired value of drain current and, thus, a proper Q-point. Three types of bias are self-bias, voltage-divider bias, and current-source bias.

After completing this section, you should be able to

- **Discuss and analyze JFET bias circuits**
    - Describe self-bias
    - Analyze a self-biased JFET circuit
    - Set the self-biased Q-point
    - Analyze a JFET circuit with voltage-divider bias
    - Use transfer characteristic curves to analyze JFET bias circuits
    - Discuss Q-point stability
    - Describe current-source bias

Self-Bias

Self-bias is the most common type of JFET bias. Recall that a JFET must be operated such that the gate-source junction is always reverse-biased. This condition requires a negative V_{GS} for an n-channel JFET and a positive V_{GS} for a p-channel JFET. This can be achieved using the self-bias arrangements shown in Figure 8–16. The gate resistor, R_G, does not affect the bias because it has essentially no voltage drop across it; and therefore the gate remains at 0 V. R_G is necessary only to force the gate to be at 0 V and to isolate an ac signal from ground in amplifier applications, as you will see later.

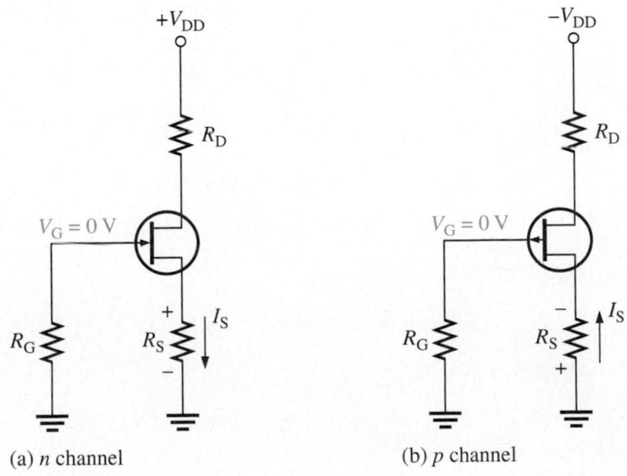

FIGURE 8–16

Self-biased JFETs ($I_S = I_D$ in all FETs).

For the n-channel JFET in Figure 8–16(a), I_S produces a voltage drop across R_S and makes the source positive with respect to ground. Since $I_S = I_D$ and $V_G = 0$, then $V_S = I_D R_S$. The gate-to-source voltage is

$$V_{GS} = V_G - V_S = 0 - I_D R_S = -I_D R_S$$

Thus,

$$V_{GS} = -I_D R_S$$

For the p-channel JFET shown in Figure 8–16(b), the current through R_S produces a negative voltage at the source, making the gate positive with respect to the source. Therefore, since $I_S = I_D$,

$$V_{GS} = +I_D R_S$$

In the following Example, the n-channel JFET in Figure 8–16(a) is used for illustration. Keep in mind that analysis of the p-channel JFET is the same except for opposite-polarity voltages. The drain voltage with respect to ground is determined as follows:

$$V_D = V_{DD} - I_D R_D$$

Since $V_S = I_D R_S$, the drain-to-source voltage is

$$V_{DS} = V_D - V_S = V_{DD} - I_D(R_D + R_S)$$

EXAMPLE 8–6

Find V_{DS} and V_{GS} in Figure 8–17. For the particular JFET in this circuit, the parameter values such as g_m, $V_{GS(off)}$, and I_{DSS} are such that a drain current (I_D) of approximately 5 mA is produced. Another JFET, even of the same type, may not produce the same results when connected in this circuit due to the variations in parameter values.

FIGURE 8–17

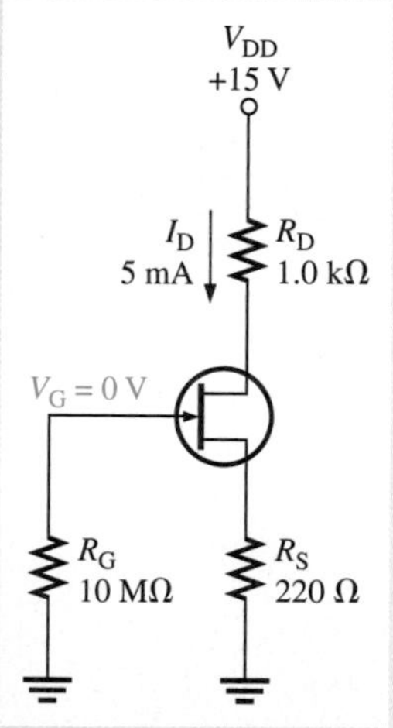

Solution

$$V_S = I_D R_S = (5\text{ mA})(220\ \Omega) = 1.1\text{ V}$$
$$V_D = V_{DD} - I_D R_D = 15\text{ V} - (5\text{ mA})(1.0\text{ k}\Omega) = 15\text{ V} - 5\text{ V} = 10\text{ V}$$

Therefore,

$$V_{DS} = V_D - V_S = 10\text{ V} - 1.1\text{ V} = \mathbf{8.9\ V}$$

Since $V_G = 0$ V,

$$V_{GS} = V_G - V_S = 0\text{ V} - 1.1\text{ V} = \mathbf{-1.1\ V}$$

Related Problem Determine V_{DS} and V_{GS} in Figure 8–17 when $I_D = 8$ mA. Assume that $R_D = 860\ \Omega$, $R_S = 390\ \Omega$, and $V_{DD} = 12$ V.

Open the Multisim file E08-06 in the Examples folder on your CD-ROM. Measure I_D, V_{GS}, and V_{DS} and compare to the calculated values from the Related Problem.

Setting the Q-Point of a Self-Biased JFET

The basic approach to establishing a JFET bias point is to determine I_D for a desired value of V_{GS} or vice versa. Then calculate the required value of R_S using the following relationship. The vertical lines indicate an absolute value.

$$R_S = \left|\frac{V_{GS}}{I_D}\right|$$

For a desired value of V_{GS}, I_D can be determined in either of two ways: from the transfer characteristic curve for the particular JFET or, more practically, from Equation 8–1 using I_{DSS} and $V_{GS(off)}$ from the JFET datasheet. The next two examples illustrate these procedures.

EXAMPLE 8–7

Determine the value of R_S required to self-bias an n-channel JFET that has the transfer characteristic curve shown in Figure 8–18 at $V_{GS} = -5$ V.

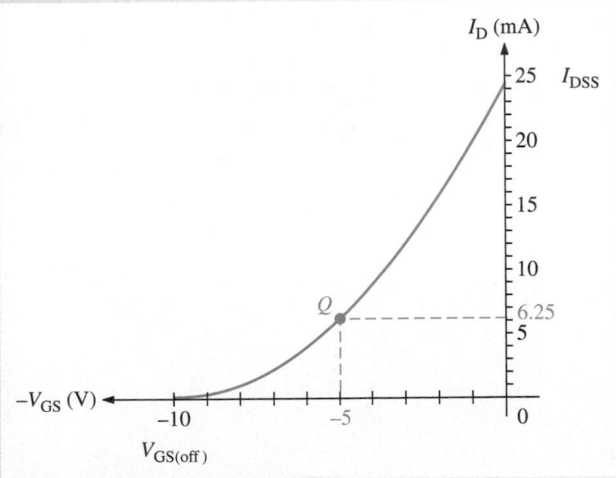

FIGURE 8–18

Solution From the graph, $I_D = 6.25$ mA when $V_{GS} = -5$ V. Calculate R_S.

$$R_S = \left|\frac{V_{GS}}{I_D}\right| = \frac{5\text{ V}}{6.25\text{ mA}} = \mathbf{800\ \Omega}$$

Related Problem Find R_S for $V_{GS} = -3$ V.

EXAMPLE 8–8

Determine the value of R_S required to self-bias a *p*-channel JFET with datasheet values of $I_{DSS} = 25$ mA and $V_{GS(off)} = 15$ V. V_{GS} is to be 5 V.

Solution Use Equation 8–1 to calculate I_D.

$$I_D \cong I_{DSS}\left(1 - \frac{V_{GS}}{V_{GS(off)}}\right)^2 = (25\text{ mA})\left(1 - \frac{5\text{ V}}{15\text{ V}}\right)^2$$

$$= (25\text{ mA})(1 - 0.333)^2 = 11.1\text{ mA}$$

Now, determine R_S.

$$R_S = \left|\frac{V_{GS}}{I_D}\right| = \frac{5\text{ V}}{11.1\text{ mA}} = \mathbf{450\ \Omega}$$

Related Problem Find the value of R_S required to self-bias a *p*-channel JFET with $I_{DSS} = 18$ mA and $V_{GS(off)} = 8$ V. $V_{GS} = 4$ V.

Midpoint Bias It is usually desirable to bias a JFET near the midpoint of its transfer characteristic curve where $I_D = I_{DSS}/2$. Under signal conditions, midpoint bias allows the maximum amount of drain current swing between I_{DSS} and 0. Using Equation 8–1, it is shown in Appendix B that I_D is approximately one-half of I_{DSS} when $V_{GS} = V_{GS(off)}/3.4$.

$$I_D \cong I_{DSS}\left(1 - \frac{V_{GS}}{V_{GS(off)}}\right)^2 = I_{DSS}\left(1 - \frac{V_{GS(off)}/3.4}{V_{GS(off)}}\right)^2 = 0.5I_{DSS}$$

So, by selecting $V_{GS} = V_{GS(off)}/3.4$, you should get a midpoint bias in terms of I_D.

To set the drain voltage at midpoint ($V_D = V_{DD}/2$), select a value of R_D to produce the desired voltage drop. Choose R_G arbitrarily large to prevent loading on the driving stage in a cascaded amplifier arrangement. Example 8–9 illustrates these concepts.

EXAMPLE 8–9

Looking at the datasheet in Figure 8–14, select resistor values for R_D and R_S in Figure 8–19 to set up an approximate midpoint bias. Use minimum datasheet values when given; otherwise, V_D should be approximately 6 V (one-half of V_{DD}).

FIGURE 8–19

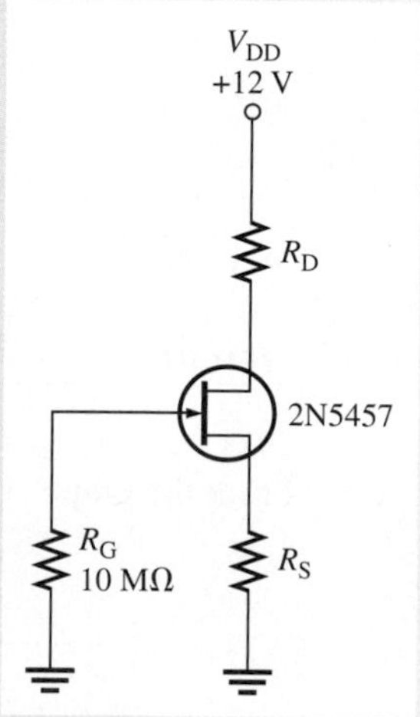

Solution For midpoint bias,

$$I_D \cong \frac{I_{DSS}}{2} = \frac{1.0\text{ mA}}{2} = 0.5\text{ mA}$$

and

$$V_{GS} \cong \frac{V_{GS(off)}}{3.4} = \frac{-0.5\text{ V}}{3.4} = -147\text{ mV}$$

Then

$$R_S = \left|\frac{V_{GS}}{I_D}\right| = \frac{147\text{ mV}}{0.5\text{ mA}} = \mathbf{294\ \Omega}$$

$$V_D = V_{DD} - I_D R_D$$

$$I_D R_D = V_{DD} - V_D$$

$$R_D = \frac{V_{DD} - V_D}{I_D} = \frac{12\text{ V} - 6\text{ V}}{0.5\text{ mA}} = \mathbf{12\ k\Omega}$$

Related Problem Select resistor values in Figure 8–19 to set up an approximate midpoint bias using a 2N5459.

Open the Multisim file E08-09 in the Examples folder on your CD-ROM. The circuit has the calculated values for R_D and R_S from the Related Problem. Verify that an approximate midpoint bias is established by measuring V_D and I_D.

Graphical Analysis of a Self-Biased JFET

You can use the transfer characteristic curve of a JFET and certain parameters to determine the Q-point (I_D and V_{GS}) of a self-biased circuit. A circuit is shown in Figure 8–20(a), and a transfer characteristic curve is shown in Figure 8–20(b). If a curve is not available from a datasheet, you can plot it from Equation 8–1 using datasheet values for I_{DSS} and $V_{GS(off)}$.

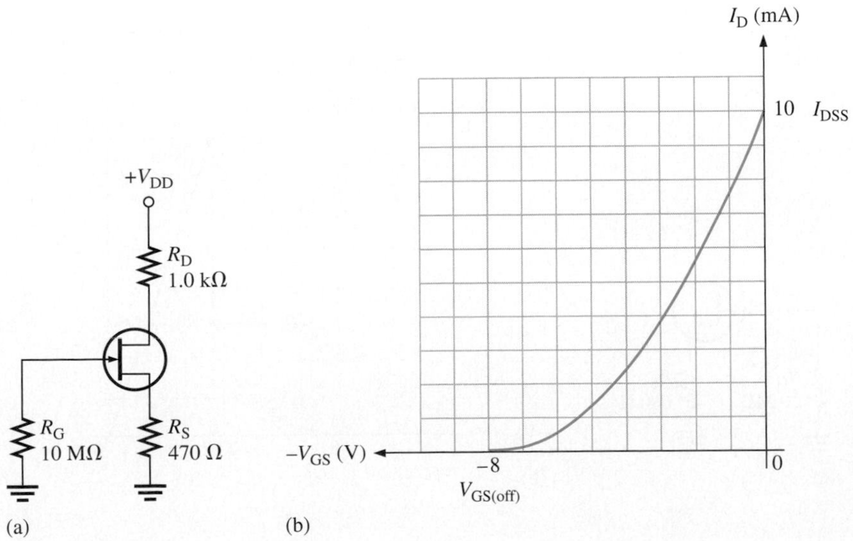

FIGURE 8–20

A self-biased JFET and its transfer characteristic curve.

To determine the Q-point of the circuit in Figure 8–20(a), a self-bias dc load line is established on the graph in part (b) as follows. First, calculate V_{GS} when I_D is zero.

$$V_{GS} = -I_D R_S = (0)(470\ \Omega) = 0\ \text{V}$$

This establishes a point at the origin on the graph ($I_D = 0$, $V_{GS} = 0$). Next, calculate V_{GS} when $I_D = I_{DSS}$. From the curve in Figure 8–20(b), $I_{DSS} = 10$ mA.

$$V_{GS} = -I_D R_S = -(10\ \text{mA})(470\ \Omega) = -4.7\ \text{V}$$

This establishes a second point on the graph ($I_D = 10$ mA, $V_{GS} = -4.7$ V). Now, with two points, the load line can be drawn on the transfer characteristic curve as shown in Figure 8–21. The point where the line intersects the transfer characteristic curve is the Q-point of the circuit as shown.

FIGURE 8–21

The intersection of the self-bias dc load line and the transfer characteristic curve is the Q-point.

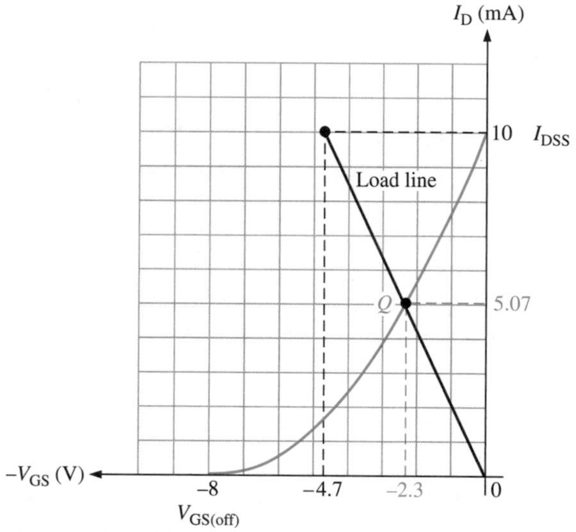

EXAMPLE 8–10

Determine the Q-point for the JFET circuit in Figure 8–22(a). The transfer characteristic curve is given in Figure 8–22(b).

FIGURE 8–22

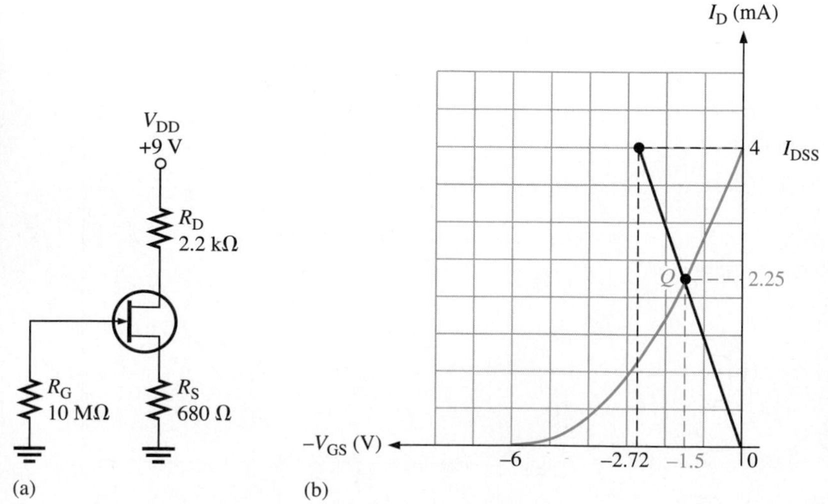

Solution For $I_D = 0$,

$$V_{GS} = -I_D R_S = (0)(680\ \Omega) = 0\text{ V}$$

This gives a point at the origin. From the curve, $I_{DSS} = 4$ mA; so $I_D = I_{DSS} = 4$ mA.

$$V_{GS} = -I_D R_S = -(4\text{ mA})(680\ \Omega) = -2.72\text{ V}$$

This gives a second point at 4 mA and −2.72 V. A line is now drawn between the two points, and the values of I_D and V_{GS} at the intersection of the line and the curve are taken from the graph, as illustrated in Figure 8–22(b). The Q-point values from the graph are

$$I_D = \mathbf{2.25\ mA}$$
$$V_{GS} = \mathbf{-1.5\ V}$$

Related Problem If R_S is increased to 1.0 kΩ in Figure 8–22(a), what is the new Q-point?

For increased Q-point stability, the value of R_S in the self-bias circuit is increased and connected to a negative supply voltage. This is sometimes called *dual-supply bias.*

Voltage-Divider Bias

An *n*-channel JFET with voltage-divider bias is shown in Figure 8–23. The voltage at the source of the JFET must be more positive than the voltage at the gate in order to keep the gate-source junction reverse-biased.

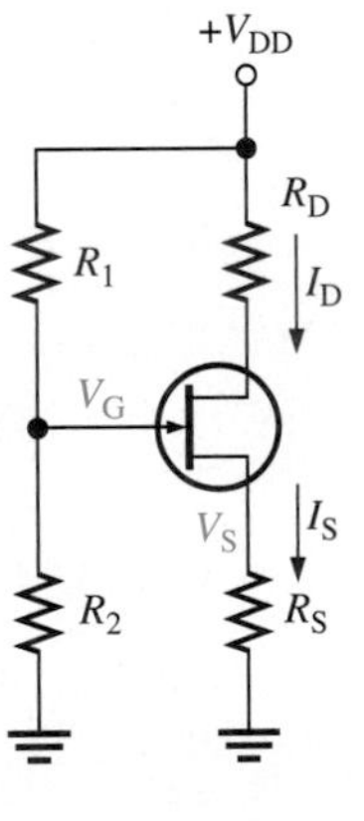

FIGURE 8–23

An *n*-channel JFET with voltage-divider bias ($I_S = I_D$).

The source voltage is

$$V_S = I_D R_S$$

The gate voltage is set by resistors R_1 and R_2 as expressed by the following equation using the voltage-divider formula:

$$V_G = \left(\frac{R_2}{R_1 + R_2}\right)V_{DD}$$

The gate-to-source voltage is

$$V_{GS} = V_G - V_S$$

and the source voltage is

$$V_S = V_G - V_{GS}$$

The drain current can be expressed as

$$I_D = \frac{V_S}{R_S}$$

Substituting for V_S,

$$I_D = \frac{V_G - V_{GS}}{R_S}$$

EXAMPLE 8–11

Determine I_D and V_{GS} for the JFET with voltage-divider bias in Figure 8–24, given that for this particular JFET the parameter values are such that $V_D \cong 7\text{ V}$.

FIGURE 8–24

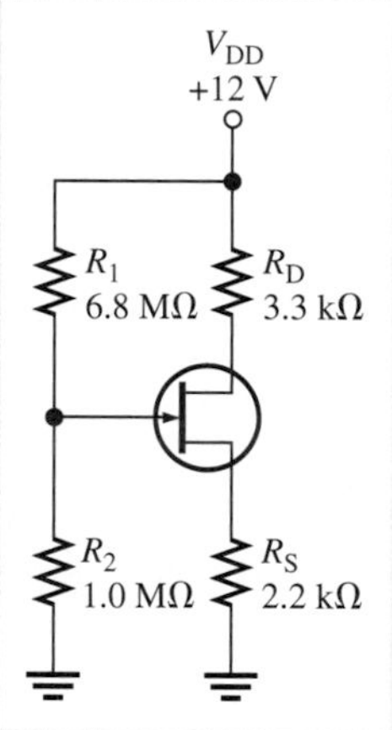

Solution

$$I_D = \frac{V_{DD} - V_D}{R_D} = \frac{12\text{ V} - 7\text{ V}}{3.3\text{ k}\Omega} = \frac{5\text{ V}}{3.3\text{ k}\Omega} = \mathbf{1.52\text{ mA}}$$

Calculate the gate-to-source voltage as follows:

$$V_S = I_D R_S = (1.52\text{ mA})(2.2\text{ k}\Omega) = 3.34\text{ V}$$

$$V_G = \left(\frac{R_2}{R_1 + R_2}\right)V_{DD} = \left(\frac{1.0\text{ M}\Omega}{7.8\text{ M}\Omega}\right)12\text{ V} = 1.54\text{ V}$$

$$V_{GS} = V_G - V_S = 1.54\text{ V} - 3.34\text{ V} = \mathbf{-1.8\text{ V}}$$

If V_D had not been given in this example, the Q-point values could not have been found without the transfer characteristic curve.

Related Problem Given that $V_D = 6\text{ V}$ when another JFET is inserted in the circuit of Figure 8–24, determine the Q-point.

Graphical Analysis of a JFET with Voltage-Divider Bias

An approach similar to the one used for self-bias can be used with voltage-divider bias to graphically determine the Q-point of a circuit on the transfer characteristic curve.

In a JFET with voltage-divider bias when $I_D = 0$, V_{GS} is not zero, as in the self-biased case, because the voltage divider produces a voltage at the gate independent of the drain current. The voltage-divider dc load line is determined as follows.

For $I_D = 0$,

$$V_S = I_D R_S = (0)R_S = 0\text{ V}$$

$$V_{GS} = V_G - V_S = V_G - 0\text{ V} = V_G$$

Therefore, one point on the line is at $I_D = 0$ and $V_{GS} = V_G$.

For $V_{GS} = 0$,

$$I_D = \frac{V_G - V_{GS}}{R_S} = \frac{V_G}{R_S}$$

A second point on the line is at $I_D = V_G/R_S$ and $V_{GS} = 0$. The generalized dc load line is shown in Figure 8–25. The point at which the load line intersects the transfer characteristic curve is the Q-point.

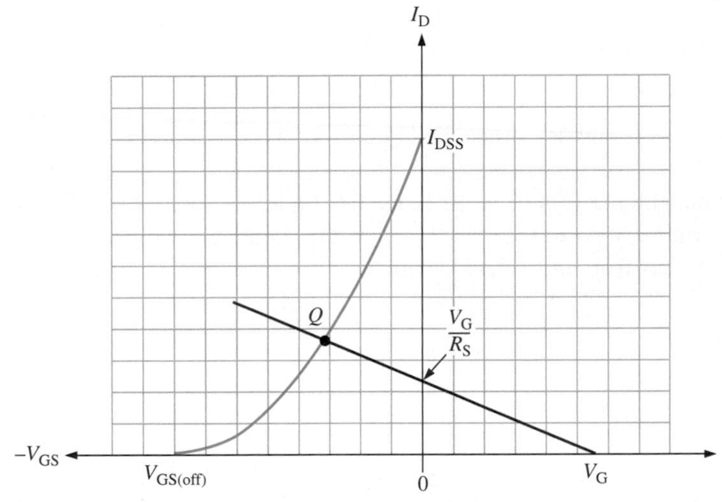

▲ FIGURE 8–25

Generalized dc load line (red) for a JFET with voltage-divider bias.

EXAMPLE 8–12

Determine the approximate Q-point for the JFET with voltage-divider bias in Figure 8–26(a), given that this particular device has a transfer characteristic curve as shown in Figure 8–26(b).

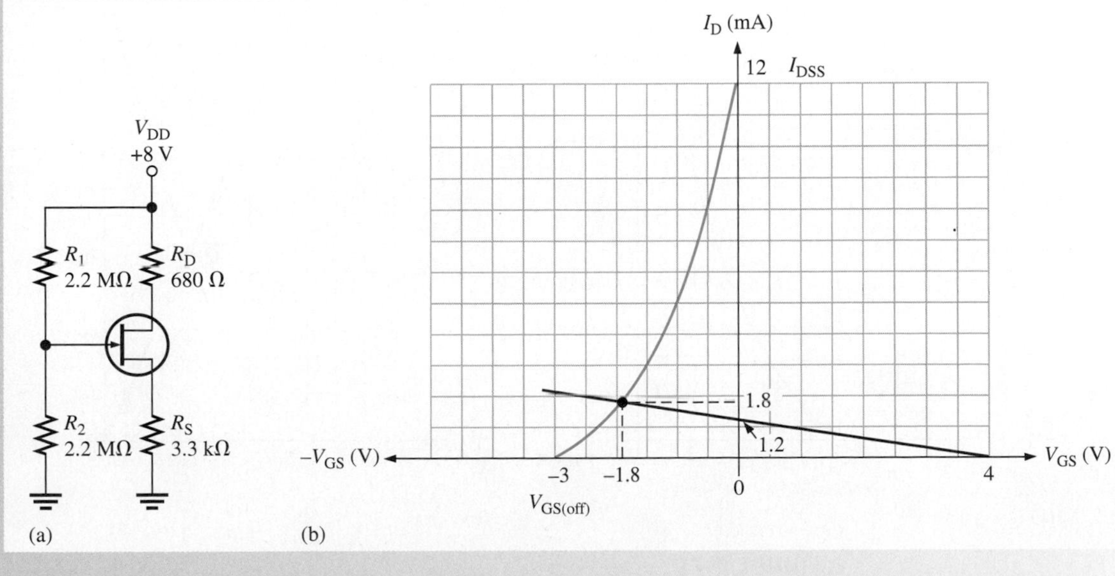

▲ FIGURE 8–26

Solution First, establish the two points for the load line. For $I_D = 0$,

$$V_{GS} = V_G = \left(\frac{R_2}{R_1 + R_2}\right)V_{DD} = \left(\frac{2.2\ \text{M}\Omega}{4.4\ \text{M}\Omega}\right)8\ \text{V} = 4\ \text{V}$$

The first point is at $I_D = 0$ and $V_{GS} = 4$ V. For $V_{GS} = 0$,

$$I_D = \frac{V_G - V_{GS}}{R_S} = \frac{V_G}{R_S} = \frac{4\ \text{V}}{3.3\ \text{k}\Omega} = 1.2\ \text{mA}$$

The second point is at $I_D = 1.2$ mA and $V_{GS} = 0$.

The load line is drawn in Figure 8–26(b), and the approximate Q-point values of $I_D \cong$ **1.8 mA** and $V_{GS} \cong$ **−1.8 V** are picked off the graph, as indicated.

Related Problem Change R_S to 4.7 kΩ and determine the Q-point for the circuit in Figure 8–26(a).

Open the Multisim file E08-12 in the Examples folder on your CD-ROM. Measure the Q-point values of I_D and V_{GS} and see how they compare to the graphically determined values from the Related Problem.

Q-Point Stability

Unfortunately, the transfer characteristic of a JFET can differ considerably from one device to another of the same type. If, for example, a 2N5459 JFET is replaced in a given bias circuit with another 2N5459, the transfer characteristic curve can vary greatly, as illustrated in Figure 8–27(a). In this case, the maximum I_{DSS} is 16 mA and the minimum I_{DSS} is 4 mA. Likewise, the maximum $V_{GS(off)}$ is −8 V and the minimum $V_{GS(off)}$ is −2 V. This means that if you have a selection of 2N5459s and you randomly pick one out, it can have values anywhere within these ranges.

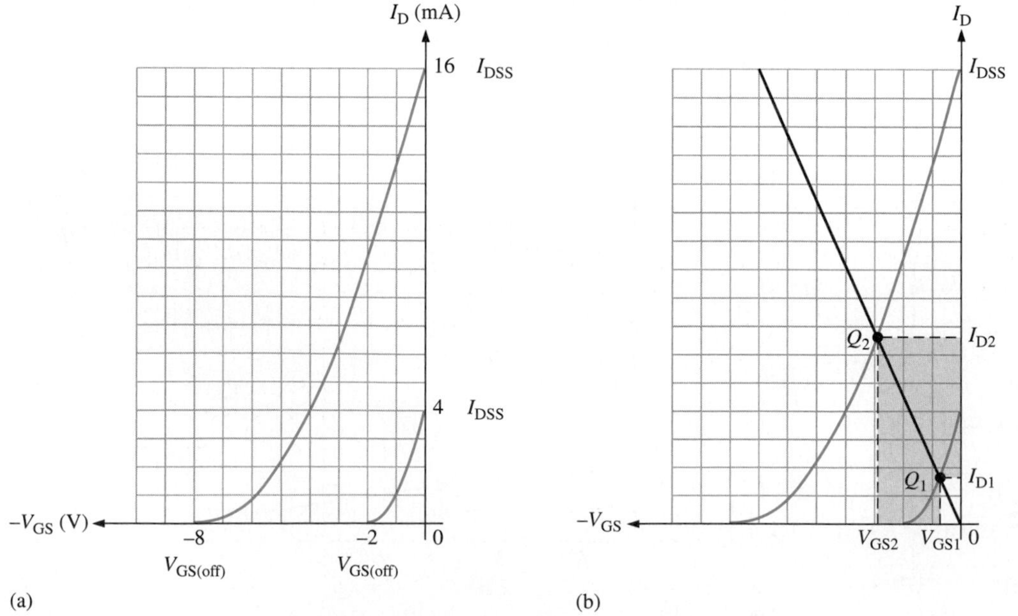

▲ **FIGURE 8–27**

Variation in the transfer characteristic of 2N5459 JFETs and the effect on the Q-point.

If a self-bias dc load line is drawn as illustrated in Figure 8–27(b), the same circuit using a 2N5459 can have a Q-point anywhere along the line from Q_1, the minimum bias point, to Q_2, the maximum bias point. Accordingly, the drain current can be any value between I_{D1} and I_{D2}, as shown by the shaded area. This means that the dc voltage at the drain can have a range of values depending on I_D. Also, the gate-to-source voltage can be any value between V_{GS1} and V_{GS2}, as indicated.

Figure 8–28 illustrates Q-point stability for a self-biased JFET and for a JFET with voltage-divider bias. With voltage-divider bias, the dependency of I_D on the range of Q-points is reduced because the slope of the load line is less than for self-bias for a given JFET. Although V_{GS} varies quite a bit for both self-bias and voltage-divider bias, I_D is much more stable with voltage-divider bias.

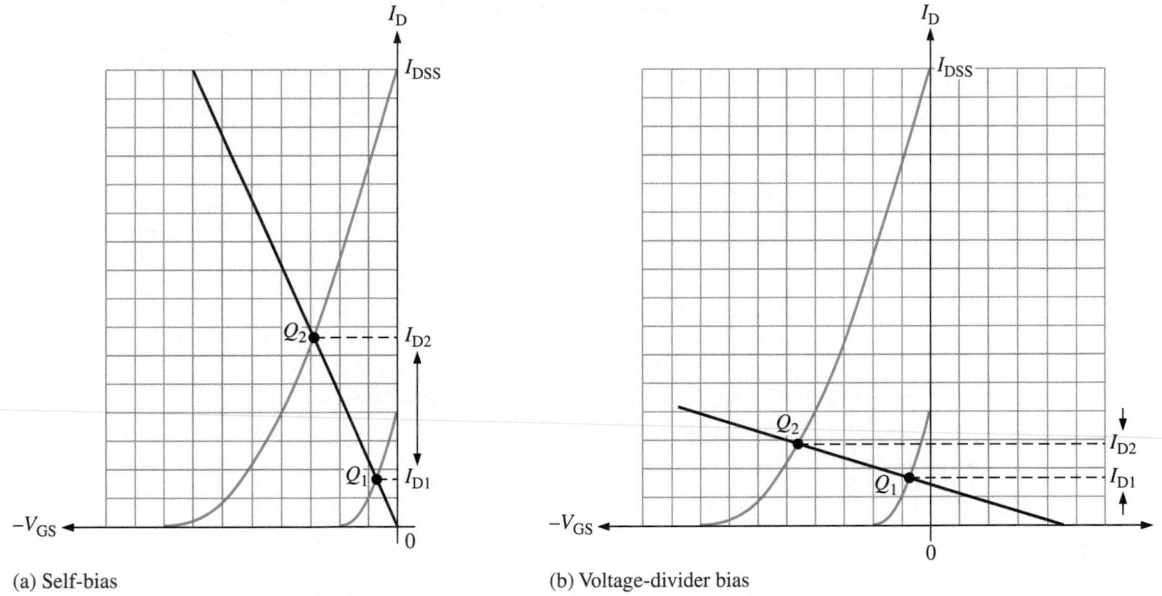

▲ **FIGURE 8–28**

The change in I_D between the minimum and the maximum Q-points is much less for a JFET with voltage-divider bias than for a self-biased JFET.

Current-Source Bias

Current-source bias is a method for increasing the Q-point stability of a self-biased JFET by making the drain current essentially independent of V_{GS}. This is accomplished by using a constant-current source in series with the JFET source, as shown in Figure 8–29(a). In this circuit, a BJT acts as the constant-current source because its emitter current is essentially constant if $V_{EE} >> V_{BE}$. A FET can also be used as a constant-current source.

$$I_E = \frac{V_{EE} - V_{BE}}{R_E} \cong \frac{V_{EE}}{R_E}$$

Since $I_E \cong I_D$,

$$I_D \cong \frac{V_{EE}}{R_E}$$

As you can see in Figure 8–29(b), I_D remains constant for any transfer characteristic curve, as indicated by the horizontal load line.

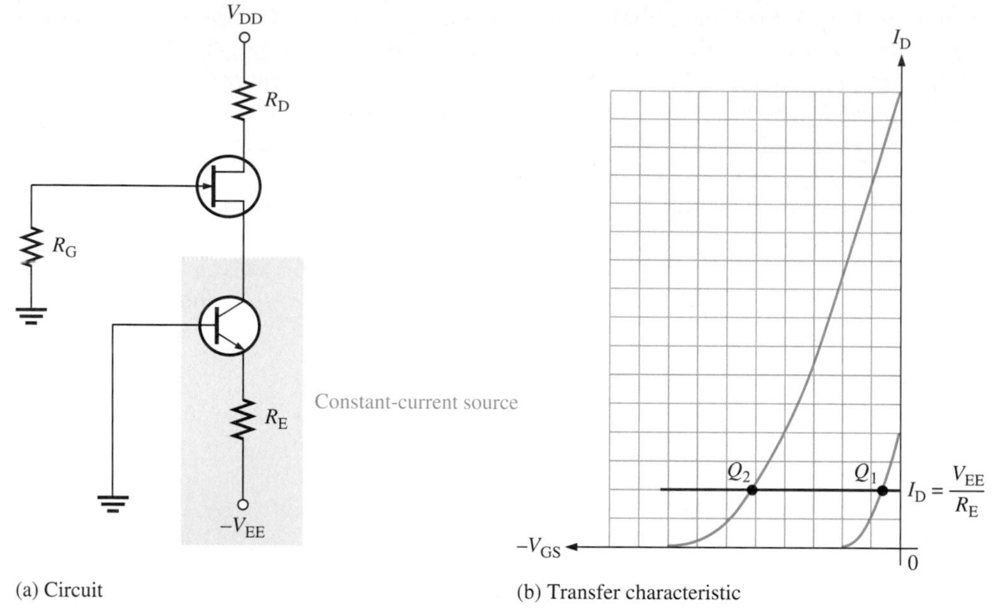

▲ **FIGURE 8–29**

Current-source bias.

EXAMPLE 8–13

A current-source bias circuit like Figure 8–29 has the following values: $V_{DD} = 9\text{ V}$, $V_{EE} = -6\text{ V}$, and $R_G = 10\text{ M}\Omega$. To produce a 10 mA drain current and a 5 V drain voltage, determine the values of R_E and R_D.

Solution

$$R_E = \frac{V_{EE}}{I_D} = \frac{6\text{ V}}{10\text{ mA}} = \mathbf{600\ \Omega}$$

$$R_D = \frac{V_{DD} - V_D}{I_D} = \frac{9\text{ V} - 5\text{ V}}{10\text{ mA}} = \mathbf{400\ \Omega}$$

Related Problem If V_{DD} is increased to 12 V, how much does I_D change?

SECTION 8–3 CHECKUP

1. Should a *p*-channel JFET have a positive or a negative V_{GS}?
2. In a certain self-biased *n*-channel JFET circuit, $I_D = 8\text{ mA}$ and $R_S = 1.0\text{ k}\Omega$. Determine V_{GS}.
3. An *n*-channel JFET with voltage-divider bias has a gate voltage of 3 V and a source voltage of 5 V. Calculate V_{GS}.

8–4 THE OHMIC REGION

The **ohmic region** is the portion of the FET characteristic curves in which Ohm's law can be applied. When properly biased in the ohmic region, a JFET exhibits the properties of a variable resistance, where the value of resistance is controlled by V_{GS}.

After completing this section, you should be able to

- **Discuss the ohmic region**
    - Explain how the slope of a characteristic curve in the ohmic region affects JFET resistance
    - Describe how a JFET can be used as a variable resistor
    - Discuss the operation of a JFET when it is biased at the origin

The ohmic region extends from the origin of the characteristic curves to the break point (where the active region begins) of the $V_{GS} = 0$ curve in a roughly parabolic shape, as shown on a typical set of curves in Figure 8–30. The characteristic curves in this region have a relatively constant slope for small values of I_D. The slope of the characteristic curve in the ohmic region is the dc drain-to-source conductance of the JFET.

$$\text{Slope} = G_{DS} \cong \frac{I_D}{V_{DS}}$$

Recall that resistance is the reciprocal of the conductance. Thus, the dc drain-to-source resistance is given by

$$R_{DS} = \frac{1}{G_{DS}} \cong \frac{V_{DS}}{I_D}$$

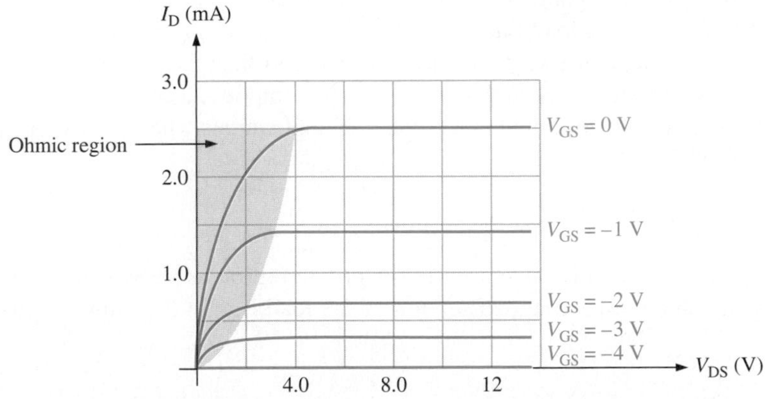

FIGURE 8–30

The ohmic region is the shaded area. The characteristic curves are straight lines with a slope of I_D/V_{DS} for small values of I_D.

The JFET as a Variable Resistance A JFET can be biased in either the active region or the ohmic region. JFETs are often biased in the ohmic region for use as a voltage-controlled variable resistor. The control voltage is V_{GS}, and it determines the resistance by varying the Q-point. To bias a JFET in the ohmic region, the dc load line must intersect the characteristic curve in the ohmic region, as illustrated in Figure 8–31. To do this in a way that allows V_{GS} to control R_{DS}, the dc saturation current is set for a value much less than I_{DSS} so that the load line intersects most of the characteristic curves in the ohmic region, as illustrated. In this case,

$$I_{D(sat)} = \frac{V_{DD}}{R_D} = \frac{12\text{ V}}{24\text{ k}\Omega} = 0.50\text{ mA}$$

Figure 8–31 shows the operating region expanded with three Q-points shown (Q_0,Q_1, and Q_2), depending on V_{GS}.

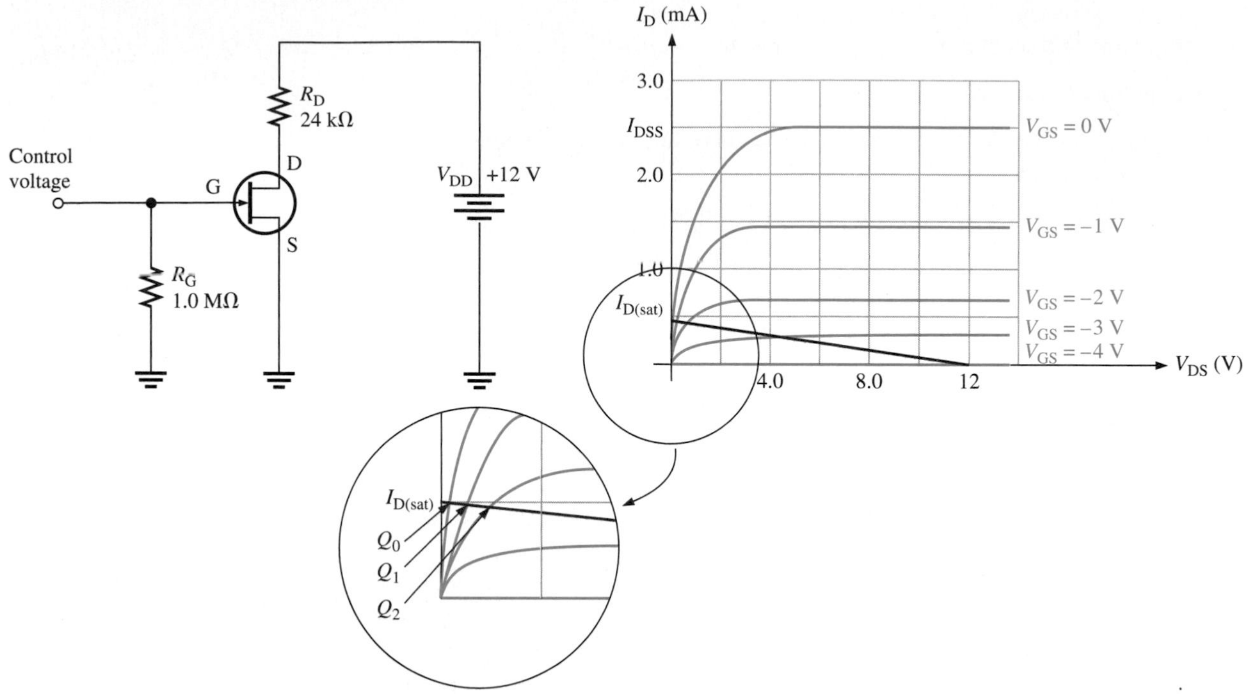

▲ FIGURE 8–31

The load line intersects the curves inside the ohmic region.

As you move along the load line in the ohmic region of Figure 8–31, the value of R_{DS} varies as the Q-point falls successively on curves with different slopes. The Q-point is moved along the load line by varying $V_{GS} = 0$ to $V_{GS} = -2$ V, in this case. As this happens, the slope of each successive curve is less than the previous one. A decrease in slope corresponds to less I_D and more V_{DS}, which implies an increase in R_{DS}. This change in resistance can be exploited in a number of applications where voltage control of a resistance is useful.

EXAMPLE 8–14

An *n*-channel JFET is biased in the ohmic region as shown in Figure 8–32. The graph shows an expanded section of the load line in the ohmic region. As V_{GS} is

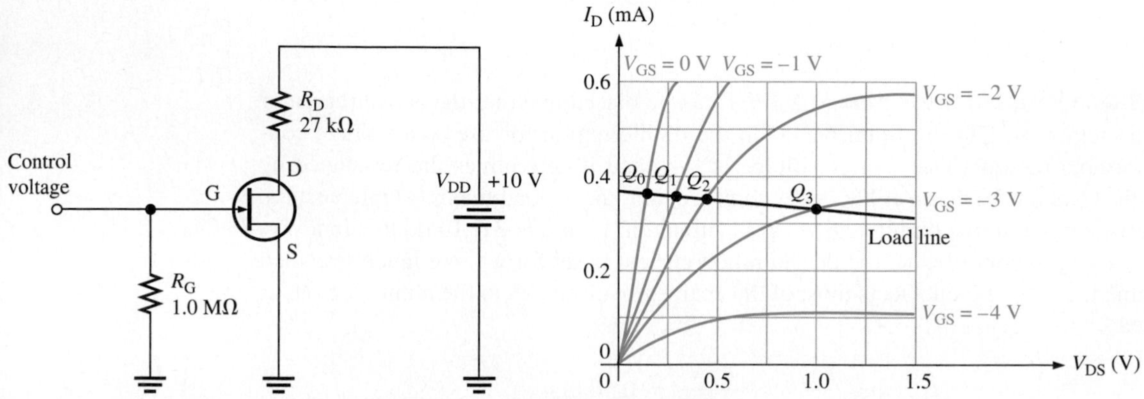

▲ FIGURE 8–32

varied from 0 V to −3 V as indicated, assume that the graph shows the following Q-point values:

$$Q_0: I_D = 0.360\text{ mA}, V_{DS} = 0.13\text{ V}$$
$$Q_1: I_D = 0.355\text{ mA}, V_{DS} = 0.27\text{ V}$$
$$Q_2: I_D = 0.350\text{ mA}, V_{DS} = 0.42\text{ V}$$
$$Q_3: I_D = 0.33\text{ mA}, V_{DS} = 0.97\text{ V}$$

Determine the range of R_{DS} as V_{GS} is varied from 0 V to −3 V.

Solution

$$Q_0: R_{DS} = \frac{V_{DS}}{I_D} = \frac{0.13\text{ V}}{0.360\text{ mA}} = 361\ \Omega$$
$$Q_1: R_{DS} = \frac{V_{DS}}{I_D} = \frac{0.27\text{ V}}{0.355\text{ mA}} = 760\ \Omega$$
$$Q_2: R_{DS} = \frac{V_{DS}}{I_D} = \frac{0.42\text{ V}}{0.27\text{ mA}} = 1.2\text{ k}\Omega$$
$$Q_3: R_{DS} = \frac{V_{DS}}{I_D} = \frac{0.6\text{ V}}{0.26\text{ mA}} = 2.9\text{ k}\Omega$$

When V_{GS} is varied from 0 V to −3 V, **R_{DS} changes from 361 Ω to 2.9 kΩ.**

Related Problem If $I_{D(sat)}$ is reduced, what happens to the range of R_{DS} values?

Q-point at the Origin In certain amplifiers, you may want to change the resistance seen by the ac signal without affecting the dc bias in order to control the gain. Sometimes you will see a JFET used as a variable resistance in a circuit where both I_D and V_{DS} are set at 0, which means that the Q-point is at the origin. A Q-point at the origin is achieved by using a capacitor in the drain circuit of the JFET. This makes the dc quantities $V_{DS} = 0$ V and $I_D = 0$ mA, so the only variables are V_{GS} and I_d, the ac drain current. At the origin you have the ac drain current controlled by V_{GS}. As you learned earlier, transconductance is defined as a change in drain current for a given change in gate-to-source voltage. So, the key factor when you bias at the origin is the transconductançe. Figure 8–33 shows the characteristic curve expanded at the origin. Notice that the ohmic region extends into the third quadrant.

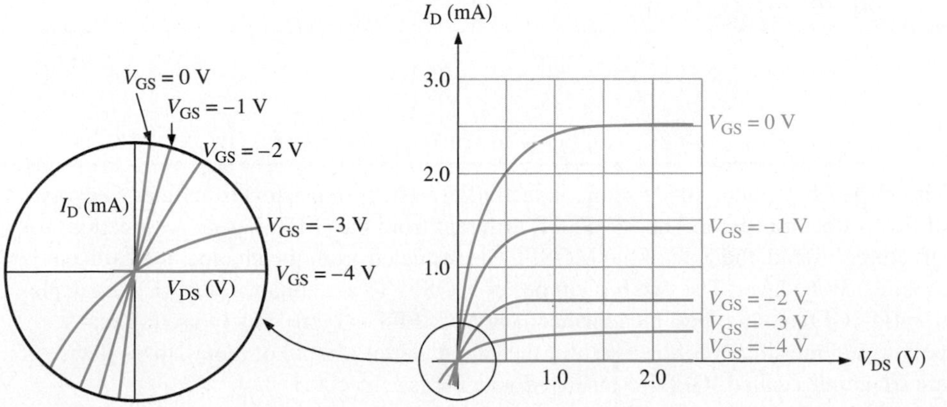

▲ FIGURE 8–33

At the origin, where $V_{DS} = 0$ V and $I_D = 0$ mA, the formula for transconductance, introduced earlier in this chapter, is

$$g_m = g_{m0}\left(1 - \frac{V_{GS}}{V_{GS(off)}}\right)$$

where g_m is transconductance and g_{m0} is transconductance for $V_{GS} = 0$ V. g_{m0} can be calculated from the following equation, which was also given earlier:

$$g_{m0} = \frac{2I_{DSS}}{|V_{GS(off)}|}$$

EXAMPLE 8–15

For the characteristic curve in Figure 8–33, calculate the ac drain-to-source resistance for a JFET biased at the origin if $V_{GS} = -2$ V. Assume $I_{DSS} = 2.5$ mA and $V_{GS(off)} = -4$ V.

Solution First, find the transconductance for $V_{GS} = 0$ V.

$$g_{m0} = \frac{2I_{DSS}}{|V_{GS(off)}|} = \frac{2(2.5\text{ mA})}{4.0\text{ V}} = 1.25\text{ mS}$$

Next, calculate g_m at $V_{DS} = -2$ V.

$$g_m = g_{m0}\left(1 - \frac{V_{GS}}{V_{GS(off)}}\right) = 1.25\text{ mS}\left(1 - \frac{-2\text{ V}}{-4\text{ V}}\right) = 0.625\text{ mS}$$

The ac drain-to-source resistance of the JFET is the reciprocal of the transconductance.

$$r'_{ds} = \frac{1}{g_m} = \frac{1}{0.625\text{ mS}} = 1.6\text{ k}\Omega$$

Related Problem What is the ac drain-to-source resistance if $V_{GS} = -1$ V?

SECTION 8–4 CHECKUP

1. For a certain Q-point in the ohmic region, $I_D = 0.3$ mA and $V_{DS} = 0.6$ V. What is the resistance of the JFET when it is biased at this Q-point?
2. How does the drain-to-source resistance change as V_{GS} becomes more negative?
3. For a JFET biased at the origin, $g_m = 0.850$ mS. Determine the corresponding ac resistance.

8–5 THE MOSFET

The **MOSFET** (metal oxide semiconductor field-effect transistor) is another category of field-effect transistor. The MOSFET, different from the JFET, has no *pn* junction structure; instead, the gate of the MOSFET is insulated from the channel by a silicon dioxide (SiO_2) layer. The two basic types of MOSFETs are enhancement (E) and depletion (D). Of the two types, the enhancement MOSFET is more widely used. Because polycrystalline silicon is now used for the gate material instead of metal, these devices are sometimes called IGFETs (insulated-gate FETs).

After completing this section, you should be able to

- **Explain the operation of MOSFETs**
  - Describe the structural difference between an *n*-channel and a *p*-channel enhancement MOSFET (E-MOSFET)
  - Identify the symbols for *n*-channel and *p*-channel E-MOSFETs
  - Explain the enhancement mode
  - Describe the structural difference between an *n*-channel and a *p*-channel depletion MOSFET (D-MOSFET)
  - Explain the depletion mode
  - Identify the symbols for *n*-channel and *p*-channel D-MOSFETs
  - Explain how D-MOSFETs and E-MOSFETs differ
  - Discuss power MOSFETs
  - Discuss dual-gate MOSFETs

Enhancement MOSFET (E-MOSFET)

The E-MOSFET operates *only* in the enhancement mode and has no depletion mode. It differs in construction from the D-MOSFET, which is discussed next, in that it has no structural channel. Notice in Figure 8–34(a) that the substrate extends completely to the SiO_2 layer. For an *n*-channel device, a positive gate voltage above a threshold value *induces* a channel by creating a thin layer of negative charges in the substrate region adjacent to the SiO_2 layer, as shown in Figure 8–34(b). The conductivity of the channel is enhanced by increasing the gate-to-source voltage and thus pulling more electrons into the channel area. For any gate voltage below the threshold value, there is no channel.

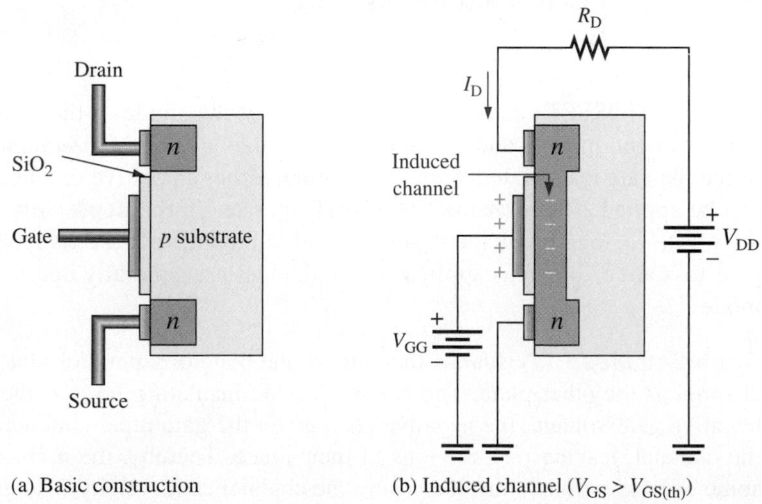

FIGURE 8–34

Representation of the basic E-MOSFET construction and operation (*n*-channel).

The schematic symbols for the *n*-channel and *p*-channel E-MOSFETs are shown in Figure 8–35. The broken lines symbolize the absence of a physical channel. An inward-pointing substrate arrow is for *n* channel, and an outward-pointing arrow is for *p* channel. Some E-MOSFET devices have a separate substrate connection.

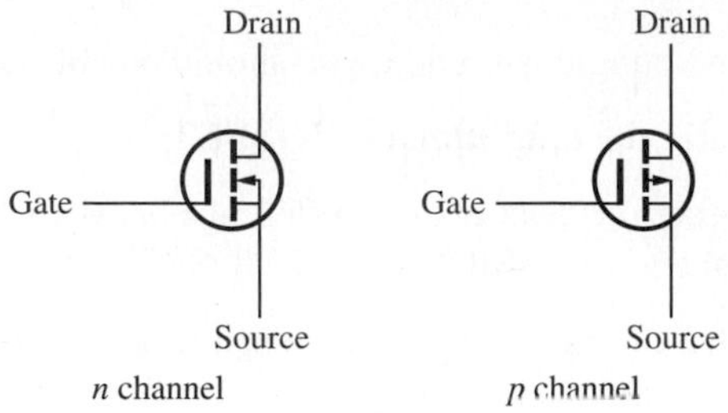

FIGURE 8–35

E-MOSFET schematic symbols.

Depletion MOSFET (D-MOSFET)

Another type of MOSFET is the depletion MOSFET (D-MOSFET), and Figure 8–36 illustrates its basic structure. The drain and source are diffused into the substrate material and then connected by a narrow channel adjacent to the insulated gate. Both *n*-channel and *p*-channel devices are shown in the figure. We will use the *n*-channel device to describe the basic operation. The *p*-channel operation is the same, except the voltage polarities are opposite those of the *n*-channel.

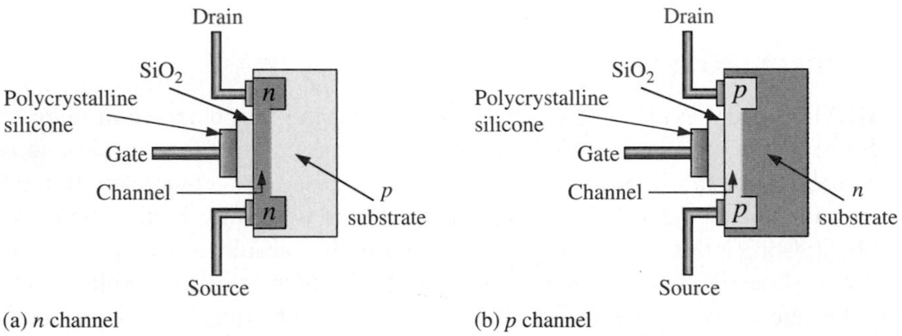

FIGURE 8–36

Representation of the basic structure of D-MOSFETs.

The D-MOSFET can be operated in either of two modes—the depletion mode or the enhancement mode—and is sometimes called a *depletion/enhancement MOSFET.* Since the gate is insulated from the channel, either a positive or a negative gate voltage can be applied. The *n*-channel MOSFET operates in the **depletion** mode when a negative gate-to-source voltage is applied and in the **enhancement** mode when a positive gate-to-source voltage is applied. These devices are generally operated in the depletion mode.

Depletion Mode Visualize the gate as one plate of a parallel-plate capacitor and the channel as the other plate. The silicon dioxide insulating layer is the dielectric. With a negative gate voltage, the negative charges on the gate repel conduction electrons from the channel, leaving positive ions in their place. Thereby, the *n* channel is depleted of some of its electrons, thus decreasing the channel conductivity. The greater the negative voltage on the gate, the greater the depletion of *n*-channel electrons. At a sufficiently negative gate-to-source voltage, $V_{GS(off)}$, the channel is totally depleted and the drain current is zero. This depletion mode is illustrated in Figure 8–37(a). Like the *n*-channel JFET, the *n*-channel D-MOSFET conducts drain current for gate-to-source voltages between $V_{GS(off)}$ and zero. In addition, the D-MOSFET conducts for values of V_{GS} above zero.

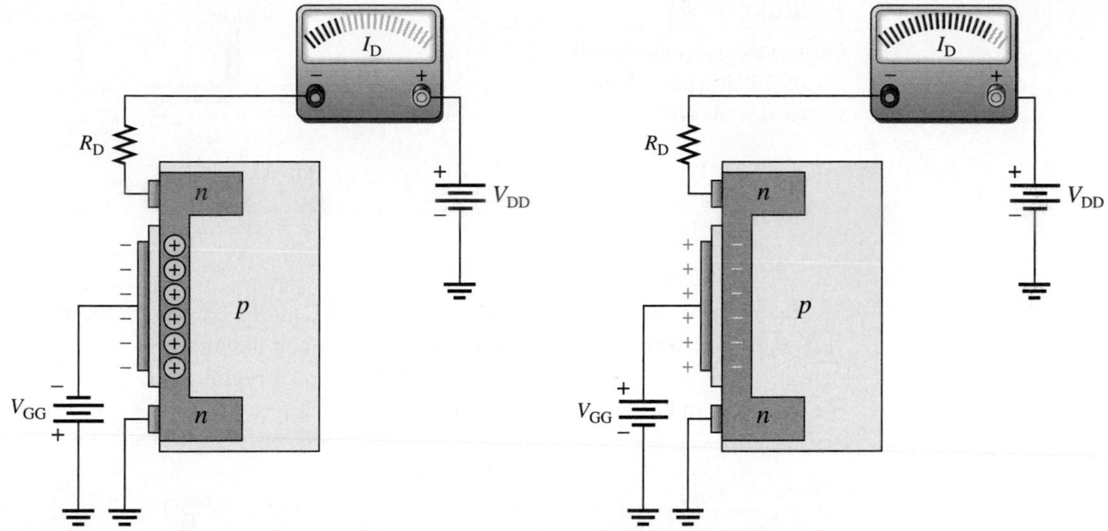

(a) Depletion mode: V_{GS} negative and less than $V_{GS(off)}$ (b) Enhancement mode: V_{GS} positive

▲ FIGURE 8–37

Operation of *n*-channel D-MOSFET.

Enhancement Mode With a positive gate voltage, more conduction electrons are attracted into the channel, thus increasing (enhancing) the channel conductivity, as illustrated in Figure 8–37(b).

D-MOSFET Symbols The schematic symbols for both the *n*-channel and the *p*-channel depletion MOSFETs are shown in Figure 8–38. The substrate, indicated by the arrow, is normally (but not always) connected internally to the source. Sometimes, there is a separate substrate pin.

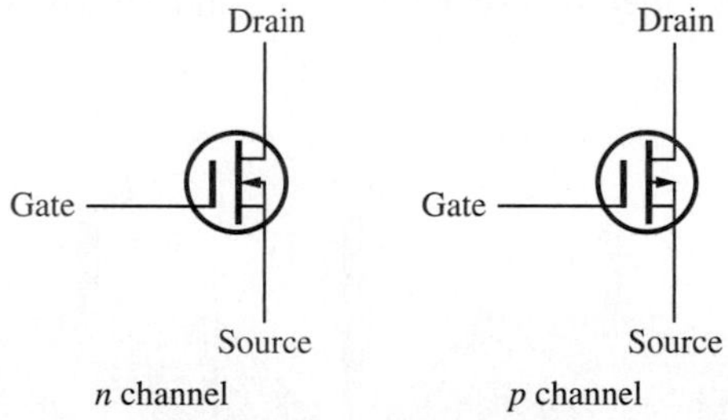

◀ FIGURE 8–38

D-MOSFET schematic symbols.

Power MOSFET Structures

The conventional enhancement MOSFETs have a long thin lateral channel as shown in the structural view in Figure 8–39. This results in a relatively high drain-to-source resistance and limits the E-MOSFET to low power applications. When the gate is positive, the channel is formed close to the gate between the source and the drain, as shown.

Laterally Diffused MOSFET (LDMOSFET) The LDMOSFET has a lateral channel structure and is a type of enhancement MOSFET designed for power applications. This device has a shorter channel between drain and source than does the conventional E-MOSFET. The shorter channel results in lower resistance, which allows higher current and voltage.

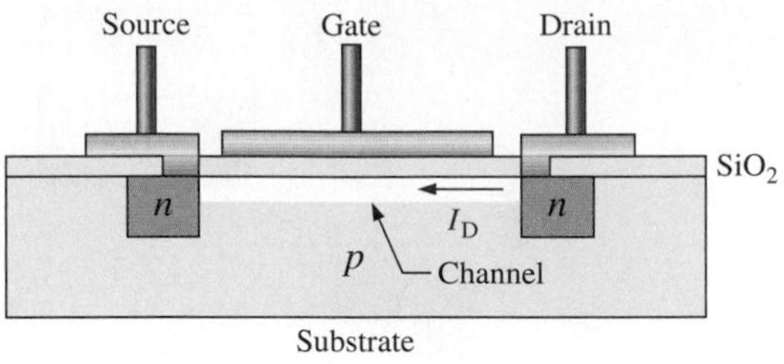

FIGURE 8–39

Cross section of conventional E-MOSFET structure. Channel is shown as white area.

Figure 8–40 shows the basic structure of an LDMOSFET. When the gate is positive, a very short n channel is induced in the p layer between the lightly doped source and the n^- region. There is current from the drain, through the n regions and the induced channel to the source as indicated.

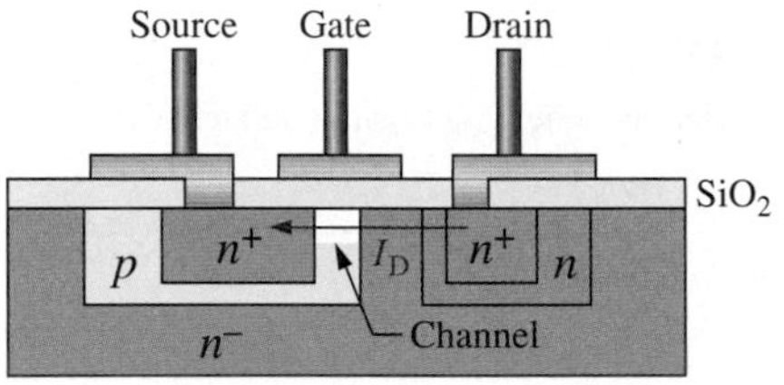

FIGURE 8–40

Cross section of LDMOSFET lateral channel structure.

VMOSFET The V-groove MOSFET is another example of the conventional E-MOSFET designed to achieve higher power capability by creating a shorter and wider channel with less resistance between the drain and source using a vertical channel structure. The shorter, wider channels allow for higher currents and, thus, greater power dissipation. Frequency response is also improved.

The VMOSFET has two source connections, a gate connection on top, and a drain connection on the bottom, as shown in Figure 8–41. The channel is induced vertically along both sides of the V-shaped groove between the drain (n^+ substrate where n^+ means a higher doping level than n^-) and the source connections. The channel length is set by the thickness of the layers, which is controlled by doping densities and diffusion time rather than by mask dimensions.

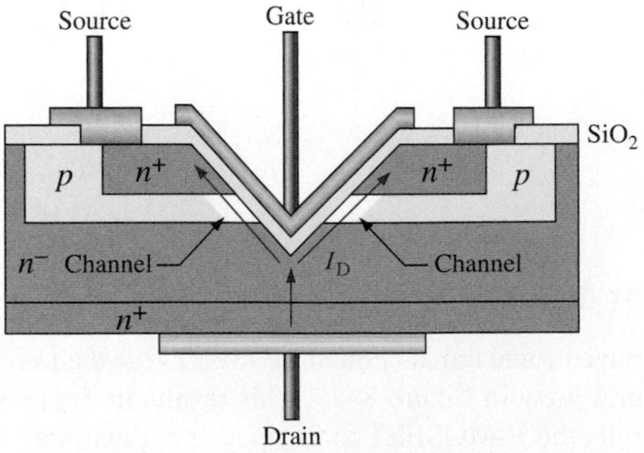

FIGURE 8–41

Cross section of VMOSFET vertical channel structure.

TMOSFET The vertical channel structure of the TMOSFET is illustrated in Figure 8–42. The gate structure is embedded in a silicon dioxide layer, and the source contact is continuous over the entire surface area. The drain is on the bottom. TMOSFET achieves greater packing density than VMOSFET, while retaining the short vertical channel advantage.

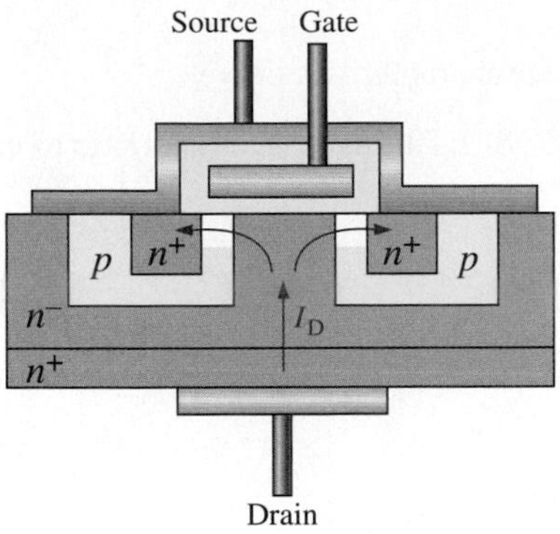

FIGURE 8–42
Cross section of TMOSFET vertical channel structure.

Dual-Gate MOSFETs

The dual-gate MOSFET can be either a depletion or an enhancement type. The only difference is that it has two gates, as shown in Figure 8–43. As previously mentioned, one drawback of a FET is its high input capacitance, which restricts its use at higher frequencies. By using a dual-gate device, the input capacitance is reduced, thus making the device useful in high-frequency RF amplifier applications. Another advantage of the dual-gate arrangement is that it allows for an automatic gain control (AGC) input in RF amplifiers. Another application is demonstrated in the Application Activity where the bias on the second gate is used to adjust the transconductance curve.

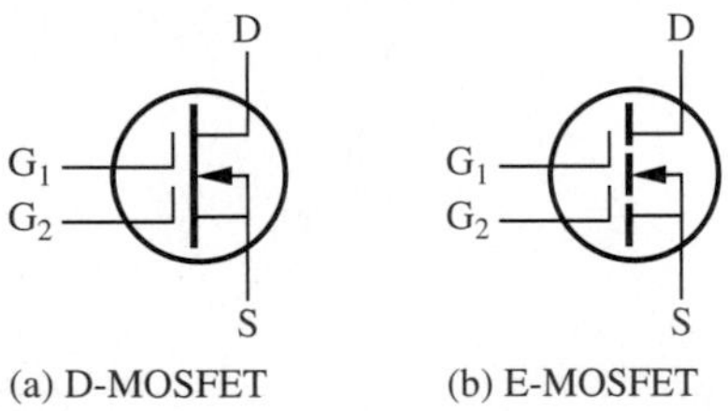

FIGURE 8–43
Dual-gate *n*-channel MOSFET symbols.

SECTION 8–5 CHECKUP

1. Name the two basic types of MOSFETs.
2. If the gate-to-source voltage in an *n*-channel E-MOSFET is made more positive, does the drain current increase or decrease?
3. If the gate-to-source voltage in an *n*-channel depletion MOSFET is made more negative, does the drain current increase or decrease?

8–6 MOSFET CHARACTERISTICS AND PARAMETERS

Much of the discussion concerning JFET characteristics and parameters applies equally to MOSFETs. In this section, MOSFET parameters are discussed.

After completing this section, you should be able to

- **Define, discuss, and apply important MOSFET parameters**
  - Analyze an E-MOSFET transfer characteristic curve
  - Use the equation for the E-MOSFET transfer characteristic to calculate I_D

- Analyze a D-MOSFET transfer characteristic curve
- Use the equation for the D-MOSFET transfer characteristic to calculate I_D
- Use a MOSFET datasheet
- Discuss handling precautions for MOS devices

E-MOSFET Transfer Characteristic

The E-MOSFET uses only channel enhancement. Therefore, an *n*-channel device requires a positive gate-to-source voltage, and a *p*-channel device requires a negative gate-to-source voltage. Figure 8–44 shows the general transfer characteristic curves for both types of E-MOSFETs. As you can see, there is no drain current when $V_{GS} = 0$. Therefore, the E-MOSFET does not have a significant I_{DSS} parameter, as do the JFET and the D-MOSFET. Notice also that there is ideally no drain current until V_{GS} reaches a certain nonzero value called the *threshold voltage,* $V_{GS(th)}$.

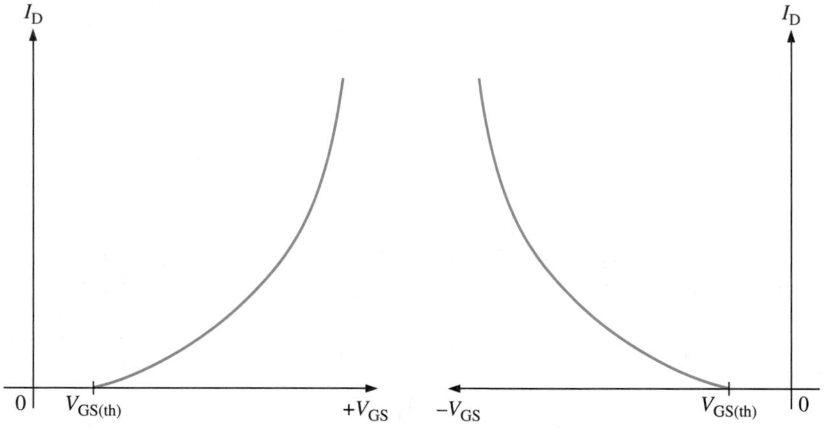

▲ **FIGURE 8–44**

E-MOSFET general transfer characteristic curves.

The equation for the parabolic transfer characteristic curve of the E-MOSFET differs from that of the JFET and the D-MOSFET because the curve starts at $V_{GS(th)}$ rather than $V_{GS(off)}$ on the horizontal axis and never intersects the vertical axis. The equation for the E-MOSFET transfer characteristic curve is

Equation 8–4

$$I_D = K(V_{GS} - V_{GS(th)})^2$$

The constant K depends on the particular MOSFET and can be determined from the datasheet by taking the specified value of I_D, called $I_{D(on)}$, at the given value of V_{GS} and substituting the values into Equation 8–4 as illustrated in Example 8–16.

EXAMPLE 8–16

The datasheet (see www.fairchild.com) for a 2N7002 E-MOSFET gives $I_{D(on)} = 500$ mA (minimum) at $V_{GS} = 10$ V and $V_{GS(th)} = 1$ V. Determine the drain current for $V_{GS} = 5$ V.

Solution First, solve for K using Equation 8–4.

$$K = \frac{I_{D(on)}}{(V_{GS} - V_{GS(th)})^2} = \frac{500\text{ mA}}{(10\text{ V} - 1\text{ V})^2} = \frac{500\text{ mA}}{81\text{ V}^2} = 6.17\text{ mA/V}^2$$

Next, using the value of K, calculate I_D for $V_{GS} = 5$ V.

$$I_D = K(V_{GS} - V_{GS(th)})^2 = (6.17\text{ mA/V}^2)(5\text{ V} - 1\text{ V})^2 = \mathbf{98.7\text{ mA}}$$

Related Problem The datasheet for an E-MOSFET gives $I_{D(on)} = 100$ mA at $V_{GS} = 8$ V and $V_{GS(th)} = 4$ V. Find I_D when $V_{GS} = 6$ V.

D-MOSFET Transfer Characteristic

As previously discussed, the D-MOSFET can operate with either positive or negative gate voltages. This is indicated on the general transfer characteristic curves in Figure 8–45 for both n-channel and p-channel MOSFETs. The point on the curves where $V_{GS} = 0$ corresponds to I_{DSS}. The point where $I_D = 0$ corresponds to $V_{GS(off)}$. As with the JFET, $V_{GS(off)} = -V_P$.

The square-law expression in Equation 8–1 for the JFET curve also applies to the D-MOSFET curve, as Example 8–17 demonstrates.

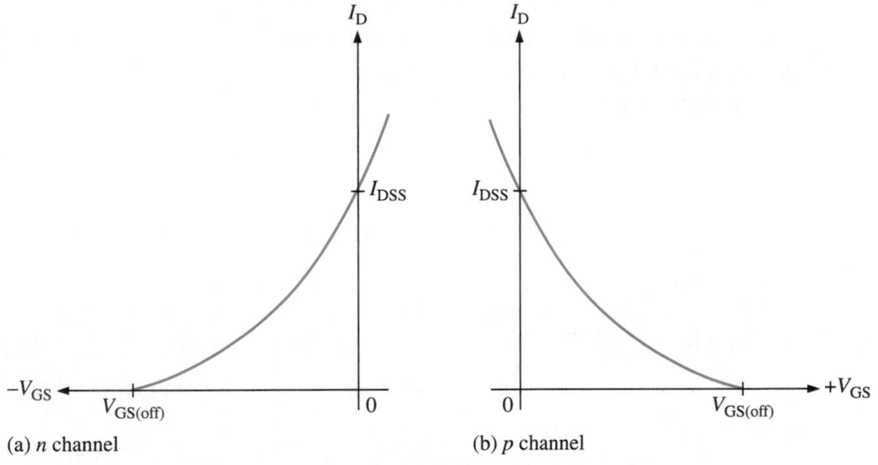

FIGURE 8–45

D-MOSFET general transfer characteristic curves.

EXAMPLE 8–17

For a certain D-MOSFET, $I_{DSS} = 10$ mA and $V_{GS(off)} = -8$ V.

(a) Is this an n-channel or a p-channel?

(b) Calculate I_D at $V_{GS} = -3$ V.

(c) Calculate I_D at $V_{GS} = +3$ V.

Solution **(a)** The device has a negative $V_{GS(off)}$; therefore, it is an ***n*-channel** MOSFET.

(b) $$I_D \cong I_{DSS}\left(1 - \frac{V_{GS}}{V_{GS(off)}}\right)^2 = (10\text{ mA})\left(1 - \frac{-3\text{ V}}{-8\text{ V}}\right)^2 = \mathbf{3.91\text{ mA}}$$

(c) $$I_D \cong (10\text{ mA})\left(1 - \frac{+3\text{ V}}{-8\text{ V}}\right)^2 = \mathbf{18.9\text{ mA}}$$

Related Problem For a certain D-MOSFET, $I_{DSS} = 18$ mA and $V_{GS(off)} = +10$ V.

(a) Is this an *n*-channel or a *p*-channel?

(b) Determine I_D at $V_{GS} = +4$ V.

(c) Determine I_D at $V_{GS} = -4$ V.

Handling Precautions

All MOS devices are subject to damage from electrostatic discharge (ESD). Because the gate of a MOSFET is insulated from the channel, the input resistance is extremely high (ideally infinite). The gate leakage current, I_{GSS}, for a typical MOSFET is in the pA range, whereas the gate reverse current for a typical JFET is in the nA range. The input capacitance results from the insulated gate structure. Excess static charge can be accumulated because the input capacitance combines with the very high input resistance and can result in damage to the device. To avoid damage from ESD, certain precautions should be taken when handling MOSFETs:

1. Carefully remove MOSFET devices from their packaging. They are shipped in conductive foam or special foil conductive bags. Usually they are shipped with a wire ring around the leads, which is removed just prior to installing the MOSFET in a circuit.
2. All instruments and metal benches used in assembly or test should be connected to earth ground (round or third prong of 110 V wall outlets).
3. The assembler's or handler's wrist should be connected to a commercial grounding strap, which has a high-value series resistor for safety. The resistor prevents accidental contact with voltage from becoming lethal.
4. Never remove a MOS device (or any other device, for that matter) from the circuit while the power is on.
5. Do not apply signals to a MOS device while the dc power supply is off.

SECTION 8–6 CHECKUP

1. What is the major difference in construction of the D-MOSFET and the E-MOSFET?
2. Name two parameters of an E-MOSFET that are not specified for D-MOSFETs?
3. What is ESD?

8–7 MOSFET BIASING

Three ways to bias a MOSFET are zero-bias, voltage-divider bias, and drain-feedback bias. Biasing is important in FET amplifiers, which you will study in the next chapter.

After completing this section, you should be able to

- **Discuss and analyze MOSFET bias circuits**
  - Describe voltage-divider bias of an E-MOSFET
  - Describe drain-feedback bias of an E-MOSFET
  - Describe zero-bias of a D-MOSFET
  - Analyze a zero-biased MOSFET circuit

E-MOSFET Bias

Because E-MOSFETs must have a V_{GS} greater than the threshold value, $V_{GS(th)}$, zero bias cannot be used. Figure 8–46 shows two ways to bias an E-MOSFET (D-MOSFETs can also be biased using these methods). An n-channel device is used for purposes of illustration. In either the voltage-divider or drain-feedback bias arrangement, the purpose is to make the gate voltage more positive than the source by an amount exceeding $V_{GS(th)}$. Equations for the analysis of the voltage-divider bias in Figure 8–46(a) are as follows:

$$V_{GS} = \left(\frac{R_2}{R_1 + R_2}\right)V_{DD}$$

$$V_{DS} = V_{DD} - I_D R_D$$

where $I_D = K(V_{GS} - V_{GS(th)})^2$ from Equation 8–4.

In the drain-feedback bias circuit in Figure 8–46(b), there is negligible gate current and, therefore, no voltage drop across R_G. This makes $V_{GS} = V_{DS}$.

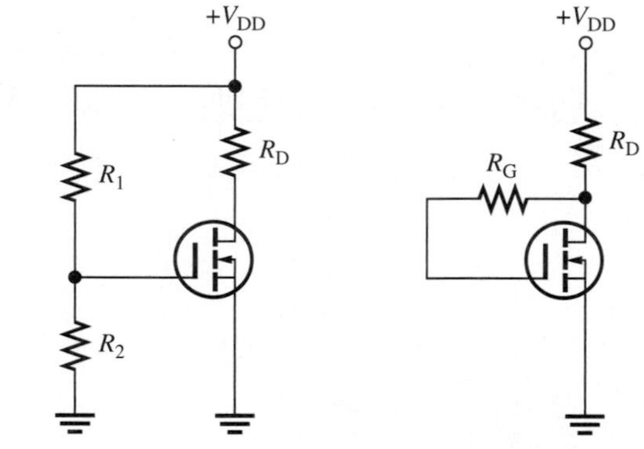

FIGURE 8–46

Common E-MOSFET biasing arrangements.

EXAMPLE 8–18

Determine V_{GS} and V_{DS} for the E-MOSFET circuit in Figure 8–47. Assume this particular MOSFET has minimum values of $I_{D(on)} = 200$ mA at $V_{GS} = 4$ V and $V_{GS(th)} = 2$ V.

FIGURE 8–47

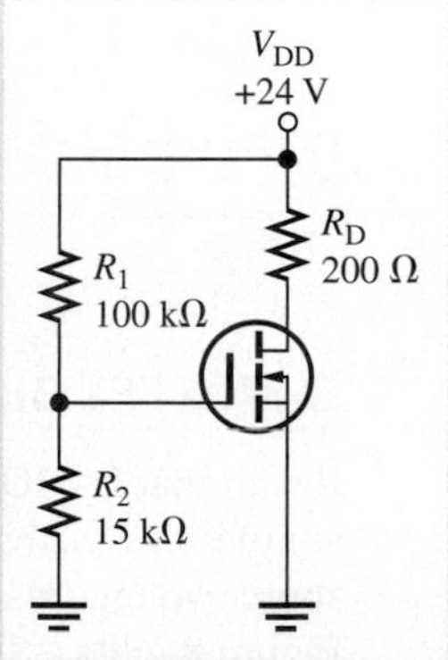

Solution For the E-MOSFET in Figure 8–47, the gate-to-source voltage is

$$V_{GS} = \left(\frac{R_2}{R_1 + R_2}\right)V_{DD} = \left(\frac{15\text{ k}\Omega}{115\text{ k}\Omega}\right)24\text{ V} = \mathbf{3.13\text{ V}}$$

To determine V_{DS}, first find K using the minimum value of $I_{D(on)}$ and the specified voltage values.

$$K = \frac{I_{D(on)}}{(V_{GS} - V_{GS(th)})^2} = \frac{200\text{ mA}}{(4\text{ V} - 2\text{ V})^2} = \frac{200\text{ mA}}{4\text{ V}^2} = 50\text{ mA/V}^2$$

Now calculate I_D for $V_{GS} = 3.13$ V.

$$I_D = K(V_{GS} - V_{GS(th)})^2 = (50\text{ mA/V}^2)(3.13\text{ V} - 2\text{ V})^2$$
$$= (50\text{ mA/V}^2)(1.13\text{ V})^2 = 63.8\text{ mA}$$

Finally, calculate V_{DS}.

$$V_{DS} = V_{DD} - I_D R_D = 24\text{ V} - (63.8\text{ mA})(200\ \Omega) = \mathbf{11.2\ V}$$

Related Problem Determine V_{GS} and V_{DS} for the circuit in Figure 8–47 given $I_{D(on)} = 100$ mA at $V_{GS} = 4$ V and $V_{GS(th)} = 3$ V.

EXAMPLE 8–19 Determine the amount of drain current in Figure 8–48. The MOSFET has a $V_{GS(th)} = 3$ V.

FIGURE 8–48

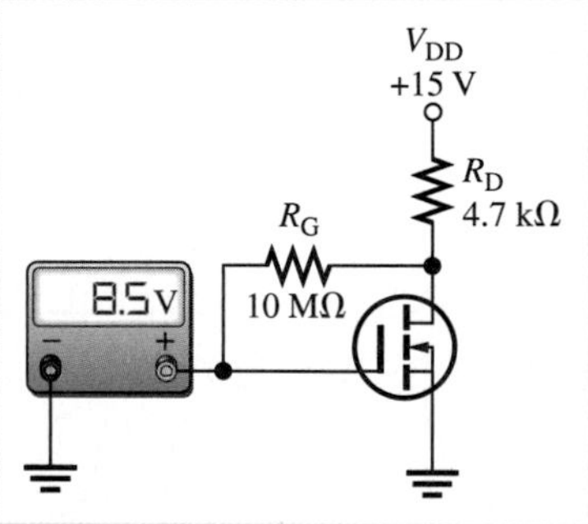

Solution The meter indicates $V_{GS} = 8.5$ V. Since this is a drain-feedback configuration, $V_{DS} = V_{GS} = 8.5$ V.

$$I_D = \frac{V_{DD} - V_{DS}}{R_D} = \frac{15\text{ V} - 8.5\text{ V}}{4.7\text{ k}\Omega} = \mathbf{1.38\ mA}$$

Related Problem Determine I_D if the meter in Figure 8–48 reads 5 V.

D-MOSFET Bias

Recall that D-MOSFETs can be operated with either positive or negative values of V_{GS}. A simple bias method is to set $V_{GS} = 0$ so that an ac signal at the gate varies the gate-to-source voltage above and below this 0 V bias point. A MOSFET with zero bias is shown in Figure 8–49(a). Since $V_{GS} = 0$, $I_D = I_{DSS}$ as indicated. The drain-to-source voltage is expressed as follows:

$$V_{DS} = V_{DD} - I_{DSS}R_D$$

The purpose of R_G is to accommodate an ac signal input by isolating it from ground, as shown in Figure 8–49(b). Since there is no dc gate current, R_G does not affect the zero gate-to-source bias.

as well as charred resistors. Next, disconnect power and measure suspected resistors for opens. If these are okay, the JFET is probably bad. Any of the following faults can produce this symptom:

1. No ground connection at R_S
2. R_S open
3. Open drain lead connection
4. Open source lead connection
5. FET internally open between drain and source

Symptom 2: V_D Significantly Less Than Normal For this condition, unless the supply voltage is lower than it should be, the drain current must be larger than normal because the drop across R_D is too much. Figure 8–54(b) indicates this situation. This symptom can be caused by any of the following:

1. Open R_G
2. Open gate lead
3. FET internally open at gate

Any of these three faults will cause the depletion region in the JFET to disappear and the channel to widen so that the drain current is limited only by R_D, R_S, and the small channel resistance.

Faults in D-MOSFET and E-MOSFET Circuits

One fault that is difficult to detect is when the gate opens in a zero-biased D-MOSFET. In a zero-biased D-MOSFET, the gate-to-source voltage ideally remains zero when an open occurs in the gate circuit; thus, the drain current doesn't change, and the bias appears normal, as indicated in Figure 8–55. However, static charge as a result of the open may cause I_D to behave irratically.

FIGURE 8–55

An open fault in the gate circuit of a MOSFET causes no change in I_D.

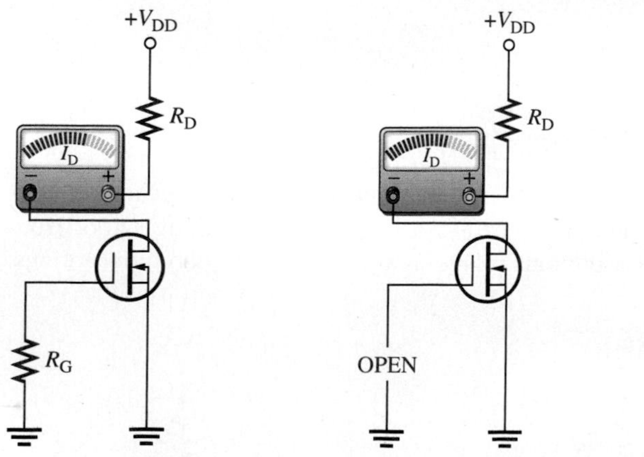

(a) Normal operation

(b) Gate circuit open (can be either external or internal)

In an E-MOSFET circuit with voltage-divider bias, an open R_1 makes the gate voltage zero. This causes the transistor to be off and act like an open switch because a gate-to-source threshold voltage greater than zero is required to turn the device on. This condition is illustrated in Figure 8–56(a). If R_2 opens, the gate is at $+V_{DD}$ and the channel resistance is very low so the device approximates a closed switch. The drain current is limited only by R_D. This condition is illustrated in Figure 8–56(b).

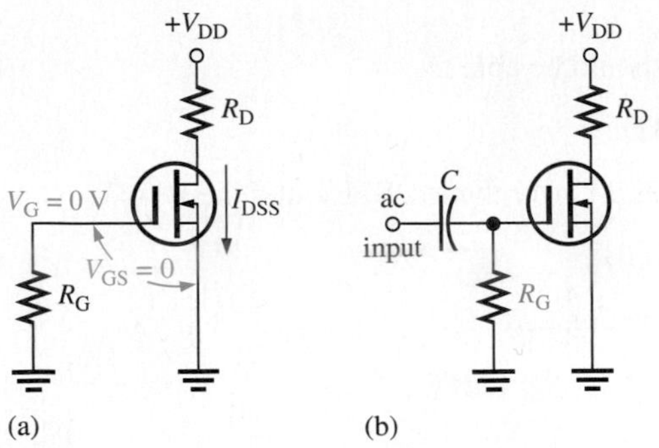

FIGURE 8–49

A zero-biased D-MOSFET.

EXAMPLE 8–20

Determine the drain-to-source voltage in the circuit of Figure 8–50. The MOSFET datasheet gives $V_{GS(off)} = -8$ V and $I_{DSS} = 12$ mA.

FIGURE 8–50

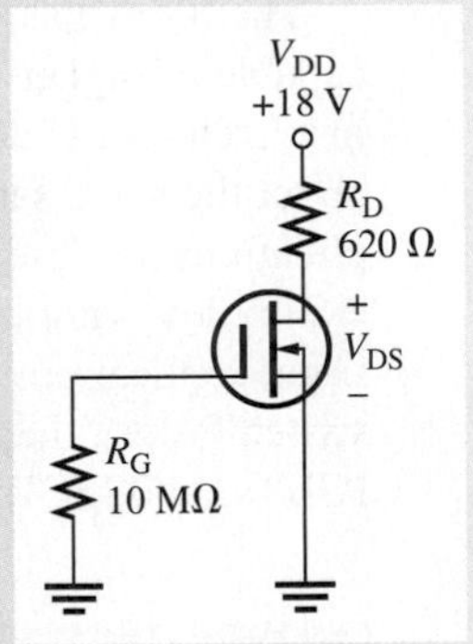

Solution Since $I_D = I_{DSS} = 12$ mA, the drain-to-source voltage is

$$V_{DS} = V_{DD} - I_{DSS}R_D = 18\text{ V} - (12\text{ mA})(620\ \Omega) = \mathbf{10.6\ V}$$

Related Problem Find V_{DS} in Figure 8–50 when $V_{GS(off)} = -10$ V and $I_{DSS} = 20$ mA.

SECTION 8–7 CHECKUP

1. For a D-MOSFET biased at $V_{GS} = 0$, is the drain current equal to zero, I_{GSS}, or I_{DSS}?
2. For an n-channel E-MOSFET with $V_{GS(th)} = 2$ V, V_{GS} must be in excess of what value in order to conduct?

8–8 THE IGBT

The IGBT (insulated-gate bipolar transistor) combines features from both the MOSFET and the BJT that make it useful in high-voltage and high-current switching applications. The IGBT has largely replaced the MOSFET and the BJT in many of these applications.

After completing this section, you should be able to

- **Explain the operation of IGBTs**
  - State the advantages of the IGBT over the MOSFET and the BJT
  - Identify the symbol for the IGBT
  - Discuss how an IGBT is turned on and off
  - Describe the equivalent circuit for the IGBT

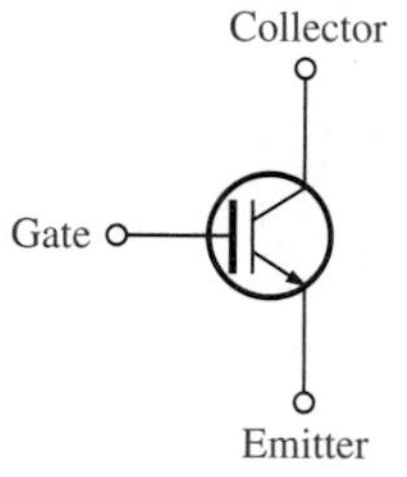

FIGURE 8–51

A symbol for the IGBT (insulated-gate bipolar transistor).

The **IGBT** is a device that has the output conduction characteristics of a BJT but is voltage controlled like a MOSFET; it is an excellent choice for many high-voltage switching applications. The IGBT has three terminals: gate, collector, and emitter. One common circuit symbol is shown in Figure 8–51. As you can see, it is similar to the BJT symbol except there is an extra bar representing the gate structure of a MOSFET rather than a base.

The IGBT has MOSFET input characteristics and BJT output characteristics. BJTs are capable of higher currents than FETs, but MOSFETs have no gate current because of the insulated gate structure. IGBTs exhibit a lower saturation voltage than MOSFETs and have about the same saturation voltage as BJTs. IGBTs are superior to MOSFETs in some applications because they can handle high collector-to-emitter voltages exceeding 200 V and exhibit less saturation voltage when they are in the *on* state. IGBTs are superior to BJTs in some applications because they can switch faster. In terms of switching speed, MOSFETs switch fastest, then IGBTs, followed by BJTs, which are slowest. A general comparison of IGBTs, MOSFETs, and BJTs is given in Table 8–1.

TABLE 8–1

Comparison of several device features for switching applications.

FEATURES	IGBT	MOSFET	BJT
Type of input drive	Voltage	Voltage	Current
Input resistance	High	High	Low
Operating frequency	Medium	High	Low
Switching speed	Medium	Fast (ns)	Slow (μs)
Saturation voltage	Low	High	Low

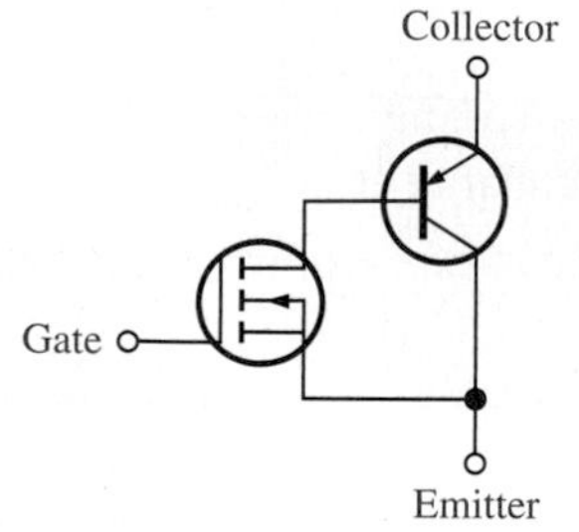

FIGURE 8–52

Simplified equivalent circuit for an IGBT.

Operation

The IGBT is controlled by the gate voltage just like a MOSFET. Essentially, an IGBT can be thought of as a voltage-controlled BJT, but with faster switching speeds. Because it is controlled by voltage on the insulated gate, the IGBT has essentially no input current and does not load the driving source. A simplified equivalent circuit for an IGBT is shown in Figure 8–52. The input element is a MOSFET, and the output element is a bipolar transistor. When the gate voltage with respect to the emitter is less than a threshold voltage, V_{thresh}, the device is turned off. The device is turned on by increasing the gate voltage to a value exceeding the threshold voltage.

The *npnp* structure of the IGBT forms a parasitic transistor and an inherent parasitic resistance within the device, as shown in red in Figure 8–53. These parasitic components have no effect during normal operation. However, if the maximum collector current is exceeded under certain conditions, the parasitic transistor, Q_p can turn on. If Q_p turns on, it effectively combines with Q_1 to form a parasitic element, as shown in Figure 8–53, in which a latchup condition can occur. In latch-up, the device will stay on and cannot be controlled by the gate voltage. Latch-up can be avoided by always operating within the specified limits of the device.

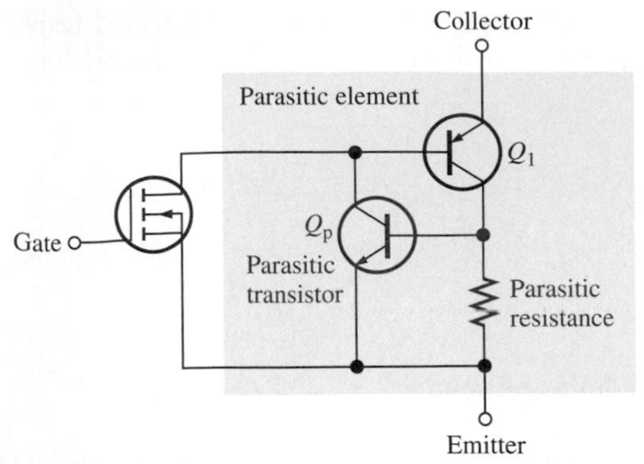

FIGURE 8–53

Parasitic components of an IGBT that can cause latch-up.

SECTION 8–8 CHECKUP

1. What does IGBT stand for?
2. What is a major application area for IGBTs?
3. Name an advantage of an IGBT over a power MOSFET.
4. Name an advantage of an IGBT over a power BJT.
5. What is latch-up?

8–9 TROUBLESHOOTING

In this section, we discuss some common faults that may be encountered in FET circuits and the probable causes for each fault.

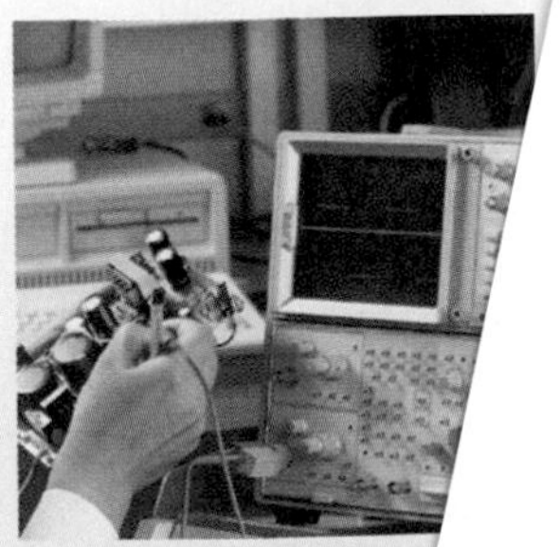

After completing this section, you should be able to

- **Troubleshoot FET circuits**
  - Troubleshoot self-biased JFET circuits
  - Troubleshoot MOSFET circuits with zero bias and voltage-divider bias

Faults in Self-Biased JFET Circuits

Symptom 1: $V_D = V_{DD}$ For this condition, the drain current must be zero because there is no voltage drop across R_D, as illustrated in Figure 8–54(a). As in any circuit, it is good troubleshooting practice to first check for obvious problems such as open or poor connections,

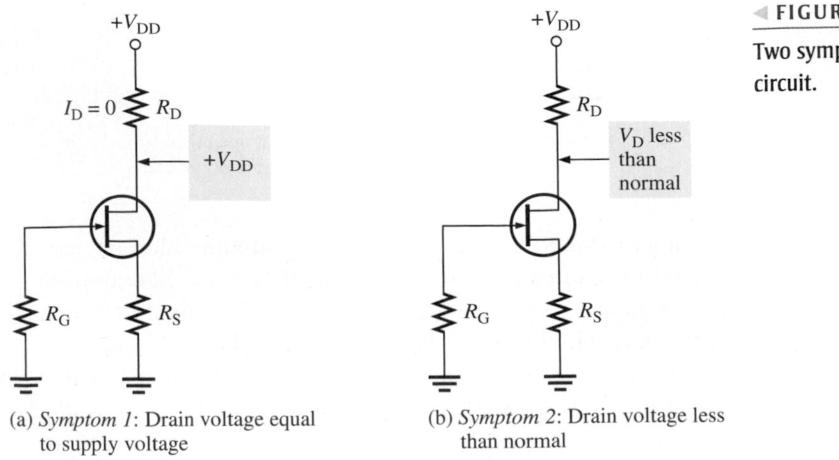

FIGURE 8–54

Two symptoms in a circuit.

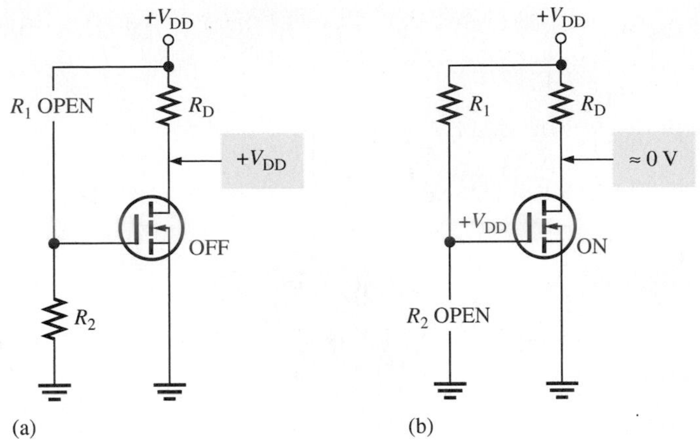

FIGURE 8–56

Failures in an E-MOSFET circuit with voltage-divider bias.

Multisim Troubleshooting Exercises

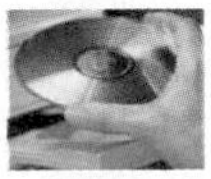

These file circuits are in the Troubleshooting Exercises folder on your CD-ROM.

1. Open file TSE08-01. Determine if the circuit is working properly and, if not, determine the fault.
2. Open file TSE08-02. Determine if the circuit is working properly and, if not, determine the fault.
3. Open file TSE08-03. Determine if the circuit is working properly and, if not, determine the fault.

SECTION 8–9 CHECKUP

1. In a self-biased JFET circuit, the drain voltage equals V_{DD}. If the JFET is okay, what are other possible faults?
2. Why doesn't the drain current change when an open occurs in the gate circuit of a zero-biased D-MOSFET circuit?
3. If the gate of an E-MOSFET becomes shorted to ground in a circuit with voltage-divider bias, what is the drain voltage?

On the Job

Application Activity: *pH Sensor Circuit*

This application involves electronic instrumentation in a waste water treatment facility. The system controls the amount of acid and base reagent added to waste water in order to neutralize it. The diagram of the waste water neutralization pH system is shown in Figure 8–57. The system measures and controls the pH of the water, which is a measure of the degree of acidity or alkalinity. The pH scale ranges from 0 for the strongest acids through 7 for neutral solutions and up to 14 for the strongest bases (caustics). Typically, the pH for waste water ranges from greater than 2 and less than 11. The pH of the water is measured by sensor probes at the inlets and outlets of the tanks. The processor and controller unit uses the inputs from the pH sensor circuits to adjust the amount of acid or base introduced into the neutralization tank. The pH should be 7 at the outlet of the smoothing tank.

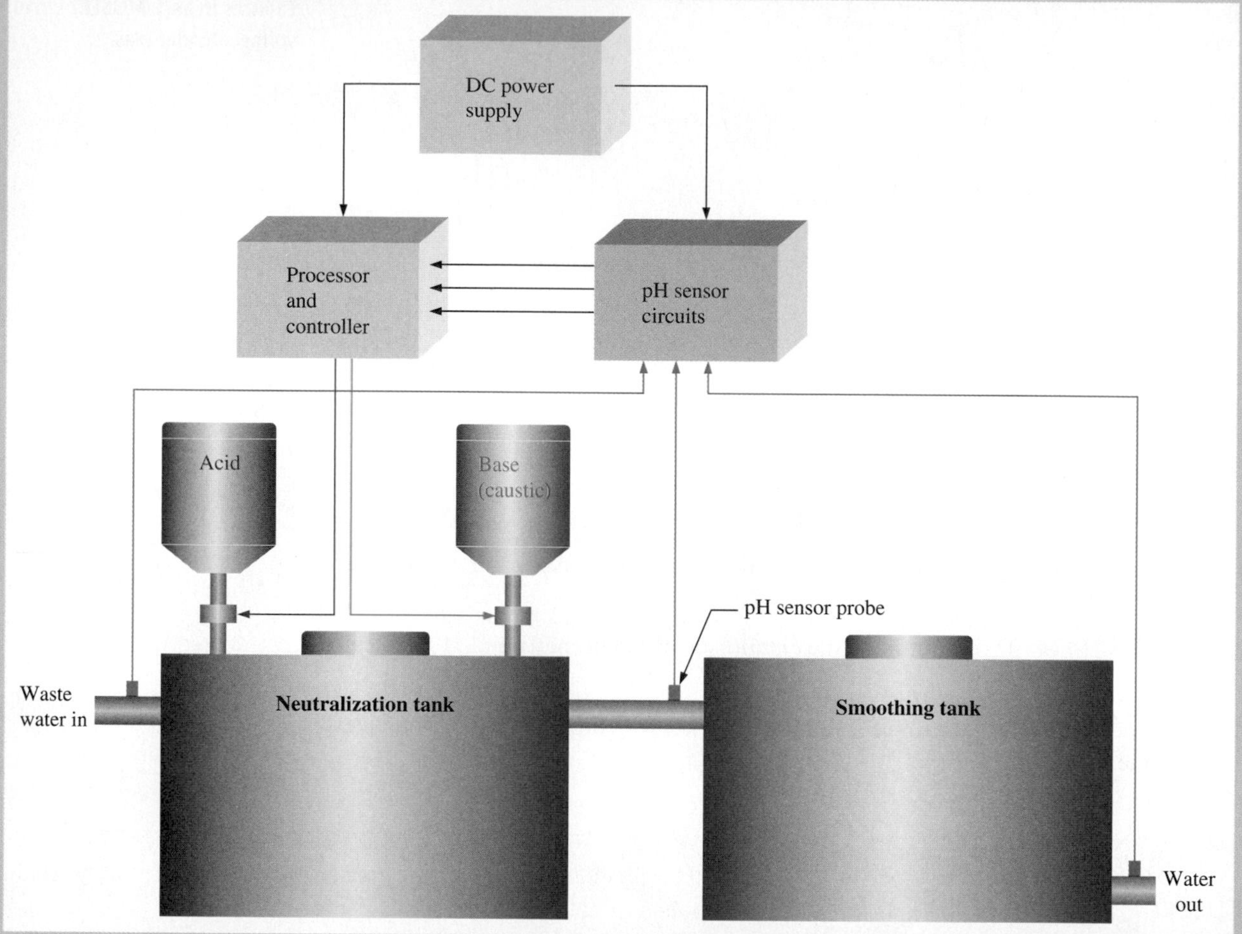

FIGURE 8–57

Simplified waste water pH neutralization system.

Generally, waste water treatment is done in three steps as follows:

- *Primary treatment* Collecting, screening, and initial storage
- *Secondary treatment* Removal of solids and the majority of contaminants using filters, coagulation, flocculation, and membranes
- *Tertiary treatment* Polishing, pH adjustment, carbon treatment to remove taste and smells, disinfection, and temporary storage to allow the disinfecting agent to work

In this application, we are focusing on the process of pH adjustment in the tertiary stage of treatment.

The Sensor Circuit

There are three identical pH sensor circuits, one for each of the inlet/outlets indicated in Figure 8–57. The pH sensor produces a small voltage (mV) proportional to the pH of the water in which it is immersed. The pH sensor produces a negative voltage if the water is acidic, no voltage if it is neutral, and a positive voltage if it is basic. The sensor output goes to the gate of a MOSFET circuit, which amplifies the sensor voltage for processing by the digital controller.

Figure 8–58 shows the pH sensor probe and a graph of output voltage versus pH. Figure 8–59 is the sensor circuit using a BF998 dual-gate *n*-channel MOSFET. A rheostat in the drain of the MOSFET is used to calibrate the circuit so that each of the three sensor circuits produce the same output voltage for a given value of pH.

1. Determine the approximate sensor voltage for a pH of 8.
2. Determine the approximate sensor voltage for a pH of 3.

The partial datasheet for the BF998 D-MOSFET is shown in Figure 8–60. In this application, the MOSFET is used as a dc amplifier. Recall that a D-MOSFET can operate

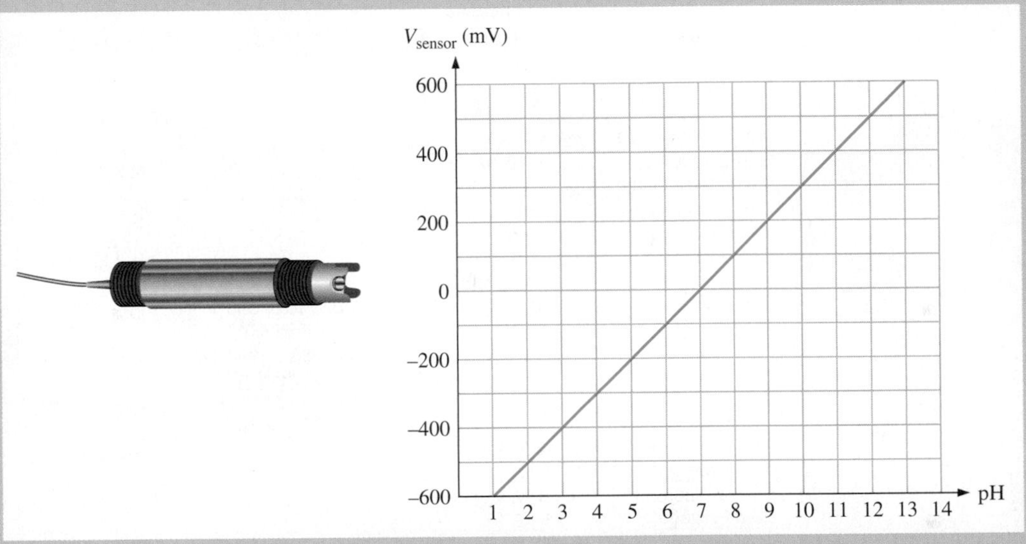

FIGURE 8–58

pH sensor and graph of pH vs. output voltage.

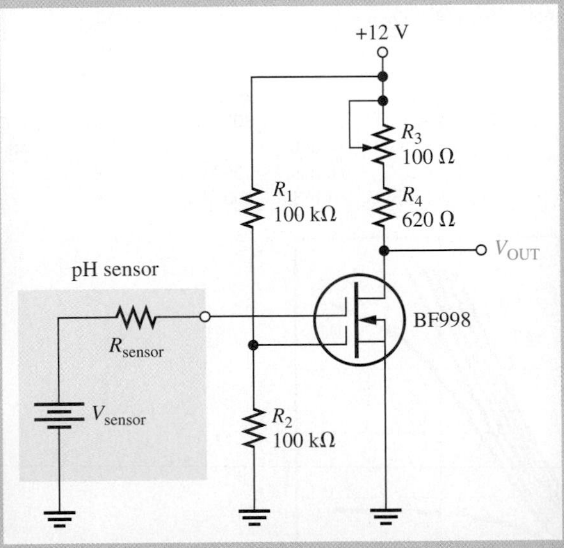

FIGURE 8–59

pH sensor circuit.

FIGURE 8–60

Partial datasheet for a BF998 MOSFET. Datasheet courtesy of Vishay Intertechnology, Inc.

Absolute Maximum Ratings

T_{amb} = 25°C, unless otherwise specified

Parameter	Test Conditions	Symbol	Value	Unit
Drain - source voltage		V_{DS}	12	V
Drain current		I_D	30	mA
Gate 1/Gate 2 - source peak current		$\pm I_{G1/G2SM}$	10	mA
Gate 1/Gate 2 - source voltage		$\pm V_{G1S/G2S}$	7	V
Total power dissipation	$T_{amb} \leq 60$ °C	P_{tot}	200	mW
Channel temperature		T_{Ch}	150	°C
Storage temperature range		T_{stg}	–65 to +150	°C

Electrical DC Characteristics

T_{amb} = 25°C, unless otherwise specified

Parameter	Test Conditions	Type	Symbol	Min	Typ	Max	Unit
Drain - source breakdown voltage	$I_D = 10\ \mu A$, $-V_{G1S} = -V_{G2S} = 4$ V		$V_{(BR)DS}$	12			V
Gate 1 - source breakdown voltage	$\pm I_{G1S} = 10$ mA, $V_{G2S} = V_{DS} = 0$		$\pm V_{(BR)G1SS}$	7		14	V
Gate 2 - source breakdown voltage	$\pm I_{G2S} = 10$ mA, $V_{G1S} = V_{DS} = 0$		$\pm V_{(BR)G2SS}$	7		14	V
Gate 1 - source leakage current	$\pm V_{G1S} = 5$ V, $V_{G2S} = V_{DS} = 0$		$\pm I_{G1SS}$			50	nA
Gate 2 - source leakage current	$\pm V_{G2S} = 5$ V, $V_{G1S} = V_{DS} = 0$		$\pm I_{G2SS}$			50	nA
Drain current	$V_{DS} = 8$ V, $V_{G1S} = 0$, $V_{G2S} = 4$ V	BF998/BF998R/ BF998RW	I_{DSS}	4		18	mA
		BF998A/BF998RA/ BF998RAW	I_{DSS}	4		10.5	mA
		BF998B/BF998RB/ BF998RBW	I_{DSS}	9.5		18	mA
Gate 1 - source cut-off voltage	$V_{DS} = 8$ V, $V_{G2S} = 4$ V, $I_D = 20\ \mu A$		$-V_{G1S(OFF)}$		1.0	2.0	V
Gate 2 - source cut-off voltage	$V_{DS} = 8$ V, $V_{G1S} = 0$, $I_D = 20\ \mu A$		$-V_{G2S(OFF)}$		0.6	1.0	V

Electrical AC Characteristics

$V_{DS} = 8$ V, $I_D = 10$ mA, $V_{G2S} = 4$ V, f = 1 MHz , T_{amb} = 25°C, unless otherwise specified

Parameter	Test Conditions	Symbol	Min	Typ	Max	Unit
Forward transadmittance		$\lvert y_{21s} \rvert$	21	24		mS
Gate 1 input capacitance		C_{issg1}		2.1	2.5	pF
Gate 2 input capacitance	$V_{G1S} = 0$, $V_{G2S} = 4$ V	C_{issg2}		1.1		pF
Feedback capacitance		C_{rss}		25		fF
Output capacitance		C_{oss}		1.05		pF
Power gain	$G_S = 2$ mS, $G_L = 0.5$ mS, f = 200 MHz	G_{ps}		28		dB
	G_S = 3,3 mS, $G_L = 1$ mS, f = 800 MHz	G_{ps}	16.5	20		dB
AGC range	V_{G2S} = 4 to –2 V, f = 800 MHz	ΔG_{ps}	40			dB
Noise figure	$G_S = 2$ mS, $G_L = 0.5$ mS, f = 200 MHz	F		1.0		dB
	G_S = 3,3 mS, $G_L = 1$ mS, f = 800 MHz	F		1.5		dB

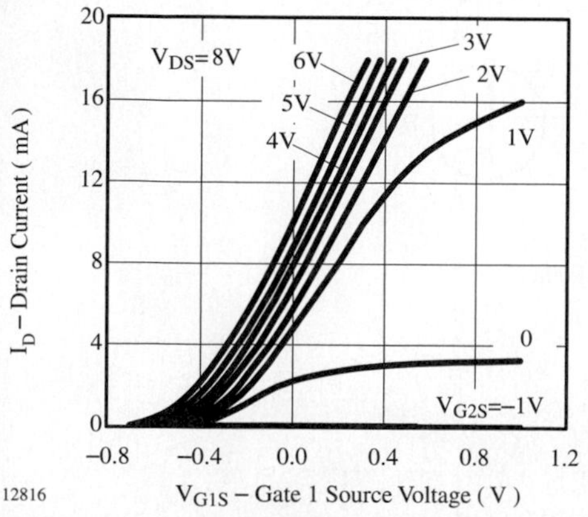

with both positive and negative gate voltages, making it ideal for this particular application where the input voltage can have either polarity. The graph in Figure 8–60 shows that the transconductance curve depends on the value of the voltage on the second gate which, in this particular design, is set at 6 V by the R_1-R_2 voltage divider. The input from the sensor is applied to the first gate.

3. What is the specified typical transconductance (transadmittance) for the BF998?
4. If the drain-to-source voltage is 10 V, determine the maximum allowable drain current.
5. If one gate is biased to 1 V, what is I_D when the other gate is 0 V?

Simulation

The pH sensor circuit is simulated in Multisim, and the results for a series of sensor input voltages are shown in Figure 8–61. The sensor is modeled as a dc source in series with an internal resistance. Notice that the output of the circuit increases as the sensor input decreases. Rheostat R_3 is used to calibrate each of the three sensor circuits so that they have an identical output voltage for a given sensor input voltage.

6. If the output of the sensor circuit is 7 V, is the solution acidic, neutral, or basic (caustic)?
7. Plot a graph of V_{OUT} vs. pH for each measurement in Figure 8–61.

Simulate the pH sensor circuit using your Multisim software. Measure the output voltage for V_{sensor} = 50 mV, V_{sensor} = 150 mV, and V_{sensor} = −25 mV.

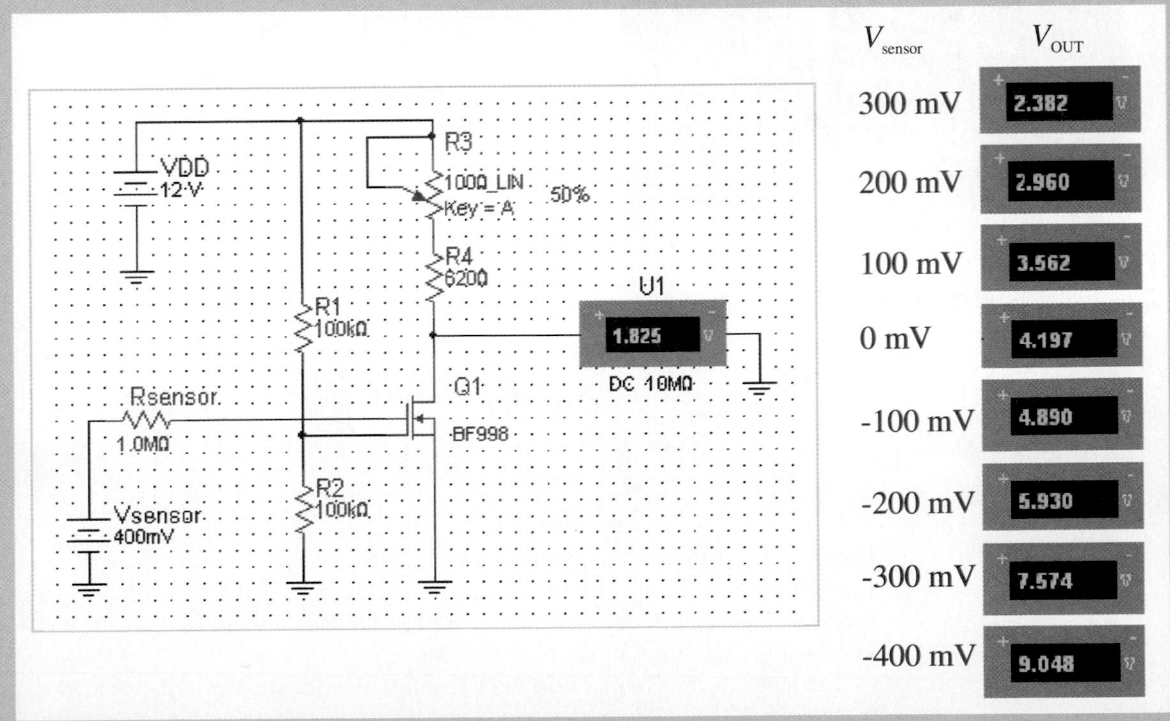

FIGURE 8–61

Simulation results for the pH sensor circuit.

Prototyping and Testing

Now that the circuit has been simulated, the prototype circuit is constructed and tested. A dc voltage source can be used to provide the sensor input voltages. After the circuit is successfully tested on a protoboard, it is ready to be finalized on a printed circuit board.

Lab Experiment

To build and test a similar circuit, go to Experiment 8 in your lab manual (*Laboratory Exercises for Electronic Devices* by David Buchla and Steven Wetterling).

Circuit Board

The pH sensor circuits are implemented on a printed circuit board as shown in Figure 8–62. Each circuit monitors one of the three pH sensors in the system. Note that a single voltage divider provides +6 V to the second gate of each transistor.

8. Check the printed circuit board for correctness by comparing with the schematic in Figure 8–59.

9. Identify the connections on the back side of the board.

10. Label each input and output pin according to function.

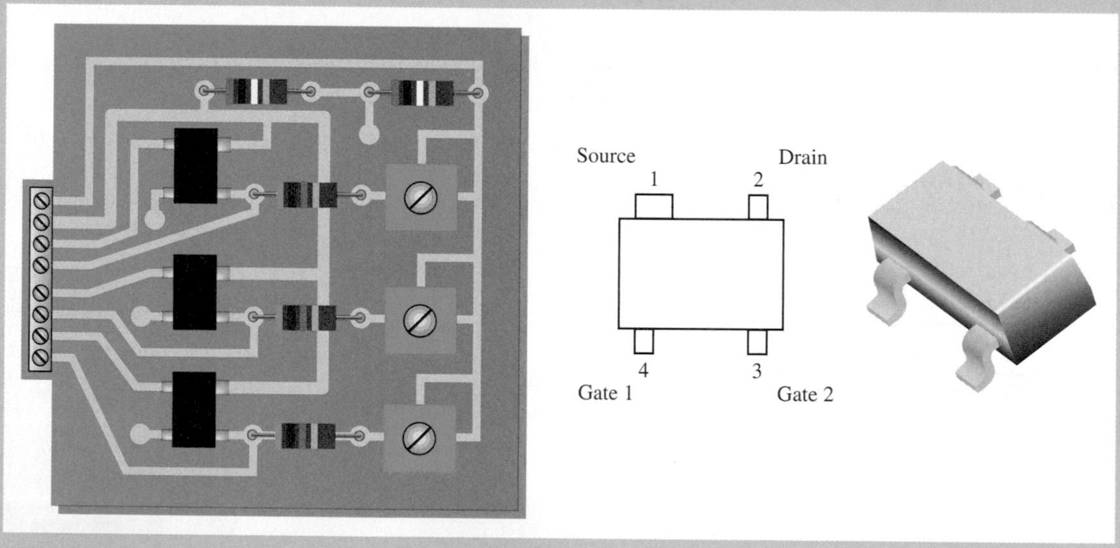

FIGURE 8–62

pH sensor circuit board.

Calibration and Testing

The first step is to calibrate each of the three circuits for a pH of 7. Using a known neutral test solution in a container into which the sensors are placed, the rheostat is adjusted (if necessary) to produce the same output voltage for each circuit. In this case it is 4.197 V, as shown in Figure 8–63.

The next step is to replace the neutral solution with one that has an acidity with a known pH. All the circuits should produce the same voltage within a specified tolerance. Finally, using a basic solution with a known pH, measure the output voltages. Again, they should all agree.

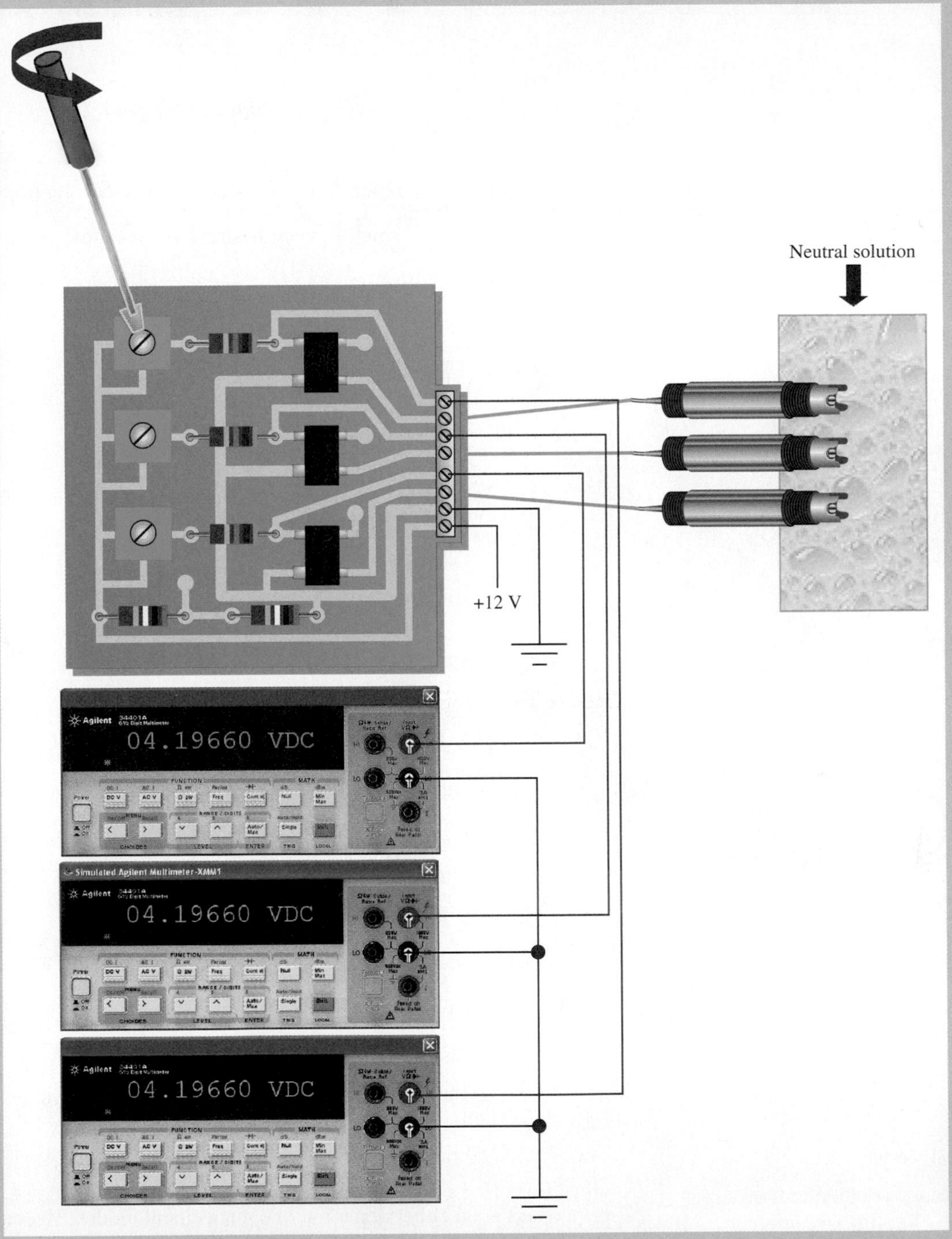

FIGURE 8–63

Calibration and testing of the pH sensor circuits.

SUMMARY OF FIELD-EFFECT TRANSISTORS

JFETs

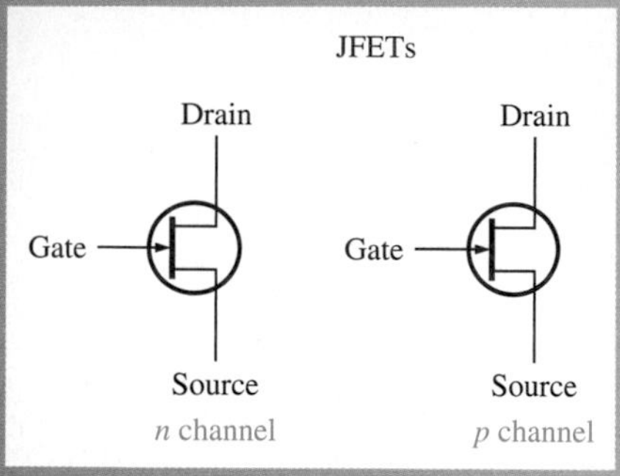

- Gate-source *pn* junction must be reverse-biased.
- V_{GS} controls I_D.
- Value of V_{DS} at which I_D becomes constant is the pinch-off voltage.
- Value of V_{GS} at which I_D becomes zero is the cutoff voltage, $V_{GS(off)}$.
- I_{DSS} is drain current when $V_{GS} = 0$.
- Transfer characteristic:

$$I_D \cong I_{DSS}\left(1 - \frac{V_{GS}}{V_{GS(off)}}\right)^2$$

- Forward transconductance:

$$g_m = g_{m0}\left(1 - \frac{V_{GS}}{V_{GS(off)}}\right)$$

$$g_{m0} = \frac{2I_{DSS}}{|V_{GS(off)}|}$$

E-MOSFETs

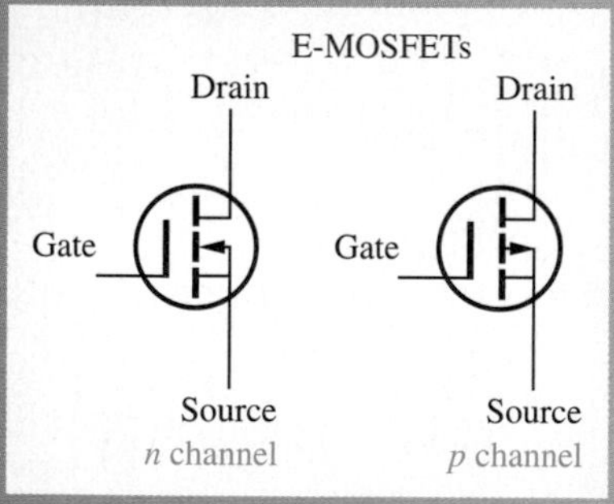

Operates in enhancement mode only.

- V_{GS} must exceed $V_{GS(th)}$.
- *Enhancement mode:*

  n channel: V_{GS} positive

  p channel: V_{GS} negative
- V_{GS} controls I_D.
- Value of V_{GS} at which I_D begins is the threshold voltage, $V_{GS(th)}$.
- Transfer characteristic:

  $$I_D = K(V_{GS} - V_{GS(th)})^2$$
- K in formula can be calculated by substituting datasheet values $I_{D(on)}$ for I_D and V_{GS} at which $I_{D(on)}$ is specified for V_{GS}.

D-MOSFETs

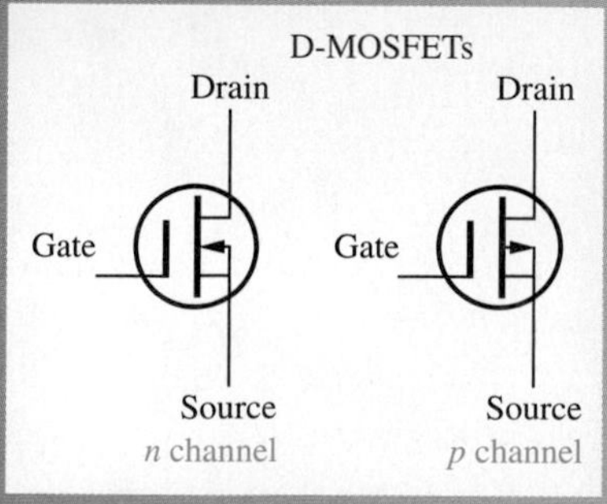

Can be operated in either depletion or enhancement modes. V_{GS} can be either polarity when biased at $V_{GS} = 0$ V.

- *Depletion mode:*

  n channel: V_{GS} negative

  p channel: V_{GS} positive
- *Enhancement mode:*

  n channel: V_{GS} positive

  p channel: V_{GS} negative

- V_{GS} controls I_D.
- Value of V_{GS} at which I_D becomes zero is the cutoff voltage, $V_{GS(off)}$.
- I_{DSS} is drain current when $V_{GS} = 0$.
- Transfer characteristic:

$$I_D \cong I_{DSS}\left(1 - \frac{V_{GS}}{V_{GS(off)}}\right)^2$$

IGBT

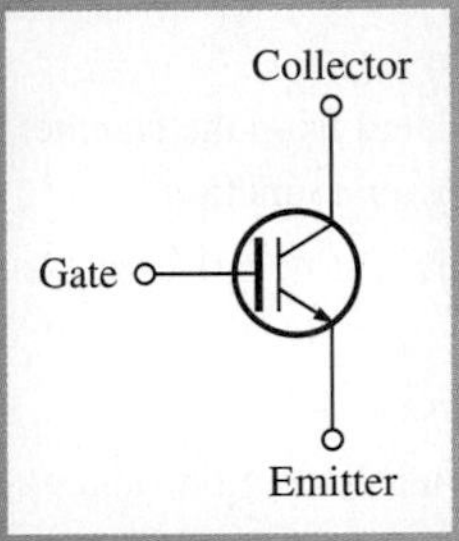

- Voltage controlled like a MOSFET
- Output characteristics like a BJT
- Three terminals: gate, collector, emitter

FET BIASING (Voltage polarities and current directions reverse for *p* channel)

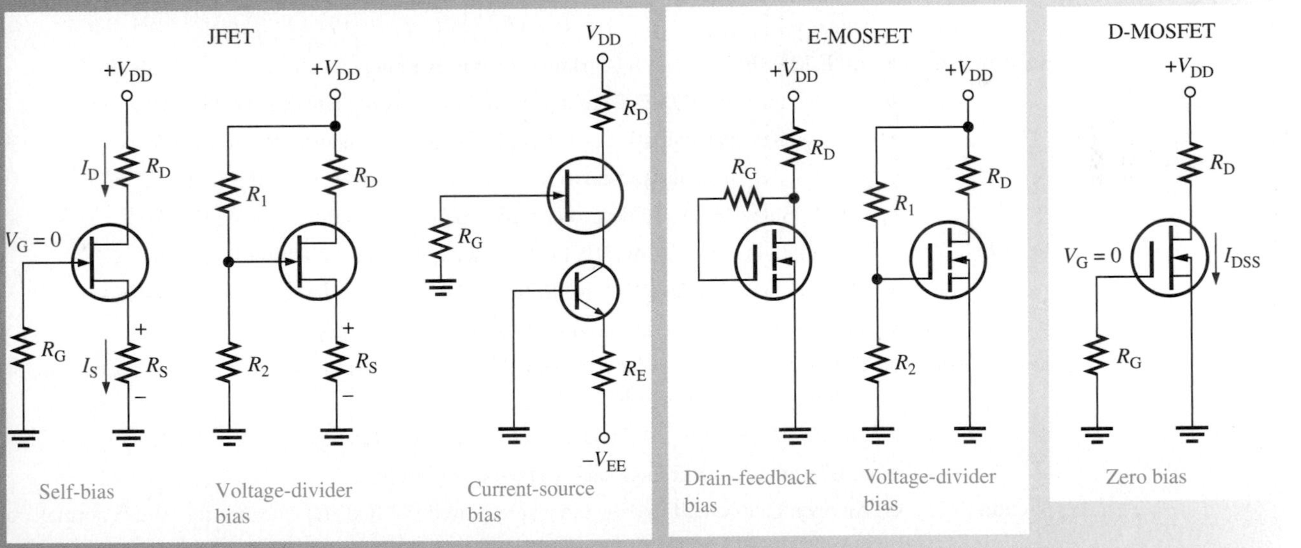

SUMMARY

Section 8–1

- ◆ Field-effect transistors are unipolar devices (one-charge carrier).
- ◆ The three FET terminals are source, drain, and gate.
- ◆ The JFET operates with a reverse-biased *pn* junction (gate-to-source).
- ◆ The high input resistance of a JFET is due to the reverse-biased gate-source junction.
- ◆ Reverse bias of a JFET produces a depletion region within the channel, thus increasing channel resistance.

Section 8–2
- For an *n*-channel JFET, V_{GS} can vary from zero negatively to cutoff, $V_{GS(off)}$. For a *p*-channel JFET, V_{GS} can vary from zero positively to $V_{GS(off)}$.
- I_{DSS} is the constant drain current when $V_{GS} = 0$. This is true for both JFETs and D-MOSFETs.
- A FET is called a *square-law device* because of the relationship of I_D to the square of a term containing V_{GS}.

Section 8–3
- Midpoint bias for a JFET is $I_D = I_{DSS}/2$, obtained by setting $V_{GS} \cong V_{GS(off)}/3.4$.
- The Q-point in a JFET with voltage-divider bias is more stable than in a self-biased JFET.
- Current-source bias increases the stability of a self-biased JFET.

Section 8–4
- A JFET used as a variable resistor is biased in the ohmic region.
- To bias in the ohmic region, I_D must be much smaller than I_{DSS}.
- The gate voltage controls R_{DS} in the ohmic region.
- When a JFET is biased at the origin ($V_{DS} = 0$, $I_D = 0$), the ac channel resistance is controlled by the gate voltage.

Section 8–5
- MOSFETs differ from JFETs in that the gate of a MOSFET is insulated from the channel by an SiO_2 layer, whereas the gate and channel in a JFET are separated by a *pn* junction.
- A depletion MOSFET (D-MOSFET) can operate with a zero, positive, or negative gate-to-source voltage.
- The D-MOSFET has a physical channel between the drain and source.
- For an *n*-channel D-MOSFET, negative values of V_{GS} produce the depletion mode and positive values produce the enhancement mode.
- The enhancement MOSFET (E-MOSFET) has no physical channel.
- Unlike JFETs and D-MOSFETs, the E-MOSFET cannot operate with $V_{GS} = 0$ V.
- A channel is induced in an E-MOSFET by the application of a V_{GS} greater than the threshold value, $V_{GS(th)}$.

Section 8–6
- An E-MOSFET has no I_{DSS} parameter. It is extremely small, if specified (ideally 0).
- An *n*-channel E-MOSFET has a positive $V_{GS(th)}$. A *p*-channel E-MOSFET has a negative $V_{GS(th)}$.
- The transfer characteristic curve for a D-MOSFET intersects the vertical I_D axis.
- The transfer characteristic curve for an E-MOSFET does not intersect the I_D axis.
- All MOS devices are subject to damage from electrostatic discharge (ESD).

Section 8–7
- Midpoint bias for a D-MOSFET is $I_D = I_{DSS}$ obtained by setting $V_{GS} = 0$.
- The gate of a zero-biased D-MOSFET is at 0 V due to a large resistor to ground.
- An E-MOSFET must have a V_{GS} greater than the threshold value.

Section 8–8
- The insulated-gate bipolar transistor (IGBT) combines the input characteristics of a MOSFET with the output characteristics of a BJT.
- The IGBT has three terminals: emitter, gate, and collector.
- IGBTs are used in high-voltage switching applications.

Section 8–9
- An open gate is hard to detect in a zero-biased D-MOSFET because the gate is normally at 0 V; however, erratic behavior may occur.
- An open gate is easy to detect in an E-MOSFET because the gate is normally at a voltage other than 0 V.

KEY TERMS

Key terms and other bold terms in the chapter are defined in the end-of-book glossary.

Depletion In a MOSFET, the process of removing or depleting the channel of charge carriers and thus decreasing the channel conductivity.

Drain One of the three terminals of a FET analogous to the collector of a BJT.

Enhancement In a MOSFET, the process of creating a channel or increasing the conductivity of the channel by the addition of charge carriers.

Gate One of the three terminals of a FET analogous to the base of a BJT.

IGBT Insulated-gate bipolar transistor; a device that combines features of the MOSFET and the BJT and used mainly for high-voltage switching applications.

JFET Junction field-effect transistor; one of two major types of field-effect transistors.

MOSFET Metal oxide semiconductor field-effect transistor; one of two major types of FETs; sometimes called IGFET for insulated-gate FET.

Ohmic region The portion of the FET characteristic curve lying below pinch-off in which Ohm's law applies.

Pinch-off voltage The value of the drain-to-source voltage of a FET at which the drain current becomes constant when the gate-to-source voltage is zero.

Source One of the three terminals of a FET analogous to the emitter of a BJT.

Transconductance (g_m) The ratio of a change in drain current to a change in gate-to-source voltage in a FET.

KEY FORMULAS

8–1	$I_D \cong I_{DSS}\left(1 - \dfrac{V_{GS}}{V_{GS(off)}}\right)^2$	JFET/D-MOSFET transfer characteristic
8–2	$g_m = g_{m0}\left(1 - \dfrac{V_{GS}}{V_{GS(off)}}\right)$	Transconductance
8–3	$g_{m0} = \dfrac{2I_{DSS}}{\lvert V_{GS(off)}\rvert}$	Transconductance at $V_{GS} = 0$
8–4	$I_D = K(V_{GS} - V_{GS(th)})^2$	E-MOSFET transfer characteristic

TRUE/FALSE QUIZ

Answers are at the end of the chapter.

1. The JFET always operates with a reverse-biased gate-to-source *pn* junction.
2. The channel resistance of a JFET is a constant.
3. The gate-to-source voltage of an *n*-channel JFET must be negative.
4. I_D becomes zero at the pinch-off voltage.
5. V_{GS} has no effect on I_D.
6. $V_{GS(off)}$ and V_P are always equal in magnitude but opposite in polarity.
7. The JFET is a square-law device because of the mathematical expression of its transfer characteristic curve.
8. Forward transconductance is the change in drain voltage for a given change in gate voltage.
9. The parameters g_m and y_{fs} are the same.
10. The D-MOSFET can be operated in two modes.
11. An E-MOSFET operates in the depletion mode.
12. A D-MOSFET has a physical channel and an E-MOSFET has an induced channel.
13. ESD means *electronic semiconductor device.*
14. MOSFETs must be handled with care.

CIRCUIT-ACTION QUIZ

Answers are at the end of the chapter.

1. If the drain current in Figure 8–17 is increased, V_{DS} will
   (a) increase **(b)** decrease **(c)** not change
2. If the drain current in Figure 8–17 is increased, V_{GS} will
   (a) increase **(b)** decrease **(c)** not change

3. If the value of R_D in Figure 8–24 is increased, I_D will
   (a) increase (b) decrease (c) not change
4. If the value of R_2 in Figure 8–24 is decreased, V_G will
   (a) increase (b) decrease (c) not change
5. If V_{GS} in Figure 8–47 is increased, I_D will
   (a) increase (b) decrease (c) not change
6. If R_2 in Figure 8–47 opens, V_{GS} will
   (a) increase (b) decrease (c) not change
7. If the value of R_G in Figure 8–50 is increased, V_G will
   (a) increase (b) decrease (c) not change
8. If the value of I_{DSS} in Figure 8–50 is increased, V_{DS} will
   (a) increase (b) decrease (c) not change

SELF-TEST

Answers are at the end of the chapter.

Section 8–1

1. The JFET is
   (a) a unipolar device
   (b) a voltage-controlled device
   (c) a current-controlled device
   (d) answers (a) and (c)
   (e) answers (a) and (b)
2. The channel of a JFET is between the
   (a) gate and drain
   (b) drain and source
   (c) gate and source
   (d) input and output
3. A JFET always operates with
   (a) the gate-to-source *pn* junction reverse-biased
   (b) the gate-to-source *pn* junction forward-biased
   (c) the drain connected to ground
   (d) the gate connected to the source

Section 8–2

4. For $V_{GS} = 0$ V, the drain current becomes constant when V_{DS} exceeds
   (a) cutoff
   (b) V_{DD}
   (c) V_P
   (d) 0 V
5. The constant-current region of a FET lies between
   (a) cutoff and saturation
   (b) cutoff and pinch-off
   (c) 0 and I_{DSS}
   (d) pinch-off and breakdown
6. I_{DSS} is
   (a) the drain current with the source shorted
   (b) the drain current at cutoff
   (c) the maximum possible drain current
   (d) the midpoint drain current

7. Drain current in the constant-current region increases when

(a) the gate-to-source bias voltage decreases

(b) the gate-to-source bias voltage increases

(c) the drain-to-source voltage increases

(d) the drain-to-source voltage decreases

8. In a certain FET circuit, $V_{GS} = 0$ V, $V_{DD} = 15$ V, $I_{DSS} = 15$ mA, and $R_D = 470\ \Omega$. If R_D is decreased to $330\ \Omega$, I_{DSS} is

(a) 19.5 mA **(b)** 10.5 mA

(c) 15 mA **(d)** 1 mA

9. At cutoff, the JFET channel is

(a) at its widest point

(b) completely closed by the depletion region

(c) extremely narrow

(d) reverse-biased

10. A certain JFET datasheet gives $V_{GS(off)} = -4$ V. The pinch-off voltage, V_P,

(a) cannot be determined

(b) is -4 V

(c) depends on V_{GS}

(d) is $+4$ V

11. The JFET in Question 10

(a) is an *n* channel

(b) is a *p* channel

(c) can be either

12. For a certain JFET, $I_{GSS} = 10$ nA at $V_{GS} = 10$ V. The input resistance is

(a) 100 MΩ **(b)** 1 MΩ

(c) 1000 MΩ **(d)** 1000 mΩ

Section 8–3

13. For a certain *p*-channel JFET, $V_{GS(off)} = 8$ V. The value of V_{GS} for an approximate midpoint bias is

(a) 4 V **(b)** 0 V

(c) 1.25 V **(d)** 2.34 V

14. In a self-biased JFET, the gate is at

(a) a positive voltage

(b) 0 V

(c) a negative voltage

(d) ground

Section 8–4

15. The drain-to-source resistance in the ohmic region depends on

(a) V_{GS}

(b) the Q-point values

(c) the slope of the curve at the Q-point

(d) all of these

16. To be used as a variable resistor, a JFET must be

(a) an *n*-channel device

(b) a *p*-channel device

(c) biased in the ohmic region

(d) biased in saturation

17. When a JFET is biased at the origin, the ac channel resistance is determined by

(a) the Q-point values **(b)** V_{GS}

(c) the transconductance **(d)** answers (b) and (c)

Section 8–5

18. A MOSFET differs from a JFET mainly because
(a) of the power rating
(b) the MOSFET has two gates
(c) the JFET has a *pn* junction
(d) MOSFETs do not have a physical channel

19. A D-MOSFET operates in
(a) the depletion mode only
(b) the enhancement mode only
(c) the ohmic region only
(d) both the depletion and enhancement modes

Section 8–6

20. An *n*-channel D-MOSFET with a positive V_{GS} is operating in
(a) the depletion mode
(b) the enhancement mode
(c) cutoff
(d) saturation

21. A certain *p*-channel E-MOSFET has a $V_{GS(th)} = -2$ V. If $V_{GS} = 0$ V, the drain current is
(a) 0 A **(b)** $I_{D(on)}$
(c) maximum **(d)** I_{DSS}

22. In an E-MOSFET, there is no drain current until V_{GS}
(a) reaches $V_{GS(th)}$ **(b)** is positive
(c) is negative **(d)** equals 0 V

23. All MOS devices are subject to damage from
(a) excessive heat
(b) electrostatic discharge
(c) excessive voltage
(d) all of these

Section 8–7

24. A certain D-MOSFET is biased at $V_{GS} = 0$ V. Its datasheet specifies $I_{DSS} = 20$ mA and $V_{GS(off)} = -5$ V. The value of the drain current
(a) is 0 A
(b) cannot be determined
(c) is 20 mA

Section 8–8

25. An IGBT is generally used in
(a) low-power applications
(b) rf applications
(c) high-voltage applications
(d) low-current applications

PROBLEMS

Answers to all odd-numbered problems are at the end of the book.

BASIC PROBLEMS

Section 8–1 **The JFET**

1. The V_{GS} of a *p*-channel JFET is increased from 1 V to 3 V.
(a) Does the depletion region narrow or widen?
(b) Does the resistance of the channel increase or decrease?

2. Why must the gate-to-source voltage of an *n*-channel JFET always be either 0 or negative?

3. Draw the schematic diagrams for a *p*-channel and an *n*-channel JFET. Label the terminals.
4. Show how to connect bias voltages between the gate and source of the JFETs in Figure 8–64.

▶ FIGURE 8–64

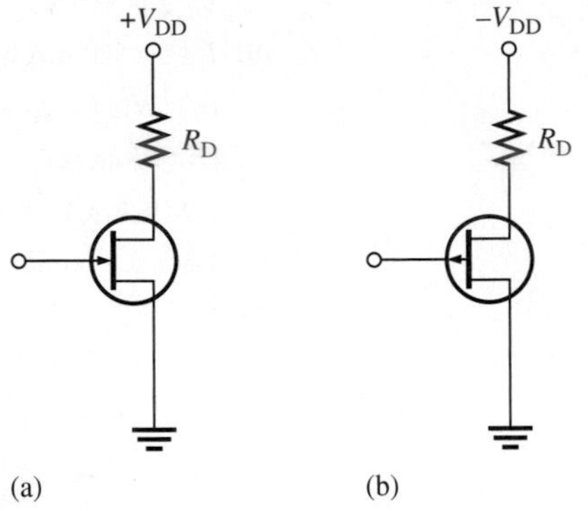

Section 8–2 JFET Characteristics and Parameters

5. A JFET has a specified pinch-off voltage of 5 V. When $V_{GS} = 0$, what is V_{DS} at the point where the drain current becomes constant?
6. A certain *n*-channel JFET is biased such that $V_{GS} = -2$ V. What is the value of $V_{GS(off)}$ if V_P is specified to be 6 V? Is the device on?
7. A certain JFET datasheet gives $V_{GS(off)} = -8$ V and $I_{DSS} = 10$ mA. When $V_{GS} = 0$, what is I_D for values of V_{DS} above pinch off? $V_{DD} = 15$ V.
8. A certain *p*-channel JFET has a $V_{GS(off)} = 6$ V. What is I_D when $V_{GS} = 8$ V?
9. The JFET in Figure 8–65 has a $V_{GS(off)} = -4$ V. Assume that you increase the supply voltage, V_{DD}, beginning at zero until the ammeter reaches a steady value. What does the voltmeter read at this point?

▶ FIGURE 8–65

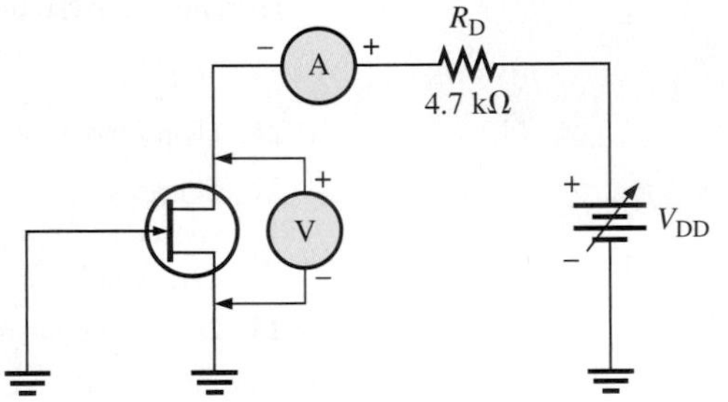

10. The following parameters are obtained from a certain JFET datasheet: $V_{GS(off)} = -8$ V and $I_{DSS} = 5$ mA. Determine the values of I_D for each value of V_{GS} ranging from 0 V to −8 V in 1 V steps. Plot the transfer characteristic curve from these data.
11. For the JFET in Problem 10, what value of V_{GS} is required to set up a drain current of 2.25 mA?
12. For a particular JFET, $g_{m0} = 3200\ \mu$S. What is g_m when $V_{GS} = -4$ V, given that $V_{GS(off)} = -8$ V?
13. Determine the forward transconductance of a JFET biased at $V_{GS} = -2$ V. From the datasheet, $V_{GS(off)} = -7$ V and $g_m = 2000\ \mu$S at $V_{GS} = 0$ V. Also determine the forward transfer conductance, g_{fs}.
14. A *p*-channel JFET datasheet shows that $I_{GSS} = 5$ nA at $V_{GS} = 10$ V. Determine the input resistance.
15. Using Equation 8–1, plot the transfer characteristic curve for a JFET with $I_{DSS} = 8$ mA and $V_{GS(off)} = -5$ V. Use at least four points.

Section 8–3 JFET Biasing

16. An *n*-channel self-biased JFET has a drain current of 12 mA and a 100 Ω source resistor. What is the value of V_{GS}?

17. Determine the value of R_S required for a self-biased JFET to produce a V_{GS} of -4 V when $I_D = 5$ mA.

18. Determine the value of R_S required for a self-biased JFET to produce $I_D = 2.5$ mA when $V_{GS} = -3$ V.

19. $I_{DSS} = 20$ mA and $V_{GS(off)} = -6$ V for a particular JFET.
   (a) What is I_D when $V_{GS} = 0$ V?
   (b) What is I_D when $V_{GS} = V_{GS(off)}$?
   (c) If V_{GS} is increased from -4 V to -1 V, does I_D increase or decrease?

20. For each circuit in Figure 8–66, determine V_{DS} and V_{GS}.

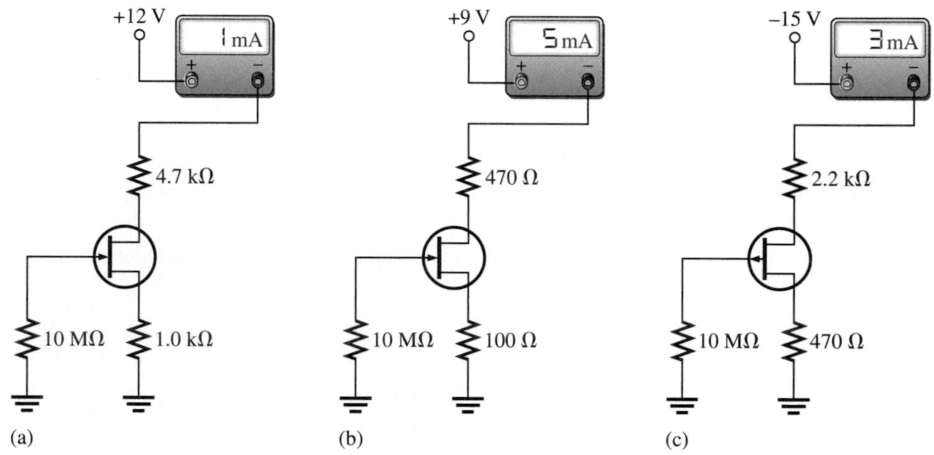

▲ FIGURE 8–66

Multisim file circuits are identified with a CD logo and are in the Problems folder on your CD-ROM. Filenames correspond to figure numbers (e.g., F08-66).

21. Using the curve in Figure 8–67, determine the value of R_S required for a 9.5 mA drain current.

22. Set up a midpoint bias for a JFET with $I_{DSS} = 14$ mA and $V_{GS(off)} = -10$ V. Use a 24 V dc source as the supply voltage. Show the circuit and resistor values. Indicate the values of I_D, V_{GS}, and V_{DS}.

23. Determine the total input resistance in Figure 8–68. $I_{GSS} = 20$ nA at $V_{GS} = -10$ V.

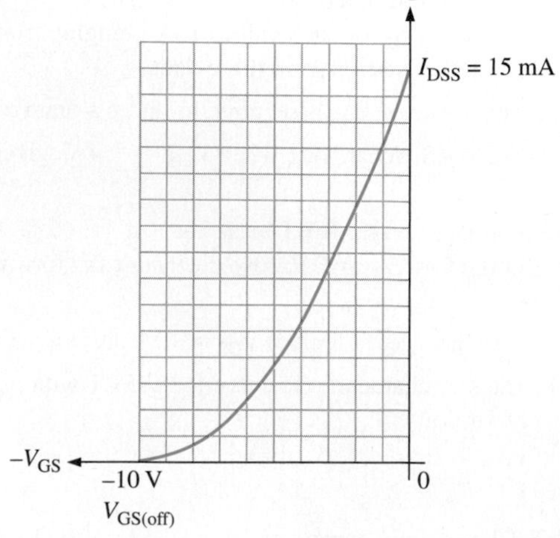

▲ FIGURE 8–67

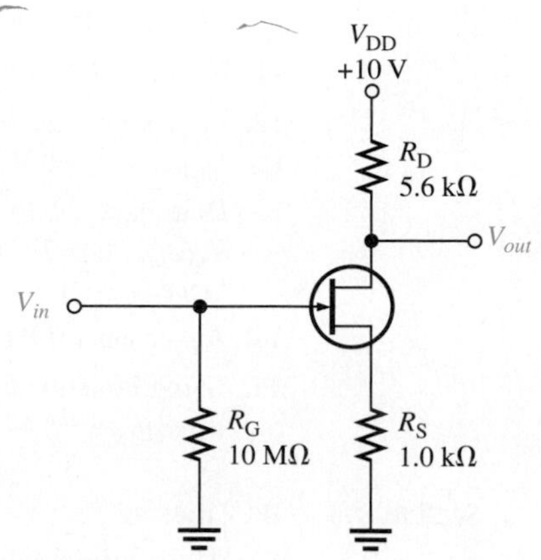

▲ FIGURE 8–68

24. Graphically determine the Q-point for the circuit in Figure 8–69(a) using the transfer characteristic curve in Figure 8–69(b).

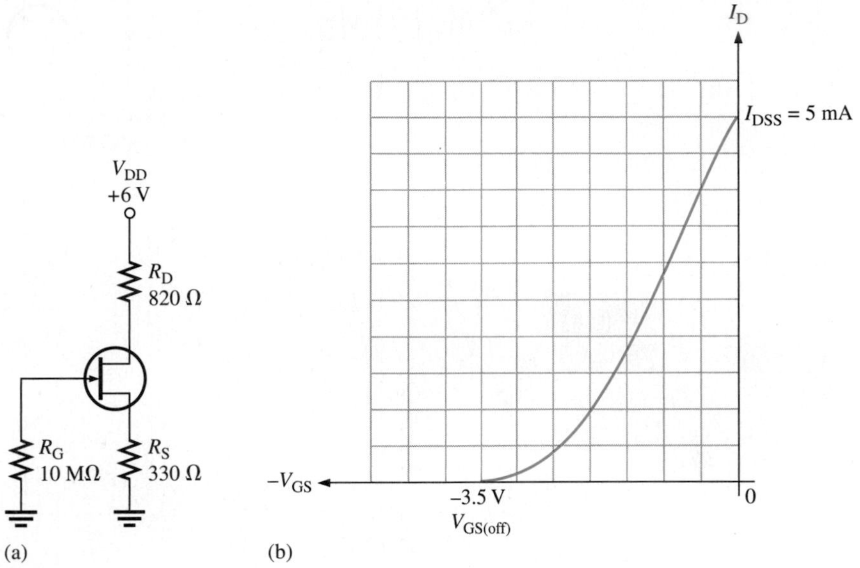

▲ FIGURE 8–69

25. Find the Q-point for the p-channel JFET circuit in Figure 8–70.

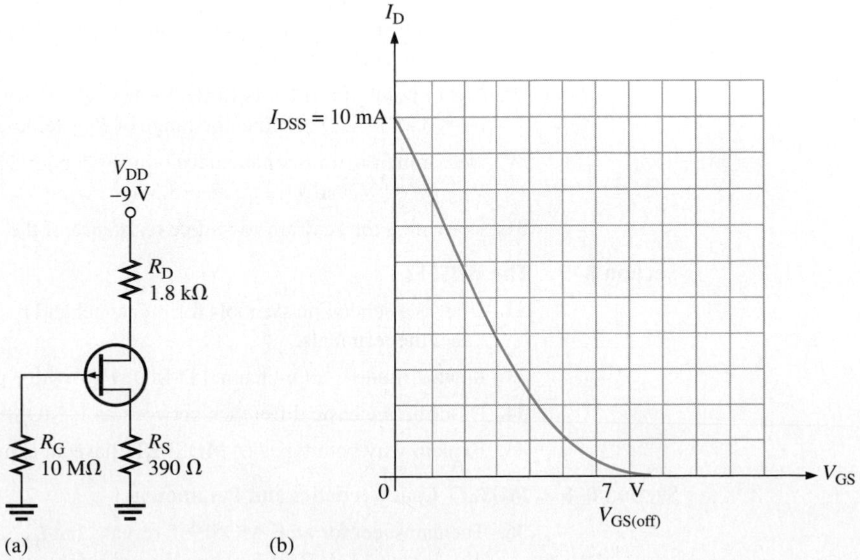

▲ FIGURE 8–70

FIGURE 8–71

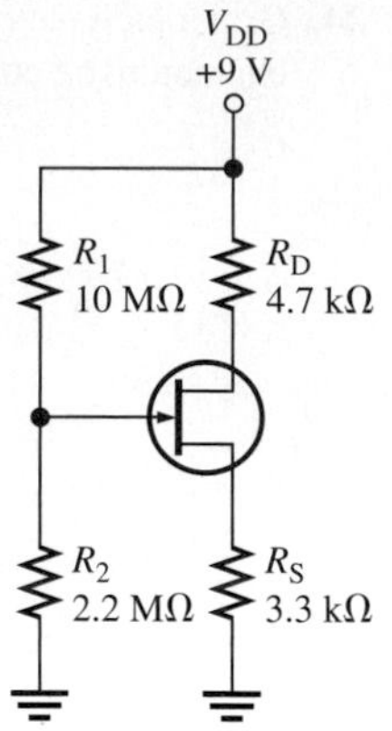

FIGURE 8–72

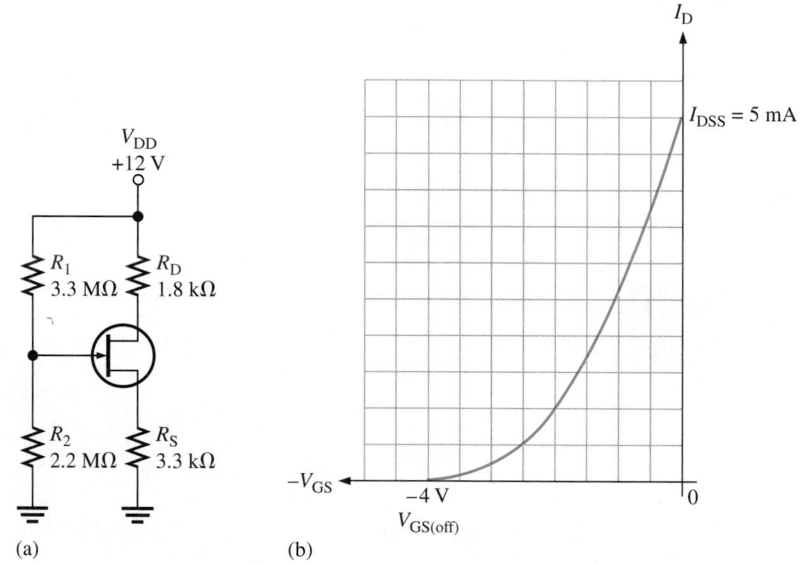

26. Given that the drain-to-ground voltage in Figure 8–71 is 5 V, determine the Q-point of the circuit.
27. Find the Q-point values for the JFET with voltage-divider bias in Figure 8–72.

Section 8–4 **The Ohmic Region**

28. A certain JFET is biased in the ohmic region at $V_{DS} = 0.8$ V and $I_D = 0.20$ mA. What is the drain-to-source resistance?
29. The Q-point of a JFET is varied from $V_{DS} = 0.4$ V and $I_D = 0.15$ mA to $V_{DS} = 0.6$ V and $I_D = 0.45$ mA. Determine the range of R_{DS} values.
30. Determine the transconductance of a JFET biased at the origin given that $g_{m0} = 1.5$ mS, $V_{GS} = -1$ V, and $V_{GS(off)} = -3.5$ V.
31. Determine the ac drain-to-source resistance of the JFET in Problem 30.

Section 8–5 **The MOSFET**

32. Draw the schematic symbols for *n*-channel and *p*-channel E-MOSFETs and D-MOSFETs. Label the terminals.
33. In what mode is an *n*-channel D-MOSFET with a positive V_{GS} operating?
34. Describe the basic difference between an E-MOSFET and a D-MOSFET.
35. Explain why both types of MOSFETs have an extremely high input resistance at the gate.

Section 8–6 **MOSFET Characteristics and Parameters**

36. The datasheet for an E-MOSFET reveals that $I_{D(on)} = 10$ mA at $V_{GS} = -12$ V and $V_{GS(th)} = -3$ V. Find I_D when $V_{GS} = -6$ V.
37. Determine I_{DSS}, given $I_D = 3$ mA, $V_{GS} = -2$ V, and $V_{GS(off)} = -10$ V.

38. The datasheet for a certain D-MOSFET gives $V_{GS(off)} = -5$ V and $I_{DSS} = 8$ mA.

(a) Is this device *p* channel or *n* channel?

(b) Determine I_D for values of V_{GS} ranging from -5 V to $+5$ V in increments of 1 V.

(c) Plot the transfer characteristic curve using the data from part (b).

Section 8–7 MOSFET Biasing

39. Determine in which mode (depletion, enhancement or neither) each D-MOSFET in Figure 8–73 is biased.

FIGURE 8–73

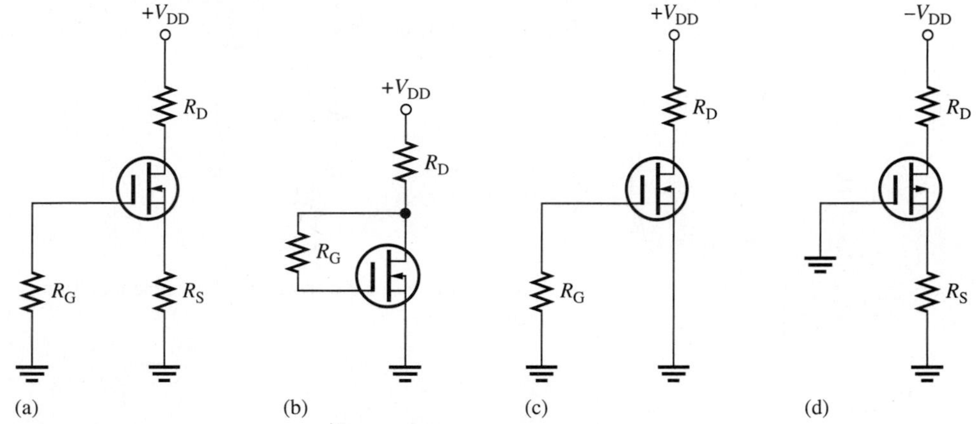

40. Each E-MOSFET in Figure 8–74 has a $V_{GS(th)}$ of $+5$ V or -5 V, depending on whether it is an *n*-channel or a *p*-channel device. Determine whether each MOSFET is on or off.

FIGURE 8–74

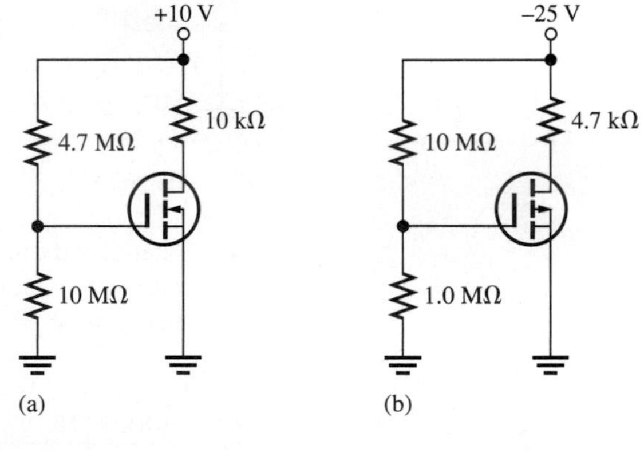

41. Determine V_{DS} for each circuit in Figure 8–75. $I_{DSS} = 8$ mA.

FIGURE 8–75

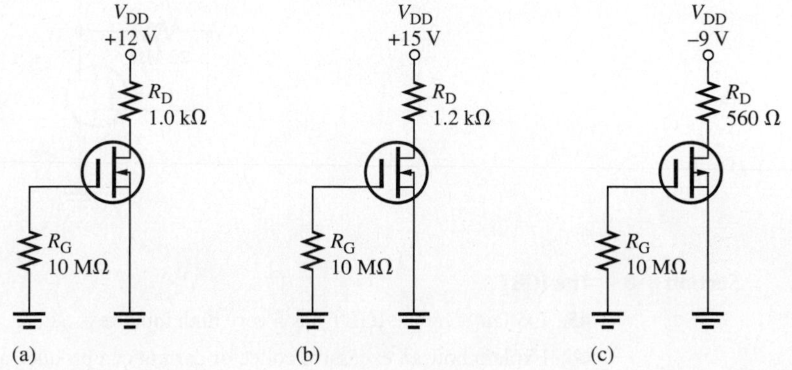

42. Find V_{GS} and V_{DS} for the E-MOSFETs in Figure 8–76. Datasheet information is listed with each circuit.

▶ FIGURE 8–76

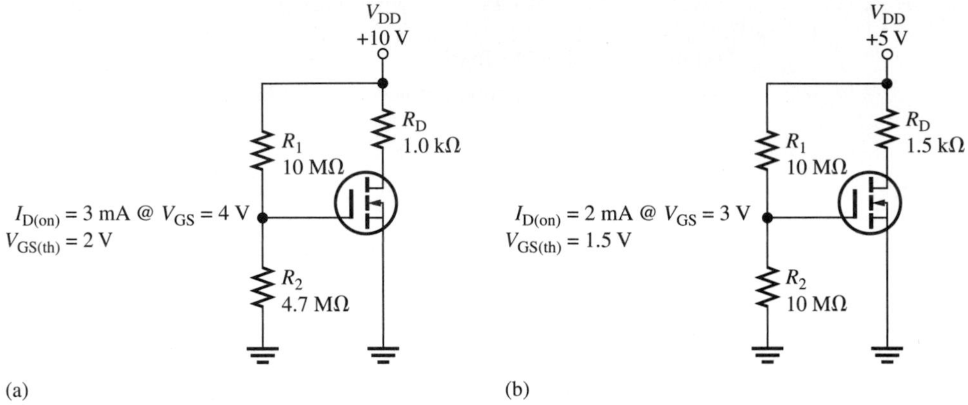

43. Based on the V_{GS} measurements, determine the drain current and drain-to-source voltage for each circuit in Figure 8–77.

▶ FIGURE 8–77

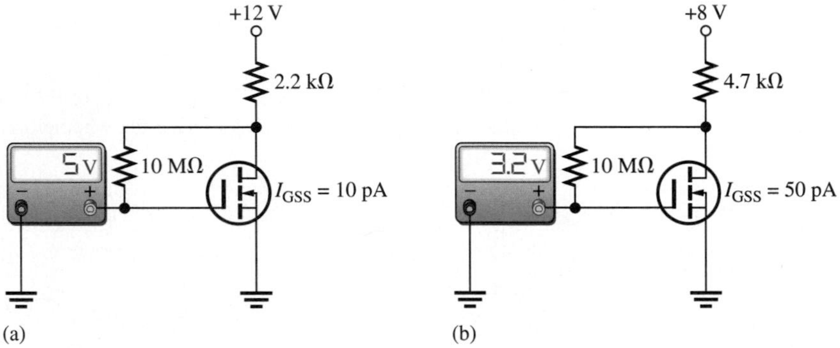

44. Determine the actual gate-to-source voltage in Figure 8–78 by taking into account the gate leakage current, I_{GSS}. Assume that I_{GSS} is 50 pA and I_D is 1 mA under the existing bias conditions.

▶ FIGURE 8–78

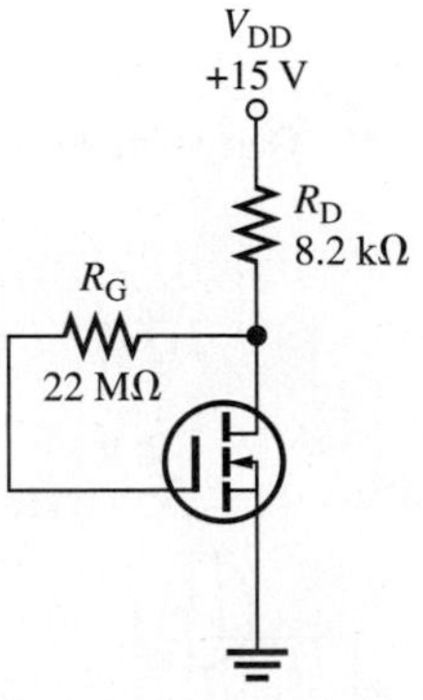

Section 8–8 The IGBT

45. Explain why the IGBT has a very high input resistance.

46. Explain how an excessive collector current can produce a latch-up condition in an IGBT.

Section 8–9 Troubleshooting

47. The current reading in Figure 8–66(a) suddenly goes to zero. What are the possible faults?

48. The current reading in Figure 8–66(b) suddenly jumps to approximately 16 mA. What are the possible faults?

49. If the supply voltage in Figure 8–66(c) is changed to -20 V, what would you see on the ammeter?

50. You measure $+10$ V at the drain of the MOSFET in Figure 8–74(a). The transistor checks good and the ground connections are okay. What can be the problem?

51. You measure approximately 0 V at the drain of the MOSFET in Figure 8–74(b). You can find no shorts and the transistor checks good. What is the most likely problem?

APPLICATION ACTIVITY PROBLEMS

52. Refer to Figure 8–58 and determine the sensor voltage for each of the following pH values.
(a) 2 **(b)** 5 **(c)** 7 **(d)** 11

53. Referring to the transconductance curves for the BF998 in Figure 8–79, determine the change in I_D when the bias on the second gate is changed from 6 V to 1 V and V_{G1S} is 0.0 V. Each curve represents a different V_{G2S} value.

FIGURE 8–79

Transconductance curves for BF998.

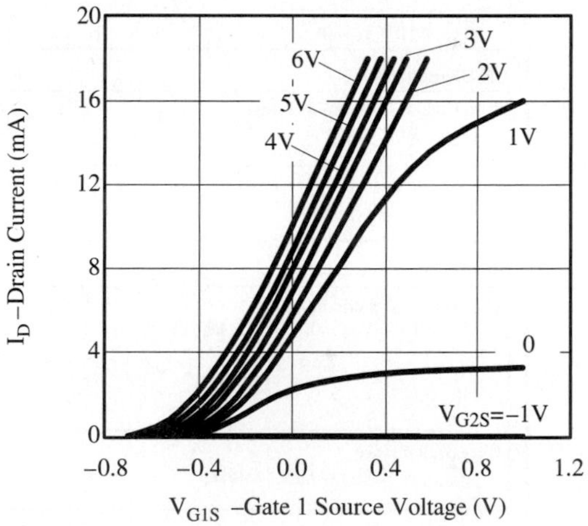

54. Refer to Figure 8–61 and plot the transconductance curve (I_D vs. V_{GS1}).

55. Refer to Figure 8–79. Determine the output voltage of the circuit in Figure 8–61 if $V_{G1S} = V_{sensor} = 0$ V and R_2 is changed to 50 kΩ.

DATASHEET PROBLEMS

56. What type of FET is the 2N5457?

57. Referring to the datasheet in Figure 8–14, determine the following:
(a) Minimum $V_{GS(off)}$ for the 2N5457.
(b) Maximum drain-to-source voltage for the 2N5457.
(c) Maximum power dissipation for the 2N5458 at an ambient temperature of 25°C.
(d) Maximum reverse gate-to-source voltage for the 2N5459.

58. Referring to Figure 8–14, determine the maximum power dissipation for a 2N5457 at an ambient temperature of 65°C.

59. Referring to Figure 8–14, determine the minimum g_{m0} for the 2N5459 at a frequency of 1 kHz.

60. Referring to Figure 8–14, what is the typical drain current in a 2N5459 for $V_{GS} = 0$ V?

61. Referring to the 2N3796 datasheet in Figure 8–80, determine the drain current for $V_{GS} = 0$ V.

62. Referring to Figure 8–80, what is the drain current for a 2N3796 when $V_{GS} = 6$ V?

63. Referring to the datasheet in Figure 8–80, determine I_D in a 2N3797 when $V_{GS} = +3$ V. Determine I_D when $V_{GS} = -2$ V.

Maximum Ratings

Rating	Symbol	Value	Unit
Drain-Source voltage 2N3796 2N3797	V_{DS}	 25 20	V dc
Gate-Source voltage	V_{GS}	±10	V dc
Drain current	I_D	20	mA dc
Total device dissipation @ T_A = 25°C Derate above 25°C	P_D	200 1.14	mW mW/°C
Junction temperature range	T_J	+175	°C
Storage channel temperature range	T_{stg}	–65 to +200	°C

2N3796
2N3797

Case 22-03, Style 2
TO-18 (TO-206AA)

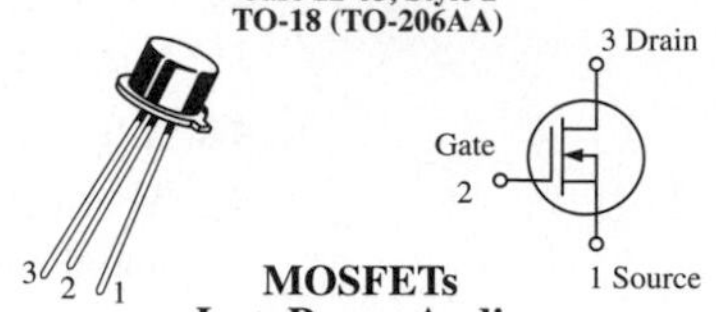

MOSFETs
Low Power Audio
***N* channel — Depletion**

Electrical Characteristics (T_A = 25°C unless otherwise noted.)

Characteristic		Symbol	Min	Typ	Max	Unit
OFF Characteristics						
Drain-Source breakdown voltage (V_{GS} = –4.0 V, I_D = 5.0 μA) (V_{GS} = –7.0 V, I_D = 5.0 μA)	 2N3796 2N3797	$V_{(BR)DSX}$	 25 20	 30 25	 – –	V dc
Gate reverse current (V_{GS} = –10 V, V_{DS} = 0) (V_{GS} = –10 V, V_{DS} = 0, T_A = 150°C)		I_{GSS}	 – –	 – –	 1.0 200	pA dc
Gate-Source cutoff voltage (I_D = 0.5 μA, V_{DS} = 10 V) (I_D = 2.0 μA, V_{DS} = 10 V)	 2N3796 2N3797	$V_{GS(off)}$	 – –	 –3.0 –5.0	 –4.0 –7.0	V dc
Drain-Gate reverse current (V_{DG} = 10 V, I_S = 0)		I_{DGO}	–	–	1.0	pA dc
ON Characteristics						
Zero-Gate-Voltage drain current (V_{DS} = 10 V, V_{GS} = 0)	2N3796 2N3797	I_{DSS}	0.5 2.0	1.5 2.9	3.0 6.0	mA dc
On-State drain current (V_{DS} = 10 V, V_{GS} = +3.5 V)	2N3796 2N3797	$I_{D(on)}$	7.0 9.0	8.3 14	14 18	mA dc
Small-Signal Characteristics						
Forward-transfer admittance (V_{DS} = 10 V, V_{GS} = 0, f = 1.0 kHz) (V_{DS} = 10 V, V_{GS} = 0, f = 1.0 MHz)	 2N3796 2N3797 2N3796 2N3797	$\|y_{fs}\|$	 900 1500 900 1500	 1200 2300 – –	 1800 3000 – –	μmhos or μS
Output admittance (V_{DS} = 10 V, V_{GS} = 0, f = 1.0 kHz)	2N3796 2N3797	$\|y_{os}\|$	– –	12 27	25 60	μmhos or μS
Input capacitance (V_{DS} = 10 V, V_{GS} = 0, f = 1.0 MHz)	2N3796 2N3797	C_{iss}	– –	5.0 6.0	7.0 8.0	pF
Reverse transfer capacitance (V_{DS} = 10 V, V_{GS} = 0, f = 1.0 MHz)		C_{rss}	–	0.5	0.8	pF
Functional Characteristics						
Noise figure (V_{DS} = 10 V, V_{GS} = 0, f = 1.0 kHz, R_S = 3 megohms)		NF	–	3.8	–	dB

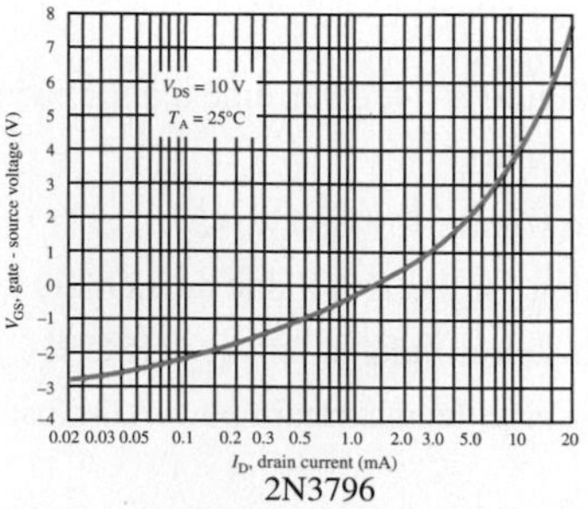

2N3796

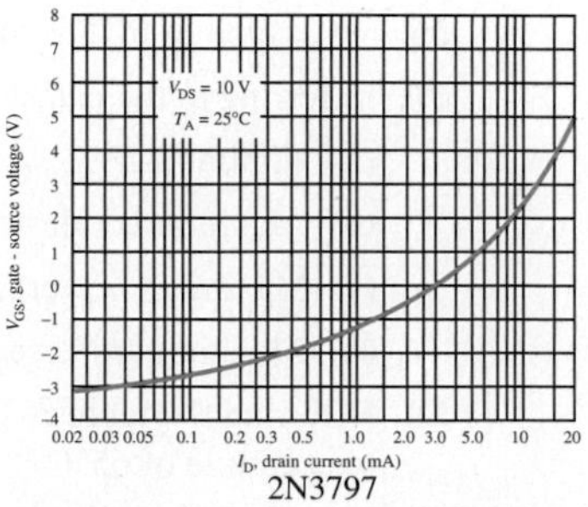

2N3797

▲ **FIGURE 8–80**

Partial datasheet for the 2N3797 D-MOSFET.

64. Referring to Figure 8–80, how much does the maximum forward transconductance of a 2N3796 change over a range of signal frequencies from 1 kHz to 1 MHz?

65. Referring to Figure 8–80, determine the typical value of gate-to-source voltage at which the 2N3796 will go into cutoff.

ADVANCED PROBLEMS

66. Find V_{DS} and V_{GS} in Figure 8–81 using minimum datasheet values.

FIGURE 8–81

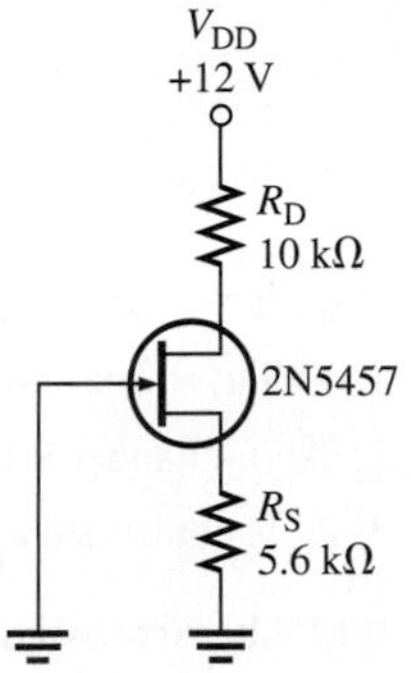

67. Determine the maximum I_D and V_{GS} for the circuit in Figure 8–82.

FIGURE 8–82

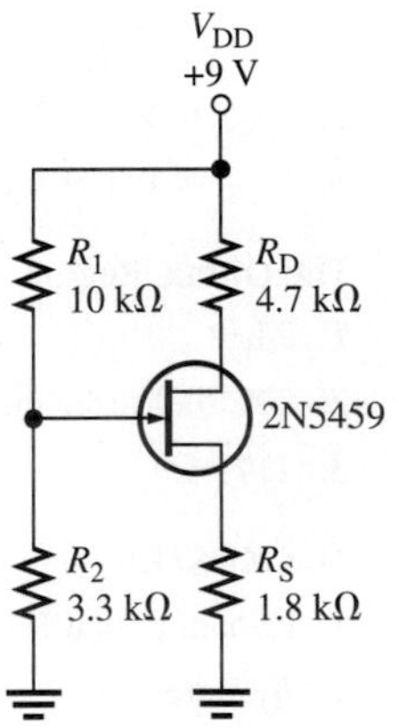

68. Determine the range of possible Q-point values from minimum to maximum for the circuit in Figure 8–81.

69. Find the drain-to-source voltage for the pH sensor circuit in Figure 8–59 when a pH of 5 is measured. Assume the rheostat is set to produce 4 V at the drain when a pH of 7 is measured.

70. Design a MOSFET circuit with zero bias using a 2N3797 that operates from a +9 V dc supply and produces a V_{DS} of 4.5 V. The maximum current drawn from the source is to be 1 mA.

71. Design a circuit using an *n*-channel E-MOSFET with the following datasheet specifications: $I_{D(on)} = 500$ mA@ $V_{GS} = 10$ V and $V_{GS(th)} = 1$ V. Use a +12 V dc supply voltage with voltage-divider bias. The voltage at the drain with respect to ground is to be +8 V. The maximum current from the supply is to be 20 mA.

MULTISIM TROUBLESHOOTING PROBLEMS

These file circuits are in the Troubleshooting Problems folder on your CD-ROM.

72. Open file TSP08-72 and determine the fault.

73. Open file TSP08-73 and determine the fault.

74. Open file TSP08-74 and determine the fault.
75. Open file TSP08-75 and determine the fault.
76. Open file TSP08-76 and determine the fault.
77. Open file TSP08-77 and determine the fault.
78. Open file TSP08-78 and determine the fault.
79. Open file TSP08-79 and determine the fault.
80. Open file TSP08-80 and determine the fault.

ANSWERS

SECTION CHECKUPS

Section 8–1 **The JFET**

1. Drain, source, and gate
2. An *n*-channel JFET requires a negative V_{GS}.
3. I_D is controlled by V_{GS}.

Section 8–2 **JFET Characteristics and Parameters**

1. When $V_{DS} = 7\text{ V}$ at pinch-off and $V_{GS} = 0\text{ V}, V_P = -7\text{ V}$.
2. As V_{GS} increases negatively, I_D decreases.
3. For $V_P = -3\text{ V}$, $V_{GS(off)} = +3\text{ V}$.

Section 8–3 **JFET Biasing**

1. A *p*-channel JFET requires a positive V_{GS}.
2. $V_{GS} = V_G - V_S = 0\text{ V} - (8\text{ mA})(1.0\text{ k}\Omega) = -8\text{ V}$
3. $V_{GS} = V_G - V_S = 3\text{ V} - 5\text{ V} = -2\text{ V}$

Section 8–4 **The Ohmic Region**

1. $2\text{ k}\Omega$
2. Channel resistance increases.
3. $1176\ \Omega$

Section 8–5 **The MOSFET**

1. Enhancement MOSFET (E-MOSFET) and depletion MOSFET (D-MOSFET)
2. I_D increases.
3. I_D decreases.

Section 8–6 **MOSFET Characteristics and Parameters**

1. The D-MOSFET has a structural channel; the E-MOSFET does not.
2. $V_{GS(th)}$ and K are not specified for D-MOSFETs.
3. ESD is ElectroStatic Discharge.

Section 8 –7 **MOSFET Biasing**

1. When $V_{GS} = 0\text{ V}$, the drain current is equal to I_{DSS}.
2. V_{GS} must exceed $V_{GS(th)} = 2\text{ V}$ for conduction to occur.

Section 8–8 **The IGBT**

1. IGBT stands for insulated-gate bipolar transistor.
2. High-voltage switching applications
3. The IGBT has a lower output saturation voltage than the MOSFET.
4. The IGBT has a very high input resistance compared to a BJT.
5. Latch-up is a condition in which the IGBT is in the *on* state and cannot be turned off by the gate voltage.

Section 8–9 Troubleshooting

1. R_S open, no ground connection
2. Because V_{GS} remains at approximately zero
3. The device is off and $V_D = V_{DD}$.

RELATED PROBLEMS FOR EXAMPLES

8–1 I_D remains at approximately 12 mA.

8–2 $V_P = -4$ V

8–3 $I_D = 3.52$ mA

8–4 $g_m = 1800\ \mu$S; $I_D = 4.32$ mA

8–5 $R_{IN} = 25{,}000$ MΩ

8–6 $V_{DS} = 2$ V; $V_{GS} = -3.12$ V

8–7 $R_S = 245\ \Omega$

8–8 $R_S = 889\ \Omega$

8–9 $R_S = 294\ \Omega$; $R_D = 3$ kΩ

8–10 $V_{GS} \cong -1.8$ V, $I_D \cong 1.8$ mA

8–11 $I_D = 1.81$ mA, $V_{GS} = -2.44$ V

8–12 $I_D \cong 1.25$ mA, $V_{GS} \cong -2.25$ V

8–13 I_D does not change because of the constant-current source.

8–14 R_{DS} values would not change because slopes are constant.

8–15 1.07 kΩ

8–16 $I_D = 25$ mA

8–17 **(a)** *p* channel **(b)** 6.48 mA **(c)** 35.3 mA

8–18 $V_{GS} = 3.13$ V; $V_{DS} = 21.4$ V

8–19 $I_D = 2.13$ mA

8–20 $V_{DS} = 5.6$ V

TRUE/FALSE QUIZ

1. T **2.** F **3.** T **4.** F **5.** F **6.** T **7.** T
8. F **9.** T **10.** T **11.** F **12.** T **13.** F **14.** T

CIRCUIT-ACTION QUIZ

1. (b) **2.** (c) **3.** (b) **4.** (b) **5.** (a) **6.** (a) **7.** (c) **8.** (b)

SELF-TEST

1. (e) **2.** (b) **3.** (a) **4.** (c) **5.** (d) **6.** (c) **7.** (a) **8.** (c)
9. (b) **10.** (d) **11.** (a) **12.** (c) **13.** (d) **14.** (b) **15.** (d) **16.** (c)
17. (d) **18.** (c) **19.** (d) **20.** (b) **21.** (a) **22.** (a) **23.** (d) **24.** (c)
25. (c)

9 FET Amplifiers and Switching Circuits

CHAPTER OUTLINE

CHAPTER OBJECTIVES

- Explain and analyze the operation of common-source FET amplifiers
- Explain and analyze the operation of common-drain FET amplifiers
- Explain and analyze the operation of common-gate FET amplifiers
- Discuss the operation of a class D amplifier
- Describe how MOSFETs are used in analog switching applications
- Describe how MOSFETs are used in digital switching applications
- Troubleshoot FET amplifiers

KEY TERMS

- Common-source
- Common-drain
- Source-follower
- Common-gate
- Class D amplifier
- Pulse-width modulation
- Analog switch
- CMOS

APPLICATION ACTIVITY PREVIEW

A JFET common-source amplifier and a common-gate amplifier are combined in a cascode arrangement for an active antenna. Cascode amplifiers are often used for RF (radio frequency) applications to achieve improved high-frequency performance. In this application, the cascode amplifier provides a high resistance input for a whip antenna, as well as high gain to amplify extremely small antenna signals.

VISIT THE COMPANION WEBSITE

Study aids for this chapter are available at http://www.prenhall.com/floyd

INTRODUCTION

Because of their extremely high input resistance and low noise, FET amplifiers are a good choice for certain applications, such as amplifying low-level signals in the first stage of a communication receiver. FETs also have the advantage in certain power amplifiers and in switching circuits because biasing is simple and more efficient. The standard amplifier configurations are common-source (CS), common-drain (CD) and common-base (CB), which are analogous to CE, CC, and CB configurations of BJTs.

FETs can be used in any of the amplifier types introduced earlier (class A, class B, and class C). In some cases, the FET circuit will perform better; in other cases, the BJT circuit is superior because BJTs have higher gain and better linearity. Another type of amplifier (class D) is introduced in this chapter because FETs are always superior to BJTs in class D and you will rarely see BJTs used in class D. The class D amplifier is a switching amplifier that is normally either in cutoff or saturation. It is used in analog power amplifiers with a circuit called a pulse-width modulator, introduced in Section 9–4.

FETs are superior to BJTs in nearly all switching applications. Various switching circuits—analog switches, analog multiplexers, and switched capacitors—are discussed. In addition, common digital switching circuits are introduced using CMOS (complementary MOS).

9–1 THE COMMON-SOURCE AMPLIFIER

When used in amplifier applications, the FET has an important advantage compared to the BJT due to the FET's extremely high input impedance. Disadvantages, however, include higher distortion and lower gain. The particular application will usually determine which type of transistor is best suited. The common-source (CS) amplifier is comparable to the common-emitter BJT amplifier that you studied in Chapter 6.

After completing this section, you should be able to

- **Explain and analyze the operation of common-source FET amplifiers**
    - Analyze JFET and MOSFET CS amplifiers
    - Determine the dc values of a CS amplifier
    - Develop an ac equivalent circuit and determine the voltage gain of a CS amplifier
    - Describe the effect of an ac load on the voltage gain
    - Discuss phase inversion in a CS amplifier
    - Determine the input resistance of a CS amplifier

An equivalent FET model is shown in Figure 9–1. In part (a), the internal resistance, r'_{gs}, appears between the gate and source, and a current source equal to $g_m V_{gs}$ appears between the drain and source. Also, the internal drain-to-source resistance, r'_{ds}, is included. In part (b), a simplified ideal model is shown. The resistance, r'_{gs}, is assumed to be extremely large so that an open circuit between the gate and source can be assumed. Also, r'_{ds} is assumed large enough to neglect.

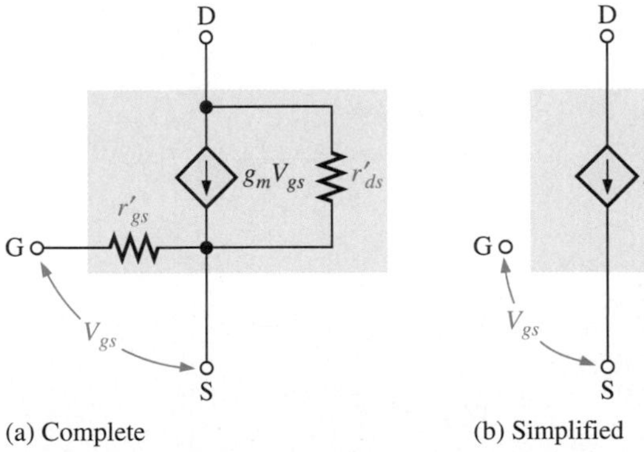

◀ FIGURE 9–1

Internal FET equivalent circuits.

An ideal FET circuit model with an external ac drain resistance is shown in Figure 9–2. The ac voltage gain of this circuit is V_{out}/V_{in}, where $V_{in} = V_{gs}$ and $V_{out} = V_{ds}$. The voltage gain expression is, therefore,

$$A_v = \frac{V_{ds}}{V_{gs}}$$

From the equivalent circuit,

$$V_{ds} = I_d R_d$$

and from the definition of transconductance,

$$V_{gs} = \frac{I_d}{g_m}$$

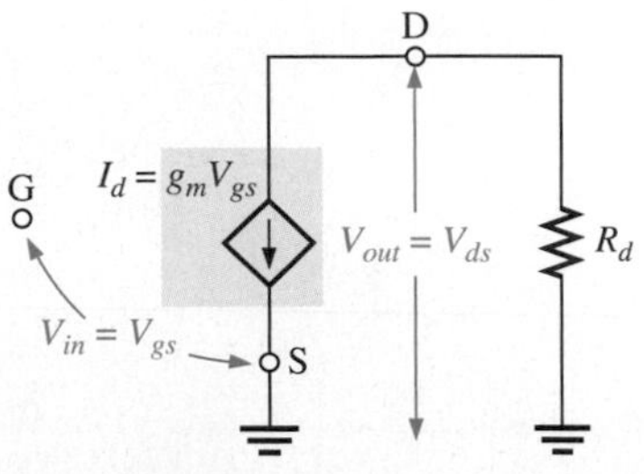

▲ FIGURE 9–2

Simplified FET equivalent circuit with an external ac drain resistance.

Substituting the two preceding expressions into the equation for voltage gain yields

$$A_v = \frac{I_d R_d}{I_d/g_m} = \frac{g_m I_d R_d}{I_d}$$

Equation 9–1

$$\mathbf{A_v = g_m R_d}$$

EXAMPLE 9–1

A certain JFET has a $g_m = 4$ mS. With an external ac drain resistance of 1.5 kΩ, what is the ideal voltage gain?

Solution $A_v = g_m R_d = (4\text{ mS})(1.5\text{ k}\Omega) = \mathbf{6}$

*Related Problem** What is the ideal voltage gain when $g_m = 6000\ \mu$S and $R_d = 2.2$ kΩ?

*Answers are at the end of the chapter.

A **common-source** JFET amplifier is one in which the ac input signal is applied to the gate and the ac output signal is taken from the drain. The source terminal is common to both the input and output signal. A common-source amplifier either has no source resistor or has a bypassed source resistor, so the source is connected to ac ground. A self-biased common-source n-channel JFET amplifier with an ac source capacitively coupled to the gate is shown in Figure 9–3(a). The resistor, R_G, serves two purposes: It keeps the gate at approximately 0 V dc (because I_{GSS} is extremely small), and its large value (usually several megohms) prevents loading of the ac signal source. A bias voltage is produced by the drop across R_S. The bypass capacitor, C_2, keeps the source of the JFET at ac ground.

FIGURE 9–3

JFET common-source amplifier.

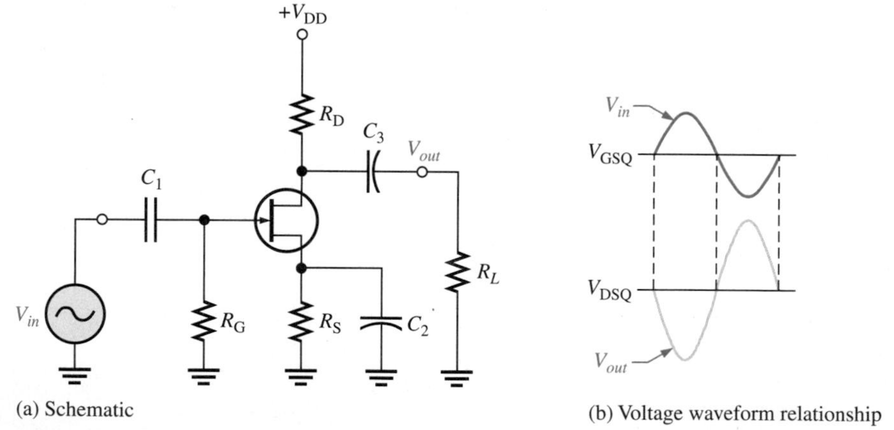

The input signal voltage causes the gate-to-source voltage to swing above and below its Q-point value (V_{GSQ}), causing a corresponding swing in drain current. As the drain current increases, the voltage drop across R_D also increases, causing the drain voltage to decrease. The drain current swings above and below its Q-point value in phase with the gate-to-source voltage. The drain-to-source voltage swings above and below its Q-point value (V_{DSQ}) and is 180° out of phase with the gate-to-source voltage, as illustrated in Figure 9–3(b).

A Graphical Picture The operation just described for an n-channel JFET is illustrated graphically on both the transfer characteristic curve and the drain characteristic curve in Figure 9–4. Part (a) shows how a sinusoidal variation, V_{gs}, produces a corresponding sinusoidal variation in I_d. As V_{gs} swings from its Q-point value to a more negative value, I_d decreases from its Q-point value. As V_{gs} swings to a less negative value, I_d increases.

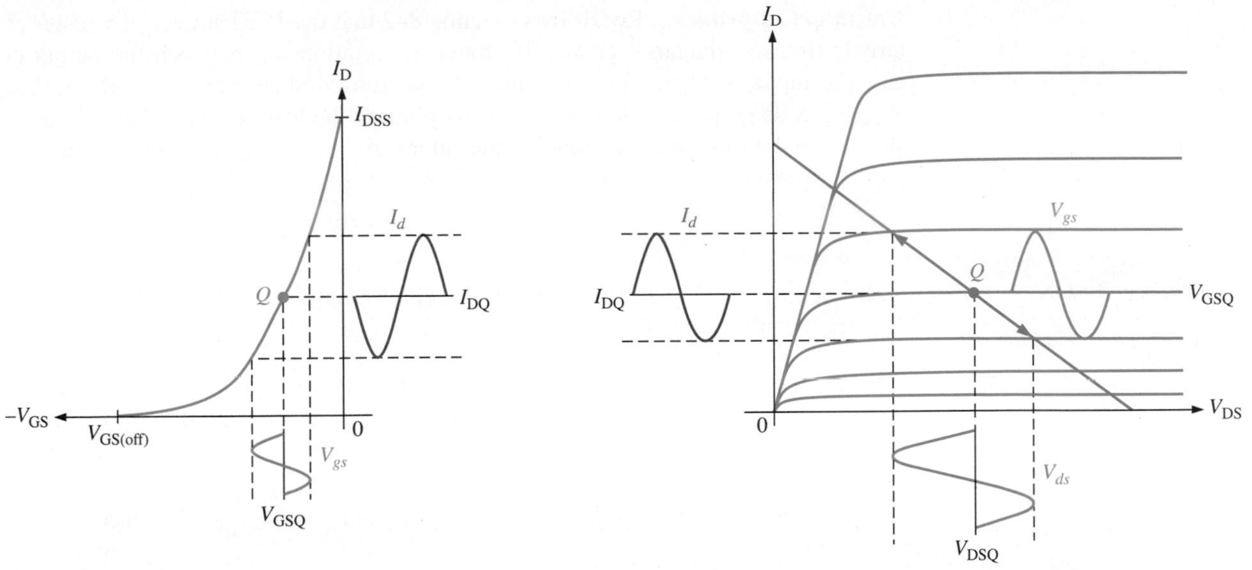

(a) JFET (n-channel) transfer characteristic curve showing signal operation

(b) JFET (n-channel) drain curves showing signal operation

▲ **FIGURE 9–4**

JFET characteristic curves.

Figure 9–4(b) shows a view of the same operation using the drain curves. The signal at the gate drives the drain current above and below the Q-point on the load line, as indicated by the arrows. Lines projected from the peaks of the gate voltage across to the I_D axis and down to the V_{DS} axis indicate the peak-to-peak variations of the drain current and drain-to-source voltage, as shown. Because the transfer characteristic curve is nonlinear, the output will have some distortion. This can be minimized if the signal swings over a limited portion of the load line.

DC Analysis

The first step in analyzing a FET amplifier is to determine the dc conditions including I_D and V_S. I_D determines the Q-point for an amplifier and enables you to calculate V_D, so it is useful to determine its value. It can be found either graphically or mathematically. The graphical approach, introduced in Chapter 8 using the transconductance curve, will be applied to an amplifier here. The same result can be obtained by expanding Equation 8–1, which is the mathematical description of the transconductance curve. The amplifier shown in Figure 9–5 will be used to illustrate both approaches. To simplify the dc analysis, the equivalent circuit is shown in Figure 9–6; capacitors appear open to dc, so they are removed.

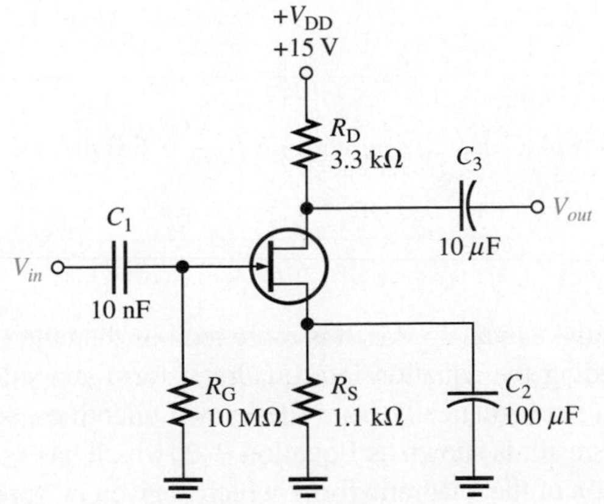

◄ **FIGURE 9–5**

JFET common-source amplifier.

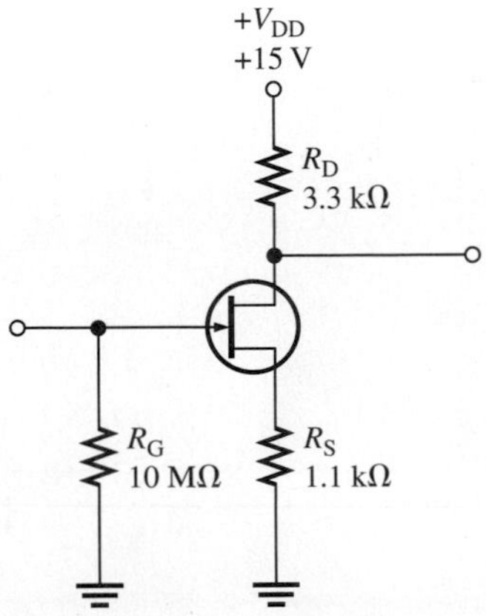

▲ **FIGURE 9–6**

DC equivalent for the amplifier in Figure 9–5.

Graphical Approach Recall from Section 8–2 that the JFET universal transfer characteristic (transconductance curve) illustrates the relationship between the output current and the input voltage. The endpoints of the transconductance curve are at I_{DSS} and $V_{GS(off)}$. A dc graphical solution is done by plotting the load line (for the self-biased case shown) on the same plot and reading the values of V_{GS} and I_D at the intersection of these plots (Q-point).

EXAMPLE 9–2

Determine I_D and V_{GS} at the Q-point for the JFET amplifier in Figure 9–6. The typical I_{DSS} for this particular JFET is 4.3 mA and $V_{GS(off)}$ is −7.7 V.

Solution Plot the transconductance curve. The end points are at I_{DSS} and $V_{GS(off)}$. You can plot two additional points quickly by noting from the universal curve in Figure 8–12, that

$$V_{GS} = 0.3V_{GS(off)} = -2.31\text{ V} \qquad \text{when } I_D = \frac{I_{DSS}}{2} = 2.15\text{ mA}$$

and

$$V_{GS} = 0.5V_{GS(off)} = -3.85\text{ V} \qquad \text{when } I_D = \frac{I_{DSS}}{4} = 1.075\text{ mA}$$

For this particular JFET, the points are plotted as shown in Figure 9–7(a). Recall from Chapter 8 that the load line starts at the origin and goes to a point where $I_D = I_{DSS}$ and $V_{GS} = I_{DSS}R_S$. Add the load line to the graph and read the I_D and V_{GS} values from the intersection (Q-point), as shown in Figure 9–7(b). For the graph shown, I_D = **2.2 mA** and V_{GS} = **−2.4 V.**

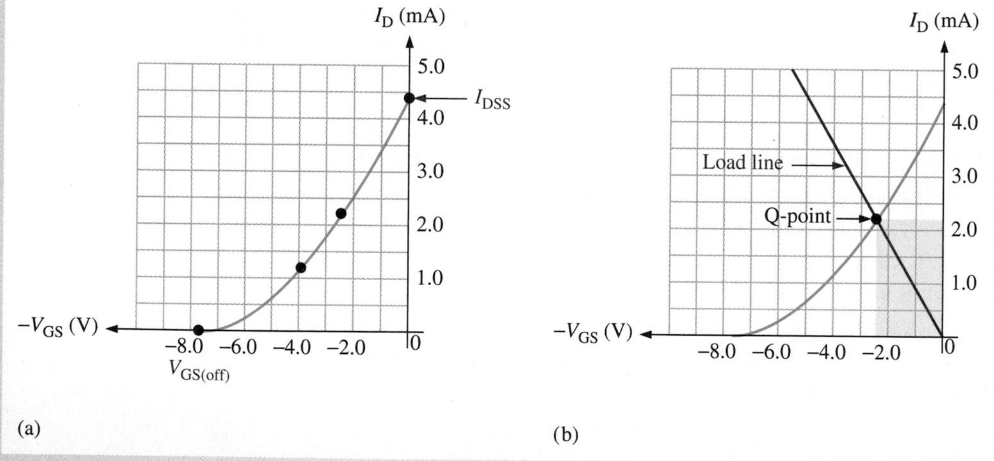

FIGURE 9–7

Related Problem Show the Q-point if the transistor is replaced with one with an I_{DSS} = 5.0 mA and a $V_{GS(off)}$ = −8 V.

Mathematical Approach The mathematical approach is more tedious than the graphical approach because it involves expanding the equation into quadratic form and solving the quadratic equation. To determine I_D mathematically from the known quantities, substitute $V_{GS} = I_DR_S$ into Equation 8–1. The result is shown as Equation 9–2, which has I_D on both sides. Isolating I_D requires the solution of the quadratic form, which is given in Appendix B.

A much easier approach is to enter Equation 9–2 into a graphing calculator such as the TI-86 or TI-89. The steps for determining I_D using the TI-86 are given in Example 9–3.

$$I_D = I_{DSS}\left(1 - \frac{I_D R_S}{V_{GS(off)}}\right)^2 \qquad \text{Equation 9–2}$$

EXAMPLE 9–3

Determine I_D and V_{GS} at the Q-point for the JFET amplifier in Figure 9–6 using the mathematical approach. The I_{DSS} for this particular JFET is 4.3 mA and $V_{GS(off)}$ is -7.7 V.

Solution First enter the equation into the calculator. For the TI-86, the keystrokes are as follows.

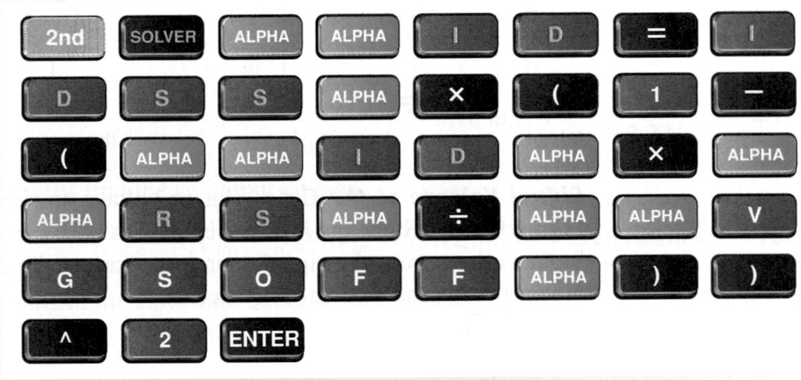

The screen will appear as follows:

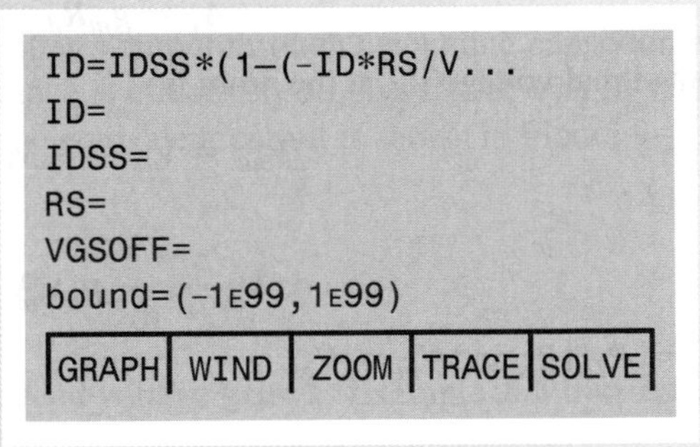

Use the up/down arrow keys to move the blinking cursor to the IDSS = and enter 4.3E–3. Next, move the cursor to the RS = and enter 1.1E3. Finally, move the cursor to the VGSOFF = and enter the absolute value 7.7 (do not include the sign if negative). Move the cursor back up to ID = but do not enter a value. Select SOLVE with the F5 key. The calculator will show that I_D = **2.104 mA**, in excellent agreement with the graphical solution. (The solution using the TI-89 is given in Appendix D.)

Calculate V_{GS}.

$$V_{GS} = -I_D R_S = -(2.1 \text{ mA})(1.1 \text{ k}\Omega) = \mathbf{-2.31\ V}$$

Related Problem Calculate the solution for the Related Problem in Example 9–2.

AC Equivalent Circuit

To analyze the signal operation of the amplifier in Figure 9–5, develop an ac equivalent circuit as follows. Replace the capacitors by effective shorts, based on the simplifying assumption that $X_C \cong 0$ at the signal frequency. Replace the dc source by a ground, based

The PWM signal is typically produced using a comparator circuit. Comparators are discussed in more detail in Chapter 13, but here is a basic explanation of how they work. A comparator has two inputs and one output, as shown by the symbol in Figure 9–27. The input labeled + is called the noninverting input, and the input labeled – is the inverting input. When the voltage on the inverting input exceeds the voltage on the noninverting input, the comparator switches to its *negative* saturated output state. When the voltage on the noninverting input exceeds the voltage on the inverting input, the comparator switches to its *positive* saturated output state. This is illustrated in Figure 9–27 for one cycle of a sine wave voltage on the noninverting input and a higher frequency triangular wave voltage on the inverting input.

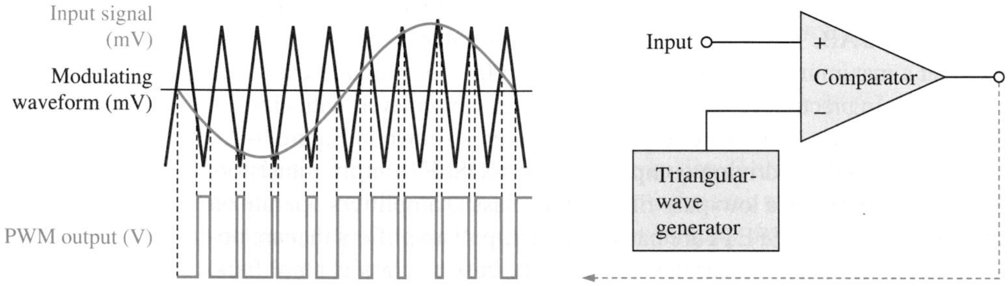

▲ **FIGURE 9–27**

A basic pulse-width modulator.

The comparator inputs are typically very small voltages (mV range); and the comparator output is "rail-to-rail," which means that the positive maximum is near the positive dc supply voltage and the negative maximum is near the negative dc supply voltage. An output of ±12 V or 24 V peak-to-peak is not unusual. From this, you can see that the gain can be quite high. For example, if the input signal is 10 mVpp, the voltage gain is 24 Vpp/10 mVpp = 2400. Since the comparator output amplitude is constant for a specified range of input voltages, the gain is dependent on the input signal voltage. If the input signal is 100 mVpp, the output is still 24 Vpp, and the gain is 240 instead of 2400.

Frequency Spectra All nonsinusoidal waveforms are made up of harmonic frequencies. The frequency content of a particular waveform is called its *spectrum.* When the triangular waveform modulates the input sine wave, the resulting spectrum contains the sine wave frequency, f_{input}, plus the fundamental frequency of the triangular modulating signal, f_m, and harmonic frequencies above and below the fundamental frequency. These harmonic frequencies are due to the fast rise and fall times of the PWM signal and the flat areas between the pulses. A simplified frequency spectrum of a PWM signal is shown in Figure 9–28. The frequency of the triangular waveform must be significantly higher than the high-

► **FIGURE 9–28**

Frequency spectrum of a PWM signal.

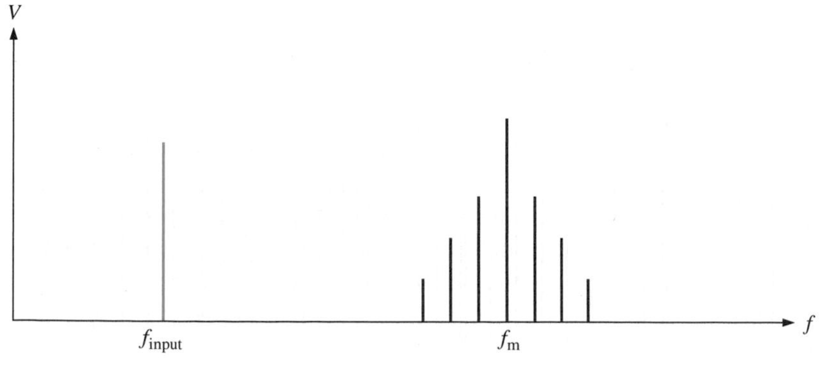

est input signal frequency so that the lowest frequency harmonic is well above the range of input signal frequencies.

The Complementary MOSFET Stage

The MOSFETs are arranged in a common-source complementary configuration to provide power gain. Each transistor switches between the *on* state and the *off* state and when one transistor is *on,* the other one is *off,* as shown in Figure 9–29. When a transistor is *on,* there is very little voltage across it and, therefore, there is very little power dissipated even though it may have a high current through it. When a transistor is *off,* there is no current through it and, therefore, there is no power dissipated. The only time power is dissipated in the transistors is during the short switching time. Power delivered to a load can be very high because a load will have a voltage across it nearly equal to the supply voltages and a high current through it.

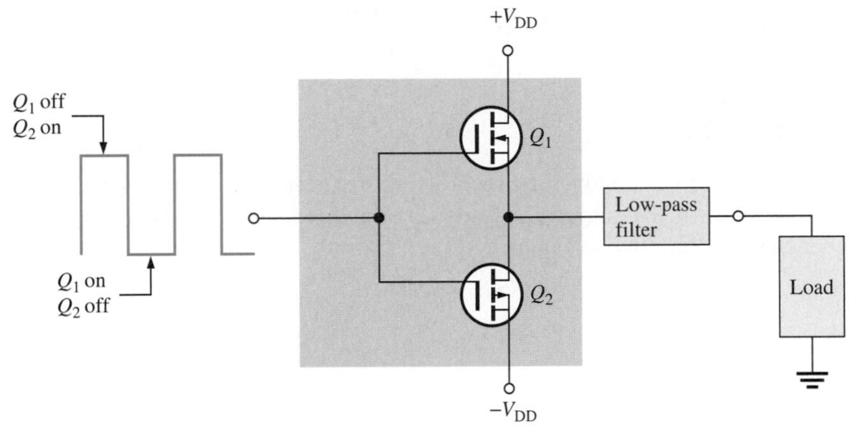

FIGURE 9–29

The complementary MOSFETs operating as switches to amplify power.

Efficiency When Q_1 is *on,* it is providing current to the load. However, ideally the voltage across it is zero so the internal power dissipated by Q_1 is

$$P_{DQ} = V_{Q1}I_L = (0\text{ V})I_L = 0\text{ W}$$

At the same time, Q_2 is *off* and the current through it is zero, so the internal power is

$$P_{DQ} = V_{Q2}I_L = V_{Q2}(0\text{ A}) = 0\text{ W}$$

Ideally, the output power to the load is $2V_QI_L$. The maximum ideal efficiency is, therefore,

$$\eta_{max} = \frac{P_{out}}{P_{tot}} = \frac{P_{out}}{P_{out} + P_{DQ}} = \frac{2V_QI_L}{2V_QI_L + 0\text{ W}} = 1$$

As a percentage, $\eta_{max} = 100\%$.

In a practical case, each MOSFET would have a few tenths of a volt across it in the *on* state. There is also a small internal power dissipation in the comparator and triangular wave generator. Also, power is dissipated during the finite switching time, so the ideal efficiency of 100% can never be reached in practice.

EXAMPLE 9–12

A certain class D amplifier dissipates an internal power of 100 mW in the comparator, triangular-wave generator, and filter combined. Each MOSFET in the complementary stage has a voltage of 0.4 V across it in the *on* state. The amplifier operates from ±15 V dc sources and provides 0.5 A to the load. Neglecting any voltage dropped across the filter, determine the output power and the overall efficiency.

Solution The output power to the load is

$$P_{out} = (V_{DD} - V_Q)I_L = (15\text{ V} - 0.4\text{ V})(0.5\text{ A}) = \mathbf{7.3\text{ W}}$$

The total internal power dissipation ($P_{tot(int)}$) is the power in the complementary stage in the *on* state (P_{DQ}) plus the internal power in the comparator, triangular-wave generator, and filter (P_{int}).

$$P_{tot(int)} = P_{DQ} + P_{int} = (400\text{ mV})(0.5\text{ A}) + 100\text{ mW}$$
$$= 200\text{ mW} + 100\text{ mW} = 300\text{ mW}$$

The efficiency is

$$\eta = \frac{P_{out}}{P_{out} + P_{tot(int)}} = \frac{7.3\text{ W}}{7.6\text{ W}} = \mathbf{0.961}$$

Related Problem There is 0.5 V across each MOSFET when it is *on* and the class D amplifier operates with ±12 V dc supply voltages. Assuming all other circuits in the amplifier dissipate 75 mW and 0.8 A is supplied to the load, determine the efficiency.

Low-Pass Filter

The low-pass filter removes the modulating frequency and harmonics and passes only the original signal to the output. The filter has a bandwidth that passes only the input signal frequencies, as illustrated in Figure 9–30.

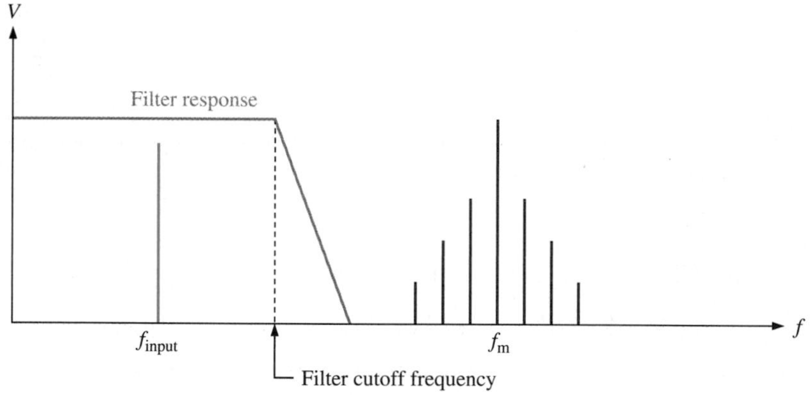

FIGURE 9–30

The low-pass filter removes all but the input signal frequency from the PWM signal.

Signal Flow

Figure 9–31 shows the signals at each point in a class D amplifier. A small audio signal is applied and pulse-width modulated to produce a PWM signal at the output of the modulator where voltage gain is achieved. The PWM drives the complementary MOSFET stage to achieve power amplification. The PWM signal is filtered and the amplified audio signal appears on the output with sufficient power to drive a speaker.

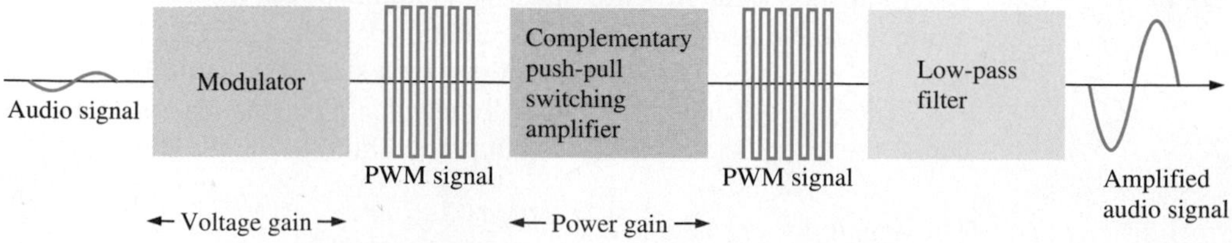

FIGURE 9–31

Signal flow in a class D amplifier.

SECTION 9–4 CHECKUP

1. Name the three stages of a class D amplifier.
2. In pulse-width modulation, to what is the pulse width proportional?
3. How is the PWM signal changed to an audio signal?

9–5 MOSFET Analog Switching

MOSFETs are widely used in analog and digital switching applications. In the preceding section, you saw how MOSFETs are used in the switching mode in Class D amplifiers. Generally, they exhibit very low *on*-resistance, very high off-resistance, and fast switching times.

After completing this section, you should be able to

- **Describe how MOSFETs are used in analog switching applications**
  - Explain how a MOSFET operates as a switch
  - Discuss analog switches
  - Discuss analog multiplexing

MOSFET Switching Operation

E-MOSFETs are generally used for switching applications because of their threshold characteristic, $V_{GS(th)}$. When the gate-to-source voltage is less than the threshold value, the MOSFET is *off.* When the gate-to-source voltage is greater than the threshold value, the MOSFET is *on.* When V_{GS} is varied between $V_{GS(th)}$ and $V_{GS(on)}$, the MOSFET is being operated as a switch, as illustrated in Figure 9–32. In the *off* state, when $V_{GS} < V_{GS(th)}$, the device is operating at the lower end of the load line and acts like an open switch (very high R_{DS}). When V_{GS} is sufficently greater than $V_{GS(th)}$, the device is operating at the upper end of the load line in the ohmic region and acts like a closed switch (very low R_{DS}).

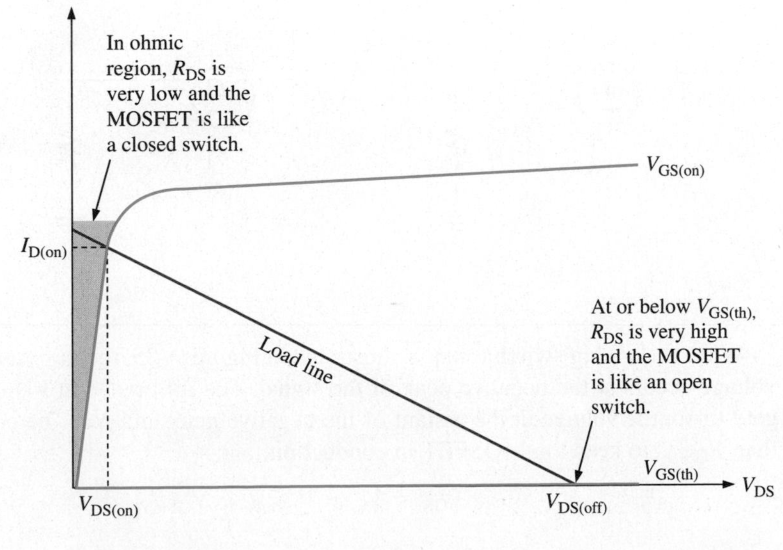

FIGURE 9–32
Switching operation on the load line.

Omair Mehdi

The Ideal Switch Refer to Figure 9–33(a). When the gate voltage of the *n*-channel MOSFET is $+V$, the gate is more positive than the source by an amount exceeding $V_{GS(th)}$. The MOSFET is *on* and appears as a closed switch between the drain and source. When the gate voltage is zero, the gate-to-source voltage is 0 V. The MOSFET is *off* and appears as an open switch between the drain and source.

Refer to Figure 9–33(b). When the gate voltage of the *p*-channel MOSFET is 0 V, the gate is less positive than the source by an amount exceeding $V_{GS(th)}$. The MOSFET is *on* and appears as a closed switch between the drain and source. When the gate voltage is $+V$, the gate-to-source voltage is 0 V. The MOSFET is *off* and appears as an open switch between the drain and source.

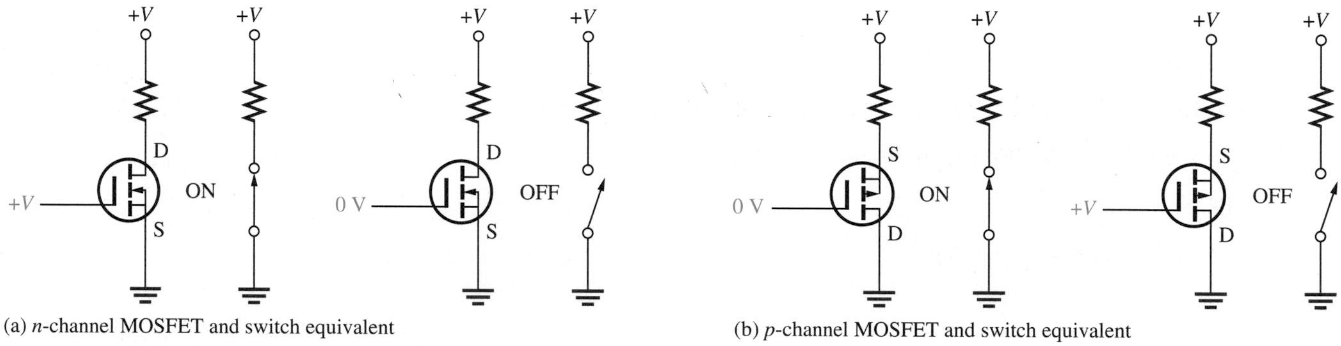

▲ **FIGURE 9–33**

The MOSFET as a switch.

The Analog Switch

MOSFETs are commonly used for switching analog signals. Basically, a signal applied to the drain can be switched through to the source by a voltage on the gate. A major restriction is that the signal level at the source must not cause the gate-to-source voltage to drop below $V_{GS(th)}$.

A basic *n*-channel MOSFET **analog switch** is shown in Figure 9–34. The signal at the drain is connected to the source when the MOSFET is turned on by a positive V_{GS} and is disconnected when V_{GS} is 0, as indicated.

▶ **FIGURE 9–34**

Operation of an *n*-channel MOSFET analog switch.

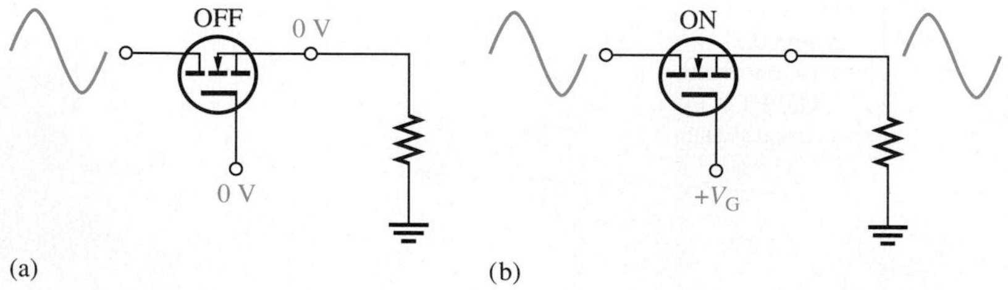

When the analog switch is *on,* as illustrated in Figure 9–35, the minimum gate-to-source voltage occurs at the negative peak of the signal. The difference in V_G and $-V_{p(out)}$ is the gate-to-source voltage at the instant of the negative peak and must be equal to or greater than $V_{GS(th)}$ to keep the MOSFET in conduction.

$$V_{GS} = V_G - V_{p(out)} \geq V_{GS(th)}$$

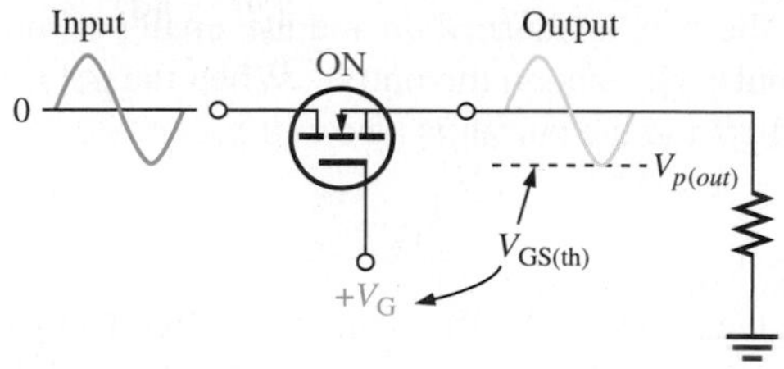

FIGURE 9–35

Signal amplitude is limited by $V_{GS(th)}$.

EXAMPLE 9–13

A certain analog switch similar to the one shown in Figure 9–35 uses an *n*-channel MOSFET with $V_{GS(th)} = 2$ V. A voltage of +5 V is applied at the gate to turn the switch *on*. Determine the maximum peak-to-peak input signal that can be applied, assuming no voltage drop across the switch.

Solution The difference between the gate voltage and the negative peak of the signal voltage must equal or exceed the threshold voltage. For maximum $V_{p(out)}$,

$$V_G - V_{p(out)} = V_{GS(th)}$$
$$V_{p(out)} = V_G - V_{GS(th)} = 5\text{ V} - 2\text{ V} = 3\text{ V}$$
$$V_{pp(in)} = 2V_{p(out)} = 2(3\text{ V}) = \mathbf{6\text{ V}}$$

Related Problem What would happen if $V_{p(in)}$ exceeded the maximum value?

Analog Switch Applications

Sampling Circuit One application of analog switches is in analog-to-digital conversion. The analog switch is used in a *sample-and-hold* circuit to sample the input signal at a certain rate. Each sampled signal value is then temporarily stored on a capacitor until it can be converted to a digital code by an analog-to-digital converter (ADC). To accomplish this, the MOSFET is turned *on* for short intervals during one cycle of the input signal by pulses applied to the gate. The basic operation, showing only a few samples for clarity, is illustrated in Figure 9–36.

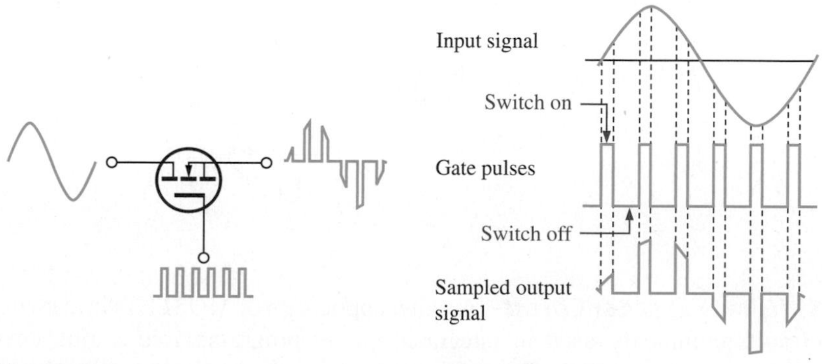

(a) Circuit action (b) Waveform diagram

FIGURE 9–36

The analog switch operating as a sampling circuit.

The minimum rate at which a signal can be sampled and reconstructed from the samples must be more than twice the maximum frequency contained in the signal. The minimum sampling frequency is called the *Nyquist frequency*.

$$f_{sample(min)} > 2f_{signal}$$

When a gate pulse is at its high level, the switch is turned *on* and the small portion of the input waveform occurring during that pulse appears on the output. When the pulse waveform is at its 0 V level, the switch is turned *off* and the output is also at 0 V.

EXAMPLE 9–14

An analog switch is used to sample an audio signal with a maximum frequency of 8 kHz. Determine the minimum frequency of the pulses applied to the MOSFET gate.

Solution

$$f_{sample(min)} > 2f_{signal} = 2(8\text{ kHz}) = \mathbf{16\text{ kHz}}$$

The sampling frequency must be greater than 16 kHz.

Related Problem What is the minimum sampling frequency if the highest frequency in the audio signal is 12 kHz?

Analog Multiplexer Analog multiplexers are used where two or more signals are to be routed to the same destination. For example, a two-channel analog sampling multiplexer is shown in Figure 9–37. The MOSFETs are alternately turned *on* and *off* so that first one signal sample is connected to the output and then the other. The pulses are applied to the gate of switch A, and the inverted pulses are applied to the gate of switch B. A digital circuit known as an *inverter* is used for this. When the pulses are high, switch A is *on* and switch B is *off*. When the pulses are low, switch B is *on* and switch A is *off*. This is called *time-division multiplexing* because signal A appears on the output during time intervals when the pulse is high and signal B appears during the time intervals when the pulse is low. That is, they are interleaved on a time basis for transmission on a single line.

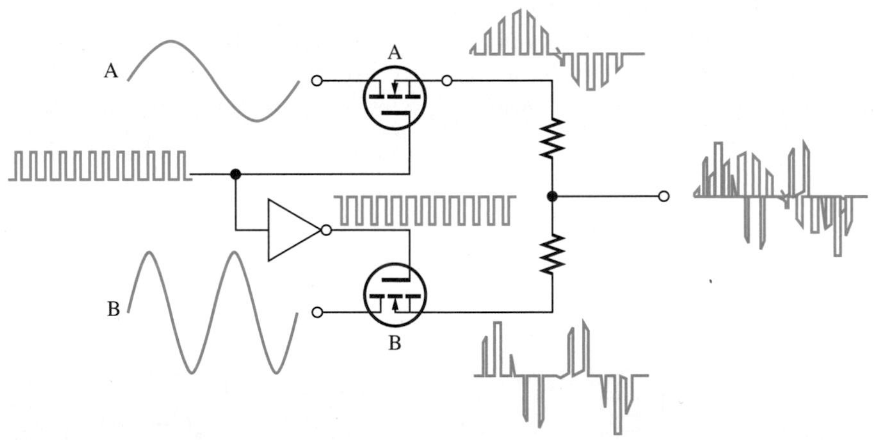

FIGURE 9–37

The analog multiplexer is alternately sampling two signals and interleaving them on a single output line.

Switched-Capacitor Circuit Another application of MOSFETs is in **switched-capacitor circuits** commonly used in integrated circuit programmable analog devices known as *analog signal processors.* Because capacitors can be implemented in ICs more easily than a resistor, they are used to emulate resistors. Capacitors also take up less space on a chip than an IC resistor and dissipate no power. Many types of analog circuits use resistors to determine voltage gain and other characteristics and by using switched capacitors to emulate resistors, dynamic programming of analog circuits can be achieved.

For example, in a certain type of IC amplifier circuit that you will study later, two external resistors are required as shown in Figure 9–38. The values of these resistors establish the voltage gain of the amplifier as $A_v = R_2/R_1$.

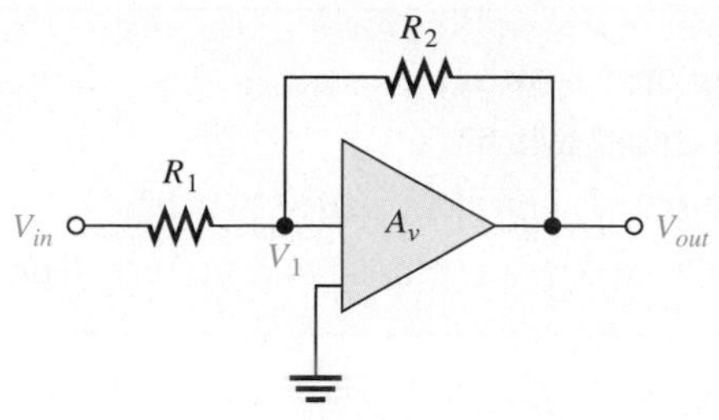

FIGURE 9–38

A type of IC amplifier.

A switched-capacitor can be used to emulate a resistor as shown in Figure 9–39 using a mechanical switch analogy (MOSFETs are actually used as the switches). Switch 1 and switch 2 are alternately turned *on* and *off* at a certain frequency to charge or discharge C, depending on the values of the voltage sources. In the case of R_1 in Figure 9–38, V_{in} and V_1 are represented by V_A and V_B, respectively. For R_2, V_1 and V_{out} are represented by V_A and V_B, respectively.

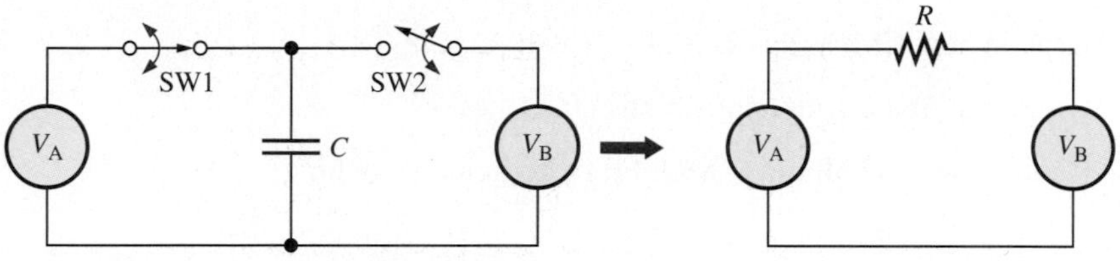

FIGURE 9–39

A switched capacitor emulates a resistance.

It can be shown (see Appendix B) that the capacitor emulates a resistance with a value that depends on the frequency at which the switches are turned *on* and *off* and the capacitance value.

$$R = \frac{1}{fC}$$

Equation 9–10

By changing the frequency, the effective resistance value can be altered.

Complementary E-MOSFETs and capacitors can be used to replace the resistors in the amplifier, as shown in Figure 9–40. When Q_1 is *on*, Q_2 is *off* and vice versa. The frequency f_1 and C_1 are selected to provide the required value of R_1. Likewise, f_2 and C_2 provide the required value of R_2. To reprogram the amplifier for a different gain, the frequencies are changed.

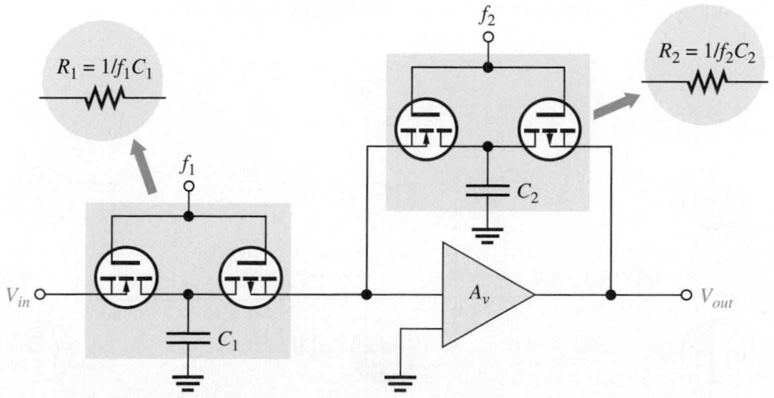

FIGURE 9–40

The IC amplifier in Figure 9–38 with switched-capacitor circuits replacing the resistors.

SECTION 9–5 CHECKUP

1. When does an E-MOSFET act as an open switch?
2. When does an E-MOSFET act as a closed switch?
3. What type of voltage is generally used to control an analog switch?
4. In a switched-capacitor circuit, on what does the emulated resistance depend?

9–6 MOSFET DIGITAL SWITCHING

In the preceding section, you saw how MOSFETs are used to switch analog signals. MOSFETs are also used in switching applications in digital integrated circuits and in power control circuits. MOSFETs used in digital ICs are low-power types, and those used in power control are high-power devices.

After completing this section, you should be able to

- **Describe how MOSFETs are used in digital switching applications**
  - Explain how CMOS operates as a switch
  - Discuss digital circuit applications
  - Describe the use of power MOSFETs in digital switching

CMOS (Complementary MOS)

CMOS combines *n*-channel and *p*-channel E-MOSFETs in a series arrangement as shown in Figure 9–41(a). The input voltage at the gates is either 0 V or V_{DD}. Notice that V_{DD} and ground are both connected to source terminals of the transistors. To avoid confusion, the term V_{DD} is used for the positive voltage, which is on the *p*-channel device's source terminal. When $V_{in} = 0$ V, Q_1 is *on* and Q_2 is *off,* as shown in part (b). Because Q_1 is acting as a closed switch, the output is approximately V_{DD}. When $V_{in} = V_{DD}$, Q_2 is on and Q_1 is off, as shown in part (c). Because Q_2 is acting as a closed switch, the output is essentially connected to ground (0 V).

A major advantage of CMOS is that it consumes very little dc power. Because the MOSFETs are in series and one of them is always off, there is essentially no current from the dc supply in the quiescent state. When the MOSFETs are switching, there is current for a very short time because both transistors are on during this very short transition from one state to the other.

Notice that the circuit in Figure 9–41 actually inverts the input because when the input is 0 V or low, the output is V_{DD} or high. When the input is V_{DD} or high, the output is 0 V or low. For this reason, this circuit is called an *inverter* in digital electronics.

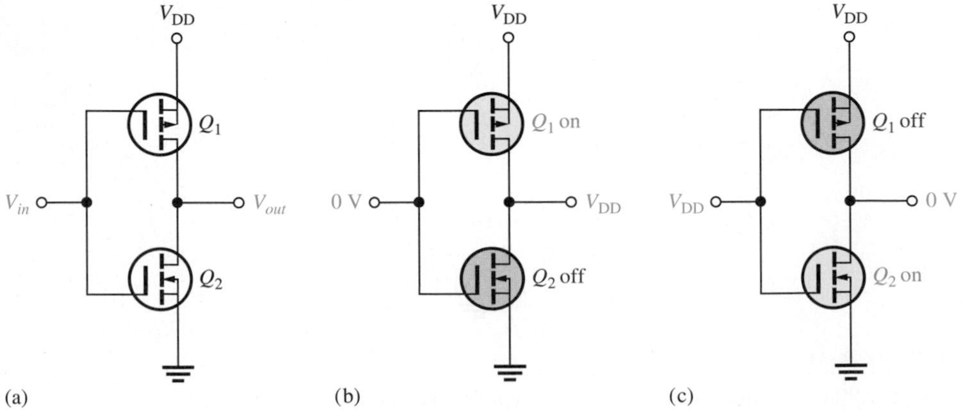

FIGURE 9–41

CMOS inverter operation.

Other Examples of CMOS Circuits

NAND Gate In Figure 9–42(a), two additional MOSFETs and a second input are added to the CMOS pair to create a digital circuit known as a NAND gate. Q_4 is connected in parallel with Q_1, and Q_3 is connected in series with Q_2. When both inputs, V_A and V_B, are 0, Q_1 and Q_4 are *on* while Q_2 and Q_3 are *off*, making $V_{out} = V_{DD}$. When both inputs are equal to V_{DD}, Q_1 and Q_4 are *off* while Q_2 and Q_3 are *on*, making $V_{out} = 0$. You can verify that when the inputs are different, one at V_{DD} and the other at 0, the output is equal to V_{DD}. The operation is summarized in the table of Figure 9–42(b) and can be stated:

When V_A *AND* V_B are high, the output is low; otherwise, the output is high.

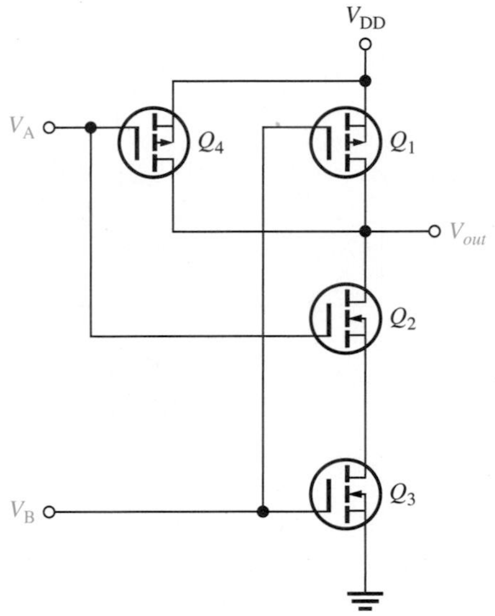

(a)

V_A	V_B	Q_1	Q_2	Q_3	Q_4	V_{out}
0	0	on	off	off	on	V_{DD}
0	V_{DD}	off	off	off	on	V_{DD}
V_{DD}	0	on	off	off	off	V_{DD}
V_{DD}	V_{DD}	off	on	on	off	0

(b)

FIGURE 9–42

CMOS NAND gate operation.

NOR Gate In Figure 9–43(a), two additional MOSFETs and a second input are added to the CMOS pair to create a digital circuit known as a NOR gate. Q_4 is connected in parallel

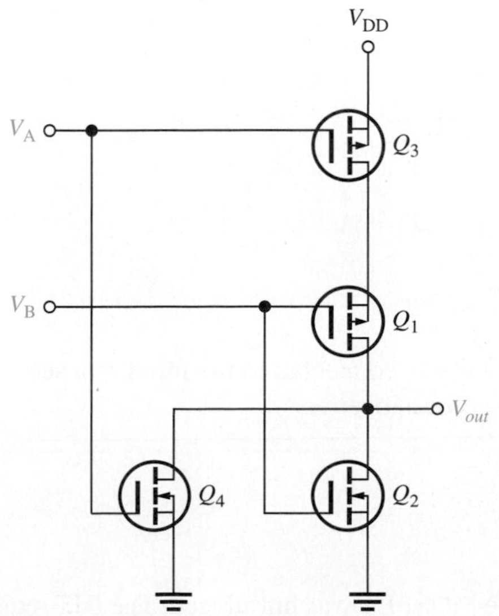

(a)

V_A	V_B	Q_1	Q_2	Q_3	Q_4	V_{out}
0	0	on	off	on	off	V_{DD}
0	V_{DD}	off	on	on	off	0
V_{DD}	0	on	off	on	off	0
V_{DD}	V_{DD}	off	on	off	on	0

(b)

FIGURE 9–43

CMOS NOR gate operation.

with Q_2, and Q_3 is connected in series with Q_1. When both inputs, V_A and V_B, are 0, Q_1 and Q_3 are *on* while Q_2 and Q_4 are *off,* making $V_{out} = V_{DD}$. When both inputs are equal to V_{DD}, Q_1 and Q_3 are *off* while Q_2 and Q_4 are *on,* making $V_{out} = 0$. You can verify that when the inputs are different, one at V_{DD} and the other at 0, the output is equal to 0. The operation is summarized in the table of Figure 9–43(b) and can be stated:

When V_A *OR* V_B *OR* both are high, the output is low; otherwise, the output is high.

EXAMPLE 9–15

A pulse waveform is applied to a CMOS inverter as shown in Figure 9–44. Determine the output waveform and explain the operation.

FIGURE 9–44

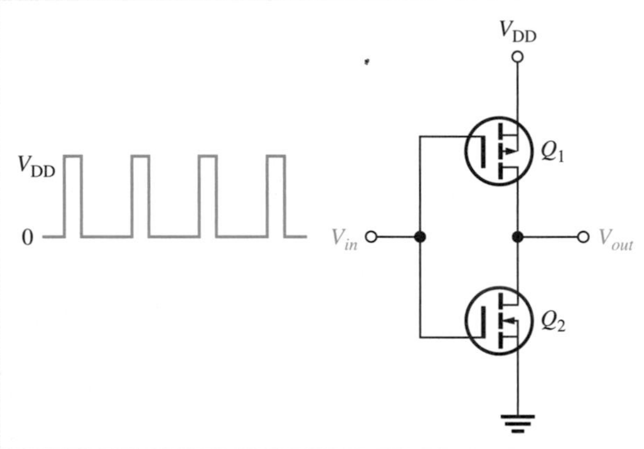

Solution The output waveform is shown in Figure 9–45 in relation to the input. When the input pulse is at V_{DD}, Q_1 is *off* and Q_2 is *on,* connecting the output to ground (0 V). When the input pulse is at 0, Q_1 is *on* and Q_2 is *off,* connecting the output to V_{DD}.

FIGURE 9–45

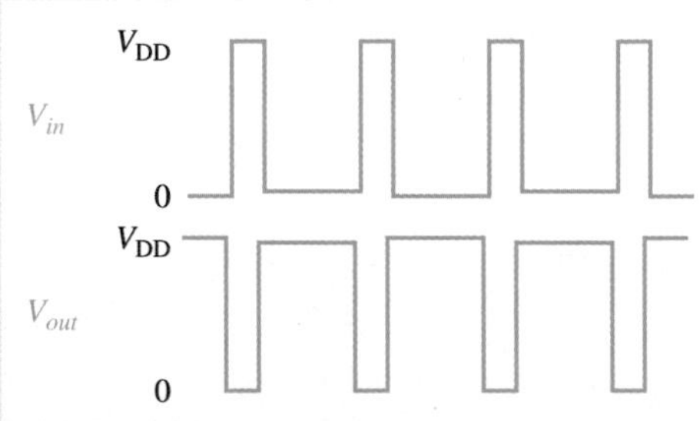

Related Problem If the output of the CMOS inverter in Figure 9–44 is connected to the input of a second CMOS inverter, what is the output of the second inverter?

MOSFETs in Power Switching

The BJT was the only power transistor until the MOSFET was introduced. The BJT requires a base current to turn on, has relatively slow turn-off characteristics, and is susceptible to thermal runaway due to a negative temperature coefficient. The MOSFET, however, is voltage controlled and has a positive temperature coefficient, which prevents thermal runaway. The MOSFET can turn *off* faster than the BJT, and the low *on*-state-resistance results in conduction power losses lower than with BJTs. Power MOSFETs are used for motor control, dc-to-ac conversion, dc-to-dc conversion, load switching, and other applications that require high current and precise digital control.

SECTION 9–6 CHECKUP

1. Describe a basic CMOS inverter.
2. What type of 2-input digital CMOS circuit has a low output only when both inputs are high?
3. What type of 2-input digital CMOS circuit has a high output only when both inputs are low?

9–7 TROUBLESHOOTING

A technician who understands the basics of circuit operation and who can, if necessary, perform basic analysis on a given circuit is much more valuable than one who is limited to carrying out routine test procedures. In this section, you will see how to test a circuit board that has only a schematic with no specified test procedure or voltage levels. In this case, basic knowledge of how the circuit operates and the ability to do a quick circuit analysis are useful.

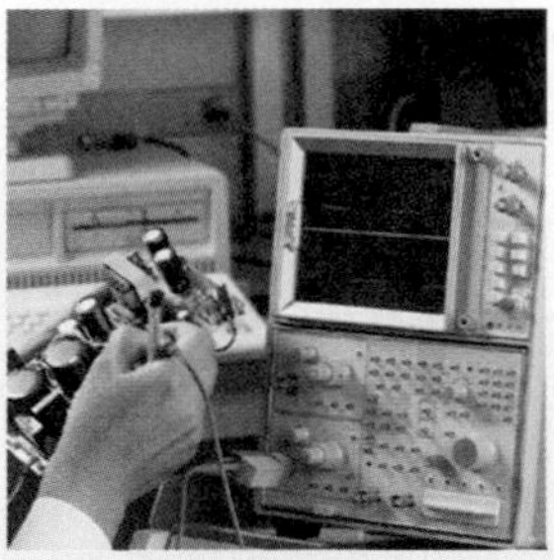

After completing this section, you should be able to

- **Troubleshoot FET amplifiers**
  - Troubleshoot a two-stage CS amplifier
  - Relate a schematic to a circuit board

Assume that you are given a circuit board containing an audio amplifier and told simply that it is not working properly. The circuit is a two-stage FET amplifier, as shown in Figure 9–46.

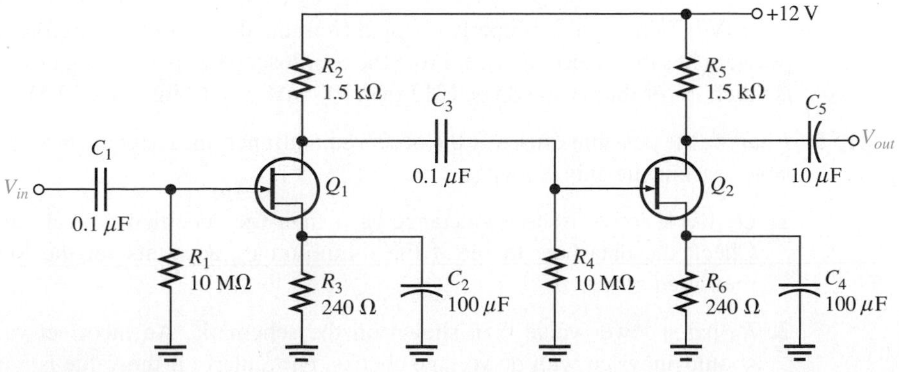

▲ FIGURE 9–46

A two-stage FET amplifier circuit.

The problem is approached in the following sequence.

Step 1: Determine what the voltage levels in the circuit should be so that you know what to look for. First, pull a datasheet on the particular transistor (assume both Q_1 and Q_2 are found to be the same type of transistor) and determine the g_m so that you can calculate the typical voltage gain. Assume that for this particular device, a typical g_m of 5000 μS is specified. Calculate the expected typical voltage gain of each stage (notice they are identical) based on the typical value of g_m. The g_m of actual devices may be any value between the specified minimum and maximum values. Because the input resistance is very high, the second stage does not significantly load the first stage, as in a BJT amplifier. So, the unloaded voltage gain for each stage is

$$A_v = g_m R_2 = (5000\,\mu\text{S})(1.5\,\text{k}\Omega) = 7.5$$

Since the stages are identical, the typical overall gain should be

$$A'_v = (7.5)(7.5) = 56.3$$

Assume the dc levels have been checked and verified. You are now ready to move to ac signal checks.

Step 2: Arrange a test setup to permit connection of an input test signal, a dc supply voltage, and ground to the circuit. The schematic shows that the dc supply voltage must be +12 V. Choose 10 mV rms as an input test signal. This value is arbitrary (although the capability of your signal source is a factor), but small enough that the expected output signal voltage is well below the absolute peak-to-peak limit of 12 V set by the supply voltage and ground (you know that the output voltage swing cannot go higher than 12 V or lower than 0 V). Set the frequency of the sinusoidal signal source to an arbitrary value in the audio range (say 10 kHz) because you know this is an audio amplifier. The audio frequency range is generally accepted as 20 Hz to 20 kHz.

Step 3: Check the input signal at the gate of Q_1 and the output signal at the drain of Q_2 with an oscilloscope. The results are shown in Figure 9–47. The measured output voltage has a peak value of 226 mV. The expected typical peak output voltage is

$$V_{out} = V_{in}A'_v = (14.14\,\text{mV})(56.3) = 796\,\text{mV peak}$$

The output is much less than it should be.

Step 4: Trace the signal from the output toward the input to determine the fault. Figure 9–47 shows the oscilloscope displays of the measured signal voltages. The voltage at the gate of Q_2 is 106 mV peak, as expected (14.14 mV × 7.5 = 106 mV). This signal is properly coupled from the drain of Q_1. Therefore, the problem lies in the second stage. From the oscilloscope displays, the gain of Q_2 is much lower than it should be (213 mV/100 mV = 2.13 instead of 7.5).

Step 5: Analyze the possible causes of the observed malfunction. There are three possible reasons the gain is low:

1. Q_2 has a lower transconductance (g_m) than the specified typical value. Check the datasheet to see if the minimum g_m accounts for the lower measured gain.

2. R_5 has a lower value than shown on the schematic. An incorrect value should show up with dc voltage checks, particularly if the value is much different than specified, so this is not the likely cause in this case.

3. The bypass capacitor C_4 is open.

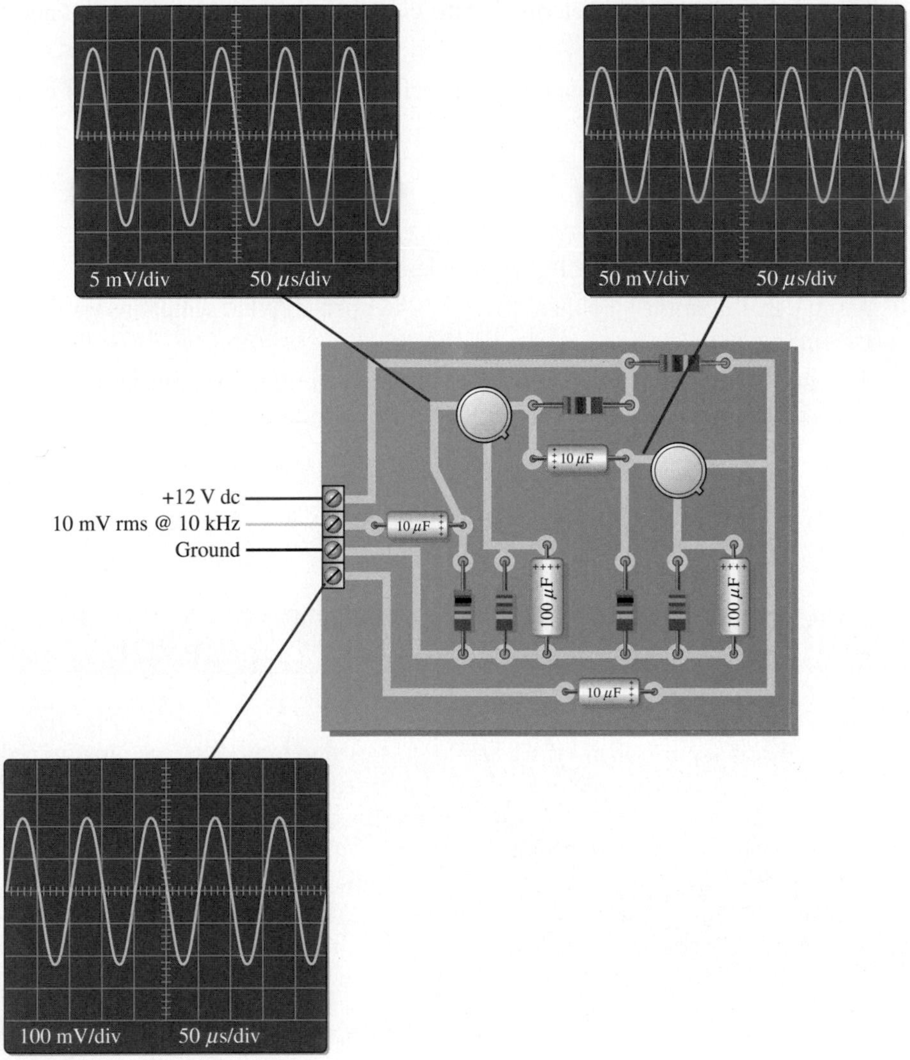

FIGURE 9–47
Oscilloscope displays of signals in the two-stage FET amplifier.

The best way to check the g_m is by replacing Q_2 with a new transistor of the same type and rechecking the output signal. You can make certain that R_5 is the proper value by removing one end of the resistor from the circuit board and measuring the resistance with an ohmmeter. To avoid having to unsolder a component, the best way to start isolating the fault is by checking the signal voltage at the source of Q_2. If the capacitor is working properly, there will be only a dc voltage at the source. The presence of a signal voltage at the source indicates that C_4 is open. With R_6 unbypassed, the gain expression is $g_mR_d/(1 + g_mR_s)$ rather than simply g_mR_d, thus resulting in less gain.

Multisim Troubleshooting Exercises

These file circuits are in the Troubleshooting Exercises folder on your CD-ROM.

1. Open file TSE09-01. Determine if the circuit is working properly and, if not, determine the fault.

2. Open file TSE09-02. Determine if the circuit is working properly and, if not, determine the fault.

3. Open file TSE09-03. Determine if the circuit is working properly and, if not, determine the fault.

4. Open file TSE09-04. Determine if the circuit is working properly and, if not, determine the fault.

SECTION 9–7 CHECKUP

1. What is the prerequisite to effective troubleshooting?
2. Assume that C_2 in the amplifier of Figure 9–46 opened. What symptoms would indicate this failure?
3. If C_3 opened in the amplifier, would the voltage gain of the first stage be affected?

On the Job

Application Activity: *Active Antenna*

In this application, a broadband JFET amplifier is used to provide a high input impedance and voltage gain for a whip antenna. When an antenna is connected to the input of a receiver or a coaxial cable, signal deterioration may be unacceptable due to a distant station, noisy conditions, or an impedance mismatch. An active antenna can alleviate this problem by providing a stronger signal. The block diagram in Figure 9–48 shows an active antenna, followed by a low impedance output buffer to drive a coaxial cable or a receiver input. The focus in this application is the active antenna. The low output impedance buffer can be a BJT emitter-follower or an impedance-matching transformer.

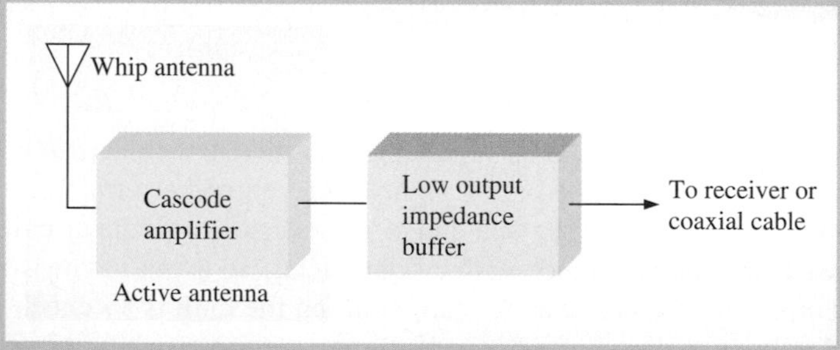

FIGURE 9–48

An active antenna driving a receiver or a coax through a buffer.

The Amplifier Circuit

Figure 9–49 is a broadband amplifier using two JFETs in a cascode arrangement commonly used in RF (radio frequency) applications. The advantage of using a JFET is that its high input impedance does not load the antenna and cause a reduction in signal voltage, resulting in poor signal reception. It also is a low noise device and can be located close to the antenna before additional noise is picked up by the system. Generally, an antenna produces signal voltages in the microvolt range, and any signal loss because of loading or noise can significantly degrade the signal. The active antenna also provides a large voltage gain that results in a stronger signal to the receiver with improved signal-to-noise ratio. The active antenna is powered by a separate 9 V battery, which also provides isolation from noise pickup in the signal lines and is located in an enclosed metal box to provide additional isolation.

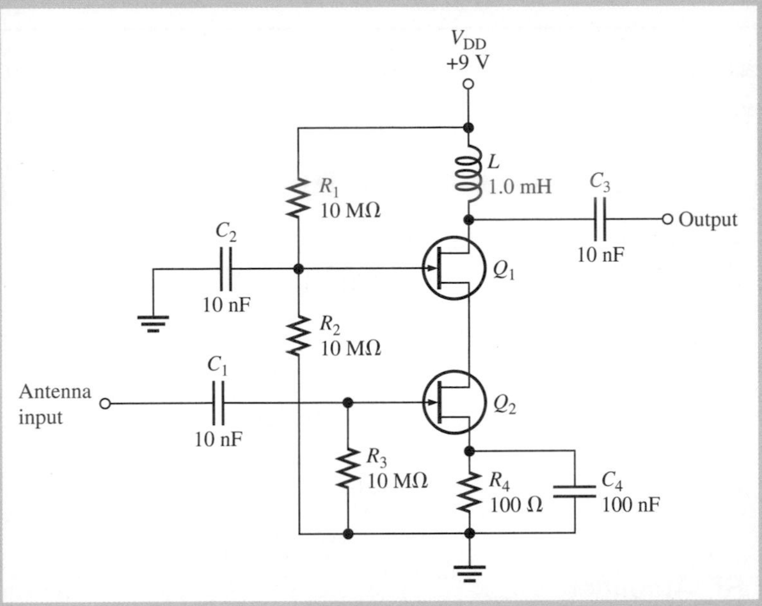

FIGURE 9–49

Cascode amplifier for active antenna.

This active antenna has a voltage gain of approximately 2000 at 88 MHz and a gain of approximately 10,000 at 1 GHz which makes it applicable for the FM broadcast band, some TV channel bands, some amateur radio (HAM) bands, cell phone bands, and many others. Also below the FM band, the gain may be adequate for other radio and TV bands as well, depending on receiver requirements. The coil can be changed to optimize gain within a specified band or to adjust the band downward.

1. Research the Internet to determine the frequency band for TV channels 7–13.
2. Research the Internet to find the frequency bands allocated for cellular telephones.
3. What is the purpose of C_2 in Figure 9–49?

The transistors used in the active antenna are 2N5484 *n*-channel JFETs. The partial datasheet is shown in Figure 9–50.

4. Using the datasheet, determine $R_{IN(gate)}$ of the JFET (Q_2).
5. What input resistance is presented to the antenna in Figure 9–49?
6. From the datasheet, what is the minimum forward transconductance?

Simulation

The active antenna circuit is simulated in Multisim with the antenna input represented by a 10 μV peak source. The output signal is shown for 88 MHz and 1 GHz inputs in Figure 9–51 on page 474.

7. What is the significance of the 88 MHz frequency?
8. Determine the rms output voltage in Figure 9–51(b) and (c), and calculate the gain for both frequencies.

Simulate the active antenna circuit using your Multisim software. Measure the output voltage at 10 MHz, 100 MHz, and 500 MHz.

Prototyping and Testing

Now that the circuit has been simulated, the prototype circuit is constructed and tested. After the circuit is successfully tested, it is ready to be finalized. Because you are working at high frequencies where stray capacitances can cause unwanted resonant conditions, the circuit layout is very critical.

FAIRCHILD
SEMICONDUCTOR™

2N5484
2N5485
2N5486

MMBF5484
MMBF5485
MMBF5486

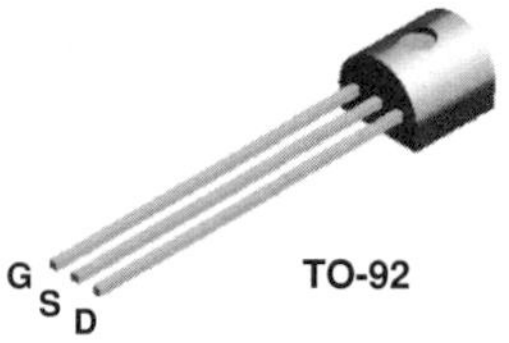

TO-92

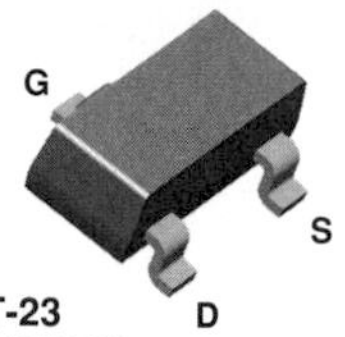

SOT-23
Mark: 6B / 6M / 6H

NOTE: Source & Drain are interchangeable

N-Channel RF Amplifier

This device is designed primarily for electronic switching applications such as low On Resistance analog switching. Sourced from Process 50.

Absolute Maximum Ratings*

TA = 25°C unless otherwise noted

Symbol	Parameter	Value	Units
V_{DG}	Drain-Gate Voltage	25	V
V_{GS}	Gate-Source Voltage	- 25	V
I_{GF}	Forward Gate Current	10	mA
T_J ,T_{stg}	Operating and Storage Junction Temperature Range	-55 to +150	°C

*These ratings are limiting values above which the serviceability of any semiconductor device may be impaired.

NOTES:
1) These ratings are based on a maximum junction temperature of 150 degrees C.
2) These are steady state limits. The factory should be consulted on applications involving pulsed or low duty cycle operations.

Thermal Characteristics

TA = 25°C unless otherwise noted

Symbol	Characteristic	Max		Units
		2N5484-5486	*MMBF5484-5486	
P_D	Total Device Dissipation Derate above 25°C	350 2.8	225 1.8	mW mW/°C
$R_{\theta JC}$	Thermal Resistance, Junction to Case	125		°C/W
$R_{\theta JA}$	Thermal Resistance, Junction to Ambient	357	556	°C/W

*Device mounted on FR-4 PCB 1.6" X 1.6" X 0.06."

FIGURE 9–50

Partial datasheet for the 2N5484 RF *n*-channel JFET. © 2003 Fairchild semiconductor corporation. Used by permission.

N-Channel RF Amplifier
(continued)

Electrical Characteristics TA = 25°C unless otherwise noted

Symbol	Parameter	Test Conditions	Min	Typ	Max	Units
OFF CHARACTERISTICS						
$V_{(BR)GSS}$	Gate-Source Breakdown Voltage	I_G = - 1.0 μA, V_{DS} = 0	- 25			V
I_{GSS}	Gate Reverse Current	V_{GS} = - 20 V, V_{DS} = 0 V_{GS}= - 20 V, V_{DS}= 0, T_A= 100°C			- 1.0 - 0.2	nA μA
$V_{GS(off)}$	Gate-Source Cutoff Voltage	V_{DS} = 15 V, I_D = 10 nA **5484** **5485** **5486**	- 0.3 - 0.5 - 2.0		- 3.0 - 4.0 - 6.0	V V V
ON CHARACTERISTICS						
I_{DSS}	Zero-Gate Voltage Drain Current*	V_{DS} = 15 V, V_{GS} = 0 **5484** **5485** **5486**	1.0 4.0 8.0		5.0 10 20	mA mA mA
SMALL SIGNAL CHARACTERISTICS						
g_{fs}	Forward Transfer Conductance	V_{DS} = 15 V, V_{GS} = 0, f = 1.0 kHz **5484** **5485** **5486**	 3000 3500 4000		 6000 7000 8000	 μmhos μmhos μmhos
$Re(y_{is})$	Input Conductance	V_{DS} = 15 V, V_{GS} = 0, f = 100 MHz **5484** V_{DS} = 15 V, V_{GS} = 0, f = 400 MHz **5485 / 5486**			 100 1000	 μmhos μmhos
g_{os}	Output Conductance	V_{DS} = 15 V, V_{GS} = 0, f = 1.0 kHz **5484** **5485** **5486**			 50 60 75	 μmhos μmhos μmhos
$Re(y_{os})$	Output Conductance	V_{DS} = 15 V, V_{GS} = 0, f = 100 MHz **5484** V_{DS} = 15 V, V_{GS} = 0, f = 400 MHz **5485 / 5486**			 75 100	 μmhos μmhos
$Re(y_{fs})$	Forward Transconductance	V_{DS} = 15 V, V_{GS} = 0, f = 100 MHz **5484** V_{DS} = 15 V, V_{GS} = 0, f = 400 MHz **5485** **5486**	 2500 3000 3500			 μmhos μmhos μmhos
C_{iss}	Input Capacitance	V_{DS} = 15 V, V_{GS} = 0, f = 1.0 MHz			5.0	pF
C_{rss}	Reverse Transfer Capacitance	V_{DS} = 15 V, V_{GS} = 0, f = 1.0 MHz			1.0	pF
C_{oss}	Output Capacitance	V_{DS} = 15 V, V_{GS} = 0, f = 1.0 MHz			2.0	pF
NF	Noise Figure	V_{DS}= 15 V, R_G = 1.0 kΩ, f = 100 MHz **5484** V_{DS}= 15 V, R_G = 1.0 kΩ, f = 400 MHz **5484** V_{DS}= 15 V , R_G = 1.0 kΩ, f = 100 MHz **5485 / 5486** V_{DS}= 15 V, R_G = 1.0 kΩ, f = 400 MHz **5485 / 5486**		 4.0 	3.0 2.0 4.0	dB dB dB dB

*Pulse Test: Pulse Width ≤ 300 ms, Duty Cycle ≤ 2%

FIGURE 9–50

(continued)

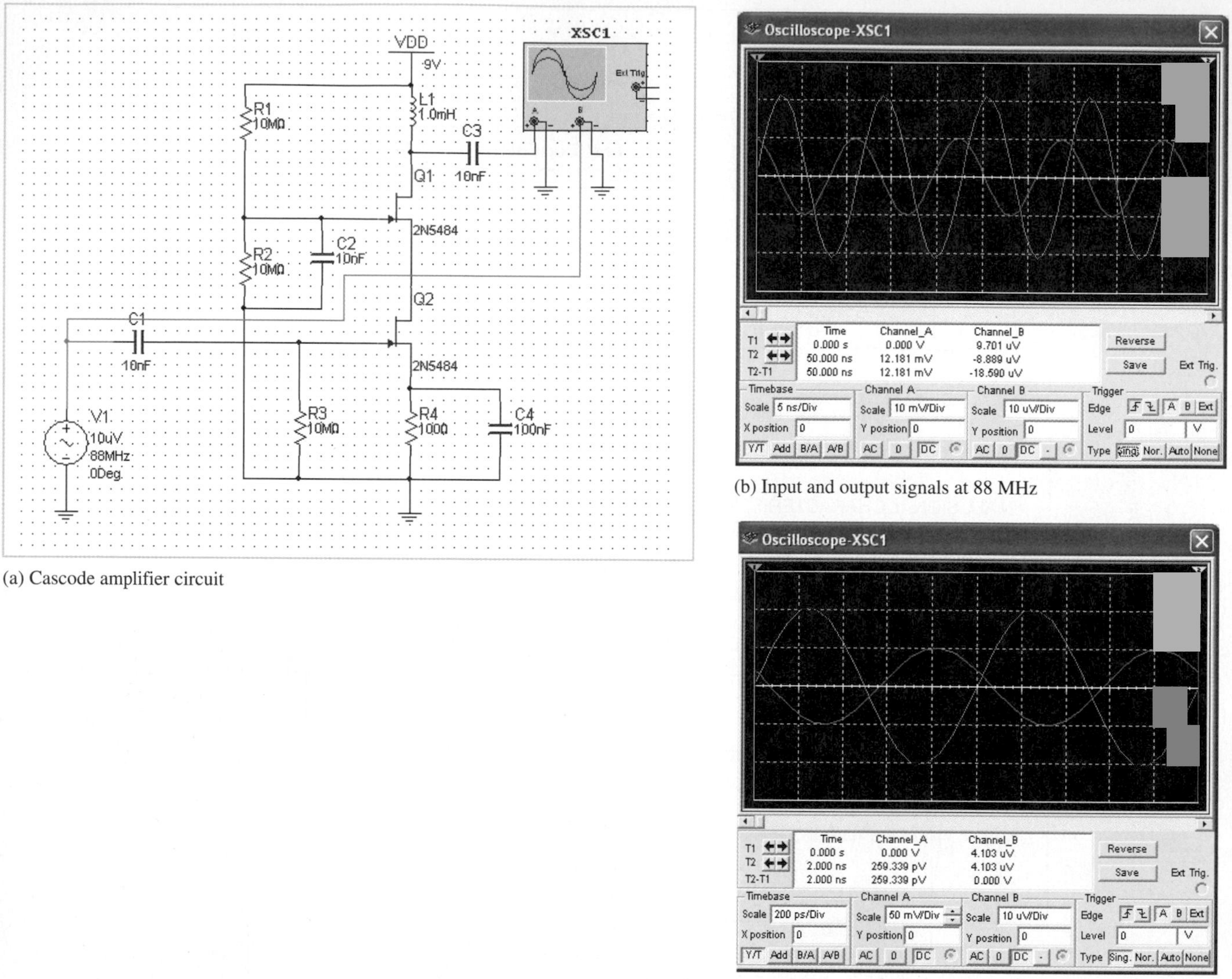

(a) Cascode amplifier circuit

(b) Input and output signals at 88 MHz

(c) Input and output signals at 1 GHz

FIGURE 9–51

Simulation results for the active antenna circuit. Input is green and output is red.

Lab Experiment

To build and test a similar circuit, go to Experiment 9 in your lab manual (*Laboratory Exercises for Electronic Devices* by David Buchla and Steven Wetterling).

Circuit Board

Certain considerations should be observed when laying out a printed circuit board for RF circuits. EMI (electromagnetic interference), line inductance, and stray capacitance all become important at high frequencies. A few basic features that should be incorporated on an RF circuit board are

- Keep traces as short and wide as possible.
- Do not run parallel signal lines that are in close proximity.
- Capacitively decouple supply voltages.
- Provide a large ground plane for shielding and to minimize noise.

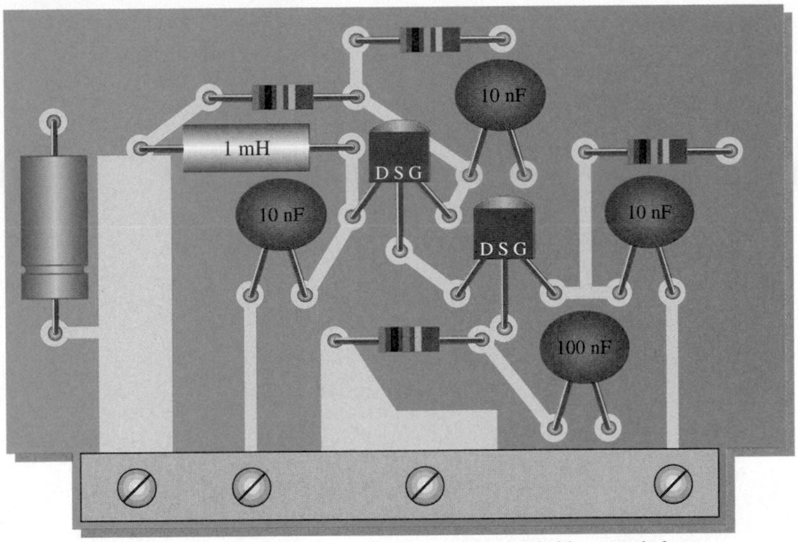

(a) Component side. Terminated pads are feedthroughs to backside ground plane.

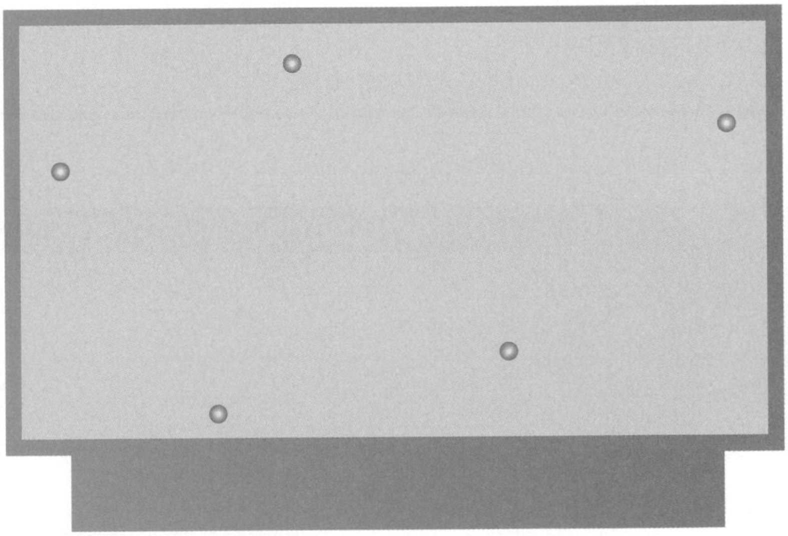

(b) Backside of board showing ground plane with feedthroughs indicated.

FIGURE 9–52
Active antenna circuit board.

The circuit board for the active antenna is shown in Figure 9–52. A large ground plane is on the back side. Components are connected to the ground plane with feedthrough connections as indicated.

9. Check the printed circuit board for correctness by comparing with the schematic in Figure 9–49.
10. State the purpose of the large blue capacitor that is not shown on the schematic.
11. Label each input and output pin according to function.

The Complete Active Antenna Unit

Typically, the active antenna circuit should be enclosed in a metal box for proper shielding, similar to that shown in Figure 9–53. The particular configuration shown includes an impedance-matching transformer connected to a type of connector (BNC) used with coaxial cables for illustration. Other interface configurations, such as an emitter-follower output may be used for interfacing. The particular antenna shown is a telescoping whip antenna.

FIGURE 9–53

A configuration of the active antenna in a metal housing (cover removed) with a 9 V battery and an impedance-matching transformer.

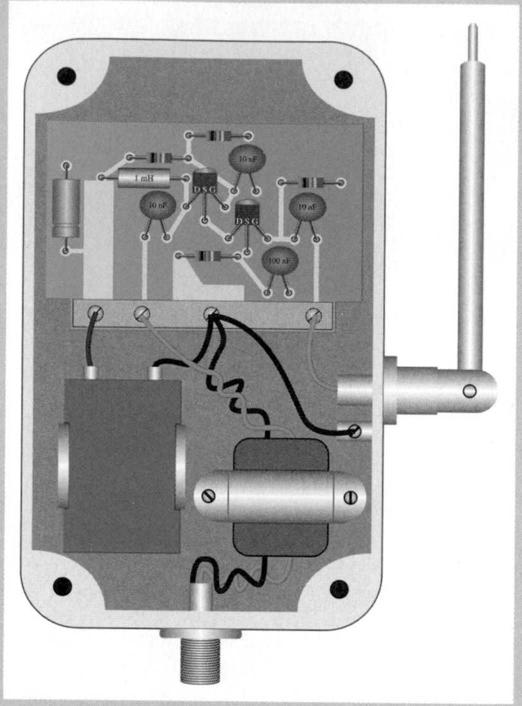

SUMMARY OF FET AMPLIFIERS

N channels are shown. V_{DD} is negative for *p* channel.

COMMON-SOURCE AMPLIFIERS

JFET ***Self-bias***

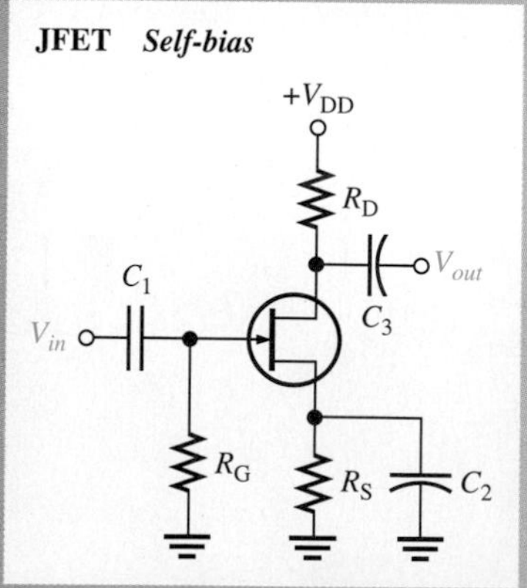

- $I_D = I_{DSS}\left(1 - \frac{I_D R_S}{V_{GS(off)}}\right)^2$
- $A_v = g_m R_d$
- $R_{in} = R_G \| \left(\frac{V_{GS}}{I_{GSS}}\right)$

D-MOSFET ***Zero-bias***

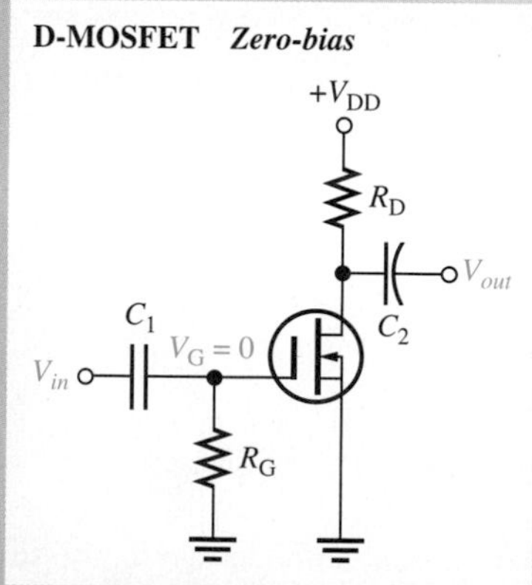

- $I_D = I_{DSS}$
- $A_v = g_m R_d$
- $R_{in} = R_G \| \left(\frac{V_{GS}}{I_{GSS}}\right)$

E-MOSFET ***Voltage-divider bias***

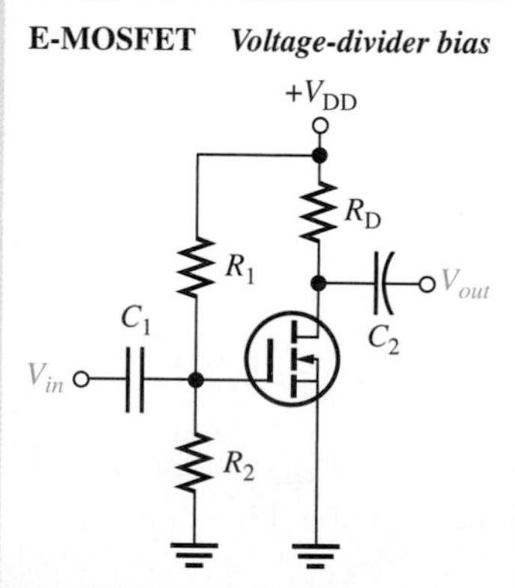

- $I_D = K(V_{GS} - V_{GS(th)})^2$
- $A_v = g_m R_d$
- $R_{in} = R_1 \| R_2 \| \left(\frac{V_{GS}}{I_{GSS}}\right)$

COMMON-DRAIN AMPLIFIER

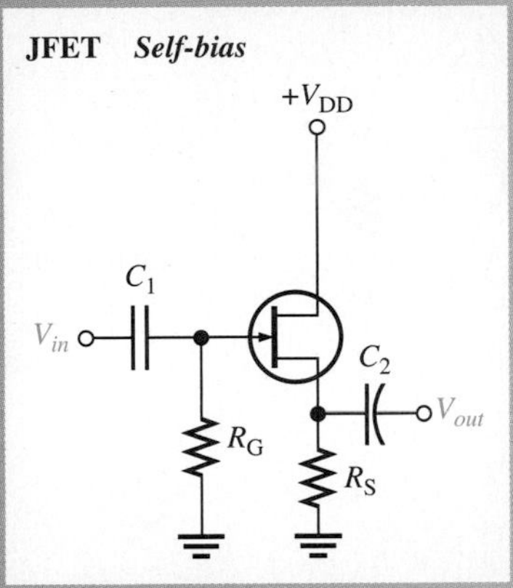

- $I_D = I_{DSS}\left(1 - \frac{I_D R_S}{V_{GS(off)}}\right)^2$
- $A_v = \frac{g_m R_s}{1 + g_m R_s}$
- $R_{in} = R_G \parallel \left(\frac{V_{GS}}{I_{GSS}}\right)$

COMMON-GATE AMPLIFIER

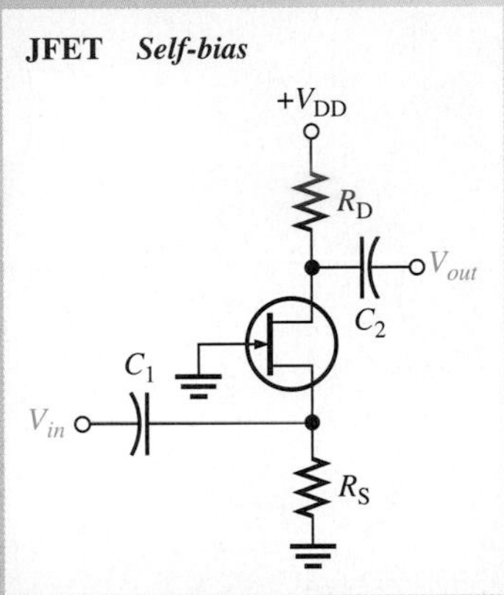

- $I_D = I_{DSS}\left(1 - \frac{I_D R_S}{V_{GS(off)}}\right)^2$
- $A_v = g_m R_d$
- $R_{in} = \left(\frac{1}{g_m}\right) \parallel R_S$

CASCODE AMPLIFIER

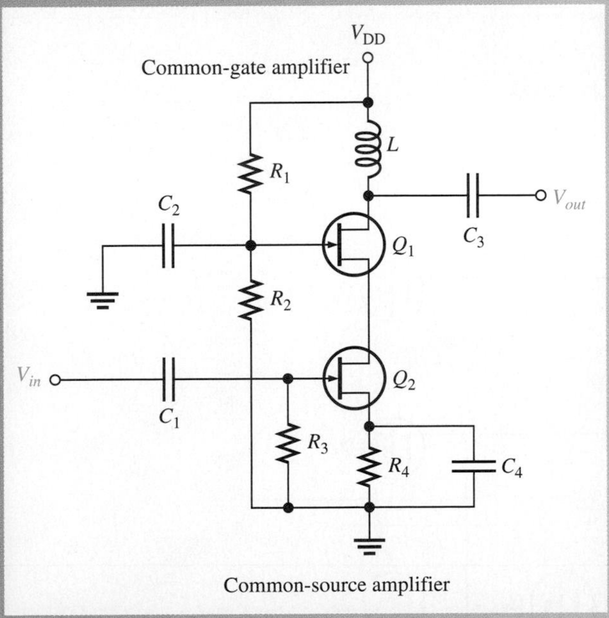

- $A_v \cong g_{m(CG)} X_L$

SUMMARY OF FET SWITCHING CIRCUITS

ANALOG SWITCH

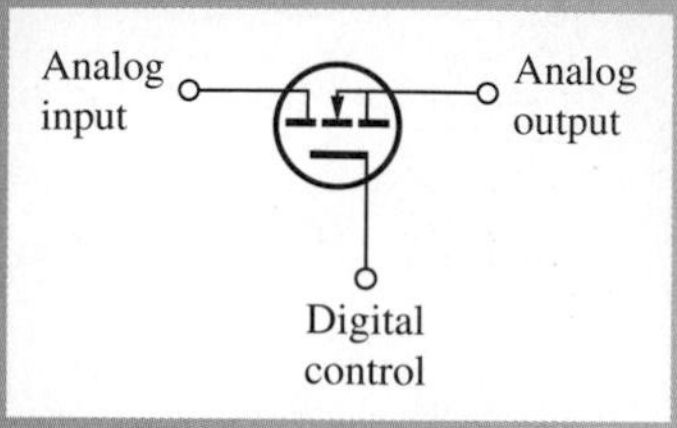

ANALOG MULTIPLEXER

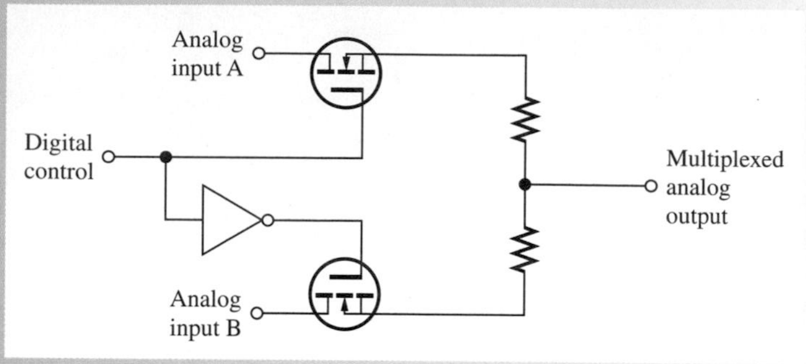

SWITCHED CAPACITOR

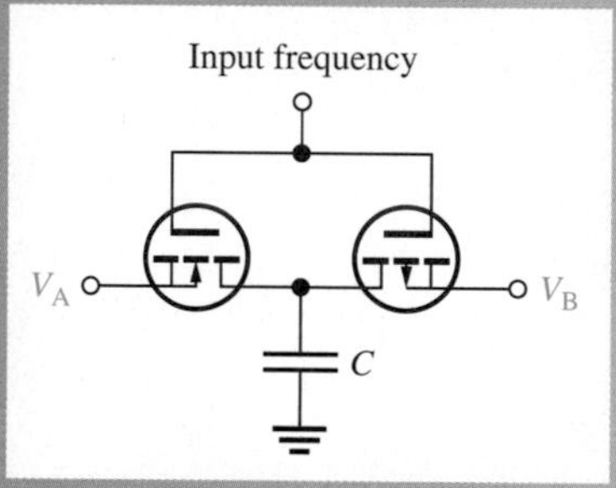

CMOS INVERTER

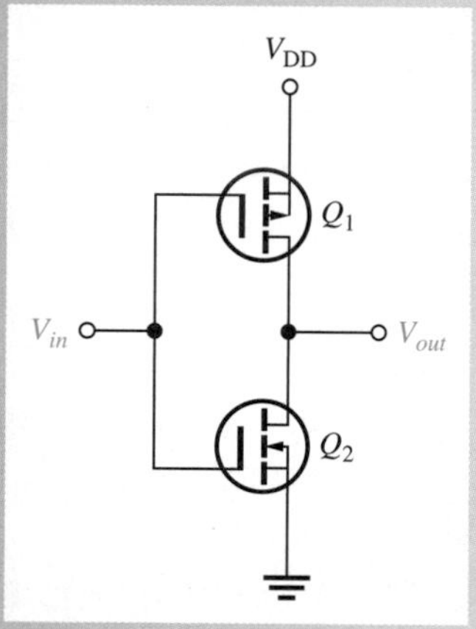

CMOS NAND GATE

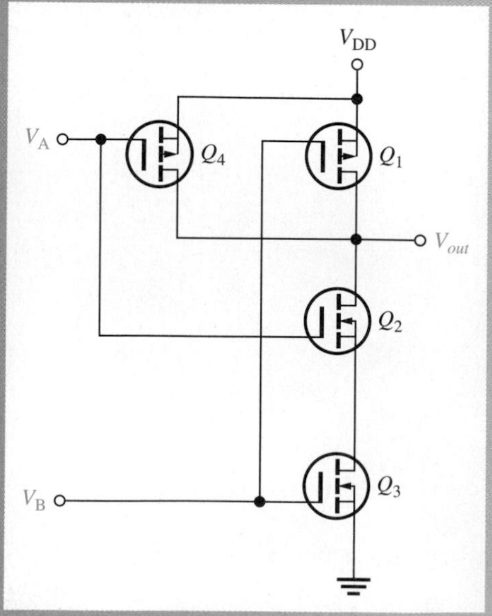

CMOS NOR GATE

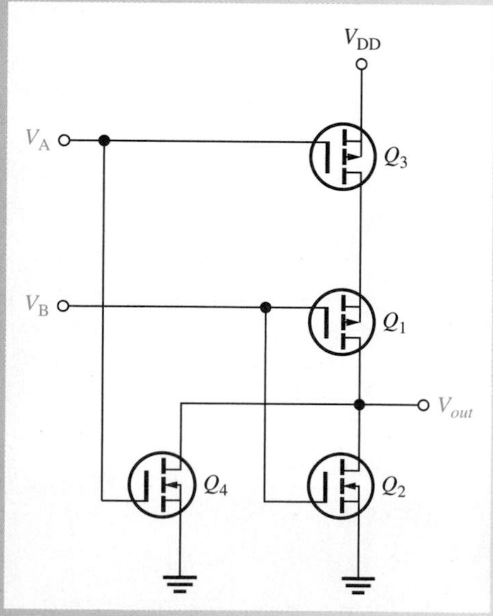

SUMMARY

Section 9–1

- The transconductance, g_m, of a FET relates the output current, I_d, to the input voltage, V_{gs}.
- The voltage gain of a common-source amplifier is determined largely by the transconductance, g_m, and the drain resistance, R_d.
- The internal drain-to-source resistance, r'_{ds}, of a FET influences (reduces) the gain if it is not sufficiently greater than R_d so that it can be neglected.
- An unbypassed resistance between source and ground (R_S) reduces the voltage gain of a FET amplifier.
- A load resistance connected to the drain of a common-source amplifier reduces the voltage gain.
- There is a 180° phase inversion between gate and drain voltages.
- The input resistance at the gate of a FET is extremely high.

Section 9–2

- The voltage gain of a common-drain amplifier (source-follower) is always slightly less than 1.
- There is no phase inversion between gate and source in a source-follower.

Section 9–3

- The input resistance of a common-gate amplifier is the reciprocal of g_m.
- The cascode amplifier combines a CS amplifier and a CG amplifier.

Section 9–4

- The class D amplifier is a nonlinear amplifier because the transistors operate as switches.
- The class D uses pulse-width modulation (PWM) to represent the input signal.
- A low-pass filter converts the PWM signal back to the original input signal.
- The efficiency of a class D amplifier approaches 100%.

Section 9–5

- An analog switch passes or blocks an analog signal when turned *on* or *off* by a digital control input.
- A sampling circuit is an analog switch that is turned *on* for short time intervals to allow a sufficient number of discrete input signal values to appear on the output so that the input signal can be accurately represented by those discrete values.
- An analog multiplexer consists of two or more analog switches that connect sampled portions of their analog input signals to a single output in a time sequence.
- Switched-capacitors are used to emulate resistance in programmable IC analog arrays.

Section 9–6

- Complementary MOS (CMOS) is used in low-power digital switching circuits.
- CMOS uses an *n*-channel MOSFET and a *p*-channel MOSFET connected in series.
- The inverter, NAND gate, and NOR gate are examples of digital logic circuits.

KEY TERMS

Key terms and other bold terms in the chapter are defined in the end-of-book glossary.

Analog switch A device that switches an analog signal on and off.

Class D A nonlinear amplifier in which the transistors are operated as switches.

CMOS Complementary MOS.

Common-drain A FET amplifier configuration in which the drain is the grounded terminal.

Common-gate A FET amplifier configuration in which the gate is the grounded terminal.

Common-source A FET amplifier configuration in which the source is the grounded terminal.

Pulse-width modulation A process in which a signal is converted to a series of pulses with widths that vary proportionally to the signal amplitude.

Source-follower The common-drain amplifier.

KEY FORMULAS

Common-Source Amplifier

9–1	$A_v = g_m R_d$	Voltage gain with source grounded or R_s bypassed
9–2	$I_D = I_{DSS}\left(1 - \dfrac{I_D R_S}{V_{GS(off)}}\right)^2$	Self-biased JFET current

9–3	$A_v = g_m R_d$	Voltage gain
9–4	$R_{in} = R_G \parallel \left(\frac{V_{GS}}{I_{GSS}}\right)$	Input resistance, self-bias and zero-bias
9–5	$R_{in} = R_1 \parallel R_2 \parallel R_{IN(gate)}$	Input resistance, voltage-divider bias

Common-Drain Amplifier

9–6	$A_v = \frac{g_m R_s}{1 + g_m R_s}$	Voltage gain
9–7	$R_{in} = R_G \parallel R_{IN(gate)}$	Input resistance

Common-Gate Amplifier

9–8	$A_v = g_m R_d$	Voltage gain
9–9	$R_{in(source)} = \frac{1}{g_m}$	Input resistance

MOSFET Analog Switching

9–10	$R = \frac{1}{fC}$	Emulated resistance

TRUE/FALSE QUIZ

Answers are at the end of the chapter.

1. A common-source (CS) amplifier has a very high input resistance.
2. The drain current in a CS amplifier can be calculated using a quadratic formula.
3. The voltage gain of a CS amplifier is the transconductance times the source resistance.
4. There is no phase inversion in a CS amplifier.
5. A CS amplifier using a D-MOSFET can operate with both positive and negative input voltages.
6. A common-drain (CD) amplifier is called a *drain-follower.*
7. The input resistance of a CD amplifier is very low.
8. The input resistance of a common-gate (CG) amplifier is very low.
9. A cascode amplifier uses both a CS and a CG amplifier.
10. The class D amplifier always operates in the linear region.
11. The class D amplifier uses pulse-width modulation.
12. An analog switch is controlled by a digital input.
13. The purpose of a switched-capacitor circuit is to emulate resistance.
14. CMOS is a device used in linear amplifiers.
15. CMOS utilizes a *pnp* MOSFET and an *npn* MOSFET connected together.

CIRCUIT-ACTION QUIZ

Answers are at the end of the chapter.

1. If the drain current is increased in Figure 9–9, V_{GS} will
   (a) increase **(b)** decrease **(c)** not change
2. If the JFET in Figure 9–9 is substituted with one having a lower value of I_{DSS}, the voltage gain will
   (a) increase **(b)** decrease **(c)** not change
3. If the JFET in Figure 9–9 is substituted with one having a lower value of $V_{GS(off)}$, the voltage gain will
   (a) increase **(b)** decrease **(c)** not change
4. If the value of R_G in Figure 9–9 is increased, V_{GS} will
   (a) increase **(b)** decrease **(c)** not change

5. If the value of R_G in Figure 9–11 is increased, the input resistance seen by the signal source will

   (a) increase (b) decrease (c) not change

6. If the value of R_1 in Figure 9–17 is increased, V_{GS} will

   (a) increase (b) decrease (c) not change

7. If the value of R_L in Figure 9–17 is decreased, the voltage gain will

   (a) increase (b) decrease (c) not change

8. If the value of R_S in Figure 9–20 is increased, the voltage gain will

   (a) increase (b) decrease (c) not change

9. If C_4 in Figure 9–46 opens, the output signal voltage will

   (a) increase (b) decrease (c) not change

SELF-TEST

Answers are at the end of the chapter.

Section 9–1

1. In a common-source amplifier, the output voltage is

   (a) 180° out of phase with the input (b) in phase with the input
   (c) taken at the source (d) taken at the drain
   (e) answers (a) and (c) (f) answers (a) and (d)

2. In a certain common-source (CS) amplifier, $V_{ds} = 3.2$ V rms and $V_{gs} = 280$ mV rms. The voltage gain is

   (a) 1 (b) 11.4 (c) 8.75 (d) 3.2

3. In a certain CS amplifier, $R_D = 1.0\ \text{k}\Omega$, $R_S = 560\ \Omega$, $V_{DD} = 10$ V, and $g_m = 4500\ \mu$S. If the source resistor is completely bypassed, the voltage gain is

   (a) 450 (b) 45 (c) 4.5 (d) 2.52

4. Ideally, the equivalent circuit of a FET contains

   (a) a current source in series with a resistance
   (b) a resistance between drain and source terminals
   (c) a current source between gate and source terminals
   (d) a current source between drain and source terminals

5. The value of the current source in Question 4 is dependent on the

   (a) transconductance and gate-to-source voltage
   (b) dc supply voltage
   (c) external drain resistance
   (d) answers (b) and (c)

6. A certain common-source amplifier has a voltage gain of 10. If the source bypass capacitor is removed,

   (a) the voltage gain will increase
   (b) the transconductance will increase
   (c) the voltage gain will decrease
   (d) the Q-point will shift

7. A CS amplifier has a load resistance of $10\ \text{k}\Omega$ and $R_D = 820\ \Omega$. If $g_m = 5$ mS and $V_{in} = 500$ mV, the output signal voltage is

   (a) 1.89 V (b) 2.05 V (c) 25 V (d) 0.5 V

8. If the load resistance in Question 7 is removed, the output voltage will

   (a) stay the same (b) decrease (c) increase (d) be zero

Section 9–2

9. A certain common-drain (CD) amplifier with $R_S = 1.0\ \text{k}\Omega$ has a transconductance of $6000\ \mu$S. The voltage gain is

   (a) 1 (b) 0.86 (c) 0.98 (d) 6

10. The datasheet for the transistor used in a CD amplifier specifies $I_{GSS} = 5$ nA at $V_{GS} = 10$ V. If the resistor from gate to ground, R_G, is 50 MΩ, the total input resistance is approximately
(a) 50 MΩ (b) 200 MΩ (c) 40 MΩ (d) 20.5 MΩ

Section 9–3

11. The common-gate (CG) amplifier differs from both the CS and CD configurations in that it has a
(a) much higher voltage gain (b) much lower voltage gain
(c) much higher input resistance (d) much lower input resistance
12. If you are looking for both good voltage gain and high input resistance, you must use a
(a) CS amplifier (b) CD amplifier (c) CG amplifier
13. A cascode amplifier consists of
(a) a CD and a CS amplifier (b) a CS and a CG amplifier
(c) a CG and a CD amplifier (d) two CG amplifiers

Section 9–4

14. The class D amplifier is similar to
(a) class C (b) class B (c) class A (d) none of these
15. The class D amplifier uses
(a) frequency modulation (b) amplitude modulation
(c) pulse-width modulation (d) duty cycle modulation

Section 9–5

16. E-MOSFETs are generally used for switching applications because of their
(a) threshold characteristic (b) high input resistance
(c) linearity (d) high gain
17. A sampling circuit must sample a signal at a minimum of
(a) one time per cycle (b) the signal frequency
(c) twice the signal frequency (d) alternate cycles
18. The value of resistance emulated by a switched-capacitor circuit is a function of
(a) voltage and capacitance (b) frequency and capacitance
(c) gain and transconductance (d) frequency and transconductance

Section 9–6

19. A basic CMOS circuit uses a combination of
(a) *n*-channel MOSFETs (b) *p*-channel MOSFETs
(c) *pnp* and *npn* BJTs (d) an *n*-channel and a *p*-channel MOSFET
20. CMOS is commonly used in
(a) digital circuits (b) linear circuits (c) RF circuits (d) power circuits

Section 9–7

21. If there is an internal open between the drain and source in a CS amplifier, the drain voltage is equal to
(a) 0 V (b) V_{DD} (c) V_{GS} (d) V_{GD}

PROBLEMS

Answers to all odd-numbered problems are at the end of the book.

BASIC PROBLEMS

Section 9–1 The Common-Source Amplifier

1. A FET has a $g_m = 6000\ \mu S$. Determine the rms drain current for each of the following rms values of V_{gs}.
(a) 10 mV (b) 150 mV (c) 0.6 V (d) 1 V
2. The gain of a certain JFET amplifier with a source resistance of zero is 20. Determine the drain resistance if the g_m is 3500 μS.
3. A certain FET amplifier has a g_m of 4.2 mS, $r'_{ds} = 12$ kΩ, and $R_D = 4.7$ kΩ. What is the voltage gain? Assume the source resistance is 0 Ω.
4. What is the gain for the amplifier in Problem 3 if the source resistance is 1.0 kΩ?
5. Identify the type of FET and its bias arrangement in Figure 9–54. Ideally, what is V_{GS}?
6. Calculate the dc voltages from each terminal to ground for the FETs in Figure 9–54.

FIGURE 9–54

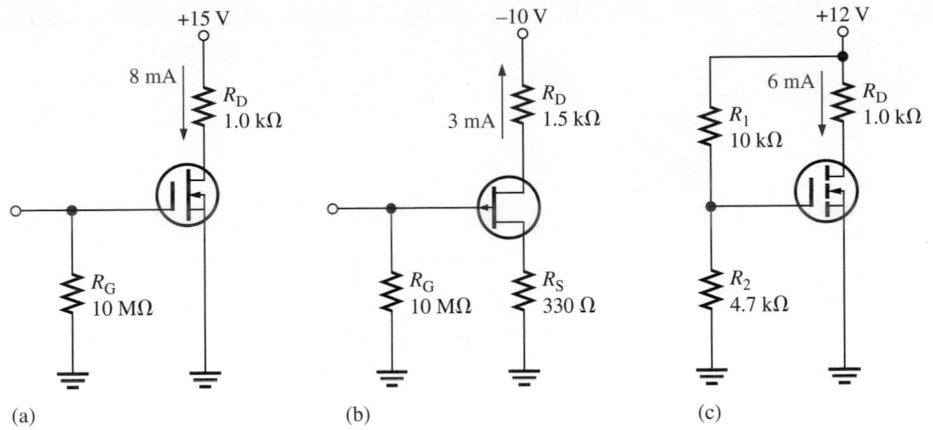

7. Identify each characteristic curve in Figure 9–55 by the type of FET that it represents.

FIGURE 9–55

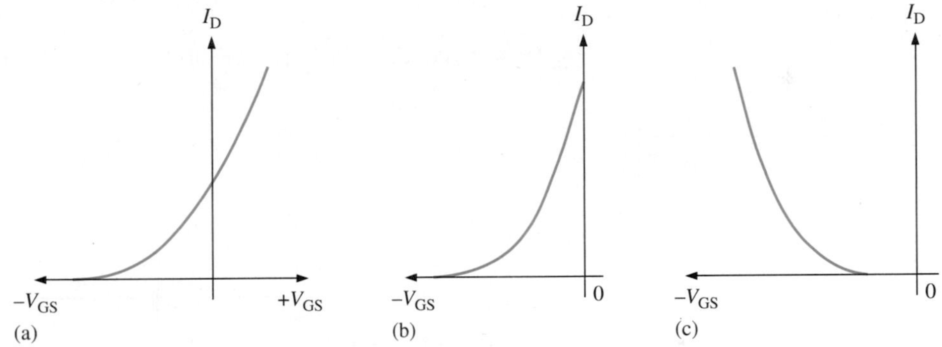

8. Refer to the JFET transfer characteristic curve in Figure 9–16(a) and determine the peak-to-peak value of I_d when V_{gs} is varied ± 1.5 V about its Q-point value.
9. Repeat Problem 8 for the curves in Figure 9–16(b) and Figure 9–16(c).
10. Given that $I_D = 2.83$ mA in Figure 9–56, find V_{DS} and V_{GS}. $V_{GS(off)} = -7$ V and $I_{DSS} = 8$ mA.

FIGURE 9–56

Multisim file circuits are identified with a CD logo and are in the Problems folder on your CD-ROM. Filenames correspond to figure numbers (e.g., F09-56).

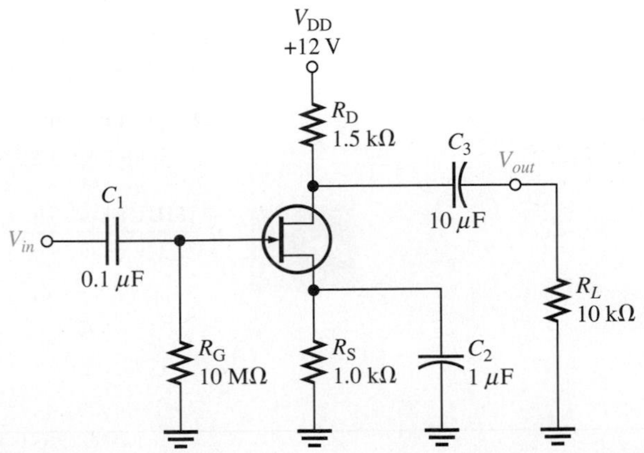

11. If a 50 mV rms input signal is applied to the amplifier in Figure 9–56, what is the peak-to-peak output voltage? $g_m = 5000\ \mu$S.
12. If a 1500 Ω load is ac coupled to the output in Figure 9–56, what is the resulting output voltage (rms) when a 50 mV rms input is applied? $g_m = 5000\ \mu$S.

13. Determine the voltage gain of each common-source amplifier in Figure 9–57.

▶ FIGURE 9–57

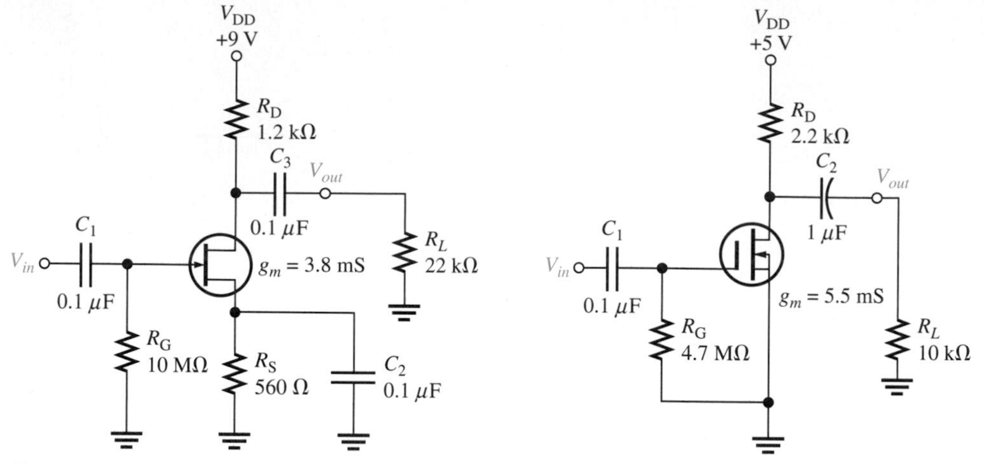

(a) (b)

14. Draw the dc and ac equivalent circuits for the amplifier in Figure 9–58.

▶ FIGURE 9–58

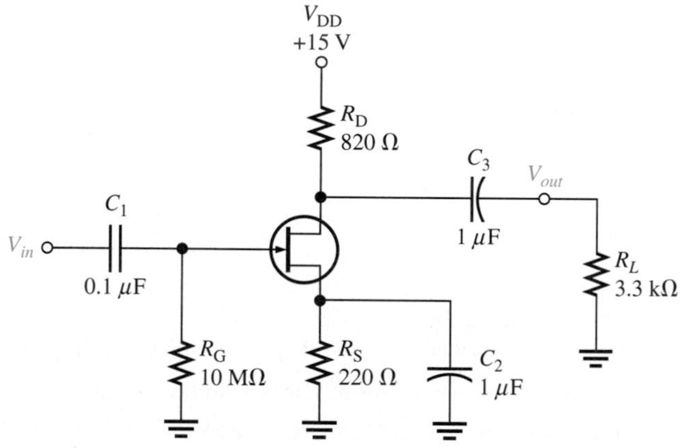

15. Determine the drain current in Figure 9–58 given that $I_{DSS} = 15$ mA and $V_{GS(off)} = -4$ V. The Q-point is centered.

16. What is the gain of the amplifier in Figure 9–58 if C_2 is removed?

17. A 4.7 kΩ resistor is connected in parallel with R_L in Figure 9–58. What is the voltage gain?

18. For the common-source amplifier in Figure 9–59, determine I_D, V_{GS}, and V_{DS} for a centered Q-point. $I_{DSS} = 9$ mA, and $V_{GS(off)} = -3$ V.

▶ FIGURE 9–59

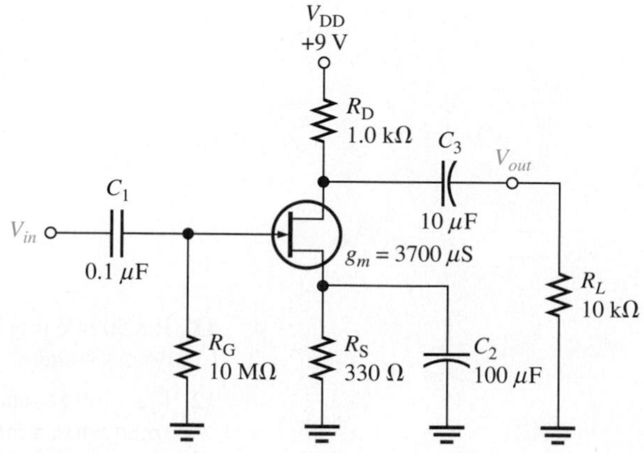

19. If a 10 mV rms signal is applied to the input of the amplifier in Figure 9–59, what is the rms value of the output signal?

20. Determine V_{GS}, I_D, and V_{DS} for the amplifier in Figure 9–60. $I_{D(on)} = 18$ mA at $V_{GS} = 10$ V, $V_{GS(th)} = 2.5$ V, and $g_m = 3000\ \mu$S.

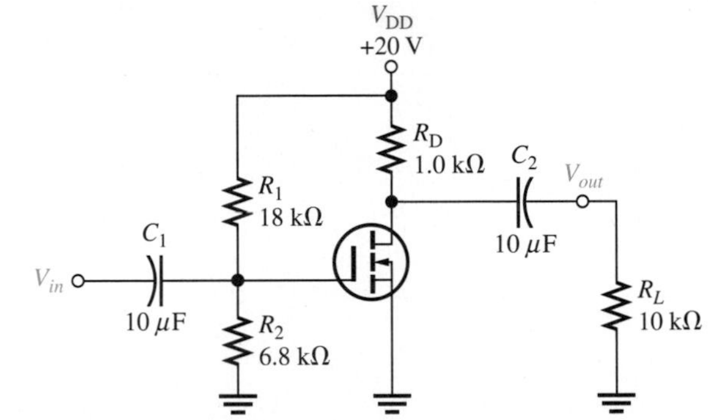

▲ FIGURE 9–60

21. Determine R_{in} seen by the signal source in Figure 9–61. $I_{GSS} = 25$ nA at $V_{GS} = -15$ V.

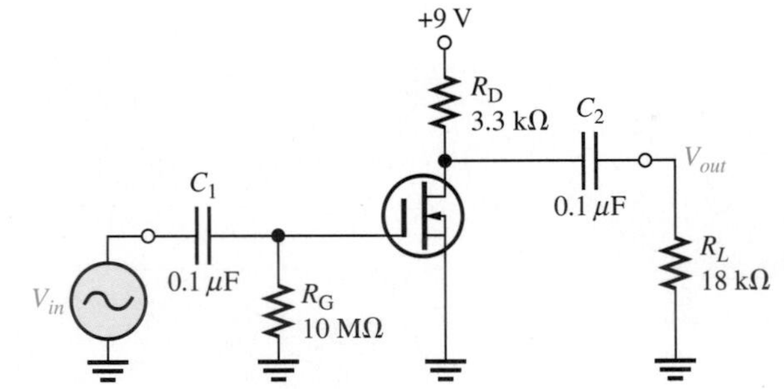

▲ FIGURE 9–61

22. Determine the total drain voltage waveform (dc and ac) and the V_{out} waveform in Figure 9–62. $g_m = 4.8$ mS and $I_{DSS} = 15$ mA. Observe that $V_{GS} = 0$.

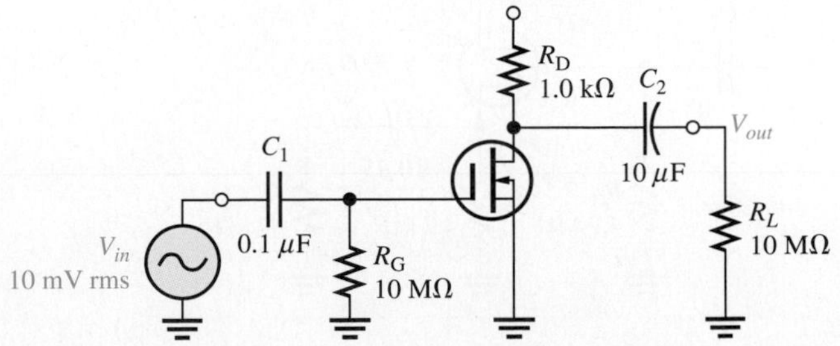

▲ FIGURE 9–62

23. For the unloaded amplifier in Figure 9–63, find V_{GS}, I_D, V_{DS}, and the rms output voltage V_{ds}. $I_{D(on)} = 8$ mA at $V_{GS} = 12$ V, $V_{GS(th)} = 4$ V, and $g_m = 4500$ S.

▶ FIGURE 9–63

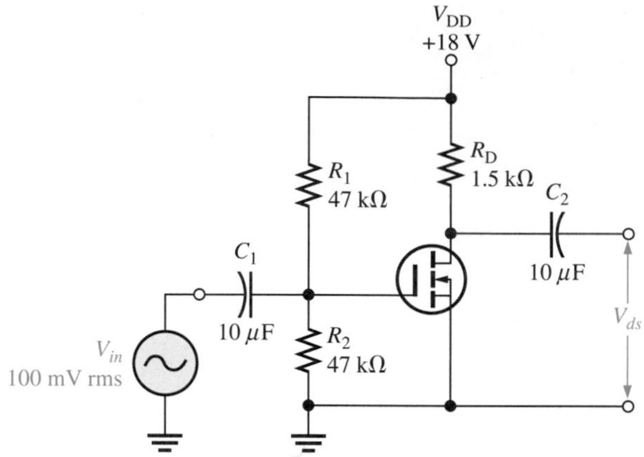

Section 9–2 The Common-Drain Amplifier

24. For the source-follower in Figure 9–64, determine the voltage gain and input resistance. $I_{GSS} = 50$ pA at $V_{GS} = -15$ V and $g_m = 5500\ \mu$S.

▶ FIGURE 9–64

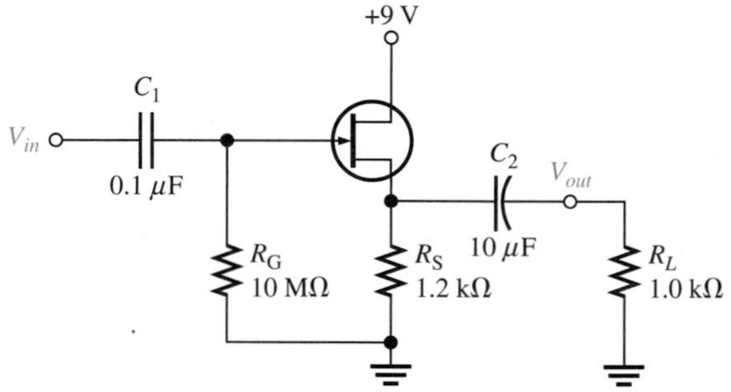

25. If the JFET in Figure 9–64 is replaced with one having a g_m of 3000 μS, what are the gain and the input resistance with all other conditions the same?

26. Find the gain of each amplifier in Figure 9–65.

27. Determine the voltage gain of each amplifier in Figure 9–65 when the capacitively coupled load is changed to 10 kΩ.

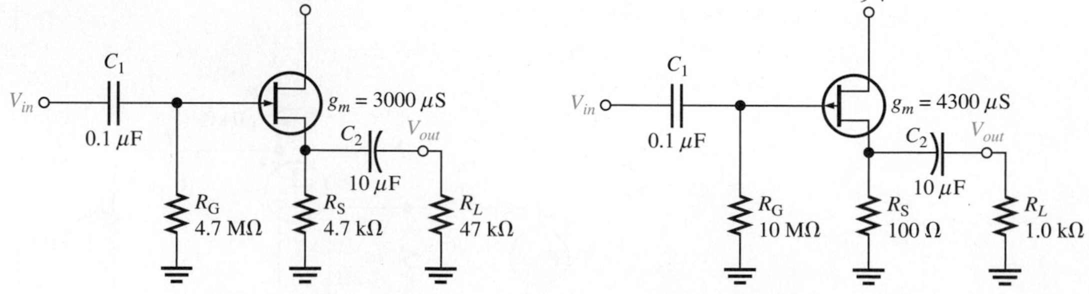

(a) (b)

▲ FIGURE 9–65

Section 9–3 The Common-Gate Amplifier

28. A common-gate amplifier has a $g_m = 4000\ \mu S$ and $R_d = 1.5\ k\Omega$. What is its gain?

29. What is the input resistance of the amplifier in Problem 28?

30. Determine the voltage gain and input resistance of the common-gate amplifier in Figure 9–66.

31. For a cascode amplifier like shown in Figure 9–24, $g_m = 2800\ \mu S$, $I_{GSS} = 2\ nA$ at $V_{GS} = 15\ V$. If $R_3 = 15\ M\Omega$ and $L = 1.5\ mH$, determine the voltage gain and the input impedance at $f = 100\ MHz$.

FIGURE 9–66

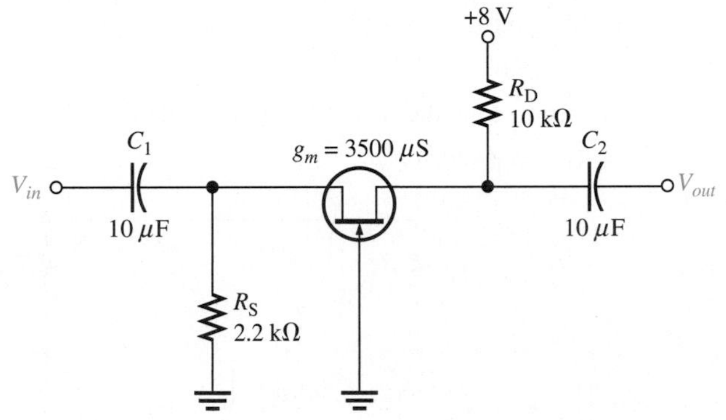

Section 9–4 The Class D Amplifier

32. A class D amplifier has an output of ±9 V. If the input signal is 5 mV, what is the voltage gain?

33. A certain class D amplifier dissipates an internal power of 140 mW in the comparator and the triangular wave generator. Each complementary MOSFET has 0.25 V drop in the *on* state. The amplifier operates from ±12 Vdc sources and provides 0.35 A to a load. Determine the efficiency.

Section 9–5 MOSFET Analog Switching

34. An analog switch uses an *n*-channel MOSFET with $V_{GS(th)} = 4\ V$. A voltage of +8 V is applied to the gate. Determine the maximum peak-to-peak input signal that can be applied if the drain-to-source voltage drop is neglected.

35. An analog switch is used to sample a signal with a maximum frequency of 15 kHz. Determine the minimum frequency of the pulses applied to the MOSFET gate.

36. A switched-capacitor circuit uses a 10 pF capacitor. Determine the frequency required to emulate a 10 kΩ resistor.

37. For a frequency of 25 kHz, what is the emulated resistance in a switched-capacitor circuit if $C = 0.001\ \mu F$?

Section 9–6 MOSFET Digital Switching

38. What is the output voltage of a CMOS inverter that operates with $V_{DD} = +5\ V$, when the input is 0 V? When the input is +5 V?

39. For each of the following input combinations, determine the output of a CMOS NAND gate that operates with $V_{DD} = +3.3\ V$.

(a) $V_A = 0\ V$, $V_B = 0\ V$

(b) $V_A = +3.3\ V$, $V_B = 0\ V$

(c) $V_A = 0\ V$, $V_B = +3.3\ V$

(d) $V_A = +3.3\ V$, $V_B = +3.3\ V$

40. Repeat Problem 39 for a CMOS NOR gate.

41. List two advantages of the MOSFET over the BJT in power switching.

Section 9–7 Troubleshooting

42. What symptom(s) would indicate each of the following failures when a signal voltage is applied to the input in Figure 9–67?

(a) Q_1 open from drain to source

(b) R_3 open

(c) C_2 shorted

(d) C_3 open

(e) Q_2 open from drain to source

43. If $V_{in} = 10$ mV rms in Figure 9–67, what is V_{out} for each of the following faults?

(a) C_1 open

(b) C_4 open

(c) a short from the source of Q_2 to ground

(d) Q_2 has an open gate

▶ FIGURE 9–67

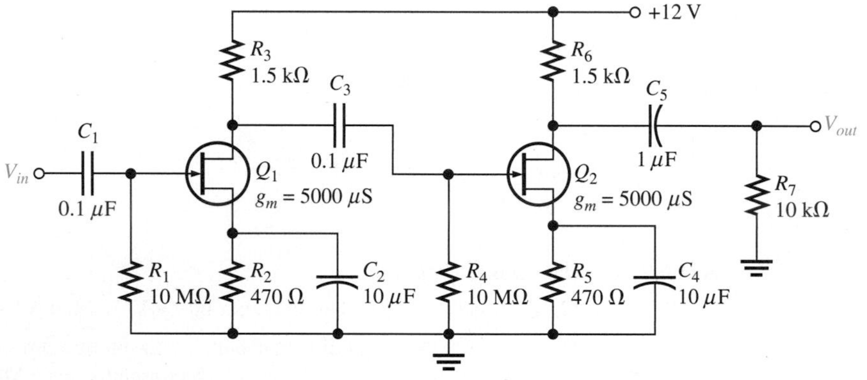

DATASHEET PROBLEMS

44. What type of FET is the 2N3796?

45. Referring to the datasheet in Figure 9–68, determine the following:

(a) typical $V_{GS(off)}$ for the 2N3796

(b) maximum drain-to-source voltage for the 2N3797

(c) maximum power dissipation for the 2N3797 at an ambient temperature of 25°C

(d) maximum gate-to-source voltage for the 2N3797

46. Referring to Figure 9–68, determine the maximum power dissipation for a 2N3796 at an ambient temperature of 55°C.

47. Referring to Figure 9–68, determine the minimum g_{m0} for the 2N3796 at a frequency of 1 kHz.

48. What is the drain current when $V_{GS} = +3.5$ V for the 2N3797?

49. Typically, what is the drain current for a zero-biased 2N3796?

50. What is the maximum possible voltage gain for a 2N3796 common-source amplifier with $R_d = 2.2\text{ k}\Omega$?

ADVANCED PROBLEMS

51. The MOSFET in a certain single-stage common-source amplifier has a range of forward transconductance values from 2.5 mS to 7.5 mS. If the amplifier is capacitively coupled to a variable load that ranges from 4 kΩ to 10 kΩ and the dc drain resistance is 1.0 kΩ, determine the minimum and maximum voltage gains.

52. Design an amplifier using a 2N3797 that operates from a 24 V supply voltage. The typical dc drain-to-source voltage should be approximately 12 V and the typical voltage gain should be approximately 9.

53. Modify the amplifier you designed in Problem 52 so that the voltage gain can be set at 9 for any randomly selected 2N3797.

2N3796
2N3797

Case 22-03, Style 2
TO-18 (TO-206AA)

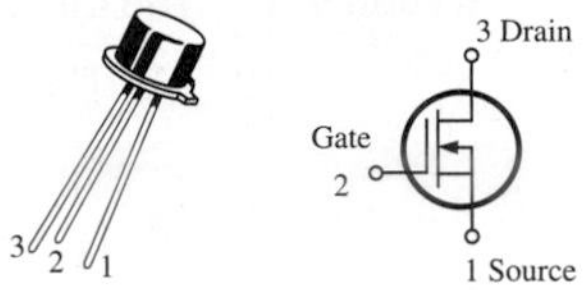

MOSFETs
Low Power Audio
***N*-channel — Depletion**

Maximim Ratings

Rating	Symbol	Value	Unit
Drain-Source voltage 2N3796 2N3797	V_{DS}	 25 20	V dc
Gate-Source voltage	V_{GS}	±10	V dc
Drain current	I_D	20	mA dc
Total device dissipation @ $T_A = 25°C$ Derate above 25°C	P_D	200 1.14	mW mW/°C
Junction temperature	T_J	+175	°C
Storage channel temperature range	T_{stg}	– 65 to + 200	°C

Electrical Characteristics ($T_A = 25°C$ unless otherwise noted.)

Characteristic		Symbol	Min	Typ	Max	Unit
OFF Characteristics						
Drain-Source breakdown voltage ($V_{GS} = -4.0$ V, $I_D = 5.0$ μA) ($V_{GS} = -7.0$ V, $I_D = 5.0$ μA)	 2N3796 2N3797	$V_{(BR)DSX}$	 25 20	 30 25	 – –	V dc
Gate reverse current ($V_{GS} = -10$ V, $V_{DS} = 0$) ($V_{GS} = -10$ V, $V_{DS} = 0$, $T_A = 150°C$)		I_{GSS}	 – –	 – –	 1.0 200	pA dc
Gate-Source cutoff voltage ($I_D = 0.5$ μA, $V_{DS} = 10$ V) ($I_D = 2.0$ μA, $V_{DS} = 10$ V)	 2N3796 2N3797	$V_{GS(off)}$	 – –	 – 3.0 – 5.0	 – 4.0 – 7.0	V dc
Drain-Gate reverse current ($V_{DG} = 10$ V, $I_S = 0$)		I_{DGO}	–	–	1.0	pA dc
ON Characteristics						
Zero-gate-voltage drain current ($V_{DS} = 10$ V, $V_{GS} = 0$)	2N3796 2N3797	I_{DSS}	0.5 2.0	1.5 2.9	3.0 6.0	mA dc
On-State drain current ($V_{DS} = 10$ V, $V_{GS} = +3.5$ V)	2N3796 2N3797	$I_{D(on)}$	7.0 9.0	8.3 14	14 18	mA dc
Small-Signal Characteristics						
Forward transfer admittance ($V_{DS} = 10$ V, $V_{GS} = 0, f = 1.0$ kHz)	2N3796 2N3797	$\vert Y_{fs} \vert$	900 1500	1200 2300	1800 3000	μmhos or μS
($V_{DS} = 10$ V, $V_{GS} = 0, f = 1.0$ MHz)	2N3796 2N3797		900 1500	– –	– –	
Output admittance ($V_{DS} = 10$ V, $V_{GS} = 0, f = 1.0$ kHz)	2N3796 2N3797	$\vert Y_{os} \vert$	– –	12 27	25 60	μmhos or μS
Input capacitance ($V_{DS} = 10$ V, $V_{GS} = 0, f = 1.0$ MHz)	2N3796 2N3797	C_{iss}	– –	5.0 6.0	7.0 8.0	pF
Reverse transfer capacitance ($V_{DS} = 10$ V, $V_{GS} = 0, f = 1.0$ MHz)		C_{rss}	–	0.5	0.8	pF

▲ FIGURE 9–68

Partial datasheet for the 2N3796 and 2N3797 D-MOSFETs.

MULTISIM TROUBLESHOOTING PROBLEMS

These file circuits are in the Troubleshooting Problems folder on your CD-ROM.

54. Open file TSP09-54 and determine the fault.

55. Open file TSP09-55 and determine the fault.

56. Open file TSP09-56 and determine the fault.

57. Open file TSP09-57 and determine the fault.

58. Open file TSP09-58 and determine the fault.

59. Open file TSP09-59 and determine the fault.

60. Open file TSP09-60 and determine the fault.

61. Open file TSP09-61 and determine the fault.

62. Open file TSP09-62 and determine the fault.

ANSWERS

SECTION CHECKUPS

Section 9–1 The Common-Source Amplifier

1. The FET with $g_m = 3.5$ mS can produce the higher gain.
2. $A_v = g_m R_d = (2500\ \mu S)(10\ k\Omega) = 25$
3. I_d is at its positive peak and V_{ds} is at its negative peak when V_{gs} is at its positive peak.
4. V_{gs} is an ac quantity, V_{GS} is a dc quantity.
5. Voltage gain of a CS amplifier is determined by g_m and R_d.
6. The gain is halved because $R_d = R_D/2$.

Section 9–2 The Common-Drain Amplifier

1. The ideal maximum voltage gain of a CD amplifier is 1.
2. The voltage gain of a CD amplifier is determined by g_m and R_s.

Section 9–3 The Common-Gate Amplifier

1. The CG amplifier has a low input resistance ($1/g_m$).
2. g_m affects both voltage gain and input resistance.
3. High input resistance, high gain, and excellent high-frequency response

Section 9–4 The Class D Amplifier

1. Modulator, push-pull switching amplifier, and low-pass filter
2. The pulse width is proportional to the amplitude of the input signal.
3. It is passed through a low-pass filter.

Section 9–5 MOSFET Analog Switching

1. When it is turned *off*
2. When it is turned *on*
3. A pulse or dc voltage
4. *R* depends inversely on frequency and capacitance.

Section 9–6 MOSFET Digital Switching

1. A CMOS inverter consists of a *p*-channel and an *n*-channel MOSFET connected in a cascade arrangement.
2. NAND gate
3. NOR gate

Section 9–7 Troubleshooting

1. To be a good troubleshooter, you must understand the circuit.
2. There would be a lower than normal first-stage gain if C_2 opens.
3. No, but there would be a loss of signal to the second stage.

RELATED PROBLEMS FOR EXAMPLES

9–1 13.2

9–2 See Figure 9–69.

9–3 2.319 mA

9–4 See Figure 9–70.

9–5 350 mV

9–6 $R_{in} = 9.99\ M\Omega$

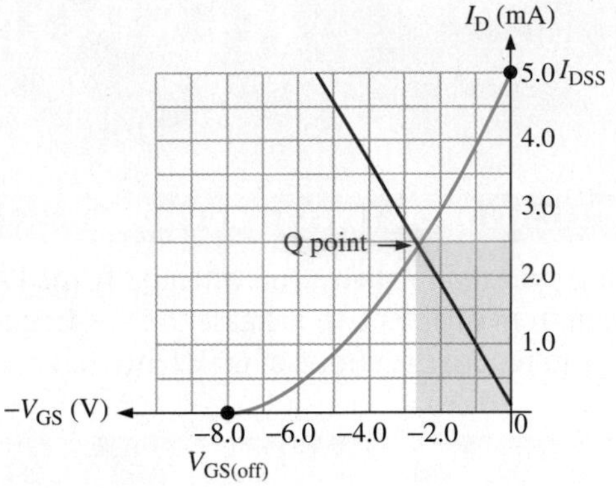

FIGURE 9–69

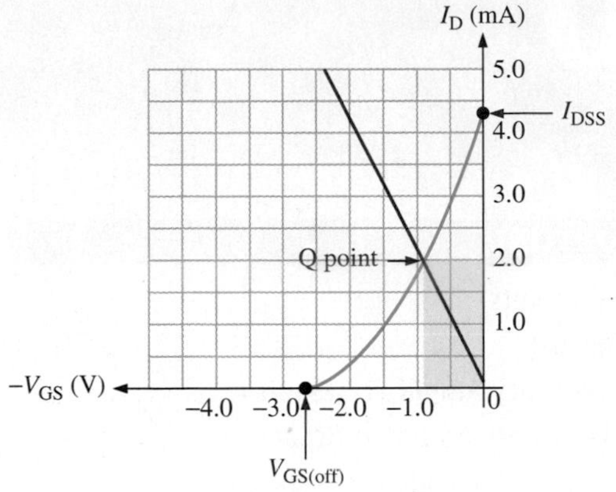

FIGURE 9–70

9–7 ΔI_D decreases; distortion and clipping at cutoff

9–8 $V_{GS} = 2.23$ V; $I_D = 1.09$ mA; $V_{DS} = 11.4$ V; $V_{out} = 750$ mV

9–9 0.976

9–10 $R_{in} = 476\ \Omega$

9–11 The gain increases.

9–12 0.951

9–13 The MOSFET switch would turn *off*.

9–14 > 24 kHz

9–15 Same as the input (no inversion)

TRUE/FALSE QUIZ

1. T **2.** T **3.** F **4.** F **5.** T **6.** F **7.** F **8.** T
9. T **10.** F **11.** T **12.** T **13.** T **14.** F **15.** F

CIRCUIT-ACTION QUIZ

1. (a) **2.** (b) **3.** (a) **4.** (c) **5.** (a) **6.** (b)
7. (b) **8.** (c) **9.** (b)

SELF-TEST

1. (f) **2.** (b) **3.** (c) **4.** (d) **5.** (a) **6.** (c) **7.** (a) **8.** (c)
9. (b) **10.** (a) **11.** (d) **12.** (a) **13.** (b) **14.** (d) **15.** (c) **16.** (a)
17. (c) **18.** (b) **19.** (d) **20.** (a) **21.** (b)

10 Amplifier Frequency Response

CHAPTER OUTLINE

CHAPTER OBJECTIVES

- Discuss the effects of circuit capacitances on the frequency response of an amplifier
- Express the gain of an amplifier in decibels (dB)
- Analyze the low-frequency response of amplifiers
- Analyze the high-frequency response of amplifiers
- Analyze an amplifier for total frequency response
- Analyze multistage amplifiers for frequency response
- Measure the frequency response of an amplifier

KEY TERMS

- Decibel
- Midrange gain
- Critical frequency
- Roll-off
- Decade
- Bode Plot
- Bandwidth

APPLICATION ACTIVITY PREVIEW

In the Application Activity, you will modify the PA system preamp from Chapter 6 to increase the low-frequency response to reduce the effects of 60 Hz interference.

VISIT THE COMPANION WEBSITE

Study aids for this chapter are available at
http://www.prenhall.com/floyd

INTRODUCTION

In the previous chapters on amplifiers, the effects of the input frequency on an amplifier's operation due to capacitive elements in the circuit were neglected in order to focus on other concepts. The coupling and bypass capacitors were considered to be ideal shorts and the internal transistor capacitances were considered to be ideal opens. This treatment is valid when the frequency is in an amplifier's midrange.

As you know, capacitive reactance decreases with increasing frequency and vice versa. When the frequency is low enough, the coupling and bypass capacitors can no longer be considered as shorts because their reactances are large enough to have a significant effect. Also, when the frequency is high enough, the internal transistor capacitances can no longer be considered as opens because their reactances become small enough to have a significant effect on the amplifier operation. A complete picture of an amplifier's response must take into account the full range of frequencies over which the amplifier can operate.

In this chapter, you will study the frequency effects on amplifier gain and phase shift. The coverage applies to both BJT and FET amplifiers, and a mix of both are included to illustrate the concepts.

10–1 BASIC CONCEPTS

In amplifiers, the coupling and bypass capacitors appear to be shorts to ac at the midband frequencies. At low frequencies the capacitive reactance of these capacitors affect the gain and phase shift of signals, so they must be taken into account. The **frequency response** of an amplifier is the change in gain or phase shift over a specified range of input signal frequencies.

After completing this section, you should be able to

- **Discuss the effects of circuit capacitances on the frequency response of an amplifier**
  - Explain the effect of coupling capacitors
  - Explain the effect of bypass capacitors
  - Discuss the internal transistor capacitances and explain their effects
  - Use Miller's theorem to determine amplifier capacitances

Effect of Coupling Capacitors

Recall from basic circuit theory that $X_C = 1/(2\pi fC)$. This formula shows that the capacitive reactance varies inversely with frequency. At lower frequencies the reactance is greater, and it decreases as the frequency increases. At lower frequencies—for example, audio frequencies below 10 Hz—capacitively coupled amplifiers such as those in Figure 10–1 have less voltage gain than they have at higher frequencies. The reason is that at lower frequencies more signal voltage is dropped across C_1 and C_3 because their reactances are higher. This higher signal voltage drop at lower frequencies reduces the voltage gain. Also, a phase shift is introduced by the coupling capacitors because C_1 forms a lead circuit with the R_{in} of the amplifier and C_3 forms a lead circuit with R_L in parallel with R_C or R_D. Recall that a *lead circuit* is an *RC* circuit in which the output voltage across *R* leads the input voltage in phase.

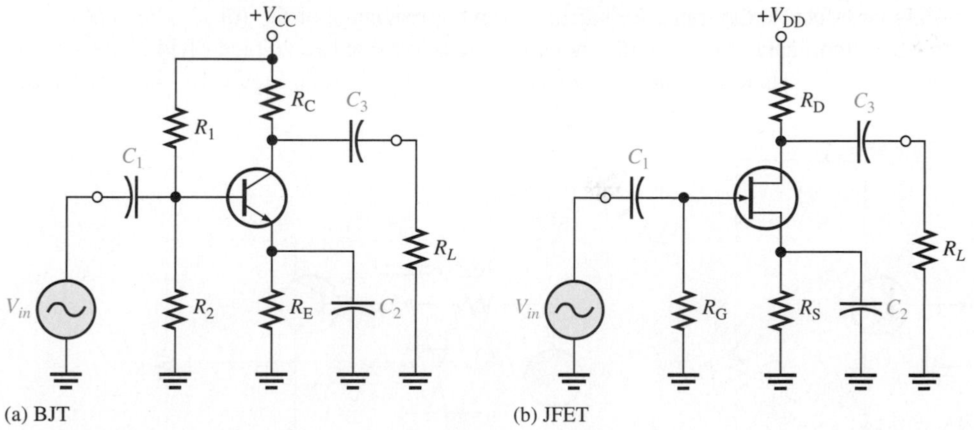

▲ FIGURE 10–1

Examples of capacitively coupled BJT and FET amplifiers.

Effect of Bypass Capacitors

At lower frequencies, the reactance of the bypass capacitor, C_2 in Figure 10–1, becomes significant and the emitter (or FET source terminal) is no longer at ac ground. The capacitive

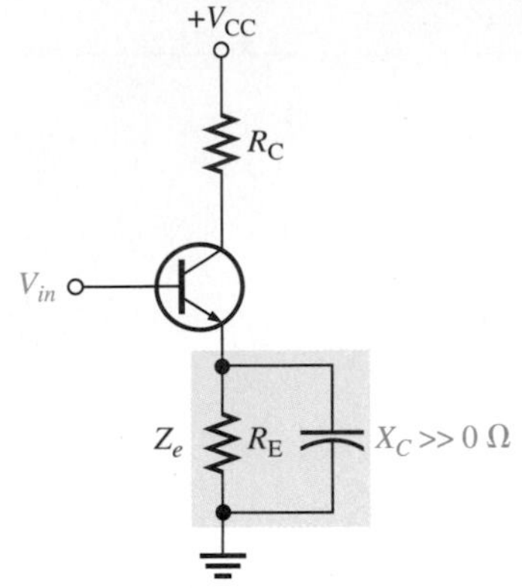

▲ **FIGURE 10–2**

Nonzero reactance of the bypass capacitor in parallel with R_E creates an emitter impedance, (Z_e), which reduces the voltage gain.

reactance X_{C2} in parallel with R_E (or R_S) creates an impedance that reduces the gain. This is illustrated in Figure 10–2.

For example, when the frequency is sufficiently high, $X_C \cong 0\ \Omega$ and the voltage gain of the CE amplifier is $A_v = R_C/r'_e$. At lower frequencies, $X_C >> 0\ \Omega$ and the voltage gain is $A_v = R_C/(r'_e + Z_e)$.

Effect of Internal Transistor Capacitances

At high frequencies, the coupling and bypass capacitors become effective ac shorts and do not affect an amplifier's response. Internal transistor junction capacitances, however, do come into play, reducing an amplifier's gain and introducing phase shift as the signal frequency increases.

Figure 10–3 shows the internal *pn* junction capacitances for both a bipolar junction transistor and a JFET. In the case of the BJT, C_{be} is the base-emitter junction capacitance and C_{bc} is the base-collector junction capacitance. In the case of the JFET, C_{gs} is the capacitance between gate and source and C_{gd} is the capacitance between gate and drain.

► **FIGURE 10–3**

Internal transistor capacitances.

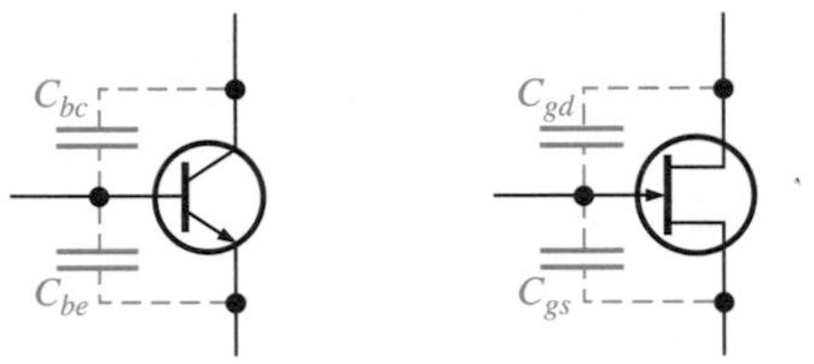

(a) BJT (b) JFET

Datasheets often refer to the BJT capacitance C_{bc} as the output capacitance, often designated C_{ob}. The capacitance C_{be} is often designated as the input capacitance C_{ib}. Datasheets for FETs normally specify input capacitance C_{iss} and reverse transfer capacitance C_{rss}. From these, C_{gs} and C_{gd} can be calculated, as you will see in Section 10–4.

At lower frequencies, the internal capacitances have a very high reactance because of their low capacitance value (usually only a few picofarads) and the low frequency value. Therefore, they look like opens and have no effect on the transistor's performance. As the frequency goes up, the internal capacitive reactances go down, and at some point they begin to have a significant effect on the transistor's gain. When the reactance of C_{be} (or C_{gs}) becomes small enough, a significant amount of the signal voltage is lost due to a voltage-divider effect of the signal source resistance and the reactance of C_{be}, as illustrated in Figure 10 –4(a). When the

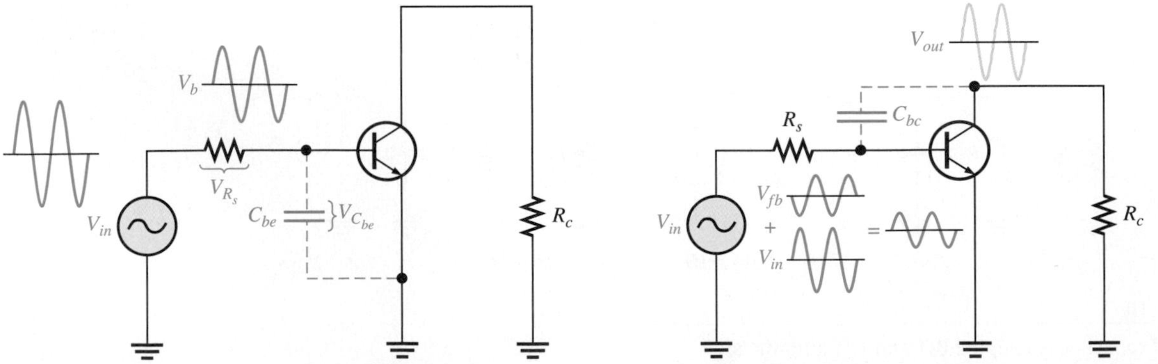

(a) Effect of C_{be}, where V_b is reduced by the voltage-divider action of R_s and $X_{C_{be}}$.

(b) Effect of C_{bc}, where part of V_{out} (V_{fb}) goes back through C_{bc} to the base and reduces the input signal because it is approximately 180° out of phase with V_{in}.

▲ **FIGURE 10–4**

AC equivalent circuit for a BJT amplifier showing effects of the internal capacitances C_{be} and C_{bc}.

reactance of C_{bc} (or C_{gd}) becomes small enough, a significant amount of output signal voltage is fed back out of phase with the input (negative feedback), thus effectively reducing the voltage gain. This is illustrated in Figure 10–4(b).

Miller's Theorem

Miller's theorem is used to simplify the analysis of inverting amplifiers at high frequencies where the internal transistor capacitances are important. The capacitance C_{bc} in BJTs (C_{gd} in FETs) between the input (base or gate) and the output (collector or drain) is shown in Figure 10–5(a) in a generalized form. A_v is the absolute voltage gain of the inverting amplifier at midrange frequencies, and C represents either C_{bc} or C_{gd}.

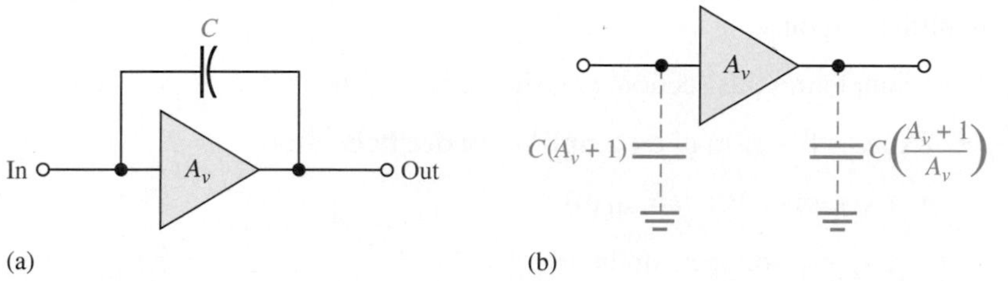

FIGURE 10–5

General case of Miller input and output capacitances. C represents C_{bc} or C_{gd}.

Miller's theorem states that C effectively appears as a capacitance from input to ground, as shown in Figure 10–5(b), that can be expressed as follows:

$$C_{in(Miller)} = C(A_v + 1)$$

Equation 10–1

This formula shows that C_{bc} (or C_{gd}) has a much greater impact on input capacitance than its actual value. For example, if $C_{bc} = 6$ pF and the amplifier gain is 50, then $C_{in(Miller)} =$ 306 pF. Figure 10–6 shows how this effective input capacitance appears in the actual ac equivalent circuit in parallel with C_{be} (or C_{gs}).

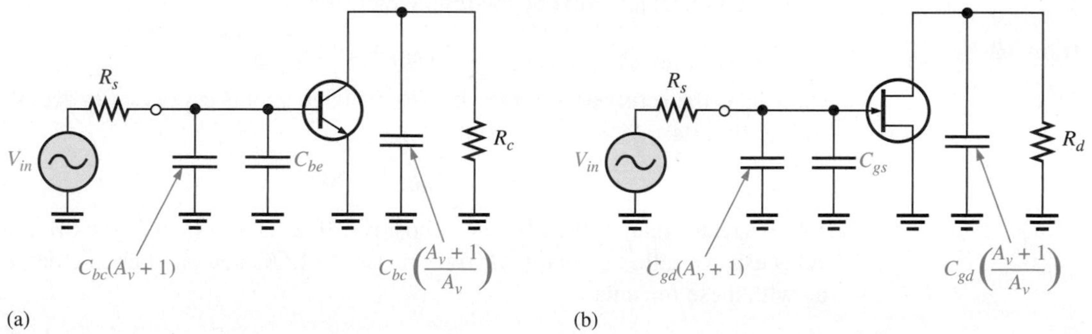

FIGURE 10–6

Amplifier ac equivalent circuits showing internal capacitances and effective Miller capacitances.

Miller's theorem also states that C effectively appears as a capacitance from output to ground, as shown in Figure 10–5(b), that can be expressed as follows:

$$C_{out(Miller)} = C\left(\frac{A_v + 1}{A_v}\right)$$

Equation 10–2

This formula indicates that if the voltage gain is 10 or greater, $C_{out(Miller)}$ is approximately equal to C_{bc} or C_{gd} because $(A_v + 1)/A_v$ is approximately equal to 1. Figure 10–6 also shows how this effective output capacitance appears in the ac equivalent circuit for BJTs and FETs. Equations 10–1 and 10–2 are derived in Appendix B.

SECTION 10–1 CHECKUP
Answers are at the end of the chapter.

1. In an ac amplifier, which capacitors affect the low-frequency gain?
2. How is the high-frequency gain of an amplifier limited?
3. When can coupling and bypass capacitors be neglected?
4. Determine $C_{in(Miller)}$ if $A_v = -50$ and $C_{bc} = 5$ pF.
5. Determine $C_{out(Miller)}$ if $A_v = -25$ and $C_{bc} = 3$ pF.

10–2 THE DECIBEL

The decibel is a unit of logarithmic gain measurement and is commonly used to express amplifier response.

After completing this section, you should be able to

- **Express the gain of an amplifier in decibels (dB)**
  - Express power gain in dB
  - Express voltage gain in dB
  - Define *critical frequency*
  - Express power in terms of dBm

The use of decibels to express gain was introduced in Chapter 6. The decibel unit is important in amplifier measurements. The basis for the decibel unit stems from the logarithmic response of the human ear to the intensity of sound. The **decibel** is a logarithmic measurement of the ratio of one power to another or one voltage to another. Power gain is expressed in decibels (dB) by the following formula:

Equation 10–3

$$A_{p(\mathrm{dB})} = 10 \log A_p$$

where A_p is the actual power gain, P_{out}/P_{in}. Voltage gain is expressed in decibels by the following formula:

Equation 10–4

$$A_{v(\mathrm{dB})} = 20 \log A_v$$

If A_v is greater than 1, the dB gain is positive. If A_v is less than 1, the dB gain is negative and is usually called *attenuation*. You can use the LOG key on your calculator when working with these formulas.

EXAMPLE 10–1

Express each of the following ratios in dB:

(a) $\dfrac{P_{out}}{P_{in}} = 250$ **(b)** $\dfrac{P_{out}}{P_{in}} = 100$ **(c)** $A_v = 10$

(d) $A_p = 0.5$ **(e)** $\dfrac{V_{out}}{V_{in}} = 0.707$

Solution

(a) $A_{p(\mathrm{dB})} = 10 \log (250) = \mathbf{24\ dB}$

(b) $A_{p(\mathrm{dB})} = 10 \log (100) = \mathbf{20\ dB}$

(c) $A_{v(dB)} = 20 \log (10) = \mathbf{20\ dB}$

(d) $A_{p(dB)} = 10 \log (0.5) = \mathbf{-3\ dB}$

(e) $A_{v(dB)} = 20 \log (0.707) = \mathbf{-3\ dB}$

*Related Problem** Express each of the following gains in dB: **(a)** $A_v = 1200$, **(b)** $A_p = 50$, **(c)** $A_v = 125{,}000$.

*Answers are at the end of the chapter.

0 dB Reference

It is often convenient in amplifiers to assign a certain value of gain as the 0 dB reference. This does not mean that the actual voltage gain is 1 (which is 0 dB); it means that the reference gain, no matter what its actual value, is used as a reference with which to compare other values of gain and is therefore assigned a 0 dB value.

Many amplifiers exhibit a maximum gain over a certain range of frequencies and a reduced gain at frequencies below and above this range. The maximum gain occurs for the range of frequencies between the upper and lower critical frequencies and is called the **midrange gain**, which is assigned a 0 dB value. Any value of gain below midrange can be referenced to 0 dB and expressed as a negative dB value. For example, if the midrange voltage gain of a certain amplifier is 100 and the gain at a certain frequency below midrange is 50, then this reduced voltage gain can be expressed as $20 \log (50/100) = 20 \log (0.5) = -6$ dB. This indicates that it is 6 dB *below* the 0 dB reference. Halving the output voltage for a steady input voltage is always a 6 dB *reduction* in the gain. Correspondingly, a doubling of the output voltage is always a 6 dB *increase* in the gain. Figure 10–7 illustrates a normalized gain-versus-frequency curve showing several dB points. The term *normalized* means that the midrange voltage gain is assigned a value of 1 or 0 dB.

TECH NOTE

The factor of 20 in Equation 10–4 is because power is proportional to voltage squared. Technically, the equation should be applied only when the voltages are measured in the same impedance. This is the case for many communication systems, such as in television or microwave systems.

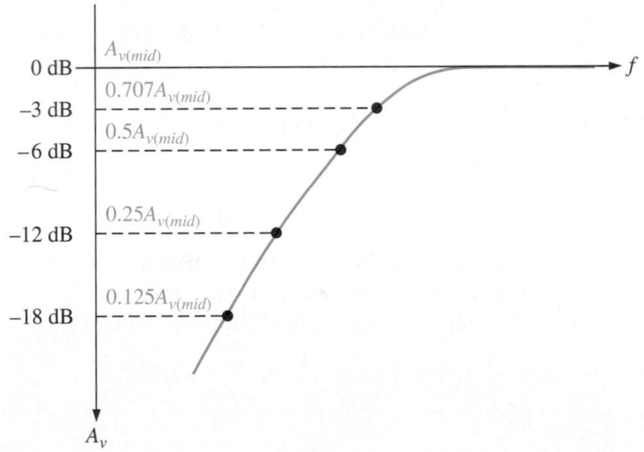

FIGURE 10–7

Normalized voltage gain versus frequency curve.

Table 10–1 shows how doubling or halving voltage gains translates into decibel values. Notice in the table that every time the voltage gain is doubled, the decibel value increases by 6 dB, and every time the gain is halved, the dB value decreases by 6 dB.

Critical Frequency

A **critical frequency** (also known as **cutoff frequency** or *corner frequency*) is a frequency at which the output power drops to one-half of its midrange value. This corresponds to a 3 dB reduction in the power gain, as expressed in dB by the following formula:

$$A_{p(dB)} = 10 \log (0.5) = -3 \text{ dB}$$

TABLE 10–1

Decibel values corresponding to doubling and halving of the voltage gain.

VOLTAGE GAIN (A_V)	DECIBEL VALUE*
32	$20 \log (32) = 30$ dB
16	$20 \log (16) = 24$ dB
8	$20 \log (8) = 18$ dB
4	$20 \log (4) = 12$ dB
2	$20 \log (2) = 6$ dB
1	$20 \log (1) = 0$ dB
0.707	$20 \log (0.707) = -3$ dB
0.5	$20 \log (0.5) = -6$ dB
0.25	$20 \log (0.25) = -12$ dB
0.125	$20 \log (0.125) = -18$ dB
0.0625	$20 \log (0.0625) = -24$ dB
0.03125	$20 \log (0.03125) = -30$ dB

*Decibel values are with respect to zero reference.

Also, at the critical frequencies the output voltage is 70.7% of its midrange value and is expressed in dB as

$$A_{v(\text{dB})} = 20 \log (0.707) = -3 \text{ dB}$$

EXAMPLE 10–2

A certain amplifier has a midrange rms output voltage of 10 V. What is the rms output voltage for each of the following dB gain reductions with a constant rms input voltage?

(a) -3 dB **(b)** -6 dB **(c)** -12 dB **(d)** -24 dB

Solution Multiply the midrange output voltage by the voltage gain corresponding to the specified decibel value in Table 10–1.

(a) At -3 dB, $V_{out} = 0.707(10 \text{ V}) = \mathbf{7.07 \text{ V}}$

(b) At -6 dB, $V_{out} = 0.5(10 \text{ V}) = \mathbf{5 \text{ V}}$

(c) At -12 dB, $V_{out} = 0.25(10 \text{ V}) = \mathbf{2.5 \text{ V}}$

(d) At -24 dB, $V_{out} = 0.0625(10 \text{ V}) = \mathbf{0.625 \text{ V}}$

Related Problem Determine the output voltage at the following decibel levels for a midrange value of 50 V:

(a) 0 dB **(b)** -18 dB **(c)** -30 dB

Power Measurement in dBm

TECH NOTE

The unit of dBmV is used in some applications such as cable TV where the reference level is 1 mV, which corresponds to 0 dB. Just as the dBm is used to indicate actual power, the dBmV unit is used to indicate actual voltage.

The **dBm** is a unit for measuring power levels referenced to 1 mW. Positive dBm values represent power levels above 1 mW, and negative dBm values represent power levels below 1 mW.

Because the decibel (dB) can be used to represent only power *ratios,* not actual power, the dBm provides a convenient way to express actual power output of an amplifier or other device. Each 3 dBm increase corresponds to a doubling of the power, and a 3 dBm decrease corresponds to a halving of the power.

To state that an amplifier has a 3 dB power gain indicates only that the output power is twice the input power and nothing about the actual output power. To indicate actual output

POWER	dBm
32 mW	15 dBm
16 mW	12 dBm
8 mW	9 dBm
4 mW	6 dBm
2 mW	3 dBm
1 mW	0 dBm
0.5 mW	−3 dBm
0.25 mW	−6 dBm
0.125 mW	−9 dBm
0.0625 mW	−12 dBm
0.03125 mW	−15 dBm

◄ **TABLE 10–2**

Power in terms of dBm.

power, the dBm can be used. For example, 3 dBm is equivalent to 2 mW because 2 mW is twice the 1 mW reference. 6 dBm is equivalent to 4 mW, and so on. Likewise, −3 dBm is the same as 0.5 mW. Table 10–2 shows several values of power in terms of dBm.

SECTION 10–2 CHECKUP

1. How much increase in actual voltage gain corresponds to +12 dB?
2. Convert a power gain of 25 to decibels.
3. What power corresponds to 0 dBm?

10–3 LOW-FREQUENCY AMPLIFIER RESPONSE

The voltage gain and phase shift of capacitively coupled amplifiers are affected when the signal frequency is below a critical value. At low frequencies, the reactance of the coupling capacitors becomes significant, resulting in a reduction in voltage gain and an increase in phase shift. Frequency responses of both BJT and FET capacitively coupled amplifiers are discussed.

After completing this section, you should be able to

- **Analyze the low-frequency response of amplifiers**
  - Determine midrange voltage gain
  - Generally describe how *RC* circuits affect the gain
  - Identify the input *RC* circuit
  - Determine the lower critical frequency of the input *RC* circuit
  - Discuss gain roll-off in terms of dB/decade and dB/octave
  - Determine phase shift of the input *RC* circuit
  - Identify the output *RC* circuit
  - Determine the lower critical frequency of the output *RC* circuit
  - Determine phase shift of the output *RC* circuit

- Identify the bypass RC circuit
- Determine the lower critical frequency of the bypass RC circuit
- Describe a Bode plot
- Analyze an amplifier for total low-frequency response

BJT Amplifiers

A typical capacitively coupled common-emitter amplifier is shown in Figure 10–8. Assuming that the coupling and bypass capacitors are ideal shorts at the midrange signal frequency, you can determine the midrange voltage gain using Equation 10–5, where $R_c = R_C \| R_L$.

Equation 10–5

$$A_{v(mid)} = \frac{R_c}{r'_e}$$

If a swamping resistor (R_{E1}) is used, it appears in series with r'_e and the equation becomes

$$A_{v(mid)} = \frac{R_c}{r'_e + R_{E1}}$$

▶ FIGURE 10–8

A capacitively coupled BJT amplifier.

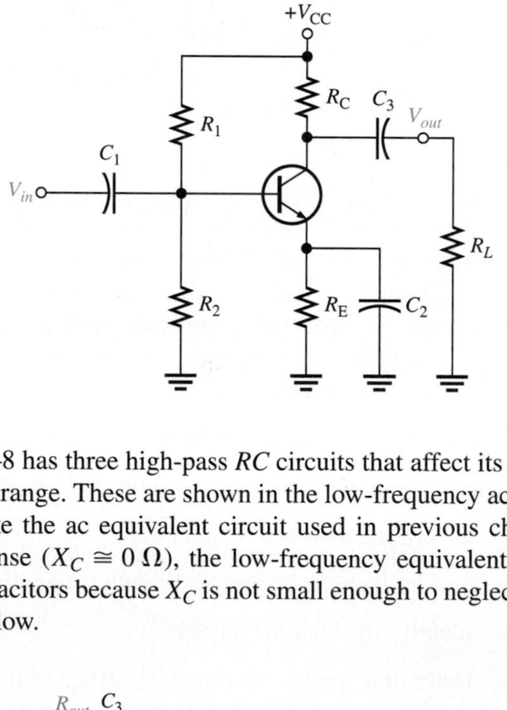

The BJT amplifier in Figure 10–8 has three high-pass RC circuits that affect its gain as the frequency is reduced below midrange. These are shown in the low-frequency ac equivalent circuit in Figure 10–9. Unlike the ac equivalent circuit used in previous chapters, which represented midrange response ($X_C \cong 0\ \Omega$), the low-frequency equivalent circuit retains the coupling and bypass capacitors because X_C is not small enough to neglect when the signal frequency is sufficiently low.

▶ FIGURE 10–9

The low-frequency ac equivalent circuit of the amplifier in Figure 10–8 consists of three high-pass RC circuits.

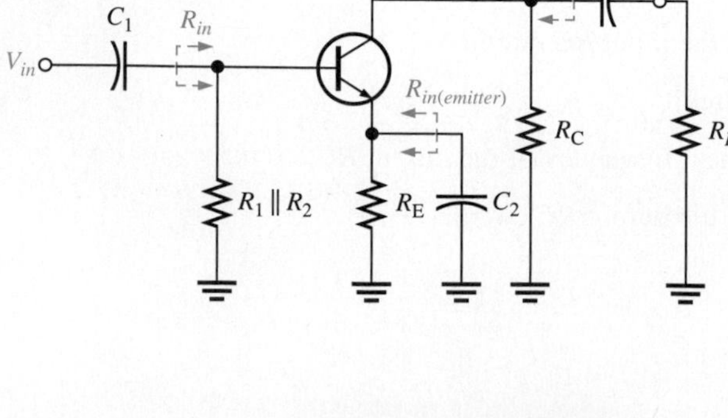

One RC circuit is formed by the input coupling capacitor C_1 and the input resistance of the amplifier. The second RC circuit is formed by the output coupling capacitor C_3, the resistance looking in at the collector (R_{out}), and the load resistance. The third RC circuit that affects the low-frequency response is formed by the emitter-bypass capacitor C_2 and the resistance looking in at the emitter.

The Input *RC* Circuit

The input RC circuit for the BJT amplifier in Figure 10–8 is formed by C_1 and the amplifier's input resistance and is shown in Figure 10–10. (Input resistance was discussed in Chapter 6.) As the signal frequency decreases, X_{C1} increases. This causes less voltage across the input resistance of the amplifier at the base because more voltage is dropped across C_1 and because of this, the overall voltage gain of the amplifier is reduced. The base voltage for the input RC circuit in Figure 10–10 (neglecting the internal resistance of the input signal source) can be stated as

$$V_{base} = \left(\frac{R_{in}}{\sqrt{R_{in}^2 + X_{C1}^2}}\right)V_{in}$$

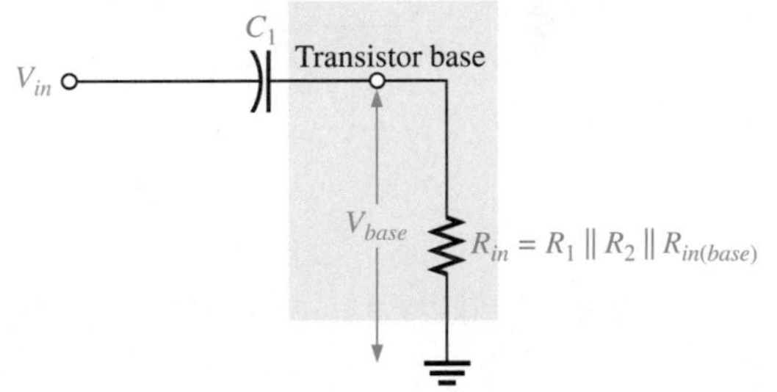

FIGURE 10–10

Input *RC* circuit formed by the input coupling capacitor and the amplifier's input resistance.

As previously mentioned, a critical point in the amplifier's response occurs when the output voltage is 70.7% of its midrange value. This condition occurs in the input RC circuit when $X_{C1} = R_{in}$.

$$V_{base} = \left(\frac{R_{in}}{\sqrt{R_{in}^2 + R_{in}^2}}\right)V_{in} = \left(\frac{R_{in}}{\sqrt{2R_{in}^2}}\right)V_{in} = \left(\frac{R_{in}}{\sqrt{2}R_{in}}\right)V_{in} = \left(\frac{1}{\sqrt{2}}\right)V_{in} = 0.707V_{in}$$

In terms of measurement in decibels,

$$20\log\left(\frac{V_{base}}{V_{in}}\right) = 20\log(0.707) = -3\text{ dB}$$

Lower Critical Frequency The condition where the gain is down 3 dB is logically called the -3 *dB point* of the amplifier response; the overall gain is 3 dB less than at midrange frequencies because of the attenuation (gain less than 1) of the input RC circuit. The frequency, f_{cl}, at which this condition occurs is called the *lower critical frequency* (also known as the *lower cutoff frequency, lower corner frequency,* or *lower break frequency*) and can be calculated as follows:

$$X_{C1} = \frac{1}{2\pi f_{cl(input)}C_1} = R_{in}$$

$$\mathbf{f_{cl(input)} = \frac{1}{2\pi R_{in}C_1}} \qquad \text{Equation 10–6}$$

If the resistance of the input source is taken into account, Equation 10–6 becomes

$$f_{cl(input)} = \frac{1}{2\pi(R_s + R_{in})C_1}$$

EXAMPLE 10–3 For the circuit in Figure 10–11, calculate the lower critical frequency due to the input *RC* circuit. Assumed $r'_e = 9.6\ \Omega$ and $\beta = 200$. Notice that a swamping resistor, R_{E1}, is used.

FIGURE 10–11

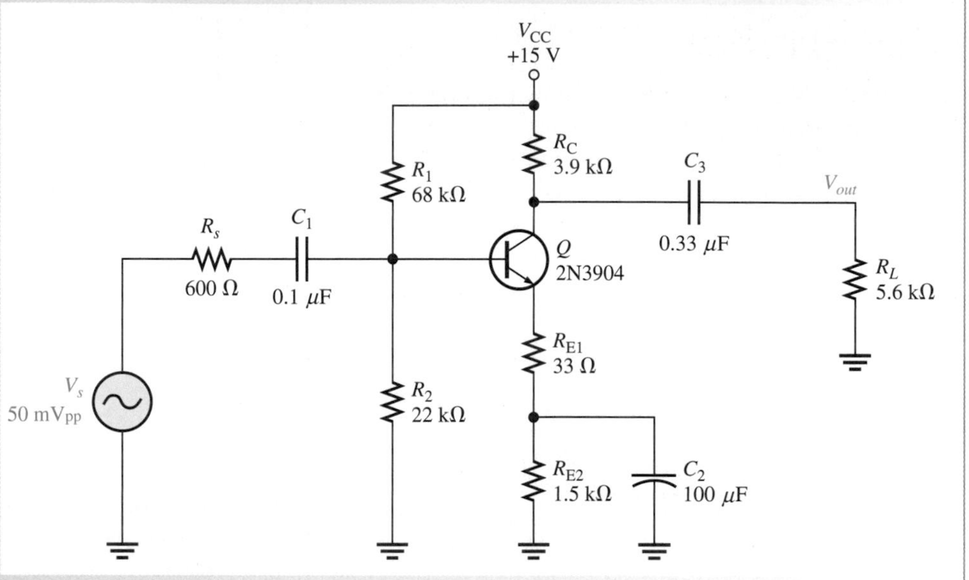

Solution The input resistance is

$$R_{in} = R_1 \| R_2 \| (\beta(r'_e + R_{E1})) = 68\ \text{k}\Omega \| 22\ \text{k}\Omega \| (200\,(9.6\ \Omega + 33\ \Omega)) = 5.63\ \text{k}\Omega$$

The lower critical frequency is

$$f_{cl(input)} = \frac{1}{2\pi R_{in} C_1} = \frac{1}{2\pi(5.63\ \text{k}\Omega)(0.1\ \mu\text{F})} = \mathbf{282\ Hz}$$

Related Problem What value of input capacitor will move the lower cutoff frequency to 130 Hz?

Open the Multisim file E10-03 and read the critical frequency on the Bode plotter. The Bode plotter is not an actual instrument available, but allows the user to see the response of a circuit in the frequency domain (frequency is the independent variable). Notice that C_2 and C_3 are taken out of the calculation by making their value huge (1 F!). While this is unrealistic, it works nicely for the computer simulation to isolate the input response.

Voltage Gain Roll-Off at Low Frequencies As you have seen, the input *RC* circuit reduces the overall voltage gain of an amplifier by 3 dB when the frequency is reduced to the critical value f_c. As the frequency continues to decrease below f_c, the overall voltage gain also continues to decrease. The rate of decrease in voltage gain with frequency is called **roll-off**. For each ten times reduction in frequency below f_c, *there is a 20 dB reduction in voltage gain.*

Let's consider a frequency that is one-tenth of the critical frequency ($f = 0.1f_c$). Since $X_{C1} = R_{in}$ at f_c, then $X_{C1} = 10R_{in}$ at $0.1f_c$ because of the inverse relationship of X_{C1} and f. The attenuation of the input *RC* circuit is, therefore,

$$\text{Attenuation} = \frac{V_{base}}{V_{in}} = \frac{R_{in}}{\sqrt{R_{in}^2 + X_{C1}^2}} = \frac{R_{in}}{\sqrt{R_{in}^2 + (10R_{in})^2}} = \frac{R_{in}}{\sqrt{R_{in}^2 + 100R_{in}^2}}$$

$$= \frac{R_{in}}{\sqrt{R_{in}^2(1 + 100)}} = \frac{R_{in}}{R_{in}\sqrt{101}} = \frac{1}{\sqrt{101}} \cong \frac{1}{10} = 0.1$$

The dB attenuation is

$$20 \log \left(\frac{V_{base}}{V_{in}} \right) = 20 \log (0.1) = -20 \text{ dB}$$

HISTORY NOTE

Hendrik Wade Bode pronounced *Boh-dee* (1905–1982) was born in Madison, Wisconsin. He received his B.A. degree in 1924, at the age of 19, from Ohio State University and his M.A. degree in 1926, both in Mathematics. He was hired by Bell Labs and completed his Ph.D. in Physics in 1935. In 1938 he developed his now well-known magnitude and phase plots. His work on automatic control systems introduced innovative methods to the study of system stability.

The Bode Plot A ten-times change in frequency is called a **decade**. So, for the input *RC* circuit, the attenuation is reduced by 20 dB for each decade that the frequency decreases below the critical frequency. This causes the overall voltage gain to drop 20 dB per decade.

A plot of dB voltage gain versus frequency on semilog graph paper (logarithmic horizontal axis scale and a linear vertical axis scale) is called a **Bode plot**. A generalized Bode plot for an input *RC* circuit appears in Figure 10–12. The ideal response curve is shown in blue. Notice that it is flat (0 dB) down to the critical frequency, at which point the gain drops at −20 dB/decade as shown. Above f_c are the midrange frequencies. The actual response curve is shown in red. Notice that it decreases gradually beginning in midrange and is down to −3 dB at the critical frequency. Often, the ideal response is used to simplify amplifier analysis. As previously mentioned, the critical frequency at which the curve "breaks" into a −20dB/decade drop is sometimes called the *lower break frequency.*

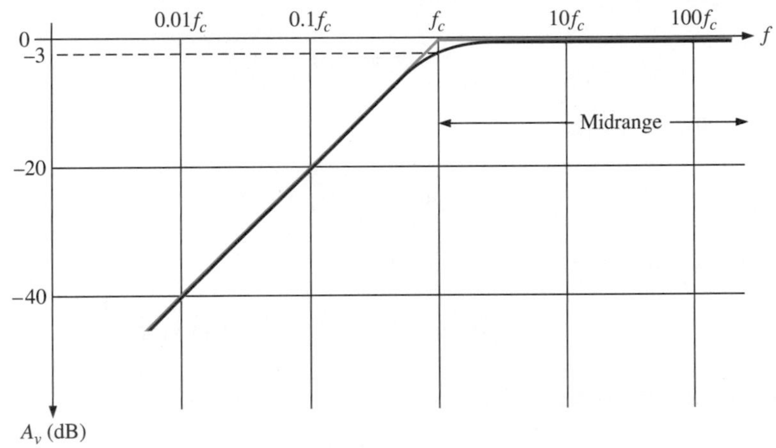

FIGURE 10–12

Bode plot. (Blue is ideal; red is actual.)

Sometimes, the voltage gain roll-off of an amplifier is expressed in dB/octave rather than dB/decade. An **octave** corresponds to a doubling or halving of the frequency. For example, an increase in frequency from 100 Hz to 200 Hz is an octave. Likewise, a decrease in frequency from 100 kHz to 50 kHz is also an octave. A rate of −20 dB/decade is approximately equivalent to −6 dB/octave, a rate of −40 dB/decade is approximately equivalent to −12 dB/octave, and so on.

EXAMPLE 10–4

The midrange voltage gain of a certain amplifier is 100. The input *RC* circuit has a lower critical frequency of 1 kHz. Determine the actual voltage gain at f = 1 kHz, f = 100 Hz, and f = 10 Hz.

Solution When f = 1 kHz, the voltage gain is 3 dB less than at midrange. At −3 dB, the voltage gain is reduced by a factor of 0.707.

$$A_v = (0.707)(100) = \mathbf{70.7}$$

When f = 100 Hz = 0.1f_c, the voltage gain is 20 dB less than at f_c. The voltage gain at −20 dB is one-tenth of that at the midrange frequencies.

$$A_v = (0.1)(100) = \mathbf{10}$$

When $f = 10\text{ Hz} = 0.01f_c$, the voltage gain is 20 dB less than at $f = 0.1f_c$ or −40 dB. The voltage gain at −40 dB is one-tenth of that at −20 dB or one-hundredth that at the midrange frequencies.

$$A_v = (0.01)(100) = \mathbf{1}$$

Related Problem The midrange voltage gain of an amplifier is 300. The lower critical frequency of the input *RC* circuit is 400 Hz. Determine the actual voltage gain at 400 Hz, 40 Hz, and 4 Hz.

Phase Shift in the Input RC Circuit In addition to reducing the voltage gain, the input *RC* circuit also causes an increasing phase shift through an amplifier as the frequency decreases. At midrange frequencies, the phase shift through the input *RC* circuit is approximately zero because the capacitive reactance, X_{C1}, is approximately 0 Ω. At lower frequencies, higher values of X_{C1} cause a phase shift to be introduced, and the output voltage of the *RC* circuit leads the input voltage. As you learned in ac circuit theory, the phase angle in an input *RC* circuit is expressed as

Equation 10–7

$$\boldsymbol{\theta = \tan^{-1}\left(\frac{X_{C1}}{R_{in}}\right)}$$

For midrange frequencies, $X_{C1} \cong 0\ \Omega$, so

$$\theta = \tan^{-1}\left(\frac{0\ \Omega}{R_{in}}\right) = \tan^{-1}(0) = 0°$$

At the critical frequency, $X_{C1} = R_{in}$, so

$$\theta = \tan^{-1}\left(\frac{R_{in}}{R_{in}}\right) = \tan^{-1}(1) = 45°$$

At a decade below the critical frequency, $X_{C1} = 10R_{in}$, so

$$\theta = \tan^{-1}\left(\frac{10R_{in}}{R_{in}}\right) = \tan^{-1}(10) = 84.3°$$

A continuation of this analysis will show that the phase shift through the input *RC* circuit approaches 90° as the frequency approaches zero. A plot of phase angle versus frequency is shown in Figure 10–13. The result is that the voltage at the base of the transistor *leads* the input signal voltage in phase below midrange, as shown in Figure 10–14.

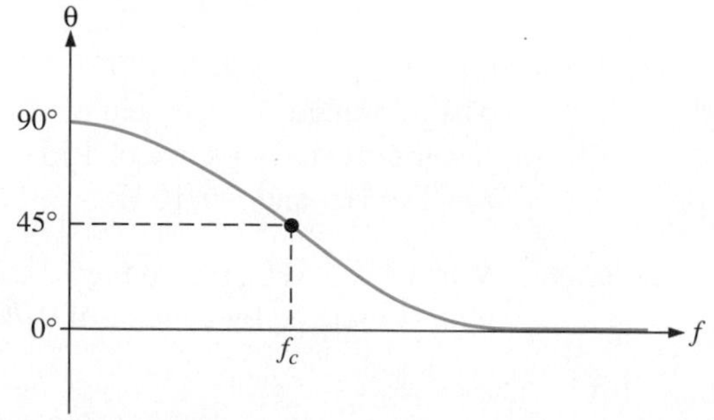

FIGURE 10–13

Phase angle versus frequency for the input *RC* circuit.

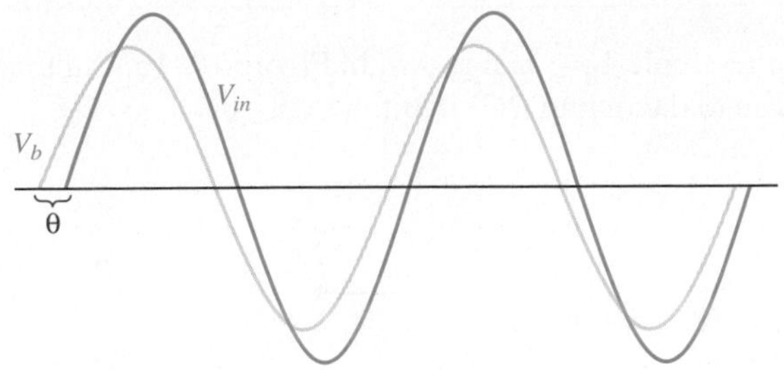

▲ FIGURE 10–14

The input *RC* circuit causes the base voltage to lead the input voltage below midrange by an amount equal to the circuit phase angle, θ.

The Output *RC* Circuit

The second high-pass *RC* circuit in the BJT amplifier of Figure 10–8 is formed by the coupling capacitor C_3, the resistance looking in at the collector, and the load resistance R_L, as shown in Figure 10–15(a). In determining the output resistance, looking in at the collector, the transistor is treated as an ideal current source (with infinite internal resistance), and the upper end of R_C is effectively at ac ground, as shown in Figure 10–15(b). Therefore, thevenizing the circuit to the left of capacitor C_3 produces an equivalent voltage source equal to the collector voltage and a series resistance equal to R_C, as shown in Figure 10–15(c). The lower critical frequency of this output *RC* circuit is

$$f_{cl(output)} = \frac{1}{2\pi(R_C + R_L)C_3}$$

Equation 10–8

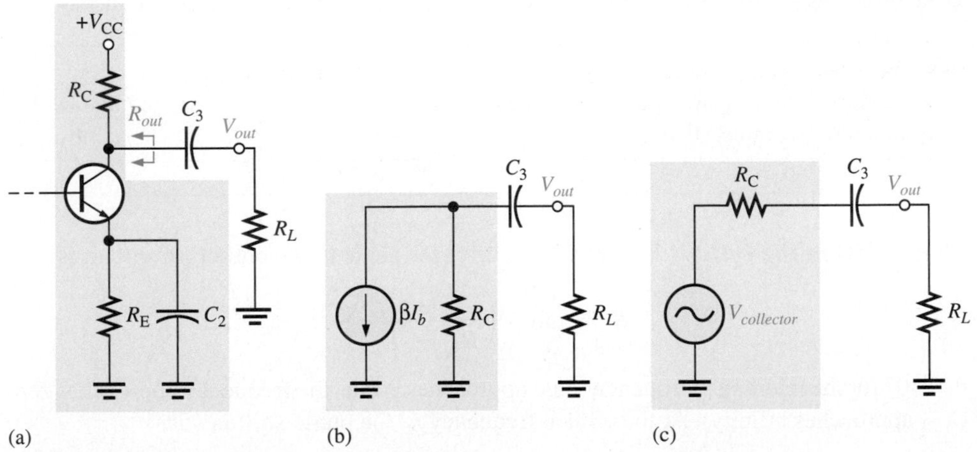

▲ FIGURE 10–15

Development of the equivalent low-frequency output *RC* circuit.

The effect of the output *RC* circuit on the amplifier voltage gain is similar to that of the input *RC* circuit. As the signal frequency decreases, X_{C3} increases. This causes less voltage across the load resistance because more voltage is dropped across C_3. The signal voltage is reduced by a factor of 0.707 when frequency is reduced to the lower critical value, f_{cl}, for the circuit. This corresponds to a 3 dB reduction in voltage gain.

EXAMPLE 10–5

For the circuit from Example 10–3 and shown in Figure 10–16, calculate the lower critical frequency due to the output *RC* circuit.

FIGURE 10–16

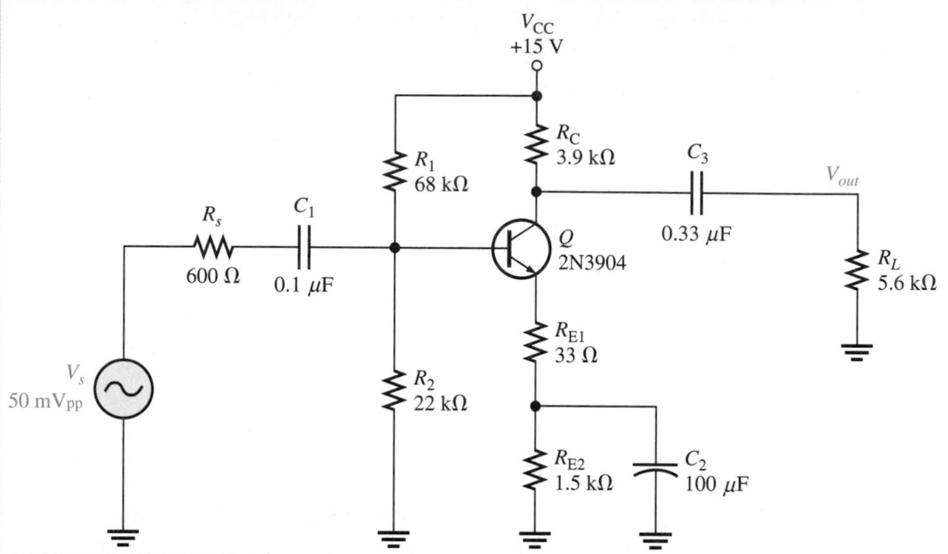

Solution The resistance in the output *RC* circuit is

$$R_C + R_L = 3.9\,\text{k}\Omega + 5.6\,\text{k}\Omega = 9.5\,\text{k}\Omega$$

The lower critical frequency is

$$f_{cl(output)} = \frac{1}{2\pi(R_C + R_L)C_3} = \frac{1}{2\pi(9.5\,\text{k}\Omega)(0.33\,\mu\text{F})} = \mathbf{50.8\,Hz}$$

Related Problem What effect does a larger load resistor have on the gain and the lower cutoff frequency?

Open the Multisim file E10-05 and read the critical frequency on the Bode plotter. Notice that C_1 and C_2 are taken out of the calculation by making their value huge as explained in Example 10–3.

Phase Shift in the Output RC Circuit The phase angle in the output *RC* circuit is

Equation 10–9

$$\theta = \tan^{-1}\left(\frac{X_{C3}}{R_C + R_L}\right)$$

$\theta \cong 0°$ for the midrange frequencies and approaches 90° as the frequency approaches zero (X_{C3} approaches infinity). At the critical frequency f_c, the phase shift is 45°.

The Bypass *RC* Circuit

The third *RC* circuit that affects the low-frequency gain of the BJT amplifier in Figure 10–8 includes the bypass capacitor C_2. As illustrated in Figure 10–17(a) for midrange frequencies, it is assumed that $X_{C2} \cong 0\,\Omega$, effectively shorting the emitter to ground so that the amplifier gain is R_c/r'_e, as you already know. As the frequency is reduced, X_{C2} increases and no longer provides a sufficiently low reactance to effectively place the emitter at ac ground, as shown in part(b). Because the impedance from emitter to ground increases, the gain decreases. In this case, R_e in the formula, $A_v = R_c/(r'_e + R_e)$, is replaced by an impedance formed by R_E in parallel with X_{C2}.

The bypass *RC* circuit is formed by C_2 and the resistance looking in at the emitter, $R_{in(emitter)}$, as shown in Figure 10–18(a). The resistance looking in at the emitter is derived

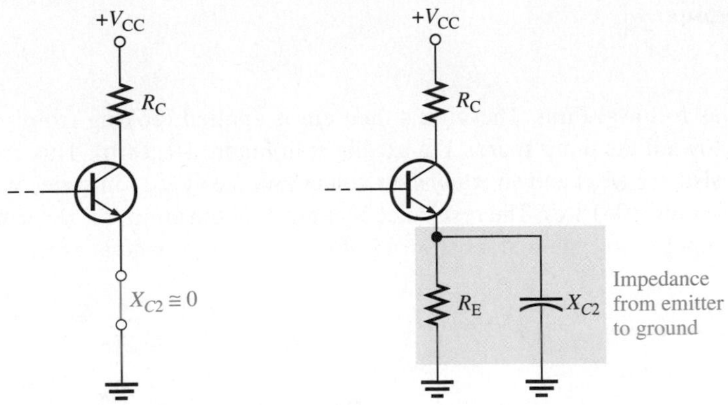

(a) For midrange frequencies, C_2 effectively shorts the emitter to ground.

(b) Below f_c, X_{C2} and R_E form an impedance between the emitter and ground.

FIGURE 10–17

At low frequencies, X_{C2} in parallel with R_E creates an impedance that reduces the voltage gain.

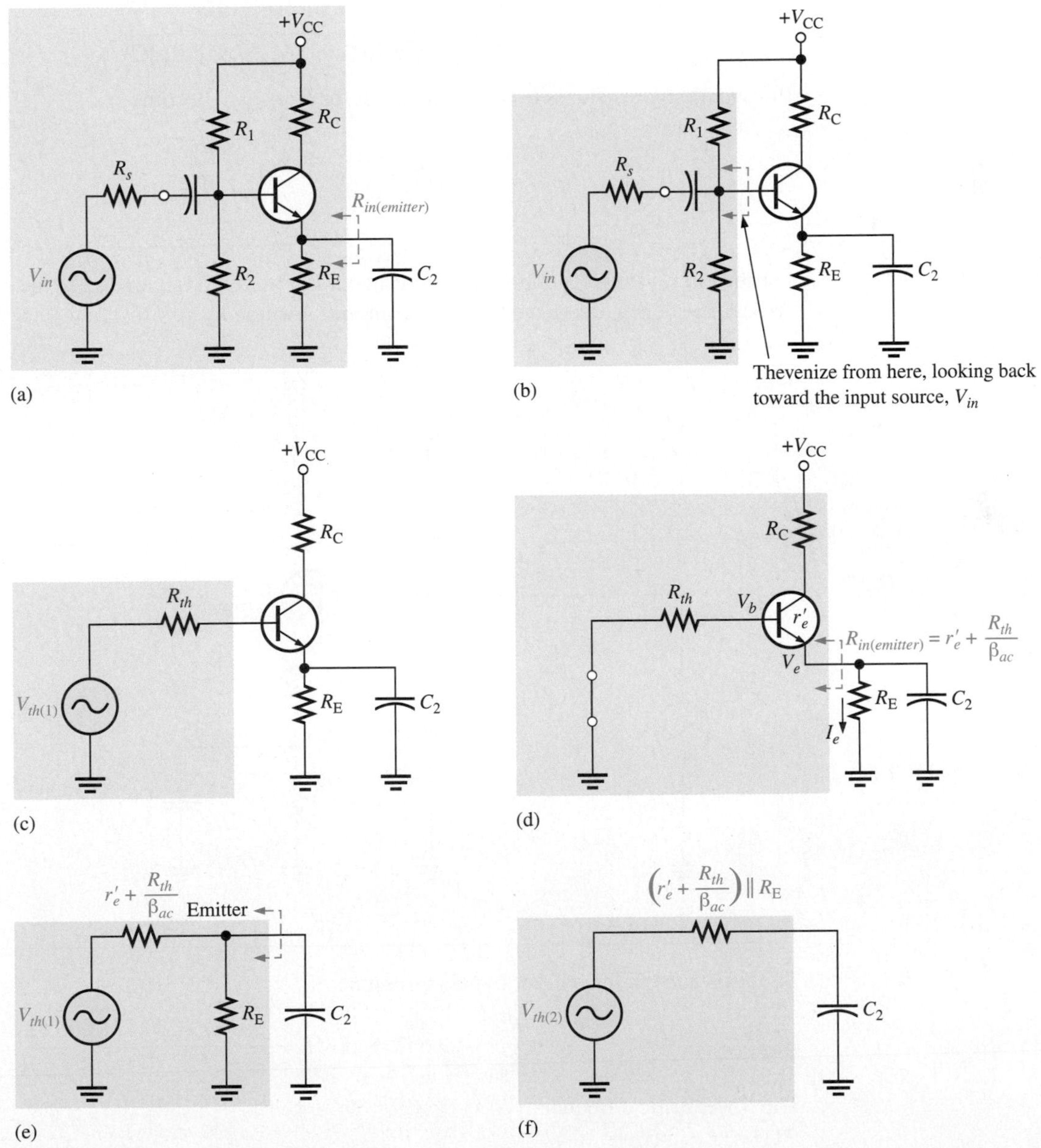

FIGURE 10–18

Development of the equivalent bypass *RC* circuit.

as follows. First, Thevenin's theorem is applied looking from the base of the transistor toward the input source V_{in}, as shown in Figure 10–18(b). This results in an equivalent resistance (R_{th}) and an equivalent voltage source ($V_{th(1)}$) in series with the base, as shown in Figure 10–18(c). The resistance looking in at the emitter is determined with the equivalent input source shorted, as shown in Figure 10–18(d), and is expressed as follows:

$$R_{in(emitter)} = r'_e + \frac{V_e}{I_e} \cong r'_e + \frac{V_b}{\beta_{ac} I_b} = r'_e + \frac{I_b R_{th}}{\beta_{ac} I_b}$$

Equation 10–10

$$\mathbf{R_{in(emitter)} = r'_e + \frac{R_{th}}{\beta_{ac}}}$$

Looking from the capacitor C_2, $r'_e + R_{th}/\beta_{ac}$ is in parallel with R_E, as shown in Figure 10–18(e). Thevenizing again, we get the equivalent RC circuit shown in Figure 10–18(f). The lower critical frequency for this equivalent bypass RC circuit is

Equation 10–11

$$\mathbf{f_{cl(bypass)} = \frac{1}{2\pi[(r'_e + R_{th}/\beta_{ac}) \parallel R_E]C_2}}$$

If a swamping resistor is used, the equation for $R_{in(emitter)}$ becomes

$$R_{in(emitter)} = r'_e + R_{E1} + \frac{R_{th}}{\beta_{ac}}$$

EXAMPLE 10–6

For the circuit from Example 10–3 and shown in Figure 10–19, calculate the lower critical frequency due to the bypass RC circuit. Assume $r'_e = 9.6\ \Omega$ and $\beta = 200$.

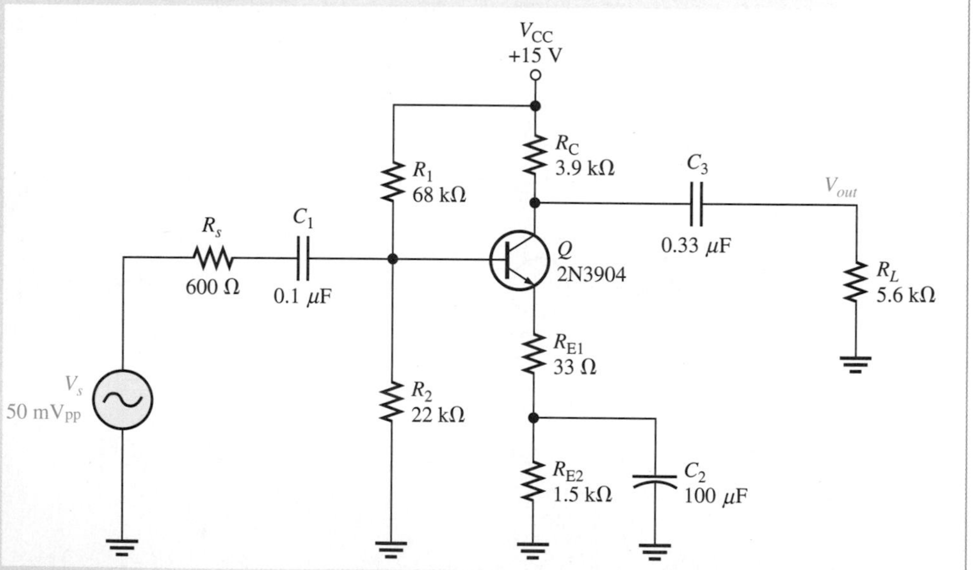

FIGURE 10–19

Solution The resistance in the emitter bypass circuit is

$$R_{in(emitter)} = r'_e + R_{E1} + \frac{R_{th}}{\beta_{ac}} = 9.6\ \Omega + 33\ \Omega + \frac{68\ \text{k}\Omega \parallel 22\ \text{k}\Omega \parallel 600\ \Omega}{200} = 45.5\ \Omega$$

The lower critical frequency is

$$f_{cl(bypass)} = \frac{1}{2\pi(R_{in(emitter)} \parallel R_{E2})C_2} = \frac{1}{2\pi(45.5\ \Omega \parallel 1.5\ \text{k}\Omega)(100\ \mu\text{F})} = \mathbf{36.0\ Hz}$$

Related Problem Explain why C_2 is larger than C_1 or C_3.

Open the Multisim file E10-06 and read the critical frequency on the Bode plotter. Notice that C_1 and C_3 are taken out of the calculation by making their value huge as before (1 F!).

FET Amplifiers

A zero-biased D-MOSFET amplifier with capacitive coupling on the input and output is shown in Figure 10–20. As you learned in Chapter 9, the midrange voltage gain of a zero-biased amplifier is

$$A_{v(mid)} = g_m R_d$$

This is the gain at frequencies high enough so that the capacitive reactances are approximately zero.

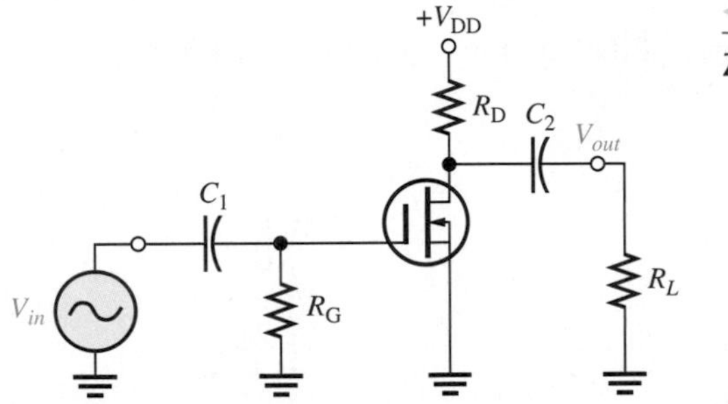

FIGURE 10–20

Zero-biased D-MOSFET amplifier.

The amplifier in Figure 10–20 has only two high-pass *RC* circuits that influence its low-frequency response. One *RC* circuit is formed by the input coupling capacitor C_1 and the input resistance. The other circuit is formed by the output coupling capacitor C_2 and the output resistance looking in at the drain.

The Input *RC* Circuit

The input *RC* circuit for the FET amplifier in Figure 10–20 is shown in Figure 10–21. As in the case for the BJT amplifier, the reactance of the input coupling capacitor increases as the frequency decreases. When $X_{C1} = R_{in}$, the gain is down 3 dB below its midrange value.

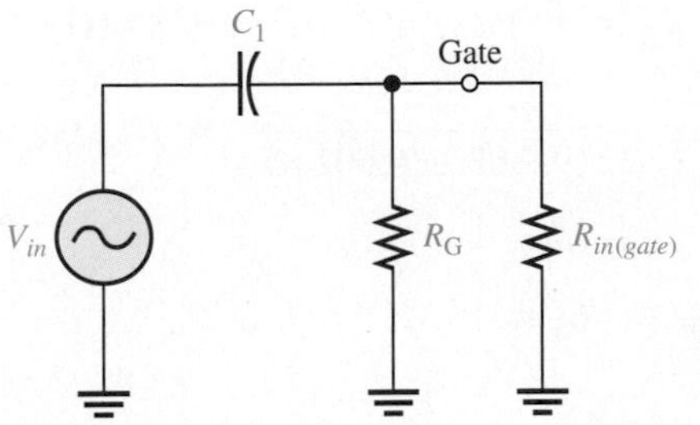

FIGURE 10–21

Input *RC* circuit.

The lower critical frequency is

$$f_{cl(input)} = \frac{1}{2\pi R_{in} C_1}$$

The input resistance is

$$R_{in} = R_G \| R_{in(gate)}$$

where $R_{in(gate)}$ is determined from datasheet information as

$$R_{in(gate)} = \left|\frac{V_{GS}}{I_{GSS}}\right|$$

Therefore, the lower critical frequency is

Equation 10–12

$$f_{cl(input)} = \frac{1}{2\pi(R_G \,\|\, R_{in(gate)})C_1}$$

For practical work, the value of $R_{in(gate)}$ is so large it can be ignored, as will be illustrated in Example 10–7.

The gain rolls off below f_c at 20 dB/decade, as previously shown. The phase angle in the low-frequency input *RC* circuit is

Equation 10–13

$$\theta = \tan^{-1}\left(\frac{X_{C1}}{R_{in}}\right)$$

EXAMPLE 10–7

What is the lower critical frequency of the input *RC* circuit in the FET amplifier of Figure 10–22?

FIGURE 10–22

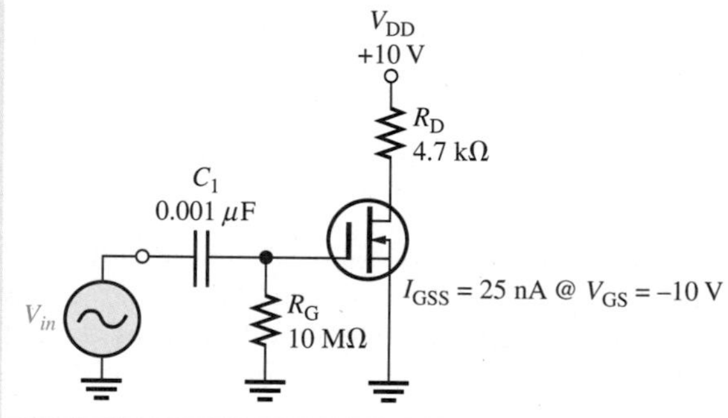

Solution First determine R_{in} and then calculate f_c.

$$R_{in(gate)} = \left|\frac{V_{GS}}{I_{GSS}}\right| = \frac{10\text{ V}}{25\text{ nA}} = 400\text{ M}\Omega$$

$$R_{in} = R_G \,\|\, R_{in(gate)} = 10\text{ M}\Omega \,\|\, 400\text{ M}\Omega = 9.8\text{ M}\Omega$$

$$f_{cl(input)} = \frac{1}{2\pi R_{in}C_1} = \frac{1}{2\pi(9.8\text{ M}\Omega)(0.001\ \mu\text{F})} = \mathbf{16.2\ Hz}$$

For all practical purposes,

$$R_{in} \cong R_G = 10\text{ M}\Omega$$

and

$$f_{cl(input)} = \frac{1}{2\pi R_G C_1} = \frac{1}{2\pi(10\text{ M}\Omega)(0.001\ \mu\text{F})} \cong 15.9\text{ Hz}$$

There is very little difference in the two results.

The critical frequency of the input *RC* circuit of a FET amplifier is usually very low because of the very high input resistance and the high value of R_G.

Related Problem How much does the lower critical frequency of the input RC circuit change if the FET in Figure 10–22 is replaced by one with $I_{GSS} = 10\text{ nA}$ @ $V_{GS} = -8\text{ V}$?

Open the Multisim file E10-07 in the Examples folder on your CD-ROM and measure the low critical frequency for the input circuit. Compare to the calculated results.

The Output *RC* Circuit

The second RC circuit that affects the low-frequency response of the amplifier in Figure 10–20 is formed by a coupling capacitor C_2 and the output resistance looking in at the drain, as shown in Figure 10–23(a). The load resistor, R_L, is also included. As in the case of the BJT, the FET is treated as a current source, and the upper end of R_D is effectively ac ground, as shown in Figure 10–23(b). The Thevenin equivalent of the circuit to the left of C_2 is shown in Figure 10–23(c). The lower critical frequency for this RC circuit is

$$f_{cl(output)} = \frac{1}{2\pi(R_D + R_L)C_2}$$ **Equation 10–14**

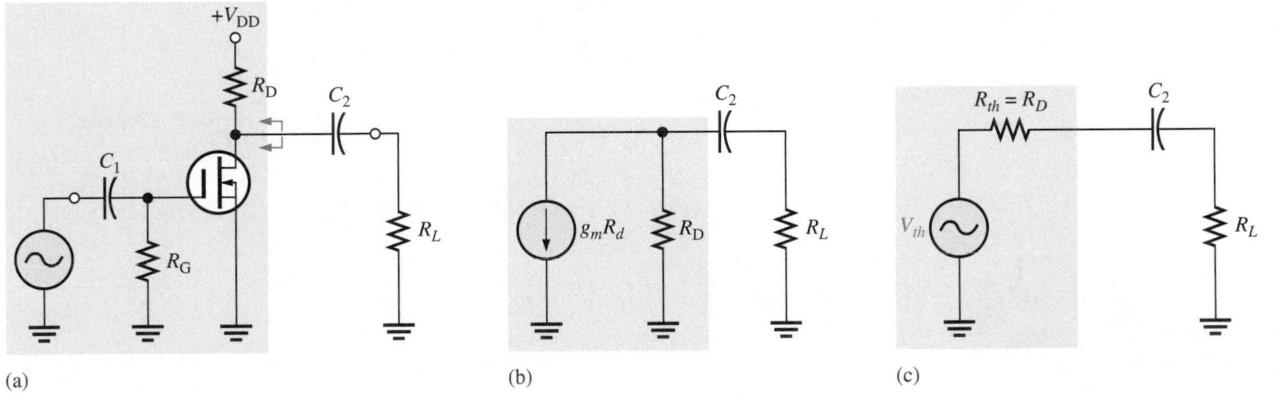

▲ FIGURE 10–23

Development of the equivalent low-frequency output *RC* circuit.

The effect of the output RC circuit on the amplifier's voltage gain below the midrange is similar to that of the input RC circuit. The circuit with the highest critical frequency dominates because it is the one that first causes the gain to roll off as the frequency drops below its midrange values. The phase angle in the low-frequency output RC circuit is

$$\theta = \tan^{-1}\left(\frac{X_{C2}}{R_D + R_L}\right)$$ **Equation 10–15**

Again, at the critical frequency, the phase angle is 45° and approaches 90° as the frequency approaches zero. However, starting at the critical frequency, the phase angle decreases from 45° and becomes very small as the frequency goes higher.

EXAMPLE 10–8 Determine the lower critical frequencies for the FET amplifier in Figure 10–24. Assume that the load is another identical amplifier with the same R_{in}. The datasheet shows $I_{GSS} = 100\text{ nA}$ at $V_{GS} = -12\text{ V}$.

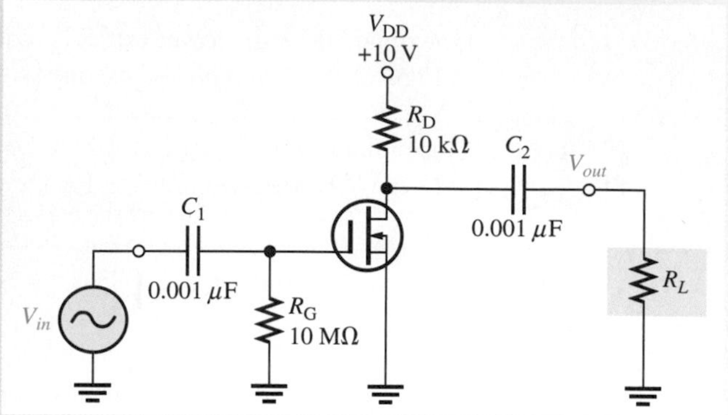

FIGURE 10–24

Solution First, find the lower critical frequency for the input *RC* circuit.

$$R_{in(gate)} = \left|\frac{V_{GS}}{I_{GSS}}\right| = \frac{12\text{ V}}{100\text{ nA}} = 120\text{ M}\Omega$$

$$R_{in} = R_G \| R_{in(gate)} = 10\text{ M}\Omega \| 120\text{ M}\Omega = 9.2\text{ M}\Omega$$

$$f_{cl(input)} = \frac{1}{2\pi R_{in}C_1} = \frac{1}{2\pi(9.2\text{ M}\Omega)(0.001\ \mu\text{F})} = \mathbf{17.3\ Hz}$$

The output *RC* circuit has a lower critical frequency of

$$f_{cl(output)} = \frac{1}{2\pi(R_D + R_L)C_2} = \frac{1}{2\pi(9.21\text{ M}\Omega)(0.001\ \mu\text{F})} \cong \mathbf{17.3\ Hz}$$

Related Problem If the circuit in Figure 10–24 were operated with no load, how is the low-frequency response affected?

Open the Multisim file E10-08 in the Examples folder on your CD-ROM. Determine the total low-frequency response of the amplifier.

Total Low-Frequency Response of an Amplifier

Now that we have individually examined the high-pass *RC* circuits that affect a BJT or FET amplifier's voltage gain at low frequencies, let's look at the combined effect of the three *RC* circuits in a BJT amplifier. Each circuit has a critical frequency determined by the *R* and *C* values. The critical frequencies of the three *RC* circuits are not necessarily all equal. If one of the *RC* circuits has a critical (break) frequency higher than the other two, then it is the *dominant RC* circuit. The dominant circuit determines the frequency at which the overall voltage gain of the amplifier begins to drop at −20 dB/decade. The other circuits each cause an additional −20 dB/decade roll-off below their respective critical (break) frequencies.

To get a better picture of what happens at low frequencies, refer to the Bode plot in Figure 10–25, which shows the superimposed ideal responses for the three *RC* circuits (green lines) of a BJT amplifier. In this example, each *RC* circuit has a different critical frequency. The input *RC* circuit is dominant (highest f_c) in this case, and the bypass *RC* circuit has the lowest f_c. The ideal overall response is shown as the blue line.

Here is what happens. As the frequency is reduced from midrange, the first "break point" occurs at the critical frequency of the input *RC* circuit, $f_{cl(input)}$, and the gain begins to drop at −20 dB/decade. This constant roll-off rate continues until the critical frequency of the output *RC* circuit, $f_{cl(output)}$, is reached. At this break point, the output *RC* circuit adds another −20 dB/decade to make a total roll-off of −40 dB/decade. This constant

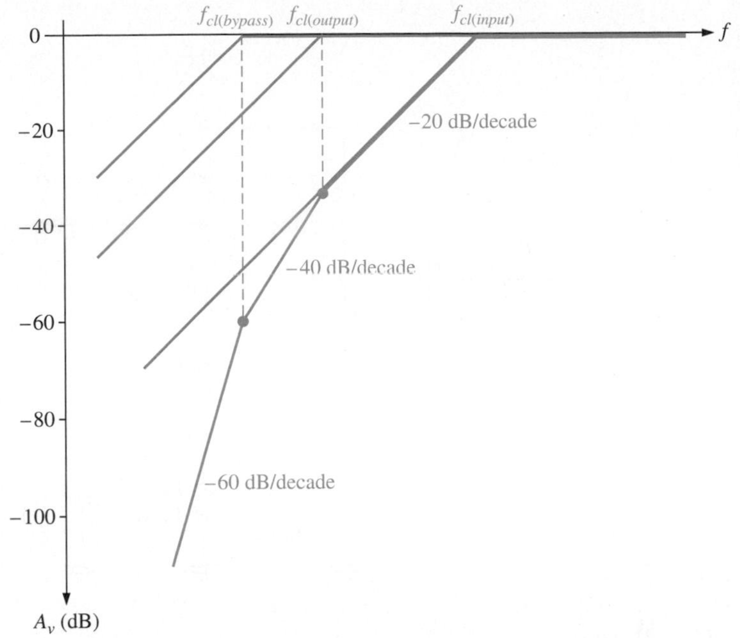

FIGURE 10–25

Composite Bode plot of a BJT amplifier response for three low-frequency *RC* circuits with different critical frequencies. Total response is shown by the blue curve.

−40 dB/decade roll-off continues until the critical frequency of the bypass *RC* circuit, $f_{cl(bypass)}$, is reached. At this break point, the bypass *RC* circuit adds still another −20 dB/decade, making the gain roll-off at −60 dB/decade.

If all *RC* circuits have the same critical frequency, the response curve has one break point at that value of f_{cl}, and the voltage gain rolls off at −60 dB/decade below that value, as shown by the ideal curve (blue) in Figure 10–26. Actually, the midrange voltage gain does not extend down to the dominant critical frequency but is really at −9 dB below the midrange voltage gain at that point (−3 dB for each *RC* circuit), as shown by the red curve.

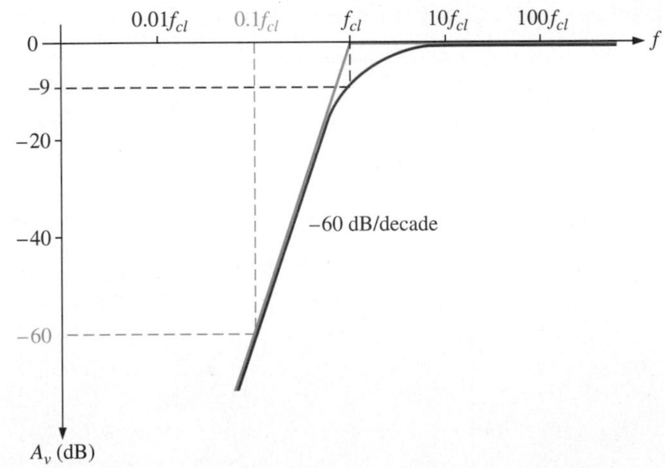

FIGURE 10–26

Composite Bode plot of an amplifier response where all *RC* circuits have the same f_{cl}. (Blue is ideal; red is actual.)

EXAMPLE 10–9

For the circuit from Example 10–3 and shown in Figure 10–27, determine the midband gain in decibels and draw the Bode plot, showing each of the lower critical frequencies. Assume $r'_e = 9.6\ \Omega$.

Solution The midband gain is

$$A_v = \frac{R_C R_L}{r'_e + R_{E1}} = \frac{(3.9\ \text{k}\Omega)(5.6\ \text{k}\Omega)}{9.6\ \Omega + 33\ \Omega} = 54.0$$

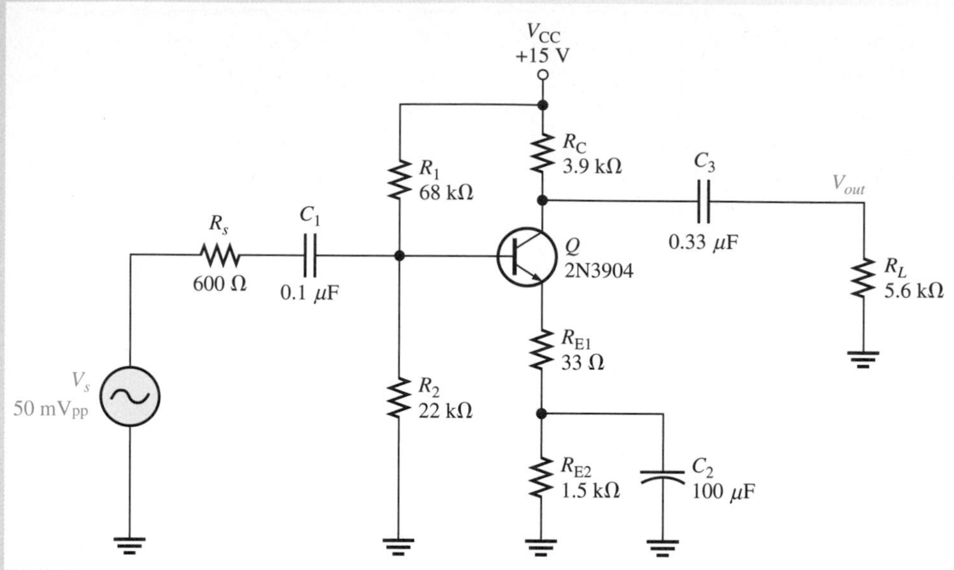

FIGURE 10–27

In decibels,

$$A_v = 20 \log (54.0) = \mathbf{34.3\ dB}$$

The critical frequency for the input circuit was found in Example 10–3 and is 282 Hz. The critical frequency for the output circuit was found in Example 10–5 and is 50.8 Hz. The critical frequency for the emitter bypass circuit was found in Example 10–6 and is 36.0 Hz.

The overall response is shown in the Bode plot of Figure 10–28. The lower critical frequency of the input circuit has the highest value and is therefore the overall or dominant critical frequency because the response first begins to roll off at this frequency.

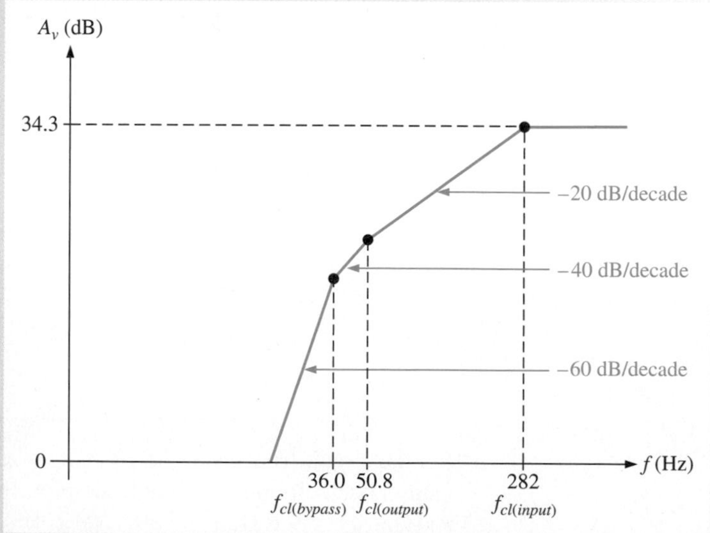

FIGURE 10–28

Ideal Bode plot for the overall low-frequency response of the amplifier in Figure 10–27.

Related Problem If the overall gain of the amplifier is reduced by increasing R_{E1}, how will the lower critical frequency be affected?

Computer Simulation of Frequency Response

As you saw in the previous example, the calculation of multiple critical frequencies is involved and each critical frequency contributes to the overall response. The ideal response shown in Example 10–9 is an excellent first approximation, but when more accuracy is required, a computer simulation is used. The computer takes into account all of the parameters for the particular device including effects such as internal capacitances that are usually ignored in manual calculations, and it can calculate in detail the interactions that occur when there are multiple breakpoints as in Example 10–9.

Multisim is based on SPICE models that can show the frequency response of circuits on the Bode plotter. As mentioned earlier, the Bode plotter is not a real instrument. It performs the same function as an instrument called the spectrum analyzer, which can also plot the frequency response of a circuit. Example 10–10 illustrates the application of computer analysis to the circuit in the previous example.

TECH NOTE

SPICE was one of the first computer programs that could simulate electronic circuits. Its origins can be traced to a program called CANCER (Computer Analysis of Nonlinear Circuits, Excluding Radiation) at the University of California. It was developed as a computer aid for designing integrated circuits in the 1960s. SPICE is an acronym for Simulation Program with Integrated Circuit Emphasis. Over the years, SPICE has been revised many times but is still the underlying software for much of today's simulations.

EXAMPLE 10–10 Use Multisim to show the overall low-frequency response of the circuit in Example 10–9.

Solution Figure 10–29 shows the circuit in Multisim with an oscilloscope display and Bode plotter. The cursor is set to the critical frequency on the Bode plotter so that the

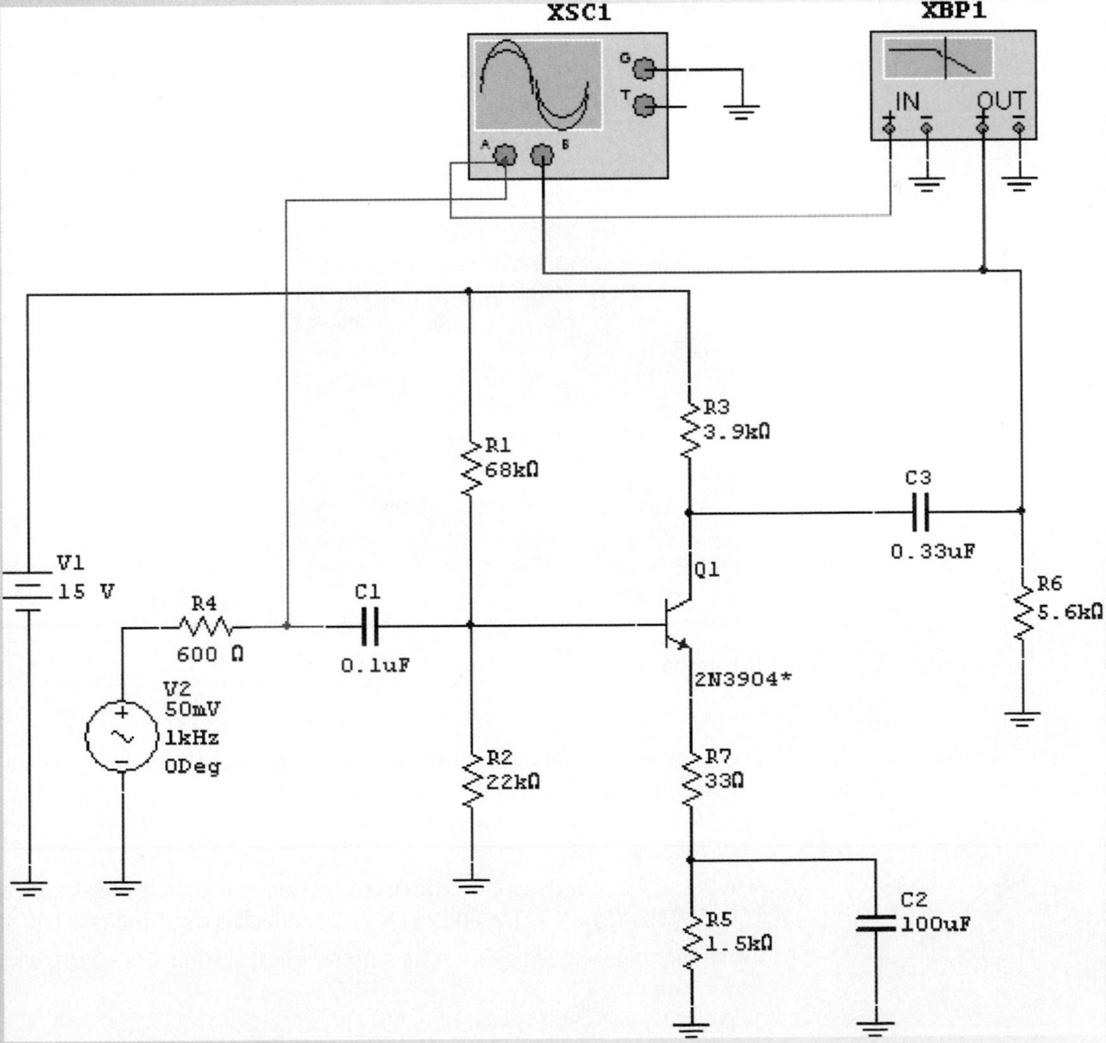

FIGURE 10–29

frequency can be read directly. The result in Figure 10–30 indicates the overall critical frequency is 328 Hz.

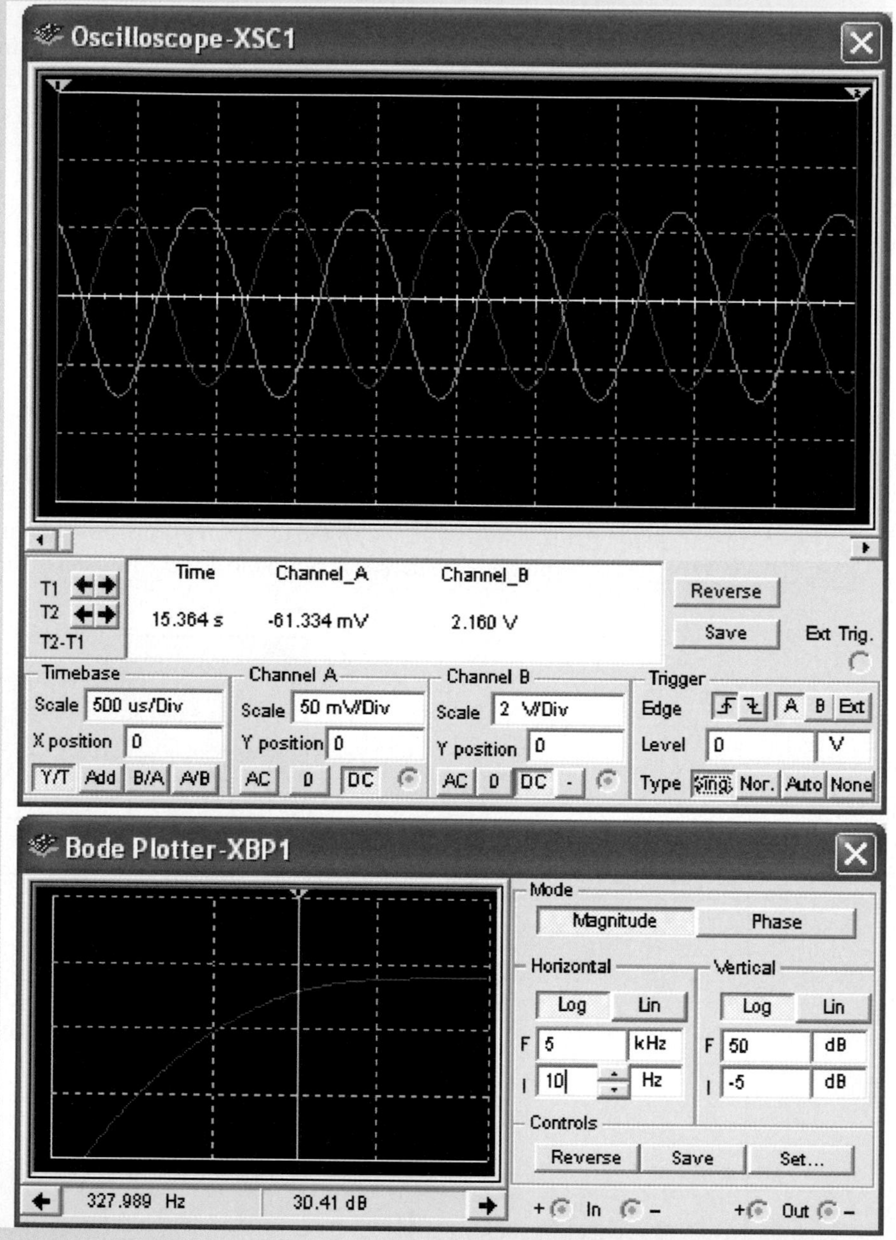

FIGURE 10–30

Related Problem What change would you make to reduce the lower critical frequency to 100 Hz?

SECTION 10–3 CHECKUP

1. A certain BJT amplifier exhibits three critical frequencies in its low-frequency response: $f_{cl1} = 130$ Hz, $f_{cl2} = 167$ Hz, and $f_{cl3} = 75$ Hz. Which is the dominant critical frequency?
2. If the midrange voltage gain of the amplifier in Question 1 is 50 dB, what is the gain at the dominant f_{cl}?

3. A certain *RC* circuit has an $f_{cl} = 235$ Hz, above which the attenuation is 0 dB. What is the dB attenuation at 23.5 Hz?
4. What is the amount of phase shift contributed by an input circuit when $X_C = 0.5R_{in}$ at a certain frequency below f_{cl1}?
5. What is the critical frequency when $R_D = 1.5\ k\Omega$, $R_L = 5\ k\Omega$, and $C_2 = 0.0022\ \mu F$ in a circuit like Figure 10–24?

10–4 HIGH-FREQUENCY AMPLIFIER RESPONSE

You have seen how the coupling and bypass capacitors affect the voltage gain of an amplifier at lower frequencies where the reactances of the coupling and bypass capacitors are significant. In the midrange of an amplifier, the effects of the capacitors are minimal and can be neglected. If the frequency is increased sufficiently, a point is reached where the transistor's internal capacitances begin to have a significant effect on the gain. The basic differences between BJTs and FETs are the specifications of the internal capacitances and the input resistance.

After completing this section, you should be able to

- **Analyze the high-frequency response of amplifiers**
  - Apply Miller's theorem
  - Generally describe how internal capacitances affect the gain
  - Identify the input *RC* circuit
  - Determine the upper critical frequency of the input *RC* circuit
  - Determine phase shift of the input *RC* circuit
  - Identify the output *RC* circuit
  - Determine the upper critical frequency of the output *RC* circuit
  - Determine phase shift of the output *RC* circuit
  - Analyze an amplifier for total high-frequency response

BJT Amplifiers

A high-frequency ac equivalent circuit for the BJT amplifier in Figure 10–31(a) is shown in Figure 10–31(b). Notice that the coupling and bypass capacitors are treated as effective shorts and do not appear in the equivalent circuit. The internal capacitances, C_{be} and C_{bc}, which are significant only at high frequencies, do appear in the diagram. As previously mentioned. C_{be} is sometimes called the input capacitance C_{ib}, and C_{bc} is sometimes called the output capacitance C_{ob}. C_{be} is specified on datasheets at a certain value of V_{BE}. Often, a datasheet will list C_{ib} as C_{ibo} and C_{ob} as C_{obo}. The *o* as the last letter in the subscript indicates the capacitance is measured with the base open. For example, a 2N2222A transistor has a C_{be} of 25 pF at $V_{BE} = 0.5$ V dc, $I_C = 0$, and $f = 1$ MHz. Also, C_{bc} is specified at a certain value of V_{BC}. The 2N2222A has a maximum C_{bc} of 8 pF at $V_{BC} = 10$ V dc.

Miller's Theorem in High-Frequency Analysis By applying Miller's theorem to the inverting amplifier in Figure 10–31(b) and using the midrange voltage gain, you have a circuit

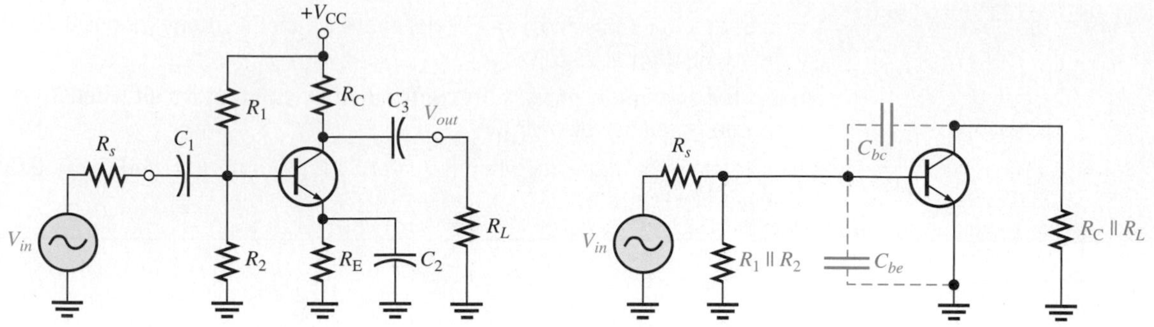

(a) Capacitively coupled amplifier

(b) High-frequency equivalent circuit

▲ FIGURE 10–31

Capacitively coupled amplifier and its high-frequency equivalent circuit.

that can be analyzed for high-frequency response. Looking in from the signal source, the capacitance C_{bc} appears in the Miller input capacitance from base to ground.

$$C_{in(Miller)} = C_{bc}(A_v + 1)$$

C_{be} simply appears as a capacitance to ac ground, as shown in Figure 10–32, in parallel with $C_{in(Miller)}$. Looking in at the collector, C_{bc} appears in the Miller output capacitance from collector to ground. As shown in Figure 10–32, the Miller output capacitance appears in parallel with R_c.

$$C_{out(Miller)} = C_{bc}\left(\frac{A_v + 1}{A_v}\right)$$

These two Miller capacitances create a high-frequency input RC circuit and a high-frequency output RC circuit. These two circuits differ from the low-frequency input and output circuits, which act as high-pass filters, because the capacitances go to ground and therefore act as low-pass filters. The equivalent circuit in Figure 10–32 is an ideal model because stray capacitances that are due to circuit interconnections are neglected.

▶ FIGURE 10–32

High-frequency equivalent circuit after applying Miller's theorem.

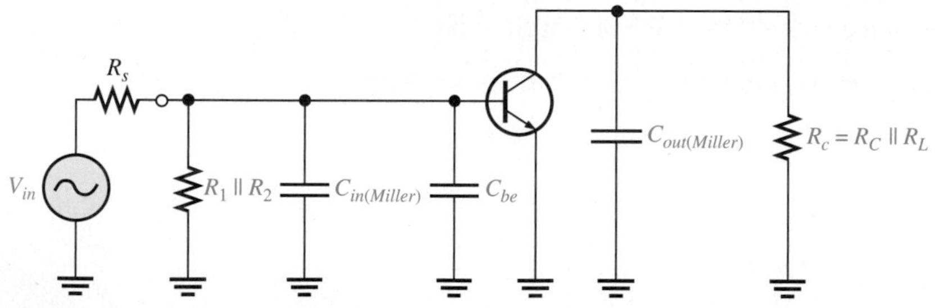

The Input *RC* Circuit

At high frequencies, the input circuit is as shown in Figure 10–33(a), where $\beta_{ac}r'_e$ is the input resistance at the base of the transistor because the bypass capacitor effectively shorts the emitter to ground. By combining C_{be} and $C_{in(Miller)}$ in parallel and repositioning, you get the simplified circuit shown in Figure 10–33(b). Next, by thevenizing the circuit to the left of the capacitor, as indicated, the input RC circuit is reduced to the equivalent form shown in Figure 10–33(c).

As the frequency increases, the capacitive reactance becomes smaller. This causes the signal voltage at the base to decrease, so the amplifier's voltage gain decreases. The reason for this is that the capacitance and resistance act as a voltage divider and, as the frequency increases, more voltage is dropped across the resistance and less across the capacitance. At the critical frequency, the gain is 3 dB less than its midrange value. The upper critical high

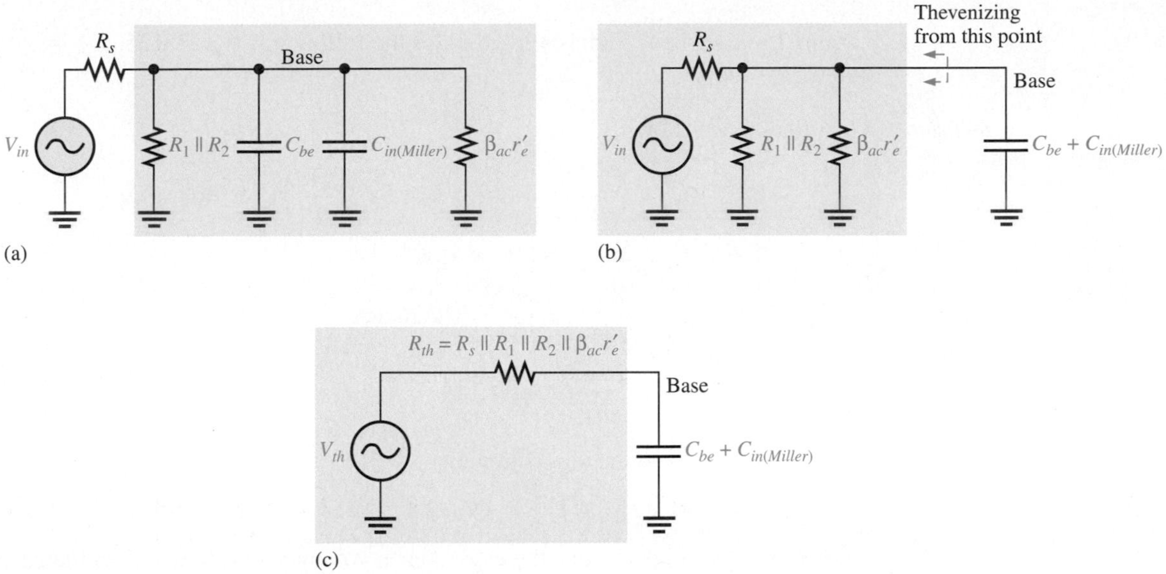

▲ **FIGURE 10–33**

Development of the equivalent high-frequency input *RC* circuit.

frequency of the input circuit, $f_{cu(input)}$, is the frequency at which the capacitive reactance is equal to the total resistance.

$$X_{C_{tot}} = R_s \parallel R_1 \parallel R_2 \parallel \beta_{ac} r'_e$$

Therefore,

$$\frac{1}{2\pi f_{cu(input)} C_{tot}} = R_s \parallel R_1 \parallel R_2 \parallel \beta_{ac} r'_e$$

and

$$\boldsymbol{f_{cu(input)} = \frac{1}{2\pi(R_s \parallel R_1 \parallel R_2 \parallel \beta_{ac} r'_e)C_{tot}}}$$

Equation 10–16

where R_s is the resistance of the signal source and $C_{tot} = C_{be} + C_{in(Miller)}$. As the frequency goes above $f_{cu(input)}$, the input *RC* circuit causes the gain to roll off at a rate of -20 dB/decade just as with the low-frequency response.

EXAMPLE 10–11

Derive the equivalent high-frequency input *RC* circuit for the BJT amplifier in Figure 10–34. Use this to determine the upper critical frequency due to the input

FIGURE 10–34

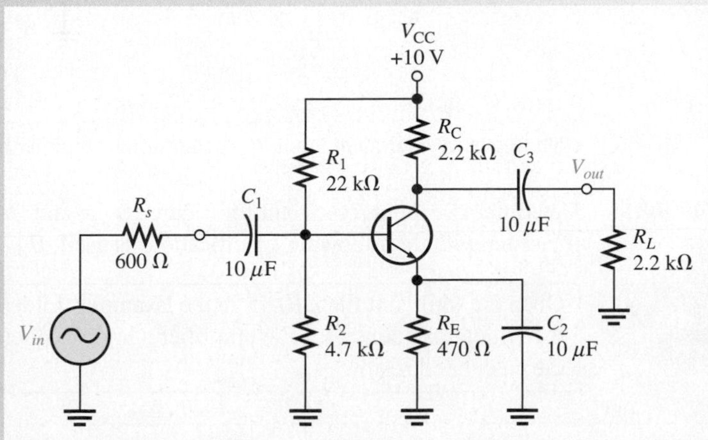

circuit. The transistor's datasheet provides the following: $\beta_{ac} = 125$, $C_{be} = 20$ pF, and $C_{bc} = 2.4$ pF.

Solution First, find r'_e as follows:

$$V_B = \left(\frac{R_2}{R_1 + R_2}\right)V_{CC} = \left(\frac{4.7\text{ k}\Omega}{26.7\text{ k}\Omega}\right)10\text{ V} = 1.76\text{ V}$$

$$V_E = V_B - 0.7\text{ V} = 1.06\text{ V}$$

$$I_E = \frac{V_E}{R_E} = \frac{1.06\text{ V}}{470\ \Omega} = 2.26\text{ mA}$$

$$r'_e = \frac{25\text{ mV}}{I_E} = 11.1\ \Omega$$

The total resistance of the input circuit is

$$R_{in(tot)} = R_s \| R_1 \| R_2 \| \beta_{ac}r'_e = 600\ \Omega \| 22\text{ k}\Omega \| 4.7\text{ k}\Omega \| 125(11.1\ \Omega) = 378\ \Omega$$

Next, in order to determine the capacitance, you must calculate the midrange gain of the amplifier so that you can apply Miller's theorem.

$$A_{v(mid)} = \frac{R_c}{r'_e} = \frac{R_C \| R_L}{r'_e} = \frac{1.1\text{ k}\Omega}{11.1\ \Omega} = 99$$

Apply Miller's theorem.

$$C_{in(Miller)} = C_{bc}(A_{v(mid)} + 1) = (2.4\text{ pF})(100) = 240\text{ pF}$$

The total input capacitance is $C_{in(Miller)}$ in parallel with C_{be}.

$$C_{in(tot)} = C_{in(Miller)} + C_{be} = 240\text{ pF} + 20\text{ pF} = 260\text{ pF}$$

The resulting high-frequency input RC circuit is shown in Figure 10–35. The upper critical frequency is

$$f_{cu(input)} = \frac{1}{2\pi(R_{in(tot)})(C_{in(tot)})} = \frac{1}{2\pi(378\ \Omega)(260\text{ pF})} = \mathbf{1.62\ MHz}$$

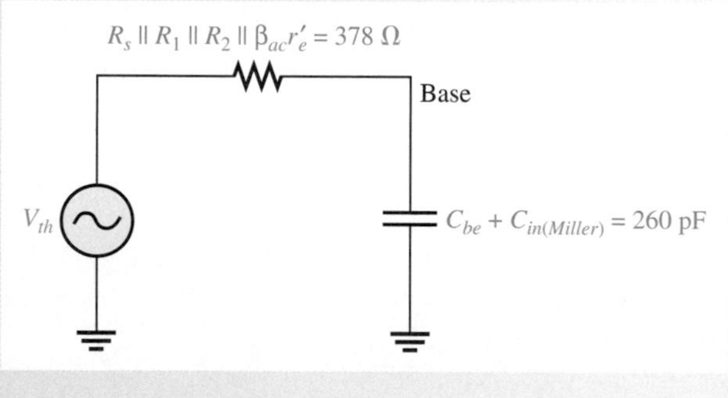

FIGURE 10–35

High-frequency equivalent input *RC* circuit for the amplifier in Figure 10–34.

Related Problem Determine the input RC circuit for Figure 10–34 and find its upper critical frequency if a transistor with the following specifications is used: $\beta_{ac} = 75$, $C_{be} = 15$ pF, $C_{bc} = 2$ pF.

Open the Multisim file E10-11 in the Examples folder on your CD-ROM. Measure the critical frequency for the amplifier's high-frequency response and compare to the calculated result.

Phase Shift of the Input RC Circuit Because the output voltage of a high-frequency input RC circuit is across the capacitor, the output of the circuit lags the input. The phase angle is expressed as

$$\theta = \tan^{-1}\left(\frac{R_s \| R_1 \| R_2 \| \beta_{ac}r'_e}{X_{C_{tot}}}\right)$$

Equation 10–17

At the critical frequency, the phase angle is 45° with the signal voltage at the base of the transistor lagging the input signal. As the frequency increases above f_c, the phase angle increases above 45° and approaches 90° when the frequency is sufficiently high.

The Output *RC* Circuit

The high-frequency output RC circuit is formed by the Miller output capacitance and the resistance looking in at the collector, as shown in Figure 10–36(a). In determining the output resistance, the transistor is treated as a current source (open) and one end of R_C is effectively ac ground, as shown in Figure 10–36(b). By rearranging the position of the capacitance in the diagram and thevenizing the circuit to the left, as shown in Figure 10–36(c), you get the equivalent circuit in Figure 10–36(d). The equivalent output RC circuit consists of a resistance equal to the parallel combination of R_C and R_L in series with a capacitance that is determined by the following Miller formula:

$$C_{out(Miller)} = C_{bc}\left(\frac{A_v + 1}{A_v}\right)$$

If the voltage gain is at least 10, this formula can be approximated as

$$C_{out(Miller)} \cong C_{bc}$$

The upper critical frequency for the output circuit is determined with the following equation, where $R_c = R_C \| R_L$.

$$f_{cu(output)} = \frac{1}{2\pi R_c C_{out(Miller)}}$$

Equation 10–18

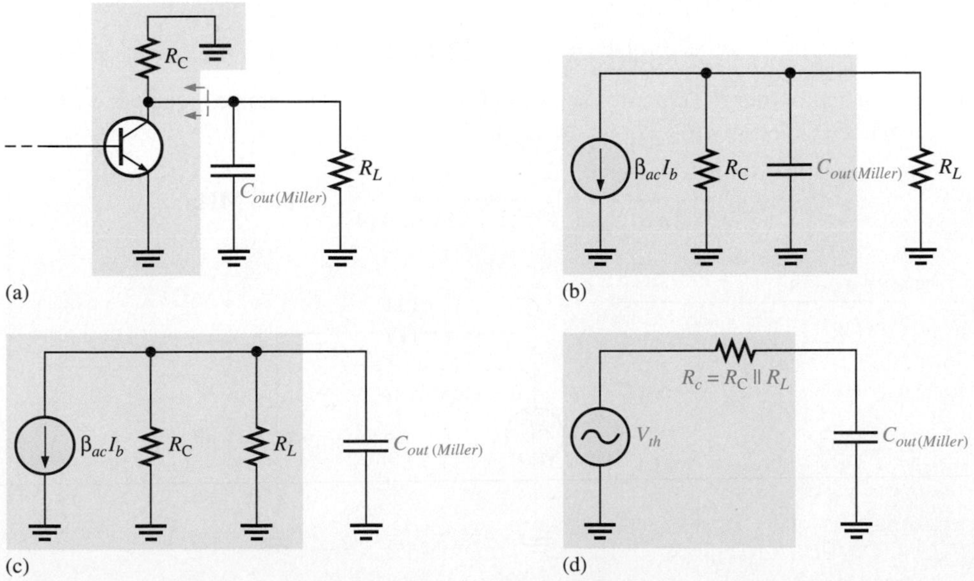

▲ FIGURE 10–36

Development of the equivalent high-frequency output RC circuit.

Just as in the input RC circuit, the output RC circuit reduces the gain by 3 dB at the critical frequency. When the frequency goes above the critical value, the gain drops at a −20 dB/decade rate. The phase angle introduced by the output RC circuit is

Equation 10–19

$$\theta = \tan^{-1}\left(\frac{R_c}{X_{C_{out(Miller)}}}\right)$$

EXAMPLE 10–12

Determine the upper critical frequency of the amplifier in Example 10–11 shown in Figure 10–37 due to its output RC circuit.

FIGURE 10–37

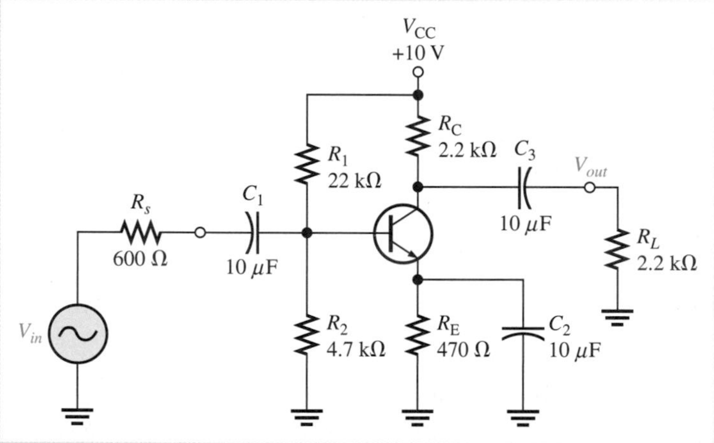

Solution Calculate the Miller output capacitance.

$$C_{out(Miller)} = C_{bc}\left(\frac{A_v + 1}{A_v}\right) = (2.4\text{ pF})\left(\frac{99 + 1}{99}\right) \cong 2.4\text{ pF}$$

The equivalent resistance is

$$R_c = R_C \| R_L = 2.2\text{ k}\Omega \| 2.2\text{ k}\Omega = 1.1\text{ k}\Omega$$

The equivalent output RC circuit is shown in Figure 10–38. Determine the upper critical frequency as follows ($C_{out(Miller)} \cong C_{bc}$):

$$f_{cu(output)} = \frac{1}{2\pi R_c C_{bc}} = \frac{1}{2\pi(1.1\text{ k}\Omega)(2.4\text{ pF})} = \mathbf{60.3\text{ MHz}}$$

FIGURE 10–38

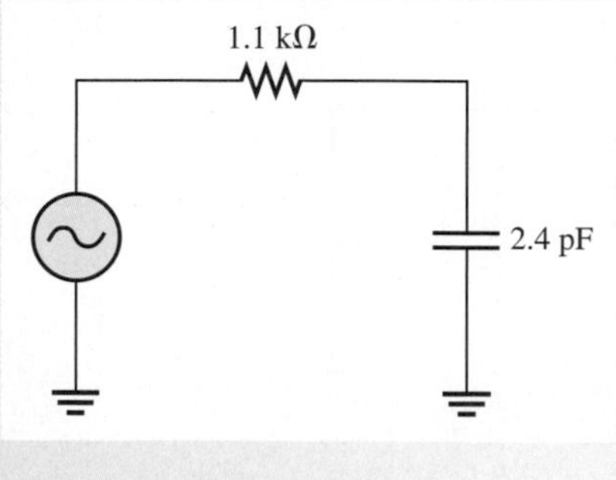

Related Problem If another transistor with $C_{bc} = 5$ pF is used in the amplifier, what is $f_{cu(output)}$?

FET Amplifiers

The approach to the high-frequency analysis of a FET amplifier is similar to that of a BJT amplifier. The basic differences are the specifications of the internal FET capacitances and the determination of the input resistance.

Figure 10–39(a) shows a JFET common-source amplifier that will be used to illustrate high-frequency analysis. A high-frequency equivalent circuit for the amplifier is shown in Figure 10–39(b). Notice that the coupling and bypass capacitors are assumed to have negligible reactances and are considered to be shorts. The internal capacitances C_{gs} and C_{gd} appear in the equivalent circuit because their reactances are significant at high frequencies.

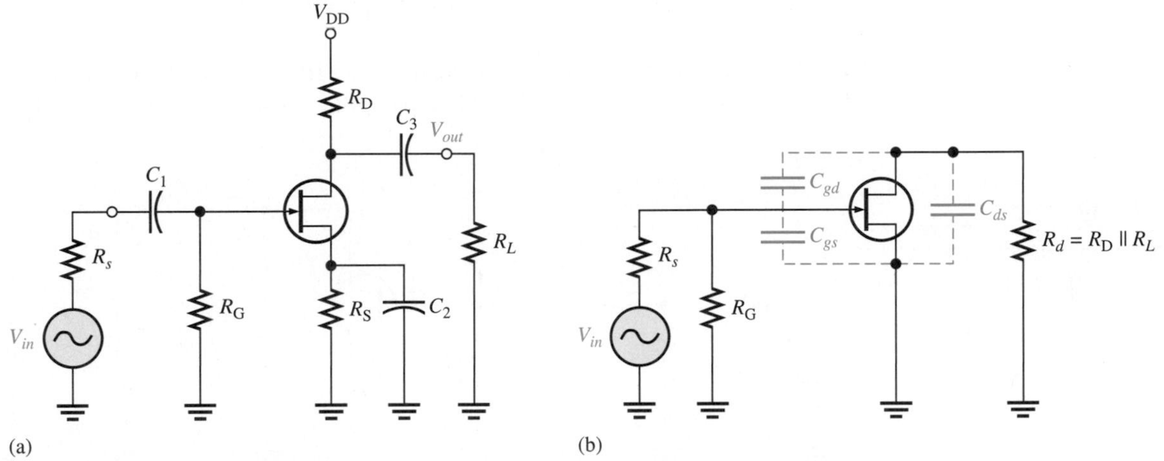

▲ **FIGURE 10–39**

Example of a JFET amplifier and its high-frequency equivalent circuit.

Values of C_{gs}, C_{gd}, and C_{ds} FET datasheets do not normally provide values for C_{gs}, C_{gd}, or C_{ds}. Instead, three other values are usually specified because they are easier to measure. These are C_{iss}, the input capacitance; C_{rss}, the reverse transfer capacitance; and C_{oss}, the output capacitance. Because of the manufacturer's method of measurement, the following relationships allow you to determine the capacitor values needed for analysis.

$$C_{gd} = C_{rss} \quad \text{Equation 10–20}$$

$$C_{gs} = C_{iss} - C_{rss} \quad \text{Equation 10–21}$$

$$C_{ds} = C_{oss} - C_{rss} \quad \text{Equation 10–22}$$

C_{oss} is not specified as often as the other values on datasheets. Sometimes, it is designated as $C_{d(sub)}$, the drain-to-substrate capacitance. In cases where a value is not available, you must either assume a value or neglect C_{ds}.

EXAMPLE 10–13

The datasheet for a 2N3823 JFET gives $C_{iss} = 6$ pF and $C_{rss} = 2$ pF. Determine C_{gd} and C_{gs}.

Solution

$$C_{gd} = C_{rss} = \mathbf{2\,pF}$$
$$C_{gs} = C_{iss} - C_{rss} = 6\,\text{pF} - 2\,\text{pF} = \mathbf{4\,pF}$$

Related Problem Although C_{oss} is not specified on the datasheet for the 2N3823 JFET, assume a value of 3 pF and determine C_{ds}.

Using Miller's Theorem Miller's theorem is applied the same way in FET inverting amplifier high-frequency analysis as was done in BJT amplifiers. Looking in from the signal source in Figure 10–39(b), C_{gd} effectively appears in the Miller input capacitance, which was given in Equation 10–1, as follows:

$$C_{in(Miller)} = C_{gd}(A_v + 1)$$

C_{gs} simply appears as a capacitance to ac ground in parallel with $C_{in(Miller)}$, as shown in Figure 10–40. Looking in at the drain, C_{gd} effectively appears in the Miller output capacitance (from Equation 10–2) from drain to ground in parallel with R_d, as shown in Figure 10–40.

$$C_{out(Miller)} = C_{gd}\left(\frac{A_v + 1}{A_v}\right)$$

These two Miller capacitances contribute to a high-frequency input *RC* circuit and a high-frequency output *RC* circuit. Both are low-pass filters, which produce phase lag.

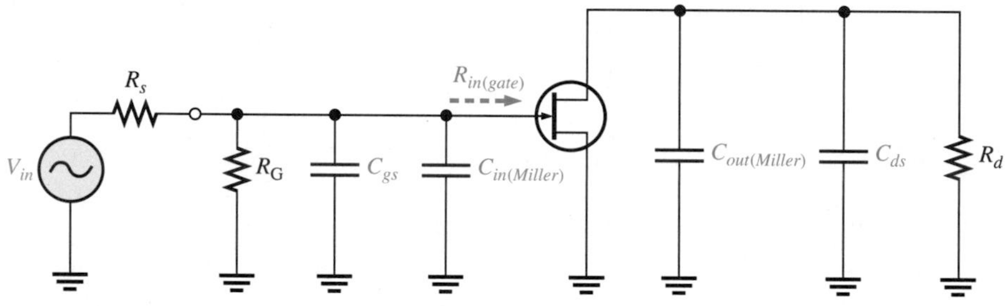

▶ **FIGURE 10–40**

High-frequency equivalent circuit after applying Miller's theorem.

The Input *RC* Circuit

The high-frequency input circuit forms a low-pass type of filter and is shown in Figure 10–41(a). Because both R_G and the input resistance at the gate of FETs are extremely high, the controlling resistance for the input circuit is the resistance of the input source as long as $R_s \ll R_{in}$. This is because R_s appears in parallel with R_{in} when Thevenin's theorem is applied. The simplified input *RC* circuit appears in Figure 10–41(b). The upper critical frequency for the input circuit is

Equation 10–23

$$f_{cu(input)} = \frac{1}{2\pi R_s C_{tot}}$$

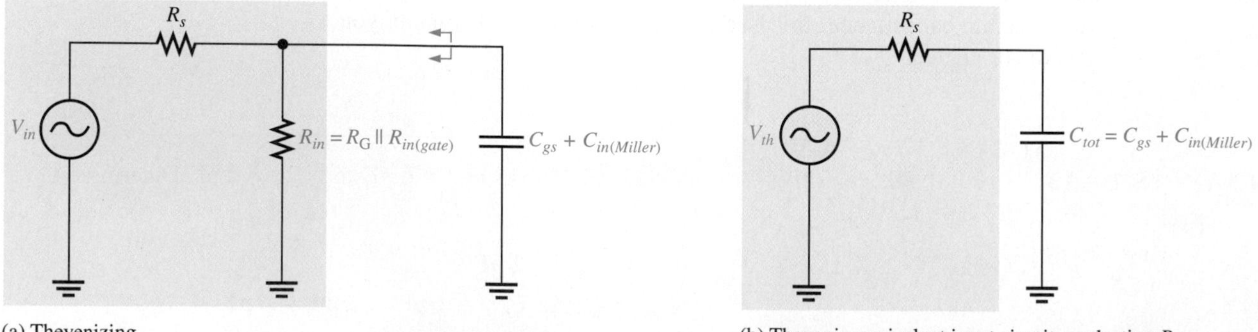

▲ **FIGURE 10–41**

Input *RC* circuit.

where $C_{tot} = C_{gs} + C_{in(Miller)}$. The input RC circuit produces a phase angle of

$$\theta = \tan^{-1}\left(\frac{R_s}{X_{C_{tot}}}\right)$$ **Equation 10–24**

The effect of the input RC circuit is to reduce the midrange gain of the amplifier by 3 dB at the critical frequency and to cause the gain to decrease at -20 dB/decade above f_c.

EXAMPLE 10–14

Find the upper critical frequency of the input RC circuit for the FET amplifier in Figure 10–42. $C_{iss} = 8$ pF, $C_{rss} = 3$ pF, and $g_m = 6500\ \mu$S.

FIGURE 10–42

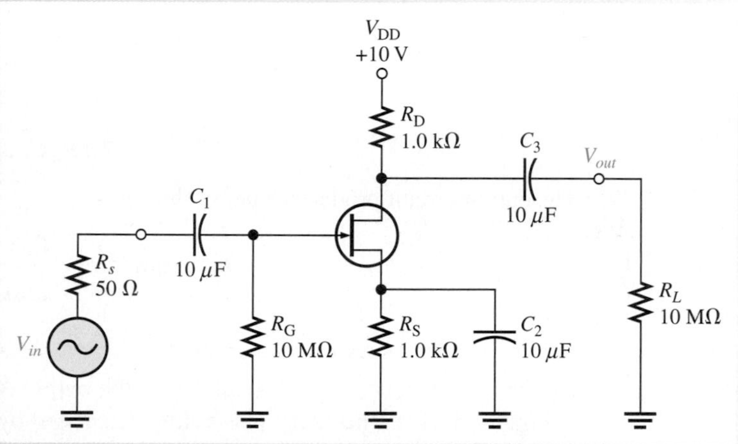

Solution Determine C_{gd} and C_{gs}.

$$C_{gd} = C_{rss} = 3\text{ pF}$$
$$C_{gs} = C_{iss} - C_{rss} = 8\text{ pF} - 3\text{ pF} = 5\text{ pF}$$

Determine the upper critical frequency for the input RC circuit as follows:

$$A_v = g_m R_d = g_m(R_D \| R_L) \cong (6500\ \mu\text{S})(1\text{ k}\Omega) = 6.5$$
$$C_{in(Miller)} = C_{gd}(A_v + 1) = (3\text{ pF})(7.5) = 22.5\text{ pF}$$

The total input capacitance is

$$C_{in(tot)} = C_{gs} + C_{in(Miller)} = 5\text{ pF} + 22.5\text{ pF} = 27.5\text{ pF}$$

The upper critical frequency is

$$f_{cu(input)} = \frac{1}{2\pi R_s C_{in(tot)}} = \frac{1}{2\pi(50\ \Omega)(27.5\text{ pF})} = \mathbf{116\ MHz}$$

Related Problem If the gain of the amplifier in Figure 10–42 is increased to 10, what happens to f_c?

The Output *RC* Circuit

The high-frequency output RC circuit is formed by the Miller output capacitance and the output resistance looking in at the drain, as shown in Figure 10–43(a). As in the case of the BJT, the FET is treated as a current source. When you apply Thevenin's theorem, you get an equivalent output RC circuit consisting of R_D in parallel with R_L and an equivalent output capacitance.

$$C_{out(Miller)} = C_{gd}\left(\frac{A_v + 1}{A_v}\right)$$

▶ FIGURE 10–43

Output *RC* circuit.

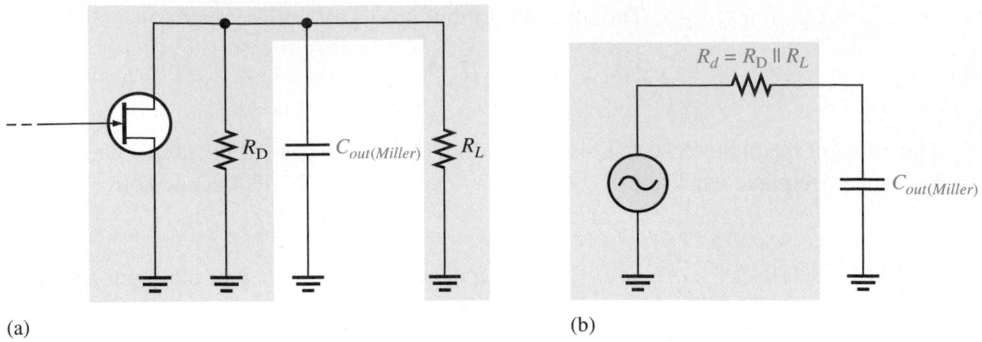

This equivalent output circuit is shown in Figure 10–43(b). The critical frequency of the output *RC* lag circuit is

Equation 10–25

$$f_{cu(output)} = \frac{1}{2\pi R_d C_{out(Miller)}}$$

The output circuit produces a phase shift of

Equation 10–26

$$\theta = \tan^{-1}\left(\frac{R_d}{X_{C_{out(Miller)}}}\right)$$

EXAMPLE 10–15

Determine the upper critical frequency of the output *RC* circuit for the FET amplifier in Figure 10–42. What is the phase shift introduced by this circuit at the critical frequency? Which *RC* circuit is dominant, that is, which one has the lower value of upper critical frequency?

Solution Since R_L is very large compared to R_D, it can be neglected, and the equivalent output resistance is

$$R_d \cong R_D = 1.0\,\text{k}\Omega$$

The equivalent output capacitance is

$$C_{out(Miller)} = C_{gd}\left(\frac{A_v + 1}{A_v}\right) = (3\,\text{pF})\left(\frac{7.5}{6.5}\right) = 3.46\,\text{pF}$$

Therefore, the upper critical frequency is

$$f_{cu(output)} = \frac{1}{2\pi R_d C_{out(Miller)}} = \frac{1}{2\pi(1.0\,\text{k}\Omega)(3.46\,\text{pF})} = \mathbf{46\,MHz}$$

Although it has been neglected, any stray wiring capacitance could significantly affect the frequency response because $C_{out(Miller)}$ is very small.

The phase angle is always **45°** at f_c for an *RC* circuit and the output lags.

In Example 10–14, the upper critical frequency of the input *RC* circuit was found to be 116 MHz. Therefore, the upper critical frequency for the output circuit is dominant because it is the lower of the two.

Related Problem If A_v of the amplifier in Figure 10–42 is increased to 10, what is the upper critical frequency of the output circuit?

Total High-Frequency Response of an Amplifier

As you have seen, the two *RC* circuits created by the internal transistor capacitances influence the high-frequency response of both BJT and FET amplifiers. As the frequency increases and reaches the high end of its midrange values, one of the *RC* circuits will cause the amplifier's gain to begin dropping off. The frequency at which this occurs is the dominant upper critical frequency; it is the lower of the two upper critical high frequencies. An ideal high-frequency Bode plot is shown in Figure 10–44(a). It shows the first break point at $f_{cu(input)}$ where the voltage gain begins to roll off at -20 dB/decade. At $f_{cu(output)}$, the gain begins dropping at -40 dB/decade because each *RC* circuit is providing a -20 dB/decade roll-off. Figure 10–44(b) shows a nonideal Bode plot where the voltage gain is actually -3 dB/decade below midrange at $f_{cu(input)}$. Other possibilities are that the output *RC* circuit is dominant or that both circuits have the same critical frequency.

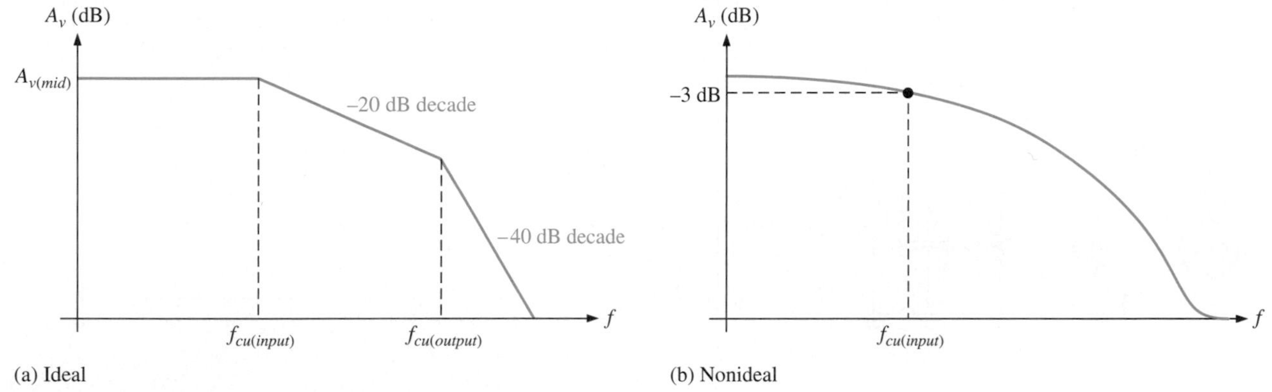

FIGURE 10–44

High-frequency Bode plots.

SECTION 10–4 CHECKUP

1. What determines the high-frequency response of an amplifier?
2. If an amplifier has a midrange voltage gain of 80, the transistor's C_{bc} is 4 pF, and $C_{be} = 8$ pF, what is the total input capacitance?
3. A certain amplifier has $f_{cu(input)} = 3.5$ MHz and $f_{cu(output)} = 8.2$ MHz. Which circuit dominates the high-frequency response?
4. What are the capacitances that are usually specified on a FET datasheet?
5. If $C_{gs} = 4$ pF and $C_{gd} = 3$ pF, what is the total input capacitance of a FET amplifier whose voltage gain is 25?

10–5 TOTAL AMPLIFIER FREQUENCY RESPONSE

In the previous sections, you learned how each *RC* circuit in an amplifier affects the frequency response. In this section, we will bring these concepts together and examine the total response of typical amplifiers and the specifications relating to their performance.

After completing this section, you should be able to

- **Analyze an amplifier for total frequency response**
  - Explain the dominant half-power frequencies
  - Determine the bandwidth
  - Define *gain-bandwidth product*

Figure 10–45(b) shows a generalized ideal response curve (Bode plot) for the BJT amplifier shown in Figure 10–45(a). As previously discussed, the three break points at the lower critical frequencies (f_{cl1}, f_{cl2}, and f_{cl3}) are produced by the three low-frequency *RC* circuits formed by the coupling and bypass capacitors. The break points at the upper critical frequencies, f_{cu1} and f_{cu2}, are produced by the two high-frequency *RC* circuits formed by the transistor's internal capacitances.

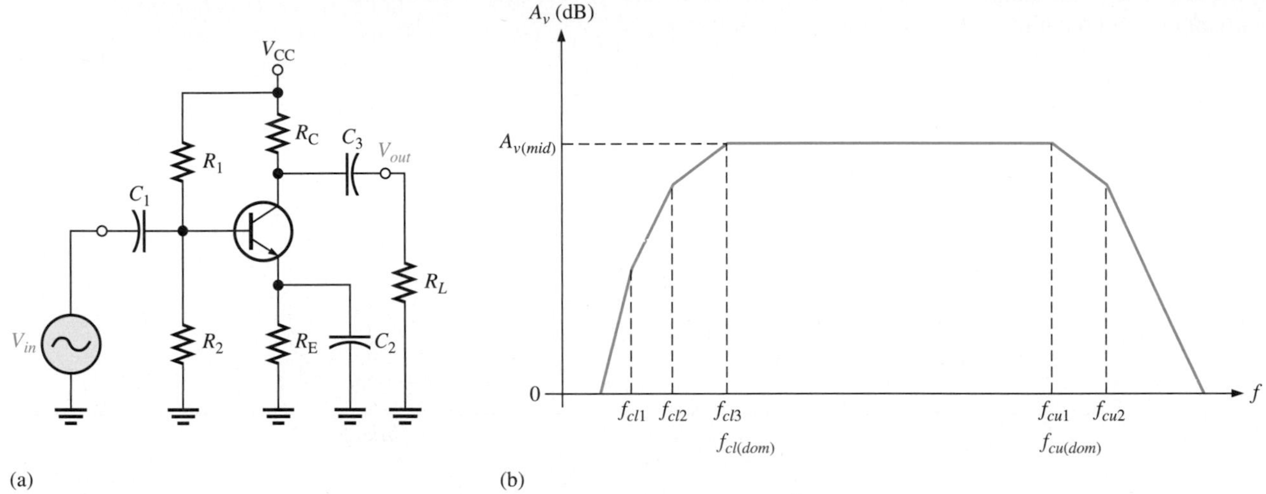

FIGURE 10–45

A BJT amplifier and its generalized ideal response curve (Bode plot).

Of particular interest are the two dominant critical frequencies, f_{cl3} and f_{cu1}, in Figure 10–45(b). These two frequencies are where the voltage gain of the amplifier is 3 dB below its midrange value. These dominant frequencies are designated $f_{cl(dom)}$ and $f_{cu(dom)}$.

The upper and lower dominant critical frequencies are sometimes called the *half-power frequencies.* This term is derived from the fact that the output power of an amplifier at its critical frequencies is one-half of its midrange power, as previously mentioned. This can be shown as follows, starting with the fact that the output voltage is 0.707 of its midrange value at the dominant critical frequencies.

$$V_{out(f_c)} = 0.707V_{out(mid)}$$

$$P_{out(f_c)} = \frac{V^2_{out(f_c)}}{R_{out}} = \frac{(0.707V_{out(mid)})^2}{R_{out}} = \frac{0.5V^2_{out(mid)}}{R_{out}} = 0.5P_{out(mid)}$$

Bandwidth

An amplifier normally operates with signal frequencies between $f_{cl(dom)}$ and $f_{cu(dom)}$. As you know, when the input signal frequency is at $f_{cl(dom)}$ or $f_{cu(dom)}$, the output signal voltage level is 70.7% of its midrange value or −3 dB. If the signal frequency drops below $f_{cl(dom)}$, the gain and thus the output signal level drops at 20 dB/decade until the next critical frequency is reached. The same occurs when the signal frequency goes above $f_{cu(dom)}$.

The range (band) of frequencies lying between $f_{cl(dom)}$ and $f_{cu(dom)}$ is defined as the **bandwidth** of the amplifier, as illustrated in Figure 10–46. Only the dominant critical frequencies appear in the response curve because they determine the bandwidth. Also, sometimes the other critical frequencies are far enough away from the dominant frequencies that they play no significant role in the total amplifier response and can be neglected. The amplifier's bandwidth is expressed in units of hertz as

Equation 10–27

$$BW = f_{cu(dom)} - f_{cl(dom)}$$

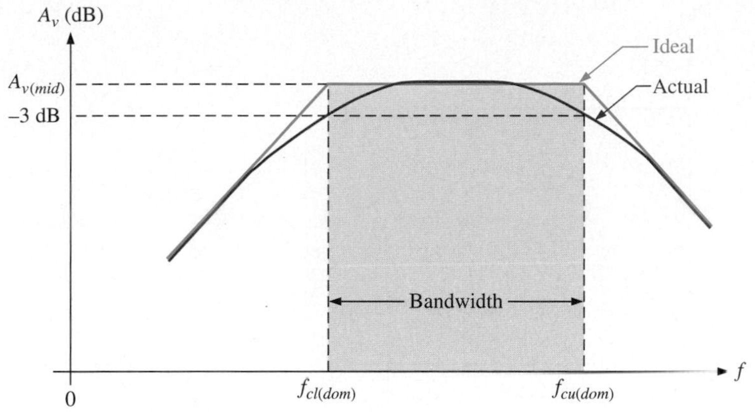

FIGURE 10–46

Response curve illustrating the bandwidth of an amplifier.

Ideally, all signal frequencies lying in an amplifier's bandwidth are amplified equally. For example, if a 10 mV rms signal is applied to an amplifier with a voltage gain of 20, it is amplified to 200 mV rms for all frequencies in the bandwidth. Actually, the gain is down 3 dB at $f_{cl(dom)}$ and $f_{cu(dom)}$.

EXAMPLE 10–16

What is the bandwidth of an amplifier having an $f_{cl(dom)}$ of 200 Hz and an $f_{cu(dom)}$ of 2 kHz?

Solution

$$BW = f_{cu(dom)} - f_{cl(dom)} = 2000\text{ Hz} - 200\text{ Hz} = \mathbf{1800\text{ Hz}}$$

Notice that bandwidth has the unit of hertz.

Related Problem If $f_{cl(dom)}$ is increased, does the bandwidth increase or decrease? If $f_{cu(dom)}$ is increased, does the bandwidth increase or decrease?

Gain-Bandwidth Product

One characteristic of amplifiers is that the product of the voltage gain and the bandwidth is always constant when the roll-off is -20 dB/decade. This characteristic is called the **gain-bandwidth product.** Let's assume that the dominant lower critical frequency of a particular amplifier is much less than the dominant upper critical frequency.

$$f_{cl(dom)} \ll f_{cu(dom)}$$

The bandwidth can then be approximated as

$$BW = f_{cu(dom)} - f_{cl(dom)} \cong f_{cu}$$

Unity-Gain Frequency The simplified Bode plot for this condition is shown in Figure 10–47. Notice that $f_{cl(dom)}$ is neglected because it is so much smaller than $f_{cu(dom)}$, and the bandwidth approximately equals $f_{cu(dom)}$. Beginning at $f_{cu(dom)}$, the gain rolls off until unity gain (0 dB) is reached. The frequency at which the amplifier's gain is 1 is called the *unity-gain frequency*, f_T. The significance of f_T is that it always equals the midrange voltage gain times the bandwidth and is constant for a given transistor.

$$\mathbf{f_T = A_{v(mid)}BW}$$ Equation 10–28

For the case shown in Figure 10–47, $f_T = A_{v(mid)}f_{cu(dom)}$. For example, if a transistor datasheet specifies $f_T = 100$ MHz, this means that the transistor is capable of producing a voltage gain of 1 up to 100 MHz, or a gain of 100 up to 1 MHz, or any combination of gain and bandwidth that produces a product of 100 MHz.

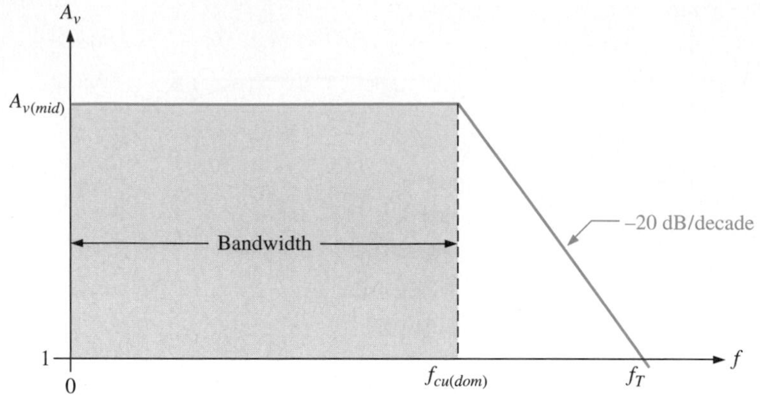

▶ FIGURE 10–47

Simplified response curve where $f_{cl(dom)}$ is negligible (assumed to be zero) compared to $f_{cu(dom)}$.

EXAMPLE 10–17

A certain transistor has an f_T of 175 MHz. When this transistor is used in an amplifier with a midrange voltage gain of 50, what bandwidth can be achieved ideally?

Solution

$$f_T = A_{v(mid)}BW$$

$$BW = \frac{f_T}{A_{v(mid)}} = \frac{175\text{ MHz}}{50} = \mathbf{3.5\ MHz}$$

Related Problem An amplifier has a midrange voltage gain of 20 and a bandwidth of 1 MHz. What is the f_T of the transistor?

SECTION 10–5 CHECKUP

1. What is the voltage gain of an amplifier at f_T?
2. What is the bandwidth of an amplifier when $f_{cu(dom)} = 25$ kHz and $f_{cl(dom)} = 100$ Hz?
3. The f_T of a certain transistor is 130 MHz. What voltage gain can be achieved with a bandwidth of 50 MHz?

10–6 FREQUENCY RESPONSE OF MULTISTAGE AMPLIFIERS

To this point, you have seen how the voltage gain of a single-stage amplifier changes over frequency. When two or more stages are cascaded to form a multistage amplifier, the overall frequency response is determined by the frequency response of each stage depending on the relationships of the critical frequencies.

After completing this section, you should be able to

- **Analyze multistage amplifiers for frequency response**
  - Determine the dominant critical frequencies when the critical frequencies of each stage are the same
  - Determine the dominant critical frequencies when the critical frequencies of each stage differ
  - Determine the bandwidth of a multistage amplifier

When amplifier stages are cascaded to form a multistage amplifier, the dominant frequency response is determined by the responses of the individual stages. There are two cases to consider:

1. Each stage has a different dominant lower critical frequency and a different dominant upper critical frequency.
2. Each stage has the same dominant lower critical frequency and the same dominant upper critical frequency.

Different Critical Frequencies

Ideally, when the dominant lower critical frequency, $f_{cl(dom)}$, of each amplifier stage is different from the other stages, the overall dominant lower critical frequency, $f'_{cl(dom)}$, equals the dominant critical frequency of the stage with the highest $f_{cl(dom)}$.

Ideally, when the dominant upper critical frequency, $f_{cu(dom)}$, of each amplifier stage is different from the other stages, the overall dominant upper critical frequency, $f'_{cu(dom)}$, equals the dominant critical frequency of the stage with the lowest $f_{cu(dom)}$.

In practice, the critical frequencies interact, so these calculated values should be considered approximations that are useful for troubleshooting or estimating the response. When more accuracy is required, a computer simulation is the best solution.

Overall Bandwidth The bandwidth of a multistage amplifier is the difference between the overall dominant lower critical frequency and the overall dominant upper critical frequency.

$$BW = f'_{cu(dom)} - f'_{cl(dom)}$$

EXAMPLE 10–18

In a certain 2-stage amplifier, one stage has a dominant lower critical frequency of 850 Hz and a dominant upper critical frequency of 100 kHz. The other has a dominant lower critical frequency of 1 kHz and a dominant upper critical frequency of 230 kHz. Determine the overall bandwidth of the 2-stage amplifier.

Solution

$$f'_{cl(dom)} = 1\text{ kHz}$$

$$f'_{cu(dom)} = 100\text{ kHz}$$

$$BW = f'_{cu(dom)} - f'_{cl(dom)} = 100\text{ kHz} - 1\text{ kHz} = \mathbf{99\text{ kHz}}$$

Related Problem A certain 3-stage amplifier has the following dominant lower critical frequencies for each stage: $f_{cl(dom)(1)} = 500$ Hz, $f_{cl(dom)(2)} = 980$ Hz, and $f_{cl(dom)(3)} = 130$ Hz. What is the overall dominant lower critical frequency?

Equal Critical Frequencies

When each amplifier stage in a multistage arrangement has equal dominant critical frequencies, you may think that the overall dominant critical frequency is equal to the critical frequency of each stage. This is not the case, however.

When the dominant lower critical frequencies of each stage in a multistage amplifier are all the same, the overall dominant lower critical frequency is increased by a factor of $1/\sqrt{2^{1/n} - 1}$ as shown by the following formula (n is the number of stages in the multistage amplifier):

$$f'_{cl(dom)} = \frac{f_{cl(dom)}}{\sqrt{2^{1/n} - 1}}$$

Equation 10–29

When the dominant upper critical frequencies of each stage are all the same, the overall dominant upper critical frequency is reduced by a factor of $\sqrt{2^{1/n} - 1}$, as shown by the following formula:

$$f'_{cu(dom)} = f_{cu(dom)}\sqrt{2^{1/n} - 1}$$

Equation 10–30

The proofs of these formulas are given in Appendix B.

EXAMPLE 10–19

Both stages in a 2-stage amplifier have a dominant lower critical frequency of 500 Hz and a dominant upper critical frequency of 80 kHz. Determine the overall bandwidth.

Solution

$$f'_{cl(dom)} = \frac{f_{cl(dom)}}{\sqrt{2^{1/n} - 1}} = \frac{500\text{ Hz}}{\sqrt{2^{0.5} - 1}} = \frac{500\text{ Hz}}{0.644} = 776\text{ Hz}$$

$$f'_{cu(dom)} = f_{cu(dom)}\sqrt{2^{1/n} - 1} = (80\text{ kHz})(0.644) = 51.5\text{ kHz}$$

$$BW = f'_{cu(dom)} - f'_{cl(dom)} = 51.5\text{ kHz} - 776\text{ Hz} = \mathbf{50.7\text{ kHz}}$$

Related Problem If a third identical stage is connected in cascade to the 2-stage amplifier in this example, what is the resulting overall bandwidth?

Computer Simulation for Multistage Amplifiers

With multistage amplifiers, the detailed calculation of the frequency response is greatly simplified by computer simulation. There are several interactions within each stage and other interactions between the stages that affect the overall response. When you need more accuracy, a computer simulation is used. This is particularly useful in design work because you can change a component and see the effect immediately on the frequency response. The following example illustrates the application of computer analysis to a multistage amplifier.

EXAMPLE 10–20

A dc coupled two-stage amplifier is simulated with Multisim in Figure 10–48 to determine the overall frequency response.

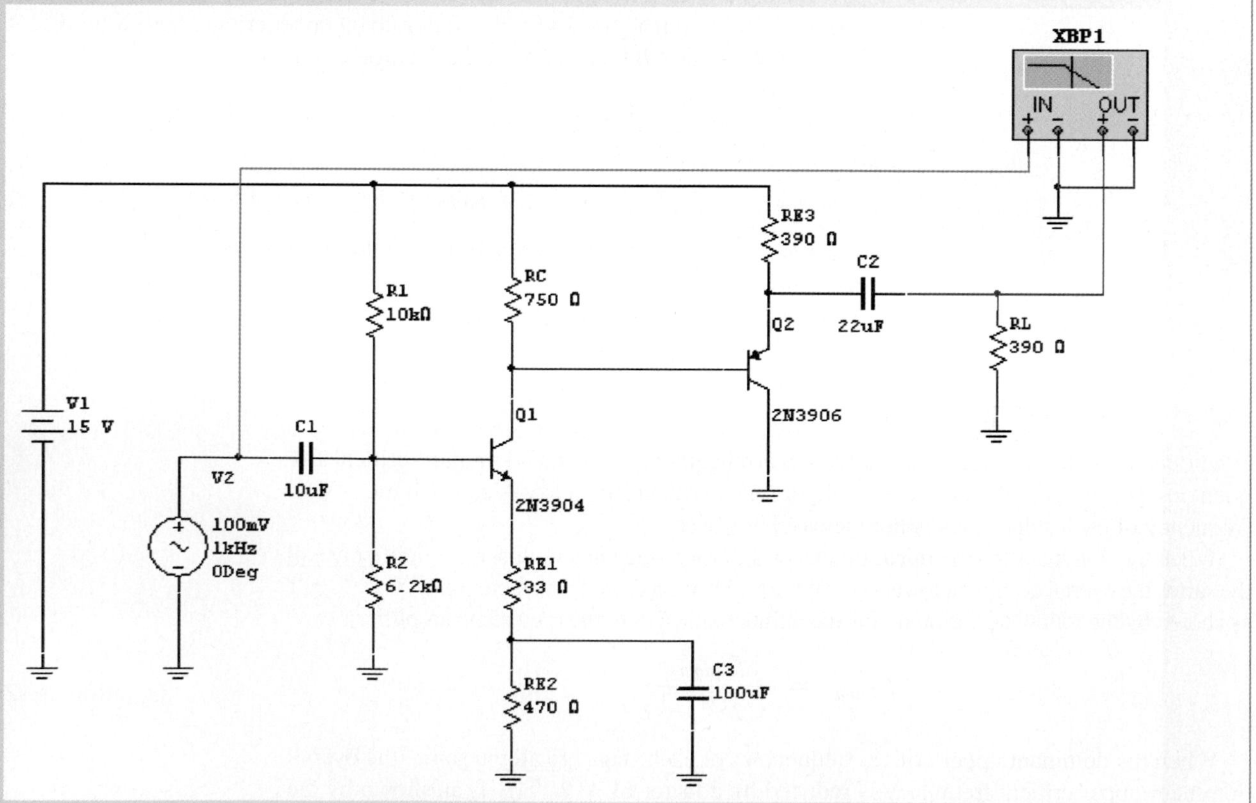

FIGURE 10–48

Solution The circuit was constructed in Multisim by dragging the parts needed onto the simulated workbench and connecting them. Connect the Bode plotter and adjust it to show the complete response curve with upper and lower critical frequencies. Figure 10–49 shows the display. When the cursor is moved to the lower critical frequency (3 dB below midrange), a reading of approximately 56 Hz is observed. When the cursor is moved to the upper critical frequency, a reading of approximately 34 MHz is observed.

FIGURE 10–49

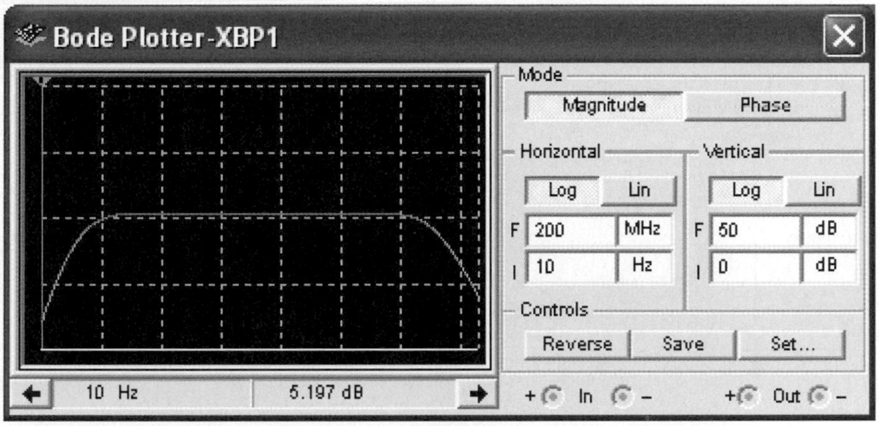

Related Problem Determine the gain of the amplifier in Figure 10–48.

SECTION 10–6 CHECKUP

1. One stage in an amplifier has f_{cl} = 1 kHz and the other stage has f_{cl} = 325 Hz. What is the dominant lower critical frequency?
2. In a certain 3-stage amplifier $f_{cu(1)}$ = 50 kHz, $f_{cu(2)}$ = 55 kHz, and $f_{cu(3)}$ = 49 kHz. What is the dominant upper critical frequency?
3. When more identical stages are added to a multistage amplifier with each stage having the same critical frequency, does the bandwidth increase or decrease?

10–7 FREQUENCY RESPONSE MEASUREMENTS

Two basic methods are used to measure the frequency response of an amplifier. The methods apply to both BJT and FET amplifiers although a BJT amplifier is used as an example. You will concentrate on determining the two dominant critical frequencies. From these values, you can get the bandwidth.

After completing this section, you should be able to

- **Measure the frequency response of an amplifier**
  - Use frequency/amplitude measurement to determine the critical frequencies of an amplifier
  - Relate pulse characteristics to frequency
  - Identify the effects of frequency response on pulse shape
  - Use step-response measurement to determine the critical frequencies of an amplifier

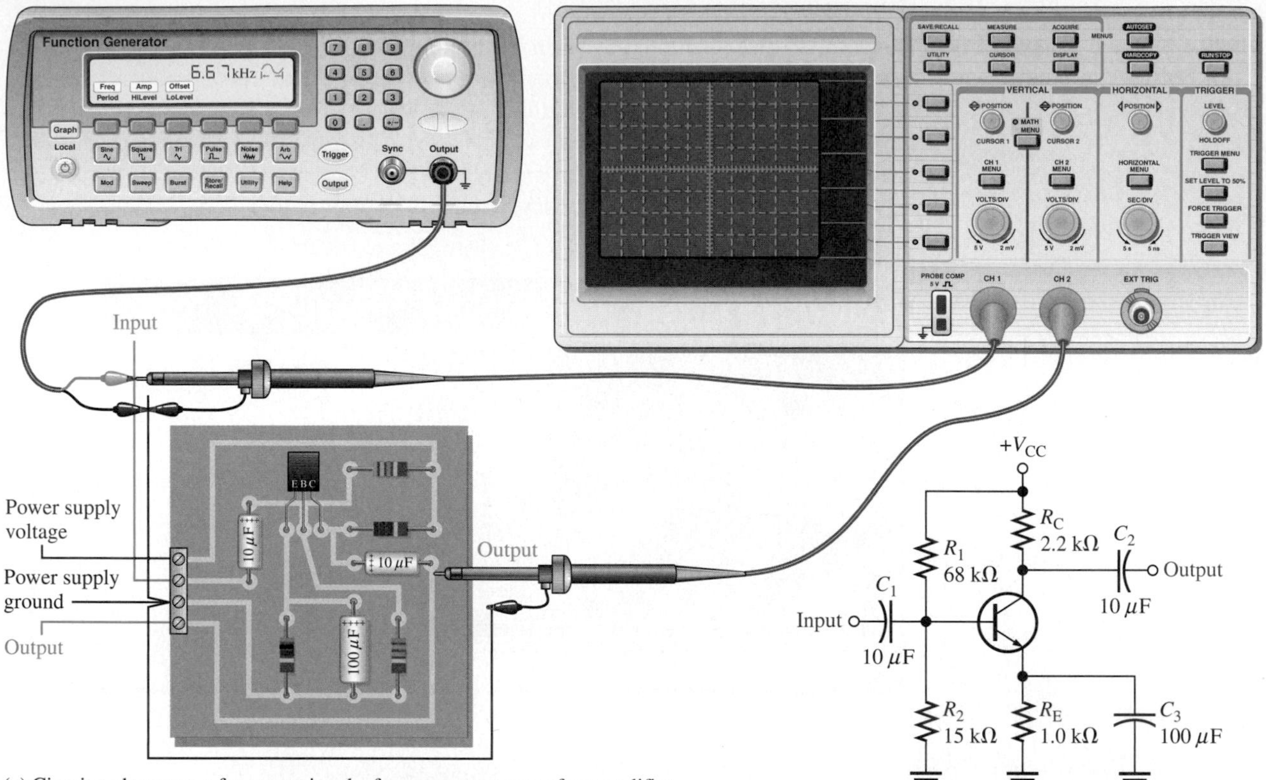

(a) Circuit and test setup for measuring the frequency response of an amplifier

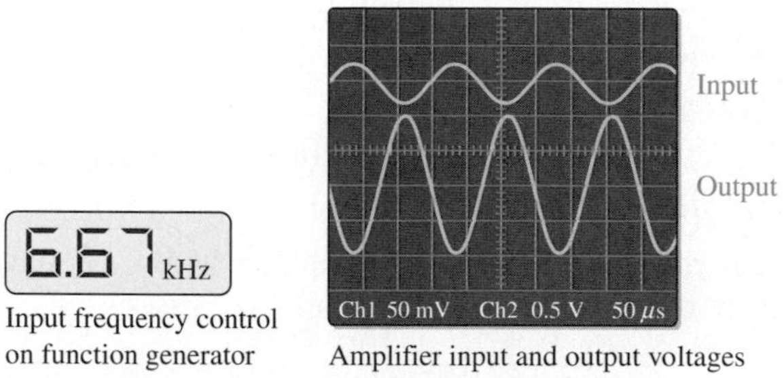

(b) Frequency is set to a midrange value (6.67 kHz in this case). Input voltage adjusted for an output of 1 V peak.

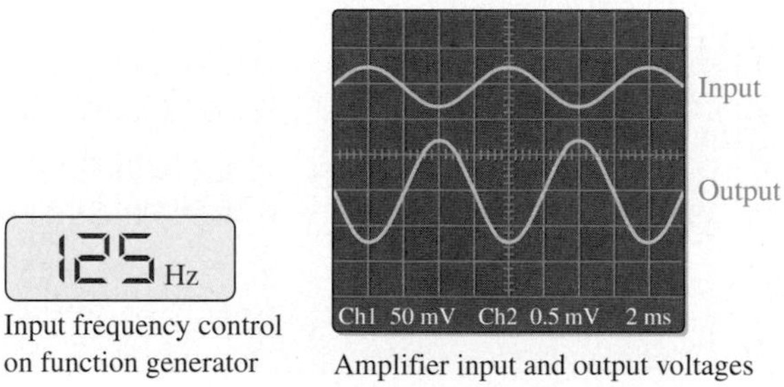

(c) Frequency is reduced until the output is 0.707 V peak. This is the lower critical frequency.

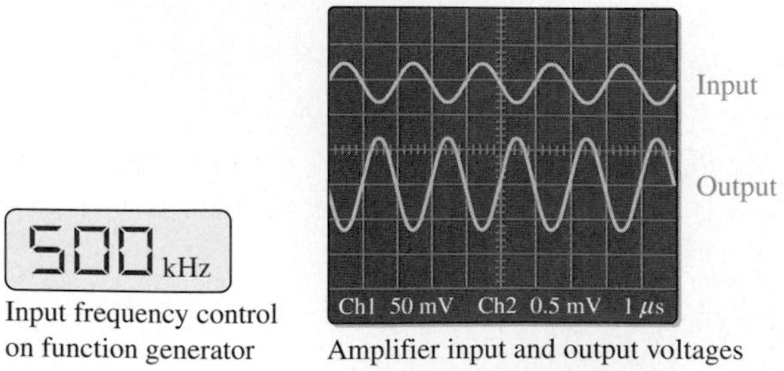

(d) Frequency is increased until the output is again 0.707 V peak. This is the upper critical frequency.

▲ **FIGURE 10–50**

A general procedure for measuring an amplifier's frequency response.

Frequency/Amplitude Measurement

Figure 10–50(a) shows the test setup for an amplifier circuit board. The schematic for the circuit board is also shown. The amplifier is driven by a sinusoidal voltage source with a dual-channel oscilloscope connected to the input and to the output. The input frequency is set to a midrange value, and its amplitude is adjusted to establish an output signal reference level, as shown in Figure 10–50(b). This output voltage reference level for midrange should be set at a convenient value within the linear operation of the amplifier: for example, 100 mV, 1 V, 10 V, and so on. In this case, set the output signal to a peak value of 1 V.

Next, the frequency of the input voltage is decreased until the peak value of the output drops to 0.707 V. The amplitude of the input voltage must be kept constant as the frequency is reduced. Readjustment may be necessary because of changes in loading of the voltage source with frequency. When the output is 0.707 V, the frequency is measured, and you have the value for f_{cl} as indicated in Figure 10–50(c).

Next, the input frequency is increased back up through midrange and beyond until the peak value of the output voltage again drops to 0.707 V. Again, the amplitude of the input must be kept constant as the frequency is increased. When the output is 0.707 V, the frequency is measured and you have the value for f_{cu} as indicated in Figure 10–50(d). From these two frequency measurements, you can find the bandwidth by the formula $BW = f_{cu} - f_{cl}$.

Step-Response Measurement

The lower and upper critical frequencies of an amplifier can be determined using the *step-response method* by applying a voltage step to the input of the amplifier and measuring the rise and fall times of the resulting output voltage. The basic test setup shown in Figure 10–50(a) is used except that the pulse output of the function generator is selected. The input step is created by the rising edge of a pulse that has a long duration compared to the rise and fall times to be measured. The rise time of the input pulse must be fast compared to the rise time you measure from the amplifier.

High-Frequency Measurement When a step input is applied, the amplifier's high-frequency *RC* circuits (internal capacitances) prevent the output from responding immediately to the step input. As a result, the output voltage has a rise time (t_r) associated with it, as shown in Figure 10–51(a). In fact, the rise time is inversely related to the upper critical frequency (f_{cu}) of the amplifier. As f_{cu} becomes lower, the rise time of the output becomes greater. The oscilloscope display illustrates how the rise time is measured from the 10% amplitude point to the 90% amplitude point. The scope must be set on a short time base so the relatively short interval of the rise time can be accurately observed. Once this measurement is made, f_{cu} can be calculated with the following formula:

$$f_{cu} = \frac{0.35}{t_r}$$

Equation 10–31

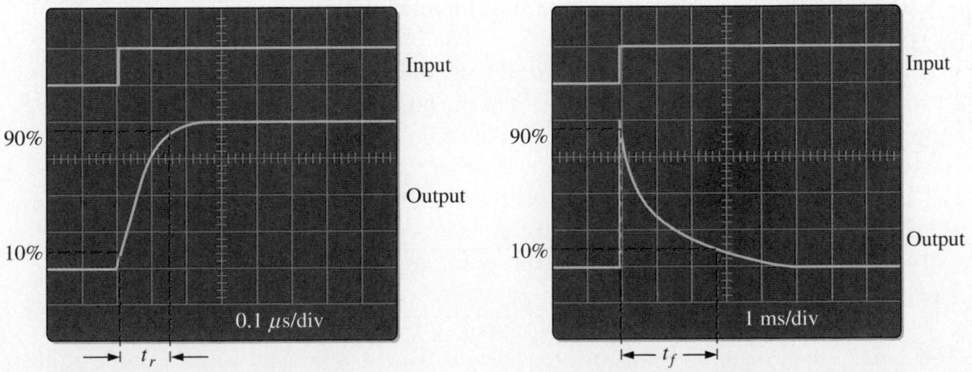

(a) Measurement of output rise time to determine the upper critical frequency

(b) Measurement of output fall time to determine the lower critical frequency

FIGURE 10–51

Measurement of the rise and fall times associated with the amplifier's step response. The outputs are inverted.

Low-Frequency Measurement To determine the lower critical frequency (f_{cl}) of the amplifier, the step input must be of sufficiently long duration to observe the full charging time of the low-frequency *RC* circuits (coupling capacitances), which cause the "sloping" of the output and which we will refer to as the fall time (t_f). This is illustrated in Figure 10–51(b). The fall time is inversely related to the low critical frequency of the amplifier. As f_{cl} becomes higher, the fall time of the output becomes less. The scope display illustrates how the fall time is measured from the 90% point to the 10% point. The scope must be set on a long time base so the complete interval of the fall time can be observed. Once this measurement is made, f_{cl} can be determined with the following formula.

Equation 10–32

$$f_{cl} = \frac{0.35}{t_f}$$

The derivations of Equations 10–31 and 10–32 are in Appendix B.

SECTION 10–7 CHECKUP

1. In Figure 10–50, what are the lower and upper critical frequencies?
2. The rise time and the fall time of an amplifier's output voltage are measured between what two points on the voltage transition?
3. In Figure 10–51, what is the rise time?
4. In Figure 10–51, what is the fall time?
5. What is the bandwidth of the amplifier whose step response is measured in Figure 10–51?

On the Job

Application Activity: *Frequency Analysis of Audio Amplifier*

A utility company is interested in purchasing a large quantity of the PA systems that were developed in the Application Activities in Chapters 6 and 7. Because the company frequently works near high-voltage power lines, where 60 Hz interference is common, it has requested that the PA systems be designed to minimize pickup from power lines. You have been assigned to analyze the frequency response of the PA system and determine the best way to avoid the 60 Hz interference. The modified PA system will be marketed only for voice communication.

The audio frequency spectrum is defined to be the range of frequencies from 20 Hz to 20 kHz. However, the range of frequencies of the human voice is generally accepted to be between 300 Hz and 3 kHz. Based on this, the audio amplifier is to be redesigned for a 300 Hz ± 10% cutoff (critical) frequency in order to minimize the 60 Hz interference. The utility company has requested that the gain at 60 Hz should be down by a minimum of −20 dB from the midrange gain for the units it is purchasing. The high-frequency response of the amplifier is no concern at this point, as long as it is greater than approximately 3 kHz.

The original audio amplifier, shown in Figure 10–52 and in the simulation of Figure 10–53, has a dominant lower critical frequency of 16 Hz as indicated on the Bode plotter in Figure 10–53(c). In order to meet the new specification for a lower critical frequency of 300 Hz, the amplifier must be modified with lower capacitance values.

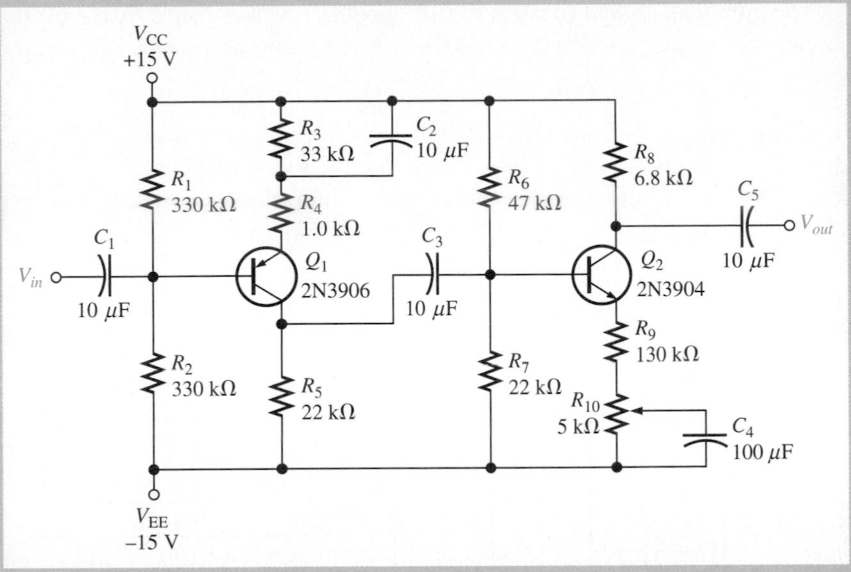

FIGURE 10–52

The audio preamplifier with original capacitor values.

A frequency analysis of the original amplifier is as follows. For the Q_1 stage, the *input circuit* consists of C_1 and $R_1 \| R_2 \| \beta_{ac}R_4$. r'_e is neglected. The critical frequency is (assuming $\beta_{ac} = 100$)

$$f_{cl(input)} = \frac{1}{2\pi(R_1 \| R_2 \| \beta_{ac}R_4)C_1} = \frac{1}{2\pi(62.3\text{ k}\Omega)10\ \mu\text{F}} = 0.255\text{ Hz}$$

The *bypass circuit* consists of C_2 and

$$\left(R_4 + \left(\frac{R_1 \| R_2 \| R_{source}}{\beta_{ac}}\right)\right) \| R_3 \cong R_4$$

The expression reduces to approximately R_4 because R_{source} is assumed to be 300 Ω (microphone impedance) and R_3 is much greater than R_4.

$$f_{cl(bypass)} = \frac{1}{2\pi R_4 C_2} = \frac{1}{2\pi(1\text{ k}\Omega)10\ \mu\text{F}} = 15.9\text{ Hz}$$

The *output circuit* consists of C_3 and $R_5 + R_6 \| R_7 \| \beta_{ac}(R_9 + R_{10})$. r'_e is neglected. Assuming that R_{10} is set at 1 kΩ,

$$f_{cl(output)} = \frac{1}{2\pi(R_5 + R_6 \| R_7 \| \beta_{ac}(R_9 + R_{10}))C_3} = \frac{1}{2\pi(35.2\text{ k}\Omega)10\ \mu\text{F}} = 0.452\text{ Hz}$$

For the Q_2 stage, the *input circuit* is the same as the output circuit of the Q_1 stage.

$$f_{cl(input)} = \frac{1}{2\pi(R_5 + R_6 \| R_7 \| \beta_{ac}(R_9 + R_{10}))C_3} = \frac{1}{2\pi(35.2\text{ k}\Omega)10\ \mu\text{F}} = 0.452\text{ Hz}$$

The *bypass circuit* consists of C_4 and approximately $R_9 + R_{10} + (R_6 \| R_7)/\beta_{ac}$. The resistance is partially dependent on the setting of R_{10}. We will assume that the gain setting is such that R_{10} has negligible effect on the frequency.

$$f_{cl(bypass)} = \frac{1}{2\pi\left(R_9 + \dfrac{R_6 \| R_7}{\beta_{ac}}\right)C_4} = \frac{1}{2\pi(280\ \Omega)100\ \mu\text{F}} = 5.68\text{ Hz}$$

The *output circuit* consists of C_5 and $R_8 + R_L$. The load is the 29 kΩ input resistance of the power amplifier.

$$f_{cl(output)} = \frac{1}{2\pi(R_8 + R_L)C_5} = \frac{1}{2\pi(35.8\text{ k}\Omega)10\ \mu\text{F}} = 0.445\text{ Hz}$$

The dominant critical frequency of the amplifier is established by the Q_1 stage bypass circuit and is $f_{cl(bypass)} = 15.9$ Hz, which is in very close agreement with the simulation.

Simulation of Original Circuit

The Multisim preamp with the original capacitor values is shown in Figure 10–53(a). A Bode plotter is connected to measure the frequency response. Figure 10–53(b) shows the logarithmic response curve with a midrange gain at 5 kHz of 33.3 dB.

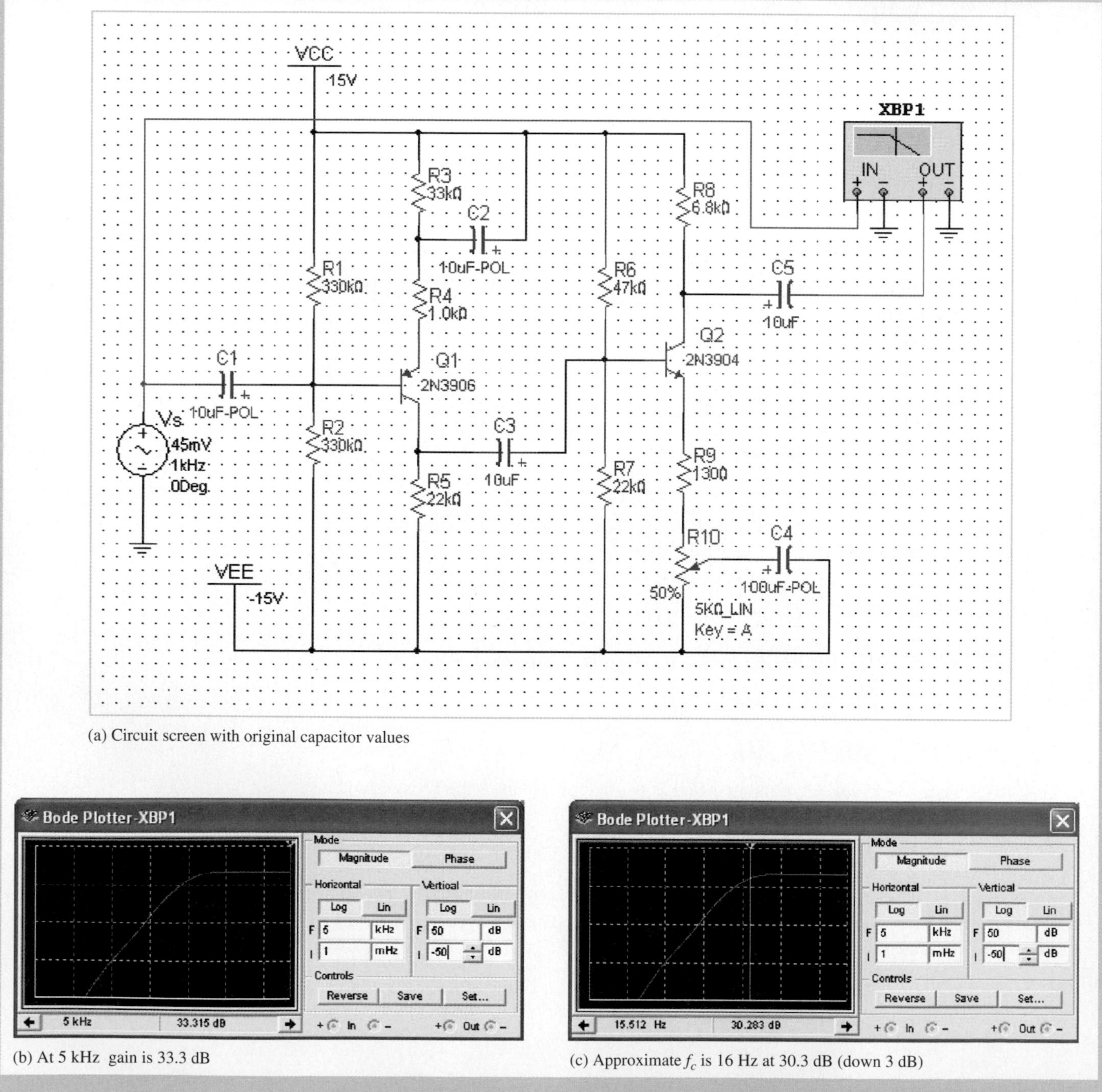

(a) Circuit screen with original capacitor values

(b) At 5 kHz gain is 33.3 dB

(c) Approximate f_c is 16 Hz at 30.3 dB (down 3 dB)

FIGURE 10–53

Preamp frequency response with original capacitor values.

Moving the Bode plotter cursor down until the gain is approximately 3 dB below midrange, or 30.3 dB, results in a lower critical frequency of 16 Hz at this gain setting (note that there is a small effect on the response for different gains due to a different path for C_4 to charge and discharge). This verifies that the response of the preamp includes the potentially troublesome 60 Hz interference.

Modification to Increase the Overall Lower Critical Frequency

Capacitor values must be reduced to achieve a critical frequency of 300 Hz $\pm$ 10%. The approach, in this case, will be to use C_1 and C_3 to set the new dominant critical frequency. C_2 and C_5 will be used to produce a faster roll-off below 60 Hz. C_4 will be left at 100 μF to avoid a change in frequency response when the gain is changed.

C_1 is part of the stage 1 input circuit, and C_3 is part of the stage 2 input circuit. These capacitor values will determine the proper dominant lower critical frequencies required to achieve an overall dominant critical frequency of 300 Hz.

Multistage Frequency Response When the lower critical frequencies of each stage are equal, Equation 10–29 applies. The overall lower critical frequency, f'_{cl}, is 300 Hz. Solving the equation for the dominant lower critical frequency of each stage; $f_{cl(dom)}$, you get

$$f_{cl(dom)} = f'_{cl}\sqrt{(2^{1/2} - 1)} = 300\text{ Hz}\sqrt{(1.414 - 1)} = 300\text{ Hz}(0.643) = 193\text{ Hz}$$

Setting the dominant critical frequency of both stages of the amplifier to 193 Hz will produce an overall lower critical frequency of 300 Hz. Using the frequency analysis that was done for the original circuit as a guide, do the following calculations.

1. Calculate the value of C_1 to produce a lower critical frequency of 193 Hz.
2. Calculate the value of C_3 to produce a lower critical frequency of 193 Hz.

The results of your calculation should agree with the values shown in Figure 10–54. The value for C_2 is the next lower available value in Multisim.

The Multisim circuit with reduced capacitor values is shown in Figure 10–54(a). As you can see in part (c), the new critical frequency is 276.604 Hz, which is within the specified 10% tolerance of 300 Hz. The gain is 9.744 dB for a frequency near 60 Hz with the volume setting at 85%, as shown in part (d).

3. From the Bode plots in Figure 10–54, determine how much the gain at 60 Hz is down from the midrange gain.

Simulate the preamp circuit using your Multisim software. Observe the operation with the Bode plotter.

Prototyping and Testing

Now that the revised circuit has been simulated and its operation verified, the modifications are made to the circuit and it is constructed and tested. After the circuit is successfully tested on a protoboard, it is ready to be finalized on a printed circuit board.

Lab Experiment

To build and test a similar circuit, go to Experiment 10 in your lab manual (*Laboratory Exercises for Electronic Devices* by David Buchla and Steven Wetterling).

Circuit Board

The capacitor values on the preamp circuit board are changed and the board is tested at 5 kHz and at 60 Hz using an oscilloscope, as shown in Figure 10–55.

4. What is the measured rms output voltage at 5 kHz in Figure 10–55?
5. What is the measured rms output voltage at 60 Hz in Figure 10–55?

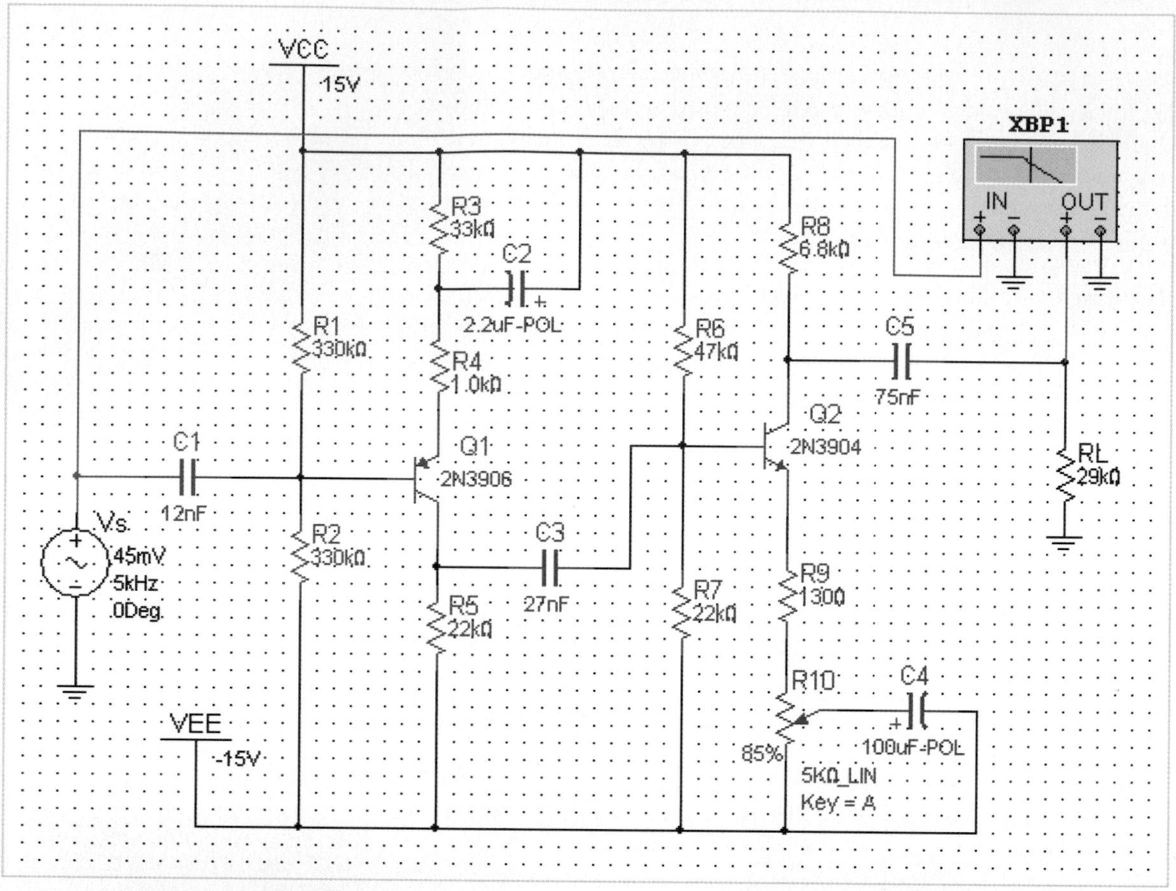

(a) Circuit screen with reduced capacitor values

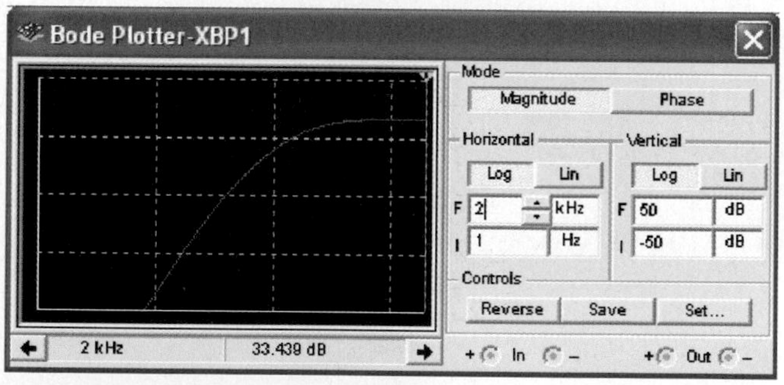

(b) Midrange gain is 33.439 Hz

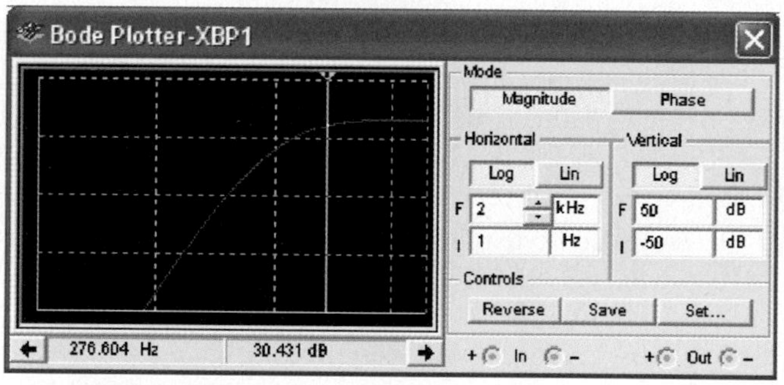

(c) f_c is 276.604 Hz at 30.431 dB (-3 dB)

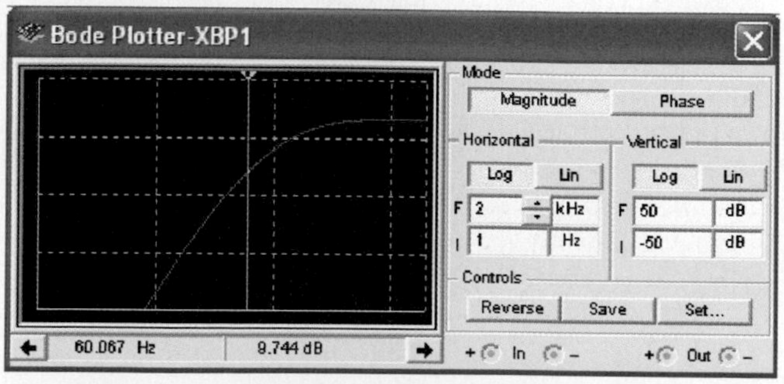

(d) At 60.067 Hz the gain is 9.744 dB (down 23.7 dB)

FIGURE 10–54

Preamp frequency response with reduced capacitor values.

6. What would be the approximate rms amplitude of the output waveform at 300 Hz?
7. Based on the oscilloscope measurement in Figure 10–55, express the voltage gain at 5 kHz in dB.
8. Based on the oscilloscope measurement in Figure 10–55, express the voltage gain at 60 Hz in dB.

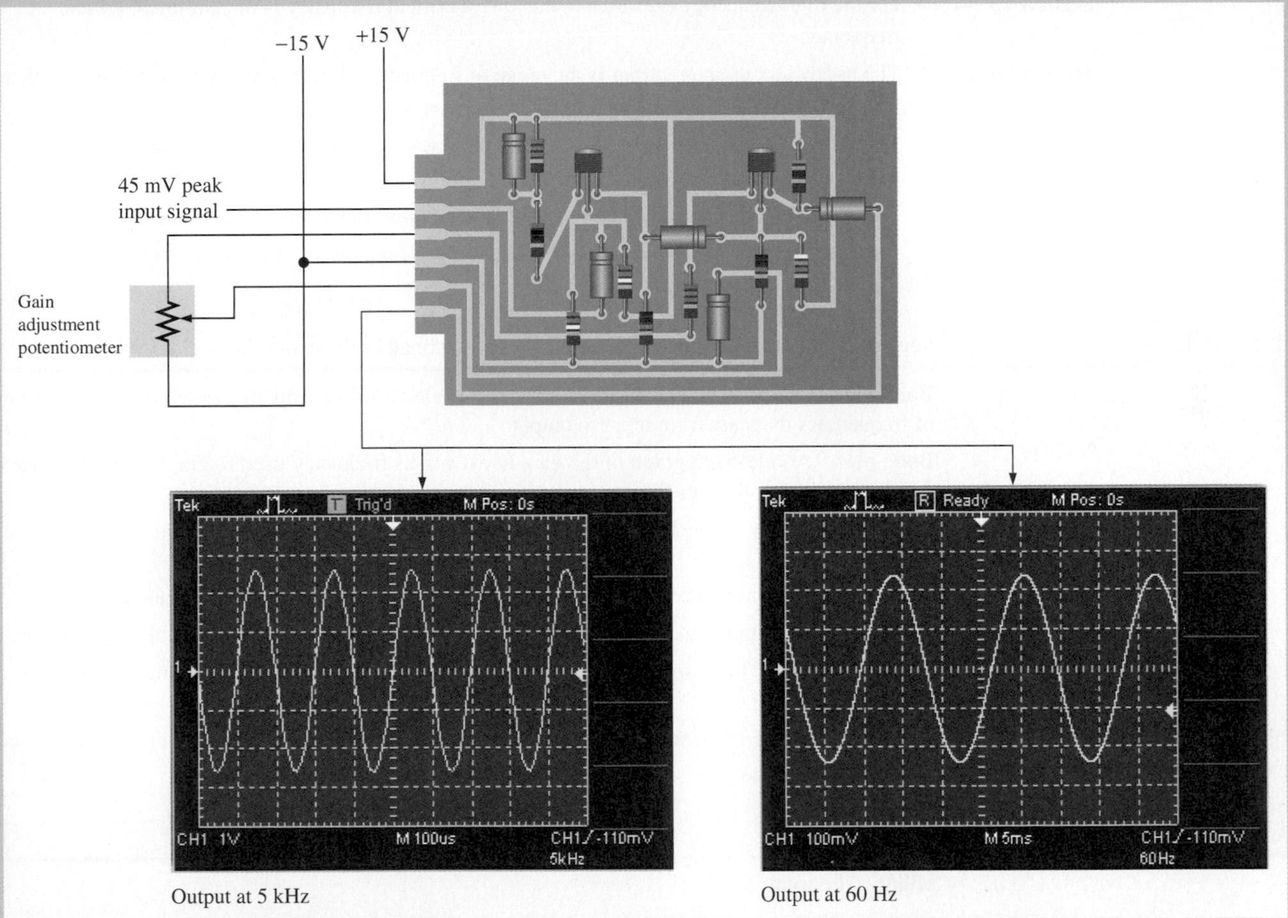

FIGURE 10–55

Frequency test of new preamp board using an oscilloscope.

SUMMARY

Section 10–1

- The coupling and bypass capacitors of an amplifier affect the low-frequency response.
- The internal transistor capacitances affect the high-frequency response.

Section 10–2

- The decibel is a logarithmic unit of measurement for power gain and voltage gain.
- A decrease in voltage gain to 70.7% of midrange value is a reduction of 3 dB.
- A halving of the voltage gain corresponds to a reduction of 6 dB.
- The dBm is a unit for measuring power levels referenced to 1 mW.

- Critical frequencies are values of frequency at which the *RC* circuits reduce the voltage gain to 70.7% of its midrange value.

Section 10–3
- Each *RC* circuit causes the gain to drop at a rate of 20 dB/decade.
- For the low-frequency *RC* circuits, the *highest* critical frequency is the dominant critical frequency.
- A decade of frequency change is a ten-times change (increase or decrease).
- An octave of frequency change is a two-times change (increase or decrease).

Section 10–4
- For the high-frequency *RC* circuits, the *lowest* critical frequency is the dominant critical frequency.

Section 10–5
- The bandwidth of an amplifier is the range of frequencies between the dominant lower critical frequency and the dominant upper critical frequency.
- The gain-bandwidth product is a transistor parameter that is constant and equal to the unity-gain frequency.

Section 10–6
- The dominant critical frequencies of a multistage amplifier establish the bandwidth.

Section 10–7
- Two frequency response measurement methods are frequency/amplitude and step.

KEY TERMS

Key terms and other bold terms in the chapter are defined in the end-of-book glossary.

Bandwidth The characteristic of certain types of electronic circuits that specifies the usable range of frequencies that pass from input to output.

Bode plot An idealized graph of the gain in dB versus frequency used to graphically illustrate the response of an amplifier or filter.

Critical frequency The frequency at which the response of an amplifier or filter is 3 dB less than at midrange.

Decade A ten-times increase or decrease in the value of a quantity such as frequency.

Decibel A logarithmic measure of the ratio of one power to another or one voltage to another.

Midrange gain The gain that occurs for the range of frequencies between the lower and upper critical frequencies.

Roll-off The rate of decrease in the gain of an amplifier above or below the critical frequencies.

KEY FORMULAS

Miller's Theorem

10–1 $C_{in(Miller)} = C(A_v + 1)$ — Miller input capacitance, where $C = C_{bc}$ or C_{gd}

10–2 $C_{out(Miller)} = C\left(\frac{A_v + 1}{A_v}\right)$ — Miller output capacitance, where $C = C_{bc}$ or C_{gd}

The Decibel

10–3 $A_{p(\text{dB})} = 10 \log A_p$ — Power gain in decibels

10–4 $A_{v(\text{dB})} = 20 \log A_v$ — Voltage gain in decibels

BJT Amplifier Low-Frequency Response

10–5 $A_{v(mid)} = \frac{R_c}{r'_e}$ — Midrange voltage gain

10–6 $f_{cl(input)} = \frac{1}{2\pi R_{in} C_1}$ — Lower critical frequency, input *RC* circuit

10–7 $\theta = \tan^{-1}\left(\frac{X_{C1}}{R_{in}}\right)$ — Phase angle, input *RC* circuit

10–8 $f_{cl(output)} = \frac{1}{2\pi(R_C + R_L)C_3}$ — Lower critical frequency, output *RC* circuit

10–9	$\theta = \tan^{-1}\left(\dfrac{X_{C3}}{R_C + R_L}\right)$	Phase angle, output *RC* circuit
10–10	$R_{in(emitter)} = r'_e + \dfrac{R_{th}}{\beta_{ac}}$	Resistance looking in at emitter
10–11	$f_{cl(bypass)} = \dfrac{1}{2\pi[(r'_e + R_{th}/\beta_{ac}) \parallel R_E]C_2}$	Lower critical frequency, bypass *RC* circuit

FET Amplifier Low-Frequency Response

10–12	$f_{cl(input)} = \dfrac{1}{2\pi(R_G \parallel R_{in(gate)})C_1}$	Lower critical frequency, input *RC* circuit
10–13	$\theta = \tan^{-1}\left(\dfrac{X_{C1}}{R_{in}}\right)$	Phase angle, input *RC* circuit
10–14	$f_{cl(output)} = \dfrac{1}{2\pi(R_D + R_L)C_2}$	Lower critical frequency, output *RC* circuit
10–15	$\theta = \tan^{-1}\left(\dfrac{X_{C2}}{R_D + R_L}\right)$	Phase angle, output *RC* circuit

BJT Amplifier High-Frequency Response

10–16	$f_{cu(input)} = \dfrac{1}{2\pi(R_s \parallel R_1 \parallel R_2 \parallel \beta_{ac}r'_e)C_{tot}}$	Upper critical frequency, input *RC* circuit
10–17	$\theta = \tan^{-1}\left(\dfrac{R_s \parallel R_1 \parallel R_2 \parallel \beta_{ac}r'_e}{X_{C_{tot}}}\right)$	Phase angle, input *RC* circuit
10–18	$f_{cu(output)} = \dfrac{1}{2\pi R_c C_{out(Miller)}}$	Upper critical frequency, output *RC* circuit
10–19	$\theta = \tan^{-1}\left(\dfrac{R_c}{X_{C_{out(Miller)}}}\right)$	Phase angle, output *RC* circuit

FET Amplifier High-Frequency Response

10–20	$C_{gd} = C_{rss}$	Gate-to-drain capacitance
10–21	$C_{gs} = C_{iss} - C_{rss}$	Gate-to-source capacitance
10–22	$C_{ds} = C_{oss} - C_{rss}$	Drain-to-source capacitance
10–23	$f_{cu(input)} = \dfrac{1}{2\pi R_s C_{tot}}$	Upper critical frequency, input *RC* circuit
10–24	$\theta = \tan^{-1}\left(\dfrac{R_s}{X_{C_{tot}}}\right)$	Phase angle, input *RC* circuit
10–25	$f_{cu(output)} = \dfrac{1}{2\pi R_d C_{out(Miller)}}$	Upper critical frequency, output *RC* circuit
10–26	$\theta = \tan^{-1}\left(\dfrac{R_d}{X_{C_{out(Miller)}}}\right)$	Phase angle, output *RC* circuit

Total Response

10–27	$BW = f_{cu} - f_{cl}$	Bandwidth
10–28	$f_T = A_{v(mid)}BW$	Unity-gain bandwidth

Multistage Response

10–29	$f'_{cl(dom)} = \dfrac{f_{cl(dom)}}{\sqrt{2^{1/n} - 1}}$	Overall dominant lower critical frequency for case of equal dominant critical frequencies

10–30	$f'_{cu(dom)} = f_{cu(dom)}\sqrt{2^{1/n} - 1}$	Overall dominant upper critical frequency for case of equal dominant critical frequencies

Measurement Techniques

10–31	$f_{cu} = \dfrac{0.35}{t_r}$	Upper critical frequency
10–32	$f_{cl} = \dfrac{0.35}{t_f}$	Lower critical frequency

TRUE/FALSE QUIZ

Answers are at the end of the chapter.

1. Coupling capacitors in an amplifier determine the low-frequency response.
2. Bypass capacitors in an amplifier determine the high-frequency response.
3. Internal transistor capacitance has no effect on an amplifier's frequency response.
4. Miller's theorem states that both gain and internal capacitances influence high-frequency response.
5. The midrange gain is between the upper and lower critical frequencies.
6. The critical frequency is where the gain is 6 dB less than the midrange gain.
7. dBm is a unit for measuring power levels.
8. A ten-times change in frequency is called a decade.
9. An octave corresponds to a doubling or halving of the frequency.
10. The input and output *RC* circuits have no effect on the frequency response.
11. A Bode plot shows the voltage gain versus frequency on a logarithmic scale.
12. Phase shift is part of an amplifier's frequency response.

CIRCUIT-ACTION QUIZ

Answers are at the end of the chapter.

1. If the value of R_1 in Figure 10–8 is increased, the signal voltage at the base will
   (a) increase (b) decrease (c) not change
2. If the value of C_1 in Figure 10–27 is decreased, the critical frequency associated with the input circuit will
   (a) increase (b) decrease (c) not change
3. If the value of R_L in Figure 10–27 is increased, the voltage gain will
   (a) increase (b) decrease (c) not change
4. If the value of R_C in Figure 10–27 is decreased, the voltage gain will
   (a) increase (b) decrease (c) not change
5. If V_{CC} in Figure 10–34 is increased, the dc emitter voltage will
   (a) increase (b) decrease (c) not change
6. If the transistor in Figure 10–34 is replaced with one having a higher β_{ac}, the critical frequency will
   (a) increase (b) decrease (c) not change
7. If the transistor in Figure 10–34 is replaced with one having a lower β_{ac}, the midrange voltage gain will
   (a) increase (b) decrease (c) not change
8. If the value of R_D in Figure 10–42 is increased, the voltage gain will
   (a) increase (b) decrease (c) not change
9. If the value of R_L in Figure 10–42 is increased, the critical frequency will
   (a) increase (b) decrease (c) not change
10. If the FET in Figure 10–42 is replaced with one having a higher g_m, the critical frequency will
    (a) increase (b) decrease (c) not change

SELF-TEST

Answers are at the end of the chapter.

Section 10–1

1. The low-frequency response of an amplifier is determined in part by
(a) the voltage gain **(b)** the type of transistor
(c) the supply voltage **(d)** the coupling capacitors

2. The high-frequency response of an amplifier is determined in part by
(a) the gain-bandwidth product **(b)** the bypass capacitor
(c) the internal transistor capacitances **(d)** the roll-off

3. The Miller input capacitance of an amplifier is dependent, in part, on
(a) the input coupling capacitor **(b)** the voltage gain
(c) the bypass capacitor **(d)** none of these

Section 10–2

4. The decibel is used to express
(a) power gain **(b)** voltage gain **(c)** attenuation **(d)** all of these

5. When the voltage gain is 70.7% of its midrange value, it is said to be
(a) attenuated **(b)** down 6 dB **(c)** down 3 dB **(d)** down 1 dB

6. In an amplifier, the gain that occurs between the lower and upper critical frequencies is called the
(a) critical gain **(b)** midrange gain **(c)** bandwidth gain **(d)** decibel gain

7. A certain amplifier has a voltage gain of 100 at midrange. If the gain decreases by 6 dB, it is equal to
(a) 50 **(b)** 70.7 **(c)** 0 **(d)** 20

Section 10–3

8. The gain of a certain amplifier decreases by 6 dB when the frequency is reduced from 1 kHz to 10 Hz. The roll-off is
(a) −3 dB/decade **(b)** −6 dB/decade **(c)** −3 dB/octave **(d)** −6 dB/octave

9. The gain of a particular amplifier at a given frequency decreases by 6 dB when the frequency is doubled. The roll-off is
(a) −12 dB/decade **(b)** −20 dB/decade
(c) −6 dB/octave **(d)** answers (b) and (c)

10. The lower critical frequency of a direct-coupled amplifier with no bypass capacitor is
(a) variable **(b)** 0 Hz **(c)** dependent on the bias **(d)** none of these

Section 10–4

11. At the upper critical frequency, the peak output voltage of a certain amplifier is 10 V. The peak voltage in the midrange of the amplifier is
(a) 7.07 V **(b)** 6.37 V **(c)** 14.14 V **(d)** 10 V

12. The high-frequency response of an amplifier is determined by the
(a) coupling capacitors **(b)** bias circuit
(c) transistor capacitances **(d)** all of these

13. The Miller input and output capacitances for a BJT inverting amplifier depend on
(a) C_{bc} **(b)** β_{ac} **(c)** A_v **(d)** answers (a) and (c)

Section 10–5

14. The bandwidth of an amplifier is determined by
(a) the midrange gain **(b)** the critical frequencies
(c) the roll-off rate **(d)** the input capacitance

15. An amplifier has the following critical frequencies: 1.2 kHz, 950 Hz, 8 kHz, and 8.5 kHz. The bandwidth is
(a) 7550 Hz **(b)** 7300 Hz **(c)** 6800 Hz **(d)** 7050 Hz

16. Ideally, the midrange gain of an amplifier
(a) increases with frequency
(b) decreases with frequency
(c) remains constant with frequency
(d) depends on the coupling capacitors

17. The frequency at which an amplifier's gain is 1 is called the
(a) unity-gain frequency **(b)** midrange frequency
(c) corner frequency **(d)** break frequency

18. When the voltage gain of an amplifier is increased, the bandwidth
(a) is not affected **(b)** increases **(c)** decreases **(d)** becomes distorted

19. If the f_T of the transistor used in a certain amplifier is 75 MHz and the bandwidth is 10 MHz, the voltage gain must be
(a) 750 **(b)** 7.5 **(c)** 10 **(d)** 1

20. In the midrange of an amplifier's bandwidth, the peak output voltage is 6 V. At the lower critical frequency, the peak output voltage is
(a) 3 V **(b)** 3.82 V **(c)** 8.48 V **(d)** 4.24 V

Section 10–6

21. The dominant lower critical frequency of a multistage amplifier is the
(a) lowest f_{cl} **(b)** highest f_{cl} **(c)** average of all the f_{cl}'s **(d)** none of these

22. When the critical frequencies of all of the stages are the same, the dominant critical frequency is
(a) higher than any individual f_{cl} **(b)** lower than any individual f_{cl}
(c) equal to the individual f_{cl}'s **(d)** the sum of all individual f_{cl}'s

Section 10–7

23. In the step response of a noninverting amplifier, a longer rise time means
(a) a narrower bandwidth **(b)** a lower f_{cl}
(c) a higher f_{cu} **(d)** answers (a) and (b)

PROBLEMS

Answers to all odd-numbered problems are at the end of the book.

BASIC PROBLEMS

Section 10–1 **Basic Concepts**

1. In a capacitively coupled amplifier, the input coupling capacitor and the output coupling capacitor form two of the circuits (along with the respective resistances) that determine the low-frequency response. Assuming that the input and output impedances are the same and neglecting the bypass circuit, which circuit will first cause the gain to drop from its midrange value as the frequency is lowered?

2. Explain why the coupling capacitors do not have a significant effect on gain at sufficiently high-signal frequencies.

3. List the capacitances that affect high-frequency gain in both BJT and FET amplifiers.

4. In the amplifier of Figure 10–56, list the capacitances that affect the low-frequency response of the amplifier and those that affect the high-frequency response.

▶ **FIGURE 10–56**

Multisim file circuits are identified with a CD logo and are in the Problems folder on your CD-ROM. Filenames correspond to figure numbers (e.g., F10-56).

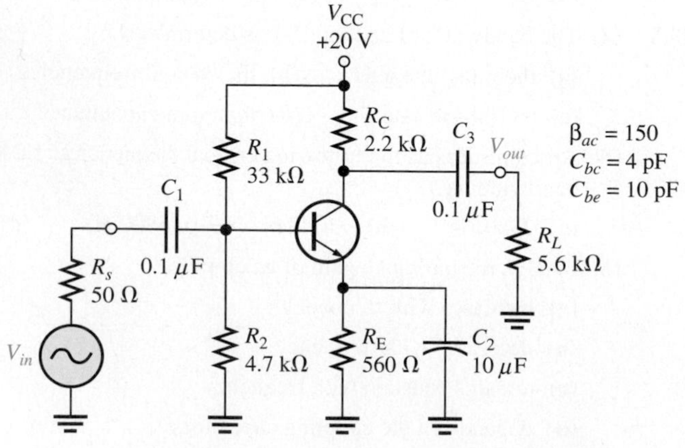

5. Determine the Miller input capacitance in Figure 10–56.
6. Determine the Miller output capacitance in Figure 10–56.
7. Determine the Miller input and output capacitances for the amplifier in Figure 10–57.

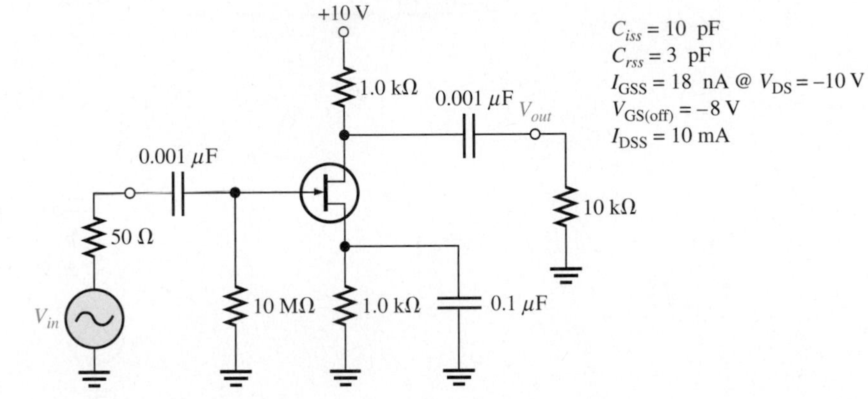

▲ FIGURE 10–57

Section 10–2 The Decibel

8. A certain amplifier exhibits an output power of 5 W with an input power of 0.5 W. What is the power gain in dB?
9. If the output voltage of an amplifier is 1.2 V rms and its voltage gain is 50, what is the rms input voltage? What is the gain in dB?
10. The midrange voltage gain of a certain amplifier is 65. At a certain frequency beyond midrange, the gain drops to 25. What is the gain reduction in dB?
11. What are the dBm values corresponding to the following power values?
    (a) 2 mW **(b)** 1 mW **(c)** 4 mW **(d)** 0.25 mW
12. Express the midrange voltage gain of the amplifier in Figure 10–56 in decibels. Also express the voltage gain in dB for the critical frequencies.

Section 10–3 Low-Frequency Amplifier Response

13. Determine the critical frequencies of each *RC* circuit in Figure 10–58.

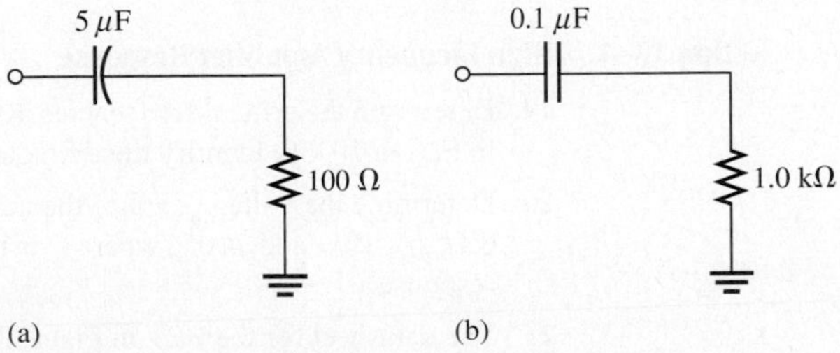

▲ FIGURE 10–58

14. Determine the critical frequencies associated with the low-frequency response of the BJT amplifier in Figure 10–59. Which is the dominant critical frequency? Sketch the Bode plot.

15. Determine the voltage gain of the amplifier in Figure 10–59 at one-tenth of the dominant critical frequency, at the dominant critical frequency, and at ten times the dominant critical frequency for the low-frequency response.

▶ FIGURE 10–59

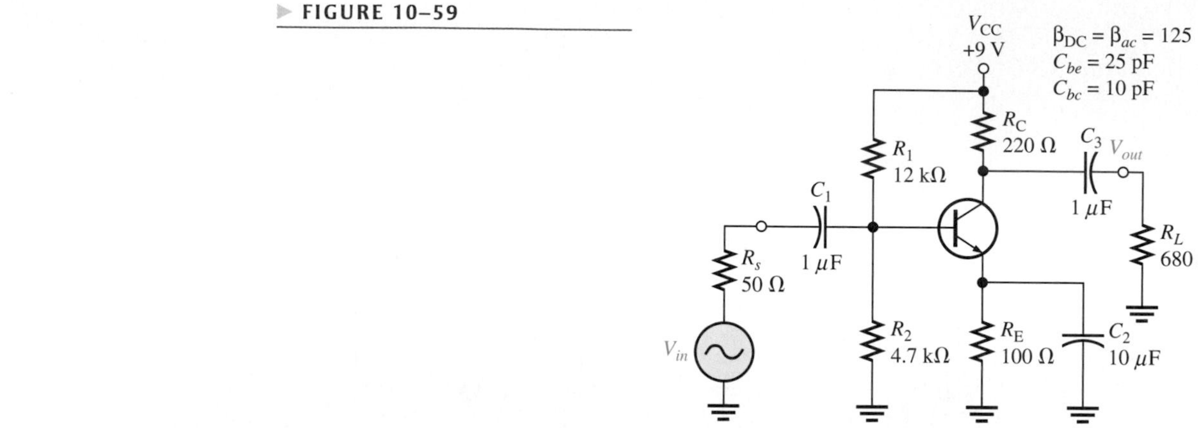

16. Determine the phase shift at each of the frequencies used in Problem 15.

17. Determine the critical frequencies associated with the low-frequency response of the FET amplifier in Figure 10–60. Indicate the dominant critical frequency and draw the Bode plot.

18. Find the voltage gain of the amplifier in Figure 10–60 at the following frequencies: f_c, $0.1f_c$, and $10f_c$, where f_c is the dominant critical frequency.

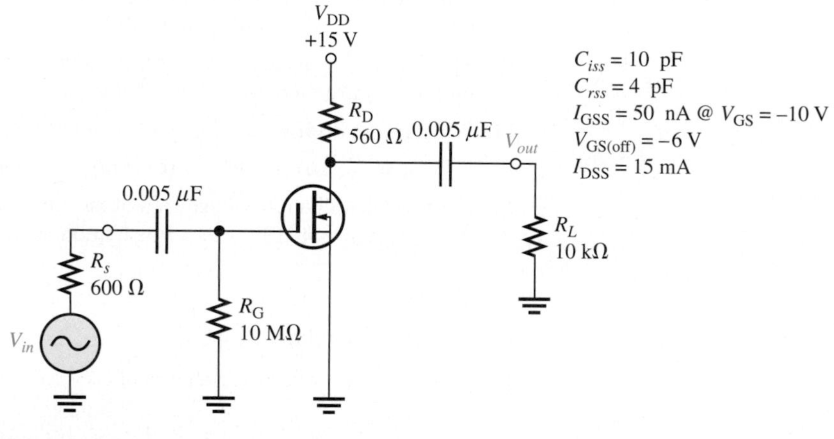

▲ FIGURE 10–60

Section 10–4 High-Frequency Amplifier Response

19. Determine the critical frequencies associated with the high-frequency response of the amplifier in Figure 10–59. Identify the dominant critical frequency and sketch the Bode plot.

20. Determine the voltage gain of the amplifier in Figure 10–59 at the following frequencies: $0.1f_c$, f_c, $10f_c$, and $100f_c$, where f_c is the dominant critical frequency in the high-frequency response.

21. The datasheet for the FET in Figure 10–60 gives $C_{rss} = 4$ pF and $C_{iss} = 10$ pF. Determine the critical frequencies associated with the high-frequency response of the amplifier, and indicate the dominant frequency.

22. Determine the voltage gain in dB and the phase shift at each of the following multiples of the dominant critical frequency in Figure 10–60 for the high-frequency response: $0.1f_c$, f_c, $10f_c$, and $100f_c$.

Section 10–5 Total Amplifier Frequency Response

23. A particular amplifier has the following low critical frequencies: 25 Hz, 42 Hz, and 136 Hz. It also has high critical frequencies of 8 kHz and 20 kHz. Determine the upper and lower critical frequencies.

24. Determine the bandwidth of the amplifier in Figure 10–59.

25. $f_T = 200$ MHz is taken from the datasheet of a transistor used in a certain amplifier. If the midrange gain is determined to be 38 and if f_{cl} is low enough to be neglected compared to f_{cu}, what bandwidth would expect? What value of f_{cu} would you expect?

26. If the midrange gain of a given amplifier is 50 dB and therefore 47 dB at f_{cu}, how much gain is there at $2f_{cu}$? At $4f_{cu}$? At $10f_{cu}$?

Section 10–6 Frequency Response of Multistage Amplifiers

27. In a certain two-stage amplifier, the first stage has critical frequencies of 230 Hz and 1.2 MHz. The second stage has critical frequencies of 195 Hz and 2 MHz. What are the dominant critical frequencies?

28. What is the bandwidth of the two-stage amplifier in Problem 27?

29. Determine the bandwidth of a two-stage amplifier in which each stage has a lower critical frequency of 400 Hz and an upper critical frequency of 800 kHz.

30. What is the dominant lower critical frequency of a three-stage amplifier in which $f_{cl} = 50$ Hz for each stage.

31. In a certain two-stage amplifier, the lower critical frequencies are $f_{cl(1)} = 125$ Hz and $f_{cl(2)} = 125$ Hz, and the upper critical frequencies are $f_{cu(1)} = 3$ MHz and $f_{cu(2)} = 2.5$ MHz. Determine the bandwidth.

Section 10–7 Frequency Response Measurements

32. In a step response test of a certain amplifier, $t_r = 20$ ns and $t_f = 1$ ms. Determine f_{cl} and f_{cu}.

33. Suppose you are measuring the frequency response of an amplifier with a signal source and an oscilloscope. Assume that the signal level and frequency are set such that the oscilloscope indicates an output voltage level of 5 V rms in the midrange of the amplifier's response. If you wish to determine the upper critical frequency, indicate what you would do and what scope indication you would look for.

34. Determine the approximate bandwidth of an amplifier from the indicated results of the step-response test in Figure 10–61.

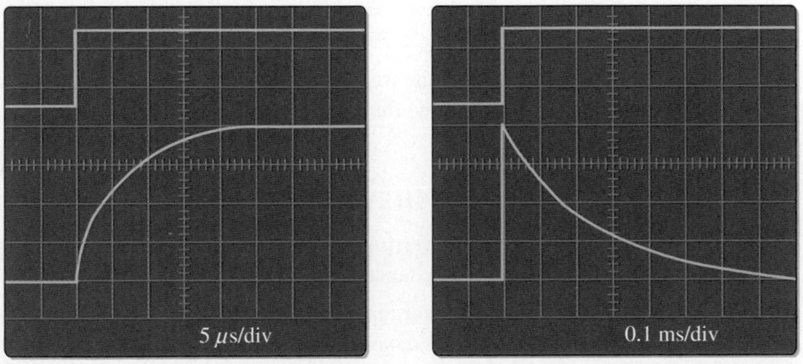

▲ **FIGURE 10–61**

APPLICATION ACTIVITY PROBLEMS

35. Determine the dominant lower critical frequency for the amplifier in Figure 10–52 if the coupling capacitors are changed to 1 μF. Assume $R_L = 29\text{ k}\Omega$ and $\beta_{ac} = 200$.

36. Does the change in Problem 35 significantly affect the overall bandwidth?

Electrical Characteristics T_A = 25°C unless otherwise noted

Symbol	Parameter	Test Conditions	Min	Max	Units
OFF CHARACTERISTICS					
$V_{(BR)CEO}$	Collector-Emitter Breakdown Voltage	I_C = 1.0 mA, I_B = 0	40		V
$V_{(BR)CBO}$	Collector-Base Breakdown Voltage	I_C = 10 μA, I_E = 0	60		V
$V_{(BR)EBO}$	Emitter-Base Breakdown Voltage	I_E = 10 μA, I_C = 0	6.0		V
I_{BL}	Base Cutoff Current	V_{CE} = 30 V, V_{EB} = 3V		50	nA
I_{CEX}	Collector Cutoff Current	V_{CE} = 30 V, V_{EB} = 3V		50	nA
ON CHARACTERISTICS*					
h_{FE}	DC Current Gain	I_C = 0.1 mA, V_{CE} = 1.0 V I_C = 1.0 mA, V_{CE} = 1.0 V I_C = 10 mA, V_{CE} = 1.0 V I_C = 50 mA, V_{CE} = 1.0 V I_C = 100 mA, V_{CE} = 1.0 V	40 70 100 60 30	 300 	
$V_{CE(sat)}$	Collector-Emitter Saturation Voltage	I_C = 10 mA, I_B = 1.0 mA I_C = 50 mA, I_B = 5.0 mA		0.2 0.3	V V
$V_{BE(sat)}$	Base-Emitter Saturation Voltage	I_C = 10 mA, I_B = 1.0 mA I_C = 50 mA, I_B = 5.0 mA	0.65	0.85 0.95	V V
SMALL SIGNAL CHARACTERISTICS					
f_T	Current Gain - Bandwidth Product	I_C = 10 mA, V_{CE} = 20 V, f = 100 MHz	300		MHz
C_{obo}	Output Capacitance	V_{CB} = 5.0 V, I_E = 0, f = 1.0 MHz		4.0	pF
C_{ibo}	Input Capacitance	V_{EB} = 0.5 V, I_C = 0, f = 1.0 MHz		8.0	pF
NF	Noise Figure	I_C = 100 μA, V_{CE} = 5.0 V, R_S =1.0kΩ,f=10 Hz to 15.7kHz		5.0	dB
SWITCHING CHARACTERISTICS					
t_d	Delay Time	V_{CC} = 3.0 V, V_{BE} = 0.5 V,		35	ns
t_r	Rise Time	I_C = 10 mA, I_{B1} = 1.0 mA		35	ns
t_s	Storage Time	V_{CC} = 3.0 V, I_C = 10mA		200	ns
t_f	Fall Time	I_{B1} = I_{B2} = 1.0 mA		50	ns

▲ FIGURE 10–62

Partial datasheet for the 2N3904. ©2003 Fairchild Semiconductor Corporation. Used by permission.

37. How does a change from 29 kΩ to 100 kΩ in load resistance on the final output of the amplifier in Figure 10–52 affect the dominant lower critical frequency?

38. If the transistors in the modified preamp in the Application Activity have a β_{ac} of 300, determine the effect on the dominant lower critical frequency.

DATASHEET PROBLEMS

39. Referring to the partial datasheet for a 2N3904 in Figure 10–62, determine the total input capacitance for an amplifier if the voltage gain is 25.

40. A certain amplifier uses a 2N3904 and has a midrange voltage gain of 50. Referring to the partial datasheet in Figure 10–62, determine its minimum bandwidth.

41. The datasheet for a 2N4351 MOSFET specifies the maximum values of internal capacitances as follows: C_{iss} = 5 pF, C_{rss} = 1.3 pF, and $C_{d(sub)}$ = 5 pF. Determine C_{gd}, C_{gs}, and C_{ds}.

ADVANCED PROBLEMS

42. Two single-stage capacitively coupled amplifiers like the one in Figure 10–56 are connected as a two-stage amplifier (R_L is removed from the first stage). Determine whether or not this configuration will operate as a linear amplifier with an input voltage of 10 mV rms. If not, modify the design to achieve maximum gain without distortion.

43. Two stages of the amplifier in Figure 10–60 are connected in cascade. Determine the overall bandwidth.

44. Redesign the amplifier in Figure 10–52 for an adjustable voltage gain of 50 to 500 and a lower critical frequency of 1 kHz.

MULTISIM TROUBLESHOOTING PROBLEMS

These file circuits are in the Troubleshooting Problems folder on your CD-ROM.

45. Open file TSP10-45 and determine the fault.

46. Open file TSP10-46 and determine the fault.

47. Open file TSP10-47 and determine the fault.

48. Open file TSP10-48 and determine the fault.

ANSWERS

SECTION CHECKUPS

Section 10–1 Basic Concepts

(a) The coupling and bypass capacitors affect the low-frequency gain.

(b) The high-frequency gain is limited by internal capacitances.

(c) Coupling and bypass capacitors can be neglected at frequencies for which their reactances are negligible.

(d) $C_{in(Miller)} = (5\text{ pF})(51) = 255\text{ pF}$

(e) $C_{out(Miller)} = (3\text{ pF})(1.04) = 3.12\text{ pF}$

Section 10–2 The Decibel

(a) +12 dB corresponds to a voltage gain of approximately 4.

(b) $A_p = 10 \log (25) = 13.98\text{ dB}$

(c) 0 dBm corresponds to 1 mW.

Section 10–3 Low-Frequency Amplifier Response

(a) $f_{cl2} = 167\text{ Hz}$ is dominant.

(b) $A_{v(\text{dB})} = 50\text{ dB} - 3\text{ dB} = 47\text{ dB}$

(c) -20 dB attenuation at one decade below f_{cl}.

(d) $\theta = \tan^{-1}(0.5) = 26.6°$

(e) $f_{cl} = 1/(2\pi(6500\ \Omega)(0.0022\ \mu\text{F})) = 11.1\text{ kHz}$

Section 10–4 High-Frequency Amplifier Response

(a) The internal transistor capacitances determine the high-frequency response.

(b) $C_{in(tot)} = C_{in(Miller)} + C_{ce} = (4\text{ pF})(81) + 8\text{ pF} = 342\text{ pF}$

(c) The input *RC* circuit dominates.

(d) C_{iss} and C_{rss} are usually specified on a FET datasheet.

(e) $C_{in(tot)} = (3\text{ pF})(26) + 4\text{ pF} = 82\text{ pF}$

Section 10–5 Total Amplifier Frequency Response

(a) The gain is 1 at f_T.

(b) $BW = 25\text{ kHz} - 100\text{ Hz} = 24.9\text{ kHz}$

(c) $A_v = 130\text{ MHz}/50\text{ MHz} = 2.6$

Section 10–6 Frequency Response of Multistage Amplifiers

(a) $f'_{cl(dom)} = 1\text{ kHz}$

(b) $f'_{cu(dom)} = 49\text{ kHz}$

(c) *BW* decreases.

Section 10–7 Frequency Response Measurements

(a) $f_{cl} = 125$ Hz; $f_{cu} = 500$ kHz

(b) Rise time is between the 10% and 90% points and fall time is between the 90% and 10% points.

(c) $t_r = 150$ ns

(d) $t_f = 2.8$ ms

(e) Since $f_{cu} >> f_{cl}$, $BW \cong f_{cu} = 2.5$ MHz.

RELATED PROBLEMS FOR EXAMPLES

10–1 **(a)** 61.6 dB **(b)** 17 dB **(c)** 102 dB

10–2 **(a)** 50 V **(b)** 6.25 V **(c)** 1.56 V

10–3 0.22 μF

10–4 212 @ 400 Hz; 30 @ 40 Hz; 3 @ 4 Hz

10–5 It will increase the gain and reduce the lower critical frequency.

10–6 C_2 "sees" a smaller resistance.

10–7 f_{cl} changes from 16.2 Hz to 16.1 Hz.

10–8 Ideally, the low-frequency response is not affected because an infinite load makes f_c of the output stage even lower, so the input stage determines the lower cutoff frequency of the amplifier.

10–9 The resistance of the input will be higher, so the critical frequency is lower.

10–10 Change C_1 to 0.68 μF.

10–11 320 Ω in series with 215 pF, $f_c = 2.31$ MHz

10–12 28.7 MHz

10–13 1 pF

10–14 f_c decreases to 83.8 MHz.

10–15 48.2 MHz

10–16 *BW* decreases; *BW* increases

10–17 20 MHz

10–18 980 Hz

10–19 39.8 kHz

10–20 26 dB

TRUE/FALSE QUIZ

1. T **2.** F **3.** F **4.** T **5.** T **6.** F
7. T **8.** T **9.** T **10.** F **11.** T **12.** T

CIRCUIT-ACTION QUIZ

1. (a) **2.** (a) **3.** (a) **4.** (b) **5.** (a) **6.** (b) **7.** (c) **8.** (a) **9.** (b) **10.** (b)

SELF-TEST

1. (d) **2.** (c) **3.** (b) **4.** (d) **5.** (c) **6.** (b) **7.** (a) **8.** (a)
9. (d) **10.** (b) **11.** (c) **12.** (c) **13.** (d) **14.** (b) **15.** (c) **16.** (c)
17. (a) **18.** (c) **19.** (b) **20.** (d) **21.** (b) **22.** (a) **23.** (a)

THYRISTORS

11

CHAPTER OUTLINE

CHAPTER OBJECTIVES

- Describe the basic structure and operation of a 4-layer diode
- Describe the basic structure and operation of an SCR
- Discuss several SCR applications
- Describe the basic structure and operation of diacs and triacs
- Describe the basic operation of an SCS
- Describe the basic structure and operation of a UJT
- Describe the structure and operation of a PUT

KEY TERMS

- 4-layer diode
- Thyristor
- Forward-breakover voltage ($V_{BR(F)}$)
- Holding current (I_H)
- SCR
- LASCR
- Diac
- Triac
- SCS
- UJT
- Standoff ratio
- PUT

APPLICATION ACTIVITY PREVIEW

The Application Activity in this chapter is a motor speed-control system for a production line conveyor. The system senses the number of parts passing a point in a specified period of time and adjusts the rate of movement of the conveyor belt to achieve a desired rate of flow of the parts. The focus is on the conveyor motor speed-control circuit.

VISIT THE COMPANION WEBSITE

Study aids for this chapter are available at http://www.prenhall.com/floyd

INTRODUCTION

In this chapter, several types of semiconductor devices are introduced. A family of devices known as thyristors are constructed of four semiconductor layers (*pnpn*). Thyristors include the 4-layer diode, the silicon-controlled rectifier (SCR), the diac, the triac, and the silicon-controlled switch (SCS). These types of thyristors share certain common characteristics in addition to their four-layer construction. They act as open circuits capable of withstanding a certain rated voltage until they are triggered. When triggered, they turn on and become low-resistance current paths and remain so, even after the trigger is removed, until the current is reduced to a certain level or until they are triggered off, depending on the type of device. Thyristors can be used to control the amount of ac power to a load and are used in lamp dimmers, motor speed controls, ignition systems, and charging circuits, to name a few.

Other devices described in this chapter include the unijunction transistor (UJT) and the programmable unijunction transistor (PUT). UJTs and PUTs are used as trigger devices for thyristors and also in oscillators and timing circuits.

11–1 THE FOUR-LAYER DIODE

The basic thyristor is a 4-layer device with two terminals, the anode and the cathode. It is constructed of four semiconductor layers that form a *pnpn* structure. The device acts as a switch and remains off until the forward voltage reaches a certain value; then it turns on and conducts. Conduction continues until the current is reduced below a specified value. Although the 4-layer diode is seldom used in new designs, the principles form the basis of other thyristors that you will study.

After completing this section, you should be able to

- **Describe the basic structure and operation of a 4-layer diode**
  - Identify the 4-layer diode symbol
  - Define *forward-breakover voltage*
  - Define *holding current*
  - Define *switching current*
  - Discuss an application

The **4-layer diode** (also known as Shockley diode and SUS) is a type of **thyristor**, which is a class of devices constructed of four semiconductor layers. The basic construction of a 4-layer diode and its schematic symbol are shown in Figure 11–1.

The *pnpn* structure can be represented by an equivalent circuit consisting of a *pnp* transistor and an *npn* transistor, as shown in Figure 11–2(a). The upper *pnp* layers form Q_1 and the lower *npn* layers form Q_2, with the two middle layers shared by both equivalent transistors. Notice that the base-emitter junction of Q_1 corresponds to *pn* junction 1 in Figure 11–1, the base-emitter junction of Q_2 corresponds to *pn* junction 3, and the base-collector junctions of both Q_1 and Q_2 correspond to *pn* junction 2.

When a positive bias voltage is applied to the anode with respect to the cathode, as shown in Figure 11–2(b), the base-emitter junctions of Q_1 and Q_2 (*pn* junctions 1 and 3 in Figure 11–1(a)) are forward-biased, and the common base-collector junction (*pn* junction 2 in Figure 11–1(a)) is reverse-biased.

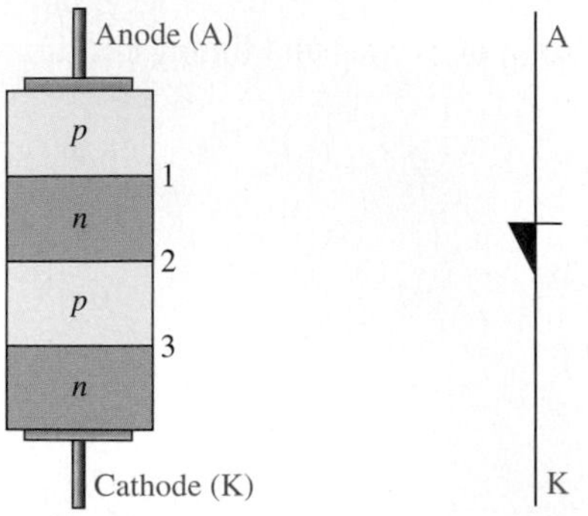

(a) Basic construction (b) Schematic symbol

▲ FIGURE 11–1

The 4-layer diode.

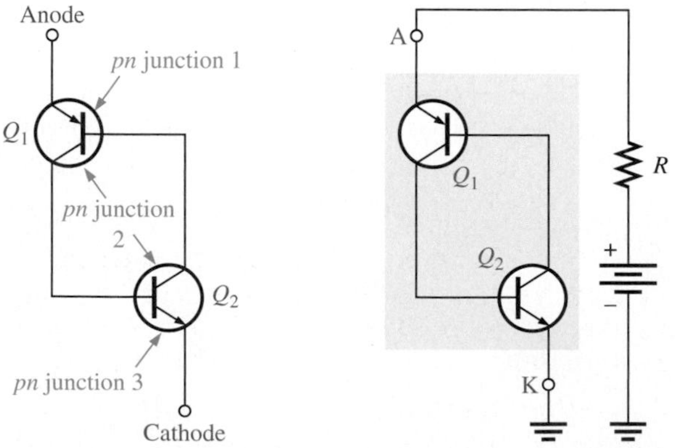

(a) (b)

▲ FIGURE 11–2

A 4-layer diode equivalent circuit.

The currents in a 4-layer diode are shown in the equivalent circuit in Figure 11–3. At low-bias levels, there is very little anode current, and thus it is in the *off* state or forward-blocking region.

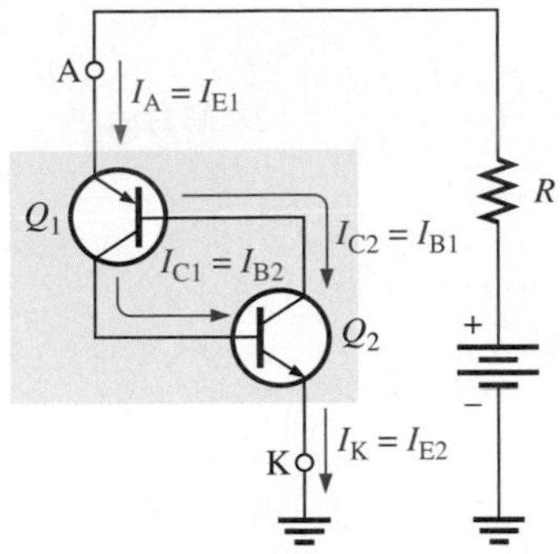

FIGURE 11–3

Currents in a 4-layer diode equivalent circuit.

HISTORY NOTE

The four-layer diode (also called a Shockley diode) was invented by William Shockley while he was at Bell Labs. Shockley believed the four-layer diode would revolutionize telephone switching circuits because it could replace the mechanical switches used by the telephone equipment of the time. In the mid-1950s, Shockley founded Shockley Semiconductor Labs with the intent of constructing silicon transistors but soon changed the emphasis of his company to the four-layer diode. It was difficult to manufacture with the technology of the time, and Shockley's company never managed to turn a profit. The four-layer diode eventually evolved into the SCR, which is essentially a four-layer diode with an added control gate.

Forward-Breakover Voltage The operation of the 4-layer diode may seem unusual because when it is forward-biased, it can act essentially as an open switch. There is a region of forward bias, called the *forward-blocking region,* in which the device has a very high forward resistance (ideally an open) and is in the *off* state. The forward-blocking region exists from $V_{AK} = 0$ V up to a value of V_{AK} called the **forward-breakover voltage,** $V_{BR(F)}$. This is indicated on the 4-layer diode characteristic curve in Figure 11–4.

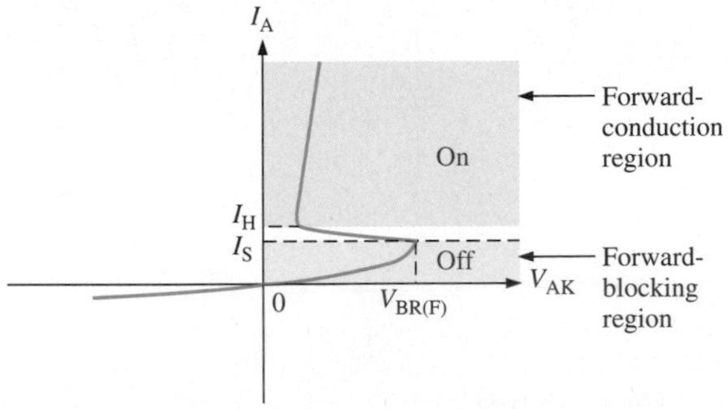

FIGURE 11–4

A 4-layer diode characteristic curve.

As V_{AK} is increased from 0, the anode current, I_A, gradually increases, as shown on the graph. As I_A increases, a point is reached where $I_A = I_S$, the switching current. At this point, $V_{AK} = V_{BR(F)}$, and the internal transistor structures become saturated. When this happens, the forward voltage drop, V_{AK}, suddenly decreases to a low value, and the 4-layer diode enters the *forward-conduction region* as indicated in Figure 11–4. Now, the device is in the *on* state and acts as a closed switch. When the anode current drops back below the holding value, I_H, the device turns off.

Holding Current Once the 4-layer diode is conducting (in the *on* state), it will continue to conduct until the anode current is reduced below a specified level, called the **holding current**, I_H. This parameter is also indicated on the characteristic curve in Figure 11–4. When I_A falls below I_H, the device rapidly switches back to the *off* state and enters the forward-blocking region.

Switching Current The value of the anode current at the point where the device switches from the forward-blocking region (off) to the forward-conduction region (on) is called the **switching current, I_S**. This value of current is always less than the holding current, I_H.

EXAMPLE 11–1

A certain 4-layer diode is biased in the forward-blocking region with an anode-to-cathode voltage of 20 V. Under this bias condition, the anode current is 1 μA. Determine the resistance of the diode in the forward-blocking region.

Solution The resistance is

$$R_{AK} = \frac{V_{AK}}{I_A} = \frac{20\text{ V}}{1\ \mu\text{A}} = \mathbf{20\ M\Omega}$$

*Related Problem** If the anode current is 2 μA and $V_{AK} = 20$ V, what is the 4-layer diode's resistance in the forward-blocking region?

*Answers are at the end of the chapter.

EXAMPLE 11–2

Determine the value of anode current in Figure 11–5 when the device is on. $V_{BR(F)} =$ 10 V. Assume the forward voltage drop is 0.9 V.

FIGURE 11–5

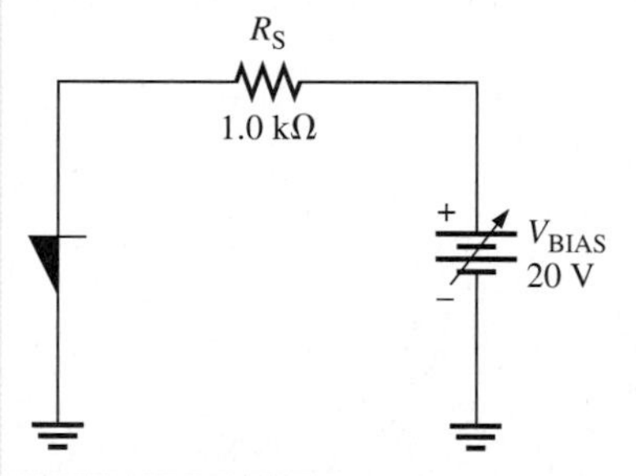

Solution The voltage at the anode, V_A, is 0.9. The voltage across R_S is

$$V_{R_S} = V_{BIAS} - V_A = 20\text{ V} - 0.9\text{ V} = 19.1\text{ V}$$

The anode current is

$$I_A = \frac{V_{R_S}}{R_S} = \frac{19.1\text{ V}}{1.0\text{ k}\Omega} = \mathbf{19.1\ mA}$$

Related Problem What is the resistance in the forward-conduction region of the 4-layer diode in Figure 11–5?

An Application

The circuit in Figure 11–6(a) is a **relaxation oscillator.** The operation is as follows. When the switch is closed, the capacitor charges through R until its voltage reaches the forward-breakover voltage of the 4-layer diode. At this point the diode switches into conduction, and the capacitor rapidly discharges through the diode. Discharging continues until the current through the diode falls below the holding value. At this point, the diode switches back

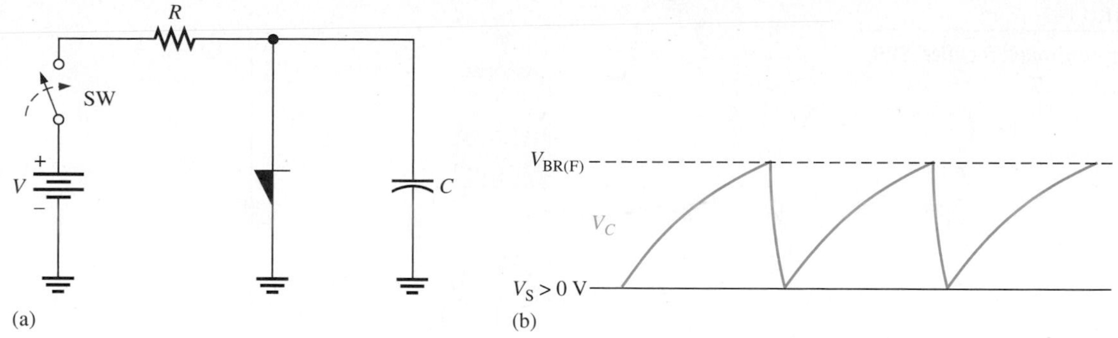

FIGURE 11–6

A 4-layer diode relaxation oscillator.

to the *off* state, and the capacitor begins to charge again. The result of this action is a voltage waveform across C like that shown in Figure 11–6(b).

SECTION 11–1 CHECKUP

Answers are at the end of the chapter.

1. Why is the 4-layer diode classified as a thyristor?
2. What is the forward-blocking region?
3. What happens when the anode-to-cathode voltage exceeds the forward-breakover voltage?
4. Once it is on, how can the 4-layer diode be turned off?

11–2 THE SILICON-CONTROLLED RECTIFIER (SCR)

Like the 4-layer diode, the SCR has two possible states of operation. In the *off* state, it acts ideally as an open circuit between the anode and the cathode; actually, rather than an open, there is a very high resistance. In the *on* state, the SCR acts ideally as a short from the anode to the cathode; actually, there is a small *on* (forward) resistance. The LASCR operates as an SCR except it is triggered by light.

After completing this section, you should be able to

- **Describe the basic structure and operation of an SCR**
  - Identify the schematic symbol and draw an equivalent circuit
  - Explain the SCR characteristic curves and define various SCR parameters
  - Define *forced commutation*
  - Describe the LASCR and its operation

An **SCR** (silicon-controlled rectifier) is a 4-layer *pnpn* device similar to the 4-layer diode except with three terminals: anode, cathode, and gate. The basic structure of an SCR is shown in Figure 11–7(a), and the schematic symbol is shown in Figure 11–7(b). Typical SCR packages are shown in Figure 11–7(c). Other types of thyristors are found in the same or similar packages.

FIGURE 11–7

The silicon-controlled rectifier (SCR).

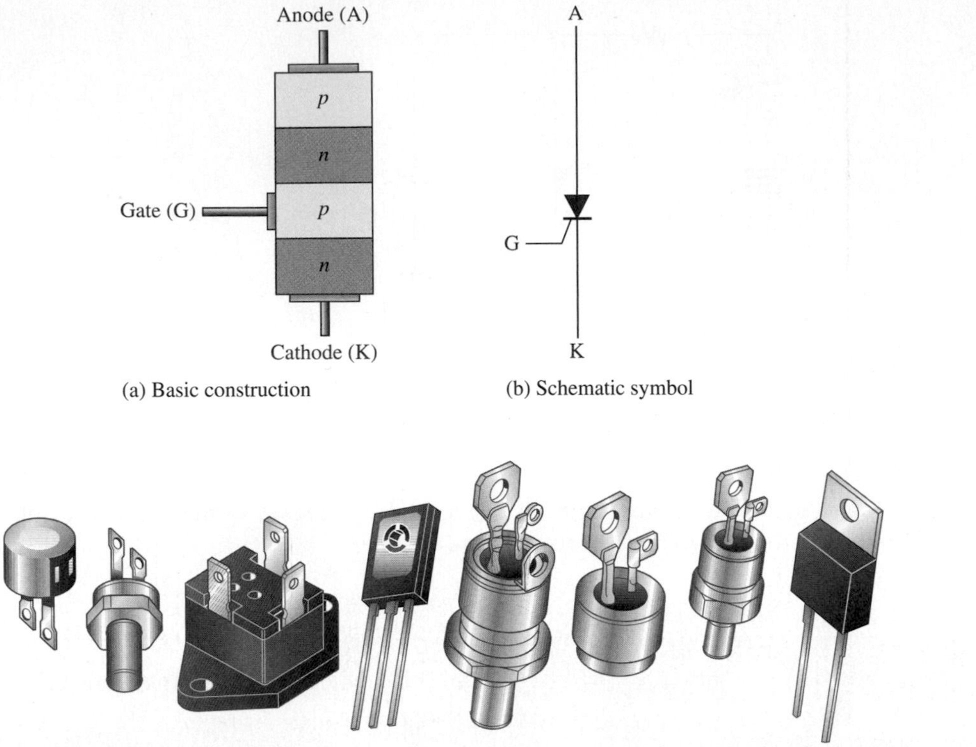

(a) Basic construction

(b) Schematic symbol

(c) Typical packages

SCR Equivalent Circuit

Like the 4-layer diode operation, the SCR operation can best be understood by thinking of its internal *pnpn* structure as a two-transistor arrangement, as shown in Figure 11–8. This structure is like that of the 4-layer diode except for the gate connection. The upper *pnp* layers act as a transistor, Q_1, and the lower *npn* layers act as a transistor, Q_2. Again, notice that the two middle layers are "shared."

FIGURE 11–8

SCR equivalent circuit.

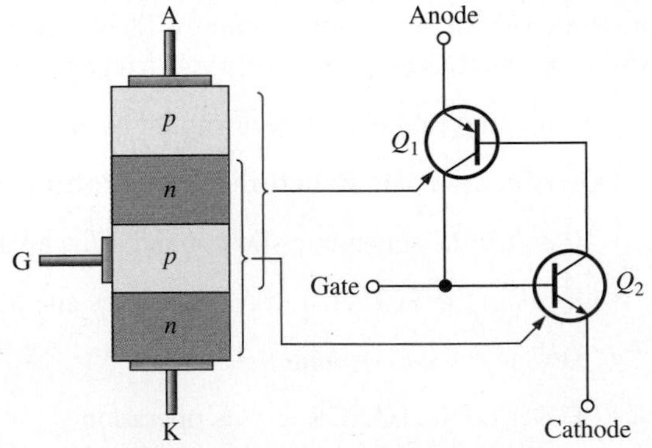

Turning the SCR On

When the gate current, I_G, is zero, as shown in Figure 11–9(a), the device acts as a 4-layer diode in the *off* state. In this state, the very high resistance between the anode and cathode can be approximated by an open switch, as indicated. When a positive pulse of current (**trigger**) is applied to the gate, both transistors turn on (the anode must be more positive than the cathode). This action is shown in Figure 11–9(b). I_{B2} turns on Q_2, providing a path for I_{B1} into the Q_2 collector, thus turning on Q_1. The collector current of Q_1 provides

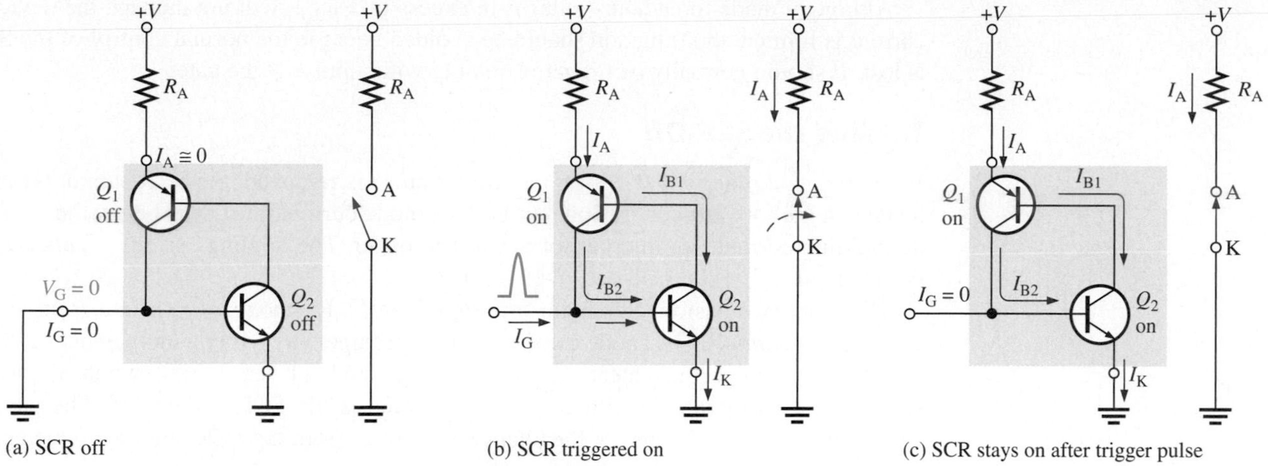

(a) SCR off (b) SCR triggered on (c) SCR stays on after trigger pulse

▲ **FIGURE 11–9**

The SCR turn-on process with the switch equivalents shown.

additional base current for Q_2 so that Q_2 stays in conduction after the trigger pulse is removed from the gate. By this regenerative action, Q_2 sustains the saturated conduction of Q_1 by providing a path for I_{B1}; in turn, Q_1 sustains the saturated conduction of Q_2 by providing I_{B2}. Thus, the device stays on (latches) once it is triggered on, as shown in Figure 11–9(c). In this state, the very low resistance between the anode and cathode can be approximated by a closed switch, as indicated.

Like the 4-layer diode, an SCR can also be turned on without gate triggering by increasing the anode-to-cathode voltage to a value exceeding the forward-breakover voltage $V_{BR(F)}$, as shown on the characteristic curve in Figure 11–10(a). The forward-breakover voltage decreases as I_G is increased above 0 V, as shown by the set of curves in Figure 11–10(b). Eventually, a value of I_G is reached at which the SCR turns on at a very low anode-to-cathode voltage. So, as you can see, the gate current controls the value of forward breakover voltage, $V_{BR(F)}$, required for turn-on.

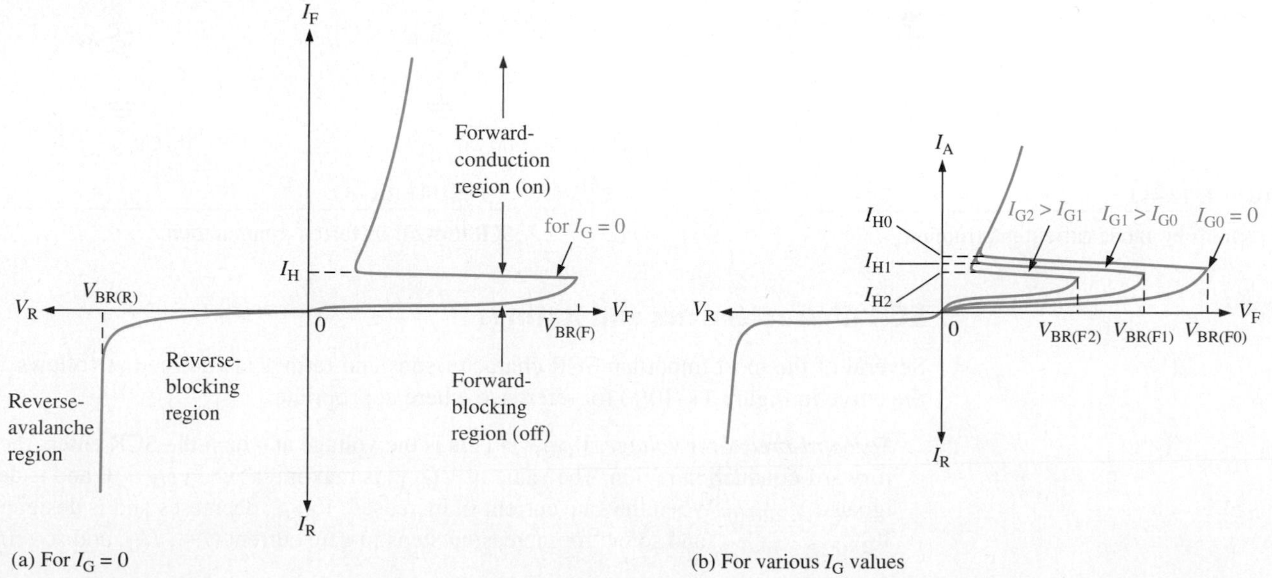

(a) For $I_G = 0$ (b) For various I_G values

▲ **FIGURE 11–10**

SCR characteristic curves.

Although anode-to-cathode voltages in excess of $V_{BR(F)}$ will not damage the device if current is limited, this situation should be avoided because the normal control of the SCR is lost. It should normally be triggered on only with a pulse at the gate.

Turning the SCR Off

When the gate returns to 0 V after the trigger pulse is removed, the SCR cannot turn off; it stays in the forward-conduction region. The anode current must drop below the value of the holding current, I_H, in order for turn-off to occur. The holding current is indicated in Figure 11–10.

There are two basic methods for turning off an SCR: *anode current interruption* and *forced commutation.* The anode current can be interrupted by either a momentary series or parallel switching arrangement, as shown in Figure 11–11. The series switch in part (a) simply reduces the anode current to zero and causes the SCR to turn off. The parallel switch in part (b) routes part of the total current away from the SCR, thereby reducing the anode current to a value less than I_H.

The **forced commutation** method basically requires momentarily forcing current through the SCR in the direction opposite to the forward conduction so that the net forward current is reduced below the holding value. The basic circuit, as shown in Figure 11–12, consists of a switch (normally a transistor switch) and a capacitor. While the SCR is conducting, the switch is open and C_c is charged to the supply voltage through R_c, as shown in part (a). To turn off the SCR, the switch is closed, placing the capacitor across the SCR and forcing current through it opposite to the forward current, as shown in part (b). Typically, turn-off times for SCRs range from a few microseconds up to about 30 μs.

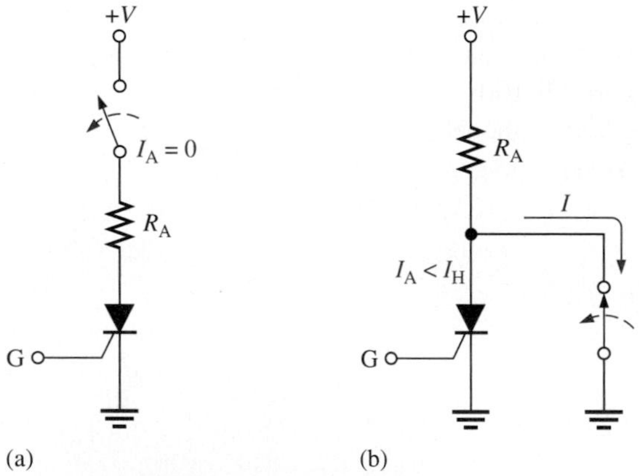

▲ FIGURE 11–11

SCR turn-off by anode current interruption.

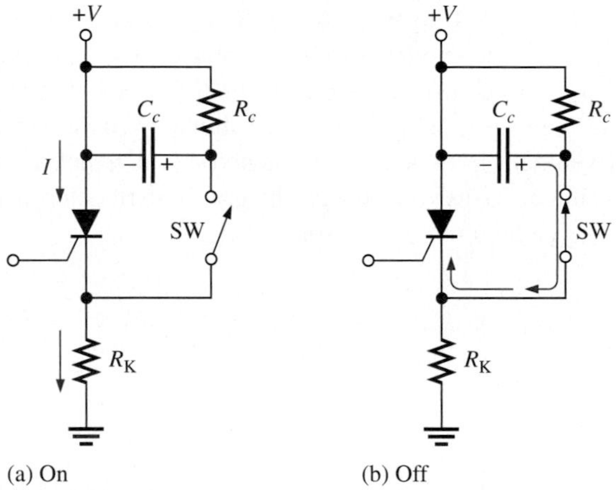

▲ FIGURE 11–12

SCR turn-off by forced commutation.

SCR Characteristics and Ratings

Several of the most important SCR characteristics and ratings are defined as follows. Use the curve in Figure 11–10(a) for reference where appropriate.

Forward-breakover voltage, $V_{BR(F)}$ This is the voltage at which the SCR enters the forward-conduction region. The value of $V_{BR(F)}$ is maximum when $I_G = 0$ and is designated $V_{BR(F0)}$. When the gate current in increased, $V_{BR(F)}$ decreases and is designated $V_{BR(F1)}$, $V_{BR(F2)}$, and so on, for increasing steps in gate current (I_{G1}, I_{G2}, and so on).

Holding current, I_H This is the value of anode current below which the SCR switches from the forward-conduction region to the forward-blocking region. The value increases with decreasing values of I_G and is maximum for $I_G = 0$.

Gate trigger current, I_{GT} This is the value of gate current necessary to switch the SCR from the forward-blocking region to the forward-conduction region under specified conditions.

Average forward current, $I_{F(avg)}$ This is the maximum continuous anode current (dc) that the device can withstand in the conduction state under specified conditions.

Forward-conduction region This region corresponds to the *on* condition of the SCR where there is forward current from anode to cathode through the very low resistance (approximate short) of the SCR.

Forward-blocking and reverse-blocking regions These regions correspond to the *off* condition of the SCR where the forward current from anode to cathode is blocked by the effective open circuit of the SCR.

Reverse-breakdown voltage, $V_{BR(R)}$ This parameter specifies the value of reverse voltage from cathode to anode at which the device breaks into the avalanche region and begins to conduct heavily (the same as in a *pn* junction diode).

The Light-Activated SCR (LASCR)

The light-activated silicon-controlled rectifier (**LASCR**) is a four-layer semiconductor device (thyristor) that operates essentially as does the conventional SCR except that it can also be light-triggered. The LASCR conducts current in one direction when activated by a sufficient amount of light and continues to conduct until the current falls below a specified value. Figure 11–13 shows a LASCR schematic symbol. The LASCR is most sensitive to light when the gate terminal is open. If necessary, a resistor from the gate to the cathode can be used to reduce the sensitivity.

Figure 11–14 shows a LASCR used to energize a latching relay. The input source turns on the lamp; the resulting incident light triggers the LASCR. The anode current energizes the relay and closes the contact. Notice that the input source is electrically isolated from the rest of the circuit.

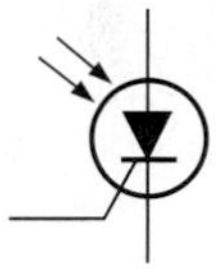

▲ FIGURE 11–13

LASCR symbol.

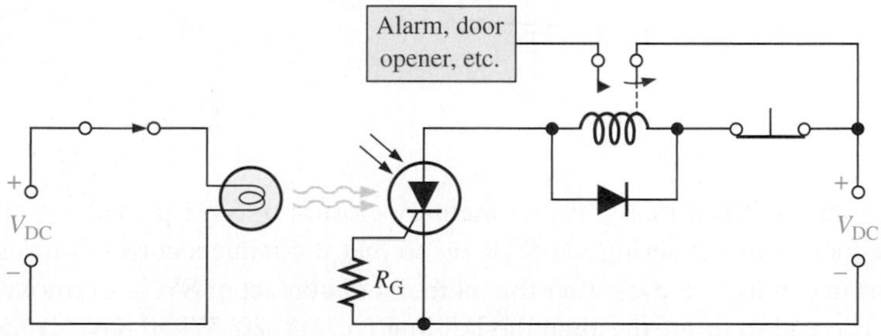

▲ FIGURE 11–14

A LASCR circuit.

SECTION 11–2 CHECKUP

1. What is an SCR?
2. Name the SCR terminals.
3. How can an SCR be turned on (made to conduct)?
4. How can an SCR be turned off?
5. What is required in Figure 11–14 to turn off the LASCR and de-energize the relay?

11–3 SCR APPLICATIONS

The SCR is used in many applications, including motor controls, time-delay circuits, heater controls, phase controls, relay controls, and sawtooth generators.

After completing this section, you should be able to

- **Discuss several SCR applications**
  - Explain how an SCR is used to control current
  - Describe half-wave power control
  - Explain a basic phase-control circuit
  - Discuss the function of an SCR in a lighting system for power interruptions
  - Explain an over-voltage protection or "crowbar" circuit

On-Off Control of Current

Figure 11–15 shows an SCR circuit that permits current to be switched to a load by the momentary closure of switch SW1 and removed from the load by the momentary closure of switch SW2.

FIGURE 11–15

On-Off SCR control circuit.

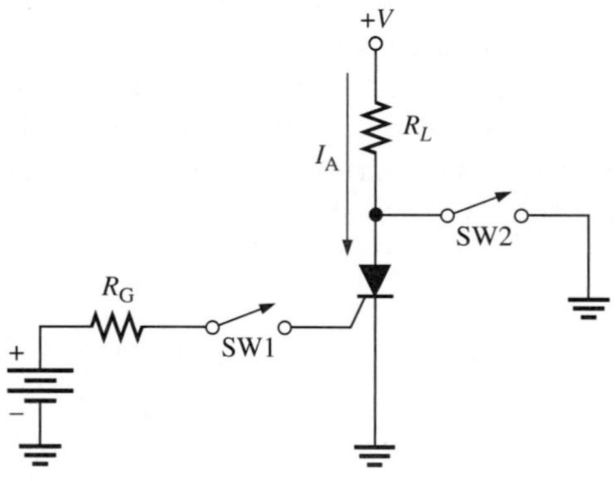

Assuming the SCR is initially off, momentary closure of SW1 provides a pulse of current into the gate, thus triggering the SCR on so that it conducts current through R_L. The SCR remains in conduction even after the momentary contact of SW1 is removed if the anode current is equal to or greater than the holding current, I_H. When SW2 is momentarily closed, current is shunted around the SCR, thus reducing its anode current below the holding value, I_H. This turns the SCR off and reduces the load current to zero.

EXAMPLE 11–3

Determine the gate trigger current and the anode current when the switch, SW1, is momentarily closed in Figure 11–16. Assume $V_{AK} = 0.2$ V, $V_{GK} = 0.7$ V, and $I_H = 5$ mA.

Solution

$$I_G = \frac{V_{TRIG} - V_{GK}}{R_G} = \frac{3\text{ V} - 0.7\text{ V}}{5.6\text{ k}\Omega} = \mathbf{410\ \mu A}$$

$$I_A = \frac{V_A - V_{AK}}{R_A} = \frac{15\text{ V} - 0.2\text{ V}}{33\ \Omega} = \mathbf{448\text{ mA}}$$

FIGURE 11–16

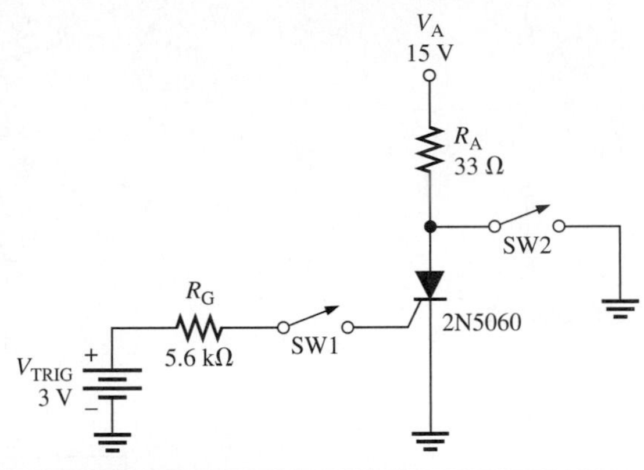

Related Problem Will the SCR turn on if V_A is reduced to 12 V? Explain.

Open the Multisim file E11-03 in the Examples folder on your CD-ROM. With SW2 open, momentarily close SW1. Compare the measured anode current with the calculated value. Notice that the anode current continues even after SW1 is opened. Close SW2 and observe the anode current. Explain your observation.

Half-Wave Power Control

A common application of SCRs is in the control of ac power for lamp dimmers, electric heaters, and electric motors. A half-wave, variable-resistance, phase-control circuit is shown in Figure 11–17; 120 V ac are applied across terminals *A* and *B*; R_L represents the resistance of the load (for example, a heating element or lamp filament). Resistor R_1 limits the current, and potentiometer R_2 sets the trigger level for the SCR.

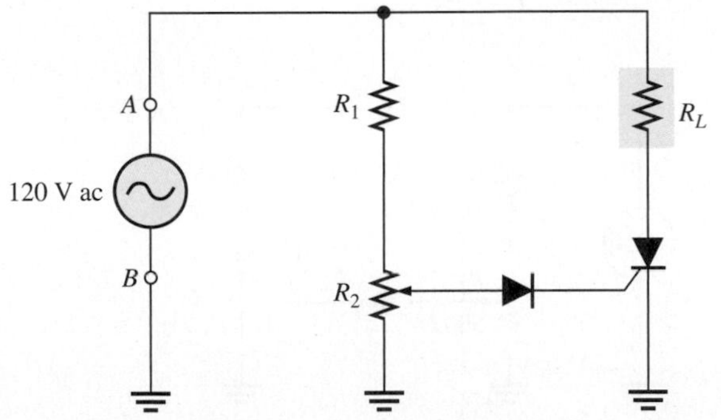

FIGURE 11–17

Half-wave, variable-resistance, phase-control circuit.

By adjusting R_2, the SCR can be made to trigger at any point on the positive half-cycle of the ac waveform between 0° and 90°, as shown in Figure 11–18.

When the SCR triggers near the beginning of the cycle (approximately 0°), as in Figure 11–18(a), it conducts for approximately 180° and maximum power is delivered to the load. When it triggers near the peak of the positive half-cycle (90°), as in Figure 11–18(b), the SCR conducts for approximately 90° and less power is delivered to the load. By adjusting R_2, triggering can be made to occur anywhere between these two extremes, and therefore, a variable amount of power can be delivered to the load. Figure 11–18(c) shows triggering at the 45° point as an example. When the ac input goes negative, the SCR turns off and does not conduct again until the trigger point on the next positive half-cycle. The diode prevents the negative ac voltage from being applied to the gate of the SCR.

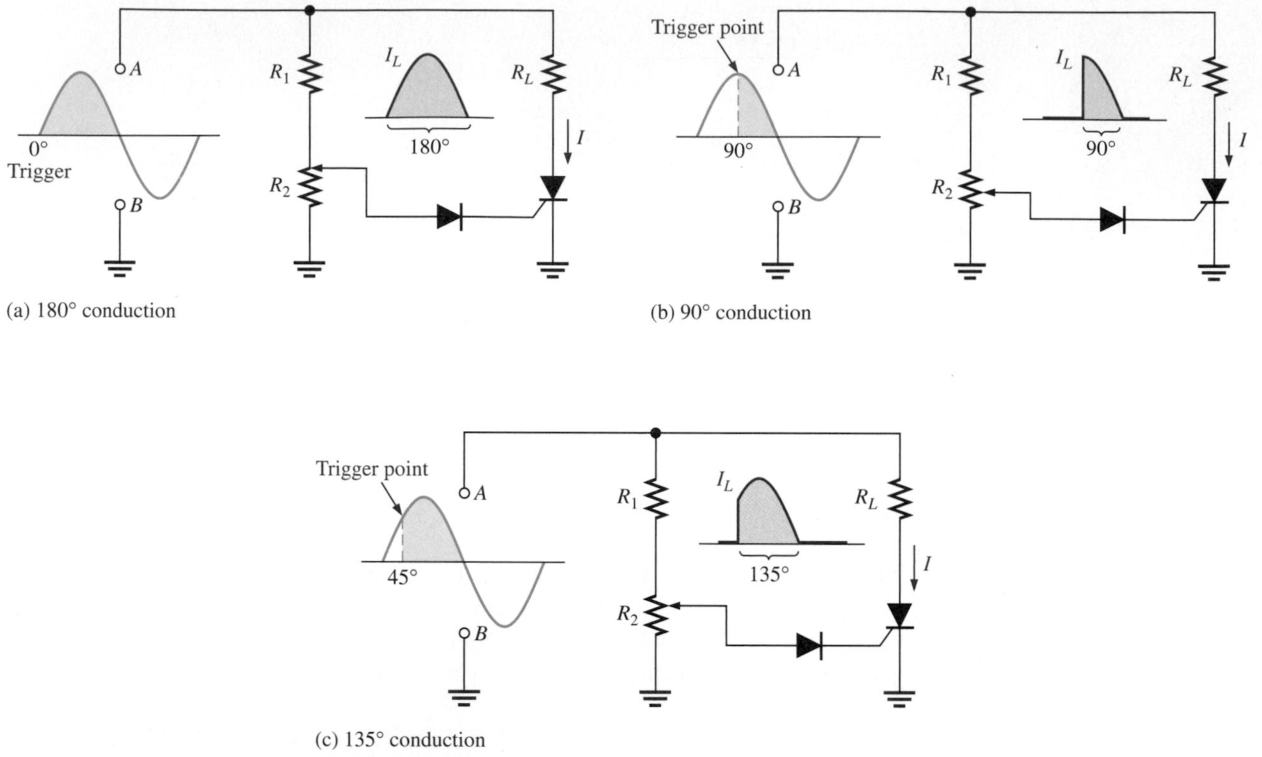

FIGURE 11–18

Operation of the phase-control circuit.

EXAMPLE 11–4

Show the voltage waveform across the SCR in Figure 11–19 from anode to cathode (ground) in relation to the load current for 180°, 45°, and 90° conduction. Assume an ideal SCR.

FIGURE 11–19

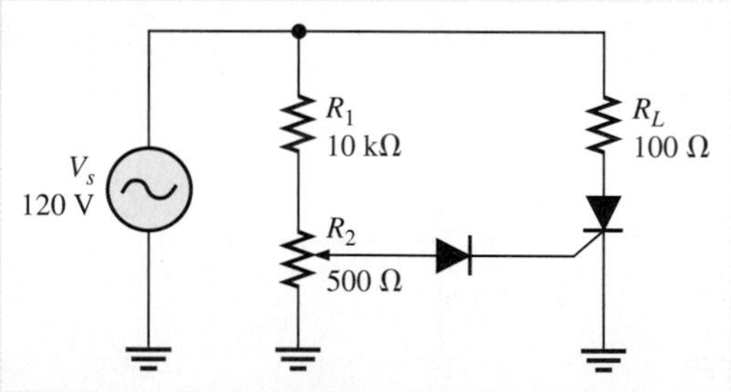

Solution When there is load current, the SCR is conducting and the voltage across it is ideally zero. When there is no load current, the voltage across the SCR is the same as the applied voltage. The waveforms are shown in Figure 11–20.

Related Problem What is the voltage across the SCR if it is never triggered?

Open the Multisim file E11-04 in the Examples folder on your CD-ROM. View the voltage across the SCR with the oscilloscope. Vary the potentiometer setting and observe how V_{AK} changes.

FIGURE 11–20

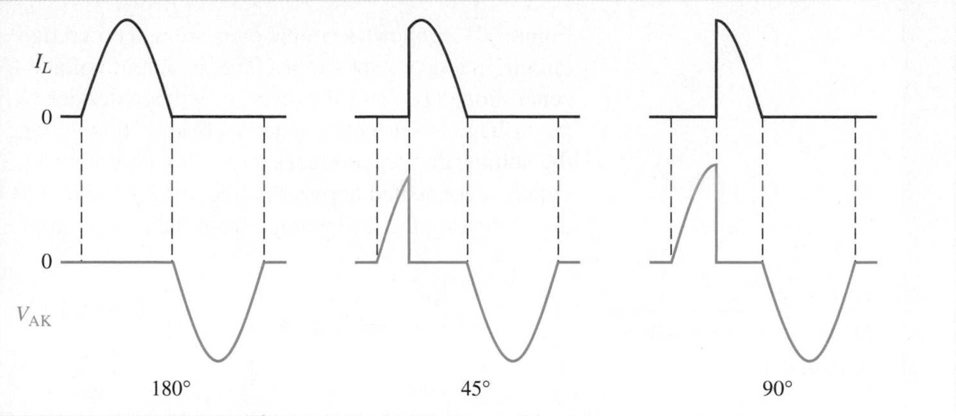

Lighting System for Power Interruptions

As another example of SCR applications, let's examine a circuit that will maintain lighting by using a backup battery when there is an ac power failure. Figure 11–21 shows a center-tapped full-wave rectifier used for providing ac power to a low-voltage lamp. As long as the ac power is available, the battery charges through diode D_3 and R_1.

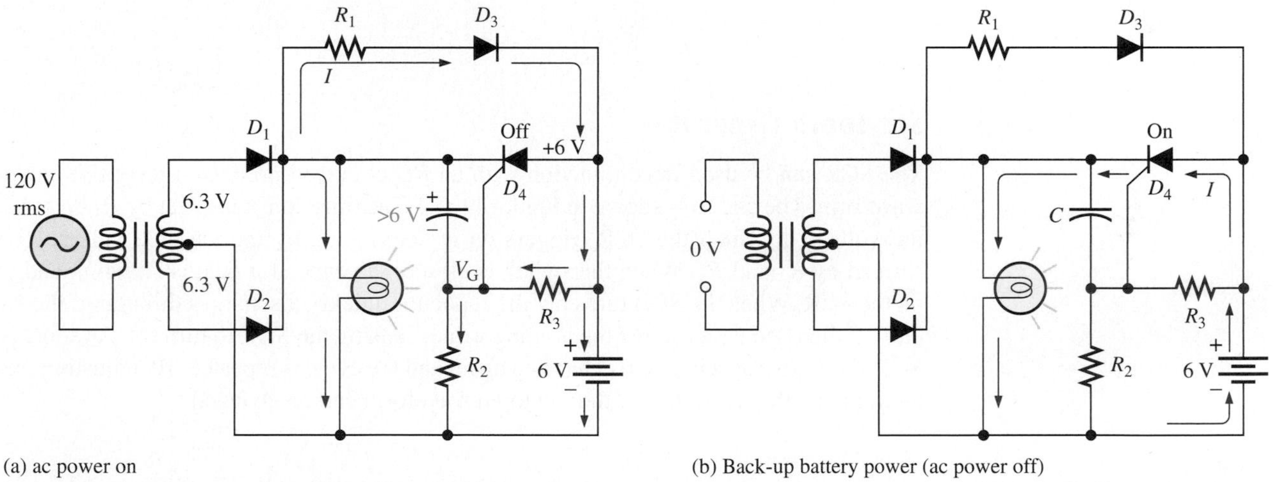

▲ FIGURE 11–21

Automatic back-up lighting circuit.

The SCR's cathode voltage is established when the capacitor charges to the peak value of the full-wave rectified ac (6.3 V rms less the drops across R_2 and D_1). The anode is at the 6 V battery voltage, making it less positive than the cathode, thus preventing conduction. The SCR's gate is at a voltage established by the voltage divider made up of R_2 and R_3. Under these conditions the lamp is illuminated by the ac input power and the SCR is off, as shown in Figure 11–21(a).

When there is an interruption of ac power, the capacitor discharges through the closed path R_1, D_3, and R_3, making the cathode less positive than the anode or the gate. This action establishes a triggering condition, and the SCR begins to conduct. Current from the battery is through the SCR and the lamp, thus maintaining illumination, as shown in Figure 11–21(b). When ac power is restored, the capacitor recharges and the SCR turns off. The battery begins recharging.

An Over-Voltage Protection Circuit

Figure 11–22 shows a simple over-voltage protection circuit, sometimes called a "crowbar" circuit, in a dc power supply. The dc output voltage from the regulator is monitored by the zener diode (D_1) and the resistive voltage divider (R_1 and R_2). The upper limit of the output voltage is set by the zener voltage. If this voltage is exceeded, the zener conducts and the voltage divider produces an SCR trigger voltage. The trigger voltage turns on the SCR, which is connected across the line voltage. The SCR current causes the fuse to blow, thus disconnecting the line voltage from the power supply.

FIGURE 11–22

A basic SCR over-voltage protection circuit (shown in blue).

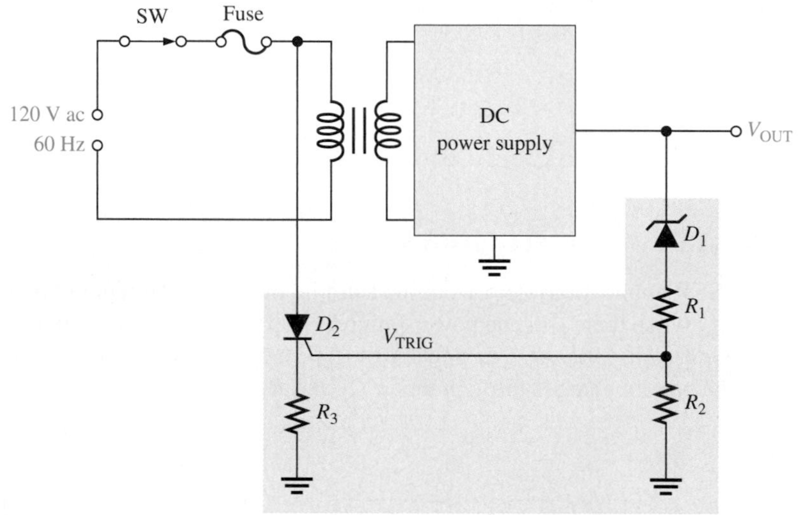

Sawtooth Generator

The SCR can be used in conjunction with an *RC* circuit to produce a repetitive sawtooth waveform. The circuit is shown in Figure 11–23. The time constant is set by R_1 and C_1, and the voltage at which the SCR triggers on is determined by the variable voltage-divider formed by R_2 and R_3. When the switch is closed, the capacitor begins charging and turns on the SCR. When the SCR turns on, the capacitor quickly discharges through it; the anode current then decreases below the holding value, causing the SCR to turn off. As soon as the SCR is off, the capacitor starts charging again and the cycle is repeated. By adjusting the potentiometer, the frequency of the sawtooth waveform can be changed.

FIGURE 11–23

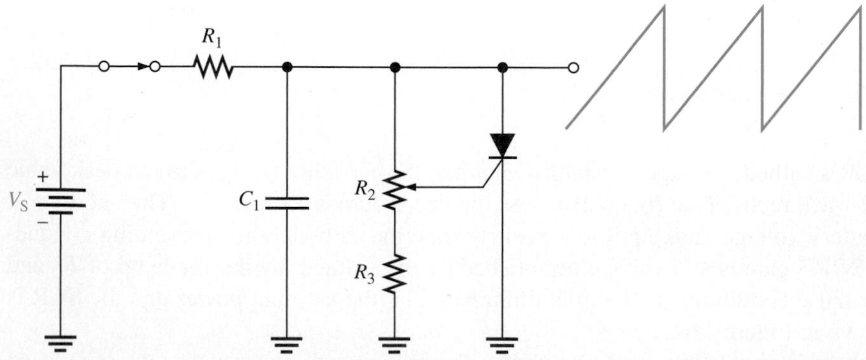

SECTION 11–3 CHECKUP

1. If the potentiometer in Figure 11–18 is set at its midpoint, during what part of the input cycle will the SCR conduct?
2. In Figure 11–21, what is the purpose of diode D_3?

11–4 THE DIAC AND TRIAC

Both the diac and the triac are types of thyristors that can conduct current in both directions (bilateral). The difference between the two devices is that a diac has two terminals, while a triac has a third terminal, which is the gate for triggering. The diac functions basically like two parallel 4-layer diodes turned in opposite directions. The triac functions basically like two parallel SCRs turned in opposite directions with a common gate terminal.

After completing this section, you should be able to

- **Describe the basic structure and operation of diacs and triacs**
    - Identify a diac or triac by the schematic symbol
    - Discuss the equivalent circuit and bias conditions
    - Explain the characteristic curve
    - Discuss an application

The Diac

A **diac** is a two-terminal four-layer semiconductor device (thyristor) that can conduct current in either direction when activated. The basic construction and schematic symbol for a diac are shown in Figure 11–24. Notice that there are two terminals, labelled A_1 and A_2. The top and bottom layers contain both *n* and *p* materials. The right side of the stack can be regarded as a *pnpn* structure with the same characteristics as a four-layer diode, while the left side is an inverted four-layer diode having an *npnp* structure.

Conduction occurs in a diac when the breakover voltage is reached with either polarity across the two terminals. The curve in Figure 11–25 illustrates this characteristic. Once breakover occurs, current is in a direction depending on the polarity of the voltage across the terminals. The device turns off when the current drops below the holding value.

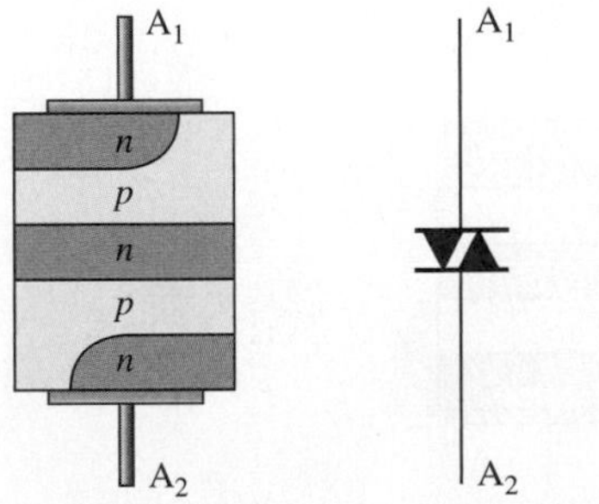

(a) Basic construction (b) Symbol

FIGURE 11–24

The diac.

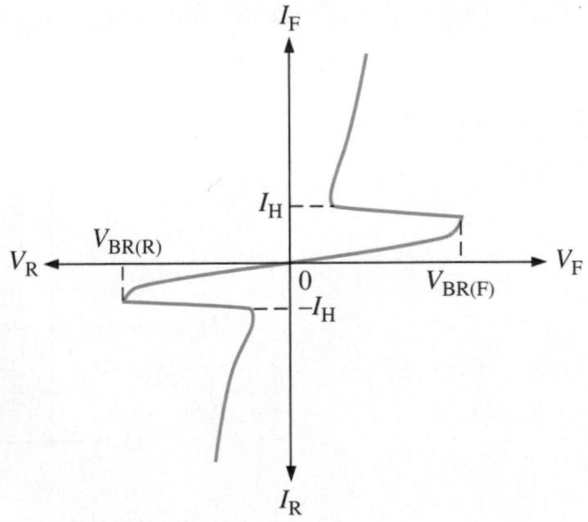

FIGURE 11–25

Diac characteristic curve.

The equivalent circuit of a diac consists of four transistors arranged as shown in Figure 11–26(a). When the diac is biased as in Figure 11–26(b), the *pnpn* structure from A_1 to A_2 provides the same operation as was described for the 4-layer diode. In the equivalent circuit, Q_1 and Q_2 are forward-biased, and Q_3 and Q_4 are reverse-biased. The device operates on the upper right portion of the characteristic curve in Figure 11–25 under this bias

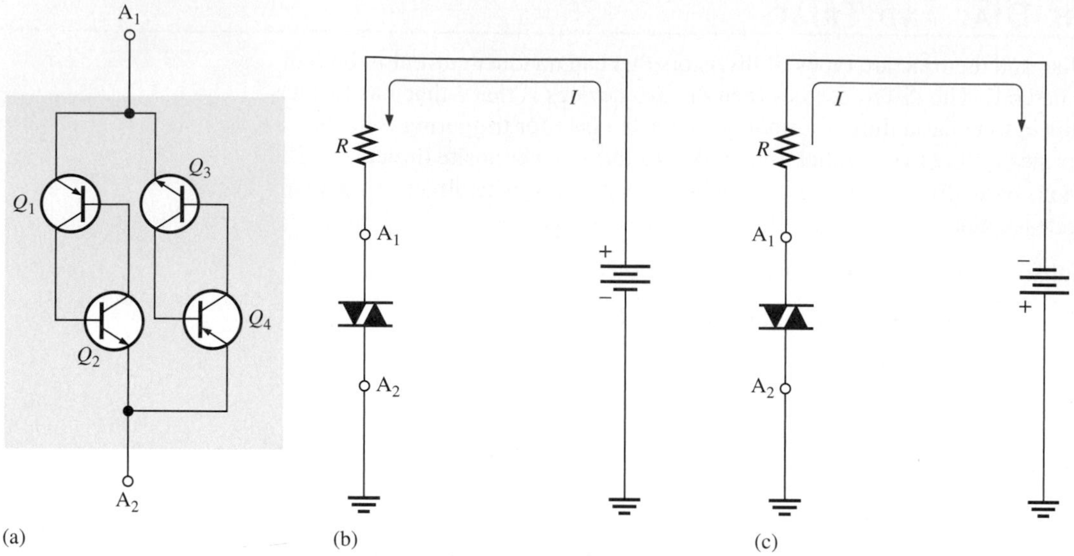

▲ **FIGURE 11–26**

Diac equivalent circuit and bias conditions.

condition. When the diac is biased as shown in Figure 11–26(c), the *pnpn* structure from A_2 and A_1 is used. In the equivalent circuit, Q_3 and Q_4 are forward-biased, and Q_1 and Q_2 are reverse-biased. Under this bias condition, the device operates on the lower left portion of the characteristic curve, as shown in Figure 11–25.

The Triac

A **triac** is like a diac with a gate terminal. A triac can be turned on by a pulse of gate current and does not require the breakover voltage to initiate conduction, as does the diac. Basically, a triac can be thought of simply as two SCRs connected in parallel and in opposite directions with a common gate terminal. Unlike the SCR, the triac can conduct current in either direction when it is triggered on, depending on the polarity of the voltage across its A_1 and A_2 terminals. Figure 11–27 shows the basic construction and schematic symbol for a triac.

► **FIGURE 11–27**

The triac.

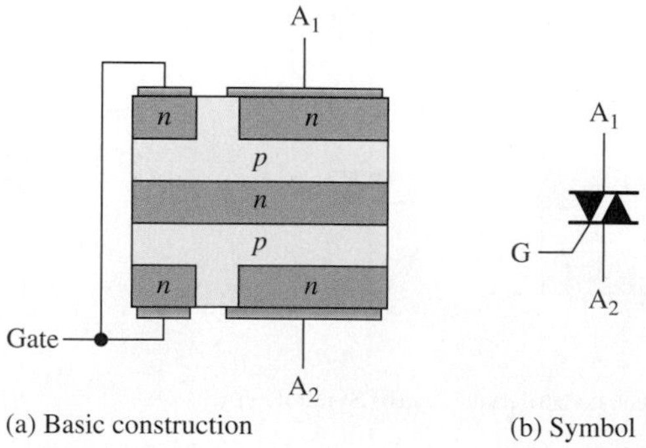

The characteristic curve is shown in Figure 11–28. Notice that the breakover potential decreases as the gate current increases, just as with the SCR. As with other thyristors, the triac ceases to conduct when the anode current drops below the specified value of the holding current, I_H. The only way to turn off the triac is to reduce the current to a sufficiently low level.

Figure 11–29 shows the triac being triggered into both directions of conduction. In part (a), terminal A_1 is biased positive with respect to A_2, so the triac conducts as shown when

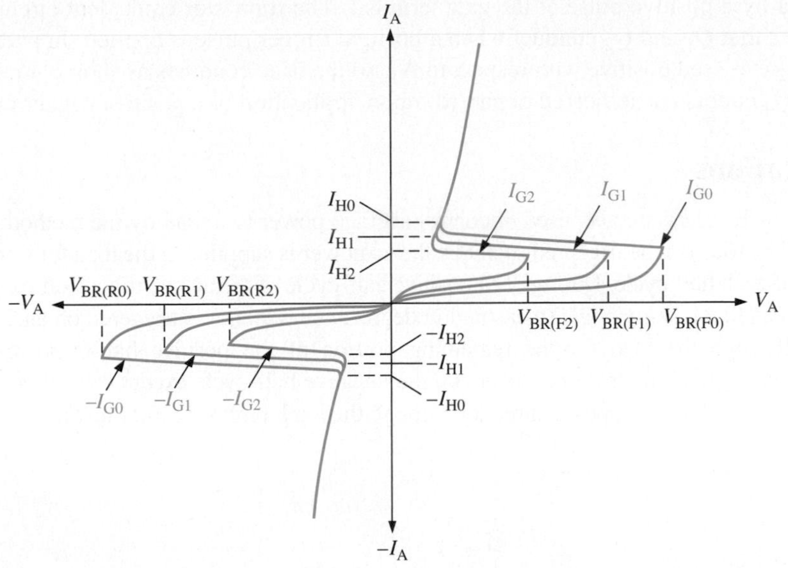

▲ **FIGURE 11–28**

Triac characteristic curves.

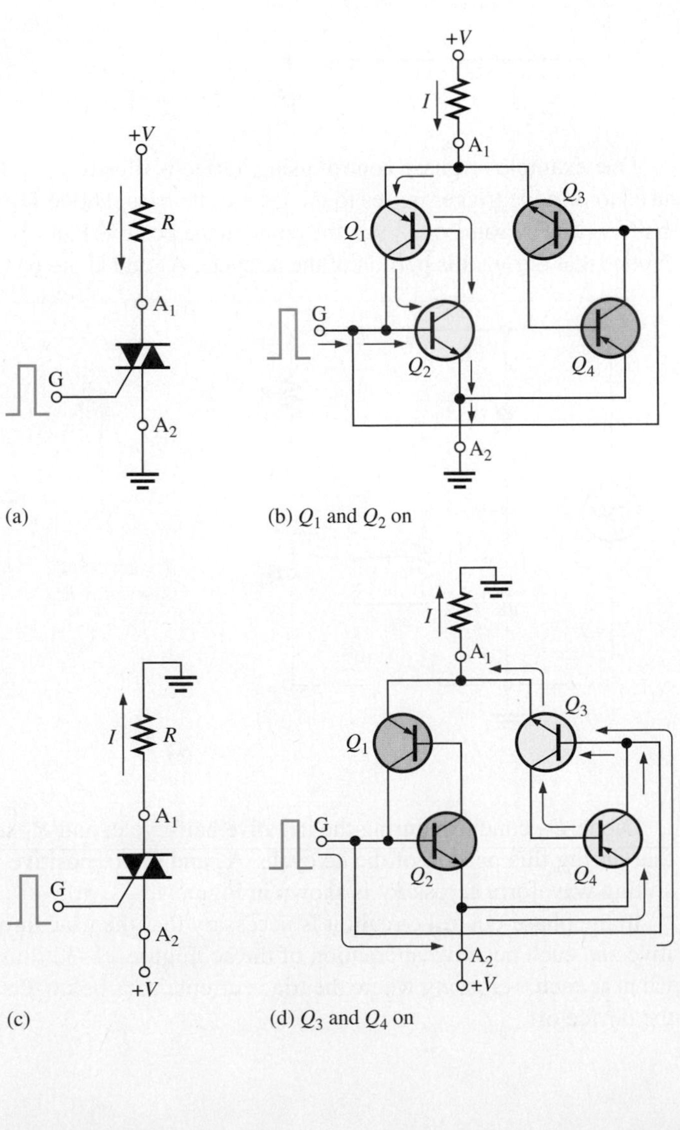

◀ **FIGURE 11–29**

Bilateral operation of a triac.

triggered by a positive pulse at the gate terminal. The transistor equivalent circuit in part (b) shows that Q_1 and Q_2 conduct when a positive trigger pulse is applied. In part (c), terminal A_2 is biased positive with respect to A_1, so the triac conducts as shown. In this case, Q_3 and Q_4 conduct as indicated in part (d) upon application of a positive trigger pulse.

Applications

Like the SCR, triacs are also used to control average power to a load by the method of phase control. The triac can be triggered such that the ac power is supplied to the load for a controlled portion of each half-cycle. During each positive half-cycle of the ac, the triac is off for a certain interval, called the *delay angle* (measured in degrees), and then it is triggered on and conducts current through the load for the remaining portion of the positive half-cycle, called the *conduction angle.* Similar action occurs on the negative half-cycle except that, of course, current is conducted in the opposite direction through the load. Figure 11–30 illustrates this action.

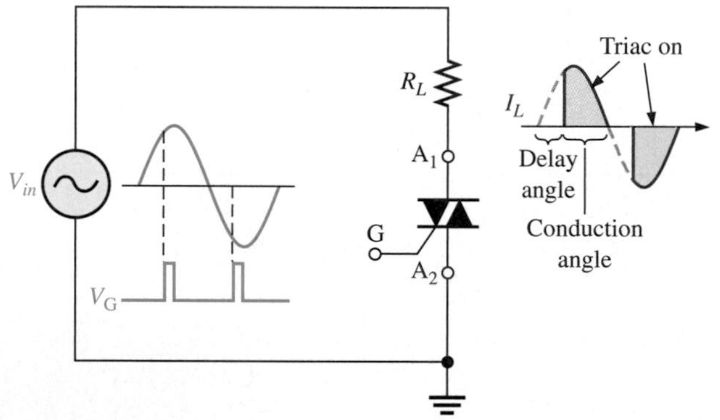

FIGURE 11–30

Basic triac phase control.

One example of phase control using a triac is illustrated in Figure 11–31(a). Diodes are used to provide trigger pulses to the gate of the triac. Diode D_1 conducts during the positive half-cycle. The value of R_1 sets the point on the positive half-cycle at which the triac triggers. Notice that during this portion of the ac cycle, A_1 and G are positive with respect to A_2.

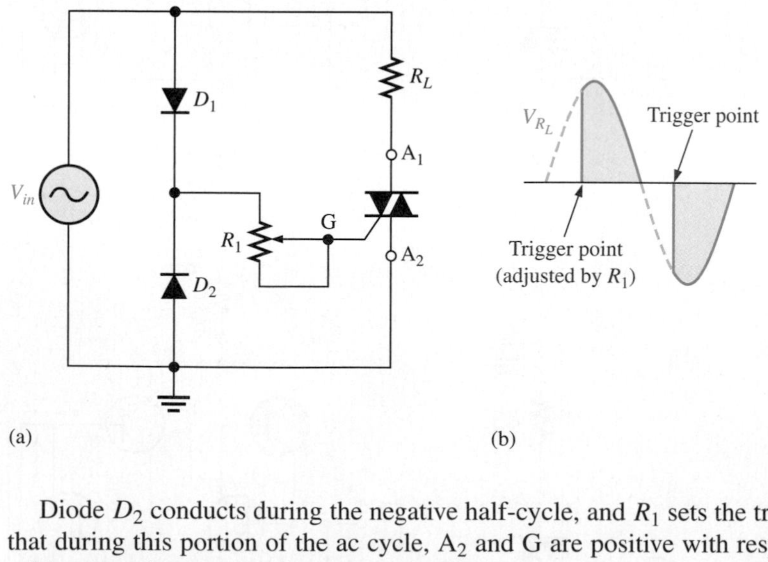

FIGURE 11–31

Triac phase-control circuit.

Diode D_2 conducts during the negative half-cycle, and R_1 sets the trigger point. Notice that during this portion of the ac cycle, A_2 and G are positive with respect to A_1. The resulting waveform across R_L is shown in Figure 11–31(b).

In the phase-control circuit, it is necessary that the triac turn off at the end of each positive and each negative alternation of the ac. Figure 11–32 illustrates that there is an interval near each 0 crossing where the triac current drops below the holding value, thus turning the device off.

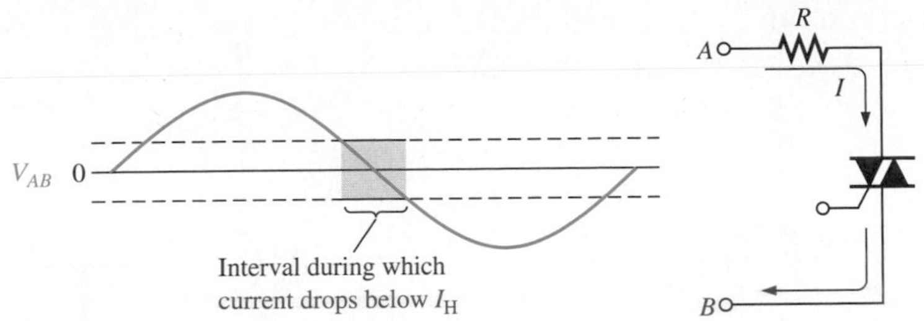

FIGURE 11–32
Triac turn-off interval.

SECTION 11–4 CHECKUP

1. Compare the diac to the 4-layer diode in terms of basic operation.
2. Compare the triac with the SCR in terms of basic operation.
3. How does a triac differ from a diac?

11–5 THE SILICON-CONTROLLED SWITCH (SCS)

The silicon-controlled switch (SCS) is similar in construction to the SCR. The SCS, however, has two gate terminals, the cathode gate and the anode gate. The SCS can be turned on and off using either gate terminal. Remember that the SCR can be only turned on using its gate terminal. Normally, the SCS is available in power ratings lower than those of the SCR.

After completing this section, you should be able to

- **Describe the basic operation of an SCS**
  - Identify an SCS by its schematic symbol
  - Use an equivalent circuit to describe SCS operation
  - Compare the SCS to the SCR

An **SCS** (silicon-controlled switch) is a four-terminal thyristor that has two gate terminals that are used to trigger the device on and off. The symbol and terminal identification for an SCS are shown in Figure 11–33.

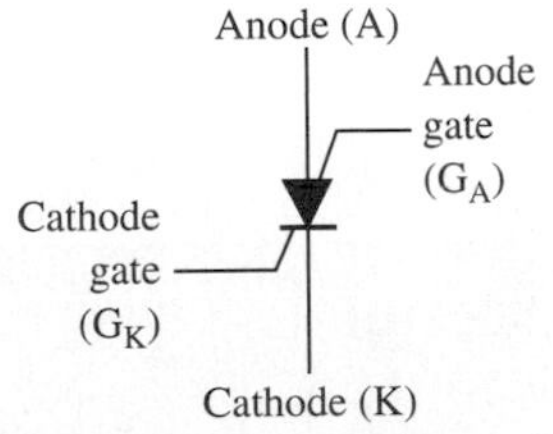

FIGURE 11–33
The silicon-controlled switch (SCS).

As with the previous thyristors, the basic operation of the SCS can be understood by referring to the transistor equivalent, shown in Figure 11–34. To start, assume that both Q_1 and Q_2 are off, and therefore that the SCS is not conducting. A positive pulse on the cathode gate drives Q_2 into conduction and thus provides a path for Q_1 base current. When Q_1 turns on, its collector current provides base current for Q_2, thus sustaining the *on* state of the device. This regenerative action is the same as in the turn-on process of the SCR and the 4-layer diode and is illustrated in Figure 11–34(a).

The SCS can also be turned on with a negative pulse on the anode gate, as indicated in Figure 11–34(a). This drives Q_1 into conduction which, in turn, provides base current for Q_2. Once Q_2 is on, it provides a path for Q_1 base current, thus sustaining the *on* state.

To turn the SCS off, a positive pulse is applied to the anode gate. This reverse-biases the base-emitter junction of Q_1 and turns it off. Q_2, in turn, cuts off and the SCS ceases conduction, as shown in Figure 11–34(b). The device can also be turned off with a negative pulse on the cathode gate, as indicated in part (b). The SCS typically has a faster turn-off time than the SCR.

▶ FIGURE 11–34

SCS operation.

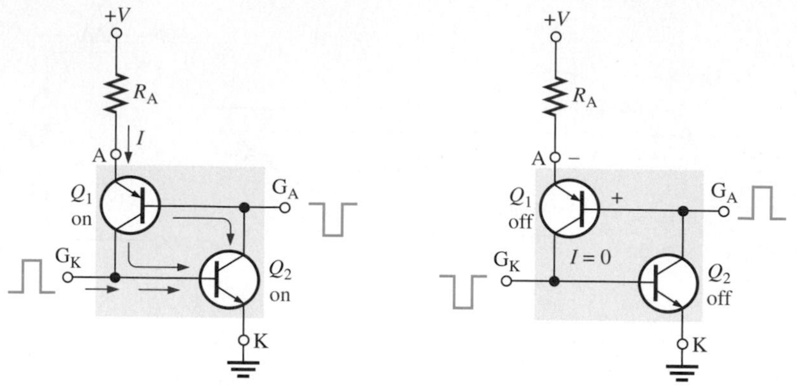

(a) *Turn-on*: Positive pulse on G_K *or* negative pulse on G_A

(b) *Turn-off*: Positive pulse on G_A *or* negative pulse on G_K

In addition to the positive pulse on the anode gate or the negative pulse on the cathode gate, there is another method for turning off an SCS. Figure 11–35(a) and (b) shows two switching methods to reduce the anode current below the holding value. In each case, the bipolar junction transistor (BJT) acts as a switch to interrupt the anode current.

▶ FIGURE 11–35

The transistor switch in both series and shunt configurations reduces I_A below I_H and turns off the SCS.

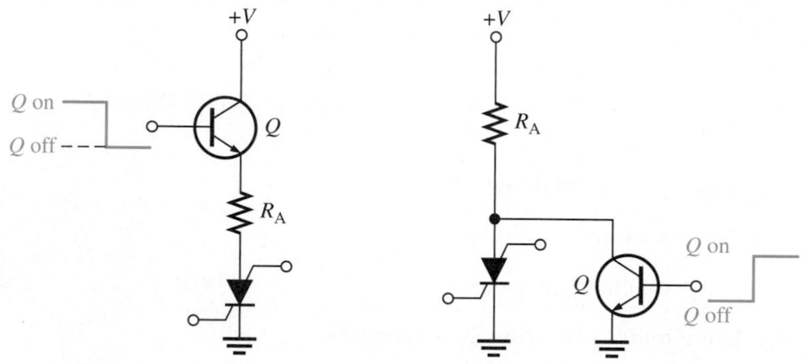

(a) Series switch turns off SCS

(b) Shunt switch turns off SCS

Applications

The SCS and SCR are used in similar applications. The SCS has the advantage of faster turn-off with pulses on either gate terminal; however, it is more limited in terms of maximum current and voltage ratings. Also, the SCS is sometimes used in digital applications such as counters, registers, and timing circuits.

SECTION 11–5 CHECKUP

1. Explain the difference between an SCS and an SCR.
2. How can an SCS be turned on?
3. Describe three ways an SCS can be turned off.

11–6 THE UNIJUNCTION TRANSISTOR (UJT)

The unijunction transistor does not belong to the thyristor family because it does not have a four-layer type of construction. The term *unijunction* refers to the fact that the UJT has a single *pn* junction. The UJT is useful in certain oscillator applications and as a triggering device in thyristor circuits.

After completing this section, you should be able to

- **Describe the basic structure and operation of a UJT**
  - Explain why a UJT is not a thyristor
  - Identify a UJT by its symbol
  - Draw the equivalent circuit
  - Define *standoff ratio*
  - Analyze the operation of a UJT relaxation oscillator

The **UJT** (unijunction transistor) is a three-terminal device whose basic construction is shown in Figure 11–36(a). The schematic symbol appears in Figure 11–36(b). Notice the terminals are labelled Emitter (E), Base 1 (B_1), and Base 2 (B_2). Do not confuse this symbol with that of a JFET; the difference is that the arrow is at an angle for the UJT. The UJT has only one *pn* junction, and therefore, the characteristics of this device are different from those of either the BJT or the FET, as you will see.

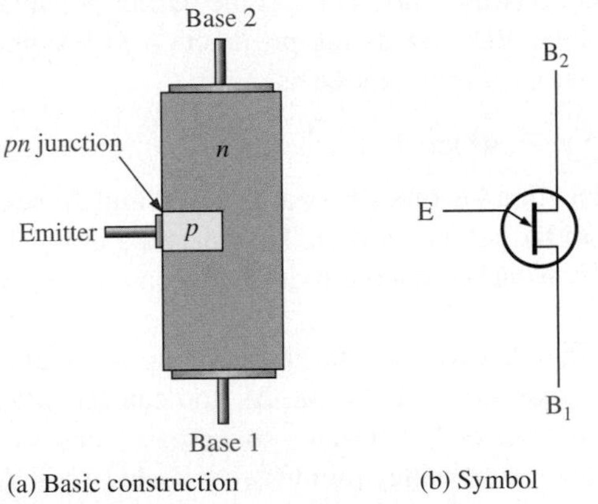

FIGURE 11–36

The unijunction transistor (UJT).

Equivalent Circuit

The equivalent circuit for the UJT, shown in Figure 11–37(a), will aid in understanding the basic operation. The diode shown in the figure represents the *pn* junction, r'_{B1} represents the

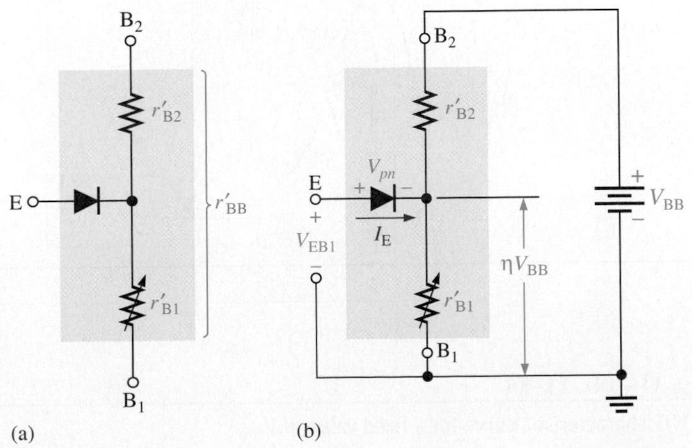

FIGURE 11–37

UJT equivalent circuit.

internal dynamic resistance of the silicon bar between the emitter and base 1, and r'_{B2} represents the dynamic resistance between the emitter and base 2. The total resistance between the base terminals is the sum of r'_{B1} and r'_{B2} and is called the *interbase resistance,* r'_{BB}.

$$r'_{BB} = r'_{B1} + r'_{B2}$$

The value of r'_{B1} varies inversely with emitter current I_E, and therefore, it is shown as a variable resistor. Depending on I_E, the value of r'_{B1} can vary from several thousand ohms down to tens of ohms. The internal resistances r'_{B1} and r'_{B2} form a voltage divider when the device is biased, as shown in Figure 11–37(b). The voltage across the resistance r'_{B1} can be expressed as

$$V_{r'_{B1}} = \left(\frac{r'_{B1}}{r'_{BB}}\right)V_{BB}$$

Standoff Ratio

The ratio r'_{B1}/r'_{BB} is a UJT characteristic called the intrinsic **standoff ratio** and is designated by η (Greek *eta*).

Equation 11–1

$$\boldsymbol{\eta = \frac{r'_{B1}}{r'_{BB}}}$$

As long as the applied emitter voltage V_{EB1} is less than $V_{r'_{B1}} + V_{pn}$, there is no emitter current because the *pn* junction is not forward-biased (V_{pn} is the barrier potential of the *pn* junction). The value of emitter voltage that causes the *pn* junction to become forward-biased is called V_P (peak-point voltage) and is expressed as

Equation 11–2

$$\boldsymbol{V_P = \eta V_{BB} + V_{pn}}$$

When V_{EB1} reaches V_P, the *pn* junction becomes forward-biased and I_E begins. Holes are injected into the *n*-type bar from the *p*-type emitter. This increase in holes causes an increase in free electrons, thus increasing the conductivity between emitter and B_1 (decreasing r'_{B1}).

After turn-on, the UJT operates in a negative resistance region up to a certain value of I_E, as shown by the characteristic curve in Figure 11–38. As you can see, after the peak point ($V_E = V_P$ and $I_E = I_P$), V_E decreases as I_E continues to increase, thus producing the negative resistance characteristic. Beyond the valley point ($V_E = V_V$ and $I_E = I_V$), the device is in saturation, and V_E increases very little with an increasing I_E.

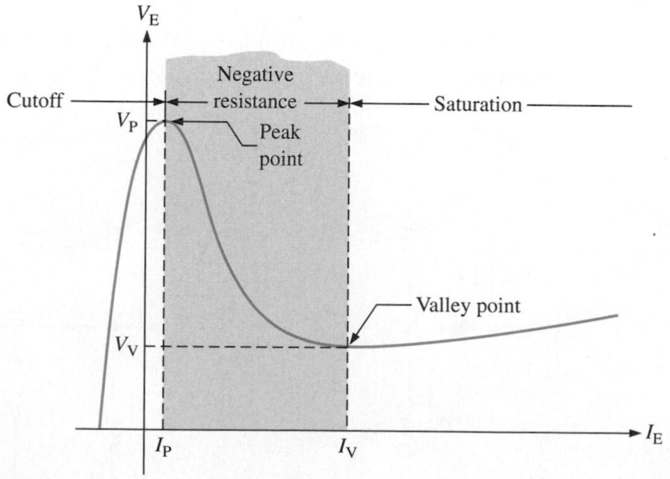

▲ **FIGURE 11–38**

UJT characteristic curve for a fixed value of V_{BB}.

EXAMPLE 11–5 The datasheet of a certain UJT gives $\eta = 0.6$. Determine the peak-point emitter voltage V_P if $V_{BB} = 20$ V.

Solution

$$V_P = \eta V_{BB} + V_{pn} = 0.6(20\text{ V}) + 0.7\text{ V} = \mathbf{12.7\text{ V}}$$

Related Problem How can the peak-point emitter voltage of a UJT be increased?

A UJT Application

The UJT can be used as a trigger device for SCRs and triacs. Other applications include nonsinusoidal oscillators, sawtooth generators, phase control, and timing circuits. Figure 11–39 shows a UJT relaxation oscillator as an example of one application.

The operation is as follows. When dc power is applied, the capacitor C charges exponentially through R_1 until it reaches the peak-point voltage V_P. At this point, the pn junction becomes forward-biased, and the emitter characteristic goes into the negative resistance region (V_E decreases and I_E increases). The capacitor then quickly discharges through the forward-biased junction, r'_B, and R_2. When the capacitor voltage decreases to the valley-point voltage V_V, the UJT turns off, the capacitor begins to charge again, and the cycle is repeated, as shown in the emitter voltage waveform in Figure 11–40 (top). During the discharge time of the capacitor, the UJT is conducting. Therefore, a voltage is developed across R_2, as shown in the waveform diagram in Figure 11–40 (bottom).

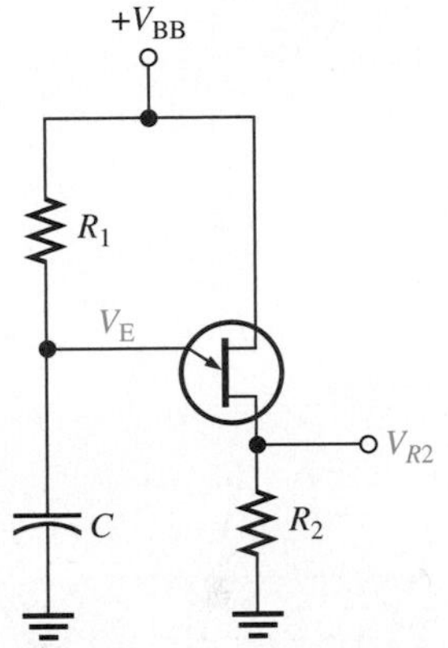

FIGURE 11–39

Relaxation oscillator.

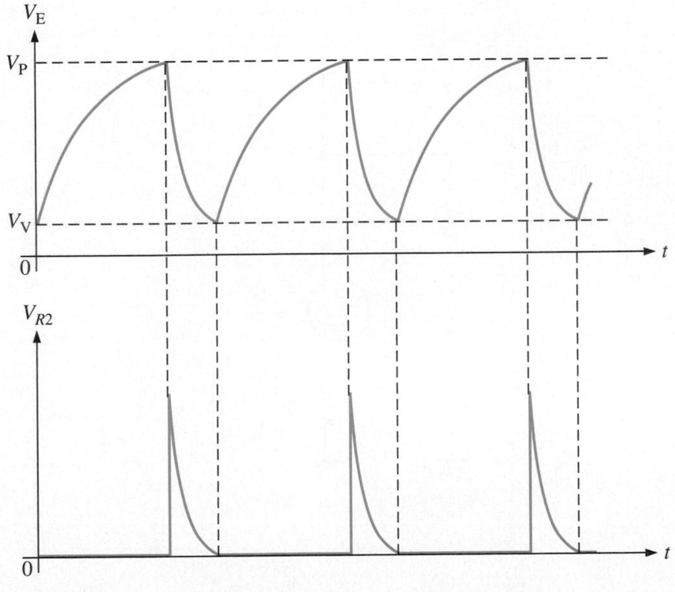

FIGURE 11–40

Waveforms for UJT relaxation oscillator.

Conditions for Turn-On and Turn-Off

In the relaxation oscillator of Figure 11–39, certain conditions must be met for the UJT to reliably turn on and turn off. First, to ensure turn-on, R_1 must not limit I_E at the peak point to less than I_P. To ensure this, the voltage drop across R_1 at the peak point should be greater than $I_P R_1$. Thus, the condition for turn-on is

$$V_{BB} - V_P > I_P R_1$$

or

$$R_1 < \frac{V_{BB} - V_P}{I_P}$$

To ensure turn-off of the UJT at the valley point, R_1 must be large enough that I_E (at the valley point) can decrease below the specified value of I_V. This means that the voltage across R_1 at the valley point must be less than $I_V R_1$. Thus, the condition for turn-off is

$$V_{BB} - V_V < I_V R_1$$

or

$$R_1 > \frac{V_{BB} - V_V}{I_V}$$

Therefore, for a proper turn-on and turn-off, R_1 must be in the range

$$\frac{V_{BB} - V_P}{I_P} > R_1 > \frac{V_{BB} - V_V}{I_V}$$

EXAMPLE 11–6

Determine a value of R_1 in Figure 11–41 that will ensure proper turn-on and turn-off of the UJT. The characteristic of the UJT exhibits the following values: $\eta = 0.5$, $V_V = 1\text{ V}$, $I_V = 10\text{ mA}$, $I_P = 20\ \mu\text{A}$, and $V_P = 14\text{ V}$.

FIGURE 11–41

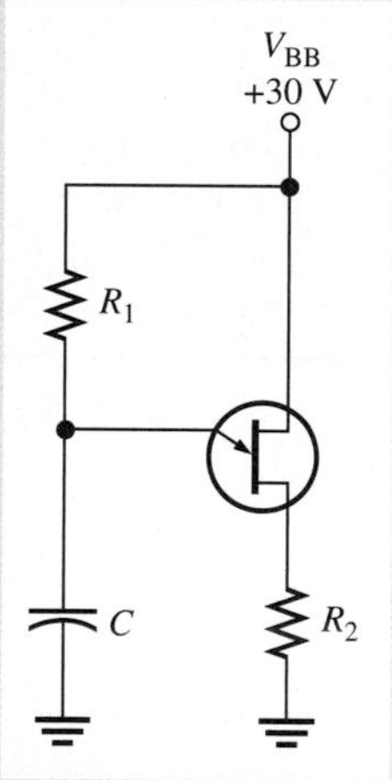

Solution

$$\frac{V_{BB} - V_P}{I_P} > R_1 > \frac{V_{BB} - V_V}{I_V}$$

$$\frac{30\text{ V} - 14\text{ V}}{20\ \mu\text{A}} > R_1 > \frac{30\text{ V} - 1\text{ V}}{10\text{ mA}}$$

$$\mathbf{800\ k\Omega} > R_1 > \mathbf{2.9\ k\Omega}$$

As you can see, R_1 has quite a wide range of possible values that will work.

Related Problem Determine a value of R_1 in Figure 11–41 that will ensure proper turn-on and turn-off for the following values: $\eta = 0.33$, $V_V = 0.8\text{ V}$, $I_V = 15\text{ mA}$, $I_P = 35\ \mu\text{A}$, and $V_P = 18\text{ V}$.

SECTION 11–6 CHECKUP

1. Name the UJT terminals.
2. What is the intrinsic standoff ratio?
3. In a basic UJT relaxation oscillator such as in Figure 11–39, what three factors determine the period of oscillation?

11–7 THE PROGRAMMABLE UNIJUNCTION TRANSISTOR (PUT)

The programmable unijunction transistor (PUT) is actually a type of thyristor and not like the UJT at all in terms of structure. The only similarity to a UJT is that the PUT can be used in some oscillator applications to replace the UJT. The PUT is similar to an SCR except that its anode-to-gate voltage can be used to both turn on and turn off the device.

After completing this section, you should be able to

- **Describe the structure and operation of a PUT**
  - Compare PUT structure to that of the SCR
  - Explain the difference between a PUT and a UJT
  - State how to set the PUT trigger voltage
  - Discuss an application

A **PUT** (programmable unijunction transistor) is a type of three-terminal thyristor that is triggered into conduction when the voltage at the anode exceeds the voltage at the gate. The structure of the PUT is more similar to that of an SCR (four-layer) than to a UJT. The exception is that the gate is brought out as shown in Figure 11–42. Notice that the gate is connected to the *n* region adjacent to the anode. This *pn* junction controls the *on* and *off* states of the device. The gate is always biased positive with respect to the cathode. When the anode voltage exceeds the gate voltage by approximately 0.7 V, the *pn* junction is forward-biased and the PUT turns on. The PUT stays on until the anode voltage falls back below this level, then the PUT turns off.

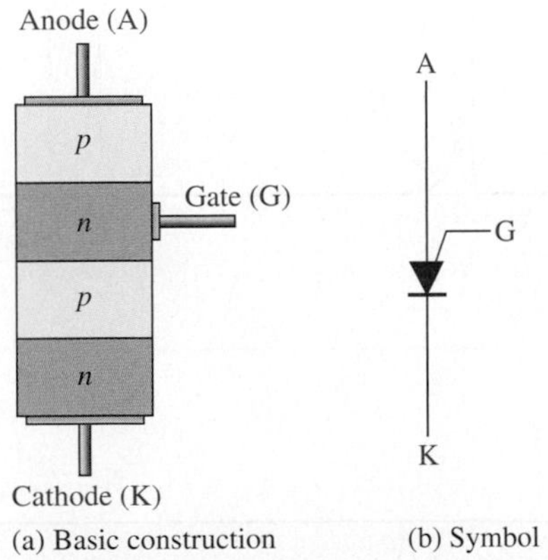

FIGURE 11–42
The programmable unijunction transistor (PUT).

Setting the Trigger Voltage

The gate can be biased to a desired voltage with an external voltage divider, as shown in Figure 11–43(a), so that when the anode voltage exceeds this "programmed" level, the PUT turns on.

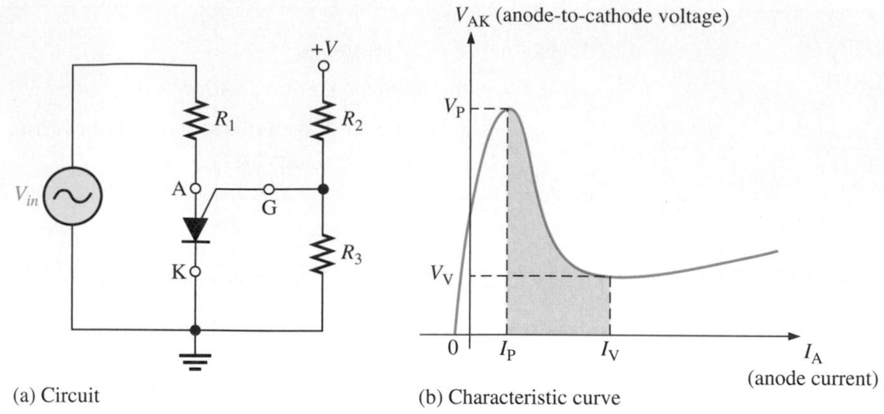

▶ FIGURE 11–43

PUT biasing.

An Application

A plot of the anode-to-cathode voltage, V_{AK}, versus anode current, I_A, in Figure 11–43(b) reveals a characteristic curve similar to that of the UJT. Therefore, the PUT replaces the UJT in many applications. One such application is the relaxation oscillator in Figure 11–44(a).

The basic operation of the PUT is as follows. The gate is biased at +9 V by the voltage divider consisting of resistors R_2 and R_3. When dc power is applied, the PUT is off and the capacitor charges toward +18 V through R_1. When the capacitor reaches $V_G + 0.7$ V, the PUT turns on and the capacitor rapidly discharges through the low *on* resistance of the PUT and R_4. A voltage spike is developed across R_4 during the discharge. As soon as the capacitor discharges, the PUT turns off and the charging cycle starts over, as shown by the waveforms in Figure 11–44(b).

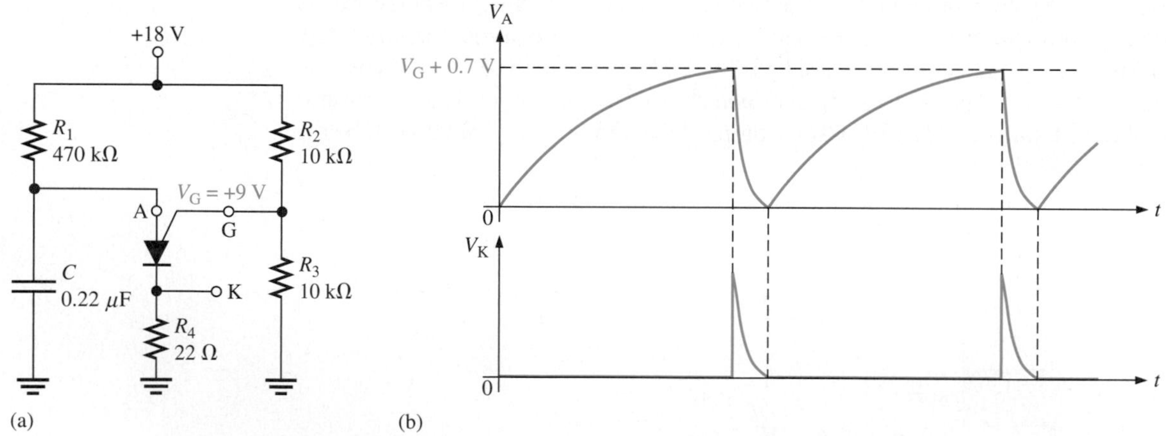

▲ FIGURE 11–44

PUT relaxation oscillator.

SECTION 11–7 CHECKUP

1. What does the term *programmable* mean as used in programmable unijunction transistor (PUT)?
2. Compare the structure and the operation of a PUT to those of other devices such as the UJT and SCR.

On the Job

Application Activity: *Motor Speed Control*

In this application, an SCR and a PUT are used to control the speed of a conveyor belt motor. The circuit controls the speed of the conveyor so that a predetermined average number of randomly spaced parts flow past a point on the production line in a specified period of time. This is to allow an adequate amount of time for the production line workers to perform certain tasks on each part. A basic diagram of the conveyor speed-control system is shown in Figure 11–45.

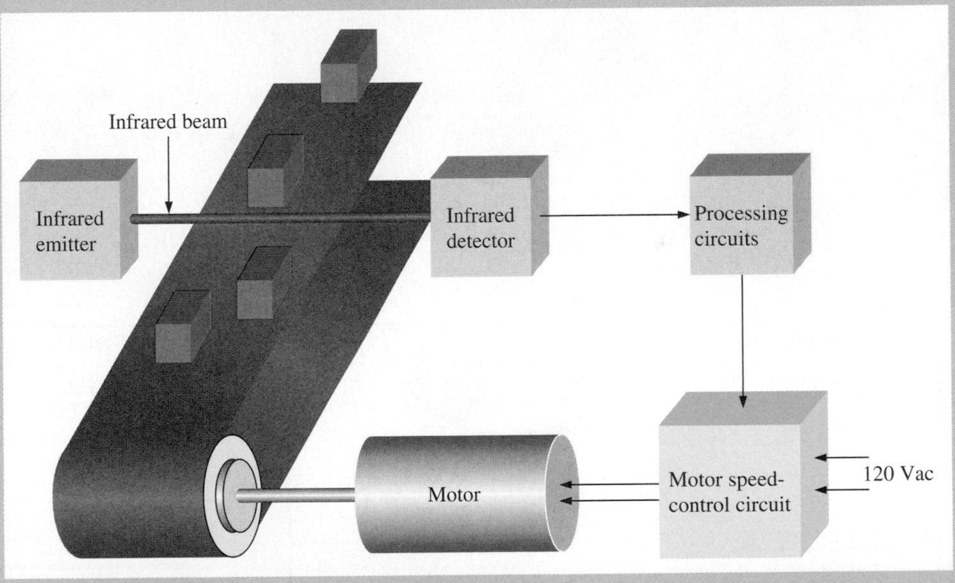

FIGURE 11–45

Block diagram of conveyor speed-control system.

Each time a part on the moving conveyor belt passes the infrared detector and interrupts the IR beam, a digital counter in the processing circuits is advanced by one. The count of the passing parts is accumulated over a specified period of time and converted to a proportional voltage by the processing circuits. The more parts that pass the IR detector during the specified time, the higher the voltage. The proportional voltage is applied to the motor speed-control circuit which, in turn, adjusts the speed of the electric motor that drives the conveyor belt in order to maintain the desired number of parts in a specified period of time.

The Motor Speed-Control Circuit

The proportional voltage from the processing circuits is applied to the gate of a PUT. This voltage determines the point in the ac cycle at which the SCR is triggered *on.* For a higher PUT gate voltage, the SCR turns on later in the half-cycle and therefore delivers less average power to the motor to decrease its speed. For a lower PUT gate voltage, the SCR turns on earlier in the half-cycle and delivers more average power to the motor to increase its speed. This process continually adjusts the motor speed to maintain the required number of parts per unit time moving on the conveyor. A potentiometer is used for calibration of the SCR trigger point. The motor speed-control circuit is shown in Figure 11–46.

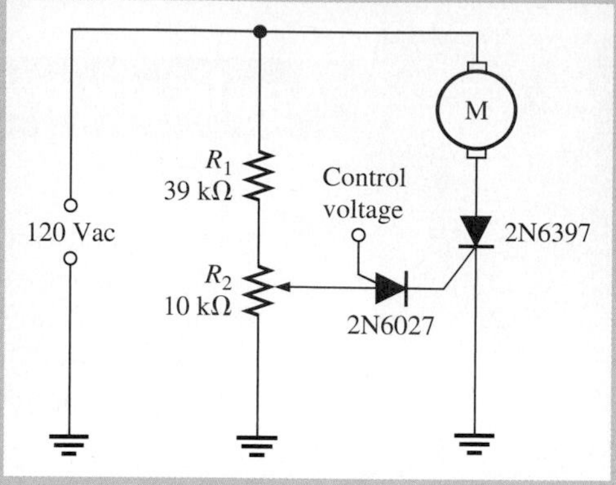

FIGURE 11–46

Motor speed-control circuit.

The SCR used in the motor speed control is the 2N6397 *n*-channel. The partial datasheet is shown in Figure 11–47. The PUT is the 2N6027 and its partial datasheet is also shown in Figure 11–47.

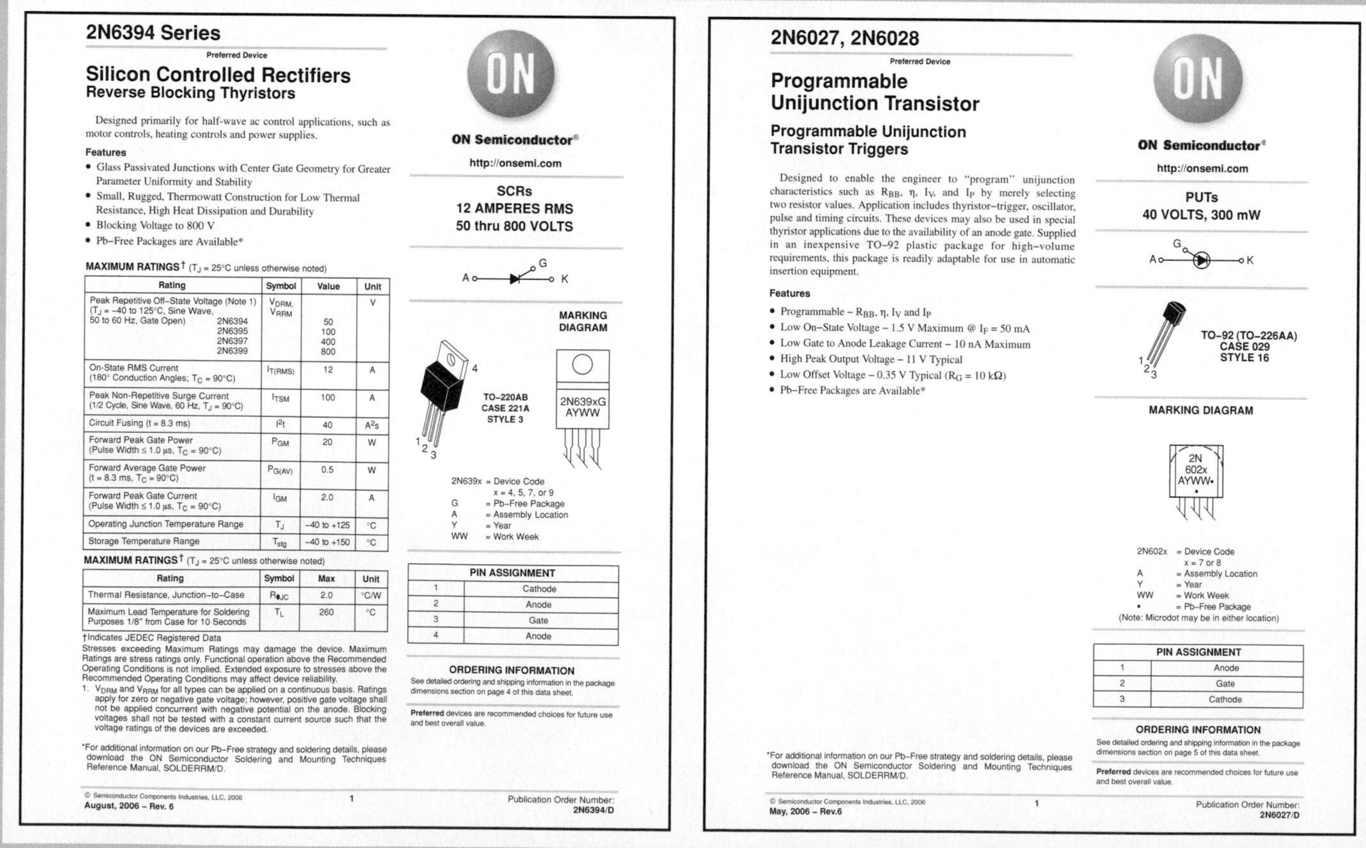

2N6394 Series

Preferred Device

Silicon Controlled Rectifiers
Reverse Blocking Thyristors

Designed primarily for half-wave ac control applications, such as motor controls, heating controls and power supplies.

Features

- Glass Passivated Junctions with Center Gate Geometry for Greater Parameter Uniformity and Stability
- Small, Rugged, Thermowatt Construction for Low Thermal Resistance, High Heat Dissipation and Durability
- Blocking Voltage to 800 V
- Pb–Free Packages are Available*

MAXIMUM RATINGS† (T_J = 25°C unless otherwise noted)

Rating	Symbol	Value	Unit
Peak Repetitive Off–State Voltage (Note 1) (T_J = −40 to 125°C, Sine Wave, 50 to 60 Hz, Gate Open) 2N6394 / 2N6395 / 2N6397 / 2N6399	V_{DRM}, V_{RRM}	50 / 100 / 400 / 800	V
On-State RMS Current (180° Conduction Angles; T_C = 90°C)	$I_{T(RMS)}$	12	A
Peak Non-Repetitive Surge Current (1/2 Cycle, Sine Wave, 60 Hz, T_J = 90°C)	I_{TSM}	100	A
Circuit Fusing (t = 8.3 ms)	I^2t	40	A^2s
Forward Peak Gate Power (Pulse Width ≤ 1.0 μs, T_C = 90°C)	P_{GM}	20	W
Forward Average Gate Power (t = 8.3 ms, T_C = 90°C)	$P_{G(AV)}$	0.5	W
Forward Peak Gate Current (Pulse Width ≤ 1.0 μs, T_C = 90°C)	I_{GM}	2.0	A
Operating Junction Temperature Range	T_J	−40 to +125	°C
Storage Temperature Range	T_{stg}	−40 to +150	°C

MAXIMUM RATINGS† (T_J = 25°C unless otherwise noted)

Rating	Symbol	Max	Unit
Thermal Resistance, Junction–to–Case	$R_{\theta JC}$	2.0	°C/W
Maximum Lead Temperature for Soldering Purposes 1/8″ from Case for 10 Seconds	T_L	260	°C

†Indicates JEDEC Registered Data

Stresses exceeding Maximum Ratings may damage the device. Maximum Ratings are stress ratings only. Functional operation above the Recommended Operating Conditions is not implied. Extended exposure to stresses above the Recommended Operating Conditions may affect device reliability.

1. V_{DRM} and V_{RRM} for all types can be applied on a continuous basis. Ratings apply for zero or negative gate voltage; however, positive gate voltage shall not be applied concurrent with negative potential on the anode. Blocking voltages shall not be tested with a constant current source such that the voltage ratings of the devices are exceeded.

*For additional information on our Pb–Free strategy and soldering details, please download the ON Semiconductor Soldering and Mounting Techniques Reference Manual, SOLDERRM/D.

ON Semiconductor®

http://onsemi.com

SCRs
12 AMPERES RMS
50 thru 800 VOLTS

A G K

MARKING DIAGRAM

TO–220AB
CASE 221A
STYLE 3

2N639xG
AYWW

2N639x = Device Code
x = 4, 5, 7, or 9
G = Pb–Free Package
A = Assembly Location
Y = Year
WW = Work Week

PIN ASSIGNMENT	
1	Cathode
2	Anode
3	Gate
4	Anode

ORDERING INFORMATION

See detailed ordering and shipping information in the package dimensions section on page 4 of this data sheet.

Preferred devices are recommended choices for future use and best overall value.

© Semiconductor Components Industries, LLC, 2006
August, 2006 – Rev. 6
1
Publication Order Number:
2N6394/D

2N6027, 2N6028

Preferred Device

Programmable Unijunction Transistor
Programmable Unijunction Transistor Triggers

Designed to enable the engineer to "program" unijunction characteristics such as R_{BB}, η, I_V, and I_P by merely selecting two resistor values. Application includes thyristor–trigger, oscillator, pulse and timing circuits. These devices may also be used in special thyristor applications due to the availability of an anode gate. Supplied in an inexpensive TO–92 plastic package for high–volume requirements, this package is readily adaptable for use in automatic insertion equipment.

Features

- Programmable – R_{BB}, η, I_V and I_P
- Low On–State Voltage – 1.5 V Maximum @ I_F = 50 mA
- Low Gate to Anode Leakage Current – 10 nA Maximum
- High Peak Output Voltage – 11 V Typical
- Low Offset Voltage – 0.35 V Typical (R_G = 10 kΩ)
- Pb–Free Packages are Available*

*For additional information on our Pb–Free strategy and soldering details, please download the ON Semiconductor Soldering and Mounting Techniques Reference Manual, SOLDERRM/D.

ON Semiconductor®

http://onsemi.com

PUTs
40 VOLTS, 300 mW

G A K

TO–92 (TO–226AA)
CASE 029
STYLE 16

MARKING DIAGRAM

2N
602x
AYWW•

2N602x = Device Code
x = 7 or 8
A = Assembly Location
Y = Year
WW = Work Week
• = Pb–Free Package
(Note: Microdot may be in either location)

PIN ASSIGNMENT	
1	Anode
2	Gate
3	Cathode

ORDERING INFORMATION

See detailed ordering and shipping information in the package dimensions section on page 5 of this data sheet.

Preferred devices are recommended choices for future use and best overall value.

© Semiconductor Components Industries, LLC, 2006
May, 2006 – Rev.6
1
Publication Order Number:
2N6027/D

FIGURE 11–47

Partial datasheets for the 2N6397 silicon-controlled rectifier and for the 2N6027 programmable unijunction transistor. Copyright of Semiconductor Component Industries, LLC. Used by permission.

Answer the following questions using the partial datasheets in Figure 11–47. If sufficient information doesn't appear on these datasheets, go to *onsemi.com* and download the complete datasheet(s).

1. How much peak voltage can the SCR withstand in the *off* state?
2. What is the maximum SCR current when it is turned on?
3. What is the maximum power dissipation of the PUT?

Simulation

The motor speed-control circuit is simulated in Multisim with a resistive/inductive load in place of the motor and a dc voltage source in place of the input from the processing circuit, as shown in Figure 11–48. The diode is placed across the motor for transient suppression.

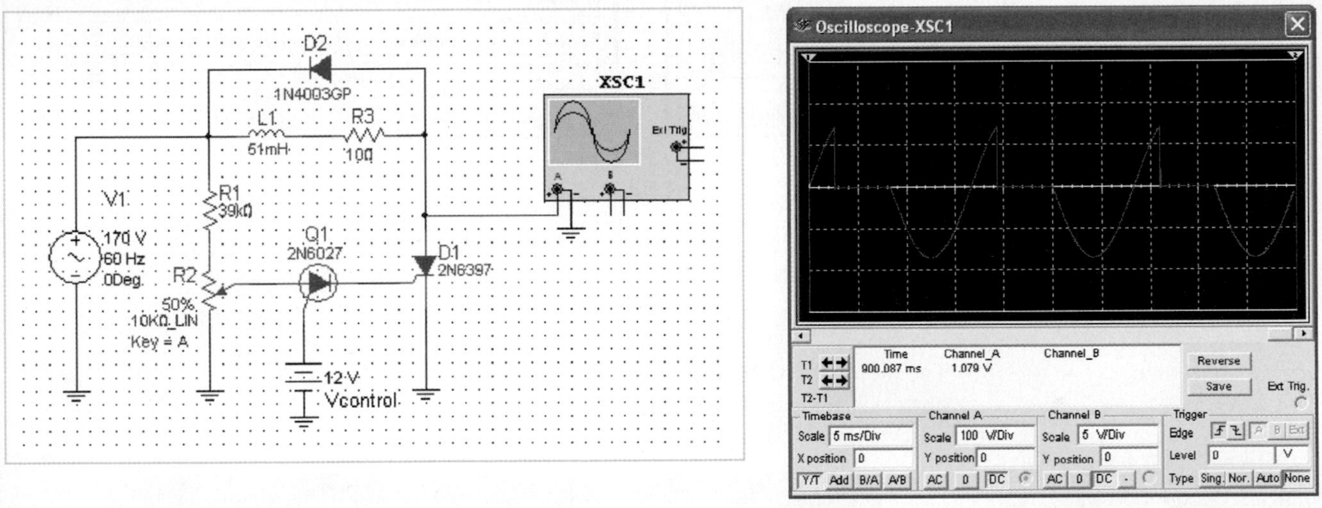

(a) Motor speed-control circuit

(b) Voltage across the SCR at $V_{cont} = 12$ V

FIGURE 11–48

Simulation results for the motor speed-control circuit.

4. On the scope display in Figure 11–48 identify when the SCR is conducting.
5. If the control voltage is reduced, will the SCR conduct more or less?
6. If the control voltage is reduced, will the motor speed increase or decrease?

Figure 11–49 shows the results of varying $V_{control}$. You can see that as the control voltage is decreased, the SCR conducts for more of the cycle and therefore delivers more power to the motor to increase its speed.

Simulate the motor speed-control circuit using your Multisim software. Observe how the SCR voltage changes with changes in $V_{control}$.

Prototyping and Testing

Now that the circuit has been simulated, the prototype circuit is constructed and tested. After the circuit is successfully tested on a protoboard, it is ready to be finalized on a printed circuit board.

Lab Experiment

To build and test a similar circuit, go to Experiment 11 in your lab manual (*Laboratory Exercises for Electronic Devices* by David Buchla and Steven Wetterling).

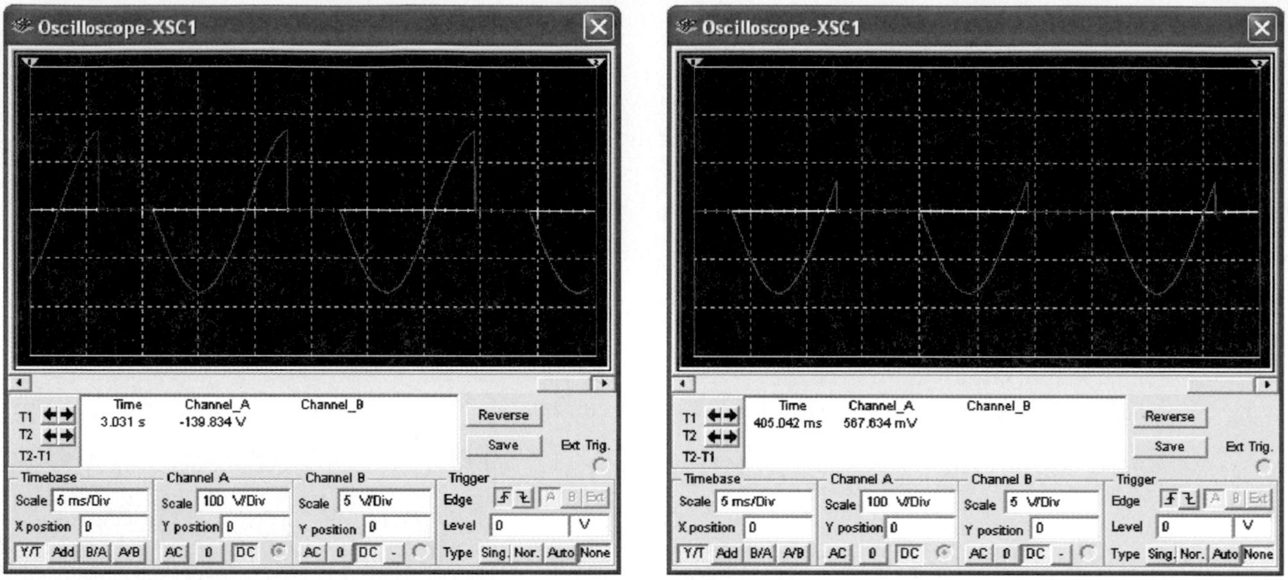

(a) Voltage across the SCR at $V_{cont} = 14.2$ V

(b) Voltage across the SCR at $V_{cont} = 4$ V

FIGURE 11–49

SCR waveforms for two control voltages.

Circuit Board

The motor speed-control circuit board is shown in Figure 11–50. The heat sink is for power dissipation in the SCR.

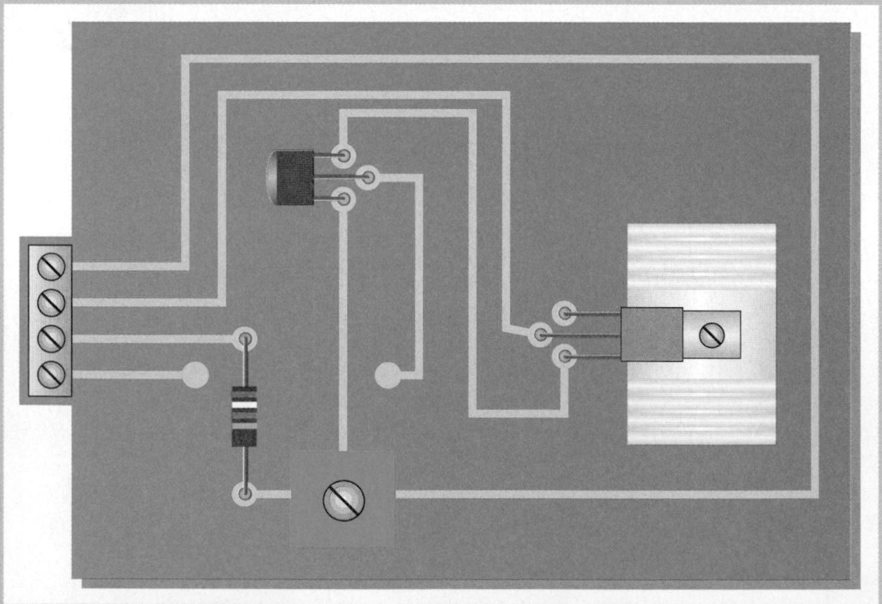

FIGURE 11–50

Motor speed-control circuit board.

7. Check the printed circuit board for correctness by comparing with the schematic in Figure 11–46.
8. Label each input and output pin according to function.

Troubleshooting

Three circuit boards are tested, and the results are shown in Figure 11–51.

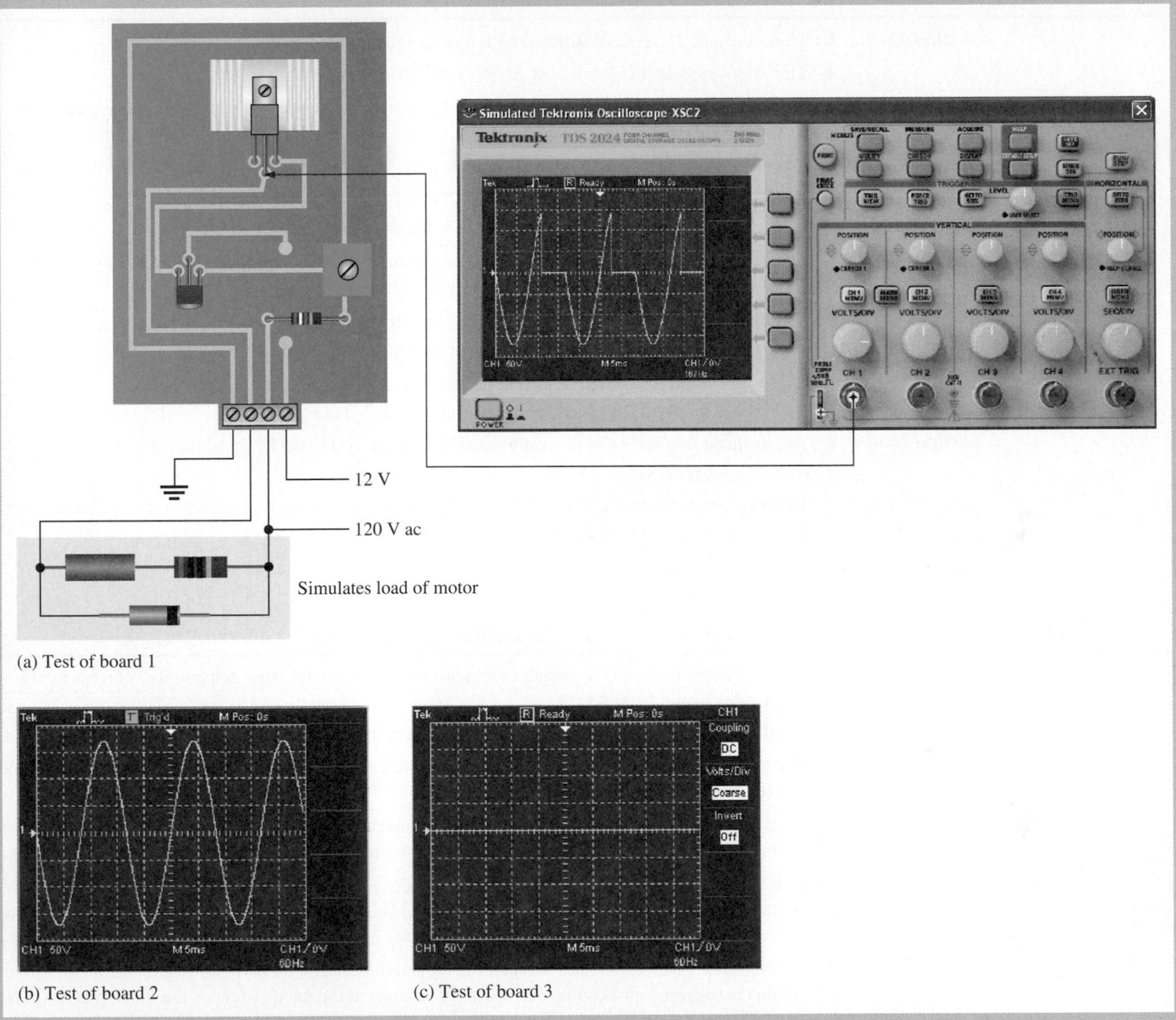

FIGURE 11–51

9. Determine the problem, if any, in each of the board tests in Figure 11–51.
10. List possible causes of any problem from item 10.

SUMMARY OF THYRISTOR SYMBOLS

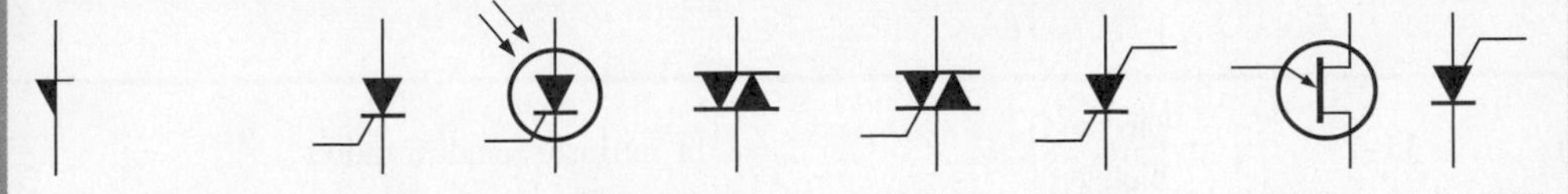

SUMMARY

Section 11–1

- Thyristors are devices constructed with four semiconductor layers (*pnpn*).
- Thyristors include 4-layer diodes, SCRs, LASCRs, diacs, triacs, SCSs, and PUTs.
- The 4-layer diode is a thyristor that conducts when the voltage across its terminals exceeds the breakover potential.

Section 11–2

- The silicon-controlled rectifier (SCR) can be triggered on by a pulse at the gate and turned off by reducing the anode current below the specified holding value.
- Light acts as the trigger source in light-activated SCRs (LASCRs).

Section 11–4

- The diac can conduct current in either direction and is turned on when a breakover voltage is exceeded. It turns off when the current drops below the holding value.
- The triac, like the diac, is a bidirectional device. It can be turned on by a pulse at the gate and conducts in a direction depending on the voltage polarity across the two anode terminals.

Section 11–5

- The silicon-controlled switch (SCS) has two gate terminals and can be turned on by a pulse at the cathode gate and turned off by a pulse at the anode gate.

Section 11–6

- The intrinsic standoff ratio of a unijunction transistor (UJT) determines the voltage at which the device will trigger on.

Section 11–7

- The programmable unijunction transistor (PUT) can be externally programmed to turn on at a desired anode-to-gate voltage level.

KEY TERMS

Key terms and other bold terms in the chapter are defined in the end-of-book glossary.

Diac A two-terminal four-layer semiconductor device (thyristor) that can conduct current in either direction when properly activated.

Forward-breakover voltage ($V_{BR(F)}$) The voltage at which a device enters the forward-blocking region.

4-layer diode The type of two-terminal thyristor that conducts current when the anode-to-cathode voltage reaches a specified "breakover" value.

Holding current (I_H) The value of the anode current below which a device switches from the forward-conduction region to the forward-blocking region.

LASCR Light-activated silicon-controlled rectifier; a four-layer semiconductor device (thyristor) that conducts current in one direction when activated by a sufficient amount of light and continues to conduct until the current falls below a specified value.

PUT Programmable unijunction transistor; a type of three-terminal thyristor (more like an SCR than a UJT) that is triggered into conduction when the voltage at the anode exceeds the voltage at the gate.

SCR Silicon-controlled rectifier; a type of three-terminal thyristor that conducts current when triggered on by a voltage at the single gate terminal and remains on until the anode current falls below a specified value.

SCS Silicon-controlled switch; a type of four-terminal thyristor that has two gate terminals that are used to trigger the device on and off.

Standoff ratio The characteristic of a UJT that determines its turn-on point.

Thyristor A class of four-layer (*pnpn*) semiconductor devices.

Triac A three-terminal thyristor that can conduct current in either direction when properly activated.

UJT Unijunction transistor; a three-terminal single *pn* junction device that exhibits a negative resistance characteristic.

KEY FORMULAS

11–1	$\eta = \dfrac{r'_{B1}}{r'_{BB}}$	UJT intrinsic standoff ratio
11–2	$V_P = \eta V_{BB} + V_{pn}$	UJT peak-point voltage

TRUE/FALSE QUIZ

Answers are at the end of the chapter.

1. A thyristor is characterized by four semiconductor layers.
2. An SCR is a silicon conduction rectifier.
3. The three terminals of an SCR are the anode, cathode, and gate.
4. One method for turning off an SCR is called forced commutation.
5. The SCR is turned on by a pulse on the anode.
6. A diac can conduct current in two directions.
7. A diac has two terminals.
8. A triac has four terminals.
9. The SCS is a silicon-controlled switch.
10. The UJT is commonly used to trigger thyristors but is not a thyristor itself.
11. The PUT is a three-terminal thyristor that can be turned on and off by a voltage on its gate.
12. PUT stands for positive unijunction transistor.

CIRCUIT-ACTION QUIZ

Answers are at the end of the chapter.

1. If the potentiometer in Figure 11–19 is adjusted from a setting near the bottom (low resistance from wiper to ground) to a setting near the top (higher resistance from wiper to ground), the average current through R_L will
   (a) increase **(b)** decrease **(c)** not change
2. If the diode in Figure 11–19 opens, the voltage across R_L will
   (a) increase **(b)** decrease **(c)** not change
3. Assume that the battery in Figure 11–21 is fully charged and the ac power goes off. If D_3 opens, the current through the lamp will immediately
   (a) increase **(b)** decrease **(c)** not change
4. If the capacitor in Figure 11–44 shorts to ground, the voltage at the cathode of the PUT will
   (a) increase **(b)** decrease **(c)** not change

SELF-TEST

Answers are at the end of the chapter.

Section 11–1

1. A thyristor has
   (a) two *pn* junctions **(b)** three *pn* junctions
   (c) four *pn* junctions **(d)** only two terminals
2. Common types of thyristors include
   (a) BJTs and SCRs **(b)** UJTs and PUTs
   (c) FETs and triacs **(d)** diacs and triacs
3. A 4-layer diode turns on when the anode-to-cathode voltage exceeds
   (a) 0.7 V **(b)** the gate voltage
   (c) the forward-breakover voltage **(d)** the forward-blocking voltage
4. Once it is conducting, a 4-layer diode can be turned off by
   (a) reducing the current below a certain value
   (b) disconnecting the anode voltage
   (c) answers (a) and (b)
   (d) neither answer (a) nor (b)

Section 11–2

5. An SCR differs from the 4-layer diode because
   (a) it has a gate terminal **(b)** it is not a thyristor
   (c) it does not have four layers **(d)** it cannot be turned on and off

6. An SCR can be turned off by
   (a) forced commutation **(b)** a negative pulse on the gate
   (c) anode current interruption **(d)** answers (a), (b), and (c)
   (e) answers (a) and (c)
7. In the forward-blocking region, the SCR is
   (a) reverse-biased **(b)** in the *off* state
   (c) in the *on* state **(d)** at the point of breakdown
8. The specified value of holding current for an SCR means that
   (a) the device will turn on when the anode current exceeds this value
   (b) the device will turn off when the anode current falls below this value
   (c) the device may be damaged if the anode current exceeds this value
   (d) the gate current must equal or exceed this value to turn the device on

Section 11–4

9. The diac is
   (a) a thyristor
   (b) a bilateral, two-terminal device
   (c) like two parallel 4-layer diodes in reverse directions
   (d) answers (a), (b), and (c)
10. The triac is
   (a) like a bidirectional SCR
   (b) a four-terminal device
   (c) not a thyristor
   (d) answers (a) and (b)

Section 11–5

11. The SCS differs from the SCR because
   (a) it does not have a gate terminal
   (b) its holding current is less
   (c) it can handle much higher currents
   (d) it has two gate terminals
12. The SCS can be turned on by
   (a) an anode voltage that exceeds forward-breakover voltage
   (b) a positive pulse on the cathode gate
   (c) a negative pulse on the anode gate
   (d) either (b) or (c)
13. The SCS can be turned off by
   (a) a negative pulse on the cathode gate and a positive pulse on the anode gate
   (b) reducing the anode current to below the holding value
   (c) answers (a) and (b)
   (d) a positive pulse on the cathode gate and a negative pulse on the anode gate

Section 11–6

14. Which of the following is *not* a characteristic of the UJT?
   (a) intrinsic standoff ratio
   (b) negative resistance
   (c) peak-point voltage
   (d) bilateral conduction

Section 11–7

15. The PUT is
   (a) much like the UJT
   (b) not a thyristor
   (c) triggered on and off by the gate-to-anode voltage
   (d) not a four-layer device

PROBLEMS

Answers to all odd-numbered problems are at the end of the book.

BASIC PROBLEMS

Section 11–1 The Four-Layer Diode

1. The 4-layer diode in Figure 11–52 is biased such that it is in the forward-conduction region. Determine the anode current for $V_{BR(F)} = 20$ V, $V_{BE} = 0.7$ V, and $V_{CE(sat)} = 0.2$ V.

▶ FIGURE 11–52

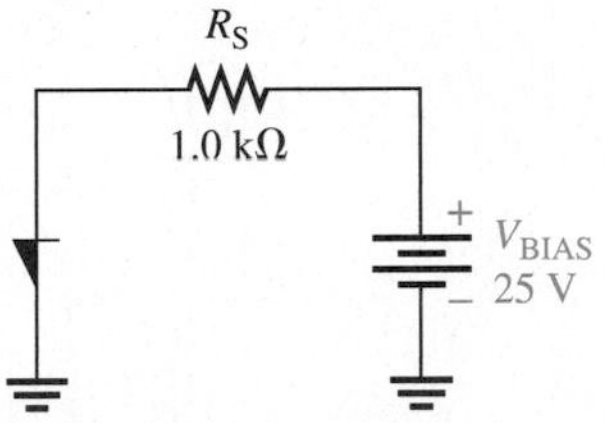

2. (a) Determine the resistance of a certain 4-layer diode in the forward-blocking region if $V_{AK} = 15$ V and $I_A = 1\ \mu$A.
   (b) If the forward-breakover voltage is 50 V, how much must V_{AK} be increased to switch the diode into the forward-conduction region?

Section 11–2 The Silicon-Controlled Rectifier (SCR)

3. Explain the operation of an SCR in terms of its transistor equivalent.
4. To what value must the variable resistor be adjusted in Figure 11–53 in order to turn the SCR off? Assume $I_H = 10$ mA and $V_{AK} = 0.7$ V.

▶ FIGURE 11–53

Multisim file circuits are identified with a CD logo and are in the Problems folder on your CD-ROM. Filenames correspond to figure numbers (e.g., F11-53).

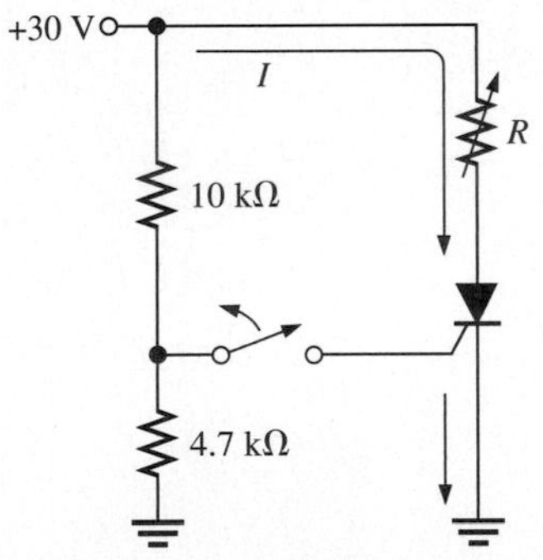

5. By examination of the circuit in Figure 11–54, explain its purpose and basic operation.

▶ FIGURE 11–54

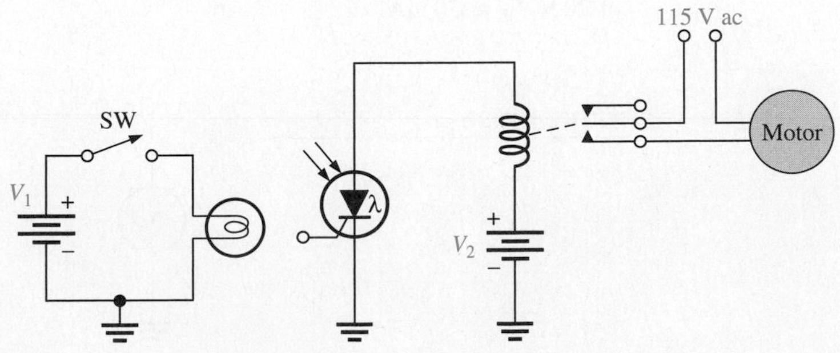

6. Determine the voltage waveform across R_K in Figure 11–55.

▶ FIGURE 11–55

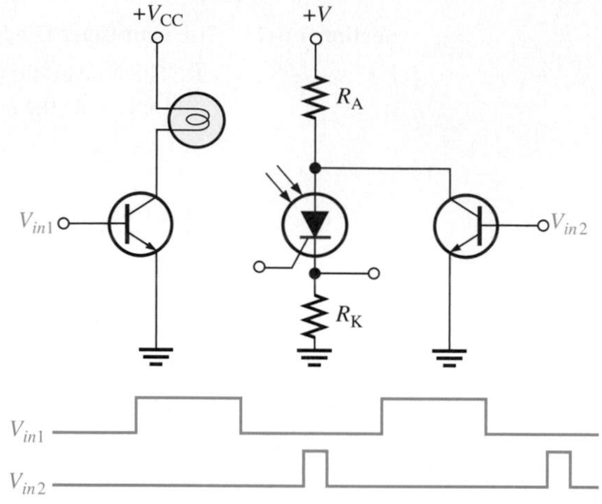

Section 11–3 SCR Applications

7. Describe how you would modify the circuit in Figure 11–17 so that the SCR triggers and conducts on the negative half-cycle of the input.

8. What is the purpose of diodes D_1 and D_2 in Figure 11–21?

9. Sketch the V_R waveform for the circuit in Figure 11–56, given the indicated relationship of the input waveforms.

▶ FIGURE 11–56

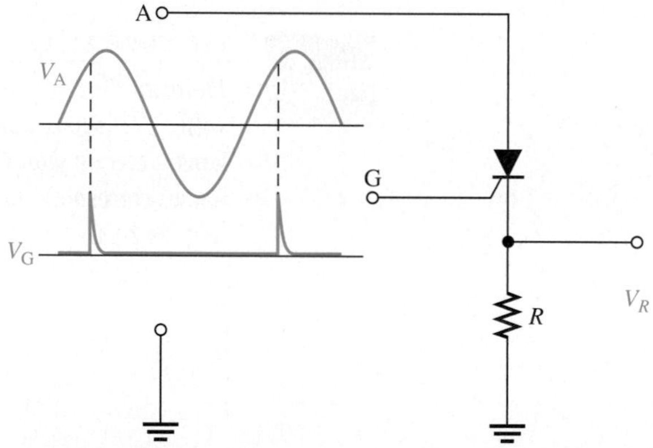

Section 11–4 The Diac and Triac

10. Sketch the current waveform for the circuit in Figure 11–57. The diac has a breakover potential of 20 V. I_H = 20 mA.

▶ FIGURE 11–57

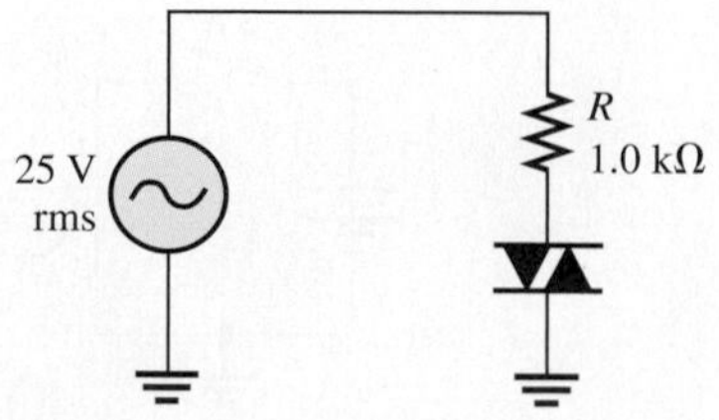

11. Repeat Problem 10 for the triac circuit in Figure 11–58. The breakover potential is 25 V and $I_H = 1$ mA.

FIGURE 11–58

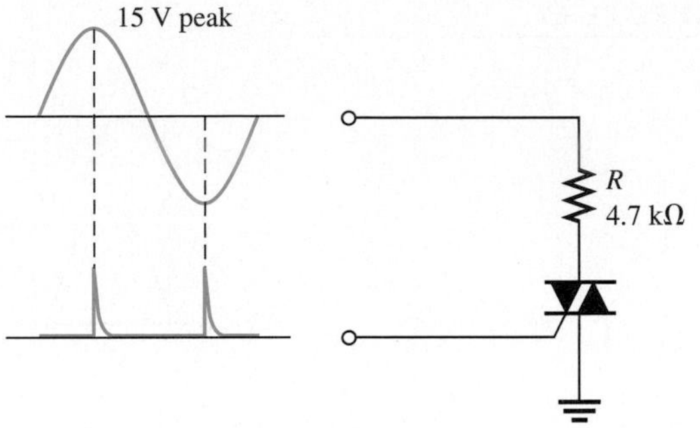

Section 11–5 The Silicon-Controlled Switch (SCS)

12. Explain the turn-on and turn-off operation of an SCS in terms of its transistor equivalent.
13. Name the terminals of an SCS.

Section 11–6 The Unijunction Transistor (UJT)

14. In a certain UJT, $r'_{B1} = 2.5\text{ k}\Omega$ and $r'_{B2} = 4\text{ k}\Omega$. What is the intrinsic standoff ratio?
15. Determine the peak-point voltage for the UJT in Problem 14 if $V_{BB} = 15$ V.
16. Find the range of values of R_1 in Figure 11–59 that will ensure proper turn-on and turn-off of the UJT. $\eta = 0.68$, $V_V = 0.8$ V, $I_V = 15$ mA, $I_P = 10\ \mu$A, and $V_P = 10$ V.

FIGURE 11–59

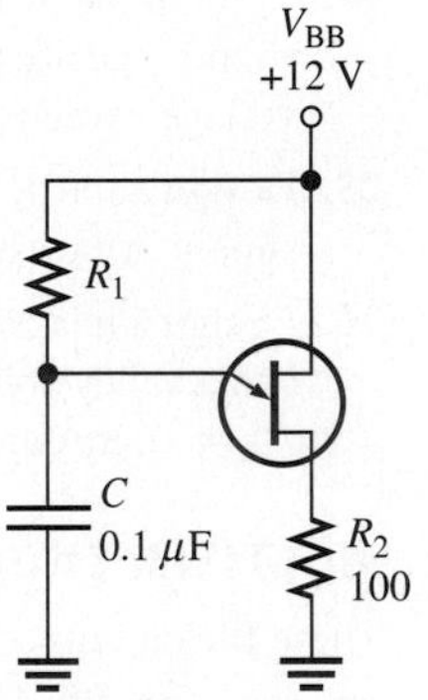

Section 11–7 The Programmable Unijunction Transistor (PUT)

17. At what anode voltage (V_A) will each PUT in Figure 11–60 begin to conduct?
18. Draw the current waveform for each circuit in Figure 11–60 when there is a 10 V peak sinusoidal voltage at the anode. Neglect the forward voltage of the PUT.

FIGURE 11–60

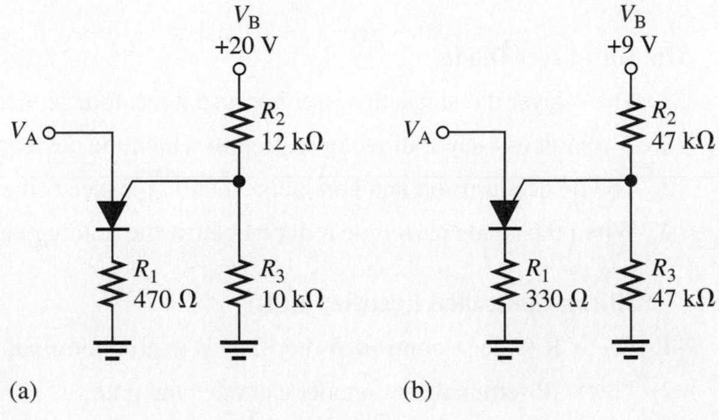

19. Sketch the voltage waveform across R_1 in Figure 11–61 in relation to the input voltage waveform.

20. Repeat Problem 19 if R_3 is increased to 15 kΩ.

FIGURE 11–61

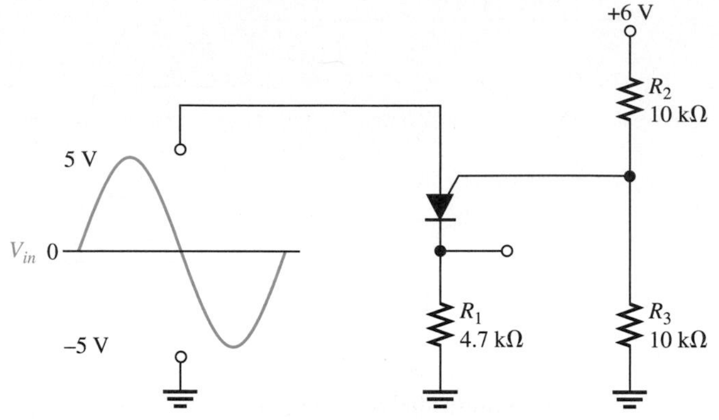

APPLICATION ACTIVITY PROBLEMS

21. In the motor speed-control circuit of Figure 11–46, at which PUT gate voltage does the electric motor run at the fastest speed: 0 V, 2 V, or 5 V?

22. Does the SCR in the motor speed-control circuit turn on earlier or later in the ac cycle if the resistance of the rheostat is reduced?

23. Describe the SCR action as the PUT gate voltage is increased in the motor speed-control circuit.

ADVANCED PROBLEMS

24. Refer to the SCR over-voltage protection circuit in Figure 11–22. For a +12 V output dc power supply, specify the component values that will provide protection for the circuit if the output voltage exceeds +15 V. Assume the fuse is rated at 1 A.

25. Design an SCR crowbar circuit to protect electronic circuits against a voltage from the power supply in excess of 6.2 V.

26. Design a relaxation oscillator to produce a frequency of 2.5 kHz using a UJT with $\eta = 0.75$ and a valley voltage of 1 V. The circuit must operate from a +12 V dc source. Design values of $I_V = 10$ mA and $I_P = 20\ \mu$A are to be used.

MULTISIM TROUBLESHOOTING PROBLEMS

These file circuits are in the Troubleshooting Problems folder on your CD-ROM.

27. Open file TSP11-27 and determine the fault.

28. Open file TSP11-28 and determine the fault.

29. Open file TSP11-29 and determine the fault.

ANSWERS

SECTION CHECKUPS

Section 11–1 The Four-Layer Diode

1. The 4-layer diode is a thyristor because it has four semiconductor layers in a *pnpn* configuration.
2. A region of 4-layer diode operation in which the device is nonconducting
3. The device turns on and conducts when V_{AK} exceeds the forward-breakover voltage.
4. When the anode current is reduced below the holding current value, the device turns off.

Section 11–2 The Silicon-Controlled Rectifier (SCR)

1. An SCR (silicon-controlled rectifier) is a three-terminal thyristor.
2. The SCR terminals are anode, cathode, and gate.

3. A positive gate pulse turns the SCR on.
4. Reduce the anode current below I_H (holding current) to turn a conducting SCR off.
5. Open the series switch

Section 11–3 SCR Applications

1. The SCR will conduct for more than 90° but less than 180°.
2. To block discharge of the battery through that path

Section 11–4 The Diac and Triac

1. The diac is like two parallel 4-layer diodes connected in opposite directions.
2. A triac is like two parallel SCRs having a common gate and connected in opposite directions.
3. A triac has a gate terminal, but a diac does not.

Section 11–5 The Silicon-Controlled Switch (SCS)

1. An SCS can be turned off with the application of a gate pulse, but an SCR cannot.
2. A positive pulse on the cathode gate or a negative pulse on the anode gate turns the SCS on.
3. An SCS can be turned off by any of the following:
   (a) positive pulse on anode gate
   (b) negative pulse on cathode gate
   (c) reduce anode current below holding value by complete interruption of the anode current

Section 11–6 The Unijunction Transistor (UJT)

1. The UJT terminals are base 1, base 2, and emitter.
2. $\eta = r'_{B1}/r'_{BB}$
3. R, C, and η determine the period.

Section 11–7 The Programmable Unijunction Transistor (PUT)

1. *Programmable* means that the turn-on voltage can be adjusted to a desired value.
2. The PUT is a thyristor, similar in structure to an SCR, but it is turned on by the anode-to-gate voltage. It has a negative resistance characteristic like the UJT.

RELATED PROBLEMS FOR EXAMPLES

11–1 10 MΩ

11–2 47.1 Ω

11–3 Yes. The current is greater than I_H.

11–4 $V_{AK} = V_s$

11–5 By increasing V_{BB}

11–6 $343\text{ k}\Omega > R_1 > 1.95\text{ k}\Omega$

TRUE/FALSE QUIZ

1. T	2. F	3. T	4. T	5. F	6. T
7. T	8. F	9. T	10. T	11. T	12. F

CIRCUIT-ACTION QUIZ

1. (b) 2. (b) 3. (c) 4. (b)

SELF-TEST

1. (b)	2. (d)	3. (c)	4. (c)	5. (a)	6. (e)	7. (b)	8. (b)
9. (d)	10. (a)	11. (d)	12. (d)	13. (c)	14. (d)	15. (c)	

12 The Operational Amplifier

CHAPTER OUTLINE

CHAPTER OBJECTIVES

- Describe the basic op-amp and its characteristics
- Discuss op-amp modes and several parameters
- Explain negative feedback in op-amp circuits
- Analyze the noninverting, voltage-follower, and inverting op-amp configurations
- Describe impedances of the three op-amp configurations
- Discuss op-amp compensation
- Analyze the open-loop response of an op-amp
- Analyze the closed-loop frequency response of an op-amp
- Troubleshoot op-amp circuits

KEY TERMS

- Operational amplifier (op-amp)
- Differential amplifier
- Differential mode
- Common mode
- CMRR
- Open-loop voltage gain
- Slew rate
- Negative feedback
- Closed-loop voltage gain
- Noninverting amplifier
- Voltage-follower
- Inverting amplifier
- Phase shift
- Gain-bandwidth product

APPLICATION ACTIVITY PREVIEW

For the Application Activity in this chapter, the audio amplifier from the PA system in Chapter 7 is modified. The two-stage preamp portion of the amplifier is replaced by an op-amp circuit. The power amplifier portion is retained in its original configuration with the exception of the drive circuit so that the new design consists of an op-amp driving a push-pull power stage. In the original system there are two PC boards—one for the preamp and one for the power amplifier. The new design will allow both the preamp and the power amplifier to be on a single PC board.

VISIT THE COMPANION WEBSITE

Study aids for this chapter are available at http://www.prenhall.com/floyd

INTRODUCTION

In the previous chapters, you have studied a number of important electronic devices. These devices, such as the diode and the transistor, are separate devices that are individually packaged and interconnected in a circuit with other devices to form a complete, functional unit. Such devices are referred to as *discrete components.*

Now you will begin the study of linear integrated circuits (ICs), where many transistors, diodes, resistors, and capacitors are fabricated on a single tiny chip of semiconductive material and packaged in a single case to form a functional circuit. An integrated circuit, such as an operational amplifier (op-amp), is treated as a single device. This means that you will be concerned with what the circuit does more from an external viewpoint than from an internal, component-level viewpoint.

In this chapter, you will learn the basics of op-amps, which are the most versatile and widely used of all linear integrated circuits. You will also learn about open-loop and closed-loop frequency responses, bandwidth, phase shift, and other frequency-related parameters. The effects of negative feedback will be examined.

12–1 INTRODUCTION TO OPERATIONAL AMPLIFIERS

Early operational amplifiers (op-amps) were used primarily to perform mathematical operations such as addition, subtraction, integration, and differentiation—thus the term *operational.* These early devices were constructed with vacuum tubes and worked with high voltages. Today's op-amps are linear integrated circuits (ICs) that use relatively low dc supply voltages and are reliable and inexpensive.

After completing this section, you should be able to

- **Describe the basic op-amp and its characteristics**
  - Recognize the op-amp symbol
  - Identify the terminals on op-amp packages
  - Describe the ideal op-amp
  - Describe the practical op-amp

The standard **operational amplifier (op-amp)** symbol is shown in Figure 12–1(a). It has two input terminals, the inverting (−) input and the noninverting (+) input, and one output terminal. Most op-amps operate with two dc supply voltages, one positive and the other negative, as shown in Figure 12–1(b), although some have a single dc supply. Usually these dc voltage terminals are left off the schematic symbol for simplicity but are understood to be there. Some typical op-amp IC packages are shown in Figure 12–1(c).

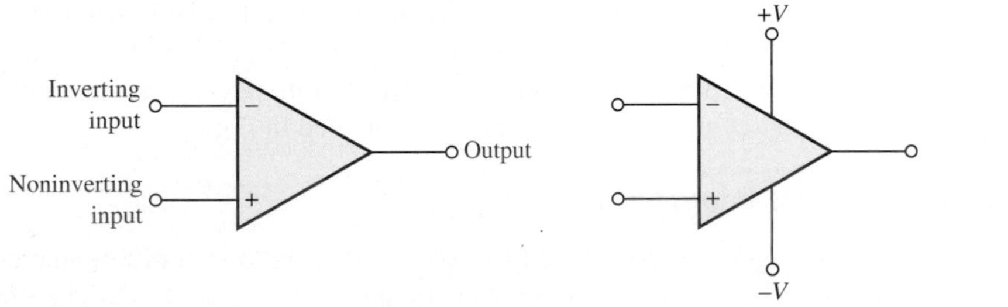

(a) Symbol

(b) Symbol with dc supply connections

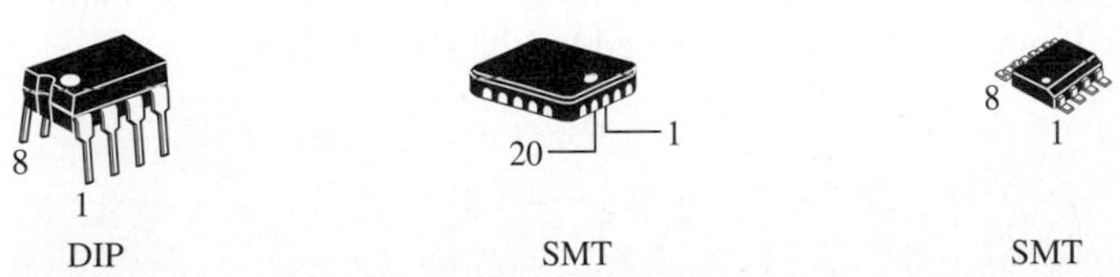

(c) Typical packages. Pin 1 is indicated by a notch or dot on dual in-line (DIP) and surface-mount technology (SMT) packages, as shown.

FIGURE 12–1

Op-amp symbols and packages.

HISTORY NOTE

The operational amplifier concept originated around 1947. It was proposed that such a device would form an extremely useful analog building block. The first commercial op-amps used vacuum tubes, but it was not until the introduction of the integrated circuit did the op-amp start to fulfill its true potential. In 1964, the first integrated circuit op-amp, designated the 702, was developed by Fairchild Semiconductor. This was later followed by the 709 and eventually the 741, which has become an industry standard.

The Ideal Op-Amp

To illustrate what an op-amp is, let's consider its ideal characteristics. A practical op-amp, of course, falls short of these ideal standards, but it is much easier to understand and analyze the device from an ideal point of view.

First, the ideal op-amp has *infinite voltage gain* and *infinite bandwidth.* Also, it has an *infinite input impedance* (open) so that it does not load the driving source. Finally, it has a *zero output impedance.* These characteristics are illustrated in Figure 12–2(a). The input voltage, V_{in}, appears between the two input terminals, and the output voltage is A_vV_{in}, as indicated by the internal voltage source symbol. The concept of infinite input impedance is a particularly valuable analysis tool for the various op-amp configurations, which will be discussed in Section 12–4.

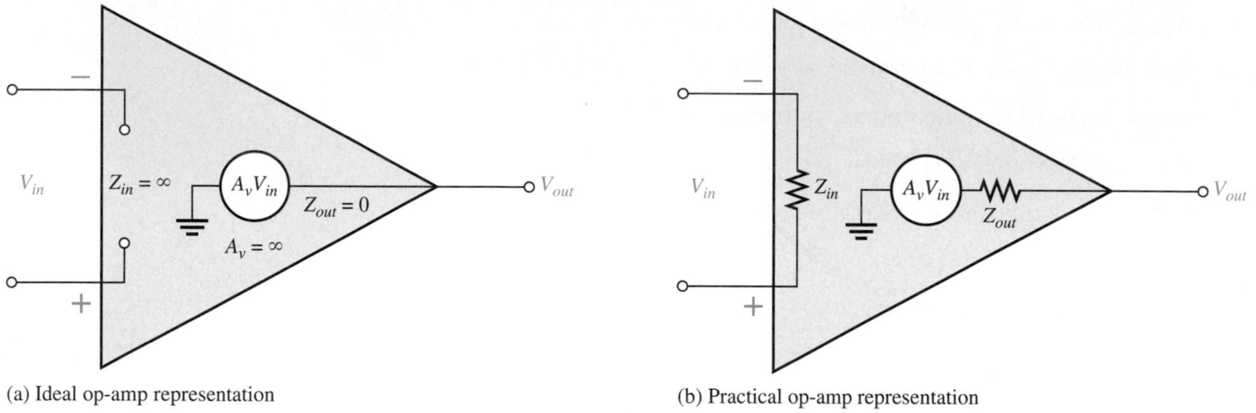

▲ FIGURE 12–2

Basic op-amp representations.

The Practical Op-Amp

Although **integrated circuit (IC)** op-amps approach parameter values that can be treated as ideal in many cases, the ideal device can never be made. Any device has limitations, and the IC op-amp is no exception. Op-amps have both voltage and current limitations. Peak-to-peak output voltage, for example, is usually limited to slightly less than the two supply voltages. Output current is also limited by internal restrictions such as power dissipation and component ratings.

Characteristics of a practical op-amp are *very high voltage gain, very high input impedance, and very low output impedance.* These are labelled in Figure 12–2(b).

Internal Block Diagram of an Op-Amp

A typical op-amp is made up of three types of amplifier circuits: a differential amplifier, a voltage amplifier, and a push-pull amplifier, as shown in Figure 12–3. The **differential amplifier** is the input stage for the op-amp. It provides amplification of the difference voltage between the two inputs. The second stage is usually a class A amplifier that provides additional gain. Some op-amps may have more than one voltage amplifier stage. A push-pull class B amplifier is typically used for the output stage.

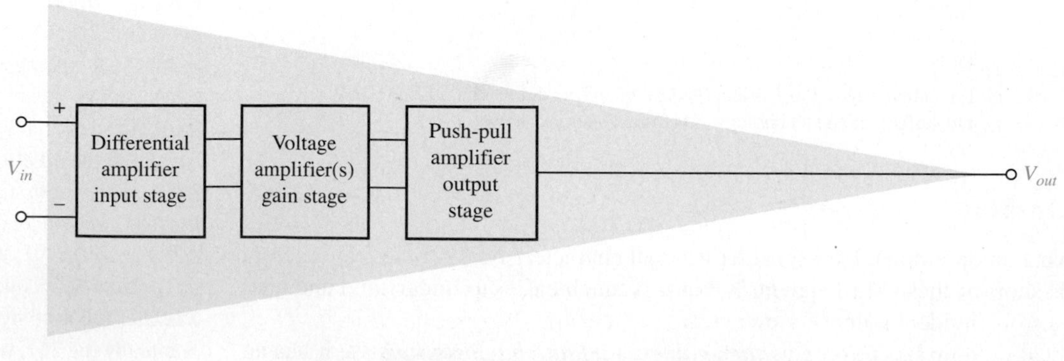

▲ FIGURE 12–3

Basic internal arrangement of an op-amp.

The differential amplifier was introduced in Chapter 6. The term *differential* comes from the amplifier's ability to amplify the difference of two input signals applied to its inputs. Only the difference in the two signals is amplified; if there is no difference, the output is zero. The differential amplifier exhibits two modes of operation based on the type of input signals. These modes are *differential* and *common,* which are described in the next section. Since the differential amplifier is the input stage of the op-amp, the op-amp exhibits the same modes.

SECTION 12–1 CHECKUP
Answers are at the end of the chapter.

1. What are the connections to a basic op-amp?
2. Describe some of the characteristics of a practical op-amp.
3. List the amplifier stages in a typical op-amp.
4. What does a differential amplifier amplify?

12–2 OP-AMP INPUT MODES AND PARAMETERS

In this section, important op-amp input modes and several parameters are defined. Also several common IC op-amps are compared in terms of these parameters.

After completing this section, you should be able to

- **Discuss op-amp modes and several parameters**
  - Explain differential-mode operation
  - Explain common-mode operation
  - Define *common-mode rejection ratio*
  - Discuss open-loop voltage gain
  - Discuss maximum output voltage swing
  - Define *input offset voltage* and discuss input offset voltage drift with temperature
  - Define *input bias current, input impedance, input offset current,* and *output impedance*
  - Define *slew rate*
  - Compare the parameters of several types of IC op-amps

Input Signal Modes

Recall that the input signal modes are determined by the differential amplifier input stage of the op-amp.

Differential Mode In the **differential mode**, either one signal is applied to an input with the other input grounded or two opposite-polarity signals are applied to the inputs. When an op-amp is operated in the single-ended differential mode, one input is grounded and a signal voltage is applied to the other input, as shown in Figure 12–4. In the case where the signal voltage is applied to the inverting input as in part (a), an inverted, amplified signal voltage appears at the output. In the case where the signal is applied to the noninverting input with the inverting input grounded, as in Figure 12–4(b), a noninverted, amplified signal voltage appears at the output.

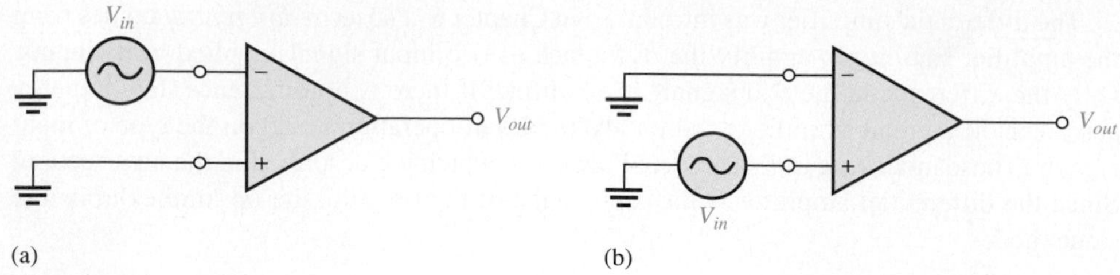

FIGURE 12–4

Single-ended differential mode.

In the double-ended differential mode, two opposite-polarity (out-of-phase) signals are applied to the inputs, as shown in Figure 12–5(a). The amplified difference between the two inputs appears on the output. Equivalently, the double-ended differential mode can be represented by a single source connected between the two inputs, as shown in Figure 12–5(b).

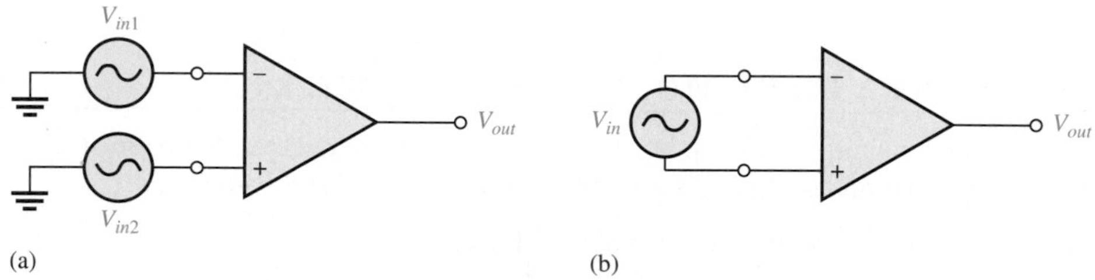

FIGURE 12–5

Double-ended differential mode.

Common Mode In the **common mode**, two signal voltages of the same phase, frequency, and amplitude are applied to the two inputs, as shown in Figure 12–6. When equal input signals are applied to both inputs, they tend to cancel, resulting in a zero output voltage.

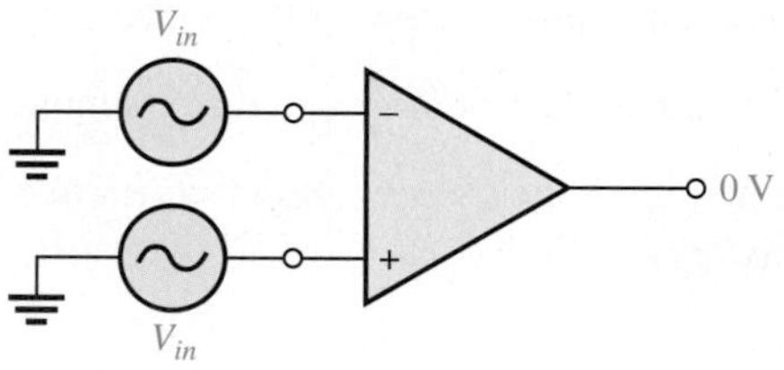

FIGURE 12–6

Common-mode operation.

This action is called *common-mode rejection.* Its importance lies in the situation where an unwanted signal appears commonly on both op-amp inputs. Common-mode rejection means that this unwanted signal will not appear on the output and distort the desired signal. Common-mode signals (noise) generally are the result of the pick-up of radiated energy on the input lines, from adjacent lines, the 60 Hz power line, or other sources.

Common-Mode Rejection Ratio

Desired signals can appear on only one input or with opposite polarities on both input lines. These desired signals are amplified and appear on the output as previously discussed. Unwanted signals (noise) appearing with the same polarity on both input lines are essentially cancelled by the op-amp and do not appear on the output. The measure of an amplifier's ability to reject common-mode signals is a parameter called the **CMRR (common-mode rejection ratio)**.

Ideally, an op-amp provides a very high gain for differential-mode signals and zero gain for common-mode signals. Practical op-amps, however, do exhibit a very small common-mode gain (usually much less than 1), while providing a high open-loop differential voltage gain (usually several thousand). The higher the open-loop gain with respect to the common-mode gain, the better the performance of the op-amp in terms of rejection of common-mode signals. This suggests that a good measure of the op-amp's performance in rejecting unwanted common-mode signals is the ratio of the open-loop differential voltage gain, A_{ol}, to the common-mode gain, A_{cm}. This ratio is the common-mode rejection ratio, CMRR.

$$\mathbf{CMRR} = \frac{A_{ol}}{A_{cm}}$$

Equation 12–1

The higher the CMRR, the better. A very high value of CMRR means that the open-loop gain, A_{ol}, is high and the common-mode gain, A_{cm}, is low.

The CMRR is often expressed in decibels (dB) as

$$\mathbf{CMRR} = 20\log\left(\frac{A_{ol}}{A_{cm}}\right)$$

Equation 12–2

The **open-loop voltage gain**, A_{ol}, of an op-amp is the internal voltage gain of the device and represents the ratio of output voltage to input voltage when there are no external components. The open-loop voltage gain is set entirely by the internal design. Open-loop voltage gain can range up to 200,000 and is not a well-controlled parameter. Datasheets often refer to the open-loop voltage gain as the *large-signal voltage gain.*

A CMRR of 100,000, for example, means that the desired input signal (differential) is amplified 100,000 times more than the unwanted noise (common-mode). If the amplitudes of the differential input signal and the common-mode noise are equal, the desired signal will appear on the output 100,000 times greater in amplitude than the noise. Thus, the noise or interference has been essentially eliminated.

EXAMPLE 12–1

A certain op-amp has an open-loop differential voltage gain of 100,000 and a common-mode gain of 0.2. Determine the CMRR and express it in decibels.

Solution $A_{ol} = 100{,}000$, and $A_{cm} = 0.2$. Therefore,

$$\text{CMRR} = \frac{A_{ol}}{A_{cm}} = \frac{100{,}000}{0.2} = \mathbf{500{,}000}$$

Expressed in decibels,

$$\text{CMRR} = 20\log(500{,}000) = \mathbf{114\ dB}$$

*Related Problem** Determine the CMRR and express it in dB for an op-amp with an open-loop differential voltage gain of 85,000 and a common-mode gain of 0.25.

*Answers are at the end of the chapter.

Maximum Output Voltage Swing ($V_{O(p\text{-}p)}$)

With no input signal, the output of an op-amp is ideally 0 V. This is called the *quiescent output voltage.* When an input signal is applied, the ideal limits of the peak-to-peak output signal is $\pm V_{CC}$. In practice, however, this ideal can be approached but never reached. $V_{O(p\text{-}p)}$ varies with the load connected to the op-amp and increases directly with load resistance. For example, the Fairchild KA741 datasheet shows a typical $V_{O(p\text{-}p)}$ of ± 13 V for $V_{CC} = \pm 15$ V when $R_L = 2\ \text{k}\Omega$. $V_{O(p\text{-}p)}$ increases to ± 14 V when $R_L = 10\ \text{k}\Omega$.

Input Offset Voltage

The ideal op-amp produces zero volts out for zero volts in. In a practical op-amp, however, a small dc voltage, $V_{OUT(error)}$, appears at the output when no differential input voltage is applied. Its primary cause is a slight mismatch of the base-emitter voltages of the differential amplifier input stage of an op-amp.

As specified on an op-amp datasheet, the *input offset voltage*, V_{OS}, is the differential dc voltage required between the inputs to force the output to zero volts. Typical values of input offset voltage are in the range of 2 mV or less. In the ideal case, it is 0 V.

The *input offset voltage drift* is a parameter related to V_{OS} that specifies how much change occurs in the input offset voltage for each degree change in temperature. Typical values range anywhere from about 5 μV per degree Celsius to about 50 μV per degree Celsius. Usually, an op-amp with a higher nominal value of input offset voltage exhibits a higher drift.

Input Bias Current

You have seen that the input terminals of a bipolar differential amplifier are the transistor bases and, therefore, the input currents are the base currents.

The *input bias current* is the dc current required by the inputs of the amplifier to properly operate the first stage. By definition, the input bias current is the *average* of both input currents and is calculated as follows:

Equation 12–3

$$I_{BIAS} = \frac{I_1 + I_2}{2}$$

The concept of input bias current is illustrated in Figure 12–7.

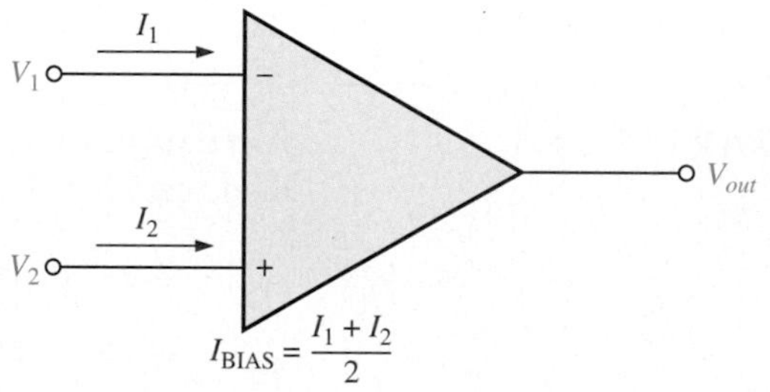

FIGURE 12–7

Input bias current is the average of the two op-amp input currents.

Input Impedance

Two basic ways of specifying the input impedance of an op-amp are the differential and the common mode. The *differential input impedance* is the total resistance between the inverting and the noninverting inputs, as illustrated in Figure 12–8(a). Differential impedance is measured by determining the change in bias current for a given change in differential input voltage. The *common-mode input impedance* is the resistance between each input and ground and is measured by determining the change in bias current for a given change in common-mode input voltage. It is depicted in Figure 12–8(b).

FIGURE 12–8

Op-amp input impedance.

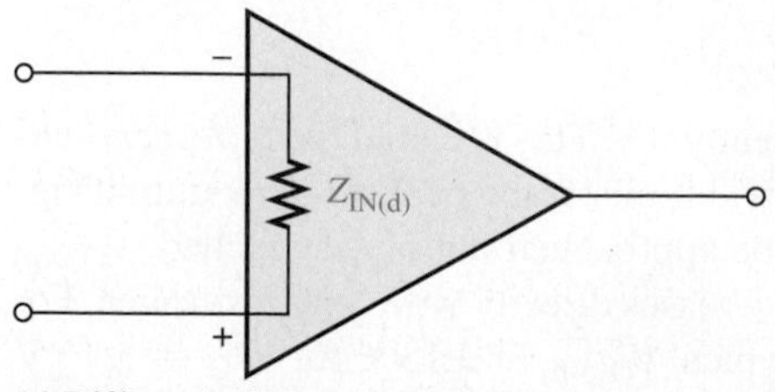

(a) Differential input impedance

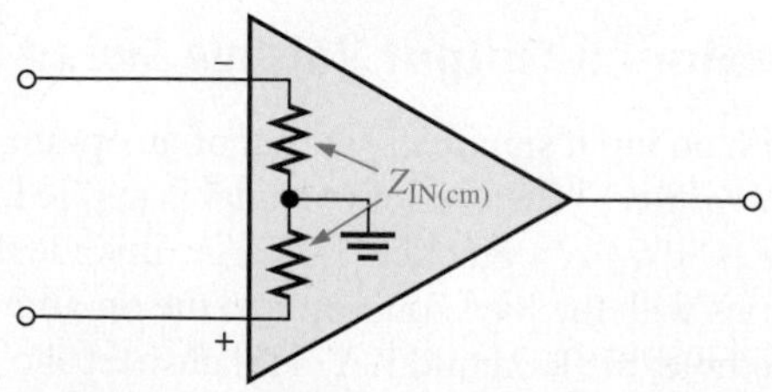

(b) Common-mode input impedance

Input Offset Current

Ideally, the two input bias currents are equal, and thus their difference is zero. In a practical op-amp, however, the bias currents are not exactly equal.

The *input offset current,* I_{OS}, is the difference of the input bias currents, expressed as an absolute value.

$$I_{OS} = |I_1 - I_2|$$

Equation 12–4

Actual magnitudes of offset current are usually at least an order of magnitude (ten times) less than the bias current. In many applications, the offset current can be neglected. However, high-gain, high-input impedance amplifiers should have as little I_{OS} as possible because the difference in currents through large input resistances develops a substantial offset voltage, as shown in Figure 12–9.

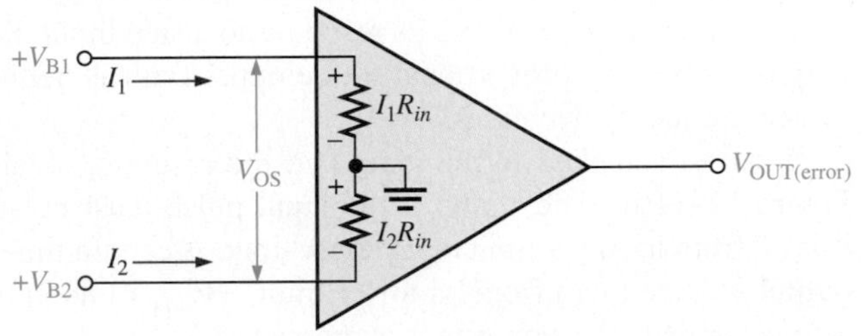

FIGURE 12–9

Effect of input offset current.

The offset voltage developed by the input offset current is

$$V_{OS} = I_1R_{in} - I_2R_{in} = (I_1 - I_2)R_{in}$$

$$V_{OS} = I_{OS}R_{in}$$

Equation 12–5

The error created by I_{OS} is amplified by the gain A_v of the op-amp and appears in the output as

$$V_{OUT(error)} = A_vI_{OS}R_{in}$$

Equation 12–6

A change in offset current with temperature affects the error voltage. Values of temperature coefficient for the offset current in the range of 0.5 nA per degree Celsius are common.

Output Impedance

The *output impedance* is the resistance viewed from the output terminal of the op-amp, as indicated in Figure 12–10.

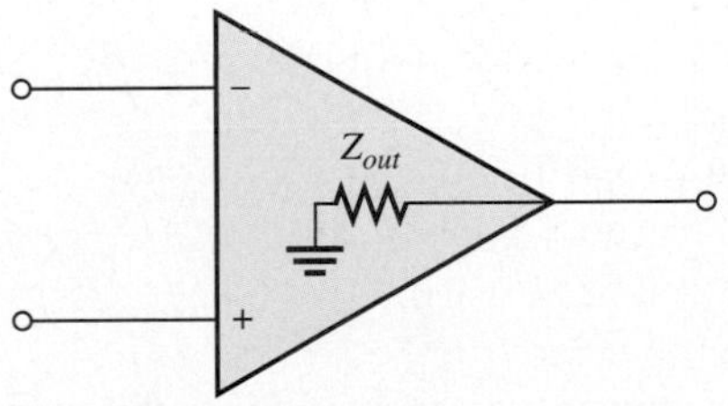

FIGURE 12–10

Op-amp output impedance.

Slew Rate

The maximum rate of change of the output voltage in response to a step input voltage is the **slew rate** of an op-amp. The slew rate is dependent upon the high-frequency response of the amplifier stages within the op-amp.

Slew rate is measured with an op-amp connected as shown in Figure 12–11(a). This particular op-amp connection is a unity-gain, noninverting configuration that will be discussed in Section 12–4. It gives a worst-case (slowest) slew rate. Recall that the high-frequency

FIGURE 12–11

Slew-rate measurement.

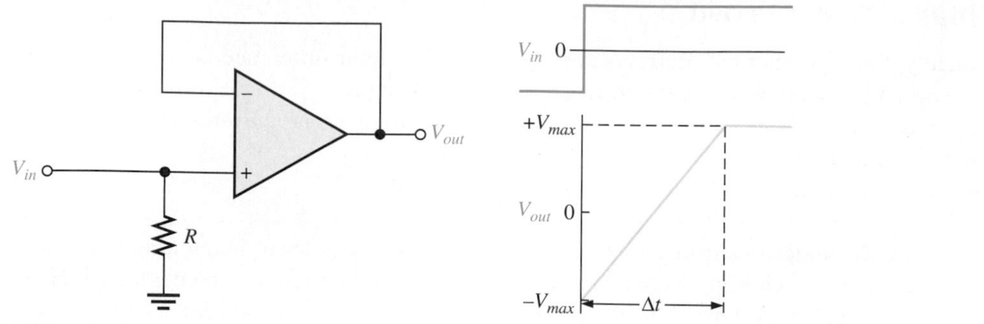

(a) Test circuit

(b) Step input voltage and the resulting output voltage

components of a voltage step are contained in the rising edge and that the upper critical frequency of an amplifier limits its response to a step input. For a step input, the slope on the output is inversely proportional to the upper critical frequency. Slope increases as upper critical frequency decreases.

A pulse is applied to the input and the resulting ideal output voltage is indicated in Figure 12–11(b). The width of the input pulse must be sufficient to allow the output to "slew" from its lower limit to its upper limit. A certain time interval, Δt, is required for the output voltage to go from its lower limit $-V_{max}$ to its upper limit $+V_{max}$, once the input step is applied. The slew rate is expressed as

Equation 12–7

$$\textbf{Slew rate} = \frac{\Delta V_{out}}{\Delta t}$$

where $\Delta V_{out} = +V_{max} - (-V_{max})$. The unit of slew rate is volts per microsecond (V/μs).

EXAMPLE 12–2

The output voltage of a certain op-amp appears as shown in Figure 12–12 in response to a step input. Determine the slew rate.

FIGURE 12–12

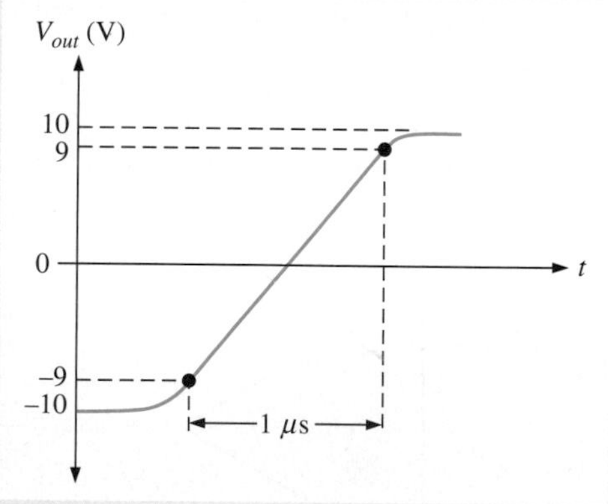

Solution The output goes from the lower to the upper limit in 1 μs. Since this response is not ideal, the limits are taken at the 90% points, as indicated. So, the upper limit is +9 V and the lower limit is −9 V. The slew rate is

$$\text{Slew rate} = \frac{\Delta V_{out}}{\Delta t} = \frac{+9\text{ V} - (-9\text{ V})}{1\ \mu\text{s}} = \mathbf{18\ V/\mu s}$$

Related Problem When a pulse is applied to an op-amp, the output voltage goes from −8 V to +7 V in 0.75 μs. What is the slew rate?

Frequency Response

The internal amplifier stages that make up an op-amp have voltage gains limited by junction capacitances, as discussed in Chapter 10. Although the differential amplifiers used in op-amps are somewhat different from the basic amplifiers discussed earlier, the same principles apply. An op-amp has no internal coupling capacitors, however; therefore, the low-frequency response extends down to dc (0 Hz).

Comparison of Op-Amp Parameters

Table 12–1 provides a comparison of values of some of the parameters just described for several common IC op-amps. Any values not listed were not given on the manufacturer's datasheet.

Most available op-amps have three important features: short-circuit protection, no latch-up, and input offset nulling. Short-circuit protection keeps the circuit from being damaged if the output becomes shorted, and the no latch-up feature prevents the op-amp from hanging up in one output state (high or low voltage level) under certain input conditions. Input offset nulling is achieved by an external potentiometer that sets the output voltage at precisely zero with zero input.

▼ **TABLE 12–1**

OP-AMP	CMRR (dB) (MIN)	OPEN-LOOP GAIN (TYP)	INPUT OFFSET VOLTAGE (mV) (MAX)	INPUT BIAS CURRENT (nA) (MAX)	INPUT IMPEDANCE (MΩ) (MIN)	SLEW RATE (V/μs) (TYP)	COMMENT
LM741C	70	200,000	6	500	0.3	0.5	Industry standard
LM101A	80	160,000	7.5	250	1.5	—	General-purpose
OP113	100	2,400,000	0.075	600	—	1.2	Low noise, low drift
OP177	130	12,000,000	0.01	1.5	26	0.3	Ultra precision
OP184	60	240,000	0.065	350	—	2.4	Precision rail-to-rail*
AD8009	50	—	5	150	—	5500	BW = 700 MHz, ultra fast, low distortion, current feedback
AD8041	74	56,000	7	2000	.16	160	BW = 160 MHz, rail-to-rail
AD8055	82	3500	5	1200	10	1400	Very fast voltage feedback

*Rail-to-rail means that the output voltage can go as high as the supply voltages.

SECTION 12–2 CHECKUP

1. Distinguish between single-ended and double-ended differential mode.
2. Define *common-mode rejection.*
3. For a given value of open-loop differential gain, does a higher common-mode gain result in a higher or lower CMRR?
4. List at least ten op-amp parameters.
5. How is slew rate measured?

12–3 NEGATIVE FEEDBACK

Negative feedback is one of the most useful concepts in electronics, particularly in op-amp applications. **Negative feedback** is the process whereby a portion of the output voltage of an amplifier is returned to the input with a phase angle that opposes (or subtracts from) the input signal.

After completing this section, you should be able to

- **Explain negative feedback in op-amp circuits**
  - Discuss why negative feedback is used
  - Describe the effects of negative feedback

Negative feedback is illustrated in Figure 12–13. The inverting (−) input effectively makes the feedback signal 180° out of phase with the input signal.

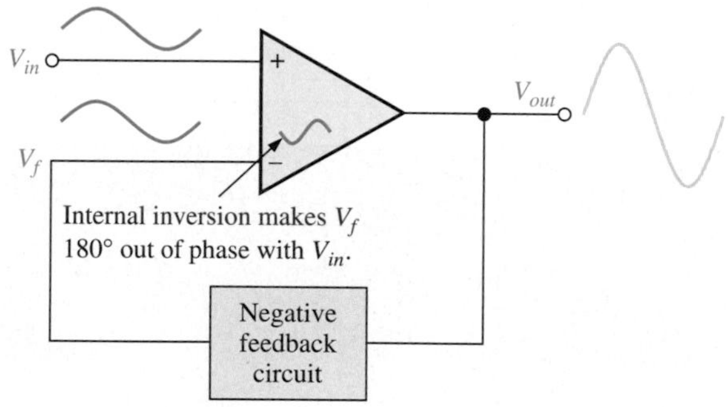

▶ FIGURE 12–13

Illustration of negative feedback.

Why Use Negative Feedback?

As you can see in Table 12–1, the inherent open-loop voltage gain of a typical op-amp is very high (usually greater than 100,000). Therefore, an extremely small input voltage drives the op-amp into its saturated output states. In fact, even the input offset voltage of the op-amp can drive it into saturation. For example, assume $V_{IN} = 1$ mV and $A_{ol} = 100{,}000$. Then,

$$V_{IN}A_{ol} = (1 \text{ mV})(100{,}000) = 100 \text{ V}$$

Since the output level of an op-amp can never reach 100 V, it is driven deep into saturation and the output is limited to its maximum output levels, as illustrated in Figure 12–14 for both a positive and a negative input voltage of 1 mV.

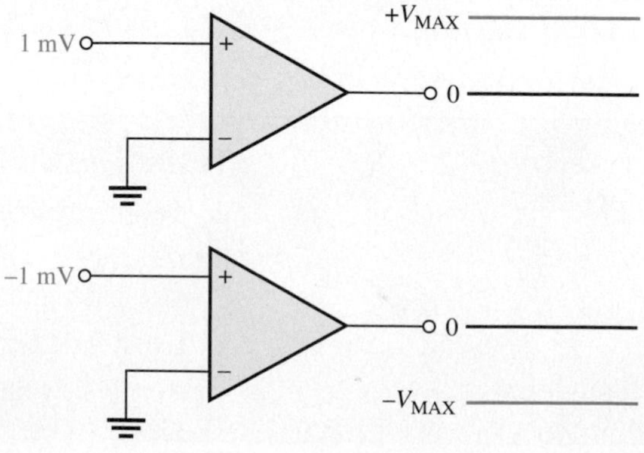

▶ FIGURE 12–14

Without negative feedback, a small input voltage drives the op-amp to its output limits and it becomes nonlinear.

The usefulness of an op-amp operated without negative feedback is generally limited to comparator applications (to be studied in Chapter 13). With negative feedback, the closed-loop voltage gain (A_{cl}) can be reduced and controlled so that the op-amp can function as a linear amplifier. In addition to providing a controlled, stable voltage gain, negative feedback also provides for control of the input and output impedances and amplifier bandwidth. Table 12–2 summarizes the general effects of negative feedback on op-amp performance.

▼ TABLE 12–2

	VOLTAGE GAIN	INPUT Z	OUTPUT Z	BANDWIDTH
Without negative feedback	A_{ol} is too high for linear amplifier applications	Relatively high (see Table 12–1)	Relatively low	Relatively narrow (because the gain is so high)
With negative feedback	A_{cl} is set to desired value by the feedback circuit	Can be increased or reduced to a desir[illegible] value depending [illegible] type of circuit	Can be reduced to a des[illegible]	Significantly wider

SECTION 12–3 CHECKUP

1. What are the benefits of negative feedback in an op-amp circuit?
2. Why is it generally necessary to reduce the gain of an op-amp from its open-loop value?

12–4 OP-AMPS WITH NEGATIVE FEEDBACK

An op-amp can be connected using negative feedback to stabilize the gain and increase frequency response. Negative feedback takes a portion of the output and applies it back out of phase with the input, creating an effective reduction in gain. This closed-loop gain is usually much less than the open-loop gain and independent of it.

After completing this section, you should be able to

- **Analyze the noninverting, voltage-follower, and inverting op-amp configurations**
  - Define *closed-loop voltage gain*
  - Identify the noninverting amplifier configuration
  - Determine the closed-loop voltage gain of a noninverting amplifier
  - Identify the voltage-follower configuration
  - Identify the inverting amplifier configuration
  - Determine the closed-loop voltage gain of an inverting amplifier

Closed-Loop Voltage Gain, A_{cl}

The **closed-loop voltage gain** is the voltage gain of an op-amp with external feedback. The amplifier configuration consists of the op-amp and an external negative feedback circuit that connects the output to the inverting input. The closed-loop voltage gain is determined by the external component values and can be precisely controlled by them.

Noninverting Amplifier

An op-amp connected in a **closed-loop** configuration as a **noninverting amplifier** with a controlled amount of voltage gain is shown in Figure 12–15. The input signal is applied to the noninverting (+) input. The output is applied back to the inverting (−) input through the feedback circuit (closed loop) formed by the input resistor R_i and the feedback resistor R_f. This creates negative feedback as follows. Resistors R_i and R_f form a voltage-divider circuit, which reduces V_{out} and connects the reduced voltage V_f to the inverting input. The feedback voltage is expressed as

$$V_f = \left(\frac{R_i}{R_i + R_f}\right)V_{out}$$

▶ FIGURE 12–15

Noninverting amplifier.

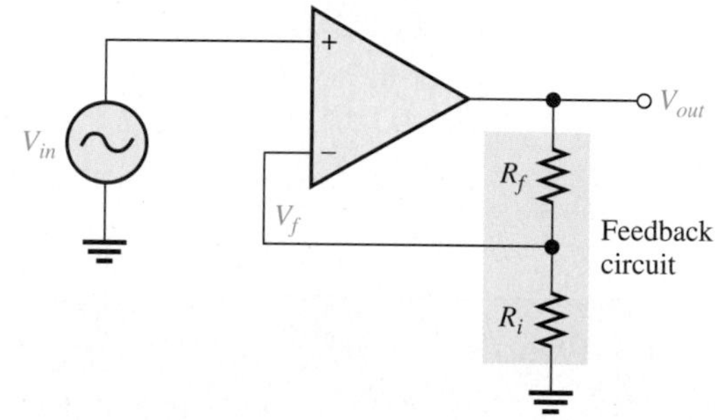

The difference of the input voltage, V_{in}, and the feedback voltage, V_f, is the differential input to the op-amp, as shown in Figure 12–16. This differential voltage is amplified by the open-loop voltage gain of the op-amp (A_{ol}) and produces an output voltage expressed as

$$V_{out} = A_{ol}(V_{in} - V_f)$$

The attenuation, B, of the feedback circuit is

$$B = \frac{R_i}{R_i + R_f}$$

Substituting BV_{out} for V_f in the V_{out} equation,

$$V_{out} = A_{ol}(V_{in} - BV_{out})$$

Then applying basic algebra,

$$V_{out} = A_{ol}V_{in} - A_{ol}BV_{out}$$
$$V_{out} + A_{ol}BV_{out} = A_{ol}V_{in}$$
$$V_{out}(1 + A_{ol}B) = A_{ol}V_{in}$$

▶ FIGURE 12–16

Differential input, $V_{in} - V_f$.

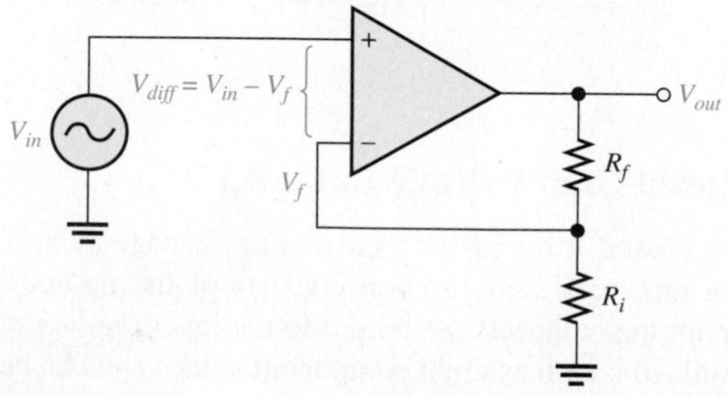

Since the overall voltage gain of the amplifier in Figure 12–15 is V_{out}/V_{in}, it can be expressed as

$$\frac{V_{out}}{V_{in}} = \frac{A_{ol}}{1 + A_{ol}B}$$

The product $A_{ol}B$ is typically much greater than 1, so the equation simplifies to

$$\frac{V_{out}}{V_{in}} \cong \frac{A_{ol}}{A_{ol}B} = \frac{1}{B}$$

The closed-loop gain of the noninverting (NI) amplifier is the reciprocal of the attenuation (B) of the feedback circuit (voltage-divider).

$$A_{cl(\text{NI})} = \frac{V_{out}}{V_{in}} \cong \frac{1}{B} = \frac{R_i + R_f}{R_i}$$

Therefore,

$$\mathbf{A_{cl(NI)} = 1 + \frac{R_f}{R_i}}$$

Equation 12–8

Notice that the closed-loop voltage gain is not at all dependent on the op-amp's open-loop voltage gain under the condition $A_{ol}B \gg 1$. The closed-loop gain can be set by selecting values of R_i and R_f.

EXAMPLE 12–3

Determine the closed-loop voltage gain of the amplifier in Figure 12–17.

FIGURE 12–17

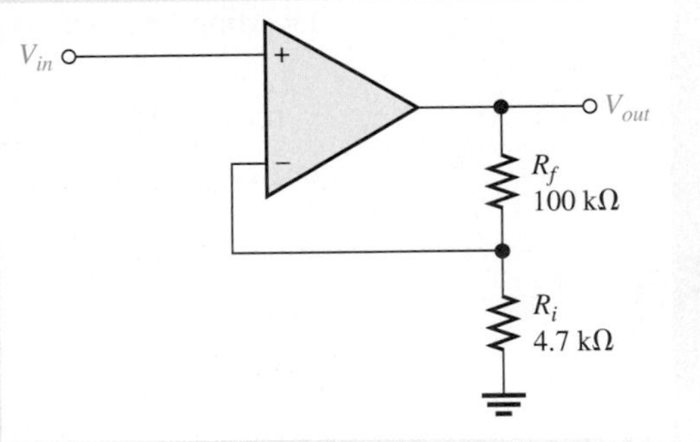

Solution This is a noninverting op-amp configuration. Therefore, the closed-loop voltage gain is

$$A_{cl(\text{NI})} = 1 + \frac{R_f}{R_i} = 1 + \frac{100\text{ k}\Omega}{4.7\text{ k}\Omega} = \mathbf{22.3}$$

Related Problem If R_f in Figure 12–17 is increased to 150 kΩ, determine the closed-loop gain.

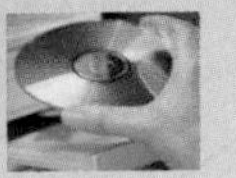

Open the Multisim file E12-03 in the Examples folder on your CD-ROM. Measure the closed-loop voltage gain of the amplifier and compare with the calculated value.

Voltage-Follower

The **voltage-follower** configuration is a special case of the noninverting amplifier where all of the output voltage is fed back to the inverting (−) input by a straight connection, as shown in Figure 12–18. As you can see, the straight feedback connection has a voltage gain of 1 (which means there is no gain). The closed-loop voltage gain of a noninverting amplifier is $1/B$ as previously derived. Since $B = 1$ for a voltage-follower, the closed-loop voltage gain of the voltage-follower is

Equation 12–9

$$A_{cl(VF)} = 1$$

▶ FIGURE 12–18

Op-amp voltage-follower.

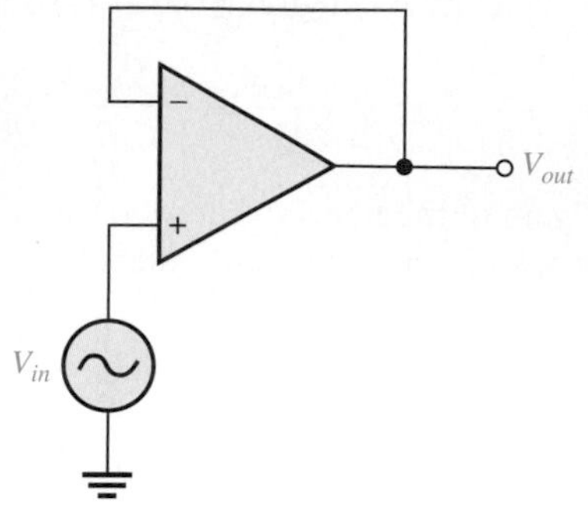

The most important features of the voltage-follower configuration are its very high input impedance and its very low output impedance. These features make it a nearly ideal buffer amplifier for interfacing high-impedance sources and low-impedance loads. This is discussed further in Section 12–5.

Inverting Amplifier

An op-amp connected as an **inverting amplifier** with a controlled amount of voltage gain is shown in Figure 12–19. The input signal is applied through a series input resistor R_i to the inverting (−) input. Also, the output is fed back through R_f to the same input. The noninverting (+) input is grounded.

▶ FIGURE 12–19

Inverting amplifier.

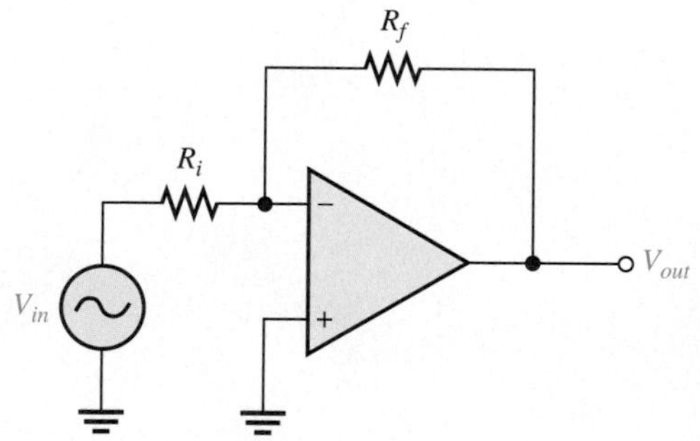

At this point, the ideal op-amp parameters mentioned earlier are useful in simplifying the analysis of this circuit. In particular, the concept of infinite input impedance is of great value. An infinite input impedance implies zero current at the inverting input. If there is zero current through the input impedance, then there must be *no* voltage drop between the inverting and noninverting inputs. This means that the voltage at the inverting (−) input is zero because the noninverting (+) input is grounded. This zero voltage at the inverting input terminal is referred to as *virtual ground.* This condition is illustrated in Figure 12–20(a).

Since there is no current at the inverting input, the current through R_i and the current through R_f are equal, as shown in Figure 12–20(b).

$$I_{in} = I_f$$

The voltage across R_i equals V_{in} because the resistor is connected to virtual ground at the inverting input of the op-amp. Therefore,

$$I_{in} = \frac{V_{in}}{R_i}$$

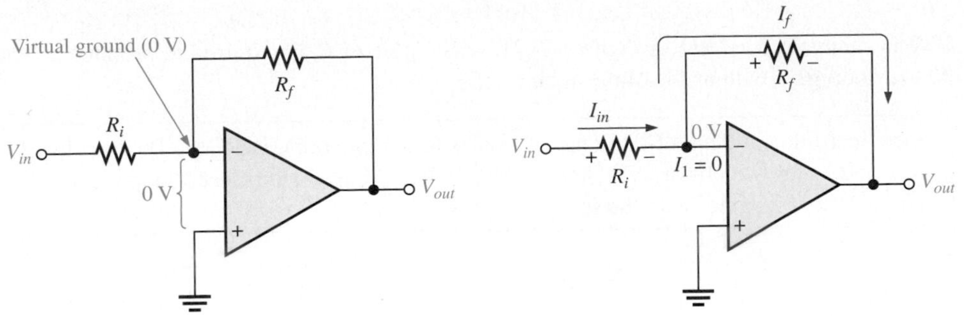

(a) Virtual ground

(b) $I_{in} = I_f$ and current at the inverting input (I_1) is 0.

FIGURE 12–20

Virtual ground concept and closed-loop voltage gain development for the inverting amplifier.

Also, the voltage across R_f equals $-V_{out}$ because of virtual ground, and therefore,

$$I_f = \frac{-V_{out}}{R_f}$$

Since $I_f = I_{in}$,

$$\frac{-V_{out}}{R_f} = \frac{V_{in}}{R_i}$$

Rearranging the terms,

$$\frac{V_{out}}{V_{in}} = -\frac{R_f}{R_i}$$

Of course, V_{out}/V_{in} is the overall gain of the inverting (I) amplifier.

$$\mathbf{A_{cl(I)} = -\frac{R_f}{R_i}}$$

Equation 12–10

Equation 12–10 shows that the closed-loop voltage gain of the inverting amplifier ($A_{cl(I)}$) is the ratio of the feedback resistance (R_f) to the input resistance (R_i). *The closed-loop gain is independent of the op-amp's internal open-loop gain.* Thus, the negative feedback stabilizes the voltage gain. The negative sign indicates inversion.

EXAMPLE 12–4

Given the op-amp configuration in Figure 12–21, determine the value of R_f required to produce a closed-loop voltage gain of -100.

FIGURE 12–21

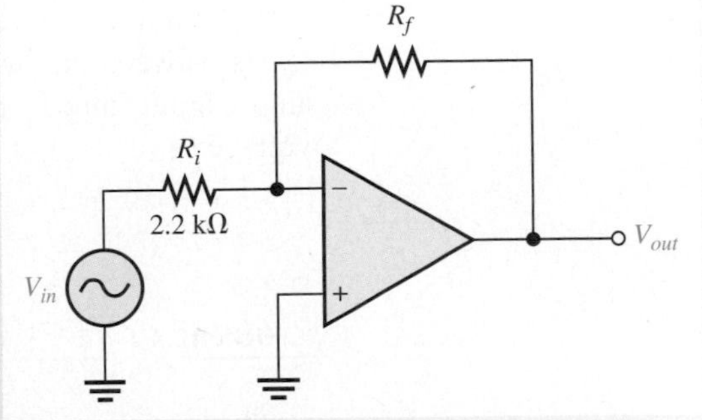

Solution Knowing that $R_i = 2.2\ \text{k}\Omega$ and the absolute value of the closed-loop gain is $|A_{cl(I)}| = 100$, calculate R_f as follows:

$$|A_{cl(I)}| = \frac{R_f}{R_i}$$

$$R_f = |A_{cl(I)}|R_i = (100)(2.2\ \text{k}\Omega) = \mathbf{220\ k\Omega}$$

Related Problem If R_i is changed to 2.7 kΩ in Figure 12–21, what value of R_f is required to produce a closed-loop gain with an absolute value of 25?

Open the Multisim file E12-04 in the Examples folder on your CD-ROM. The circuit has a value of R_f which was calculated to be 220 kΩ. Measure the closed-loop voltage gain and see if it agrees with the specified value.

SECTION 12–4 CHECKUP

1. What is the main purpose of negative feedback?
2. The closed-loop voltage gain of each of the op-amp configurations discussed is dependent on the internal open-loop voltage gain of the op-amp. (True or False)
3. The attenuation of the negative feedback circuit of a noninverting op-amp configuration is 0.02. What is the closed-loop gain of the amplifier?

12–5 EFFECTS OF NEGATIVE FEEDBACK ON OP-AMP IMPEDANCES

Negative feedback affects the input and output impedances of an op-amp. The effects on both inverting and noninverting amplifiers are examined in this section.

After completing this section, you should be able to

- **Describe impedances of the three op-amp configurations**
  - Determine input and output impedances of a noninverting amplifier
  - Determine input and output impedances of a voltage-follower
  - Determine input and output impedances of an inverting amplifier

Impedances of a Noninverting Amplifier

Input Impedance The input impedance of a noninverting amplifier can be developed with the aid of Figure 12–22. For this analysis, assume a small differential voltage, V_d, exists between the two inputs, as indicated. This means that you cannot assume the op-amp's input impedance to be infinite or the input current to be zero. Express the input voltage as

$$V_{in} = V_d + V_f$$

FIGURE 12–22

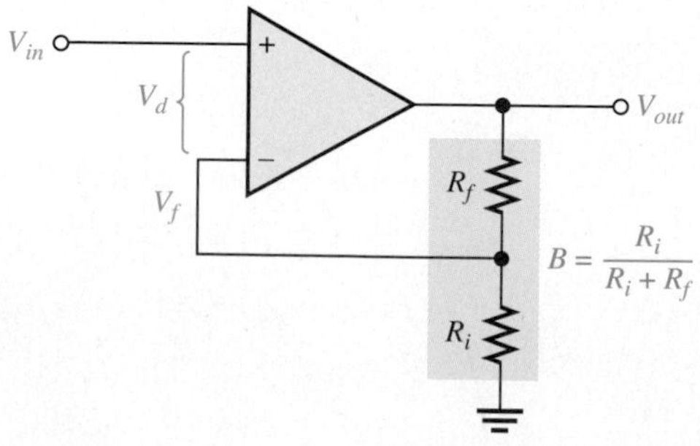

Substituting BV_{out} for the feedback voltage, V_f, yields

$$V_{in} = V_d + BV_{out}$$

Remember, B is the attenuation of the negative feedback circuit and is equal to $R_i/(R_i + R_f)$.

Since $V_{out} \cong A_{ol}V_d$ (A_{ol} is the open-loop gain of the op-amp),

$$V_{in} = V_d + A_{ol}BV_d = (1 + A_{ol}B)V_d$$

Now substituting $I_{in}Z_{in}$ for V_d,

$$V_{in} = (1 + A_{ol}B)I_{in}Z_{in}$$

where Z_{in} is the open-loop input impedance of the op-amp (without feedback connections).

$$\frac{V_{in}}{I_{in}} = (1 + A_{ol}B)Z_{in}$$

V_{in}/I_{in} is the overall input impedance of a closed-loop noninverting amplifier configuration.

$$\mathbf{Z_{in(NI)} = (1 + A_{ol}B)Z_{in}}$$

Equation 12–11

This equation shows that the input impedance of the noninverting amplifier configuration with negative feedback is much greater than the internal input impedance of the op-amp itself (without feedback).

Output Impedance An expression for output impedance of a noninverting amplifier can be developed with the aid of Figure 12–23.

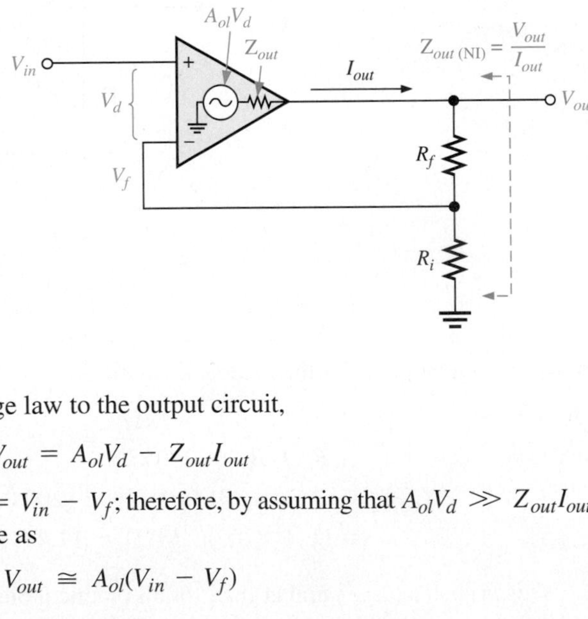

FIGURE 12–23

By applying Kirchhoff's voltage law to the output circuit,

$$V_{out} = A_{ol}V_d - Z_{out}I_{out}$$

The differential input voltage is $V_d = V_{in} - V_f$; therefore, by assuming that $A_{ol}V_d \gg Z_{out}I_{out}$, you can express the output voltage as

$$V_{out} \cong A_{ol}(V_{in} - V_f)$$

Substituting BV_{out} for V_f,

$$V_{out} \cong A_{ol}(V_{in} - BV_{out})$$

Expanding and factoring yields

$$V_{out} \cong A_{ol}V_{in} - A_{ol}BV_{out}$$

$$A_{ol}V_{in} \cong V_{out} + A_{ol}BV_{out} \cong (1 + A_{ol}B)V_{out}$$

Since the output impedance of the noninverting amplifier configuration is $Z_{out(NI)} = V_{out}/I_{out}$, you can substitute $I_{out}Z_{out(NI)}$ for V_{out}; therefore,

$$A_{ol}V_{in} = (1 + A_{ol}B)I_{out}Z_{out(NI)}$$

Dividing both sides of the previous expression by I_{out},

$$\frac{A_{ol}V_{in}}{I_{out}} = (1 + A_{ol}B)Z_{out(NI)}$$

The term on the left is the internal output impedance of the op-amp (Z_{out}) because, without feedback, $A_{ol}V_{in} = V_{out}$. Therefore,

$$Z_{out} = (1 + A_{ol}B)Z_{out(NI)}$$

Thus,

Equation 12–12

$$\mathbf{Z_{out(NI)} = \frac{Z_{out}}{1 + A_{ol}B}}$$

This equation shows that the output impedance of the noninverting amplifier configuration with negative feedback is much less than the internal output impedance, Z_{out}, of the op-amp itself (without feedback) because Z_{out} is divided by the factor $1 + A_{ol}B$.

EXAMPLE 12–5

(a) Determine the input and output impedances of the amplifier in Figure 12–24. The op-amp datasheet gives $Z_{in} = 2\ \text{M}\Omega$, $Z_{out} = 75\ \Omega$, and $A_{ol} = 200{,}000$.

(b) Find the closed-loop voltage gain.

FIGURE 12–24

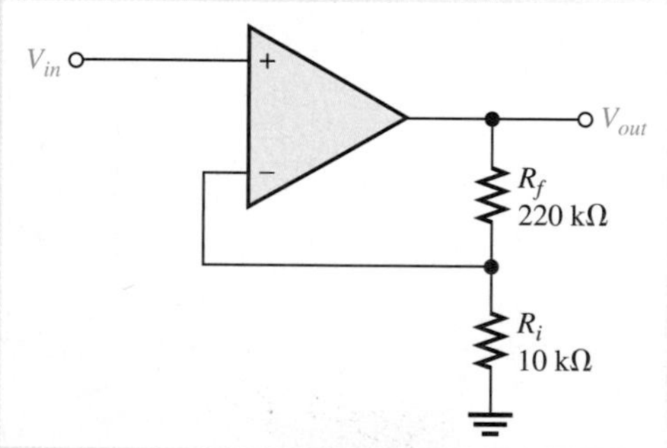

Solution **(a)** The attenuation, B, of the feedback circuit is

$$B = \frac{R_i}{R_i + R_f} = \frac{10\ \text{k}\Omega}{230\ \text{k}\Omega} = 0.0435$$

$$Z_{in(NI)} = (1 + A_{ol}B)Z_{in} = [1 + (200{,}000)(0.0435)](2\ \text{M}\Omega)$$
$$= (1 + 8700)(2\ \text{M}\Omega) = \mathbf{17.4\ G\Omega}$$

This is such a large number that, for all practical purposes, it can be assumed to be infinite as in the ideal case.

$$Z_{out(NI)} = \frac{Z_{out}}{1 + A_{ol}B} = \frac{75\ \Omega}{1 + 8700} = \mathbf{8.6\ m\Omega}$$

This is such a small number that, for all practical purposes, it can be assumed to be zero as in the ideal case.

(b) $$A_{cl(NI)} = 1 + \frac{R_f}{R_i} = 1 + \frac{220\ \text{k}\Omega}{10\ \text{k}\Omega} = \mathbf{23.0}$$

Related Problem **(a)** Determine the input and output impedances in Figure 12–24 for op-amp datasheet values of $Z_{in} = 3.5\ \text{M}\Omega$, $Z_{out} = 82\ \Omega$, and $A_{ol} = 135{,}000$.

(b) Find A_{cl}.

Open the Multisim file E12-05 in the Examples folder on your CD-ROM. Measure the closed-loop voltage gain and compare with the calculated value.

Voltage-Follower Impedances

Since a voltage-follower is a special case of the noninverting amplifier configuration, the same impedance formulas are used but with $B = 1$.

$$\mathbf{Z_{in(VF)} = (1 + A_{ol})Z_{in}}$$ Equation 12–13

$$\mathbf{Z_{out(VF)} = \frac{Z_{out}}{1 + A_{ol}}}$$ Equation 12–14

As you can see, the voltage-follower input impedance is greater for a given A_{ol} and Z_{in} than for the noninverting amplifier configuration with the voltage-divider feedback circuit. Also, its output impedance is much smaller.

EXAMPLE 12–6

The op-amp in Example 12–5 is used in a voltage-follower configuration. Determine the input and output impedances.

Solution Since $B = 1$,

$$Z_{in(VF)} = (1 + A_{ol})Z_{in} = (1 + 200{,}000)(2\ \text{M}\Omega) \cong \mathbf{400\ G\Omega}$$

$$Z_{out(VF)} = \frac{Z_{out}}{1 + A_{ol}} = \frac{75\ \Omega}{1 + 200{,}000} = \mathbf{375\ \mu\Omega}$$

Notice that $Z_{in(VF)}$ is much greater than $Z_{in(NI)}$, and $Z_{out(VF)}$ is much less than $Z_{out(NI)}$ from Example 12–5. Again for all practical purposes, the ideal values can be assumed.

Related Problem If the op-amp in this example is replaced with one having a higher open-loop gain, how are the input and output impedances affected?

Impedances of an Inverting Amplifier

The input and output impedances of an inverting op-amp configuration are developed with the aid of Figure 12–25. Both the input signal and the negative feedback are applied, through resistors, to the inverting (−) terminal as shown.

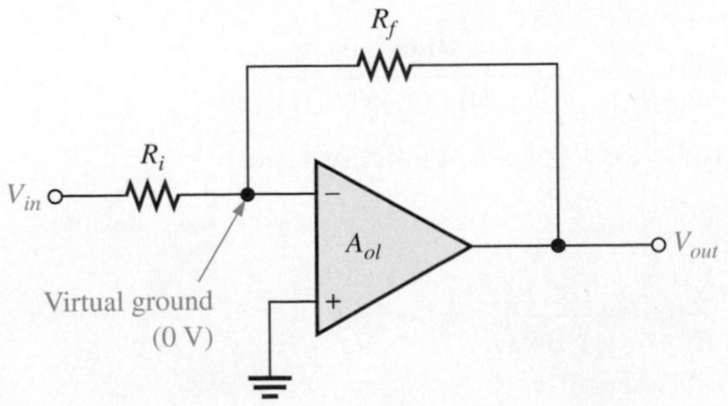

FIGURE 12–25

Inverting amplifier.

Input Impedance The input impedance for an inverting amplifier is

Equation 12–15

$$Z_{in(I)} \cong R_i$$

This is because the inverting input of the op-amp is at virtual ground (0 V), and the input source simply sees R_i to ground, as shown in Figure 12–26.

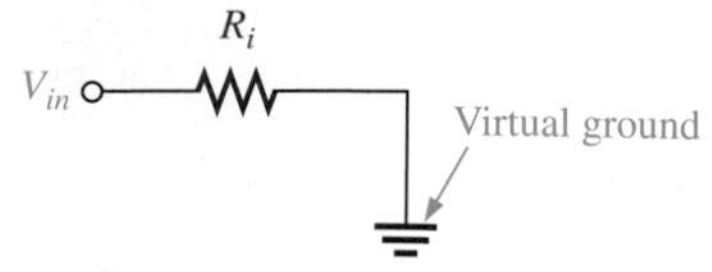

▲ FIGURE 12–26

Output Impedance As with a noninverting amplifier, the output impedance of an inverting amplifier is decreased by the negative feedback. In fact, the expression is the same as for the noninverting case.

Equation 12–16

$$Z_{out(I)} = \frac{Z_{out}}{1 + A_{ol}B}$$

The output impedance of both the noninverting and the inverting amplifier configurations is very low; in fact, it is almost zero in practical cases. Because of this near zero output impedance, any load impedance within limits can be connected to the op-amp output and not change the output voltage. The limits for the load impedance are determined by the maximum peak-to-peak swing of the output ($V_{O(p\text{-}p)}$) and the current limit of the op-amp.

EXAMPLE 12–7

Find the values of the input and output impedances in Figure 12–27. Also, determine the closed-loop voltage gain. The op-amp has the following parameters: $A_{ol} = 50{,}000$; $Z_{in} = 4\ \text{M}\Omega$; and $Z_{out} = 50\ \Omega$.

▶ FIGURE 12–27

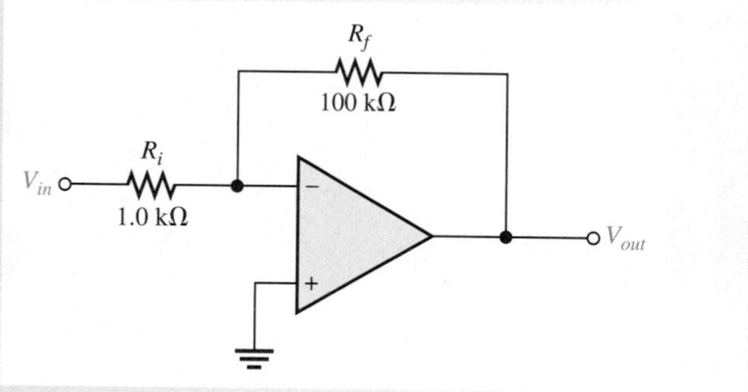

Solution

$$Z_{in(I)} \cong R_i = \mathbf{1.0\ k\Omega}$$

The feedback attenuation, B, is

$$B = \frac{R_i}{R_i + R_f} = \frac{1.0\ \text{k}\Omega}{101\ \text{k}\Omega} = 0.001$$

Then

$$Z_{out(I)} = \frac{Z_{out}}{1 + A_{ol}B} = \frac{50\ \Omega}{1 + (50{,}000)(0.001)}$$
$$= \mathbf{980\ m\Omega} \text{ (zero for all practical purposes)}$$

The closed-loop voltage gain is

$$A_{cl(I)} = -\frac{R_f}{R_i} = -\frac{100\ \text{k}\Omega}{1.0\ \text{k}\Omega} = \mathbf{-100}$$

Related Problem Determine the input and output impedances and the closed-loop voltage gain in Figure 12–27. The op-amp parameters and circuit values are as follows: $A_{ol} = 100{,}000$; $Z_{in} = 5\ \text{M}\Omega$; $Z_{out} = 75\ \Omega$; $R_i = 560\ \Omega$; and $R_f = 82\ \text{k}\Omega$.

Open the Multisim file E12-07 in the Examples folder on your CD-ROM and measure the closed-loop voltage gain. Compare to the calculated result.

SECTION 12–5 CHECKUP

1. How does the input impedance of a noninverting amplifier configuration compare to the input impedance of the op-amp itself?
2. When an op-amp is connected in a voltage-follower configuration, does the input impedance increase or decrease?
3. Given that $R_f = 100\ \text{k}\Omega$; $R_i = 2\ \text{k}\Omega$; $A_{ol} = 120{,}000$; $Z_{in} = 2\ \text{M}\Omega$; and $Z_{out} = 60\ \Omega$, what are $Z_{in(I)}$ and $Z_{out(I)}$ for an inverting amplifier configuration?

12–6 BIAS CURRENT AND OFFSET VOLTAGE

Certain deviations from the ideal op-amp must be recognized because of their effects on its operation. Transistors within the op-amp must be biased so that they have the correct values of base and collector currents and collector-to-emitter voltages. The ideal op-amp has no input current at its terminals; but in fact, the practical op-amp has small input bias currents typically in the nA range. Also, small internal imbalances in the transistors effectively produce a small offset voltage between the inputs. These nonideal parameters were described in Section 12–2.

After completing this section, you should be able to

- **Discuss op-amp compensation**
  - Describe the effect of input bias current
  - Explain bias current compensation
  - Describe the effect of input offset voltage
  - Explain input offset voltage compensation

Effect of an Input Bias Current

Figure 12–28 is an inverting amplifier with zero input voltage. Ideally, the current through R_i is zero because the input voltage is zero and the voltage at the inverting ($-$) terminal is zero. The small input bias current, I_1, is through R_f from the output terminal. I_1 creates a voltage drop across R_f, as indicated. The positive side of R_f is the output terminal, and therefore, the output error voltage is I_1R_f when it should be zero.

Figure 12–29 is a voltage-follower with zero input voltage and a source resistance, R_s. In this case, an input bias current, I_1, produces a drop across R_s and creates an output voltage error as shown. The voltage at the inverting input terminal decreases to $-I_1R_s$ because the negative feedback tends to maintain a differential voltage of zero, as indicated. Since

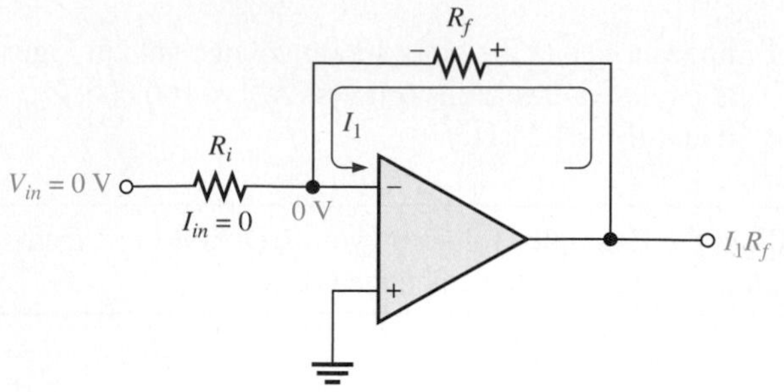

FIGURE 12–28

Input bias current creates output error voltage (I_1R_f) in an inverting amplifier.

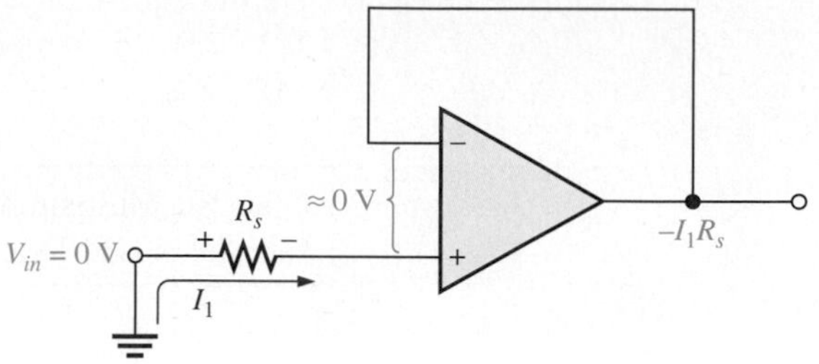

FIGURE 12–29

Input bias current creates output error voltage in a voltage-follower.

the inverting terminal is connected directly to the output terminal, the output error voltage is $-I_1R_s$.

Figure 12–30 is a noninverting amplifier with zero input voltage. Ideally, the voltage at the inverting terminal is also zero, as indicated. The input bias current, I_1, produces a voltage drop across R_f and thus creates an output error voltage of I_1R_f, just as with the inverting amplifier.

FIGURE 12–30

Input bias current creates output error voltage in a noninverting amplifier.

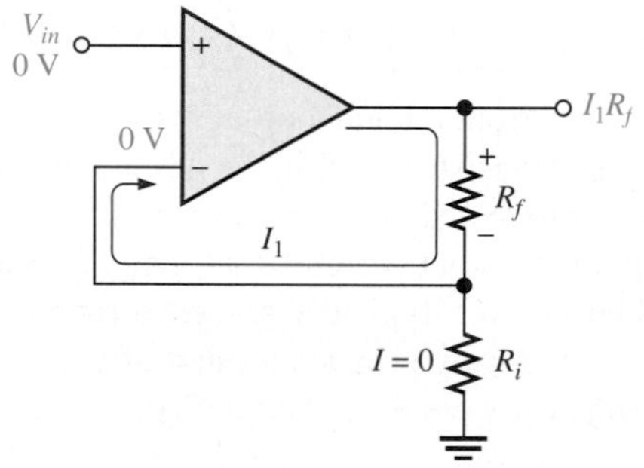

Bias Current Compensation in a Voltage-Follower

The output error voltage due to bias currents in a voltage-follower can be sufficiently reduced by adding a resistor, R_f, equal to the source resistance, R_s, in the feedback path, as shown in Figure 12–31. The voltage drop created by I_1 across the added resistor subtracts from the $-I_2R_s$ output error voltage. If $I_1 = I_2$, then the output voltage is zero. Usually I_1 does not quite equal I_2; but even in this case, the output error voltage is reduced as follows because I_{OS} is less than I_2.

$$V_{OUT(error)} = |I_1 - I_2|R_s = I_{OS}R_s$$

where I_{OS} is the input offset current.

FIGURE 12–31

Bias current compensation in a voltage-follower.

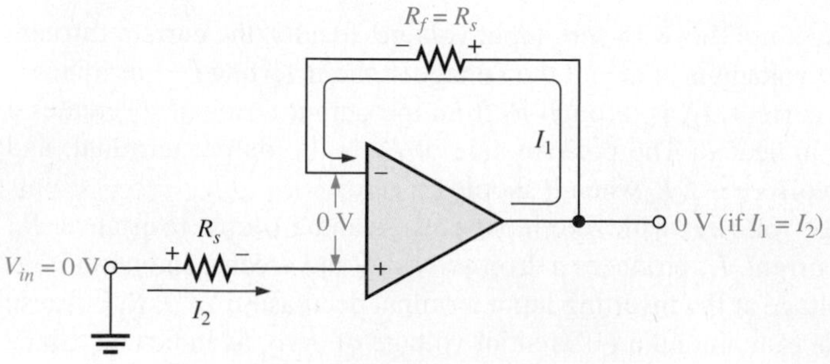

Bias Current Compensation in Other Op-Amp Configurations

To compensate for the effect of bias current in the noninverting amplifier, a resistor R_c is added, as shown in Figure 12–32(a). The compensating resistor value equals the parallel combination of R_i and R_f. The input current creates a voltage drop across R_c that offsets the voltage across the combination of R_i and R_f, thus sufficiently reducing the output error voltage. The inverting amplifier is similarly compensated, as shown in Figure 12–32(b).

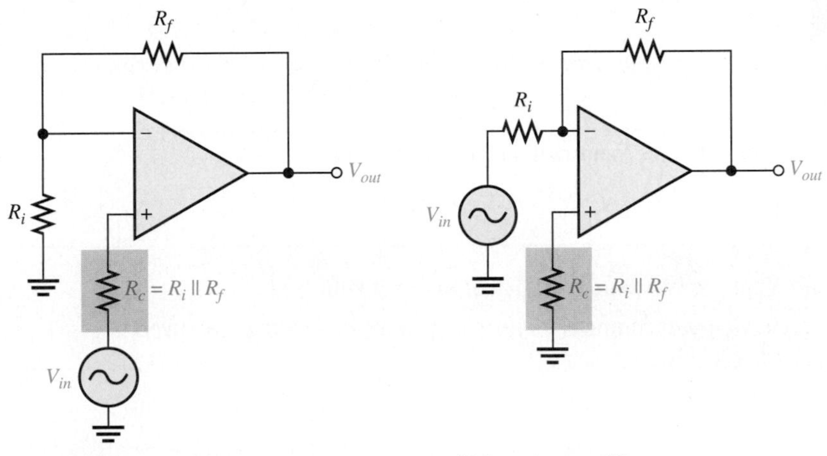

FIGURE 12–32

Bias current compensation in the noninverting and inverting amplifier configurations.

Use of a BIFET Op-Amp to Eliminate the Need for Bias Current Compensation The BIFET op-amp uses both BJTs and JFETs in its internal circuitry. The JFETs are used as the input devices to achieve a higher input impedance than is possible with standard BJT amplifiers. Because of their very high input impedance, BIFETs typically have input bias currents that are much smaller than in BJT op-amps, thus reducing or eliminating the need for bias current compensation.

Effect of Input Offset Voltage

The output voltage of an op-amp should be zero when the differential input is zero. However, there is always a small output error voltage present whose value typically ranges from microvolts to millivolts. This is due to unavoidable imbalances within the internal op-amp transistors aside from the bias currents previously discussed. In a negative feedback configuration, the input offset voltage V_{IO} can be visualized as an equivalent small dc voltage source, as illustrated in Figure 12–33 for a voltage-follower. Generally, the output error voltage due to the input offset voltage is

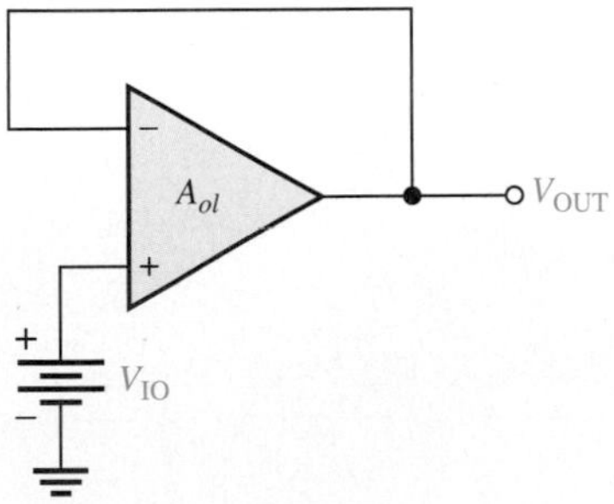

FIGURE 12–33

Input offset voltage equivalent.

$$V_{OUT(error)} = A_{cl}V_{IO}$$

For the case of the voltage-follower, $A_{cl} = 1$, so

$$V_{OUT(error)} = V_{IO}$$

Input Offset Voltage Compensation

Most integrated circuit op-amps provide a means of compensating for offset voltage. This is usually done by connecting an external potentiometer to designated pins on the IC package, as illustrated in Figure 12–34(a) and (b) for a 741 op-amp. The two terminals are labelled *offset null.* With no input, the potentiometer is simply adjusted until the output voltage reads 0, as shown in Figure 12–34(c).

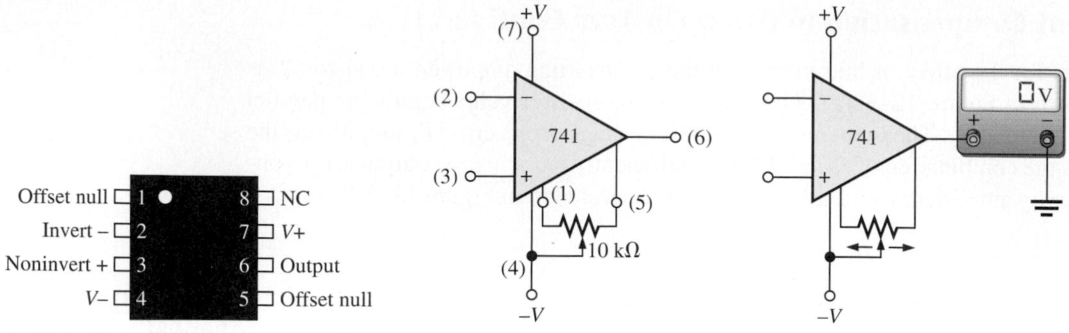

(a) 8-pin DIP or SMT package (b) External potentiometer (c) Adjust for zero output

▲ FIGURE 12–34

Input offset voltage compensation for a 741 op-amp.

SECTION 12–6 CHECKUP

1. What are two sources of dc output error voltages?
2. How do you compensate for bias current in a voltage-follower?

12–7 OPEN-LOOP RESPONSE

In this section, the open-loop frequency response and the open-loop phase response of an op-amp are covered. Open-loop responses relate to an op-amp with no external feedback. The frequency response indicates how the voltage gain changes with frequency, and the phase response indicates how the phase shift between the input and output signal changes with frequency. The open-loop gain, like the β of a transistor, varies greatly from one device to the next of the same type and cannot be depended upon to have a constant value.

After completing this section, you should be able to

- **Analyze the open-loop response of an op-amp**
  - Discuss the frequency dependency of gain
  - Explain the open-loop bandwidth
  - Explain the unity-gain bandwidth
  - Determine phase shift
  - Discuss how internal stages affect the overall response
  - Discuss critical frequencies and roll-off rates
  - Determine overall phase response

Review of Op-Amp Gains

Figure 12–35 illustrates the open-loop and closed-loop amplifier configurations. As shown in part (a), the open-loop voltage gain, A_{ol}, of an op-amp is the internal voltage gain of the device and represents the ratio of output voltage to input voltage. Notice that there are no external components, so the open-loop voltage gain is set entirely by the internal design. In the

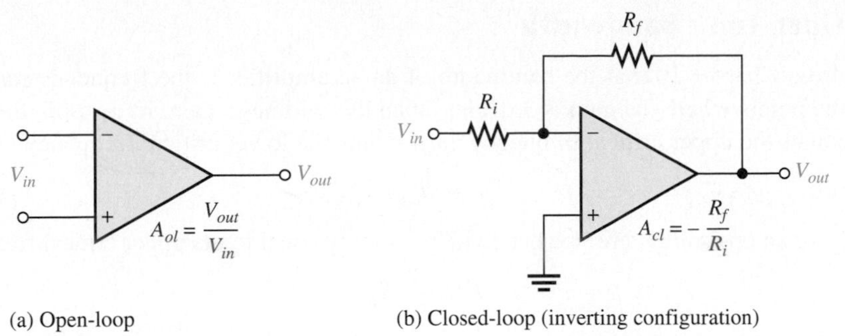

▲ **FIGURE 12–35**

Open-loop and closed-loop op-amp configurations.

closed-loop op-amp configuration shown in part (b), the closed-loop voltage gain, A_{cl}, is the voltage gain of an op-amp with external feedback. The closed-loop voltage gain is determined by the external component values for an inverting amplifier configuration and is always less than the open-loop gain. The closed-loop voltage gain can be precisely controlled by external component values. The closed-loop response of op-amps is covered in Section 12–8.

Bandwidth Limitations

In the previous sections, all of the voltage gain expressions were based on the midrange gain and were considered independent of the frequency. The midrange open-loop gain of an op-amp extends from zero frequency (dc) up to a critical frequency at which the gain is 3 dB less than the midrange value. This concept should be familiar from your study of Chapter 10. Op-amps are dc amplifiers (no capacitive coupling between stages), and therefore, there is no lower critical frequency. This means that the midrange gain extends down to zero frequency (dc), and dc voltages are amplified the same as midrange signal frequencies.

An open-loop response curve (Bode plot) for a certain op-amp is shown in Figure 12–36. Most op-amp datasheets show this type of curve or specify the midrange open-loop gain. Notice that the curve rolls off (decreases) at −20 dB per decade (−6 dB per octave). The midrange gain is 200,000, which is 106 dB, and the critical (cutoff) frequency is approximately 10 Hz.

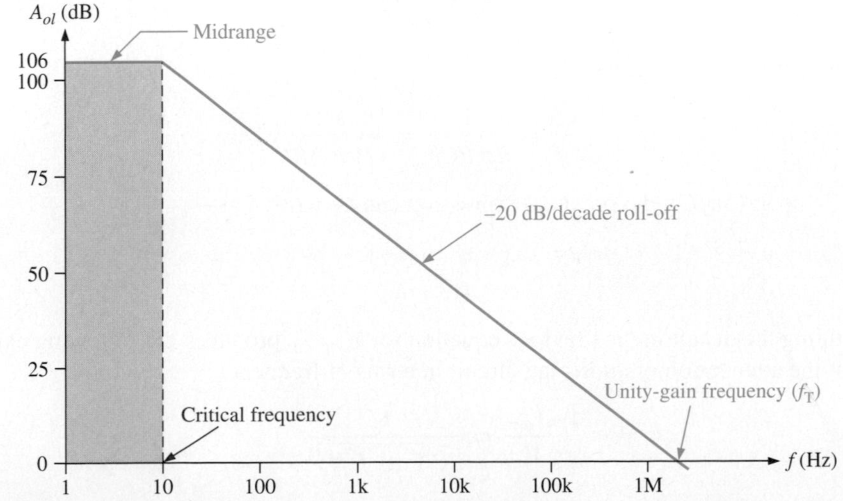

▲ **FIGURE 12–36**

Ideal plot of open-loop voltage gain versus frequency for a typical op-amp. The frequency scale is logarithmic.

3 dB Open-Loop Bandwidth

Recall from Chapter 10 that the bandwidth of an ac amplifier is the frequency range between the points where the gain is 3 dB less than the midrange gain. In general, the bandwidth equals the upper critical frequency (f_{cu}) minus the lower critical frequency (f_{cl}).

$$BW = f_{cu} - f_{cl}$$

Since f_{cl} for an op-amp is zero, the bandwidth is simply equal to the upper critical frequency.

Equation 12–17

$$\mathbf{BW = f_{cu}}$$

From now on, we will refer to f_{cu} as simply f_c; and we will use open-loop (ol) or closed-loop (cl) subscript designators, for example, $f_{c(ol)}$.

Unity-Gain Bandwidth

Notice in Figure 12–36 that the gain steadily decreases to a point where it is equal to unity (1 or 0 dB). The value of the frequency at which this unity gain occurs is the *unity-gain frequency* designated f_T. f_T is also called the *unity-gain bandwidth.*

Gain-Versus-Frequency Analysis

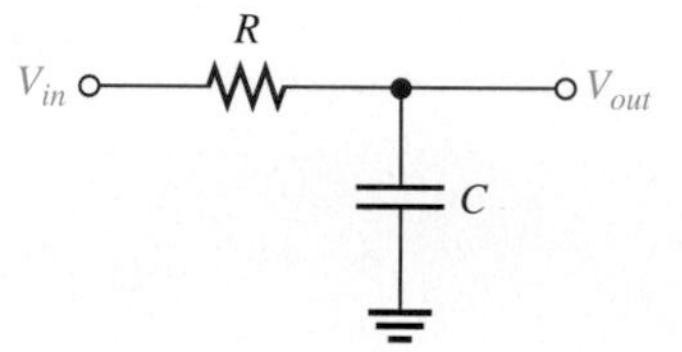

FIGURE 12–37

RC lag circuit.

The *RC* lag (low-pass) circuits within an op-amp are responsible for the roll-off in gain as the frequency increases, just as was discussed for the discrete amplifiers in Chapter 10. From basic ac circuit theory, the attenuation of an *RC* lag circuit, such as in Figure 12–37, is expressed as

$$\frac{V_{out}}{V_{in}} = \frac{X_C}{\sqrt{R^2 + X_C^2}}$$

Dividing both the numerator and denominator to the right of the equals sign by X_C,

$$\frac{V_{out}}{V_{in}} = \frac{1}{\sqrt{1 + R^2/X_C^2}}$$

The critical frequency of an *RC* circuit is

$$f_c = \frac{1}{2\pi RC}$$

Dividing both sides by f gives

$$\frac{f_c}{f} = \frac{1}{2\pi RCf} = \frac{1}{(2\pi fC)R}$$

Since $X_C = 1/(2\pi fC)$, the previous expression can be written as

$$\frac{f_c}{f} = \frac{X_C}{R}$$

Substituting this result in the previous equation for V_{out}/V_{in} produces the following expression for the attenuation of an *RC* lag circuit in terms of frequency:

Equation 12–18

$$\mathbf{\frac{V_{out}}{V_{in}} = \frac{1}{\sqrt{1 + f^2/f_c^2}}}$$

If an op-amp is represented by a voltage gain element with a gain of $A_{ol(mid)}$ plus a single *RC* lag circuit, as shown in Figure 12–38, it is known as a compensated op-amp. The total open-loop gain of the op-amp is the product of the midrange open-loop gain, $A_{ol(mid)}$, and the attenuation of the *RC* circuit.

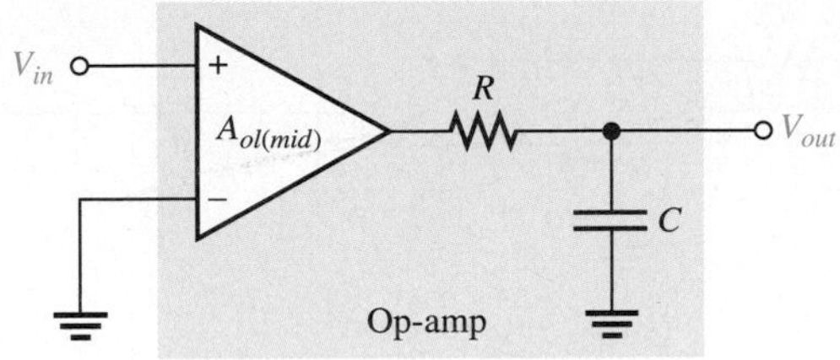

FIGURE 12–38

Op-amp represented by a gain element and an internal *RC* circuit.

$$A_{ol} = \frac{A_{ol(mid)}}{\sqrt{1 + f^2/f_c^2}}$$

Equation 12–19

As you can see from Equation 12–19, the open-loop gain equals the midrange gain when the signal frequency f is much less than the critical frequency f_c and drops off as the frequency increases. Since f_c is part of the open-loop response of an op-amp, we will refer to it as $f_{c(ol)}$.

The following example demonstrates how the open-loop gain decreases as the frequency increases above $f_{c(ol)}$.

EXAMPLE 12–8

Determine A_{ol} for the following values of f. Assume $f_{c(ol)} = 100$ Hz and $A_{ol(mid)} = 100{,}000$.

(a) $f = 0$ Hz **(b)** $f = 10$ Hz **(c)** $f = 100$ Hz **(d)** $f = 1000$ Hz

Solution

(a) $A_{ol} = \frac{A_{ol(mid)}}{\sqrt{1 + f^2/f_{c(ol)}^2}} = \frac{100{,}000}{\sqrt{1 + 0}} = \mathbf{100{,}000}$

(b) $A_{ol} = \frac{100{,}000}{\sqrt{1 + (0.1)^2}} = \mathbf{99{,}503}$

(c) $A_{ol} = \frac{100{,}000}{\sqrt{1 + (1)^2}} = \frac{100{,}000}{\sqrt{2}} = \mathbf{70{,}710}$

(d) $A_{ol} = \frac{100{,}000}{\sqrt{1 + (10)^2}} = \mathbf{9950}$

Related Problem Find A_{ol} for the following frequencies. Assume $f_{c(ol)} = 200$ Hz and $A_{ol(mid)} = 80{,}000$.

(a) $f = 2$ Hz **(b)** $f = 10$ Hz **(c)** $f = 2500$ Hz

Phase Shift

As you know from Chapter 10, an *RC* circuit causes a propagation delay from input to output, thus creating a **phase shift** between the input signal and the output signal. An *RC* lag circuit such as found in an op-amp stage causes the output signal voltage to lag the input, as shown in Figure 12–39. From basic ac circuit theory, the phase shift, θ, is

$$\theta = -\tan^{-1}\left(\frac{R}{X_C}\right)$$

Since $R/X_C = f/f_c$,

$$\theta = -\tan^{-1}\left(\frac{f}{f_c}\right)$$

Equation 12–20

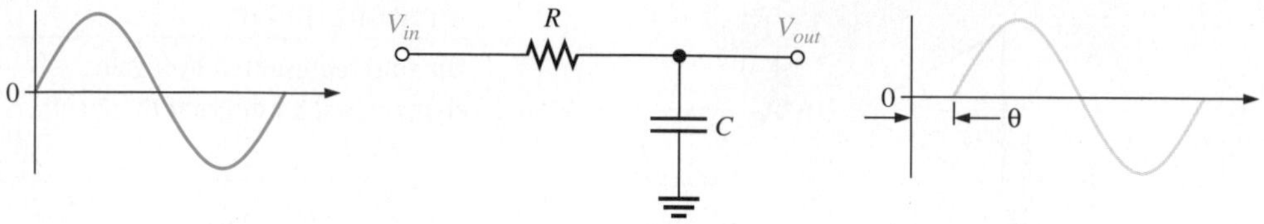

FIGURE 12–39

Output voltage lags input voltage.

The negative sign indicates that the output lags the input. This equation shows that the phase shift increases with frequency and approaches $-90°$ as f becomes much greater than f_c.

EXAMPLE 12–9

Calculate the phase shift for an *RC* lag circuit for each of the following frequencies, and then plot the curve of phase shift versus frequency. Assume $f_c = 100$ Hz.

(a) $f = 1$ Hz **(b)** $f = 10$ Hz **(c)** $f = 100$ Hz
(d) $f = 1000$ Hz **(e)** $f = 10{,}000$ Hz

Solution

(a) $\theta = -\tan^{-1}\left(\frac{f}{f_c}\right) = -\tan^{-1}\left(\frac{1\text{ Hz}}{100\text{ Hz}}\right) = \mathbf{-0.573°}$

(b) $\theta = -\tan^{-1}\left(\frac{10\text{ Hz}}{100\text{ Hz}}\right) = \mathbf{-5.71°}$

(c) $\theta = -\tan^{-1}\left(\frac{100\text{ Hz}}{100\text{ Hz}}\right) = \mathbf{-45°}$

(d) $\theta = -\tan^{-1}\left(\frac{1000\text{ Hz}}{100\text{ Hz}}\right) = \mathbf{-84.3°}$

(e) $\theta = -\tan^{-1}\left(\frac{10{,}000\text{ Hz}}{100\text{ Hz}}\right) = \mathbf{-89.4°}$

The phase shift-versus-frequency curve is plotted in Figure 12–40. Note that the frequency axis is logarithmic.

FIGURE 12–40

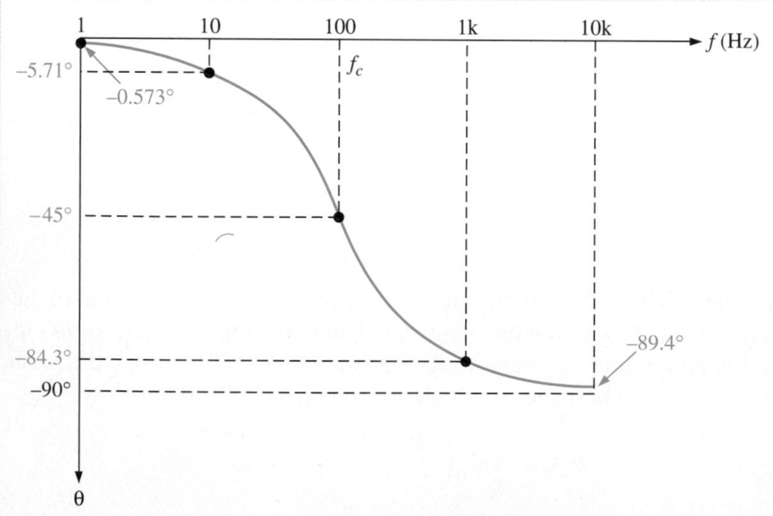

Related Problem At what frequency, in this example, is the phase shift 60°?

Overall Frequency Response

Previously, an op-amp was defined to have a constant roll-off of −20 dB/decade above its critical frequency. For most op-amps this is the case; for some, however, the situation is more complex. The more complex IC operational amplifier may consist of two or more cascaded amplifier stages. The gain of each stage is frequency dependent and rolls off at −20 dB/decade above its critical frequency. Therefore, the total response of an op-amp is a composite of the individual responses of the internal stages. As an example, a three-stage op-amp is represented in Figure 12–41(a), and the frequency response of each stage is shown in Figure 12–41(b). As you know, dB gains are added so that the total op-amp frequency response is as shown in Figure 12–41(c). Since the roll-off rates are additive, the total roll-off rate increases by −20 dB/decade (−6 dB/decade) as each critical frequency is reached.

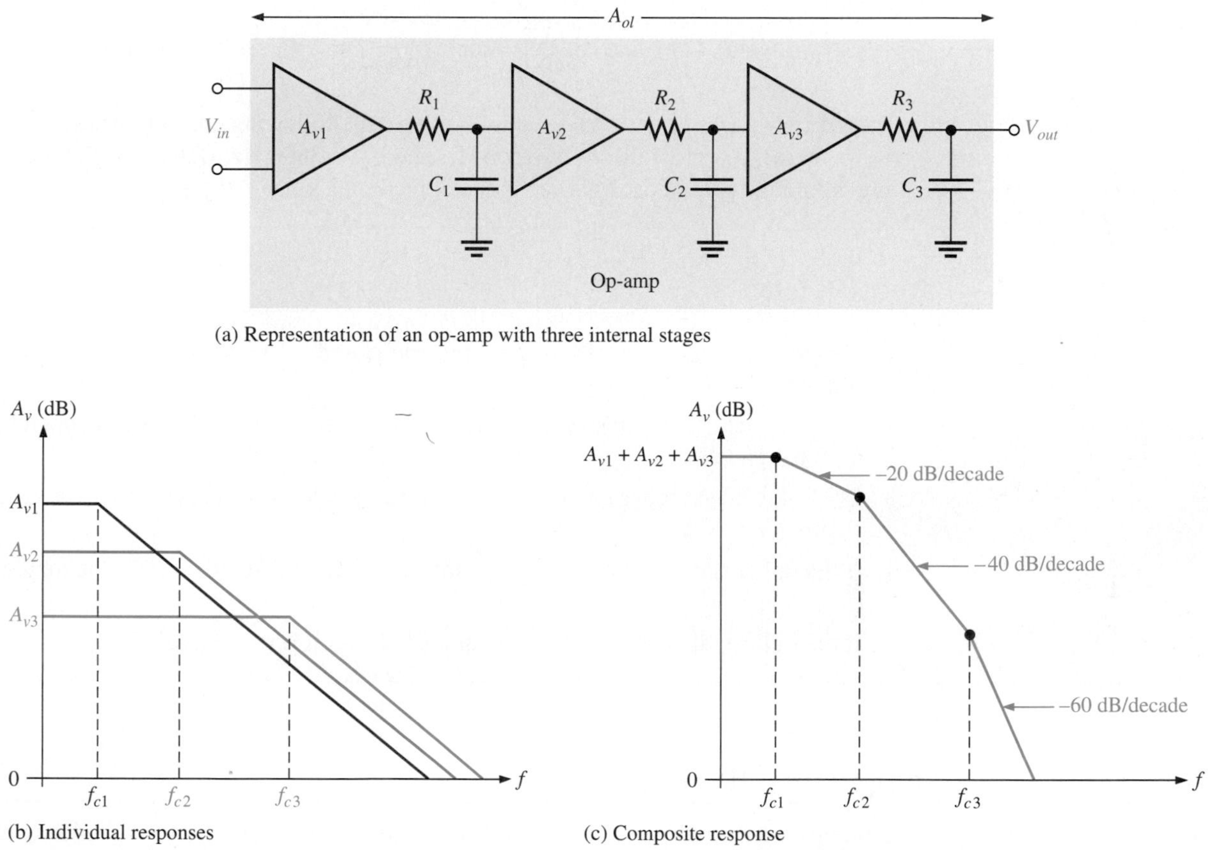

▲ **FIGURE 12–41**

Op-amp open-loop frequency response.

Overall Phase Response

In a multistage amplifier, each stage contributes to the total phase lag. As you have seen, each *RC* lag circuit can produce up to a −90° phase shift. Since each stage in an op-amp includes an *RC* lag circuit, a three-stage op-amp, for example, can have a maximum phase lag of −270°. Also, the phase lag of each stage is less than −45° when the frequency is below the critical frequency, equal to −45° at the critical frequency, and greater than −45° when the frequency is above the critical frequency. The phase lags of the stages of an op-amp are added to produce a total phase lag, according to the following formula for three stages:

$$\theta_{tot} = -\tan^{-1}\left(\frac{f}{f_{c1}}\right) - \tan^{-1}\left(\frac{f}{f_{c2}}\right) - \tan^{-1}\left(\frac{f}{f_{c3}}\right)$$

EXAMPLE 12–10

A certain op-amp has three internal amplifier stages with the following gains and critical frequencies:

Stage 1: $A_{v1} = 40\text{ dB}, f_{c1} = 2\text{ kHz}$

Stage 2: $A_{v2} = 32\text{ dB}, f_{c2} = 40\text{ kHz}$

Stage 3: $A_{v3} = 20\text{ dB}, f_{c3} = 150\text{ kHz}$

Determine the open-loop midrange gain in decibels and the total phase lag when $f = f_{c1}$.

Solution

$$A_{ol(mid)} = A_{v1} + A_{v2} + A_{v3} = 40\text{ dB} + 32\text{ dB} + 20\text{ dB} = 92\text{ dB}$$

$$\theta_{tot} = -\tan^{-1}\left(\frac{f}{f_{c1}}\right) - \tan^{-1}\left(\frac{f}{f_{c2}}\right) - \tan^{-1}\left(\frac{f}{f_{c3}}\right)$$

$$= -\tan^{-1}(1) - \tan^{-1}\left(\frac{2}{40}\right) - \tan^{-1}\left(\frac{2}{150}\right) = -45^\circ - 2.86^\circ - 0.76^\circ = \mathbf{-48.6^\circ}$$

Related Problem

The internal stages of a two-stage amplifier have the following characteristics: $A_{v1} = 50\text{ dB}$, $A_{v2} = 25\text{ dB}$, $f_{c1} = 1500\text{ Hz}$, and $f_{c2} = 3000\text{ Hz}$. Determine the open-loop midrange gain in decibels and the total phase lag when $f = f_{c1}$.

SECTION 12–7 CHECKUP

1. How do the open-loop voltage gain and the closed-loop voltage gain of an op-amp differ?
2. The upper critical frequency of a particular op-amp is 100 Hz. What is its open-loop 3 dB bandwidth?
3. Does the open-loop gain increase or decrease with frequency above the critical frequency?
4. If the individual stage gains of an op-amp are 20 dB and 30 dB, what is the total gain in decibels?
5. If the individual phase lags are -49° and -5.2°, what is the total phase lag?

12–8 CLOSED-LOOP FREQUENCY RESPONSE

Op-amps are normally used in a closed-loop configuration with negative feedback in order to achieve precise control of the gain and bandwidth. In this section, you will see how feedback affects the gain and frequency response of an op-amp.

After completing this section, you should be able to

- **Analyze the closed-loop frequency response of an op-amp**
  - Determine the closed-loop gain
  - Explain the effect of negative feedback on bandwidth
  - Explain gain-bandwidth product

Recall that midrange gain of an op-amp is reduced by negative feedback, as indicated by the following closed-loop gain expressions for the three amplifier configurations previously covered, where B is the feedback attenuation. For a noninverting amplifier,

$$A_{cl(NI)} = \frac{A_{ol}}{1 + A_{ol}B} \cong \frac{1}{B} = 1 + \frac{R_f}{R_i}$$

For an inverting amplifier,

$$A_{cl(I)} \cong -\frac{R_f}{R_i}$$

For a voltage-follower,

$$A_{cl(VF)} = 1$$

Effect of Negative Feedback on Bandwidth

You know how negative feedback affects the gain; now you will learn how it affects the amplifier's bandwidth. The closed-loop critical frequency of an op-amp is

$$f_{c(cl)} = f_{c(ol)}(1 + BA_{ol(mid)})$$ **Equation 12–21**

This expression shows that the closed-loop critical frequency, $f_{c(cl)}$, is higher than the open-loop critical frequency $f_{c(ol)}$ by the factor $1 + BA_{ol(mid)}$. You will find a derivation of Equation 12–21 in Appendix B.

Since $f_{c(cl)}$ equals the bandwidth for the closed-loop amplifier, the closed-loop bandwidth (BW_{cl}) is also increased by the same factor.

$$BW_{cl} = BW_{ol}(1 + BA_{ol(mid)})$$ **Equation 12–22**

EXAMPLE 12–11

A certain amplifier has an open-loop midrange gain of 150,000 and an open-loop 3 dB bandwidth of 200 Hz. The attenuation (B) of the feedback loop is 0.002. What is the closed-loop bandwidth?

Solution $BW_{cl} = BW_{ol}(1 + BA_{ol(mid)}) = 200\text{ Hz}[1 + (0.002)(150{,}000)] = \mathbf{60.2\ kHz}$

Related Problem If $A_{ol(mid)} = 200{,}000$ and $B = 0.05$, what is the closed-loop bandwidth?

Figure 12–42 graphically illustrates the concept of closed-loop response. When the open-loop gain of an op-amp is reduced by negative feedback, the bandwidth is increased.

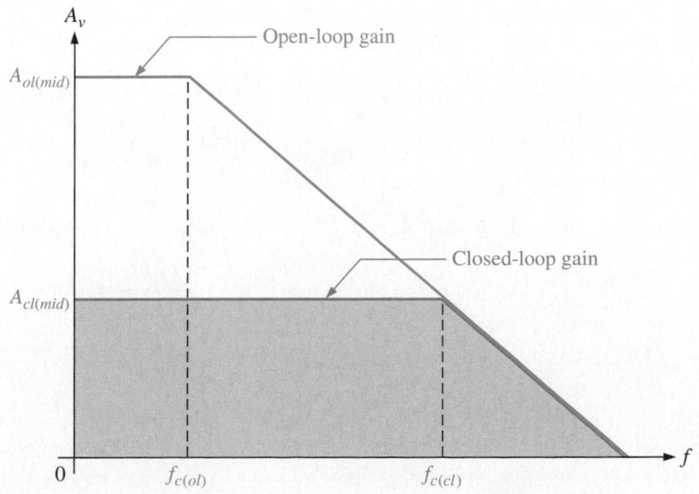

FIGURE 12–42

Closed-loop gain compared to open-loop gain.

The closed-loop gain is independent of the open-loop gain up to the point of intersection of the two gain curves. This point of intersection is the critical frequency, $f_{c(cl)}$, for the closed-loop response. Notice that the closed-loop gain has the same roll-off rate as the open-loop gain, beyond the closed-loop critical frequency.

Gain-Bandwidth Product

An increase in closed-loop gain causes a decrease in the bandwidth and vice versa, such that the product of gain and bandwidth is a constant. This is true as long as the roll-off rate is fixed, as in the case of a compensated op-amp. If you let A_{cl} represent the gain of any of the closed-loop configurations and $f_{c(cl)}$ represent the closed-loop critical frequency (same as the bandwidth), then

$$A_{cl}f_{c(cl)} = A_{ol}f_{c(ol)}$$

The **gain-bandwidth product** is always equal to the frequency at which the op-amp's open-loop gain is unity or 0 dB (unity-gain bandwidth, f_T).

Equation 12–23

$$\boldsymbol{f_T = A_{cl}f_{c(cl)}}$$

EXAMPLE 12–12

Determine the bandwidth of each of the amplifiers in Figure 12–43. Both op-amps have an open-loop gain of 100 dB and a unity-gain bandwidth (f_T) of 3 MHz.

FIGURE 12–43

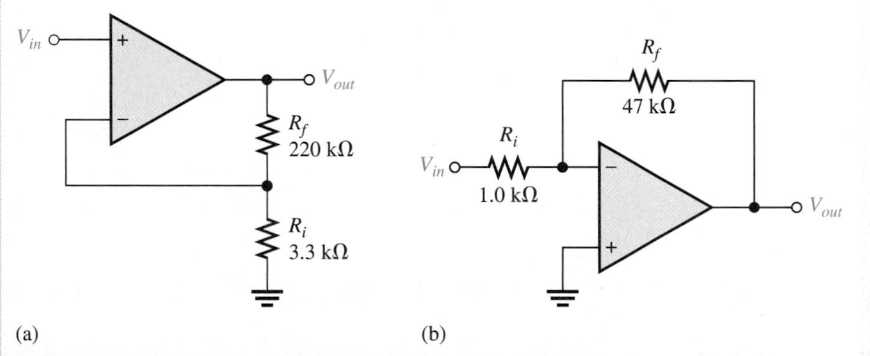

Solution **(a)** For the noninverting amplifier in Figure 12–43(a), the closed-loop gain is

$$A_{cl} = 1 + \frac{R_f}{R_i} = 1 + \frac{220\ \text{k}\Omega}{3.3\ \text{k}\Omega} = 67.7$$

Use Equation 12–23 and solve for $f_{c(cl)}$ (where $f_{c(cl)} = BW_{cl}$).

$$f_{c(cl)} = BW_{cl} = \frac{f_T}{A_{cl}}$$

$$BW_{cl} = \frac{3\ \text{MHz}}{67.7} = \mathbf{44.3\ kHz}$$

(b) For the inverting amplifier in Figure 12–43(b), the closed-loop gain is

$$A_{cl} = -\frac{R_f}{R_i} = -\frac{47\ \text{k}\Omega}{1.0\ \text{k}\Omega} = -47$$

Using the absolute value of A_{cl}, the closed-loop bandwidth is

$$BW_{cl} = \frac{3\ \text{MHz}}{47} = \mathbf{63.8\ kHz}$$

Related Problem Determine the bandwidth of each of the amplifiers in Figure 12–43. Both op-amps have an A_{ol} of 90 dB and a unity-gain bandwidth of 2 MHz.

Open the Multisim file E12-12 in the Examples folder on your CD-ROM. Measure the bandwidth of each amplifier using the Bode plotter and compare with the calculated values.

SECTION 12–8 CHECKUP

1. Is the closed-loop gain always less than the open-loop gain?
2. A certain op-amp is used in a feedback configuration having a gain of 30 and a bandwidth of 100 kHz. If the external resistor values are changed to increase the gain to 60, what is the new bandwidth?
3. What is the unity-gain bandwidth of the op-amp in Question 2?

12–9 TROUBLESHOOTING

As a technician, you may encounter situations in which an op-amp or its associated circuitry has malfunctioned. The op-amp is a complex integrated circuit with many types of internal failures possible. However, since you cannot troubleshoot the op-amp internally, treat it as a single device with only a few connections to it. If it fails, replace it just as you would a resistor, capacitor, or transistor.

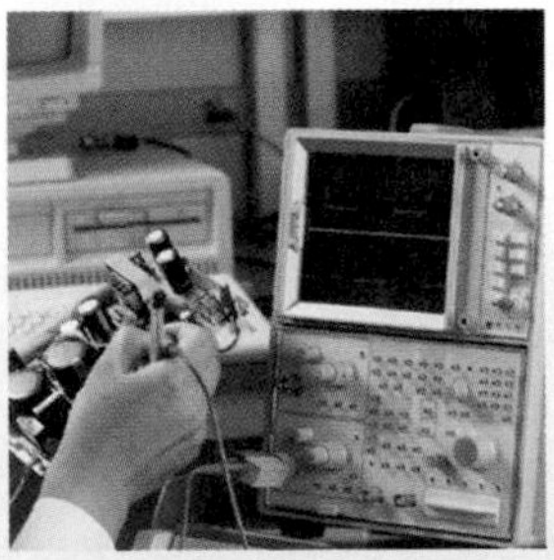

After completing this section, you should be able to

- **Troubleshoot op-amp circuits**
  - Analyze faults in a noninverting amplifier
  - Analyze faults in a voltage-follower
  - Analyze faults in an inverting amplifier

In the basic op-amp configurations, there are only a few external components that can fail. These are the feedback resistor, the input resistor, and the potentiometer used for offset voltage compensation. Also, of course, the op-amp itself can fail or there can be faulty contacts in the circuit. Let's examine the three basic configurations for possible faults and the associated symptoms.

The first thing to do when you suspect a faulty circuit is to check for the proper supply voltage and ground at the pins of the op-amp. Having done that, several other possible faults are as follows. A visual inspection should also be done.

Faults in the Noninverting Amplifier

Open Feedback Resistor If the feedback resistor, R_f, in Figure 12–44 opens, the op-amp is operating with its very high open-loop gain, which causes the input signal to drive the device into nonlinear operation and results in a severely clipped output signal as shown in part (a).

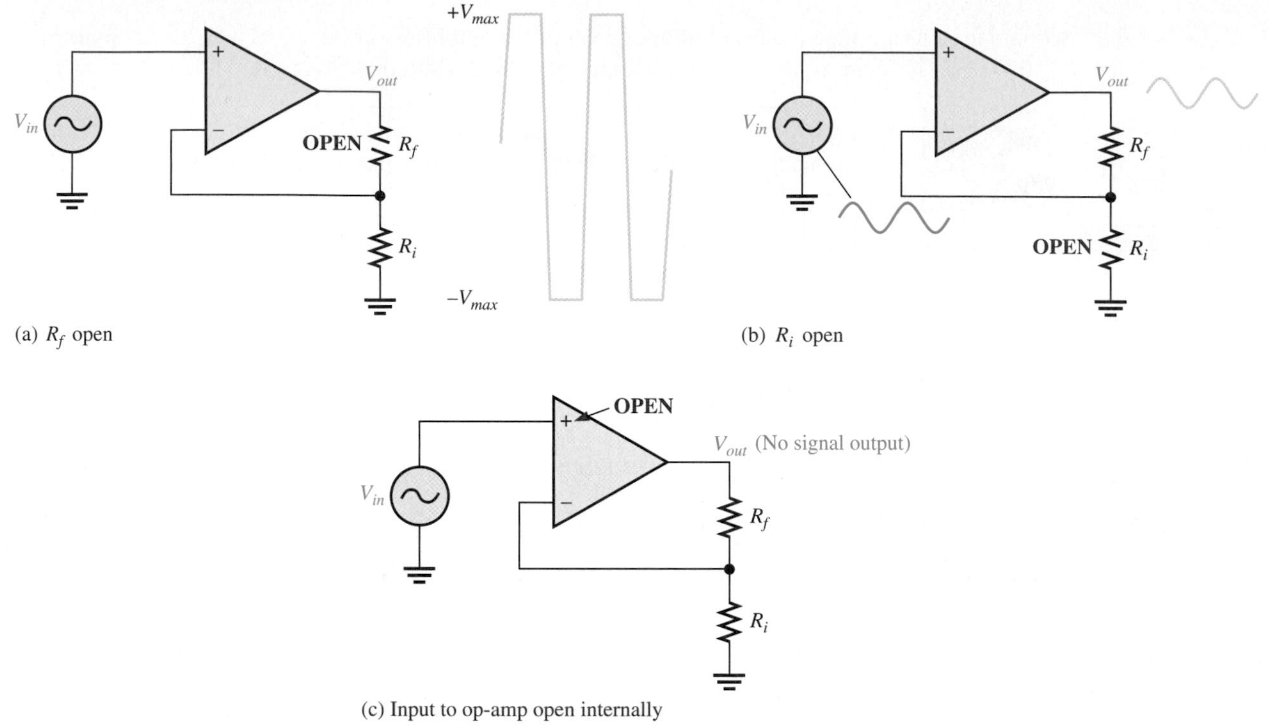

▲ FIGURE 12–44

Faults in the noninverting amplifier.

Open Input Resistor In this case, you still have a closed-loop configuration. Since R_i is open and effectively equal to infinity (∞), the closed-loop gain from Equation 12–8 is

$$A_{cl(\mathrm{NI})} = 1 + \frac{R_f}{R_i} = 1 + \frac{R_f}{\infty} = 1 + 0 = 1$$

This shows that the amplifier acts like a voltage-follower. You would observe an output signal that is the same as the input, as indicated in Figure 12–44(b).

Internally Open Noninverting Op-Amp Input In this situation, because the input voltage is not applied to the op-amp, the output is zero. This is indicated in Figure 12–44(c).

Other Op-Amp Faults In general, an internal failure will result in a loss or distortion of the output signal. The best approach is to first make sure that there are no external failures or faulty conditions. If everything else is good, then the op-amp must be bad.

Faults in the Voltage-Follower

The voltage-follower is a special case of the noninverting amplifier. Except for a faulty op-amp, an open or shorted external connection, or a problem with the offset null potentiometer, about the only thing that can happen in a voltage-follower circuit is an open feedback loop. This would have the same effect as an open feedback resistor as previously discussed.

Faults in the Inverting Amplifier

Open Feedback Resistor If R_f opens, as indicated in Figure 12–45(a), the input signal still feeds through the input resistor and is amplified by the high open-loop gain of the op-amp. This forces the device to be driven into nonlinear operation, and you will see an output something like that shown. This is a similar result as in the noninverting amplifier configuration.

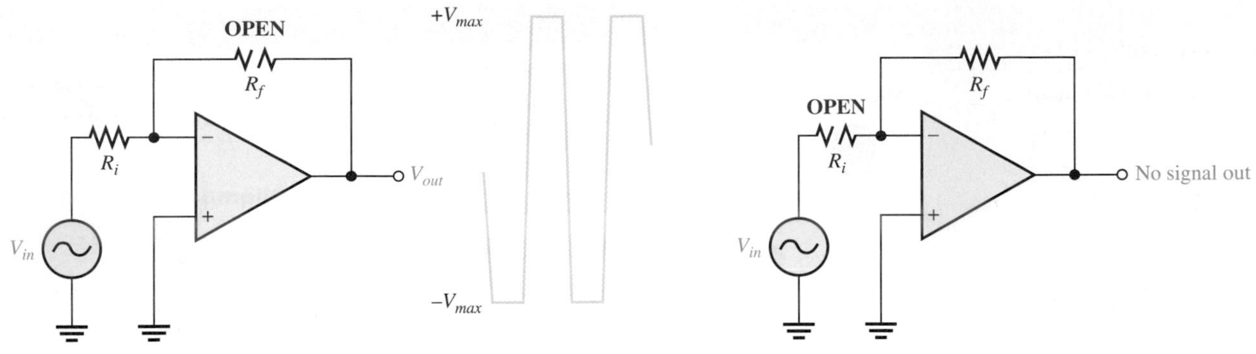

(a) (b)

FIGURE 12–45

Faults in the inverting amplifier.

Open Input Resistor This prevents the input signal from getting to the op-amp input, so there will be no output signal, as indicated in Figure 12–45(b).

Failures in the op-amp itself or the offset null potentiometer have the same effects as previously discussed for the noninverting amplifier configuration.

Multisim Troubleshooting Exercises

These file circuits are in the Troubleshooting Exercises folder on your CD-ROM.

1. Open file TSE12-01. Determine if the circuit is working properly, and, if not, determine the fault.
2. Open file TSE12-02. Determine if the circuit is working properly, and, if not, determine the fault.
3. Open file TSE12-03. Determine if the circuit is working properly, and, if not, determine the fault.
4. Open file TSE12-04. Determine if the circuit is working properly, and, if not, determine the fault.

SECTION 12–9 CHECKUP

1. **If you notice that the op-amp output signal is beginning to clip on one peak as you increase the input signal, what should you check?**
2. **For a noninverting amplifier, if there is no op-amp output signal when there is a verified input signal at the input pin, what would you suspect as being faulty?**

On the Job

Application Activity: *Op-Amp Audio Amplifier*

The company that manufactures the PA system developed in Chapters 6 and 7 wants to replace the audio amplifier with a new design using an op-amp instead of the discrete transistor preamp circuit to reduce parts and cost. The power amplifier will still retain its basic design with only a few changes. The block diagram of the PA system is shown in Figure 12–46.

Simulation

The audio amplifier is simulated with an input signal of 50 mV using Multisim. The results are shown in Figure 12–49 where an 8.2 Ω resistor is used to simulate the speaker.

FIGURE 12–49

Simulation of the audio amplifier.

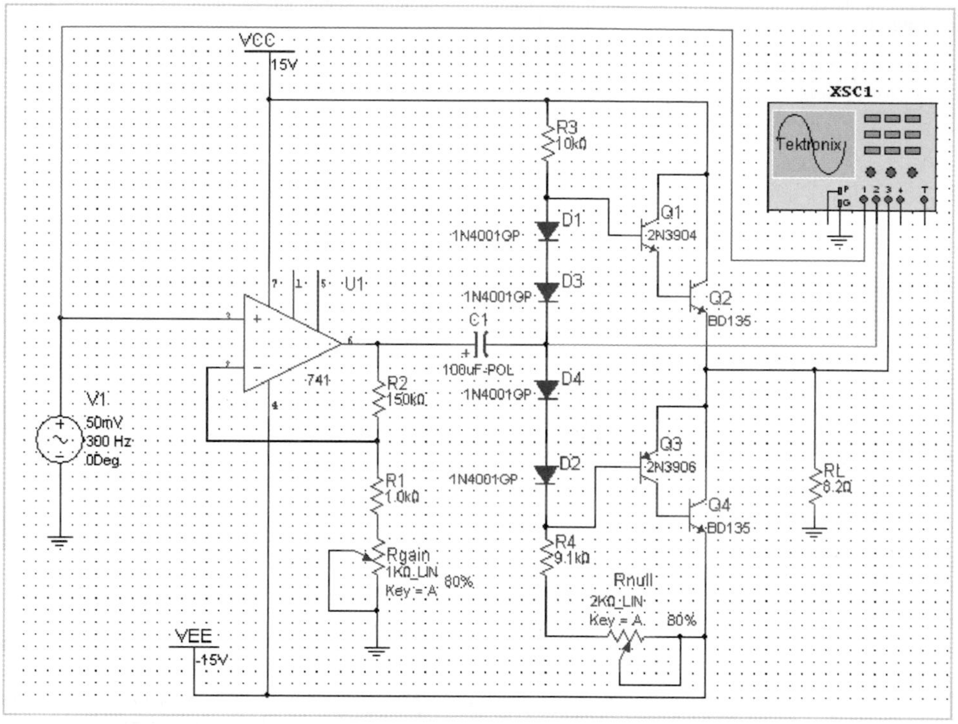

(a) Circuit screen

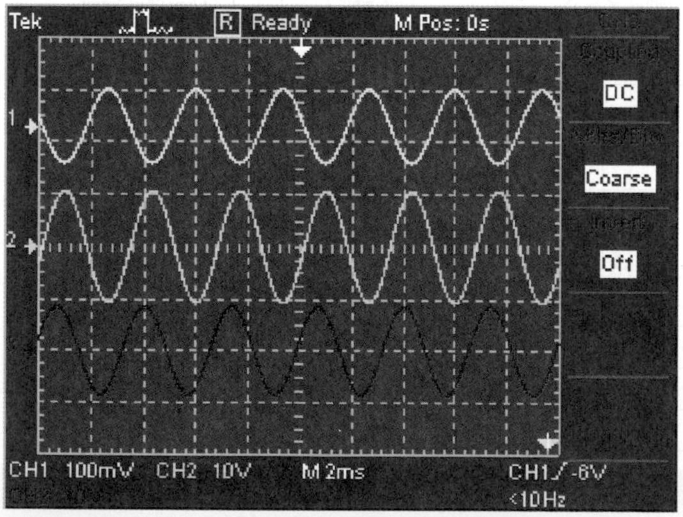

(b) Input signal, op-amp output signal, and final output signal

8. From the scope display in Figure 12–49, determine the rms value of each voltage.
9. Determine the voltage gain of the op-amp stage from the measured signals.
10. Determine the overall voltage gain from the measured signals.

Simulate the op-amp audio amplifier using your Multisim software. Observe the signal voltages with the oscilloscope.

Prototyping and Testing

Now that the circuit has been simulated, the prototype circuit is constructed and tested. After the circuit is successfully tested on a protoboard, it is ready to be finalized on a printed circuit board.

To build and test a similar circuit, go to Experiment 12–A in your lab manual (*Laboratory Exercises for Electronic Devices* by David Buchla and Steven Wetterling).

Lab Experiment

The original system had two boards, the preamp board and the power amp board. By using the new op-amp design, the audio amplifier is simplified to one board, as shown in Figure 12–50.

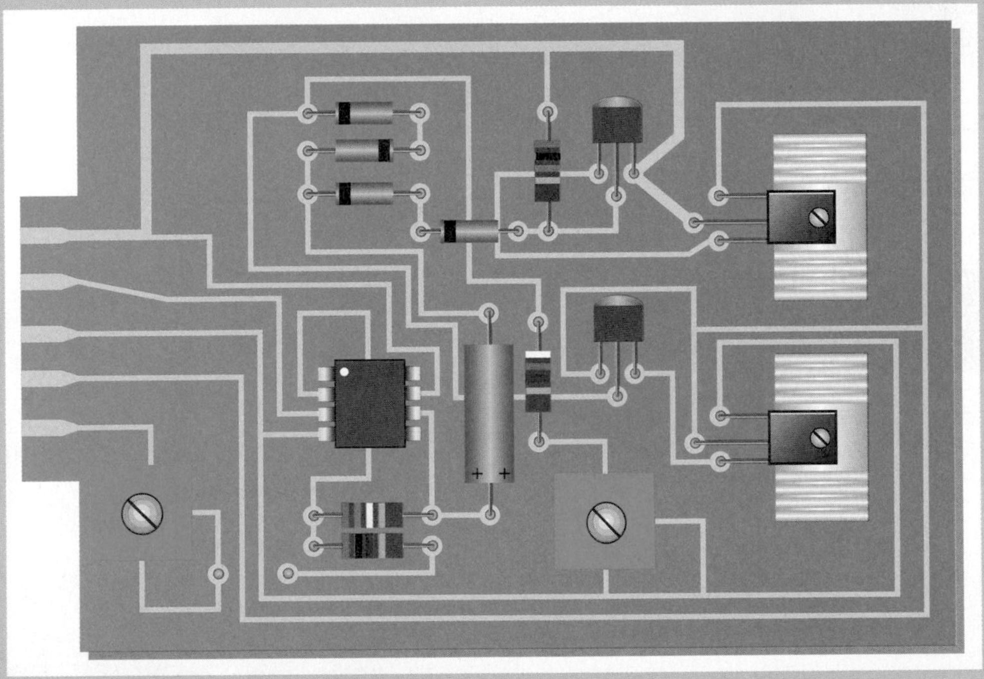

FIGURE 12–50

Audio amplifier board.

11. Check the printed circuit board for correctness by comparing it with the schematic in Figure 12–47.
12. Label each input and output pin according to function.

Troubleshooting

Four circuit boards are tested, and the results are shown in Figure 12–51.

13. Determine the problem, if any, in each of the board tests in Figure 12–51.

14. List possible causes of any problem from item 13.

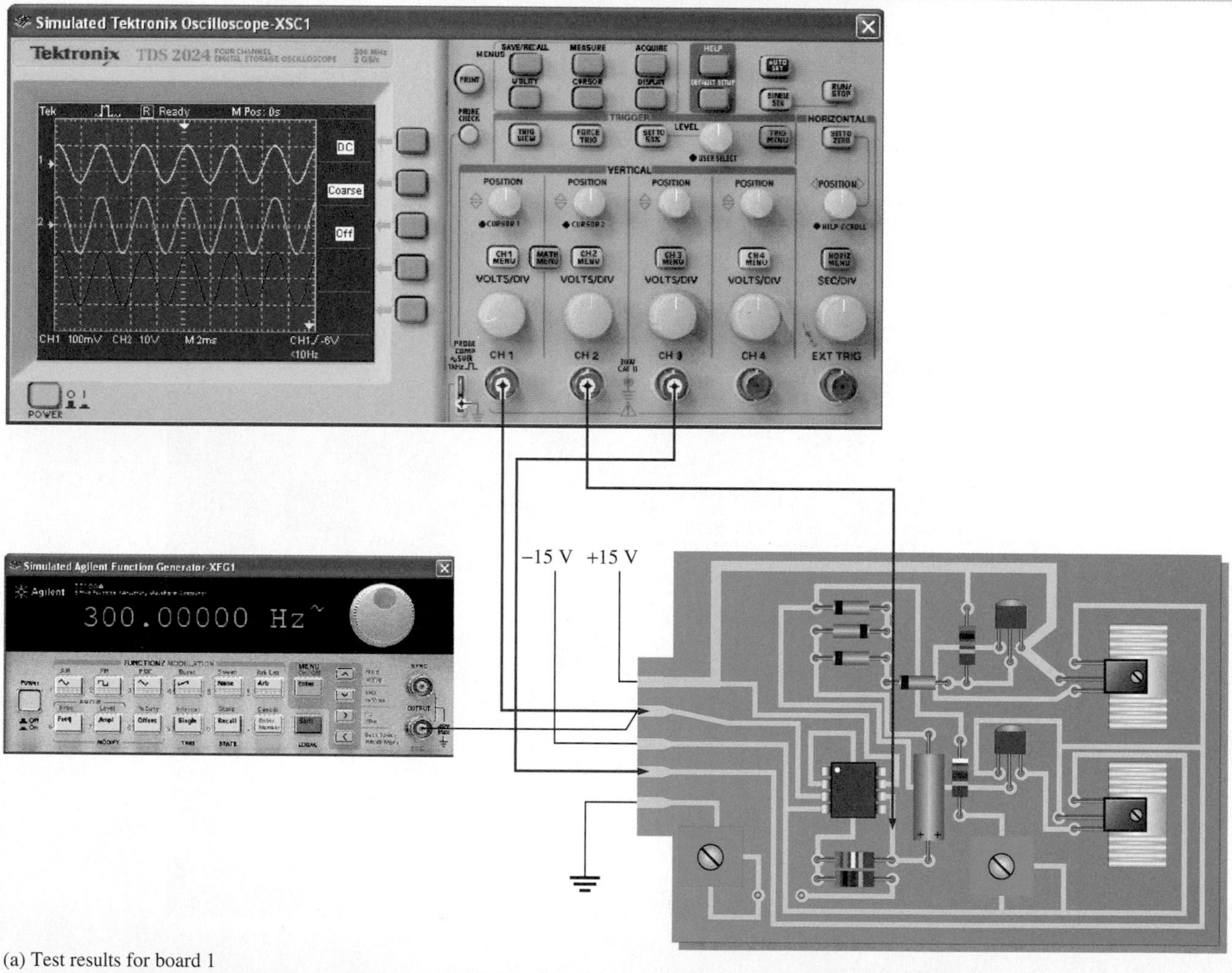

(a) Test results for board 1

FIGURE 12–51

Results of audio amplifier board tests.

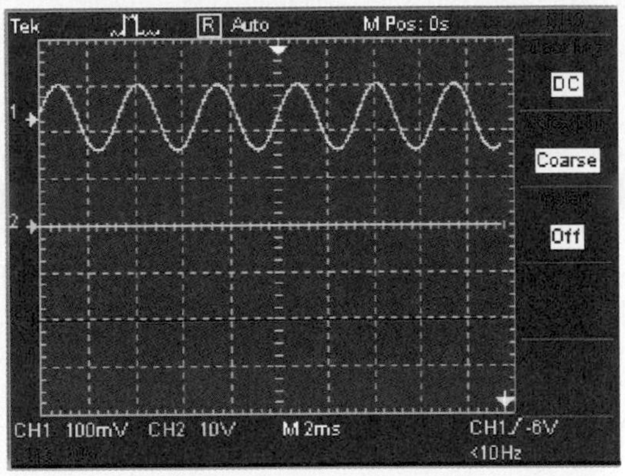

(b) Test results for board 2

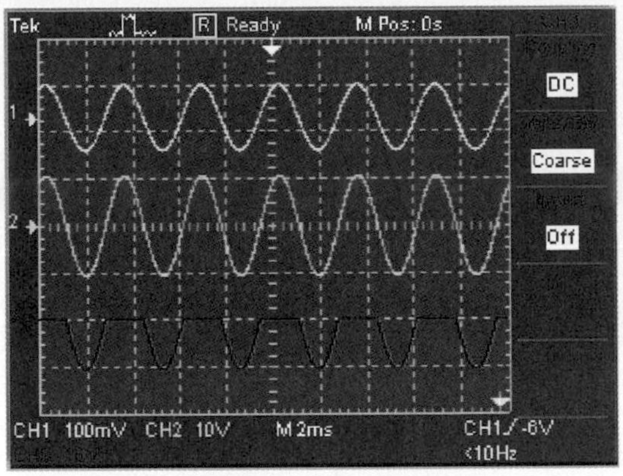

(c) Test results for board 3

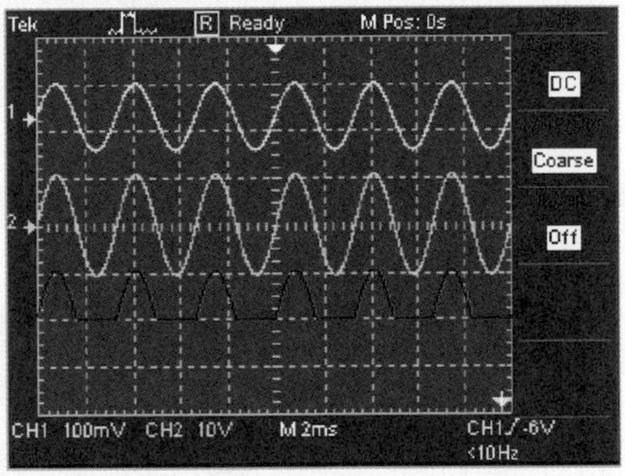

(d) Test results for board 4

FIGURE 12–51

Continued

Anadigm.com

Programmable Analog Design

FPAAs (field-programmable analog arrays) and dpASPs (dynamically programmable analog signal processors) are based on switched-capacitor technology that was covered in Chapter 9. These are integrated circuit devices that can be programmed with software for various types of analog functions and designs. Both the FPAA and the dpASP can be reprogrammed, but the FPAA is statically reconfigurable and must first be reset whereas the dpASP can be **dynamically reconfigured** "on-the-fly" while operating in a system. The software allows you to experiment and design with various analog devices that are covered in this textbook by specifying parameters, observing operation, interconnecting devices for more complex circuits, and preparing the software for **downloading** to an actual device. Refer to the tutorial available with the AnadigmDesigner®2 software or the tutorial in the *Laboratory Exercises for Electronic Devices* lab manual.

The Designer2 software differs in its primary purpose from the Multisim software you have been using. Electronic Workbench Multisim is basically a simulation software that allows you to test circuits on the computer using simulated discrete and integrated components. The Multisim software is useful for verifying that a circuit actually works as intended, but it has limited hardware interface capability. AnadigmDesigner2 is both a simulation and a hardware interface tool that allows you to custom program an analog design and implement it in an integrated circuit chip. It is based on an extensive library of analog functions called configurable analog modules (**CAMs**) that can be connected using a simple "drag-and drop" format and tested with virtual instruments on the computer. The design can then be converted to hardware by downloading it to an actual FPAA or dpASP IC chip.

A free trial version of the AnadigmDesigner2 software is available for downloading at www.anadigm.com. The basic steps in implementing a design are as follows using a specific CAM for illustration. These steps apply to any CAM or multiple CAMs.

CAM Selection

The first step in implementing a programmable analog design is to open the AnadigmDesigner2 software. The representation of a blank FPAA or dpASP chip will appear as shown in Figure 12–52. Select the *Configure* icon to open a list of available CAMs, as shown in Figure 12–53.

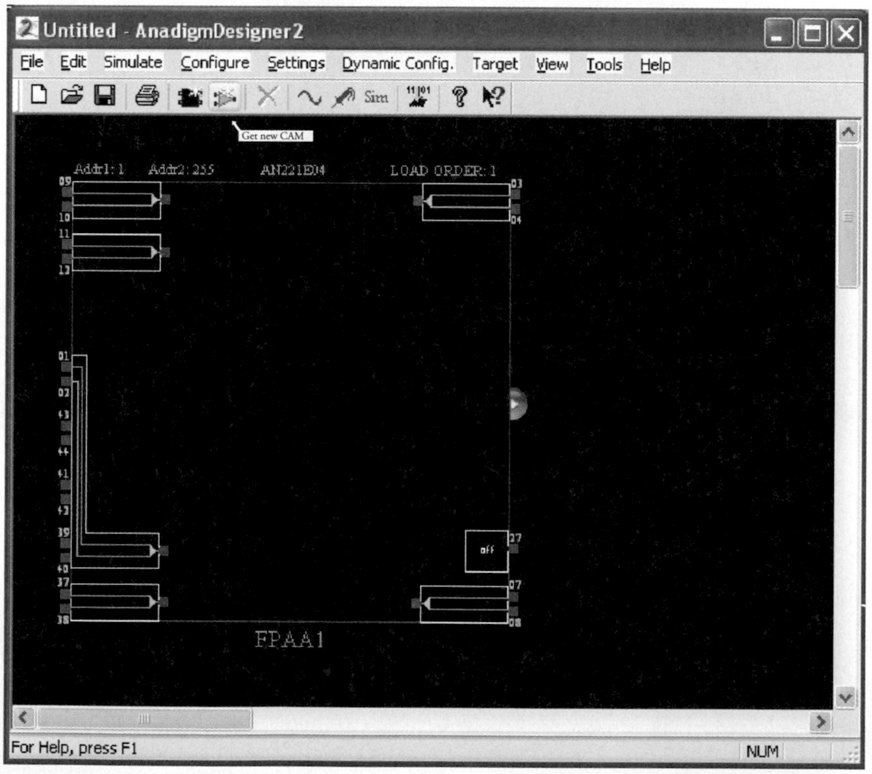

FIGURE 12–52

Click on *Configure* icon.

Configure and Place CAM

Select the desired CAM and set up the parameters in the *Set CAM Parameters* screen shown in Figure 12–54. Next place the CAM in the chip outline and connect to an input and output, as shown in Figure 12–55. Several CAMs can be placed in one chip and interconnected.

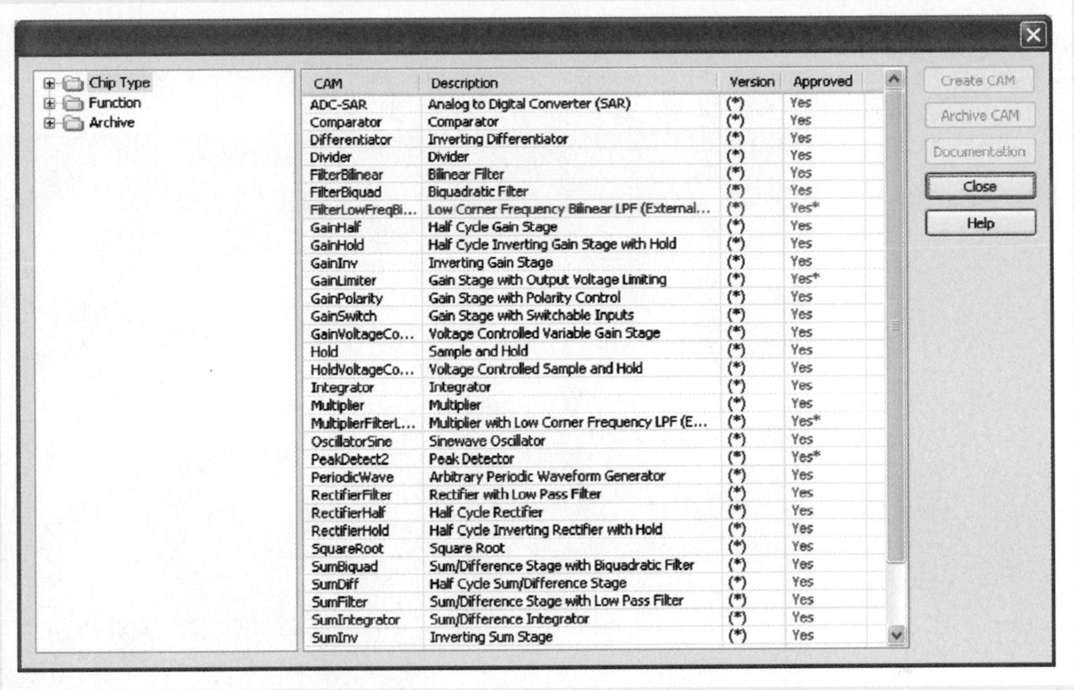

▲ **FIGURE 12–53**

List of available CAMs.

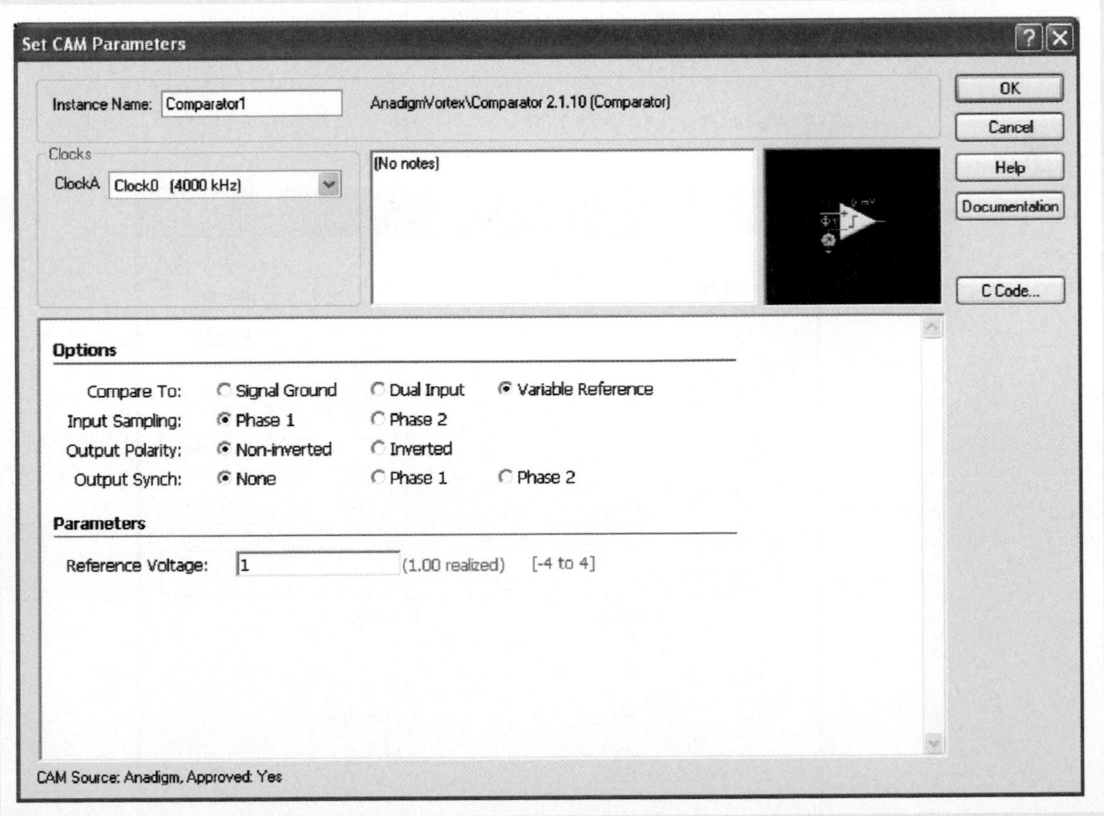

▲ **FIGURE 12–54**

Set CAM parameters.

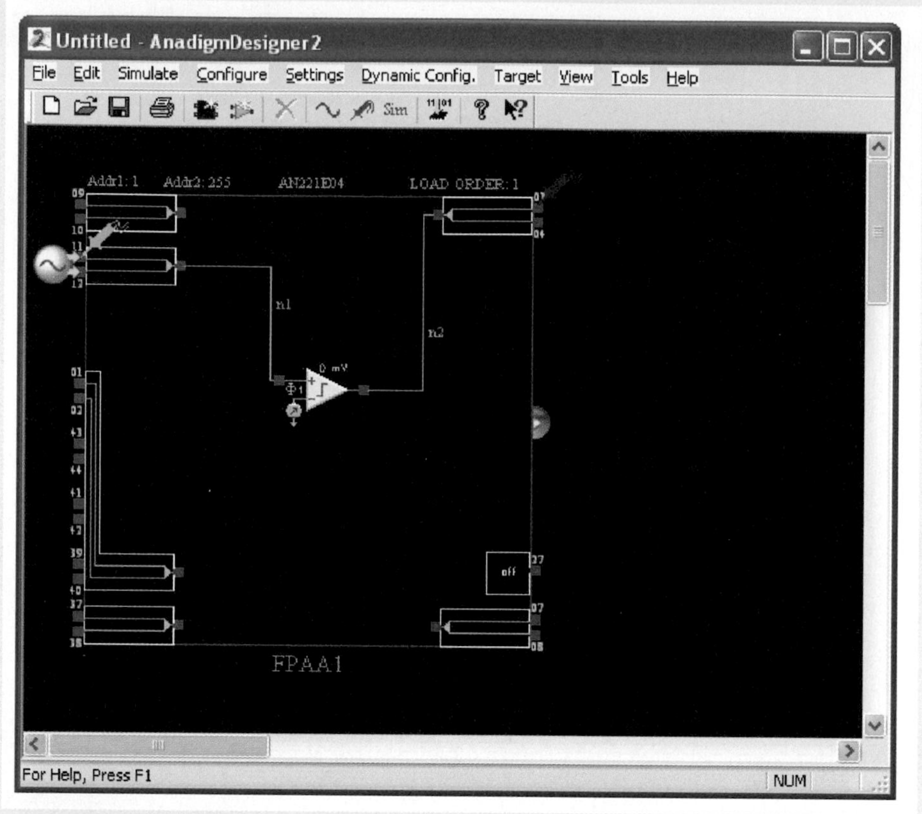

FIGURE 12–55

CAM placed and connected. Signal source and scope probes placed.

Test the Design

Place a signal source on the input by clicking on the sine wave icon and set its function and parameters using the *Signal Generator Control* window, as shown in Figure 12–56. Place probes at

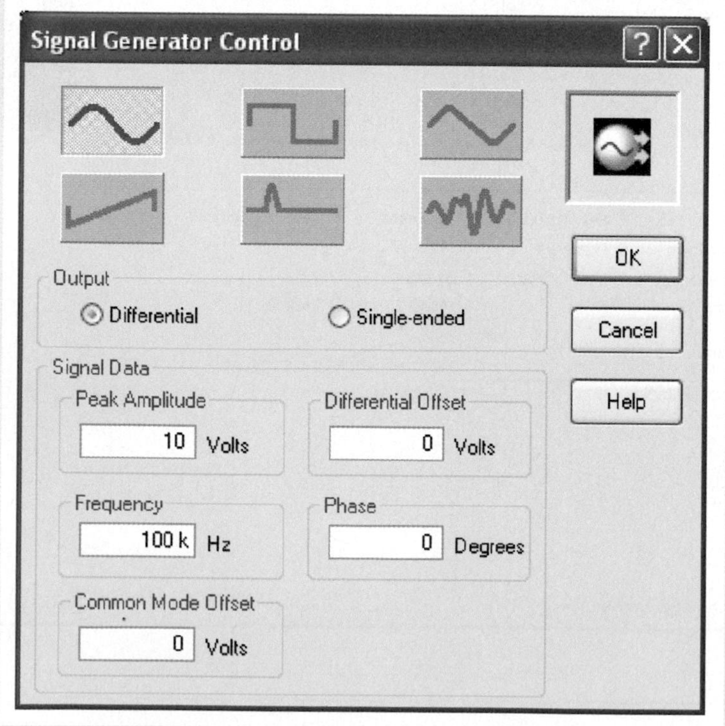

FIGURE 12–56

Set signal generator parameters.

appropriate points by clicking on the probe icon and then measure the waveforms on the virtual oscilloscope, as shown in Figure 12–57, by clicking on *Sim*.

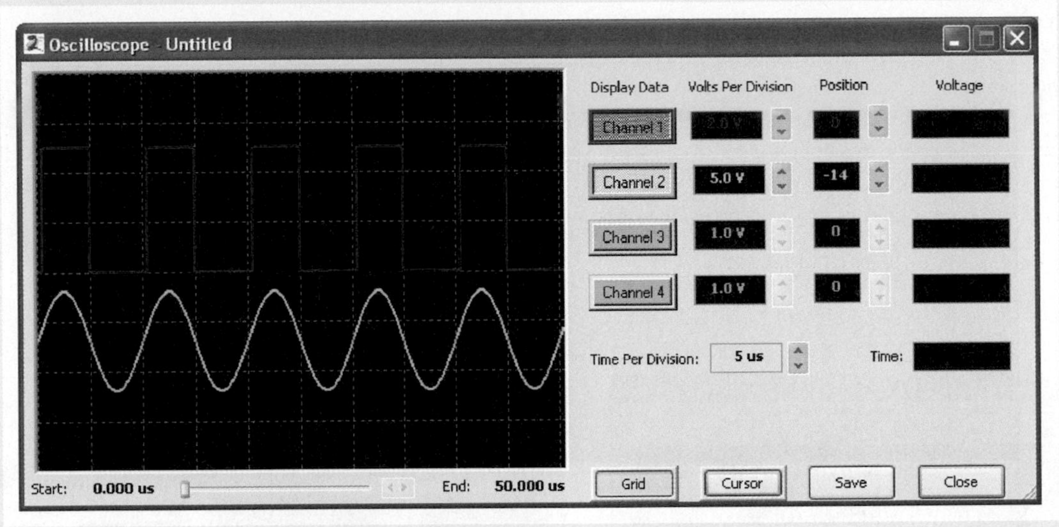

FIGURE 12–57

Measure waveforms at probe locations.

Download the Design

Once the design is finalized and tested on your computer, it can be downloaded to an actual FPAA or dpASP chip, known as a target device, by selecting *Target* on the screen. The chip is normally mounted on a special PC board with peripheral devices, test points, and connectors so that the downloaded design can be fully tested in hardware. One type of board for this purpose is the programmable analog module (PAM) board available from Servenger LLC (www.servenger.com). Figure 12–58 illustrates the setup.

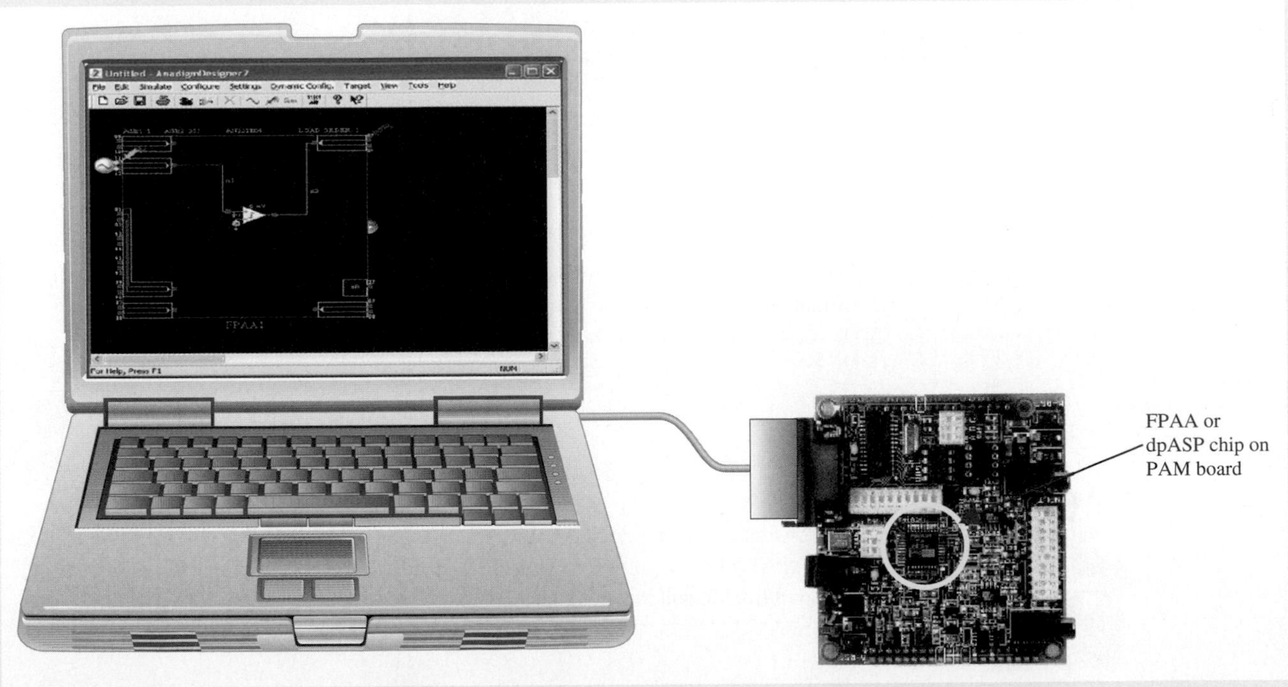

FIGURE 12–58

EXAMPLE

Simulate an inverting amplifier with a gain of −1. Apply a sinusoidal input and check both its input and output with the oscilloscope.

Step 1: Open the software and the outline of the blank FPAA chip appears with input and output pins.
Step 2: Select the *Inverting Gain Stage* and drop into the chip outline.
Step 3: Set the gain.
Step 4: Connect a signal generator to an input and set its parameters.
Step 5: Connect a scope probe to an output.
Step 6: Select "begin simulation."
Step 7: Observe input and output waveforms on oscilloscope.

The results of these steps are shown in Figure 12–59.

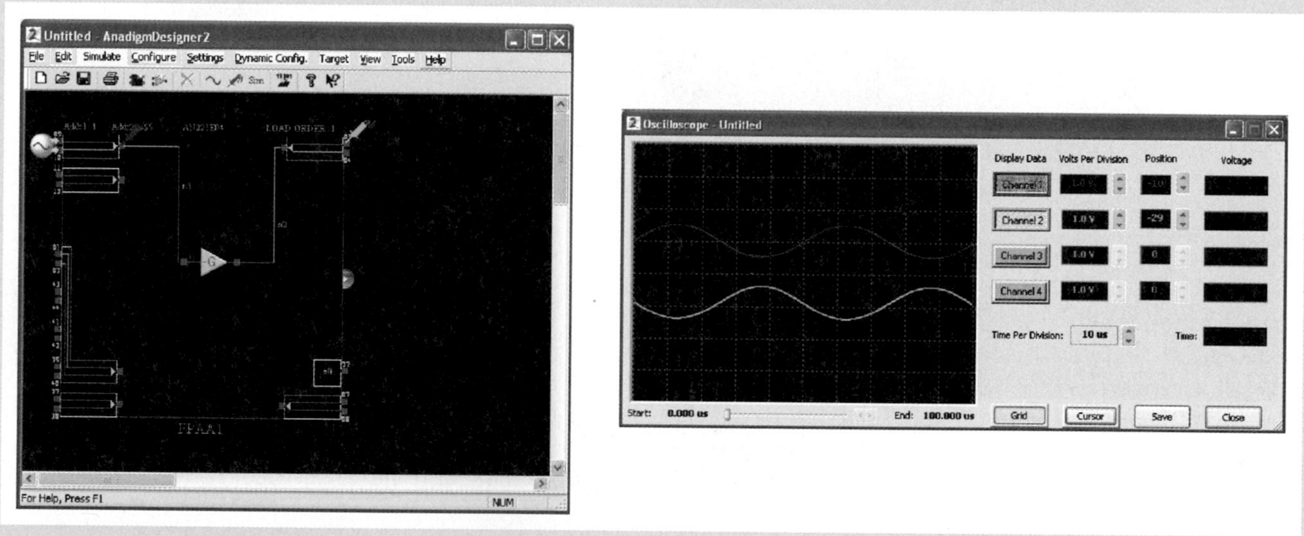

FIGURE 12–59

Programming Exercises

1. Simulate and test an inverting amplifier with a gain of −10.
2. Simulate and test a 2-stage noninverting amplifier with a gain of 50.
3. If you have an evaluation board, download each of the simulated circuits.

PAM Experiment

To program, download, and test circuits using AnadigmDesigner2 software and the programmable analog module (PAM) board, go to Experiment 12–B in *Laboratory Exercises for Electronic Devices* by David Buchla and Steve Wetterling.

SUMMARY OF OP-AMP CONFIGURATIONS

BASIC OP-AMP

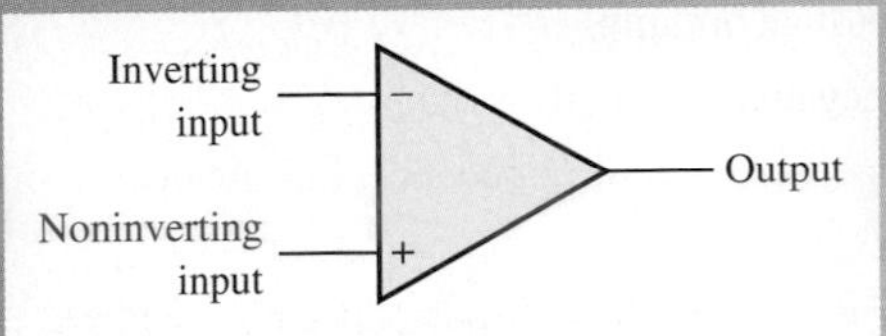

Symbol

- Very high open-loop voltage gain
- Very high input impedance
- Very low output impedance

NONINVERTING AMPLIFIER

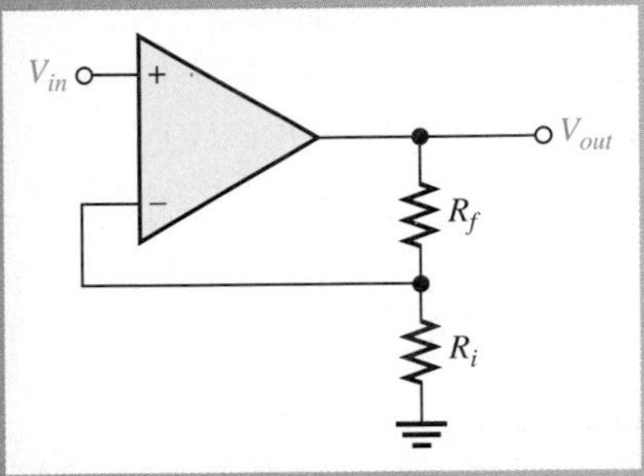

- Voltage gain:

$$A_{cl(NI)} = 1 + \frac{R_f}{R_i}$$

- Input impedance:

$$Z_{in(NI)} = (1 + A_{ol}B)Z_{in}$$

- Output impedance:

$$Z_{out(NI)} = \frac{Z_{out}}{1 + A_{ol}B}$$

VOLTAGE-FOLLOWER

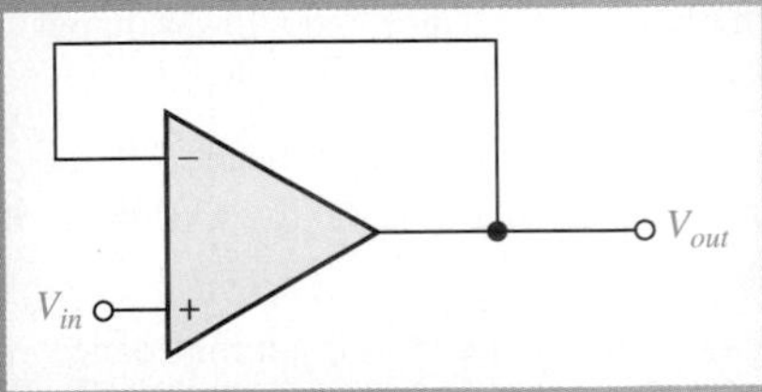

- Voltage gain:

$$A_{cl(VF)} = 1$$

- Input impedance:

$$Z_{in(VF)} = (1 + A_{ol})Z_{in}$$

- Output impedance:

$$Z_{out(VF)} = \frac{Z_{out}}{1 + A_{ol}}$$

INVERTING AMPLIFIER

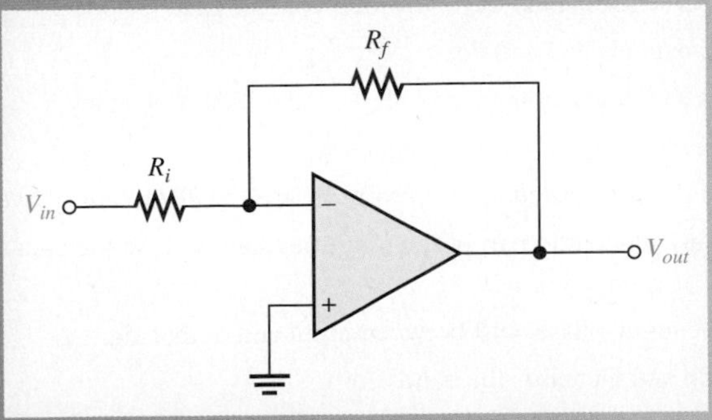

- Voltage gain:

$$A_{cl(I)} = -\frac{R_f}{R_i}$$

- Input impedance:

$$Z_{in(I)} \cong R_i$$

- Output impedance:

$$Z_{out(I)} = \frac{Z_{out}}{1 + A_{ol}B}$$

SUMMARY

Section 12–1

- The basic op-amp has three terminals not including power and ground: inverting (−) input, noninverting (+) input, and output.
- A differential amplifier forms the input stage of an op-amp.
- Most op-amps require both a positive and a negative dc supply voltage.
- The ideal op-amp has infinite input impedance, zero output impedance, infinite open-loop voltage gain, and infinite bandwidth.
- A practical op-amp has very high input impedance, very low output impedance, and very high open-loop voltage gain.

Section 12–2

- Two types of op-amp input operation are the differential mode and the common mode.
- Common mode occurs when equal in-phase voltages are applied to both input terminals.
- The common-mode rejection ratio (CMRR) is a measure of an op-amp's ability to reject common-mode inputs.
- Open-loop voltage gain is the gain of an op-amp with no external feedback connections.
- Input offset voltage produces an output error voltage (with no input voltage).
- Input bias current also produces an output error voltage (with no input voltage).
- Input offset current is the difference between the two bias currents.
- Slew rate is the rate in volts per microsecond at which the output voltage of an op-amp can change in response to a step input.

Section 12–3

- Negative feedback occurs when a portion of the output voltage is connected back to the inverting input such that it subtracts from the input voltage, thus reducing the voltage gain but increasing the stability and bandwidth.

Section 12–4

- There are three basic op-amp configurations: inverting, noninverting, and voltage-follower.
- The three basic op-amp configurations employ negative feedback.
- Closed-loop voltage gain is the gain of an op-amp with external feedback.

Section 12–5

- A noninverting amplifier configuration has a higher input impedance and a lower output impedance than the op-amp itself (without feedback).
- An inverting amplifier configuration has an input impedance approximately equal to the input resistor R_i and an output impedance approximately equal to the output impedance of the op-amp itself.
- The voltage-follower has the highest input impedance and the lowest output impedance of the three amplifier configurations.

Section 12–6

- All practical op-amps have small input bias currents and input offset voltages that produce small output error voltages.
- The input bias current effect can be compensated for with external resistors.
- The input offset voltage can be compensated for with an external potentiometer between the two offset null pins provided on the IC op-amp package and as recommended by the manufacturer.

Section 12–7

- The closed-loop voltage gain is always less than the open-loop voltage gain.
- The midrange gain of an op-amp extends down to dc.
- The gain of an op-amp decreases as frequency increases above the critical frequency.
- The bandwidth of an op-amp equals the upper critical frequency.
- The open-loop response curve of a compensated op-amp rolls off at −20 dB/decade above f_c.

Section 12–8

- The internal RC lag circuits that are inherently part of the amplifier stages cause the gain to roll off as frequency goes up.
- The internal RC lag circuits also cause a phase shift between input and output signals.
- Negative feedback lowers the gain and increases the bandwidth.
- The product of gain and bandwidth is constant for a given op-amp.
- The gain-bandwidth product equals the frequency at which unity voltage gain occurs.

KEY TERMS

Key terms and other bold terms in the chapter are defined in the end-of-book glossary.

Closed-loop voltage gain (A_{cl}) The voltage gain of an op-amp with external feedback.

CMRR Common-mode rejection ratio; the ratio of open-loop gain to common-mode gain; a measure of an op-amp's ability to reject common-mode signals.

Common mode A condition characterized by the presence of the same signal on both op-amp inputs.

Differential amplifier A type of amplifier with two inputs and two outputs that is used as the input stage of an op-amp.

Differential mode A mode of op-amp operation in which two opposite-polarity signal voltages are applied to the two inputs (double-ended) or in which a signal is applied to one input and ground to the other input (single-ended).

Gain-bandwidth product A constant parameter which is always equal to the frequency at which the op-amp's open-loop gain is unity (1).

Inverting amplifier An op-amp closed-loop configuration in which the input signal is applied to the inverting input.

Negative feedback The process of returning a portion of the output signal to the input of an amplifier such that it is out of phase with the input signal.

Noninverting amplifier An op-amp closed-loop configuration in which the input signal is applied to the noninverting input.

Open-loop voltage gain (A_{ol}) The voltage gain of an op-amp without external feedback.

Operational amplifier (op-amp) A type of amplifier that has very high voltage gain, very high input impedance, very low output impedance, and good rejection of common-mode signals.

Phase shift The relative angular displacement of a time-varying function relative to a reference.

Slew rate The rate of change of the output voltage of an op-amp in response to a step input.

Voltage-follower A closed-loop, noninverting op-amp with a voltage gain of 1.

KEY FORMULAS

Op-Amp Input Modes and Parameters

12–1	$\mathrm{CMRR} = \dfrac{A_{ol}}{A_{cm}}$	Common-mode rejection ratio
12–2	$\mathrm{CMRR} = 20\log\left(\dfrac{A_{ol}}{A_{cm}}\right)$	Common-mode rejection ratio (dB)
12–3	$I_{\mathrm{BIAS}} = \dfrac{I_1 + I_2}{2}$	Input bias current
12–4	$I_{\mathrm{OS}} = \lvert I_1 - I_2 \rvert$	Input offset current
12–5	$V_{\mathrm{OS}} = I_{\mathrm{OS}}R_{in}$	Offset voltage
12–6	$V_{\mathrm{OUT(error)}} = A_v I_{\mathrm{OS}} R_{in}$	Output error voltage
12–7	$\text{Slew rate} = \dfrac{\Delta V_{out}}{\Delta t}$	Slew rate

Op-Amp Configurations

12–8	$A_{cl(\mathrm{NI})} = 1 + \dfrac{R_f}{R_i}$	Voltage gain (noninverting)
12–9	$A_{cl(\mathrm{VF})} = 1$	Voltage gain (voltage-follower)
12–10	$A_{cl(\mathrm{I})} = -\dfrac{R_f}{R_i}$	Voltage gain (inverting)

Op-Amp Impedances

12–11	$Z_{in(NI)} = (1 + A_{ol}B)Z_{in}$	Input impedance (noninverting)
12–12	$Z_{out(NI)} = \dfrac{Z_{out}}{1 + A_{ol}B}$	Output impedance (noninverting)
12–13	$Z_{in(VF)} = (1 + A_{ol})Z_{in}$	Input impedance (voltage-follower)
12–14	$Z_{out(VF)} = \dfrac{Z_{out}}{1 + A_{ol}}$	Output impedance (voltage-follower)
12–15	$Z_{in(I)} \cong R_i$	Input impedance (inverting)
12–16	$Z_{out(I)} = \dfrac{Z_{out}}{1 + A_{ol}B}$	Output impedance (inverting)

Op-Amp Frequency Responses

12–17	$BW = f_{cu}$	Op-amp bandwidth
12–18	$\dfrac{V_{out}}{V_{in}} = \dfrac{1}{\sqrt{1 + f^2/f_c^2}}$	*RC* attenuation
12–19	$A_{ol} = \dfrac{A_{ol(mid)}}{\sqrt{1 + f^2/f_c^2}}$	Open-loop voltage gain
12–20	$\theta = -\tan^{-1}\left(\dfrac{f}{f_c}\right)$	*RC* phase shift
12–21	$f_{c(cl)} = f_{c(ol)}(1 + BA_{ol(mid)})$	Closed-loop critical frequency
12–22	$BW_{cl} = BW_{ol}(1 + BA_{ol(mid)})$	Closed-loop bandwidth
12–23	$f_T = A_{cl}f_{c(cl)}$	Unity-gain bandwidth

TRUE/FALSE QUIZ

Answers are at the end of the chapter.

1. An ideal op-amp has an infinite input impedance.
2. An ideal op-amp has a very high output impedance.
3. The op-amp can operate in both the differential mode or the common mode.
4. Common-mode rejection means that a signal appearing on both inputs is effectively cancelled.
5. CMRR stands for common-mode rejection reference.
6. Slew rate determines how fast the output can change in response to a step input.
7. Negative feedback reduces the gain of an op-amp from its open-loop value.
8. Negative feedback reduces the bandwidth of an op-amp from its open-loop value.
9. A noninverting amplifier uses negative feedback.
10. The gain of a voltage-follower is very high.
11. Negative feedback affects the input and output impedances of an op-amp.
12. A compensated op-amp has a gain roll-off of −20 dB/decade above the critical frequency.
13. The gain-bandwidth product equals the unity-gain frequency.
14. If the feedback resistor in an inverting amplifier opens, the gain becomes zero.

CIRCUIT-ACTION QUIZ

Answers are at the end of the chapter.

1. If R_f is decreased in the circuit of Figure 12–17, the voltage gain will
   (a) increase **(b)** decrease **(c)** not change
2. If $V_{in} = 1$ mV and R_f opens in the circuit of Figure 12–17, the output voltage will
   (a) increase **(b)** decrease **(c)** not change

3. If R_i is increased in the circuit of Figure 12–17, the voltage gain will
   (a) increase (b) decrease (c) not change
4. If 10 mV are applied to the input to the op-amp circuit of Figure 12–21 and R_f is increased, the output voltage will
   (a) increase (b) decrease (c) not change
5. In Figure 12–27, if R_f is changed from 100 kΩ to 68 kΩ, the feedback attenuation will
   (a) increase (b) decrease (c) not change
6. If the closed-loop gain in Figure 12–43(a) is increased by increasing the value of R_f, the closed-loop bandwidth will
   (a) increase (b) decrease (c) not change
7. If R_f is changed to 470 kΩ and R_i is changed to 10 kΩ in Figure 12–43(b), the closed-loop bandwidth will
   (a) increase (b) decrease (c) not change
8. If R_i in Figure 12–43(b) opens, the output voltage will
   (a) increase (b) decrease (c) not change

SELF-TEST

Answers are at the end of the chapter.

Section 12–1

1. An integrated circuit (IC) op-amp has
   (a) two inputs and two outputs (b) one input and one output
   (c) two inputs and one output
2. Which of the following characteristics does not necessarily apply to an op-amp?
   (a) High gain (b) Low power
   (c) High input impedance (d) Low output impedance
3. A differential amplifier
   (a) is part of an op-amp (b) has one input and one output
   (c) has two outputs (d) answers (a) and (c)

Section 12–2

4. When an op-amp is operated in the single-ended differential mode,
   (a) the output is grounded
   (b) one input is grounded and a signal is applied to the other
   (c) both inputs are connected together
   (d) the output is not inverted
5. In the double-ended differential mode,
   (a) a signal is applied between the two inputs (b) the gain is 1
   (c) the outputs are different amplitudes (d) only one supply voltage is used
6. In the common mode,
   (a) both inputs are grounded (b) the outputs are connected together
   (c) an identical signal appears on both inputs (d) the output signals are in-phase
7. Common-mode gain is
   (a) very high (b) very low
   (c) always unity (d) unpredictable
8. If $A_{ol} = 3500$ and $A_{cm} = 0.35$, the CMRR is
   (a) 1225 (b) 10,000
   (c) 80 dB (d) answers (b) and (c)
9. With zero volts on both inputs, an op-amp ideally should have an output equal to
   (a) the positive supply voltage (b) the negative supply voltage
   (c) zero (d) the CMRR

10. Of the values listed, the most realistic value for open-loop gain of an op-amp is
    (a) 1 **(b)** 2000 **(c)** 80 dB **(d)** 100,000
11. A certain op-amp has bias currents of 50 μA and 49.3 μA. The input offset current is
    (a) 700 nA **(b)** 99.3 μA **(c)** 49.7 μA **(d)** none of these
12. The output of a particular op-amp increases 8 V in 12 μs. The slew rate is
    (a) 96 V/μs **(b)** 0.67 V/μs **(c)** 1.5 V/μs **(d)** none of these

Section 12–3

13. The purpose of offset nulling is to
    (a) reduce the gain **(b)** equalize the input signals
    (c) zero the output error voltage **(d)** answers (b) and (c)
14. The use of negative feedback
    (a) reduces the voltage gain of an op-amp **(b)** makes the op-amp oscillate
    (c) makes linear operation possible **(d)** answers (a) and (c)

Section 12–4

15. For an op-amp with negative feedback, the output is
    (a) equal to the input
    (b) increased
    (c) fed back to the inverting input
    (d) fed back to the noninverting input
16. A certain noninverting amplifier has an R_i of 1.0 kΩ and an R_f of 100 kΩ. The closed-loop gain is
    (a) 100,000 **(b)** 1000 **(c)** 101 **(d)** 100
17. If the feedback resistor in Question 16 is open, the voltage gain
    (a) increases **(b)** decreases **(c)** is not affected **(d)** depends on R_i
18. A certain inverting amplifier has a closed-loop gain of 25. The op-amp has an open-loop gain of 100,000. If another op-amp with an open-loop gain of 200,000 is substituted in the configuration, the closed-loop gain
    (a) doubles **(b)** drops to 12.5 **(c)** remains at 25 **(d)** increases slightly
19. A voltage-follower
    (a) has a gain of 1 **(b)** is noninverting
    (c) has no feedback resistor **(d)** has all of these

Section 12–5

20. Negative feedback
    (a) increases the input and output impedances
    (b) increases the input impedance and the bandwidth
    (c) decreases the output impedance and the bandwidth
    (d) does not affect impedances or bandwidth

Section 12–6

21. Bias current compensation
    (a) reduces gain **(b)** reduces output error voltage
    (c) increases bandwidth **(d)** has no effect

Section 12–7

22. The midrange open-loop gain of an op-amp
    (a) extends from the lower critical frequency to the upper critical frequency
    (b) extends from 0 Hz to the upper critical frequency
    (c) rolls off at 20 dB/decade beginning at 0 Hz
    (d) answers (b) and (c)
23. The frequency at which the open-loop gain is equal to 1 is called
    (a) the upper critical frequency **(b)** the cutoff frequency
    (c) the notch frequency **(d)** the unity-gain frequency
24. Phase shift through an op-amp is caused by
    (a) the internal *RC* circuits **(b)** the external *RC* circuits
    (c) the gain roll-off **(d)** negative feedback

25. Each RC circuit in an op-amp
(a) causes the gain to roll off at −6 dB/octave
(b) causes the gain to roll off at −20 dB/decade
(c) reduces the midrange gain by 3 dB
(d) answers (a) and (b)

26. If a certain op-amp has a midrange open-loop gain of 200,000 and a unity-gain frequency of 5 MHz, the gain-bandwidth product is
(a) 200,000 Hz **(b)** 5,000,000 Hz
(c) 1×10^{12} Hz **(d)** not determinable from the information

Section 12–8

27. The bandwidth of an ac amplifier having a lower critical frequency of 1 kHz and an upper critical frequency of 10 kHz is
(a) 1 kHz **(b)** 9 kHz **(c)** 10 kHz **(d)** 11 kHz

28. The bandwidth of a dc amplifier having an upper critical frequency of 100 kHz is
(a) 100 kHz **(b)** unknown **(c)** infinity **(d)** 0 kHz

29. When negative feedback is used, the gain-bandwidth product of an op-amp
(a) increases **(b)** decreases **(c)** stays the same **(d)** fluctuates

30. If a certain op-amp has a closed-loop gain of 20 and an upper critical frequency of 10 MHz, the gain-bandwidth product is
(a) 200 MHz **(b)** 10 MHz **(c)** the unity-gain frequency **(d)** answers (a) and (c)

PROBLEMS

Answers to all odd-numbered problems are at the end of the book.

BASIC PROBLEMS

Section 12–1

Introduction to Operational Amplifiers

1. Compare a practical op-amp to an ideal op-amp.

2. Two IC op-amps are available to you. Their characteristics are listed below. Choose the one you think is more desirable.
Op-amp 1: $Z_{in} = 5\ \text{M}\Omega$, $Z_{out} = 100\ \Omega$, $A_{ol} = 50{,}000$
Op-amp 2: $Z_{in} = 10\ \text{M}\Omega$, $Z_{out} = 75\ \Omega$, $A_{ol} = 150{,}000$

Section 12–2

Op-Amp Input Modes and Parameters

3. Identify the type of input mode for each op-amp in Figure 12–60.

4. A certain op-amp has a CMRR of 250,000. Convert this to decibels.

5. The open-loop gain of a certain op-amp is 175,000. Its common-mode gain is 0.18. Determine the CMRR in decibels.

6. An op-amp datasheet specifies a CMRR of 300,000 and an A_{ol} of 90,000. What is the common-mode gain?

7. Determine the bias current, I_{BIAS}, given that the input currents to an op-amp are 8.3 μA and 7.9 μA.

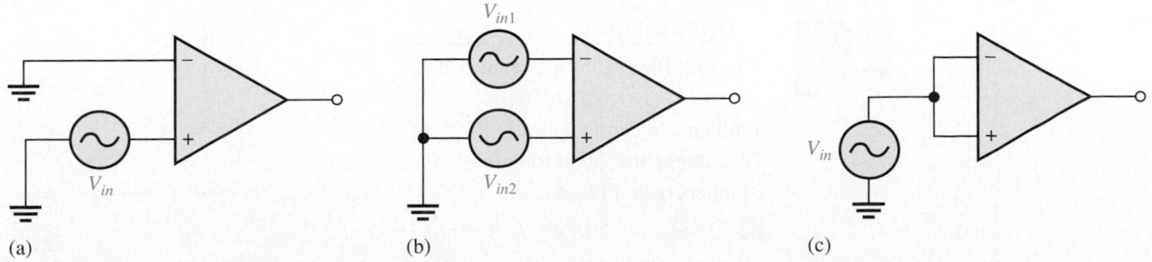

▲ FIGURE 12–60

8. Distinguish between input bias current and input offset current, and then calculate the input offset current in Problem 7.
9. Figure 12–61 shows the output voltage of an op-amp in response to a step input. What is the slew rate?
10. How long does it take the output voltage of an op-amp to go from −10 V to +10 V if the slew rate is 0.5 V/μS?

▶ FIGURE 12–61

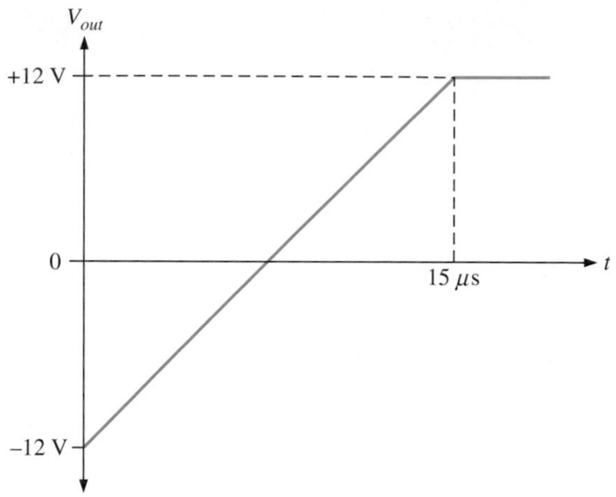

Section 12–4 Op-Amps with Negative Feedback

11. Identify each of the op-amp configurations in Figure 12–62.

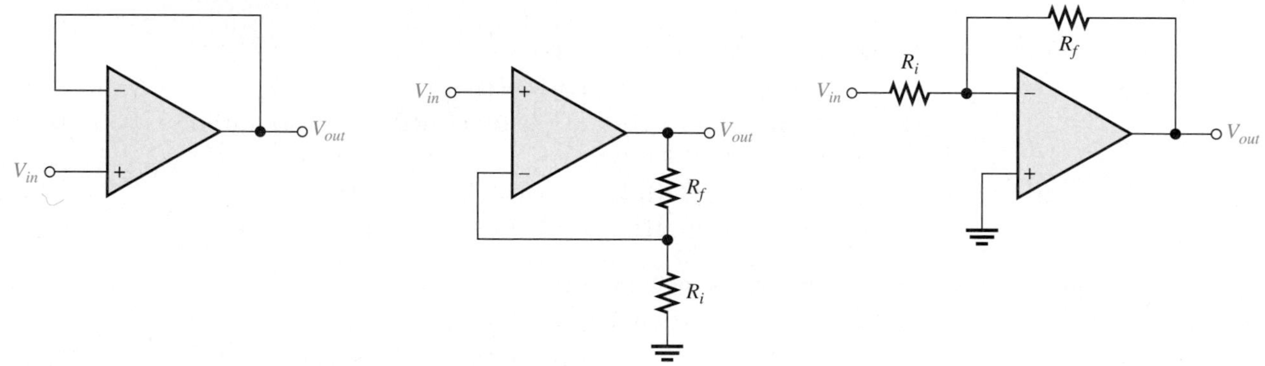

(a) (b) (c)

▲ FIGURE 12–62

12. A noninverting amplifier has an R_i of 1.0 kΩ and an R_f of 100 kΩ. Determine V_f and B if $V_{out} = 5$ V.
13. For the amplifier in Figure 12–63, determine the following:
    (a) $A_{cl(NI)}$ **(b)** V_{out} **(c)** V_f

▶ FIGURE 12–63

Multisim file circuits are identified with a CD logo and are in the Problems folder on your CD-ROM. Filenames correspond to figure numbers (e.g., F12-63).

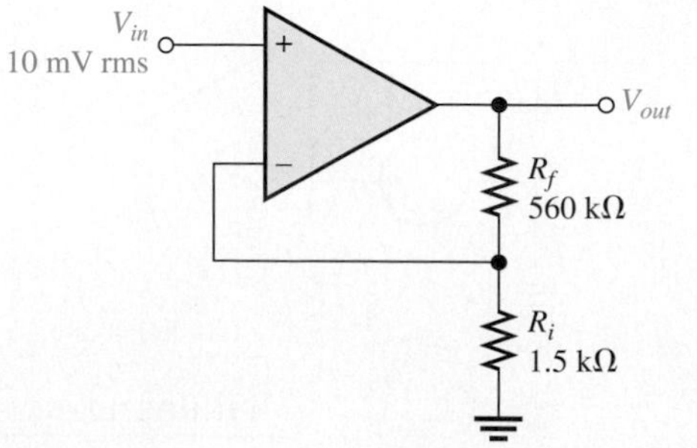

14. Determine the closed-loop gain of each amplifier in Figure 12–64.

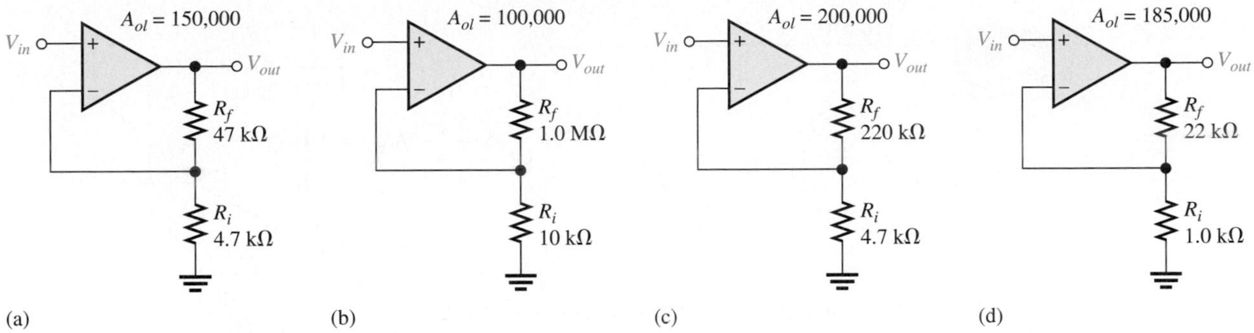

▲ FIGURE 12–64

15. Find the value of R_f that will produce the indicated closed-loop gain in each amplifier in Figure 12–65.

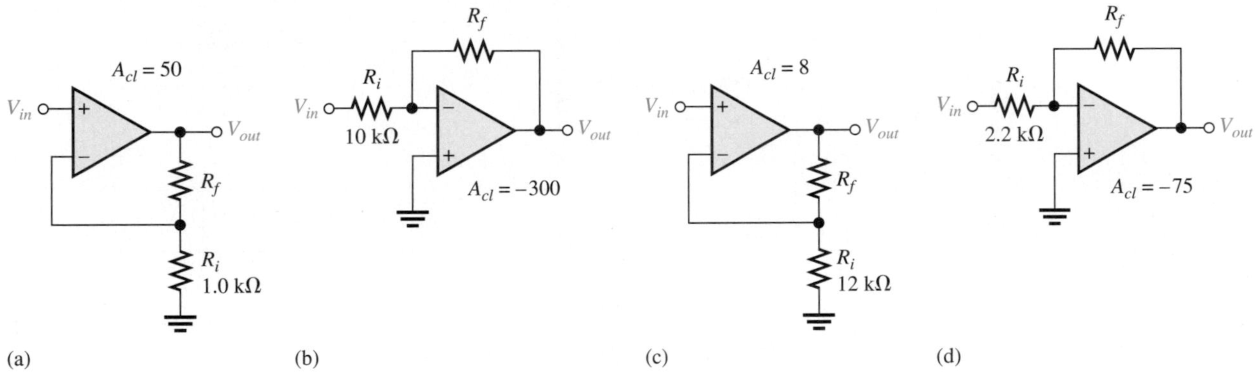

▲ FIGURE 12–65

16. Find the gain of each amplifier in Figure 12–66.

17. If a signal voltage of 10 mV rms is applied to each amplifier in Figure 12–66, what are the output voltages and what is their phase relationship with inputs?

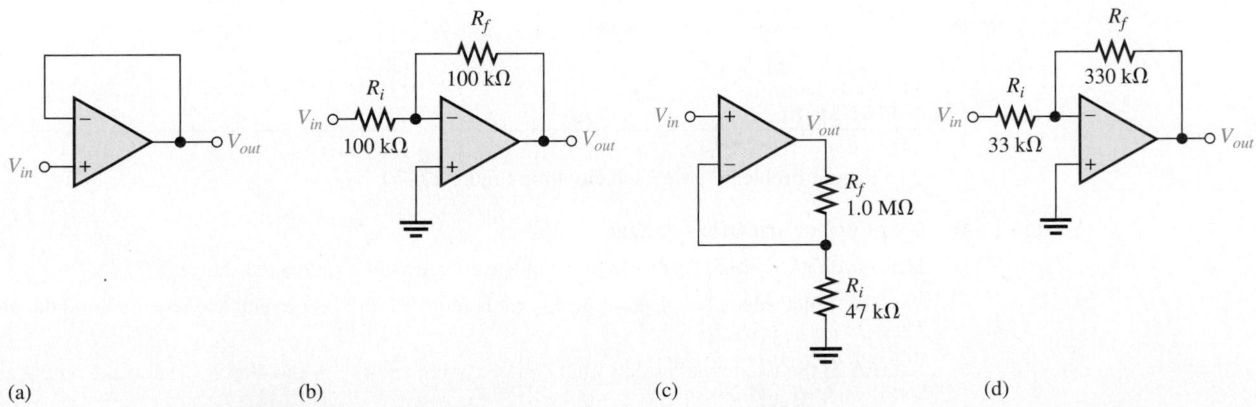

▲ FIGURE 12–66

18. Determine the approximate values for each of the following quantities in Figure 12–67.
(a) I_{in} **(b)** I_f **(c)** V_{out} **(d)** closed-loop gain

▶ FIGURE 12–67

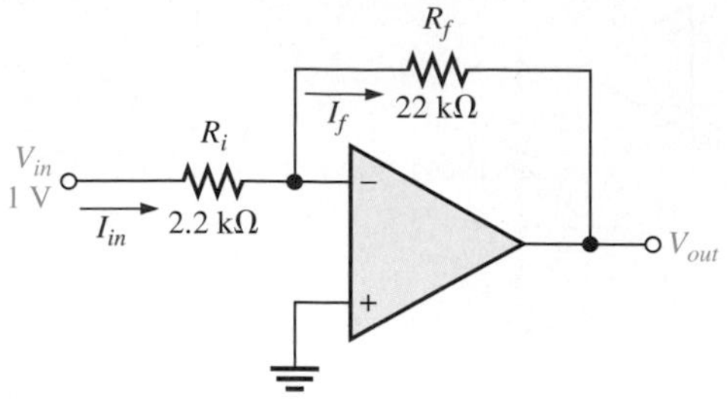

Section 12–5 Effects of Negative Feedback on Op-Amp Impedances

19. Determine the input and output impedances for each amplifier configuration in Figure 12–68.

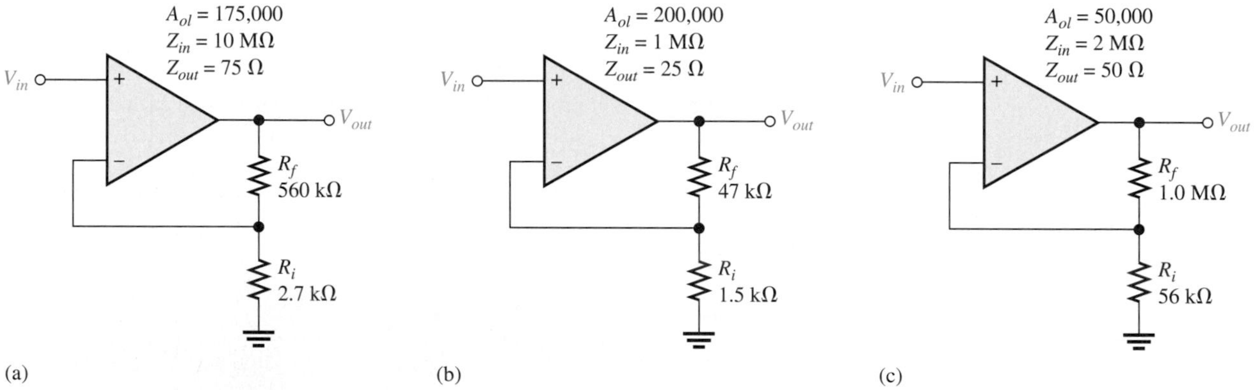

▲ FIGURE 12–68

20. Repeat Problem 19 for each circuit in Figure 12–69.

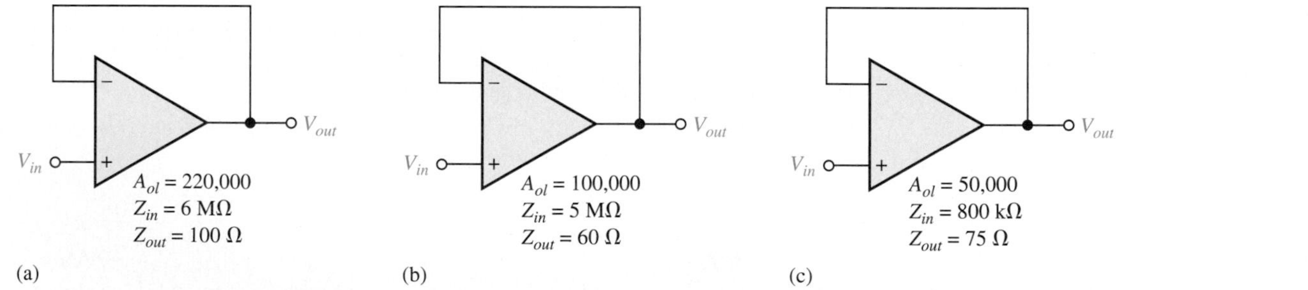

▲ FIGURE 12–69

21. Repeat Problem 19 for each circuit in Figure 12–70.

Section 12–6 Bias Current and Offset Voltage

22. A voltage-follower is driven by a voltage source with a source resistance of 75 Ω.
(a) What value of compensating resistor is required for bias current, and where should the resistor be placed?
(a) If the two input currents after compensation are 42 μA and 40 μA, what is the output error voltage?

23. Determine the compensating resistor value for each amplifier configuration in Figure 12–68, and indicate the placement of the resistor.

24. A particular op-amp voltage-follower has an input offset voltage of 2 nV. What is the output error voltage?

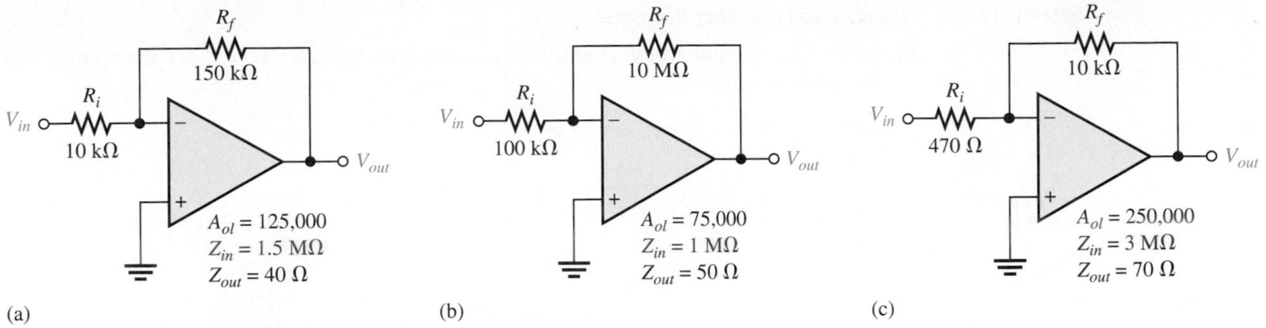

▲ FIGURE 12–70

25. What is the input offset voltage of an op-amp if a dc output voltage of 35 mV is measured when the input voltage is zero? The op-amp's open-loop gain is specified to be 200,000.

Section 12–7 Open-Loop Response

26. The midrange open-loop gain of a certain op-amp is 120 dB. Negative feedback reduces this gain by 50 dB. What is the closed-loop gain?

27. The upper critical frequency of an op-amp's open-loop response is 200 Hz. If the midrange gain is 175,000, what is the ideal gain at 200 Hz? What is the actual gain? What is the op-amp's open-loop bandwidth?

28. An *RC* lag circuit has a critical frequency of 5 kHz. If the resistance value is 1.0 kΩ, what is X_C when $f = 3$ kHz?

29. Determine the attenuation of an *RC* lag circuit with $f_c = 12$ kHz for each of the following frequencies.

(a) 1 kHz **(b)** 5 kHz **(c)** 12 kHz **(d)** 20 kHz **(e)** 100 kHz

30. The midrange open-loop gain of a certain op-amp is 80,000. If the open-loop critical frequency is 1 kHz, what is the open-loop gain at each of the following frequencies?

(a) 100 Hz **(b)** 1 kHz **(c)** 10 kHz **(d)** 1 MHz

31. Determine the phase shift through each circuit in Figure 12–71 at a frequency of 2 kHz.

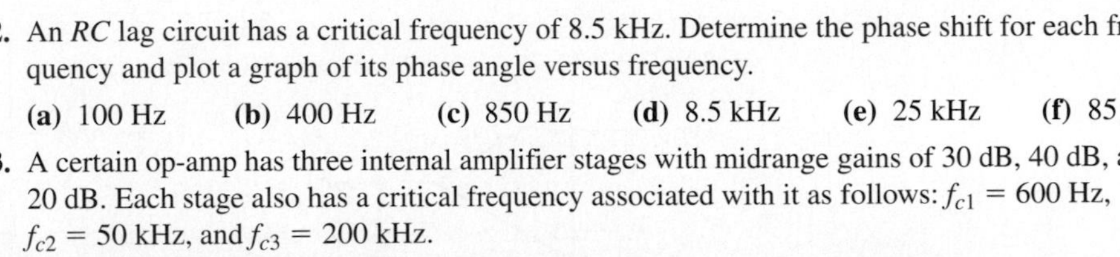

▲ FIGURE 12–71

32. An *RC* lag circuit has a critical frequency of 8.5 kHz. Determine the phase shift for each frequency and plot a graph of its phase angle versus frequency.

(a) 100 Hz **(b)** 400 Hz **(c)** 850 Hz **(d)** 8.5 kHz **(e)** 25 kHz **(f)** 85 kHz

33. A certain op-amp has three internal amplifier stages with midrange gains of 30 dB, 40 dB, and 20 dB. Each stage also has a critical frequency associated with it as follows: $f_{c1} = 600$ Hz, $f_{c2} = 50$ kHz, and $f_{c3} = 200$ kHz.

(a) What is the midrange open-loop gain of the op-amp, expressed in dB?

(b) What is the total phase shift through the amplifier, including inversion, when the signal frequency is 10 kHz?

34. What is the gain roll-off rate in Problem 33 between the following frequencies?

(a) 0 Hz and 600 Hz **(b)** 600 Hz and 50 kHz

(c) 50 kHz and 200 kHz **(d)** 200 kHz and 1 MHz

Section 12–8 Closed-Loop Frequency Response

35. Determine the midrange gain in dB of each amplifier in Figure 12–72. Are these open-loop or closed-loop gains?

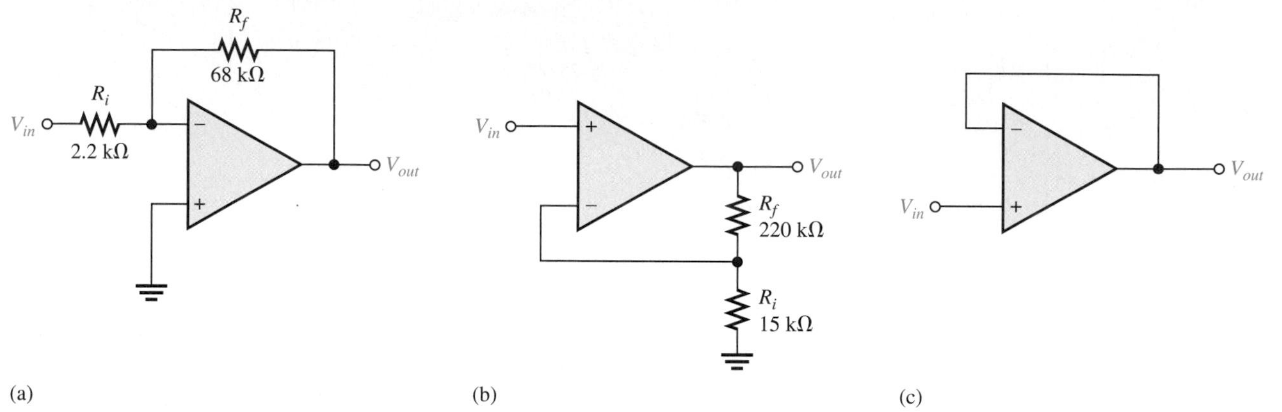

▲ FIGURE 12–72

36. A certain amplifier has an open-loop gain in midrange of 180,000 and an open-loop critical frequency of 1500 Hz. If the attenuation of the feedback path is 0.015, what is the closed-loop bandwidth?

37. Given that $f_{c(ol)} = 750$ Hz, $A_{ol} = 89$ dB, and $f_{c(cl)} = 5.5$ kHz, determine the closed-loop gain in decibels.

38. What is the unity-gain bandwidth in Problem 37?

39. For each amplifier in Figure 12–73, determine the closed-loop gain and bandwidth. The op-amps in each circuit exhibit an open-loop gain of 125 dB and a unity-gain bandwidth of 2.8 MHz.

▶ FIGURE 12–73

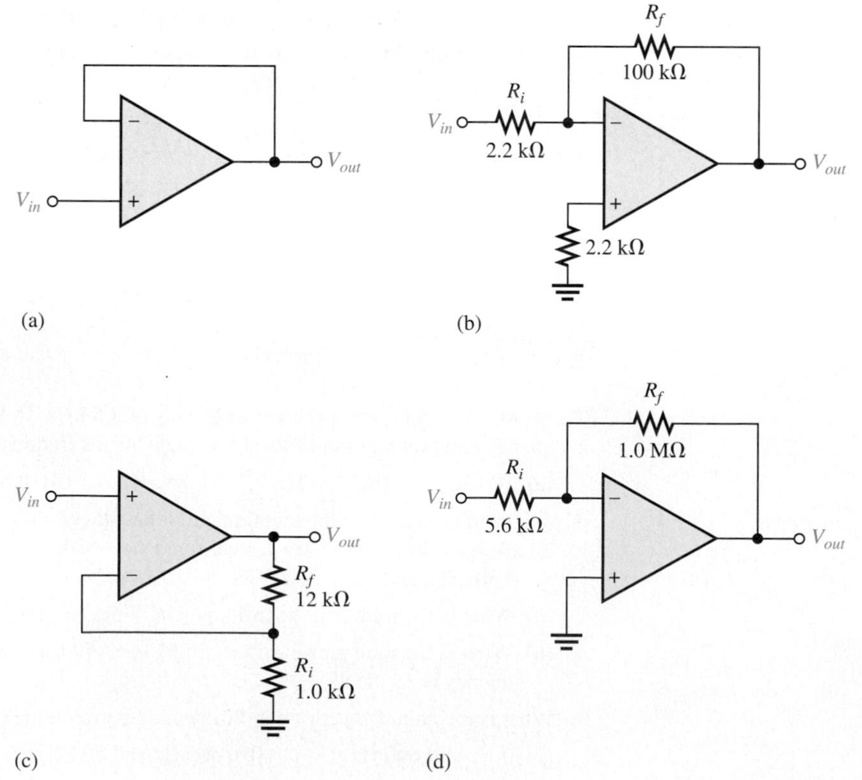

40. Which of the amplifiers in Figure 12–74 has the smaller bandwidth?

FIGURE 12–74

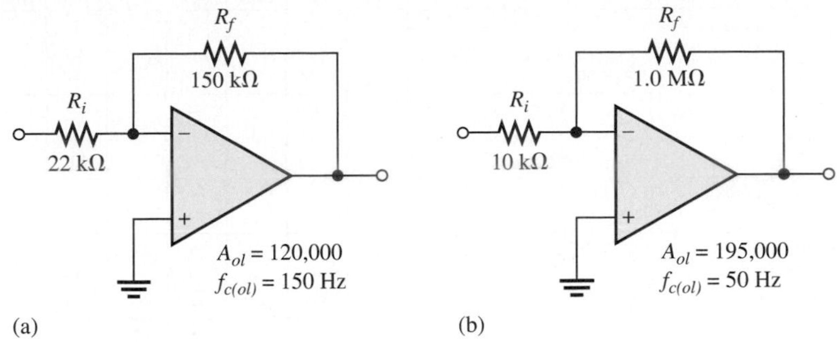

Section 12–9 Troubleshooting

41. Determine the most likely fault(s) for each of the following symptoms in Figure 12–75 with a 100 mV signal applied.

(a) No output signal. **(b)** Output severely clipped on both positive and negative swings.

FIGURE 12–75

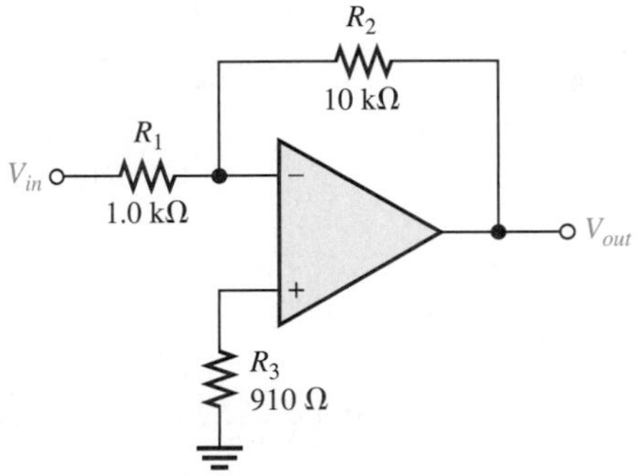

42. Determine the effect on the output if the circuit in Figure 12–75 has the following fault (one fault at a time).

(a) Output pin is shorted to the inverting input. **(b)** R_3 is open.

(c) R_3 is 10 kΩ instead of 910 Ω. **(d)** R_1 and R_2 are swapped.

43. On the circuit board in Figure 12–76, what happens if the middle lead (wiper) of the 100 kΩ potentiometer is broken?

FIGURE 12–76

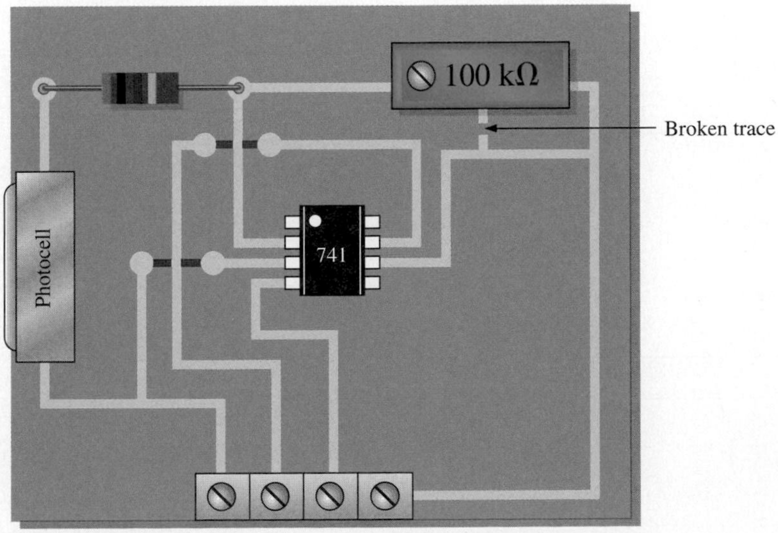

Electrical Characteristics

Parameter	Conditions	LM741A			LM741			LM741C			Units
		Min	Typ	Max	Min	Typ	Max	Min	Typ	Max	
Input offset voltage	$T_A = 25°C$	—	—	—	—	—	—	—	—	—	
	$R_S \le 10\ k\Omega$	—	—	—	—	—	5.0	—	2.0	6.0	mV
	$R_S \le 50\ \Omega$	—	0.8	3.0	—	—	—	—	—	—	mV
	$T_{AMIN} \le T_A \le T_{AMAX}$	—	—	—	—	—	—	—	—	—	
	$R_S \le 50\ \Omega$	—	—	4.0	—	—	—	—	—	—	mV
	$R_S \le 10\ k\Omega$	—	—	—	—	—	6.0	—	—	7.5	mV
Average input offset voltage drift		—	—	15	—	—	—	—	—	—	$\mu V/°C$
		—	—	—	—	—	—	—	—	—	
Input offset voltage adjustment range	$T_A = 25°C, V_S = \pm 20\ V$	±10	—	—	—	±15	—	—	±15	—	mV
		—	—	—	—	—	—	—	—	—	
Input offset current	$T_A = 25°C$	—	3.0	30	—	20	200	—	20	200	nA
	$T_{AMIN} \le T_A \le T_{AMAX}$	—	—	70	—	85	500	—	—	300	nA
Average input offset current drift		—	—	0.5	—	—	—	—	—	—	nA/°C
		—	—	—	—	—	—	—	—	—	
Input bias current	$T_A = 25°C$	—	30	80	—	80	500	—	80	500	nA
	$T_{AMIN} \le T_A \le T_{AMAX}$	—	—	0.210	—	—	1.5	—	—	0.8	μA
Input resistance	$T_A = 25°C, V_S = \pm 20\ V$	1.0	6.0	—	0.3	2.0	—	0.3	2.0	—	MΩ
	$T_{AMIN} \le T_A \le T_{AMAX}$	0.5	—	—	—	—	—	—	—	—	MΩ
	$V_S = \pm 20\ V$	—	—	—	—	—	—	—	—	—	
Input voltage range	$T_A = 25°C$	—	—	—	—	—	—	—	±12	±13	V
	$T_{AMIN} \le T_A \le T_{AMAX}$	—	—	—	±12	±13	—	—	—	—	V
Large-signal voltage gain	$T_A = 25°C, R_L \ge 2\ k\Omega$	—	—	—	—	—	—	—	—	—	
	$V_S = \pm 20\ V, V_O = \pm 15\ V$	50	—	—	—	—	—	—	—	—	V/mV
	$V_S = \pm 15\ V, V_O = \pm 10\ V$	—	—	—	50	200		20	200		V/mV
	$T_{AMIN} \le T_A \le T_{AMAX}$	—	—	—	—	—	—	—	—	—	
	$R_L \ge 2\ k\Omega$	—	—	—	—	—	—	—	—	—	
	$V_S = \pm 20\ V, V_O = \pm 15\ V$	32	—	—	—	—	—	—	—	—	V/mV
	$V_S = \pm 15\ V, V_O = \pm 10\ V$	—	—	—	25	—	—	15	—	—	V/mV
	$V_S = \pm 5\ V, V_O = \pm 2\ V$	10	—	—	—	—	—	—	—	—	V/mV
Output voltage swing	$V_S = \pm 20\ V$	—	—	—	—	—	—	—	—	—	
	$R_L \ge 10\ k\Omega$	±16	—	—	—	—	—	—	—	—	V
	$R_L \ge 2\ k\Omega$	±15	—	—	—	—	—	—	—	—	V
	$V_S = \pm 15\ V$	—	—	—	—	—	—	—	—	—	
	$R_L \ge 10\ k\Omega$	—	—	—	±12	±14	—	±12	±14	—	V
	$R_L \ge 2\ k\Omega$	—	—	—	±10	±13	—	±10	±13	—	V
Output short circuit Current	$T_A = 25°C$	10	25	35	—	25	—	—	25	—	mA
	$T_{AMIN} \le T_A \le T_{AMAX}$	10	—	40	—	—	—	—	—	—	mA
Common-mode rejection ratio	$T_{AMIN} \le T_A \le T_{AMAX}$	—	—	—	—	—	—	—	—	—	
	$R_S \le 10\ k\Omega, V_{CM} = \pm 12\ V$	—	—	—	70	90	—	70	90	—	dB
	$R_S \le 50\ \Omega, V_{CM} = \pm 12\ V$	80	95	—	—	—	—	—	—	—	dB
Supply voltage rejection ratio	$T_{AMIN} \le T_A \le T_{AMAX}$,	—	—	—	—	—	—	—	—	—	
	$V_S = \pm 20\ V$ to $V_S = \pm 5\ V$	—	—	—	—	—	—	—	—	—	
	$R_S \le 50\ \Omega$	86	96	—	—	—	—	—	—	—	dB
	$R_S \le 10\ k\Omega$	—	—	—	77	96	—	77	96	—	dB
Transient response	$T_A = 25°C$, Unity gain	—	—	—	—	—	—	—	—	—	
Rise time		—	0.25	0.8	—	0.3	—	—	0.3	—	μs
Overshoot		—	6.0	20	—	5	—	—	5	—	%
Bandwidth	$T_A = 25°C$	0.437	1.5	—	—	—	—	—	—	—	MHz
Slew rate	$T_A = 25°C$, Unity gain	0.3	0.7	—	—	0.5	—	—	0.5	—	$V/\mu s$
Supply current	$T_A = 25°C$			—	—	1.7	2.8	—	1.7	2.8	mA
Power consumption	$T_A = 25°C$	—	—	—	—	—	—	—	—	—	
	$V_S = \pm 20\ V$	—	80	150	—	—	—	—	—	—	mW
	$V_S = \pm 15\ V$	—	—	—	—	50	85	—	50	85	mW
LM741A	$V_S = \pm 20\ V$	—	—	—	—	—	—	—	—	—	
	$T_A = T_{AMIN}$	—	—	165	—	—	—	—	—	—	mW
	$T_A = T_{AMAX}$	—	—	135	—	—	—	—	—	—	mW
LM741	$V_S = \pm 15\ V$	—	—	—	—	—	—	—	—	—	
	$T_A = T_{AMIN}$	—	—	—	—	60	100	—	—	—	mW
	$T_A = T_{AMAX}$	—	—	—	—	45	75	—	—	—	mW

▲ **FIGURE 12–77**

APPLICATION ACTIVITY PROBLEMS

44. In the amplifier circuit of Figure 12–47, list the possible faults that will cause the push-pull stage to operate nonlinearly.

45. What indication would you observe if a 100 kΩ resistor is incorrectly installed for R_2 in Figure 12–47?

46. What voltage will you measure on the output of the amplifier in Figure 12–47 if diode D_1 opens?

DATASHEET PROBLEMS

47. Refer to the partial 741 datasheet (LM741) in Figure 12–77. Determine the input resistance (impedance) of a noninverting amplifier which uses a 741 op-amp with $R_f = 47\ \text{k}\Omega$ and $R_i = 470\ \Omega$. Use typical values.

48. Refer to the partial datasheet in Figure 12–77. Determine the input impedances of a LM741 op-amp connected as an inverting amplifier with a closed-loop voltage gain of 100 and $R_f = 100\ \text{k}\Omega$.

49. Refer to Figure 12–77 and determine the minimum open-loop voltage gain for an LM741 expressed as a ratio of output volts to input volts.

50. Refer to Figure 12–77. How long does it typically take the output voltage of an LM741 to make a transition from −8 V to +8 V in response to a step input?

ADVANCED PROBLEMS

51. Design a noninverting amplifier with an appropriate closed-loop voltage gain of 150 and a minimum input impedance of 100 MΩ using a 741 op-amp. Include bias current compensation.

52. Design an inverting amplifier using a 741 op-amp. The voltage gain must be 68 ± 5% and the input impedance must be approximately 10 kΩ. Include bias current compensation.

53. Design a noninverting amplifier with an upper critical frequency, f_{cu}, of 10 kHz using a 741 op-amp. The dc supply voltages are ±15 V. Refer to Figure 12–78. Include bias current compensation.

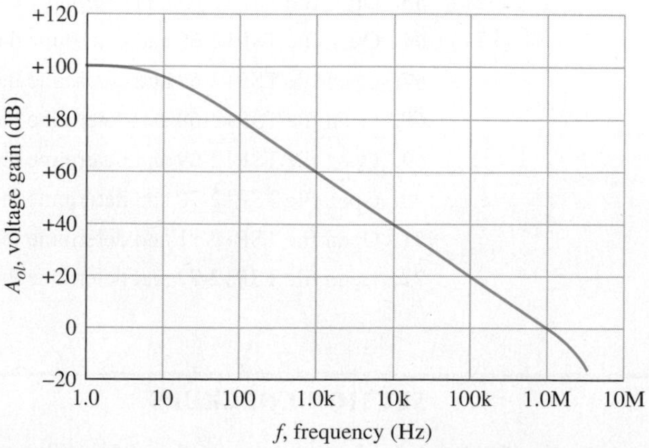

▲ FIGURE 12–78

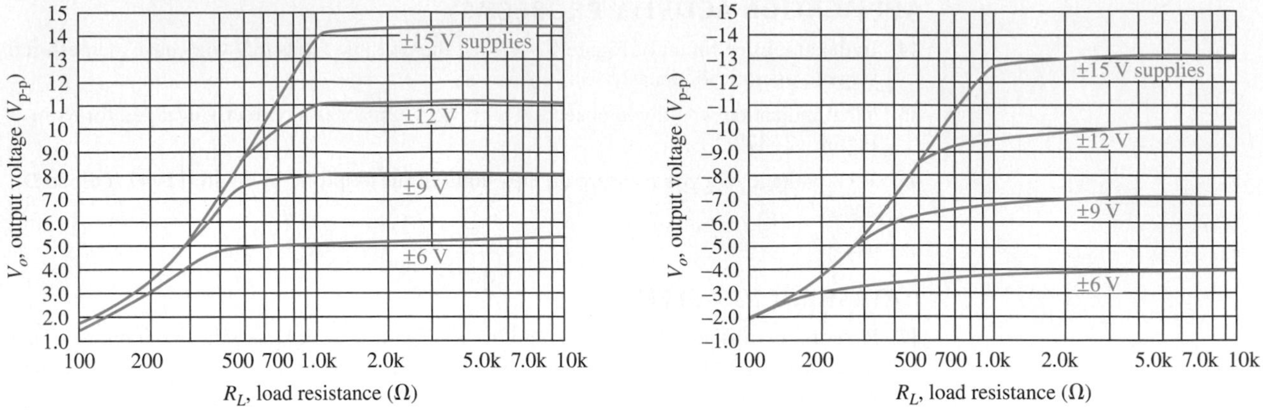

(a) Positive output voltage swing versus load resistance

(b) Negative output voltage swing versus load resistance

▲ **FIGURE 12–79**

54. For the circuit you designed in Problem 53, determine the minimum load resistance if the minimum output voltage swing is to be ± 10 V. Refer to the datasheet graphs in Figure 12–79.

55. Design an inverting amplifier using a 741 op-amp if a midrange voltage gain of 50 and a bandwidth of 20 kHz is required. Include bias current compensation.

56. What is the maximum closed-loop voltage gain that can be achieved with a 741 op-amp if the bandwidth must be no less than 5 kHz?

MULTISIM TROUBLESHOOTING PROBLEMS

These file circuits are in the Troubleshooting Problems folder on your CD-ROM.

57. Open file TSP12-57 and determine the fault.

58. Open file TSP12-58 and determine the fault.

59. Open file TSP12-59 and determine the fault.

60. Open file TSP12-60 and determine the fault.

61. Open file TSP12-61 and determine the fault.

62. Open file TSP12-62 and determine the fault.

63. Open file TSP12-63 and determine the fault.

64. Open file TSP12-64 and determine the fault.

65. Open file TSP12-65 and determine the fault.

66. Open file TSP12-66 and determine the fault.

67. Open file TSP12-67 and determine the fault.

68. Open file TSP12-68 and determine the fault.

69. Open file TSP12-69 and determine the fault.

70. Open file TSP12-70 and determine the fault.

71. Open file TSP12-71 and determine the fault.

72. Open file TSP12-72 and determine the fault.

ANSWERS

SECTION CHECKUPS

Section 12–1 Introduction to Operational Amplifiers

1. Inverting input, noninverting input, output, positive and negative supply voltages

2. A practical op-amp has very high input impedance, very low output impedance, and very high voltage gain.

3. Differential amplifier, voltage amplifier, and push-pull amplifier

4. The difference between its two input voltages

Section 12–2 Op-Amp Input Modes and Parameters

1. Double-ended input is between two input terminals. Single-ended input is from one input terminal to ground (with other input grounded).
2. Common-mode rejection is the ability of an op-amp to produce very little output when the same signal is applied to both inputs.
3. A higher common-mode gain results in a lower CMRR.
4. Input bias current, input offset voltage, drift, input offset current, input impedance, output impedance, maximum output voltage swing, CMRR, open-loop voltage gain, slew rate, frequency response.
5. A fast pulse is used as the input and the rate of change of the output is measured.

Section 12–3 Negative Feedback

1. Negative feedback provides a stable controlled voltage gain, control of impedances, and wider bandwidth.
2. The open-loop gain is so high that a very small signal on the input will drive the op-amp into saturation.

Section 12–4 Op-Amps with Negative Feedback

1. The main purpose of negative feedback is to stabilize the gain.
2. False
3. $A_{cl} = 1/0.02 = 50$

Section 12–5 Effects of Negative Feedback on Op-Amp Impedances

1. The noninverting configuration has a higher Z_{in} than the op-amp alone.
2. Z_{in} increases in a voltage-follower.
3. $Z_{in(I)} \cong R_i = 2\text{ k}\Omega, Z_{out(I)} = Z_{out}/(1 + A_{ol}B) = 25\text{ m}\Omega$

Section 12–6 Bias Current and Offset Voltage

1. Input bias current and input offset voltage are sources of output error.
2. Add a resistor in the feedback path equal to the input source resistance.

Section 12–7 Open-Loop Response

1. Open-loop voltage gain is without feedback, and closed-loop voltage gain is with negative feedback. Open-loop voltage gain is larger.
2. $BW = 100$ Hz
3. A_{ol} decreases.
4. $A_{v(tot)} = 20\text{ dB} + 30\text{ dB} = 50\text{ dB}$
5. $\theta_{tot} = -49° + (-5.2°) = -54.2°$

Section 12–8 Closed-Loop Frequency Response

1. Yes, A_{cl} is always less than A_{ol}.
2. $BW = 3{,}000\text{ kHz}/60 = 50\text{ kHz}$
3. $f_T = 3{,}000\text{ kHz}/1 = 3\text{ MHz}$

Section 12–9 Troubleshooting

1. Check the output null adjustment.
2. After a verification that there is power supply voltage to the op-amp, then the absence of an output signal probably indicates a bad op-amp.

RELATED PROBLEMS FOR EXAMPLES

12–1 340,000; 111 dB

12–2 20 V/μs

12–3 32.9

12–4 67.5 kΩ

12–5 **(a)** 20.6 GΩ, 14 mΩ **(b)** 23

12–6 Z_{in} increases, Z_{out} decreases.

12–7 $Z_{in(I)} = 560\ \Omega$; $Z_{out(I)} = 110\ \text{m}\Omega$; $A_{cl} = -146$

12–8 **(a)** 79,996 **(b)** 79,900 **(c)** 6380

12–9 173 Hz

12–10 75 dB; −71.6°

12–11 2 MHz

12–12 **(a)** 29.6 kHz **(b)** 42.6 kHz

TRUE/FALSE QUIZ

1. T	**2.** F	**3.** T	**4.** T	**5.** F	**6.** T	**7.** T
8. F	**9.** T	**10.** F	**11.** T	**12.** T	**13.** T	**14.** F

CIRCUIT-ACTION QUIZ

1. (b)	**2.** (a)	**3.** (b)	**4.** (a)
5. (a)	**6.** (b)	**7.** (c)	**8.** (b)

SELF-TEST

1. (c)	**2.** (b)	**3.** (d)	**4.** (b)	**5.** (a)	**6.** (c)
7. (b)	**8.** (d)	**9.** (c)	**10.** (d)	**11.** (a)	**12.** (b)
13. (c)	**14.** (d)	**15.** (c)	**16.** (c)	**17.** (a)	**18.** (c)
19. (d)	**20.** (b)	**21.** (b)	**22.** (b)	**23.** (d)	**24.** (a)
25. (d)	**26.** (b)	**27.** (b)	**28.** (a)	**29.** (c)	**30.** (d)

BASIC OP-AMP CIRCUITS

13

CHAPTER OUTLINE

CHAPTER OBJECTIVES

- Analyze the operation of several basic comparator circuits
- Analyze the operation of several types of summing amplifiers
- Analyze the operation of integrators and differentiators
- Troubleshoot basic op-amp circuits

KEY TERMS

- Comparator
- Hysteresis
- Schmitt trigger
- Bounding
- Summing amplifier
- Integrator
- Differentiator

APPLICATION ACTIVITY PREVIEW

In the Application Activity in this chapter, an audio signal generator manufactured by a certain company is modified to include a pulse generator for the purpose of providing a signal source for digital circuits. A voltage comparator is used to generate the pulse waveform from the sine wave output of the audio generator. The duty cycle of the pulse waveform can be varied and is compatible with +5 V logic circuits.

VISIT THE COMPANION WEBSITE

Study aids for this chapter are available at http://www.prenhall.com/floyd

INTRODUCTION

In the last chapter, you learned about the principles, operation, and characteristics of the operational amplifier. Op-amps are used in such a wide variety of circuits and applications that it is impossible to cover all of them in one chapter, or even in one book. Therefore, in this chapter, four fundamentally important circuits are covered to give you a foundation in op-amp circuits.

13–1 COMPARATORS

Operational amplifiers are often used as comparators to compare the amplitude of one voltage with another. In this application, the op-amp is used in the open-loop configuration, with the input voltage on one input and a reference voltage on the other.

After completing this section, you should be able to

- **Analyze the operation of several basic comparator circuits**
  - Describe the operation of a zero-level detector
  - Describe the operation of a nonzero-level detector
  - Discuss how input noise affects comparator operation
  - Define *hysteresis*
  - Explain how hysteresis reduces noise effects
  - Describe a Schmitt trigger circuit
  - Describe the operation of bounded comparators
  - Discuss two comparator applications including analog-to-digital conversion

A **comparator** is a specialized op-amp circuit that compares two input voltages and produces an output that is always at either one of two states, indicating the greater or less than relationship between the inputs. Comparators provide very fast switching times, and many have additional capabilities (such as fast propagation delay or internal reference voltages) to optimize the comparison function. For example, some ultra-high-speed comparators can have propagation delays of as little as 500 ps. Because the output is always in one of two states, comparators are often used to interface between an analog and digital circuit.

For less critical applications, an op-amp running without negative feedback (open-loop) is often used as a comparator. Although op-amps are much slower and lack other special features, they have very high open-loop gain, which enables them to detect very tiny differences in the inputs. In general, comparators cannot be used as op-amps, but op-amps can be used as comparators in noncritical applications. Because an op-amp without negative feedback is essentially a comparator, we will look at the comparison function using a typical op-amp.

Zero-Level Detection

One application of an op-amp used as a comparator is to determine when an input voltage exceeds a certain level. Figure 13–1(a) shows a zero-level detector. Notice that the inverting (−) input is grounded to produce a zero level and that the input signal voltage is applied to the noninverting (+) input. Because of the high open-loop voltage gain, a very small difference voltage between the two inputs drives the amplifier into saturation, causing the output voltage to go to its limit. For example, consider an op-amp having $A_{ol} = 100{,}000$. A voltage difference of only 0.25 mV between the inputs could produce an output voltage of $(0.25\text{ mV})(100{,}000) = 25\text{ V}$ *if* the op-amp were capable. However, since most op-amps have maximum output voltage limitations near the value of their dc supply voltages, the device would be driven into saturation.

Figure 13–1(b) shows the result of a sinusoidal input voltage applied to the noninverting (+) input of the zero-level detector. When the sine wave is positive, the output is at its maximum positive level. When the sine wave crosses 0, the amplifier is driven to its opposite state and the output goes to its maximum negative level, as shown. As you can see, the zero-level detector can be used as a squaring circuit to produce a square wave from a sine wave.

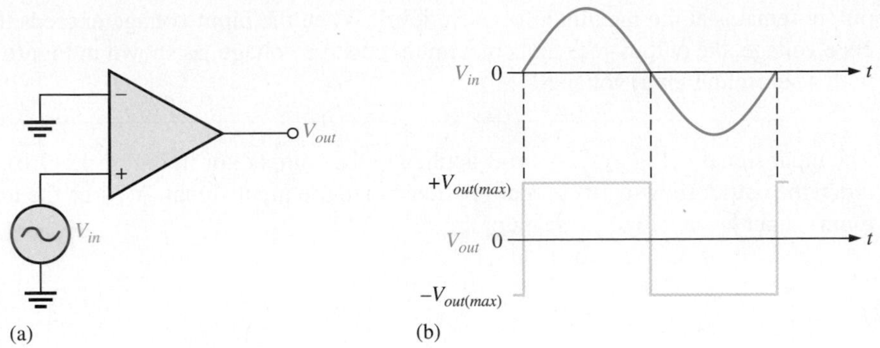

◀ FIGURE 13–1

The op-amp as a zero-level detector.

Nonzero-Level Detection

The zero-level detector in Figure 13–1 can be modified to detect positive and negative voltages by connecting a fixed reference voltage source to the inverting (−) input, as shown in Figure 13–2(a). A more practical arrangement is shown in Figure 13–2(b) using a voltage divider to set the reference voltage, V_{REF}, as follows:

$$V_{\text{REF}} = \frac{R_2}{R_1 + R_2}(+V)$$

where $+V$ is the positive op-amp dc supply voltage. The circuit in Figure 13–2(c) uses a zener diode to set the reference voltage ($V_{\text{REF}} = V_Z$). As long as V_{in} is less than V_{REF}, the

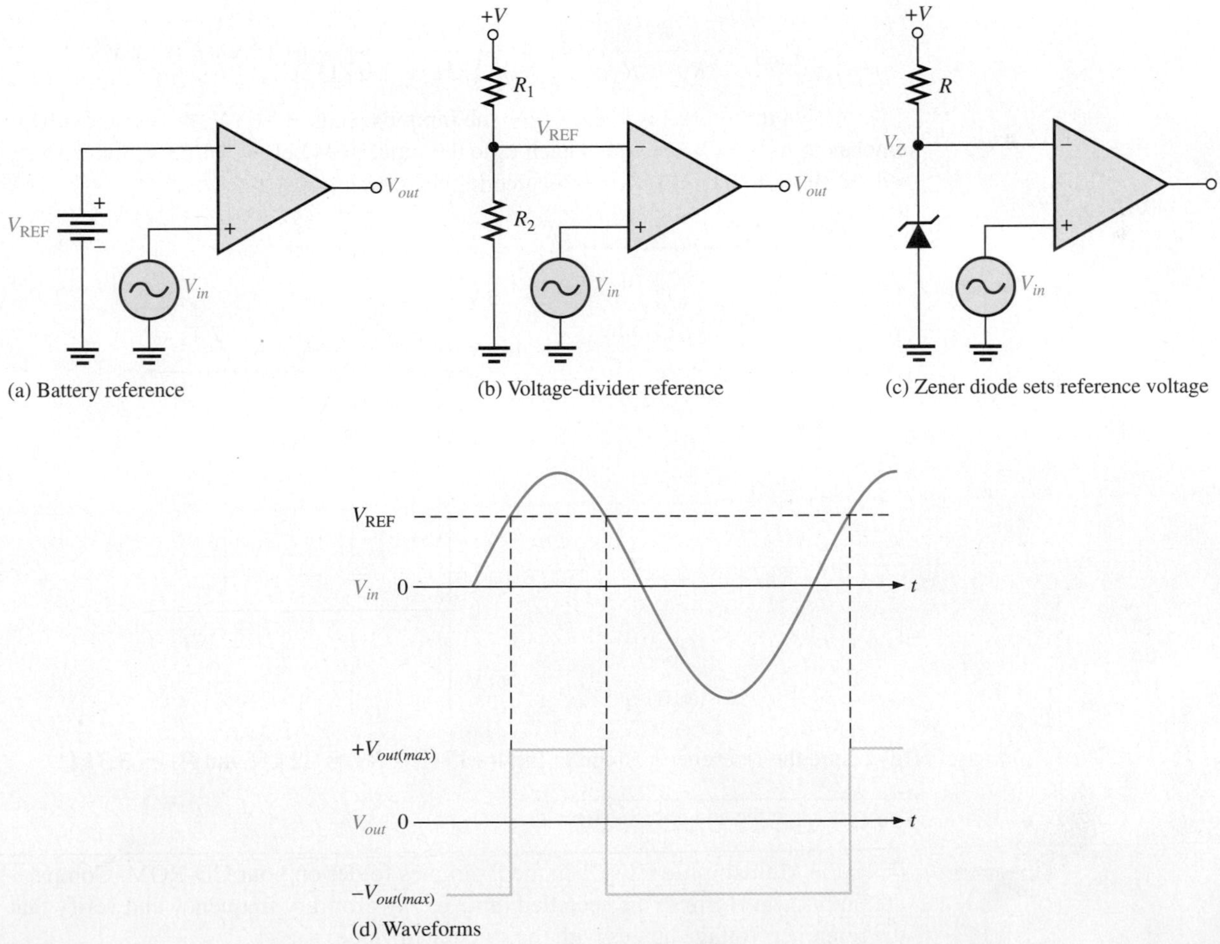

▲ FIGURE 13–2

Nonzero-level detectors.

output remains at the maximum negative level. When the input voltage exceeds the reference voltage, the output goes to its maximum positive voltage, as shown in Figure 13–2(d) with a sinusoidal input voltage.

EXAMPLE 13–1

The input signal in Figure 13–3(a) is applied to the comparator in Figure 13–3(b). Draw the output showing its proper relationship to the input signal. Assume the maximum output levels of the comparator are ± 14 V.

FIGURE 13–3

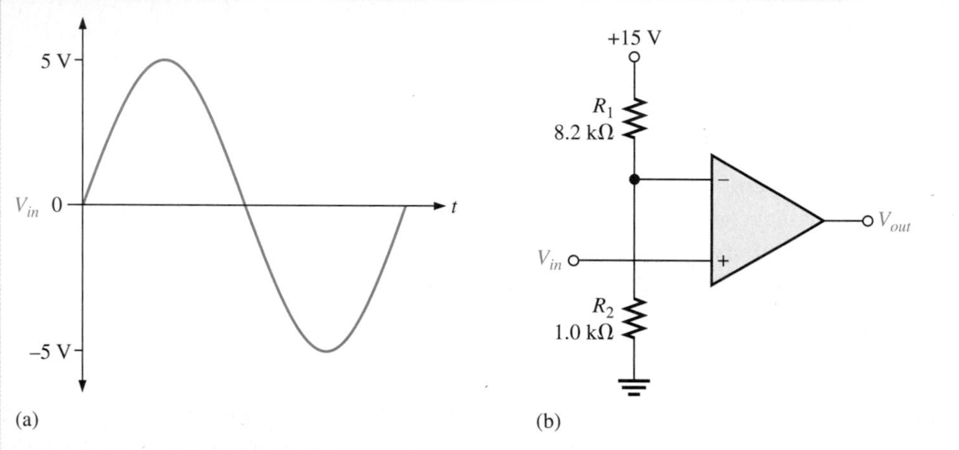

Solution The reference voltage is set by R_1 and R_2 as follows:

$$V_{REF} = \frac{R_2}{R_1 + R_2}(+V) = \frac{1.0\,\text{k}\Omega}{8.2\,\text{k}\Omega + 1.0\,\text{k}\Omega}(+15\text{ V}) = 1.63\text{ V}$$

As shown in Figure 13–4, each time the input exceeds $+1.63$ V, the output voltage switches to its $+14$ V level, and each time the input goes below $+1.63$ V, the output switches back to its -14 V level, neglecting hysteresis.

FIGURE 13–4

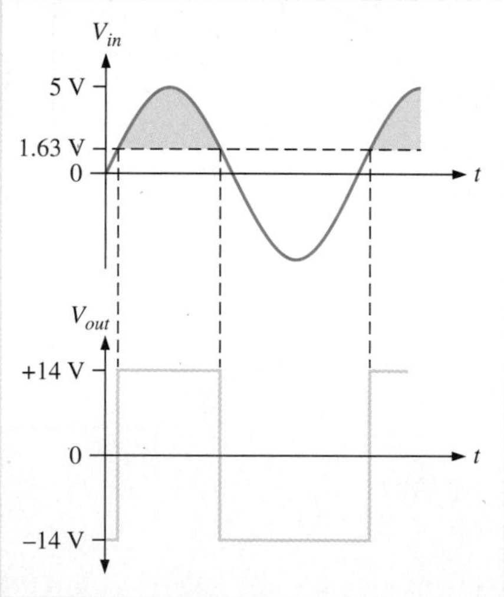

*Related Problem** Determine the reference voltage in Figure 13–3 if $R_1 = 22\,\text{k}\Omega$ and $R_2 = 3.3\,\text{k}\Omega$.

*Answers are at the end of the chapter.

Open the Multisim file E13-01 in the Examples folder on your CD-ROM. Compare the output waveform to the specified input at any arbitrary frequency and verify that the reference voltage agrees with the calculated value.

Effects of Input Noise on Comparator Operation

In many practical situations, **noise** (unwanted voltage fluctuations) appears on the input line. This noise voltage becomes superimposed on the input voltage, as shown in Figure 13–5 for the case of a sine wave, and can cause a comparator to erratically switch output states.

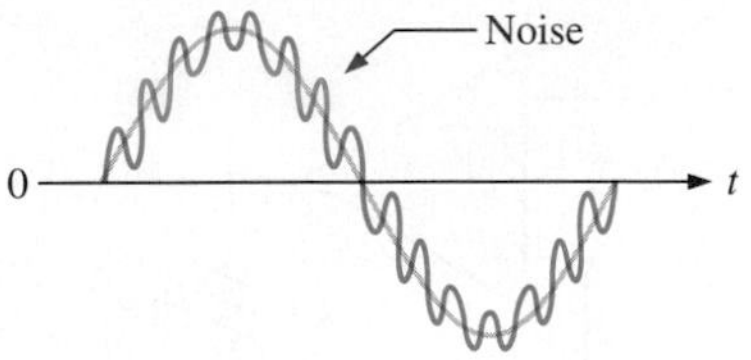

FIGURE 13–5

Sine wave with superimposed noise.

In order to understand the potential effects of noise voltage, consider a low-frequency sinusoidal voltage applied to the noninverting (+) input of an op-amp comparator used as a zero-level detector, as shown in Figure 13–6(a). Part (b) of the figure shows the input sine wave plus noise and the resulting output. When the sine wave approaches 0, the fluctuations due to noise may cause the total input to vary above and below 0 several times, thus producing an erratic output voltage.

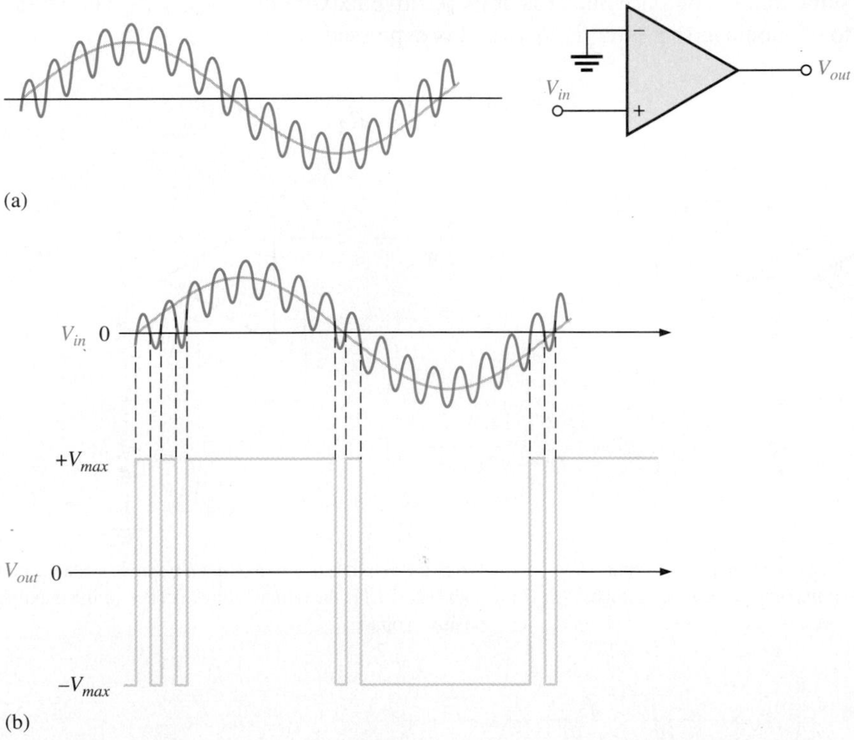

FIGURE 13–6

Effects of noise on comparator circuit.

Reducing Noise Effects with Hysteresis An erratic output voltage caused by noise on the input occurs because the op-amp comparator switches from its negative output state to its positive output state at the same input voltage level that causes it to switch in the opposite direction, from positive to negative. This unstable condition occurs when the input voltage hovers around the reference voltage, and any small noise fluctuations cause the comparator to switch first one way and then the other.

In order to make the comparator less sensitive to noise, a technique incorporating positive feedback, called **hysteresis**, can be used. Basically, hysteresis means that there is a higher reference level when the input voltage goes from a lower to higher value than when it goes from a higher to a lower value. A good example of hysteresis is a common household thermostat that turns the furnace on at one temperature and off at another.

The two reference levels are referred to as the upper trigger point (UTP) and the lower trigger point (LTP). This two-level hysteresis is established with a positive feedback arrangement, as shown in Figure 13–7. Notice that the noninverting (+) input is connected to a resistive voltage divider such that a portion of the output voltage is fed back to the input. The input signal is applied to the inverting (−) input in this case.

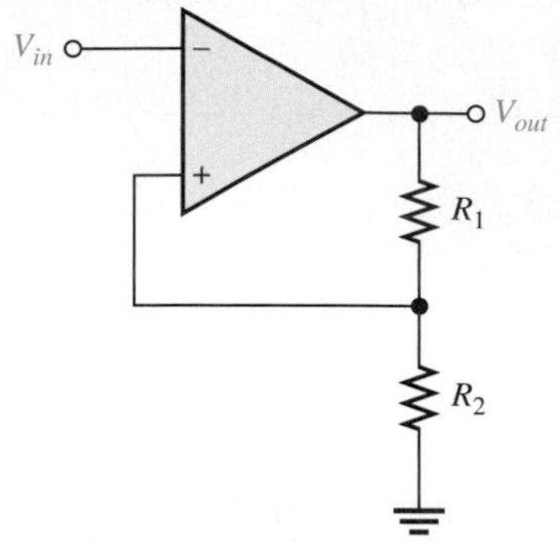

▶ FIGURE 13–7

Comparator with positive feedback for hysteresis.

The basic operation of the comparator with hysteresis is illustrated in Figure 13–8. Assume that the output voltage is at its positive maximum, $+V_{out(max)}$. The voltage fed back to the noninverting input is V_{UTP} and is expressed as

Equation 13–1

$$V_{UTP} = \frac{R_2}{R_1 + R_2}(+V_{out(max)})$$

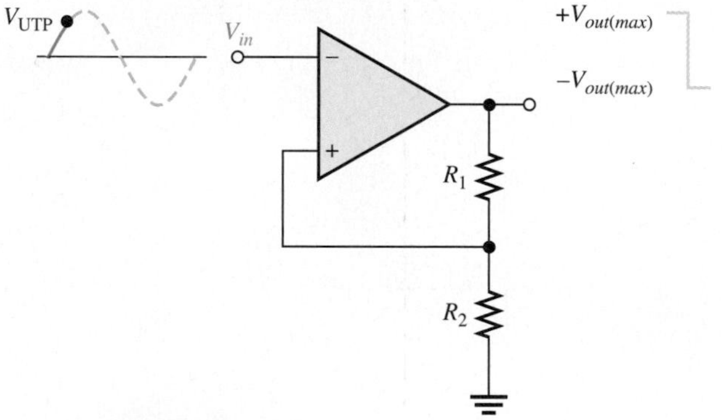

(a) When the output is at the maximum positive voltage and the input exceeds UTP, the output switches to the maximum negative voltage.

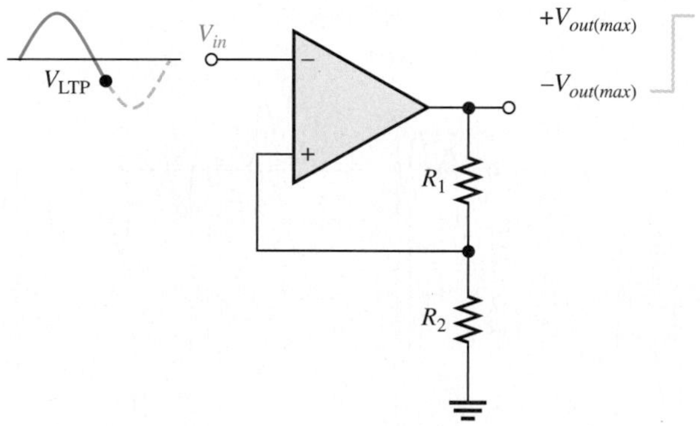

(b) When the output is at the maximum negative voltage and the input goes below LTP, the output switches back to the maximum positive voltage.

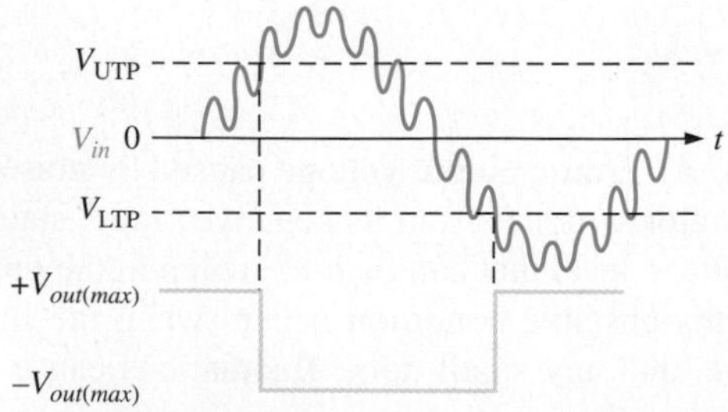

(c) Device triggers only once when UTP or LTP is reached; thus, there is immunity to noise that is riding on the input signal.

▲ FIGURE 13–8

Operation of a comparator with hysteresis.

When V_{in} exceeds V_{UTP}, the output voltage drops to its negative maximum, $-V_{out(max)}$, as shown in part (a). Now the voltage fed back to the noninverting input is V_{LTP} and is expressed as

$$V_{LTP} = \frac{R_2}{R_1 + R_2}(-V_{out(max)})$$ Equation 13–2

The input voltage must now fall below V_{LTP}, as shown in part (b), before the device will switch from the maximum negative voltage back to the maximum positive voltage. This means that a small amount of noise voltage has no effect on the output, as illustrated by Figure 13–8(c).

A comparator with built-in hysteresis is sometimes known as a **Schmitt trigger**. The amount of hysteresis is defined by the difference of the two trigger levels.

$$V_{HYS} = V_{UTP} - V_{LTP}$$ Equation 13–3

EXAMPLE 13–2 Determine the upper and lower trigger points for the comparator circuit in Figure 13–9. Assume that $+V_{out(max)} = +5$ V and $-V_{out(max)} = -5$ V.

FIGURE 13–9

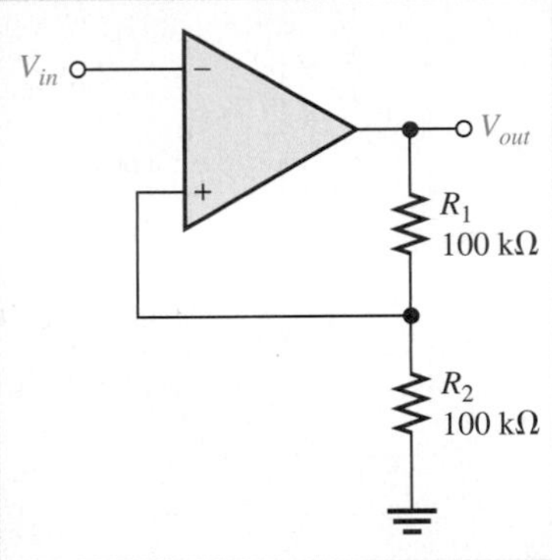

Solution

$$V_{UTP} = \frac{R_2}{R_1 + R_2}(+V_{out(max)}) = 0.5(5\text{ V}) = \mathbf{+2.5\text{ V}}$$

$$V_{LTP} = \frac{R_2}{R_1 + R_2}(-V_{out(max)}) = 0.5(-5\text{ V}) = \mathbf{-2.5\text{ V}}$$

Related Problem Determine the upper and lower trigger points in Figure 13–9 for $R_1 = 68$ kΩ and $R_2 = 82$ kΩ. Also assume the maximum output voltage levels are now ±7 V.

Open the Multisim file E13-02 in the Examples folder on your CD-ROM. Determine the upper and lower trigger points and compare with the calculated values using a 5 V rms, 60 Hz sine wave for the input.

Output Bounding

In some applications, it is necessary to limit the output voltage levels of a comparator to a value less than that provided by the saturated op-amp. A single zener diode can be used, as shown in Figure 13–10, to limit the output voltage to the zener voltage in one direction and to the forward diode drop in the other. This process of limiting the output range is called **bounding**.

▶ FIGURE 13–10

Comparator with output bounding.

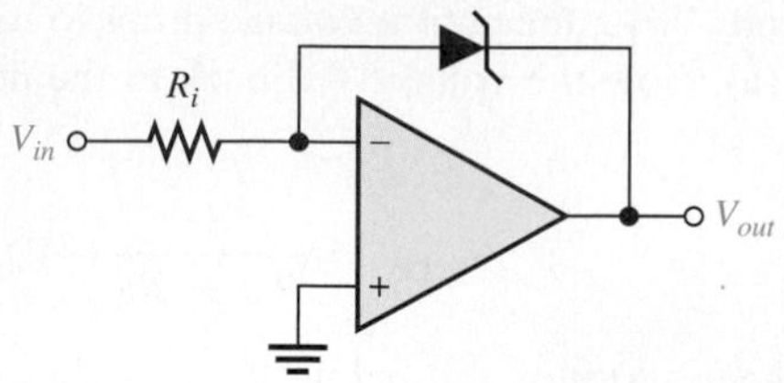

The operation is as follows. Since the anode of the zener is connected to the inverting (−) input, it is at virtual ground (≅ 0 V). Therefore, when the output voltage reaches a positive value equal to the zener voltage, it limits at that value, as illustrated in Figure 13–11(a). When the output switches negative, the zener acts as a regular diode and becomes forward-biased at 0.7 V, limiting the negative output voltage to this value, as shown in part (b). Turning the zener around limits the output voltage in the opposite direction.

▶ FIGURE 13–11

Operation of a bounded comparator.

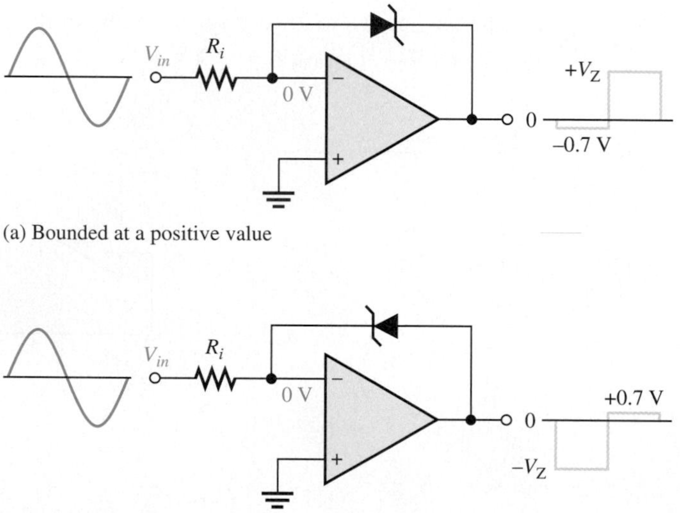

(a) Bounded at a positive value

(b) Bounded at a negative value

Two zener diodes arranged as in Figure 13–12 limit the output voltage to the zener voltage plus the forward voltage drop (0.7 V) of the forward-biased zener, both positively and negatively, as shown.

▶ FIGURE 13–12

Double-bounded comparator.

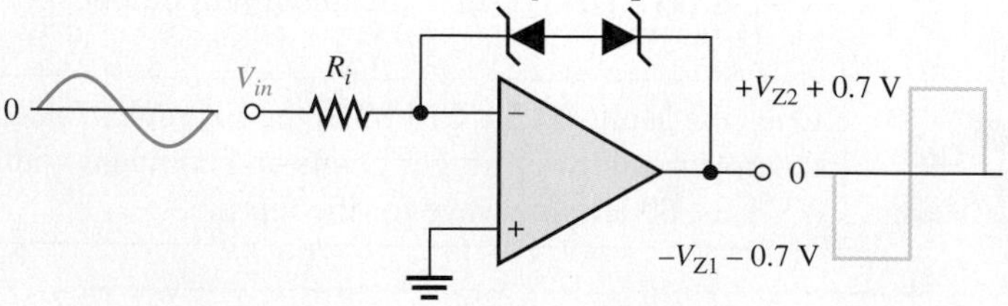

EXAMPLE 13–3

Determine the output voltage waveform for Figure 13–13.

Solution This comparator has both hysteresis and zener bounding. The voltage across D_1 and D_2 in either direction is 4.7 V + 0.7 V = 5.4 V. This is because one zener is always forward-biased with a drop of 0.7 V when the other one is in breakdown.

FIGURE 13–13

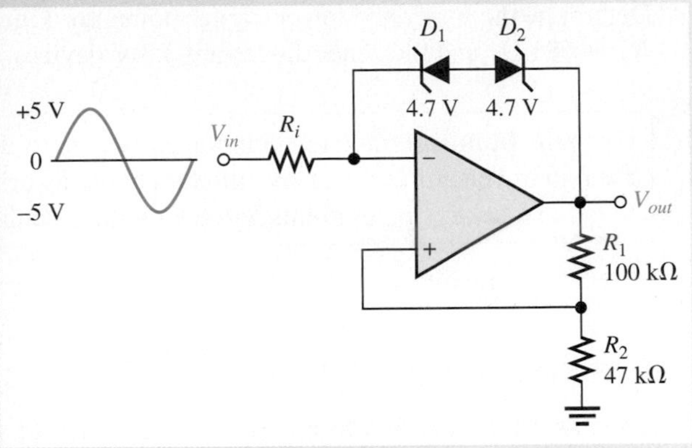

The voltage at the inverting (−) op-amp input is $V_{out} \pm 5.4$ V. Since the differential voltage is negligible, the voltage at the noninverting (+) op-amp input is also approximately $V_{out} \pm 5.4$ V. Thus,

$$V_{R1} = V_{out} - (V_{out} \pm 5.4\text{ V}) = \pm 5.4\text{ V}$$

$$I_{R1} = \frac{V_{R1}}{R_1} = \frac{\pm 5.4\text{ V}}{100\text{ k}\Omega} = \pm 54\ \mu\text{A}$$

Since the noninverting input current is negligible,

$$I_{R2} = I_{R1} = \pm 54\ \mu\text{A}$$

$$V_{R2} = R_2 I_{R2} = (47\text{ k}\Omega)(\pm 54\ \mu\text{A}) = \pm 2.54\text{ V}$$

$$V_{out} = V_{R1} + V_{R2} = \pm 5.4\text{ V} \pm 2.54\text{ V} = \pm 7.94\text{ V}$$

The upper trigger point (UTP) and the lower trigger point (LTP) are as follows:

$$V_{\text{UTP}} = \left(\frac{R_2}{R_1 + R_2}\right)(+V_{out}) = \left(\frac{47\text{ k}\Omega}{147\text{ k}\Omega}\right)(+7.94\text{ V}) = +2.54\text{ V}$$

$$V_{\text{LTP}} = \left(\frac{R_2}{R_1 + R_2}\right)(-V_{out}) = \left(\frac{47\text{ k}\Omega}{147\text{ k}\Omega}\right)(-7.94\text{ V}) = -2.54\text{ V}$$

The output waveform for the given input voltage is shown in Figure 13–14.

FIGURE 13–14

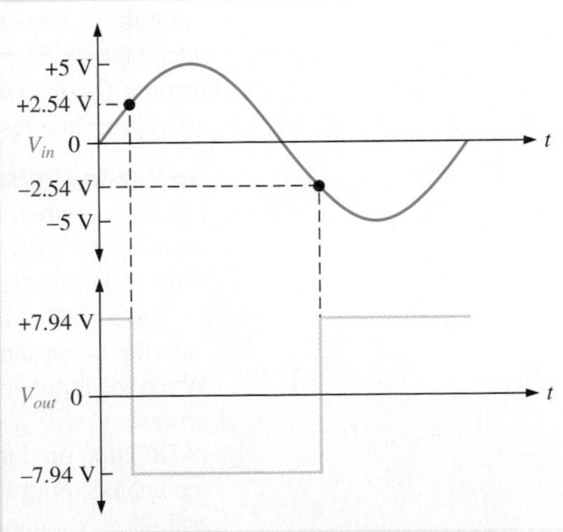

Related Problem Determine the upper and lower trigger points for Figure 13–13 if $R_1 = 150\text{ k}\Omega$, $R_2 = 68\text{ k}\Omega$, and the zener diodes are 3.3 V devices.

Open the Multisim file E13-03 in the Examples folder on your CD-ROM. Compare the output waveform to the specified input at any arbitrary frequency and see if the upper and lower trigger points agree with the calculated values.

Comparator Applications

Over-Temperature Sensing Circuit Figure 13–15 shows an op-amp comparator used in a precision over-temperature sensing circuit to determine when the temperature reaches a certain critical value. The circuit consists of a Wheatstone bridge with the op-amp used to detect when the bridge is balanced. One leg of the bridge contains a thermistor (R_1), which is a temperature-sensing resistor with a negative temperature coefficient (its resistance decreases as temperature increases). The potentiometer (R_2) is set at a value equal to the resistance of the thermistor at the critical temperature. At normal temperatures (below critical), R_1 is greater than R_2, thus creating an unbalanced condition that drives the op-amp to its low saturated output level and keeps transistor Q_1 off.

FIGURE 13–15

An over-temperature sensing circuit.

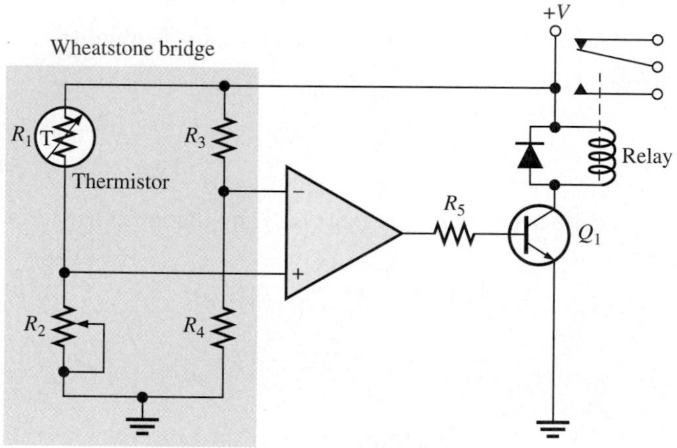

As the temperature increases, the resistance of the thermistor decreases. When the temperature reaches the critical value, R_1 becomes equal to R_2, and the bridge becomes balanced (since $R_3 = R_4$). At this point the op-amp switches to its high saturated output level, turning Q_1 on. This energizes the relay, which can be used to activate an alarm or initiate an appropriate response to the over-temperature condition.

Analog-to-Digital (A/D) Conversion **A/D conversion** is a common interfacing process often used when a linear **analog** system must provide inputs to a **digital** system. Many methods for A/D conversion are available. However, in this discussion, only one type is used to demonstrate the concept.

The *simultaneous,* or *flash,* method of A/D conversion uses parallel comparators to compare the linear input signal with various reference voltages developed by a voltage divider. When the input voltage exceeds the reference voltage for a given comparator, a high level is produced on that comparator's output. Figure 13–16 shows an analog-to-digital converter (ADC) that produces three-digit binary numbers on its output, which represent the values of the analog input voltage as it changes. This converter requires seven comparators. In general, $2^n - 1$ comparators are required for conversion to an n-digit binary number. The large number of comparators necessary for a reasonably sized binary number is one of the drawbacks of the simultaneous ADC, but newer IC technology has reduced the problem somewhat by combining multiple comparators and associated circuits on a single IC chip. For example, 6- or 8-bit flash converters are readily available. These ADCs are useful in applications that require the fastest possible conversion times, such as video processing.

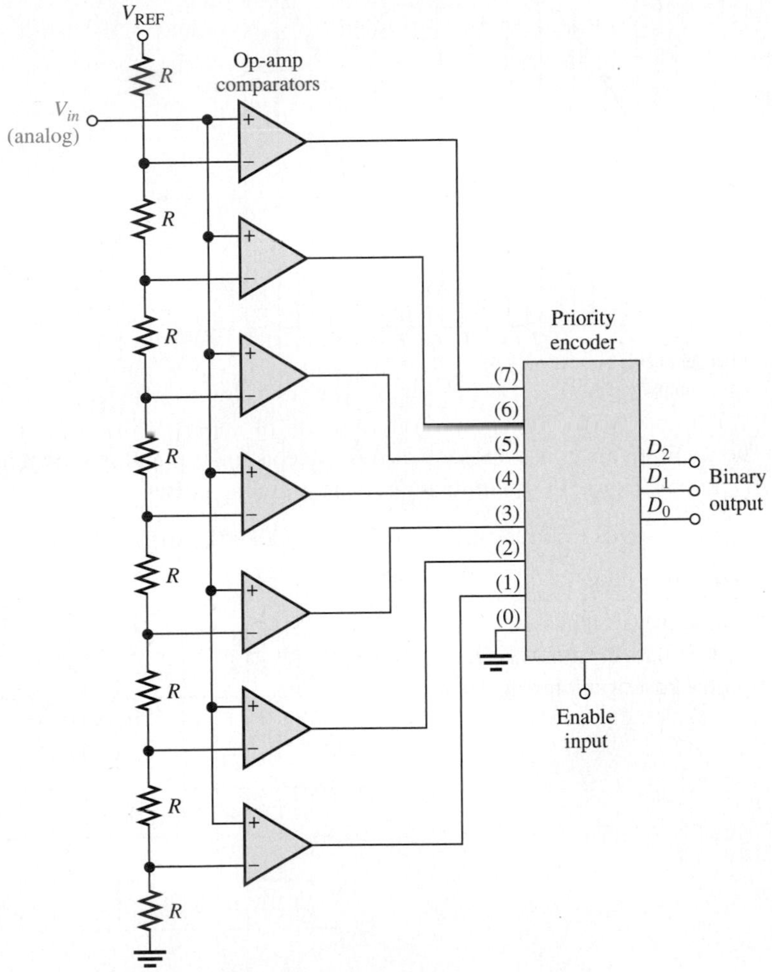

FIGURE 13–16

A simplified simultaneous (flash) analog-to-digital converter (ADC) using op-amps as comparators.

In Figure 13–16, the reference voltage for each comparator is set by the resistive voltage-divider circuit and V_{REF}. The output of each comparator is connected to an input of the priority encoder. The *priority encoder* is a digital device that produces a binary number on its output representing the highest value input.

The encoder samples its input when a pulse occurs on the enable line (sampling pulse), and a three-digit binary number proportional to the value of the analog input signal appears on the encoder's outputs.

The sampling rate determines the accuracy with which the sequence of binary numbers represents the changing input signal. The more samples taken in a given unit of time, the more accurately the analog signal is represented in digital form.

The following example illustrates the basic operation of the simultaneous ADC in Figure 13–16.

EXAMPLE 13–4

Determine the binary number sequence of the three-digit simultaneous ADC in Figure 13–16 for the input signal in Figure 13–17 and the sampling pulses (encoder enable) shown. Draw the resulting digital output waveforms.

FIGURE 13–17

Sampling of values on analog waveform for conversion to digital.

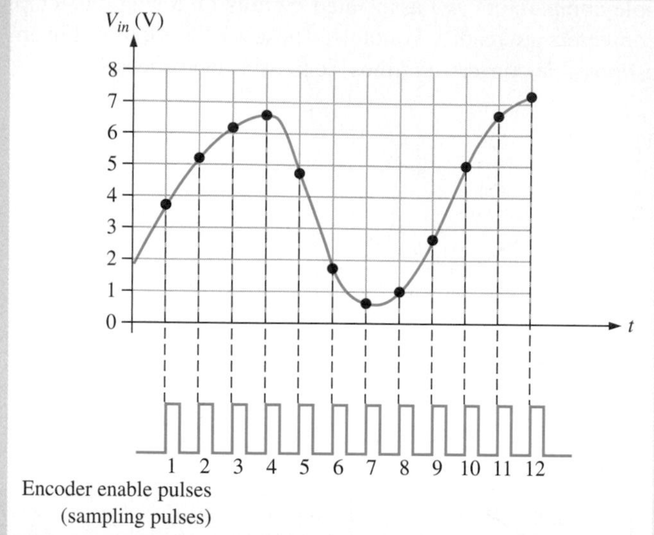

Solution The resulting binary output sequence is listed as follows and is shown in the waveform diagram of Figure 13–18 in relation to the sampling pulses.

011, 101, 110, 110, 100, 001, 000, 001, 010, 101, 110, 111

FIGURE 13–18

Resulting digital outputs for sampled values in Figure 13–17. D_0 is the least significant digit.

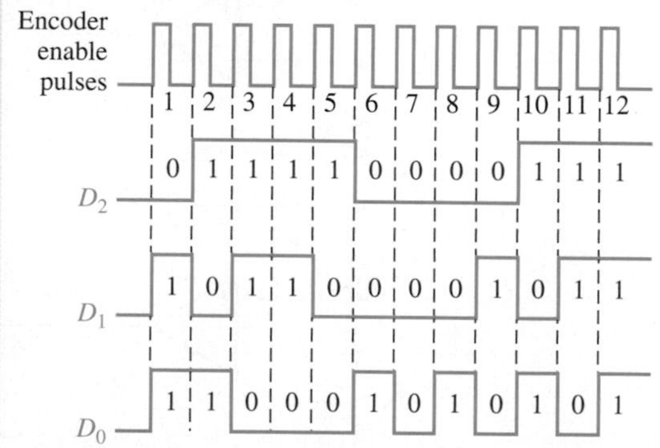

Related Problem If the frequency of the enable pulses in Figure 13–17 is doubled, does the resulting binary output sequence represent the analog waveform more or less accurately?

Specific Comparators

The LM111, LM211, and LM311 are examples of specific comparators that exhibit high switching speeds and other features not normally found on the general type of op-amp. These comparators can operate with supply voltages from ±15 V to a single +5 V. The open collector output provides the capability of driving loads that require voltages up to 50 V referenced to ground or to the supply voltages. An offset balancing input and a strobe input allow the output to be turned on or off regardless of the differential input.

SECTION 13–1 CHECKUP
Answers are at the end of the chapter.

1. What is the reference voltage for each comparator in Figure 13–19?
2. What is the purpose of hysteresis in a comparator?
3. Define the term *bounding* in relation to a comparator's output.

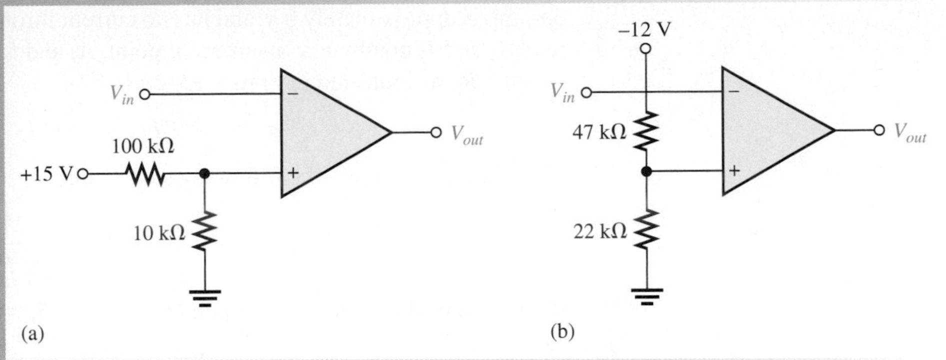

FIGURE 13–19

13–2 SUMMING AMPLIFIERS

The summing amplifier is an application of the inverting op-amp configuration covered in Chapter 12. The averaging amplifier and the scaling amplifier are variations of the basic summing amplifier.

After completing this section, you should be able to

- **Analyze the operation of several types of summing amplifiers**
  - Describe the operation of a unity-gain summing amplifier
  - Discuss how to achieve any specified gain greater than unity
  - Describe the operation of an averaging amplifier
  - Describe the operation of a scaling adder
  - Discuss a scaling adder used as a digital-to-analog converter

Summing Amplifier with Unity Gain

A **summing amplifier** has two or more inputs, and its output voltage is proportional to the negative of the algebraic sum of its input voltages. A two-input summing amplifier is shown in Figure 13–20, but any number of inputs can be used. The operation of the circuit and derivation of the output expression are as follows. Two voltages, V_{IN1} and V_{IN2}, are applied to the inputs and produce currents I_1 and I_2, as shown. Using the concepts of infinite

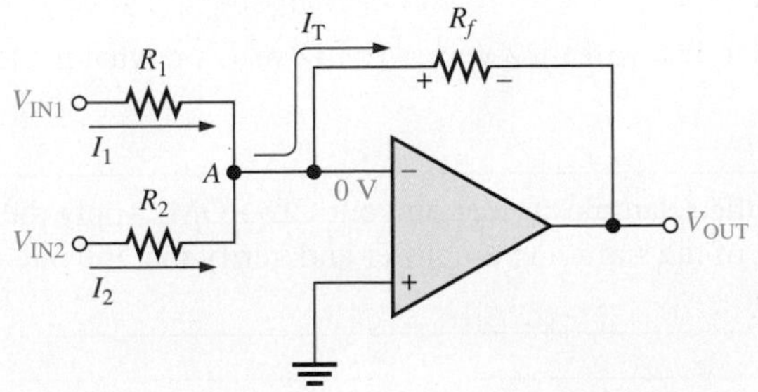

FIGURE 13–20

Two-input inverting summing amplifier.

input impedance and virtual ground, you can determine that the inverting (−) input of the op-amp is approximately 0 V and has no current through it. This means that both input currents I_1 and I_2 combine at a summing point, A, and form the total current (I_T), which goes through R_f, as indicated in Figure 13–20.

$$I_T = I_1 + I_2$$

Since $V_{OUT} = -I_T R_f$, the following steps apply:

$$V_{OUT} = -(I_1 + I_2)R_f = -\left(\frac{V_{IN1}}{R_1} + \frac{V_{IN2}}{R_2}\right)R_f$$

If all three of the resistors are equal ($R_1 = R_2 = R_f = R$), then

$$V_{OUT} = -\left(\frac{V_{IN1}}{R} + \frac{V_{IN2}}{R}\right)R = -(V_{IN1} + V_{IN2})$$

The previous equation shows that the output voltage has the same magnitude as the sum of the two input voltages but with a negative sign, indicating inversion.

A general expression is given in Equation 13–4 for a unity-gain summing amplifier with n inputs, as shown in Figure 13–21 where all resistors are equal in value.

Equation 13–4

$$\mathbf{V_{OUT} = -(V_{IN1} + V_{IN2} + V_{IN3} + \cdots + V_{INn})}$$

FIGURE 13–21

Summing amplifier with n inputs.

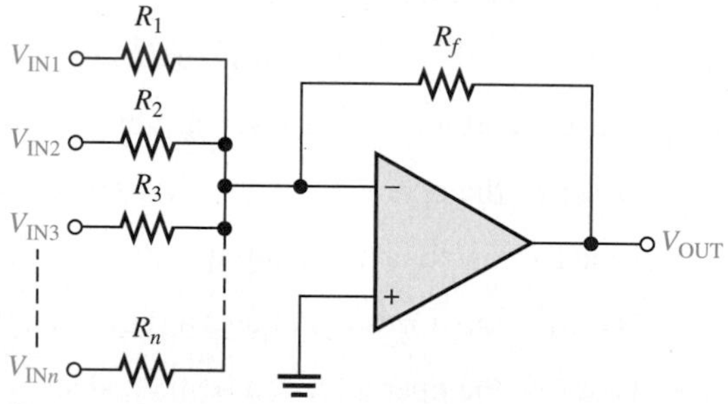

EXAMPLE 13–5

Determine the output voltage in Figure 13–22.

FIGURE 13–22

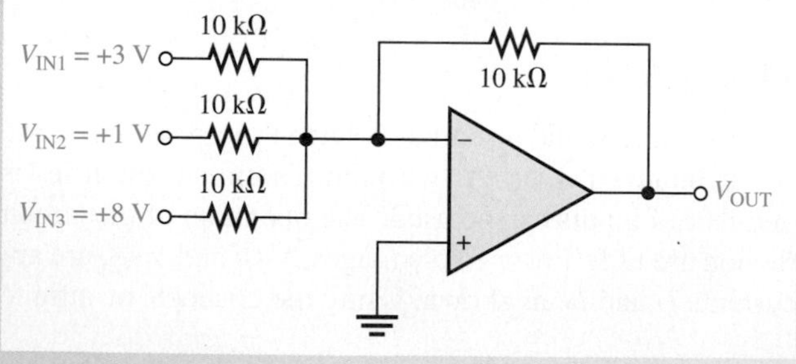

Solution

$$V_{OUT} = -(V_{IN1} + V_{IN2} + V_{IN3}) = -(3\text{ V} + 1\text{ V} + 8\text{ V}) = \mathbf{-12\text{ V}}$$

Related Problem If a fourth input of −0.5 V is added to Figure 13–22 with a 10 kΩ resistor, what is the output voltage?

Open the Multisim file E13-05 in the Examples folder on your CD-ROM. Apply the indicated dc voltages to the inputs of the summing amplifier and verify that the output is the inverted sum of the inputs.

Summing Amplifier with Gain Greater Than Unity

When R_f is larger than the input resistors, the amplifier has a gain of R_f/R, where R is the value of each equal-value input resistor. The general expression for the output is

$$V_{OUT} = -\frac{R_f}{R}(V_{IN1} + V_{IN2} + \cdots + V_{INn})$$

Equation 13–5

As you can see, the output voltage has the same magnitude as the sum of all the input voltages multiplied by a constant determined by the ratio $-(R_f/R)$.

EXAMPLE 13–6 Determine the output voltage for the summing amplifier in Figure 13–23.

FIGURE 13–23

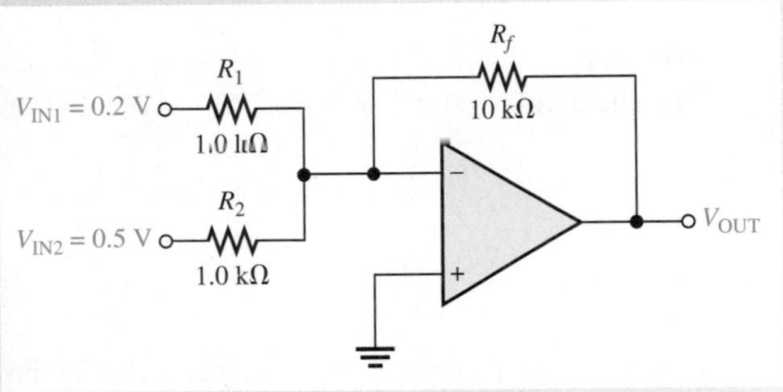

Solution $R_f = 10\text{ k}\Omega$ and $R = R_1 = R_2 = 1.0\text{ k}\Omega$. Therefore,

$$V_{OUT} = -\frac{R_f}{R}(V_{IN1} + V_{IN2}) = -\frac{10\text{ k}\Omega}{1.0\text{ k}\Omega}(0.2\text{ V} + 0.5\text{ V}) = -10(0.7\text{ V}) = \mathbf{-7\text{ V}}$$

Related Problem Determine the output voltage in Figure 13–23 if the two input resistors are 2.2 kΩ and the feedback resistor is 18 kΩ.

Open the Multisim file E13-06 in the Examples folder on your CD-ROM. Apply the indicated dc voltages to the inputs of the summing amplifier and verify that the output is the inverted sum of the inputs times a gain of 10.

Averaging Amplifier

A summing amplifier can be made to produce the mathematical average of the input voltages. This is done by setting the ratio R_f/R equal to the reciprocal of the number of inputs (n).

$$\frac{R_f}{R} = \frac{1}{n}$$

You obtain the average of several numbers by first adding the numbers and then dividing by the quantity of numbers you have. Examination of Equation 13–5 and a little thought will convince you that a summing amplifier can be designed to do this. The next example will illustrate.

EXAMPLE 13–7

Show that the amplifier in Figure 13–24 produces an output whose magnitude is the mathematical average of the input voltages.

FIGURE 13–24

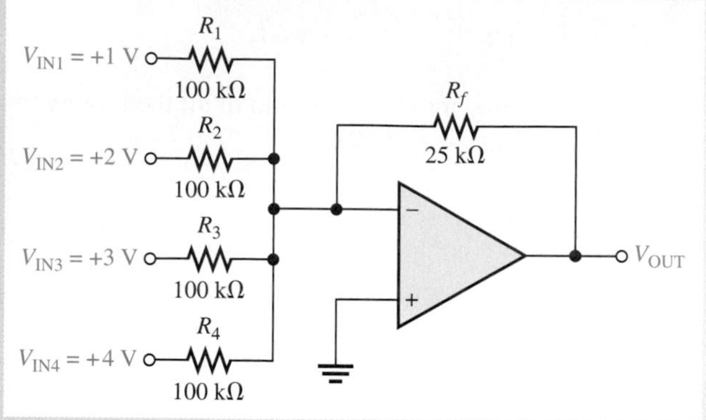

Solution Since the input resistors are equal, $R = 100\text{ k}\Omega$. The output voltage is

$$V_{OUT} = -\frac{R_f}{R}(V_{IN1} + V_{IN2} + V_{IN3} + V_{IN4})$$

$$= -\frac{25\text{ k}\Omega}{100\text{ k}\Omega}(1\text{ V} + 2\text{ V} + 3\text{ V} + 4\text{ V}) = -\frac{1}{4}(10\text{ V}) = \mathbf{-2.5\text{ V}}$$

A simple calculation shows that the average of the input values is the same magnitude as V_{OUT} but of opposite sign.

$$V_{IN(avg)} = \frac{1\text{ V} + 2\text{ V} + 3\text{ V} + 4\text{ V}}{4} = \frac{10\text{ V}}{4} = 2.5\text{ V}$$

Related Problem Specify the changes required in the averaging amplifier in Figure 13–24 in order to handle five inputs.

Open the Multisim file E13-07 in the Examples folder on your CD-ROM. Apply the indicated dc voltages to the inputs of the summing amplifier and verify that the output is the inverted average of the inputs.

Scaling Adder

A different weight can be assigned to each input of a summing amplifier by simply adjusting the values of the input resistors. As you have seen, the output voltage can be expressed as

Equation 13–6

$$V_{OUT} = -\left(\frac{R_f}{R_1}V_{IN1} + \frac{R_f}{R_2}V_{IN2} + \cdots + \frac{R_f}{R_n}V_{INn}\right)$$

The weight of a particular input is set by the ratio of R_f to the resistance, R_x, for that input ($R_x = R_1, R_2, \cdots R_n$). For example, if an input voltage is to have a weight of 1, then $R_x = R_f$. Or, if a weight of 0.5 is required, $R_x = 2R_f$. The smaller the value of input resistance R_x, the greater the weight, and vice versa.

EXAMPLE 13–8

Determine the weight of each input voltage for the scaling adder in Figure 13–25 and find the output voltage.

FIGURE 13–25

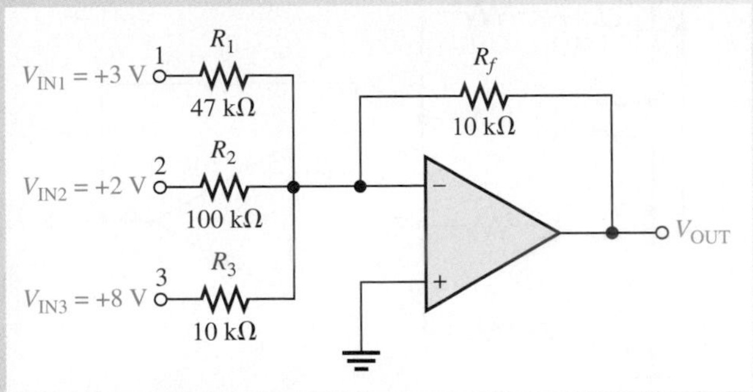

Solution Weight of input 1: $\dfrac{R_f}{R_1} = \dfrac{10\text{ k}\Omega}{47\text{ k}\Omega} = \mathbf{0.213}$

Weight of input 2: $\dfrac{R_f}{R_2} = \dfrac{10\text{ k}\Omega}{100\text{ k}\Omega} = \mathbf{0.100}$

Weight of input 3: $\dfrac{R_f}{R_3} = \dfrac{10\text{ k}\Omega}{10\text{ k}\Omega} = \mathbf{1.00}$

The output voltage is

$$V_{OUT} = -\left(\frac{R_f}{R_1}V_{IN1} + \frac{R_f}{R_2}V_{IN2} + \frac{R_f}{R_3}V_{IN3}\right)$$
$$= -[0.213(3\text{ V}) + 0.100(2\text{ V}) + 1.00(8\text{ V})]$$
$$= -(0.639\text{ V} + 0.2\text{ V} + 8\text{ V}) = \mathbf{-8.84\text{ V}}$$

Related Problem Determine the weight of each input voltage in Figure 13–25 if $R_1 = 22\text{ k}\Omega$, $R_2 = 82\text{ k}\Omega$, $R_3 = 56\text{ k}\Omega$, and $R_f = 10\text{ k}\Omega$. Also find V_{OUT}.

Open the Multisim file E13-08 in the Examples folder on your CD-ROM. Apply the indicated dc voltages to the inputs of the summing amplifier and verify that the output agrees with the calculated value.

Applications

D/A conversion is an important interface process for converting digital signals to analog (linear) signals. An example is a voice signal that is digitized for storage, processing, or transmission and must be changed back into an approximation of the original audio signal in order to drive a speaker.

One method of D/A conversion uses a scaling adder with input resistor values that represent the binary weights of the digital input code. Although this is not the most widely used method, it serves to illustrate how a scaling adder can be applied. A more common method for D/A conversion is known as the *R*/2*R* ladder method. The *R*/2*R* ladder is introduced here for comparison although it does not use a scaling adder.

Figure 13–26 shows a four-digit digital-to-analog converter (DAC) of this type (called a *binary-weighted resistor DAC*). The switch symbols represent transistor switches for applying each of the four binary digits to the inputs. The inverting (−) input is at virtual ground, and so the output voltage is proportional to the current through the feedback resistor R_f (sum of input currents). The lowest-value resistor *R* corresponds to the highest weighted binary input (2^3). All of the other resistors are multiples of *R* and correspond to the binary weights 2^2, 2^1, and 2^0.

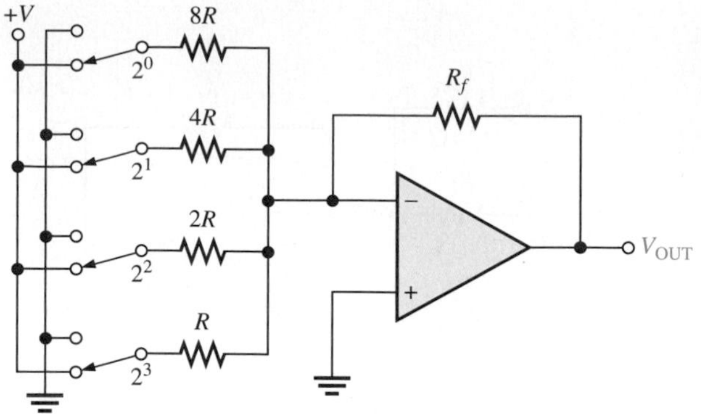

FIGURE 13–26

A scaling adder as a four-digit digital-to-analog converter (DAC).

EXAMPLE 13–9

Determine the output voltage of the DAC in Figure 13–27(a). The sequence of four-digit binary codes represented by the waveforms in Figure 13–27(b) are applied to the inputs. A high level is a binary 1, and low level is a binary 0. The least significant binary digit is D_0.

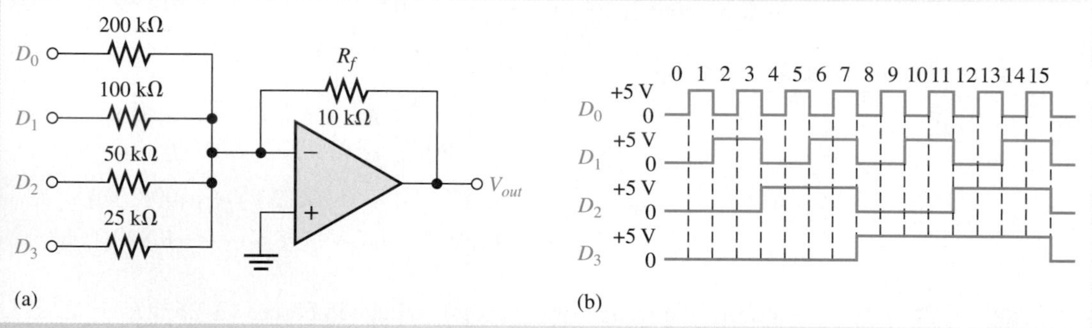

FIGURE 13–27

Solution First, determine the current for each of the weighted inputs. Since the inverting input of the op-amp is at 0 V (virtual ground) and a binary 1 corresponds to a high level (+5 V), the current through any of the input resistors equals 5 V divided by the resistance value.

$$I_0 = \frac{5\text{ V}}{200\text{ k}\Omega} = 0.025\text{ mA}$$

$$I_1 = \frac{5\text{ V}}{100\text{ k}\Omega} = 0.05\text{ mA}$$

$$I_2 = \frac{5\text{ V}}{50\text{ k}\Omega} = 0.1\text{ mA}$$

$$I_3 = \frac{5\text{ V}}{25\text{ k}\Omega} = 0.2\text{ mA}$$

There is almost no current at the inverting op-amp input because of its extremely high impedance. Therefore, assume that all of the input current is through R_f. Since one end of R_f is at 0 V (virtual ground), the drop across R_f equals the output voltage, which is negative with respect to virtual ground.

$$V_{OUT(D0)} = -R_f I_0 = -(10\text{ k}\Omega)(0.025\text{ mA}) = \mathbf{-0.25\text{ V}}$$

$$V_{OUT(D1)} = -R_f I_1 = -(10\text{ k}\Omega)(0.05\text{ mA}) = \mathbf{-0.5\text{ V}}$$

$$V_{OUT(D2)} = -R_f I_2 = -(10\text{ k}\Omega)(0.1\text{ mA}) = \mathbf{-1\text{ V}}$$

$$V_{OUT(D3)} = -R_f I_3 = -(10\text{ k}\Omega)(0.2\text{ mA}) = \mathbf{-2\text{ V}}$$

From Figure 13–27(b), the first binary input code is 0000, which produces an output voltage of 0 V. The next input code is 0001 (it stands for decimal 1). For this, the output voltage is −0.25 V. The next code is 0010, which produces an output voltage of −0.5 V. The next code is 0011, which produces an output voltage of −0.25 V + (−0.5 V) = −0.75 V. Each successive binary code increases the output voltage by −0.25 V. So, for this particular straight binary sequence on the inputs, the output is a stairstep waveform going from 0 V to −3.75 V in −0.25 V steps, as shown in Figure 13–28. If the steps are very small, the output approximates a straight line (linear).

FIGURE 13–28

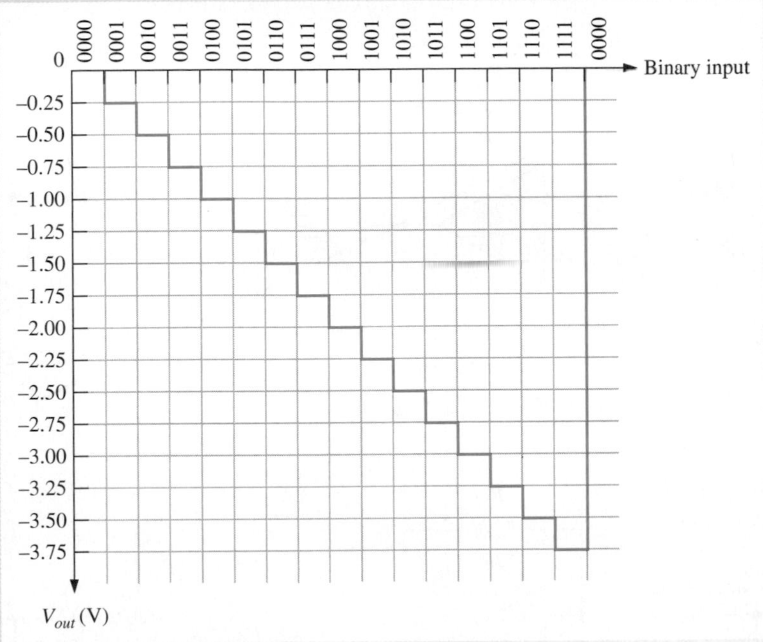

Related Problem If the 200 kΩ resistor in Figure 13–27(a) is changed to 400 kΩ, would the other resistor values have to be changed? If so, specify the values.

As mentioned before, the *R*/2*R* ladder is more commonly used for D/A conversion than the scaling adder and is shown in Figure 13–29 for four bits. It overcomes one of the disadvantages of the binary-weighted-input DAC because it requires only two resistor values.

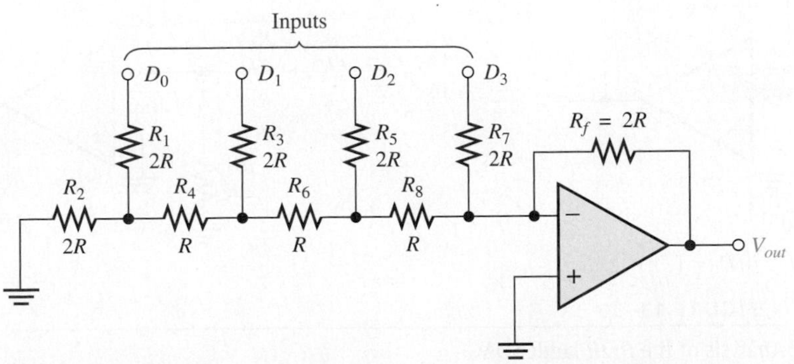

FIGURE 13–29

An *R*/2*R* ladder DAC.

Assume that the D_3 input is HIGH (+5 V) and the others are LOW (ground, 0 V). This condition represents the binary number 1000. A circuit analysis will show that this reduces to the equivalent form shown in Figure 13–30(a). Essentially no current goes through the

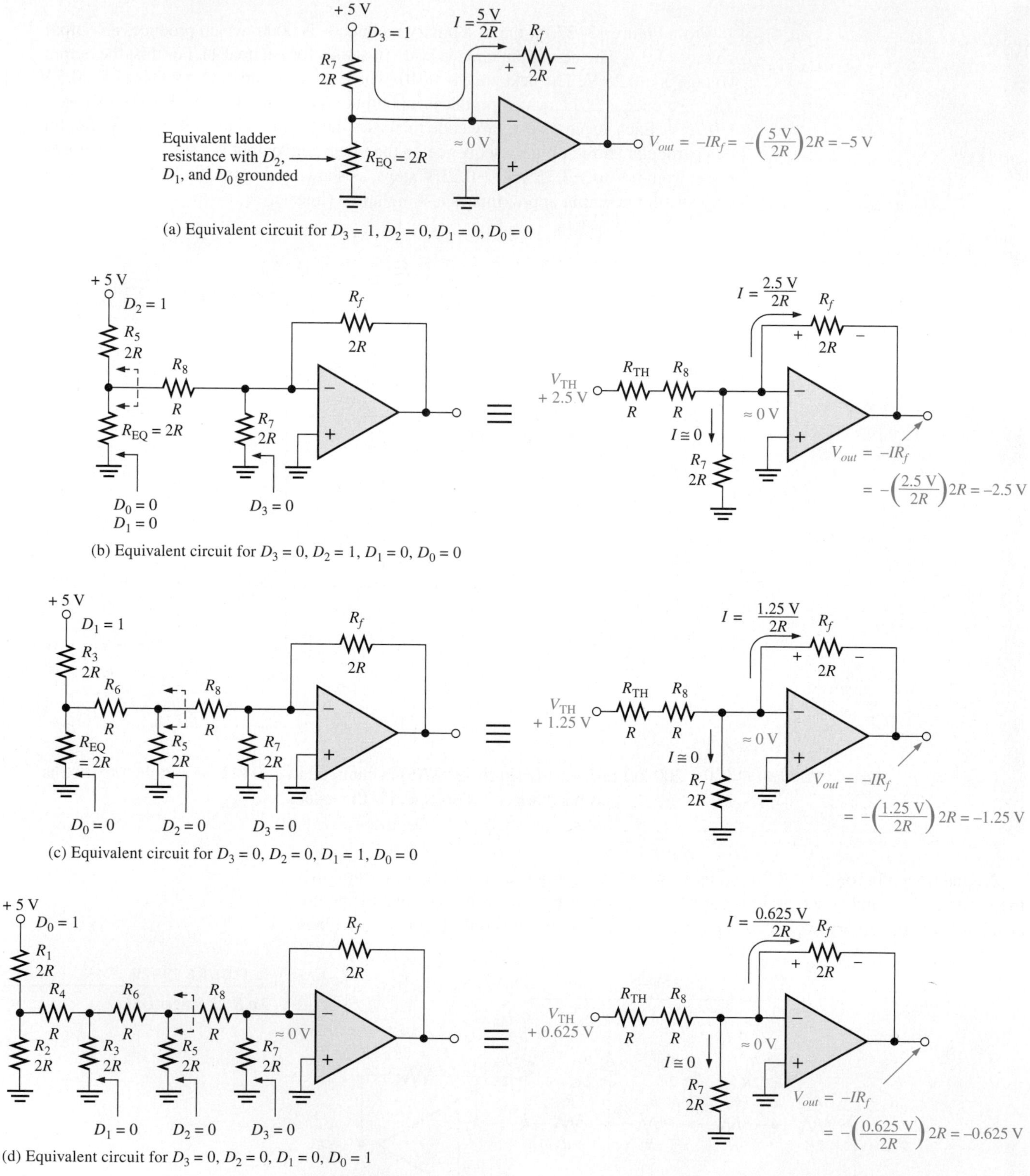

FIGURE 13–30

Analysis of the $R/2R$ ladder DAC.

2R equivalent resistance because the inverting input is at virtual ground. Thus, all of the current ($I = 5\text{ V}/2R$) through R_7 is also through R_f, and the output voltage is -5 V. The operational amplifier keeps the inverting ($-$) input near zero volts (≈ 0 V) because of negative feedback. Therefore, all current is through R_f rather than into the inverting input.

Figure 13–30(b) shows the equivalent circuit when the D_2 input is at +5 V and the others are at ground. This condition represents 0100. If we thevenize looking from R_8, we get 2.5 V in series with R, as shown. This results in a current through R_f of $I = 2.5\text{ V}/2R$, which gives an output voltage of -2.5 V. Keep in mind that there is no current into the op-amp inverting input and that there is no current through R_7 because it has 0 V across it, due to the virtual ground.

Figure 13–30(c) shows the equivalent circuit when the D_1 input is at +5 V and the others are at ground. This condition represents 0010. Again thevenizing looking from R_8, you get 1.25 V in series with R as shown. This results in a current through R_f of $I = 1.25\text{ V}/2R$, which gives an output voltage of -1.25 V.

In part (d) of Figure 13–30, the equivalent circuit representing the case where D_0 is at +5 V and the other inputs are at ground is shown. This condition represents 0001. Thevenizing from R_8 gives an equivalent of 0.625 V in series with R as shown. The resulting current through R_f is $I = 0.625\text{ V}/2R$, which gives an output voltage of -0.625 V.

Notice that each successively lower-weighted input produces an output voltage that is halved, so that the output voltage is proportional to the binary weight of the input bits.

SECTION 13–2 CHECKUP

1. Define *summing point.*
2. What is the value of R_f/R for a five-input averaging amplifier?
3. A certain scaling adder has two inputs, one having twice the weight of the other. If the resistor value for the lower-weighted input is 10 kΩ, what is the value of the other input resistor?

13–3 INTEGRATORS AND DIFFERENTIATORS

An op-amp **integrator** simulates mathematical integration, which is basically a summing process that determines the total area under the curve of a function. An op-amp **differentiator** simulates mathematical differentiation, which is a process of determining the instantaneous rate of change of a function. It is not necessary for you to understand mathematical integration or differentiation, at this point, in order to learn how an integrator and differentiator work. Ideal integrators and differentiators are used to show basic principles. Practical integrators often have an additional resistor in parallel with the feedback capacitor to prevent saturation. Practical differentiators may include a resistor in series with the comparator to reduce high frequency noise.

After completing this section, you should be able to

- **Analyze the operation of integrators and differentiators**
  - Identify an integrator
  - Discuss how a capacitor charges
  - Determine the rate of change of an integrator's output
  - Identify a differentiator
  - Determine the output voltage of a differentiator

The Op-Amp Integrator

The Ideal Integrator An ideal integrator is shown in Figure 13–31. Notice that the feedback element is a capacitor that forms an *RC* circuit with the input resistor.

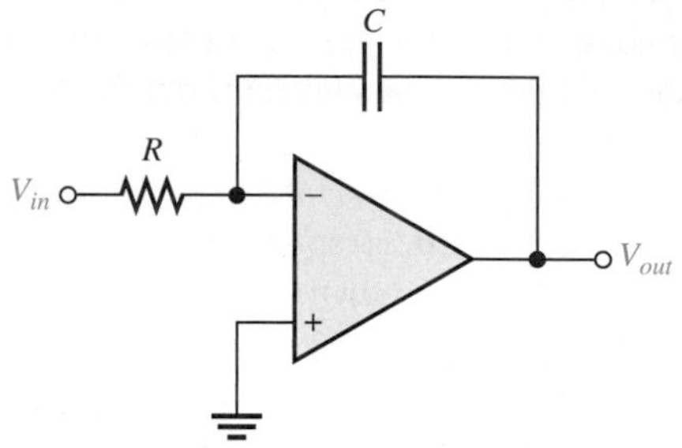

FIGURE 13–31
An op-amp integrator.

How a Capacitor Charges To understand how an integrator works, it is important to review how a capacitor charges. Recall that the charge Q on a capacitor is proportional to the charging current (I_C) and the time (t).

$$Q = I_C t$$

Also, in terms of the voltage, the charge on a capacitor is

$$Q = CV_C$$

From these two relationships, the capacitor voltage can be expressed as

$$V_C = \left(\frac{I_C}{C}\right)t$$

This expression has the form of an equation for a straight line that begins at zero with a constant slope of I_C/C. Remember from algebra that the general formula for a straight line is $y = mx + b$. In this case, $y = V_C$, $m = I_C/C$, $x = t$, and $b = 0$.

Recall that the capacitor voltage in a simple *RC* circuit is not linear but is exponential. This is because the charging current continuously decreases as the capacitor charges and causes the rate of change of the voltage to continuously decrease. The key thing about using an op-amp with an *RC* circuit to form an integrator is that the capacitor's charging current is made constant, thus producing a straight-line (linear) voltage rather than an exponential voltage. Now let's see why this is true.

In Figure 13–32, the inverting input of the op-amp is at virtual ground (0 V), so the voltage across R_i equals V_{in}. Therefore, the input current is

$$I_{in} = \frac{V_{in}}{R_i}$$

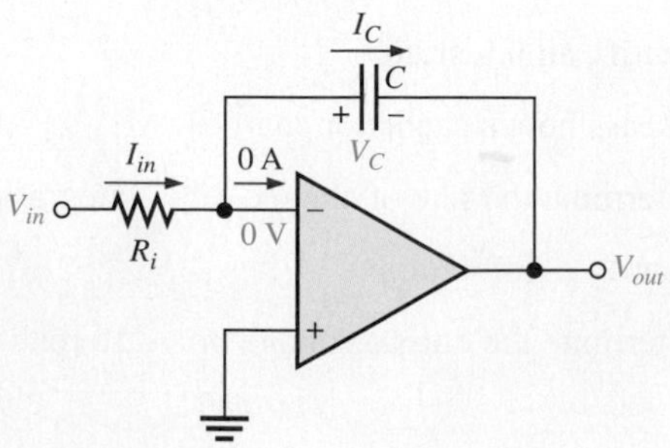

FIGURE 13–32
Currents in an integrator.

If V_{in} is a constant voltage, then I_{in} is also a constant because the inverting input always remains at 0 V, keeping a constant voltage across R_i. Because of the very high input impedance of the op-amp, there is negligible current at the inverting input. This makes all of the input current go through the capacitor, as indicated in Figure 13–32, so

$$I_C = I_{in}$$

The Capacitor Voltage Since I_{in} is constant, so is I_C. The constant I_C charges the capacitor linearly and produces a linear voltage across C. The positive side of the capacitor is held at 0 V by the virtual ground of the op-amp. The voltage on the negative side of the capacitor, which is the op-amp output voltage, decreases linearly from zero as the capacitor charges, as shown in Figure 13–33. This voltage, V_C, is called a *negative ramp* and is the consequence of a constant positive input.

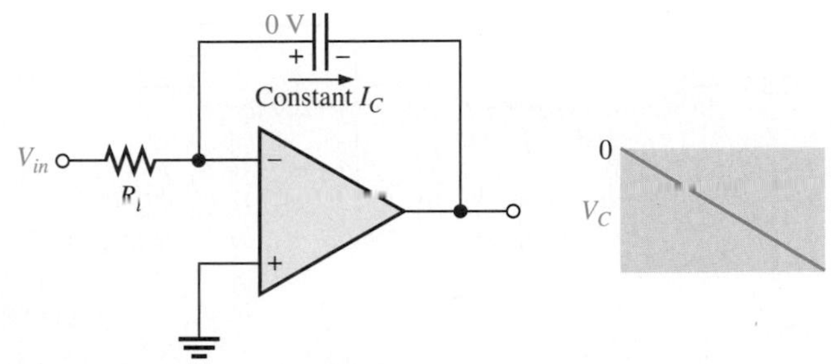

FIGURE 13–33

A linear ramp voltage is produced across the capacitor by the constant charging current.

The Output Voltage V_{out} is the same as the voltage on the negative side of the capacitor. When a constant positive input voltage in the form of a step or pulse (a pulse has a constant amplitude when high) is applied, the output ramp decreases negatively until the op-amp saturates at its maximum negative level. This is indicated in Figure 13–34.

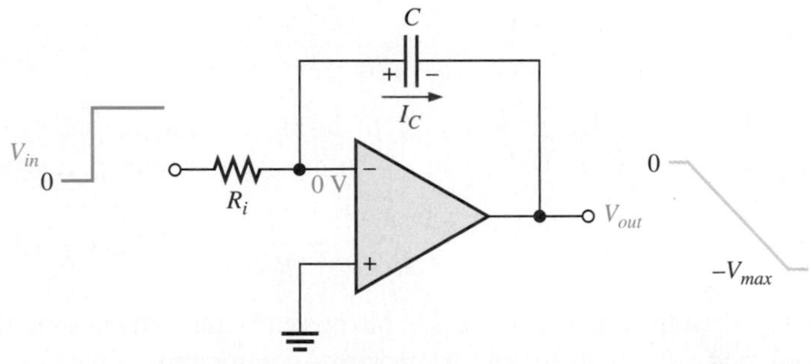

FIGURE 13–34

A constant input voltage produces a ramp on the output of the integrator.

Rate of Change of the Output The rate at which the capacitor charges, and therefore the slope of the output ramp, is set by the ratio I_C/C, as you have seen. Since $I_C = V_{in}/R_i$, the rate of change or slope of the integrator's output voltage is $\Delta V_{out}/\Delta t$.

$$\frac{\Delta V_{out}}{\Delta t} = -\frac{V_{in}}{R_i C}$$

Equation 13–7

Integrators are especially useful in triangular-wave oscillators as you will see in Chapter 16.

EXAMPLE 13–10

(a) Determine the rate of change of the output voltage in response to the input square wave, as shown for the ideal integrator in Figure 13–35(a). The output voltage is initially zero. The pulse width is 100 μs.

FIGURE 13–35

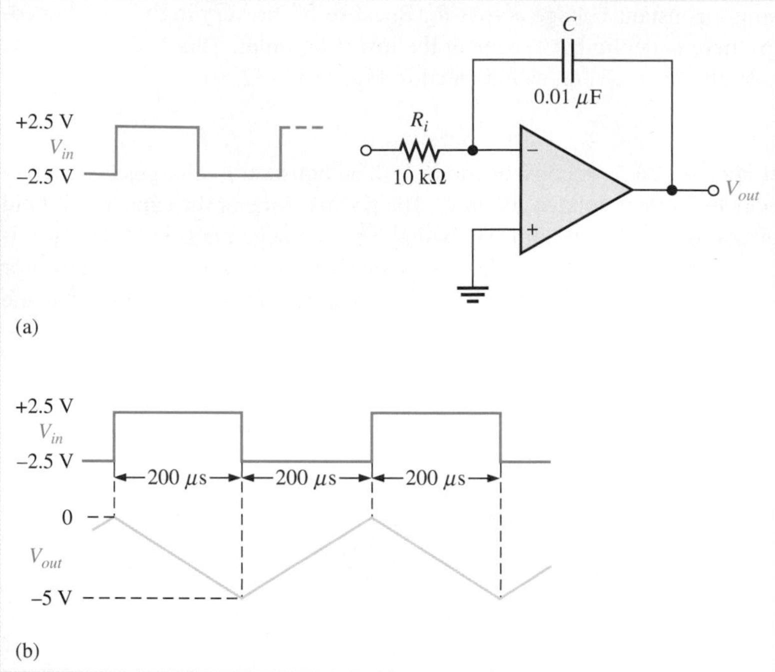

(b) Describe the output and draw the waveform.

Solution **(a)** The rate of change of the output voltage during the time that the input is at +2.5 V (capacitor charging) is

$$\frac{\Delta V_{out}}{\Delta t} = -\frac{V_{in}}{R_iC} = -\frac{2.5\text{ V}}{(10\text{ k}\Omega)(0.01\ \mu\text{F})} = -25\text{ kV/s} = \mathbf{-25\ mV/\mu s}$$

The rate of change of the output during the time that the input is negative (capacitor discharging) is the same as during charging except it is positive.

$$\frac{\Delta V_{out}}{\Delta t} = +\frac{V_{in}}{R_iC} = \mathbf{+25\ mV/\mu s}$$

(b) When the input is at +2.5 V, the output is a negative-going ramp. When the input is at −2.5 V, the output is a positive-going ramp.

$$\Delta V_{out} = (25\text{ mV}/\mu\text{s})(200\ \mu\text{s}) = 5\text{ V}$$

During the time the input is at +2.5 V, the output will go from 0 to −5 V. During the time the input is at −2.5 V, the output will go from −5 V to 0 V. Therefore, the output is a triangular wave with peaks at 0 V and −5 V, as shown in Figure 13–35(b).

Related Problem Modify the integrator in Figure 13–35 to make the output change from 0 to −5 V in 100 μs with the same input.

Open the Multisim file E13-10 in the Examples folder on your CD-ROM. Using the function generator, apply the indicated pulse waveform to the input and verify that the output voltage is a 5 V peak-to-peak triangular waveform.

The Practical Integrator The ideal integrator uses a capacitor in the feedback path, which is open to dc. This implies that the gain at dc is the open-loop gain of the op-amp. In a practical integrator, any dc error voltage due to offset error will cause the output to produce a ramp that moves toward either positive or negative saturation (depending on the offset), even when no signal is present.

Practical integrators must have some means of overcoming the effects of offset and bias current. Various solutions are available, such as chopper stabilized amplifiers; however, the simplest solution is to use a resistor in parallel with the capacitor in the feedback path, as shown in Figure 13–36. The feedback resistor, R_f, should be large compared to the input resistor R_{in}, in order to have a negligible effect on the output waveform. In addition, a compensating resistor, R_c, may be added to the noninverting input to balance the effects of bias current.

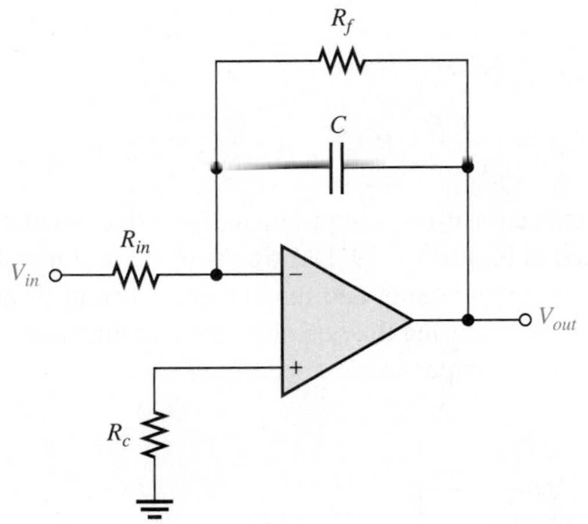

FIGURE 13–36

The Op-Amp Differentiator

The Ideal Differentiator An ideal differentiator is shown in Figure 13–37. Notice how the placement of the capacitor and resistor differ from the integrator. The capacitor is now the input element, and the resistor is the feedback element. A differentiator produces an output that is proportional to the rate of change of the input voltage.

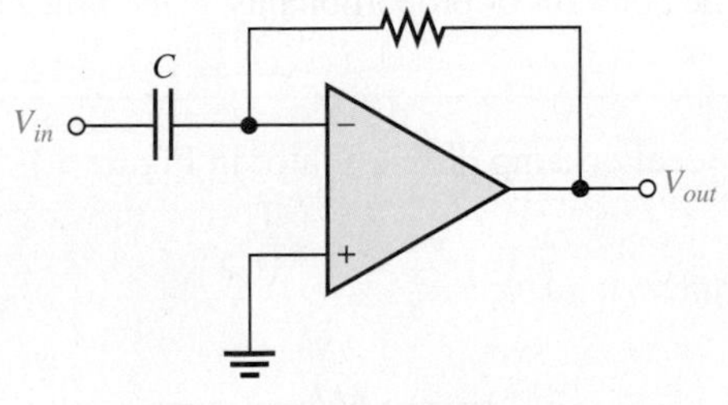

FIGURE 13–37

An op-amp differentiator.

To see how the differentiator works, apply a positive-going ramp voltage to the input as indicated in Figure 13–38. In this case, $I_C = I_{in}$ and the voltage across the capacitor is equal to V_{in} at all times ($V_C = V_{in}$) because of virtual ground on the inverting input.

From the basic formula, $V_C = (I_C/C)t$, the capacitor current is

$$I_C = \left(\frac{V_C}{t}\right)C$$

FIGURE 13–38

A differentiator with a ramp input.

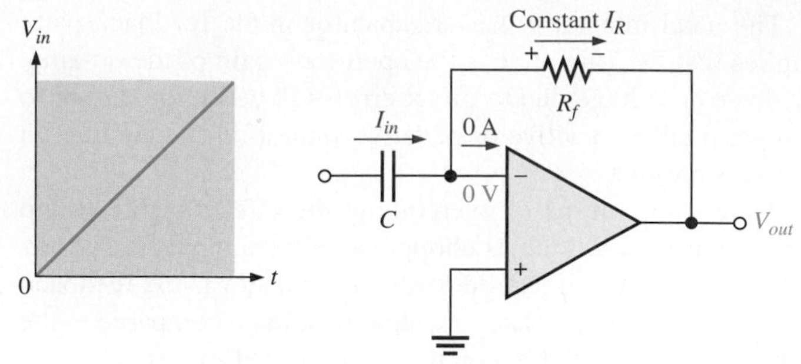

Since the current at the inverting input is negligible, $I_R = I_C$. Both currents are constant because the slope of the capacitor voltage (V_C/t) is constant. The output voltage is also constant and equal to the voltage across R_f because one side of the feedback resistor is always 0 V (virtual ground).

$$V_{out} = I_R R_f = I_C R_f$$

Equation 13–8

$$V_{out} = -\left(\frac{V_C}{t}\right)R_f C$$

The output is negative when the input is a positive-going ramp and positive when the input is a negative-going ramp, as illustrated in Figure 13–39. During the positive slope of the input, the capacitor is charging from the input source and the constant current through the feedback resistor is in the direction shown. During the negative slope of the input, the current is in the opposite direction because the capacitor is discharging.

FIGURE 13–39

Output of a differentiator with a series of positive and negative ramps (triangle wave) on the input.

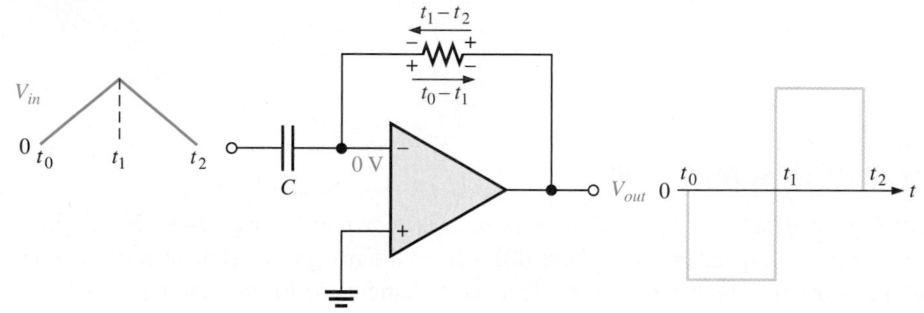

Notice in Equation 13–8 that the term V_C/t is the slope of the input. If the slope increases, V_{out} increases. If the slope decreases, V_{out} decreases. The output voltage is proportional to the slope (rate of change) of the input. The constant of proportionality is the time constant, R_fC.

EXAMPLE 13–11

Determine the output voltage of the ideal op-amp differentiator in Figure 13–40 for the triangular-wave input shown.

FIGURE 13–40

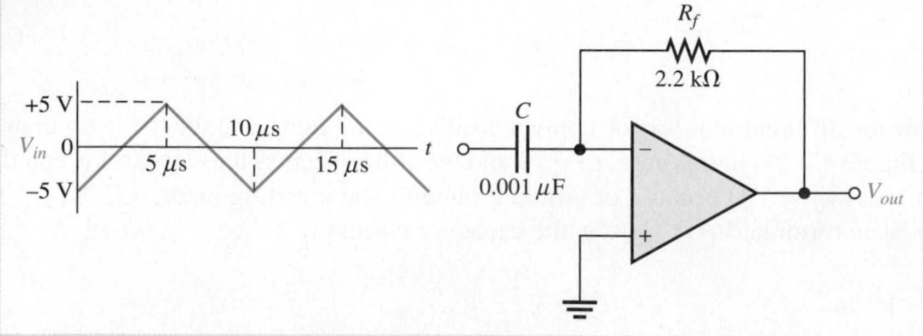

Solution Starting at $t = 0$, the input voltage is a positive-going ramp ranging from -5 V to $+5$ V (a $+10$ V change) in 5 μs. Then it changes to a negative-going ramp ranging from $+5$ V to -5 V (a -10 V change) in 5 μs.

The time constant is

$$R_fC = (2.2\,\text{k}\Omega)(0.001\,\mu\text{F}) = 2.2\,\mu\text{s}$$

Determine the slope or rate of change (V_C/t) of the positive-going ramp and calculate the output voltage as follows:

$$\frac{V_C}{t} = \frac{10\text{ V}}{5\,\mu\text{s}} = 2\text{ V}/\mu\text{s}$$

$$V_{out} = -\left(\frac{V_C}{t}\right)R_fC = -(2\text{ V}/\mu\text{s})2.2\,\mu\text{s} = \mathbf{-4.4\text{ V}}$$

Likewise, the slope of the negative-going ramp is -2 V/μs, and the output voltage is

$$V_{out} = -(-2\text{ V}/\mu\text{s})2.2\,\mu\text{s} = \mathbf{+4.4\text{ V}}$$

Figure 13–41 shows a graph of the output voltage waveform relative to the input.

FIGURE 13–41

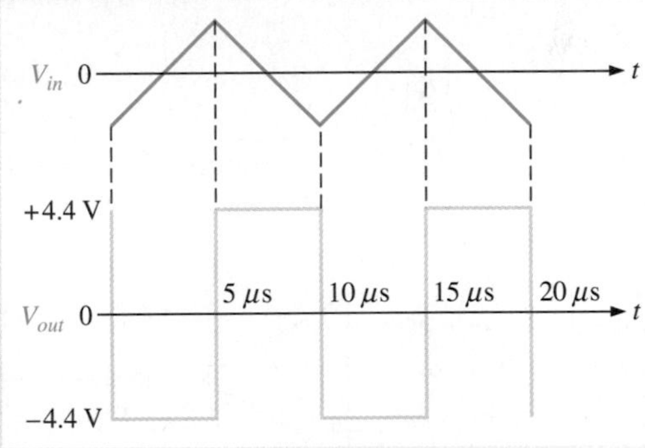

Related Problem What would the output voltage be if the feedback resistor in Figure 13–40 is changed to 3.3 kΩ?

The Practical Differentiator The ideal differentiator uses a capacitor in series with the inverting input. Because a capacitor has very low impedance at high frequencies, the combination of R_f and C form a very high gain amplifier at high frequencies. This means that a differentiator circuit tends to be noisy because electrical noise mainly consists of high frequencies. The solution to this problem is simply to add a resistor, R_{in}, in series with the capacitor to act as a low-pass filter and reduce the gain at high frequencies. The resistor should be small compared to the feedback resistor in order to have a negligible effect on the desired signal. Figure 13–42 shows a practical differentiator. A bias compensating resistor may also be used on the noninverting input.

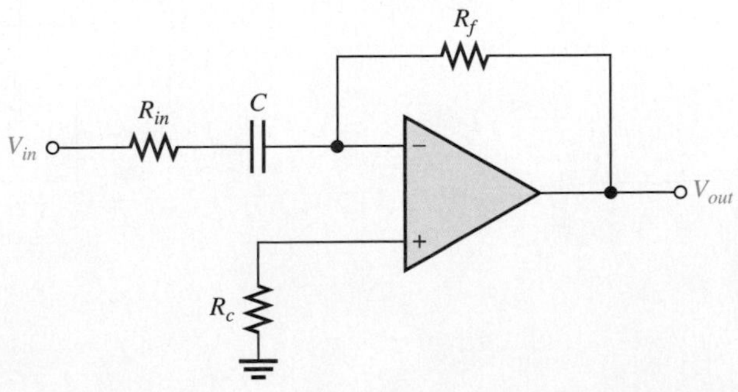

FIGURE 13–42

SECTION 13–3 CHECKUP

1. What is the feedback element in an ideal op-amp integrator?
2. For a constant input voltage to an integrator, why is the voltage across the capacitor linear?
3. What is the feedback element in an op-amp differentiator?
4. How is the output of a differentiator related to the input?

13–4 TROUBLESHOOTING

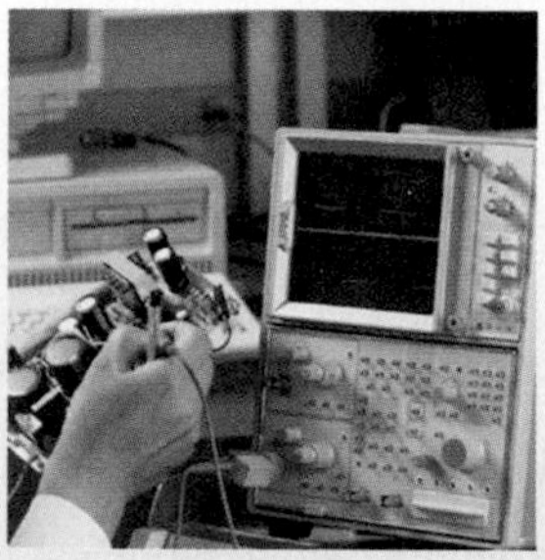

Although integrated circuit op-amps are extremely reliable and trouble-free, failures do occur from time to time. One type of internal failure mode is a condition where the op-amp output is in a saturated state, resulting in a constant high or constant low level, regardless of the input. Also, external component failures will produce various types of failure modes in op-amp circuits. Some examples are presented in this section.

After completing this section, you should be able to

- **Troubleshoot basic op-amp circuits**
  - Identify failures in comparator circuits
  - Identify failures in summing amplifiers

Figure 13–43 illustrates an internal failure of a comparator circuit that results in a "stuck" output.

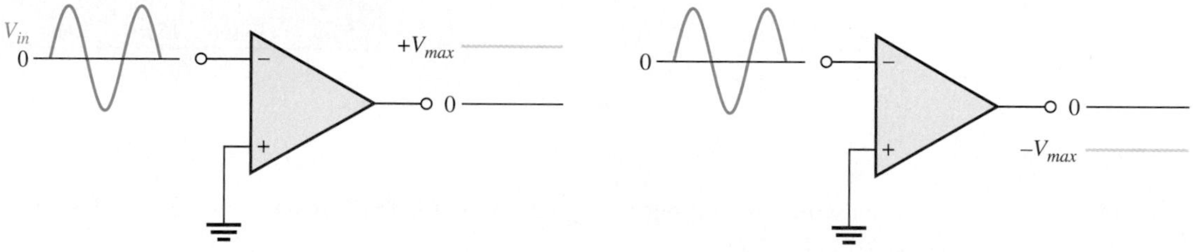

(a) Output failed in the HIGH state

(b) Output failed in the LOW state

▲ **FIGURE 13–43**

Internal comparator failures typically result in the output being "stuck" in the HIGH or LOW state.

Symptoms of External Component Failures in Comparator Circuits

A comparator with zener-bounding and hysteresis is shown in Figure 13–44. In addition to a failure of the op-amp itself, a zener diode or one of the resistors could be faulty. For example,

► **FIGURE 13–44**

A bounded comparator with hysteresis.

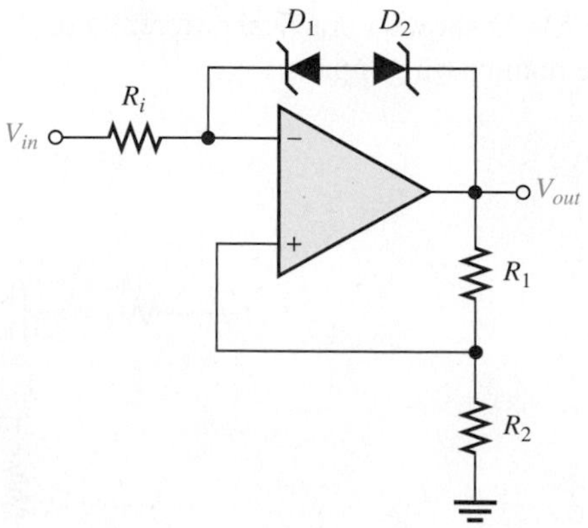

suppose one of the zener diodes opens. This effectively eliminates both zeners, and the circuit operates as an unbounded comparator, as indicated in Figure 13–45(a). With a shorted diode, the output is limited to the zener voltage (bounded) only in one direction, depending on which diode remains operational, as illustrated in Figure 13–45(b). In the other direction, the output is held at the forward diode voltage.

Recall that R_1 and R_2 set the UTP and LTP for the hysteresis comparator. Now, suppose that R_2 opens. Essentially all of the output voltage is fed back to the noninverting (+) input,

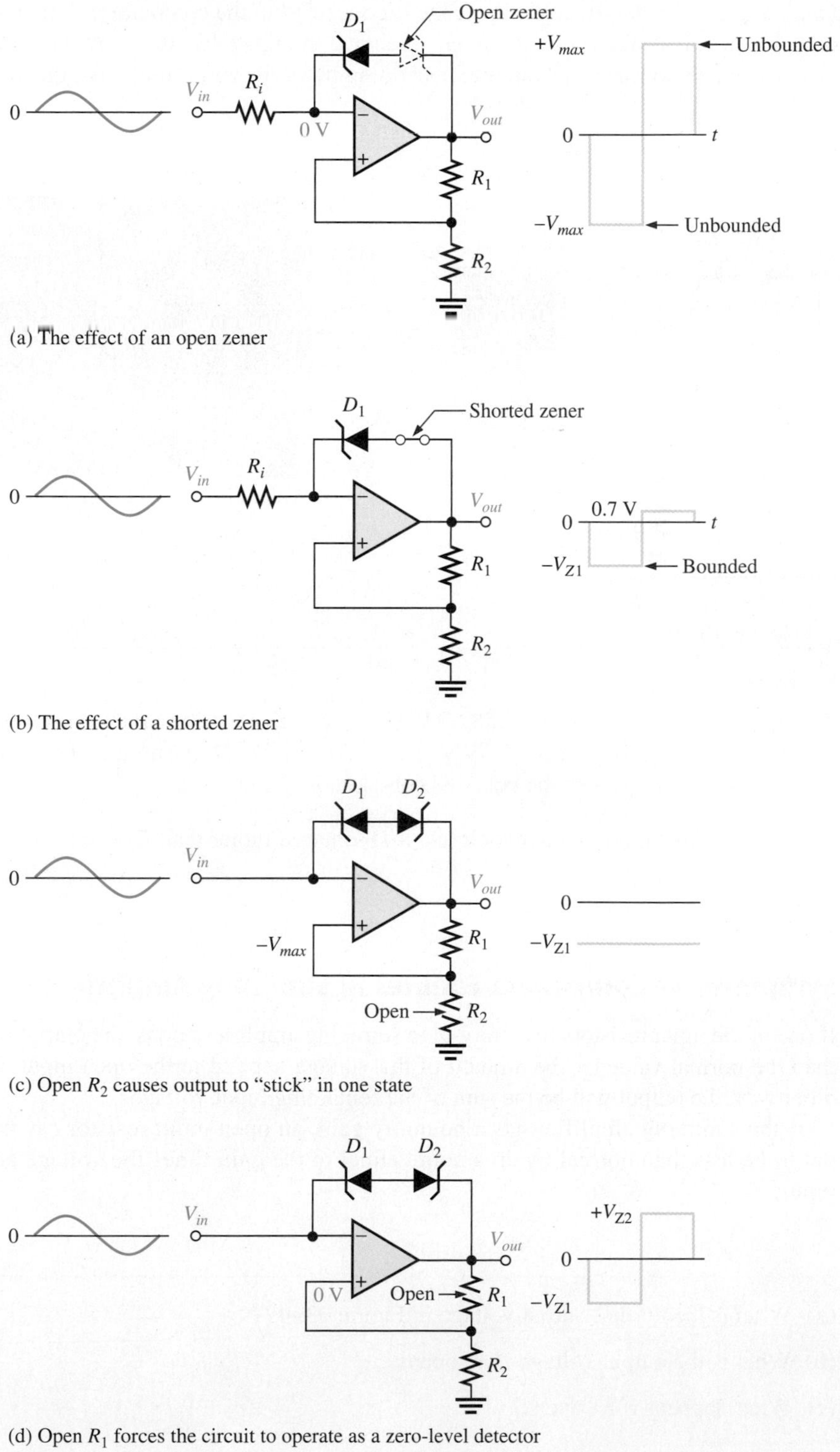

FIGURE 13–45

Examples of comparator circuit failures and their effects.

and, since the input voltage will never exceed the output, the device will remain in one of its bounded states. This symptom can also indicate a faulty op-amp, as mentioned before. Now, assume that R_1 opens. This leaves the noninverting input near ground potential and causes the circuit to operate as a zero-level detector. These conditions are shown in parts (c) and (d) of Figure 13–45.

EXAMPLE 13–12

One channel of a dual-trace oscilloscope is connected to the comparator output and the other channel is connected to the input, as shown in Figure 13–46. From the observed waveforms, determine if the circuit is operating properly, and if not, what the most likely failure is.

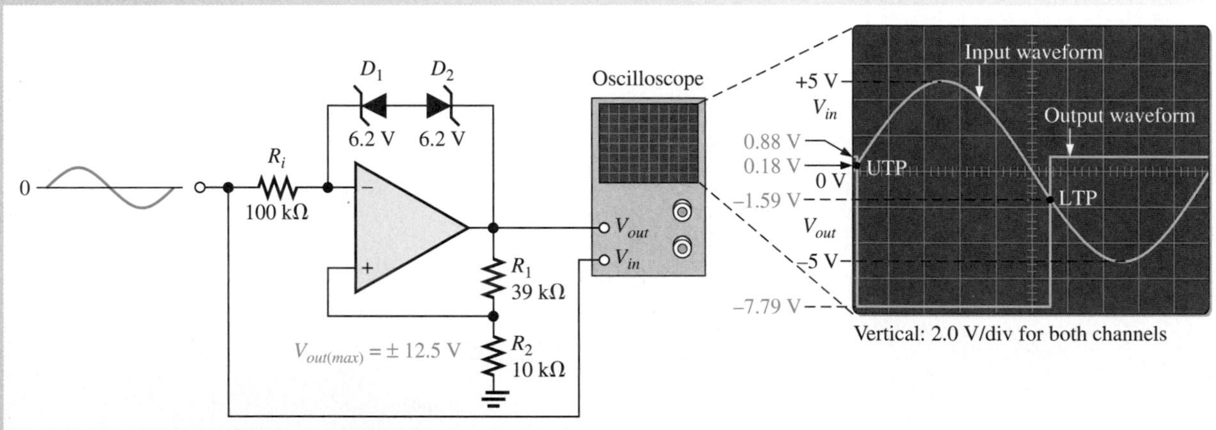

FIGURE 13–46

Solution The output should be limited to ± 8.67 V. However, the positive maximum is $+0.88$ V and the negative maximum is -7.79 V. This indicates that D_2 is shorted. Refer to Example 13–3 for analysis of the bounded comparator.

Related Problem What would the output voltage look like if D_1 shorted rather than D_2?

Symptoms of Component Failures in Summing Amplifiers

If one of the input resistors in a unity-gain summing amplifier opens, the output will be less than the normal value by the amount of the voltage applied to the open input. Stated another way, the output will be the sum of the remaining input voltages.

If the summing amplifier has a nonunity gain, an open input resistor causes the output to be less than normal by an amount equal to the gain times the voltage at the open input.

EXAMPLE 13–13

(a) What is the normal output voltage in Figure 13–47?

(b) What is the output voltage if R_2 opens?

(c) What happens if R_5 opens?

FIGURE 13–47

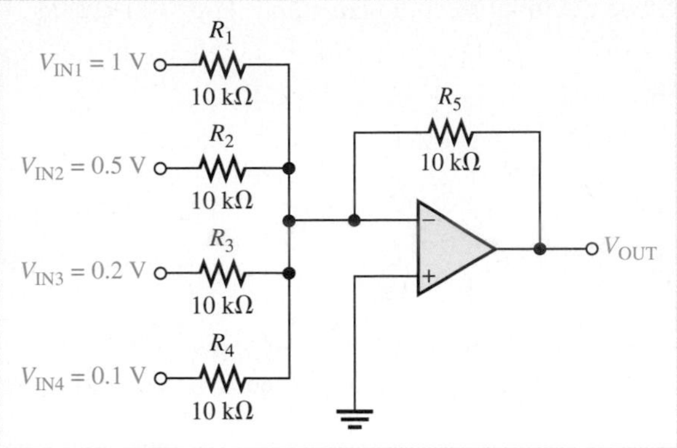

Solution

(a) $V_{OUT} = -(V_{IN1} + V_{IN2} + \cdots + V_{INn})$

$= -(1\text{ V} + 0.5\text{ V} + 0.2\text{ V} + 0.1\text{ V}) = \mathbf{-1.8\ V}$

(b) $V_{OUT} = -(1\text{ V} + 0.2\text{ V} + 0.1\text{ V}) = \mathbf{-1.3\ V}$

(c) If the feedback resistor opens, the circuit becomes a comparator and the output goes to $-V_{max}$.

Related Problem In Figure 13–47, $R_5 = 47\text{ k}\Omega$. What is the output voltage if R_1 opens?

As another example, let's look at an averaging amplifier. An open input resistor will result in an output voltage that is the average of all the inputs with the open input averaged in as a zero.

EXAMPLE 13–14

(a) What is the normal output voltage for the averaging amplifier in Figure 13–48?

(b) If R_4 opens, what is the output voltage? What does the output voltage represent?

FIGURE 13–48

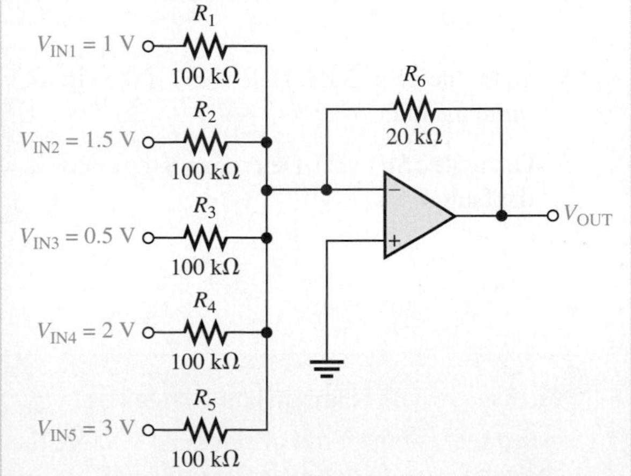

Solution Since the input resistors are equal, $R = 100\text{ k}\Omega$. $R_f = R_6$.

$$\textbf{(a)}\ V_{OUT} = -\frac{R_f}{R}(V_{IN1} + V_{IN2} + \cdots + V_{INn})$$

$$= -\frac{20\text{ k}\Omega}{100\text{ k}\Omega}(1\text{ V} + 1.5\text{ V} + 0.5\text{ V} + 2\text{ V} + 3\text{ V}) = -\frac{1}{5}(8\text{ V}) = \mathbf{-1.6\ V}$$

$$\textbf{(b)}\ V_{OUT} = -\frac{20\text{ k}\Omega}{100\text{ k}\Omega}(1\text{ V} + 1.5\text{ V} + 0.5\text{ V} + 3\text{ V}) = -\frac{1}{5}(6\text{ V}) = \mathbf{-1.2\ V}$$

1.2 V is the average of five voltages with the 2 V input replaced by 0 V. Notice that the output is not the average of the four remaining input voltages.

Related Problem If R_4 is open, as was the case in this example, what would you have to do to make the output equal to the average of the remaining four input voltages?

Multisim Troubleshooting Exercises

These file circuits are in the Troubleshooting Exercises folder on your CD-ROM.

1. Open file TSE13-01. Determine if the circuit is working properly and, if not, determine the fault.
2. Open file TSE13-02. Determine if the circuit is working properly and, if not, determine the fault.
3. Open file TSE13-03. Determine if the circuit is working properly and, if not, determine the fault.
4. Open file TSE13-04. Determine if the circuit is working properly and, if not, determine the fault.
5. Open file TSE13-05. Determine if the circuit is working properly and, if not, determine the fault.
6. Open file TSE13-06. Determine if the circuit is working properly and, if not, determine the fault.
7. Open file TSE13-07. Determine if the circuit is working properly and, if not, determine the fault.
8. Open file TSE13-08. Determine if the circuit is working properly and, if not, determine the fault.
9. Open file TSE13-09. Determine if the circuit is working properly and, if not, determine the fault.

SECTION 13–4 CHECKUP

1. Describe one type of internal op-amp failure.
2. If a certain malfunction is attributable to more than one possible component failure, what would you do to isolate the problem?

On the Job

Application Activity: *Sine/Pulse Waveform Generator*

Your company manufactures a battery-operated audio signal generator that produces a sinusoidal output with a variable frequency and amplitude and operates from ±12 V dc voltages. The frequency can be varied from 20 Hz to 20 kHz, and the peak amplitude can be varied from 10 mV to 10 V with front panel controls.

A new version of the signal generator is being designed that adds a pulse waveform generator to the audio signal generator in a single unit. The pulse generator will produce an output with a variable duty cycle that can be used to drive 5 V digital logic circuits. The sine wave generator will remain the same, but the frequency control and the output terminals will be common to both the sine wave generator and the pulse generator. The output function will be switch-selectable, and the pulse waveform will require an additional front panel control for adjusting the duty cycle. The minimum specifications are given in Table 13–1. The front panel for the sine/pulse generator is shown in Figure 13–49.

TABLE 13–1

	OUTPUT VOLTAGE RANGE	FREQUENCY RANGE	DUTY CYCLE RANGE
Sine	0.1 V–20 V p-p	20 Hz–20 kHz	——
Pulse	5 V amplitude	20 Hz–20 kHz	15%–85%

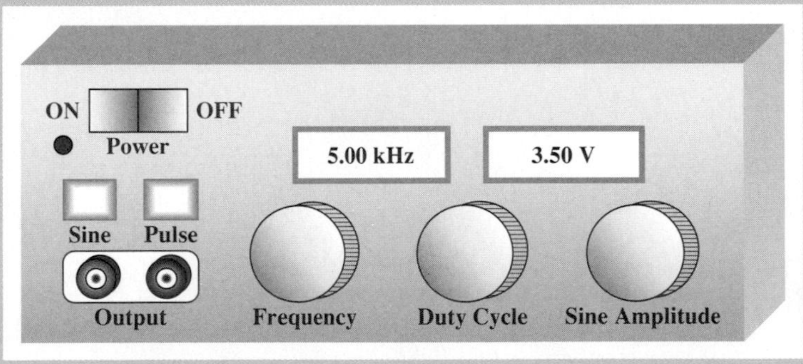

FIGURE 13–49

Front panel of the sine/pulse generator.

The Circuit

The schematic of the new design is shown in Figure 13–50. The pulse waveform is derived from the 10 V peak sine wave that is available internally in the existing signal generator. An LM111 comparator is used for producing the pulse waveform using the sine wave as the driving source. The variable reference voltage at the inverting input of the comparator provides the duty cycle control. The duty cycle adjustment range is from 10% to 90%.

The LM111 comparator has an open collector output that is pulled up to +5 V with a 1 kΩ resistor, and the emitter of the output transistor is connected to ground, as shown. As a result, the output pulses vary between 0 V and +5 V.

1. Which components determine the comparator's variable reference voltage?
2. Calculate the minimum reference voltage.
3. Calculate the maximum reference voltage.
4. What sets the amplitude of the output pulses?
5. Explain how the duty cycle control works.

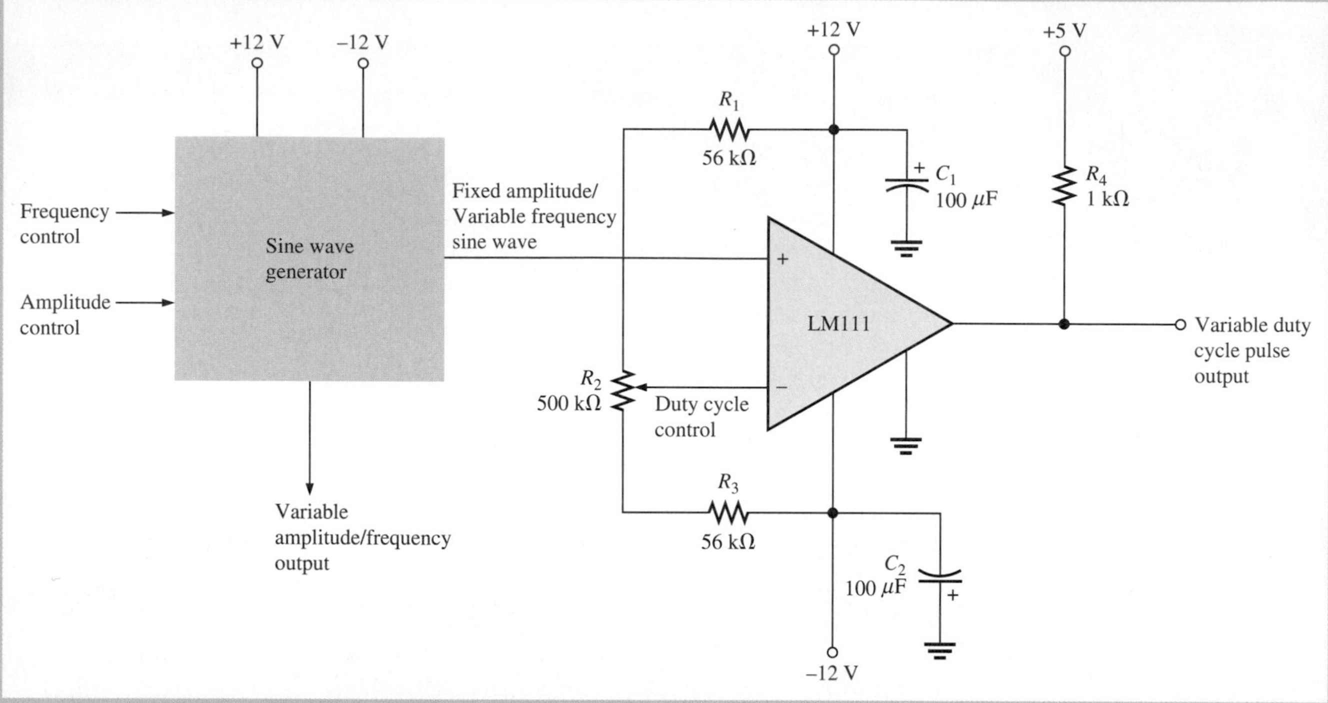

FIGURE 13–50

Sine/Pulse generator.

The pin diagram from the LM111 datasheet is shown in Figure 13–51. Pins 5 and 6 are unused in this application.

FIGURE 13–51

Pin diagram for the LM111 comparator.

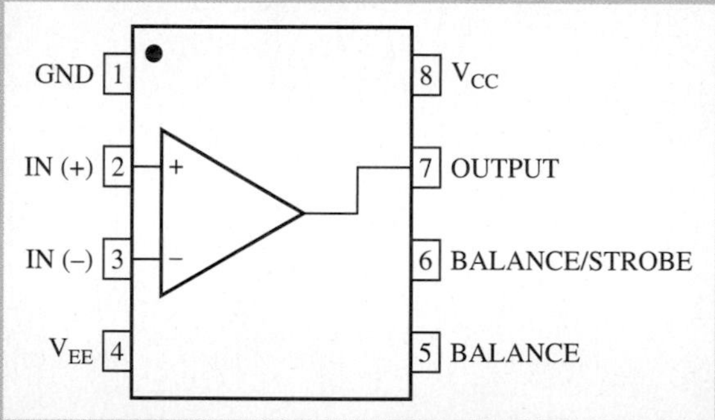

6. Referencing the pin diagram, assign pin numbers to the comparator in Figure 13–50.

Simulation

The sine/pulse generator is simulated using Multisim with an input signal of 7.07 V rms to represent the existing sine wave generator output. The results are shown in Figure 13–52 where the duty cycle of the pulse waveform is set to 50%.

7. From the scope display in Figure 13–52, verify the rms value of the sine wave.
8. Measure the amplitude of the pulse waveform on the display.
9. Verify the frequency of the waveforms on the display.

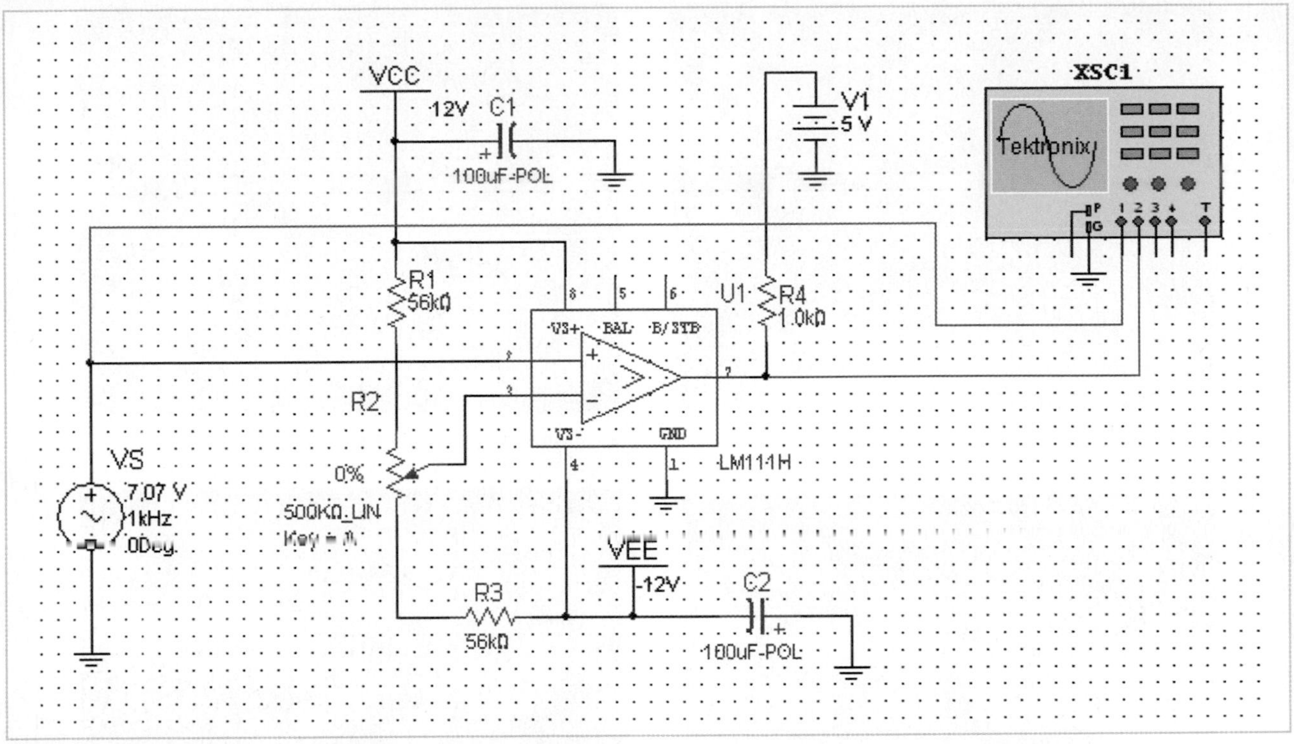

(a) Circuit screen

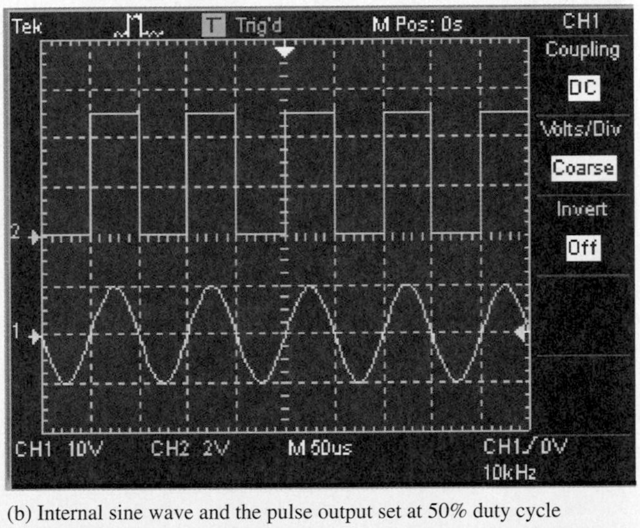

(b) Internal sine wave and the pulse output set at 50% duty cycle

FIGURE 13–52

Simulation of the sine/pulse generator at a frequency of 10 kHz.

Figure 13–53 shows the simulation results for the pulse duty cycle measurement at test frequencies of 1 kHz and 10 kHz.

10. In Figure 13–53, determine if the minimum and maximum duty cycles meet or exceed specifications for the frequencies shown.

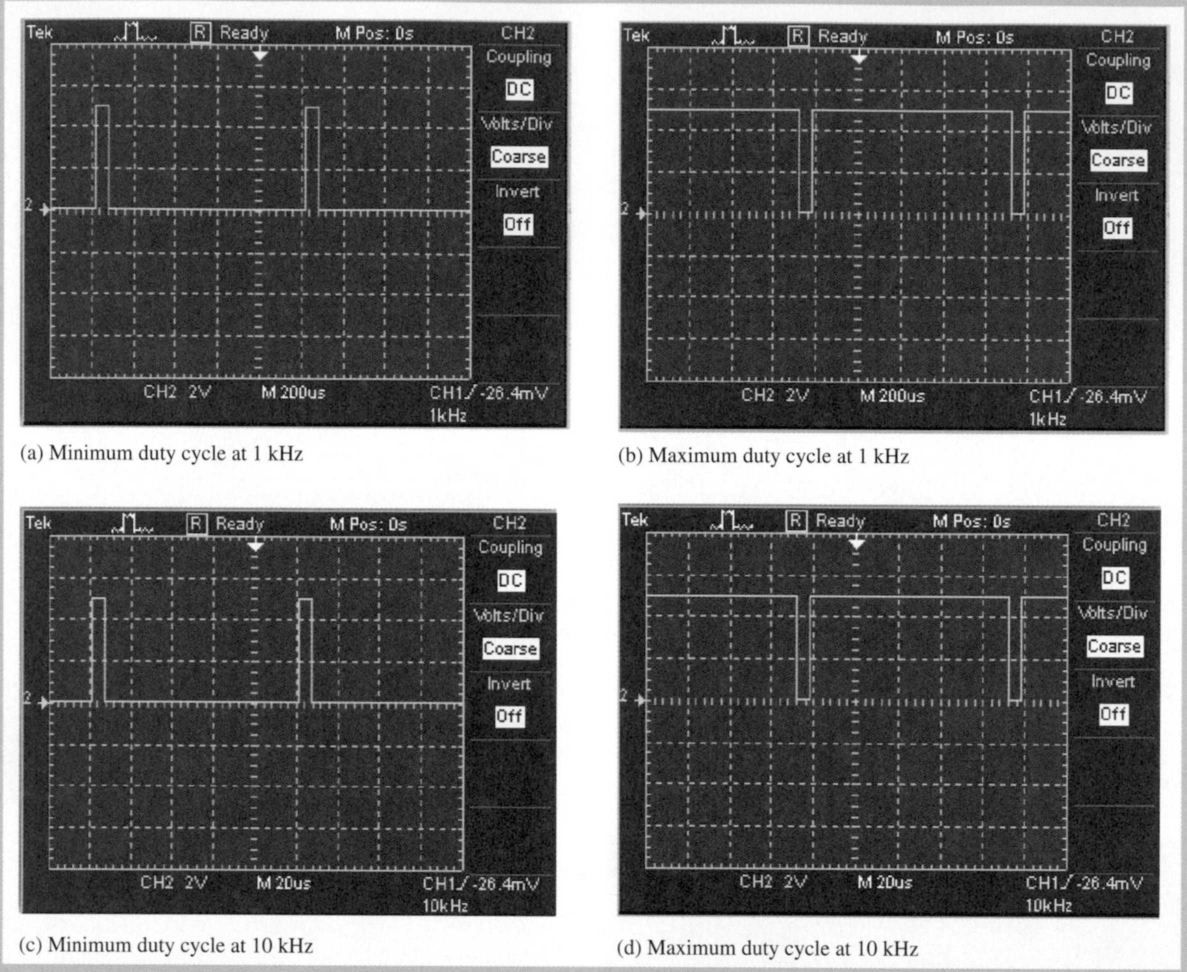

(a) Minimum duty cycle at 1 kHz

(b) Maximum duty cycle at 1 kHz

(c) Minimum duty cycle at 10 kHz

(d) Maximum duty cycle at 10 kHz

▲ FIGURE 13–53

Simulation results.

Simulate the sine/pulse generator using your Multisim software. Observe the output voltages with the oscilloscope as the duty cycle control is varied.

Hint: Before running a simulation that includes a comparator, it may be necessary to reset the default value of the relative error tolerance to avoid interpolation error, which results in slow transitions particularly at higher frequencies. To do this, select *Simulate* and click on *Interactive Simulation Settings.* Navigate to the *Analysis Options* tab and choose *Customize.* Change the *Relative Tolerance Error* to be le-005.

Prototyping and Testing

Now that the circuit has been simulated, the prototype circuit is constructed and tested. After the circuit is successfully tested on a protoboard, it is ready to be finalized on a printed circuit board.

Lab Experiment

To build and test a similar circuit, go to Experiment 13–A in your lab manual (*Laboratory Exercises for Electronic Devices* by David Buchla and Steven Wetterling).

Circuit Board

The pulse generator board is shown in Figure 13–54. This board will be added to the existing audio generator and connected to the front panel controls to complete the sine/pulse generator.

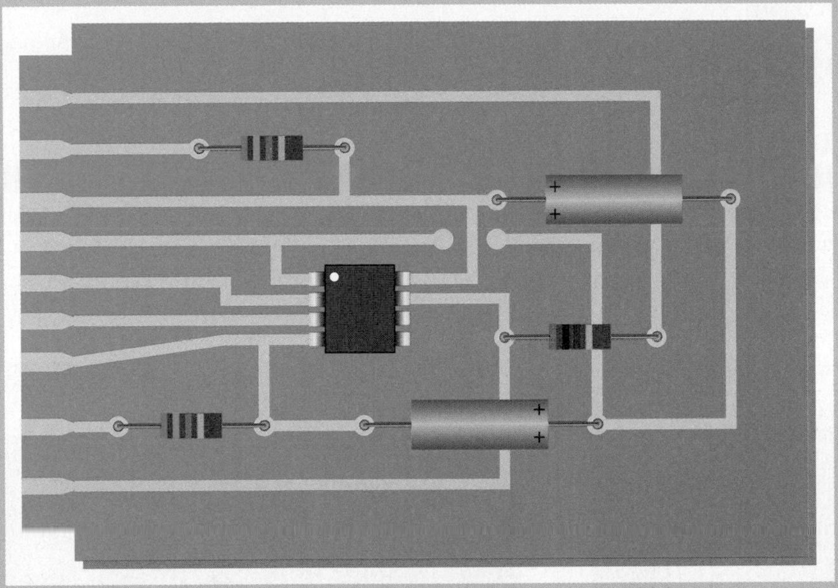

FIGURE 13–54

Pulse generator board.

11. Check the printed circuit board for correctness by comparing with the schematic in Figure 13–50.
12. Label each input and output pin according to function.

The Sine/Pulse Generator System Diagram

The complete generator unit consists of the sine wave generator, the pulse generator, the front panel controls, and the battery power supply, as shown in Figure 13–55.

13. Verify the connection from the pulse generator board to the various system components.

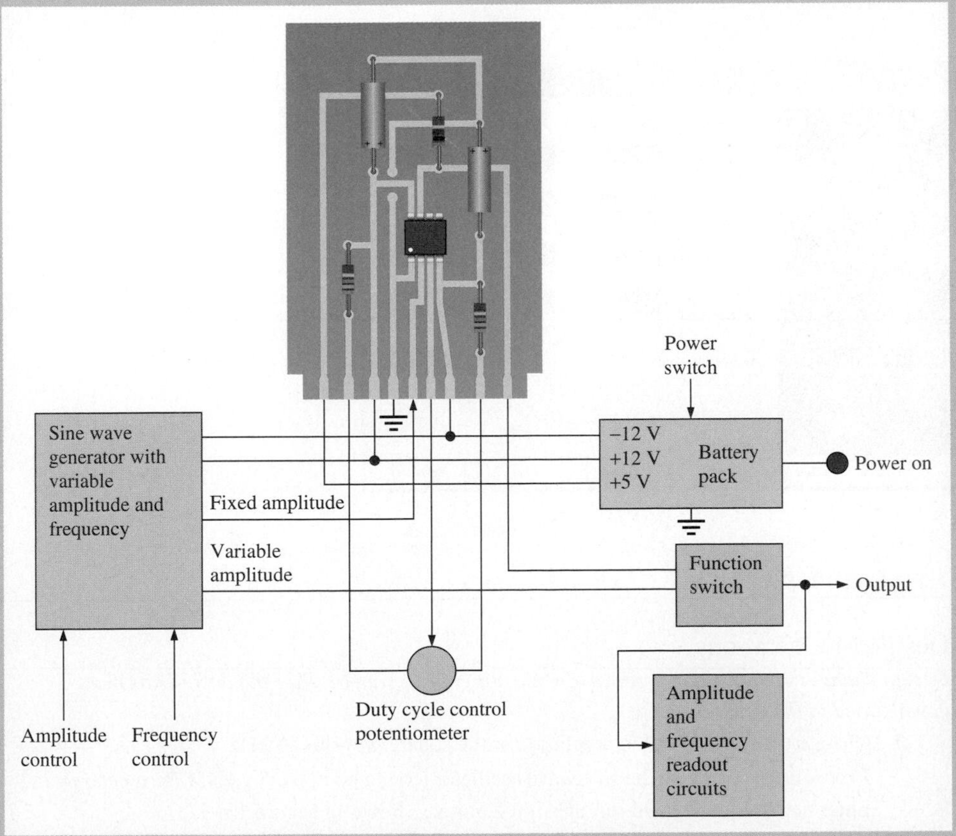

FIGURE 13–55

System diagram.

Programmable Analog Design

Design Assignment

Create a square-wave generator using a sine wave oscillator and a comparator.

Procedure: Open your Designer2 software and configure and test the FPAA using the default settings as shown in Figure 13–56.

Analysis: Determine the amplitude and frequency of the output in Figure 13–56.

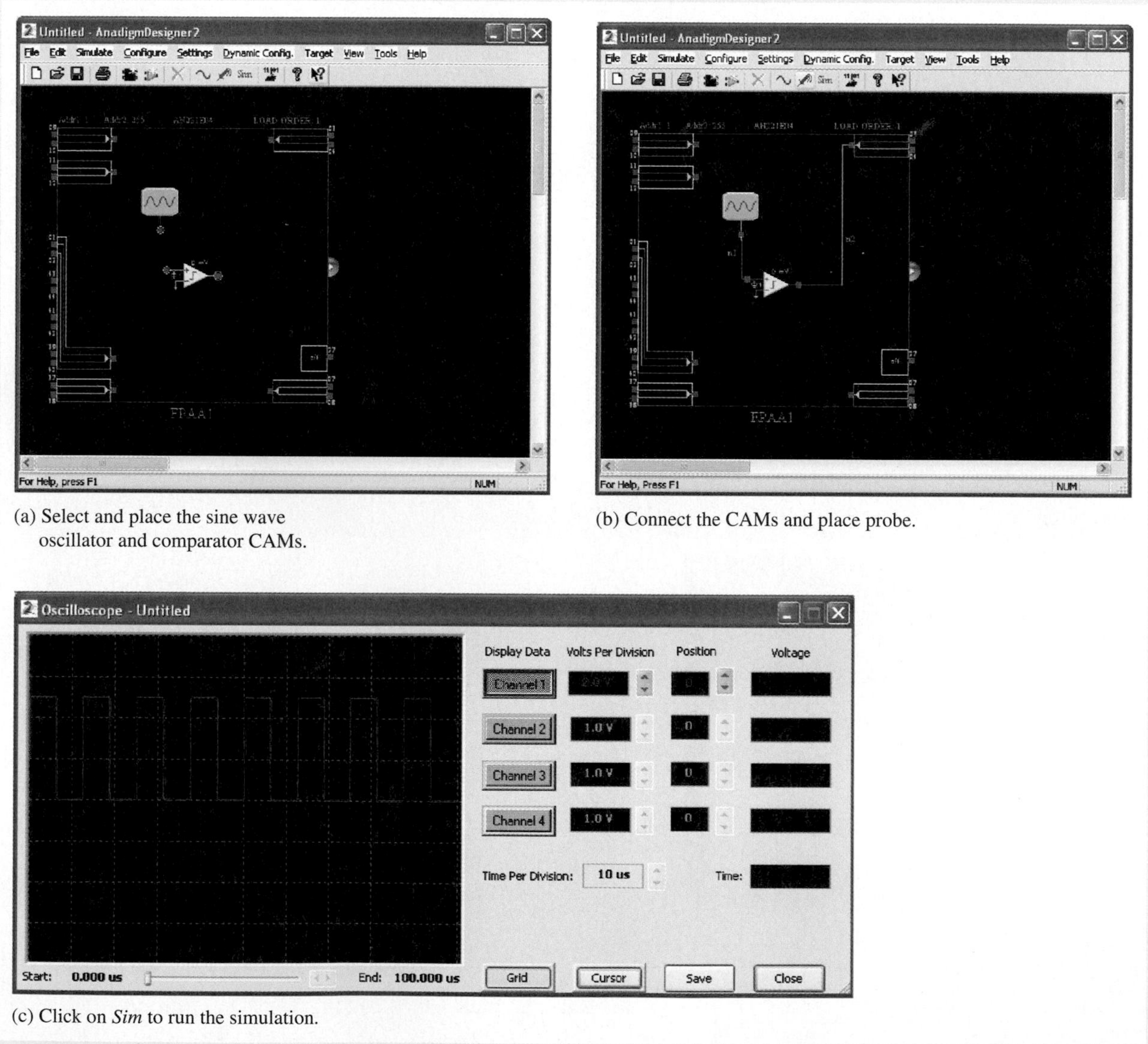

(a) Select and place the sine wave oscillator and comparator CAMs.

(b) Connect the CAMs and place probe.

(c) Click on *Sim* to run the simulation.

▲ FIGURE 13–56

Design Modifications

Design changes can be made prior to downloading the design or after the design has been downloaded to the chip.

1. Change the frequency and amplitude of the square wave to 50 kHz.

   Procedure: Click on the sine wave oscillator icon to open the *Set CAM Parameters* window. Enter new frequency and run the simulation, as shown in Figure 13–57.

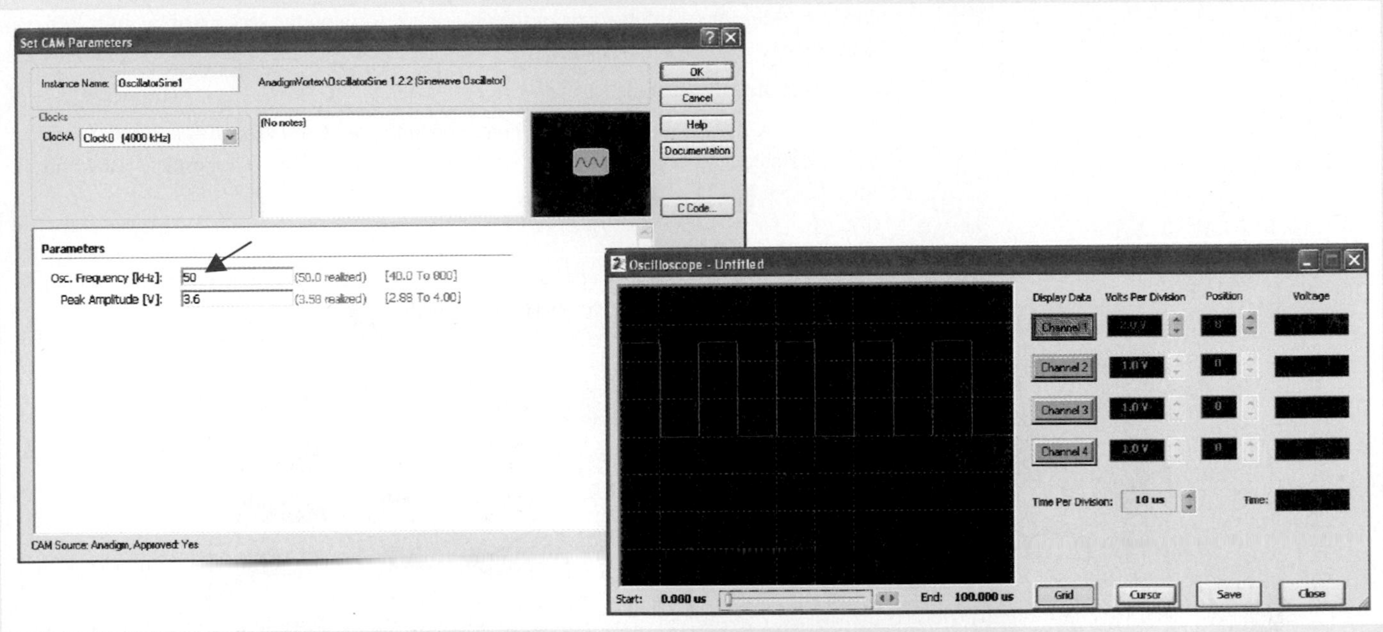

FIGURE 13–57

Analysis: Verify the frequency of the square wave on the scope display in Figure 13–57.

2. Decrease the duty cycle of the square wave.

   Procedure: Click on the comparator icon to open the *Set CAM Parameters* window. Choose variable reference, and set it to +3 V, as shown in Figure 13–58. Run the simulation. (The reference can be set to any voltage within the specified range to achieve a desired duty cycle.)

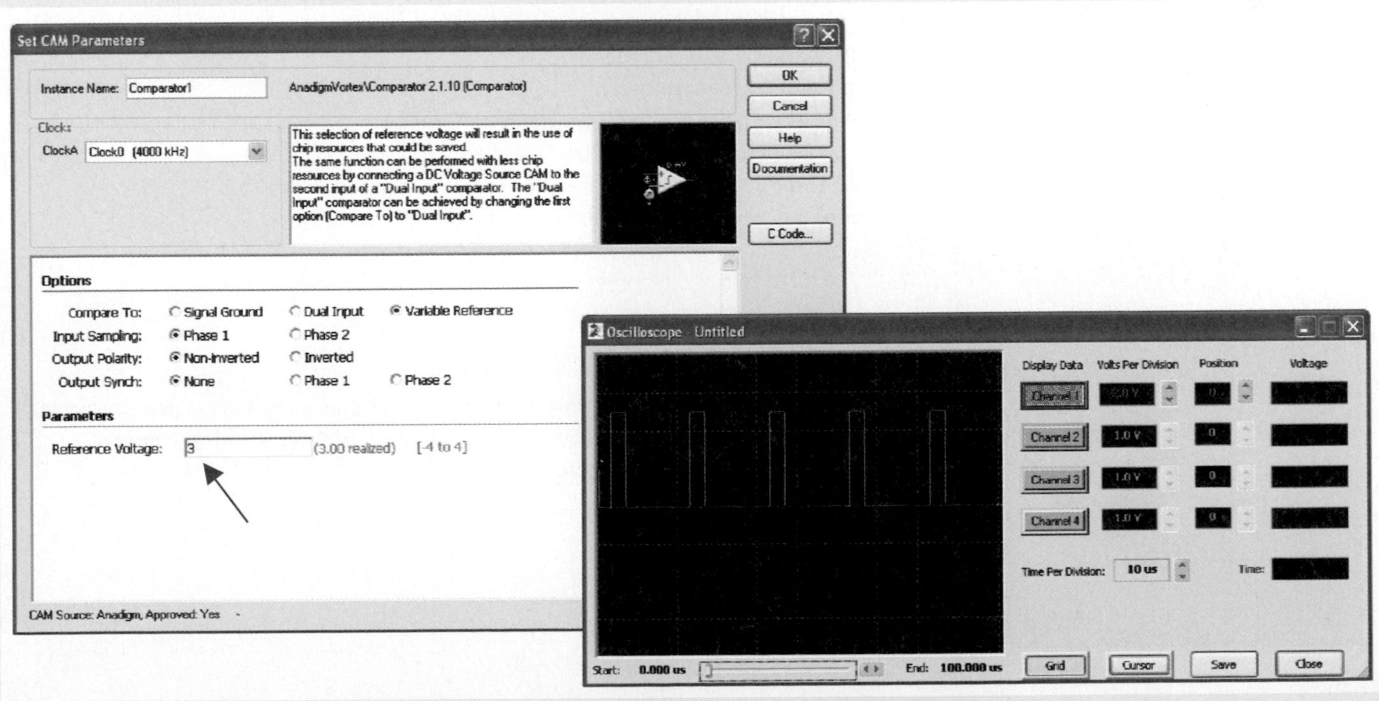

FIGURE 13–58

Analysis: Measure the duty cycle of the pulse waveform on the scope display in Figure 13–58.

3. Increase the duty cycle of the square wave.

Procedure: Click on the comparator icon to open the *Set CAM Parameters* window. Choose variable reference, and set it to −3 V. Run the simulation. The result is shown in Figure 13–59.

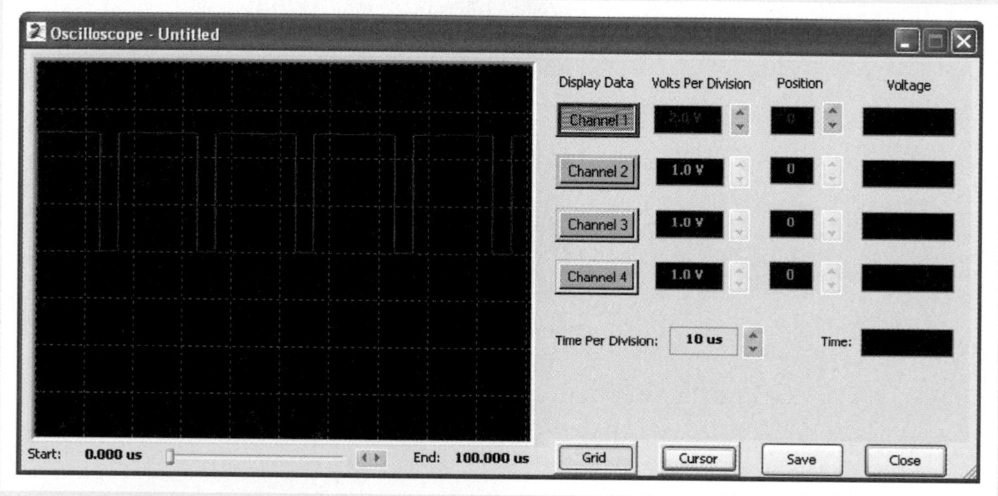

▲ FIGURE 13–59

Analysis: Measure the duty cycle of the pulse waveform in Figure 13–59.

Programming Exercises

1. Open your Designer2 software.
2. Implement the pulse waveform generator described.
3. Change the frequency to 100 kHz.
4. Decrease the duty cycle to less than that shown in Figure 13–58.
5. Increase the duty cycle to greater than that shown in Figure 13–59.

PAM Experiment

To program, download, and test a circuit using AnadigmDesigner2 software and the programmable analog module (PAM) board, go to Experiment 13–B in *Laboratory Exercises for Electronic Devices* by David Buchla and Steven Wetterling.

SUMMARY OF COMPARATORS AND OP-AMP CIRCUITS

COMPARATORS

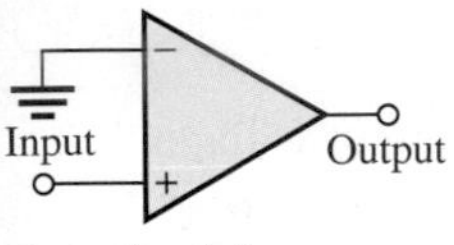

Zero-level detector

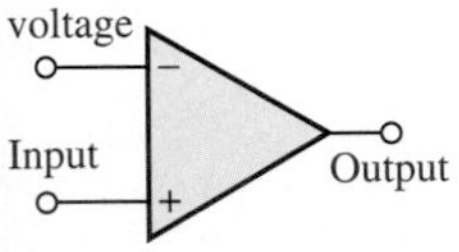

Nonzero-level detector

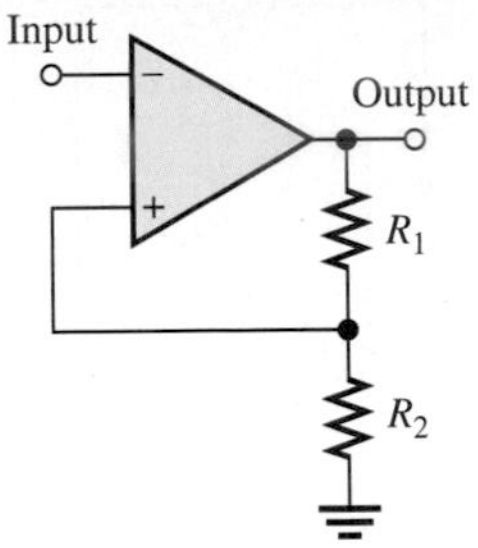

$$V_{UTP} = \frac{R_2}{R_1 + R_2}(+V_{out(max)})$$

$$V_{LTP} = \frac{R_2}{R_1 + R_2}(-V_{out(max)})$$

Comparator with hysteresis

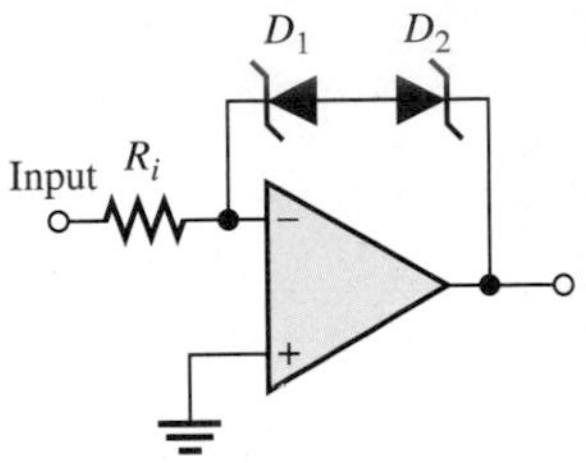

$$+V_{out(max)} = V_{Z1} + 0.7\text{ V}$$

$$-V_{out(max)} = -(V_{Z2} + 0.7\text{ V})$$

Bounded comparator

SUMMING AMPLIFIER

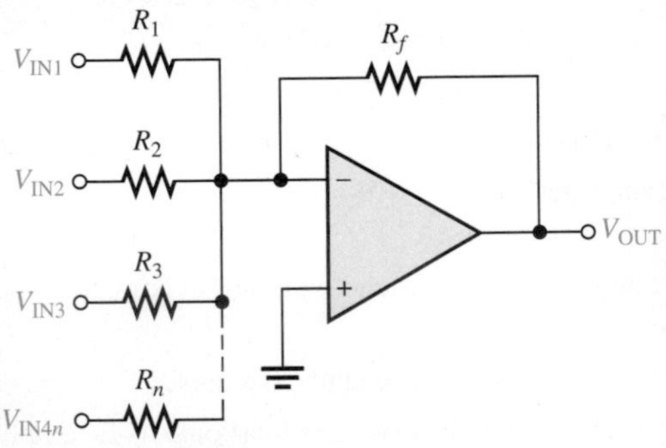

Summing amplifier

- Unity-gain amplifier.

$$R_f = R_1 = R_2 = R_3 = \cdots = R_n$$

$$V_{OUT} = -(V_{IN1} + V_{IN2} + V_{IN3} + \cdots + V_{INn})$$

- Greater than unity-gain amplifier:

$$R_f > R$$

$$R = R_1 = R_2 = R_3 = \cdots = R_n$$

$$V_{OUT} = -\frac{R_f}{R}(V_{IN1} + V_{IN2} + V_{IN3} + \cdots + V_{INn})$$

- Averaging amplifier:

$$\frac{R_f}{R} = \frac{1}{n}$$

$$R = R_1 = R_2 = R_3 = \cdots = R_n$$

$$V_{OUT} = -\frac{R_f}{R}(V_{IN1} + V_{IN2} + V_{IN3} + \cdots + V_{INn})$$

- Scaling adder:

$$V_{OUT} = -\left(\frac{R_f}{R_1}V_{IN1} + \frac{R_f}{R_2}V_{IN2} + \frac{R_f}{R_3}V_{IN3} + \cdots + \frac{R_f}{R_n}V_{INn}\right)$$

INTEGRATOR AND DIFFERENTIATOR

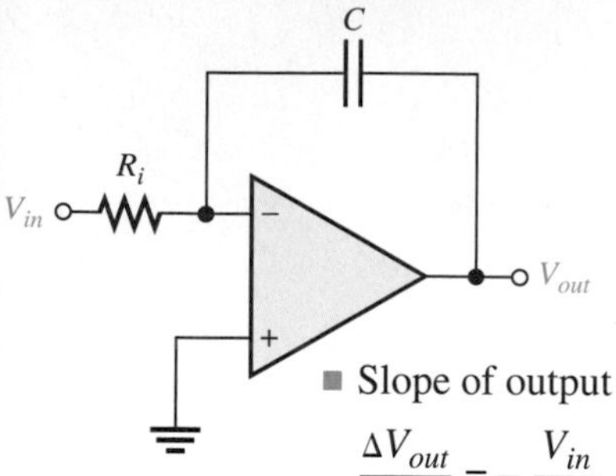

- Slope of output voltage:

$$\frac{\Delta V_{out}}{\Delta t} = -\frac{V_{in}}{R_iC}$$

Integrator

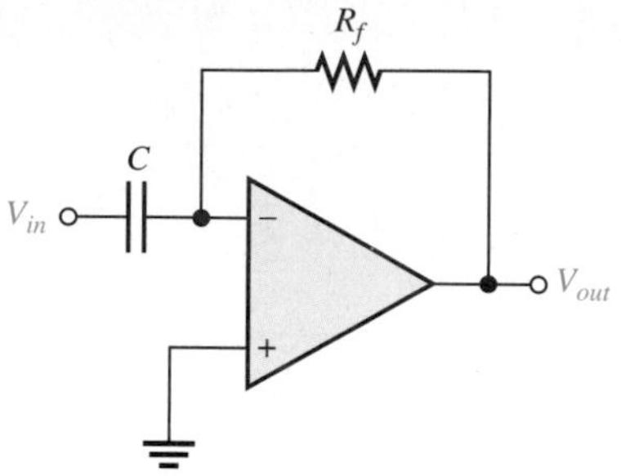

- Output voltage:

$$V_{out} = -\left(\frac{V_C}{t}\right)R_fC$$

$$V_C = V_{in}$$

Differentiator

SUMMARY

Section 13–1

- In an op-amp comparator, when the input voltage exceeds a specified reference voltage, the output changes state.
- Hysteresis gives an op-amp noise immunity.
- A comparator switches to one state when the input reaches the upper trigger point (UTP) and back to the other state when the input drops below the lower trigger point (LTP).
- The difference between the UTP and the LTP is the hysteresis voltage.
- Bounding limits the output amplitude of a comparator.

Section 13–2

- The output voltage of a summing amplifier is proportional to the sum of the input voltages.
- An averaging amplifier is a summing amplifier with a closed-loop gain equal to the reciprocal of the number of inputs.
- In a scaling adder, a different weight can be assigned to each input, thus making the input contribute more or contribute less to the output.

Section 13–3

- Integration is a mathematical process for determining the area under a curve.
- Integration of a step input produces a ramp output with a slope proportional to the amplitude.
- Differentiation is a mathematical process for determining the rate of change of a function.
- Differentiation of a ramp input produces a step output with an amplitude proportional to the slope.

KEY TERMS

Key terms and other bold terms in the chapter are defined in the end-of-book glossary.

Bounding The process of limiting the output range of an amplifier or other circuit.

Comparator A circuit that compares two input voltages and produces an output in either of two states indicating the greater than or less than relationship of the inputs.

Differentiator A circuit that produces an output which approximates the instantaneous rate of change of the input function.

Hysteresis Characteristic of a circuit in which two different trigger levels create an offset or lag in the switching action.

Integrator A circuit that produces an output which approximates the area under the curve of the input function.

Schmitt trigger A comparator with built-in hysteresis.

Summing amplifier An op-amp configuration with two or more inputs that produces an output voltage that is proportional to the negative of the algebraic sum of its input voltages.

KEY FORMULAS

Comparator

13–1	$V_{UTP} = \frac{R_2}{R_1 + R_2}(+V_{out(max)})$	Upper trigger point
13–2	$V_{LTP} = \frac{R_2}{R_1 + R_2}(-V_{out(max)})$	Lower trigger point
13–3	$V_{HYS} = V_{UTP} - V_{LTP}$	Hysteresis voltage

Summing Amplifier

13–4	$V_{OUT} = -(V_{IN1} + V_{IN2} + \cdots + V_{INn})$	n-input adder
13–5	$V_{OUT} = -\frac{R_f}{R}(V_{IN1} + V_{IN2} + \cdots + V_{INn})$	Adder with gain
13–6	$V_{OUT} = -\left(\frac{R_f}{R_1}V_{IN1} + \frac{R_f}{R_2}V_{IN2} + \cdots + \frac{R_f}{R_n}V_{INn}\right)$	Scaling adder with gain

Integrator and Differentiator

13–7	$\frac{\Delta V_{out}}{\Delta t} = -\frac{V_{in}}{R_iC}$	Integrator output rate of change
13–8	$V_{out} = -\left(\frac{V_C}{t}\right)R_fC$	Differentiator output voltage with ramp input

TRUE/FALSE QUIZ

Answers are at the end of the chapter.

1. The output of a comparator has two states.
2. The reference voltage on a comparator input establishes the gain.
3. Hysteresis incorporates positive feedback.
4. A comparator with hyteresis has two trigger points.
5. A summing amplifier can have more than two inputs.
6. The gain of a summing amplifier must always be unity (1).
7. DAC stands for digital-to-analog comparator.
8. An *R*/2*R* ladder circuit is one form of DAC.
9. An integrator produces a ramp when a step input is applied.
10. In a practical integrator, a resistor is connected across the capacitor.
11. When a triangular waveform is applied to a differential, a sine wave appears on the output.
12. In a practical differentiator, a resistor is connected in series with the capacitor.

CIRCUIT-ACTION QUIZ

Answers are at the end of the chapter.

1. If R_2 opens in the comparator of Figure 13–3, the output voltage amplitude will
   (a) increase **(b)** decrease **(c)** not change
2. In the trigger circuit of Figure 13–9, if R_1 is decreased to 50 kΩ, the upper trigger-point voltage will
   (a) increase **(b)** decrease **(c)** not change

3. If the zener diodes in Figure 13–13 are changed to ones with a rating of 5.6 V, the output voltage amplitude will
   (a) increase (b) decrease˜ (c) not change
4. If the top resistor in Figure 13–22 opens, the output voltage will
   (a) increase (b) decrease (c) not change
5. If V_{IN2} is changed to -1 V in Figure 13–22, the output voltage will
   (a) increase (b) decrease (c) not change
6. If V_{IN1} is increased to 0.4 V and V_{IN2} is reduced to 0.3 V in Figure 13–23, the output voltage will
   (a) increase (b) decrease (c) not change
7. If V_{IN3} is changed to -7 V in Figure 13–24, the output voltage will
   (a) increase (b) decrease (c) not change
8. If R_f in Figure 13–25 opens, the output voltage will
   (a) increase (b) decrease (c) not change
9. If the value of C in Figure 13–35 is reduced, the frequency of the output waveform will
   (a) increase (b) decrease (c) not change
10. If the frequency of the input waveform in Figure 13–40 is increased, the amplitude of the output voltage will
   (a) increase (b) decrease (c) not change

SELF-TEST

Answers are at the end of the chapter.

Section 13–1

1. In a zero-level detector, the output changes state when the input
   (a) is positive (b) is negative (c) crosses zero (d) has a zero rate of change
2. The zero-level detector is one application of a
   (a) comparator (b) differentiator (c) summing amplifier (d) diode
3. Noise on the input of a comparator can cause the output to
   (a) hang up in one state
   (b) go to zero
   (c) change back and forth erratically between two states
   (d) produce the amplified noise signal
4. The effects of noise can be reduced by
   (a) lowering the supply voltage (b) using positive feedback
   (c) using negative feedback (d) using hysteresis
   (e) answers (b) and (d)
5. A comparator with hysteresis
   (a) has one trigger point (b) has two trigger points
   (c) has a variable trigger point (d) is like a magnetic circuit
6. In a comparator with hysteresis,
   (a) a bias voltage is applied between the two inputs
   (b) only one supply voltage is used
   (c) a portion of the output is fed back to the inverting input
   (d) a portion of the output is fed back to the noninverting input
7. Using output bounding in a comparator
   (a) makes it faster (b) keeps the output positive
   (c) limits the output levels (d) stabilizes the output

Section 13–2

8. A summing amplifier can have
   a. only one input (b) only two inputs (c) any number of inputs

9. If the voltage gain for each input of a summing amplifier with a 4.7 kΩ feedback resistor is unity, the input resistors must have a value of
   (a) 4.7 kΩ
   (b) 4.7 kΩ divided by the number of inputs
   (c) 4.7 kΩ times the number of inputs

10. An averaging amplifier has five inputs. The ratio R_f/R_i must be
    (a) 5 (b) 0.2 (c) 1

11. In a scaling adder, the input resistors are
    (a) all the same value (b) all of different values
    (c) each proportional to the weight of its input (d) related by a factor of two

Section 13–3

12. In an ideal integrator, the feedback element is a
    (a) resistor (b) capacitor (c) zener diode (d) voltage divider

13. For a step input, the output of an integrator is
    (a) a pulse (b) a triangular waveform (c) a spike (d) a ramp

14. The rate of change of an integrator's output voltage in response to a step input is set by
    (a) the *RC* time constant (b) the amplitude of the step input
    (c) the current through the capacitor (d) all of these

15. In a differentiator, the feedback element is a
    (a) resistor (b) capacitor (c) zener diode (d) voltage divider

16. The output of a differentiator is proportional to
    (a) the *RC* time constant (b) the rate at which the input is changing
    (c) the amplitude of the input (d) answers (a) and (b)

17. When you apply a triangular waveform to the input of a differentiator, the output is
    (a) a dc level (b) an inverted triangular waveform
    (c) a square waveform (d) the first harmonic of the triangular waveform

PROBLEMS

Answers to all odd-numbered problems are at the end of the book.

BASIC PROBLEMS

Section 13–1 **Comparators**

1. A certain op-amp has an open-loop gain of 80,000. The maximum saturated output levels of this particular device are ±12 V when the dc supply voltages are ±15 V. If a differential voltage of 0.15 mV rms is applied between the inputs, what is the peak-to-peak value of the output?

2. Determine the output level (maximum positive or maximum negative) for each comparator in Figure 13–60.

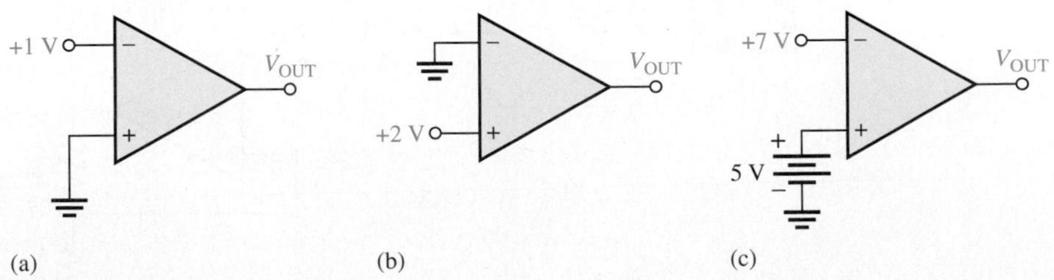

▲ FIGURE 13–60

▶ FIGURE 13–61

Multisim file circuits are identified with a CD logo and are in the Problems folder on your CD-ROM. Filenames correspond to figure numbers (e.g., F13–61).

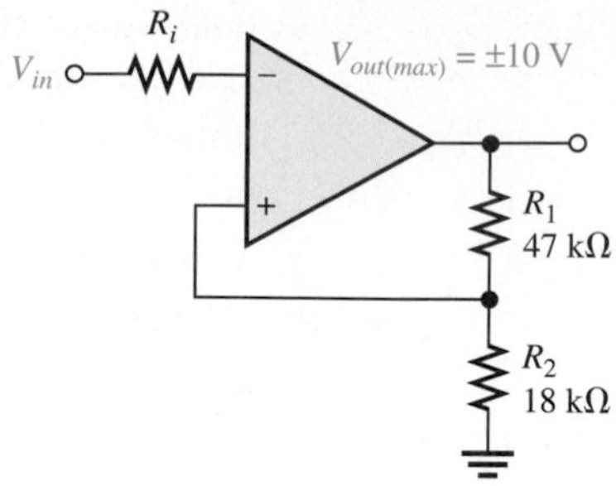

3. Calculate the V_{UTP} and V_{LTP} in Figure 13–61. $V_{out(max)} = \pm 10$ V.
4. What is the hysteresis voltage in Figure 13–61?
5. Draw the output voltage waveform for each circuit in Figure 13–62 with respect to the input. Show voltage levels.

▶ FIGURE 13–62

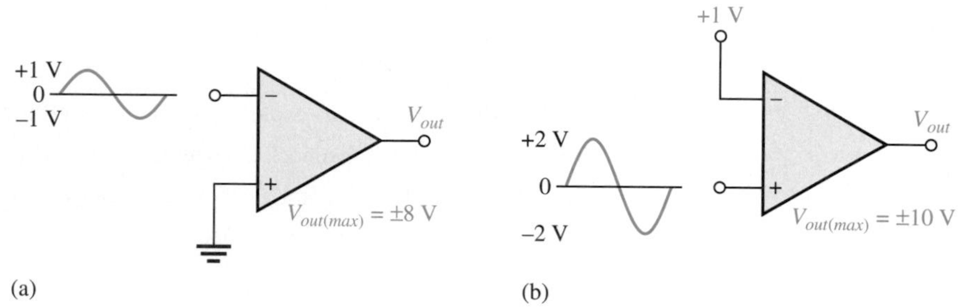

6. Determine the hysteresis voltage for each comparator in Figure 13–63. The maximum output levels are ±11 V.

▶ FIGURE 13–63

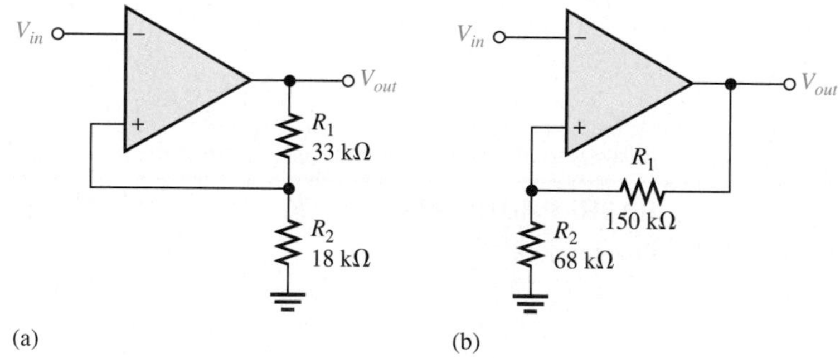

7. A 6.2 V zener diode is connected from the output to the inverting input in Figure 13–61 with the cathode at the output. What are the positive and negative output levels?
8. Determine the output voltage waveform in Figure 13–64.

▶ FIGURE 13–64

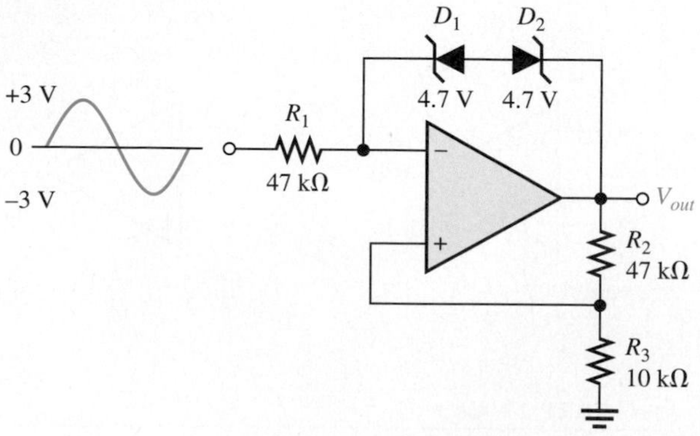

Section 13–2 Summing Amplifiers

9. Determine the output voltage for each circuit in Figure 13–65.

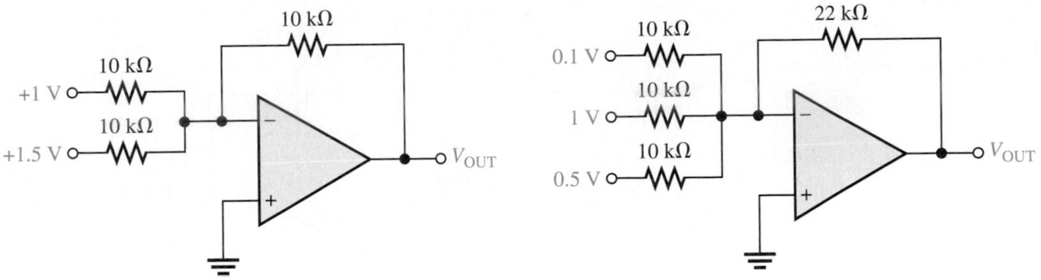

(a) (b)

▲ **FIGURE 13–65**

10. Refer to Figure 13–66. Determine the following.

(a) V_{R1} and V_{R2} **(b)** Current through R_f **(c)** V_{OUT}

▶ **FIGURE 13–66**

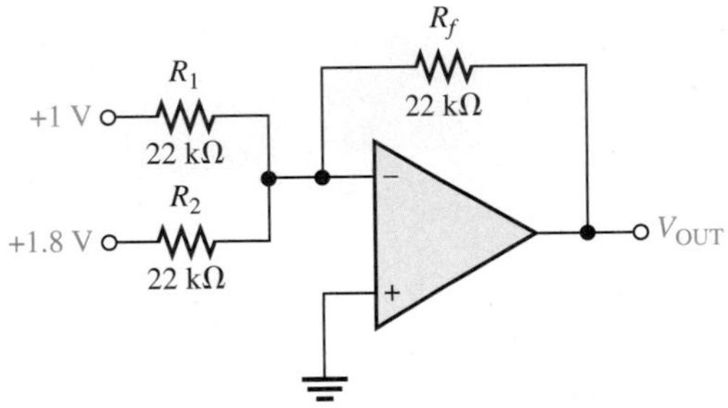

11. Find the value of R_f necessary to produce an output that is five times the sum of the inputs in Figure 13–66.

12. Show a summing amplifier that will average eight input voltages. Use input resistances of 10 kΩ each.

13. Find the output voltage when the input voltages shown in Figure 13–67 are applied to the scaling adder. What is the current through R_f?

14. Determine the values of the input resistors required in a six-input scaling adder so that the lowest weighted input is 1 and each successive input has a weight twice the previous one. Use $R_f = 100\text{ k}\Omega$.

▶ **FIGURE 13–67**

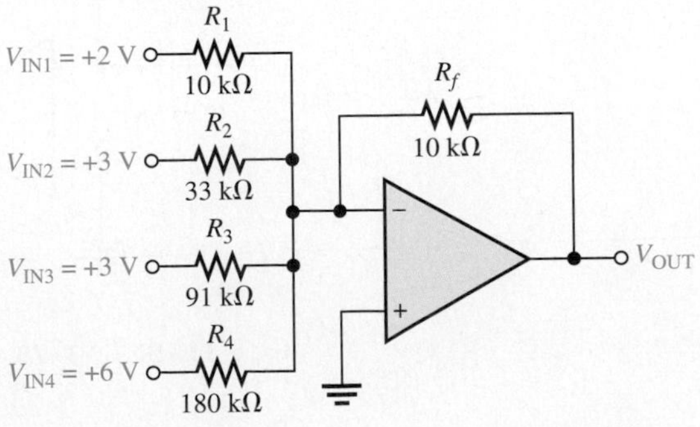

Section 13–3 Integrators and Differentiators

15. Determine the rate of change of the output voltage in response to the step input to the integrator in Figure 13–68.

▶ FIGURE 13–68

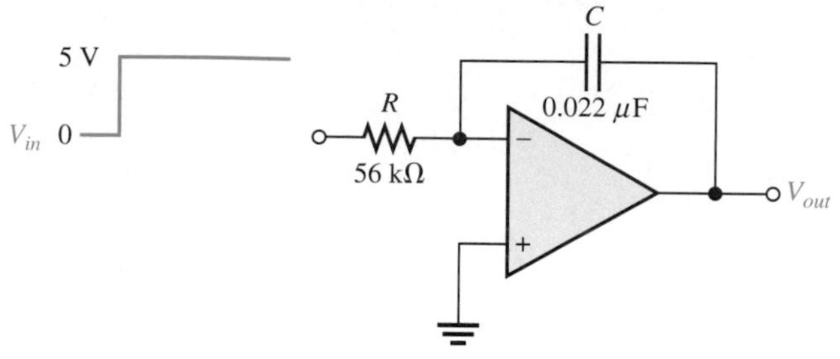

16. A triangular waveform is applied to the input of the circuit in Figure 13–69 as shown. Determine what the output should be and sketch its waveform in relation to the input.

▶ FIGURE 13–69

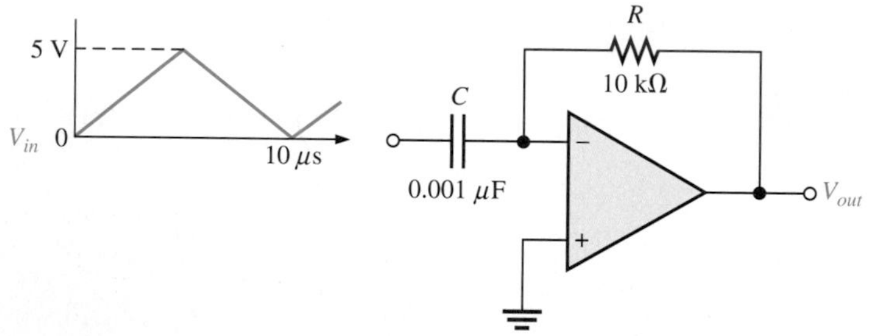

17. What is the magnitude of the capacitor current in Problem 16?

18. A triangular waveform with a peak-to-peak voltage of 2 V and a period of 1 ms is applied to the differentiator in Figure 13–70(a). What is the output voltage?

19. Beginning in position 1 in Figure 13–70(b), the switch is thrown into position 2 and held there for 10 ms, then back to position 1 for 10 ms, and so forth. Sketch the resulting output waveform if its initial value is 0 V. The saturated output levels of the op-amp are ±12 V.

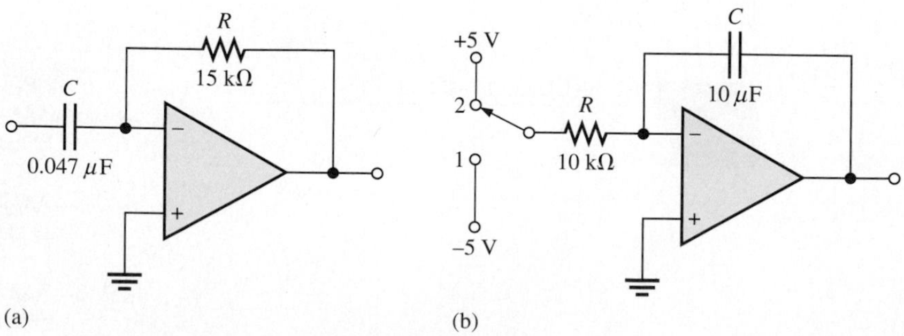

▲ FIGURE 13–70

Section 13–4 **Troubleshooting**

20. The waveforms given in Figure 13–71(a) are observed at the indicated points in Figure 13–71(b). Is the circuit operating properly? If not, what is a likely fault?

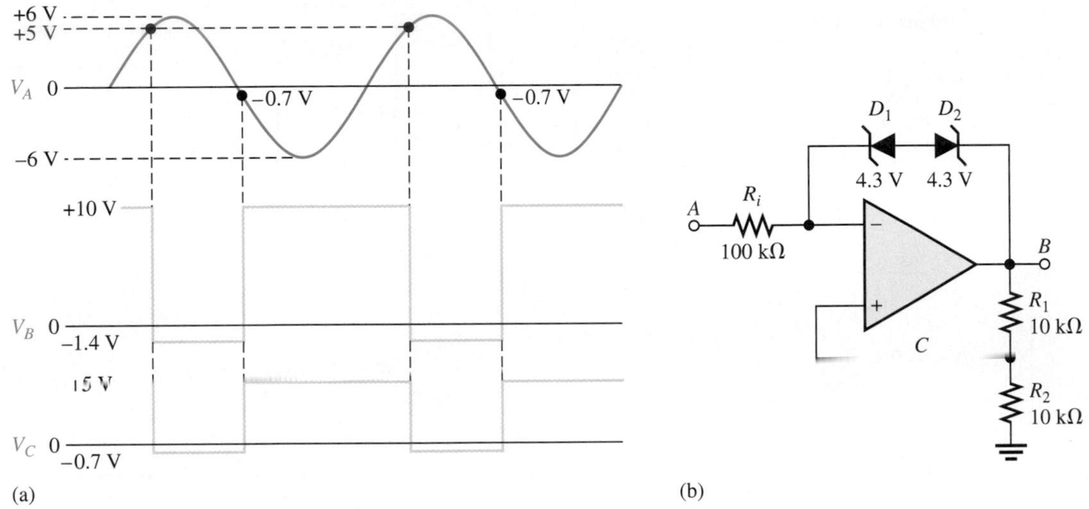

▲ FIGURE 13–71

21. The sequences of voltage levels shown in Figure 13–72 are applied to the summing amplifier and the indicated output is observed. First, determine if this output is correct. If it is not correct, determine the fault.

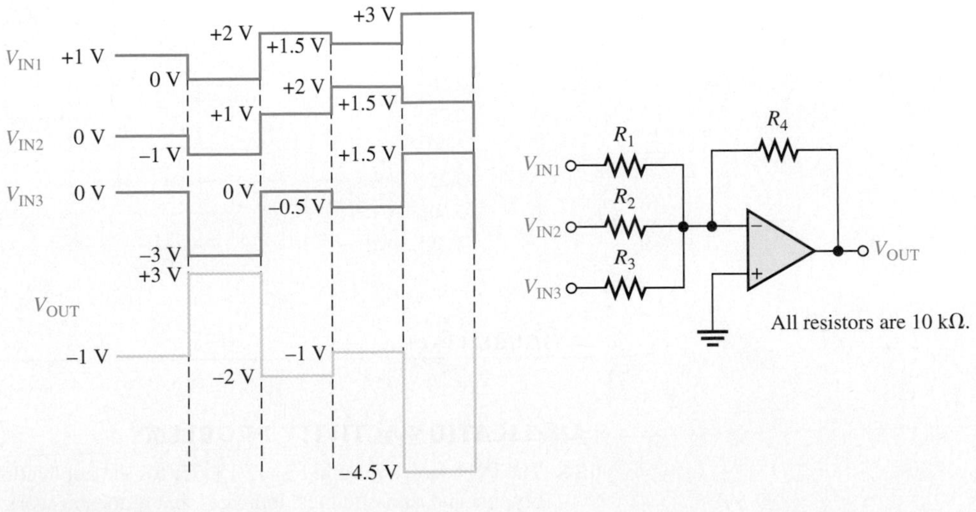

▲ FIGURE 13–72

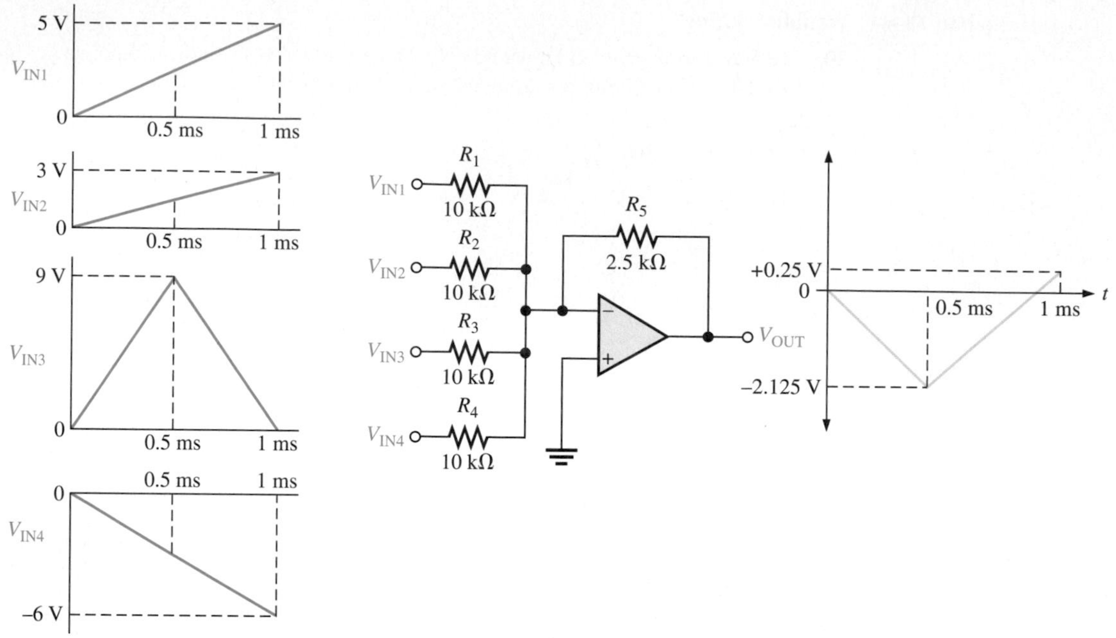

FIGURE 13–73

22. The given ramp voltages are applied to the op-amp circuit in Figure 13–73. Is the given output correct? If it isn't, what is the problem?

23. The DAC with inputs as shown in Figure 13–27 produces the output shown in Figure 13–74. Determine the fault in the circuit.

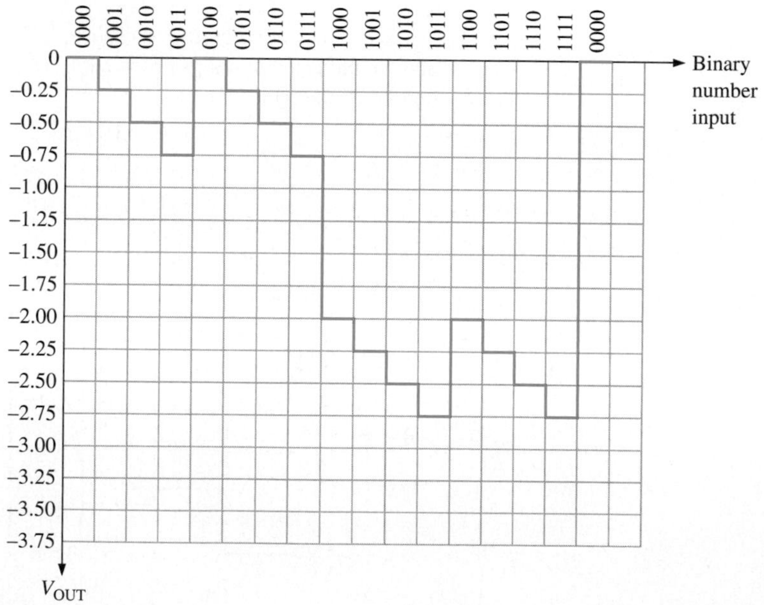

FIGURE 13–74

APPLICATION ACTIVITY PROBLEMS

24. The PC board, shown in Figure 13–75, for the application activity has just come off the assembly line and a pass/fail test indicates that it doesn't work. The board now comes to you for troubleshooting. What is the very first thing you should do? Can you isolate the problem(s) by this first step in this case?

FIGURE 13–75

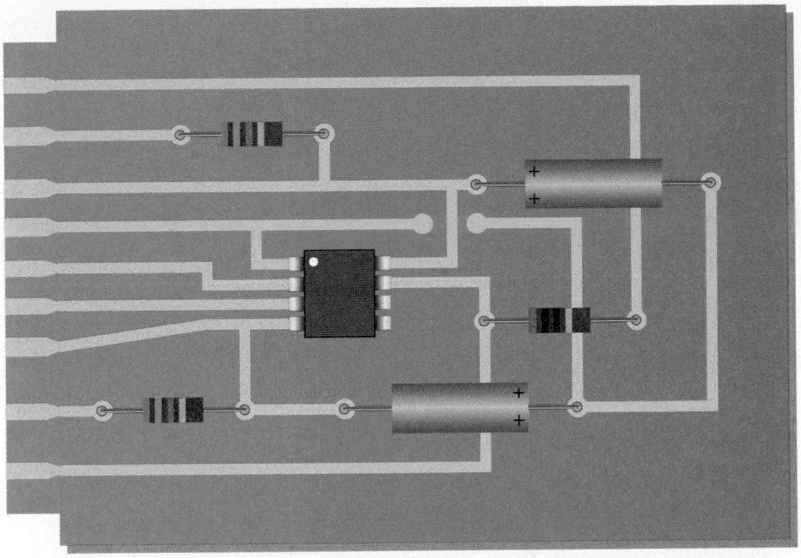

25. Describe the effect of an open decoupling capacitor on the PC board in Figure 13–75.
26. Assume that a 1.0 kΩ resistor is inadvertently used for R_1 in Figure 13–50. What effect does this have on the circuit operation?

ADVANCED PROBLEMS

27. Calculate the percent duty cycle in Figure 13–50 for minimum and maximum settings of R_2. A 10 V peak sine wave is applied to the noninverting input of the comparator with no hysteresis.
28. Redesign the circuit in Figure 13–50 for a 5 V peak sine wave.
29. Design an integrator that will produce an output voltage with a slope of 100 mV/μs when the input voltage is a constant 5 V. Specify the input frequency of a square wave with an amplitude of 5 V that will result in a 5 V peak-to-peak triangular wave output.

MULTISIM TROUBLESHOOTING PROBLEMS

These file circuits are in the Troubleshooting Problems folder on your CD-ROM.

30. Open file TSP13-30 and determine the fault.
31. Open file TSP13-31 and determine the fault.
32. Open file TSP13-32 and determine the fault.
33. Open file TSP13-33 and determine the fault.
34. Open file TSP13-34 and determine the fault.
35. Open file TSP13-35 and determine the fault.
36. Open file TSP13-36 and determine the fault.
37. Open file TSP13-37 and determine the fault.
38. Open file TSP13-38 and determine the fault.
39. Open file TSP13-39 and determine the fault.

ANSWERS

SECTION CHECKUPS

Section 13–1 **Comparators**

1. **(a)** $V = (10\,\text{k}\Omega/110\,\text{k}\Omega)15\,\text{V} = 1.36\,\text{V}$
   (b) $V = (22\,\text{k}\Omega/69\,\text{k}\Omega)(-12\,\text{V}) = -3.83\,\text{V}$

2. Hysteresis makes the comparator less susceptable to noise.
3. Bounding limits the output amplitude to a specified level.

Section 13–2 Summing Amplifiers

1. The summing point is the point where the input resistors are commonly connected.
2. $R_f/R = 1/5 = 0.2$
3. 5 kΩ

Section 13–3 Integrators and Differentiators

1. The feedback element in an ideal integrator is a capacitor.
2. The capacitor voltage is linear because the capacitor current is constant.
3. The feedback element in a differentiator is a resistor.
4. The output of a differentiator is proportional to the rate of change of the input.

Section 13–4 Troubleshooting

1. An op-amp can fail with a shorted output.
2. Replace suspected components one by one.

RELATED PROBLEMS FOR EXAMPLES

13–1 1.96 V
13–2 +3.83 V; −3.83 V
13–3 +1.81 V; −1.81 V
13–4 More accurately
13–5 −11.5 V
13–6 −5.73 V
13–7 Changes require an additional 100 kΩ input resistor and a change of R_f to 20 kΩ.
13–8 0.45, 0.12, 0.18; $V_{OUT} = -3.03$ V
13–9 Yes. All should be doubled.
13–10 Change C to 5000 pF.
13–11 Same waveform but with an amplitude of 6.6 V
13–12 A pulse from −0.88 V to +7.79 V
13–13 −3.76 V
13–14 Change R_6 to 25 kΩ.

TRUE/FALSE QUIZ

1. T **2.** F **3.** T **4.** T **5.** T **6.** F **7.** F **8.** T
9. T **10.** T **11.** F **12.** T

CIRCUIT-ACTION QUIZ

1. (b) **2.** (a) **3.** (a) **4.** (b) **5.** (b) **6.** (c) **7.** (b)
8. (a) **9.** (c) **10.** (a)

SELF-TEST

1. (c) **2.** (a) **3.** (c) **4.** (e) **5.** (b) **6.** (d)
7. (c) **8.** (c) **9.** (a) **10.** (b) **11.** (c) **12.** (b)
13. (d) **14.** (d) **15.** (a) **16.** (d) **17.** (c)

Special-Purpose Op-Amp Circuits

14

Chapter Outline

Chapter Objectives

- Analyze and explain the operation of an instrumentation amplifier
- Analyze and explain the operation of an isolation amplifier
- Analyze and explain the operation of an OTA
- Analyze and explain the operation of log and antilog amplifiers
- Analyze and explain several special types of op-amp circuits

Key Terms

- Instrumentation amplifier
- Isolation amplifier
- Operational transconductance amplifier (OTA)
- Transconductance
- Natural Logarithm

Application Activity Preview

The Application Activity in this chapter describes a liquid-level control system for an industrial storage tank. A pressure sensor, which is a type of transducer, is used to detect a change in pressure in a tube inserted in the liquid. The voltage from the pressure sensor is sent to the control circuit that consists of an instrumentation amplifier and a comparator. When the liquid in the tank reaches a predetermined minimum level, the circuit causes a pump to turn on and refill the tank to a predetermined maximum level. Also, a method for minimizing the effects of noise in an industrial environment is introduced in this chapter and used in the application.

Visit the Companion Website

Study aids for this chapter are available at http://www.prenhall.com/floyd

Introduction

A general-purpose op-amp, such as the 741, is a versatile and widely used device. However, some specialized IC amplifiers are available that have certain features or characteristics oriented to special applications. Most of these devices are actually derived from the basic op-amp. These special circuits include the instrumentation amplifier that is used in high-noise environments, the isolation amplifier that is used in high-voltage and medical applications, the operational transconductance amplifier (OTA) that is used as a voltage-to-current amplifier, and the logarithmic amplifiers that are used for linearizing certain types of inputs and for mathematical operations. Log amplifiers are also used in communication systems, including fiber optics.

14–1 INSTRUMENTATION AMPLIFIERS

Instrumentation amplifiers are commonly used in environments with high common-mode noise such as in data acquisition systems where remote sensing of input variables is required.

After completing this section, you should be able to

- **Analyze and explain the operation of an instrumentation amplifier**
  - Explain how op-amps are connected to form an instrumentation amplifier
  - Describe how the voltage gain is set
  - Describe the features of the AD622 instrumentation amplifier
  - Discuss noise in instrumentation amplifiers
  - Describe the AD522 instrumentation amplifier

An **instrumentation amplifier** is a differential voltage-gain device that amplifies the difference between the voltages existing at its two input terminals. The main purpose of an instrumentation amplifier is to amplify small signals that may be riding on large common-mode voltages. The key characteristics are high input impedance, high common-mode rejection, low output offset, and low output impedance. The basic instrumentation amplifier is an integrated circuit that internally has three operational amplifiers and several resistors. The voltage gain is usually set with an external resistor.

A basic instrumentation amplifier is shown in Figure 14–1. Op-amps A1 and A2 are noninverting configurations that provide high input impedance and voltage gain. Op-amp A3 is used as a unity-gain differential amplifier with high-precision resistors that are all equal in value ($R_3 = R_4 = R_5 = R_6$).

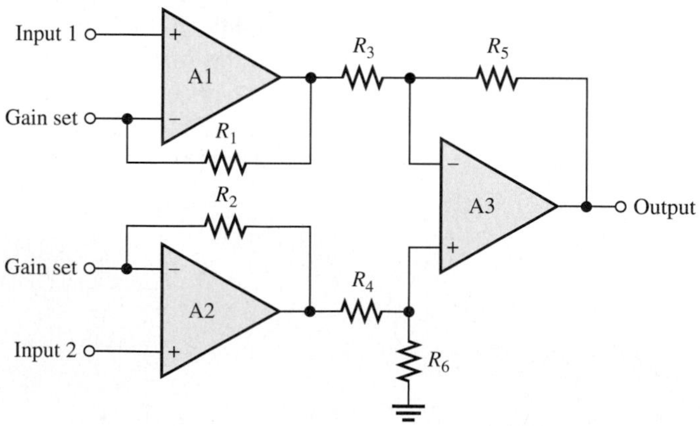

FIGURE 14–1

The basic instrumentation amplifier using three op-amps.

The gain-setting resistor, R_G, is connected externally as shown in Figure 14–2. Op-amp A1 receives the differential input signal V_{in1} on its noninverting (+) input and amplifies this signal with a voltage gain of

$$A_v = 1 + \frac{R_1}{R_G}$$

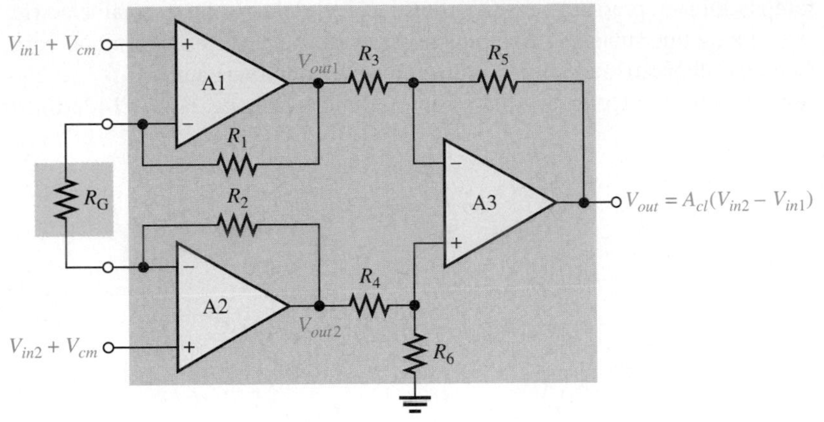

◀ **FIGURE 14–2**

The basic instrumentation amplifier with an external gain-setting resistor R_G. Differential and common-mode signals are indicated.

Op-amp A1 also has V_{in2} as an input signal to its inverting (−) input through op-amp A2 and the path formed by R_2 and R_G. The input signal V_{in2} is amplified by op-amp A1 with a voltage gain of

$$A_v = \frac{R_1}{R_G}$$

The overall closed-loop gain of the instrumentation amplifier is

$$\mathbf{A_{cl} = 1 + \frac{2R}{R_G}}$$

Equation 14–1

where $R_1 = R_2 = R$. Equation 14–1 shows that the gain of the instrumentation amplifier can be set by the value of the external resistor R_G when R_1 and R_2 have a known fixed value. See Appendix B for the derivation.

The external gain-setting resistor R_G can be calculated for a desired voltage gain by using Equation 14–1.

$$\mathbf{R_G = \frac{2R}{A_{cl} - 1}}$$

Equation 14–2

Instrumentation amplifiers in which the gain is set to specific values using a binary input instead of a resistor are also available.

EXAMPLE 14–1

Determine the value of the external gain-setting resistor R_G for a certain IC instrumentation amplifier with $R_1 = R_2 = 25\text{ k}\Omega$. The closed-loop voltage gain is to be 500.

Solution

$$R_G = \frac{2R}{A_{cl} - 1} = \frac{50\text{ k}\Omega}{500 - 1} \cong \mathbf{100\ \Omega}$$

*Related Problem** What value of external gain-setting resistor is required for an instrumentation amplifier with $R_1 = R_2 = 39\text{ k}\Omega$ to produce a gain of 325?

*Answers are at the end of the chapter.

Applications

The instrumentation amplifier is normally used to measure small differential signal voltages that are superimposed on a common-mode voltage often much larger than the signal voltage. Applications include situations where a quantity is sensed by a remote device, such as a temperature- or pressure-sensitive transducer, and the resulting small electrical signal is sent over a long line subject to electrical noise that produces common-mode voltages in the line. The instrumentation amplifier at the end of the line must amplify the small signal from the remote sensor and reject the large common-mode voltage. Figure 14–3 illustrates this.

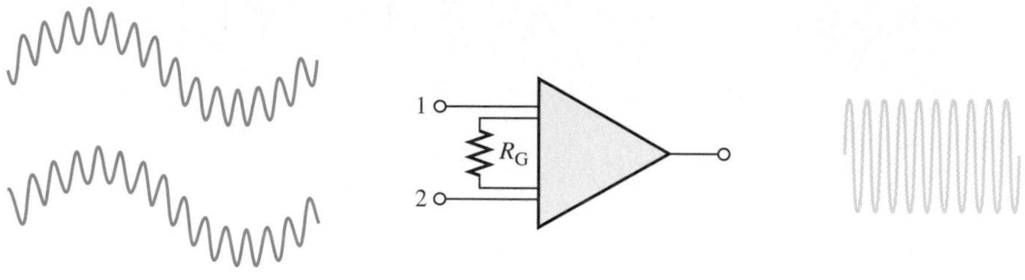

FIGURE 14–3

Illustration of the rejection of large common-mode voltages and the amplification of smaller signal voltages by an instrumentation amplifier.

A Specific Instrumentation Amplifier

Now that you have the basic idea of how an instrumentation amplifier works, let's look at a specific device. A representative device, the AD622, is shown in Figure 14–4 where IC pin numbers are given for reference. This instrumentation amplifier is based on the design using three op-amps that was shown in Figure 14–1.

FIGURE 14–4

The AD622 instrumentation amplifier.

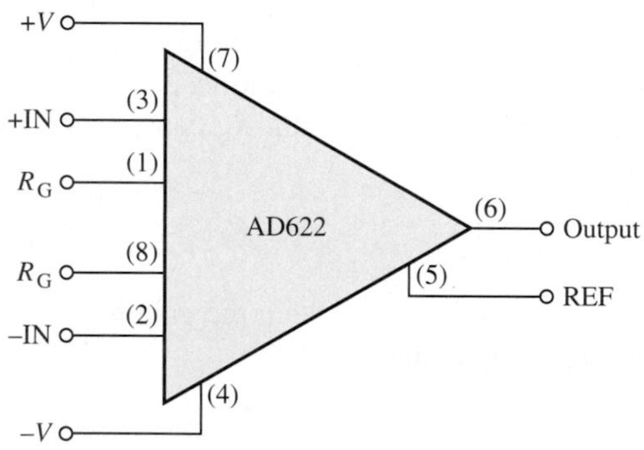

Some of the features of the AD622 are as follows. The voltage gain can be adjusted from 2 to 1000 with an external resistor R_G. There is unity gain with no external resistor. The input impedance is $10\,\text{G}\Omega$. The common-mode rejection ratio (CMRR) has a minimum value of 66 dB. Recall that a higher CMRR means better rejection of common-mode voltages. The AD622 has a bandwidth of 800 kHz at a gain of 10 and a slew rate of $1.2\ \text{V}/\mu\text{s}$.

Setting the Voltage Gain For the AD622, an external resistor must be used to achieve a voltage gain greater than unity, as indicated in Figure 14–5. Resistor R_G is connected between the R_G terminals (pins 1 and 8). No resistor is required for unity. R_G is selected for the desired gain based on the following formula:

$$R_G = \frac{50.5\ \text{k}\Omega}{A_v - 1}$$

Notice that this formula is the same as Equation 14–2 for the three-op-amp configuration with an external R_G where the internal resistors R_1 and R_2 are each $25.25\ \text{k}\Omega$.

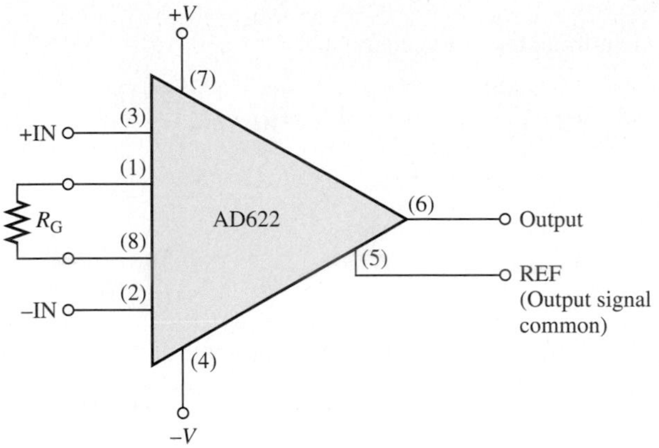

FIGURE 14–5

The AD622 with a gain-setting resistor.

Gain versus Frequency The graph in Figure 14–6 shows how the gain varies with frequency for gains of 1, 10, 100, and 1000. As you can see, the bandwidth decreases as the gain increases.

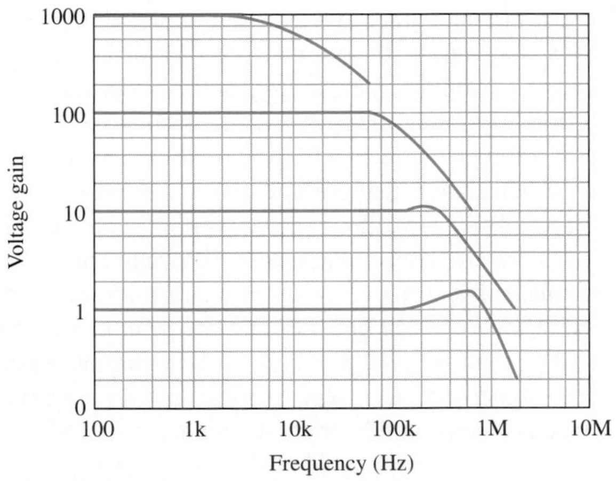

FIGURE 14–6

Gain versus frequency for the AD622 instrumentation amplifier.

EXAMPLE 14–2

Calculate the voltage gain and determine the bandwidth using the graph in Figure 14–6 for the instrumentation amplifier in Figure 14–7.

FIGURE 14–7

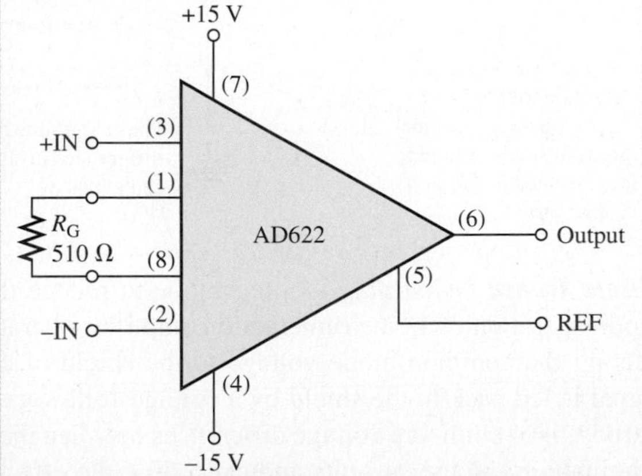

Solution Determine the voltage gain as follows:

$$R_G = \frac{50.5\ \text{k}\Omega}{A_v - 1}$$

$$A_v - 1 = \frac{50.5\ \text{k}\Omega}{R_G}$$

$$A_v = \frac{50.5\ \text{k}\Omega}{510\ \Omega} + 1 = \mathbf{100}$$

Determine the approximate bandwidth from the graph in Figure 14–6.

$$BW \approx \mathbf{60\ kHz}$$

Related Problem Modify the circuit in Figure 14–7 for a gain of approximately 45.

Noise Effects in Instrumentation Amplifier Applications

Various types of transducers are used to sense temperature, strain, pressure, and other parameters in many types of applications. Instrumentation amplifiers are generally used to process the small voltages produced by a transducer and often are used in noisy industrial environments where long cables connect the transducer output to the amplifier inputs. Noise in the form of common-mode signals picked up from external sources can be minimized, but not totally eliminated, by using coaxial cable in which the differential signal wires are surrounded by a metal mesh sheathing called a *shield.* As you know, in an electrically noisy environment any common-mode signals that are induced on the signal lines are rejected because both inputs to the amplifier have the same common-mode signal. However, when a shielded cable is used, there are stray capacitances distributed along its length between each signal line and the shield. The differences in these stray capacitances, particularly at higher frequencies, result in a phase shift between the two common-mode signals, as illustrated in Figure 14–8. The result is a degradation in the common-mode rejection of the amplifier because the two signals are no longer in phase and do not completely cancel so that a differential voltage is created at the amplifier inputs.

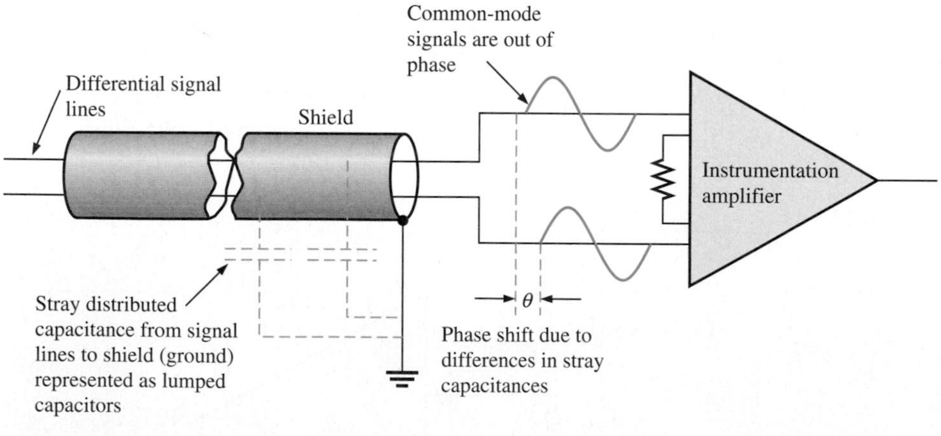

▶ **FIGURE 14–8**

Degradation of common-mode rejection in a shielded cable connection due to unwanted phase shifts.

Shield Guard *Guarding* is a technique to reduce the effects of noise on the common-mode operation of an instrumentation amplifier operating in critical environments by connecting the common-mode voltage to the shield of a coaxial cable. The common-mode signal is fed back to the shield by a voltage-follower stage, as shown in Figure 14–9. The purpose is to eliminate voltage differences between the signal lines and the shield, virtually eliminating leakage currents and cancelling the effects of the distributed capacitances so that the common-mode voltages are the same in both lines.

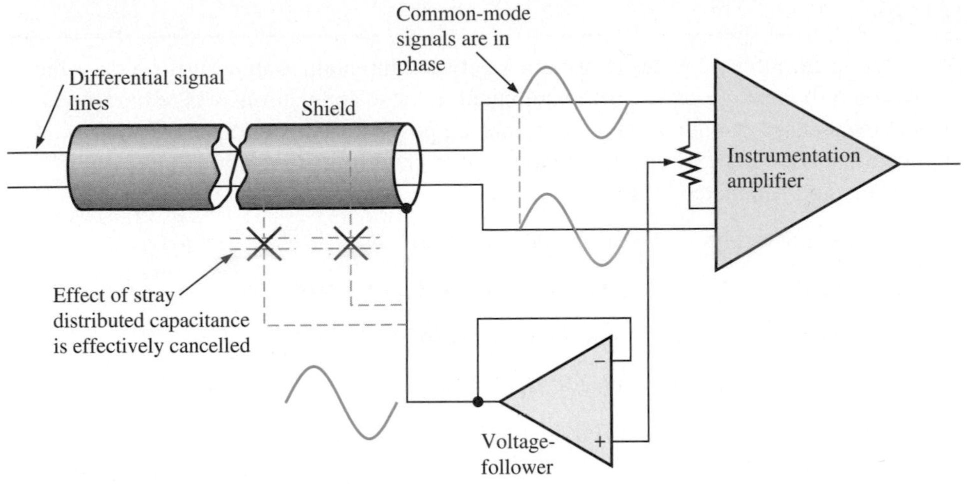

FIGURE 14–9

Instrumentation amplifier with shield guard to prevent degradation of the common-mode rejection.

The voltage-follower is a low-impedance source that drives the common-mode signal onto the shield to eliminate the voltage difference between the signal lines and the shield. When the voltage between each signal line and the shield is zero, the leakage currents are also zero and the capacitive reactances become infinitely large. An infinitely large X_C implies a zero capacitance.

A Specific Instrumentation Amplifier with a Guard Output

Most instrumentation amplifiers can be configured externally to provide a shield guard driver. Certain IC amplifiers, however, provide an internally generated guard output that is intended for very critical environments. An example is the AD522, shown in Figure 14–10, which is a precision IC instrumentation amplifier designed for applications requiring high accuracy under worst-case operating conditions and with very small signals. The pin labeled DATA GUARD is the shield-guard output.

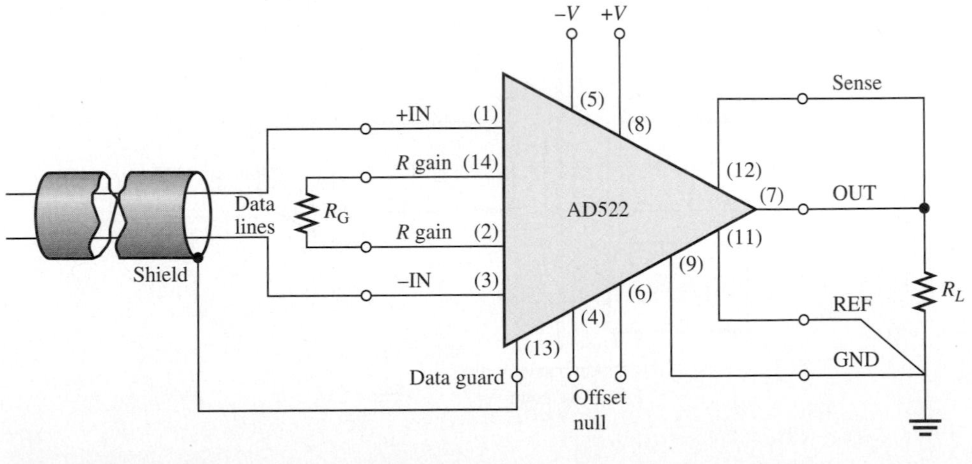

FIGURE 14–10

The AD522 instrumentation amplifier in a typical configuration.

SECTION 14–1 CHECKUP

Answers are at the end of the chapter.

1. What is the main purpose of an instrumentation amplifier and what are three of its key characteristics?
2. What components do you need to construct a basic instrumentation amplifier?
3. How is the gain determined in an instrumentation amplifier?
4. In a certain AD622 configuration, $R_G = 10\text{ k}\Omega$. What is the voltage gain?
5. Describe the purpose of a shield guard.

14–2 ISOLATION AMPLIFIERS

An isolation amplifier provides dc isolation between input and output. It is used for the protection of human life or sensitive equipment in those applications where hazardous power-line leakage or high-voltage transients are possible. The principal areas of application are in medical instrumentation, power plant instrumentation, industrial processing, and automated testing.

After completing this section, you should be able to

- **Analyze and explain the operation of an isolation amplifier**
  - Explain the basic configuration of an isolation amplifier
  - Discuss an application in medical electronics
  - Discuss the IS0124, 3656KG, and AD208 isolation amplifiers

A Basic Capacitor-Coupled Isolation Amplifier

An **isolation amplifier** is a device that consists of two electrically isolated stages. The input stage and the output stage are separated from each other by an isolation barrier so that a signal must be processed in order to be coupled across the isolation barrier. Some isolation amplifiers use optical coupling or transformer coupling to provide isolation between the stages. However, many modern isolation amplifiers use capacitive coupling for isolation. Each stage has separate supply voltages and grounds so that there are no common electrical paths between them. A simplified block diagram for a typical isolation amplifier is shown in Figure 14–11. Notice two different ground symbols are used to reinforce the concept of stage separation.

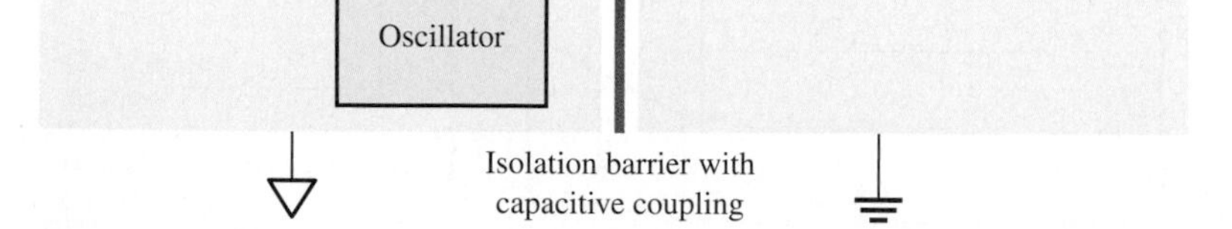

FIGURE 14–11

Simplified block diagram of a typical isolation amplifier.

The input stage consists of an amplifier, an oscillator, and a modulator. **Modulation** is the process of allowing a signal containing information to modify a characteristic of another signal, such as amplitude, frequency, or pulse width, so that the information in the first signal is also contained in the second. In this case, the modulator uses a high-frequency square-wave oscillator to modify the original signal. A small-value capacitor in the isolation barrier is used to couple the lower-frequency modulated signal or dc voltage from the input to the output. Without modulation, prohibitively high-value capacitors would be necessary with a resulting degradation in the isolation between the stages.

The output stage consists of a demodulator that extracts the original input signal from the modulated signal so that the original signal from the input stage is back to its original form.

The high-frequency oscillator output in Figure 14–11 can be either amplitude or pulse-width modulated by the signal from the input amplifier (oscillators are covered in Chapter 16). In amplitude modulation, the amplitude of the oscillator output is varied corresponding to the variations of the input signal, as indicated in Figure 14–12(a), which uses one cycle of a sine wave for illustration. In pulse-width modulation, the duty cycle of the oscillator output is varied by changing the pulse width corresponding to the variations of the input signal. An isolation amplifier using pulse-width modulation is represented in Figure 14–12(b).

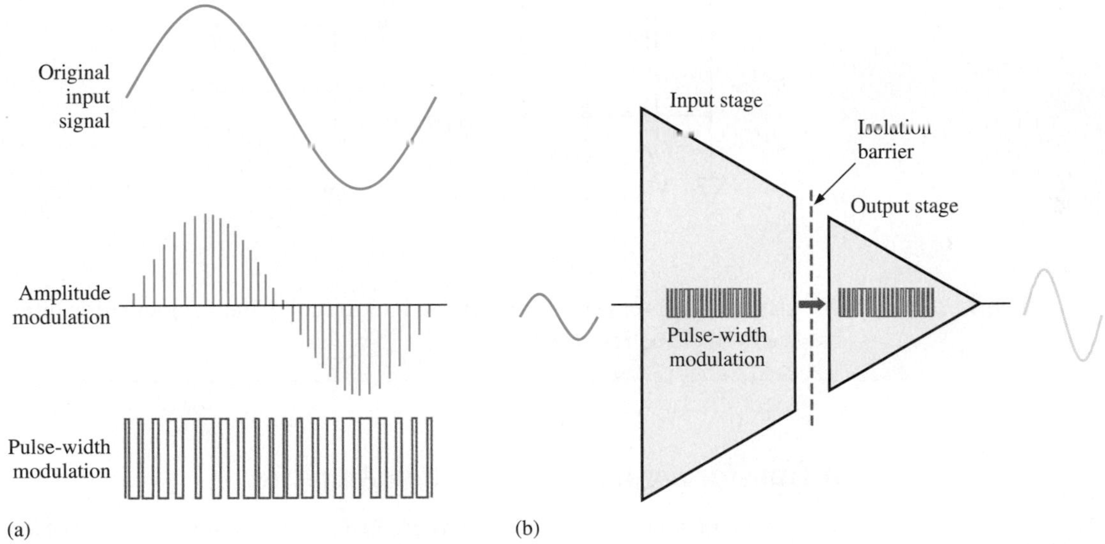

▲ FIGURE 14–12

Modulation.

Although it uses a relatively complex process internally, the isolation amplifier is still just an amplifier and is simple to use. When separate dc supply voltages and an input signal are applied, an amplified output signal is the result. The isolation function itself is an unseen process.

EXAMPLE 14–3

The IS0124 is an integrated circuit isolation amplifier. It has a voltage gain of 1 and operates on positive and negative dc supply voltages for both stages. This device uses pulse-width modulation (sometimes called duty cycle modulation) with a frequency of 500 kHz. It is recommended that the supply voltages be decoupled with external capacitors to reduce noise. Show the appropriate connections.

Solution The manufacturer recommends a 1 μF tantalum capacitor (for low leakage) from each dc power supply pin to ground. This is shown in Figure 14–13 where the supply voltages are ± 15 V.

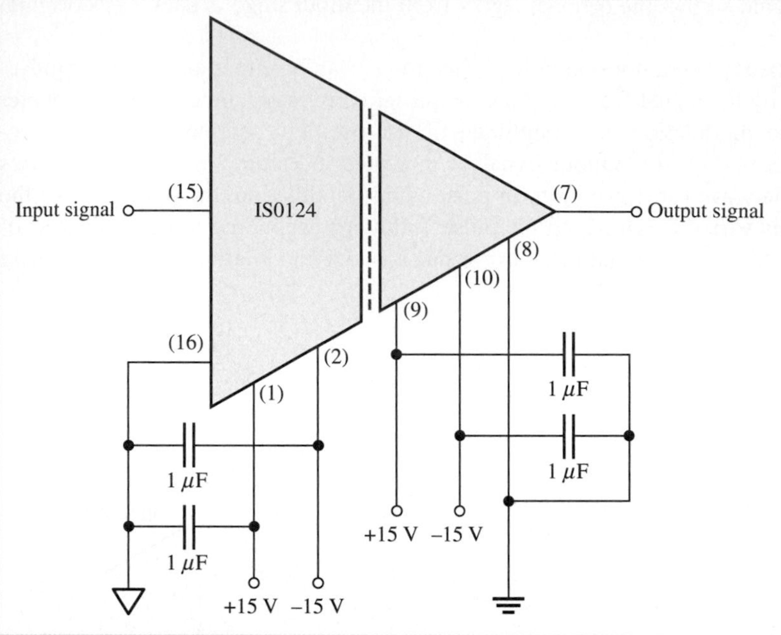

FIGURE 14–13

Basic signal and power connections for an IS0124 isolation amplifier.

Related Problem The output signal may have some ripple introduced by the demodulation process. How could this ripple be removed?

A Transformer-Coupled Isolation Amplifier

The Texas Instruments (Burr-Brown) 3656KG is an example of an isolation amplifier that uses transformer coupling to isolate the two stages. Unlike the IS0124, which has a fixed unity gain, the 3656KG provides for external gain adjustment of both stages. A diagram of the 3656KG with external gain resistors and decoupling capacitors is shown in Figure 14–14.

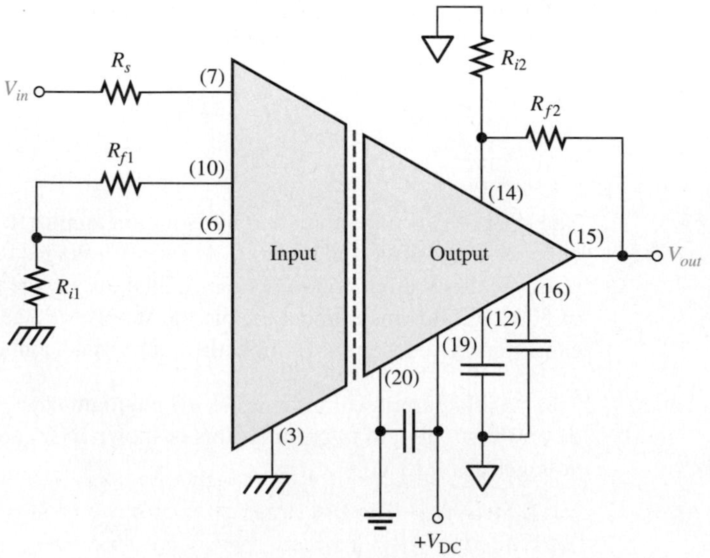

FIGURE 14–14

The 3656KG isolation amplifier.

The voltage gains of both the input stage and the output stage can be set with external resistors connected as shown in the figure. The gain of the input stage is

$$A_{v1} = \frac{R_{f1}}{R_{i1}} + 1 \qquad \text{Equation 14–3}$$

The gain of the output stage is

$$A_{v2} = \frac{R_{f2}}{R_{i2}} + 1 \qquad \text{Equation 14–4}$$

The total amplifier gain is the product of the gains of the input and output stages.

$$A_{v(tot)} = A_{v1}A_{v2}$$

EXAMPLE 14–4 Determine the total voltage gain of the 3656KG isolation amplifier in Figure 14–15.

FIGURE 14–15

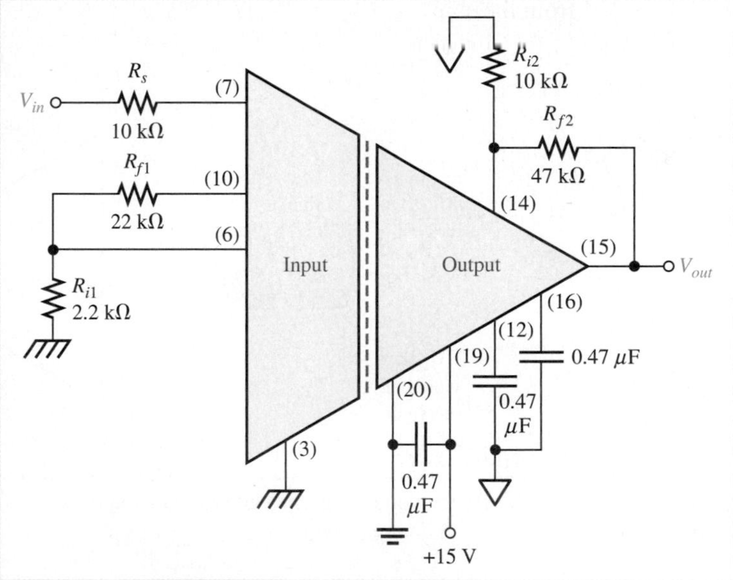

Solution The voltage gain of the input stage is

$$A_{v1} = \frac{R_{f1}}{R_{i1}} + 1 = \frac{22\ \text{k}\Omega}{2.2\ \text{k}\Omega} + 1 = 10 + 1 = 11$$

The voltage gain of the output stage is

$$A_{v2} = \frac{R_{f2}}{R_{i2}} + 1 = \frac{47\ \text{k}\Omega}{10\ \text{k}\Omega} + 1 = 4.7 + 1 = 5.7$$

The total voltage gain of the isolation amplifier is

$$A_{v(tot)} = A_{v1}A_{v2} = (11)(5.7) = \mathbf{62.7}$$

Related Problem Select resistor values in Figure 14–15 that will produce a total voltage gain of approximately 100.

Another example of a transformer-coupled isolation amplifier is the AD208. It differs from the 3656KG in that it has only an input stage gain adjustment and requires an external square wave to drive an output stage power converter.

Applications

As previously mentioned, the isolation amplifier is used in applications that require no common grounds between a transducer and the processing circuits where interfacing to sensitive equipment is required. In chemical, nuclear, and metal-processing industries, for example, millivolt signals typically exist in the presence of large common-mode voltages that can be in the kilovolt range. In this type of environment, the isolation amplifier can amplify small signals from very noisy equipment and provide a safe output to sensitive equipment such as computers.

Another important application is in various types of medical equipment. In medical applications where body functions such as heart rate and blood pressure are monitored, the very small monitored signals are combined with large common-mode signals, such as 60 Hz power-line pickup from the skin. In these situations, without isolation, dc leakage or equipment failure could be fatal. Figure 14–16 shows a simplified diagram of an isolation amplifier in a cardiac-monitoring application. In this situation, heart signals, which are very small, are combined with much larger common-mode signals caused by muscle noise, electrochemical noise, residual electrode voltage, and 60 Hz power-line pickup from the skin.

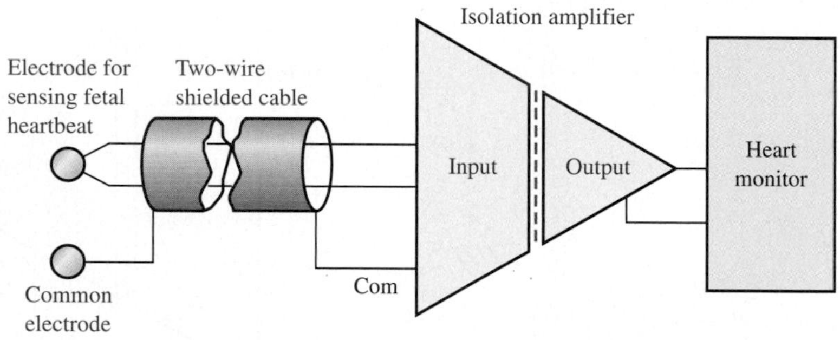

▲ FIGURE 14–16

Fetal heartbeat monitoring using an isolation amplifier.

The monitoring of fetal heartbeat, as illustrated, is the most demanding type of cardiac monitoring because in addition to the fetal heartbeat that typically generates 50 μV, there is also the mother's heartbeat that typically generates 1 mV. The common-mode voltages can run from about 1 mV to about 100 mV. The CMR (common-mode rejection) of the isolation amplifier separates the signal of the fetal heartbeat from that of the mother's heartbeat and from those common-mode signals. Therefore, the signal from the fetal heartbeat is essentially all that the amplifier sends to the monitoring equipment.

SECTION 14–2 CHECKUP

1. In what types of applications are isolation amplifiers used?
2. What are the two stages in a typical isolation amplifier and what is the purpose of having two stages?
3. How are the stages in an isolation amplifier connected?
4. What is the purpose of the oscillator in an isolation amplifier?

14–3 OPERATIONAL TRANSCONDUCTANCE AMPLIFIERS (OTAS)

Conventional op-amps are, as you know, primarily voltage amplifiers in which the output voltage equals the gain times the input voltage. The **operational transconductance amplifier (OTA)** is primarily a voltage-to-current amplifier in which the output current equals the gain times the input voltage.

After completing this section, you should be able to

- **Analyze and explain the operation of an OTA**
  - Identify the OTA symbol
  - Define *transconductance*
  - Discuss the relationship between transconductance and bias current
  - Describe the features of the LM13700 OTA
  - Discuss OTA applications

Figure 14–17 shows the symbol for an OTA. The double circle symbol at the output represents an output current source that is dependent on a bias current. Like the conventional op-amp, the OTA has two differential input terminals, a high input impedance, and a high CMRR. Unlike the conventional op-amp, the OTA has a bias-current input terminal, a high output impedance, and no fixed open-loop voltage gain.

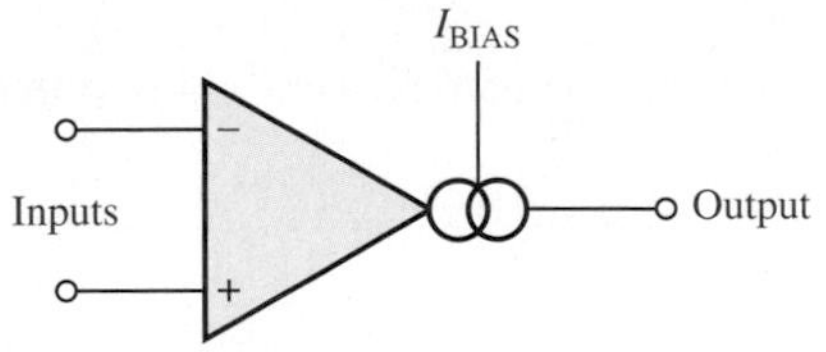

FIGURE 14–17

Symbol for an operational transconductance amplifier (OTA).

Transconductance Is the Gain of an OTA

The **transconductance** of an electronic device is the ratio of the output current to the input voltage. For an OTA, voltage is the input variable and current is the output variable; therefore, the ratio of output current to input voltage is also its gain. Consequently, the voltage-to-current gain of an OTA is the transconductance, g_m.

$$g_m = \frac{I_{out}}{V_{in}}$$

Equation 14–5

In an OTA, the transconductance is dependent on a constant (K) times the bias current (I_{BIAS}), as indicated in Equation 14–6. The value of the constant is dependent on the internal circuit design.

$$g_m = KI_{BIAS}$$

Equation 14–6

The output current is controlled by the input voltage and the bias current as shown by the following formula:

$$I_{out} = g_mV_{in} = KI_{BIAS}V_{in}$$

Transconductance Is a Function of Bias Current

The relationship of the transconductance and the bias current in an OTA is an important characteristic. The graph in Figure 14–18 illustrates a typical relationship. Notice that the transconductance increases linearly with the bias current. The constant of proportionality, K, is the slope of the line. In this case, K is approximately 16 μS/μA.

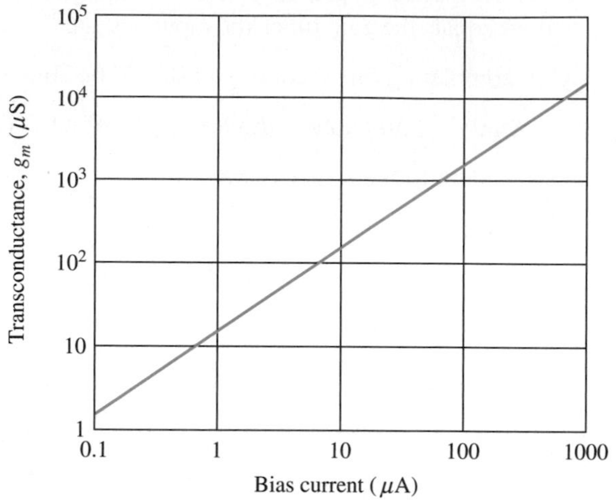

FIGURE 14–18

Example of a transconductance versus bias current graph for a typical OTA.

EXAMPLE 14–5

If an OTA has a $g_m = 1000\ \mu\text{S}$, what is the output current when the input voltage is 25 mV?

Solution

$$I_{out} = g_m V_{in} = (1000\ \mu\text{S})(25\ \text{mV}) = \mathbf{25\ \mu A}$$

Related Problem Based on $K \cong 16\ \mu\text{S}/\mu\text{A}$, calculate the approximate bias current required to produce $g_m = 1000\ \mu\text{S}$.

Basic OTA Circuits

Figure 14–19 shows the OTA used as an inverting amplifier with a fixed voltage gain. The voltage gain is set by the transconductance and the load resistance as follows.

$$V_{out} = I_{out}R_L$$

Dividing both sides by V_{in},

$$\frac{V_{out}}{V_{in}} = \left(\frac{I_{out}}{V_{in}}\right)R_L$$

FIGURE 14–19

An OTA as an inverting amplifier with a fixed voltage gain.

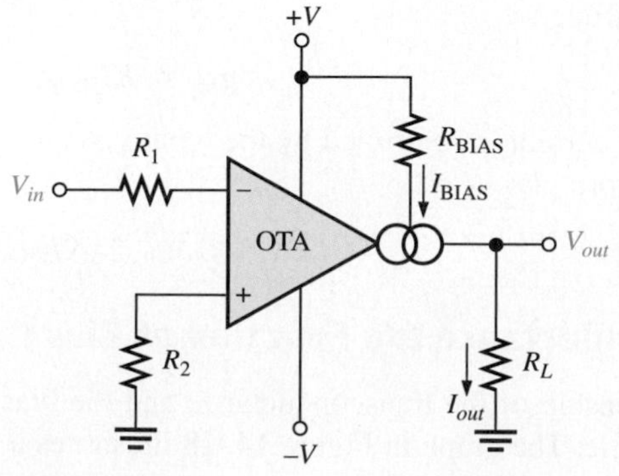

Since V_{out}/V_{in} is the voltage gain and $I_{out}/V_{in} = g_m$,

$$A_v = g_m R_L$$

The transconductance of the amplifier in Figure 14–19 is determined by the amount of bias current, which is set by the dc supply voltages and the bias resistor R_{BIAS}.

One of the most useful features of an OTA is that the voltage gain can be controlled by the amount of bias current. This can be done manually, as shown in Figure 14–20(a), by using a variable resistor in series with R_{BIAS} in the circuit of Figure 14–19. By changing the resistance, you can produce a change in I_{BIAS}, which changes the transconductance. A change in the transconductance changes the voltage gain. The voltage gain can also be controlled with an externally applied variable voltage, as shown in Figure 14–20(b). A variation in the applied bias voltage causes a change in the bias current.

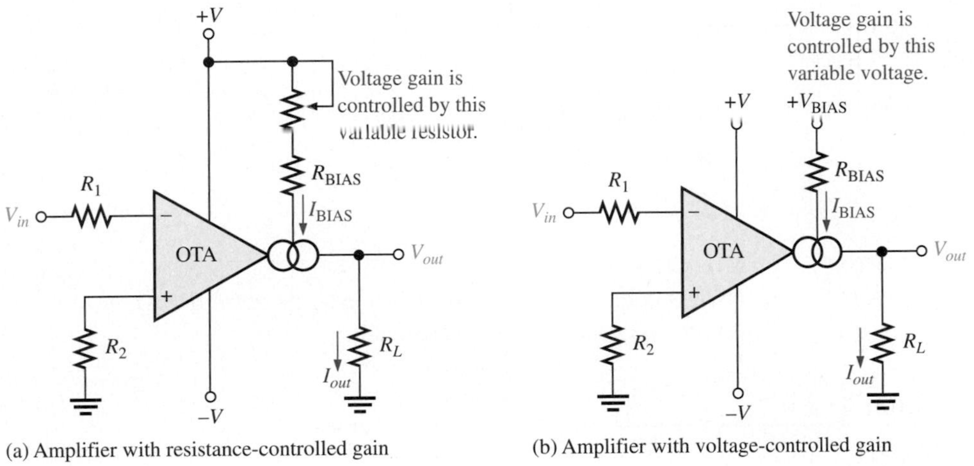

FIGURE 14–20

An OTA as an inverting amplifier with a variable-voltage gain.

A Specific OTA

The LM13700 is a typical OTA and serves as a representative device. The LM13700 is a dual-device package containing two OTAs and buffer circuits. Figure 14–21 shows the pin configuration using a single OTA in the package. The maximum dc supply voltages are ±18 V, and its transconductance characteristic happens to be the same as indicated by the graph in Figure 14–18. For an LM13700, the bias current is determined by the following formula:

$$I_{BIAS} = \frac{+V_{BIAS} - (-V) - 1.4\text{ V}}{R_{BIAS}}$$

The 1.4 V is due to the internal circuit where a base-emitter junction and a diode connect the external R_{BIAS} with the negative supply voltage $(-V)$. The positive bias voltage, $+V_{BIAS}$, may be obtained from the positive supply voltage, $+V$.

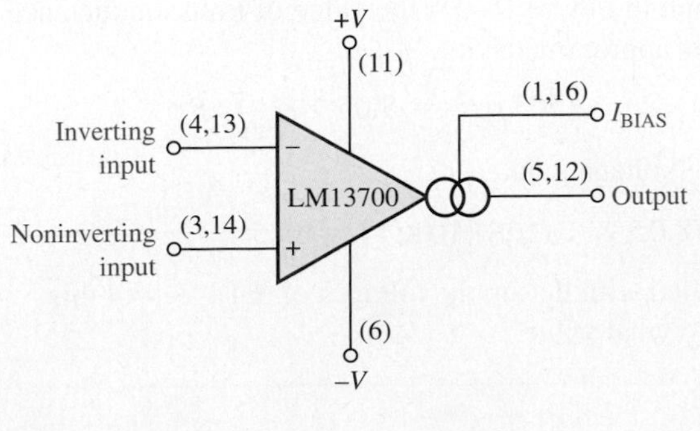

FIGURE 14–21

An LM13700 OTA. There are two in an IC package. The buffer transistors are not shown. Pin numbers for both OTAs are given in parentheses.

Not only does the transconductance of an OTA vary with bias current, but so do the input and output resistances. Both the input and output resistances decrease as the bias current increases, as shown in Figure 14–22.

FIGURE 14–22

Example of input and output resistances versus bias current.

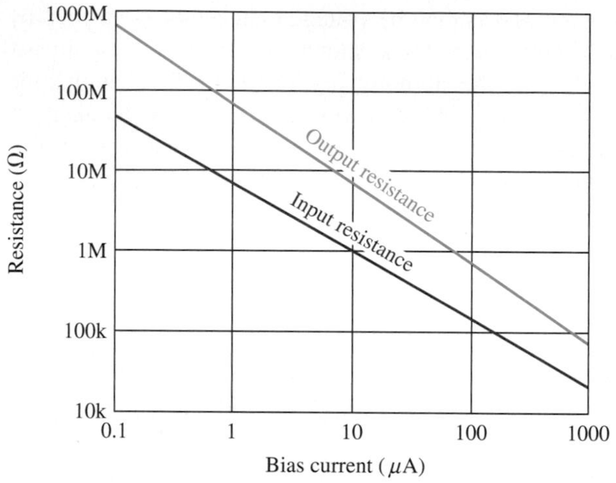

EXAMPLE 14–6

The OTA in Figure 14–23 is connected as an inverting fixed-gain amplifier where $+V_{BIAS} = +V$. Determine the approximate voltage gain.

FIGURE 14–23

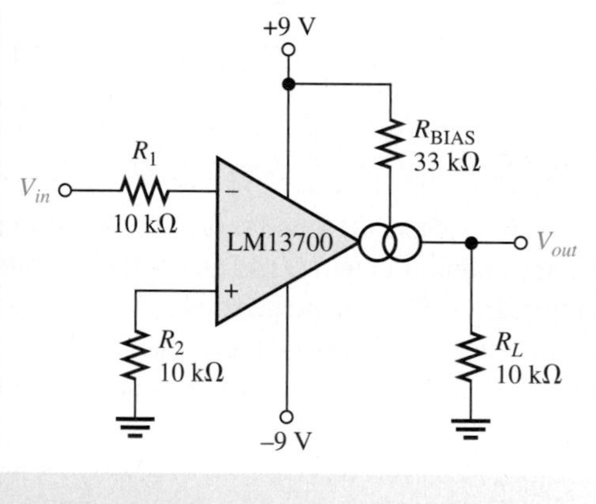

Solution Calculate the bias current as follows:

$$I_{BIAS} = \frac{+V_{BIAS} - (-V) - 1.4\text{ V}}{R_{BIAS}} = \frac{9\text{ V} - (-9\text{ V}) - 1.4\text{ V}}{33\text{ k}\Omega} = 503\ \mu\text{A}$$

Using $K \cong 16\mu\text{S}/\mu\text{A}$ from the graph in Figure 14–18, the value of transconductance corresponding to $I_{BIAS} = 503\ \mu\text{A}$ is approximately

$$g_m = KI_{BIAS} \cong (16\ \mu\text{S}/\mu\text{A})(503\ \mu\text{A}) = 8.05 \times 10^3\ \mu\text{S}$$

Using this value of g_m, calculate the voltage gain.

$$A_v = g_m R_L \cong (8.05 \times 10^3\ \mu\text{S})(10\text{ k}\Omega) = \mathbf{80.5}$$

Related Problem If the OTA in Figure 14–23 is operated with dc supply voltages of ±12 V, will this change the voltage gain and, if so, to what value?

Two OTA Applications

Amplitude Modulator Figure 14–24 illustrates an OTA connected as an amplitude modulator. The voltage gain is varied by applying a modulation voltage to the bias input. When a constant-amplitude input signal is applied, the amplitude of the output signal will vary according to the modulation voltage on the bias input. The gain is dependent on bias current, and bias current is related to the modulation voltage by the following relationship:

$$I_{BIAS} = \frac{V_{MOD} - (-V) - 1.4\text{ V}}{R_{BIAS}}$$

This modulating action is shown in Figure 14–24 for a higher-frequency sinusoidal input voltage and a lower-frequency sinusoidal modulating voltage.

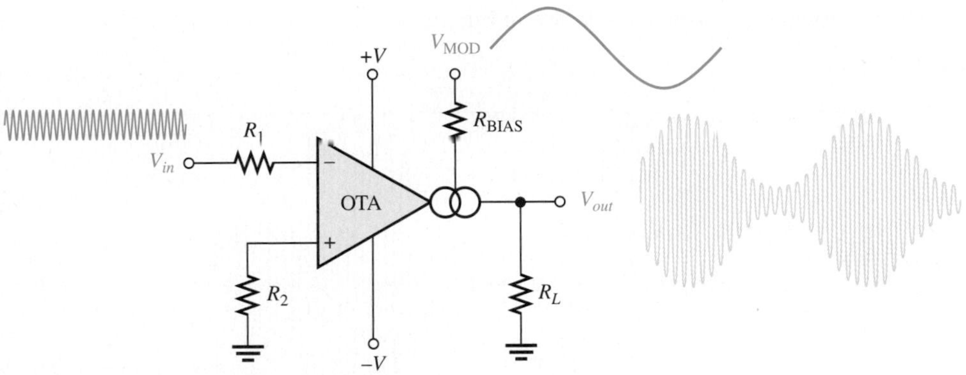

FIGURE 14–24

The OTA as an amplitude modulator.

EXAMPLE 14–7

The input to the OTA amplitude modulator in Figure 14–25 is a 50 mV peak-to-peak, 1 MHz sine wave. Determine the output signal, given the modulation voltage shown is applied to the bias input.

FIGURE 14–25

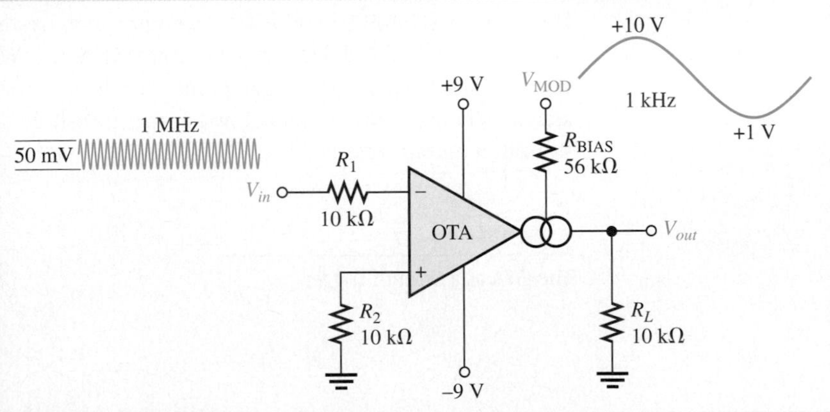

Solution The maximum voltage gain is when I_{BIAS}, and thus g_m, is maximum. This occurs at the maximum peak of the modulating voltage, V_{MOD}.

$$I_{BIAS(max)} = \frac{V_{MOD(max)} - (-V) - 1.4\text{ V}}{R_{BIAS}} = \frac{10\text{ V} - (-9\text{ V}) - 1.4\text{ V}}{56\text{ k}\Omega} = 314\ \mu\text{A}$$

From the graph in Figure 14–18, the constant K is approximately 16 μS/μA.

$$g_m = KI_{\text{BIAS(max)}} \cong (16\ \mu\text{S}/\mu\text{A})(314\ \mu\text{A}) = 5.02\ \text{mS}$$
$$A_{v(max)} = g_m R_L \cong (5.02\ \text{mS})(10\ \text{k}\Omega) = 50.2$$
$$V_{out(max)} = A_{v(max)} V_{in} \cong (50.2)(50\ \text{mV}) = 2.51\ \text{V}$$

The minimum output voltage is calculated as follows:

$$I_{\text{BIAS(min)}} = \frac{V_{\text{MOD(min)}} - (-V) - 1.4\ \text{V}}{R_{\text{BIAS}}} = \frac{1\ \text{V} - (-9\ \text{V}) - 1.4\ \text{V}}{56\ \text{k}\Omega} = 154\ \mu\text{A}$$
$$g_m = KI_{\text{BIAS(min)}} \cong (16\ \mu\text{S}/\mu\text{A})(154\ \mu\text{A}) = 2.46\ \text{mS}$$
$$A_{v(min)} = g_m R_L \cong (2.46\ \text{mS})(10\ \text{k}\Omega) = 24.6$$
$$V_{out(min)} = A_{v(min)} V_{in} \cong (24.6)(50\ \text{mV}) = 1.23\ \text{V}$$

The resulting output voltage is shown in Figure 14–26.

FIGURE 14–26

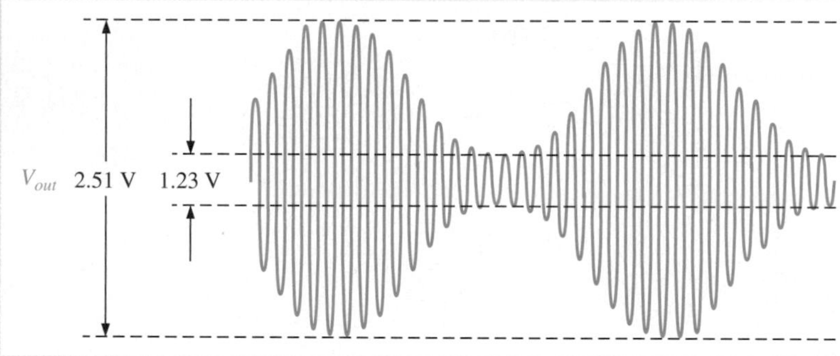

Related Problem Repeat this example with the sinusoidal modulating signal replaced by a square wave with the same maximum and minimum levels and a bias resistor of 39 kΩ.

Schmitt Trigger Figure 14–27 shows an OTA used in a Schmitt-trigger configuration. Basically, a Schmitt trigger is a comparator with hysteresis where the input voltage is large enough to drive the device into its saturated states. When the input voltage exceeds a certain threshold value or trigger point, the device switches to one of its saturated output states. When the input falls below another threshold value, the device switches to its other saturated output state.

FIGURE 14–27

The OTA as a Schmitt trigger.

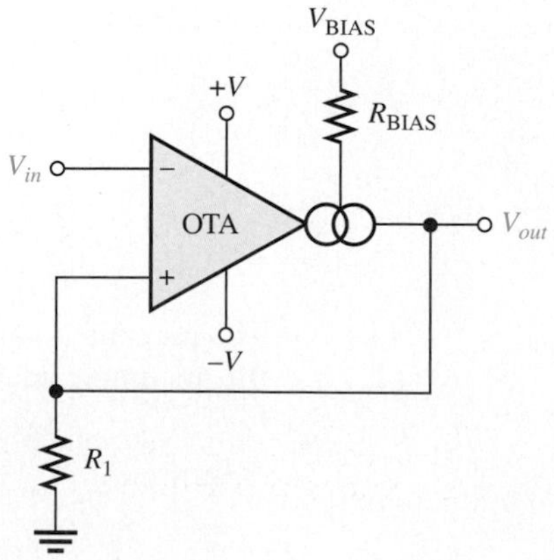

In the case of the OTA Schmitt trigger, the threshold levels are set by the current through resistor R_1. The maximum output current in an OTA equals the bias current. Therefore, in the saturated output states, $I_{out} = I_{BIAS}$. The maximum positive output voltage is $I_{out}R_1$, and this voltage is the positive threshold value or upper trigger point. When the input voltage exceeds this value, the output switches to its maximum negative voltage, which is $-I_{out}R_1$. Since $I_{out} = I_{BIAS}$, the trigger points can be controlled by the bias current. Figure 14–28 illustrates this operation.

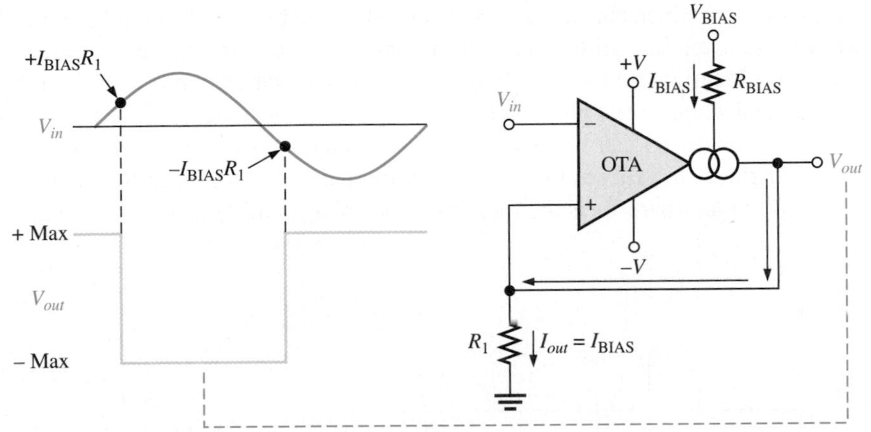

FIGURE 14–28

Basic operation of the OTA Schmitt trigger.

SECTION 14–3 CHECKUP

1. What does OTA stand for?
2. If the bias current in an OTA is increased, does the transconductance increase or decrease?
3. What happens to the voltage gain if the OTA is connected as a fixed-voltage amplifier and the supply voltages are increased?
4. What happens to the voltage gain if the OTA is connected as a variable-gain voltage amplifier and the voltage at the bias terminal is decreased?

14–4 LOG AND ANTILOG AMPLIFIERS

Log and antilog amplifiers are used in applications that require compression of analog input data, linearization of transducers that have exponential outputs, and analog multiplication and division. They are often used in high-frequency communication systems, including fiber optics, for processing wide dynamic range signals.

After completing this section, you should be able to

- **Analyze and explain the operation of log and antilog amplifiers**
  - Describe the feedback configurations
  - Define *logarithm, antilogarithm,* and *natural logarithm*
  - Discuss signal compression with logarithmic amplifiers

The **logarithm** of a number is the power to which the base must be raised to get that number. A logarithmic (log) amplifier produces an output that is proportional to the logarithm of the input, and an antilogarithmic (antilog) amplifier takes the antilog or inverse log of the input.

The Basic Logarithmic Amplifier

The key element in a log amplifier is a device that exhibits a logarithmic characteristic that, when placed in the feedback loop of an op-amp, produces a logarithmic response. This means that the output voltage is a function of the logarithm of the input voltage, as expressed by the following general equation:

Equation 14–7

$$V_{out} = -K\ln(V_{in})$$

where K is a constant and ln is the natural logarithm to the base e. A **natural logarithm** is the exponent to which the base e must be raised in order to equal a given quantity. Although we will use natural logarithms in the formulas in this section, each expression can be converted to a logarithm to the base 10 ($\log_{10}$) using the relationship $\ln x = 2.3\log_{10}x$.

The semiconductor *pn* junction in the form of either a diode or the base-emitter junction of a BJT provides a logarithmic characteristic. You may recall that a diode has a nonlinear characteristic up to a forward voltage of approximately 0.7 V. Figure 14–29 shows the characteristic curve, where V_F is the forward diode voltage and I_F is the forward diode current.

FIGURE 14–29

A portion of a diode (*pn* junction) characteristic curve (V_F versus I_F).

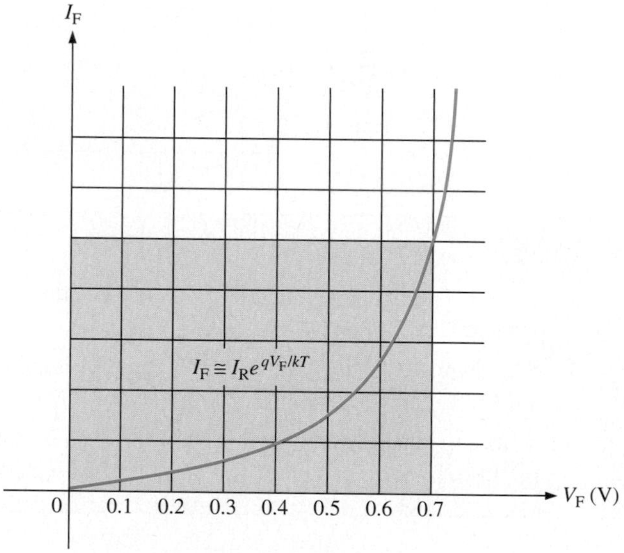

As you can see on the graph, the diode curve is nonlinear. Not only is the characteristic curve nonlinear, it is logarithmic and is specifically defined by the following formula:

$$I_F \cong I_R e^{qV_F/kT}$$

where I_R is the reverse leakage current, q is the charge on an electron, k is Boltzmann's constant, and T is the absolute temperature in Kelvin. From the previous equation, the diode forward voltage, V_F, can be determined as follows. Take the natural logarithm (*ln* is the logarithm to the base *e*) of both sides.

$$\ln I_F = \ln I_R e^{qV_F/kT}$$

The ln of a product of two terms equals the sum of the ln of each term.

$$\ln I_F = \ln I_R + \ln e^{qV_F/kT} = \ln I_R + \frac{qV_F}{kT}$$

$$\ln I_F - \ln I_R = \frac{qV_F}{kT}$$

The difference of two ln terms equals the ln of the quotient of the terms.

$$\ln\left(\frac{I_F}{I_R}\right) = \frac{qV_F}{kT}$$

Solving for V_F,

$$V_F = \left(\frac{kT}{q}\right)\ln\left(\frac{I_F}{I_R}\right)$$

Log Amplifier with a Diode When you place a diode in the feedback loop of an op-amp circuit, as shown in Figure 14–30, you have a basic log amplifier. Since the inverting input is at virtual ground (0 V), the output is at $-V_F$ when the input is positive. Since V_F is logarithmic, so is V_{out}. The output is limited to a maximum value of approximately -0.7 V because the diode's logarithmic characteristic is restricted to voltages below 0.7 V. Also, the input must be positive when the diode is connected in the direction shown in the figure. To handle negative inputs, you must turn the diode around.

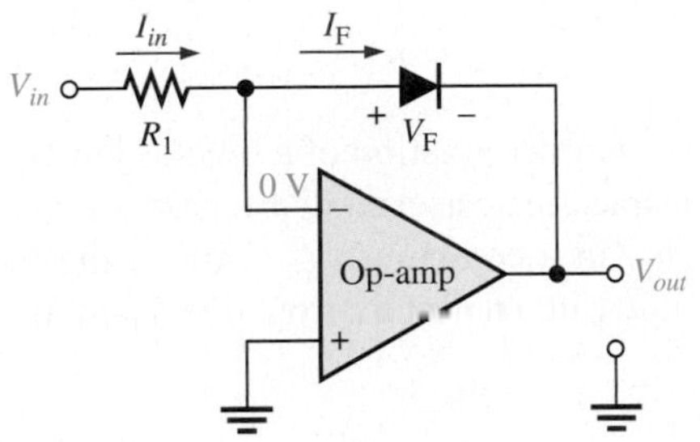

FIGURE 14–30

A basic log amplifier using a diode as the feedback element.

An analysis of the circuit in Figure 14–30 is as follows, beginning with the facts that $V_{out} = -V_F$ and $I_F = I_{in}$ because there is no current at the inverting input.

$$V_{out} = -V_F$$

$$I_F = I_{in} = \frac{V_{in}}{R_1}$$

Substituting into the formula for V_F,

$$V_{out} = -\left(\frac{kT}{q}\right)\ln\left(\frac{V_{in}}{I_R R_1}\right)$$

The term kT/q is a constant equal to approximately 25 mV at 25°C. Therefore, the output voltage can be expressed as

$$\mathbf{V_{out} \cong -(0.025\ V)\ln\left(\frac{V_{in}}{I_R R_1}\right)} \qquad \text{Equation 14–8}$$

From Equation 14–8, you can see that the output voltage is the negative of a logarithmic function of the input voltage. The value of the output is controlled by the value of the input voltage and the value of the resistor R_1. The other factor, I_R, is a constant for a given diode.

EXAMPLE 14–8

Determine the output voltage for the log amplifier in Figure 14–31. Assume $I_R = 50$ nA.

FIGURE 14–31

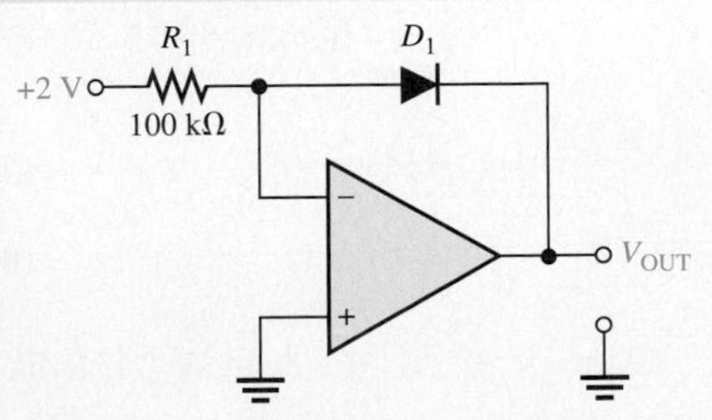

Solution The input voltage and the resistor value are given in Figure 14–31.

$$V_{OUT} = -(0.025\ \text{V})\ln\left(\frac{V_{in}}{I_R R_1}\right) = -(0.025\ \text{V})\ln\left(\frac{2\ \text{V}}{(50\ \text{nA})(100\ \text{k}\Omega)}\right)$$
$$= -(0.025\ \text{V})\ln(400) = -(0.025\ \text{V})(5.99) = \mathbf{-0.150\ V}$$

Related Problem Calculate the output voltage of the log amplifier with a +4 V input.

Open the Multisim file E14-08 in the Examples folder on your CD-ROM. Apply the specified input voltage and measure the output voltage.

Log Amplifier with a BJT The base-emitter junction of a bipolar junction transistor exhibits the same type of logarithmic characteristic as a diode because it is also a *pn* junction. A log amplifier with a BJT connected in a common-base form in the feedback loop is shown in Figure 14–32. Notice that V_{out} with respect to ground is equal to $-V_{BE}$.

▶ FIGURE 14–32

A basic log amplifier using a transistor as the feedback element.

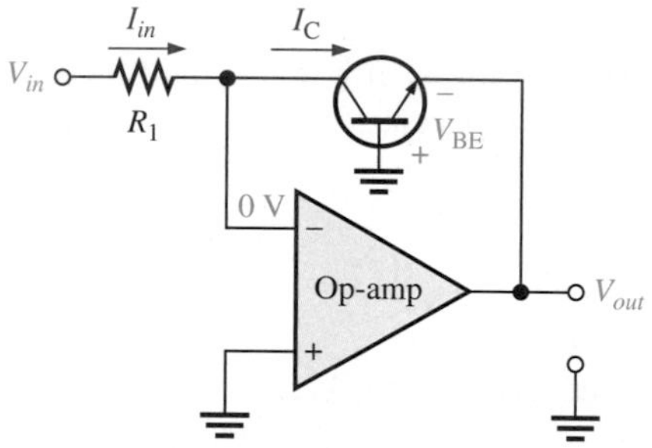

The analysis for this circuit is the same as for the diode log amplifier except that V_{BE} replaces V_F, I_C replaces I_F, and I_{EBO} replaces I_R. The expression for the V_{BE} versus I_C characteristic curve is

$$I_C = I_{EBO}e^{qV_{BE}/kT}$$

where I_{EBO} is the emitter-to-base leakage current. The expression for the output voltage is

Equation 14–9

$$\mathbf{V_{out} = -(0.025\ V)\ln\left(\frac{V_{in}}{I_{EBO}R_1}\right)}$$

EXAMPLE 14–9

What is V_{out} for a transistor log amplifier with $V_{in} = 3$ V and $R_1 = 68$ kΩ? Assume $I_{EBO} = 40$ nA.

Solution

$$V_{out} = -(0.025\ \text{V})\ln\left(\frac{V_{in}}{I_{EBO}R_1}\right) = -(0.025\ \text{V})\ln\left(\frac{3\ \text{V}}{(40\ \text{nA})(68\ \text{k}\Omega)}\right)$$
$$= -(0.025\ \text{V})\ln(1103) = \mathbf{-175.1\ mV}$$

Related Problem Calculate V_{out} if R_1 is changed to 33 kΩ.

The Basic Antilog Amplifier

The **antilogarithm** of a number is the result obtained when the base is raised to a power equal to the logarithm of that number. To get the antilogarithm, you must take the exponential of the logarithm (antilogarithm of $x = e^{\ln x}$).

An antilog amplifier is formed by connecting a transistor (or diode) as the input element as shown in Figure 14–33. The exponential formula still applies to the base-emitter *pn* junction. The output voltage is determined by the current (equal to the collector current) through the feedback resistor.

$$V_{out} = -R_f I_C$$

The characteristic equation of the *pn* junction is

$$I_C = I_{EBO} e^{qV_{BE}/kT}$$

Substituting into the equation for V_{out},

$$V_{out} = -R_f I_{EBO} e^{qV_{BE}/kT}$$

As you can see in Figure 14–33, $V_{in} = V_{BE}$.

$$V_{out} = -R_f I_{EBO} e^{qV_{in}/kT}$$

The exponential term can be expressed as an antilogarithm as follows:

$$V_{out} = -R_f I_{EBO} \text{antilog}\left(\frac{V_{in}q}{kT}\right)$$

Since kT/q is approximately 25 mV,

$$\boldsymbol{V_{out} = -R_f I_{EBO} \textbf{antilog}\left(\frac{V_{in}}{25 \text{ mV}}\right)}$$ Equation 14–10

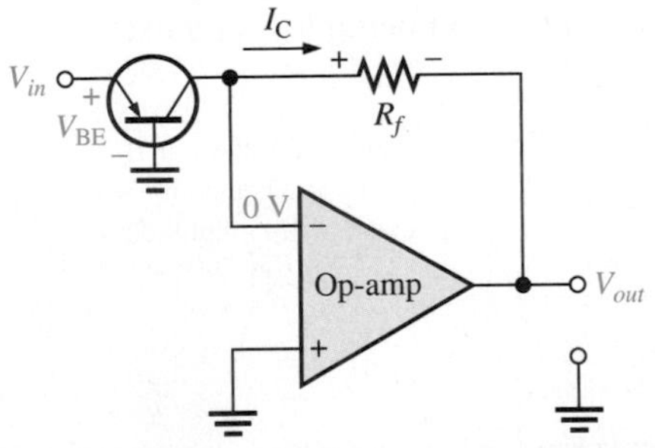

FIGURE 14–33

A basic antilog amplifier.

EXAMPLE 14–10

For the antilog amplifier in Figure 14–34, find the output voltage. Assume $I_{EBO} = 40$ nA.

FIGURE 14–34

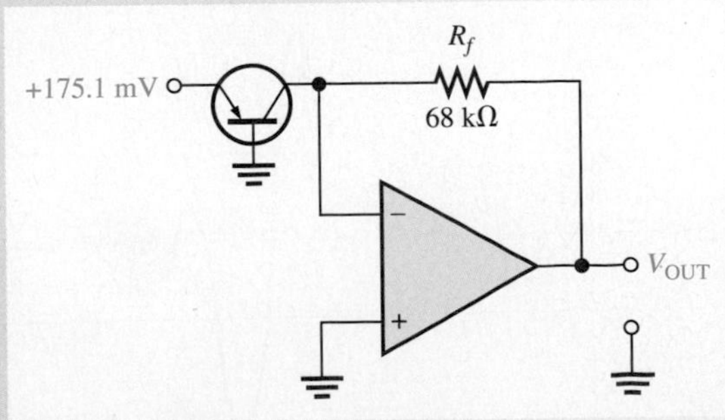

Solution First of all, notice that the input voltage in Figure 14–34 is the inverted output voltage of the log amplifier in Example 14–9, where the output voltage is proportional to the logarithm of the input voltage. In this case, the antilog amplifier reverses the process and produces an output that is proportional to the antilog of the input. Stated another way, the input of an antilog amplifier is proportional to the logarithm of the output. So, the output voltage of the antilog amplifier in Figure 14–34 should have the same magnitude as the input voltage of the log amplifier in Example 14–9 because all the constants are the same. Let's see if it does.

$$V_{OUT} = -R_f I_{EBO}\text{antilog}\left(\frac{V_{in}}{25\text{ mV}}\right) = -(68\text{ k}\Omega)(40\text{ nA})\text{antilog}\left(\frac{175.1\text{ mV}}{25\text{ mV}}\right)$$
$$= -(68\text{ k}\Omega)(40\text{ nA})(1101) = \mathbf{-3\text{ V}}$$

Related Problem Determine V_{OUT} for the amplifier in Figure 14–34 if the feedback resistor is changed to 100 kΩ.

Signal Compression with Logarithmic Amplifiers

In certain applications, a signal may be too large in magnitude for a particular system to handle. The term *dynamic range* is often used to describe the range of voltages contained in a signal. In these cases, the signal voltage must be scaled down by a process called **signal compression** so that it can be properly handled by the system. If a linear circuit is used to scale a signal down in amplitude, the lower voltages are reduced by the same percentage as the higher voltages. Linear signal compression often results in the lower voltages becoming obscured by noise and difficult to accurately distinguish, as illustrated in Figure 14–35(a). To overcome this problem, a signal with a large dynamic range can be compressed using a logarithmic response, as shown in Figure 14–35(b). In logarithmic signal compression, the higher voltages are reduced by a greater percentage than the lower voltages, thus keeping the lower voltage signals from being lost in noise.

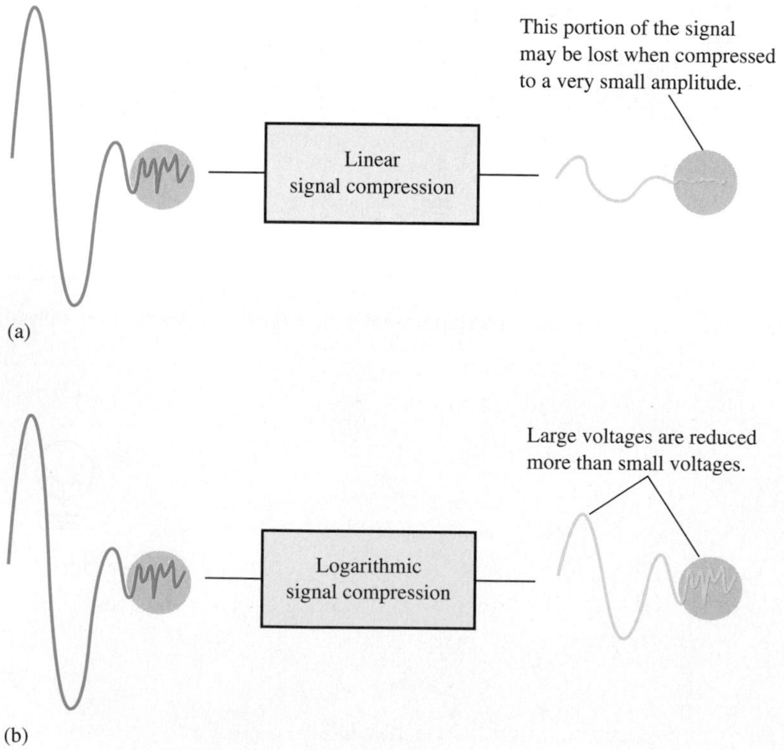

FIGURE 14–35

The basic concept of signal compression with a logarithmic amplifier.

SECTION 14–4 CHECKUP

1. What purpose does the diode or transistor perform in the feedback loop of a log amplifier?
2. Why is the output of a log amplifier limited to about 0.7 V?
3. What are the factors that determine the output voltage of a basic log amplifier?
4. In terms of implementation, how does a basic antilog amplifier differ from a basic log amplifier?

14–5 CONVERTERS AND OTHER OP-AMP CIRCUITS

This section introduces a few more op-amp circuits that represent basic applications of the op-amp. You will learn about the constant-current source, the current-to-voltage converter, the voltage-to-current converter, and the peak detector. This is, of course, not a comprehensive coverage of all possible op-amp circuits but is intended only to introduce you to some common basic uses.

After completing this section, you should be able to

- **Analyze and explain several special types of op-amp circuits**
  - Describe how the op-amp is used as a constant-current source
  - Explain the operation of a current-to-voltage converter
  - Explain the operation of a voltage-to-current converter
  - Describe the operation of a peak detector

Constant-Current Source

A constant-current source delivers a load current that remains constant when the load resistance changes. Figure 14–36 shows a basic circuit in which a stable voltage source (V_{IN}) provides a constant current (I_i) through the input resistor (R_i). Since the inverting (−) input of the op-amp is at virtual ground (0 V), the value of I_i is determined by V_{IN} and R_i as

$$I_i = \frac{V_{IN}}{R_i}$$

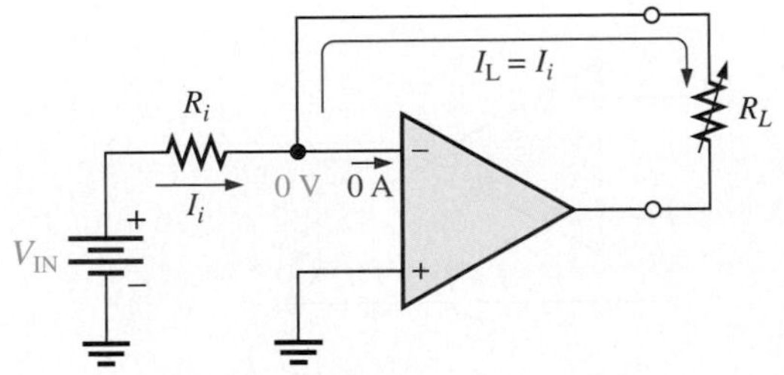

FIGURE 14–36
A basic constant-current source.

Now, since the internal input impedance of the op-amp is extremely high (ideally infinite), practically all of I_i is through R_L, which is connected in the feedback path. Since $I_i = I_L$,

$$I_L = \frac{V_{IN}}{R_i} \qquad \text{Equation 14–11}$$

If R_L changes, I_L remains constant as long as V_{IN} and R_i are held constant.

Current-to-Voltage Converter

A current-to-voltage converter converts a variable input current to a proportional output voltage. A basic circuit that accomplishes this is shown in Figure 14–37(a). Since practically all of I_i is through the feedback path, the voltage dropped across R_f is I_iR_f. Because the left side of R_f is at virtual ground (0 V), the output voltage equals the voltage across R_f, which is proportional to I_i.

Equation 14–12

$$V_{out} = I_iR_f$$

FIGURE 14–37

Current-to-voltage converter.

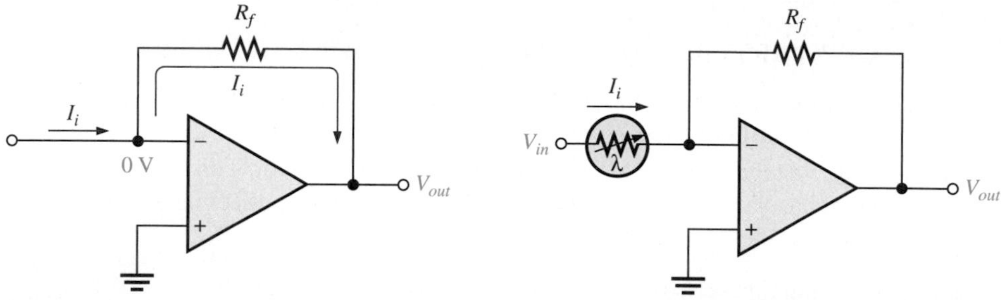

(a) Basic circuit

(b) Circuit for sensing light level and converting it to a proportional output voltage

A specific application of this circuit is illustrated in Figure 14–37(b), where a photoconductive cell is used to sense changes in light level. As the amount of light changes, the current through the photoconductive cell varies because of the cell's change in resistance. This change in resistance produces a proportional change in the output voltage ($\Delta V_{out} = \Delta I_iR_f$).

Voltage-to-Current Converter

A basic voltage-to-current converter is shown in Figure 14–38. This circuit is used in applications where it is necessary to have an output (load) current that is controlled by an input voltage.

Neglecting the input offset voltage, both inverting and noninverting input terminals of the op-amp are at the same voltage, V_{in}. Therefore, the voltage across R_1 equals V_{in}. Since there is negligible current at the inverting input, the current through R_1 is the same as the current through R_L; thus

Equation 14–13

$$I_L = \frac{V_{in}}{R_1}$$

FIGURE 14–38

Voltage-to-current converter.

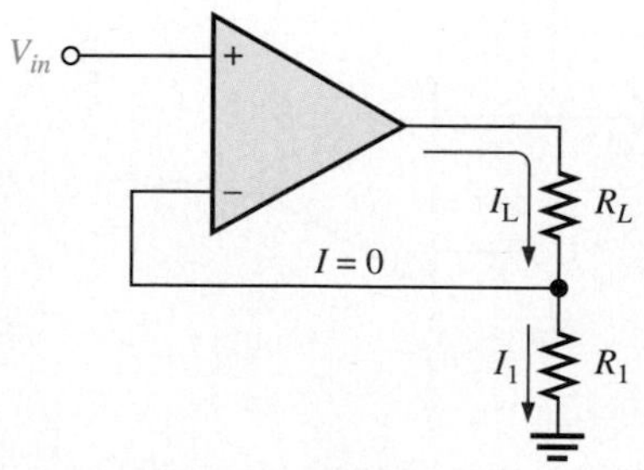

Peak Detector

An interesting application of the op-amp is in a peak detector circuit such as the one shown in Figure 14–39. In this case the op-amp is used as a comparator. This circuit is used to detect the peak of the input voltage and store that peak voltage on a capacitor. For example,

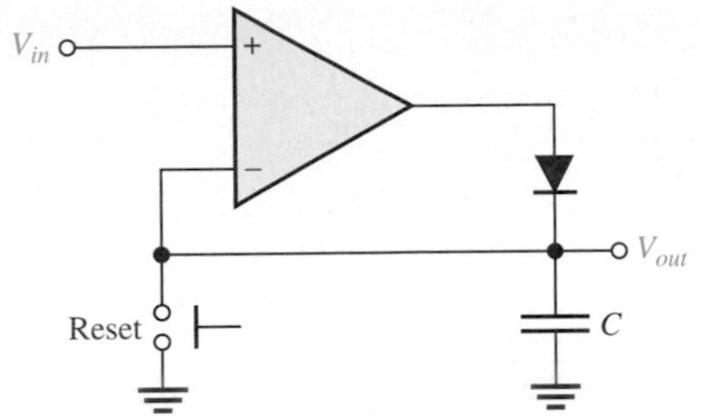

◀ FIGURE 14–39

A basic peak detector.

this circuit can be used to detect and store the maximum value of a voltage surge; this value can then be measured at the output with a voltmeter or recording device. The basic operation is as follows. When a positive voltage is applied to the noninverting input of the op-amp, the high-level output voltage of the op-amp forward-biases the diode and charges the capacitor. The capacitor continues to charge until its voltage reaches a value equal to the input voltage and thus both op-amp inputs are at the same voltage. At this point, the op-amp comparator switches, and its output goes to the low level. The diode is now reverse-biased, and the capacitor stops charging. It has reached a voltage equal to the peak of V_{in} and will hold this voltage until the charge eventually leaks off or until it is reset with a switch as indicated. If a greater input peak occurs, the capacitor charges to the new peak.

SECTION 14–5 CHECKUP

1. For the constant-current source in Figure 14–36, the input reference voltage is 6.8 V and R_i is 10 kΩ. What value of constant current does the circuit supply to a 1.0 kΩ load? To a 5 kΩ load?
2. What element determines the constant of proportionality that relates input current to output voltage in the current-to-voltage converter?

On the Job

Application Activity: *Liquid Level Control*

The system in this application is designed to maintain a constant liquid level in a tank. The level is kept constant by an electric pump and a pressure sensor (transducer) that detects a change in the level of the liquid by sensing the pressure in a tube.

Level-Sensing Method

A tube with both ends open is placed vertically in a liquid so that one end is above the surface of the liquid. The level of liquid in the tube will be the same as the level in the tank. Now, if the upper end is closed, the pressure of the air trapped in the tube will vary proportional to a change in level of the liquid. For example, if the liquid is water and it rises in the tank by 20 mm, then the pressure in the tube will increase by 20 mm of water. A pressure sensor is placed on the upper end of the tube when the liquid is at its reference level, and the other side is exposed to atmospheric pressure. When the water level decreases, a negative change in pressure is measured by the pressure sensor and a small proportional voltage is produced. The voltage from the pressure sensor is connected to an instrumentation amplifier, which amplifies the small voltage to drive a comparator with hysteresis (Schmitt trigger). The comparator reference voltage is adjusted to the desired value; and when the level falls below the reference, the comparator switches states and turns the

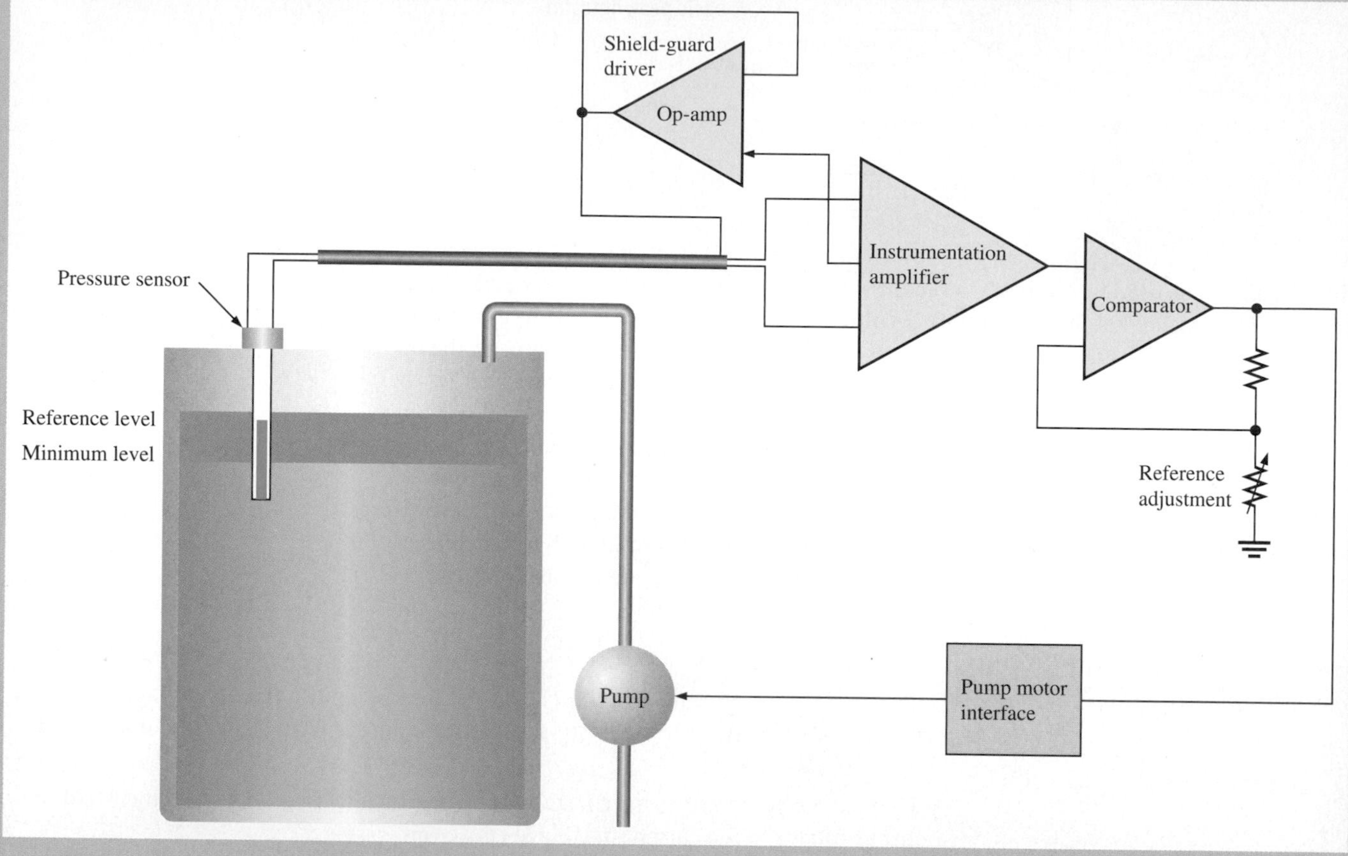

FIGURE 14–40

Block diagram of the liquid-level control system.

pump *on* to refill the tank to the reference level. The pressure sensor detects when the reference level of the liquid is reached, and the comparator switches back to its other state, turning the pump *off.* A basic diagram of the system is shown in Figure 14–40.

The Circuit

This system will operate in an industrial environment with exposure to mostly 60 Hz electrical noise. Also, the circuit will be located some distance from the tank and connected to the pressure sensor with a long coaxial cable. The output voltage of the pressure sensor is very small ($100\ \mu V - 200\ \mu V$). For these reasons, a shield-guard driver is incorporated to minimize the effects of noise on the small signal. The AD624 instrumentation amplifier is used to drive an LM111 comparator with hysteresis controlled by a rheostat in the feedback circuit. An AD741 op-amp connected as a voltage-follower is used for the guard driver. The circuit diagram is shown in Figure 14–41. Power supply connections are omitted to simplify the drawing. Resistors R_1 and R_2 provide a return path for bias currents to prevent output drift. R_3 is a pull-up resistor for the comparator output, and R_4 and R_5 provide for the adjustable reference levels by varying the hysteresis. R_6 provides a resistance in series with the shield-guard driver to limit current.

As the liquid level in the tank decreases, the pressure in the tube decreases. This decrease in pressure is translated into a proportional decrease in voltage by the pressure sensor. This decrease in voltage is processed by the circuit to trigger the comparator to its

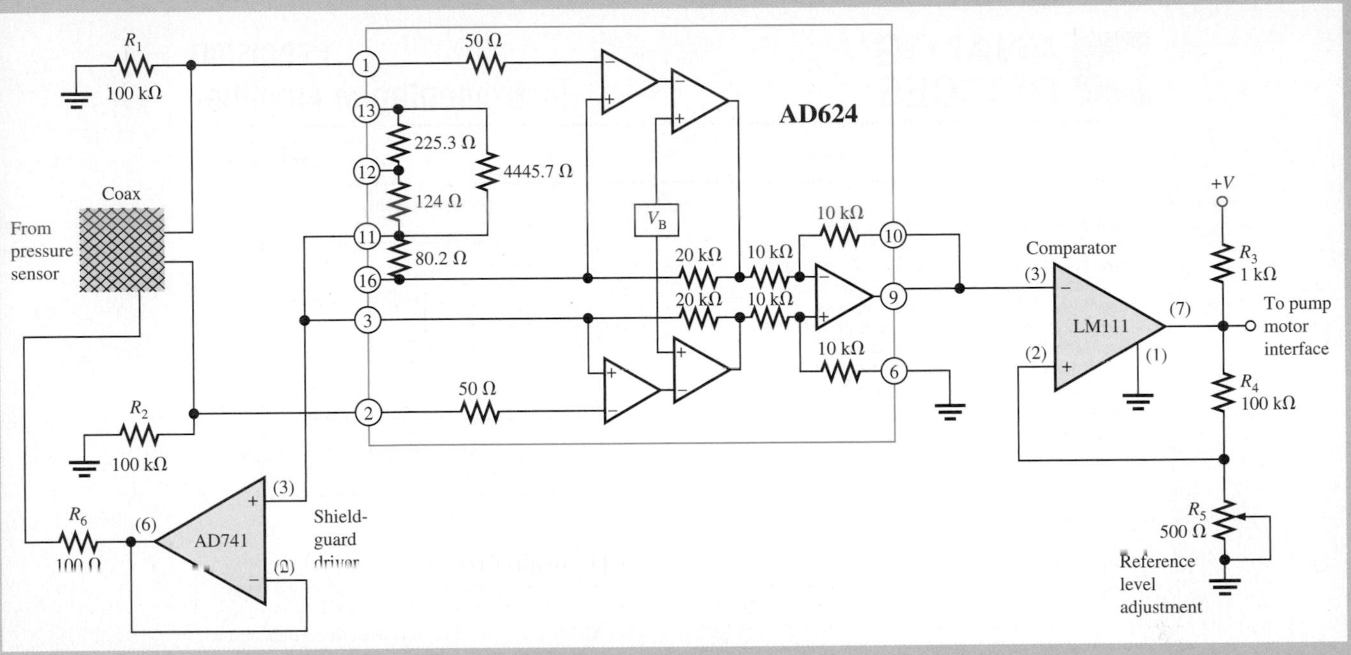

FIGURE 14–41

Schematic diagram of the liquid-level control circuit.

HIGH state to turn the pump *on* when a desired minimum level is reached. An increase in level occurs while the pump is running, causing a proportional increase in pressure. When the maximum level is reached, the circuit triggers the comparator to its LOW state to turn the pump *off*. This process is illustrated in Figure 14–42.

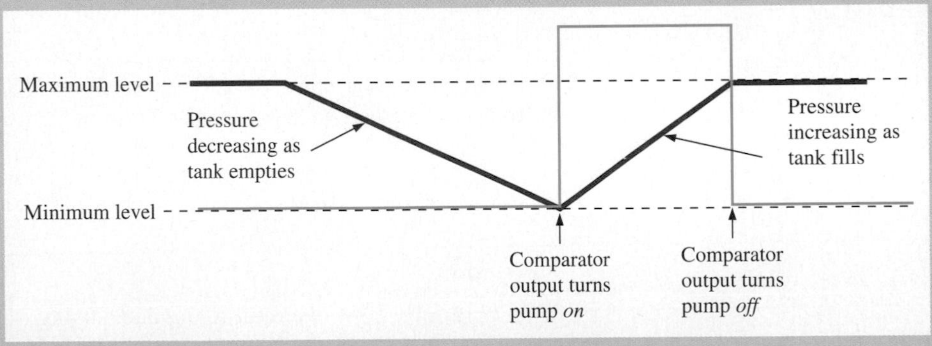

FIGURE 14–42

System operation.

AD624 Datasheet

The front page of the datasheet for the AD624 instrumentation amplifier is shown in Figure 14–43.

1. Use the datasheet to determine the voltage gain of the AD624 in Figure 14–41 based on the connections. You will have to go on-line.
2. How would you change the gain?

ANALOG DEVICES

Precision Instrumentation Amplifier

AD624

FEATURES

Low Noise: 0.2 μV p-p 0.1 Hz to 10 Hz
Low Gain TC: 5 ppm max (G = 1)
Low Nonlinearity: 0.001% max (G = 1 to 200)
High CMRR: 130 dB min (G = 500 to 1000)
Low Input Offset Voltage: 25 μV, max
Low Input Offset Voltage Drift: 0.25 μV/°C max
Gain Bandwidth Product: 25 MHz
Pin Programmable Gains of 1, 100, 200, 500, 1000
No External Components Required
Internally Compensated

FUNCTIONAL BLOCK DIAGRAM

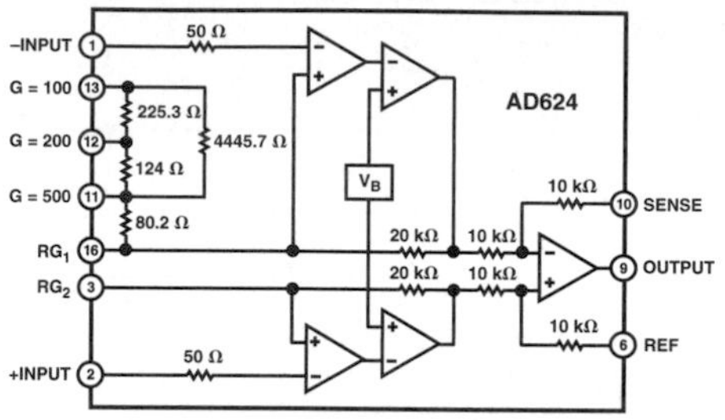

PRODUCT DESCRIPTION

The AD624 is a high precision, low noise, instrumentation amplifier designed primarily for use with low level transducers, including load cells, strain gauges and pressure transducers. An outstanding combination of low noise, high gain accuracy, low gain temperature coefficient and high linearity make the AD624 ideal for use in high resolution data acquisition systems.

The AD624C has an input offset voltage drift of less than 0.25 μV/°C, output offset voltage drift of less than 10 μV/°C, CMRR above 80 dB at unity gain (130 dB at G = 500) and a maximum nonlinearity of 0.001% at G = 1. In addition to these outstanding dc specifications, the AD624 exhibits superior ac performance as well. A 25 MHz gain bandwidth product, 5 V/μs slew rate and 15 μs settling time permit the use of the AD624 in high speed data acquisition applications.

The AD624 does not need any external components for pretrimmed gains of 1, 100, 200, 500 and 1000. Additional gains such as 250 and 333 can be programmed within one percent accuracy with external jumpers. A single external resistor can also be used to set the 624's gain to any value in the range of 1 to 10,000.

PRODUCT HIGHLIGHTS

1. The AD624 offers outstanding noise performance. Input noise is typically less than 4 nV/√Hz at 1 kHz.
2. The AD624 is a functionally complete instrumentation amplifier. Pin programmable gains of 1, 100, 200, 500 and 1000 are provided on the chip. Other gains are achieved through the use of a single external resistor.
3. The offset voltage, offset voltage drift, gain accuracy and gain temperature coefficients are guaranteed for all pretrimmed gains.
4. The AD624 provides totally independent input and output offset nulling terminals for high precision applications. This minimizes the effect of offset voltage in gain ranging applications.
5. A sense terminal is provided to enable the user to minimize the errors induced through long leads. A reference terminal is also provided to permit level shifting at the output.

FIGURE 14–43

AD624 datasheet (page l). Datasheet is courtesy of Analog Devices and is subject to revisions. To see the complete datasheet go to www.analog.com.

Simulation

The liquid-level control circuit is simulated using Multisim with an input signal of 100 mV at 100 Hz to represent the pressure sensor output. Although the sensor output will change very slowly (almost dc), we are using a higher frequency signal in order to observe the circuit operation. The simulated circuit is shown in Figure 14–44(a) for a differential input. The resulting outputs are shown in part (b). The comparator is triggered at two difference points, as indicated. Notice that there is no signal on the output of the shield-guard driver because there is no common-mode signal on the inputs.

3. Refer to Figure 14–44(b) to determine the voltage gain of the instrumentation amplifier and compare to the gain indicated by the pin connections. The input is 100 μV rms.
4. Determine the difference in mV (hysteresis) between the trigger points on the IA output waveform in Figure 14–44(b).

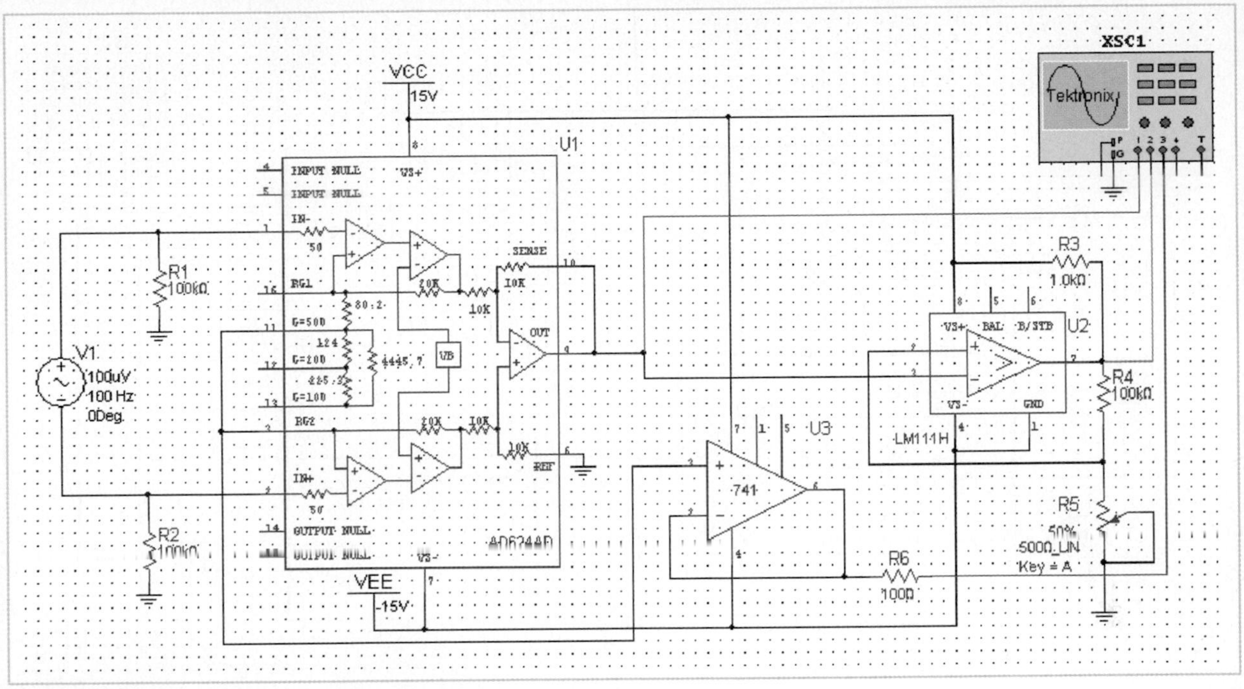

(a) Circuit screen

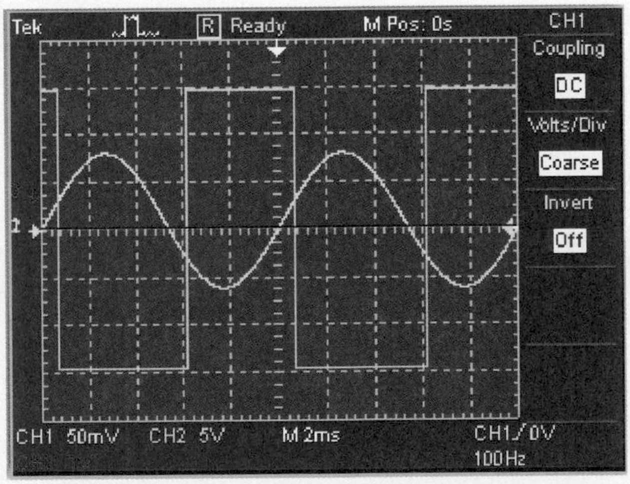

(b) Output of IA (yellow), output of comparator (blue), and output of guard driver (pink)

FIGURE 14–44

Simulation with a differential input.

Next, the input is changed to a 100 mV common-mode signal at a frequency of 60 Hz, and the simulation is run as shown in Figure 14–45. This simulates a low-frequency noise environment. Notice on the scope display that there is no output signal from the instrumentation amplifier, which indicates that it is rejecting the common-mode signal. The scope display also shows that the shield-guard driver correctly produces the common-mode signal.

5. Verify that the shield-guard driver output is equal to the common-mode signal.

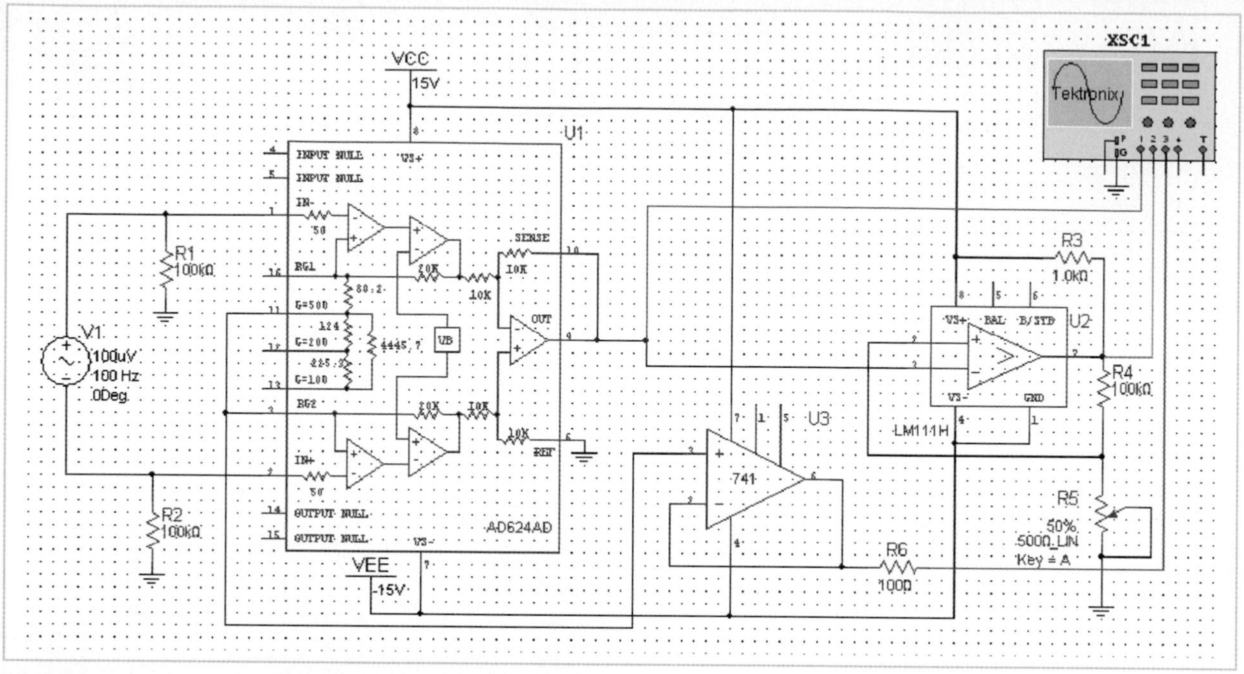

(a) Circuit screen

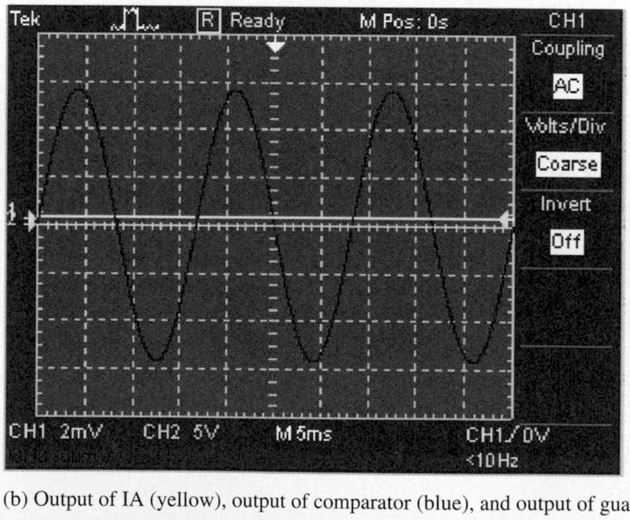

(b) Output of IA (yellow), output of comparator (blue), and output of guard driver (pink)

FIGURE 14–45

Simulation with a common-mode input.

Simulate the liquid-level control circuit using your Multisim software. Observe the operation with the virtual oscilloscope.

Prototyping and Testing

Now that the circuit has been simulated, the prototype circuit is constructed and tested. After the circuit is successfully tested on a protoboard, it is ready to be finalized on a printed circuit board.

Lab Experiment

To build and test a similar circuit, go to Experiment 14–A in your lab manual (*Laboratory Exercises for Electronic Devices* by David Buchla and Steven Wetterling).

Circuit Board

The liquid-level control circuit is implemented on a printed circuit board as shown in Figure 14–46. The dark gray lines represent backside connections.

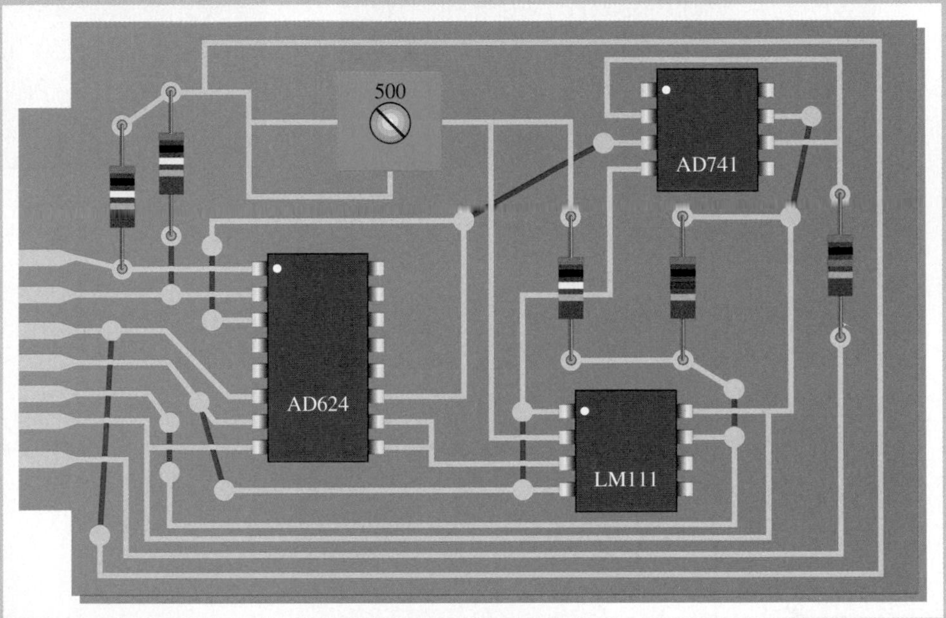

FIGURE 14–46

Liquid-level control board.

6. Check the printed circuit board and verify that it agrees with the schematic.
7. Label each input and output pin according to function.

Anadigm.com

Programmable Analog Design

Design Assignment

Create a circuit to provide a function similar to that of the level-control circuit in the Application Activity.

Procedure: Open your Designer2 software and configure the CAMs as shown in Figure 14–47.

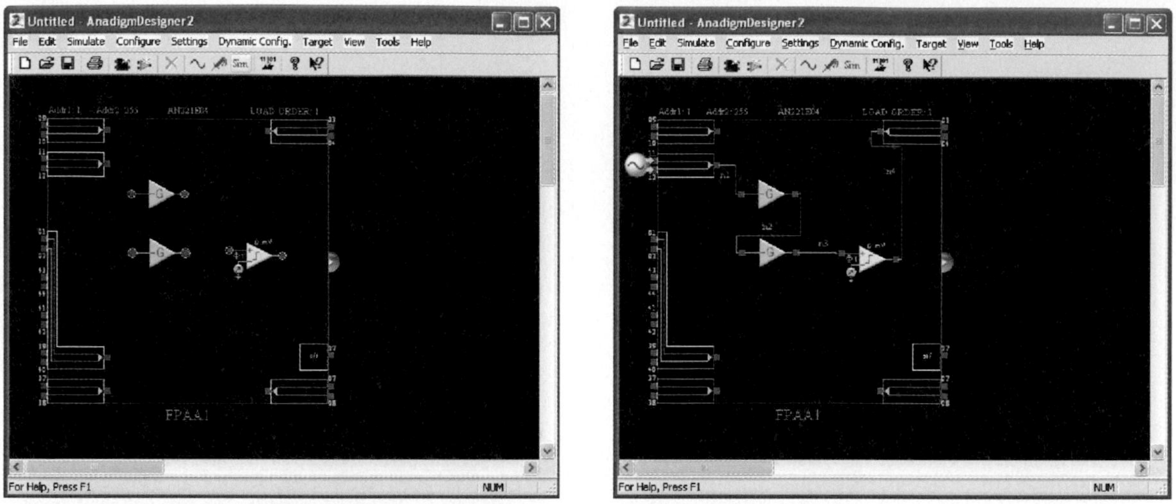

(a) Select and place two inverting gain stage CAMs and one comparator CAM.

(b) Connect the CAMs and add a differential signal source.

▲ FIGURE 14–47

Configure the signal generator as shown in Figure 14–48. Set the signal generator to represent a pressure sensor with a differential output and an amplitude of 100 μV. Note that the frequency is selected only to facilitate viewing.

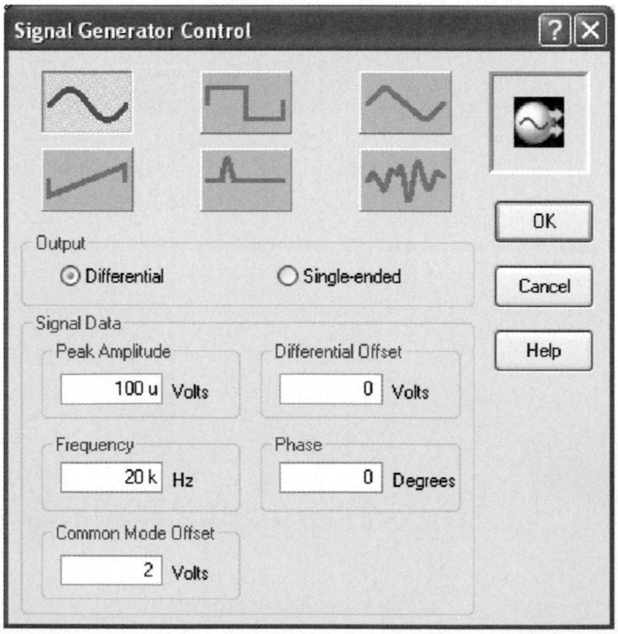

▲ FIGURE 14–48

Configure the gain stages for a total gain of 500, as shown in Figure 14–49.

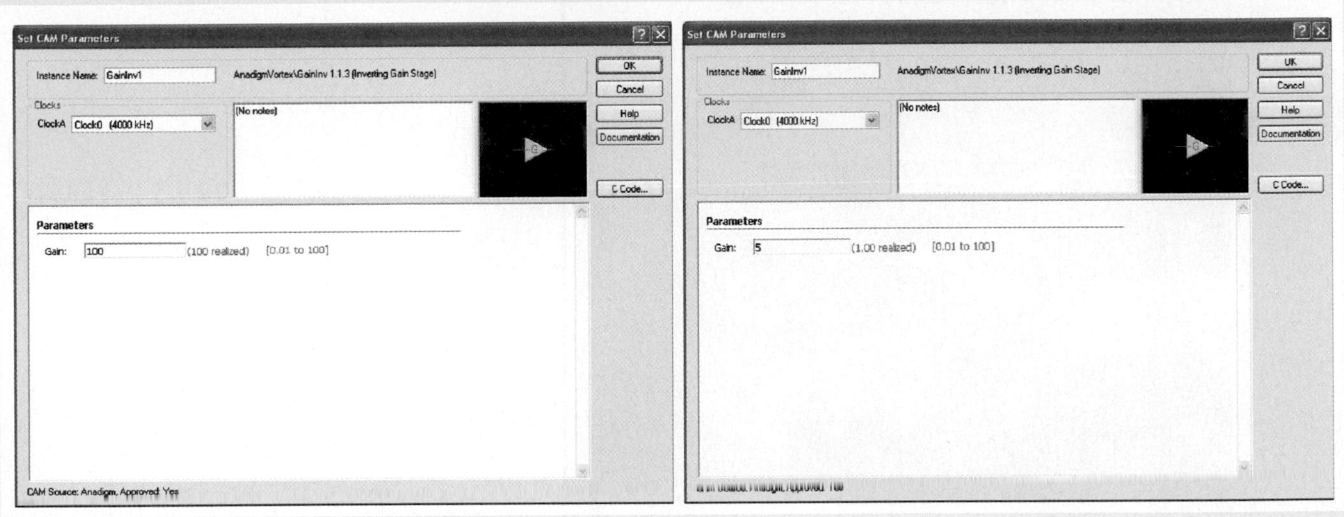

FIGURE 14–49

First stage has gain of 100 and second stage has gain of 5.

Configure the comparator for a hysteresis of 40 mV, as shown in Figure 14–50.

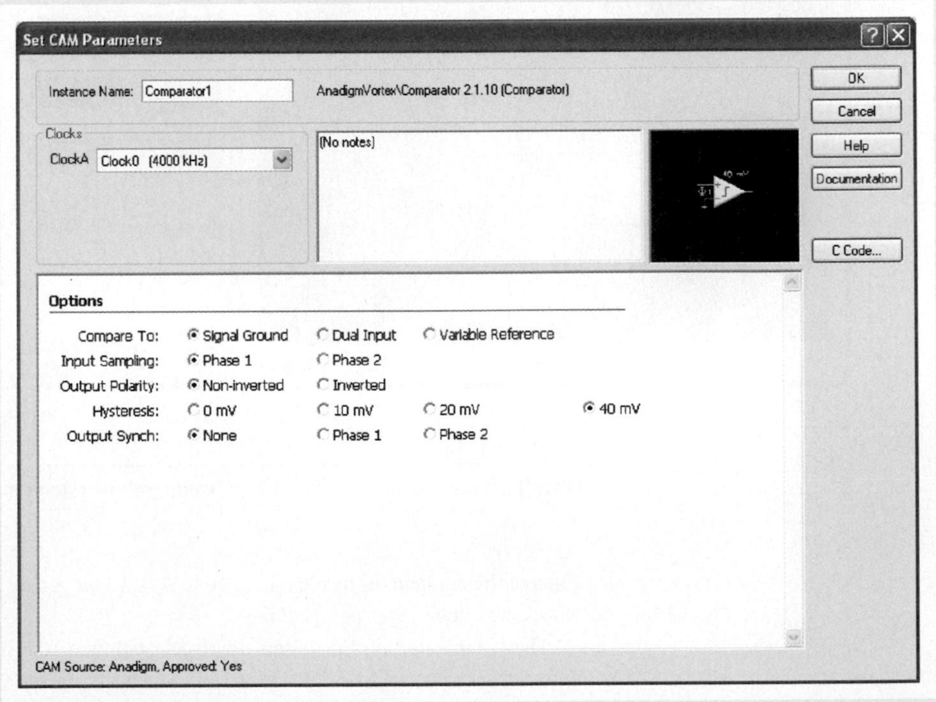

FIGURE 14–50

Analysis: Place probes as shown in Figure 14–51 (top) and run a simulation. The results are shown in Figure 14–51 (bottom).

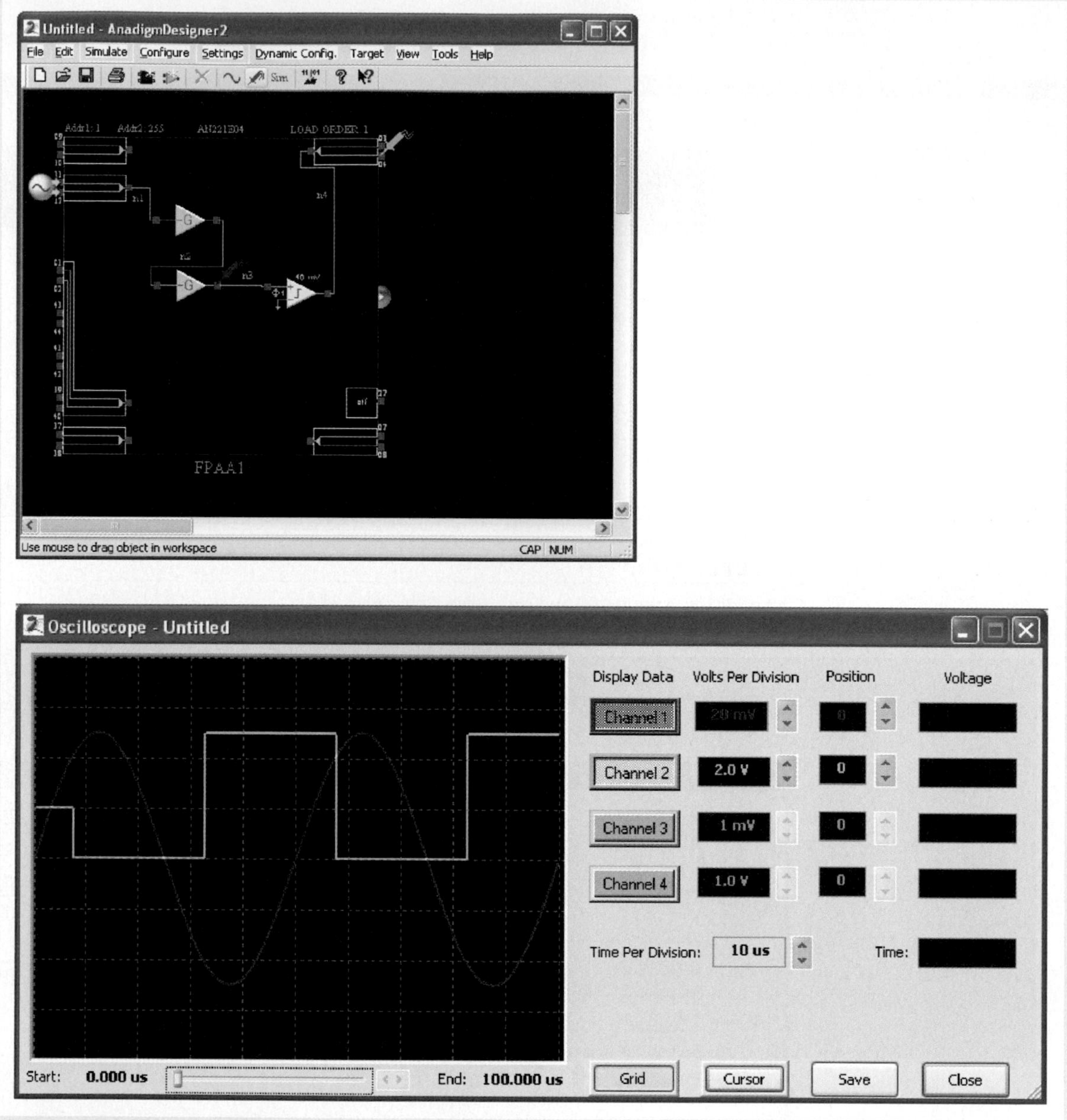

▲ **FIGURE 14–51**

Waveform measurement with a 40 mV comparator hysteresis.

Change the comparator hysteresis to 10 mV, and you get the waveform shown in Figure 14–52. Notice how the trigger points change.

The comparator hysteresis sets the trigger points on the signal so that a minimum and a maximum level can be set to control the level in a tank. Once an FPAA/dpASP is programmed with this design, the levels can be changed by programming a different hysteresis.

Programming Exercises

1. Open your Designer2 software.
2. Implement the level control circuit described.
3. Observe the output for a comparator hyteresis of 0, 10 mV, 20 mV, and 40 mV.

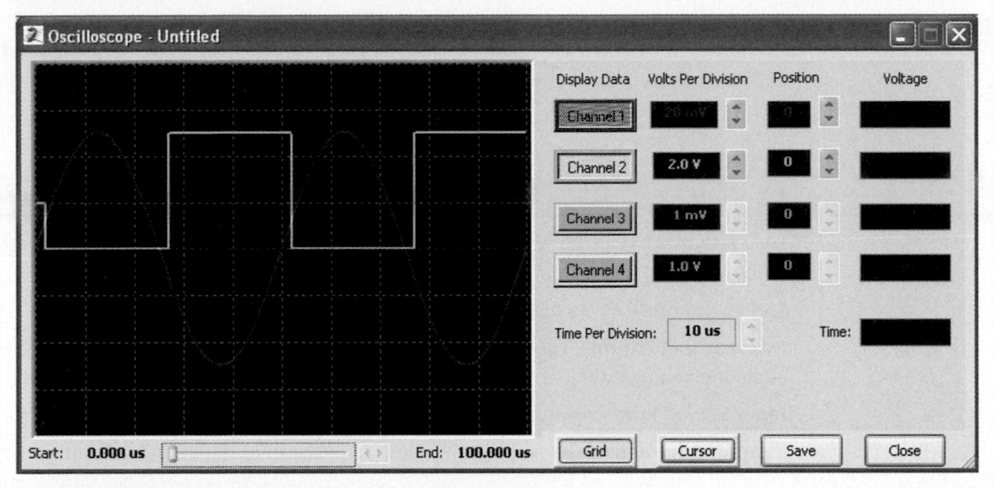

▲ FIGURE 14–52

Waveform measurement with a 10 mV comparator hysteresis.

PAM Experiment

To program, download, and test a circuit using AnadigmDesigner2 software and the programmable analog module (PAM) board, go to Experiment 14–B in *Laboratory Exercises for Electronic Devices* by David Buchla and Steven Wetterling.

SUMMARY

Section 14–1

- A basic instrumentation amplifier is formed by three op-amps and seven resistors, including the gain-setting resistor R_G.
- An instrumentation amplifier has high input impedance, high CMRR, low output offset, and low output impedance.
- The voltage gain of a basic instrumentation amplifier is set by a single external resistor.
- An instrumentation amplifier is useful in applications where small signals are embedded in large common-mode noise.

Section 14–2

- A basic isolation amplifier has electrically isolated input and output stages.
- Isolation amplifiers use capacitive, optical, or transformer coupling for isolation.
- Isolation amplifiers are used to interface sensitive equipment with high-voltage environments and to provide protection from electrical shock in certain medical applications.

Section 14–3

- The operational transconductance amplifier (OTA) is a voltage-to-current amplifier.
- The output current of an OTA is the input voltage times the transconductance.
- In an OTA, transconductance varies with bias current; therefore, the gain of an OTA can be varied with a bias voltage or a variable resistor.

Section 14–4

- The operation of log and antilog amplifiers is based on the nonlinear (logarithmic) characteristics of a *pn* junction.
- A log amplifier has a *pn* junction in the feedback loop, and an antilog amplifier has a *pn* junction in series with the input.

Section 14–5

- A constant-current source delivers the same load current regardless of load resistance (within limits).
- In a peak detector, an op-amp is used as a comparator to charge a capacitor through a diode to the peak value of the input voltage. It is useful in measuring peak voltage surges.

KEY TERMS

Key terms and other bold terms in the chapter are defined in the end-of-book glossary.

Instrumentation amplifier An amplifier used for amplifying small signals riding on large common-mode voltages.

Isolation amplifier An amplifier with electrically isolated internal stages.

Natural logarithm The exponent to which the base e ($e = 2.71828$) must be raised in order to equal a given quantity.

Operational transconductance amplifier (OTA) A voltage-to-current amplifier.

Transconductance In an electronic device, the ratio of the output current to the input voltage.

KEY FORMULAS

Instrumentation Amplifier

14–1 $$A_{cl} = 1 + \frac{2R}{R_G}$$

14–2 $$R_G = \frac{2R}{A_{cl} - 1}$$

Isolation Amplifier

14–3 $$A_{v1} = \frac{R_{f1}}{R_{i1}} + 1$$

14–4 $$A_{v2} = \frac{R_{f2}}{R_{i2}} + 1$$

Operational Transconductance Amplifier (OTA)

14–5 $$g_m = \frac{I_{out}}{V_{in}}$$

14–6 $$g_m = KI_{BIAS}$$

Log and Antilog Amplifiers

14–7 $$V_{out} = -K\ln(V_{in})$$

14–8 $$V_{out} \cong -(0.025\text{ V})\ln\left(\frac{V_{in}}{I_R R_1}\right)$$

14–9 $$V_{out} = -(0.025\text{ V})\ln\left(\frac{V_{in}}{I_{EBO} R_1}\right)$$

14–10 $$V_{out} = -R_f I_{EBO}\,\text{antilog}\left(\frac{V_{in}}{25\text{ mV}}\right)$$

Converters and Other Op-Amp Circuits

14–11 $$I_L = \frac{V_{IN}}{R_i}$$ Constant-current source

14–12 $$V_{out} = I_i R_f$$ Current-to-voltage converter

14–13 $$I_L = \frac{V_{in}}{R_1}$$ Voltage-to-current converter

TRUE/FALSE QUIZ

Answers are at the end of the chapter.

1. Instrumentation amplifiers are particularly useful for amplifying small signals in a noisy environment.
2. The gain of an instrumentation amplifier cannot be changed.
3. A basic instrumentation amplifier consists of three op-amps.
4. An isolation amplifier prefers to operate alone.
5. An isolation amplifier consists of two electrically isolated stages.
6. All isolation amplifiers use transformer coupling.
7. OTA stands for operational transistor amplifier.
8. The transconductance of an OTA is dependent on a bias current.
9. A log amplifier can be used for compression of large dynamic range signals.
10. A peak detector is a circuit that uses a diode and a capacitor to produce a dc voltage equal to the peak of the input signal voltage.

CIRCUIT-ACTION QUIZ

Answers are at the end of the chapter.

1. If the value of R_G in Figure 14–7 is increased, the voltage gain will
   (a) increase **(b)** decrease **(c)** not change
2. If the voltage gain of the instrumentation amplifier in Figure 14–7 is set to 10 at 1 kHz and the frequency is increased to 100 kHz, the gain will
   (a) increase **(b)** decrease **(c)** not change
3. If the voltage gain of the instrumentation amplifier in Figure 14–7 is increased from 10 to 100, the bandwidth will
   (a) increase **(b)** decrease **(c)** not change
4. If R_{f1} in the isolation amplifier of Figure 14–15 is increased to 33 kΩ, the total voltage gain will
   (a) increase **(b)** decrease **(c)** not change
5. If the values of all the capacitors in Figure 14–15 are changed to 0.68 μF, the gain of the output stage will
   (a) increase **(b)** decrease **(c)** not change
6. If the value of R_L in the OTA of Figure 14–23 is reduced, the voltage gain will
   (a) increase **(b)** decrease **(c)** not change
7. If the bias current in the OTA of Figure 14–23 is increased, the voltage gain will
   (a) increase **(b)** decrease **(c)** not change
8. In the log amplifier of Figure 14–31, when the value of R_1 is decreased, the output voltage will
   (a) increase **(b)** decrease **(c)** not change

SELF-TEST

Answers are at the end of the chapter.

Section 14–1

1. To make a basic instrumentation amplifier, it takes
   (a) one op-amp with a certain feedback arrangement
   (b) two op-amps and seven resistors
   (c) three op-amps and seven capacitors
   (d) three op-amps and seven resistors
2. Typically, an instrumentation amplifier has an external resistor used for
   (a) establishing the input impedance **(b)** setting the voltage gain
   (c) setting the current gain **(d)** interfacing with an instrument

3. Instrumentation amplifiers are used primarily in
   (a) high-noise environments (b) medical equipment
   (c) test instruments (d) filter circuits

Section 14–2

4. Isolation amplifiers are used primarily in
   (a) remote, isolated locations
   (b) systems that isolate a single signal from many different signals
   (c) applications where there are high voltages and sensitive equipment
   (d) applications where human safety is a concern
   (e) answers (c) and (d)
5. The two parts of a basic isolation amplifier are
   (a) amplifier and filter (b) input stage and coupling stage
   (c) input stage and output stage (d) gain stage and offset stage
6. The stages of many isolation amplifiers are connected by
   (a) copper strips (b) a capacitor (c) microwave links (d) current loops
7. The characteristic that allows an isolation amplifier to amplify small signal voltages in the presence of much greater noise voltages is its
   (a) CMRR (b) high gain
   (c) high input impedance (d) magnetic coupling between input and output

Section 14–3

8. The term *OTA* means
   (a) operational transistor amplifier (b) operational transformer amplifier
   (c) operational transconductance amplifier (d) output transducer amplifier
9. In an OTA, the transconductance is controlled by
   (a) the dc supply voltage (b) the input signal voltage
   (c) the manufacturing process (d) a bias current
10. The voltage gain of an OTA circuit is set by
    (a) a feedback resistor (b) the transconductance only
    (c) the transconductance and the load resistor (d) the bias current and supply voltage
11. An OTA is basically a
    (a) voltage-to-current amplifier (b) current-to-voltage amplifier
    (c) current-to-current amplifier (d) voltage-to-voltage amplifier

Section 14–4

12. The operation of a logarithmic amplifier is based on
    (a) the nonlinear operation of an op-amp
    (b) the logarithmic characteristic of a *pn* junction
    (c) the reverse breakdown characteristic of a *pn* junction
    (d) the logarithmic charge and discharge of an *RC* circuit
13. If the input to a log amplifier is x, the output is proportional to
    (a) e^x (b) $\ln x$ (c) $\log_{10}x$
    (d) $2.3 \log_{10}x$ (e) answers (a) and (c) (f) answers (b) and (d)
14. If the input to an antilog amplifier is x, the output is proportional to
    (a) $e^{\ln x}$ (b) e^x (c) $\ln x$ (d) e^{-x}

Section 14–5

15. A constant-current source provides a nonchanging current to a load
    (a) for all values of current
    (b) for all values of load resistance
    (c) for all values of load resistance within defined limits
16. A peak detector consists of
    (a) a comparator, a transistor, and a capacitor (b) a comparator, a diode, and a capacitor
    (c) a comparator, a diode, and an inductor (d) an integrator, a diode, and a capacitor

PROBLEMS

Answers to all odd-numbered problems are at the end of the book.

BASIC PROBLEMS

Section 14–1 Instrumentation Amplifiers

1. Determine the voltage gains of op-amps A1 and A2 for the instrumentation amplifier configuration in Figure 14–53.

FIGURE 14–53

Multisim file circuits are identified with a CD logo and are in the Problems folder on your CD-ROM. Filenames correspond to Figure numbers (e.g., F14–53).

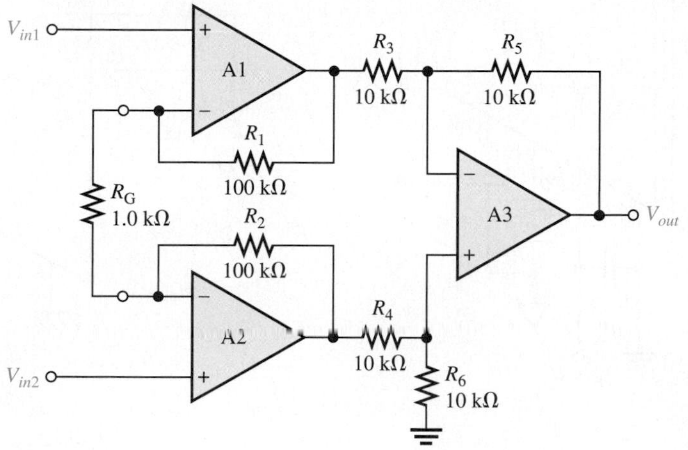

2. Find the overall voltage gain of the instrumentation amplifier in Figure 14–53.
3. The following voltages are applied to the instrumentation amplifier in Figure 14–53: $V_{in1} = 5$ mV, $V_{in2} = 10$ mV, and $V_{cm} = 225$ mV. Determine the final output voltage.
4. What value of R_G must be used to change the gain of the instrumentation amplifier in Figure 14–53 to 1000?
5. What is the voltage gain of the AD622 instrumentation amplifier in Figure 14–54?

FIGURE 14–54

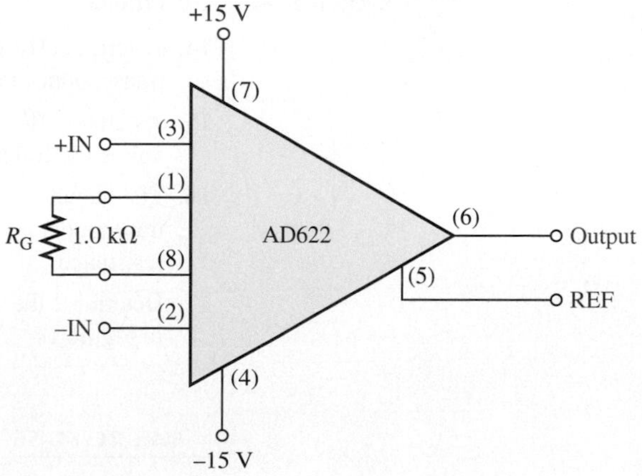

6. Determine the approximate bandwidth of the amplifier in Figure 14–54 if the voltage gain is set to 10. Use the graph in Figure 14–6.
7. Specify what you must do to change the gain of the amplifier in Figure 14–54 to approximately 24.
8. Determine the value of R_G in Figure 14–54 for a voltage gain of 20.

Section 14–2 Isolation Amplifiers

9. The op-amp in the input stage of a certain isolation amplifier has a voltage gain of 30. The output stage is set for a gain of 10. What is the total voltage gain of this device?

10. Determine the total voltage gain of each 3656KG in Figure 14–55.

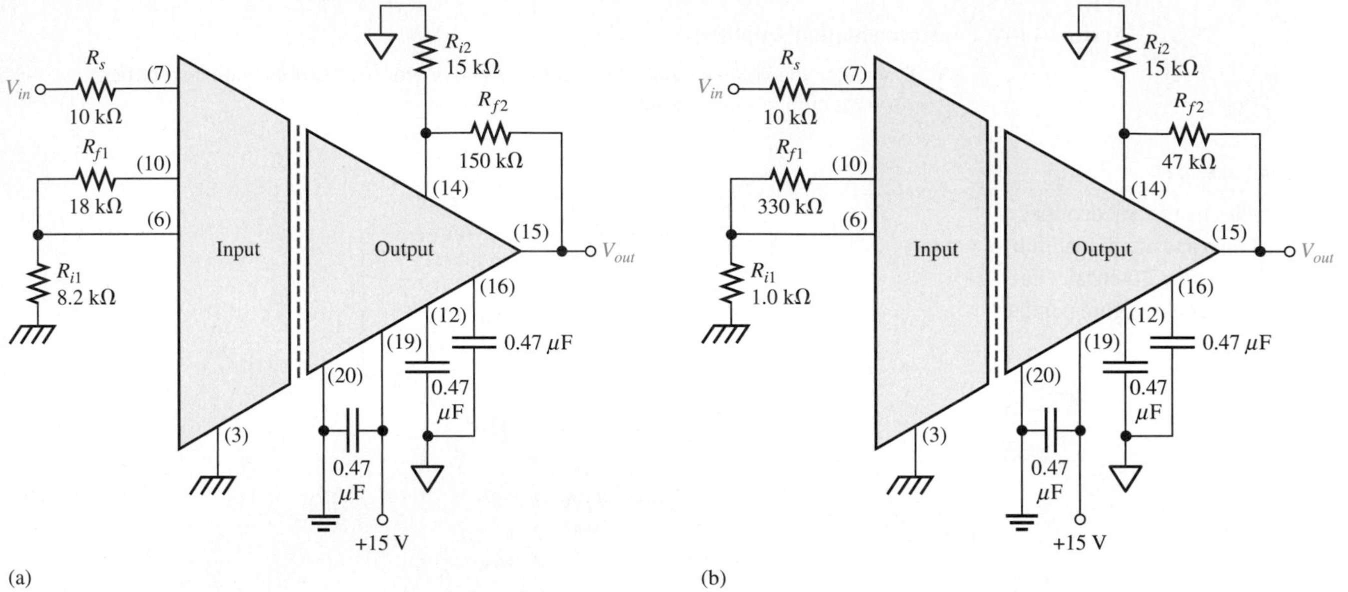

▲ FIGURE 14–55

11. Specify how you would change the total gain of the amplifier in Figure 14–55(a) to approximately 100 by changing only the gain of the input stage.

12. Specify how you would change the total gain in Figure 14–55(b) to approximately 440 by changing only the gain of the output stage.

13. Specify how you would connect each amplifier in Figure 14–55 for unity gain.

Section 14–3 Operational Transconductance Amplifiers (OTAs)

14. A certain OTA has an input voltage of 10 mV and an output current of 10 μA. What is the transconductance?

15. A certain OTA with a transconductance of 5000 μS has a load resistance of 10 kΩ. If the input voltage is 100 mV, what is the output current? What is the output voltage?

16. The output voltage of a certain OTA with a load resistance is determined to be 3.5 V. If its transconductance is 4000 μS and the input voltage is 100 mV, what is the value of the load resistance?

17. Determine the voltage gain of the OTA in Figure 14–56. Assume $K = 16\ \mu S/\mu A$ for the graph in Figure 14–57.

▶ FIGURE 14–56

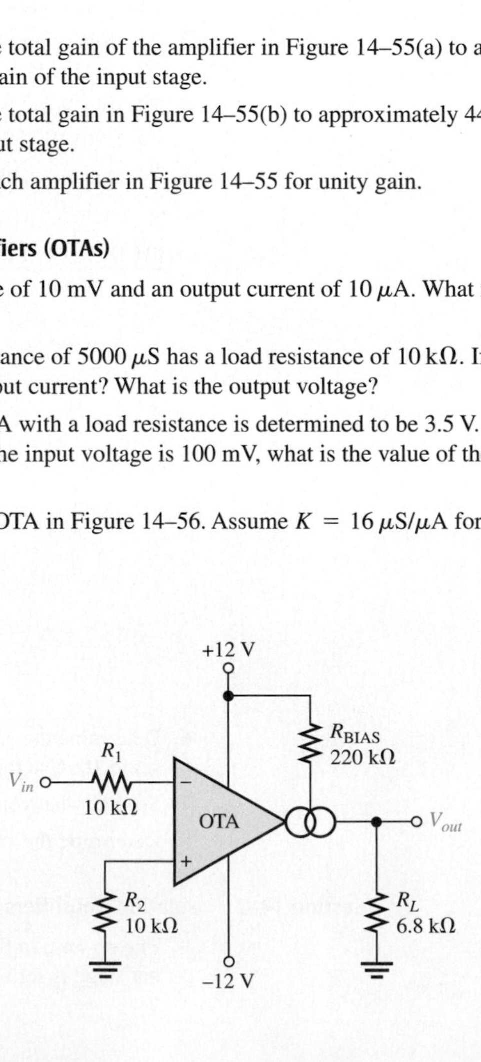

FIGURE 14–57

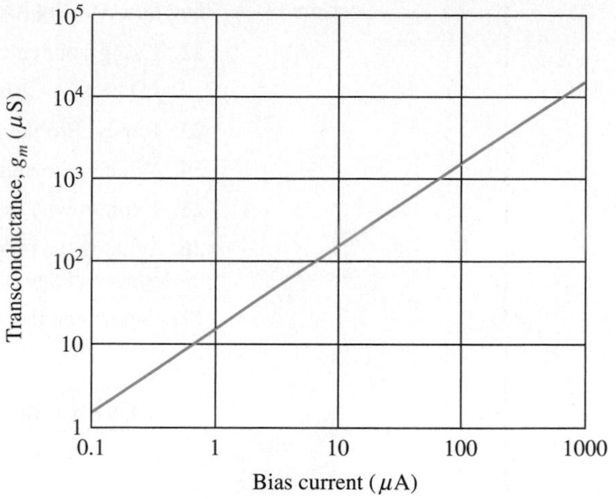

18. If a 10 kΩ rheostat is added in series with the bias resistor in Figure 14–56, what are the minimum and maximum voltage gains?

19. The OTA in Figure 14–58 functions as an amplitude modulation circuit. Determine the output voltage waveform for the given input waveforms assuming $K = 16\ \mu\text{S}/\mu\text{A}$.

FIGURE 14–58

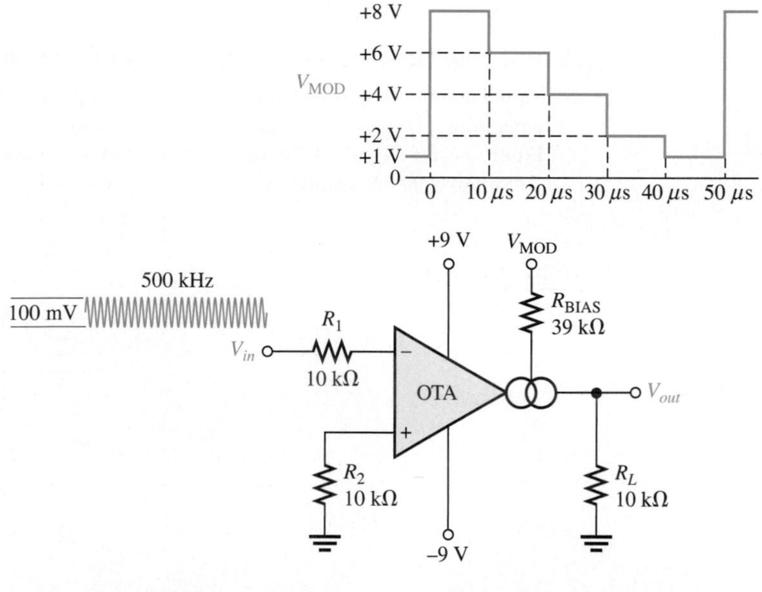

20. Determine the trigger points for the Schmitt trigger in Figure 14–59.

21. Determine the output voltage waveform for the Schmitt trigger in Figure 14–59 in relation to a 1 kHz sine wave with peak values of ±10 V.

FIGURE 14–59

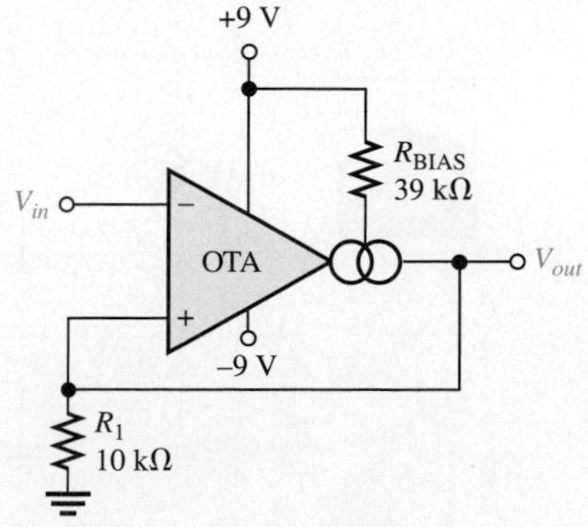

Section 14–4 Log and Antilog Amplifiers

22. Using your calculator, find the natural logarithm (ln) of each of the following numbers:

(a) 0.5 **(b)** 2 **(c)** 50 **(d)** 130

23. Repeat Problem 22 for $\log_{10}$.

24. What is the antilog of 1.6?

25. Explain why the output of a log amplifier is limited to approximately 0.7 V.

26. What is the output voltage of a certain log amplifier with a diode in the feedback path when the input voltage is 3 V? The input resistor is 82 kΩ and the reverse leakage current is 100 nA.

27. Determine the output voltage for the log amplifier in Figure 14–60. Assume $I_{EBO} = 60$ nA.

FIGURE 14–60

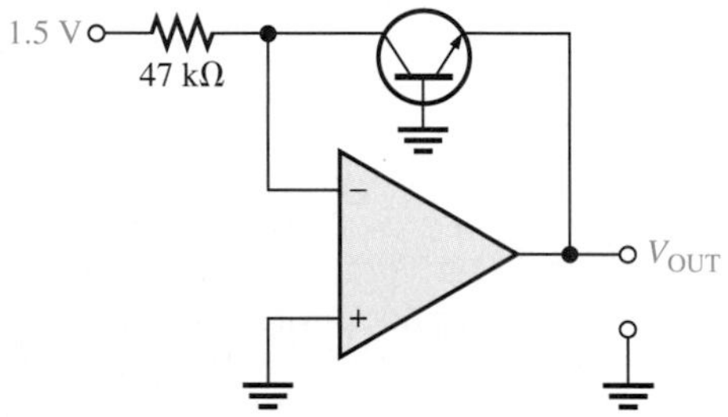

28. Determine the output voltage for the antilog amplifier in Figure 14–61. Assume $I_{EBO} = 60$ nA.

29. Signal compression is one application of logarithmic amplifiers. Suppose an audio signal with a maximum voltage of 1 V and a minimum voltage of 100 mV is applied to the log amplifier in Figure 14–60. What will be the maximum and minimum output voltages? What conclusion can you draw from this result?

FIGURE 14–61

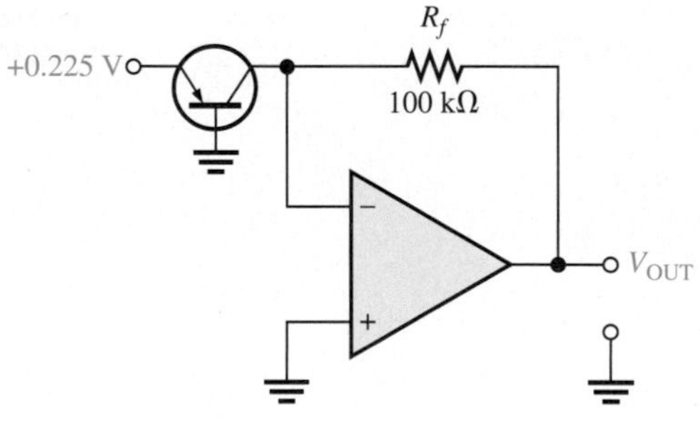

Section 14–5 Converters and Other Op-Amp Circuits

30. Determine the load current in each circuit of Figure 14–62.

FIGURE 14–62

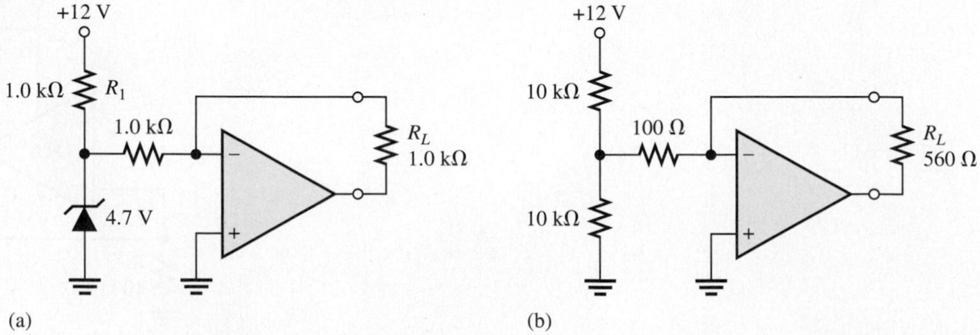

31. Devise a circuit for remotely sensing temperature and producing a proportional voltage that can then be converted to digital form for display. A thermistor can be used as the temperature-sensing element.

MULTISIM TROUBLESHOOTING PROBLEMS

These file circuits are in the Troubleshooting Problems folder on your CD-ROM.

32. Open file TSP14-32 and determine the fault.
33. Open file TSP14-33 and determine the fault.
34. Open file TSP14-34 and determine the fault.
35. Open file TSP14-35 and determine the fault.
36. Open file TSP14-36 and determine the fault.

ANSWERS

SECTION CHECKUPS

Section 14–1 Instrumentation Amplifiers

1. The main purpose of an instrumentation amplifier is to amplify small signals that occur on large common-mode voltages. The key characteristics are high input impedance, high CMRR, low output impedance, and low output offset.
2. Three op-amps and seven resistors including the gain resistor are required to construct a basic instrumentation amplifier (see Figure 14–2).
3. The gain is set by the external resistor R_G.
4. The gain is approximately 6.
5. To reduce effects of noise on the common-mode operation of an IA

Section 14–2 Isolation Amplifiers

1. Isolation amplifiers are used in medical equipment, power plant instrumentation, industrial processing, and automated testing.
2. The two stages of an isolation amplifier are input and output and their purpose is isolation.
3. The stages are connected by capacitive, optical, or transformer coupling.
4. The oscillator is used to provide the signal to be modulated.

Section 14–3 Operational Transconductance Amplifiers (OTAs)

1. OTA stands for Operational Transconductance Amplifier.
2. Transconductance increases with bias current.
3. Assuming that the bias input is connected to the supply voltage, the voltage gain increases when the supply voltage is increased because this increases the bias current.
4. The voltage gain decreases as the bias voltage decreases.

Section 14–4 Log and Antilog Amplifiers

1. A diode or transistor in the feedback loop provides the exponential (nonlinear) characteristic.
2. The output of a log amplifier is limited to the barrier potential of the *pn* junction (about 0.7 V).
3. The output voltage is determined by the input voltage, the input resistor, and the emitter-to-base leakage current.
4. The transistor in an antilog amplifier is in series with the input rather than in the feedback loop.

Section 14–5 Converters and Other Op-Amp Circuits

1. $I_L = 6.8\text{ V}/10\text{ k}\Omega = 0.68\text{ mA}$; same value to 5 kΩ load.
2. The feedback resistor is the constant of proportionality.

RELATED PROBLEMS FOR EXAMPLES

14–1 241 Ω

14–2 Make $R_G = 1.1\ k\Omega$.

14–3 The ripple could be removed by an output low-pass filter.

14–4 Many combinations are possible. Here is one: $R_{f1} = 10\ k\Omega$, $R_{i1} = 1.0\ k\Omega$, $R_{f2} = 10\ k\Omega$, and $R_{i2} = 1.0\ k\Omega$

14–5 $I_{BIAS} \cong 62.5\ \mu A$

14–6 Yes. The gain will change to approximately 110.

14–7 The output is a square-wave modulated signal with a maximum amplitude of approximately 3.6 V and a minimum amplitude of approximately 1.76 V.

14–8 −0.167 V

14–9 −0.193 V

14–10 −4.39 V

TRUE/FALSE QUIZ

1. T	**2.** F	**3.** T	**4.** F	**5.** T
6. F	**7.** F	**8.** T	**9.** T	**10.** T

CIRCUIT-ACTION QUIZ

1. (b)	**2.** (c)	**3.** (b)	**4.** (a)
5. (c)	**6.** (b)	**7.** (a)	**8.** (a)

SELF-TEST

1. (d)	**2.** (b)	**3.** (a)	**4.** (e)	**5.** (c)	**6.** (b)	**7.** (a)	**8.** (c)
9. (d)	**10.** (c)	**11.** (a)	**12.** (b)	**13.** (f)	**14.** (b)	**15.** (c)	**16.** (b)

ACTIVE FILTERS

15

CHAPTER OUTLINE

CHAPTER OBJECTIVES

- Describe the gain-versus-frequency responses of the basic filters
- Describe the three basic filter response characteristics and other filter parameters
- Analyze active low-pass filters
- Analyze active high-pass filters
- Analyze active band-pass filters
- Analyze active band-stop filters
- Discuss two methods for measuring frequency response

KEY TERMS

- Filter
- Low-pass filter
- Pole
- Roll-off
- High-pass filter
- Band-pass filter
- Band-stop filter
- Damping factor

APPLICATION ACTIVITY PREVIEW

RFID stands for Radio Frequency Identification and is a technology that enables the tracking and/or identification of objects. Typically, an RFID system consists of an RF tag containing an IC chip that transmits data about the object, a reader that receives transmitted data from the tag, and a data-processing system that processes and stores the data passed to it by the reader. In this application, you will focus on the RFID reader. RFID systems are used in metering applications such as electronic toll collection, inventory control and tracking, merchandise control, asset tracking and recovery, tracking parts moving through a manufacturing process, and tracking goods in a supply chain.

VISIT THE COMPANION WEBSITE

Study aids for this chapter are available at http://www.prenhall.com/floyd

INTRODUCTION

Power supply filters were introduced in Chapter 2. In this chapter, active filters that are used for signal processing are introduced. Filters are circuits that are capable of passing signals with certain selected frequencies while rejecting signals with other frequencies. This property is called *selectivity.*

Active filters use transistors or op-amps combined with passive *RC*, *RL*, or *RLC* circuits. The active devices provide voltage gain, and the passive circuits provide frequency selectivity. In terms of general response, the four basic categories of active filters are low-pass, high-pass, band-pass, and band-stop. In this chapter, you will study active filters using op-amps and *RC* circuits.

15–1 BASIC FILTER RESPONSES

Filters are usually categorized by the manner in which the output voltage varies with the frequency of the input voltage. The categories of active filters are low-pass, high-pass, band-pass, and band-stop. Each of these general responses are examined.

After completing this section, you should be able to

- **Describe the gain-versus-frequency responses of the basic filters**
  - Explain the low-pass filter response
  - Determine the critical frequency and bandwidth of a low-pass filter
  - Explain the high-pass filter response
  - Determine the critical frequency of a high-pass filter
  - Explain the band-pass filter response
  - Explain the significance of the quality factor
  - Determine the critical frequency, bandwidth, quality factor, and damping factor of a band-pass filter
  - Explain the band-stop filter response

Low-Pass Filter Response

A **filter** is a circuit that passes certain frequencies and attenuates or rejects all other frequencies. The **passband** of a filter is the range of frequencies that are allowed to pass through the filter with minimum attenuation (usually defined as less than -3 dB of attenuation). The **critical frequency,** f_c, (also called the *cutoff frequency*) defines the end of the passband and is normally specified at the point where the response drops -3 dB (70.7%) from the passband response. Following the passband is a region called the *transition region* that leads into a region called the *stopband.* There is no precise point between the transition region and the stopband.

A **low-pass filter** is one that passes frequencies from dc to f_c and significantly attenuates all other frequencies. The passband of the ideal low-pass filter is shown in the blue-shaded area of Figure 15–1(a); the response drops to zero at frequencies beyond the passband. This ideal response is sometimes referred to as a "brick-wall" because nothing gets through beyond the wall. The bandwidth of an ideal low-pass filter is equal to f_c.

Equation 15–1

$$BW = f_c$$

The ideal response shown in Figure 15–1(a) is not attainable by any practical filter. Actual filter responses depend on the number of **poles**, a term used with filters to describe the number of *RC* circuits contained in the filter. The most basic low-pass filter is a simple *RC* circuit consisting of just one resistor and one capacitor; the output is taken across the capacitor as shown in Figure 15–1(b). This basic *RC* filter has a single pole, and it rolls off at -20 dB/decade beyond the critical frequency. The actual response is indicated by the blue line in Figure 15–1(a). The response is plotted on a standard log plot that is used for filters to show details of the curve as the gain drops. Notice that the gain drops off slowly until the frequency is at the critical frequency; after this, the gain drops rapidly.

The -20 dB/decade **roll-off** rate for the gain of a basic *RC* filter means that at a frequency of $10f_c$, the output will be -20 dB (10%) of the input. This roll-off rate is not a particularly good filter characteristic because too much of the unwanted frequencies (beyond the passband) are allowed through the filter.

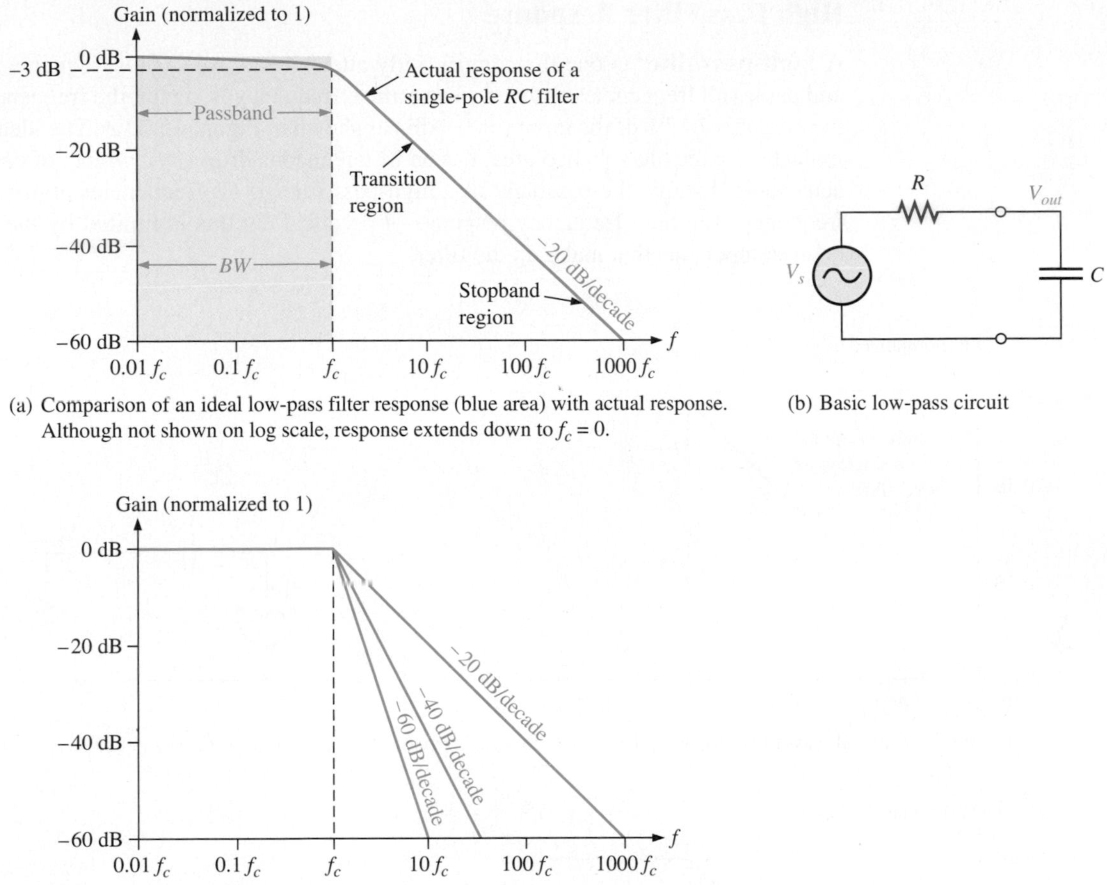

(a) Comparison of an ideal low-pass filter response (blue area) with actual response. Although not shown on log scale, response extends down to $f_c = 0$.

(b) Basic low-pass circuit

(c) Idealized low-pass filter responses

▲ **FIGURE 15–1**

Low-pass filter responses.

The critical frequency of a low-pass *RC* filter occurs when $X_C = R$, where

$$f_c = \frac{1}{2\pi RC}$$

Recall from your basic dc/ac studies that the output at the critical frequency is 70.7% of the input. This response is equivalent to an attenuation of −3 dB.

Figure 15–1(c) illustrates three idealized low-pass response curves including the basic one-pole response (−20 dB/decade). The approximations show a flat response to the cutoff frequency and a roll-off at a constant rate after the cutoff frequency. Actual filters do not have a perfectly flat response up to the cutoff frequency but drop to −3 dB at this point as described previously.

In order to produce a filter that has a steeper transition region (and hence form a more effective filter), it is necessary to add additional circuitry to the basic filter. Responses that are steeper than −20 dB/decade in the transition region cannot be obtained by simply cascading identical *RC* stages (due to loading effects). However, by combining an op-amp with frequency-selective feedback circuits, filters can be designed with roll-off rates of −40, −60, or more dB/decade. Filters that include one or more op-amps in the design are called **active filters.** These filters can optimize the roll-off rate or other attribute (such as phase response) with a particular filter design. In general, the more poles the filter uses, the steeper its transition region will be. The exact response depends on the type of filter and the number of poles.

High-Pass Filter Response

A **high-pass filter** is one that significantly attenuates or rejects all frequencies below f_c and passes all frequencies above f_c. The critical frequency is, again, the frequency at which the output is 70.7% of the input (or −3 dB) as shown in Figure 15–2(a). The ideal response, indicated by the blue-shaded area, has an instantaneous drop at f_c, which, of course, is not achievable. Ideally, the passband of a high-pass filter is all frequencies above the critical frequency. The high-frequency response of practical circuits is limited by the op-amp or other components that make up the filter.

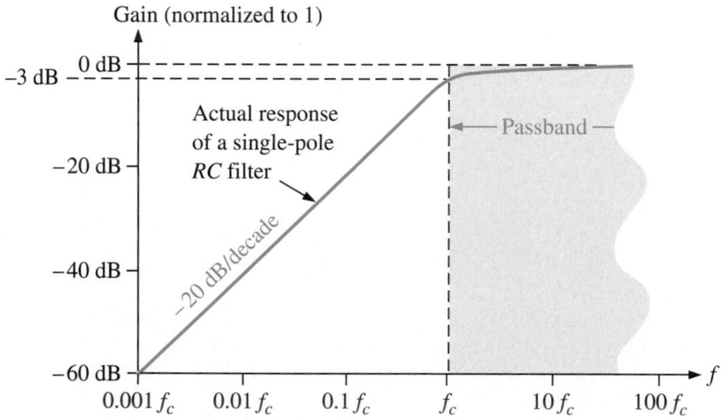

(a) Comparison of an ideal high-pass filter response (blue area) with actual response

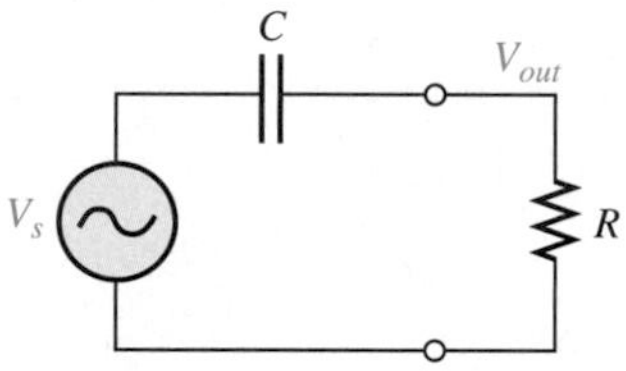

(b) Basic high-pass circuit

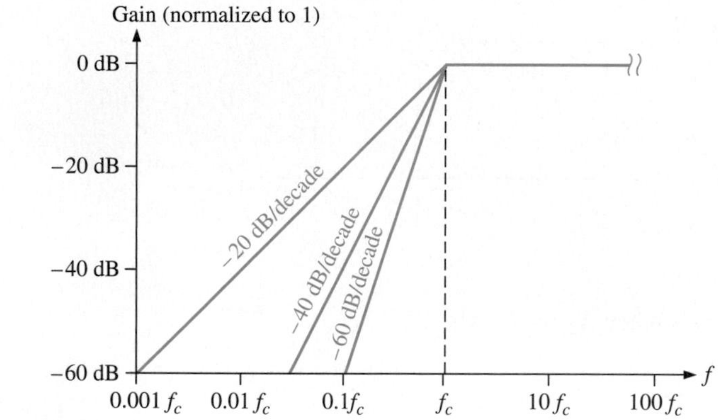

(c) Idealized high-pass filter responses

▲ **FIGURE 15–2**

High-pass filter responses.

A simple *RC* circuit consisting of a single resistor and capacitor can be configured as a high-pass filter by taking the output across the resistor as shown in Figure 15–2(b). As in the case of the low-pass filter, the basic *RC* circuit has a roll-off rate of −20 dB/decade, as indicated by the blue line in Figure 15–2(a). Also, the critical frequency for the basic high-pass filter occurs when $X_C = R$, where

$$f_c = \frac{1}{2\pi RC}$$

Figure 15–2(c) illustrates three idealized high-pass response curves including the basic one-pole response (−20 dB/decade) for a high-pass *RC* circuit. As in the case of the low-pass filter, the approximations show a flat response to the cutoff frequency and a roll-off at a constant rate after the cutoff frequency. Actual high-pass filters do not have the perfectly flat response indicated or the precise roll-off rate shown. Responses that are steeper than −20 dB/decade in the transition region are also possible with active high-pass filters; the particular response depends on the type of filter and the number of poles.

Band-Pass Filter Response

A **band-pass filter** passes all signals lying within a band between a lower-frequency limit and an upper-frequency limit and essentially rejects all other frequencies that are outside this specified band. A generalized band-pass response curve is shown in Figure 15–3. The bandwidth (*BW*) is defined as the difference between the upper critical frequency (f_{c2}) and the lower critical frequency (f_{c1}).

$$BW = f_{c2} - f_{c1}$$

Equation 15–2

The critical frequencies are, of course, the points at which the response curve is 70.7% of its maximum. Recall from Chapter 12 that these critical frequencies are also called *3 dB frequencies.* The frequency about which the passband is centered is called the *center frequency,* f_0, defined as the geometric mean of the critical frequencies.

$$f_0 = \sqrt{f_{c1}f_{c2}}$$

Equation 15–3

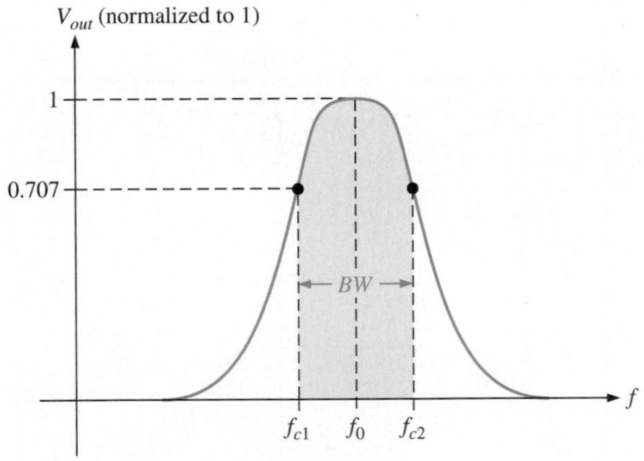

FIGURE 15–3

General band-pass response curve.

Quality Factor The **quality factor (*Q*)** of a band-pass filter is the ratio of the center frequency to the bandwidth.

$$Q = \frac{f_0}{BW}$$

Equation 15–4

The value of Q is an indication of the selectivity of a band-pass filter. The higher the value of Q, the narrower the bandwidth and the better the selectivity for a given value of f_0. Band-pass filters are sometimes classified as narrow-band ($Q > 10$) or wide-band ($Q < 10$). The quality factor (Q) can also be expressed in terms of the damping factor (DF) of the filter as

$$Q = \frac{1}{DF}$$

You will study the damping factor in Section 15–2.

EXAMPLE 15–1

A certain band-pass filter has a center frequency of 15 kHz and a bandwidth of 1 kHz. Determine Q and classify the filter as narrow-band or wide-band.

Solution

$$Q = \frac{f_0}{BW} = \frac{15\text{ kHz}}{1\text{ kHz}} = \mathbf{15}$$

Because $Q > 10$, this is a narrow-band filter.

*Related Problem** If the quality factor of the filter is doubled, what will the bandwidth be?

*Answers are at the end of the chapter.

Band-Stop Filter Response

Another category of active filter is the **band-stop filter**, also known as *notch, band-reject,* or *band-elimination* filter. You can think of the operation as opposite to that of the band-pass filter because frequencies within a certain bandwidth are rejected, and frequencies outside the bandwidth are passed. A general response curve for a band-stop filter is shown in Figure 15–4. Notice that the bandwidth is the band of frequencies between the 3 dB points, just as in the case of the band-pass filter response.

FIGURE 15–4

General band-stop filter response.

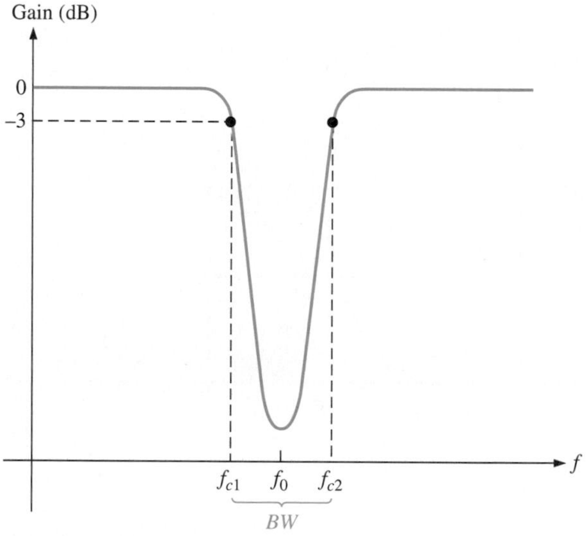

SECTION 15–1 CHECKUP
Answers are at the end of the chapter.

1. What determines the bandwidth of a low-pass filter?
2. What limits the passband of an active high-pass filter?
3. How are the Q and the bandwidth of a band-pass filter related? Explain how the selectivity is affected by the Q of a filter.

15–2 FILTER RESPONSE CHARACTERISTICS

Each type of filter response (low-pass, high-pass, band-pass, or band-stop) can be tailored by circuit component values to have either a Butterworth, Chebyshev, or Bessel characteristic. Each of these characteristics is identified by the shape of the response curve, and each has an advantage in certain applications.

After completing this section, you should be able to

- **Describe the three basic filter response characteristics and other filter parameters**
    - Describe the Butterworth characteristic
    - Describe the Chebyshev characteristic
    - Describe the Bessel characteristic
    - Define *damping factor* and discuss its significance
    - Calculate the damping factor of a filter
    - Discuss the order of a filter and its affect on the roll-off rate

Butterworth, Chebyshev, or Bessel response characteristics can be realized with most active filter circuit configurations by proper selection of certain component values. A general comparison of the three response characteristics for a low-pass filter response curve is shown in Figure 15–5. High-pass and band-pass filters can also be designed to have any one of the characteristics.

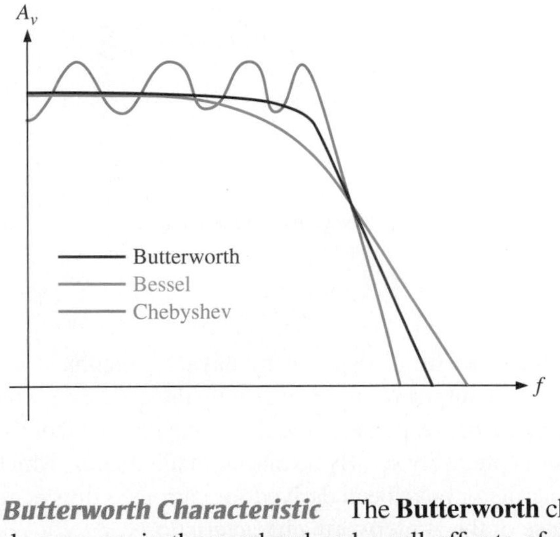

FIGURE 15–5

Comparative plots of three types of filter response characteristics.

The Butterworth Characteristic The **Butterworth** characteristic provides a very flat amplitude response in the passband and a roll-off rate of −20 dB/decade/pole. The phase response is not linear, however, and the phase shift (thus, time delay) of signals passing through the filter varies nonlinearly with frequency. Therefore, a pulse applied to a filter with a Butterworth response will cause overshoots on the output because each frequency component of the pulse's rising and falling edges experiences a different time delay. Filters with the Butterworth response are normally used when all frequencies in the passband must have the same gain. The Butterworth response is often referred to as a maximally flat response.

The Chebyshev Characteristic Filters with the **Chebyshev** response characteristic are useful when a rapid roll-off is required because it provides a roll-off rate greater than −20 dB/decade/pole. This is a greater rate than that of the Butterworth, so filters can be implemented with the Chebyshev response with fewer poles and less complex circuitry for a given roll-off rate. This type of filter response is characterized by overshoot or ripples in the passband (depending on the number of poles) and an even less linear phase response than the Butterworth.

The Bessel Characteristic The **Bessel** response exhibits a linear phase characteristic, meaning that the phase shift increases linearly with frequency. The result is almost no overshoot on the output with a pulse input. For this reason, filters with the Bessel response are used for filtering pulse waveforms without distorting the shape of the waveform.

The Damping Factor

As mentioned, an active filter can be designed to have either a Butterworth, Chebyshev, or Bessel response characteristic regardless of whether it is a low-pass, high-pass, band-pass, or band-stop type. The **damping factor (*DF*)** of an active filter circuit determines which response characteristic the filter exhibits. To explain the basic concept, a generalized active filter is shown in Figure 15–6. It includes an amplifier, a negative feedback circuit, and a filter section. The amplifier and feedback are connected in a noninverting configuration. The damping factor is determined by the negative feedback circuit and is defined by the following equation:

Equation 15–5

$$DF = 2 - \frac{R_1}{R_2}$$

FIGURE 15–6

General diagram of an active filter.

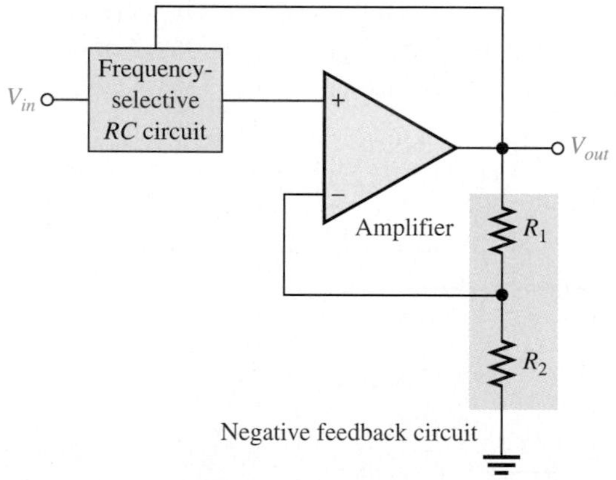

Basically, the damping factor affects the filter response by negative feedback action. Any attempted increase or decrease in the output voltage is offset by the opposing effect of the negative feedback. This tends to make the response curve flat in the passband of the filter if the value for the damping factor is precisely set. By advanced mathematics, which we will not cover, values for the damping factor have been derived for various orders of filters to achieve the maximally flat response of the Butterworth characteristic.

The value of the damping factor required to produce a desired response characteristic depends on the **order** (number of poles) of the filter. A *pole,* for our purposes, is simply a circuit with one resistor and one capacitor. The more poles a filter has, the faster its roll-off rate is. To achieve a second-order Butterworth response, for example, the damping factor must be 1.414. To implement this damping factor, the feedback resistor ratio must be

$$\frac{R_1}{R_2} = 2 - DF = 2 - 1.414 = 0.586$$

This ratio gives the closed-loop gain of the noninverting amplifier portion of the filter, $A_{cl(\mathrm{NI})}$, a value of 1.586, derived as follows:

$$A_{cl(\mathrm{NI})} = \frac{1}{B} = \frac{1}{R_2/(R_1 + R_2)} = \frac{R_1 + R_2}{R_2} = \frac{R_1}{R_2} + 1 = 0.586 + 1 = 1.586$$

EXAMPLE 15–2

If resistor R_2 in the feedback circuit of an active two-pole filter of the type in Figure 15–6 is 10 kΩ, what value must R_1 be to obtain a maximally flat Butterworth response?

Solution

$$\frac{R_1}{R_2} = 0.586$$

$$R_1 = 0.586R_2 = 0.586(10\text{ k}\Omega) = \mathbf{5.86\text{ k}\Omega}$$

Using the nearest standard 5 percent value of 5.6 kΩ will get very close to the ideal Butterworth response.

Related Problem What is the damping factor for $R_2 = 10\text{ k}\Omega$ and $R_1 = 5.6\text{ k}\Omega$?

Critical Frequency and Roll-Off Rate

The critical frequency is determined by the values of the resistors and capacitors in the frequency-selective RC circuit shown in Figure 15–6. For a single-pole (first-order) filter, as shown in Figure 15–7, the critical frequency is

$$f_c = \frac{1}{2\pi RC}$$

Although we show a low-pass configuration, the same formula is used for the f_c of a single-pole high-pass filter. The number of poles determines the roll-off rate of the filter. A Butterworth response produces −20 dB/decade/pole. So, a first-order (one-pole) filter has a roll-off of −20 dB/decade; a second-order (two-pole) filter has a roll-off rate of −40 dB/decade; a third-order (three-pole) filter has a roll-off rate of −60 dB/decade; and so on.

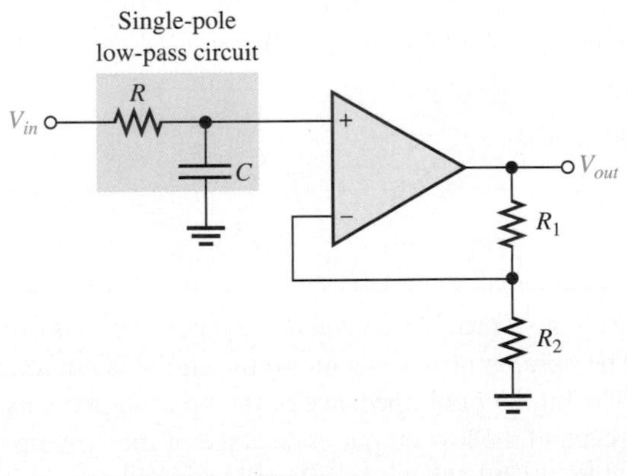

FIGURE 15–7

First-order (one-pole) low-pass filter.

Generally, to obtain a filter with three poles or more, one-pole or two-pole filters are cascaded, as shown in Figure 15–8. To obtain a third-order filter, for example, cascade a second-order and a first-order filter; to obtain a fourth-order filter, cascade two second-order filters; and so on. Each filter in a cascaded arrangement is called a *stage* or *section.*

Because of its maximally flat response, the Butterworth characteristic is the most widely used. Therefore, we will limit our coverage to the Butterworth response to illustrate basic filter concepts. Table 15–1 lists the roll-off rates, damping factors, and feedback resistor ratios for up to sixth-order Butterworth filters. Resistor designations correspond to the gain-setting resistors in Figure 15–8 and may be different on other circuit diagrams.

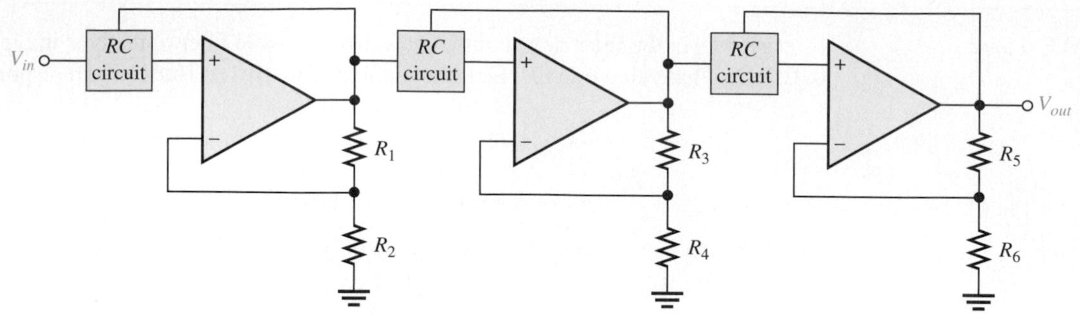

▲ FIGURE 15–8

The number of filter poles can be increased by cascading.

▼ TABLE 15–1

Values for the Butterworth response.

		1ST STAGE			2ND STAGE			3RD STAGE		
ORDER	ROLL-OFF DB/DECADE	POLES	*DF*	R_1/R_2	POLES	*DF*	R_3/R_4	POLES	*DF*	R_5/R_6
1	−20	1	Optional							
2	−40	2	1.414	0.586						
3	−60	2	1.00	1	1	1.00	1			
4	−80	2	1.848	0.152	2	0.765	1.235			
5	−100	2	1.00	1	2	1.618	0.382	1	0.618	1.382
6	−120	2	1.932	0.068	2	1.414	0.586	2	0.518	1.482

SECTION 15–2 CHECKUP

1. Explain how Butterworth, Chebyshev, and Bessel responses differ.
2. What determines the response characteristic of a filter?
3. Name the basic parts of an active filter.

15–3 ACTIVE LOW-PASS FILTERS

Filters that use op-amps as the active element provide several advantages over passive filters (*R*, *L*, and *C* elements only). The op-amp provides gain, so the signal is not attenuated as it passes through the filter. The high input impedance of the op-amp prevents excessive loading of the driving source, and the low output impedance of the op-amp prevents the filter from being affected by the load that it is driving. Active filters are also easy to adjust over a wide frequency range without altering the desired response.

After completing this section, you should be able to

- **Analyze active low-pass filters**
  - Identify a single-pole filter and determine its gain and critical frequency
  - Identify a two-pole Sallen-Key low-pass filter and determine its gain and critical frequency
  - Explain how a higher roll-off rate is achieved by cascading low-pass filters

A Single-Pole Filter

Figure 15–9(a) shows an active filter with a single low-pass *RC* frequency-selective circuit that provides a roll-off of −20 dB/decade above the critical frequency, as indicated by the response curve in Figure 15–9(b). The critical frequency of the single-pole filter is $f_c = 1/(2\pi RC)$. The op-amp in this filter is connected as a noninverting amplifier with the closed-loop voltage gain in the passband set by the values of R_1 and R_2.

$$A_{cl(NI)} = \frac{R_1}{R_2} + 1$$

Equation 15–6

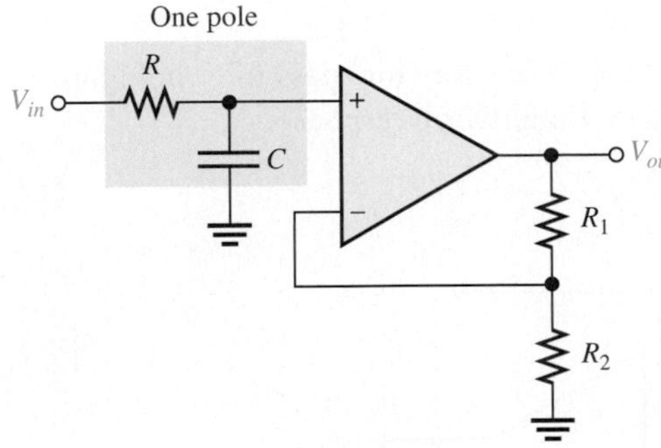

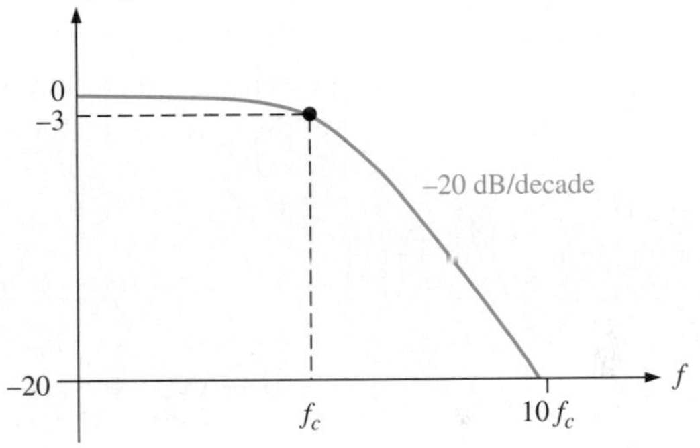

FIGURE 15–9

Single-pole active low-pass filter and response curve.

The Sallen-Key Low-Pass Filter

The Sallen-Key is one of the most common configurations for a second-order (two-pole) filter. It is also known as a VCVS (voltage-controlled voltage source) filter. A low-pass version of the Sallen-Key filter is shown in Figure 15–10. Notice that there are two low-pass *RC* circuits that provide a roll-off of −40 dB/decade above the critical frequency (assuming a Butterworth characteristic). One *RC* circuit consists of R_A and C_A, and the second circuit consists of R_B and C_B. A unique feature of the Sallen-Key low-pass filter is the capacitor C_A that provides feedback for shaping the response near the edge of the passband. The critical frequency for the Sallen-Key filter is

$$f_c = \frac{1}{2\pi\sqrt{R_A R_B C_A C_B}}$$

Equation 15–7

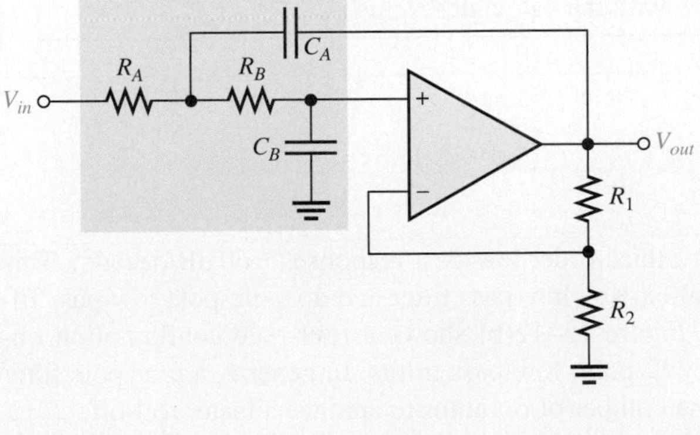

FIGURE 15–10

Basic Sallen-Key low-pass filter.

The component values can be made equal so that $R_A = R_B = R$ and $C_A = C_B = C$. In this case, the expression for the critical frequency simplifies to

$$f_c = \frac{1}{2\pi RC}$$

As in the single-pole filter, the op-amp in the second-order Sallen-Key filter acts as a noninverting amplifier with the negative feedback provided by resistors R_1 and R_2. As you have learned, the damping factor is set by the values of R_1 and R_2, thus making the filter response either Butterworth, Chebyshev, or Bessel. For example, from Table 15–1, the R_1/R_2 ratio must be 0.586 to produce the damping factor of 1.414 required for a second-order Butterworth response.

EXAMPLE 15–3

Determine the critical frequency of the Sallen-Key low-pass filter in Figure 15–11, and set the value of R_1 for an approximate Butterworth response.

FIGURE 15–11

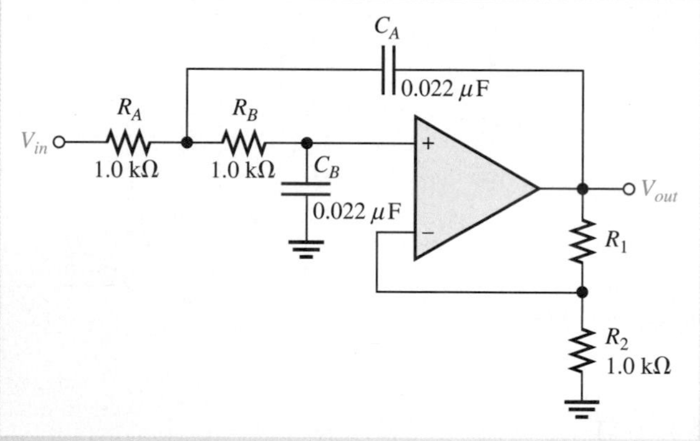

Solution Since $R_A = R_B = R = 1.0\,\text{k}\Omega$ and $C_A = C_B = C = 0.022\,\mu\text{F}$,

$$f_c = \frac{1}{2\pi RC} = \frac{1}{2\pi(1.0\,\text{k}\Omega)(0.022\,\mu\text{F})} = \mathbf{7.23\,kHz}$$

For a Butterworth response, $R_1/R_2 = 0.586$.

$$R_1 = 0.586R_2 = 0.586(1.0\,\text{k}\Omega) = \mathbf{586\,\Omega}$$

Select a standard value as near as possible to this calculated value.

Related Problem Determine f_c for Figure 15–11 if $R_A = R_B = R_2 = 2.2\,\text{k}\Omega$ and $C_A = C_B = 0.01\,\mu\text{F}$. Also determine the value of R_1 for a Butterworth response.

Open the Multisim file E15-03 in the Examples folder on your CD-ROM. Determine the critical frequency and compare with the calculated value.

Cascaded Low-Pass Filters

A three-pole filter is required to get a third-order low-pass response (−60 dB/decade). This is done by cascading a two-pole Sallen-Key low-pass filter and a single-pole low-pass filter, as shown in Figure 15–12(a). Figure 15–12(b) shows a four-pole configuration obtained by cascading two Sallen-Key (2-pole) low-pass filters. In general, a four-pole filter is preferred because it uses the same number of op-amps to achieve a faster roll-off.

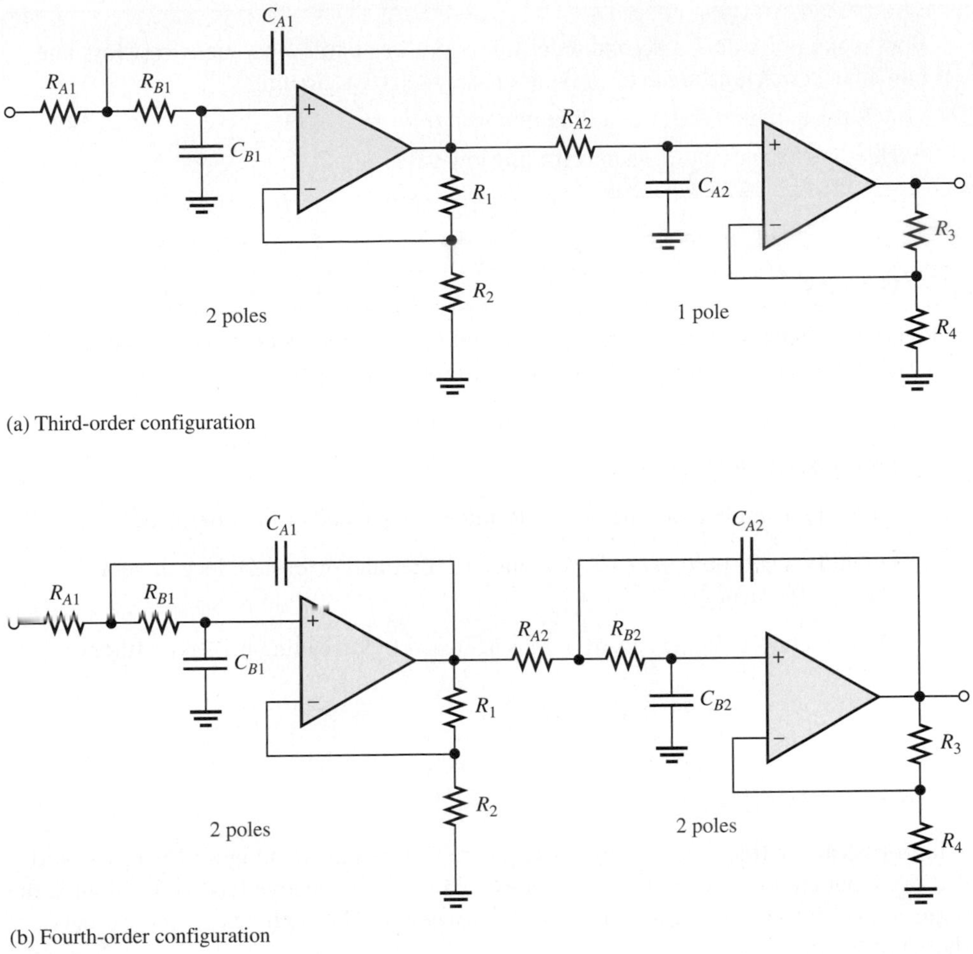

FIGURE 15–12

Cascaded low-pass filters.

EXAMPLE 15–4

For the four-pole filter in Figure 15–12(b), determine the capacitance values required to produce a critical frequency of 2680 Hz if all the resistors in the *RC* low-pass circuits are 1.8 kΩ. Also select values for the feedback resistors to get a Butterworth response.

Solution Both stages must have the same f_c. Assuming equal-value capacitors,

$$f_c = \frac{1}{2\pi RC}$$

$$C = \frac{1}{2\pi Rf_c} = \frac{1}{2\pi(1.8\text{ k}\Omega)(2680\text{ Hz})} = 0.033\ \mu\text{F}$$

$$C_{A1} = C_{B1} = C_{A2} = C_{B2} = \mathbf{0.033\ \mu F}$$

Also select $R_2 = R_4 = 1.8\text{ k}\Omega$ for simplicity. Refer to Table 15–1. For a Butterworth response in the first stage, $DF = 1.848$ and $R_1/R_2 = 0.152$. Therefore,

$$R_1 = 0.152R_2 = 0.152(1800\ \Omega) = \mathbf{274\ \Omega}$$

Choose $R_1 = 270\ \Omega$.

In the second stage, $DF = 0.765$ and $R_3/R_4 = 1.235$. Therefore,

$$R_3 = 1.235R_4 = 1.235(1800\ \Omega) = \mathbf{2.22\ k\Omega}$$

Choose $R_3 = 2.2\text{ k}\Omega$.

Related Problem For the filter in Figure 15–12(b), determine the capacitance values for $f_c = 1$ kHz if all the filter resistors are 680 Ω. Also specify the values for the feedback resistors to produce a Butterworth response.

SECTION 15–3 CHECKUP

1. How many poles does a second-order low-pass filter have? How many resistors and how many capacitors are used in the frequency-selective circuit?
2. Why is the damping factor of a filter important?
3. What is the primary purpose of cascading low-pass filters?

15–4 ACTIVE HIGH-PASS FILTERS

In high-pass filters, the roles of the capacitor and resistor are reversed in the *RC* circuits. Otherwise, the basic parameters are the same as for the low-pass filters.

After completing this section, you should be able to

- **Analyze active high-pass filters**
  - Identify a single-pole filter and determine its gain and critical frequency
  - Identify a two-pole Sallen-Key high-pass filter and determine its gain and critical frequency
  - Explain how a higher roll-off rate is achieved by cascading high-pass filters

A Single-Pole Filter

A high-pass active filter with a −20 dB/decade roll-off is shown in Figure 15–13(a). Notice that the input circuit is a single high-pass *RC* circuit. The negative feedback circuit is the same as for the low-pass filters previously discussed. The high-pass response curve is shown in Figure 15–13(b).

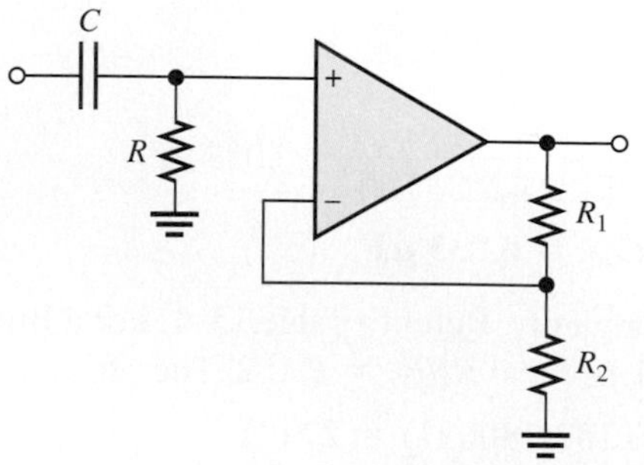

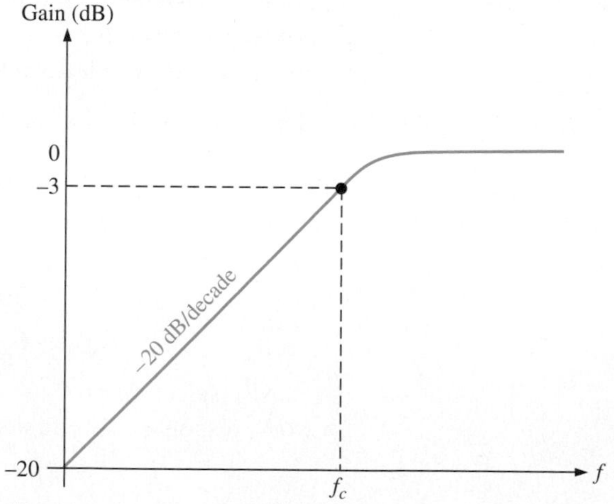

FIGURE 15–13

Single-pole active high-pass filter and response curve.

Ideally, a high-pass filter passes all frequencies above f_c without limit, as indicated in Figure 15–14(a), although in practice, this is not the case. As you have learned, all op-amps inherently have internal *RC* circuits that limit the amplifier's response at high frequencies.

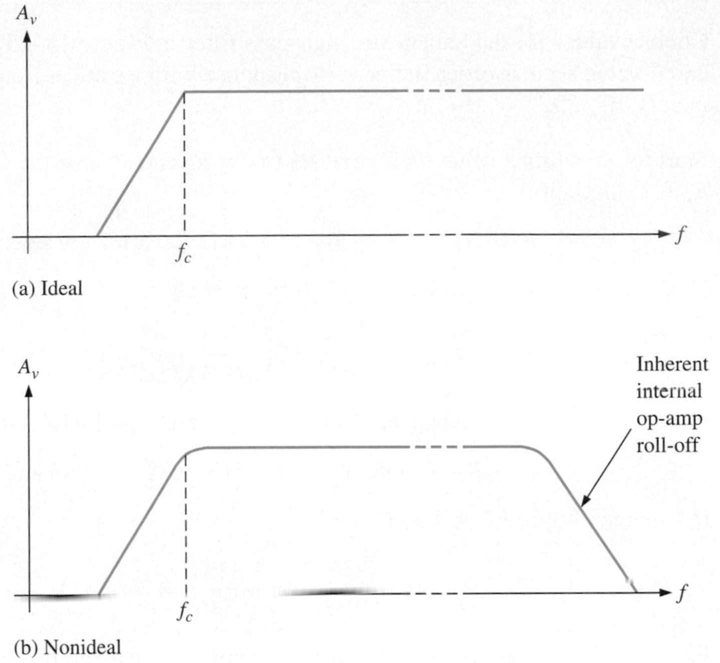

FIGURE 15–14

High-pass filter response.

Therefore, there is an upper-frequency limit on the high-pass filter's response which, in effect, makes it a band-pass filter with a very wide bandwidth. In the majority of applications, the internal high-frequency limitation is so much greater than that of the filter's critical frequency that the limitation can be neglected. In some applications, discrete transistors are used for the gain element to increase the high-frequency limitation beyond that realizable with available op-amps.

The Sallen-Key High-Pass Filter

A high-pass Sallen-Key configuration is shown in Figure 15–15. The components R_A, C_A, R_B, and C_B form the two-pole frequency-selective circuit. Notice that the positions of the resistors and capacitors in the frequency-selective circuit are opposite to those in the low-pass configuration. As with the other filters, the response characteristic can be optimized by proper selection of the feedback resistors, R_1 and R_2.

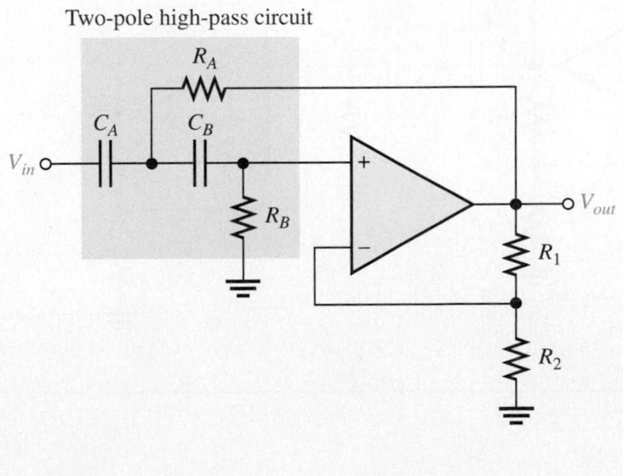

FIGURE 15–15

Basic Sallen-Key high-pass filter.

EXAMPLE 15–5

Choose values for the Sallen-Key high-pass filter in Figure 15–15 to implement an equal-value second-order Butterworth response with a critical frequency of approximately 10 kHz.

Solution Start by selecting a value for R_A and R_B (R_1 or R_2 can also be the same value as R_A and R_B for simplicity).

$$R = R_A = R_B = R_2 = \mathbf{3.3\,k\Omega} \text{ (an arbitrary selection)}$$

Next, calculate the capacitance value from $f_c = 1/(2\pi RC)$.

$$C = C_A = C_B = \frac{1}{2\pi R f_c} = \frac{1}{2\pi(3.3\,\text{k}\Omega)(10\,\text{kHz})} = \mathbf{0.0048\,\mu F}$$

For a Butterworth response, the damping factor must be 1.414 and $R_1/R_2 = 0.586$.

$$R_1 = 0.586R_2 = 0.586(3.3\,\text{k}\Omega) = \mathbf{1.93\,k\Omega}$$

If you had chosen $R_1 = 3.3\,\text{k}\Omega$, then

$$R_2 = \frac{R_1}{0.586} = \frac{3.3\,\text{k}\Omega}{0.586} = 5.63\,\text{k}\Omega$$

Either way, an approximate Butterworth response is realized by choosing the nearest standard values.

Related Problem Select values for all the components in the high-pass filter of Figure 15–15 to obtain an $f_c = 300$ Hz. Use equal-value components with $R = 10\,\text{k}\Omega$ and optimize for a Butterworth response.

Cascading High-Pass Filters

As with the low-pass configuration, first- and second-order high-pass filters can be cascaded to provide three or more poles and thereby create faster roll-off rates. Figure 15–16 shows a six-pole high-pass filter consisting of three Sallen-Key two-pole stages. With this configuration optimized for a Butterworth response, a roll-off of −120 dB/decade is achieved.

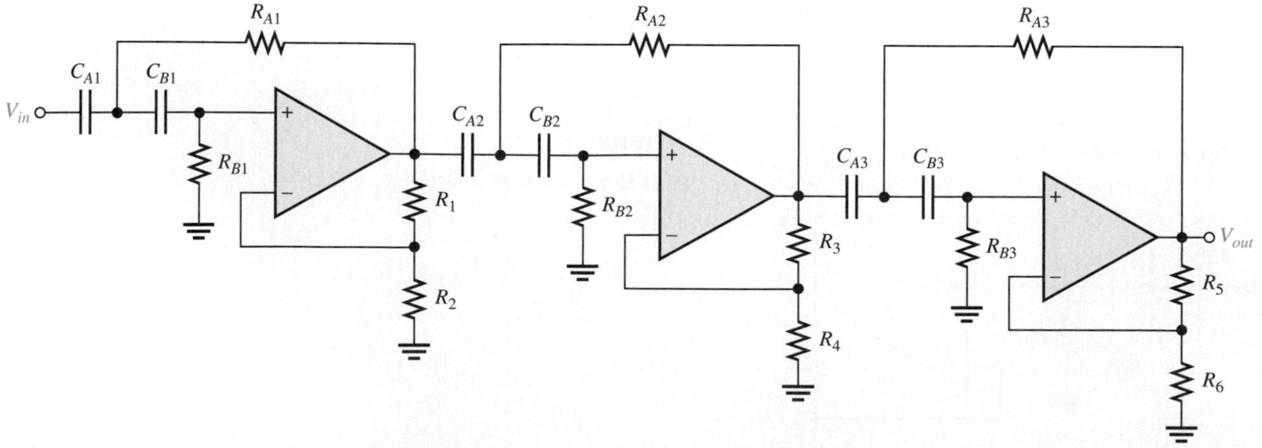

▲ **FIGURE 15–16**

Sixth-order high-pass filter.

SECTION 15–4 CHECKUP

1. How does a high-pass Sallen-Key filter differ from the low-pass configuration?
2. To increase the critical frequency of a high-pass filter, would you increase or decrease the resistor values?
3. If three two-pole high-pass filters and one single-pole high-pass filter are cascaded, what is the resulting roll-off?

15–5 ACTIVE BAND-PASS FILTERS

As mentioned, band-pass filters pass all frequencies bounded by a lower-frequency limit and an upper-frequency limit and reject all others lying outside this specified band. A band-pass response can be thought of as the overlapping of a low-frequency response curve and a high-frequency response curve.

After completing this section, you should be able to

- **Analyze active band-pass filters**
  - Describe a band-pass filter composed of a low-pass filter and a high-pass filter
  - Determine the critical frequencies and center frequency of a cascaded band-pass filter
  - Analyze a multiple-feedback band-pass filter to determine center frequency, bandwidth, and gain
  - Analyze a state-variable filter
  - Identify a biquad filter

Cascaded Low-Pass and High-Pass Filters

One way to implement a band-pass filter is a cascaded arrangement of a high-pass filter and a low-pass filter, as shown in Figure 15–17(a), as long as the critical frequencies are sufficiently separated. Each of the filters shown is a Sallen-Key Butterworth configuration so that the roll-off rates are -40 dB/decade, indicated in the composite response curve of Figure 15–17(b). The critical frequency of each filter is chosen so that the response curves overlap sufficiently, as indicated. The critical frequency of the high-pass filter must be sufficiently lower than that of the low-pass stage. This filter is generally limited to wide bandwidth applications.

The lower frequency f_{c1} of the passband is the critical frequency of the high-pass filter. The upper frequency f_{c2} is the critical frequency of the low-pass filter. Ideally, as discussed earlier, the center frequency f_0 of the passband is the geometric mean of f_{c1} and f_{c2}. The following formulas express the three frequencies of the band-pass filter in Figure 15–17.

$$f_{c1} = \frac{1}{2\pi\sqrt{R_{A1}R_{B1}C_{A1}C_{B1}}}$$

$$f_{c2} = \frac{1}{2\pi\sqrt{R_{A2}R_{B2}C_{A2}C_{B2}}}$$

$$f_0 = \sqrt{f_{c1}f_{c2}}$$

Of course, if equal-value components are used in implementing each filter, the critical frequency equations simplify to the form $f_c = 1/(2\pi RC)$.

▶ **FIGURE 15–17**

Band-pass filter formed by cascading a two-pole high-pass and a two-pole low-pass filter (it does not matter in which order the filters are cascaded).

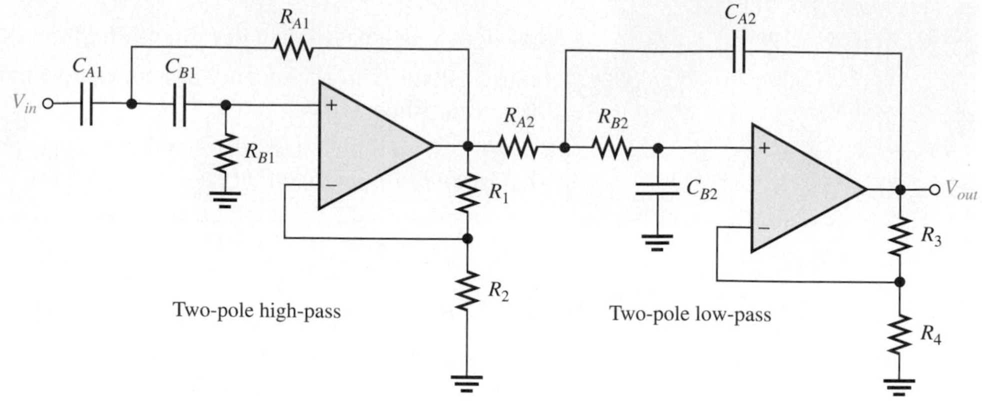

(a)

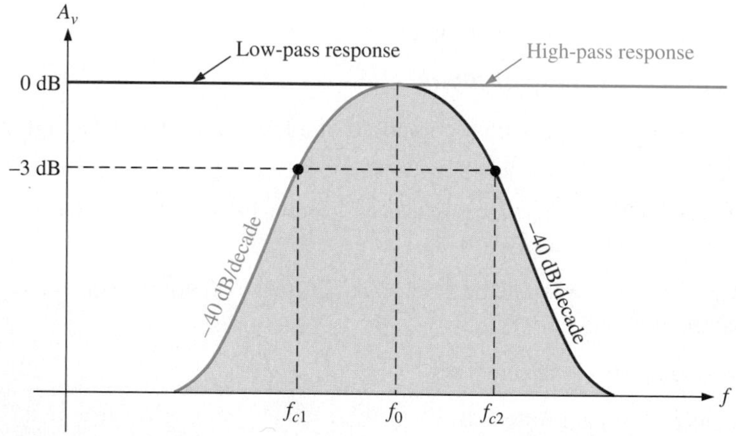

(b)

Multiple-Feedback Band-Pass Filter

Another type of filter configuration, shown in Figure 15–18, is a multiple-feedback band-pass filter. The two feedback paths are through R_2 and C_1. Components R_1 and C_1 provide the low-pass response, and R_2 and C_2 provide the high-pass response. The maximum gain, A_0, occurs at the center frequency. Q values of less than 10 are typical in this type of filter.

▶ **FIGURE 15–18**

Multiple-feedback band-pass filter.

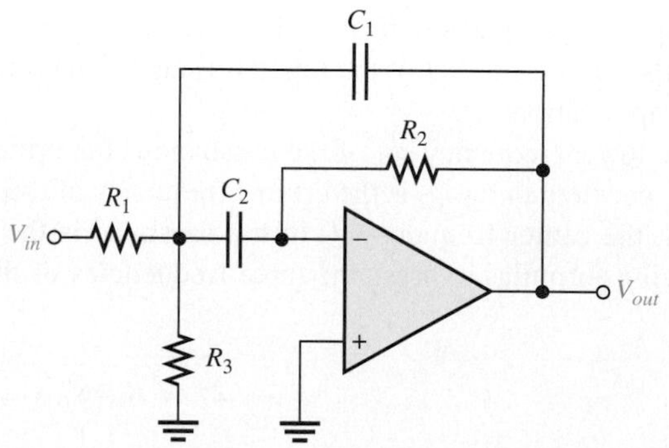

An expression for the center frequency is developed as follows, recognizing that R_1 and R_3 appear in parallel as viewed from the C_1 feedback path (with the V_{in} source replaced by a short).

$$f_0 = \frac{1}{2\pi\sqrt{(R_1 \parallel R_3)R_2C_1C_2}}$$

Making $C_1 = C_2 = C$ yields

$$f_0 = \frac{1}{2\pi\sqrt{(R_1 \| R_3)R_2C^2}} = \frac{1}{2\pi C\sqrt{(R_1 \| R_3)R_2}}$$

$$= \frac{1}{2\pi C}\sqrt{\frac{1}{R_2(R_1 \| R_3)}} = \frac{1}{2\pi C}\sqrt{\left(\frac{1}{R_2}\right)\left(\frac{1}{R_1R_3 / R_1 + R_3}\right)}$$

$$\boldsymbol{f_0 = \frac{1}{2\pi C}\sqrt{\frac{R_1 + R_3}{R_1R_2R_3}}}$$

Equation 15–8

A value for the capacitors is chosen and then the three resistor values are calculated to achieve the desired values for f_0, BW, and A_0. As you know, the Q can be determined from the relation $Q = f_0/BW$. The resistor values can be found using the following formulas (stated without derivation):

$$R_1 = \frac{Q}{2\pi f_0 C A_0}$$

$$R_2 = \frac{Q}{\pi f_0 C}$$

$$R_3 = \frac{Q}{2\pi f_0 C(2Q^2 - A_0)}$$

To develop a gain expression, solve for Q in the R_1 and R_2 formulas as follows:

$$Q = 2\pi f_0 A_0 C R_1$$
$$Q = \pi f_0 C R_2$$

Then,

$$2\pi f_0 A_0 C R_1 = \pi f_0 C R_2$$

Cancelling yields

$$2A_0R_1 = R_2$$

$$\boldsymbol{A_0 = \frac{R_2}{2R_1}}$$

Equation 15–9

In order for the denominator of the equation $R_3 = Q/[2\pi f_0 C(2Q^2 - A_0)]$ to be positive, $A_0 < 2Q^2$, which imposes a limitation on the gain.

EXAMPLE 15–6

Determine the center frequency, maximum gain, and bandwidth for the filter in Figure 15–19.

FIGURE 15–19

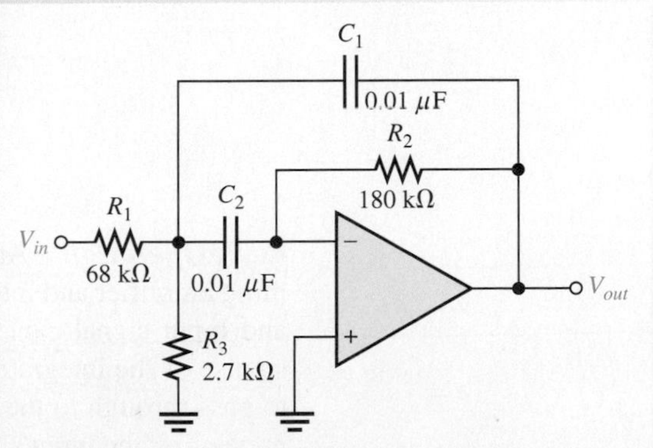

Solution

$$f_0 = \frac{1}{2\pi C}\sqrt{\frac{R_1 + R_3}{R_1 R_2 R_3}} = \frac{1}{2\pi(0.01\ \mu\text{F})}\sqrt{\frac{68\ \text{k}\Omega + 2.7\ \text{k}\Omega}{(68\ \text{k}\Omega)(180\ \text{k}\Omega)(2.7\ \text{k}\Omega)}} = \mathbf{736\ Hz}$$

$$A_0 = \frac{R_2}{2R_1} = \frac{180\ \text{k}\Omega}{2(68\ \text{k}\Omega)} = \mathbf{1.32}$$

$$Q = \pi f_0 C R_2 = \pi(736\ \text{Hz})(0.01\ \mu\text{F})(180\ \text{k}\Omega) = 4.16$$

$$BW = \frac{f_0}{Q} = \frac{736\ \text{Hz}}{4.16} = \mathbf{177\ Hz}$$

Related Problem If R_2 in Figure 15–19 is increased to 330 kΩ, determine the gain, center frequency, and bandwidth of the filter?

Open the Multisim file E15-06 in the Examples folder on your CD-ROM. Measure the center frequency and the bandwidth and compare to the calculated values.

State-Variable Filter

The state-variable or universal active filter is widely used for band-pass applications. As shown in Figure 15–20, it consists of a summing amplifier and two op-amp integrators (which act as single-pole low-pass filters) that are combined in a cascaded arrangement to form a second-order filter. Although used primarily as a band-pass (BP) filter, the state-variable configuration also provides low-pass (LP) and high-pass (HP) outputs. The center frequency is set by the *RC* circuits in both integrators. When used as a band-pass filter, the critical frequencies of the integrators are usually made equal, thus setting the center frequency of the passband.

▶ FIGURE 15–20

State-variable filter.

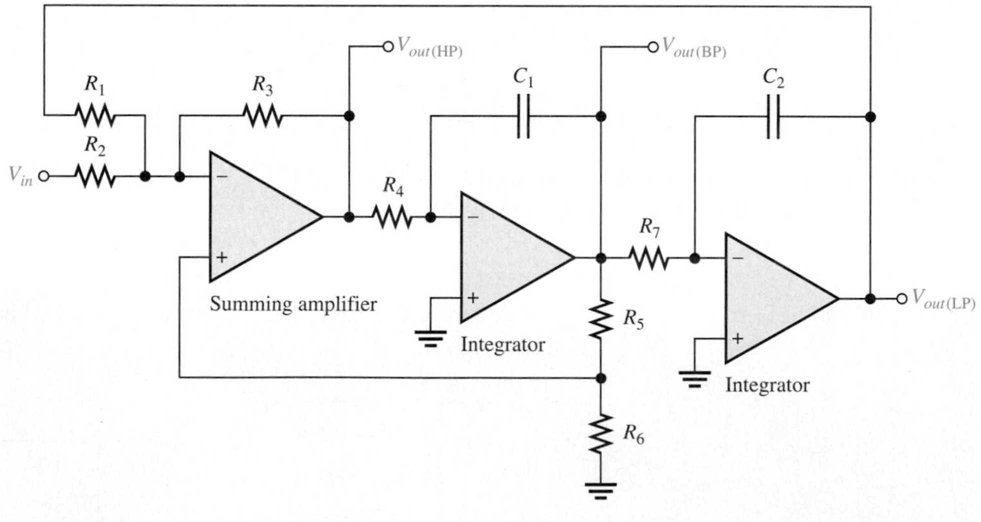

Basic Operation At input frequencies below f_c, the input signal passes through the summing amplifier and integrators and is fed back 180° out of phase. Thus, the feedback signal and input signal cancel for all frequencies below approximately f_c. As the low-pass response of the integrators rolls off, the feedback signal diminishes, thus allowing the input to pass through to the band-pass output. Above f_c, the low-pass response disappears, thus preventing the input signal from passing through the integrators. As a result, the band-pass output peaks sharply at f_c, as indicated in Figure 15–21. Stable Qs up to 100 can be

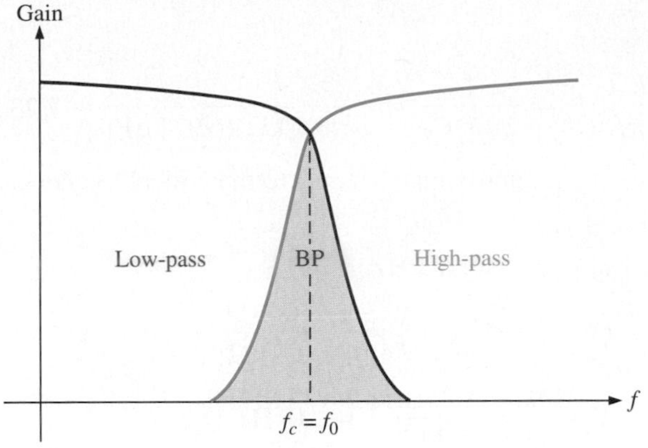

▲ **FIGURE 15–21**

General state-variable response curves. BP is for band-pass.

obtained with this type of filter. The Q is set by the feedback resistors R_5 and R_6 according to the following equation:

$$Q = \frac{1}{3}\left(\frac{R_5}{R_6} + 1\right)$$

The state-variable filter cannot be optimized for low-pass, high-pass, and narrow band-pass performance simultaneously for this reason: To optimize for a low-pass or a high-pass Butterworth response, DF must equal 1.414. Since $Q = 1/DF$, a Q of 0.707 will result. Such a low Q provides a very wide band-pass response (large BW and poor selectivity). For optimization as a narrow band-pass filter, the Q must be set high.

EXAMPLE 15–7

Determine the center frequency, Q, and BW for the band-pass output of the state-variable filter in Figure 15–22.

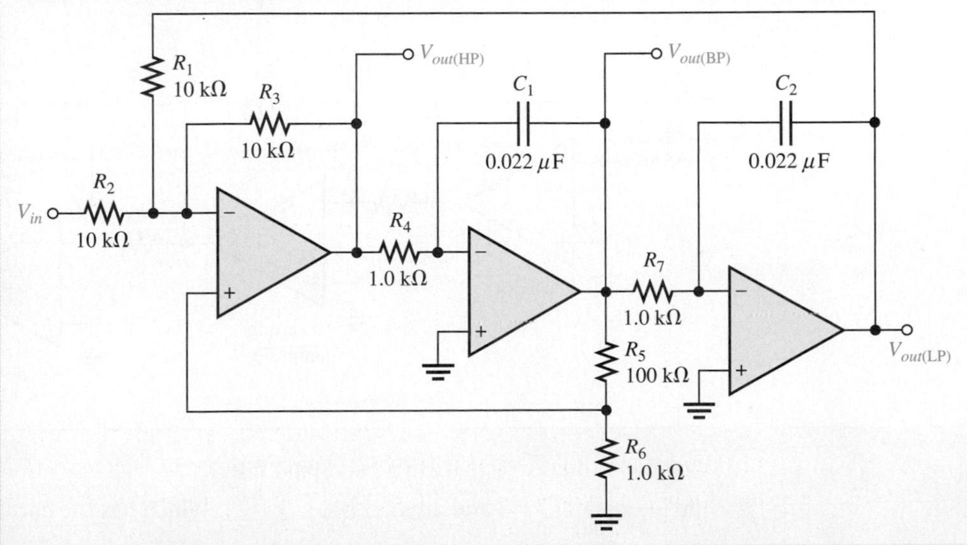

▲ **FIGURE 15–22**

Solution For each integrator,

$$f_c = \frac{1}{2\pi R_4 C_1} = \frac{1}{2\pi R_7 C_2} = \frac{1}{2\pi(1.0\,\text{k}\Omega)(0.022\,\mu\text{F})} = 7.23\text{ kHz}$$

The center frequency is approximately equal to the critical frequencies of the integrators.

$$f_0 = f_c = \mathbf{7.23\,kHz}$$

$$Q = \frac{1}{3}\left(\frac{R_5}{R_6} + 1\right) = \frac{1}{3}\left(\frac{100\,\text{k}\Omega}{1.0\,\text{k}\Omega} + 1\right) = \mathbf{33.7}$$

$$BW = \frac{f_0}{Q} = \frac{7.23\text{ kHz}}{33.7} = \mathbf{215\,Hz}$$

Related Problem Determine f_0, Q, and BW for the filter in Figure 15–22 if $R_4 = R_6 = R_7 = 330\ \Omega$ with all other component values the same as shown on the schematic.

Open the Multisim file E15-07 in the Examples folder on your CD-ROM. Measure the center frequency and the bandwidth and compare to the calculated values.

The Biquad Filter

The biquad filter is similar to the state-variable filter except that it consists of an integrator, followed by an inverting amplifier, and then another integrator, as shown in Figure 15–23. These differences in the configuration between a biquad and a state-variable filter result in some operational differences although both allow a very high Q value. In a biquad filter, the bandwidth is independent and the Q is dependent on the critical frequency; however, in the state-variable filter it is just the opposite: the bandwidth is dependent and the Q is independent on the critical frequency. Also, the biquad filter provides only band-pass and low-pass outputs.

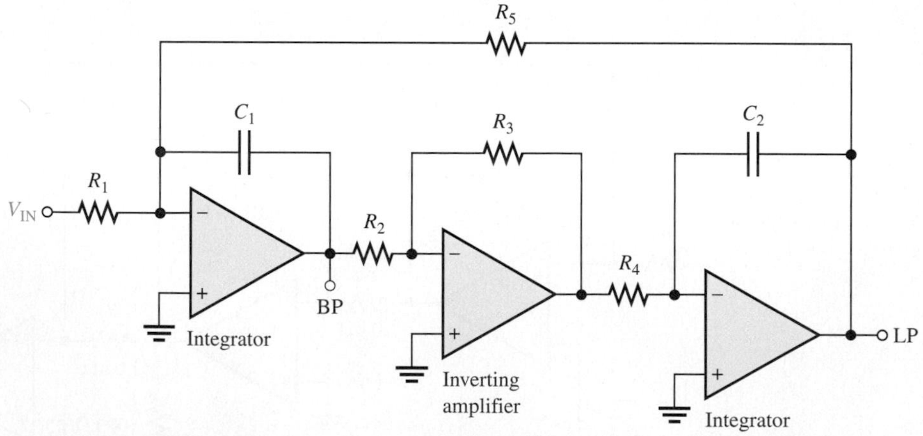

▶ **FIGURE 15–23**

A biquad filter.

SECTION 15–5 CHECKUP

1. What determines selectivity in a band-pass filter?
2. One filter has a $Q = 5$ and another has a $Q = 25$. Which has the narrower bandwidth?
3. List the active elements that make up a state-variable filter.
4. List the active elements that make up a biquad filter.

15–6 ACTIVE BAND-STOP FILTERS

Band-stop filters reject a specified band of frequencies and pass all others. The response is opposite to that of a band-pass filter. Band-stop filters are sometimes referred to as notch filters.

After completing this section, you should be able to

- **Analyze active band-stop filters**
  - Identify a multiple-feedback band-stop filter
  - Analyze a state-variable band-stop filter

Multiple-Feedback Band-Stop Filter

Figure 15–24 shows a multiple-feedback band-stop filter. Notice that this configuration is similar to the band-pass version in Figure 15–18 except that R_3 has been moved and R_4 has been added.

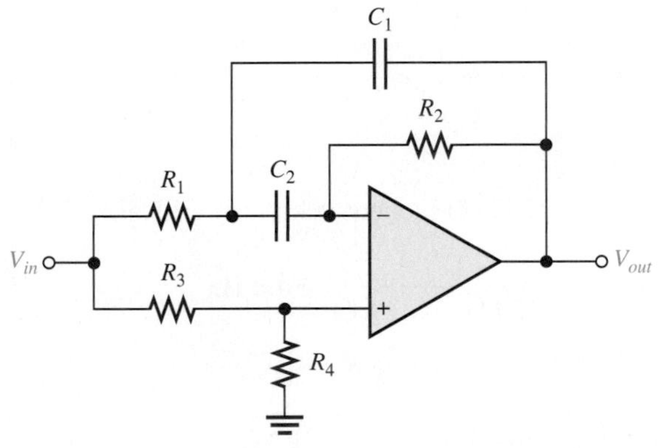

FIGURE 15–24

Multiple-feedback band-stop filter.

State-Variable Band-Stop Filter

Summing the low-pass and the high-pass responses of the state-variable filter covered in Section 15–5 with a summing amplifier creates a band-stop filter, as shown in Figure 15–25. One important application of this filter is minimizing the 60 Hz "hum" in audio systems by setting the center frequency to 60 Hz.

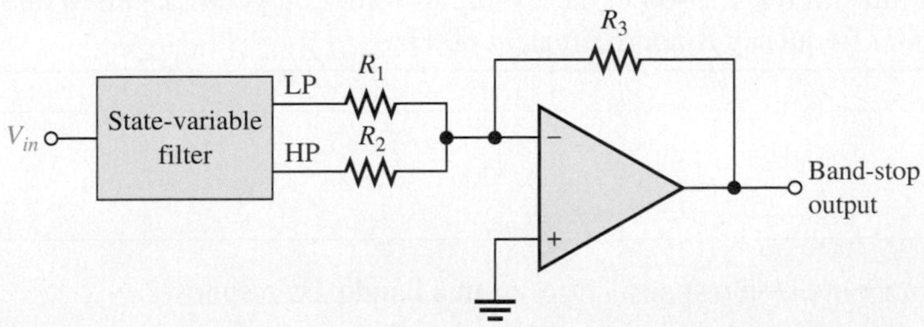

FIGURE 15–25

State-variable band-stop filter.

EXAMPLE 15–8

Verify that the band-stop filter in Figure 15–26 has a center frequency of 60 Hz, and optimize the filter for a Q of 10.

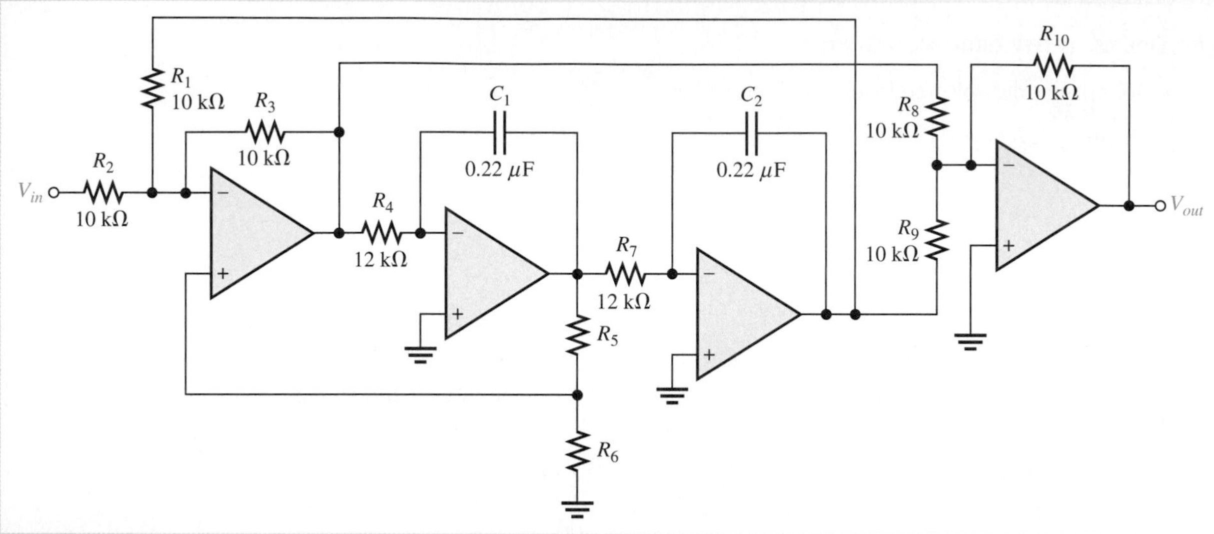

▲ FIGURE 15–26

Solution f_0 equals the f_c of the integrator stages. (In practice, component values are critical.)

$$f_0 = \frac{1}{2\pi R_4 C_1} = \frac{1}{2\pi R_7 C_2} = \frac{1}{2\pi(12\text{ k}\Omega)(0.22\ \mu\text{F})} = \mathbf{60\ Hz}$$

You can obtain a $Q = 10$ by choosing R_6 and then calculating R_5.

$$Q = \frac{1}{3}\left(\frac{R_5}{R_6} + 1\right)$$

$$R_5 = (3Q - 1)R_6$$

Choose $R_6 = \mathbf{3.3\ k\Omega}$. Then

$$R_5 = [3(10) - 1]3.3\text{ k}\Omega = \mathbf{95.7\ k\Omega}$$

Use the nearest standard value of 100 kΩ.

Related Problem How would you change the center frequency to 120 Hz in Figure 15–26?

Open the Multisim file E15-08 in the Examples folder on your CD-ROM and verify that the center frequency is approximately 60 Hz.

SECTION 15–6 CHECKUP

1. How does a band-stop response differ from a band-pass response?
2. How is a state-variable band-pass filter converted to a band-stop filter?

15–7 FILTER RESPONSE MEASUREMENTS

Two methods of determining a filter's response by measurement are discrete point measurement and swept frequency measurement.

After completing this section, you should be able to

- **Discuss two methods for measuring frequency response**
  - Explain the discrete point measurement method
  - Explain the swept frequency measurement method

Discrete Point Measurement

Figure 15–27 shows an arrangement for taking filter output voltage measurements at discrete values of input frequency using common laboratory instruments. The general procedure is as follows:

1. Set the amplitude of the sine wave generator to a desired voltage level.
2. Set the frequency of the sine wave generator to a value well below the expected critical frequency of the filter under test. For a low-pass filter, set the frequency as near as possible to 0 Hz. For a band-pass filter, set the frequency well below the expected lower critical frequency.
3. Increase the frequency in predetermined steps sufficient to allow enough data points for an accurate response curve.
4. Maintain a constant input voltage amplitude while varying the frequency.
5. Record the output voltage at each value of frequency.
6. After recording a sufficient number of points, plot a graph of output voltage versus frequency.

If the frequencies to be measured exceed the frequency response of the DMM, an oscilloscope may have to be used instead.

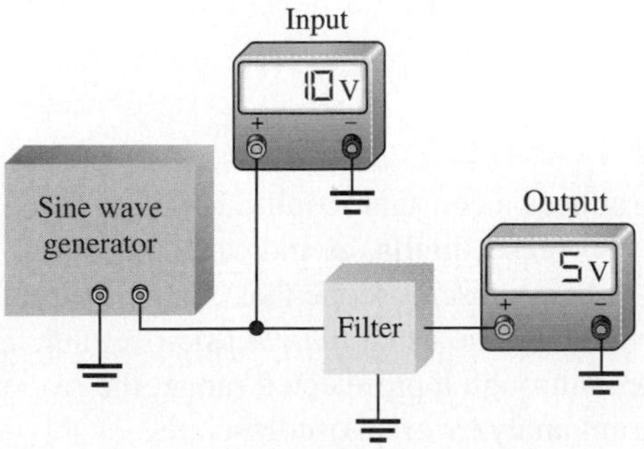

FIGURE 15–27

Test setup for discrete point measurement of the filter response. (Readings are arbitrary and for display only.)

Swept Frequency Measurement

The swept frequency method requires more elaborate test equipment than does the discrete point method, but it is much more efficient and can result in a more accurate response curve. A general test setup is shown in Figure 15–28(a) using a swept frequency generator and a spectrum analyzer. Figure 15–28(b) shows how the test can be made with an oscilloscope.

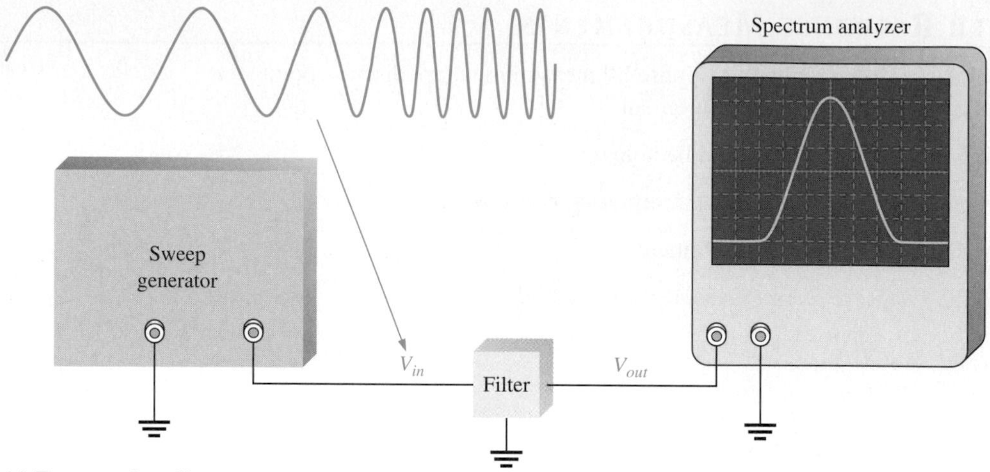

(a) Test setup for a filter response using a spectrum analyzer

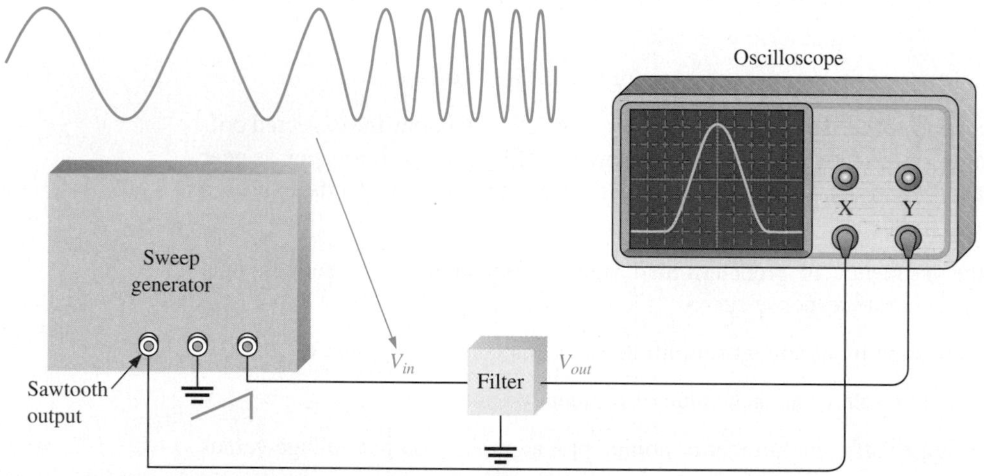

(b) Test setup for a filter response using an oscilloscope. The scope is placed in X-Y mode. The sawtooth waveform from the sweep generator drives the X-channel of the oscilloscope.

▲ **FIGURE 15–28**

Test setup for swept frequency measurement of the filter response.

The swept frequency generator produces a constant amplitude output signal whose frequency increases linearly between two preset limits, as indicated in Figure 15–28. The spectrum analyzer is essentially an elaborate oscilloscope that can be calibrated for a desired *frequency span/division* rather than for the usual *time/division* setting. Therefore, as the input frequency to the filter sweeps through a preselected range, the response curve is traced out on the screen of the spectrum analyzer or an oscilloscope.

SECTION 15–7 CHECKUP

1. What is the purpose of the two tests discussed in this section?
2. Name one disadvantage and one advantage of each test method.

Application Activity: *RFID System*

RFID (radio frequency identification) is a technology that enables the tracking and/or identification of objects. Typically, an RFID system contains an *RFID tag* that consists of an IC chip that transmits data about the object, an *RFID reader* that receives transmitted data from the tag, and a *data-processing system* that processes and stores the data passed to it by the reader. A basic block diagram is shown in Figure 15–29.

FIGURE 15–29

Basic block diagram of an RFID system.

The RFID Tag

RFID tags are tiny, very thin microchips with memory and a coil antenna. The tags listen for a radio signal sent by an RFID reader. When a tag receives a signal, it responds by transmitting its unique ID code and other data back to the reader.

Passive RFID Tag This type of tag does not require batteries. The tag is inactive until powered by the energy from the electromagnetic field of an RFID reader. Passive tags can be read from distances up to about 20 feet and are generally read-only, meaning the data they contain cannot be altered or written over.

Active RFID Tag This type of tag is powered by a battery and is capable of communicating up to 100 feet or more from the RFID reader. Generally, the active tag is larger and more expensive than a passive tag, but can hold more data about the product and is commonly used for identification of high-value assets. Active tags may be read-write, meaning the data they contain can be written over.

Tags are available in a variety of shapes. Depending on the application, they may be embedded in glass or epoxy, or they may be in label or card form. Another type of tag, often called the *smart label,* is a paper (or similar material) label with printing, but also with the RF circuitry and antenna embedded in it.

Some advantages of RFID tags compared to bar codes are

- Non-line-of-sight identification
- More information can be stored
- Coverage at greater distances
- Unattended operations are possible
- Ability to identify moving objects that have tags embedded
- Can be used in diverse environments

Disadvantages of RFID tags are that they are expensive compared to the bar code and they are bulkier because the electronics is embedded in the tag.

RFID tags and readers must be tuned to the same frequency to communicate. RFID systems use many different frequencies, but generally the most common are low frequency

(125 kHz), high frequency (13.56 MHz), and ultra-high frequency, or UHF (850–900 MHz). Microwave (2.45 GHz) is also used in some applications. The frequency used depends on the particular type of application.

Low-frequency systems are the least expensive and have the shortest range. They are most commonly used in security access, asset tracking, and animal identification applications. High-frequency systems are used for applications such as railroad car tracking and automated toll collection.

Some typical RFID application areas are

- Metering applications such as electronic toll collection
- Inventory control and tracking such as merchandise control
- Asset tracking and recovery
- Tracking parts moving through a manufacturing process
- Tracking goods in a supply chain

The RFID Reader

Data is stored on the RFID tag in digital form and is transmitted to the reader as a modulated signal. Many RFID systems use ASK (amplitude shift keying) or FSK (frequency shift keying). In ASK, the amplitude of a carrier signal is varied by the digital data. In FSK, the frequency of a carrier signal is varied by the digital data. Examples of these forms of modulation are shown Figure 15–30. In this system, the carrier is 125 kHz, and the modulating signal is a digital waveform at the rate of 10 kHz, representing a stream of 1s and 0s.

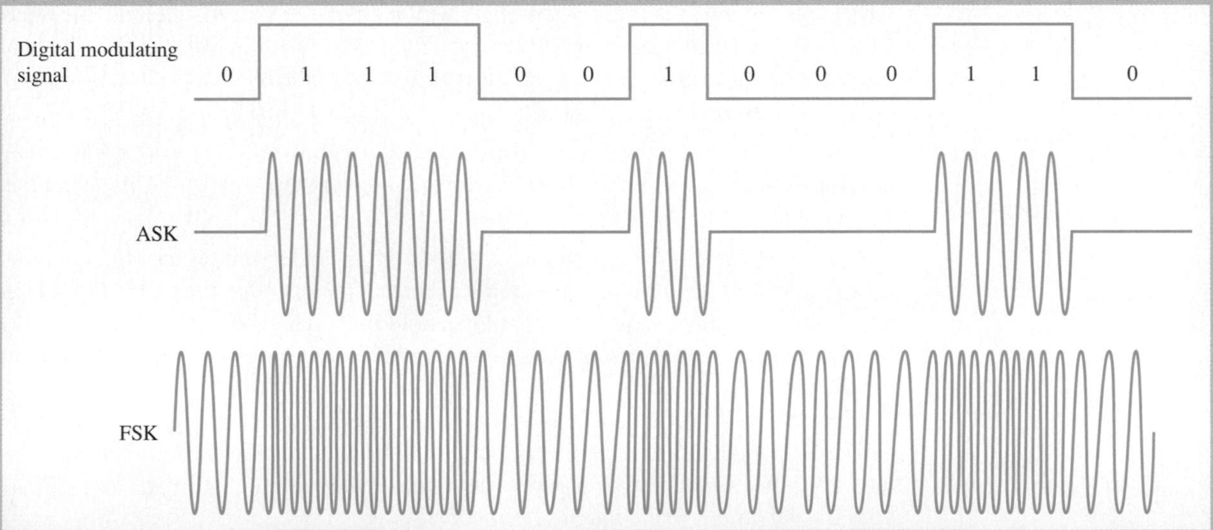

FIGURE 15–30

Examples of ASK and FSK modulation transmitted by an RFID tag.

Project

Your company is developing a new RFID reader using ASK modulation at a carrier frequency of 125 kHz. A block diagram is shown in Figure 15–31. The purpose of each block is as follows. The band-pass filter passes the 125 kHz signal and reduces signals and noise from other sources; the 2-stage amplifier increases the very small signal from the tag to a usable level; the rectifier eliminates the negative portions of the modulated signal; the low-pass filter eliminates the 125 kHz carrier frequency but passes the 10 kHz

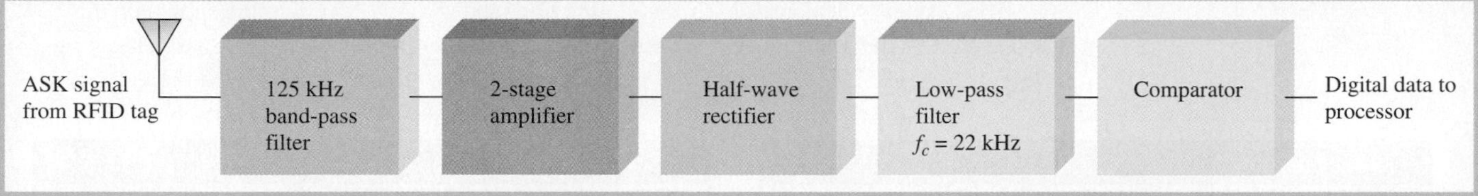

FIGURE 15–31

Block diagram of RFID reader.

modulating signal; and the comparator restores the digital signal to a usable stream of digital data.

1. In general, what are RFID systems used for?
2. Name the three basic components of an RFID system.
3. Explain the purpose of an RFID tag.
4. Explain the purpose of an RFID reader.

Simulation

The RFID reader is simulated with Multisim using an input signal of 1 mV at 125 kHz to represent the output of the RFID tag. For purposes of simulation, the 125 kHz carrier is modulated with a 10 kHz sine wave although the actual modulating signal will be a pulse waveform containing digital data. In Multisim it is difficult to produce a sinusoidal carrier signal modulated with a pulse signal, so the sinusoidal modulating signal serves to verify system operation. The simulated circuit is shown in Figure 15–32. The band-pass filter is U1, the amplifier stages are U2 and U3, the half-wave rectifier is D1, the low-pass filter is U4, and the comparator is U5. Datasheets for the OP27AH op-amp and the LM111H comparator are available at www.analog.com.

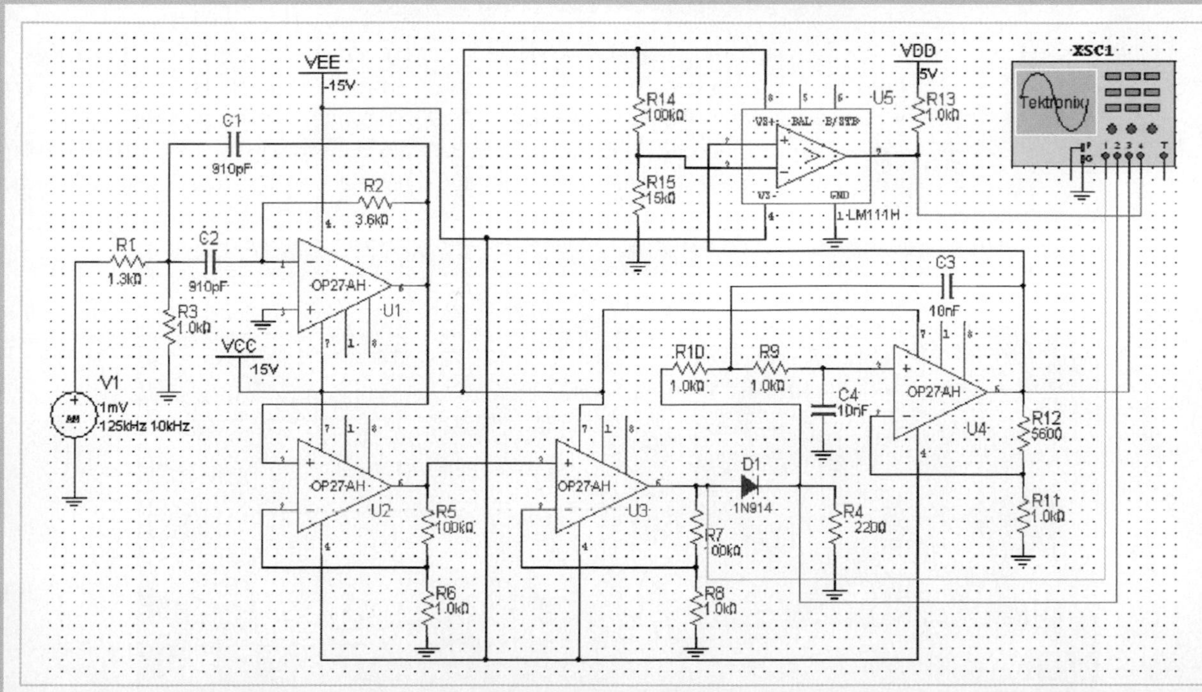

FIGURE 15–32

Multisim circuit screen for the RFID reader.

The frequency responses of the band-pass filter and the low-pass filter are shown on the Bode plotters in Figure 15–33. As you can see, the peak response of the band-pass filter is approximately 125 kHz and the critical frequency of the low-pass filter is approximately 16 kHz.

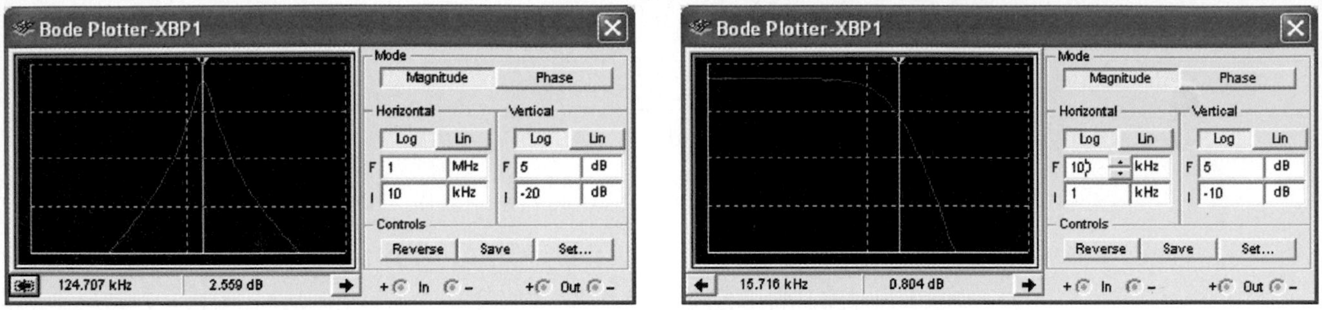

(a) Band-pass response

(b) Low-pass response

FIGURE 15–33

Bode plots for the RFID reader filters.

5. What is the purpose of the band-pass filter in the RFID reader?
6. What is the purpose of the low-pass filter in the RFID reader?
7. Calculate the gain of each amplifier in the reader in Figure 15–32.
8. Use the formula for a multiple-feedback band-pass filter to verify the center frequency of the band-pass filter in the reader.
9. What type of response characteristic is the low-pass filter set up for?
10. Calculate the critical frequency of the low-pass filter and compare to the measured value.
11. Calculate the reference voltage for the comparator and explain why a reference above ground is necessary.

Measurements at points on the reader circuit are shown on the oscilloscope in Figure 15–34. The top waveform is the modulated carrier at the output of amplifier U3. The second waveform is the output of the rectifier D1. The third waveform is the output of the low-pass filter (notice that the carrier frequency has been removed by the filter). The bottom waveform is the output of the comparator and represents the digital data sent to the processor.

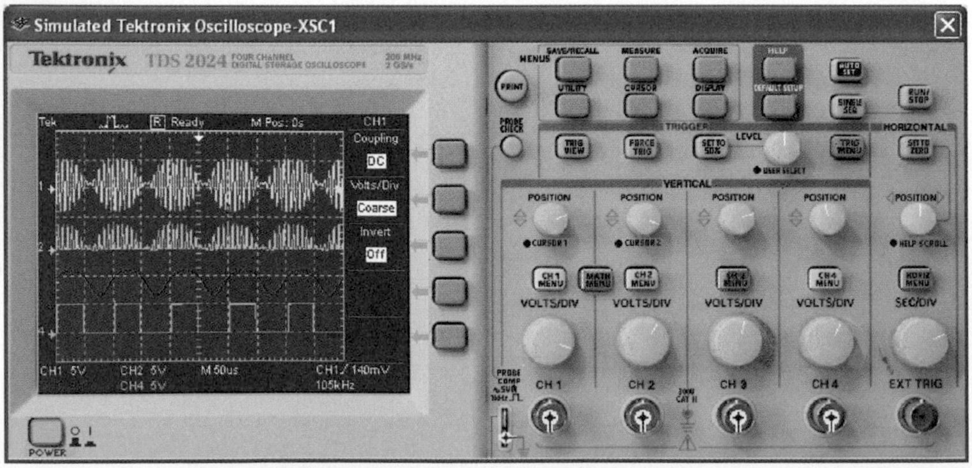

FIGURE 15–34

RFID reader waveforms.

Simulate the RFID reader circuit using your Multisim software. Observe the operation with the oscilloscope and Bode plotter.

Prototyping and Testing

Now that the circuit has been simulated, the prototype circuit is constructed and tested. After the circuit is successfully tested on a protoboard, it is ready to be finalized on a printed circuit board.

Lab Experiment

To build and test a low-pass filter similar to one used in the RFID reader, go to Experiment 15–A in your lab manual (*Laboratory Exercises for Electronic Devices* by David Buchla and Steven Wetterling).

Circuit Board

The RFID reader circuit is implemented on a printed circuit board as shown in Figure 15–35. The dark gray lines represent backside traces.

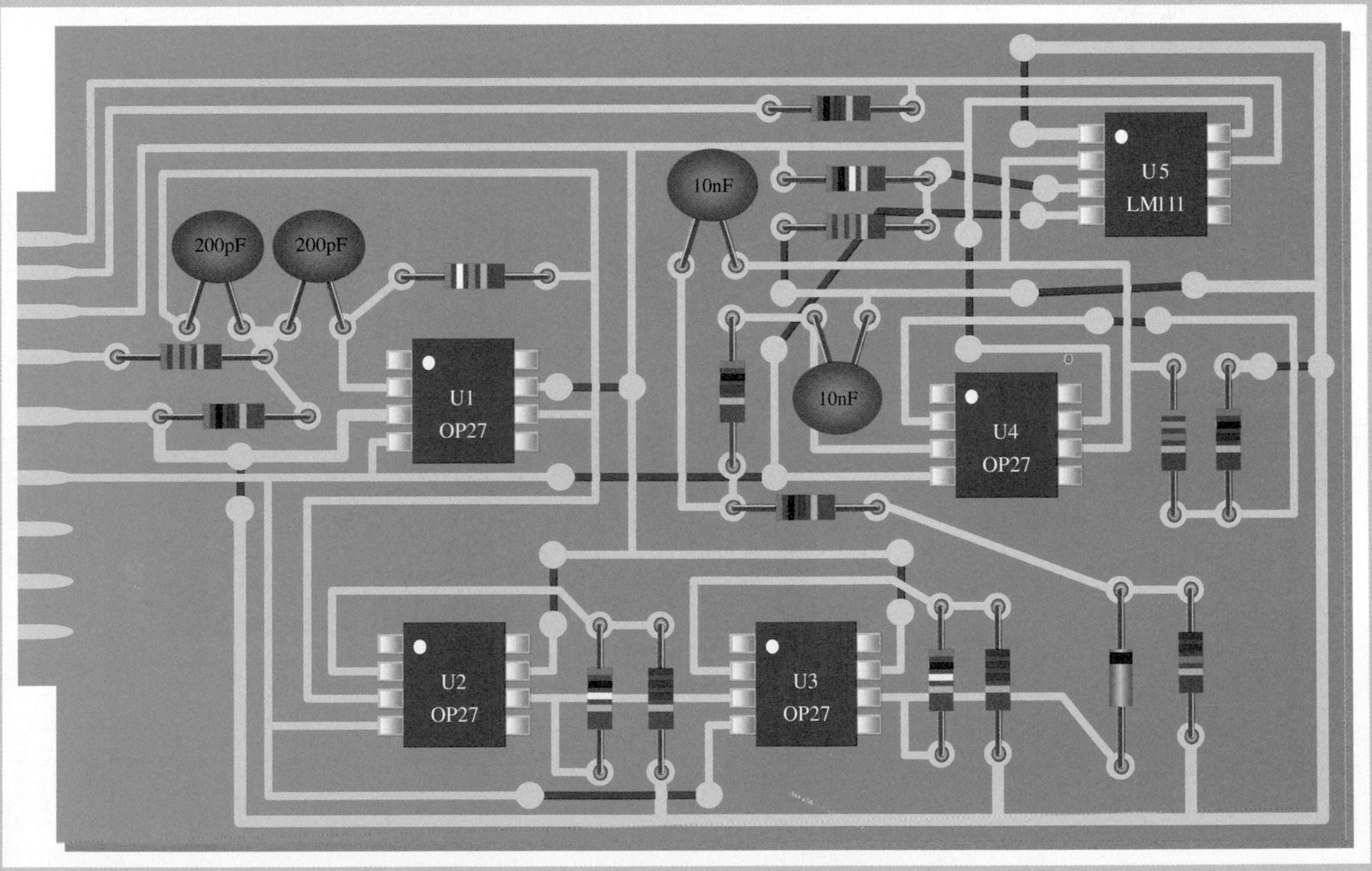

FIGURE 15–35

RFID reader board.

12. Check the printed circuit board and verify that it agrees with the simulation schematic in Figure 15–32.
13. Label each input and output pin according to function.

Anadigm.com

Programmable Analog Design

The material you have learned in this chapter is necessary to give you a basic understanding of active filters. However, filter design can be quite complex mathematically. To avoid tedious calculations and trial-and-error breadboarding, the preferred method for development of many filters is to use computer software and then download the design to a programmable analog array. AnadigmDesigner2 software is used in this section to illustrate the ease with which active filters can be developed and implemented in hardware. If you have checked out the optional *Programmable Analog Design* feature, which appeared first in Chapter 12, you are aware that this software is available and can be downloaded free from www.anadigm.com. You can easily implement a filter design in an FPAA or dpASP chip if you have an evaluation board and interface cable connected to your computer.

Filter Specification

Once you have downloaded the AnadigmDesigner2 software, the first thing you see when opening it is an outline representation of the blank FPAA chip, as shown in Figure 15–36. Under the Tools menu, select *AnadigmFilter,* as shown, and you will get the screen shown in Figure 15–37.

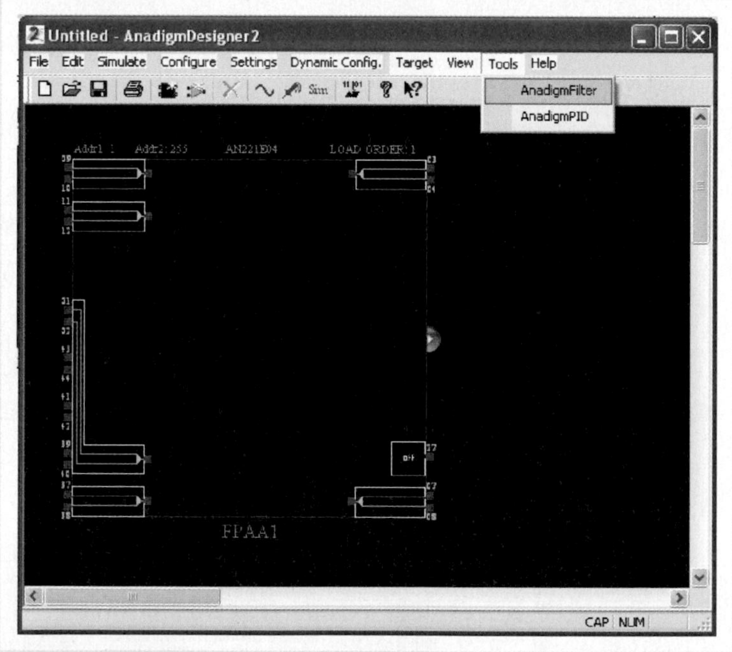

FIGURE 15–36

FPAA chip outline screen showing AnadigmFilter selection.

You are now ready to specify a filter. For example, select a filter type and approximation and enter the desired parameters, as shown in Figure 15–38, for a band-pass Butterworth filter. Note that you can use your mouse to drag the limits, shown in red and blue on the screen, to set the desired response.

When the filter has been completely specified, click on "To AnadigmDesigner2" and the filter components will be placed in the FPAA chip screen, as shown in Figure 15–39(a). Notice that the filter consists of three stages, in this case. Now use the connection tool to connect the filter to an input and output, as shown in part (b).

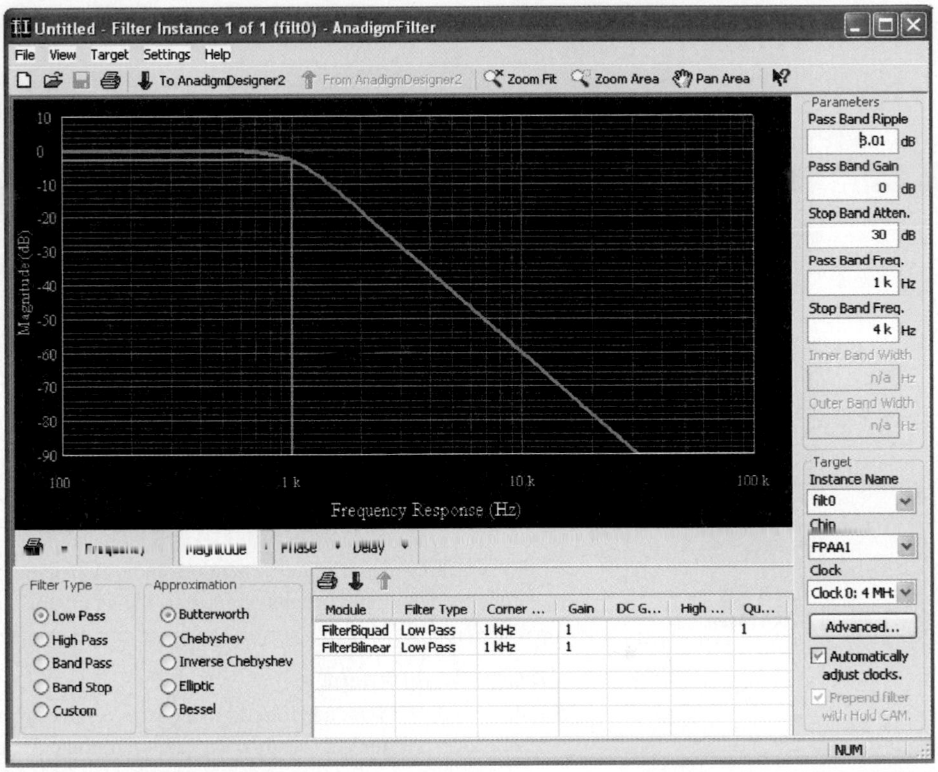

▶ FIGURE 15–37

Filter default screen.

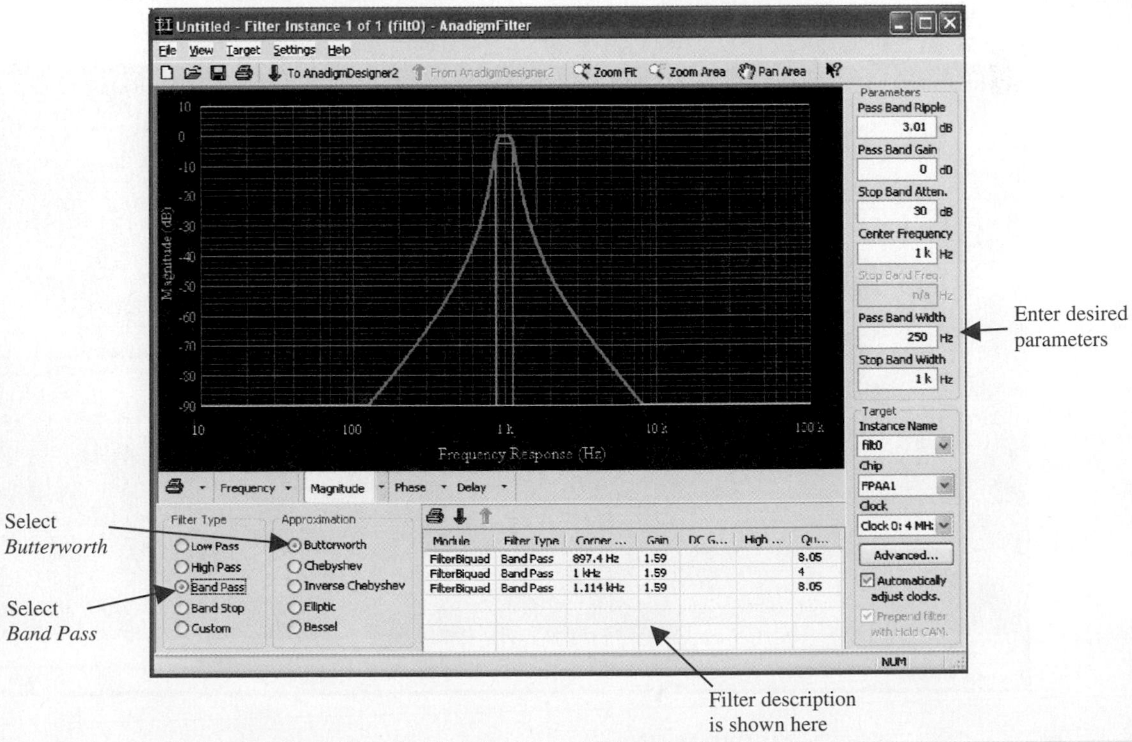

▲ FIGURE 15–38

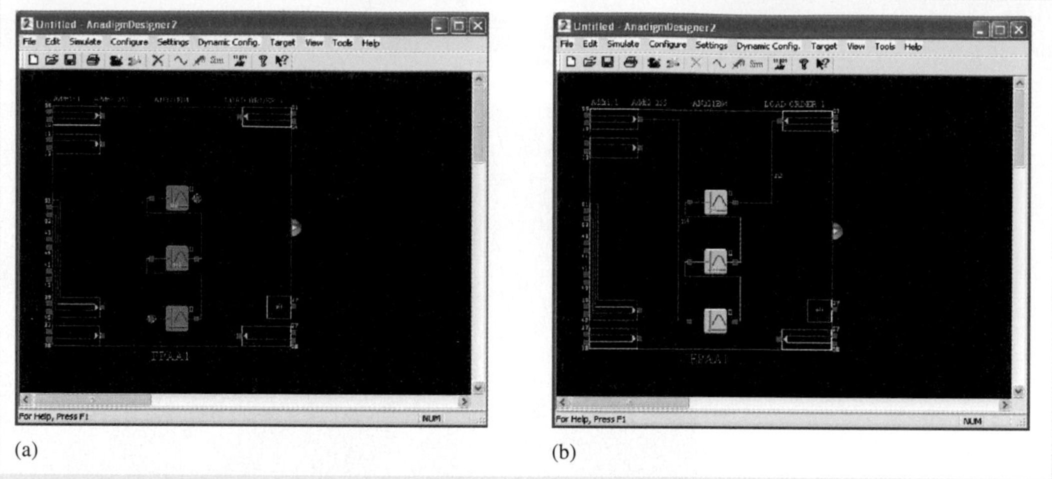

FIGURE 15–39

By attaching actual signal generators and oscilloscope probes to the board, you can verify that the downloaded circuit is behaving just as the simulator indicated it would. Note that an FPAA or dpASP is reprogrammable so you can make circuit changes, download, and test indefinitely.

Design Assignment

Implement the RFID reader circuit using AnadigmDesigner2 software.

Procedure: Figure 15–40 shows a version of the circuit implemented in FPAA1. Because of limitations on implementing the ASK input signal, modifications have been made. Since the input cell

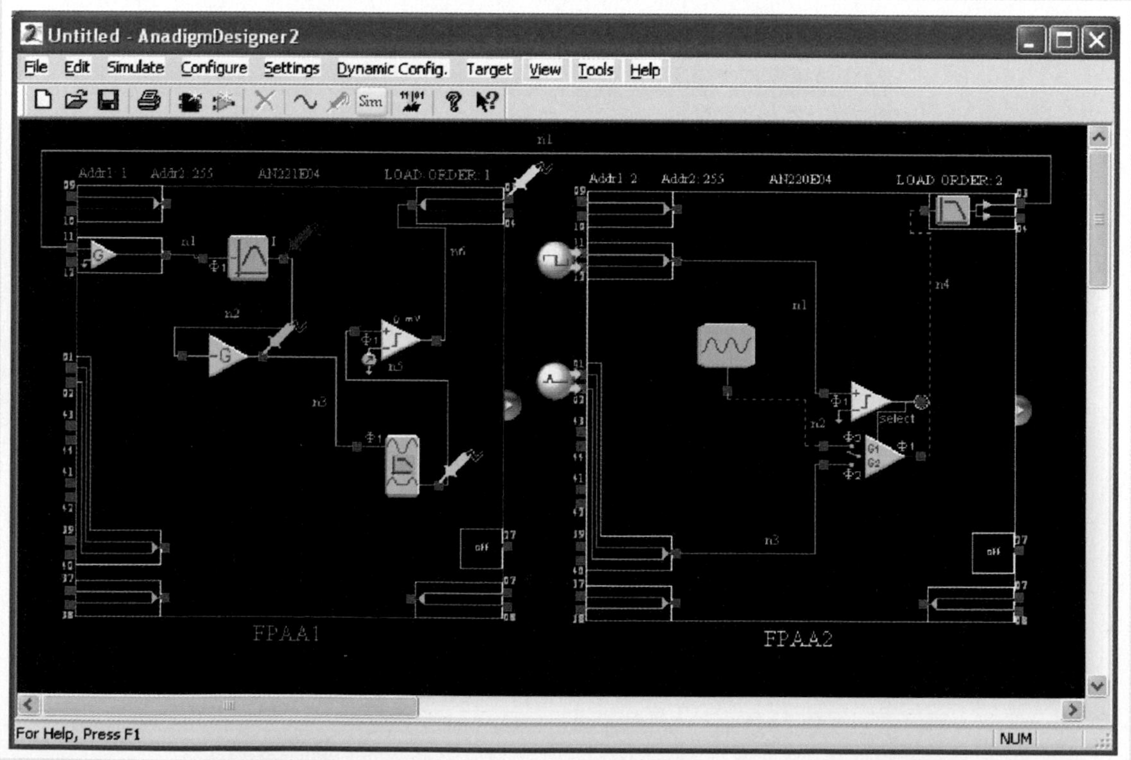

FIGURE 15–40

Design screen showing the RFID reader in FPAA1 and an ASK generator representing the RFID tag in FPAA2.

contains an amplifier with gain, the amplifier in the RFID reader circuit has less gain than if a 1 mV ASK signal were available. Also, the rectifier and low-pass filter are combined in one CAM. FPAA2 is used as a signal source to replicate a 125 kHz carrier modulated with a 10 kHz square wave. This chip is for test purposes only and is not part of the RFID reader.

Analysis: The simulation of the RFID reader is shown in Figure 15–41. The top waveform is the output of the 125 kHz band-pass filter CAM and is an ASK input signal representing a digital 1 followed by a 0. The second waveform is the output of the inverting gain stage CAM with a unity gain. The third waveform is the output of the half-wave rectifier/low-pass filter CAM. The bottom output is the digital signal from the comparator.

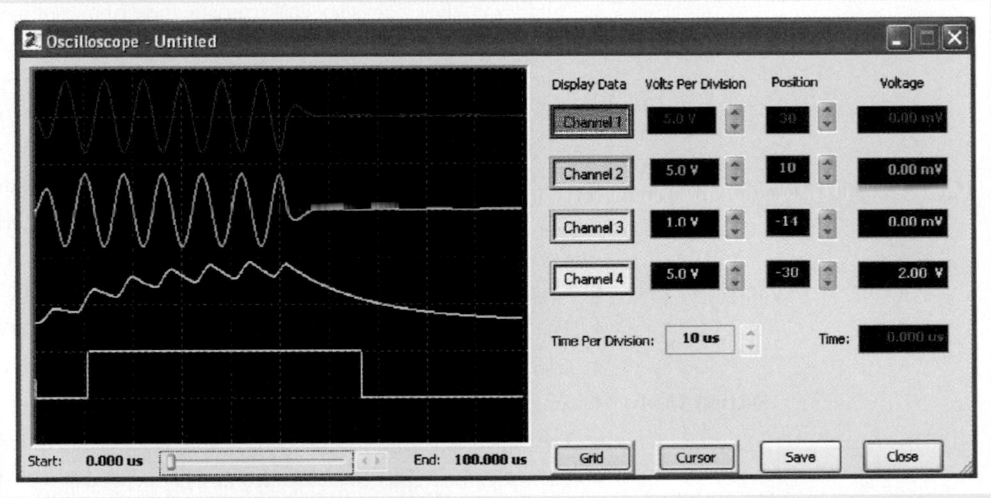

▲ **FIGURE 15–41**

Simulation waveforms for the RFID reader.

Programming Exercises

1. Why is a software program the best way to specify and implement active filters?
2. List the filter types available in the AnadigmFilter software.
3. List the filter approximations available in the AnadigmFilter software.

PAM Experiment

To program, download, and test a circuit using AnadigmDesigner2 software and the programmable analog module (PAM) board, go to Experiment 15–B in *Laboratory Exercises for Electronic Devices* by David Buchla and Steven Wetterling.

SUMMARY

Section 15–1

- ◆ In filter terminology, a single *RC* circuit is called a *pole.*
- ◆ The bandwidth in a low-pass filter equals the critical frequency because the response extends to 0 Hz.
- ◆ The passband of a high-pass filter extends above the critical frequency and is limited only by the inherent frequency limitation of the active circuit.

- A band-pass filter passes all frequencies within a band between a lower and an upper critical frequency and rejects all others outside this band.
- The bandwidth of a band-pass filter is the difference between the upper critical frequency and the lower critical frequency.
- The quality factor Q of a band-pass filter determines the filter's selectivity. The higher the Q, the narrower the bandwidth and the better the selectivity.
- A band-stop filter rejects all frequencies within a specified band and passes all those outside this band.

Section 15–2

- Filters with the Butterworth response characteristic have a very flat response in the passband, exhibit a roll-off of −20 dB/decade/pole, and are used when all the frequencies in the passband must have the same gain.
- Filters with the Chebyshev characteristic have ripples or overshoot in the passband and exhibit a faster roll-off per pole than filters with the Butterworth characteristic.
- Filters with the Bessel characteristic are used for filtering pulse waveforms. Their linear phase characteristic results in minimal waveshape distortion. The roll-off rate per pole is slower than for the Butterworth.
- Each pole in a Butterworth filter causes the output to roll off at a rate of −20 dB/decade.
- The damping factor determines the filter response characteristic (Butterworth, Chebyshev, or Bessel).

Section 15–3

- Single-pole low-pass filters have a −20 dB/decade roll-off.
- The Sallen-Key low-pass filter has two poles (second order) and has a −40 dB/decade roll-off.
- Each additional filter in a cascaded arrangement adds −20 dB to the roll-off rate.

Section 15–4

- Single-pole high-pass filters have a −20 dB/decade roll-off.
- The Sallen-Key high-pass filter has two poles (second order) and has a −40 dB/deacde roll-off.
- Each additional filter in a cascaded arrangement adds −20 dB to the roll-off rate.
- The response of an active high-pass filter is limited by the internal op-amp roll-off.

Section 15–5

- Band-pass filters pass a specified band of frequencies.
- A band-pass filter can be achieved by cascading a low-pass and a high-pass filter.
- The multiple-feedback band-pass filter uses two feedback paths to achieve its response characteristic.
- The state-variable band-pass filter uses a summing amplifier and two integrators.
- The biquad filter consists of an integrator followed by an inverting amplifier and a second integrator.

Section 15–6

- Band-stop filters reject a specified band of frequencies.
- Multiple-feedback and state-variable are common types of band-stop filters.

Section 15–7

- Filter response can be measured using discrete point measurement or swept frequency measurement.

KEY TERMS

Key terms and other bold terms in the chapter are defined in the end-of-book glossary.

Band-pass filter A type of filter that passes a range of frequencies lying between a certain lower frequency and a certain higher frequency.

Band-stop filter A type of filter that blocks or rejects a range of frequencies lying between a certain lower frequency and a certain higher frequency.

Damping factor A filter characteristic that determines the type of response.

Filter A circuit that passes certain frequencies and attenuates or rejects all other frequencies.

High-pass filter A type of filter that passes frequencies above a certain frequency while rejecting lower frequencies.

Low-pass filter A type of filter that passes frequencies below a certain frequency while rejecting higher frequencies.

Pole A circuit containing one resistor and one capacitor that contributes −20 dB/decade to a filter's roll-off rate.

Roll-off The rate of decrease in gain, below or above the critical frequencies of a filter.

KEY FORMULAS

15–1	$BW = f_c$	Low-pass bandwidth
15–2	$BW = f_{c2} - f_{c1}$	Filter bandwidth of a band-pass filter
15–3	$f_0 = \sqrt{f_{c1}f_{c2}}$	Center frequency of a band-pass filter
15–4	$Q = \dfrac{f_0}{BW}$	Quality factor of a band-pass filter
15–5	$DF = 2 - \dfrac{R_1}{R_2}$	Damping factor
15–6	$A_{cl(NI)} = \dfrac{R_1}{R_2} + 1$	Closed-loop voltage gain
15–7	$f_c = \dfrac{1}{2\pi\sqrt{R_AR_BC_AC_B}}$	Critical frequency for a second-order Sallen-Key filter
15–8	$f_0 = \dfrac{1}{2\pi C}\sqrt{\dfrac{R_1 + R_3}{R_1R_2R_3}}$	Center frequency of a multiple-feedback filter
15–9	$A_0 = \dfrac{R_2}{2R_1}$	Gain of a multiple-feedback filter

TRUE/FALSE QUIZ

Answers are at the end of the chapter.

1. The response of a filter can be identified by its passband.
2. A filter pole is the cutoff frequency of a filter.
3. A single-pole filter has one *RC* circuit.
4. A single-pole filter produces a roll-off of −25 dB/decade.
5. A low-pass filter can pass a dc voltage.
6. A high-pass filter passes any frequency above dc.
7. The critical frequency of a filter depends only on *R* and *C* values.
8. The band-pass filter has two critical frequencies.
9. The quality factor of a band-pass filter is the ratio of bandwidth to the center frequency.
10. The higher the *Q*, the narrower the bandwidth of a band-pass filter.
11. The Butterworth characteristic provides a flat response in the passband.
12. Filters with a Chebyshev response have a slow roll-off.
13. A Chebyshev response has ripples in the passband.
14. Bessel filters are useful in filtering pulse waveforms.
15. The order of a filter is the number of poles it contains.
16. A Sallen-Key filter is also known as a VCVS filter.
17. Multiple feedback is used in low-pass filters.
18. A state-variable filter uses differentiators.
19. A band-stop filter rejects certain frequencies.
20. Filter response can be measured using a sweep generator.

CIRCUIT-ACTION QUIZ

Answers are at the end of the chapter.

1. If the critical frequency of a low-pass filter is increased, the bandwidth will
   (a) increase (b) decrease (c) not change
2. If the critical frequency of a high-pass filter is increased, the bandwidth will
   (a) increase (b) decrease (c) not change
3. If the Q of a band-pass filter is increased, the bandwidth will
   (a) increase (b) decrease (c) not change
4. If the value of C_A and C_B in Figure 15–11 are increased by the same amount, the critical frequency will
   (a) increase (b) decrease (c) not change
5. If the the value of R_2 in Figure 15–11 is increased, the bandwidth will
   (a) increase (b) decrease (c) not change
6. If two filters like the one in Figure 15–15 are cascaded, the roll-off rate of the frequency response will
   (a) increase (b) decrease (c) not change
7. If the value of R_2 in Figure 15–19 is decreased, the Q will
   (a) increase (b) decrease (c) not change
8. If the capacitors in Figure 15–19 are changed to 0.022 μF, the center frequency will
   (a) increase (b) decrease (c) not change

SELF-TEST

Answers are at the end of the chapter.

Section 15–1

1. The term *pole* in filter terminology refers to
   (a) a high-gain op-amp (b) one complete active filter
   (c) a single *RC* circuit (d) the feedback circuit
2. A single resistor and a single capacitor can be connected to form a filter with a roll-off rate of
   (a) −20 dB/decade (b) −40 dB/decade
   (c) −6 dB/octave (d) answers (a) and (c)
3. A band-pass response has
   (a) two critical frequencies (b) one critical frequency
   (c) a flat curve in the passband (d) a wide bandwidth
4. The lowest frequency passed by a low-pass filter is
   (a) 1 Hz (b) 0 Hz (c) 10 Hz (d) dependent on the critical frequency
5. The quality factor (Q) of a band-pass filter depends on
   (a) the critical frequencies (b) only the bandwidth
   (c) the center frequency and the bandwidth (d) only the center frequency

Section 15–2

6. The damping factor of an active filter determines
   (a) the voltage gain (b) the critical frequency
   (c) the response characteristic (d) the roll-off rate
7. A maximally flat frequency response is known as
   (a) Chebyshev (b) Butterworth (c) Bessel (d) Colpitts
8. The damping factor of a filter is set by
   (a) the negative feedback circuit (b) the positive feedback circuit
   (c) the frequency-selective circuit (d) the gain of the op-amp
9. The number of poles in a filter affect the
   (a) voltage gain (b) bandwidth
   (c) center frequency (d) roll-off rate

Section 15–3

10. Sallen-Key low-pass filters are
    (a) single-pole filters (b) second-order filters
    (c) Butterworth filters (d) band-pass filters
11. When low-pass filters are cascaded, the roll-off rate
    (a) increases (b) decreases (c) does not change

Section 15–4

12. In a high-pass filter, the roll-off occurs
    (a) above the critical frequency (b) below the critical frequency
    (c) during the mid range (d) at the center frequency
13. A two-pole Sallen-Key high-pass filter contains
    (a) one capacitor and two resistors (b) two capacitors and two resistors
    (c) a feedback circuit (d) answers (b) and (c)

Section 15–5

14. When a low-pass and a high-pass filter are cascaded to get a band-pass filter, the critical frequency of the low-pass filter must be
    (a) equal to the critical frequency of the high-pass filter
    (b) less than the critical frequency of the high-pass filter
    (c) greater than the critical frequency of the high-pass filter
15. A state-variable filter consists of
    (a) one op-amp with multiple-feedback paths
    (b) a summing amplifier and two integrators
    (c) a summing amplifier and two differentiators
    (d) three Butterworth stages

Section 15–6

16. When the gain of a filter is minimum at its center frequency, it is
    (a) a band-pass filter (b) a band-stop filter
    (c) a notch filter (d) answers (b) and (c)

PROBLEMS

Answers to all odd-numbered problems are at the end of the book.

BASIC PROBLEMS

Section 15–1

Basic Filter Responses

1. Identify each type of filter response (low-pass, high-pass, band-pass, or band-stop) in Figure 15–42.

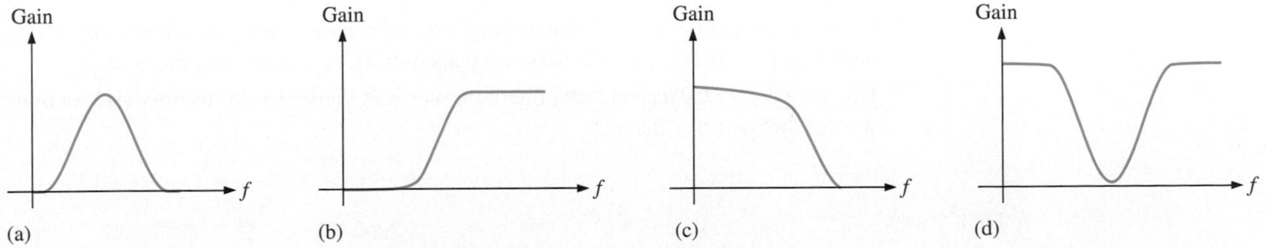

▲ FIGURE 15–42

2. A certain low-pass filter has a critical frequency of 800 Hz. What is its bandwidth?
3. A single-pole high-pass filter has a frequency-selective circuit with $R = 2.2\text{ k}\Omega$ and $C = 0.0015\ \mu\text{F}$. What is the critical frequency? Can you determine the bandwidth from the available information?
4. What is the roll-off rate of the filter described in Problem 3?
5. What is the bandwidth of a band-pass filter whose critical frequencies are 3.2 kHz and 3.9 kHz? What is the Q of this filter?
6. What is the center frequency of a filter with a Q of 15 and a bandwidth of 1 kHz?

Section 15–2 Filter Response Characteristics

7. What is the damping factor in each active filter shown in Figure 15–43? Which filters are approximately optimized for a Butterworth response characteristic?

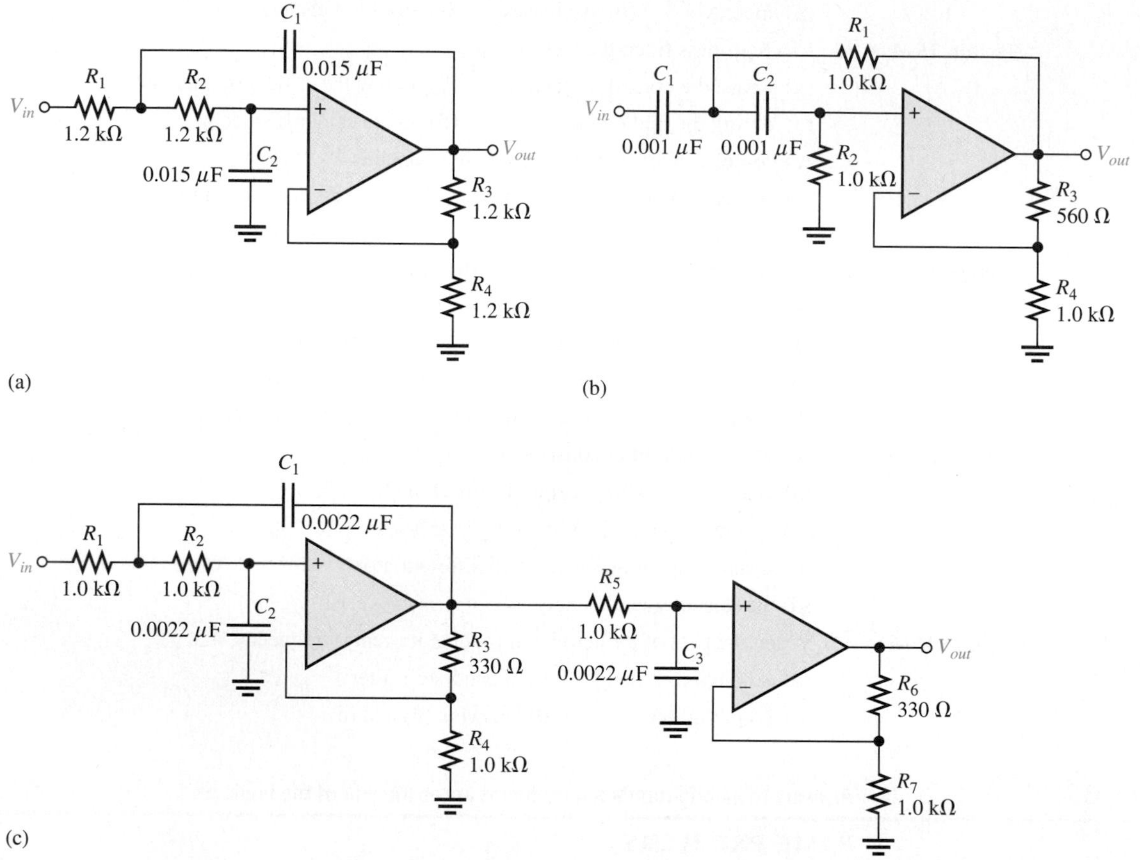

▲ **FIGURE 15–43**

Multisim file circuits are identified with a CD logo and are in the Problems folder on your CD-ROM. Filenames correspond to figure numbers (e.g., F15-43).

8. For the filters in Figure 15–43 that do not have a Butterworth response, specify the changes necessary to convert them to Butterworth responses. (Use nearest standard values.)

9. Response curves for second-order filters are shown in Figure 15–44. Identify each as Butterworth. Chebyshev, or Bessel.

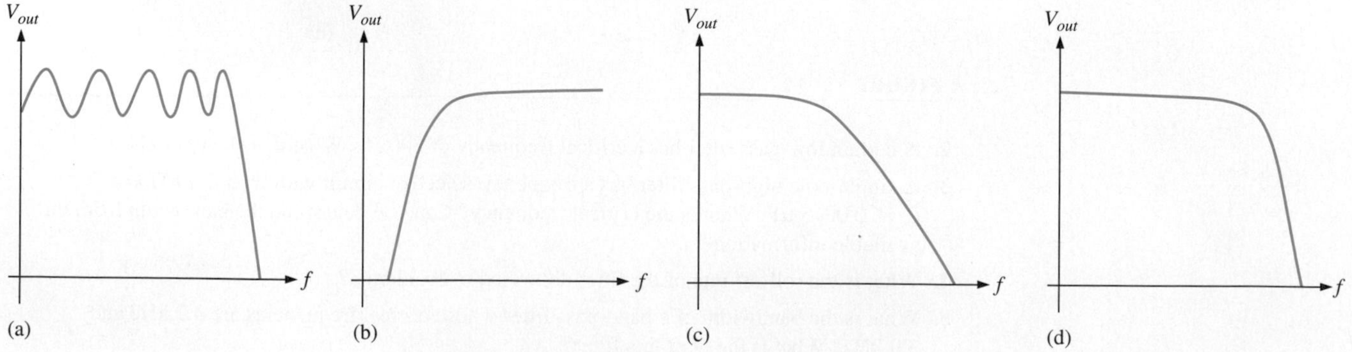

▲ **FIGURE 15–44**

Section 15–3 Active Low-Pass Filters

10. Is the four-pole filter in Figure 15–45 approximately optimized for a Butterworth response? What is the roll-off rate?

11. Determine the critical frequency in Figure 15–45.

12. Without changing the response curve, adjust the component values in the filter of Figure 15–45 to make it an equal-value filter. Select $C = 0.22\ \mu F$ for both stages.

13. Modify the filter in Figure 15–45 to increase the roll-off rate to -120 dB/decade while maintaining an approximate Butterworth response.

14. Using a block diagram format, show how to implement the following roll-off rates using single-pole and two-pole low-pass filters with Butterworth responses.

(a) -40 dB/decade **(b)** -20 dB/decade

(c) -60 dB/decade **(d)** -100 dB/decade

(e) -120 dB/decade

FIGURE 15–45

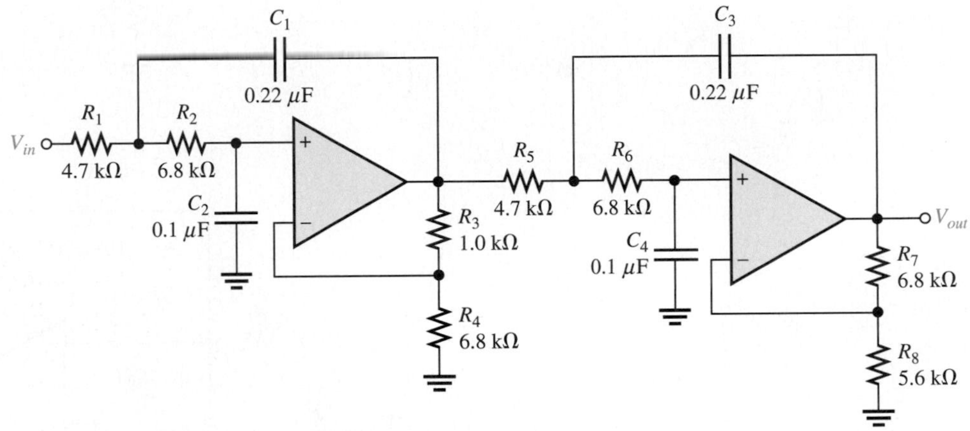

Section 15–4 Active High-Pass Filters

15. Convert the filter in Problem 12 to a high-pass with the same critical frequency and response characteristic.

16. Make the necessary circuit modification to reduce by half the critical frequency in Problem 15.

17. For the filter in Figure 15–46, **(a)** how would you increase the critical frequency? **(b)** How would you increase the gain?

FIGURE 15–46

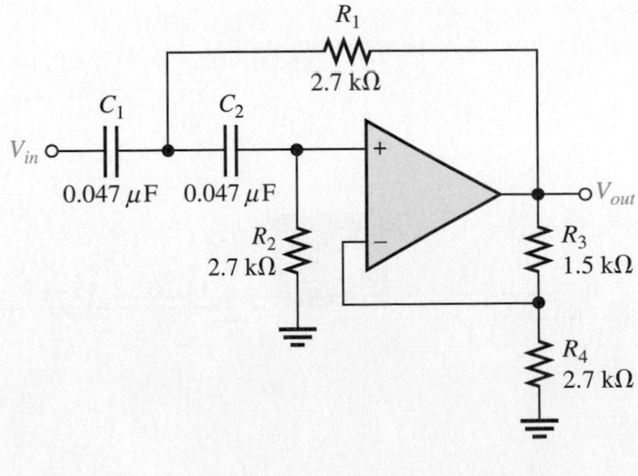

Section 15–5 Active Band-Pass Filters

18. Identify each band-pass filter configuration in Figure 15–47.

19. Determine the center frequency and bandwidth for each filter in Figure 15–47.

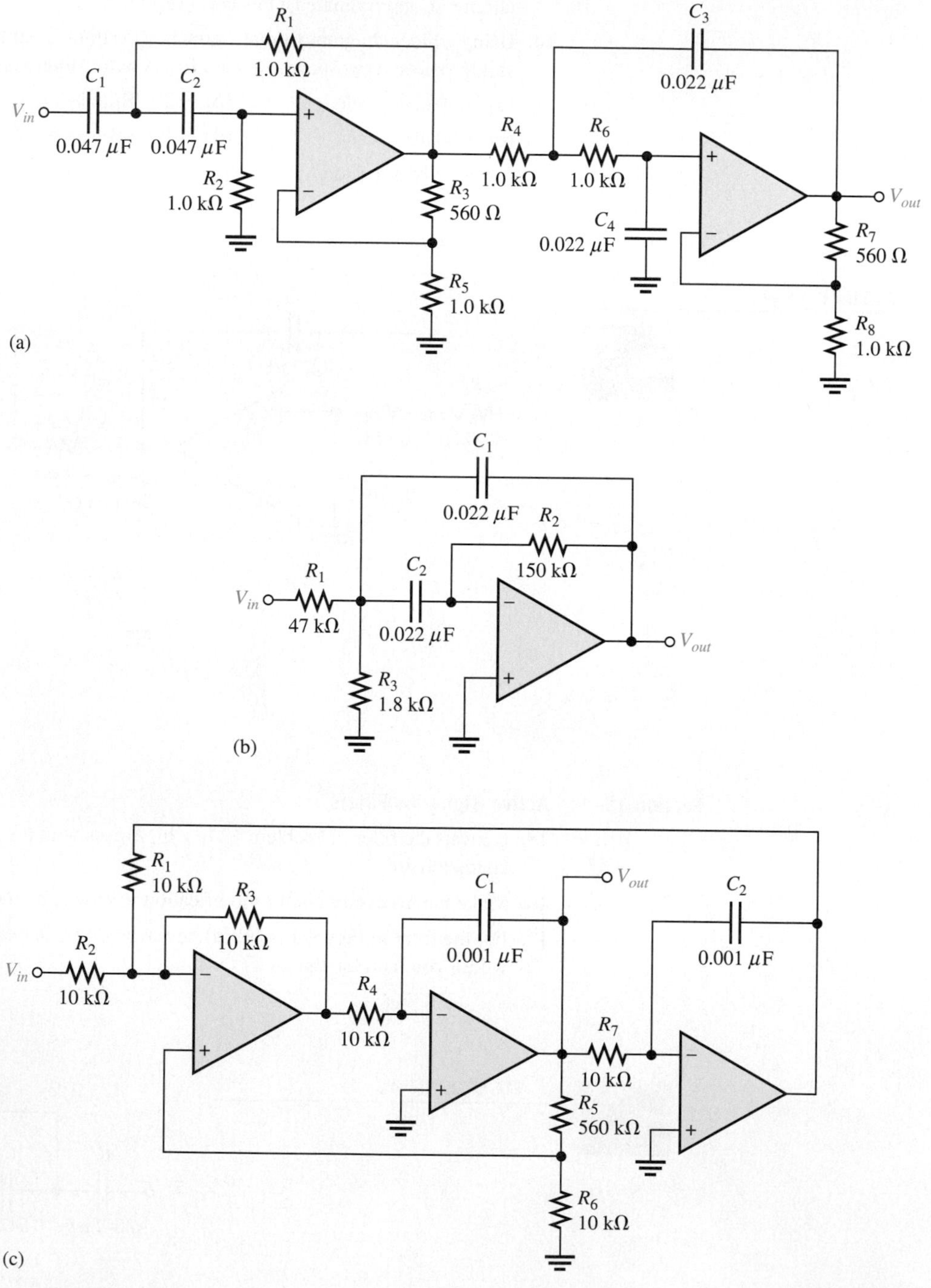

▲ FIGURE 15–47

▶ FIGURE 15–48

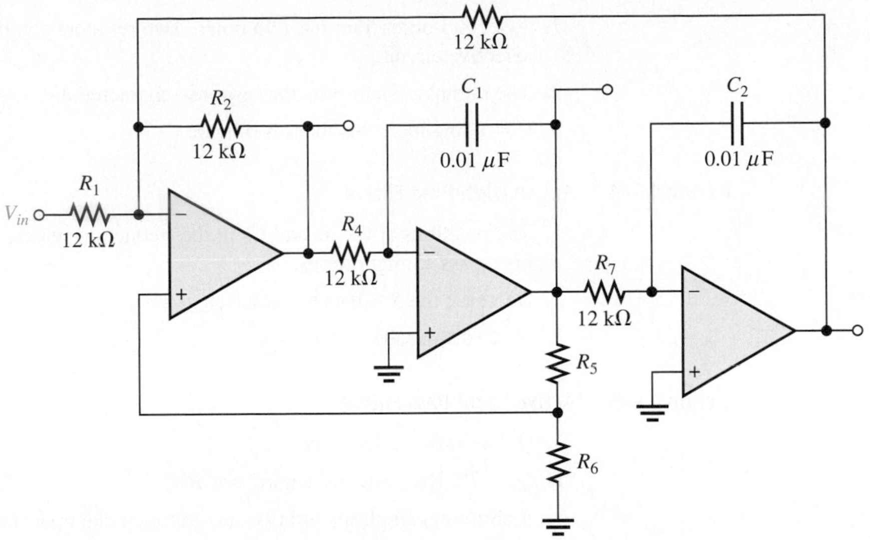

20. Optimize the state-variable filter in Figure 15–48 for $Q = 50$. What bandwidth is achieved?

Section 15–6 **Active Band-Stop Filters**

21. Show how to make a notch (band-stop) filter using the basic circuit in Figure 15–48.

22. Modify the band-stop filter in Problem 21 for a center frequency of 120 Hz.

MULTISIM TROUBLESHOOTING PROBLEMS

These file circuits are in the Troubleshooting Problems folder on your CD-ROM.

23. Open file TSP15-23 and determine the fault.

24. Open file TSP15-24 and determine the fault.

25. Open file TSP15-25 and determine the fault.

26. Open file TSP15-26 and determine the fault.

27. Open file TSP15-27 and determine the fault.

28. Open file TSP15-28 and determine the fault.

29. Open file TSP15-29 and determine the fault.

30. Open file TSP15-30 and determine the fault.

31. Open file TSP15-31 and determine the fault.

ANSWERS

SECTION CHECKUPS

Section 15–1 **Basic Filter Responses**

1. The critical frequency determines the passband.
2. The inherent frequency limitation of the op-amp limits the bandwidth.
3. Q and BW are inversely related. The higher the Q, the better the selectivity, and vice versa.

Section 15–2 **Filter Response Characteristics**

1. Butterworth is very flat in the passband and has a −20 dB/decade/pole roll-off. Chebyshev has ripples in the passband and has greater than −20 dB/decade/pole roll-off. Bessel has a linear phase characteristic and less than −20 dB/decade/pole roll-off.
2. The damping factor
3. Frequency-selective circuit, gain element, and negative feedback circuit are the parts of an active filter.

Section 15–3 **Active Low-Pass Filters**

1. A second-order filter has two poles. Two resistors and two capacitors make up the frequency-selective circuit.
2. The damping factor sets the response characteristic.
3. Cascading increases the roll-off rate.

Section 15–4 **Active High-Pass Filters**

1. The positions of the *R*s and *C*s in the frequency-selective circuit are opposite for low-pass and high-pass configurations.
2. Decrease the R values to increase f_c.
3. -140 dB/decade

Section 15–5 **Active Band-Pass Filters**

1. Q determines selectivity.
2. $Q = 25$. Higher Q gives narrower BW.
3. A summing amplifier and two integrators make up a state-variable filter.
4. An inverting amplifier and two integrators make up a biquad filter.

Section 15–6 **Active Band-Stop Filters**

1. A band-stop rejects frequencies within the stopband. A band-pass passes frequencies within the passband.
2. The low-pass and high-pass outputs are summed.

Section 15–7 **Filter Response Measurements**

1. To check the frequency response of a filter
2. Discrete point measurement: tedious and less complete; simpler equipment. Swept frequency measurement: uses more expensive equipment; more efficient, can be more accurate and complete.

RELATED PROBLEMS FOR EXAMPLES

15–1 500 Hz

15–2 1.44

15–3 7.23 kHz; 1.29 kΩ

15–4 $C_{A1} = C_{A2} = C_{B1} = C_{B2} = 0.234\ \mu F$; $R_2 = R_4 = 680\ \Omega$; $R_1 = 103\ \Omega$; $R_3 = 840\ \Omega$

15–5 $R_A = R_B = R_2 = 10\ k\Omega$; $C_A = C_B = 0.053\ \mu F$; $R_1 = 5.86\ k\Omega$

15–6 Gain increases to 2.43, frequency decreases to 544 Hz, and bandwidth decreases to 96.5 Hz.

15–7 $f_0 = 21.9$ kHz; $Q = 101$; $BW = 217$ Hz

15–8 Decrease the input resistors or the feedback capacitors of the two integrator stages by half.

TRUE/FALSE QUIZ

1. T **2.** F **3.** T **4.** F **5.** T **6.** F **7.** T
8. T **9.** F **10.** T **11.** T **12.** F **13.** T **14.** T
15. T **16.** T **17.** F **18.** F **19.** T **20.** T

CIRCUIT-ACTION QUIZ

1. (a) **2.** (b) **3.** (b) **4.** (b) **5.** (c) **6.** (a) **7.** (b) **8.** (b)

SELF-TEST

1. (c) **2.** (d) **3.** (a) **4.** (b) **5.** (c) **6.** (c) **7.** (b) **8.** (a)
9. (d) **10.** (b) **11.** (a) **12.** (a) **13.** (d) **14.** (c) **15.** (b) **16.** (d)

OSCILLATORS

16

CHAPTER OUTLINE

CHAPTER OBJECTIVES

- Describe the basic operating principles of an oscillator
- Discuss the main principle on which feedback oscillators is based
- Describe and analyze the basic operation of three *RC* feedback oscillators
- Describe and analyze the basic operation of *LC* feedback oscillators
- Describe and analyze the basic operation of relaxation oscillators
- Use a 555 timer in an oscillator application

KEY TERMS

- Feedback oscillator
- Relaxation oscillator
- Positive feedback
- Voltage-controlled oscillator (VCO)
- Astable

APPLICATION ACTIVITY PREVIEW

The application in this chapter is a circuit that produces an ASK signal for testing the RFID reader developed in the last chapter. The ASK test generator uses an oscillator, a 555 timer, and a JFET analog switch to produce a 125 kHz carrier signal modulated at 10 kHz by a digital signal. The output amplitude is adjustable down to a low level to simulate the RFID tag signal.

VISIT THE COMPANION WEBSITE

Study aids for this chapter are available at
http://www.prenhall.com/floyd

INTRODUCTION

Oscillators are electronic circuits that generate an output signal without the necessity of an input signal. They are used as signal sources in all sorts of applications. Different types of oscillators produce various types of outputs including sine waves, square waves, triangular waves, and sawtooth waves. In this chapter, several types of basic oscillator circuits using both discrete transistors and op-amps as the gain element are introduced. Also, a popular integrated circuit, the 555 timer, is discussed in relation to its oscillator applications.

Sinusoidal oscillator operation is based on the principle of positive feedback, where a portion of the output signal is fed back to the input in a way that causes it to reinforce itself and thus sustain a continuous output signal. Oscillators are widely used in most communications systems as well as in digital systems, including computers, to generate required frequencies and timing signals. Also, oscillators are found in many types of test instruments like those used in the laboratory.

16–1 THE OSCILLATOR

An **oscillator** is a circuit that produces a periodic waveform on its output with only the dc supply voltage as an input. A repetitive input signal is not required except to synchronize oscillations in some applications. The output voltage can be either sinusoidal or nonsinusoidal, depending on the type of oscillator. Two major classifications for oscillators are feedback oscillators and relaxation oscillators.

After completing this section, you should be able to

- **Describe the basic operating principles of an oscillator**
  - Explain the purpose of an oscillator
  - List the basic elements of an oscillator
  - Discuss two important oscillator classifications

Essentially, an oscillator converts electrical energy from the dc power supply to periodic waveforms. A basic oscillator is shown in Figure 16–1.

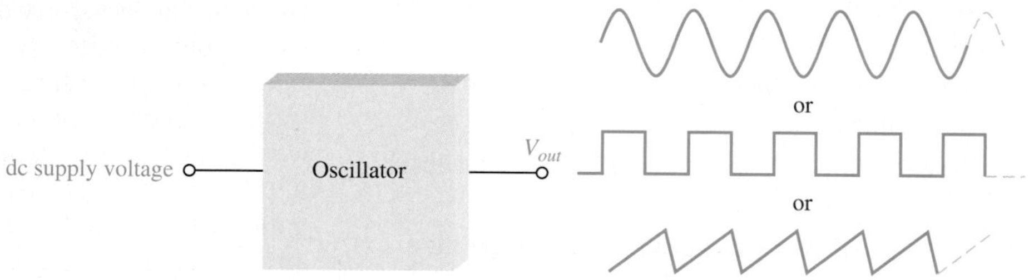

FIGURE 16–1

The basic oscillator concept showing three common types of output waveforms: sine wave, square wave, and sawtooth.

Feedback Oscillators One type of oscillator is the **feedback oscillator**, which returns a fraction of the output signal to the input with no net phase shift, resulting in a reinforcement of the output signal. After oscillations are started, the loop gain is maintained at 1.0 to maintain oscillations. A feedback oscillator consists of an amplifier for gain (either a discrete transistor or an op-amp) and a positive feedback circuit that produces phase shift and provides attenuation, as shown in Figure 16–2.

FIGURE 16–2

Basic elements of a feedback oscillator.

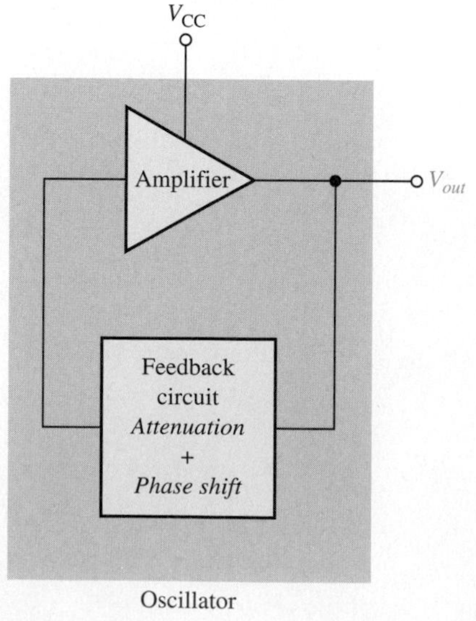

Relaxation Oscillators A second type of oscillator is the **relaxation oscillator**. A relaxation oscillator uses an *RC* timing circuit to generate a waveform that is generally a square wave or other nonsinusoidal waveform. Typically, a relaxation oscillator uses a Schmitt trigger or other device that changes states to alternately charge and discharge a capacitor through a resistor. Relaxation oscillators are discussed in Section 16–5.

SECTION 16–1 CHECKUP
Answers are at the end of the chapter.

1. What is an oscillator?
2. What type of feedback does a feedback oscillator require?
3. What is the purpose of the feedback circuit?
4. Name the two types of oscillators.

16–2 FEEDBACK OSCILLATOR PRINCIPLES

Feedback oscillator operation is based on the principle of positive feedback. In this section, we will examine this concept and look at the general conditions required for oscillation to occur. Feedback oscillators are widely used to generate sinusoidal waveforms.

After completing this section, you should be able to

- **Discuss the main principle on which feedback oscillators is based**
  - Explain positive feedback
  - Describe the conditions for oscillation
  - Discuss the start-up conditions

Positive Feedback

Positive feedback is characterized by the condition wherein a portion of the output voltage of an amplifier is fed back to the input with no net phase shift, resulting in a reinforcement of the output signal. This basic idea is illustrated in Figure 16–3(a). As you can see, the in-phase feedback voltage, V_f, is amplified to produce the output voltage, which in turn produces the feedback voltage. That is, a loop is created in which the signal sustains itself and

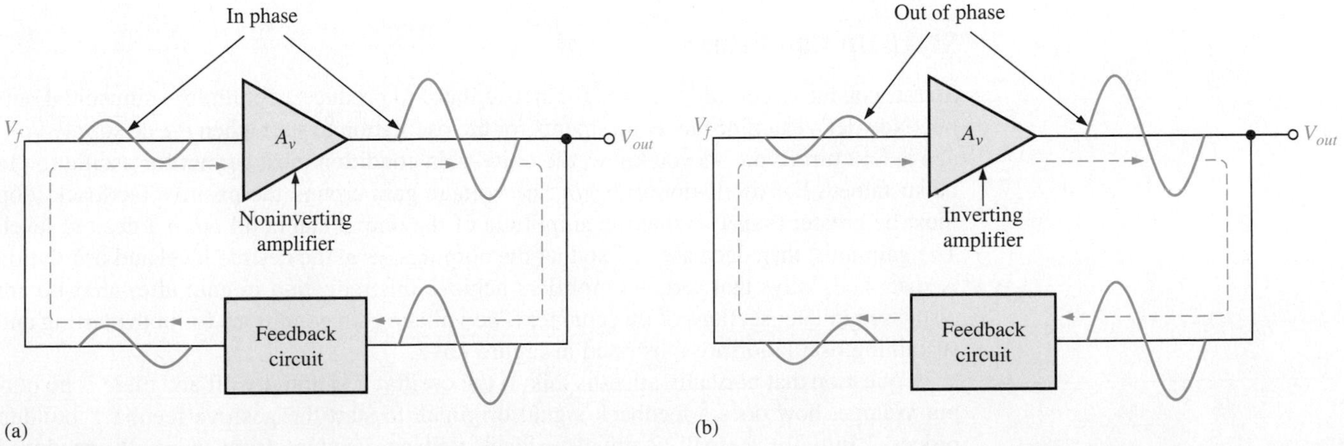

▲ FIGURE 16–3

Positive feedback produces oscillation.

a continuous sinusoidal output is produced. This phenomenon is called *oscillation.* In some types of amplifiers, the feedback circuit shifts the phase 180° and an inverting amplifier is required to provide another 180° phase shift so that there is no net phase shift. This is illustrated in Figure 16–3(b).

Conditions for Oscillation

Two conditions, illustrated in Figure 16–4, are required for a sustained state of oscillation:

1. The phase shift around the feedback loop must be effectively 0°.
2. The voltage gain, A_{cl}, around the closed feedback loop (loop gain) must equal 1 (unity).

FIGURE 16–4

General conditions to sustain oscillation.

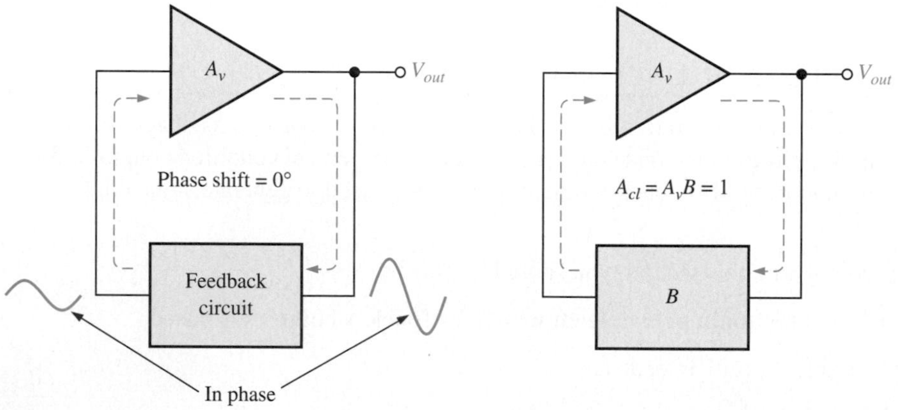

(a) The phase shift around the loop is 0°.

(b) The closed loop gain is 1.

The voltage gain around the closed feedback loop, A_{cl}, is the product of the amplifier gain, A_v, and the attenuation, B, of the feedback circuit.

$$A_{cl} = A_vB$$

If a sinusoidal wave is the desired output, a loop gain greater than 1 will rapidly cause the output to saturate at both peaks of the waveform, producing unacceptable distortion. To avoid this, some form of gain control must be used to keep the loop gain at exactly 1 once oscillations have started. For example, if the attenuation of the feedback circuit is 0.01, the amplifier must have a gain of exactly 100 to overcome this attenuation and not create unacceptable distortion ($0.01 \times 100 = 1$). An amplifier gain of greater than 100 will cause the oscillator to limit both peaks of the waveform.

Start-Up Conditions

So far, you have seen what it takes for an oscillator to produce a continuous sinusoidal output. Now let's examine the requirements for the oscillation to start when the dc supply voltage is first turned on. As you know, the unity-gain condition must be met for oscillation to be sustained. For oscillation to *begin,* the voltage gain around the positive feedback loop must be greater than 1 so that the amplitude of the output can build up to a desired level. The gain must then decrease to 1 so that the output stays at the desired level and oscillation is sustained. Ways that certain amplifiers achieve this reduction in gain after start-up are discussed in later sections of this chapter. The voltage gain conditions for both starting and sustaining oscillation are illustrated in Figure 16–5.

A question that normally arises is this: If the oscillator is initially off and there is no output voltage, how does a feedback signal originate to start the positive feedback buildup process? Initially, a small positive feedback voltage develops from thermally produced broad-band noise in the resistors or other components or from power supply turn-on transients. The feedback circuit permits only a voltage with a frequency equal to the selected oscillation frequency to appear in phase on the amplifier's input. This initial feedback volt-

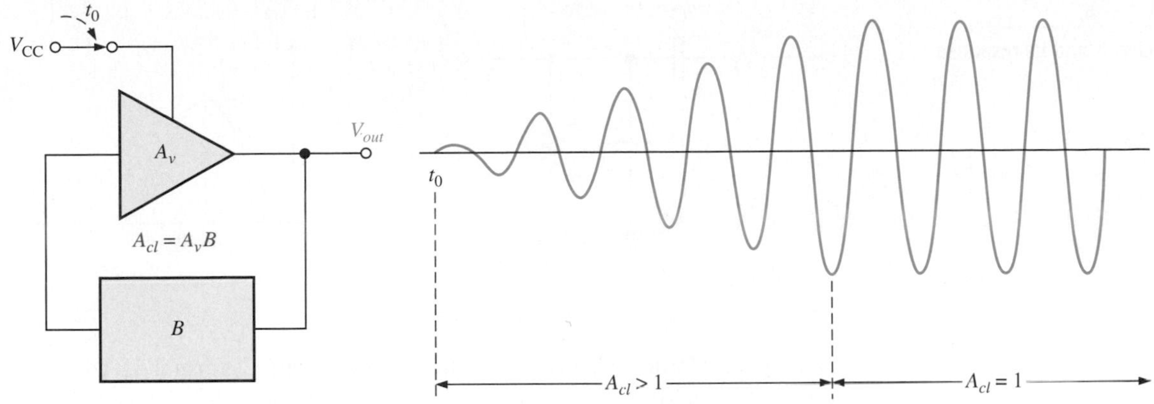

▲ **FIGURE 16–5**

When oscillation starts at t_0, the condition $A_{cl} > 1$ causes the sinusoidal output voltage amplitude to build up to a desired level. Then A_{cl} decreases to 1 and maintains the desired amplitude.

age is amplified and continually reinforced, resulting in a buildup of the output voltage as previously discussed.

SECTION 16–2 CHECKUP

1. What are the conditions required for a circuit to oscillate?
2. Define *positive feedback.*
3. What is the voltage gain condition for oscillator start-up?

16–3 OSCILLATORS WITH *RC* FEEDBACK CIRCUITS

Three types of feedback oscillators that use *RC* circuits to produce sinusoidal outputs are the Wien-bridge oscillator, the phase-shift oscillator, and the twin-T oscillator. Generally, *RC* feedback oscillators are used for frequencies up to about 1 MHz. The Wien-bridge is by far the most widely used type of *RC* feedback oscillator for this range of frequencies.

After completing this section, you should be able to

- **Describe and analyze the basic operation of three *RC* feedback oscillators**
  - Identify a Wien-bridge oscillator
  - Calculate the resonant frequency of a Wien-bridge oscillator
  - Analyze oscillator feedback conditions
  - Analyze oscillator start-up conditions
  - Describe a self-starting Wien-bridge oscillator
  - Identify a phase-shift oscillator
  - Calculate the resonant frequency and analyze the feedback conditions for a phase-shift oscillator
  - Identify a twin-T oscillator and describe its operation

The Wien-Bridge Oscillator

One type of sinusoidal feedback oscillator is the **Wien-bridge oscillator**. A fundamental part of the Wien-bridge oscillator is a lead-lag circuit like that shown in Figure 16–6(a). R_1

FIGURE 16–6

A lead-lag circuit and its response curve.

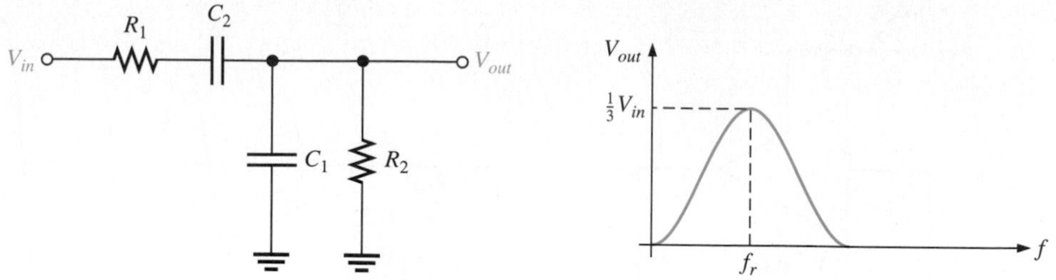

(a) Circuit

(b) Response curve

and C_1 together form the lag portion of the circuit; R_2 and C_2 form the lead portion. The operation of this lead-lag circuit is as follows. At lower frequencies, the lead circuit dominates due to the high reactance of C_2. As the frequency increases, X_{C2} decreases, thus allowing the output voltage to increase. At some specified frequency, the response of the lag circuit takes over, and the decreasing value of X_{C1} causes the output voltage to decrease.

The response curve for the lead-lag circuit shown in Figure 16–6(b) indicates that the output voltage peaks at a frequency called the resonant frequency, f_r. At this point, the attenuation (V_{out}/V_{in}) of the circuit is $1/3$ if $R_1 = R_2$ and $X_{C1} = X_{C2}$ as stated by the following equation (derived in Appendix B):

Equation 16–1

$$\frac{V_{out}}{V_{in}} = \frac{1}{3}$$

The formula for the resonant frequency (also derived in Appendix B) is

Equation 16–2

$$f_r = \frac{1}{2\pi RC}$$

To summarize, the lead-lag circuit in the Wien-bridge oscillator has a resonant frequency, f_r, at which the phase shift through the circuit is 0° and the attenuation is 1/3. Below f_r, the lead circuit dominates and the output leads the input. Above f_r, the lag circuit dominates and the output lags the input.

The Basic Circuit The lead-lag circuit is used in the positive feedback loop of an op-amp, as shown in Figure 16–7(a). A voltage divider is used in the negative feedback loop.

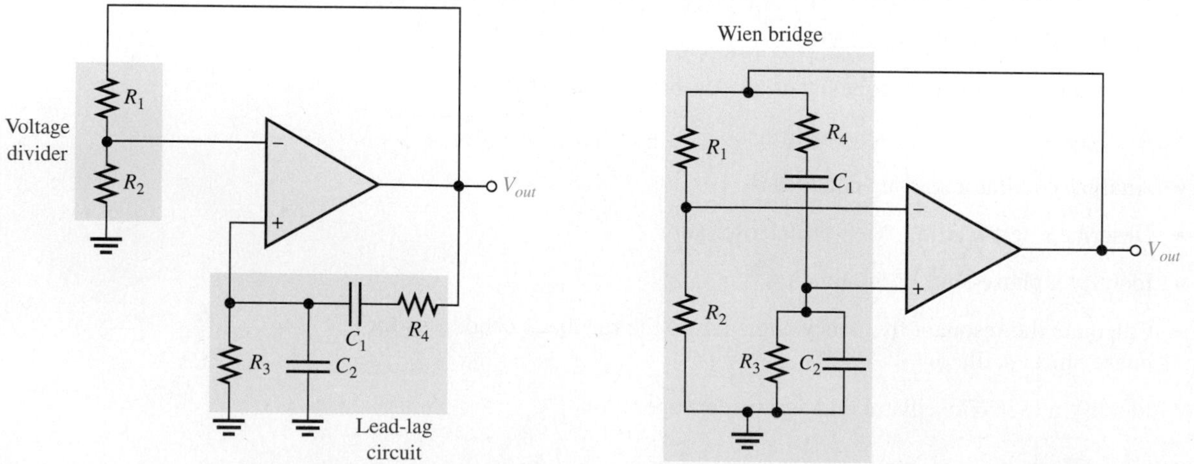

(a)

(b) Wien bridge circuit combines a voltage divider and a lead-lag circuit.

FIGURE 16–7

The Wien-bridge oscillator schematic drawn in two different but equivalent ways.

The Wien-bridge oscillator circuit can be viewed as a noninverting amplifier configuration with the input signal fed back from the output through the lead-lag circuit. Recall that the voltage divider determines the closed-loop gain of the amplifier.

$$A_{cl} = \frac{1}{B} = \frac{1}{R_2/(R_1 + R_2)} = \frac{R_1 + R_2}{R_2}$$

The circuit is redrawn in Figure 16–7(b) to show that the op-amp is connected across the bridge circuit. One leg of the bridge is the lead-lag circuit, and the other is the voltage divider.

Positive Feedback Conditions for Oscillation As you know, for the circuit to produce a sustained sinusoidal output (oscillate), the phase shift around the positive feedback loop must be 0° and the gain around the loop must equal unity (1). The 0° phase-shift condition is met when the frequency is f_r because the phase shift through the lead-lag circuit is 0° and there is no inversion from the noninverting (+) input of the op-amp to the output. This is shown in Figure 16–8(a).

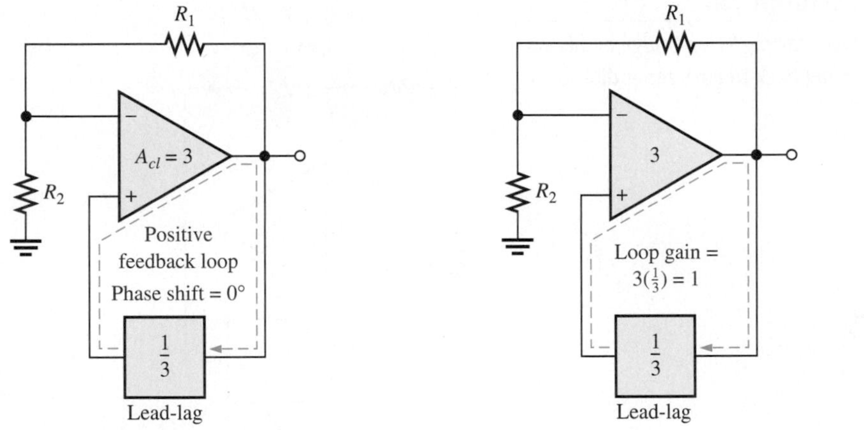

(a) The phase shift around the loop is 0°.
(b) The voltage gain around the loop is 1.

FIGURE 16–8

Conditions for sustained oscillation.

The unity-gain condition in the feedback loop is met when

$$A_{cl} = 3$$

This offsets the 1/3 attenuation of the lead-lag circuit, thus making the total gain around the positive feedback loop equal to 1, as depicted in Figure 16–8(b). To achieve a closed-loop gain of 3,

$$R_1 = 2R_2$$

Then

$$A_{cl} = \frac{R_1 + R_2}{R_2} = \frac{2R_2 + R_2}{R_2} = \frac{3R_2}{R_2} = 3$$

Start-Up Conditions Initially, the closed-loop gain of the amplifier itself must be more than 3 ($A_{cl} > 3$) until the output signal builds up to a desired level. Ideally, the gain of the amplifier must then decrease to 3 so that the total gain around the loop is 1 and the output signal stays at the desired level, thus sustaining oscillation. This is illustrated in Figure 16–9.

The circuit in Figure 16–10 illustrates a method for achieving sustained oscillations. Notice that the voltage-divider circuit has been modified to include an additional resistor R_3 in parallel with a back-to-back zener diode arrangement. When dc power is first applied,

HISTORY NOTE

Max Wien (1866–1938) was a German physicist. He theoretically developed the concept of the Wien-bridge oscillator in 1891. At that time, Wien did not have a means of developing electronic gain, so a workable oscillator could not be achieved. Based on Wien's work, William Hewlett, co-founder of Hewlett-Packard, was successful in building a practical Wien-bridge oscillator in 1939.

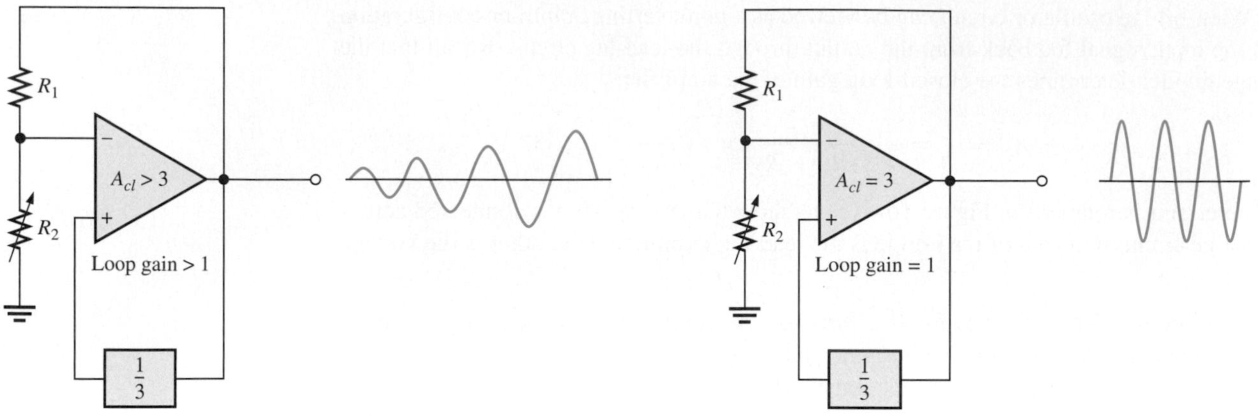

(a) Loop gain greater than 1 causes output to build up.

(b) Loop gain of 1 causes a sustained constant output.

▲ **FIGURE 16–9**

Conditions for start-up and sustained oscillations.

▶ **FIGURE 16–10**

Self-starting Wien-bridge oscillator using back-to-back zener diodes.

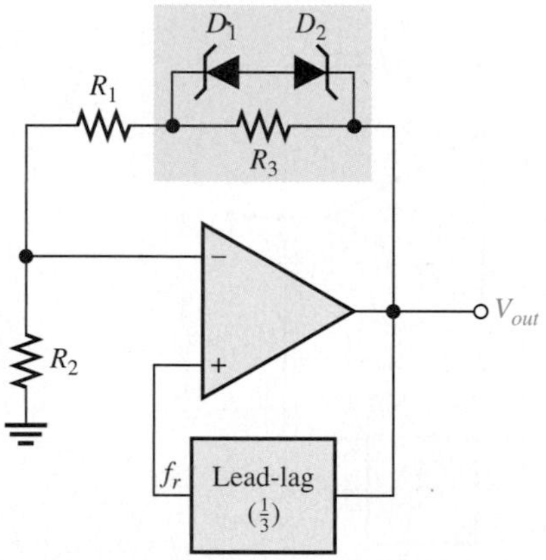

both zener diodes appear as opens. This places R_3 in series with R_1, thus increasing the closed-loop gain of the amplifier as follows ($R_1 = 2R_2$):

$$A_{cl} = \frac{R_1 + R_2 + R_3}{R_2} = \frac{3R_2 + R_3}{R_2} = 3 + \frac{R_3}{R_2}$$

Initially, a small positive feedback signal develops from noise or turn-on transients. The lead-lag circuit permits only a signal with a frequency equal to f_r to appear in phase on the noninverting input. This feedback signal is amplified and continually reinforced, resulting in a buildup of the output voltage. When the output signal reaches the zener breakdown voltage, the zeners conduct and effectively short out R_3. This lowers the amplifier's closed-loop gain to 3. At this point, the total loop gain is 1 and the output signal levels off and the oscillation is sustained.

All practical methods to achieve stability for feedback oscillators require the gain to be self-adjusting. This requirement is a form of automatic gain control (AGC). The zener diodes in Figure 16–10 limit the gain at the onset of nonlinearity, in this case, zener conduction. Although the zener feedback is simple, it suffers from the nonlinearity of the zener diodes that occurs in order to control gain. It is difficult to achieve an undistorted sinusoidal output waveform. In some older designs, a tungsten lamp was used in the feedback circuit to achieve stability.

A better method to control the gain uses a JFET as a voltage-controlled resistor in a negative feedback path. This method can produce an excellent sinusoidal waveform that is stable. A JFET operating with a small or zero V_{DS} is operating in the ohmic region. As the gate voltage increases, the drain-source resistance increases. If the JFET is placed in the negative feedback path, automatic gain control can be achieved because of this voltage-controlled resistance.

A JFET stabilized Wien bridge is shown in Figure 16–11. The gain of the op-amp is controlled by the components shown in the green box, which include the JFET. The JFET's drain-source resistance depends on the gate voltage. With no output signal, the gate is at zero volts, causing the drain-source resistance to be at the minimum. With this condition, the loop gain is greater than 1. Oscillations begin and rapidly build to a large output signal. Negative excursions of the output signal forward-bias D_1, causing capacitor C_3 to charge to a negative voltage. This voltage increases the drain-source resistance of the JFET and reduces the gain (and hence the output). This is classic negative feedback at work. With the proper selection of components, the gain can be stabilized at the required level. Example 16–1 illustrates a JFET stabilized Wien-bridge oscillator.

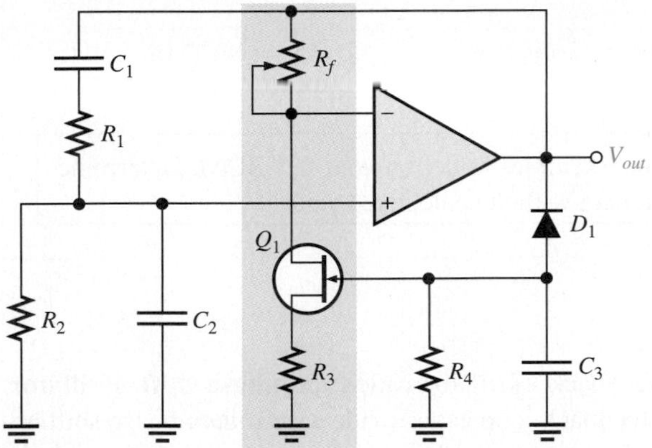

FIGURE 16–11

Self-starting Wien-bridge oscillator using a JFET in the negative feedback loop.

EXAMPLE 16–1

Determine the resonant frequency for the Wien-bridge oscillator in Figure 16–12. Also, calculate the setting for R_f assuming the internal drain-source resistance, r'_{ds}, of the JFET is 500 Ω when oscillations are stable.

FIGURE 16–12

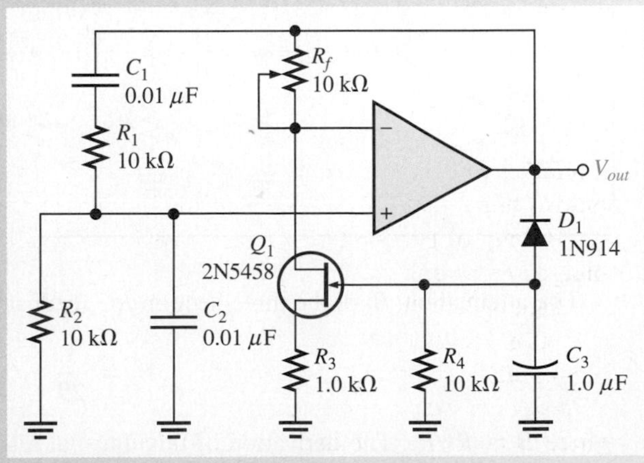

Solution For the lead-lag circuit, $R_1 = R_2 = R = 10\,\text{k}\Omega$ and $C_1 = C_2 = C = 0.01\,\mu\text{F}$. The frequency is

$$f_r = \frac{1}{2\pi RC} = \frac{1}{2\pi(10\,\text{k}\Omega)(0.01\,\mu\text{F})} = \mathbf{1.59\,kHz}$$

The closed-loop gain must be 3.0 for oscillations to be sustained. For an inverting amplifier, the gain expression is the same as for a noninverting amplifier.

$$A_v = \frac{R_f}{R_i} + 1$$

R_i is composed of R_3 (the source resistor) and r'_{ds}. Substituting,

$$A_v = \frac{R_f}{R_3 + r'_{ds}} + 1$$

Rearranging and solving for R_f,

$$R_f = (A_v - 1)(R_3 + r'_{ds}) = (3 - 1)(1.0\text{ k}\Omega + 500\ \Omega) = \mathbf{3.0\ k\Omega}$$

*Related Problem** What happens to the oscillations if the setting of R_f is too high? What happens if the setting is too low?

*Answers are at the end of the chapter.

Open the Multisim file E16-01 in the Examples folder on your CD-ROM. Determine the frequency of oscillation and compare with the calculated value.

The Phase-Shift Oscillator

Figure 16–13 shows a sinusoidal feedback oscillator called the **phase-shift oscillator.** Each of the three *RC* circuits in the feedback loop can provide a *maximum* phase shift approaching 90°. Oscillation occurs at the frequency where the total phase shift through the three *RC* circuits is 180°. The inversion of the op-amp itself provides the additional 180° to meet the requirement for oscillation of a 360° (or 0°) phase shift around the feedback loop.

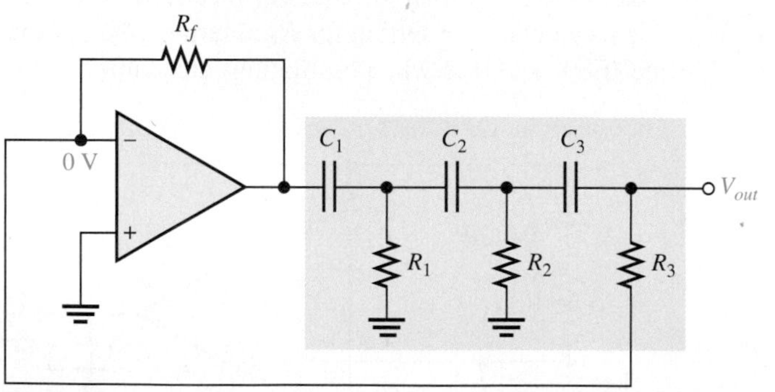

FIGURE 16–13

Phase-shift oscillator.

The attenuation, *B*, of the three-section *RC* feedback circuit is

Equation 16–3

$$B = \frac{1}{29}$$

where $B = R_3/R_f$. The derivation of this unusual result is given in Appendix B. To meet the greater-than-unity loop gain requirement, the closed-loop voltage gain of the op-amp must be greater than 29 (set by R_f and R_3). The frequency of oscillation (f_r) is also derived in Appendix B and is stated in the following equation, where $R_1 = R_2 = R_3 = R$ and $C_1 = C_2 = C_3 = C$.

Equation 16–4

$$f_r = \frac{1}{2\pi\sqrt{6}RC}$$

EXAMPLE 16–2

(a) Determine the value of R_f necessary for the circuit in Figure 16–14 to operate as an oscillator.

(b) Determine the frequency of oscillation.

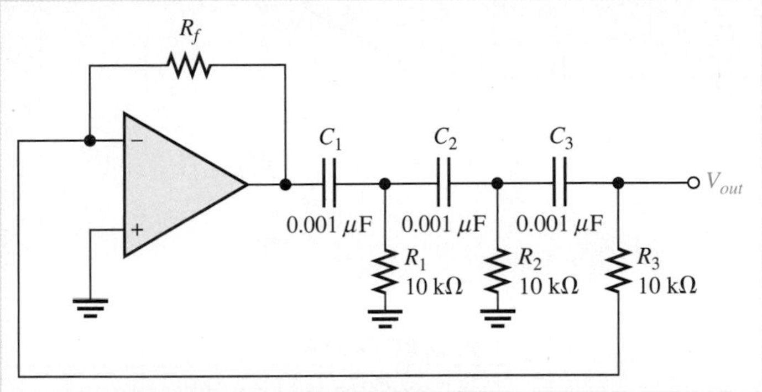

FIGURE 16–14

Solution **(a)** $A_{cl} = 29$, and $B = 1/29 = R_3/R_f$. Therefore,

$$\frac{R_f}{R_3} = 29$$

$$R_f = 29R_3 = 29(10\,\text{k}\Omega) = \mathbf{290\,k\Omega}$$

(b) $R_1 = R_2 = R_3 = R$ and $C_1 = C_2 = C_3 = C$. Therefore,

$$f_r = \frac{1}{2\pi\sqrt{6}RC} = \frac{1}{2\pi\sqrt{6}(10\,\text{k}\Omega)(0.001\,\mu\text{F})} \cong \mathbf{6.5\,kHz}$$

Related Problem **(a)** If R_1, R_2, and R_3 in Figure 16–14 are changed to 8.2 kΩ, what value must R_f be for oscillation?

(b) What is the value of f_r?

Open the Multisim file E16-02 in the Examples folder on your CD-ROM. Measure the frequency of oscillation and compare to the calculated value.

Twin-T Oscillator

Another type of *RC* feedback oscillator is called the *twin-T* because of the two T-type *RC* filters used in the feedback loop, as shown in Figure 16–15(a). One of the twin-T filters has a low-pass response, and the other has a high-pass response. The combined parallel filters produce a band-stop or notch response with a center frequency equal to the desired frequency of oscillation, f_r, as shown in Figure 16–15(b).

Oscillation cannot occur at frequencies above or below f_r because of the negative feedback through the filters. At f_r, however, there is negligible negative feedback; thus, the positive feedback through the voltage divider (R_1 and R_2) allows the circuit to oscillate.

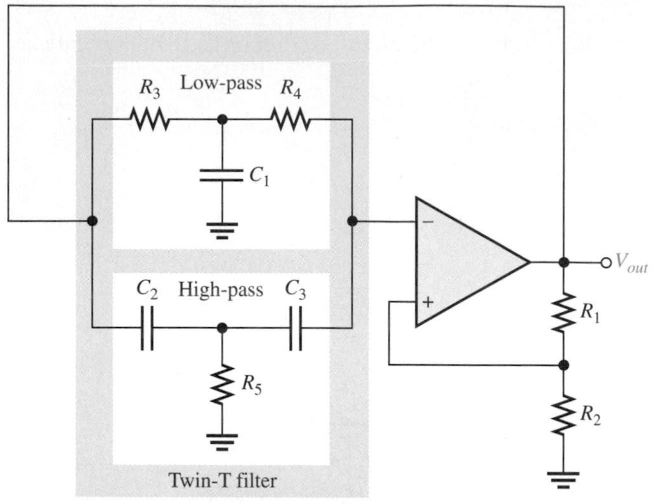

(a) Oscillator circuit

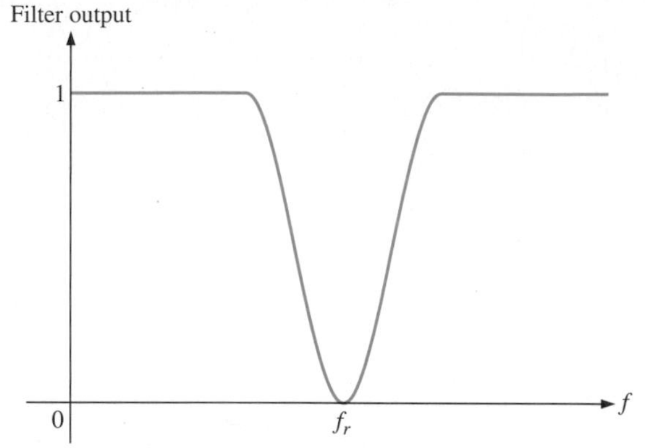

(b) Twin-T filter's frequency response curve

▲ FIGURE 16–15

Twin-T oscillator and twin-T filter response.

SECTION 16–3 CHECKUP

1. There are two feedback loops in the Wien-bridge oscillator. What is the purpose of each?
2. A certain lead-lag circuit has $R_1 = R_2$ and $C_1 = C_2$. An input voltage of 5 V rms is applied. The input frequency equals the resonant frequency of the circuit. What is the rms output voltage?
3. Why is the phase shift through the *RC* feedback circuit in a phase-shift oscillator 180°?

16–4 OSCILLATORS WITH *LC* FEEDBACK CIRCUITS

Although the *RC* feedback oscillators, particularly the Wien bridge, are generally suitable for frequencies up to about 1 MHz, *LC* feedback elements are normally used in oscillators that require higher frequencies of oscillation. Also, because of the frequency limitation (lower unity-gain frequency) of most op-amps, discrete transistors (BJT or FET) are often used as the gain element in *LC* oscillators. This section introduces several types of resonant *LC* feedback oscillators: the Colpitts, Clapp, Hartley, Armstrong, and crystal-controlled oscillators.

After completing this section, you should be able to

- **Describe and analyze the basic operation of *LC* feedback oscillators**
  - Identify and analyze a Colpitts oscillator
  - Identify and analyze a Clapp oscillator
  - Identify and analyze a Hartley oscillator
  - Identify and analyze an Armstrong oscillator
  - Discuss the operation of crystal-controlled oscillators

The Colpitts Oscillator

One basic type of resonant circuit feedback oscillator is the Colpitts, named after its inventor—as are most of the others we cover here. As shown in Figure 16–16, this type of oscillator uses an *LC* circuit in the feedback loop to provide the necessary phase shift and to act as a resonant filter that passes only the desired frequency of oscillation.

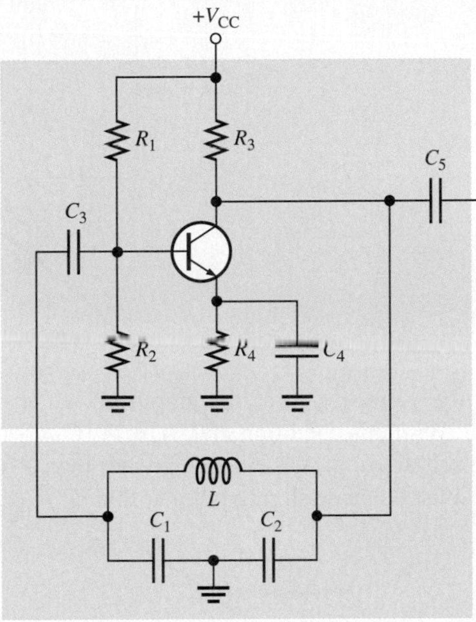

◀ **FIGURE 16–16**

A basic Colpitts oscillator with a BJT as the gain element.

The approximate frequency of oscillation is the resonant frequency of the *LC* circuit and is established by the values of C_1, C_2, and L according to this familiar formula:

$$f_r \cong \frac{1}{2\pi\sqrt{LC_T}}$$

Equation 16–5

where C_T is the total capacitance. Because the capacitors effectively appear in series around the tank circuit, the total capacitance (C_T) is

$$C_T = \frac{C_1C_2}{C_1 + C_2}$$

Conditions for Oscillation and Start-Up The attenuation, B, of the resonant feedback circuit in the Colpitts oscillator is basically determined by the values of C_1 and C_2.

Figure 16–17 shows that the circulating tank current is through both C_1 and C_2 (they are effectively in series). The voltage developed across C_2 is the oscillator's output voltage (V_{out}) and the voltage developed across C_1 is the feedback voltage (V_f), as indicated. The expression for the attenuation (B) is

$$B = \frac{V_f}{V_{out}} \cong \frac{IX_{C1}}{IX_{C2}} = \frac{X_{C1}}{X_{C2}} = \frac{1/(2\pi f_r C_1)}{1/(2\pi f_r C_2)}$$

Cancelling the $2\pi f_r$ terms gives

$$B = \frac{C_2}{C_1}$$

As you know, a condition for oscillation is $A_vB = 1$. Since $B = C_2/C_1$,

$$\mathbf{A_v = \frac{C_1}{C_2}}$$

Equation 16–6

FIGURE 16–17

The attenuation of the tank circuit is the output of the tank (V_f) divided by the input to the tank (V_{out}). $B = V_f/V_{out} = C_2/C_1$. For $A_vB > 1$, A_v must be greater than C_1/C_2.

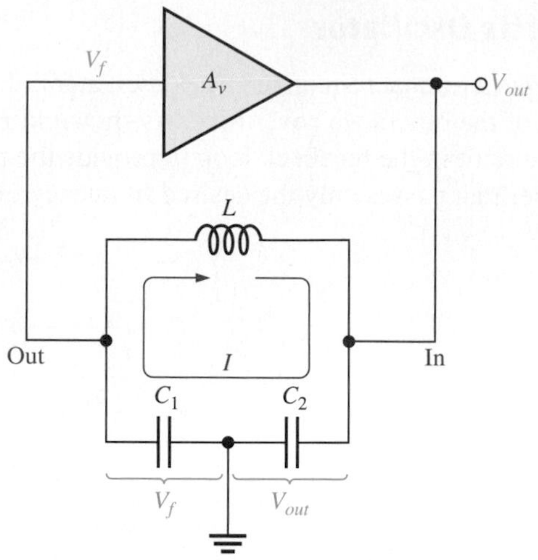

HISTORY NOTE

Edwin H. Colpitts was involved in the development of oscillators and vacuum tube push-pull amplifiers at Western Electric in the early 1900s. Western Electric research laboratories became part of Bell Laboratories in 1925, and Colpitts became vice-president of Bell Labs before retirement. The Colpitts oscillator is named in his honor.

where A_v is the voltage gain of the amplifier, which is represented by the triangle in Figure 16–17. With this condition met, $A_vB = (C_1/C_2)(C_2/C_1) = 1$. Actually, for the oscillator to be self-starting, A_vB must be greater than 1 (that is, $A_vB > 1$). Therefore, the voltage gain must be made slightly greater than C_1/C_2.

$$A_v > \frac{C_1}{C_2}$$

Loading of the Feedback Circuit Affects the Frequency of Oscillation As indicated in Figure 16–18, the input impedance of the amplifier acts as a load on the resonant feedback circuit and reduces the Q of the circuit. The resonant frequency of a parallel resonant circuit depends on the Q, according to the following formula:

Equation 16–7

$$f_r = \frac{1}{2\pi\sqrt{LC_T}}\sqrt{\frac{Q^2}{Q^2 + 1}}$$

As a rule of thumb, for a Q greater than 10, the frequency is approximately $1/(2\pi\sqrt{LC_T})$, as stated in Equation 16–5. When Q is less than 10, however, f_r is reduced significantly.

FIGURE 16–18

Z_{in} of the amplifier loads the feedback circuit and lowers its Q, thus lowering the resonant frequency.

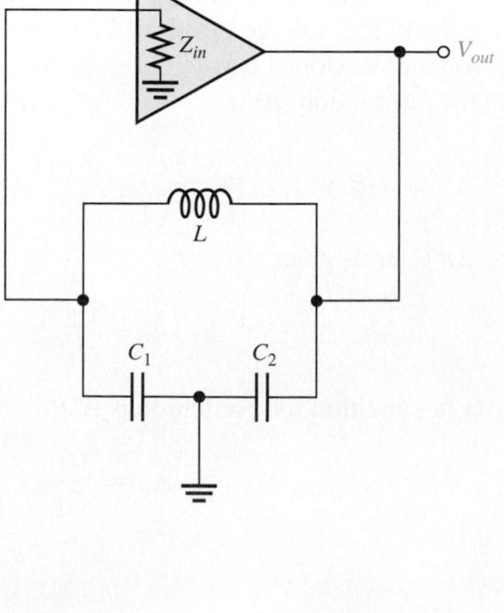

A FET can be used in place of a BJT, as shown in Figure 16–19, to minimize the loading effect of the transistor's input impedance. Recall that FETs have much higher input impedances than do bipolar junction transistors. Also, when an external load is connected to the oscillator output, as shown in Figure 16–20(a), f_r may decrease, again because of a reduction in Q. This happens if the load resistance is too small. In some cases, one way to eliminate the effects of a load resistance is by transformer coupling, as indicated in Figure 16–20(b).

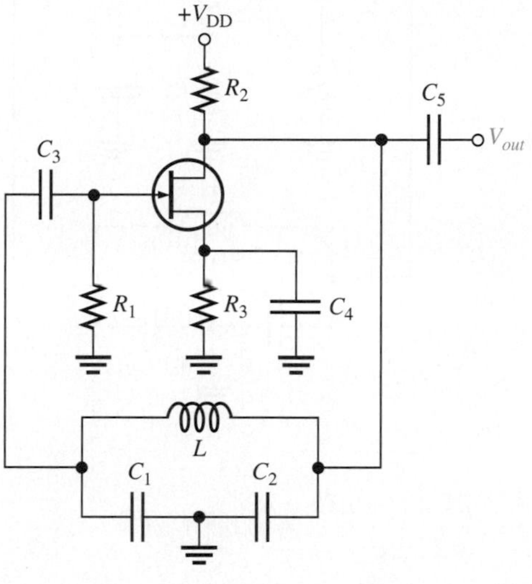

▲ **FIGURE 16–19**

A basic FET Colpitts oscillator.

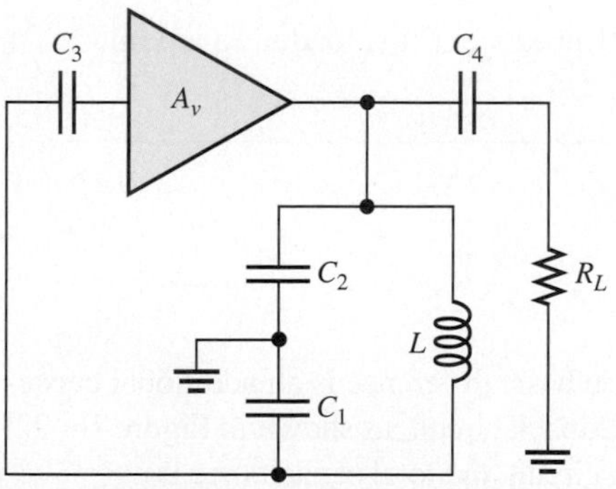

(a) A load capacitively coupled to oscillator output can reduce circuit Q and f_r.

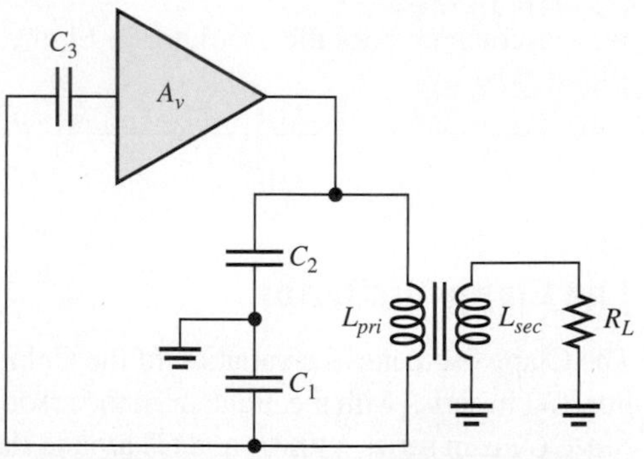

(b) Transformer coupling of load can reduce loading effect by impedance transformation.

▲ **FIGURE 16–20**

Oscillator loading.

EXAMPLE 16–3

(a) Determine the frequency for the oscillator in Figure 16–21. Assume there is negligible loading on the feedback circuit and that its Q is greater than 10.

(b) Find the frequency if the oscillator is loaded to a point where the Q drops to 8.

FIGURE 16–21

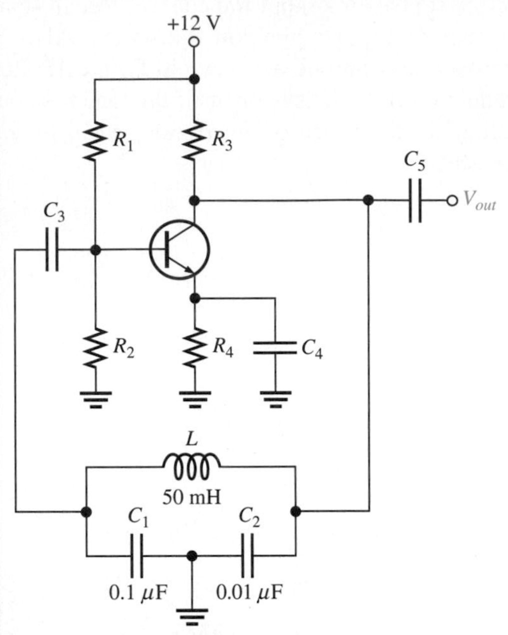

Solution (a) $C_T = \dfrac{C_1C_2}{C_1 + C_2} = \dfrac{(0.1\ \mu\text{F})(0.01\ \mu\text{F})}{0.11\ \mu\text{F}} = 0.0091\ \mu\text{F}$

$$f_r \cong \frac{1}{2\pi\sqrt{LC_T}} = \frac{1}{2\pi\sqrt{(50\text{ mH})(0.0091\ \mu\text{F})}} = \mathbf{7.46\ kHz}$$

(b) $f_r = \dfrac{1}{2\pi\sqrt{LC_T}}\sqrt{\dfrac{Q^2}{Q^2 + 1}} = (7.46\text{ kHz})(0.9923) = \mathbf{7.40\ kHz}$

Related Problem What frequency does the oscillator in Figure 16–21 produce if it is loaded to a point where $Q = 4$?

The Clapp Oscillator

The Clapp oscillator is a variation of the Colpitts. The basic difference is an additional capacitor, C_3, in series with the inductor in the resonant feedback circuit, as shown in Figure 16–22. Since C_3 is in series with C_1 and C_2 around the tank circuit, the total capacitance is

$$C_T = \frac{1}{\dfrac{1}{C_1} + \dfrac{1}{C_2} + \dfrac{1}{C_3}}$$

and the approximate frequency of oscillation ($Q > 10$) is

$$f_r \cong \frac{1}{2\pi\sqrt{LC_T}}$$

If C_3 is much smaller than C_1 and C_2, then C_3 almost entirely controls the resonant frequency ($f_r \cong 1/(2\pi\sqrt{LC_3})$). Since C_1 and C_2 are both connected to ground at one end, the junction capacitance of the transistor and other stray capacitances appear in parallel with C_1 and C_2 to ground, altering their effective values. C_3 is not affected, however, and thus provides a more accurate and stable frequency of oscillation.

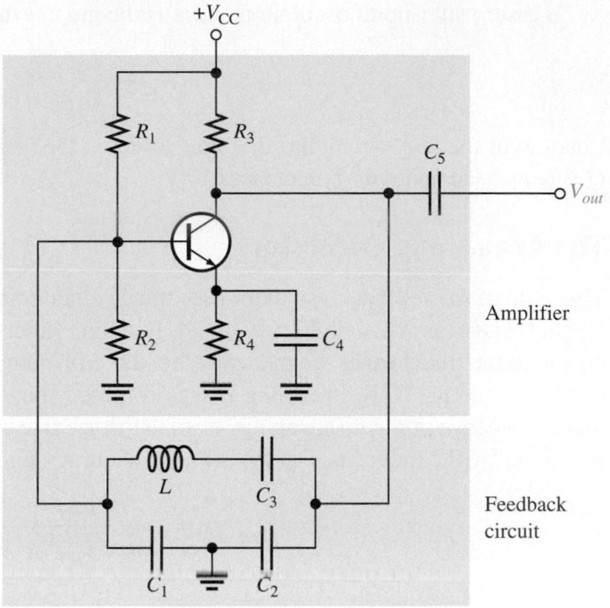

FIGURE 16–22

A basic Clapp oscillator.

The Hartley Oscillator

The Hartley oscillator is similar to the Colpitts except that the feedback circuit consists of two series inductors and a parallel capacitor as shown in Figure 16–23.

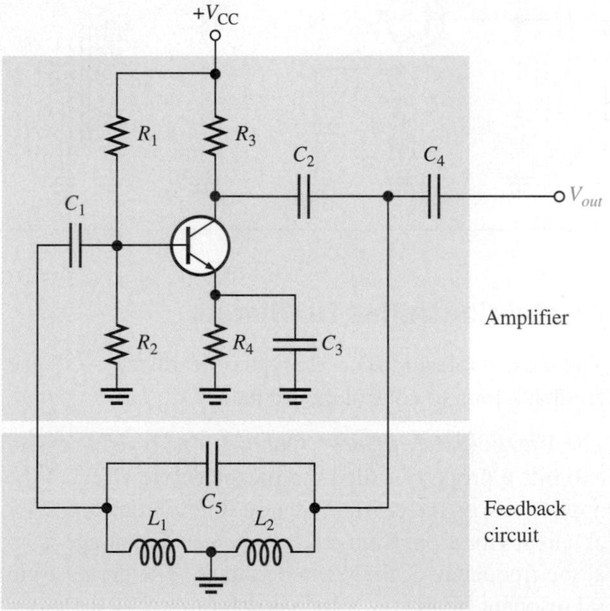

FIGURE 16–23

A basic Hartley oscillator.

In this circuit, the frequency of oscillation for $Q > 10$ is

$$f_r \cong \frac{1}{2\pi\sqrt{L_T C}}$$

where $L_T = L_1 + L_2$. The inductors act in a role similar to C_1 and C_2 in the Colpitts to determine the attenuation, B, of the feedback circuit.

$$B \cong \frac{L_1}{L_2}$$

HISTORY NOTE

Ralph Vinton Lyon Hartley (1888–1970) invented the Hartley oscillator and the Hartley transform, a mathematical analysis method, which contributed to the foundations of information theory. In 1915 he was in charge of radio receiver development for the Bell System transatlantic radiotelephone tests. During this time he developed the Hartley oscillator. A patent for the oscillator was filed in 1915 and awarded in 1920.

To assure start-up of oscillation, A_v must be greater than $1/B$.

Equation 16–8

$$A_v \cong \frac{L_2}{L_1}$$

Loading of the tank circuit has the same effect in the Hartley as in the Colpitts; that is, the Q is decreased and thus f_r decreases.

The Armstrong Oscillator

This type of *LC* feedback oscillator uses transformer coupling to feed back a portion of the signal voltage, as shown in Figure 16–24. It is sometimes called a "tickler" oscillator in reference to the transformer secondary or "tickler coil" that provides the feedback to keep the oscillation going. The Armstrong is less common than the Colpitts, Clapp, and Hartley, mainly because of the disadvantage of transformer size and cost. The frequency of oscillation is set by the inductance of the primary winding (L_{pri}) in parallel with C_1.

Equation 16–9

$$f_r = \frac{1}{2\pi\sqrt{L_{pri}C_1}}$$

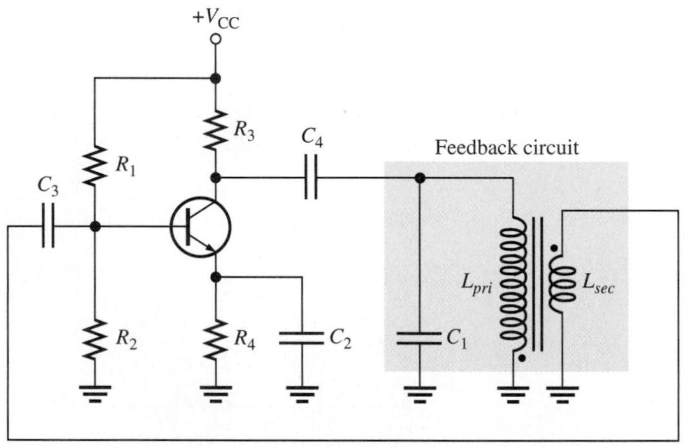

FIGURE 16–24

A basic Armstrong oscillator.

Crystal-Controlled Oscillators

The most stable and accurate type of feedback oscillator uses a piezoelectric **crystal** in the feedback loop to control the frequency.

The Piezoelectric Effect Quartz is one type of crystalline substance found in nature that exhibits a property called the **piezoelectric effect**. When a changing mechanical stress is applied across the crystal to cause it to vibrate, a voltage develops at the frequency of mechanical vibration. Conversely, when an ac voltage is applied across the crystal, it vibrates at the frequency of the applied voltage. The greatest vibration occurs at the crystal's natural resonant frequency, which is determined by the physical dimensions and by the way the crystal is cut.

Crystals used in electronic applications typically consist of a quartz wafer mounted between two electrodes and enclosed in a protective "can" as shown in Figure 16–25(a) and (b). A schematic symbol for a crystal is shown in Figure 16–25(c), and an equivalent *RLC* circuit for the crystal appears in Figure 16–25(d). As you can see, the crystal's equivalent circuit is a series-parallel *RLC* circuit and can operate in either series resonance or parallel resonance. At the series resonant frequency, the inductive reactance is cancelled by the reactance of C_s. The remaining series resistor, R_s, determines the impedance of the crystal. Parallel resonance occurs when the inductive reactance and the reactance of the parallel capacitance, C_p, are equal. The parallel resonant frequency is usually at least 1 kHz higher

HISTORY NOTE

Edwin Howard Armstrong (1890–1954) was an American electrical engineer and inventor. He was the inventor of the FM radio. Armstrong also invented the regenerative circuit (patented 1914), the superheterodyne receiver (patented 1918) and the superregenerative circuit (patented 1922). Many of Armstrong's inventions were ultimately claimed by others in patent lawsuits. The Armstrong oscillator is named in his honor.

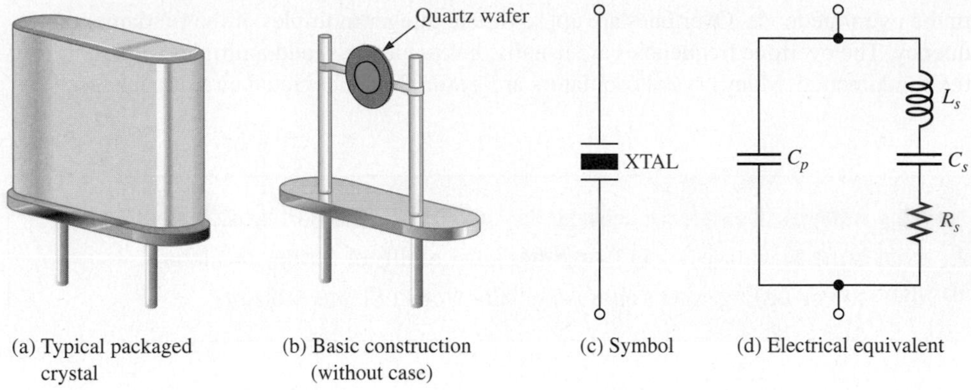

FIGURE 16–25

A quartz crystal.

than the series resonant frequency. A great advantage of the crystal is that it exhibits a very high Q (Qs with values of several thousand are typical).

An oscillator that uses a crystal as a series resonant tank circuit is shown in Figure 16–26(a). The impedance of the crystal is minimum at the series resonant frequency, thus providing maximum feedback. The crystal tuning capacitor, C_C, is used to "fine tune" the oscillator frequency by "pulling" the resonant frequency of the crystal slightly up or down.

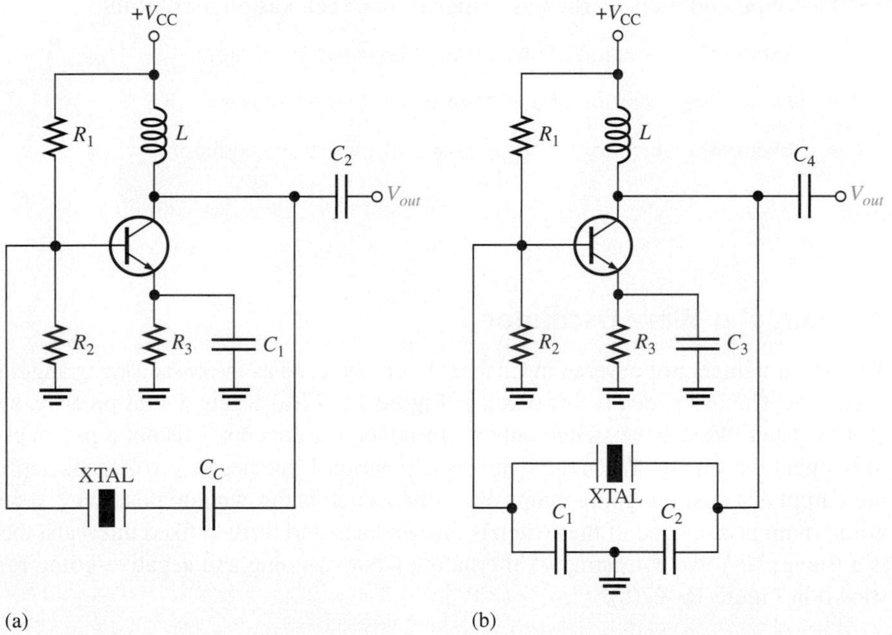

FIGURE 16–26

Basic crystal oscillators.

A modified Colpitts configuration is shown in Figure 16–26(b) with a crystal acting as a parallel resonant tank circuit. The impedance of the crystal is maximum at parallel resonance, thus developing the maximum voltage across the capacitors. The voltage across C_1 is fed back to the input.

Modes of Oscillation in the Crystal Piezoelectric crystals can oscillate in either of two modes—fundamental or overtone. The fundamental frequency of a crystal is the lowest frequency at which it is naturally resonant. The fundamental frequency depends on the crystal's mechanical dimensions, type of cut, and other factors, and is inversely proportional to the thickness of the crystal slab. Because a slab of crystal cannot be cut too thin without fracturing, there is an upper limit on the fundamental frequency. For most crystals, this upper limit is less than 20 MHz. For higher frequencies, the crystal must be operated in the overtone mode. Overtones are approximate integer multiples of the fundamental frequency. The overtone frequencies are usually, but not always, odd multiples (3, 5, 7, . . .) of the fundamental. Many crystal oscillators are available in integrated circuit packages.

SECTION 16–4 CHECKUP

1. What is the basic difference between the Colpitts and the Hartley oscillators?
2. What is the advantage of a FET amplifier in a Colpitts or Hartley oscillator?
3. How can you distinguish a Colpitts oscillator from a Clapp oscillator?

16–5 RELAXATION OSCILLATORS

The second major category of oscillators is the relaxation oscillator. Relaxation oscillators use an *RC* timing circuit and a device that changes states to generate a periodic waveform. In this section, you will learn about several circuits that are used to produce nonsinusoidal waveforms.

After completing this section, you should be able to

- **Describe and analyze the basic operation of relaxation oscillators**
  - Discuss the operation of basic triangular-wave oscillators
  - Discuss the operation of a voltage-controlled oscillator (VCO)
  - Discuss the operation of a square-wave relaxation oscillator

A Triangular-Wave Oscillator

The op-amp integrator covered in Chapter 13 can be used as the basis for a triangular-wave oscillator. The basic idea is illustrated in Figure 16–27(a) where a dual-polarity, switched input is used. We use the switch only to introduce the concept; it is not a practical way to implement this circuit. When the switch is in position 1, the negative voltage is applied, and the output is a positive-going ramp. When the switch is thrown into position 2, a negative-going ramp is produced. If the switch is thrown back and forth at fixed intervals, the output is a triangular wave consisting of alternating positive-going and negative-going ramps, as shown in Figure 16–27(b).

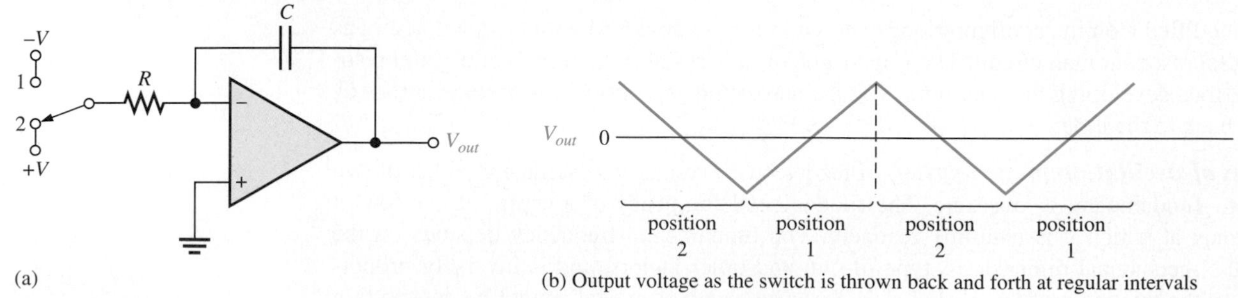

(a)

(b) Output voltage as the switch is thrown back and forth at regular intervals

▲ FIGURE 16–27

Basic triangular-wave oscillator.

A Practical Triangular-Wave Oscillator One practical implementation of a triangular-wave oscillator utilizes an op-amp comparator with hysteresis to perform the switching function, as shown in Figure 16–28. The operation is as follows. To begin, assume that the output voltage of the comparator is at its maximum negative level. This output is connected to the inverting input of the integrator through R_1, producing a positive-going ramp on the output of the integrator. When the ramp voltage reaches the upper trigger point (UTP), the comparator switches to its maximum positive level. This positive level causes the integrator ramp to change to a negative-going direction. The ramp continues in this direction until the lower trigger point (LTP) of the comparator is reached. At this point, the comparator output switches back to the maximum negative level and the cycle repeats. This action is illustrated in Figure 16–29.

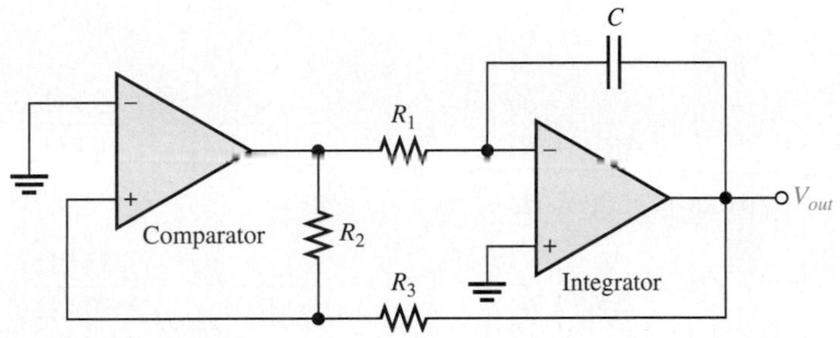

FIGURE 16–28

A triangular-wave oscillator using two op-amps.

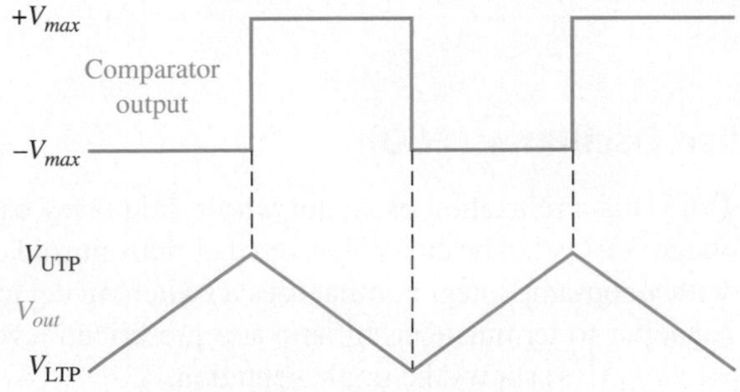

FIGURE 16–29

Waveforms for the circuit in Figure 16–28.

Since the comparator produces a square-wave output, the circuit in Figure 16–28 can be used as both a triangular-wave oscillator and a square-wave oscillator. Devices of this type are commonly known as *function generators* because they produce more than one output function. The output amplitude of the square wave is set by the output swing of the comparator, and the resistors R_2 and R_3 set the amplitude of the triangular output by establishing the UTP and LTP voltages according to the following formulas:

$$V_{UTP} = +V_{max}\left(\frac{R_3}{R_2}\right)$$

$$V_{LTP} = -V_{max}\left(\frac{R_3}{R_2}\right)$$

where the comparator output levels, $+V_{max}$ and $-V_{max}$, are equal. The frequency of both waveforms depends on the R_1C time constant as well as the amplitude-setting resistors, R_2 and R_3. By varying R_1, the frequency of oscillation can be adjusted without changing the output amplitude.

$$f_r = \frac{1}{4R_1C}\left(\frac{R_2}{R_3}\right)$$

Equation 16–10

EXAMPLE 16–4

Determine the frequency of oscillation of the circuit in Figure 16–30. To what value must R_1 be changed to make the frequency 20 kHz?

FIGURE 16–30

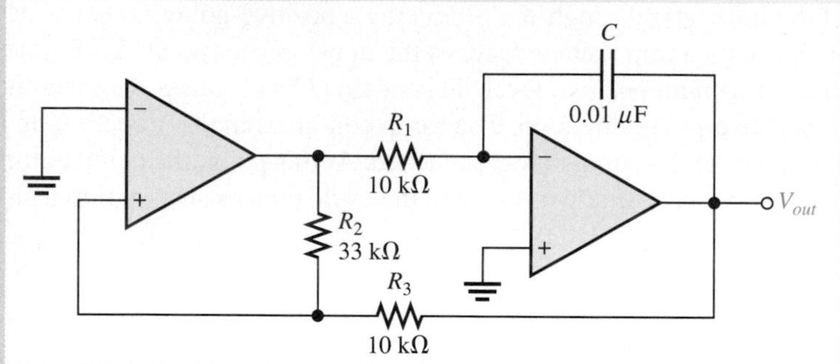

Solution

$$f_r = \frac{1}{4R_1C}\left(\frac{R_2}{R_3}\right) = \left(\frac{1}{4(10\text{ k}\Omega)(0.01\ \mu\text{F})}\right)\left(\frac{33\text{ k}\Omega}{10\text{ k}\Omega}\right) = \mathbf{8.25\ kHz}$$

To make $f = 20$ kHz,

$$R_1 = \frac{1}{4fC}\left(\frac{R_2}{R_3}\right) = \left(\frac{1}{4(20\text{ kHz})(0.01\ \mu\text{F})}\right)\left(\frac{33\text{ k}\Omega}{10\text{ k}\Omega}\right) = \mathbf{4.13\ k\Omega}$$

Related Problem What is the amplitude of the triangular wave in Figure 16–30 if the comparator output is ± 10 V?

A Sawtooth Voltage-Controlled Oscillator (VCO)

The **voltage-controlled oscillator (VCO)** is a relaxation oscillator whose frequency can be changed by a variable dc control voltage. VCOs can be either sinusoidal or nonsinusoidal. One way to build a sawtooth VCO is with an op-amp integrator that uses a switching device (PUT) in parallel with the feedback capacitor to terminate each ramp at a prescribed level and effectively "reset" the circuit. Figure 16–31(a) shows the implementation.

As you learned in Chapter 11, the PUT is a programmable unijunction transistor with an anode, a cathode, and a gate terminal. The gate is always biased positively with respect to

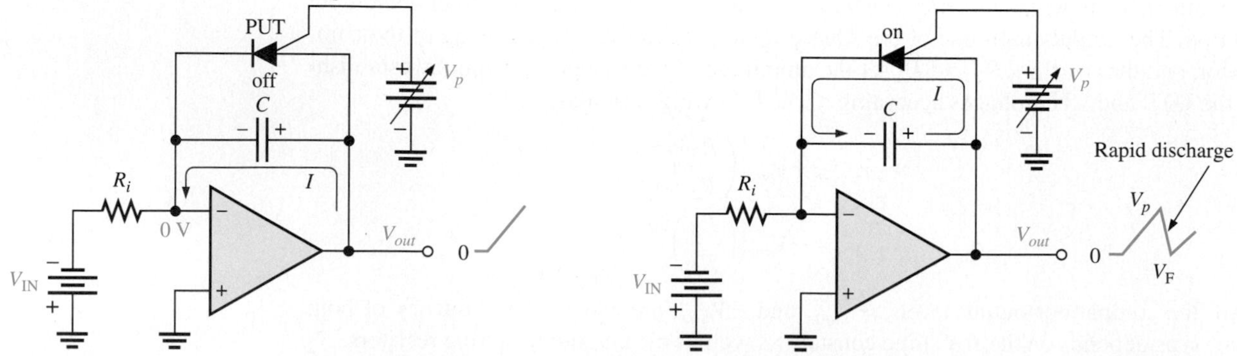

(a) Initially, the capacitor charges, the output ramp begins, and the PUT is off.

(b) The capacitor rapidly discharges when the PUT momentarily turns on.

▲ FIGURE 16–31

Sawtooth VCO operation.

the cathode. When the anode voltage exceeds the gate voltage by approximately 0.7 V, the PUT turns on and acts as a forward-biased diode. When the anode voltage falls below this level, the PUT turns off. Also, the current must be above the holding value to maintain conduction.

The operation of the sawtooth VCO begins when the negative dc input voltage, $-V_{IN}$, produces a positive-going ramp on the output. During the time that the ramp is increasing, the circuit acts as a regular integrator. The PUT triggers on when the output ramp (at the anode) exceeds the gate voltage by 0.7 V. The gate is set to the approximate desired sawtooth peak voltage. When the PUT turns on, the capacitor rapidly discharges, as shown in Figure 16–31(b). The capacitor does not discharge completely to zero because of the PUT's forward voltage, V_F. Discharge continues until the PUT current falls below the holding value. At this point, the PUT turns off and the capacitor begins to charge again, thus generating a new output ramp. The cycle continually repeats, and the resulting output is a repetitive sawtooth waveform, as shown. The sawtooth amplitude and period can be adjusted by varying the PUT gate voltage.

The frequency of oscillation is determined by the R_iC time constant of the integrator and the peak voltage set by the PUT. Recall that the charging rate of a capacitor is V_{IN}/R_iC. The time it takes a capacitor to charge from V_F to V_p is the period, T, of the sawtooth waveform (neglecting the rapid discharge time).

$$T = \frac{V_p - V_F}{|V_{IN}|/R_iC}$$

From $f = 1/T$,

$$f = \frac{|V_{IN}|}{R_iC}\left(\frac{1}{V_p - V_F}\right)$$

Equation 16–11

EXAMPLE 16–5

(a) Find the amplitude and frequency of the sawtooth output in Figure 16–32. Assume that the forward PUT voltage, V_F, is approximately 1 V.

(b) Sketch the output waveform.

FIGURE 16–32

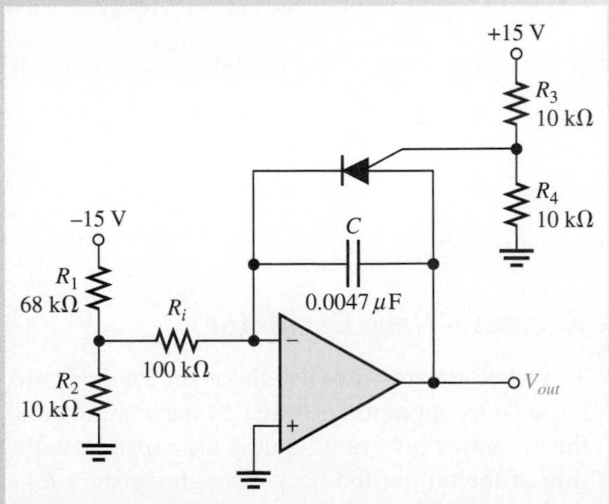

Solution **(a)** First, find the gate voltage in order to establish the approximate voltage at which the PUT turns on.

$$V_G = \frac{R_4}{R_3 + R_4}(+V) = \frac{10\text{ k}\Omega}{20\text{ k}\Omega}(15\text{ V}) = 7.5\text{ V}$$

This voltage sets the approximate maximum peak value of the sawtooth output (neglecting the 0.7 V).

$$V_p \cong 7.5\text{ V}$$

The minimum peak value (low point) is

$$V_F \cong 1\text{ V}$$

So the peak-to-peak amplitude is

$$V_{pp} = V_p - V_F = 7.5\text{ V} - 1\text{ V} = \mathbf{6.5\text{ V}}$$

Determine the frequency as follows:

$$V_{IN} = \frac{R_2}{R_1 + R_2}(-V) = \frac{10\text{ k}\Omega}{78\text{ k}\Omega}(-15\text{ V}) = -1.92\text{ V}$$

$$f = \frac{|V_{IN}|}{R_iC}\left(\frac{1}{V_p - V_F}\right) = \left(\frac{1.92\text{ V}}{(100\text{ k}\Omega)(0.0047\mu\text{F})}\right)\left(\frac{1}{7.5\text{ V} - 1\text{ V}}\right) = \mathbf{628\text{ Hz}}$$

(b) The output waveform is shown in Figure 16–33, where the period is determined as follows:

$$T = \frac{1}{f} = \frac{1}{628\text{ Hz}} = 1.59\text{ ms}$$

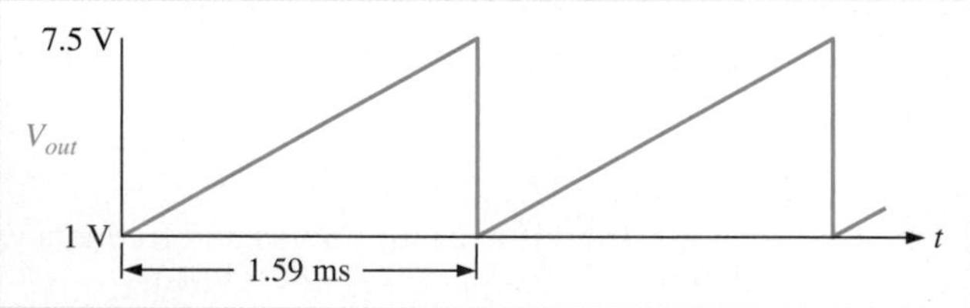

FIGURE 16–33

Output of the circuit in Figure 16–32.

Related Problem If R_i is changed to 56 kΩ in Figure 16–32, what is the frequency?

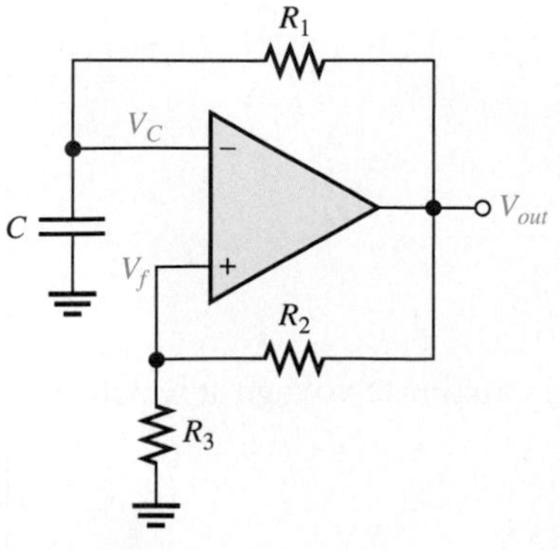

FIGURE 16–34

A square-wave relaxation oscillator.

A Square-Wave Oscillator

The basic square-wave oscillator shown in Figure 16–34 is a type of relaxation oscillator because its operation is based on the charging and discharging of a capacitor. Notice that the op-amp's inverting input is the capacitor voltage and the noninverting input is a portion of the output fed back through resistors R_2 and R_3 to provide hysteresis. When the circuit is first turned on, the capacitor is uncharged, and thus the inverting input is at 0 V. This makes the output a positive maximum, and the capacitor begins to charge toward V_{out} through R_1. When the capacitor voltage (V_C) reaches a value equal to the feedback voltage (V_f) on the noninverting input, the op-amp switches to the maximum negative state. At this point, the capacitor begins to discharge from $+V_f$ toward $-V_f$. When the capacitor voltage reaches $-V_f$, the op-amp switches back to the maximum positive state. This action continues to repeat, as shown in Figure 16–35, and a square-wave output voltage is obtained.

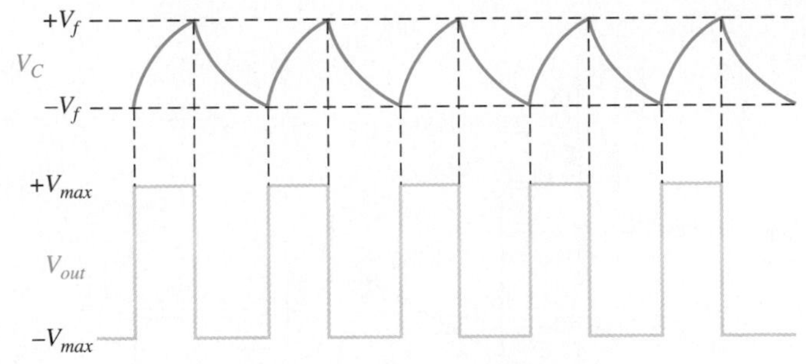

FIGURE 16–35

Waveforms for the square-wave relaxation oscillator.

SECTION 16–5 CHECKUP

1. What is a VCO, and basically, what does it do?
2. Upon what principle does a relaxation oscillator operate?

16–6 THE 555 TIMER AS AN OSCILLATOR

The 555 timer is a versatile integrated circuit with many applications. In this section, you will see how the 555 is configured as an astable or free-running multivibrator, which is essentially a square-wave oscillator. The use of the 555 timer as a voltage-controlled oscillator (VCO) is also discussed.

After completing this section, you should be able to

- **Use a 555 timer in an oscillator application**
  - Explain what the 555 timer is
  - Discuss astable operation of the 555 timer
  - Explain how to use the 555 timer as a VCO

The 555 timer consists basically of two comparators, a flip-flop, a discharge transistor, and a resistive voltage divider, as shown in Figure 16–36. The flip-flop (bistable multivibrator) is a digital device that may be unfamiliar to you at this point unless you already have taken a digital fundamentals course. Briefly, it is a two-state device whose output can be at either a high voltage level (set, *S*) or a low voltage level (reset, *R*). The state of the output can be changed with proper input signals.

The resistive voltage divider is used to set the voltage comparator levels. All three resistors are of equal value; therefore, the upper comparator has a reference of $\frac{2}{3}V_{CC}$, and the lower comparator has a reference of $\frac{1}{3}V_{CC}$. The comparators' outputs control the state of the flip-flop. When the trigger voltage goes below $\frac{1}{3}V_{CC}$, the flip-flop sets and the output jumps to its high level. The threshold input is normally connected to an external *RC* timing circuit. When the external capacitor voltage exceeds $\frac{2}{3}V_{CC}$, the upper comparator resets the flip-flop, which in turn switches the output back to its low level. When the device output is low, the discharge transistor (Q_d) is turned on and provides a path for rapid discharge of the external timing capacitor. This basic operation allows the timer to be configured with external components as an oscillator, a one-shot, or a time-delay element.

▶ FIGURE 16–36

Internal diagram of a 555 integrated circuit timer. (IC pin numbers are in parentheses.)

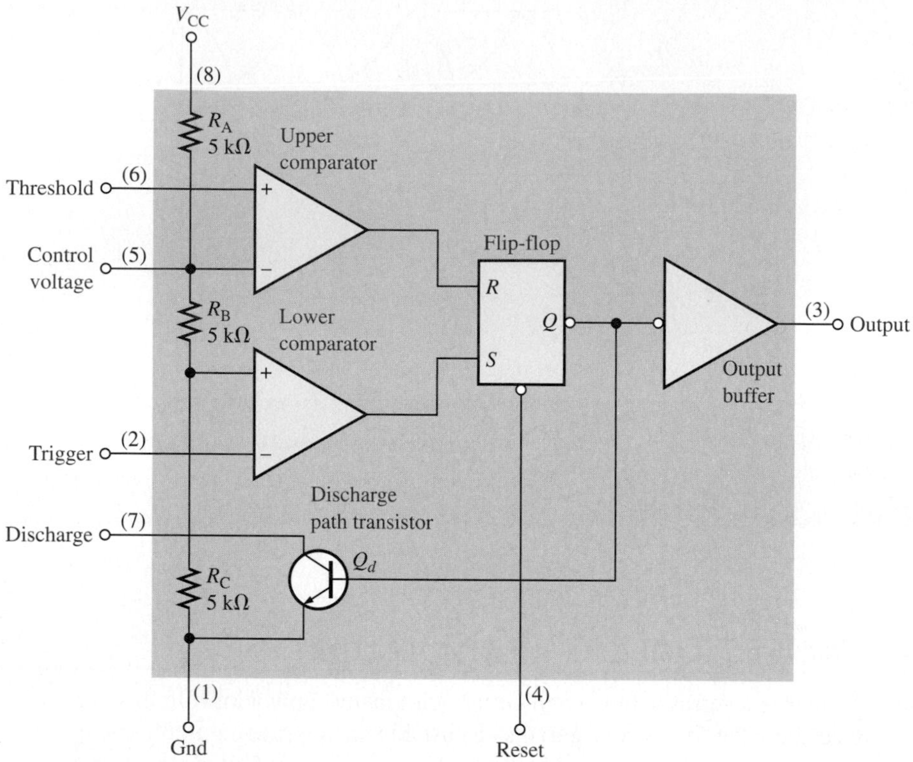

Astable Operation

A 555 timer connected to operate in the **astable** mode as a free-running relaxation oscillator (astable multivibrator) is shown in Figure 16–37. Notice that the threshold input (THRESH) is now connected to the trigger input (TRIG). The external components R_1, R_2, and C_{ext} form the timing circuit that sets the frequency of oscillation. The 0.01 μF capacitor connected to the control (CONT) input is strictly for decoupling and has no effect on the operation.

▶ FIGURE 16–37

The 555 timer connected as an astable multivibrator.

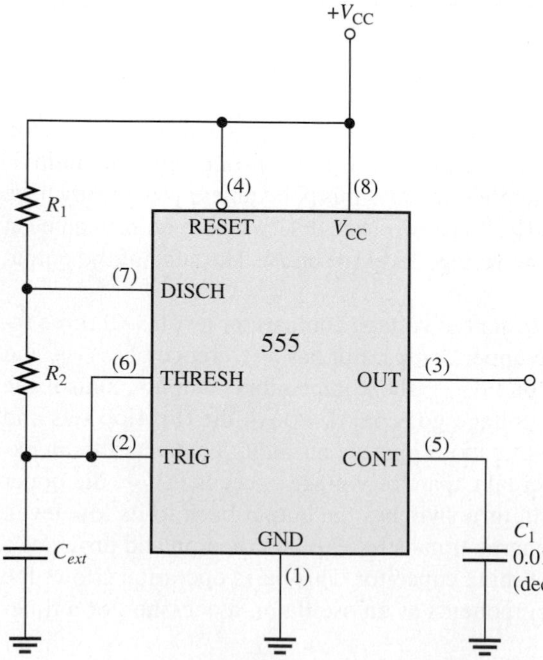

Initially, when the power is turned on, the capacitor C_{ext} is uncharged and thus the trigger voltage (pin 2) is at 0 V. This causes the output of the lower comparator to be high and the output of the upper comparator to be low, forcing the output of the flip-flop, and thus the base of Q_d, low and keeping the transistor off. Now, C_{ext} begins charging through R_1 and R_2 as indicated in Figure 16–38. When the capacitor voltage reaches $\frac{1}{3}V_{CC}$, the lower comparator switches to its low output state, and when the capacitor voltage reaches $\frac{2}{3}V_{CC}$, the upper comparator switches to its high output state. This resets the flip-flop, causes the base of Q_d to go high, and turns on the transistor. This sequence creates a discharge path for the capacitor through R_2 and the transistor, as indicated. The capacitor now begins to discharge, causing the upper comparator to go low. At the point where the capacitor discharges down to $\frac{1}{3}V_{CC}$, the lower comparator switches high, setting the flip-flop, which makes the base of Q_d low and turns off the transistor. Another charging cycle begins, and the entire process repeats. The result is a rectangular wave output whose duty cycle depends on the values of R_1 and R_2.

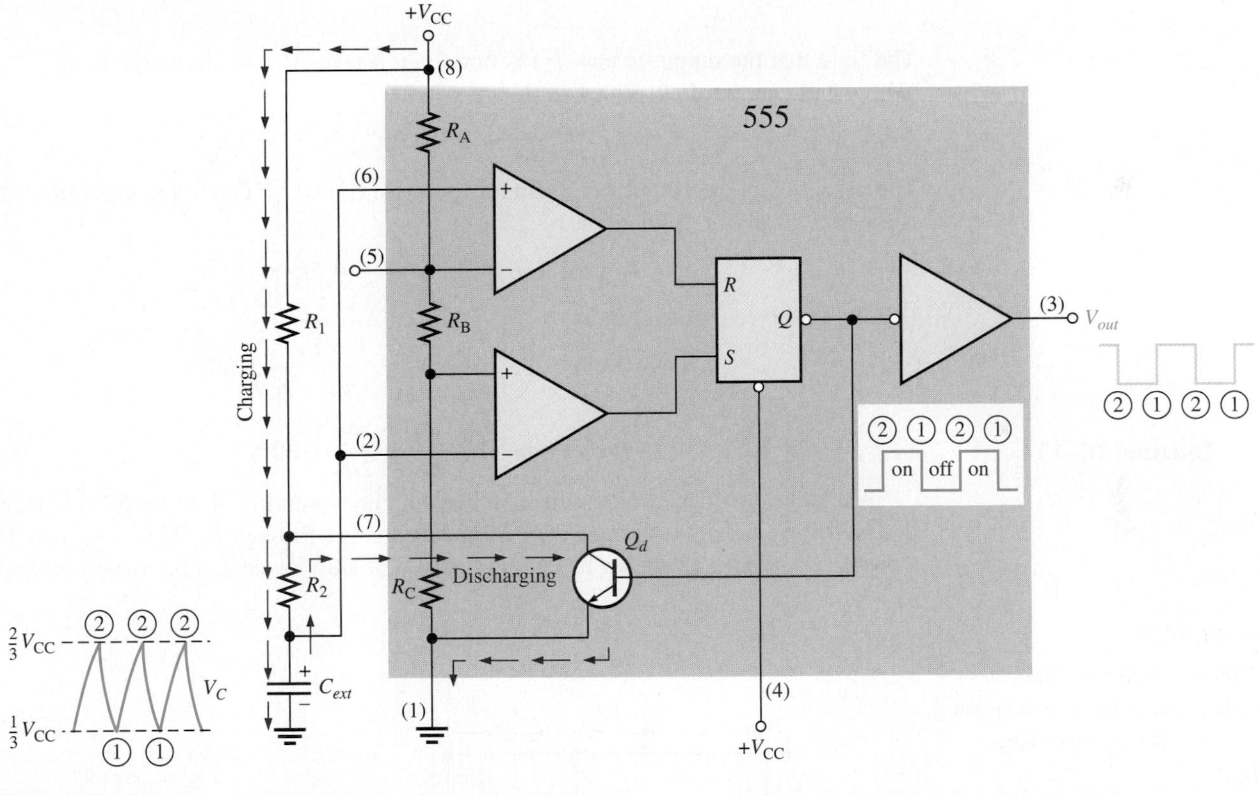

▲ **FIGURE 16–38**

Operation of the 555 timer in the astable mode.

The frequency of oscillation is given by Equation 16–12, or it can be found using the graph in Figure 16–39.

$$f_r = \frac{\mathbf{1.44}}{(\boldsymbol{R_1} + \boldsymbol{2R_2})\boldsymbol{C_{ext}}}$$

Equation 16–12

By selecting R_1 and R_2, the duty cycle of the output can be adjusted. Since C_{ext} charges through $R_1 + R_2$ and discharges only through R_2, duty cycles approaching a minimum of 50 percent can be achieved if $R_2 >> R_1$ so that the charging and discharging times are approximately equal.

A formula to calculate the duty cycle is developed as follows. The time that the output is high (t_H) is how long it takes C_{ext} to charge from $\frac{1}{3}V_{CC}$ to $\frac{2}{3}V_{CC}$. It is expressed as

$$t_H = 0.694(R_1 + R_2)C_{ext}$$

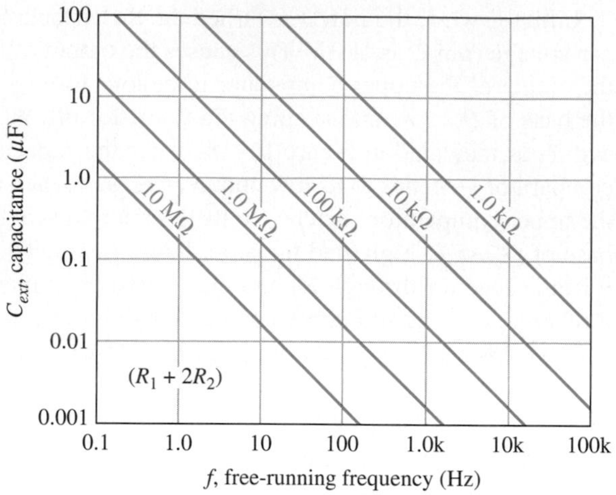

▶ FIGURE 16–39

Frequency of oscillation (free-running frequency) of a 555 timer in the astable mode as a function of C_{ext} and $R_1 + 2R_2$. The sloped lines are values of $R_1 + 2R_2$.

The time that the output is low (t_L) is how long it takes C_{ext} to discharge from $\frac{2}{3}V_{CC}$ to $\frac{1}{3}V_{CC}$. It is expressed as

$$t_L = 0.694R_2C_{ext}$$

The period, T, of the output waveform is the sum of t_H and t_L. The following formula for T is the reciprocal of f in Equation 16–12.

$$T = t_H + t_L = 0.694(R_1 + 2R_2)C_{ext}$$

Finally, the percent duty cycle is

$$\text{Duty cycle} = \left(\frac{t_H}{T}\right)100\% = \left(\frac{t_H}{t_H + t_L}\right)100\%$$

Equation 16–13

$$\textbf{Duty cycle} = \left(\frac{\mathbf{R_1 + R_2}}{\mathbf{R_1 + 2R_2}}\right)\mathbf{100\%}$$

To achieve duty cycles of less than 50 percent, the circuit in Figure 16–37 can be modified so that C_{ext} charges through only R_1 and discharges through R_2. This is achieved with a diode, D_1, placed as shown in Figure 16–40. The duty cycle can be made less than 50

▶ FIGURE 16–40

The addition of diode D_1 allows the duty cycle of the output to be adjusted to less than 50 percent by making $R_1 < R_2$.

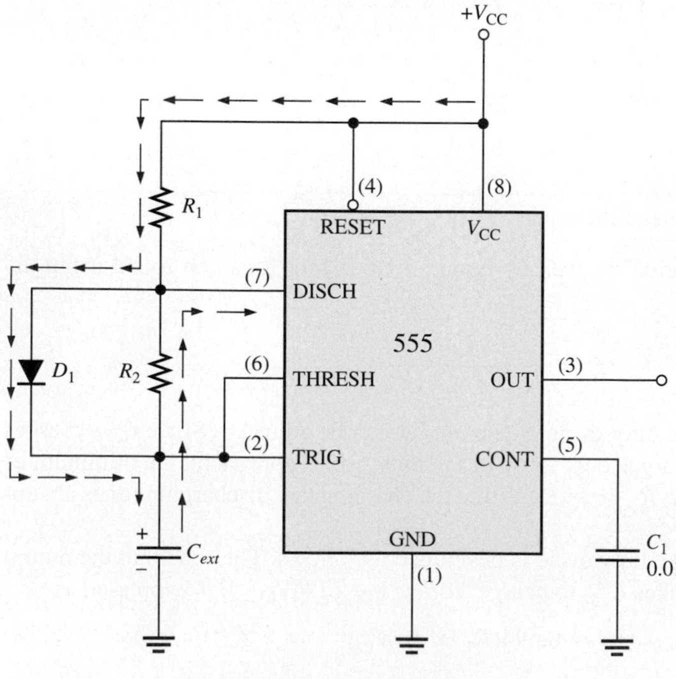

percent by making R_1 less than R_2. Under this condition, the formulas for the frequency and percent duty cycle are (assuming an ideal diode)

$$f_r \cong \frac{1.44}{(R_1 + R_2)C_{ext}}$$

$$\text{Duty cycle} \cong \left(\frac{R_1}{R_1 + R_2}\right)100\%$$

EXAMPLE 16–6

A 555 timer configured to run in the astable mode (oscillator) is shown in Figure 16–41. Determine the frequency of the output and the duty cycle.

FIGURE 16–41

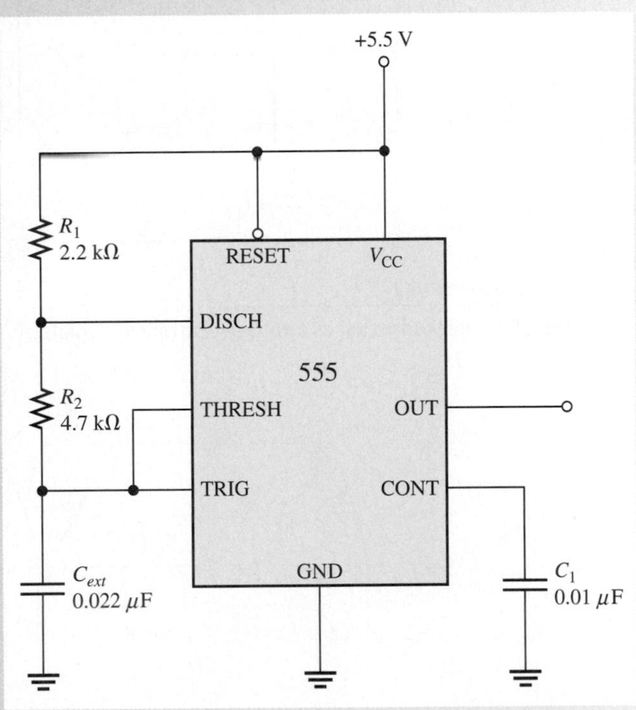

Solution

$$f_r = \frac{1.44}{(R_1 + 2R_2)C_{ext}} = \frac{1.44}{(2.2\text{ k}\Omega + 9.4\text{ k}\Omega)0.022\ \mu\text{F}} = \mathbf{5.64\ kHz}$$

$$\text{Duty cycle} = \left(\frac{R_1 + R_2}{R_1 + 2R_2}\right)100\% = \left(\frac{2.2\text{ k}\Omega + 4.7\text{ k}\Omega}{2.2\text{ k}\Omega + 9.4\text{ k}\Omega}\right)100\% = \mathbf{59.5\%}$$

Related Problem Determine the duty cycle in Figure 16–41 if a diode is connected across R_2 as indicated in Figure 16–40.

Operation as a Voltage-Controlled Oscillator (VCO)

A 555 timer can be set up to operate as a VCO by using the same external connections as for astable operation, with the exception that a variable control voltage is applied to the CONT input (pin 5), as indicated in Figure 16–42.

As shown in Figure 16–43, the control voltage (V_{CONT}) changes the threshold values of ⅓V_{CC} and ⅔V_{CC} for the internal comparators. With the control voltage, the upper value is V_{CONT} and the lower value is ½V_{CONT}, as you can see by examining the internal diagram of the 555 timer. When the control voltage is varied, the output frequency also varies. An

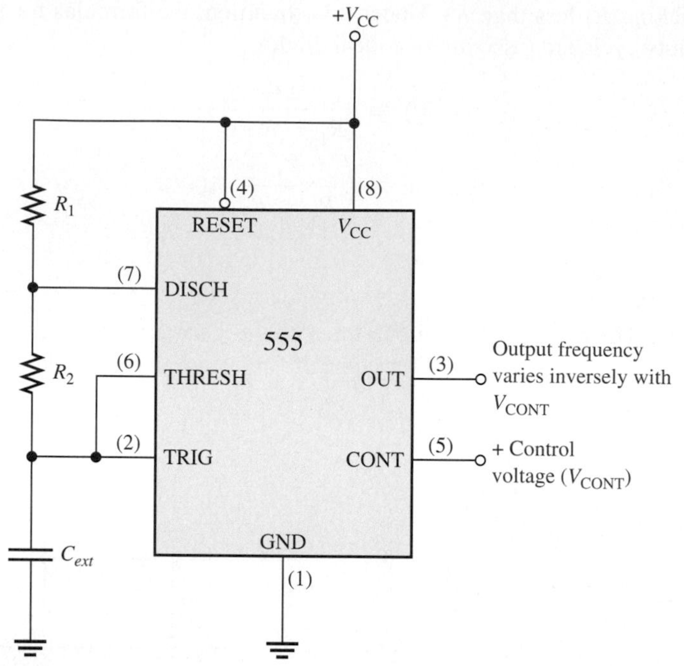

▲ **FIGURE 16–42**

The 555 timer connected as a voltage-controlled oscillator (VCO). Note the variable control voltage input on pin 5.

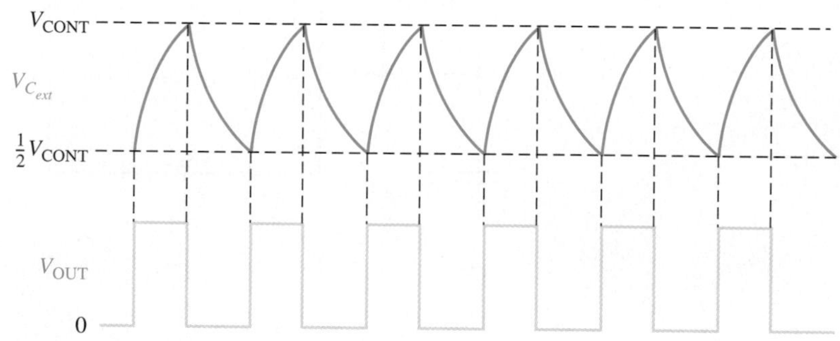

▲ **FIGURE 16–43**

The VCO output frequency varies inversely with V_{CONT} because the charging and discharging time of C_{ext} is directly dependent on the control voltage.

increase in V_{CONT} increases the charging and discharging time of the external capacitor and causes the frequency to decrease. A decrease in V_{CONT} decreases the charging and discharging time of the capacitor and causes the frequency to increase.

An interesting application of the VCO is in phase-locked loops, which are used in various types of communication receivers to track variations in the frequency of incoming signals. You will learn about the basic operation of a phase-locked loop in Section 18–8.

SECTION 16–6 CHECKUP

1. Name the five basic elements in a 555 timer IC.
2. When the 555 timer is configured as an astable multivibrator, how is the duty cycle determined?

On the Job

Application Activity: *ASK Test Generator*

The RFID reader board that was developed in the Chapter 15 Application Activity requires an ASK modulated source to test it. Recall that the RFID tag transmits a 125 kHz ASK (amplitude shift keyed) signal modulated with coded information represented by a digital waveform. The basic block diagram is shown in Figure 16–44.

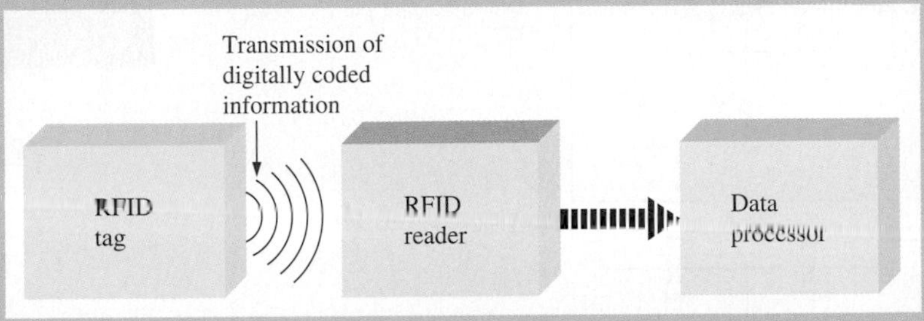

FIGURE 16–44

Basic block diagram of an RFID system.

The ASK Test Generator

The purpose of this application is to develop a signal source for testing the RFID reader circuit board. The source must produce a 125 kHz signal that is modulated with a 10 kHz pulse signal to simulate the RFID tag. An oscillator is used to generate the 125 kHz carrier signal, and a 555 timer produces the modulating pulse signal. The modulating device is an analog switch that allows the carrier signal to be turned on and off by the modulating pulse signal. A basic block diagram is shown in Figure 16–45.

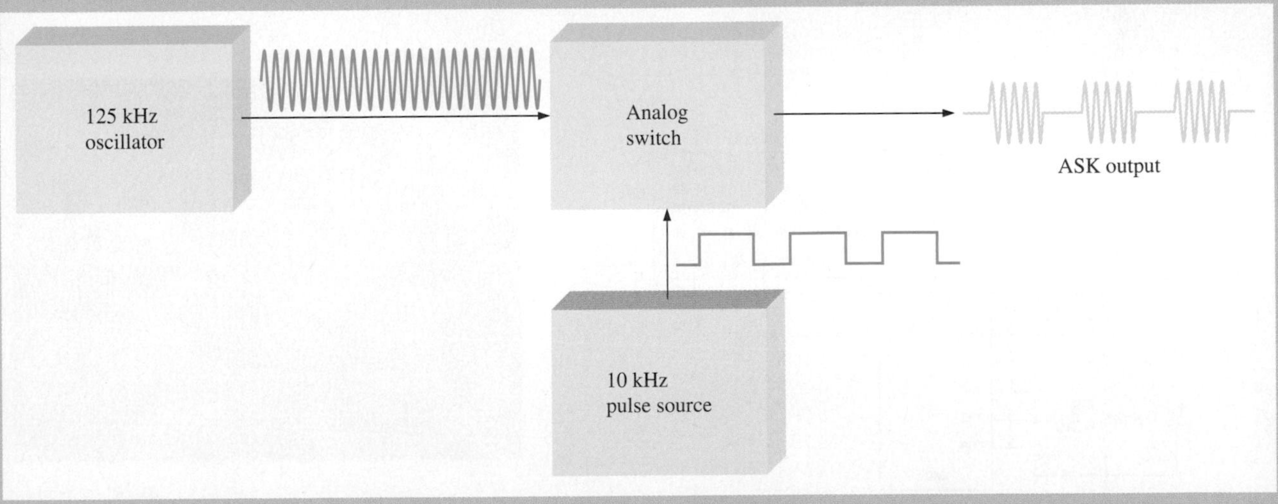

FIGURE 16–45

Basic block diagram of the ASK test generator.

Simulation

The first step is to design the 125 kHz oscillator circuit. The type of oscillator chosen for this application is the Colpitts oscillator. The simulated circuit is shown in Figure 16–46 (a), and the output waveform is shown in part (b).

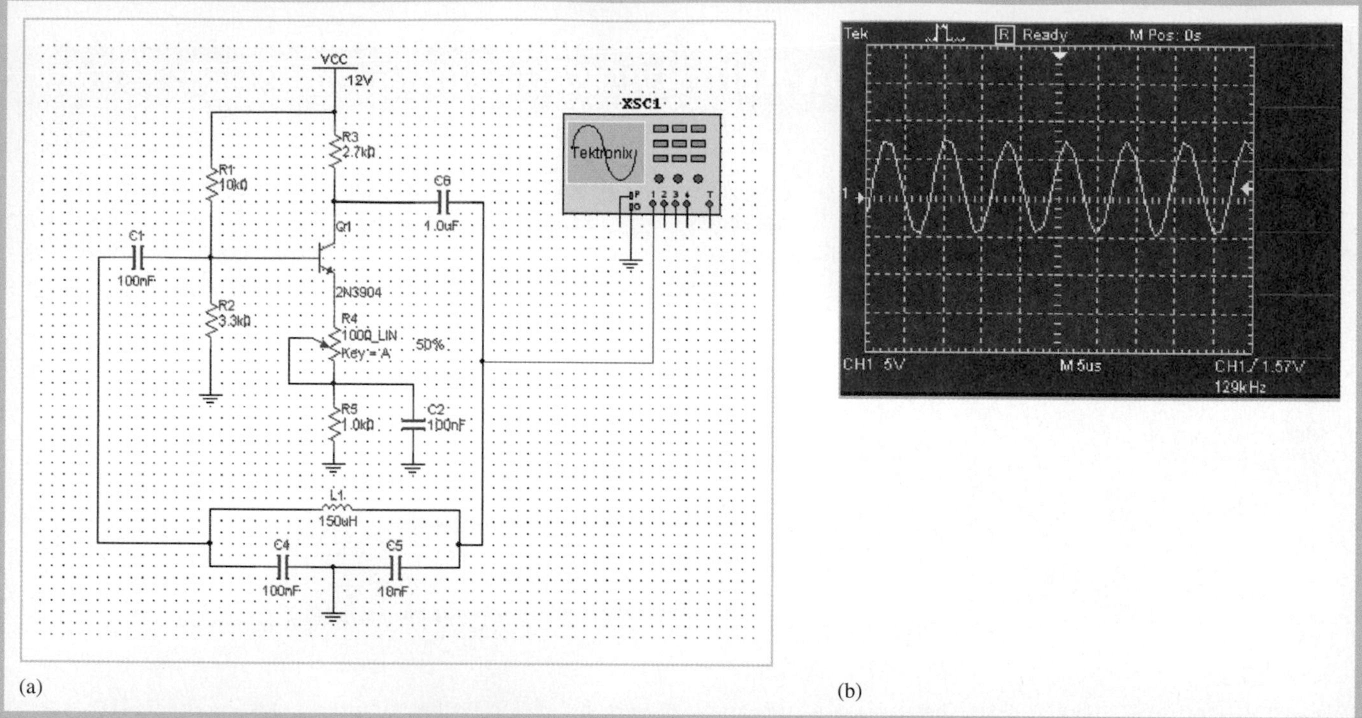

FIGURE 16–46

Colpitts oscillator for generating the 125 kHz carrier signal.

1. Calculate the gain of the Colpitts oscillator in Figure 16–46.
2. Calculate the frequency of the Colpitts oscillator and compare to the frequency measured in the simulation.

In the second step, the 10 kHz pulse oscillator is designed using a 555 timer. The simulated circuit and output waveform are shown in Figure 16–47.

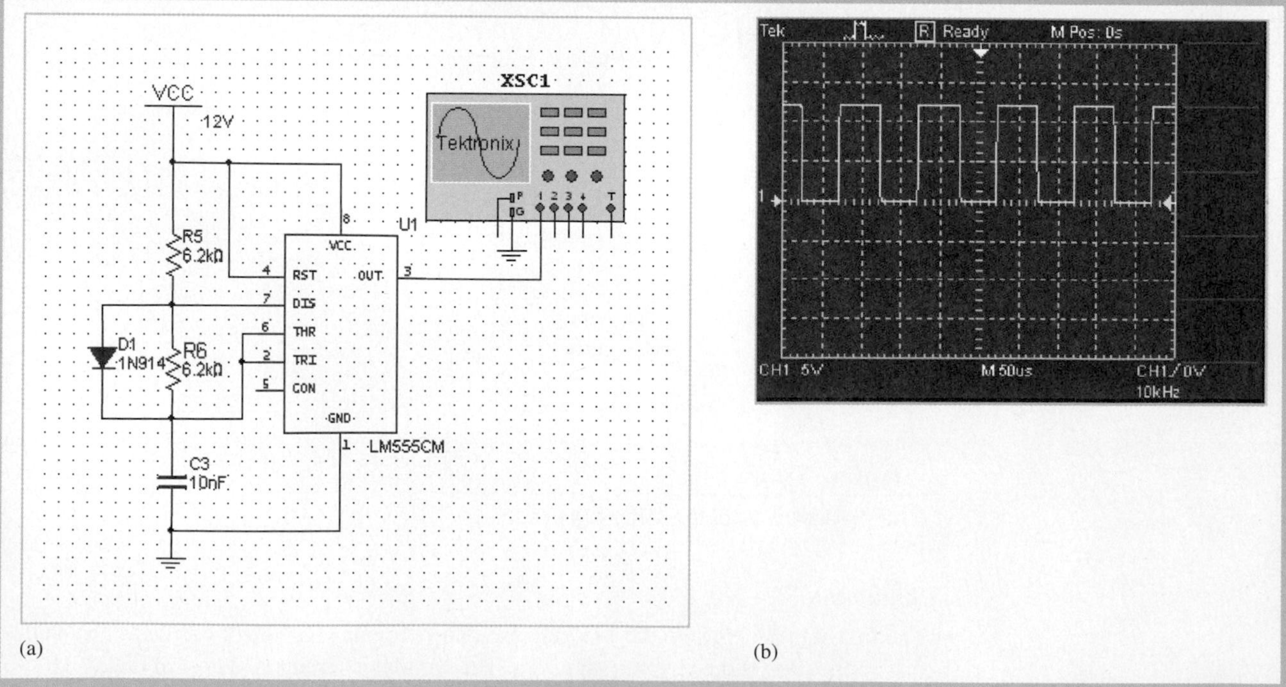

FIGURE 16–47

555 timer configured for generating a 10 kHz square wave.

3. Calculate the frequency of the pulse oscillator in Figure 16–47 and compare to the measured frequency in the simulation.
4. Describe a possible reason for the difference in the calculated and the simulated value of frequency.

The third step in the simulation of the ASK test generator is to combine the Colpitts oscillator with the 555 timer and add an analog switch. For the purpose of switching the carrier signal on and off, a *p*-channel JFET is used. When the timer output is low, the JFET turns on and passes the carrier signal to the ASK output. When the timer output is high, the JFET turns off and blocks the signal from the output. The complete circuit is shown in Figure 16–48, and the resulting waveforms are shown in Figure 16–49.

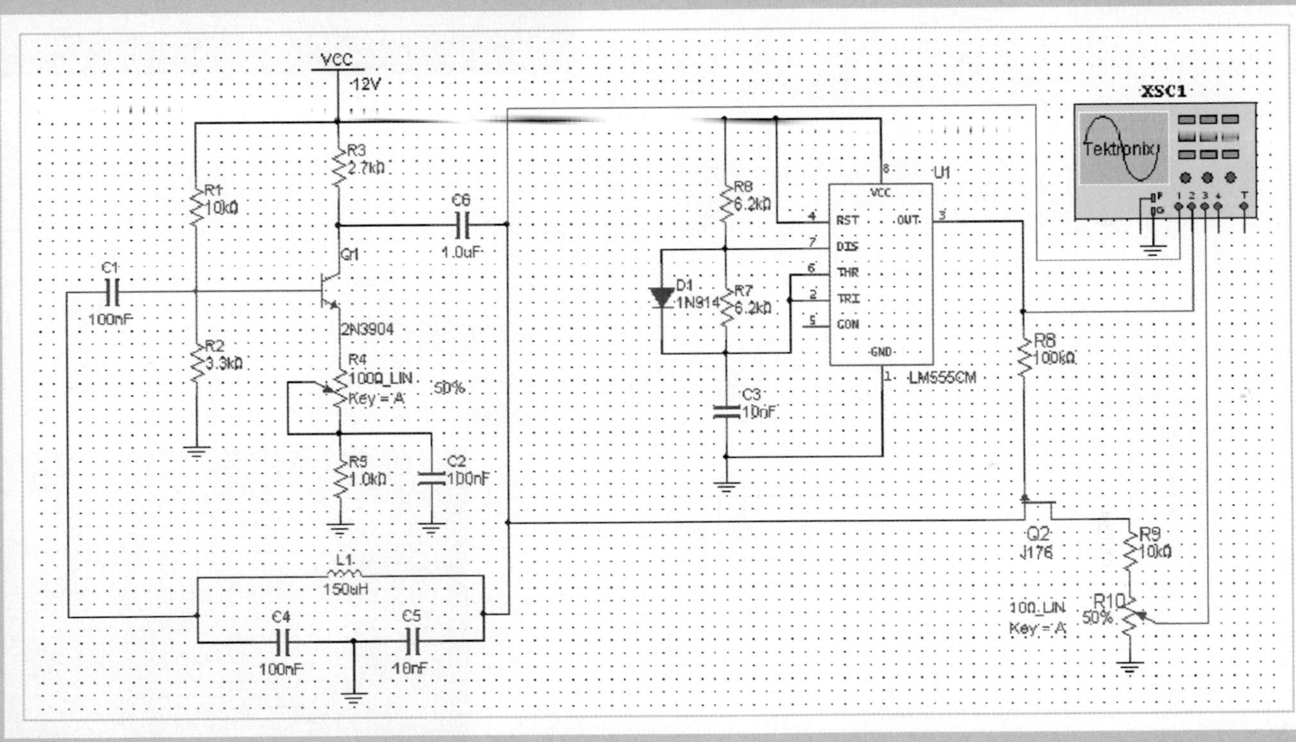

FIGURE 16–48

The complete ASK test generator.

5. What is the purpose of Q_2 in the ASK test generator circuit?

Finally, a simulation is run with the ASK test generator driving the RFID reader. This is shown in Figure 16–50.

6. Identify each waveform in Figure 16–50.

Simulate the ASK test generator using your Multisim software. Observe the operation with the oscilloscope.

Prototyping and Testing

Now that the circuit has been simulated, the prototype circuit is constructed and tested. After the circuit is successfully tested on a protoboard, it is ready to be finalized on a printed circuit board.

FIGURE 16–49

Waveforms for the ASK test generator.

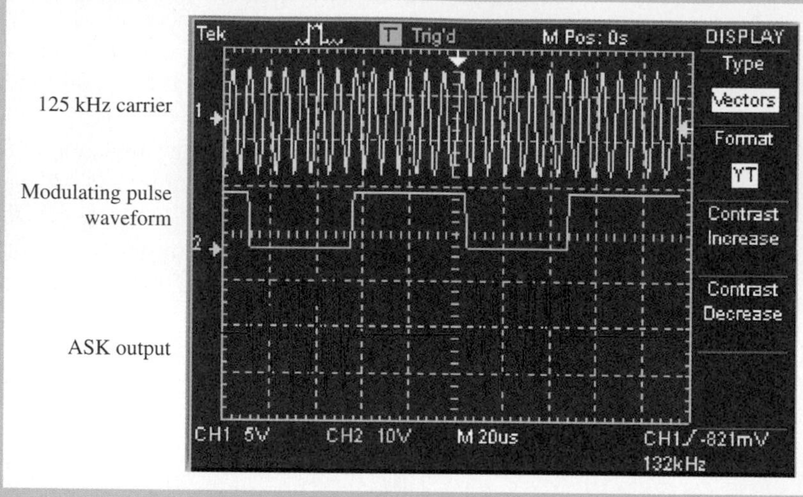

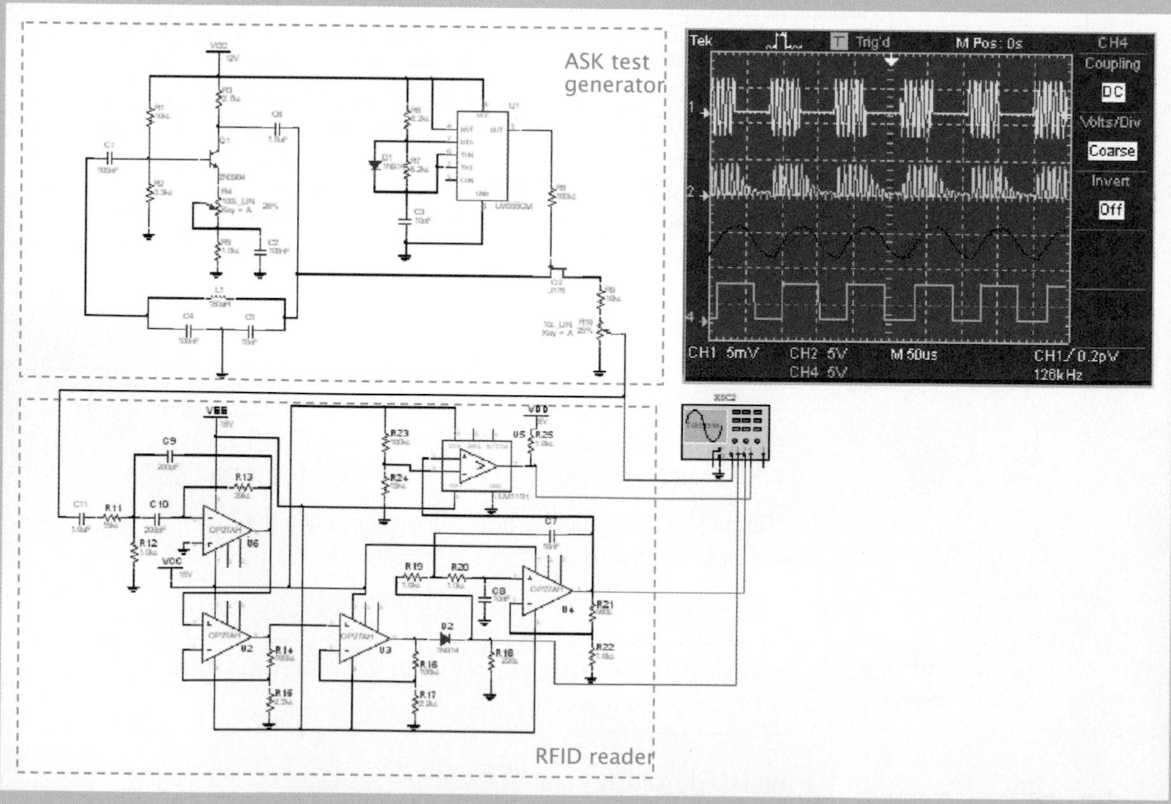

FIGURE 16–50

Simulation of the ASK test generator driving the RFID reader.

Lab Experiment

To build and test a circuit similar to one used in the ASK test generator, go to Experiment 16–A in your lab manual (*Laboratory Exercises for Electronic Devices* by David Buchla and Steven Wetterling).

Circuit Board

The ASK test generator is implemented on a printed circuit board as shown in Figure 16–51 and will be housed in a unit for use in testing RFID readers on the assembly line. The dark gray lines represent backside connections.

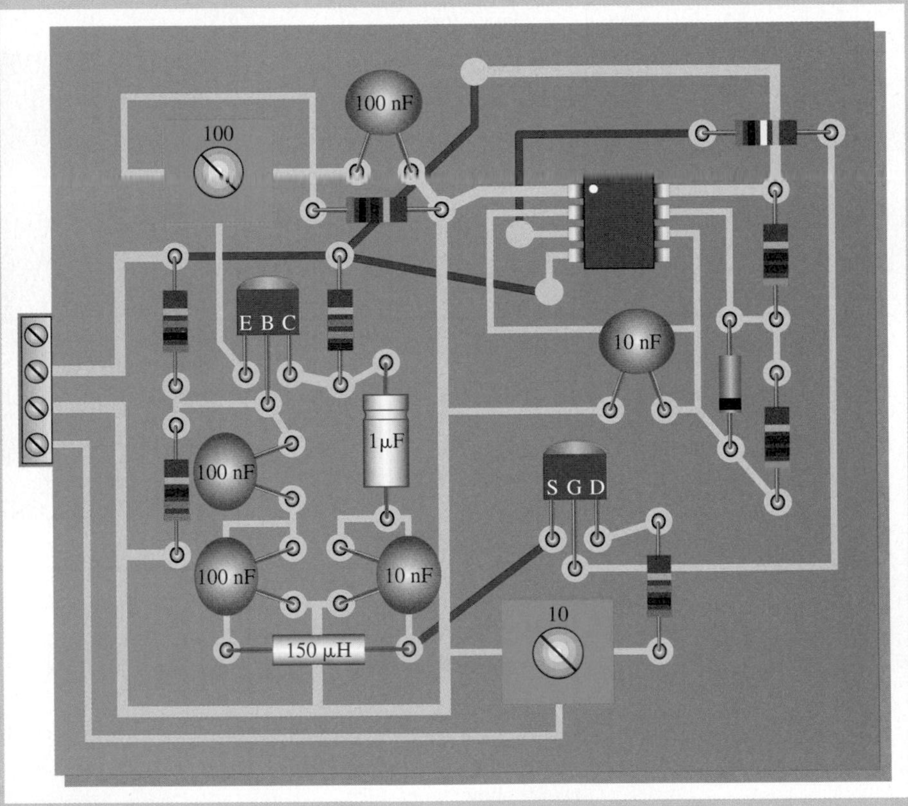

FIGURE 16–51
ASK test generator board.

7. Check the printed circuit board and verify that it agrees with the simulation schematic in Figure 16–48.
8. Label each input and output pin according to function.

Anadigm.com

Programmable Analog Design

Oscillators with various types of outputs can be programmed into an FPAA or a dpASP. These are described as follows.

Sine-Wave Oscillator

Oscillators can be implemented in programmable analog arrays using software. A sine-wave oscillator is shown in Figure 16–52 using AnadigmDesigner2.

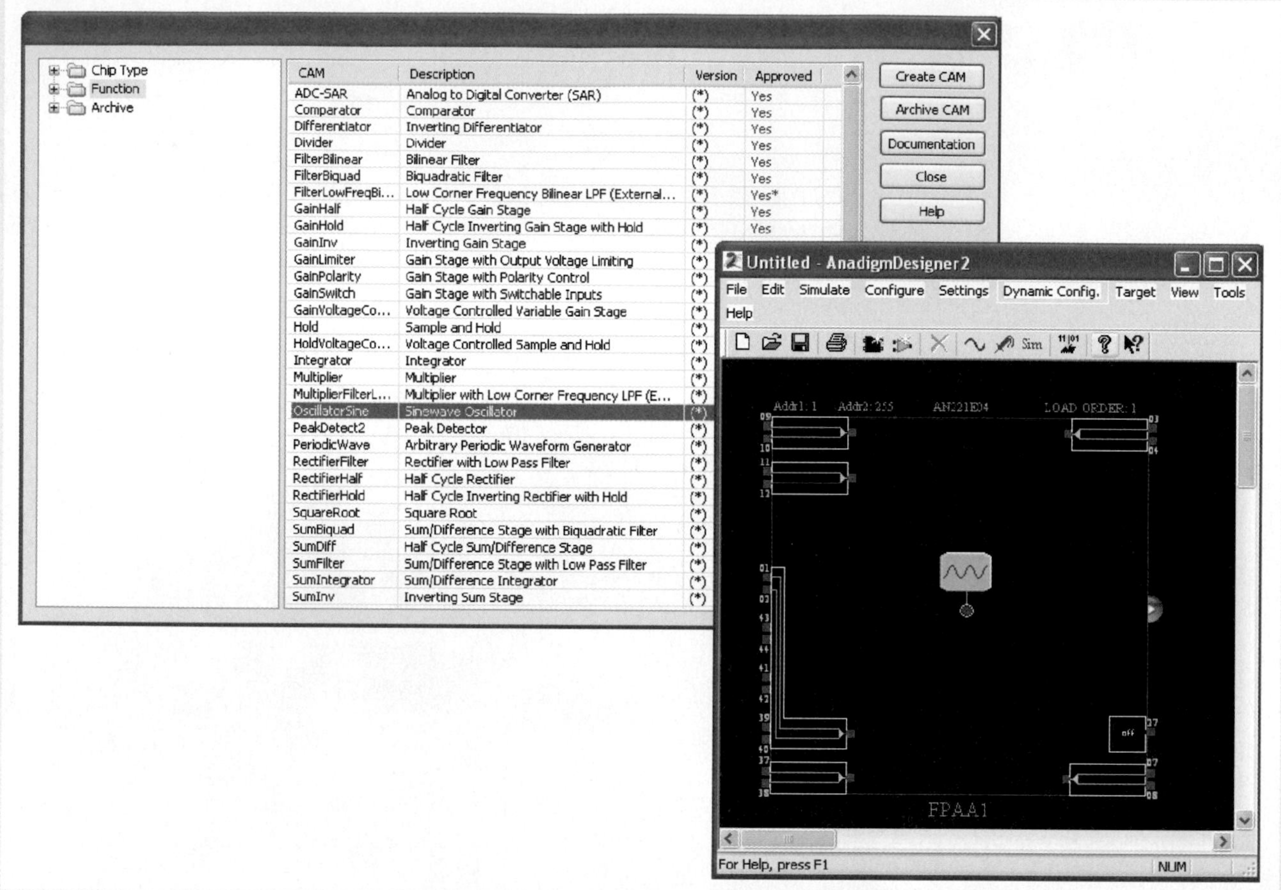

FIGURE 16–52

Selection and placement of the sine-wave oscillator CAM.

The frequency and peak amplitude of the oscillator can be programmed, and the oscillator CAM can be connected to an output as shown in Figure 16–53(a). Running the simulation produces the results shown in part (b).

Square-Wave Oscillator

A square-wave oscillator can be programmed using the sine-wave oscillator CAM and the comparator CAM, as illustrated in Figure 16–54. The frequency of the square wave can be changed by reprogramming the frequency of the sine-wave oscillator.

Variable Duty Cycle Pulse Oscillator

By adding a variable reference to the comparator CAM and by changing its value, the duty cycle as well as the frequency of the pulse waveform can be varied, as shown in Figure 16–55.

Triangular-Wave Oscillator

One way to program a triangular-wave oscillator is shown in Figure 16–56. A sine-wave oscillator is used to drive an inverting gain stage into nonlinear operation. This is followed by an integrator with a properly selected integration constant. A comparator could have been used instead of the over-driven gain stage except that Designer2 does not allow the output of a comparator to be connected to anything but a chip output.

Programming Exercises

1. How do you adjust the duty cycle of the variable duty cycle pulse oscillator?
2. To change the frequency of the triangular-wave oscillator, what parameters must be changed by programming?

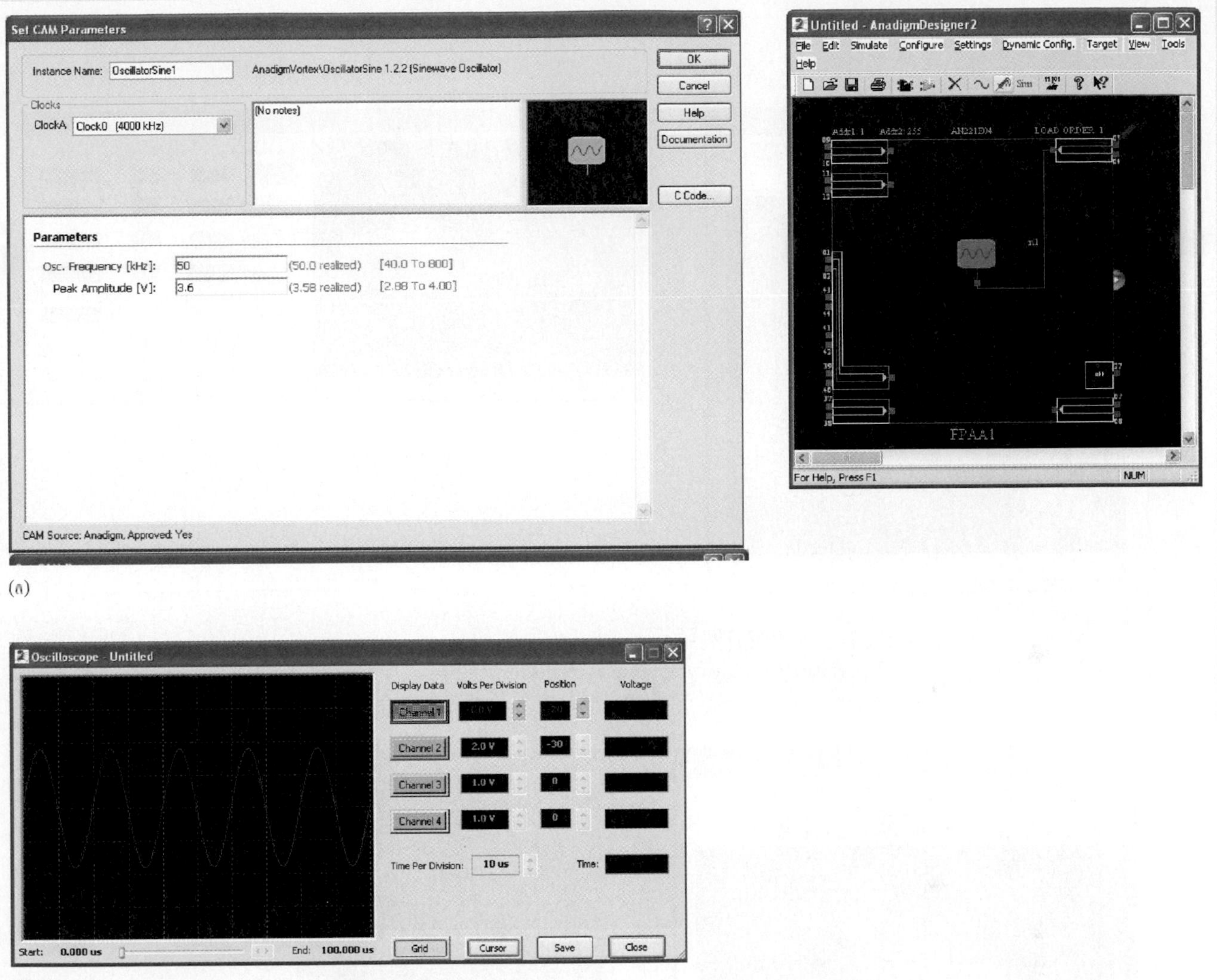

(a)

(b)

▲ **FIGURE 16–53**

Programming a chip as a sine-wave oscillator.

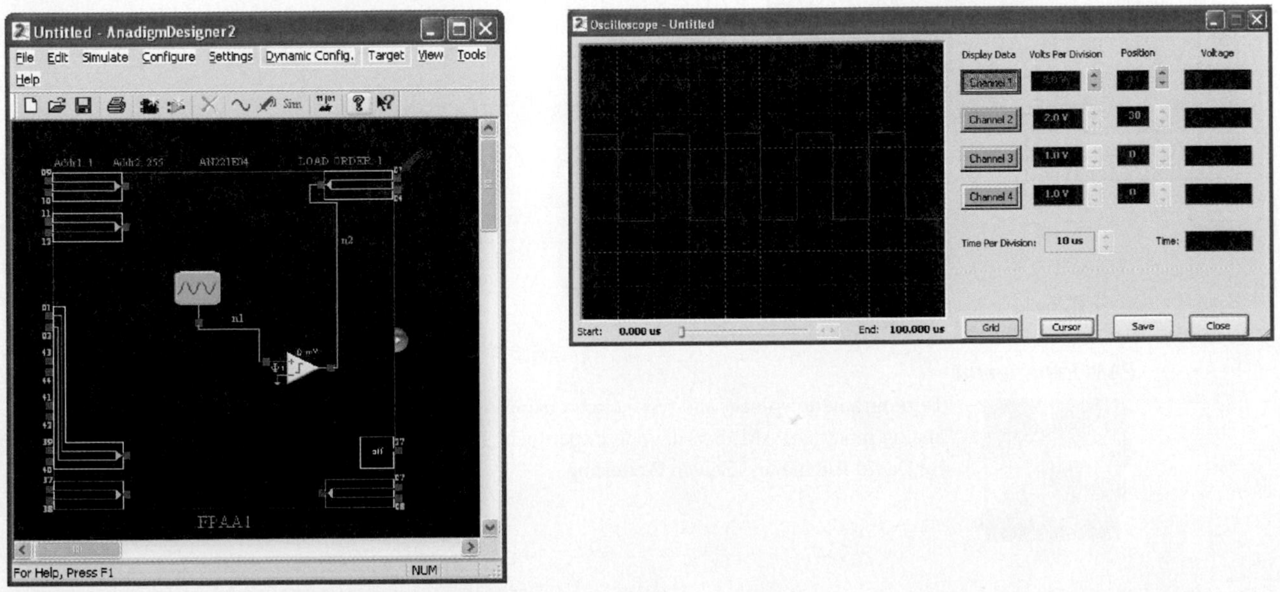

▲ **FIGURE 16–54**

A square-wave oscillator.

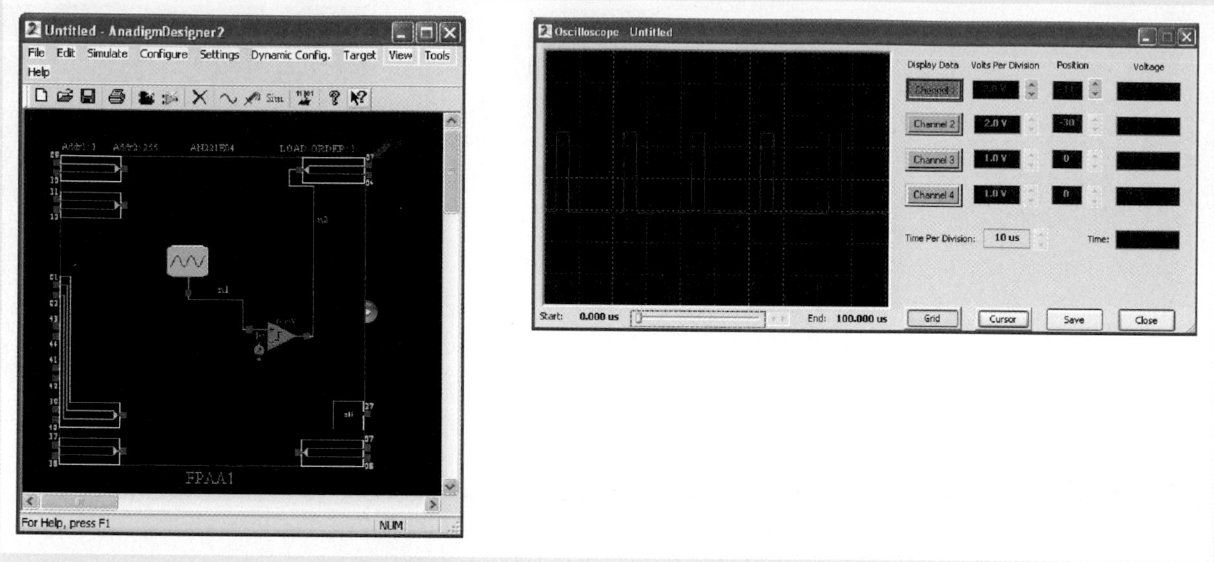

FIGURE 16–55

A variable duty cycle pulse oscillator.

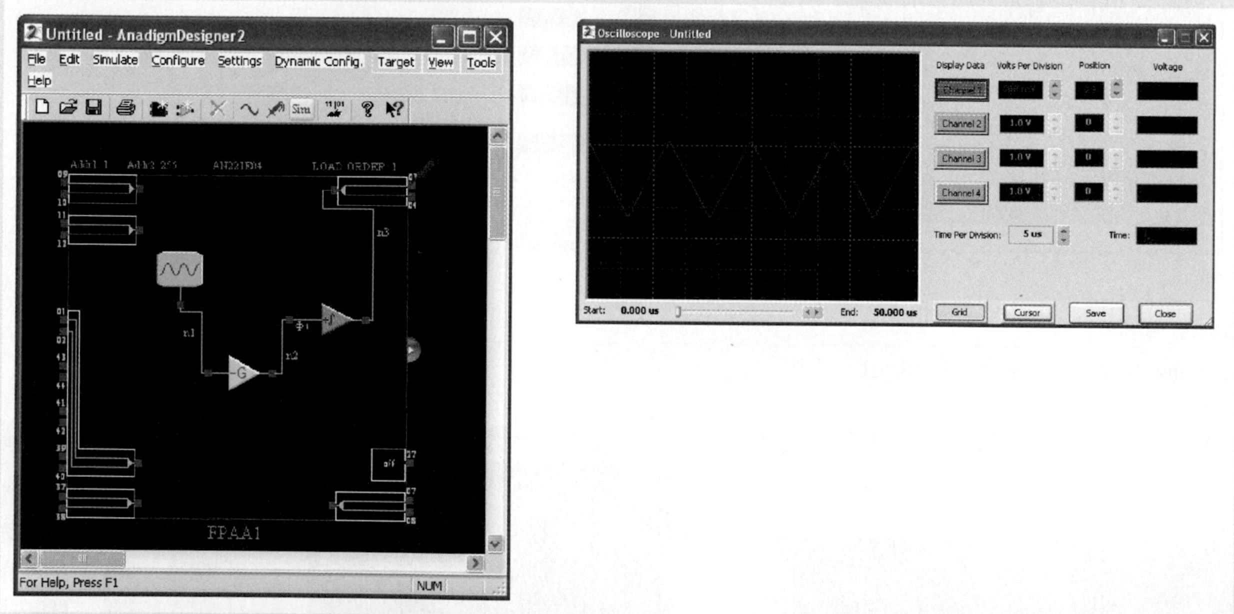

FIGURE 16–56

A triangular-wave oscillator.

PAM Experiment

To program, download, and test a circuit using AnadigmDesigner2 software and the programmable analog module (PAM) board, go to Experiment 16–B in *Laboratory Exercises for Electronic Devices* by David Buchla and Steven Wetterling.

SUMMARY

Section 16–1
- Sinusoidal feedback oscillators operate with positive feedback.
- Relaxation oscillators use an *RC* timing circuit.

Section 16–2
- The two conditions for positive feedback are the phase shift around the feedback loop must be 0° and the voltage gain around the feedback loop must equal 1.
- For initial start-up, the voltage gain around the feedback loop must be greater than 1.

Section 16–3
- Sinusoidal *RC* oscillators include the Wien-bridge, phase-shift, and twin-T.

Section 16–4
- Sinusoidal *LC* oscillators include the Colpitts, Clapp, Hartley, Armstrong, and crystal-controlled.
- The feedback signal in a Colpitts oscillator is derived from a capacitive voltage divider in the *LC* circuit.
- The Clapp oscillator is a variation of the Colpitts with a capacitor added in series with the inductor.
- The feedback signal in a Hartley oscillator is derived from an inductive voltage divider in the *LC* circuit.
- The feedback signal in an Armstrong oscillator is derived by transformer coupling.
- Crystal oscillators are the most stable type of feedback oscillator.

Section 16–5
- A relaxation oscillator uses an *RC* timing circuit and a device that changes states to generate a periodic waveform.
- The frequency in a voltage-controlled oscillator (VCO) can be varied with a dc control voltage.

Section 16–6
- The 555 timer is an integrated circuit that can be used as an oscillator, in addition to many other applications.

KEY TERMS

Key terms and other bold terms in the chapter are defined in the end-of-book glossary.

Astable Characterized by having no stable states.

Feedback oscillator An electronic circuit that operates with positive feedback and produces a time-varying output signal without an external input signal.

Positive feedback The return of a portion of the output signal to the input such that it reinforces and sustains the output.

Relaxation oscillator An electronic circuit that uses an *RC* timing circuit to generate a nonsinusoidal waveform without an external input signal.

Voltage-controlled oscillator (VCO) A type of relaxation oscillator whose frequency can be varied by a dc control voltage.

KEY FORMULAS

16–1	$\frac{V_{out}}{V_{in}} = \frac{1}{3}$	Wien-bridge positive feedback attenuation
16–2	$f_r = \frac{1}{2\pi RC}$	Wien-bridge resonant frequency
16–3	$B = \frac{1}{29}$	Phase-shift feedback attenuation
16–4	$f_r = \frac{1}{2\pi\sqrt{6}RC}$	Phase-shift oscillator frequency
16–5	$f_r \cong \frac{1}{2\pi\sqrt{LC_T}}$	Colpitts, Clapp, and Hartley approximate resonant frequency
16–6	$A_v = \frac{C_1}{C_2}$	Colpitts amplifier gain

16–7	$f_r = \dfrac{1}{2\pi\sqrt{LC_T}}\sqrt{\dfrac{Q^2}{Q^2 + 1}}$	Colpitts resonant frequency
16–8	$A_v > \dfrac{L_2}{L_1}$	Hartley self-starting gain
16–9	$f_r = \dfrac{1}{2\pi\sqrt{L_{pri}C_1}}$	Armstrong resonant frequency
16–10	$f_r = \dfrac{1}{4R_1C}\left(\dfrac{R_2}{R_3}\right)$	Triangular-wave oscillator frequency
16–11	$f = \dfrac{\lvert V_{IN}\rvert}{R_iC}\left(\dfrac{1}{V_p - V_F}\right)$	Sawtooth VCO frequency
16–12	$f_r = \dfrac{1.44}{(R_1 + 2R_2)C_{ext}}$	555 astable frequency
16–13	$\text{Duty cycle} = \left(\dfrac{R_1 + R_2}{R_1 + 2R_2}\right)100\%$	555 astable

TRUE/FALSE QUIZ

Answers are at the end of the chapter.

1. Two categories of oscillators are feedback and relaxation.
2. A feedback oscillator uses only negative feedback.
3. Positive feedback is never used in an oscillator.
4. The net phase shift around the oscillator feedback loop must be zero.
5. The voltage gain around the closed feedback loop must be greater than 1 to sustain oscillations.
6. For start-up, the loop gain must be greater than 1.
7. A Wien-bridge oscillator uses an *RC* circuit in the positive feedback loop.
8. The phase-shift oscillator utilizes *RC* circuits.
9. The twin-T oscillator contains an *LC* feedback circuit.
10. Colpitts, Clapp, Hartley, and Armstrong are examples of *LC* oscillators.
11. The crystal oscillator is based on the photoelectric effect.
12. A relaxation oscillator uses no positive feedback.
13. Most relaxation oscillators produce sinusoidal outputs.
14. VCO stands for variable-capacitance oscillator.
15. The 555 timer can be used as an oscillator.

CIRCUIT-ACTION QUIZ

Answers are at the end of the chapter.

1. If R_1 and R_2 are increased to 18 kΩ in Figure 16–12, the frequency of oscillation will
   (a) increase (b) decrease (c) not change
2. If the feedback potentiometer R_f is adjusted to a higher value, the voltage gain in Figure 16–12 will
   (a) increase (b) decrease (c) not change
3. In Figure 16–14, if the R_f is decreased, the feedback attenuation will
   (a) increase (b) decrease (c) not change
4. If the capacitors in Figure 16–14 are increased to 0.01 μF, the frequency of oscillation will
   (a) increase (b) decrease (c) not change
5. In order to increase V_{UTP} in Figure 16–30, R_3 must
   (a) increase (b) decrease (c) not change
6. If the capacitor in Figure 16–30 opens, the frequency of oscillation will
   (a) increase (b) decrease (c) not change

7. If the value of R_1 in Figure 16–32 is decreased, the peak value of the sawtooth output will
   (a) increase **(b)** decrease **(c)** not change
8. If the diode in Figure 16–40 opens, the duty cycle will
   (a) increase **(b)** decrease **(c)** not change

SELF-TEST

Answers are at the end of the chapter.

Section 16–1

1. An oscillator differs from an amplifier because the oscillator
   (a) has more gain **(b)** requires no input signal
   (c) requires no dc supply **(d)** always has the same output

Section 16–2

2. One condition for oscillation is
   (a) a phase shift around the feedback loop of 180°
   (b) a gain around the feedback loop of one-third
   (c) a phase shift around the feedback loop of 0°
   (d) a gain around the feedback loop of less than 1
3. A second condition for oscillation is
   (a) no gain around the feedback loop
   (b) a gain of 1 around the feedback loop
   (c) the attenuation of the feedback circuit must be one-third
   (d) the feedback circuit must be capacitive
4. In a certain oscillator, $A_v = 50$. The attenuation of the feedback circuit must be
   (a) 1 **(b)** 0.01 **(c)** 10 **(d)** 0.02
5. For an oscillator to properly start, the gain around the feedback loop must initially be
   (a) 1 **(b)** less than 1 **(c)** greater than 1 **(d)** equal to *B*

Section 16–3

6. Wien-bridge oscillators are based on
   (a) positive feedback **(b)** negative feedback
   (c) the piezoelectric effect **(d)** high gain
7. In a Wien-bridge oscillator, if the resistances in the positive feedback circuit are decreased, the frequency
   (a) decreases **(b)** increases **(c)** remains the same
8. The Wien-bridge oscillator's positive feedback circuit is
   (a) an *RL* circuit **(b)** an *LC* circuit
   (c) a voltage divider **(d)** a lead-lag circuit
9. A phase-shift oscillator has
   (a) three *RC* circuits **(b)** three *LC* circuits
   (c) a T-type circuit **(d)** a π-type circuit

Section 16–4

10. Colpitts, Clapp, and Hartley are names that refer to
    (a) types of *RC* oscillators **(b)** inventors of the transistor
    (c) types of *LC* oscillators **(d)** types of filters
11. The main feature of a crystal oscillator is
    (a) economy **(b)** reliability **(c)** stability **(d)** high frequency

Section 16–5

12. An oscillator whose frequency is changed by a variable dc voltage is known as
    (a) a crystal oscillator **(b)** a VCO
    (c) an Armstrong oscillator **(d)** a piezoelectric device
13. The operation of a relaxation oscillator is based on
    (a) the charging and discharging of a capacitor **(b)** a highly selective resonant circuit
    (c) a very stable supply voltage **(d)** low power consumption

Section 16–6

14. Which one of the following is *not* an input or output of the 555 timer?
    (a) Threshold (b) Control voltage (c) Clock
    (d) Trigger (e) Discharge (f) Reset

PROBLEMS

Answers to odd-numbered problems are at the end of the book.

BASIC PROBLEMS

Section 16–1

The Oscillator

1. What type of input is required for an oscillator?
2. What are the basic components of an oscillator circuit?

Section 16–2

Feedback Oscillator Principles

3. If the voltage gain of the amplifier portion of an oscillator is 75, what must be the attenuation of the feedback circuit to sustain the oscillation?
4. Generally describe the change required in the oscillator of Problem 3 in order for oscillation to begin when the power is initially turned on.

Section 16–3

Oscillators with *RC* Feedback Circuits

5. A certain lead-lag circuit has a resonant frequency of 3.5 kHz. What is the rms output voltage if an input signal with a frequency equal to f_r and with an rms value of 2.2 V is applied to the input?
6. Calculate the resonant frequency of a lead-lag circuit with the following values: $R_1 = R_2 = 6.2\ \text{k}\Omega$, and $C_1 = C_2 = 0.02\ \mu\text{F}$.
7. Determine the necessary value of R_2 in Figure 16–57 so that the circuit will oscillate. Neglect the forward resistance of the zener diodes. (*Hint:* The total gain of the circuit must be 3 when the zener diodes are conducting.)
8. Explain the purpose of R_3 in Figure 16–57.

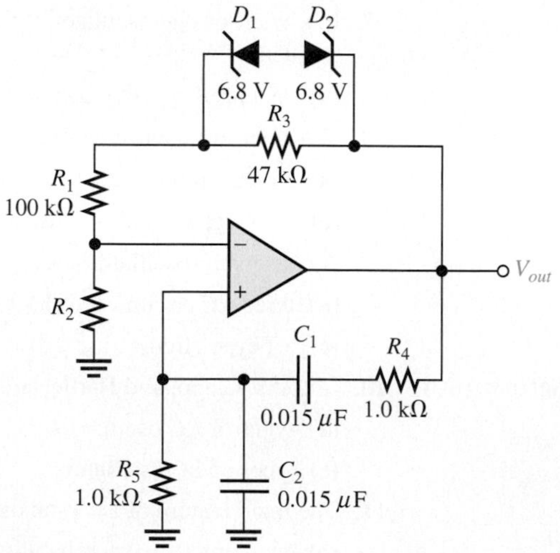

▲ FIGURE 16–57

9. For the Wien-bridge oscillator in Figure 16–58, calculate the setting for R_f, assuming the internal drain-source resistance, r'_{ds}, of the JFET is 350 Ω when oscillations are stable.
10. Find the frequency of oscillation for the Wien-bridge oscillator in Figure 16–58.

FIGURE 16–58

Multisim file circuits are identified with a CD logo and are in the Problems folder on your CD-ROM. Filenames correspond to figure numbers (e.g., F16-58).

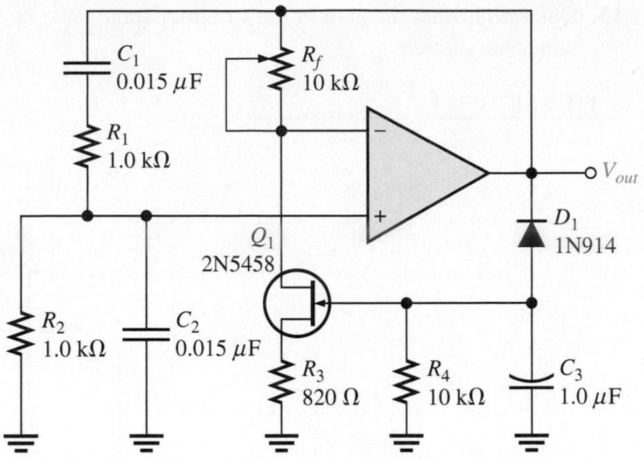

11. What value of R_f is required in Figure 16–59? What is f_r?

FIGURE 16–59

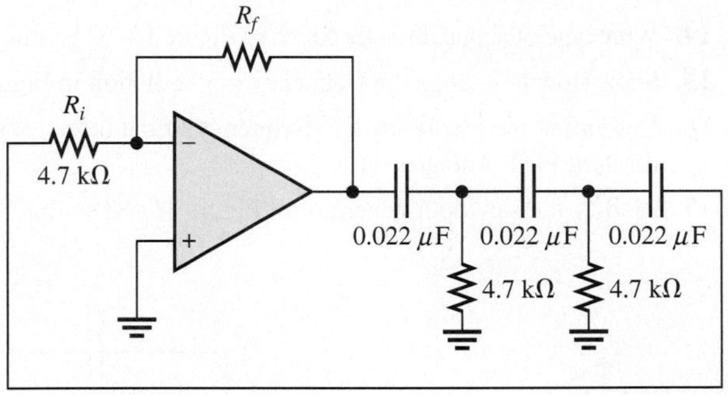

Section 16–4 **Oscillators with LC Feedback Circuits**

12. Calculate the frequency of oscillation for each circuit in Figure 16–60 and identify the type of oscillator. Assume $Q > 10$ in each case.

FIGURE 16–60

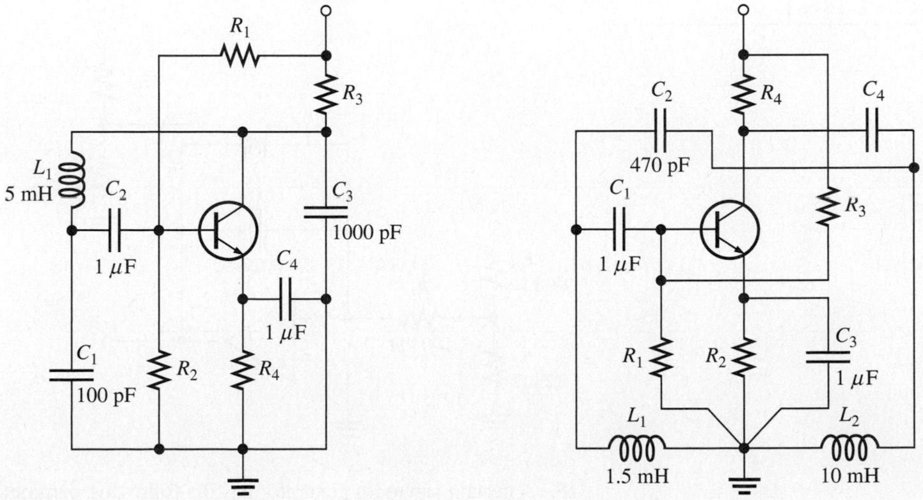

(a) (b)

13. Determine what the gain of the amplifier stage must be in Figure 16–61 in order to have sustained oscillation.

▶ FIGURE 16–61

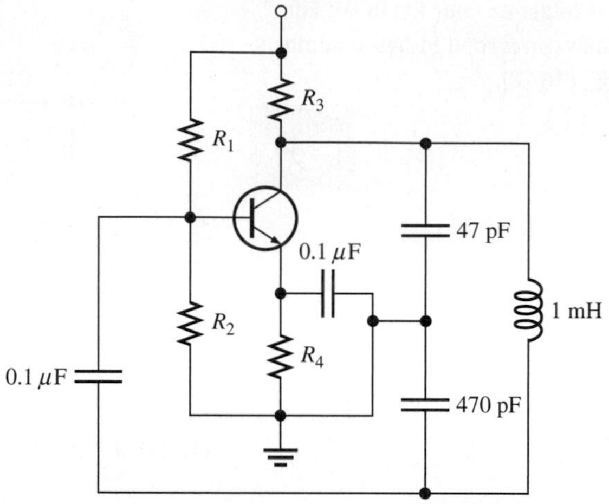

Section 16–5 **Relaxation Oscillators**

14. What type of signal does the circuit in Figure 16–62 produce? Determine the frequency of the output.

15. Show how to change the frequency of oscillation in Figure 16–62 to 10 kHz.

16. Determine the amplitude and frequency of the output voltage in Figure 16–63. Use 1 V as the forward PUT voltage.

17. Modify the sawtooth generator in Figure 16–63 so that its peak-to-peak output is 4 V.

▶ FIGURE 16–62

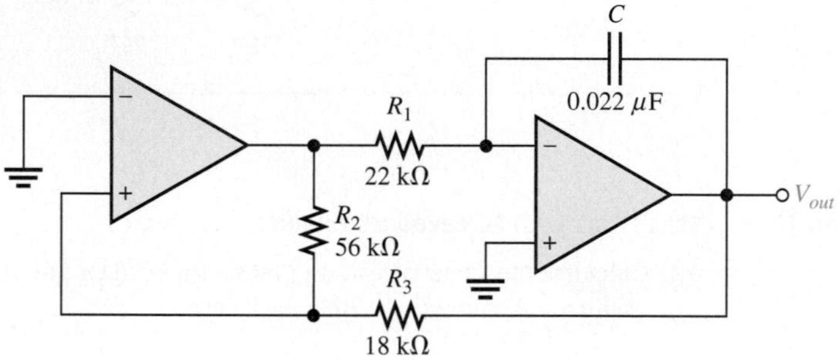

▶ FIGURE 16–63

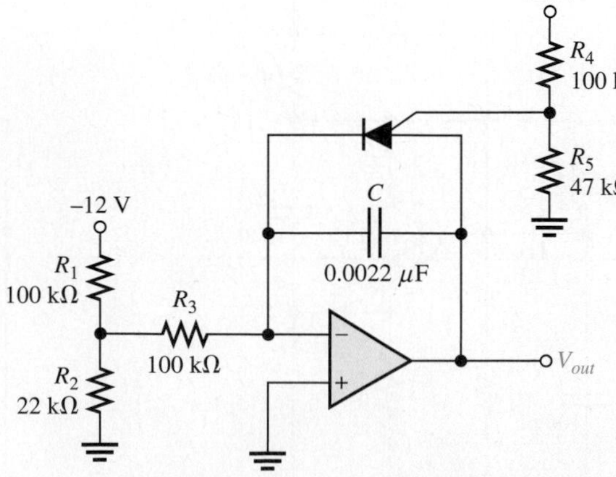

18. A certain sawtooth generator has the following parameter values: $V_{IN} = 3$ V, $R = 4.7$ kΩ, $C = 0.001$ μF. Determine its peak-to-peak output voltage if the period is 10 μs.

Section 16–6 **The 555 Timer as an Oscillator**

19. What are the two comparator reference voltages in a 555 timer when $V_{CC} = 10$ V?

20. Determine the frequency of oscillation for the 555 astable oscillator in Figure 16–64.

21. To what value must C_{ext} be changed in Figure 16–64 to achieve a frequency of 25 kHz?

22. In an astable 555 configuration, the external resistor $R_1 = 3.3$ kΩ. What must R_2 equal to produce a duty cycle of 75 percent?

▶ FIGURE 16–64

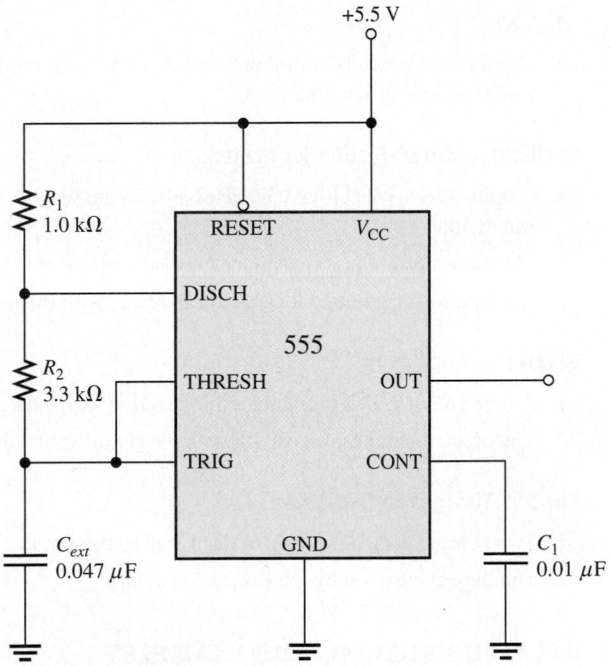

MULTISIM TROUBLESHOOTING PROBLEMS

These file circuits are in the Troubleshooting Problems folder on your CD-ROM.

23. Open file TSP16-23 and determine the fault.

24. Open file TSP16-24 and determine the fault.

25. Open file TSP16-25 and determine the fault.

26. Open file TSP16-26 and determine the fault.

27. Open file TSP16-27 and determine the fault.

28. Open file TSP16-28 and determine the fault.

ANSWERS

SECTION CHECKUPS

Section 16–1 **The Oscillator**

1. An oscillator is a circuit that produces a repetitive output waveform with only the dc supply voltage as an input.

2. Positive feedback

3. The feedback circuit provides attenuation and phase shift.

4. Feedback and relaxation

Section 16–2 Feedback Oscillator Principles

1. Zero phase shift and unity voltage gain around the closed feedback
2. Positive feedback is when a portion of the output signal is fed back to the input of the amplifier such that it reinforces itself.
3. Loop gain greater than 1

Section 16–3 Oscillators with *RC* Feedback Circuits

1. The negative feedback loop sets the closed-loop gain; the positive feedback loop sets the frequency of oscillation.
2. 1.67 V
3. The three *RC* circuits contribute a total of 180° and the inverting amplifier contributes 180° for a total of 360° around the loop.

Section 16–4 Oscillators with *LC* Feedback Circuits

1. Colpitts uses a capacitive voltage divider in the feedback circuit; Hartley uses an inductive voltage divider.
2. The higher FET input impedance has less loading effect on the resonant feedback circuit.
3. A Clapp has an additional capacitor in series with the inductor in the feedback circuit.

Section 16–5 Relaxation Oscillators

1. A voltage-controlled oscillator exhibits a frequency that can be varied with a dc control voltage.
2. The basis of a relaxation oscillator is the charging and discharging of a capacitor.

Section 16–6 The 555 Timer as an Oscillator

1. Two comparators, a flip-flop, a discharge transistor, and a resistive voltage divider
2. The duty cycle is set by the external resistors.

RELATED PROBLEMS FOR EXAMPLES

16–1 R_f too high causes clipping. R_f too low causes oscillations to die out.

16–2 **(a)** 238 kΩ **(b)** 7.92 kHz

16–3 7.24 kHz

16–4 6.06 V peak-to-peak

16–5 1122 Hz

16–6 31.9%

TRUE/FALSE QUIZ

1. T **2.** F **3.** F **4.** T **5.** F **6.** T **7.** T **8.** T
9. F **10.** T **11.** F **12.** T **13.** F **14.** F **15.** T

CIRCUIT-ACTION QUIZ

1. (b) **2.** (a) **3.** (a) **4.** (b) **5.** (a) **6.** (b) **7.** (c) **8.** (a)

SELF-TEST

1. (b) **2.** (c) **3.** (b) **4.** (d) **5.** (c) **6.** (a) **7.** (b)
8. (d) **9.** (a) **10.** (c) **11.** (c) **12.** (b) **13.** (a) **14.** (c)

VOLTAGE REGULATORS

17

CHAPTER OUTLINE

CHAPTER OBJECTIVES

- Describe the basic concept of voltage regulation
- Discuss the principles of series voltage regulators
- Discuss the principles of shunt voltage regulators
- Discuss the principles of switching regulators
- Discuss integrated circuit voltage regulators
- Discuss applications of IC voltage regulators

KEY TERMS

- Regulator
- Line regulation
- Load regulation
- Linear regulator
- Switching regulator
- Thermal overload

APPLICATION ACTIVITY PREVIEW

In the Application Activity, the dc power supply from Chapter 3 is redesigned to produce a regulated variable output voltage from 9 V to 30 V. The company that manufactures the power supply will introduce it as a new line that can be preset to a specified voltage at the factory or adjusted by the customer to any desired value in the output voltage range.

VISIT THE COMPANION WEBSITE

Study aids for this chapter are available at http://www.prenhall.com/floyd

INTRODUCTION

A voltage regulator provides a constant dc output voltage that is essentially independent of the input voltage, output load current, and temperature. The voltage regulator is one part of a power supply. Its input voltage comes from the filtered output of a rectifier derived from an ac voltage or from a battery in the case of portable systems.

Most voltage regulators fall into two broad categories: linear regulators and switching regulators. In the linear regulator category, two general types are the series regulator and the shunt regulator. These are normally available for either positive or negative output voltages. A dual regulator provides both positive and negative outputs. In the switching regulator category, three general configurations are step-down, step-up, and inverting.

Many types of integrated circuit (IC) regulators are available. The most popular types of linear regulator are the three-terminal fixed voltage regulator and the three-terminal adjustable voltage regulator. Switching regulators are also widely used. In this chapter, specific IC devices are introduced as representative of the wide range of available devices.

17–1 VOLTAGE REGULATION

Two basic categories of voltage regulation are line regulation and load regulation. The purpose of line regulation is to maintain a nearly constant output voltage when the input voltage varies. The purpose of load regulation is to maintain a nearly constant output voltage when the load varies.

After completing this section, you should be able to

- **Describe the basic concept of voltage regulation**
  - Explain line regulation
  - Calculate line regulation
  - Explain load regulation
  - Calculate load regulation

Line Regulation

When the ac input (line) voltage of a power supply changes, an electronic circuit called a **regulator** maintains a nearly constant output voltage, as illustrated in Figure 17–1. **Line regulation** can be defined as the percentage change in the output voltage for a given change in the input voltage. When taken over a range of input voltage values, line regulation is expressed as a percentage by the following formula:

Equation 17–1

$$\textbf{Line regulation} = \left(\frac{\Delta V_{OUT}}{\Delta V_{IN}}\right)100\%$$

Line regulation can also be expressed in units of %/V. For example, a line regulation of 0.05%/V means that the output voltage changes 0.05 percent when the input voltage increases or decreases by one volt. Line regulation can be calculated using the following formula (Δ means "a change in"):

Equation 17–2

$$\textbf{Line regulation} = \frac{(\Delta V_{OUT}/V_{OUT})100\%}{\Delta V_{IN}}$$

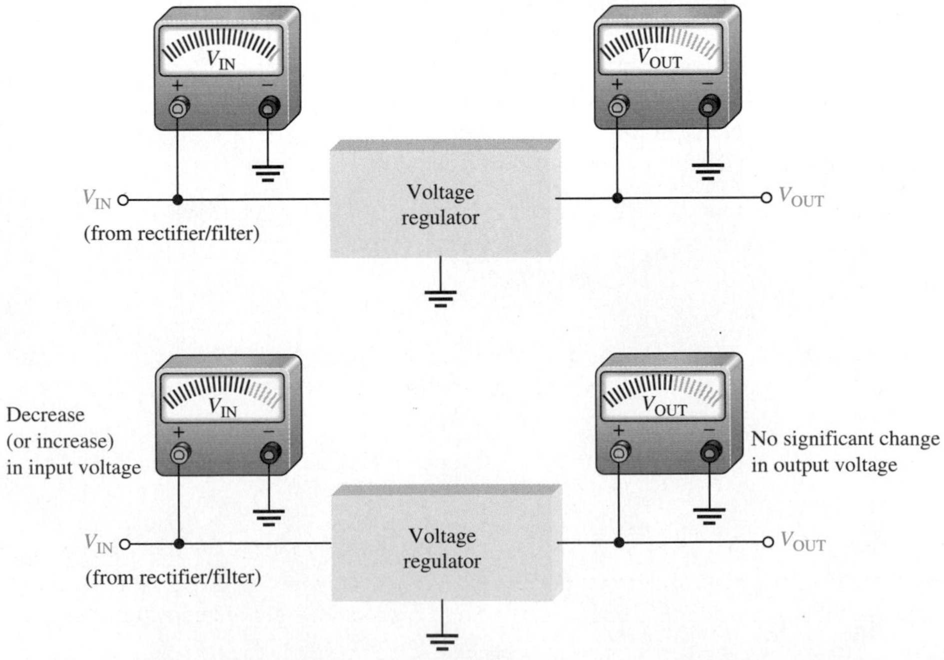

FIGURE 17–1

Line regulation. A change in input (line) voltage does not significantly affect the output voltage of a regulator (within certain limits).

EXAMPLE 17–1

When the ac input voltage of a certain power supply changes, the input to the voltage regulator decreases by 5 V as a result, and the output of the regulator decreases by 0.25 V. The nominal output is 15 V. Determine the line regulation in %/V.

Solution The line regulation as a percentage change per volt is

$$\text{line regulation} = \frac{(\Delta V_{OUT}/V_{OUT})100\%}{\Delta V_{IN}} = \frac{(0.25\text{ V}/15\text{ V})100\%}{5\text{ V}} = \mathbf{0.333\%/V}$$

*Related Problem** The input of a certain regulator increases by 3.5 V. As a result, the output voltage increases by 0.42 V. The nominal output is 20 V. Determine the regulation in %/V.

*Answers are at the end of the chapter.

Load Regulation

When the amount of current through a load changes due to a varying load resistance, the voltage regulator must maintain a nearly constant output voltage across the load, as illustrated in Figure 17–2.

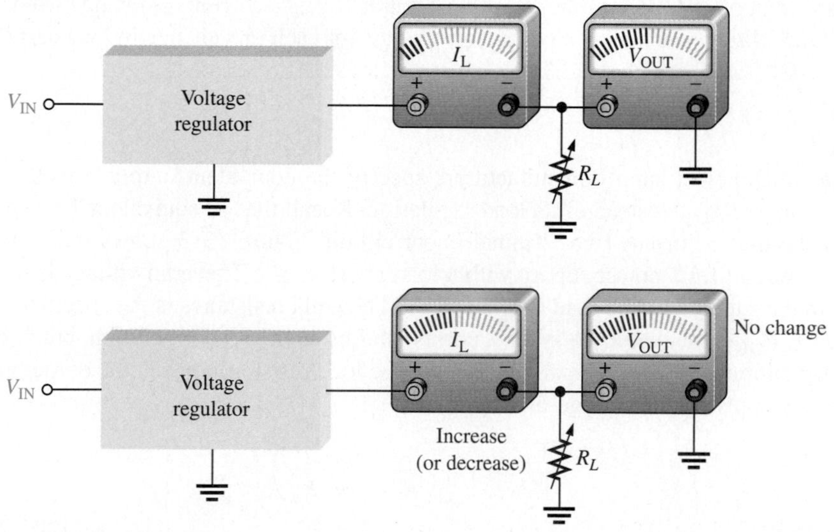

▲ **FIGURE 17–2**

Load regulation. A change in load current has practically no effect on the output voltage of a regulator (within certain limits).

Load regulation can be defined as the percentage change in output voltage for a given change in load current. One way to express load regulation is as a percentage change in output voltage from no-load (NL) to full-load (FL).

$$\textbf{Load regulation} = \left(\frac{V_{NL} - V_{FL}}{V_{FL}}\right)100\%$$

Equation 17–3

Alternately, the load regulation can be expressed as a percentage change in output voltage for each mA change in load current. For example, a load regulation of 0.01%/mA means that the output voltage changes 0.01 percent when the load current increases or decreases 1 mA.

EXAMPLE 17–2 A certain voltage regulator has a 12 V output when there is no load ($I_L = 0$). When there is a full-load current of 10 mA, the output voltage is 11.9 V. Express the voltage regulation as a percentage change from no-load to full-load and also as a percentage change for each mA change in load current.

Solution The no-load output voltage is

$$V_{NL} = 12\text{ V}$$

The full-load output voltage is

$$V_{FL} = 11.9\text{ V}$$

The load regulation as a percentage change from no-load to full-load is

$$\text{load regulation} = \left(\frac{V_{NL} - V_{FL}}{V_{FL}}\right)100\% = \left(\frac{12\text{ V} - 11.9\text{ V}}{11.9\text{ V}}\right)100\% = \mathbf{0.840\%}$$

The load regulation can also be expressed as a percentage change per milliamp as

$$\text{load regulation} = \frac{0.840\%}{10\text{ mA}} = \mathbf{0.084\%/mA}$$

where the change in load current from no-load to full-load is 10 mA.

Related Problem A regulator has a no-load output voltage of 18 V and a full-load output of 17.8 V at a load current of 50 mA. Determine the voltage regulation as a percentage change from no-load to full-load and also as a percentage change for each mA change in load current.

Sometimes power supply manufacturers specify the equivalent output resistance of a power supply (R_{OUT}) instead of its load regulation. Recall that an equivalent Thevenin circuit can be drawn for any two-terminal linear circuit. Figure 17–3 shows the equivalent Thevenin circuit for a power supply with a load resistor. The Thevenin voltage is the voltage from the supply with no load (V_{NL}), and the Thevenin resistance is the specified output resistance, R_{OUT}. Ideally, R_{OUT} is zero, corresponding to 0% load regulation, but in practical power supplies R_{OUT} is a small value. With the load resistor in place, the output voltage is found by applying the voltage-divider rule:

$$V_{OUT} = V_{NL}\left(\frac{R_L}{R_{OUT} + R_L}\right)$$

FIGURE 17–3

Thevenin equivalent circuit for a power supply with a load resistor.

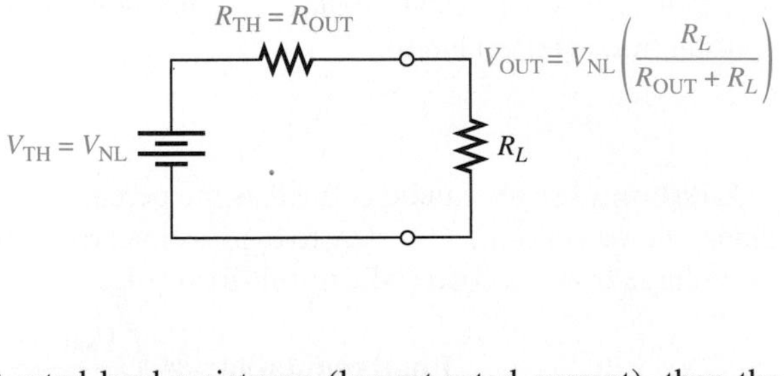

If we let R_{FL} equal the smallest-rated load resistance (largest-rated current), then the full-load output voltage (V_{FL}) is

$$V_{FL} = V_{NL}\left(\frac{R_{FL}}{R_{OUT} + R_{FL}}\right)$$

By rearranging and substituting into Equation 17–3,

$$V_{NL} = V_{FL}\left(\frac{R_{OUT} + R_{FL}}{R_{FL}}\right)$$

$$\text{Load regulation} = \frac{V_{FL}\left(\frac{R_{OUT} + R_{FL}}{R_{FL}}\right) - V_{FL}}{V_{FL}} \times 100\%$$

$$= \left(\frac{R_{OUT} + R_{FL}}{R_{FL}} - 1\right)100\%$$

$$\textbf{Load regulation} = \left(\frac{R_{OUT}}{R_{FL}}\right)\mathbf{100\%} \qquad \text{Equation 17–4}$$

Equation 17–4 is a useful way of finding the percent load regulation when the output resistance and minimum load resistance are specified.

SECTION 17–1 CHECKUP
Answers are at the end of the chapter.

1. Define *line regulation.*
2. Define *load regulation.*
3. The input of a certain regulator increases by 3.5 V. As a result, the output voltage increases by 0.042 V. The nominal output is 20 V. Determine the line regulation in both % and in %/V.
4. If a 5.0 V power supply has an output resistance of 80 mΩ and a specified maximum output current of 1.0 A, what is the load regulation? Give the result as a % and as a %/mA.

17–2 BASIC LINEAR SERIES REGULATORS

The fundamental classes of voltage regulators are linear regulators and switching regulators. Both of these are available in integrated circuit form. Two basic types of linear regulator are the series regulator and the shunt regulator.

After completing this section, you should be able to

- **Discuss the principles of series voltage regulators**
  - Explain regulating action
  - Calculate output voltage of an op-amp series regulator
  - Discuss overload protection and explain how to use current limiting
  - Describe a regulator with fold-back current limiting

A simple representation of a series type of **linear regulator** is shown in Figure 17–4(a), and the basic components are shown in the block diagram in Figure 17–4(b). The control element is a pass transistor in series with the load between the input and output. The output sample circuit senses a change in the output voltage. The error detector compares the sample voltage with a reference voltage and causes the control element to com-

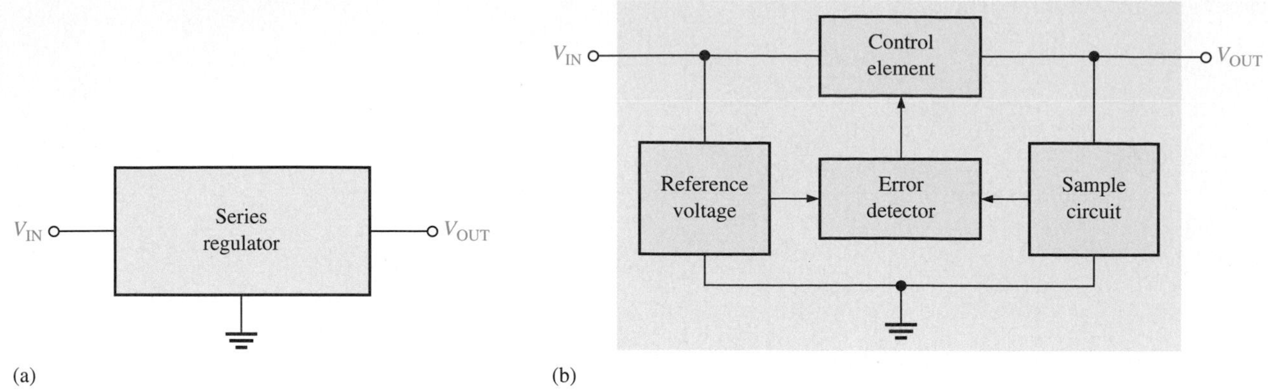

▲ **FIGURE 17–4**

Simple series voltage regulator and block diagram.

pensate in order to maintain a constant output voltage. A basic op-amp series regulator is shown in Figure 17–5.

▶ **FIGURE 17–5**

Basic op-amp series regulator.

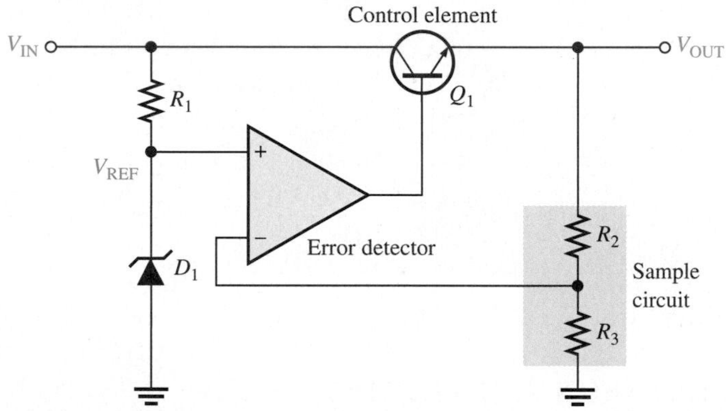

Regulating Action

The operation of the series regulator is illustrated in Figure 17–6 and is as follows. The resistive voltage divider formed by R_2 and R_3 senses any change in the output voltage. When the output tries to decrease, as indicated in Figure 17–6(a), because of a decrease in V_{IN} or because of an increase in I_L caused by a decrease in R_L, a proportional voltage decrease is applied to the op-amp's inverting input by the voltage divider. Since the zener diode (D_1) holds the other op-amp input at a nearly constant reference voltage, V_{REF}, a small difference voltage (error voltage) is developed across the op-amp's inputs. This difference voltage is amplified, and the op-amp's output voltage, V_B, increases. This increase is applied to the base of Q_1, causing the emitter voltage V_{OUT} to increase until the voltage to the inverting input again equals the reference (zener) voltage. This action offsets the attempted decrease in output voltage, thus keeping it nearly constant. The power transistor, Q_1, is usually used with a heat sink because it must handle all of the load current.

The opposite action occurs when the output tries to increase, as indicated in Figure 17–6(b). The op-amp in the series regulator is actually connected as a noninverting amplifier where the reference voltage V_{REF} is the input at the noninverting terminal, and the R_2/R_3 voltage divider forms the negative feedback circuit. The closed-loop voltage gain is

$$A_{cl} = 1 + \frac{R_2}{R_3}$$

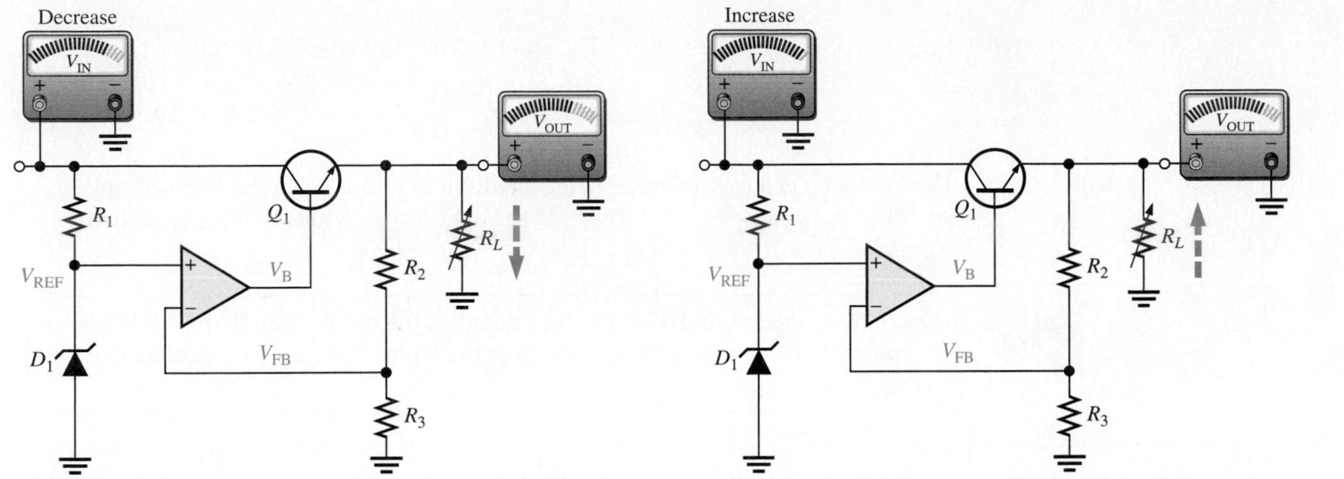

(a) When V_{IN} or R_L decreases, V_{OUT} attempts to decrease. The feedback voltage, V_{FB}, also attempts to decrease, and as a result, the op-amp's output voltage V_B attempts to increase, thus compensating for the attempted decrease in V_{OUT} by increasing the Q_1 emitter voltage. Changes in V_{OUT} are exaggerated for illustration.

When V_{IN} (or R_L) stabilizes at its new lower value, the voltages return to their original values, thus keeping V_{OUT} constant as a result of the negative feedback.

(b) When V_{IN} or R_L increases, V_{OUT} attempts to increase. The feedback voltage, V_{FB}, also attempts to increase, and as a result, V_B, applied to the base of the control transistor, attempts to decrease, thus compensating for the attempted increase in V_{OUT} by decreasing the Q_1 emitter voltage.

When V_{IN} (or R_L) stabilizes at its new higher value, the voltages return to their original values, thus keeping V_{OUT} constant as a result of the negative feedback.

▲ FIGURE 17–6

Illustration of series regulator action that keeps V_{OUT} constant when V_{IN} or R_L changes.

Therefore, the regulated output voltage of the series regulator (neglecting the base-emitter voltage of Q_1) is

$$V_{OUT} \cong \left(1 + \frac{R_2}{R_3}\right)V_{REF}$$

Equation 17–5

From this analysis, you can see that the output voltage is determined by the zener voltage and the resistors R_2 and R_3. It is relatively independent of the input voltage, and therefore, regulation is achieved (as long as the input voltage and load current are within specified limits).

EXAMPLE 17–3

Determine the output voltage for the regulator in Figure 17–7.

FIGURE 17–7

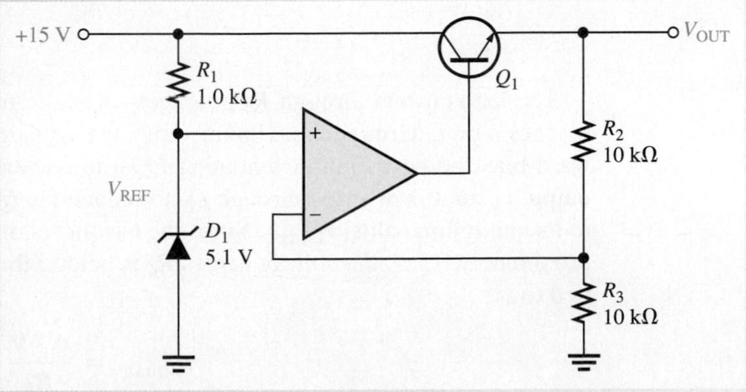

Solution $V_{REF} = 5.1$ V, the zener voltage. The regulated output voltage is therefore

$$V_{OUT} = \left(1 + \frac{R_2}{R_3}\right)V_{REF} = \left(1 + \frac{10\text{ k}\Omega}{10\text{ k}\Omega}\right)5.1\text{ V} = (2)5.1\text{ V} = \mathbf{10.2\ V}$$

Related Problem The following changes are made in the circuit in Figure 17–7: A 3.3 V zener replaces the 5.1 V zener, $R_1 = 1.8\text{ k}\Omega$, $R_2 = 22\text{ k}\Omega$, and $R_3 = 18\text{ k}\Omega$. What is the output voltage?

Open the Multisim file E17-03 in the Examples folder on your CD-ROM. Measure the output voltage with 15 V dc applied to the input. Compare to the calculated value.

Short-Circuit or Overload Protection

If an excessive amount of load current is drawn, the series-pass transistor can be quickly damaged or destroyed. Most regulators use some type of excess current protection in the form of a current-limiting mechanism. Figure 17–8 shows one method of current limiting to prevent overloads called *constant-current limiting.* The current-limiting circuit consists of transistor Q_2 and resistor R_4.

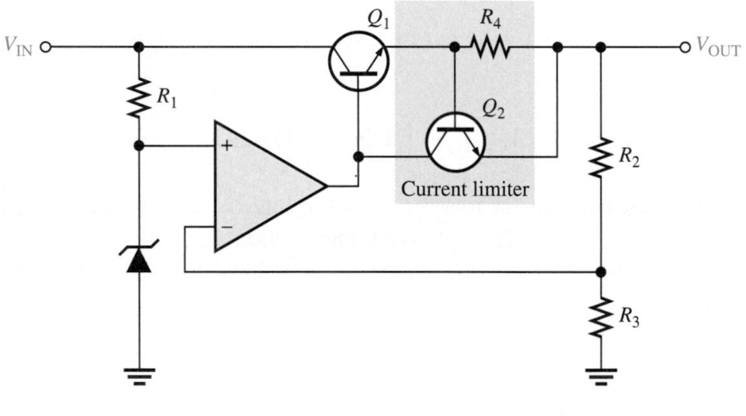

▲ FIGURE 17–8

Series regulator with constant-current limiting.

The load current through R_4 produces a voltage from base to emitter of Q_2. When I_L reaches a predetermined maximum value, the voltage drop across R_4 is sufficient to forward-bias the base-emitter junction of Q_2, thus causing it to conduct. Enough op-amp output current is diverted through Q_2 to reduce the Q_1 base current, so that I_L is limited to its maximum value, $I_{L(max)}$. Since the base-to-emitter voltage of Q_2 cannot exceed approximately 0.7 V, the voltage across R_4 is held to this value, and the load current is limited to

Equation 17–6

$$I_{L(max)} = \frac{0.7\text{ V}}{R_4}$$

EXAMPLE 17–4 Determine the maximum current that the regulator in Figure 17–9 can provide to a load.

FIGURE 17–9

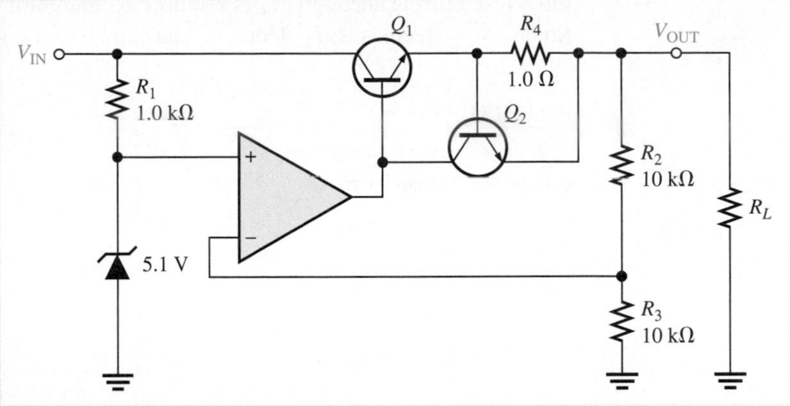

Solution

$$I_{L(max)} = \frac{0.7\text{ V}}{R_4} = \frac{0.7\text{ V}}{1.0\ \Omega} = \mathbf{0.7\ A}$$

Related Problem If the output of the regulator in Figure 17–9 is shorted, what is the current?

Regulator with Fold-Back Current Limiting

In the previous current-limiting technique, the current is restricted to a maximum constant value. **Fold-back current limiting** is a method used particularly in high-current regulators whereby the output current under overload conditions drops to a value well below the peak load current capability to prevent excessive power dissipation.

The basic concept of fold-back current limiting is as follows, with reference to Figure 17–10. The circuit in the green-shaded area is similar to the constant current-limiting arrangement in Figure 17–8, with the exception of resistors R_5 and R_6. The voltage drop developed across R_4 by the load current must not only overcome the base-emitter voltage required to turn on Q_2, but it must also overcome the voltage across R_5. That is, the voltage across R_4 must be

$$V_{R4} = V_{R5} + V_{BE}$$

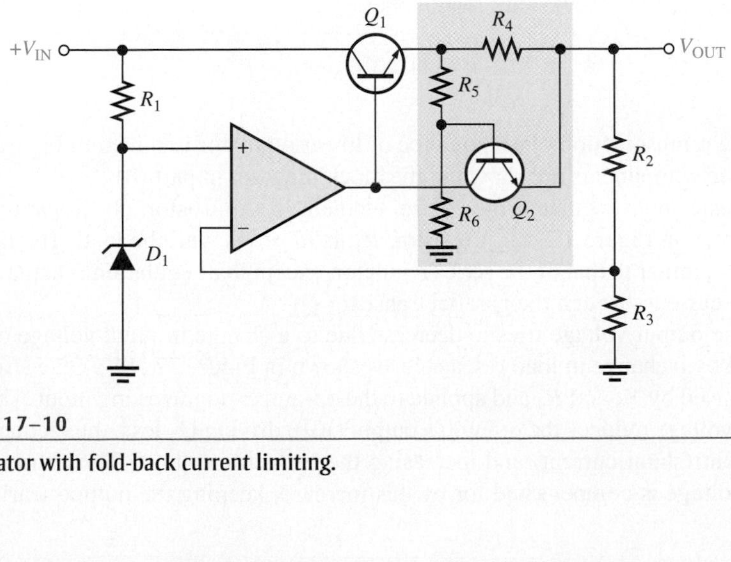

▲ FIGURE 17–10

Series regulator with fold-back current limiting.

In an overload or short-circuit condition, the load current increases to a value, $I_{L(max)}$, that is sufficient to cause Q_2 to conduct. At this point the current can increase no further. The decrease in output voltage results in a proportional decrease in the voltage across R_5; thus, less current through R_4 is required to maintain the forward-biased condition of Q_1. So, as V_{OUT} decreases, I_L decreases, as shown in the graph of Figure 17–11.

▶ FIGURE 17–11

Fold-back current limiting (output voltage versus load current).

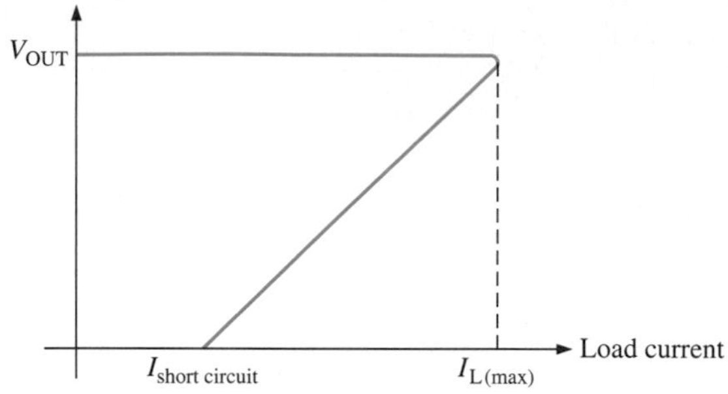

The advantage of this technique is that the regulator is allowed to operate with peak load current up to $I_{L(max)}$; but when the output becomes shorted, the current drops to a lower value to prevent overheating of the device.

SECTION 17–2 CHECKUP

1. What are the basic components in a series regulator?
2. A certain series regulator has an output voltage of 8 V. If the op-amp's closed loop gain is 4, what is the value of the reference voltage?

17–3 BASIC LINEAR SHUNT REGULATORS

The second basic type of linear voltage regulator is the shunt regulator. As you have learned, the control element in the series regulator is the series-pass transistor. In the shunt regulator, the control element is a transistor in parallel (shunt) with the load.

After completing this section, you should be able to

- **Discuss the principles of shunt voltage regulators**
  - Describe the operation of a basic op-amp shunt regulator
  - Compare series and shunt regulators

A simple representation of a shunt type of linear regulator is shown in Figure 17–12(a), and the basic components are shown in the block diagram in part (b).

In the basic shunt regulator, the control element is a transistor, Q_1, in parallel with the load, as shown in Figure 17–13. A resistor, R_1, is in series with the load. The operation of the circuit is similar to that of the series regulator, except that regulation is achieved by controlling the current through the parallel transistor Q_1.

When the output voltage tries to decrease due to a change in input voltage or load current caused by a change in load resistance, as shown in Figure 17–14(a), the attempted decrease is sensed by R_3 and R_4 and applied to the op amp's noninverting input. The resulting difference voltage reduces the op-amp's output (V_B), driving Q_1 less, thus reducing its collector current (shunt current) and increasing the collector voltage. Thus, the original decrease in voltage is compensated for by this increase, keeping the output nearly constant.

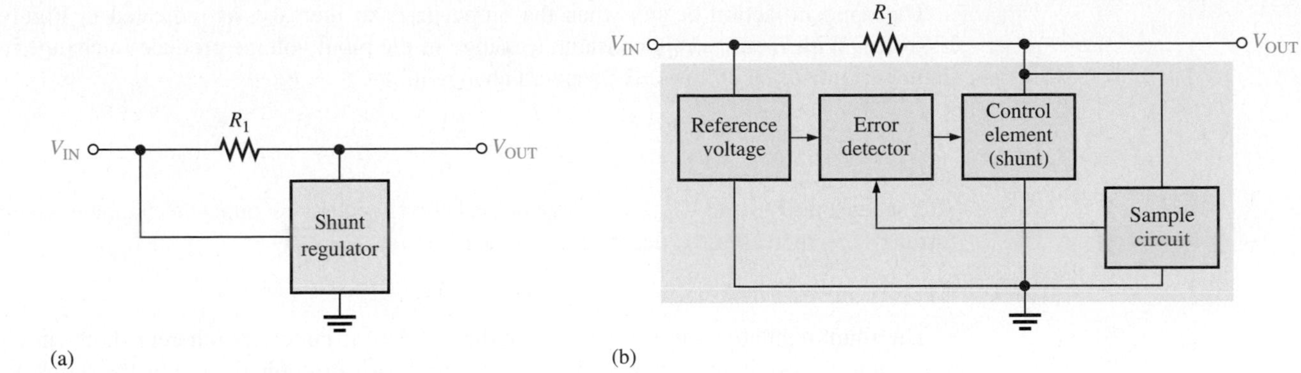

▲ **FIGURE 17–12**

Simple shunt regulator and block diagram.

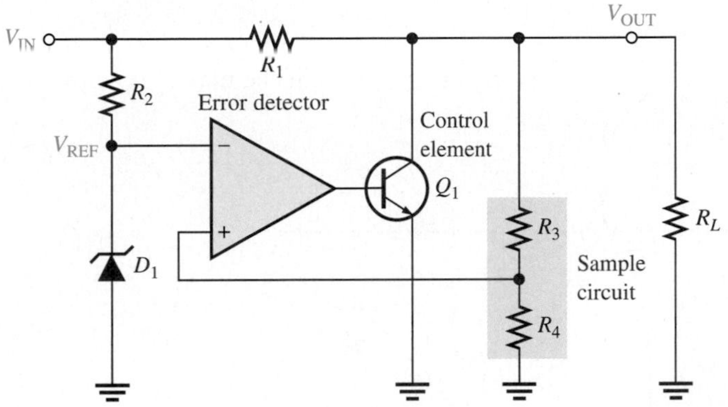

▲ **FIGURE 17–13**

Basic op-amp shunt regulator with load resistor.

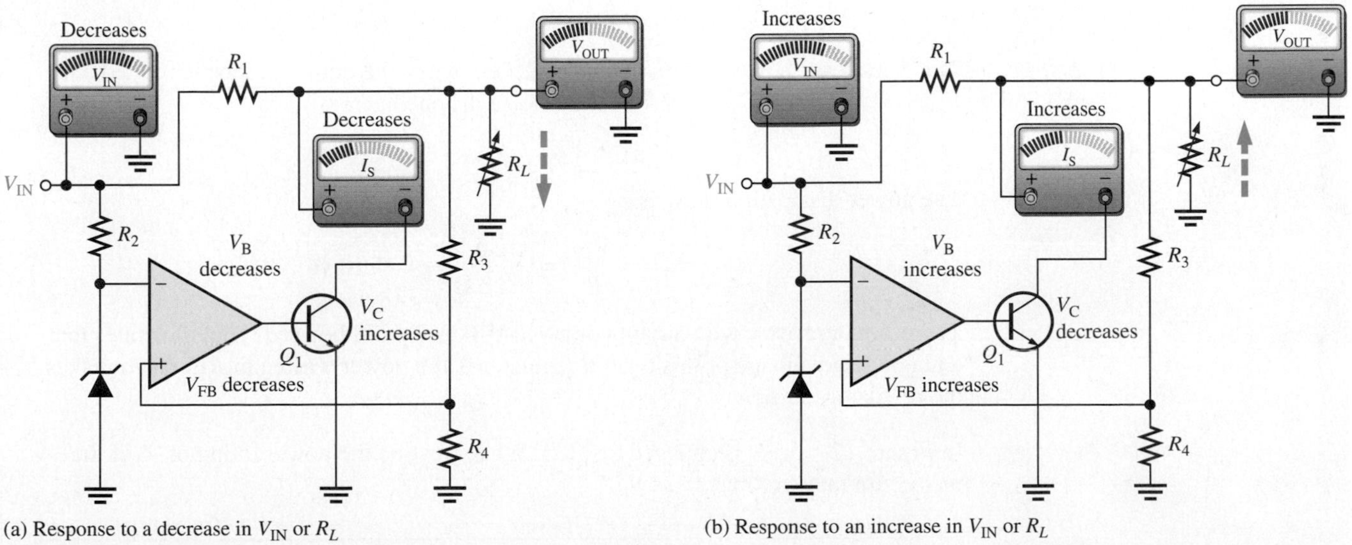

▲ **FIGURE 17–14**

Sequence of responses when V_{OUT} tries to decrease as a result of a decrease in R_L or V_{IN} (opposite responses for an attempted increase).

The opposite action occurs when the output tries to increase, as indicated in Figure 17–14(b). With I_L and V_{OUT} constant, a change in the input voltage produces a change in shunt current (I_S) as follows (Δ means "a change in"):

$$\Delta I_S = \frac{\Delta V_{IN}}{R_1}$$

With a constant V_{IN} and V_{OUT}, a change in load current causes an opposite change in shunt current. If I_L increases, I_S decreases, and vice versa.

$$\Delta I_S = -\Delta I_L$$

The shunt regulator is less efficient than the series type but offers inherent short-circuit protection. If the output is shorted ($V_{OUT} = 0$), the load current is limited by the series resistor R_1 to a maximum value as follows ($I_S = 0$).

Equation 17–7

$$\mathbf{I_{L(max)} = \frac{V_{IN}}{R_1}}$$

EXAMPLE 17–5

In Figure 17–15, what power rating must R_1 have if the maximum input voltage is 12.5 V?

FIGURE 17–15

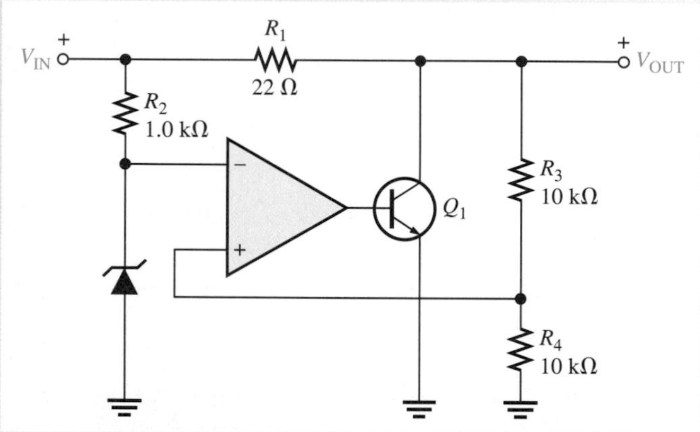

Solution The worst-case power dissipation in R_1 occurs when the output is short-circuited and $V_{OUT} = 0$. When $V_{IN} = 12.5$ V, the voltage dropped across R_1 is

$$V_{R1} = V_{IN} - V_{OUT} = 12.5 \text{ V}$$

The power dissipation in R_1 is

$$P_{R1} = \frac{V_{R1}^2}{R_1} = \frac{(12.5 \text{ V})^2}{22 \ \Omega} = \mathbf{7.10 \ W}$$

Therefore, a resistor with a rating of at least 10 W should be used. This illustrates that a major disadvantage of this type of regulator is the power wasted in R_1, which makes the regulator inefficient.

Related Problem In Figure 17–15, R_1 is changed to 33 Ω. What must be the power rating of R_1 if the maximum input voltage is 24 V?

Open the Multisim file E17 05 in the Examples folder on your CD-ROM. Measure the output voltage with 15 V applied to the input.

SECTION 17–3 CHECKUP

1. How does the control element in a shunt regulator differ from that in a series regulator?
2. What is one advantage of a shunt regulator over a series type? What is a disadvantage?

17–4 BASIC SWITCHING REGULATORS

The two types of linear regulators, series and shunt, have control elements (transistors) that are conducting all the time, with the amount of conduction varied as demanded by changes in the output voltage or current. The switching regulator is different because the control element operates as a switch.

After completing this section, you should be able to

- **Discuss the principles of switching regulators**
  - Describe the step-down configuration of a switching regulator
  - Determine the output voltage of the step-down configuration
  - Describe the step-up configuration of a switching regulator
  - Describe the voltage-inverter configuration

A much greater efficiency can be realized with a switching type of voltage regulator than with the linear types because the transistor is not always conducting. Switching regulator efficiencies can be greater than 90%. Therefore, switching regulators can provide greater load currents at low voltage than linear regulators because the control transistor doesn't dissipate as much power. Three basic configurations of switching regulators are step-down, step-up, and inverting.

Step-Down Configuration

In the step-down configuration, the output voltage is always less than the input voltage. A basic step-down **switching regulator** is shown in Figure 17–16(a), and its simplified equivalent is shown in Figure 17–16(b). Transistor Q_1 is used to switch the input voltage at

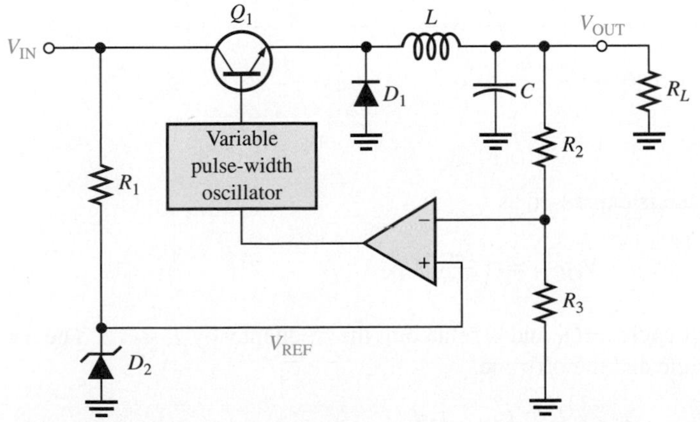

(a) Typical circuit

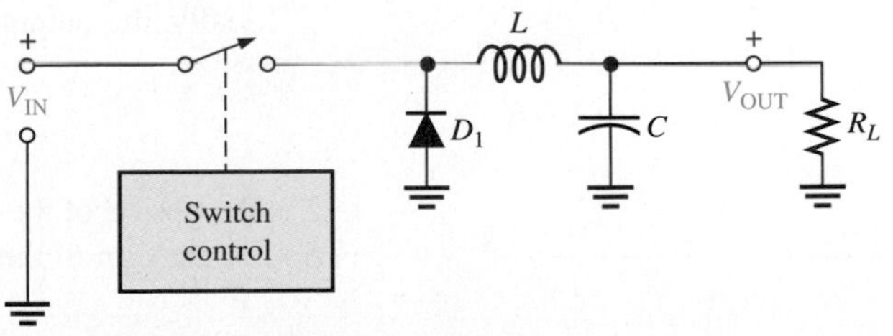

(b) Simplified equivalent circuit

▲ FIGURE 17–16

Basic step-down switching regulator.

a duty cycle that is based on the regulator's load requirement. The LC filter is then used to average the switched voltage. Since Q_1 is either *on* (saturated) or *off,* the power lost in the control element is relatively small. Therefore, the switching regulator is useful primarily in higher power applications or in applications where efficiency is of utmost concern.

The *on* and *off* intervals of Q_1 are shown in the waveform of Figure 17–17(a). The capacitor charges during the on-time (t_{on}) and discharges during the off-time (t_{off}). When the on-time is increased relative to the off-time, the capacitor charges more, thus increasing the output voltage, as indicated in Figure 17–17(b). When the on-time is decreased relative to the off-time, the capacitor discharges more, thus decreasing the output voltage, as in Figure 17–17(c). The inductor further smooths the fluctuations of the output voltage caused by the charging and discharging action.

FIGURE 17–17

Switching regulator waveforms. The V_C waveform is shown for no inductive filtering to illustrate the charge and discharge action (ripple). L and C smooth V_C to a nearly constant level, as indicated by the dashed line for V_{OUT}.

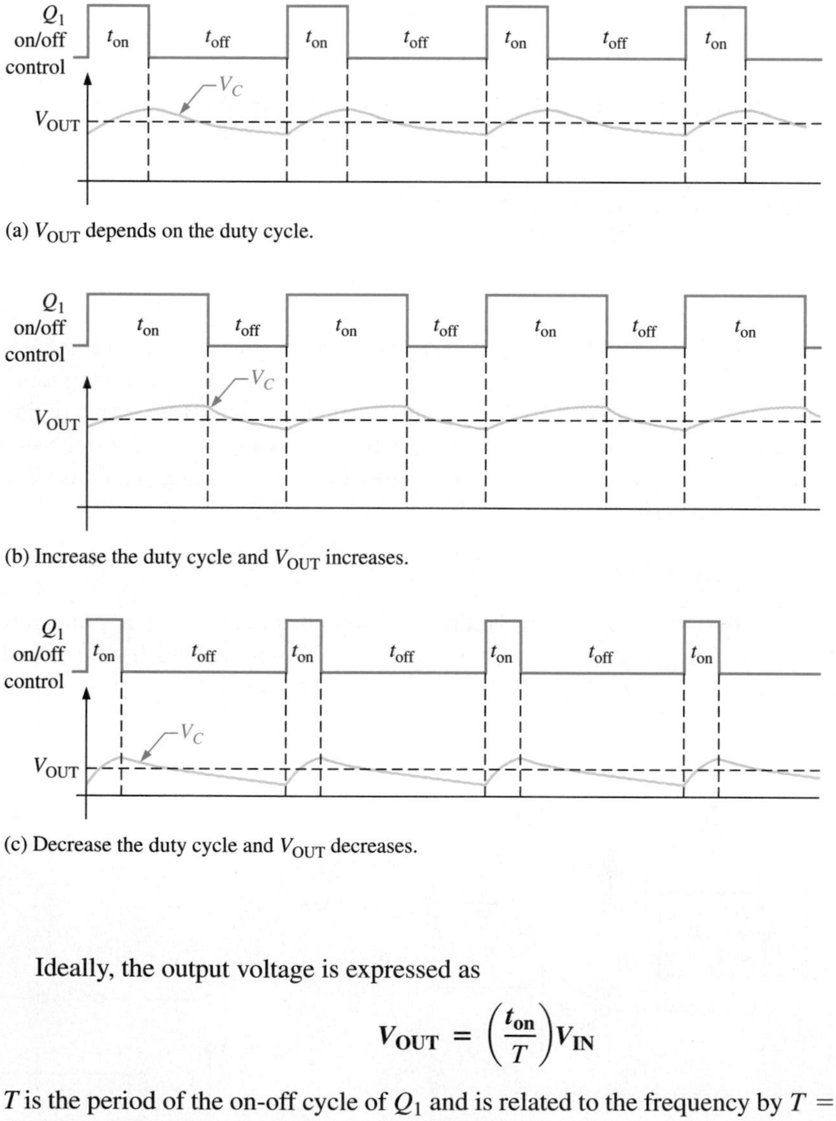

(a) V_{OUT} depends on the duty cycle.

(b) Increase the duty cycle and V_{OUT} increases.

(c) Decrease the duty cycle and V_{OUT} decreases.

Ideally, the output voltage is expressed as

Equation 17–8

$$V_{OUT} = \left(\frac{t_{on}}{T}\right)V_{IN}$$

T is the period of the on-off cycle of Q_1 and is related to the frequency by $T = 1/f$. The period is the sum of the on-time and the off-time.

$$T = t_{on} + t_{off}$$

As you know, the ratio t_{on}/T is called the *duty cycle.*

The regulating action is as follows and is illustrated in Figure 17–18. When V_{OUT} tries to decrease, the on-time of Q_1 is increased, causing an additional charge on C to offset the

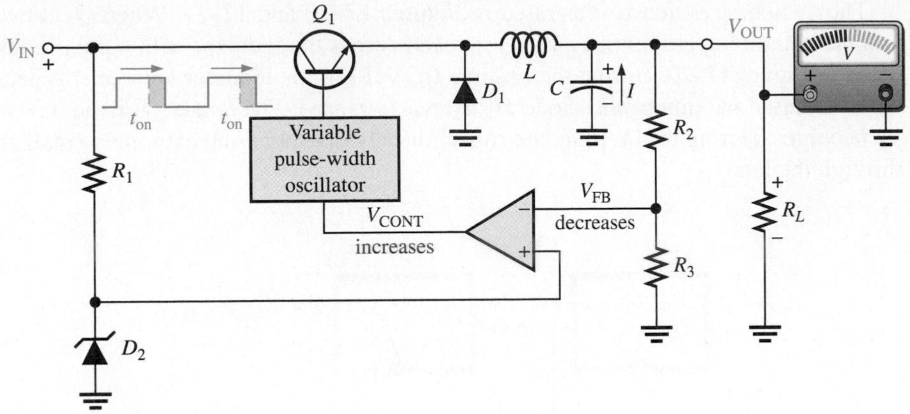

(a) When V_{OUT} attempts to decrease, the on-time of Q_1 increases.

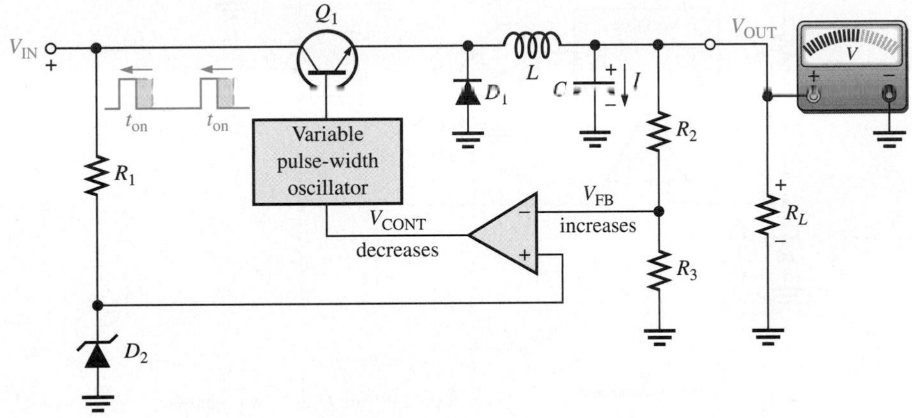

(b) When V_{OUT} attempts to increase, the on-time of Q_1 decreases.

FIGURE 17–18

Basic regulating action of a step-down switching regulator.

attempted decrease. When V_{OUT} tries to increase, the on-time of Q_1 is decreased, causing the capacitor to discharge enough to offset the attempted increase.

Step-Up Configuration

A basic step-up type of switching regulator is shown in Figure 17–19, where transistor Q_1 operates as a switch to ground.

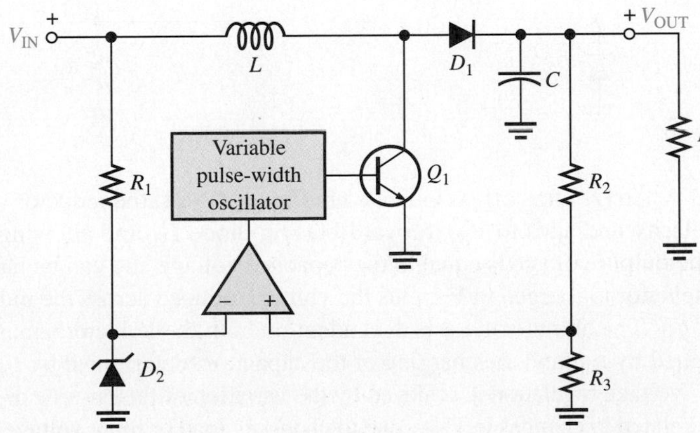

FIGURE 17–19

Basic step-up switching regulator.

The switching action is illustrated in Figures 17–20 and 17–21. When Q_1 turns on, a voltage equal to approximately V_{IN} is induced across the inductor with a polarity as indicated in Figure 17–20. During the on-time (t_{on}) of Q_1, the inductor voltage, V_L, decreases from its initial maximum and diode D_1 is reverse-biased. The longer Q_1 is on, the smaller V_L becomes. During the on-time, the capacitor only discharges an extremely small amount through the load.

FIGURE 17–20

Basic action of a step-up regulator when Q_1 is on.

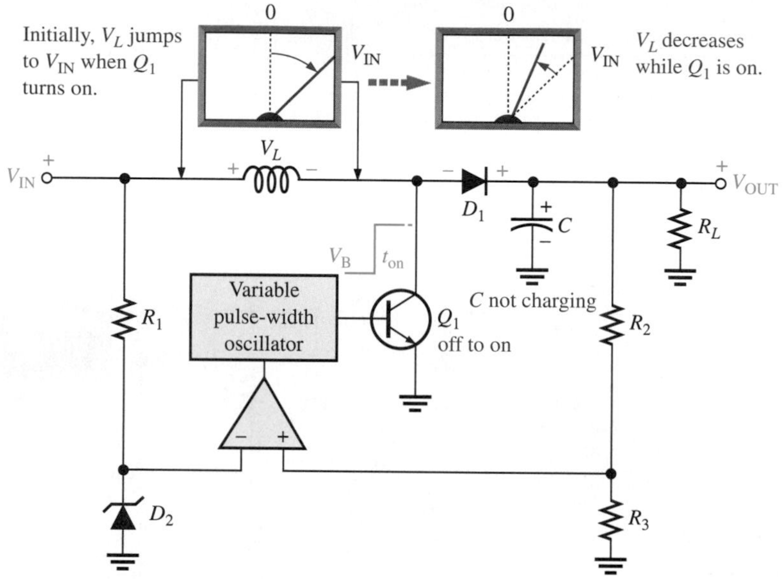

FIGURE 17–21

Basic switching action of a step-up regulator when Q_1 turns off.

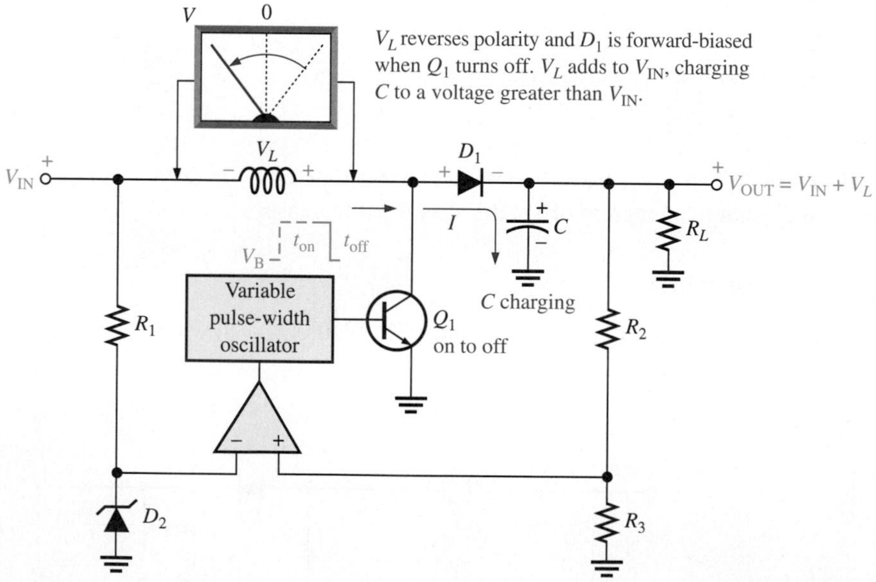

When Q_1 turns off, as indicated in Figure 17–21, the inductor voltage suddenly reverses polarity and adds to V_{IN}, forward-biasing diode D_1 and allowing the capacitor to charge. The output voltage is equal to the capacitor voltage and can be larger than V_{IN} because the capacitor is charged to V_{IN} plus the voltage induced across the inductor during the off-time of Q_1. The output voltage is dependent on both the inductor's magnetic field action (determined by t_{on}) and the charging of the capacitor (determined by t_{off}).

Voltage regulation is achieved by the variation of the on-time of Q_1 (within certain limits) as related to changes in V_{OUT} due to changing load or input voltage. If V_{OUT} tries to increase, the on-time of Q_1 will decrease, resulting in a decrease in the amount that C will charge. If

V_{OUT} tries to decrease, the on-time of Q_1 will increase, resulting in an increase in the amount that C will charge. This regulating action maintains V_{OUT} at an essentially constant level.

Voltage-Inverter Configuration

A third type of switching regulator produces an output voltage that is opposite in polarity to the input. A basic diagram is shown in Figure 17–22.

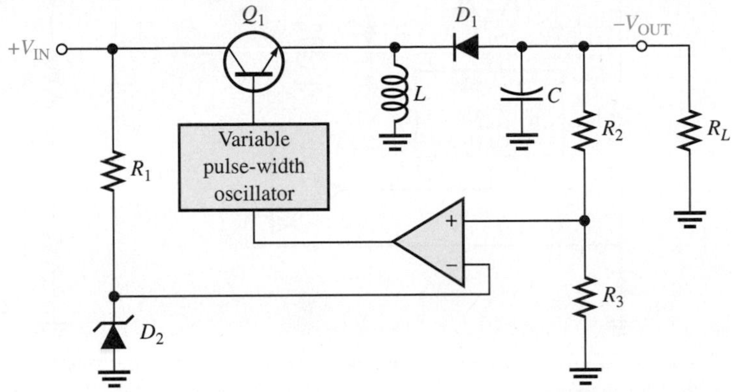

FIGURE 17–22
Basic inverting switching regulator.

When Q_1 turns on, the inductor voltage jumps to approximately $V_{IN} - V_{CE(sat)}$ and the magnetic field rapidly expands, as shown in Figure 17–23(a). While Q_1 is on, the diode is reverse-biased and the inductor voltage decreases from its initial maximum. When Q_1 turns off, the magnetic field collapses and the inductor's polarity reverses, as shown in Figure 17–23(b). This forward-biases the diode, charges C, and produces a negative output voltage, as indicated. The repetitive on-off action of Q_1 produces a repetitive charging and discharging that is smoothed by the LC filter action.

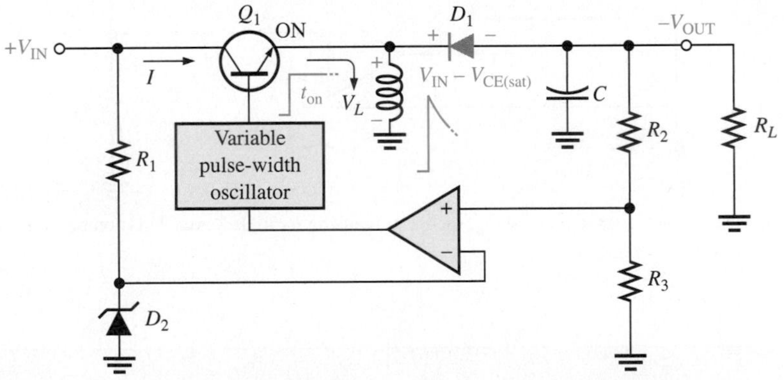

(a) When Q_1 is on, D_1 is reverse-biased.

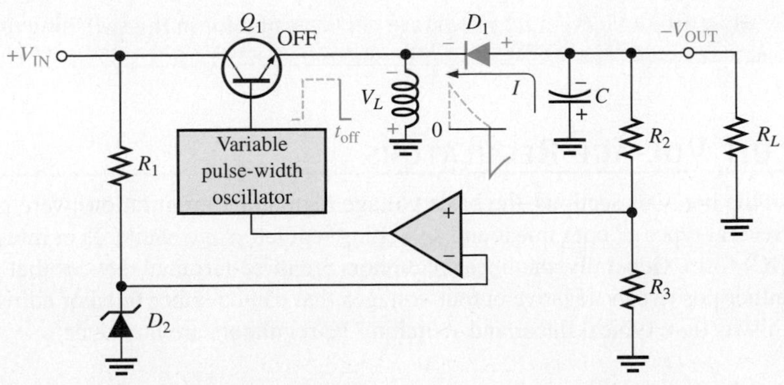

(b) When Q_1 turns off, D_1 forward biases.

FIGURE 17–23
Basic inverting action of an inverting switching regulator.

As with the step-up regulator, the less time Q_1 is on, the greater the output voltage is, and vice versa. This regulating action is illustrated in Figure 17–24.

▶ FIGURE 17–24

Basic regulating action of an inverting switching regulator.

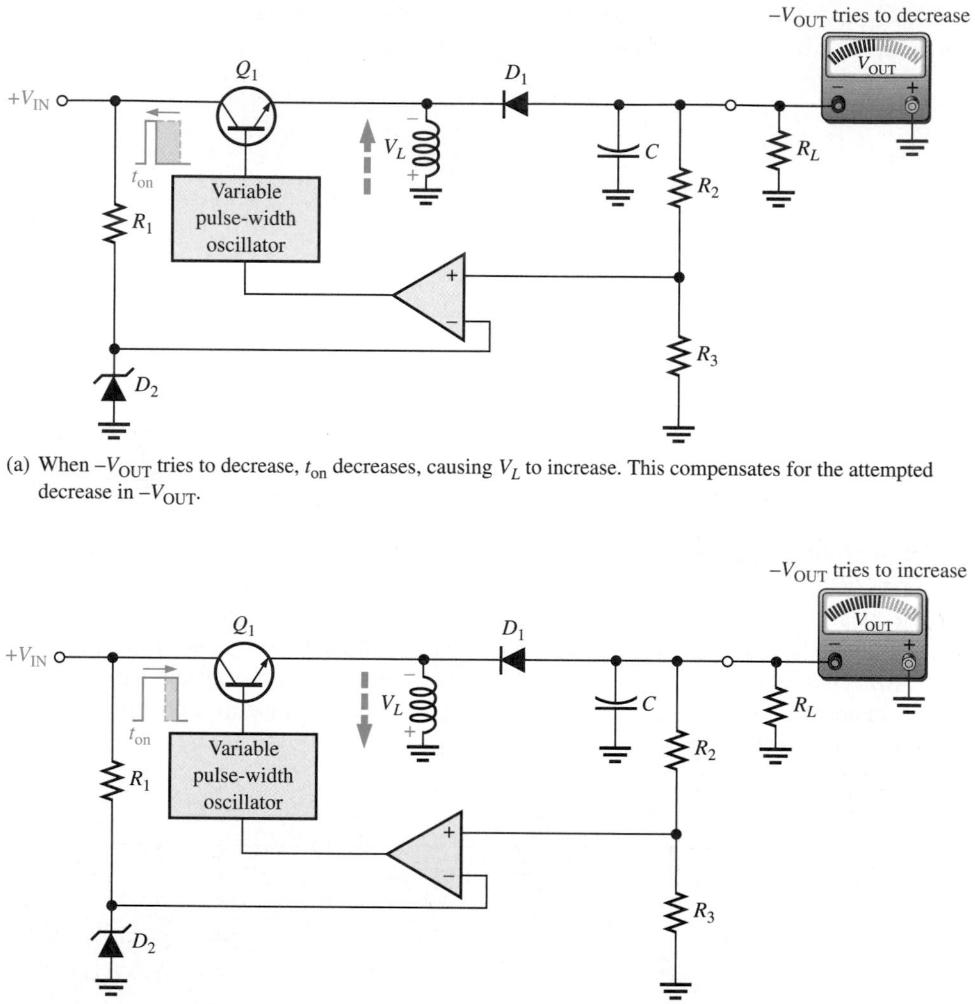

(a) When $-V_{OUT}$ tries to decrease, t_{on} decreases, causing V_L to increase. This compensates for the attempted decrease in $-V_{OUT}$.

(b) When $-V_{OUT}$ tries to increase, t_{on} increases, causing V_L to decrease. This compensates for the attempted increase in $-V_{OUT}$.

SECTION 17-4 CHECKUP

1. What are three types of switching regulators?
2. What is the primary advantage of switching regulators over linear regulators?
3. How are changes in output voltage compensated for in the switching regulator?

17–5 INTEGRATED CIRCUIT VOLTAGE REGULATORS

In the previous sections, the basic voltage regulator configurations were presented. Several types of both linear and switching regulators are available in integrated circuit (IC) form. Generally, the linear regulators are three-terminal devices that provide either positive or negative output voltages that can be either fixed or adjustable. In this section, typical linear and switching IC regulators are introduced.

After completing this section, you should be able to

- **Discuss integrated circuit voltage regulators**
  - Describe the 78XX series of positive regulators
  - Describe the 79XX series of negative regulators
  - Describe the LM317 adjustable positive regulator
  - Describe the LM337 adjustable negative regulator
  - Describe IC switching regulators

Fixed Positive Linear Voltage Regulators

Although many types of IC regulators are available, the 78XX series of IC regulators is representative of three-terminal devices that provide a fixed positive output voltage. The three terminals are input, output, and ground as indicated in the standard fixed voltage configuration in Figure 17–25(a). The last two digits in the part number designate the output voltage. For example, the 7805 is a +5.0 V regulator. For any given regulator, the output voltage can be as much as ±4% of the nominal output. Thus, a 7805 may have an output from 4.8 V to 5.2 V but will remain constant in that range. Other available output voltages are given in Figure 17–25(b) and common packages are shown in part (c).

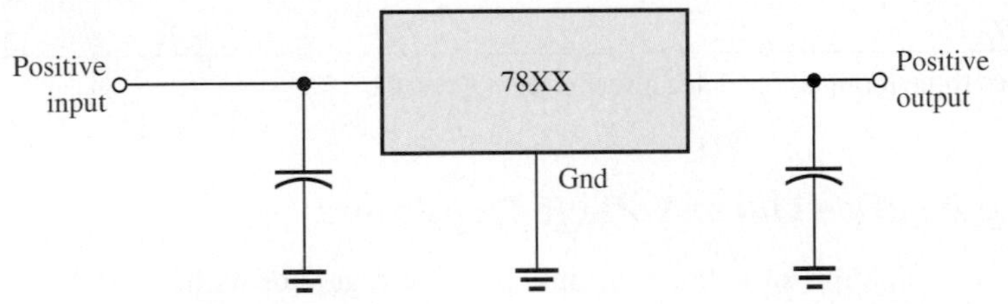

(a) Standard configuration

Type number	Output voltage
7805	+5.0 V
7806	+6.0 V
7808	+8.0 V
7809	+9.0 V
7812	+12.0 V
7815	+15.0 V
7818	+18.0 V
7824	+24.0 V

(b) The 78XX series

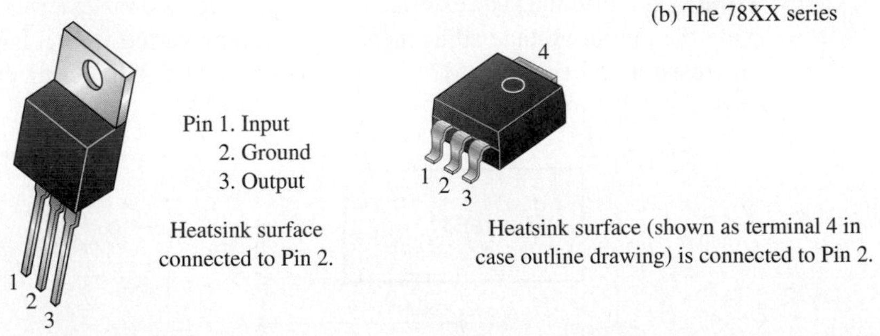

(c) Typical packages

▲ **FIGURE 17–25**

The 78XX series three-terminal fixed positive voltage regulators.

Capacitors, although not always necessary, are sometimes used on the input and output as indicated in Figure 17–25(a). The output capacitor acts basically as a line filter to improve transient response. The input capacitor filters the input and prevents unwanted oscillations when the regulator is some distance from the power supply filter such that the line has a significant inductance.

The 78XX series can produce output currents up to in excess of 1 A when used with an adequate heat sink. The input voltage must be approximately 2.5 V above the output voltage in order to maintain regulation. The circuits have internal thermal overload protection and short-circuit current-limiting features. **Thermal overload** occurs when the internal power dissipation becomes excessive and the temperature of the device exceeds a certain value. Almost all applications of regulators require that the device be secured to a heat sink to prevent thermal overload.

Fixed Negative Linear Voltage Regulators

The 79XX series is typical of three-terminal IC regulators that provide a fixed negative output voltage. This series is the negative-voltage counterpart of the 78XX series and shares most of the same features and characteristics except the pin numbers are different than the positive regulators. Figure 17–26 indicates the standard configuration and part numbers with corresponding output voltages that are available.

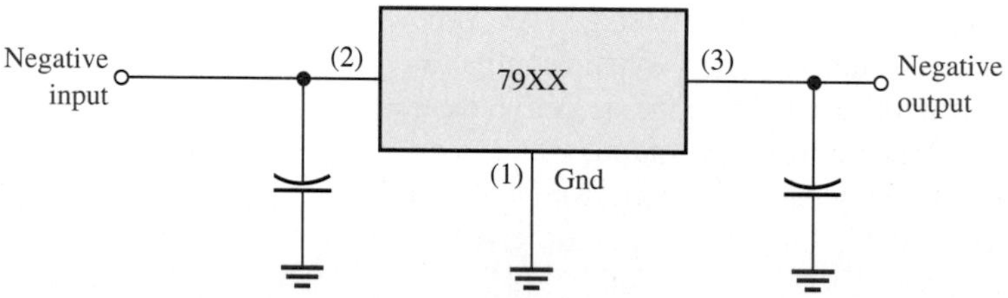

(a) Standard configuration

Type number	Output voltage
7905	−5.0 V
7905.2	−5.2 V
7906	−6.0 V
7908	−8.0 V
7912	−12.0 V
7915	−15.0 V
7918	−18.0 V
7924	−24.0 V

(b) The 79XX series

▲ **FIGURE 17–26**

The 79XX series three-terminal fixed negative voltage regulators.

Adjustable Positive Linear Voltage Regulators

The LM317 is an example of a three-terminal positive regulator with an adjustable output voltage. The standard configuration is shown in Figure 17–27. The capacitors are for decoupling and do not affect the dc operation. Notice that there is an input, an output, and an adjustment terminal. The external fixed resistor R_1 and the external variable resistor R_2 provide the output voltage adjustment. V_{OUT} can be varied from 1.2 V to 37 V depending on the resistor values. The LM317 can provide over 1.5 A of output current to a load.

▶ **FIGURE 17–27**

The LM317 three-terminal adjustable positive voltage regulator.

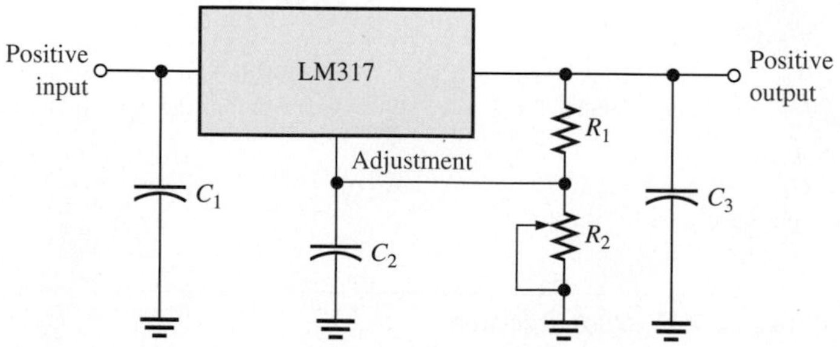

The LM317 is operated as a "floating" regulator because the adjustment terminal is not connected to ground, but floats to whatever voltage is across R_2. This allows the output voltage to be much higher than that of a fixed-voltage regulator.

Basic Operation As indicated in Figure 17–28, a constant 1.25 V reference voltage (V_{REF}) is maintained by the regulator between the output terminal and the adjustment terminal. This

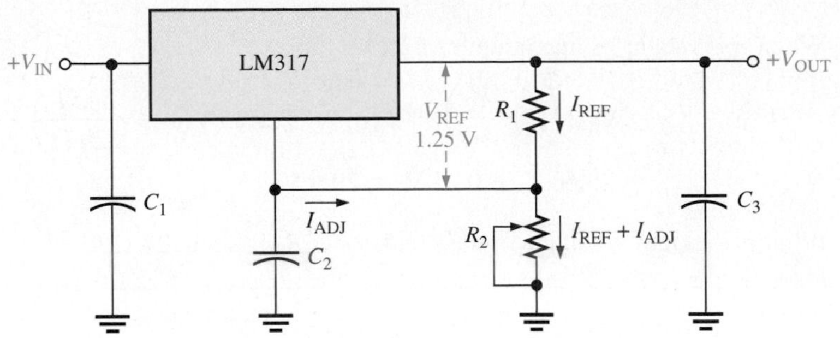

FIGURE 17–28
Operation of the LM317 adjustable positive voltage regulator.

constant reference voltage produces a constant current (I_{REF}) through R_1, regardless of the value of R_2. I_{REF} is also through R_2.

$$I_{REF} = \frac{V_{REF}}{R_1} = \frac{1.25\text{ V}}{R_1}$$

There is a very small constant current at the adjustment terminal of approximately 50 μA called I_{ADJ}, which is through R_2. A formula for the output voltage is developed as follows.

$$V_{OUT} = V_{R1} + V_{R2} = I_{REF}R_1 + I_{REF}R_2 + I_{ADJ}R_2$$

$$= I_{REF}(R_1 + R_2) + I_{ADJ}R_2 = \frac{V_{REF}}{R_1}(R_1 + R_2) + I_{ADJ}R_2$$

$$\mathbf{V_{OUT} = V_{REF}\left(1 + \frac{R_2}{R_1}\right) + I_{ADJ}R_2}$$ Equation 17–9

As you can see, the output voltage is a function of both R_1 and R_2. Once the value of R_1 is set, the output voltage is adjusted by varying R_2.

EXAMPLE 17–6

Determine the minimum and maximum output voltages for the voltage regulator in Figure 17–29. Assume $I_{ADJ} = 50\ \mu$A.

FIGURE 17–29

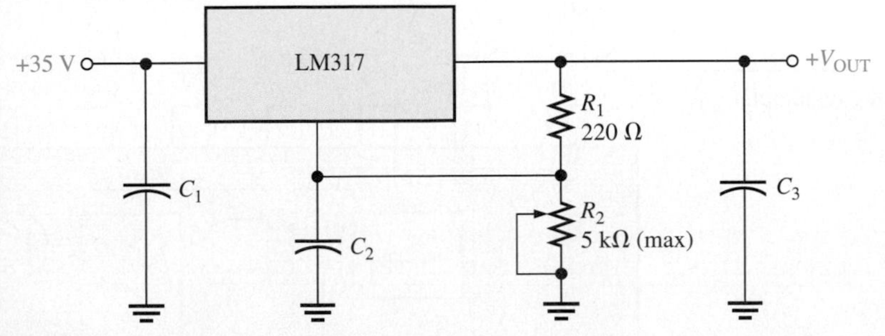

Solution

$$V_{R1} = V_{REF} = 1.25\text{ V}$$

When R_2 is set at its minimum of 0 Ω,

$$V_{OUT(min)} = V_{REF}\left(1 + \frac{R_2}{R_1}\right) + I_{ADJ}R_2 = 1.25\text{ V}(1) = \mathbf{1.25\text{ V}}$$

When R_2 is set at its maximum of 5 kΩ,

$$V_{OUT(max)} = V_{REF}\left(1 + \frac{R_2}{R_1}\right) + I_{ADJ}R_2 = 1.25\text{ V}\left(1 + \frac{5\text{ k}\Omega}{220\ \Omega}\right) + (50\ \mu\text{A})5\text{ k}\Omega$$

$$= 29.66\text{ V} + 0.25\text{ V} = \mathbf{29.9\ V}$$

Related Problem What is the output voltage of the regulator if R_2 is set at 2 kΩ?

Adjustable Negative Linear Voltage Regulators

The LM337 is the negative output counterpart of the LM317 and is a good example of this type of IC regulator. Like the LM317, the LM337 requires two external resistors for output voltage adjustment as shown in Figure 17–30. The output voltage can be adjusted from −1.2 V to −37 V, depending on the external resistor values. The capacitors are for decoupling and do not affect the dc operation.

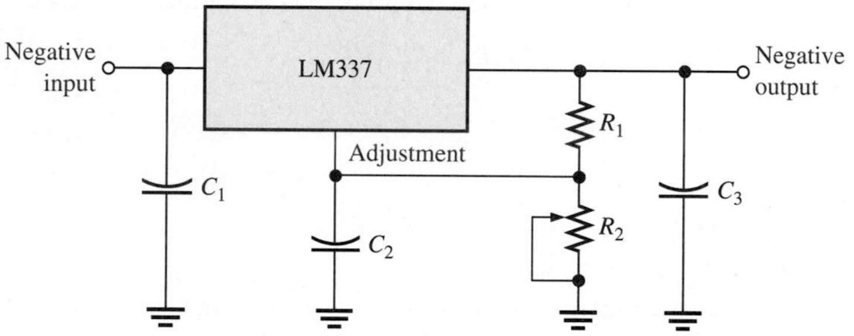

FIGURE 17–30

The LM337 three-terminal adjustable negative voltage regulator.

Switching Voltage Regulators

As an example of an IC switching voltage regulator, let's look at the 78S40. This is a universal device that can be used with external components to provide step-up, step-down, and inverting operation.

The internal circuitry of the 78S40 is shown in Figure 17–31. This circuit can be compared to the basic switching regulators that were covered in Section 17–4. For example,

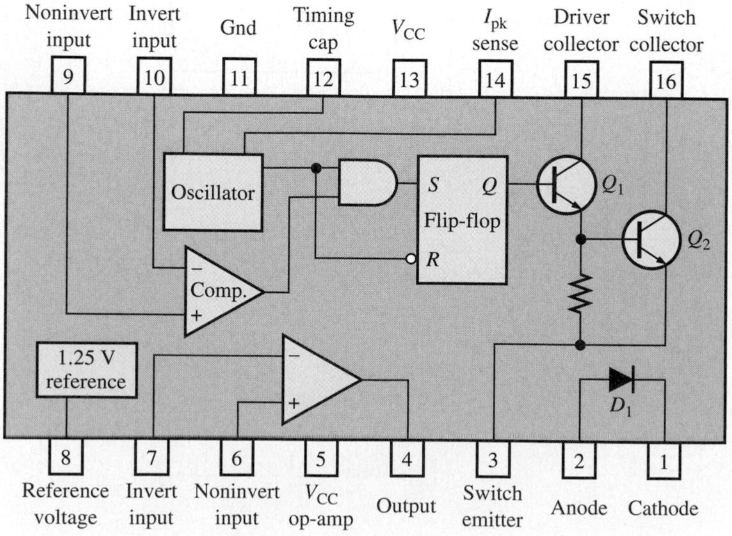

FIGURE 17–31

The 78S40 switching regulator.

look back at Figure 17–16(a). The oscillator and comparator functions are directly comparable. The gate and flip-flop in the 78S40 were not included in the basic circuit of Figure 17–16(a), but they provide additional regulating action. Transistors Q_1 and Q_2 effectively perform the same function as Q_1 in the basic circuit. The 1.25 V reference block in the 78S40 has the same purpose as the zener diode in the basic circuit, and diode D_1 in the 78S40 corresponds to D_1 in the basic circuit.

The 78S40 also has an "uncommited" op-amp thrown in for good measure. It is not used in any of the regulator configurations. External circuitry is required to make this device operate as a regulator, as you will see in Section 17–6.

SECTION 17–5 CHECKUP

1. What are the three terminals of a fixed-voltage regulator?
2. What is the output voltage of a 7809? Of a 7915?
3. What are the three terminals of an adjustable-voltage regulator?
4. What external components are required for a basic LM317 configuration?

17–6 APPLICATIONS OF IC VOLTAGE REGULATORS

In the last section, you saw several devices that are representative of the general types of IC voltage regulators. Now, several different ways these devices can be modified with external circuitry to improve or alter their performance are examined.

After completing this section, you should be able to

- **Discuss applications of IC voltage regulators**
  - Explain the use of an external pass transistor
  - Explain the use of current limiting
  - Explain how to use a voltage regulator as a constant-current source
  - Discuss some application considerations for switching regulators

The External Pass Transistor

As you know, an IC voltage regulator is capable of delivering only a certain amount of output current to a load. For example, the 78XX series regulators can handle a peak output current of 1.3 A (more under certain conditions). If the load current exceeds the maximum allowable value, there will be thermal overload and the regulator will shut down. A thermal overload condition means that there is excessive power dissipation inside the device.

If an application requires more than the maximum current that the regulator can deliver, an external pass transistor Q_{ext}, can be used. Figure 17–32 illustrates a three-terminal

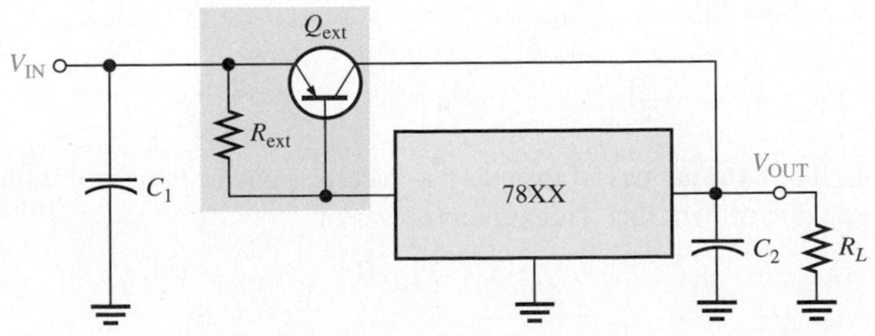

FIGURE 17–32

A 78XX-series three-terminal regulator with an external pass transistor to increase power dissipation.

regulator with an external pass transistor for handling currents in excess of the output current capability of the basic regulator.

The value of the external current-sensing resistor, R_{ext}, determines the value of current at which Q_{ext} begins to conduct because it sets the base-to-emitter voltage of the transistor. As long as the current is less than the value set by R_{ext}, the transistor Q_{ext} is off, and the regulator operates normally as shown in Figure 17–33(a). This is because the voltage drop across R_{ext} is less than the 0.7 V base-to-emitter voltage required to turn Q_{ext} on. R_{ext} is determined by the following formula, where I_{max} is the highest current that the voltage regulator is to handle internally.

Equation 17–10

$$\mathbf{R_{ext} = \frac{0.7\ V}{I_{max}}}$$

When the current is sufficient to produce at least a 0.7 V drop across R_{ext}, the external pass transistor Q_{ext} turns on and conducts any current in excess of I_{max}, as indicated in Figure 17–33(b). Q_{ext} will conduct more or less, depending on the load requirements. For example, if the total load current is 3 A and I_{max} was selected to be 1 A, the external pass transistor will conduct 2 A, which is the excess over the internal voltage regulator current I_{max}.

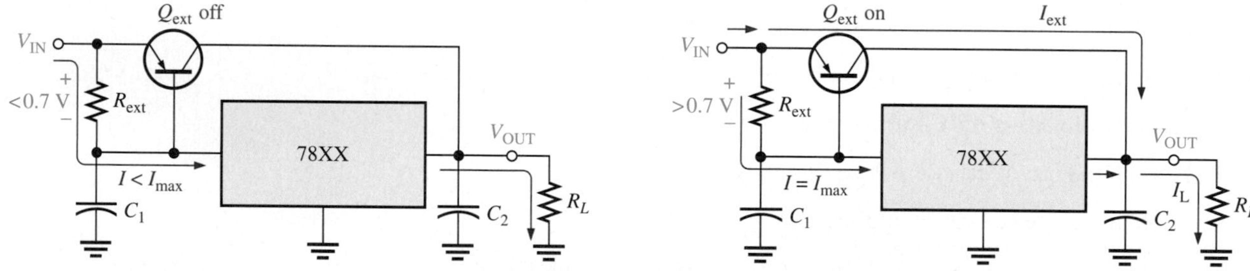

(a) When the regulator current is less than I_{max}, the external pass transistor is off and the regulator is handling all of the current.

(b) When the load current exceeds I_{max}, the drop across R_{ext} turns Q_{ext} on and it conducts the excess current.

▲ **FIGURE 17–33**

Operation of the regulator with an external pass transistor.

EXAMPLE 17–7

What value is R_{ext} if the maximum current to be handled internally by the voltage regulator in Figure 17–32 is set at 700 mA?

Solution

$$R_{ext} = \frac{0.7\ V}{I_{max}} = \frac{0.7\ V}{0.7\ A} = \mathbf{1\ \Omega}$$

Related Problem If R_{ext} is changed to 1.5 Ω, at what current value will Q_{ext} turn on?

The external pass transistor is typically a power transistor with a heat sink that must be capable of handling a maximum power of

$$P_{ext} = I_{ext}(V_{IN} - V_{OUT})$$

EXAMPLE 17–8

What must be the minimum power rating for the external pass transistor used with a 7824 regulator in a circuit such as that shown in Figure 17–32? The input voltage is 30 V and the load resistance is 10 Ω. The maximum internal current is to be 700 mA. Assume that there is no heat sink for this calculation. Keep in mind that the use of a heat sink increases the effective power rating of the transistor and you can use a lower rated transistor.

Solution The load current is

$$I_L = \frac{V_{OUT}}{R_L} = \frac{24\text{ V}}{10\ \Omega} = 2.4\text{ A}$$

The current through Q_{ext} is

$$I_{ext} = I_L - I_{max} = 2.4\text{ A} - 0.7\text{ A} = 1.7\text{ A}$$

The power dissipated by Q_{ext} is

$$P_{ext(min)} = I_{ext}(V_{IN} - V_{OUT}) = (1.7\text{ A})(30\text{ V} - 24\text{ V}) = (1.7\text{ A})(6\text{ V}) = \mathbf{10.2\ W}$$

For a safety margin, choose a power transistor with a rating greater than 10.2 W, say at least 15 W.

Related Problem Rework this example using a 7815 regulator.

Current Limiting

A drawback of the circuit in Figure 17–32 is that the external transistor is not protected from excessive current, such as would result from a shorted output. An additional current-limiting circuit (Q_{lim} and R_{lim}) can be added as shown in Figure 17–34 to protect Q_{ext} from excessive current and possible burn out.

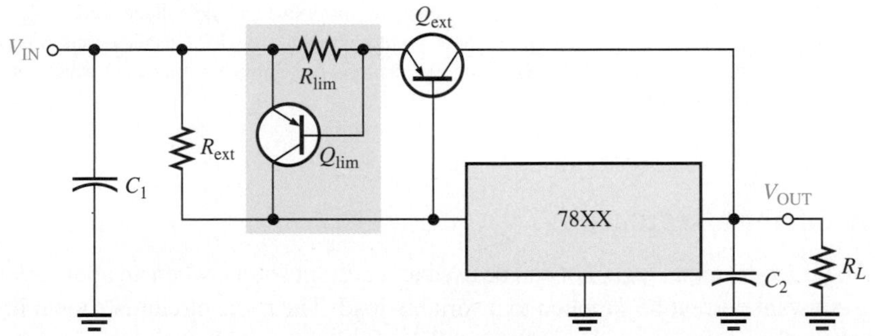

◀ **FIGURE 17–34**

Regulator with current limiting.

The following describes the way the current-limiting circuit works. The current-sensing resistor R_{lim} sets the V_{BE} of transistor Q_{lim}. The base-to-emitter voltage of Q_{ext} is now determined by $V_{R_{ext}} - V_{R_{lim}}$ because they have opposite polarities. So, for normal operation, the drop across R_{ext} must be sufficient to overcome the opposing drop across R_{lim}. If the current through Q_{ext} exceeds a certain maximum ($I_{ext(max)}$) because of a shorted output or a faulty load, the voltage across R_{lim} reaches 0.7 V and turns Q_{lim} on. Q_{lim} now conducts current through the regulator and away from Q_{ext}, forcing a thermal overload to occur and shut down the regulator. Remember, the IC regulator is internally protected from thermal overload as part of its design.

This action is illustrated in Figure 17–35. In part (a), the circuit is operating normally with Q_{ext} conducting less than the maximum current that it can handle with Q_{lim} off. Part (b) shows what happens when there is a short across the load. The current through Q_{ext} suddenly increases and causes the voltage drop across R_{lim} to increase, which turns Q_{lim} on. The current is now diverted through the regulator, which causes it to shut down due to thermal overload.

▶ FIGURE 17–35

The current-limiting action of the regulator circuit.

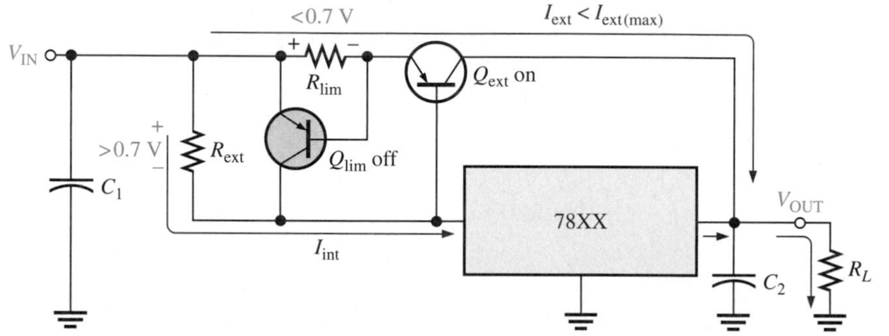

(a) During normal operation, when the load current is not excessive, Q_{lim} is off.

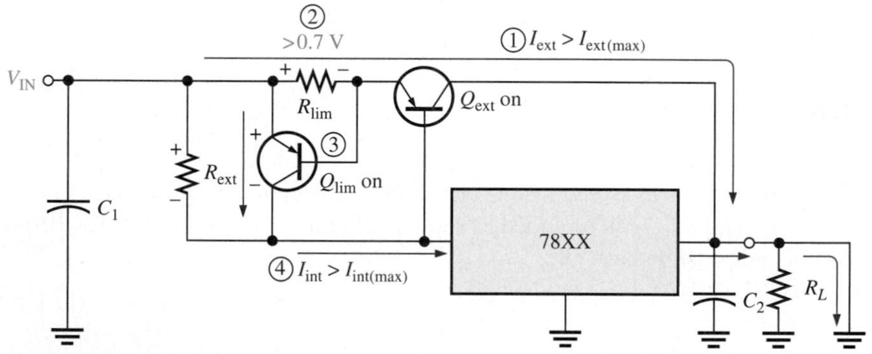

(b) When short occurs ①, the external current becomes excessive and the voltage across R_{lim} increases ② and turns on Q_{lim} ③, which then routes current through the regulator and conducts it away from Q_{ext}, causing the internal regulator current to become excessive ④ which forces the regulator into thermal shut down.

A Current Regulator

The three-terminal regulator can be used as a current source when an application requires that a constant current be supplied to a variable load. The basic circuit is shown in Figure 17–36 where R_1 is the current-setting resistor. The regulator provides a fixed constant voltage, V_{OUT},

▶ FIGURE 17–36

The three-terminal regulator as a current source.

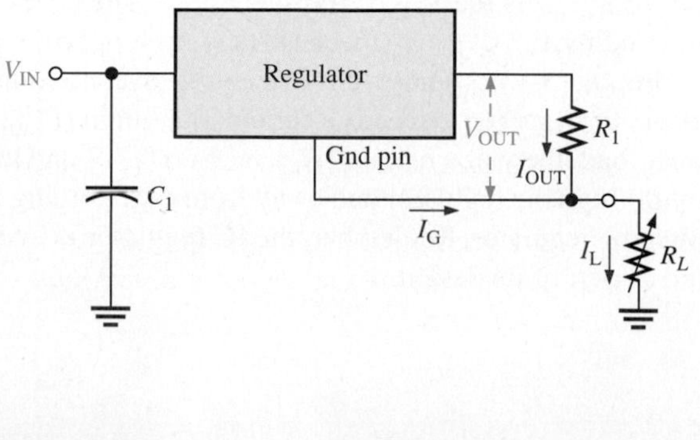

between the ground terminal (not connected to ground in this case) and the output terminal. This determines the constant current supplied to the load.

$$I_L = \frac{V_{OUT}}{R_1} + I_G$$

Equation 17–11

The current, I_G, from the ground pin is very small compared to the output current and can often be neglected.

EXAMPLE 17–9

What value of R_1 is necessary in a 7805 regulator to provide a constant current of 0.5 A to a variable load that can be adjusted from 1 Ω to 10 Ω?

Solution The 7805 produces 5 V between its ground terminal and its output terminal. Therefore, if you want 0.5 A of current, the current-setting resistor must be (neglecting I_G)

$$R_1 = \frac{V_{OUT}}{I_L} = \frac{5\text{ V}}{0.5\text{ A}} = \mathbf{10\ \Omega}$$

The circuit is shown in Figure 17–37.

FIGURE 17–37

A constant-current source of 0.5 A.

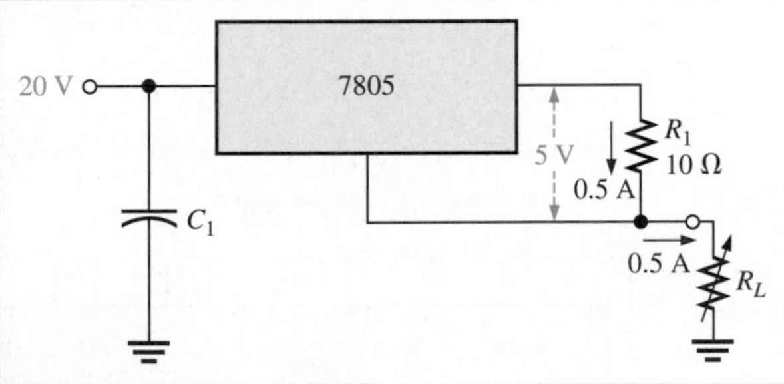

Related Problem If a 7808 regulator is used instead of the 7805, to what value would you change R_1 to maintain a constant current of 0.5 A?

Switching Regulator Configurations

In Section 17–5, the 78S40 was introduced as an example of an IC switching voltage regulator. Figure 17–38 shows the external connections for a step-down configuration where the output voltage is less than the input voltage, and Figure 17–39 shows a step-up configuration in which the output voltage is greater than the input voltage. An inverting configuration is also possible, but it is not shown here.

The timing capacitor, C_T, controls the pulse width and frequency of the oscillator and thus establishes the on-time of transistor Q_2. The voltage across the current-sensing resistor, R_{CS}, is used internally by the oscillator to vary the duty cycle based on the desired peak load current. The voltage divider, made up of R_1 and R_2, reduces the output voltage to a nominal value equal to the reference voltage. If V_{OUT} exceeds its set value, the output of the comparator switches to its low state, disabling the gate to turn Q_2 off until the output decreases. This regulating action is in addition to that produced by the duty cycle variation of the oscillator as described in Section 17–4 in relation to the basic switching regulator.

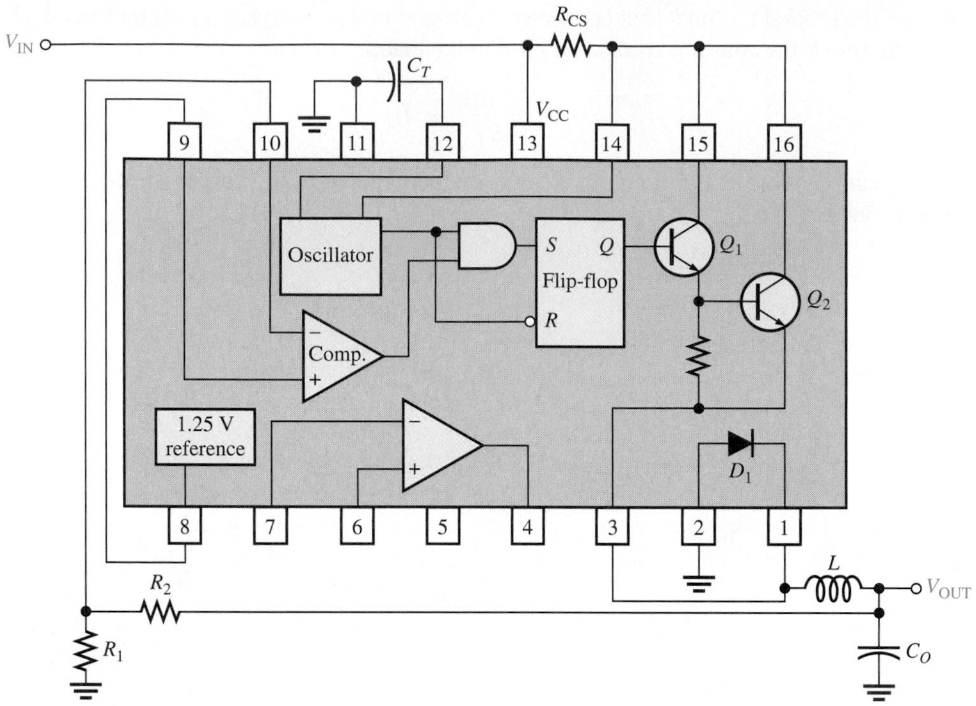

▶ FIGURE 17–38

The step-down configuration of the 78S40 switching regulator.

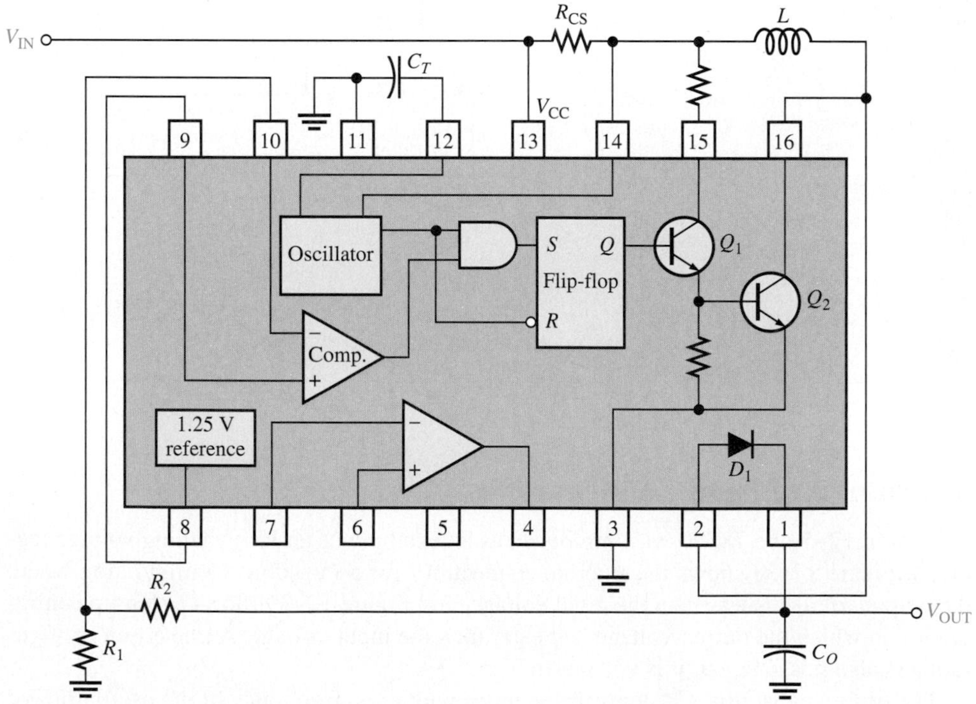

▶ FIGURE 17–39

The step-up configuration of the 78S40 switching regulator.

SECTION 17–6 CHECKUP

1. What is the purpose of using an external pass transistor with an IC voltage regulator?
2. What is the advantage of current limiting in a voltage regulator?
3. What does *thermal overload* mean?

On the Job

Application Activity: *Variable DC Power Supply*

A regulated power supply with a fixed output voltage of +12 V was developed in Chapter 3. The company that manufactures this power supply plans to offer a new line of variable power supplies for which a specified voltage can be preset at the factory or can be adjusted by the user. In this application, the power supply with a variable regulator is developed to provide an output voltage from +9 V to +30 V and a maximum load current of 250 mA.

The Circuit

Recall that in the original power supply, a 7812 provided a +12 V regulated output. In this new power supply, a 7809 is used to produce that variable output voltage. As in the earlier design, it is recommended by the manufacturer that a 0.33 μF capacitor be connected from the input terminal to ground and a 0.1 μF capacitor be connected from the output terminal to ground, as shown in Figure 17–40, to prevent high frequency oscillations and improve the performance. The reason for a small-value capacitor in parallel with a large one is that the large filter capacitor has an internal equivalent series resistance, which affects the high frequency response of the system. The effect is cancelled with the small capacitor.

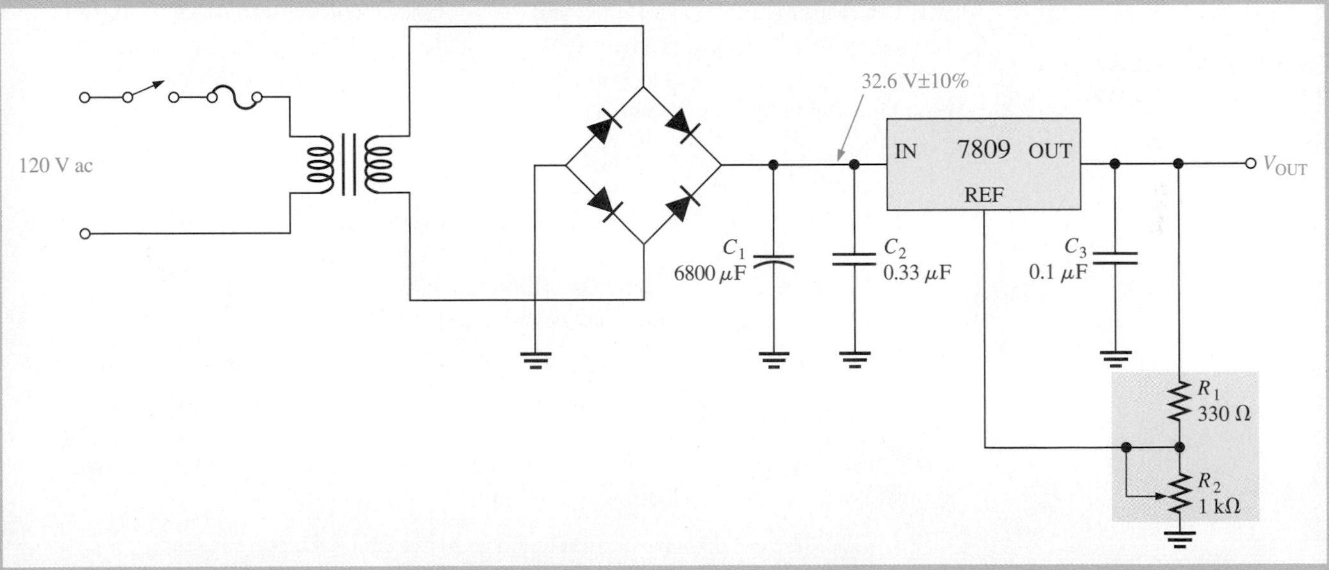

FIGURE 17–40

Variable output power supply.

The Transformer The transformer must convert the 120 V rms line voltage to an ac voltage that will result in a rectified voltage that will produce 34 V ± 10% when filtered.

The Voltage Regulator A partial datasheet for a 7809 is shown in Figure 17–41. Notice that there is a range of nominal output voltages, but it is typically 9.0 V. The line and load regulations specify how much the output can vary about the nominal output value. For example, the typical 9.0 V output will change no more than 12 mV (typical) as the load current changes from 5 mA to 1.5 A. The output voltage of the regulator is the voltage

Electrical Characteristics (LM7809) (Continued)
Refer to the test circuits. -40°C < T_J < 125°C, I_O = 500mA, V_I = 15V, C_I = 0.33μF, C_O = 0.1μF, unless otherwise specified.

Symbol	Parameter	Conditions		Min.	Typ.	Max.	Unit
V_O	Output Voltage	T_J = +25°C		8.65	9.0	9.35	V
		5mA ≤ I_O ≤ 1A, P_O ≤ 15W, V_I = 11.5V to 24V		8.6	9.0	9.4	
Regline	Line Regulation(7)	T_J = +25°C	V_I = 11.5V to 25V	–	6.0	180	mV
			V_I = 12V to 17V	–	2.0	90.0	
Regload	Load Regulation(7)	T_J = +25°C	I_O = 5mA to 1.5A	–	12.0	180	mV
			I_O = 250mA to 750mA	–	4.0	90.0	
I_Q	Quiescent Current	T_J = +25°C		–	5.0	8.0	mA
ΔI_Q	Quiescent Current Change	I_O = 5mA to 1A		–	–	0.5	mA
		V_I = 11.5V to 26V		–	–	1.3	
$\Delta V_O/\Delta T$	Output Voltage Drift(8)	I_O = 5mA		–	-1.0	–	mV/°C
V_N	Output Noise Voltage	f = 10Hz to 100kHz, T_A = +25°C		–	58.0	–	μV/V_O
RR	Ripple Rejection(8)	f = 120Hz, V_O = 13V to 23V		56.0	71.0	–	dB
V_{DROP}	Dropout Voltage	I_O = 1A, T_J = +25°C		–	2.0	–	V
r_O	Output Resistance(8)	f = 1kHz		–	17.0	–	mΩ
I_{SC}	Short Circuit Current	V_I = 35V, T_A = +25°C		–	250	–	mA
I_{PK}	Peak Current(8)	T_J = +25°C		–	2.2	–	A

FIGURE 17–41

Partial datasheet for a 7809 regulator. © 2003 Fairchild Semiconductor Corporation. Used by permission.

between the output (OUT) terminal and the reference (REF) terminal. The voltage divider formed by R_1 and R_2 provides a reference voltage other than ground and increases the output voltage with respect to ground above the 9 V nominal regulator output by an amount equal to the voltage across R_2.

1. What are the minimum and maximum nominal output voltages specified on the datasheet when I_O is 500 mA.
2. From the datasheet, determine the maximum change in the output voltage when the load current changes from 5 mA to 1.5 A.
3. Calculate the maximum power dissipation in R_1.
4. Calculate the maximum power dissipation in R_2.

The Fuse The fuse will be in series with the primary winding of the transformer, as shown in Figure 17–40. The fuse should be calculated based on the maximum allowable primary current. Recall from your dc/ac circuits course that if the voltage is stepped down, the current is stepped up. From the specifications for the unregulated power supply, the maximum load current is 100 mA.

5. Calculate the primary current and use this value to select a fuse rating for the circuit in Figure 17–40.

Simulation

Multisim is used to simulate this power supply circuit. Figure 17–42 shows the simulated regulated power supply circuit adjusted to show that it meets or exceeds the specified minimum and maximum output voltages.

Build and simulate the circuit using your Multisim software. Verify the operation.

Prototyping and Testing

Now that all the components have been selected and the circuit has been simulated, the circuit is breadboarded and tested.

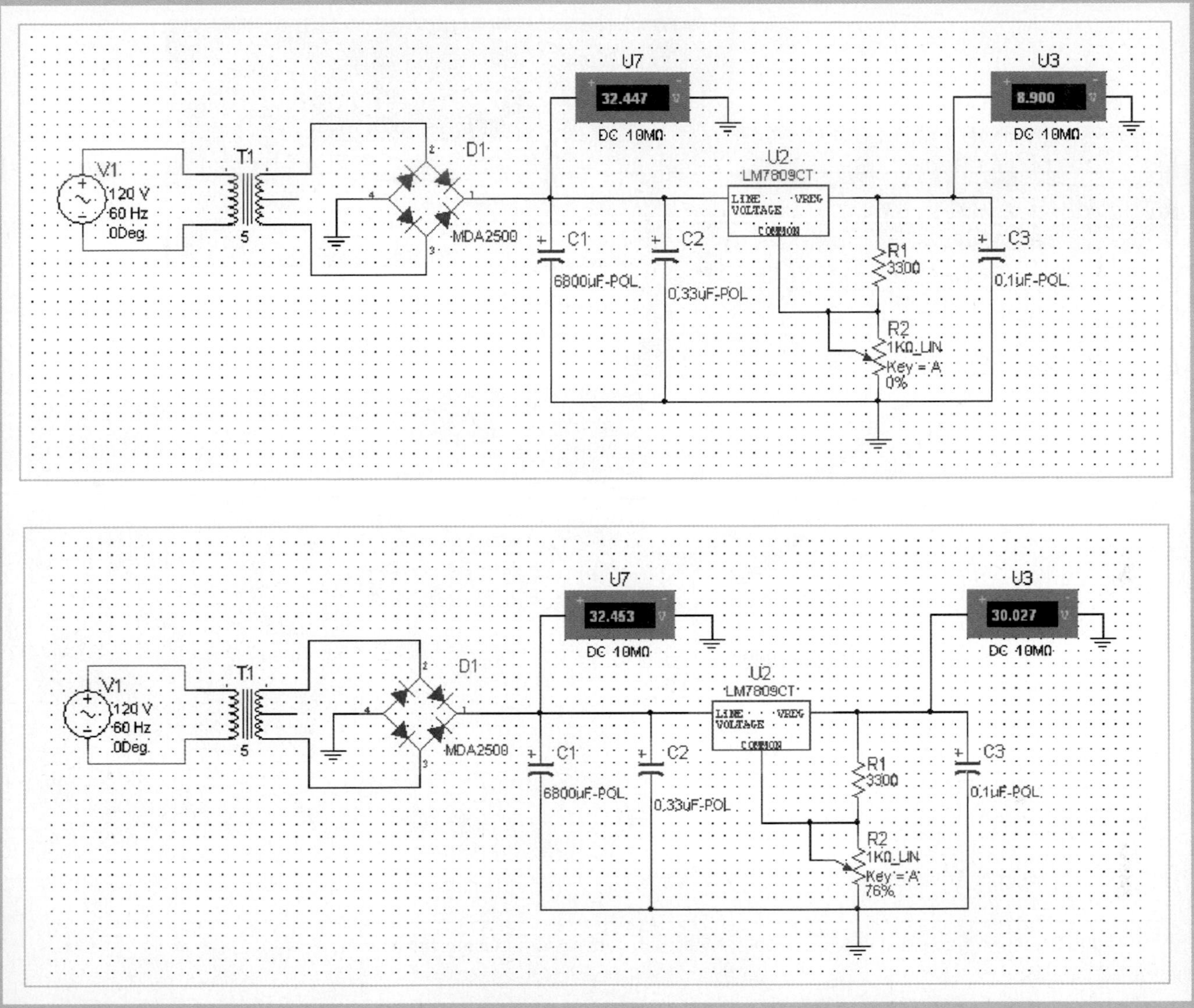

FIGURE 17–42

Simulation of the regulated variable power supply circuit at the minimum and maximum specified output voltages.

Lab Experiment

To build and test a similar circuit, go to Experiment 17 in your lab manual (*Laboratory Exercises for Electronic Devices* by David Buchla and Steven Wetterling).

Printed Circuit Board

The variable regulated power supply prototype has been built and tested. It is now committed to a printed circuit layout, as shown in Figure 17–43. Notice that a heat sink is used with the regulator IC to increase its ability to dissipate power. The output voltage is measured at the potentiometer.

6. Compare the printed circuit board to the schematic in Figure 17–40.
7. Calculate the power dissipated by the regulator for an output of 9 V and $I_L = 100$ mA.
8. Calculate the power dissipated by the regulator for an output of 30 V and $I_L = 100$ mA.

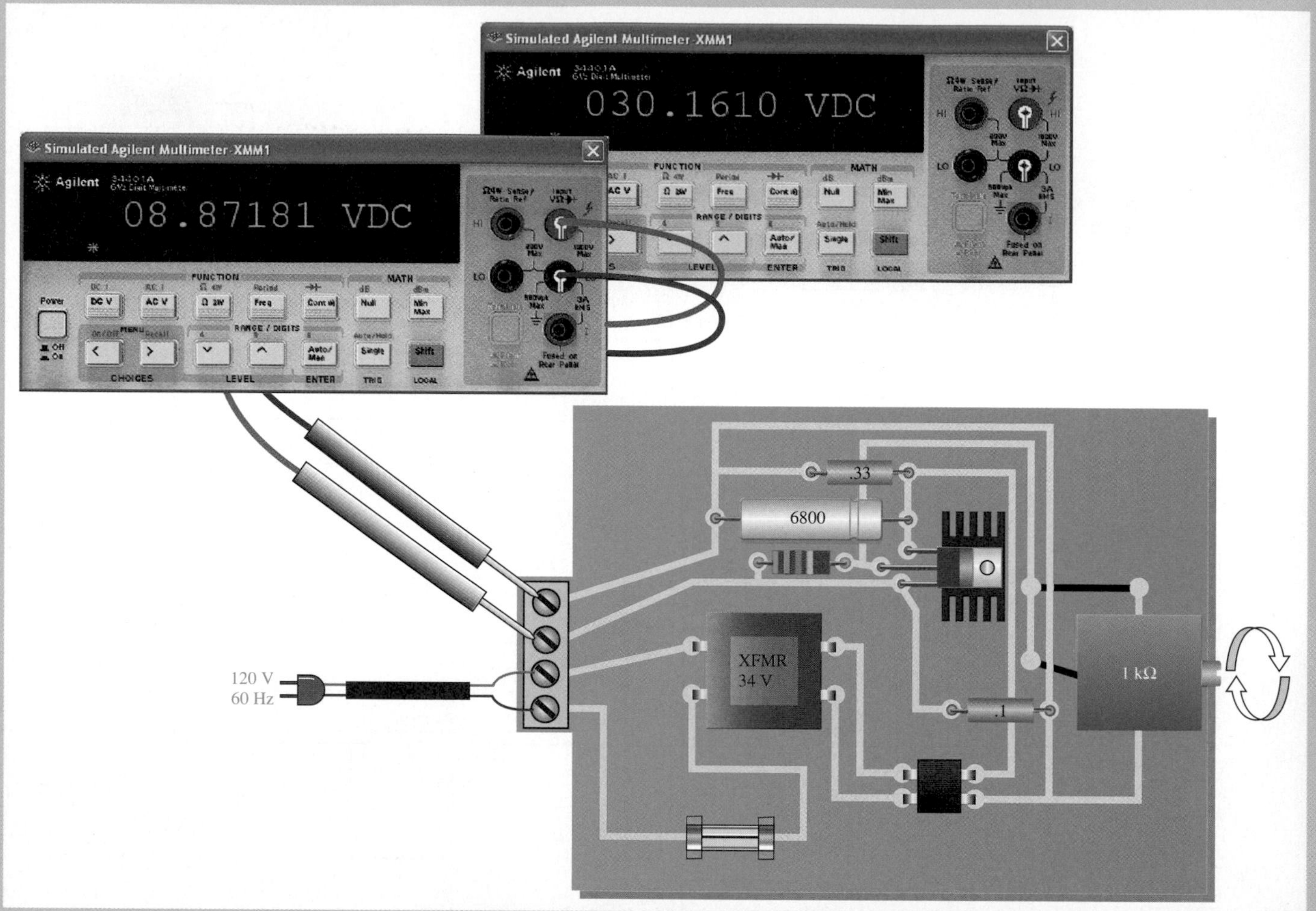

FIGURE 17–43

Regulated power supply PC board adjusted for output voltages that meet the minimum and maximum specifications.

SUMMARY

Section 17–1

- Voltage regulators keep a constant dc output voltage when the input or load varies within limits.
- Line regulation is the percentage change in the output voltage for a given change in the input voltage of a regulator.
- Load regulation is the percentage change in output voltage for a given change in load current.

Section 17–2

- A basic voltage regulator consists of a reference voltage source, an error detector, a sampling element, and a control device. Protection circuitry is also found in most regulators.
- Two basic categories of voltage regulators are linear and switching.
- Two basic types of linear regulators are series and shunt.
- In a linear series regulator, the control element is a transistor in series with the load.

Section 17–3

- In a linear shunt regulator, the control element is a transistor in parallel with the load.

Section 17–4

- Three configurations for switching regulators are step-down, step-up, and inverting.
- Switching regulators are more efficient than linear regulators and are particularly useful in low-voltage, high-current applications.

Section 17–5

- Three-terminal linear IC regulators are available for either fixed output or variable output voltages of positive or negative polarities.
- The 78XX series are three-terminal IC regulators with fixed positive output voltage.
- The 79XX series are three-terminal IC regulators with fixed negative output voltage.
- The LM317 is a three-terminal IC regulator with a positive variable output voltage.
- The LM337 is a three-terminal IC regulator with a negative variable output voltage.
- The 78S40 is a switching voltage regulator.

Section 17–6

- An external pass transistor increases the current capability of a regulator.

KEY TERMS

Key terms and other bold terms in the chapter are defined in the end-of-book glossary.

Linear regulator A voltage regulator in which the control element operates in the linear region.

Line regulation The percentage change in output voltage for a given change in input (line) voltage.

Load regulation The percentage change in output voltage for a given change in load current.

Regulator An electronic circuit that maintains an essentially constant output voltage with a changing input voltage or load current.

Switching regulator A voltage regulator in which the control element operates as a switch.

Thermal overload A condition in a rectifier where the internal power dissipation of the circuit exceeds a certain maximum due to excessive current.

KEY FORMULAS

Voltage Regulation

17–1 $$\text{Line regulation} = \left(\frac{\Delta V_{OUT}}{\Delta V_{IN}}\right)100\%$$ Line regulation as a percentage

17–2 $$\text{Line regulation} = \frac{(\Delta V_{OUT}/V_{OUT})100\%}{\Delta V_{IN}}$$ Line regulation in %/V

17–3 $$\text{Load regulation} = \left(\frac{V_{NL} - V_{FL}}{V_{FL}}\right)100\%$$ Percent load regulation

17–4 $$\text{Load regulation} = \left(\frac{R_{OUT}}{R_{FL}}\right)100\%$$ Load regulation in terms of output resistance and full-load resistance

Basic Series Regulator

17–5 $$V_{OUT} \cong \left(1 + \frac{R_2}{R_3}\right)V_{REF}$$ Regulator output

17–6 $$I_{L(max)} = \frac{0.7\text{ V}}{R_4}$$ For constant-current limiting (silicon)

Basic Shunt Regulator

17–7 $$I_{L(max)} = \frac{V_{IN}}{R_1}$$ Maximum load current

Basic Switching Regulators

17–8 $$V_{OUT} = \left(\frac{t_{on}}{T}\right)V_{IN}$$ For step-down switching regulator

IC Voltage Regulators

17–9 $$V_{OUT} = V_{REF}\left(1 + \frac{R_2}{R_1}\right) + I_{ADJ}R_2$$ IC regulator

17–10 $R_{ext} = \dfrac{0.7\text{ V}}{I_{max}}$ For external pass circuit

17–11 $I_L = \dfrac{V_{OUT}}{R_1} + I_G$ Regulator as a current source

TRUE/FALSE QUIZ

Answers are at the end of the chapter.

1. Line regulation is a measure of how constant the output voltage is for a given change in the input voltage.
2. Load regulation depends on the amount of power dissipated in the load.
3. Linear and switching are two main categories of voltage regulators.
4. Two types of linear regulator are series and bypass.
5. Three types of switching regulator are step-down, step-up, and inverting.
6. The three terminals of a 78XX series regulator are input, output, and control.
7. An external bypass transistor is sometimes used to increase the current capability of a regulator.
8. Current limiting is used to protect the external bypass transistor.
9. The purpose of a heat sink is to help the regulator dissipate excessive heat.
10. A variable pulse-width oscillator is part of a linear voltage regulator.

CIRCUIT-ACTION QUIZ

Answers are at the end of the chapter.

1. If the input voltage in Figure 17–7 is increased by 1 V, the output voltage will
   (a) increase (b) decrease (c) not change
2. If the zener diode in Figure 17–7 is changed to one with a zener voltage of 6.8 V, the output voltage will
   (a) increase (b) decrease (c) not change
3. If R_3 in Figure 17–7 is increased in value, the output voltage will
   (a) increase (b) decrease (c) not change
4. If R_4 in Figure 17–9 is reduced, the amount of current that the regulator can supply to the load will
   (a) increase (b) decrease (c) not change
5. If R_2 in Figure 17–15 is increased, the power dissipation in R_1 will
   (a) increase (b) decrease (c) not change
6. If the duty cycle of the variable pulse-width oscillator in Figure 17–16(a) is increased, the output voltage will
   (a) increase (b) decrease (c) not change
7. If R_2 in Figure 17–29 is adjusted to a lower value, the output voltage will
   (a) increase (b) decrease (c) not change
8. To increase the maximum current that the regulator in Figure 17–33 can supply, the value of R_{ext} must
   (a) increase (b) decrease (c) not change

SELF-TEST

Answers are at the end of the chapter.

Section 17–1

1. In the case of line regulation,
   (a) when the temperature varies, the output voltage stays constant
   (b) when the output voltage changes, the load current stays constant
   (c) when the input voltage changes, the output voltage stays constant
   (d) when the load changes, the output voltage stays constant

2. In the case of load regulation,
   (a) when the temperature varies, the output voltage stays constant
   (b) when the input voltage changes, the load current stays constant
   (c) when the load changes, the load current stays constant
   (d) when the load changes, the output voltage stays constant
3. All of the following are parts of a basic voltage regulator *except*
   (a) control element (b) sampling circuit
   (c) voltage-follower (d) error detector (e) reference voltage

Section 17–2

4. The basic difference between a series regulator and a shunt regulator is
   (a) the amount of current that can be handled (b) the position of the control element
   (c) the type of sample circuit (d) the type of error detector
5. In a basic series regulator, V_{OUT} is determined by
   (a) the control element (b) the sample circuit
   (c) the reference voltage (d) answers (b) and (c)
6. The main purpose of current limiting in a regulator is
   (a) protection of the regulator from excessive current
   (b) protection of the load from excessive current
   (c) to keep the power supply transformer from burning up
   (d) to maintain a constant output voltage
7. In a linear regulator, the control transistor is conducting
   (a) a small part of the time (b) half the time
   (c) all of the time (d) only when the load current is excessive

Section 17–3

8. In a basic shunt regulator, V_{OUT} is determined by
   (a) the control element (b) the sample circuit
   (c) the reference voltage (d) answers (b) and (c)

Section 17–4

9. In a switching regulator, the control transistor is conducting
   (a) part of the time (b) all of the time
   (c) only when the input voltage exceeds a set limit (d) only when there is an overload

Section 17–5

10. The LM317 is an example of an IC
    (a) three-terminal negative voltage regulator (b) fixed positive voltage regulator
    (c) switching regulator (d) linear regulator
    (e) variable positive voltage regulator (f) answers (b) and (d) only
    (g) answers (d) and (e) only

Section 17–6

11. An external pass transistor is used for
    (a) increasing the output voltage (b) improving the regulation
    (c) increasing the current that the regulator can handle (d) short-circuit protection

PROBLEMS

Answers to all odd-numbered problems are at the end of the book.

BASIC PROBLEMS

Section 17–1 **Voltage Regulation**

1. The nominal output voltage of a certain regulator is 8 V. The output changes 2 mV when the input voltage goes from 12 V to 18 V. Determine the line regulation and express it as a percentage change over the entire range of V_{IN}.
2. Express the line regulation found in Problem 1 in units of %/V.
3. A certain regulator has a no-load output voltage of 10 V and a full-load output voltage of 9.90 V. What is the percent load regulation?
4. In Problem 3, if the full-load current is 250 mA, express the load regulation in %/mA.

Section 17–2 Basic Linear Series Regulators

5. Label the functional blocks for the voltage regulator in Figure 17–44.

▶ FIGURE 17–44

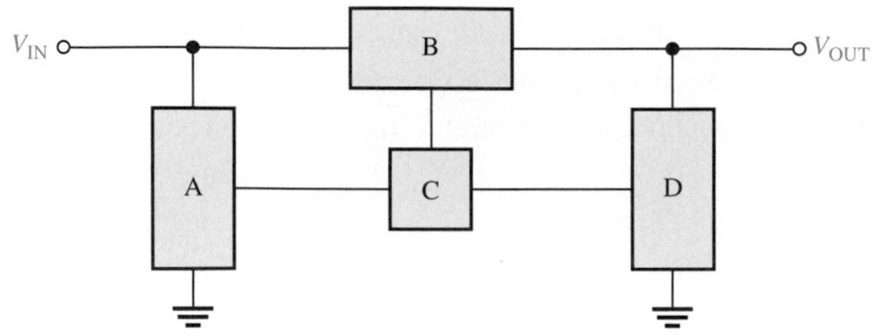

6. Determine the output voltage for the regulator in Figure 17–45.

▶ FIGURE 17–45

Multisim file circuits are identified with a CD logo and are in the Problems folder on your CD-ROM. Filenames correspond to figure numbers (e.g., F17–45).

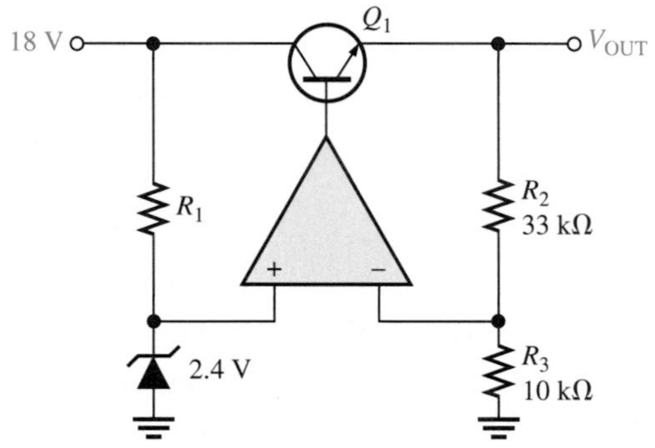

7. Determine the output voltage for the series regulator in Figure 17–46.

▶ FIGURE 17–46

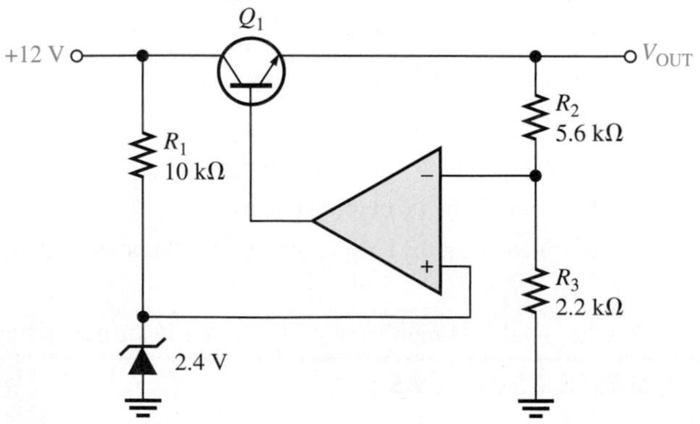

8. If R_3 in Figure 17–46 is increased to 4.7 kΩ, what happens to the output voltage?

9. If the zener voltage is 2.7 V instead of 2.4 V in Figure 17–46, what is the output voltage?

10. A series voltage regulator with constant-current limiting is shown in Figure 17–47. Determine the value of R_4 if the load current is to be limited to a maximum value of 250 mA. What power rating must R_4 have?

11. If the R_4 determined in Problem 10 is halved, what is the maximum load current?

FIGURE 17–47

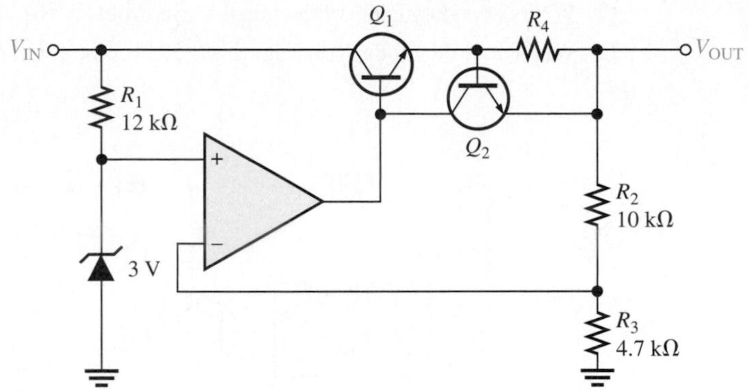

Section 17–3 Basic Linear Shunt Regulators

12. In the shunt regulator of Figure 17–48, when the load current increases, does Q_1 conduct more or less? Why?

FIGURE 17–48

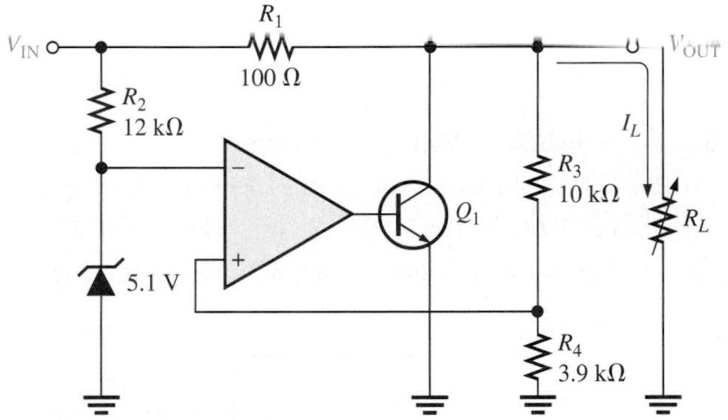

13. Assume I_L remains constant and V_{IN} changes by 1 V in Figure 17–48. What is the change in the collector current of Q_1?
14. With a constant input voltage of 17 V, the load resistance in Figure 17–48 is varied from 1 kΩ to 1.2 kΩ. Neglecting any change in output voltage, how much does the shunt current through Q_1 change?
15. If the maximum allowable input voltage in Figure 17–48 is 25 V, what is the maximum possible output current when the output is short-circuited? What power rating should R_1 have?

Section 17–4 Basic Switching Regulators

16. A basic switching regulator is shown in Figure 17–49. If the switching frequency of the transistor is 100 Hz with an off-time of 6 ms, what is the output voltage?

FIGURE 17–49

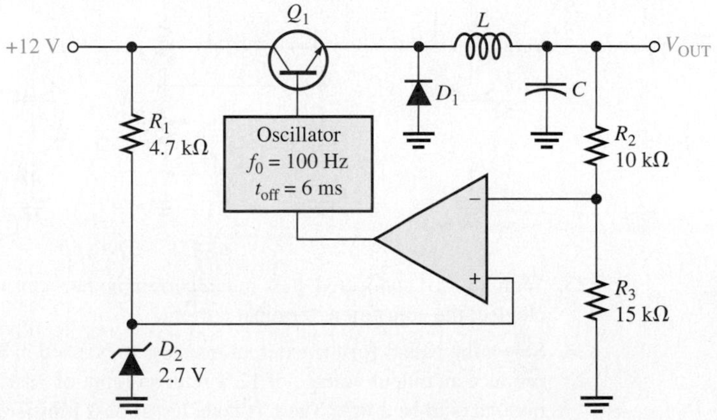

17. What is the duty cycle of the transistor in Problem 16?

18. When does the diode D_1 in Figure 17–50 become forward-biased?

FIGURE 17–50

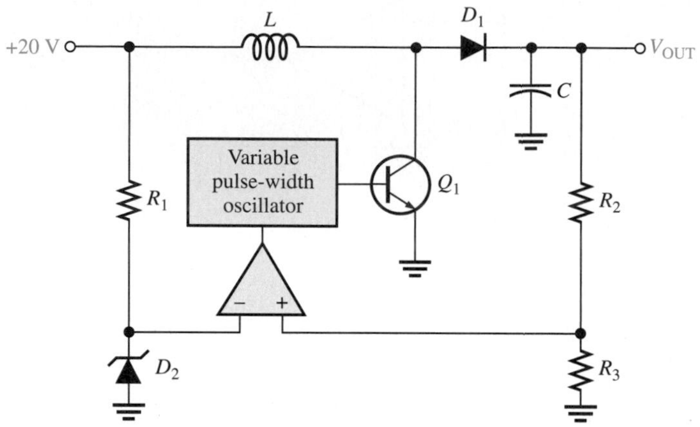

19. If the on-time of Q_1 in Figure 17–50 is decreased, does the output voltage increase or decrease?

Section 17–5 Integrated Circuit Voltage Regulators

20. What is the output voltage of each of the following IC regulators?
    (a) 7806 **(b)** 7905.2 **(c)** 7818 **(d)** 7924

21. Determine the output voltage of the regulator in Figure 17–51. $I_{ADJ} = 50\ \mu A$.

FIGURE 17–51

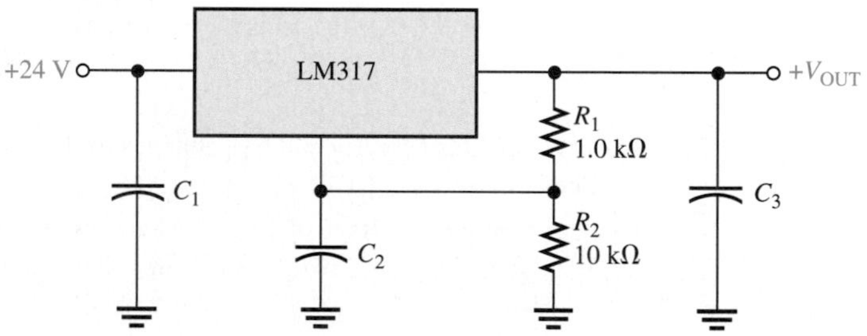

22. Determine the minimum and maximum output voltages for the circuit in Figure 17–52. $I_{ADJ} = 50\ \mu A$.

FIGURE 17–52

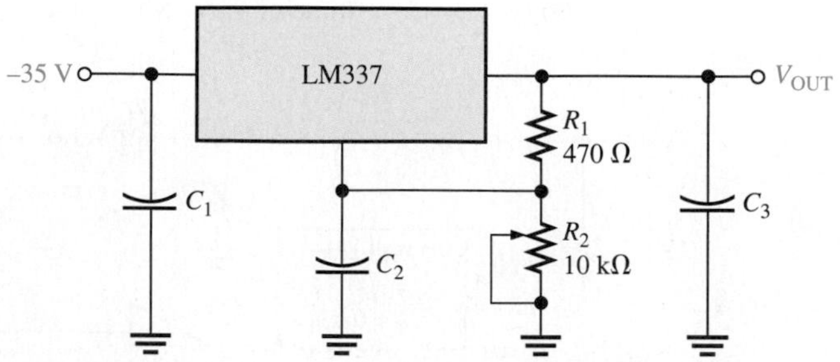

23. With no load connected, how much current is there through the regulator in Figure 17–51? Neglect the adjustment terminal current.

24. Select the values for the external resistors to be used in an LM317 circuit that is required to produce an output voltage of 12 V with an input of 18 V. The maximum regulator current with no load is to be 2 mA. There is no external pass transistor.

Section 17–6 **Applications of IC Voltage Regulators**

25. In the regulator circuit of Figure 17–53, determine R_{ext} if the maximum internal regulator current is to be 250 mA.

▶ FIGURE 17–53

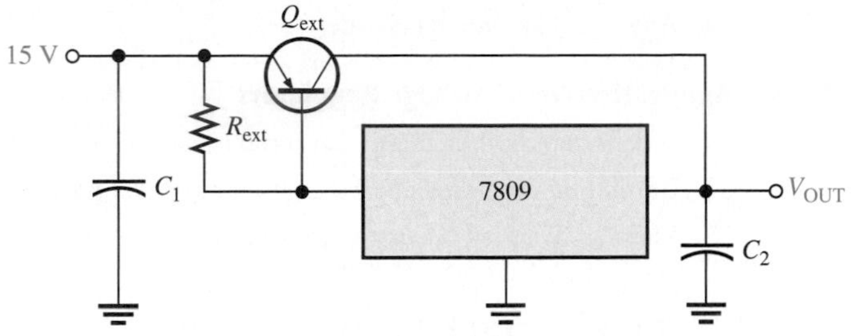

26. Using a 7812 voltage regulator and a 10 Ω load in Figure 17–53, how much power will the external pass transistor have to dissipate? The maximum internal regulator current is set at 500 mA by R_{ext}.

27. Show how to include current limiting in the circuit of Figure 17–53. What should the value of the limiting resistor be if the external current is to be limited to 2 A?

28. Using an LM317, design a circuit that will provide a constant current of 500 mA to a load.

29. Repeat Problem 28 using a 7908.

30. If a 78S40 switching regulator is to be used to regulate a 12 V input down to a 6 V output, calculate the values of the external voltage-divider resistors.

MULTISIM TROUBLESHOOTING PROBLEMS

These file circuits are in the Troubleshooting Problems folder on your CD-ROM.

31. Open file TSP17-31 and determine the fault.

32. Open file TSP17-32 and determine the fault.

33. Open file TSP17-33 and determine the fault.

34. Open file TSP17-34 and determine the fault.

ANSWERS

SECTION CHECKUPS

Section 17–1 **Voltage Regulation**

1. The percentage change in the output voltage for a given change in input voltage.
2. The percentage change in output voltage for a given change in load current.
3. 1.2%; 0.06%/V
4. 1.6%; 0.0016%/mA

Section 17–2 **Basic Linear Series Regulators**

1. Control element, error detector, sampling element, reference voltage
2. 2 V

Section 17–3 **Basic Linear Shunt Regulators**

1. In a shunt regulator, the control element is in parallel with the load rather than in series.
2. A shunt regulator has inherent current limiting. A disadvantage is that a shunt regulator is less efficient than a series regulator.

Section 17–4 **Basic Switching Regulators**

1. Step-down, step-up, inverting
2. Switching regulators operate at a higher efficiency.
3. The duty cycle varies to regulate the output.

Section 17–5 Integrated Circuit Voltage Regulators

1. Input, output, and ground
2. A 7809 has a +9 V output; A 7915 has a −15 V output.
3. Input, output, adjustment
4. A two-resistor voltage divider

Section 17–6 Applications of IC Voltage Regulators

1. A pass transistor increases the current that can be handled.
2. Current limiting prevents excessive current and prevents damage to the regulator.
3. Thermal overload occurs when the internal power dissipation becomes excessive.

RELATED PROBLEMS FOR EXAMPLES

17–1 0.6%/V

17–2 1.12%, 0.0224%/mA

17–3 7.33 V

17–4 0.7 A

17–5 17.5 W

17–6 12.7 V

17–7 467 mA

17–8 12 W

17–9 16 Ω

TRUE/FALSE QUIZ

1. T	**2.** F	**3.** T	**4.** F	**5.** T
6. F	**7.** T	**8.** T	**9.** T	**10.** F

CIRCUIT-ACTION QUIZ

1. (c) **2.** (a) **3.** (b) **4.** (a) **5.** (c) **6.** (a) **7.** (b) **8.** (b)

SELF-TEST

1. (c)	**2.** (d)	**3.** (c)	**4.** (b)	**5.** (d)	**6.** (a)	**7.** (c)	**8.** (d)	**9.** (a)
10. (g)	**11.** (c)							

COMMUNICATIONS

18

CHAPTER OUTLINE

CHAPTER OBJECTIVES

- Describe basic superheterodyne receivers
- Discuss the function of a linear multiplier
- Discuss the fundamentals of amplitude modulation
- Discuss the basic function of a mixer
- Describe AM demodulation
- Describe IF and audio amplifiers
- Describe frequency modulation
- Describe a phase-locked loop (PLL)
- Discuss fiber optics

KEY TERMS

- Amplitude modulation (AM)
- Frequency modulation (FM)
- Four-quadrant multiplier
- Balanced modulation
- Mixer
- Phase-locked loop (PLL)
- Lock range
- Capture range
- Fiber optics
- Angle of incidence
- Critical angle
- Index of refraction

VISIT THE COMPANION WEBSITE

Study aids for this chapter are available at http://www.prenhall.com/floyd

INTRODUCTION

Communications electronics encompasses a wide range of systems, including both analog and digital. Any system that sends information from one point to another over relatively long distances can be classified as a communications system. Some of the categories of communications systems are radio (broadcast, ham, CB, marine), television, telephony, radar, navigation, satellite, data (digital), and telemetry.

Many communications systems use either amplitude modulation (AM) or frequency modulation (FM) to send information. Other modulation methods include pulse modulation, phase modulation, and frequency shift keying (FSK) as well as more specialized techniques. By necessity, the scope of this chapter is limited and is intended to introduce you to basic AM and FM communications systems and circuits and to fiber optics.

18–1 BASIC RECEIVERS

Receivers based on the superheterodyne principle are standard in one form or another in most types of communications systems and are found in familiar systems such as standard broadcast radio, stereo, and television. This section provides a basic introduction to amplitude modulation and frequency modulation and an overview of the complete AM and FM receiver.

After completing this section, you should be able to

- **Describe basic superheterodyne receivers**
  - Define *AM* and *FM*
  - Discuss the major functional blocks of an AM receiver
  - Discuss the major functional blocks of an FM receiver

Amplitude Modulation

Amplitude modulation (AM) is a method for sending audible information, such as voice and music, by electromagnetic waves that are broadcast through the atmosphere. In AM, the amplitude of a signal with a specific frequency (f_c), called the **carrier,** is varied according to a modulating signal, which can be an audio signal (voice or music), as shown in Figure 18–1. The carrier frequency permits the receiver to be tuned to a specific known frequency. The resulting AM waveform contains the carrier frequency, an upper-side frequency equal to the carrier frequency plus the modulation frequency ($f_c + f_m$), and a lower-side frequency equal to the carrier frequency minus the modulation frequency ($f_c - f_m$). For example, if a 1 MHz carrier is amplitude modulated with a 5 kHz audio signal, the frequency components in the AM waveform are 1 MHz (carrier), 1 MHz + 5 kHz = 1,005,000 Hz (upper side), and 1 MHz − 5 kHz = 995,000 Hz (lower side). Harmonics of these frequencies are also present.

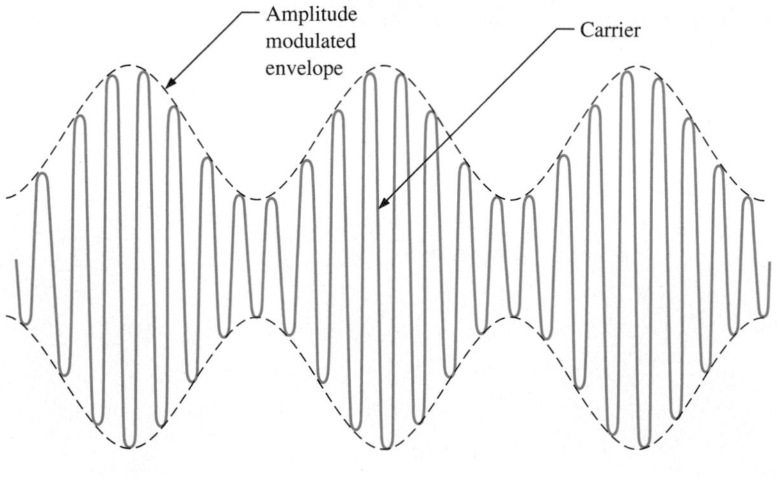

FIGURE 18–1

An example of an amplitude modulated signal. In this case, the higher-frequency carrier is modulated by a lower-frequency sinusoidal signal.

The frequency band for AM broadcast receivers is 540 kHz to 1640 kHz. This means that an AM receiver can be tuned to pick up a specific carrier frequency that lies in the broadcast band. Each AM radio station transmits at a specific carrier frequency that is different from any other station in the area, so you can tune the receiver to pick up any desired station.

The Superheterodyne AM Receiver

A block diagram of a superheterodyne AM receiver is shown in Figure 18–2. The receiver shown consists of an antenna, an RF (radio frequency) amplifier, a mixer, a local oscillator (LO), an IF (intermediate frequency) amplifier, a detector, an audio amplifier, a power amplifier, and a speaker.

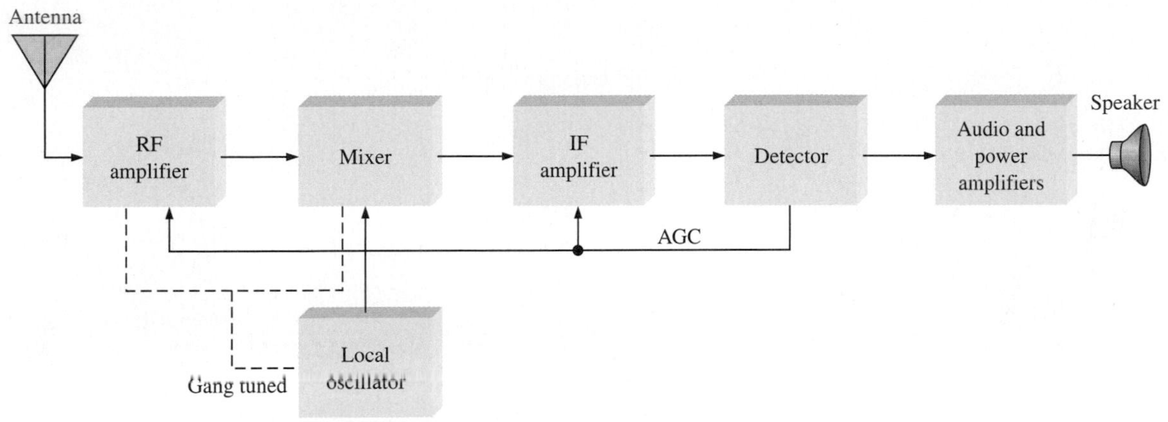

▲ **FIGURE 18–2**

Superheterodyne AM receiver block diagram.

Antenna The antenna picks up all radiated signals and feeds them into the RF amplifier. These signals are very small (usually only a few microvolts).

RF Amplifier This circuit can be adjusted (tuned) to select and amplify any carrier frequency within the AM broadcast band. Only the selected frequency and its two side bands pass through the amplifier. (Some AM receivers do not have a separate RF amplifier stage.)

Local Oscillator This circuit generates a steady sine wave at a frequency 455 kHz above the selected RF frequency.

Mixer This circuit accepts two inputs, the amplitude modulated RF signal from the output of the RF amplifier (or the antenna when there is no RF amplifier) and the sinusoidal output of the local oscillator (LO). These two signals are then "mixed" by a nonlinear process called *heterodyning* to produce sum and difference frequencies. For example, if the RF carrier has a frequency of 1000 kHz, the LO frequency is 1455 kHz and the sum and difference frequencies out of the mixer are 2455 kHz and 455 kHz, respectively. The difference frequency is always 455 kHz no matter what the RF carrier frequency.

IF Amplifier The input to the IF amplifier is the 455 kHz AM signal, a replica of the original AM carrier signal except that the frequency has been lowered to 455 kHz. The IF amplifier significantly increases the level of this signal. The advantage of the IF stage is that it can be designed for a single frequency, simplifying the receiver.

Detector This circuit recovers the modulating signal (audio signal) from the 455 kHz intermediate frequency (IF). At this point the IF is no longer needed, so the output of the detector consists of only the audio signal.

Audio and Power Amplifiers This circuit amplifies the detected audio signal and drives the speaker to produce sound.

AGC The automatic gain control (AGC) provides a dc level out of the detector that is proportional to the strength of the received signal. This level is fed back to the IF amplifier, and sometimes to the mixer and RF amplifier, to adjust the gains so as to maintain constant signal levels throughout the system over a wide range of incoming carrier signal strengths.

Figure 18–3 shows the signal flow through an AM superheterodyne receiver. The receiver can be tuned to accept any frequency in the AM band. The RF amplifier, mixer, and local oscillator are tuned simultaneously so that the LO frequency is always 455 kHz above the incoming RF signal frequency. This is called *gang tuning*.

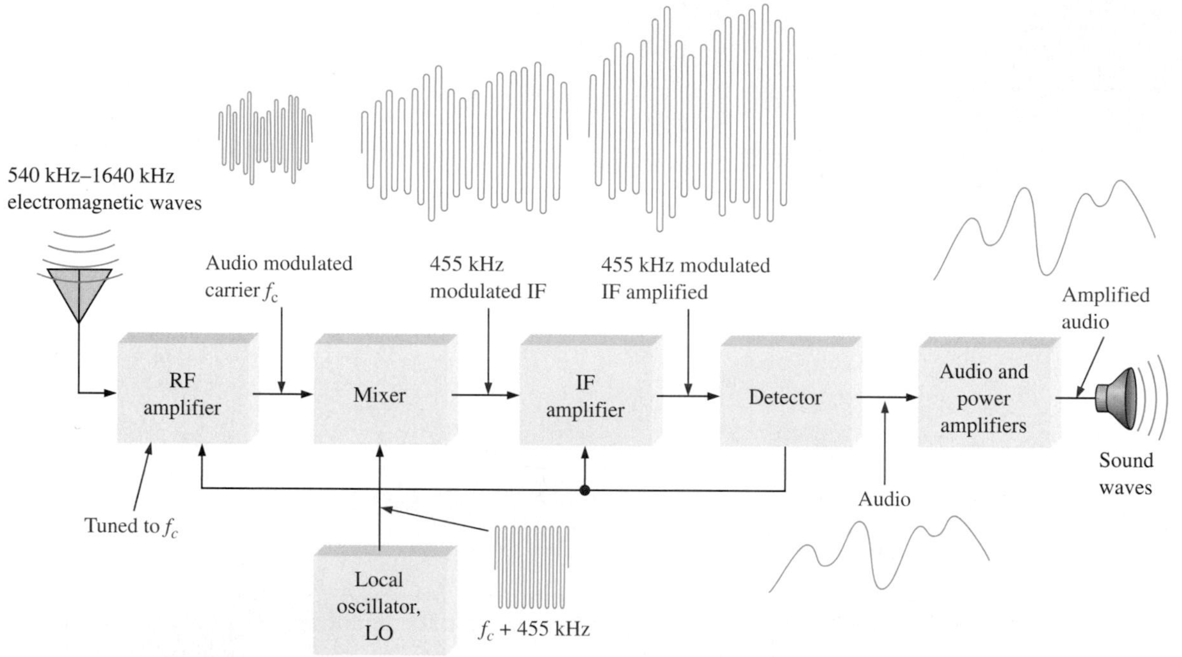

▲ **FIGURE 18–3**

Illustration of signal flow through an AM receiver.

Frequency Modulation

In **frequency modulation (FM)**, the modulating signal (audio) varies the frequency of a carrier as opposed to the amplitude, as in the case of AM. Figure 18–4 illustrates basic frequency modulation. The standard FM broadcast band consists of carrier frequencies from 88 MHz to 108 MHz, which is significantly higher than AM.

▶ **FIGURE 18–4**

An example of frequency modulation.

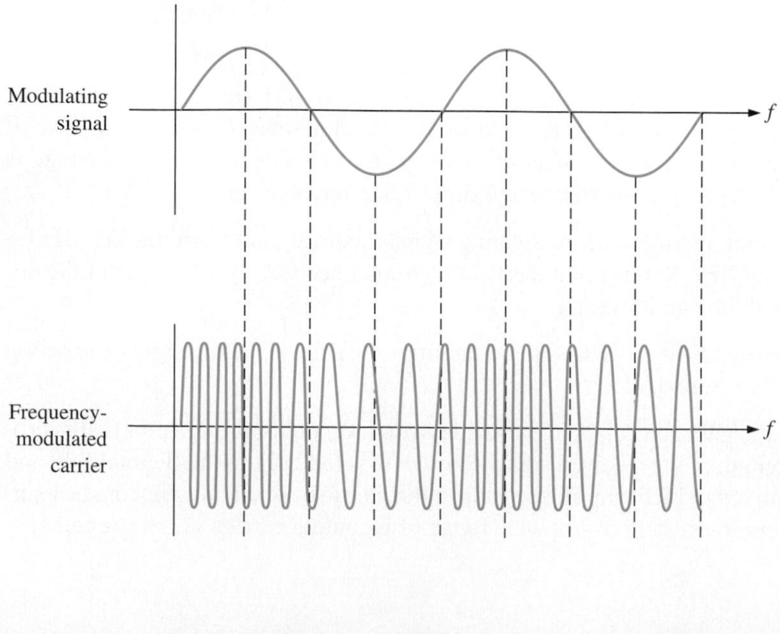

The Superheterodyne FM Receiver

The FM receiver is similar to the AM receiver in many ways, but there are several differences. A block diagram of a superheterodyne FM receiver is shown in Figure 18–5. Notice that it includes an RF amplifier, mixer, local oscillator, and IF amplifier just as in the AM receiver. These circuits operate at higher frequencies than in a commercial AM system. A significant difference in FM is the way the audio signal must be recovered from the modulated IF. This is accomplished by the limiter, discriminator, and de-emphasis network. Figure 18–6 depicts the signal flow through an FM receiver.

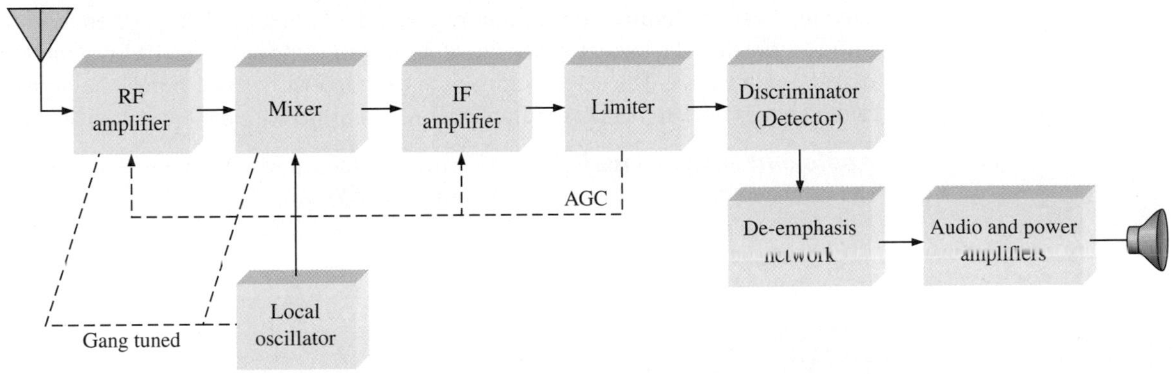

▲ **FIGURE 18–5**

Superheterodyne FM receiver block diagram.

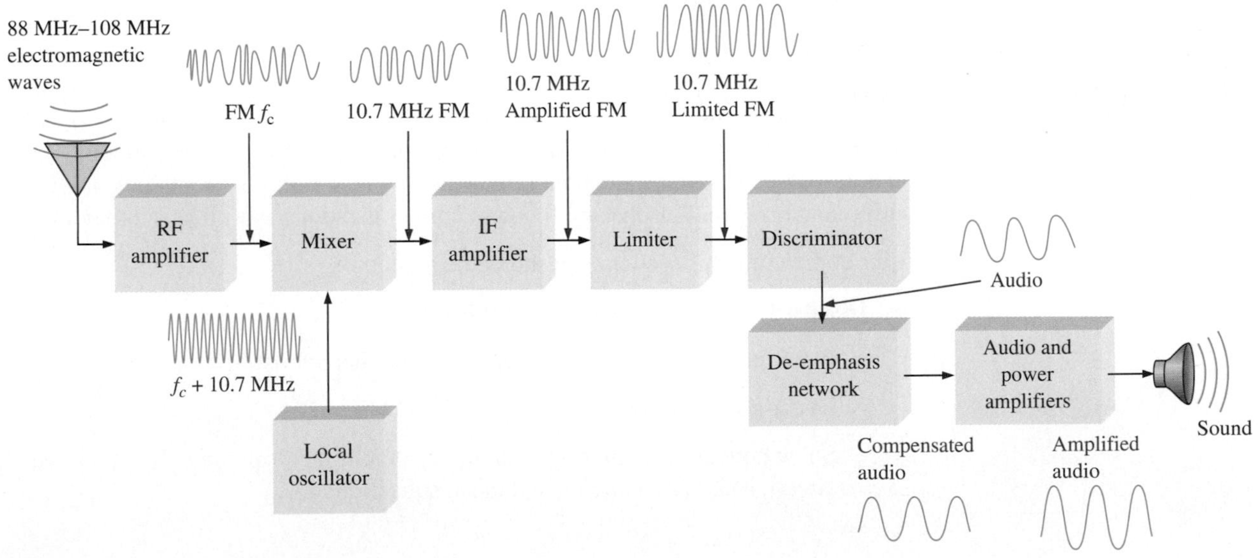

▲ **FIGURE 18–6**

Example of signal flow through an FM receiver.

RF Amplifier This circuit must be capable of amplifying any frequency between 88 MHz and 108 MHz. It is highly selective so that it passes only the selected carrier frequency and significant side-band frequencies that contain the audio.

Local Oscillator This circuit produces a sine wave at a frequency 10.7 MHz above the selected RF frequency.

Mixer This circuit performs the same function as in the AM receiver, except that its output is a 10.7 MHz FM signal regardless of the RF carrier frequency.

IF Amplifier This circuit amplifies the 10.7 MHz FM signal.

Limiter The limiter removes any unwanted variations in the amplitude of the FM signal as it comes out of the IF amplifier and produces a constant amplitude FM output at the 10.7 MHz intermediate frequency.

Discriminator This circuit performs the equivalent function of the detector in an AM system and is sometimes called a detector rather than a discriminator. The **discriminator** recovers the audio from the FM signal.

De-emphasis Network For certain reasons, the higher modulating frequencies are amplified more than the lower frequencies at the transmitting end of an FM system by a process called *preemphasis*. The de-emphasis circuit in the FM receiver brings the high-frequency audio signals back to the proper amplitude relationship with the lower frequencies.

Audio and Power Amplifiers This circuit is the same as in the AM system and can be shared when there is a dual AM/FM configuration.

SECTION 18–1 CHECKUP
Answers are at the end of the chapter.

1. What do *AM* and *FM* mean?
2. How do *AM* and *FM* differ?
3. What are the standard broadcast frequency bands for AM and FM?

18–2 THE LINEAR MULTIPLIER

The linear multiplier is used in many types of communications systems. In this section, you will examine the basic principles of linear multipliers and look at a few multiplier configurations that are found in communications as well as in other areas.

After completing this section, you should be able to

- **Discuss the function of a linear multiplier**
    - Describe multiplier quadrants and transfer characteristic
    - Discuss scale factor
    - Show how to use a multiplier circuit as a multiplier, squaring circuit, divide circuit, square root circuit, and mean square circuit

Multiplier Quadrants

There are one-quadrant, two-quadrant, and four-quadrant multipliers. The quadrant classification indicates the number of input polarity combinations that the multiplier can handle. A graphical representation of the quadrants is shown in Figure 18–7. A **four-quadrant multiplier** can accept any of the four possible input polarity combinations and produce an output with the corresponding polarity.

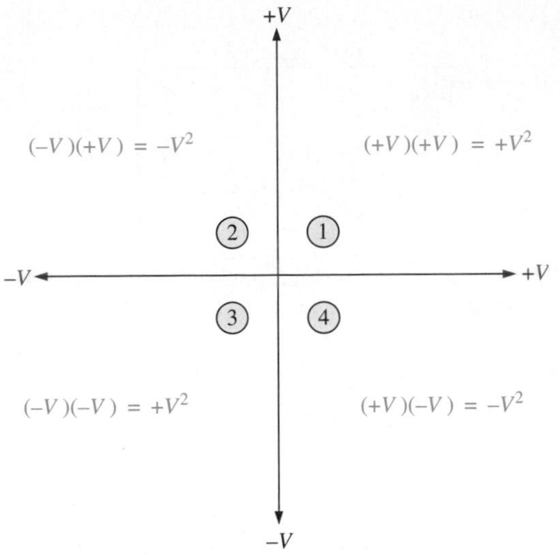

FIGURE 18–7

Four-quadrant polarities and their products.

The Multiplier Transfer Characteristic

Figure 18–8 shows the transfer characteristic for a typical linear multiplier of two input voltages V_X and V_Y. Values of V_X run along the horizontal axis and values of V_Y are the sloped lines. To find the output voltage from the transfer characteristic graph, find the intersection of the two input voltages V_X and V_Y. Then find the output voltage by projecting the point of intersection over to the vertical axis. An example will illustrate this.

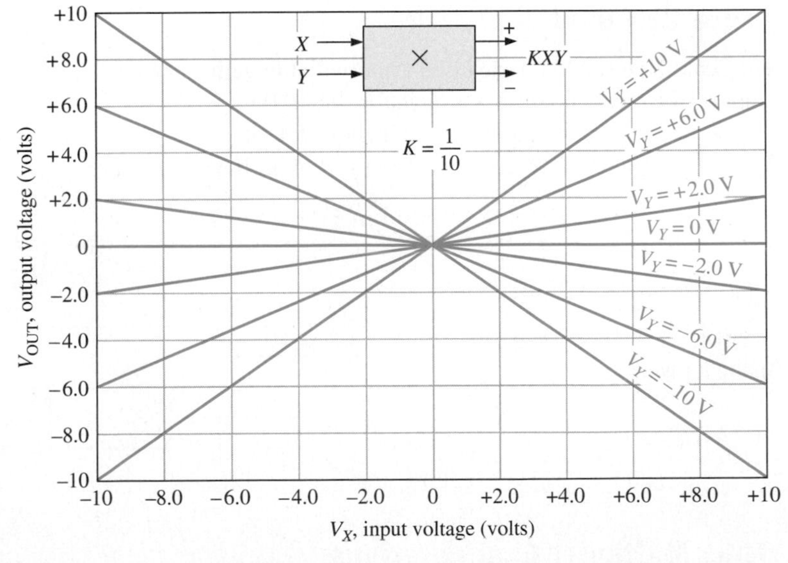

FIGURE 18–8

A four-quadrant multiplier transfer characteristic.

EXAMPLE 18–1

Determine the output voltage for a four-quadrant linear multiplier whose transfer characteristic is given in Figure 18–8. The input voltages are $V_X = -4$ V and $V_Y = +10$ V.

Solution The output voltage is **−4 V** as illustrated in Figure 18–9. For this transfer characteristic, the output voltage is a factor of ten smaller than the actual product of the two input voltages. This is due to the scale factor of the multiplier, which is discussed next.

FIGURE 18–9

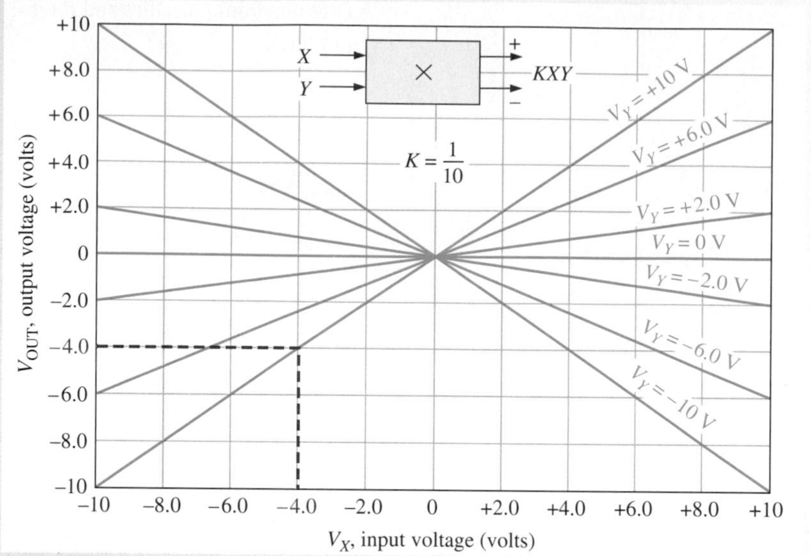

*Related Problem** Find V_{OUT} if $V_X = -6$ V and $V_Y = +6$ V.

*Answers are at the end of the chapter.

The Scale Factor, *K*

The scale factor, *K*, is basically an internal attenuation that reduces the output by a fixed amount. The scale factor has a typical value of 0.1.

The expression for the output voltage of the linear multiplier includes the scale factor, *K*, as indicated in Equation 18–1. The symbol is shown in Figure 18–10.

Equation 18–1

$$V_{OUT} = KV_XV_Y$$

FIGURE 18–10

Multiplier symbol.

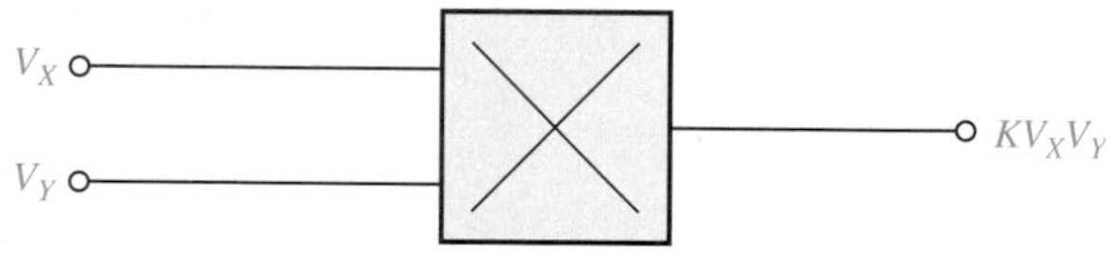

Other Multiplier Configurations

Squaring Circuit A special case of the multiplier is a squaring circuit that is realized by simply applying the same voltage to both inputs by connecting the inputs together as shown in Figure 18–11.

FIGURE 18–11

Squaring circuit.

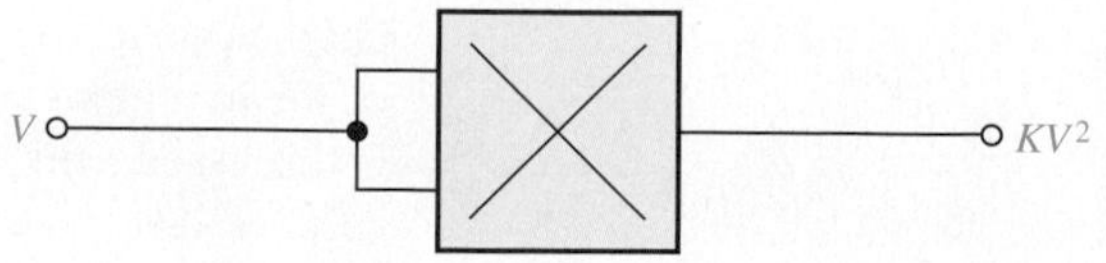

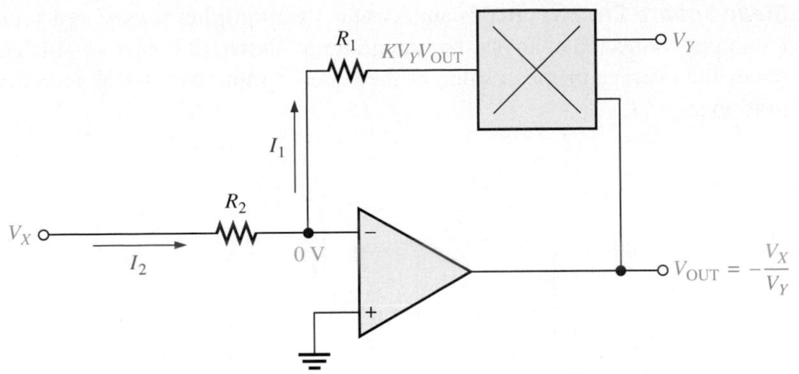

FIGURE 18–12
Divide circuit.

Divide Circuit The circuit in Figure 18–12 shows the multiplier placed in the feedback loop of an op-amp. The basic operation is as follows. There is a virtual ground at the inverting (–) input of the op-amp and therefore the current at the inverting input is negligible. Therefore, I_1 and I_2 are equal. Since the inverting input voltage is 0 V, the voltage across R_1 is KV_YV_{OUT} and the current through R_1 is

$$I_1 = \frac{KV_YV_{OUT}}{R_1}$$

The voltage across R_2 is V_X, so the current through R_2 is

$$I_2 = \frac{V_X}{R_2}$$

Since $I_1 = -I_2$,

$$\frac{KV_YV_{OUT}}{R_1} = -\frac{V_X}{R_2}$$

Solving for V_{OUT},

$$V_{OUT} = -\frac{V_XR_1}{KV_YR_2}$$

If $R_1 = KR_2$,

$$V_{OUT} = -\frac{V_X}{V_Y}$$

Square Root Circuit The square root circuit is a special case of the divide circuit where V_{OUT} is applied to both inputs of the multiplier as shown in Figure 18–13.

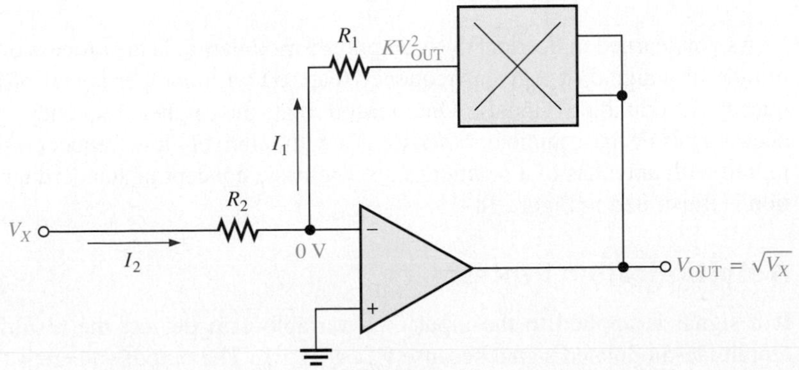

FIGURE 18–13
Square root circuit.

Mean Square Circuit In this application, the multiplier is used as a squaring circuit with its output connected to an op-amp integrator as shown in Figure 18–14. The integrator produces the average or mean value of the squared input over time, as indicated by the integration sign ($\int$).

FIGURE 18–14

Mean square circuit.

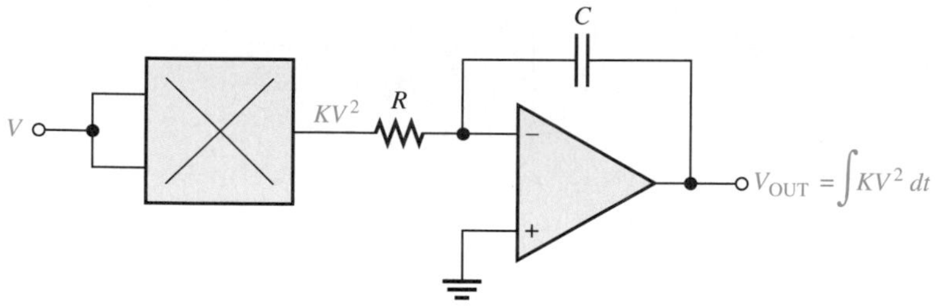

SECTION 18–2 CHECKUP

1. Compare a four-quadrant multiplier to a one-quadrant multiplier in terms of the inputs that can be handled.
2. How do you convert a basic multiplier to a squaring circuit?

18–3 AMPLITUDE MODULATION

Amplitude modulation (AM) is an important method for transmitting information. Of course, the AM superheterodyne receiver is designed to receive transmitted AM signals. In this section, we take a further look at amplitude modulation and show how the linear multiplier can be used as an amplitude-modulated device.

After completing this section, you should be able to

- **Discuss the fundamentals of amplitude modulation**
  - Explain how AM is basically a multiplication process
  - Describe sum and difference frequencies
  - Discuss balanced modulation
  - Describe the frequency spectra
  - Explain standard AM

As you learned in Section 18–1, amplitude modulation is the process of varying the amplitude of a signal of a given frequency (carrier) with another signal of much lower frequency (modulating signal). One reason that the higher-frequency carrier signal is necessary is because audio or other signals with relatively low frequencies cannot be transmitted with antennas of a practical size. The basic concept of standard amplitude modulation is illustrated in Figure 18–15.

A Multiplication Process

If a signal is applied to the input of a variable-gain device, the resulting output is an amplitude-modulated signal because $V_{out} = A_v V_{in}$. The output voltage is the input voltage

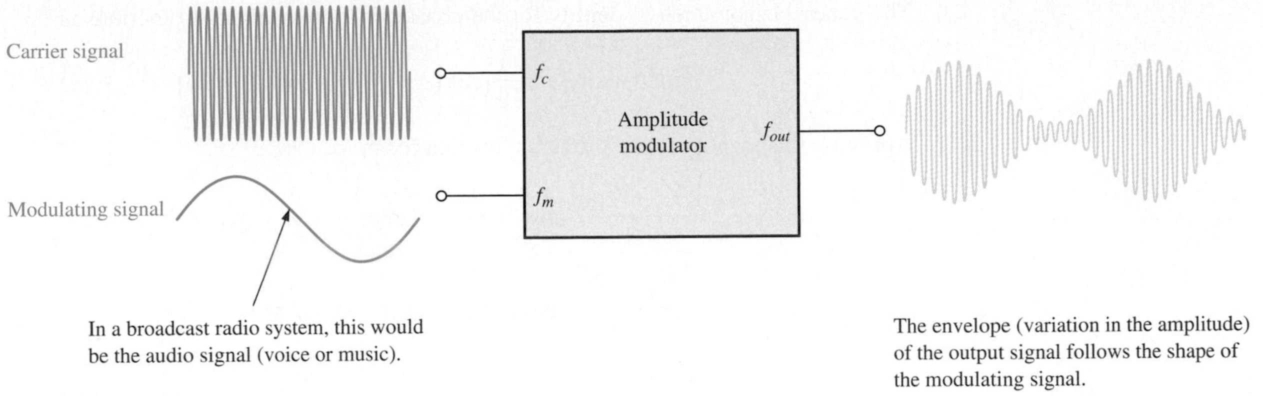

▲ FIGURE 18–15

Basic concept of amplitude modulation.

multiplied by the voltage gain. For example, if the gain of an amplifier is made to vary sinusoidally at a certain frequency and an input signal is applied at a higher frequency, the output signal will have the higher frequency. However, its amplitude will vary according to the variation in gain as illustrated in Figure 18–16. Amplitude modulation is basically a multiplication process (input voltage multiplied by a variable gain).

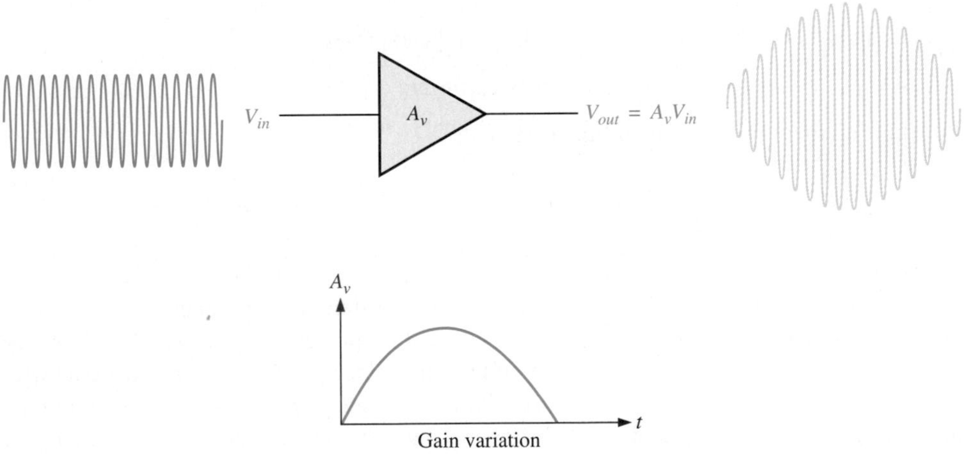

◀ FIGURE 18–16

The amplitude of the output voltage varies according to the gain and is the product of voltage gain and input voltage.

Sum and Difference Frequencies

If the expressions for two sinusoidal signals of different frequencies are multiplied mathematically, a term containing both the difference and the sum of the two frequencies is produced. Recall from ac circuit theory that a sinusoidal voltage can be expressed as

$$v = V_p \sin 2\pi ft$$

where V_p is the peak voltage and f is the frequency. Two different sinusoidal signals can be expressed as follows:

$$v_1 = V_{1(p)}\sin 2\pi f_1 t$$

$$v_2 = V_{2(p)}\sin 2\pi f_2 t$$

Multiplying these two sinusoidal wave terms,

$$v_1 v_2 = (V_{1(p)}\sin 2\pi f_1 t)(V_{2(p)}\sin 2\pi f_2 t) = V_{1(p)}V_{2(p)}(\sin 2\pi f_1 t)(\sin 2\pi f_2 t)$$

The general trigonometric identity for the product of two sinusoidal functions is

$$(\sin A)(\sin B) = \frac{1}{2}[\cos(A - B) - \cos(A + B)]$$

Applying this identity to the previous formula for v_1v_2,

$$v_1v_2 = \frac{V_{1(p)}V_{2(p)}}{2}[\cos(2\pi f_1t - 2\pi f_2t) - \cos(2\pi f_1t + 2\pi f_2t)]$$

$$= \frac{V_{1(p)}V_{2(p)}}{2}[\cos 2\pi(f_1 - f_2)t - \cos 2\pi(f_1 + f_2)t]$$

Equation 18–2

$$\boldsymbol{v_1v_2 = \frac{V_{1(p)}V_{2(p)}}{2}\cos 2\pi(f_1 - f_2)t - \frac{V_{1(p)}V_{2(p)}}{2}\cos 2\pi(f_1 + f_2)t}$$

You can see in Equation 18–2 that the product of the two sinusoidal voltages V_1 and V_2 contains a difference frequency $(f_1 - f_2)$ and a sum frequency $(f_1 + f_2)$. The fact that the product terms are cosine simply indicates a 90° phase shift in the multiplication process.

Analysis of Balanced Modulation

Since amplitude modulation is simply a multiplication process, the preceding analysis is now applied to carrier and modulating signals. The expression for the sinusoidal carrier signal can be written as

$$v_c = V_{c(p)}\sin 2\pi f_ct$$

Assuming a sinusoidal modulating signal, it can be expressed as

$$v_m = V_{m(p)}\sin 2\pi f_mt$$

Substituting these two signals in Equation 18–2,

$$v_cv_m = \frac{V_{c(p)}V_{m(p)}}{2}\cos 2\pi(f_c - f_m)t - \frac{V_{c(p)}V_{m(p)}}{2}\cos 2\pi(f_c + f_m)t$$

An output signal described by this expression for the product of two sinusoidal signals is produced by a linear multiplier. Notice that there is a difference frequency term $(f_c - f_m)$ and a sum frequency term $(f_c + f_m)$, but the original frequencies, f_c and f_m, do not appear alone in the expression. Thus, the product of two sinusoidal signals contains no signal with the carrier frequency, f_c, or with the modulating frequency, f_m. This form of amplitude modulation is called **balanced modulation** because there is no carrier frequency in the output. The carrier frequency is "balanced out."

The Frequency Spectra of a Balanced Modulator

A graphical picture of the frequency content of a signal is called its frequency spectrum. A frequency spectrum shows voltage on a frequency base rather than on a time base as a waveform diagram does. The frequency spectra of the product of two sinusoidal signals are shown in Figure 18–17. Part (a) shows the two input frequencies and part (b) shows the output frequencies. In communications terminology, the sum frequency is called the **upper-side frequency** and the difference frequency is called the **lower-side frequency** because the frequencies appear on each side of the missing carrier frequency.

The Linear Multiplier as a Balanced Modulator

As mentioned, the linear multiplier acts as a balanced modulator when a carrier signal and a modulating signal are applied to its inputs, as illustrated in Figure 18–18. A balanced modulator produces an upper-side frequency and a lower-side frequency, but it does not

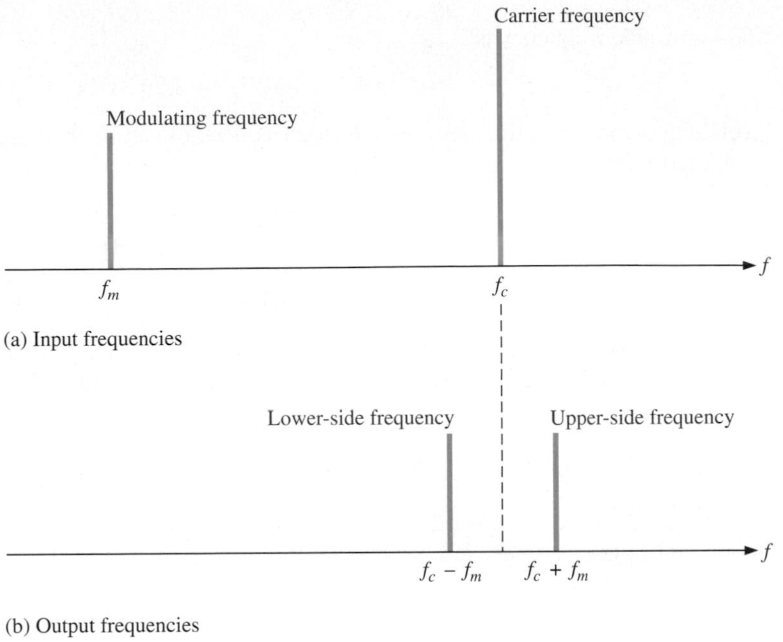

FIGURE 18–17

Illustration of the input and output frequency spectra for a linear multiplier.

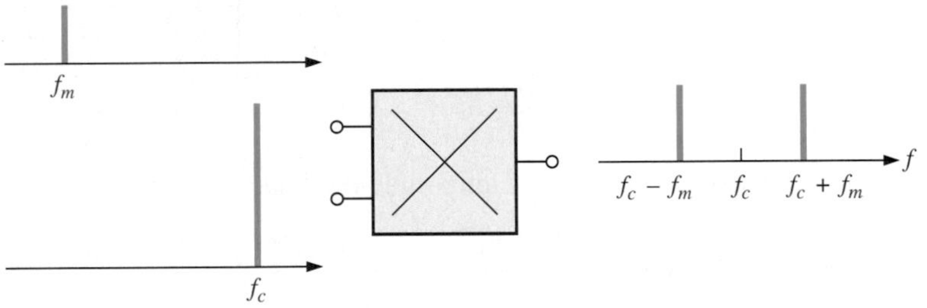

FIGURE 18–18

The linear multiplier as a balanced modulator.

produce a carrier frequency. Since there is no carrier signal, balanced modulation is sometimes known as **suppressed-carrier modulation.** Balanced modulation is used in certain types of communications such as single side-band systems, but it is not used in standard AM broadcast systems.

EXAMPLE 18–2

Determine the frequencies contained in the output signal of the balanced modulator in Figure 18–19.

FIGURE 18–19

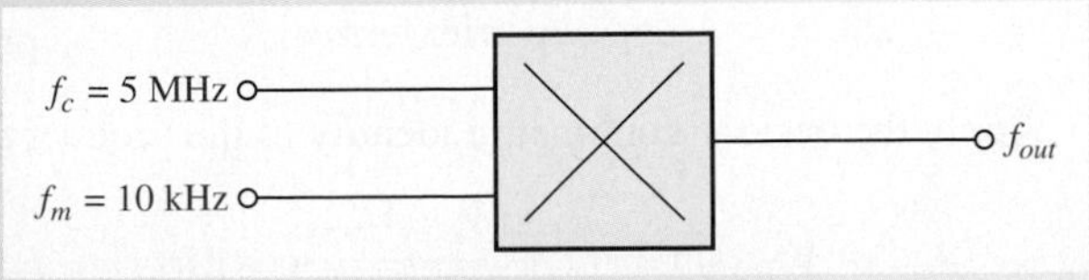

Solution The upper-side frequency is

$$f_c + f_m = 5\text{ MHz} + 10\text{ kHz} = \mathbf{5.01\text{ MHz}}$$

The lower-side frequency is

$$f_c - f_m = 5\text{ MHz} - 10\text{ kHz} = \mathbf{4.99\text{ MHz}}$$

Related Problem Explain how the separation between the side frequencies can be increased using the same carrier frequency.

Standard Amplitude Modulation (AM)

In standard AM systems, the output signal contains the carrier frequency as well as the sum and difference side frequencies. The frequency spectrum in Figure 18–20 illustrates standard amplitude modulation.

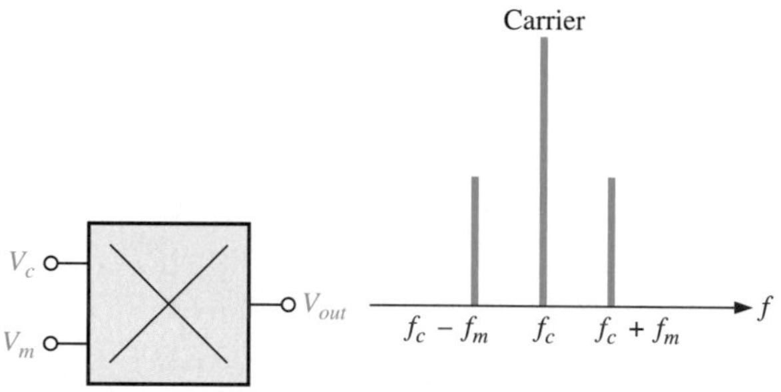

FIGURE 18–20

The output frequency spectrum of a standard amplitude modulator.

The expression for a standard amplitude-modulated signal is

Equation 18–3

$$V_{out} = V_{c(p)}^2 \sin 2\pi f_c t + \frac{V_{c(p)}V_{m(p)}}{2}\cos 2\pi(f_c - f_m)t - \frac{V_{c(p)}V_{m(p)}}{2}\cos 2\pi(f_c + f_m)t$$

Notice in Equation 18–3 that the first term is for the carrier frequency and the other two terms are for the side frequencies. Let's see how the carrier-frequency term gets into the equation.

If a dc voltage equal to the peak of the carrier voltage is added to the modulating signal before the modulating signal is multiplied by the carrier signal, a carrier-signal term appears in the final result as shown in the following steps. Add the peak carrier voltage to the modulating signal, and you get the following expression:

$$V_{c(p)} + V_{m(p)}\sin 2\pi f_m t$$

Multiply by the carrier signal.

$$\begin{aligned} V_{out} &= (V_{c(p)}\sin 2\pi f_c t)(V_{c(p)} + V_{m(p)}\sin 2\pi f_m t) \\ &= \underbrace{V_{c(p)}^2\sin 2\pi f_c t}_{\text{carrier term}} + \underbrace{V_{c(p)}V_{m(p)}(\sin 2\pi f_c t)(\sin 2\pi f_m t)}_{\text{product term}} \end{aligned}$$

Apply the basic trigonometric identity to the product term.

$$V_{out} = V_{c(p)}^2 \sin 2\pi f_c t + \frac{V_{c(p)}V_{m(p)}}{2}\cos 2\pi(f_c - f_m)t - \frac{V_{c(p)}V_{m(p)}}{2}\cos 2\pi(f_c + f_m)t$$

This result shows that the output of the multiplier contains a carrier term and two side-frequency terms. Figure 18–21 illustrates how a standard amplitude modulator can be implemented by a summing circuit followed by a linear multiplier. Figure 18–22 shows a possible implementation of the summing circuit.

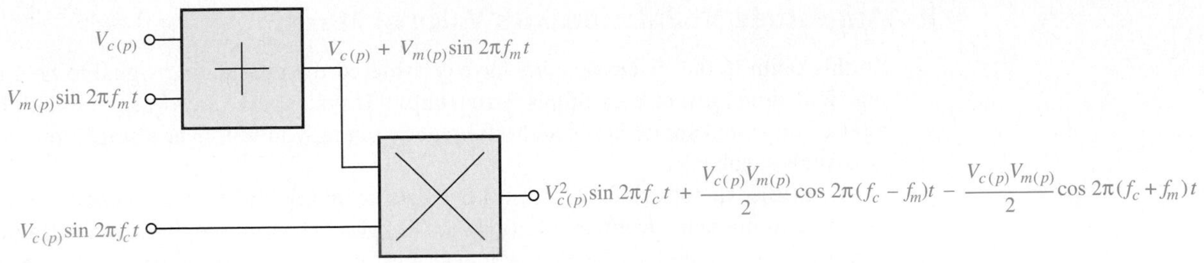

FIGURE 18–21

Basic block diagram of an amplitude modulator.

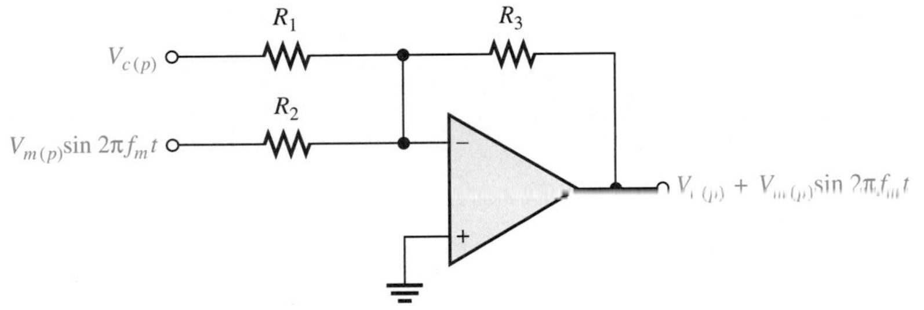

FIGURE 18–22

Implementation of the summing circuit in the amplitude modulator.

EXAMPLE 18–3

A carrier frequency of 1200 kHz is modulated by a sinusoidal wave with a frequency of 25 kHz by a standard amplitude modulator. Determine the output frequency spectrum.

Solution The lower-side frequency is

$$f_c - f_m = 1200\text{ kHz} - 25\text{ kHz} = \mathbf{1175\text{ kHz}}$$

The upper-side frequency is

$$f_c + f_m = 1200\text{ kHz} + 25\text{ kHz} = \mathbf{1225\text{ kHz}}$$

The output contains the carrier frequency and the two side frequencies as shown in Figure 18–23.

FIGURE 18–23

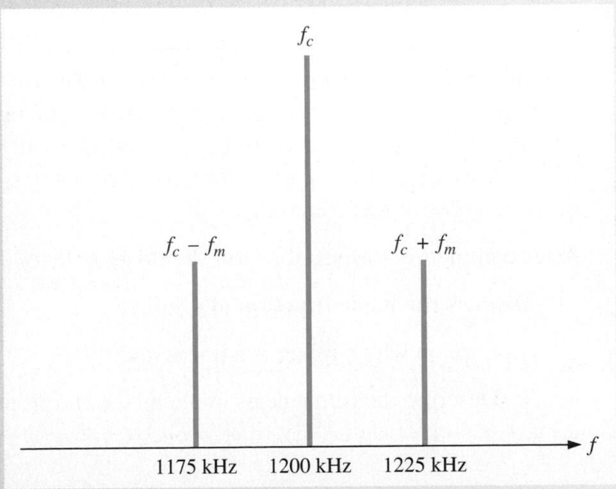

Related Problem Compare the output frequency spectrum in this example to that of a balanced modulator having the same inputs.

Amplitude Modulation with Voice or Music

To this point in our discussion, we have considered the modulating signal to be a pure sinusoidal signal just to keep things fairly simple. If you receive an AM signal modulated by a pure sinusoidal signal in the audio frequency range, you will hear a single tone from the receiver's speaker.

An **audio** signal (voice or music) consists of many sinusoidal components within a range of frequencies from about 20 Hz to 20 kHz. For example, if a carrier frequency is amplitude modulated with voice or music with frequencies from 100 Hz to 10 kHz, the frequency spectrum is as shown in Figure 18–24. Instead of one lower-side and one upper-side frequency as in the case of a single-frequency modulating signal, a band of lower-side frequencies and a band of upper-side frequencies correspond to the sum and difference frequencies of each sinusoidal component of the voice or music signal.

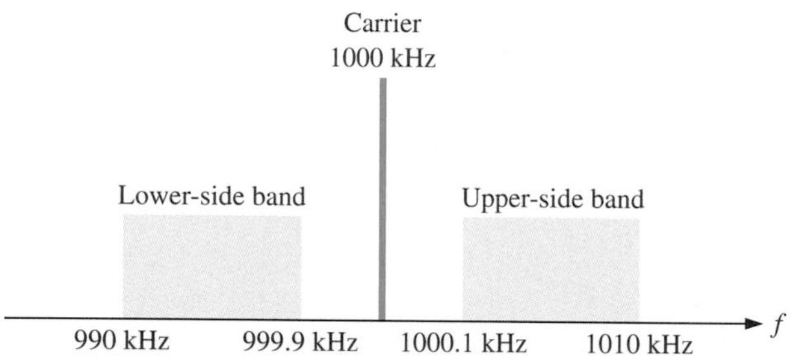

FIGURE 18–24

Example of a frequency spectrum for a voice or music signal.

SECTION 18–3 CHECKUP

1. What is amplitude modulation?
2. What is the difference between balanced modulation and standard AM?
3. What two input signals are used in amplitude modulation? Explain the purpose of each signal.
4. What are the upper-side frequency and the lower-side frequency?
5. How can a balanced modulator be changed to a standard amplitude modulator?

18–4 THE MIXER

The mixer in the receiver system discussed in Section 18–1 can be implemented with a linear multiplier as you will see in this section. The basic principles of linear multiplication of sinusoidal signals are covered, and you will see how sum and difference frequencies are produced. The difference frequency is a critical part of the operation of many types of receiver systems.

After completing this section, you should be able to

- **Discuss the basic function of a mixer**
  - Explain why a mixer is a linear multiplier
  - Describe the frequencies in the mixer and IF portion of a receiver

The **mixer** is basically a frequency converter because it changes the frequency of a signal to another value. The mixer in a receiver system takes the incoming modulated RF signal (which is sometimes amplified by an RF amplifier and sometimes not) along with the signal from the local oscillator and produces a modulated signal with a frequency equal to the difference of its two input frequencies (RF and LO). The mixer also produces a frequency equal to the sum of the input frequencies. The mixer function is illustrated in Figure 18–25.

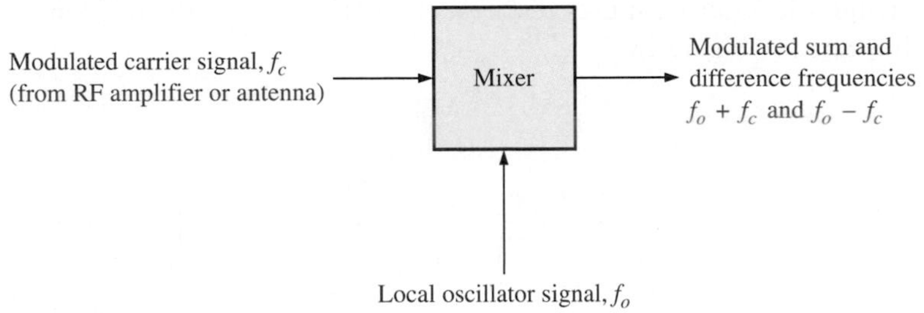

FIGURE 18–25

The mixer function.

In the case of receiver applications, the mixer must produce an output that has a frequency component equal to the difference of its input frequencies. From the mathematical analysis in Section 18–3, you can see that if two sinusoidal signals are multiplied, the product contains the difference frequency and the sum frequency. Thus, the mixer is actually a linear multiplier as indicated in Figure 18–26.

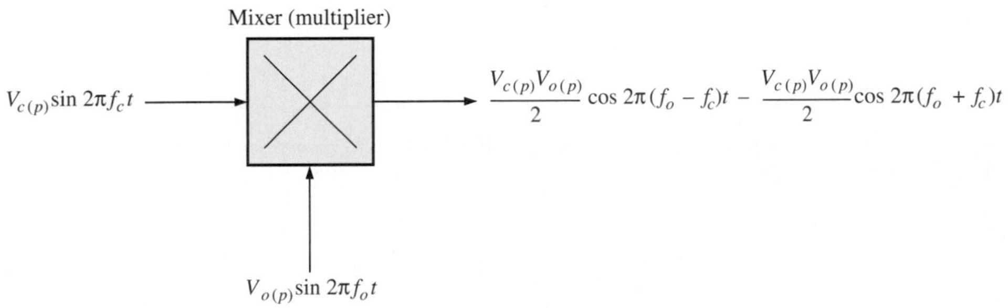

FIGURE 18–26

The mixer as a linear multiplier.

EXAMPLE 18–4

Determine the output expression for a multiplier with one sinusoidal input having a peak voltage of 5 mV and a frequency of 1200 kHz and the other input having a peak voltage of 10 mV and a frequency of 1655 kHz.

Solution The two input expressions are

$$v_1 = (5\text{ mV})\sin 2\pi(1200\text{ kHz})t$$
$$v_2 = (10\text{ mV})\sin 2\pi(1655\text{ kHz})t$$

Multiplying,

$$v_1v_2 = (5\text{ mV})(10\text{ mV})[\sin 2\pi(1200\text{ kHz})t][\sin 2\pi(1655\text{ kHz})t]$$

Applying the trigonometric identity, $(\sin A)(\sin B) = \frac{1}{2}[\cos(A - B) - \cos(A + B)]$,

$$V_{out} = \frac{(5\text{ mV})(10\text{ mV})}{2}\cos 2\pi(1655\text{ kHz} - 1200\text{ kHz})t - \frac{(5\text{ mV})(10\text{ mV})}{2}\cos 2\pi(1655\text{ kHz} + 1200\text{ kHz})t$$

$$\mathbf{V_{out} = (25\ \mu V)\cos 2\pi(455\text{ kHz})t - (25\ \mu V)\cos 2\pi(2855\text{ kHz})t}$$

Related Problem What is the value of the peak amplitude and frequency of the difference frequency component in this example?

In the receiver system, both the sum and difference frequencies from the mixer are applied to the IF (intermediate frequency) amplifier. The IF amplifier is actually a tuned amplifier that is designed to respond to the difference frequency while rejecting the sum frequency. You can think of the IF amplifier section of a receiver as a band-pass filter plus an amplifier because it uses resonant circuits to provide the frequency selectivity. This is illustrated in Figure 18–27.

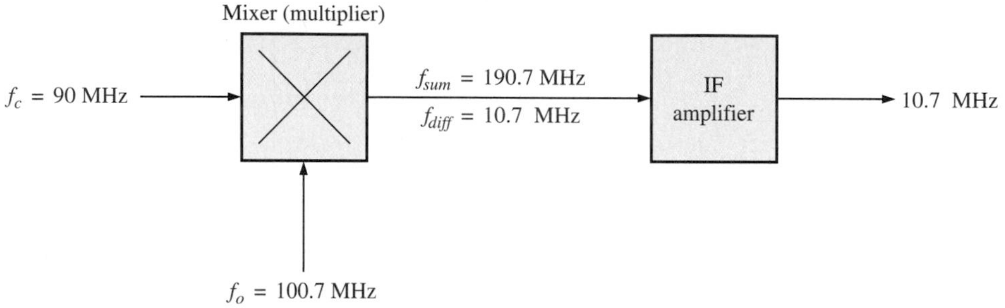

▲ **FIGURE 18–27**

Example of frequencies in the mixer and IF portion of a receiver.

EXAMPLE 18–5

Determine the output frequency of the IF amplifier for the conditions shown in Figure 18–28.

FIGURE 18–28

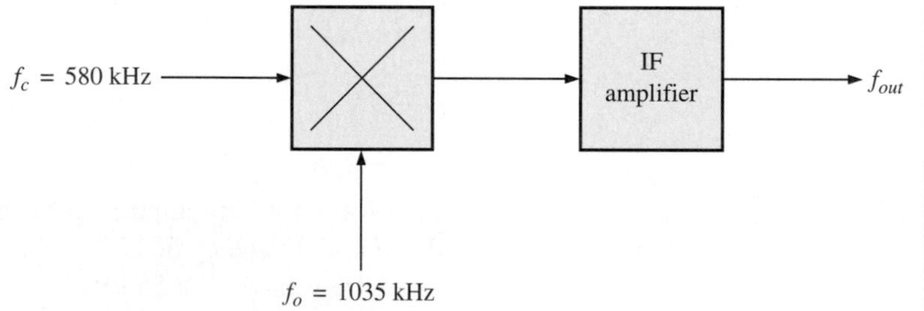

Solution The IF amplifier produces only the difference frequency signal on its output.

$$f_{out} = f_{diff} = f_o - f_c = 1035\text{ kHz} - 580\text{ kHz} = \mathbf{455\text{ kHz}}$$

Related Problem Based on your basic knowledge of the superheterodyne receiver from Section 18–1, determine the IF output frequency when the incoming RF signal changes to 1550 kHz.

SECTION 18–4 CHECKUP

1. What is the purpose of the mixer in a superheterodyne receiver?
2. How does the mixer produce its output?
3. If a mixer has 1000 kHz on one input and 350 kHz on the other, what frequencies appear on the output?

18–5 AM DEMODULATION

The linear multiplier can be used to demodulate or detect an AM signal as well as to perform the modulation process that was discussed in Section 18–3. **Demodulation** can be thought of as reverse modulation. The purpose is to get back the original modulating signal (sound in the case of standard AM receivers). The detector in the AM receiver can be implemented using a multiplier although another method using peak envelope detection is common.

After completing this section, you should be able to

- **Describe AM demodulation**
    - Discuss a basic AM demodulator
    - Discuss the frequency spectra

The Basic AM Demodulator

An AM demodulator can be implemented with a linear multiplier followed by a low-pass filter, as shown in Figure 18–29. The critical frequency of the filter is the highest audio frequency that is required for a given application (15 kHz, for example).

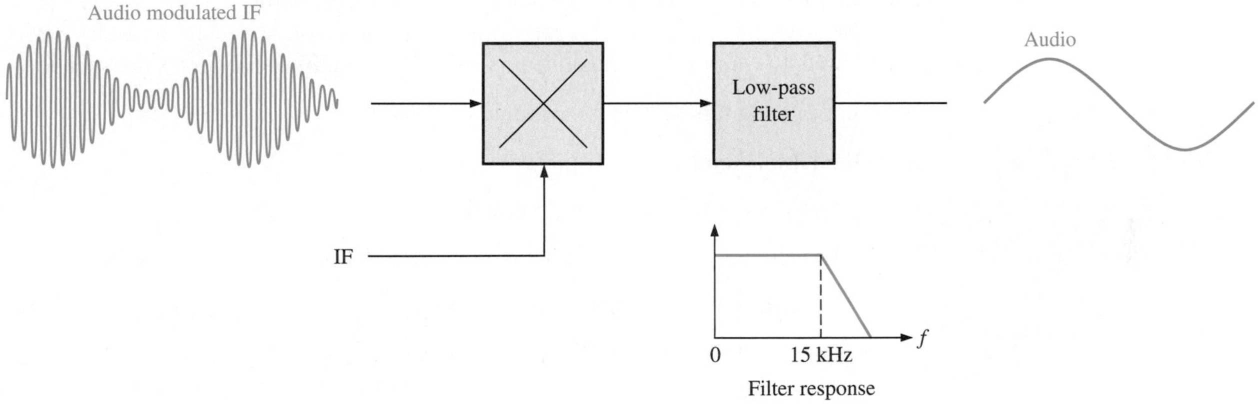

FIGURE 18–29

Basic AM demodulator.

Operation in Terms of the Frequency Spectra

Let's assume a carrier modulated by a single tone with a frequency of 10 kHz is received and converted to a modulated intermediate frequency of 455 kHz, as indicated by the frequency spectra in Figure 18–30. Notice that the upper-side and lower-side frequencies are separated from both the carrier and the IF by 10 kHz.

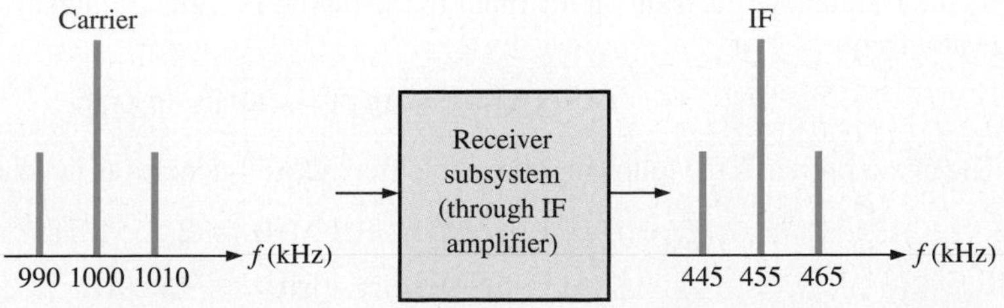

FIGURE 18–30

An AM signal converted to IF.

When the modulated output of the IF amplifier is applied to the demodulator along with the IF, sum and difference frequencies for each input frequency are produced as shown in Figure 18–31. Only the 10 kHz audio frequency is passed by the filter. A drawback to this type of AM detection is that a pure IF must be produced to mix with the modulated IF.

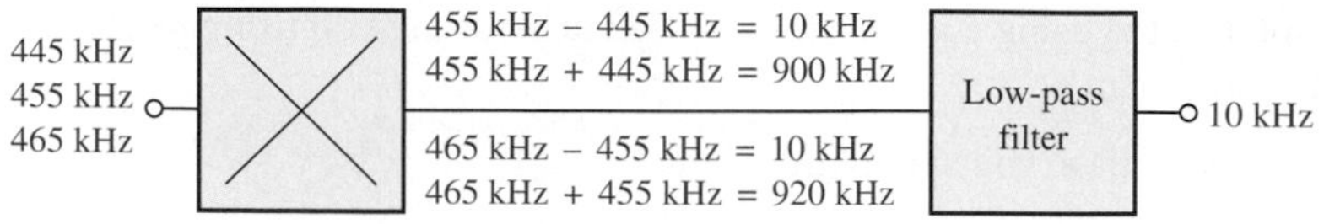

FIGURE 18–31

Example of demodulation.

SECTION 18–5 CHECKUP

1. What is the purpose of the filter in the linear multiplier demodulator?
2. If a 455 kHz IF modulated by a 1 kHz audio frequency is demodulated, what frequency or frequencies appear on the output of the demodulator?

18–6 IF AND AUDIO AMPLIFIERS

In this section, amplifiers for intermediate and audio frequencies are introduced. A typical IF amplifier is discussed and audio preamplifiers and power amplifiers are covered. As you have learned, the IF amplifier in a communications receiver provides amplification of the modulated IF signal out of the mixer before it is applied to the detector. After the audio signal is recovered by the detector, it goes to the audio preamp where it is amplified and applied to the power amplifier that drives the speaker.

After completing this section, you should be able to

- **Describe IF and audio amplifiers**
  - Discuss the function of an IF amplifier
  - Explain how the local oscillator and mixer operate with the IF amplifier
  - State the purpose of the audio amplifier
  - Discuss the LM386 audio power amplifier

IF Amplifiers

The IF amplifier in a receiver is a tuned amplifier with a specified bandwidth operating at a center frequency of 455 kHz for AM and 10.7 MHz for FM. The IF amplifier is one of the key features of a superheterodyne receiver because it is set to operate at a single resonant frequency that remains the same over the entire band of carrier frequencies that can be received. Figure 18–32 illustrates the basic function of an IF amplifier in terms of the frequency spectra.

Assume, for example, that the received carrier frequency of $f_c = 1$ MHz is modulated by an audio signal with a maximum frequency of $f_m = 5$ kHz, indicated in Figure 18–32 by the frequency spectrum on the input to the mixer. For this frequency, the local oscillator is at a frequency of

$$f_o = 1\text{ MHz} + 455\text{ kHz} = 1.455\text{ MHz}$$

The mixer produces the following sum and difference frequencies as indicated in Figure 18–32.

$$f_o + f_c = 1.455\text{ MHz} + 1\text{ MHz} = 2.455\text{ MHz}$$
$$f_o - f_c = 1.455\text{ MHz} - 1\text{ MHz} = 455\text{ kHz}$$

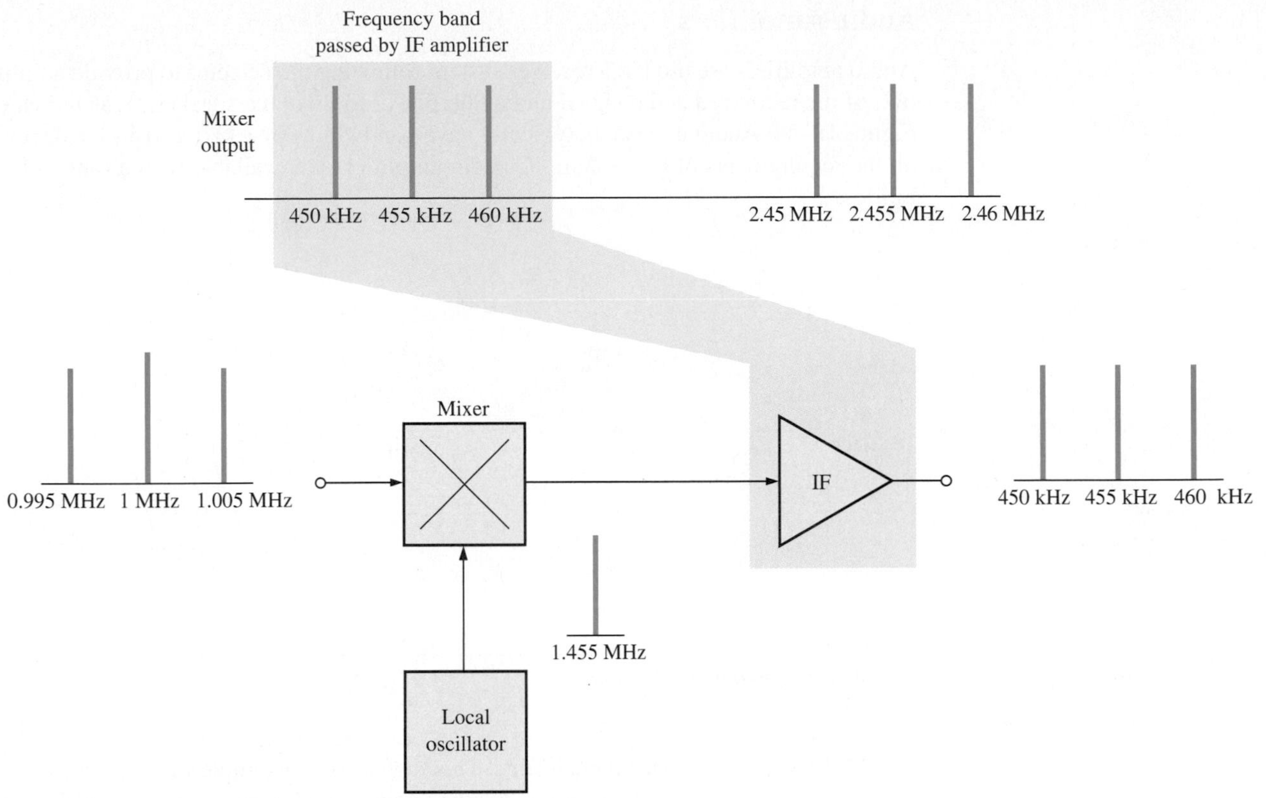

▲ FIGURE 18–32

An illustration of the basic function of the IF amplifier in an AM receiver.

$$f_o + (f_c + f_m) = 1.455\text{ MHz} + 1.005\text{ MHz} = 2.46\text{ MHz}$$
$$f_o + (f_c - f_m) = 1.455\text{ MHz} + 0.995\text{ MHz} = 2.45\text{ MHz}$$
$$f_o - (f_c + f_m) = 1.455\text{ MHz} - 1.005\text{ MHz} = 450\text{ kHz}$$
$$f_o - (f_c - f_m) = 1.455\text{ MHz} - 0.995\text{ MHz} = 460\text{ kHz}$$

Since the IF amplifier is a frequency-selective circuit, it responds only to 455 kHz and any side frequencies lying in the 10 kHz band centered at 455 kHz. All of the frequencies out of the mixer are rejected except the 455 kHz IF, all lower-side frequencies down to 450 kHz, and all upper-side frequencies up to 460 kHz. This frequency spectrum is the audio modulated intermediate frequency.

Although the detailed circuitry of the IF amplifier may differ from one system to another, it always has a tuned (resonant) circuit on the input or on the output or on both. Figure 18–33(a) shows a basic IF amplifier with tuned transformer coupling at the input and output. The general frequency response curve is shown in Figure 18–33(b).

◀ FIGURE 18–33

A basic IF amplifier with a tuned circuit on the input and output.

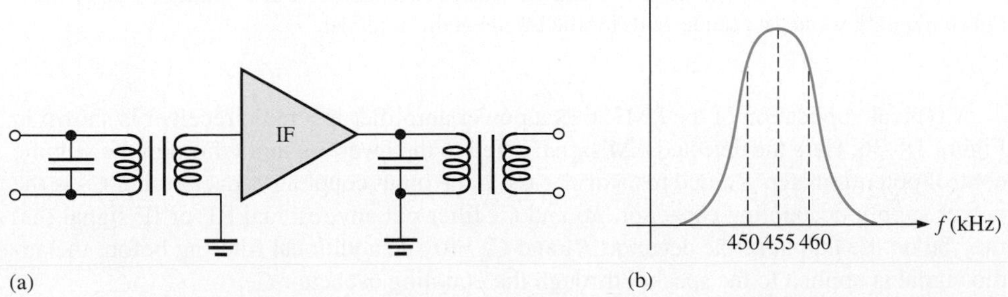

Audio Amplifiers

Audio amplifiers are used in a receiver system following the detector to provide amplification of the recovered audio signal and audio power to drive the speaker(s), as indicated in Figure 18–34. Audio amplifiers typically have bandwidths of 3 kHz to 15 kHz depending on the requirements of the system. IC audio amplifiers are available with a range of capabilities.

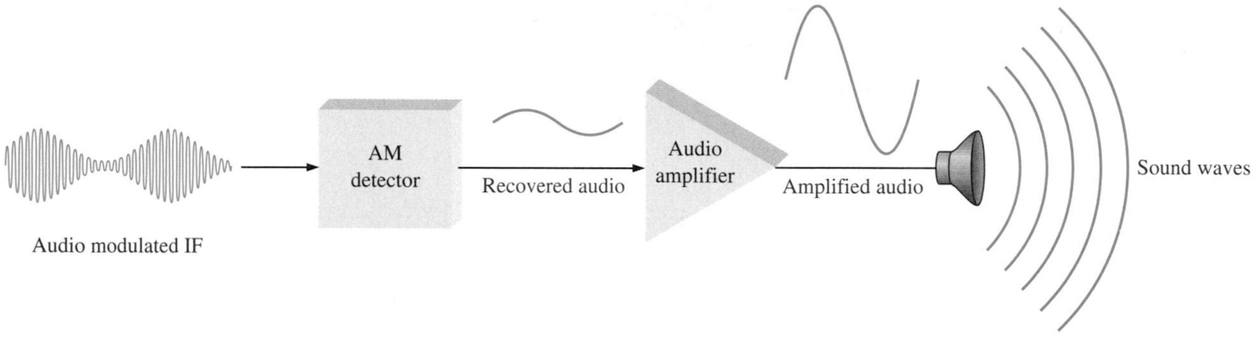

▲ FIGURE 18–34

The audio amplifier in a receiver system.

The LM386 Audio Power Amplifier This device is an example of a low-power audio amplifier that is capable of providing several hundred milliwatts to an 8 Ω speaker. It operates from any dc supply voltage in the 4 V to 12 V range, making it a good choice for battery operation. The pin configuration of the LM386 is shown in Figure 18–35(a). The voltage gain of the LM386 is 20 without external connections to the gain terminals, as shown in Figure 18–35(b). A voltage gain of 200 is achieved by connecting a 10 μF capacitor from pin 1 to pin 8, as shown in Figure 18–35(c). Voltage gains between 20 and 200 can be realized by a resistor (R_G) and capacitor (C_G) connected in series from pin 1 to pin 8 as shown in Figure 18–35(d). These external components are effectively placed in parallel with an internal gain-setting resistor.

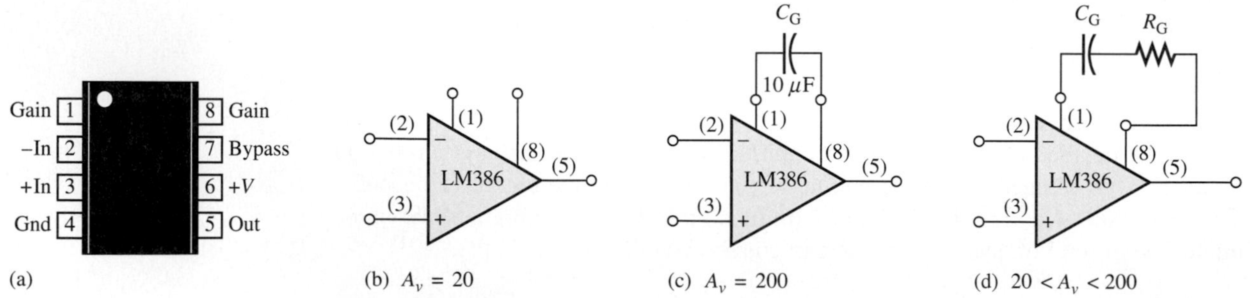

▲ FIGURE 18–35

Pin configuration and gain connections for the LM386 audio amplifier.

A typical application of the LM386 as a power amplifier in a radio receiver is shown in Figure 18–36. Here the detected AM signal is fed to the inverting input through the volume control potentiometer, R_1, and resistor R_2. C_1 is the input coupling capacitor and C_2 is the power supply decoupling capacitor. R_2 and C_3 filter out any residual RF or IF signal that may be on the output of the detector. R_3 and C_6 provide additional filtering before the audio signal is applied to the speaker through the coupling capacitor C_7.

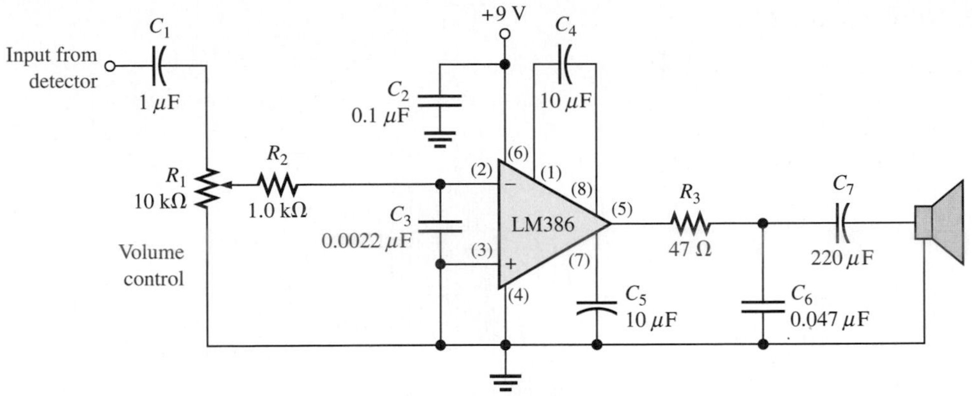

FIGURE 18–36

The LM386 as an AM audio power amplifier.

SECTION 18–6 CHECKUP

1. What is the purpose of the IF amplifier in an AM receiver?
2. What is the center frequency of an AM IF amplifier?
3. Why is the bandwidth of an AM receiver IF amplifier 10 kHz?
4. Why must the audio amplifier follow the detector in a receiver system?
5. Compare the frequency response of the IF amplifier to that of the audio amplifier.

18–7 FREQUENCY MODULATION

As you have seen, modulation is the process of varying a parameter of a carrier signal with an information signal. In amplitude modulation (AM) the parameter of amplitude is varied. In frequency modulation (FM), the frequency of a carrier is varied above and below its normal or at-rest value by a modulating signal. This section provides a further look into FM and discusses the differences between an AM and an FM receiver.

After completing this section, you should be able to

- **Describe frequency modulation**
  - Discuss the voltage-controlled oscillator
  - Explain frequency demodulation

In a frequency-modulated (FM) signal, the carrier frequency is increased or decreased according to the modulating signal. The amount of deviation above or below the carrier frequency depends on the amplitude of the modulating signal. The rate at which the frequency deviation occurs depends on the frequency of the modulating signal.

Figure 18–37 illustrates both a square wave and a sine wave modulating the frequency of a carrier. The carrier frequency is highest when the modulating signal is at its maximum positive amplitude and is lowest when the modulating signal is at its maximum negative amplitude.

A Basic Frequency Modulator

Frequency modulation is achieved by varying the frequency of an oscillator with the modulating signal. A voltage-controlled oscillator (VCO) is typically used for this purpose, as illustrated in Figure 18–38.

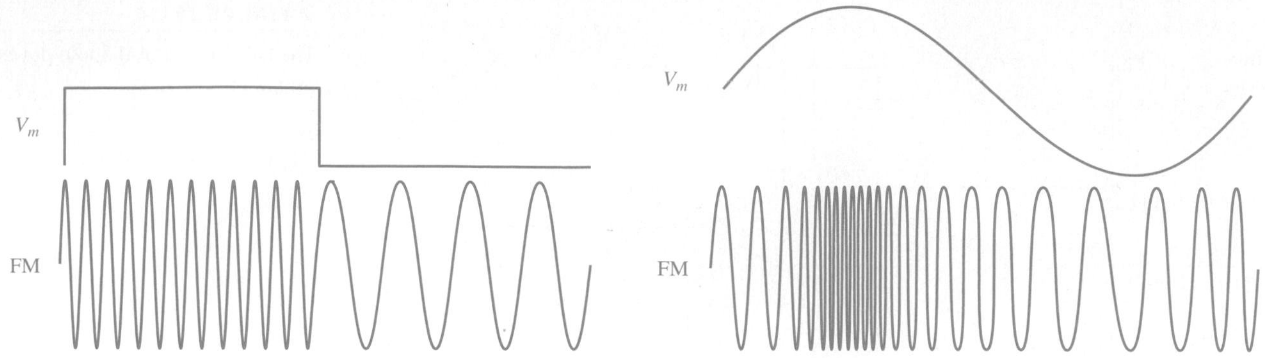

(a) Frequency modulation with a square wave

(b) Frequency modulation with a sine wave

▲ **FIGURE 18–37**

Examples of frequency modulation.

▶ **FIGURE 18–38**

Frequency modulation with a voltage-controlled oscillator.

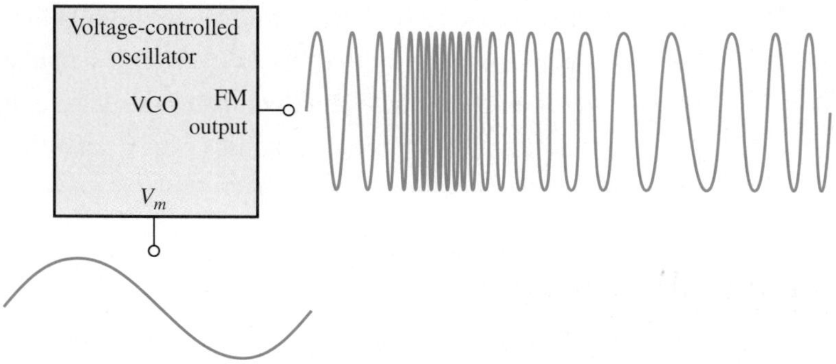

Generally, a variable-reactance type of voltage-controlled oscillator is used in FM applications. The variable-reactance VCO uses the varactor diode as a voltage-variable capacitance, as illustrated in Figure 18–39, where the capacitance is varied with the modulating voltage, V_m.

▶ **FIGURE 18–39**

Basic variable-reactance VCO.

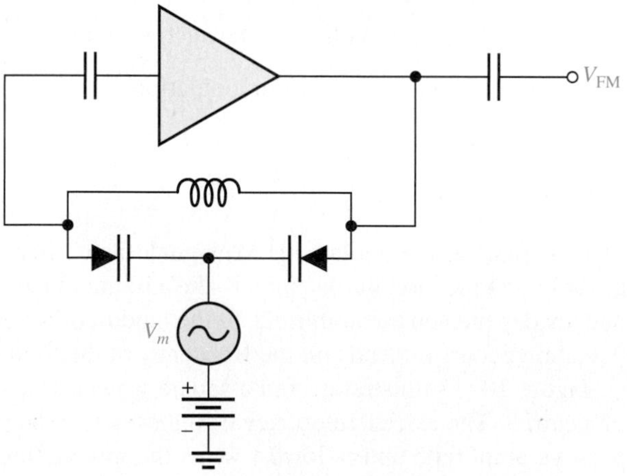

FM Demodulation

Except for the higher frequencies, the standard broadcast FM receiver is basically the same as the AM receiver up through the IF amplifier. The main difference between an FM receiver and an AM receiver, other than the frequency band, is the method used to recover the audio signal from the modulated IF.

There are several methods for demodulating an FM signal. These include slope detection, phase-shift discrimination, ratio detection, quadrature detection, and phase-locked loop demodulation. Most of these methods are covered in detail in communications courses. However, because of its importance in many types of applications, we will cover the phase-locked loop (PLL) demodulation in the next section.

SECTION 18–7 CHECKUP

1. How does an FM signal carry information?
2. On what principle are most VCOs used in FM based?

18–8 THE PHASE-LOCKED LOOP (PLL)

In the last section, the PLL was mentioned as a way to demodulate an FM signal. In addition to FM demodulation, PLLs are used in a wide variety of communications applications, which include TV receivers, tone decoders, telemetry receivers, modems, and data synchronizers, to name a few. Many of these applications are covered in electronic communications courses. In fact, entire books have been written on the finer points of PLL operation, analysis, and applications. The approach in this section is intended only to present the basic concept and give you an intuitive idea of how PLLs work and how they are used in FM demodulation. A specific PLL integrated circuit is also introduced.

After completing this section, you should be able to

- **Describe a phase-locked loop (PLL)**
  - Draw a basic block diagram for the PLL
  - Discuss the phase detector and state its purpose
  - State the purpose of the VCO
  - State the purpose of the low-pass filter
  - Explain lock range and capture range
  - Discuss the LM565 PLL and explain how it can be used as an FM demodulator

The Basic PLL Concept

The **phase-locked loop (PLL)** is a feedback circuit consisting of a phase detector, a low-pass filter, and a voltage-controlled oscillator (VCO). Some PLLs also include an amplifier in the loop, and in some applications the filter is not used.

The PLL is capable of locking onto or synchronizing with an incoming signal. When the phase of the incoming signal changes, indicating a change in frequency, the phase detector's output increases or decreases just enough to keep the VCO frequency the same as the frequency of the incoming signal. A basic PLL block diagram is shown in Figure 18–40.

The general operation of a PLL is as follows. The phase detector compares the phase difference between the incoming signal, V_i, and the VCO signal, V_o. When the frequency of the incoming signal, f_i, is different from that of the VCO frequency, f_o, the phase angle between the two signals is also different. The output of the phase detector and the filter is proportional to the phase difference of the two signals. This proportional voltage is fed to the VCO, forcing its frequency to move toward the frequency of the incoming signal until the two frequencies are equal. At this point, the PLL is locked onto the incoming frequency. If f_i changes, the phase difference also changes, forcing the VCO to track the incoming frequency.

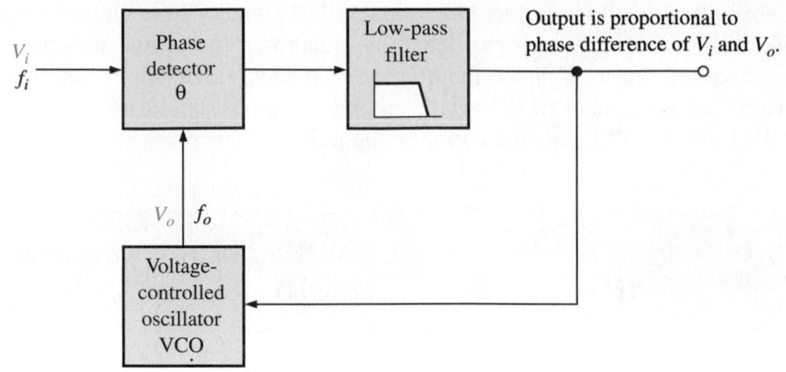

FIGURE 18–40

Basic PLL block diagram.

The Phase Detector

The phase-detector circuit in a PLL is basically a linear multiplier. The following analysis illustrates how it works in a PLL application. The incoming signal, V_i, and the VCO signal, V_o, applied to the phase detector can be expressed as

$$v_i = V_i \sin(2\pi f_i t + \theta_i)$$
$$v_o = V_o \sin(2\pi f_o t + \theta_o)$$

where θ_i and θ_o are the relative phase angles of the two signals. The phase detector multiplies these two signals and produces a sum and difference frequency output, V_d, as follows:

$$V_d = V_i \sin(2\pi f_i t + \theta_i) \times V_o \sin(2\pi f_o t + \theta_o)$$
$$= \frac{V_i V_o}{2}\cos[2\pi f_i t + \theta_i) - (2\pi f_o t + \theta_o)] - \frac{V_i V_o}{2}\cos[(2\pi f_i t + \theta_i) + (2\pi f_o t + \theta_o)]$$

When the PLL is locked,

$$f_i = f_o$$

and

$$2\pi f_i t = 2\pi f_o t$$

Therefore, the detector output voltage is

$$V_d = \frac{V_i V_o}{2}[\cos(\theta_i - \theta_o) - \cos(4\pi f_i t + \theta_i + \theta_o)]$$

The second cosine term in the above equation is a second harmonic term ($2 \times 2\pi f_i t$) and is filtered out by the low-pass filter.

The control voltage on the output of the filter is expressed as

Equation 18–4

$$V_c = \frac{V_i V_o}{2}\cos\theta_e$$

where the phase error $\theta_e = \theta_i - \theta_o$. The filter output voltage is proportional to the phase difference between the incoming signal and the VCO signal and is used as the control voltage for the VCO. This operation is illustrated in Figure 18–41.

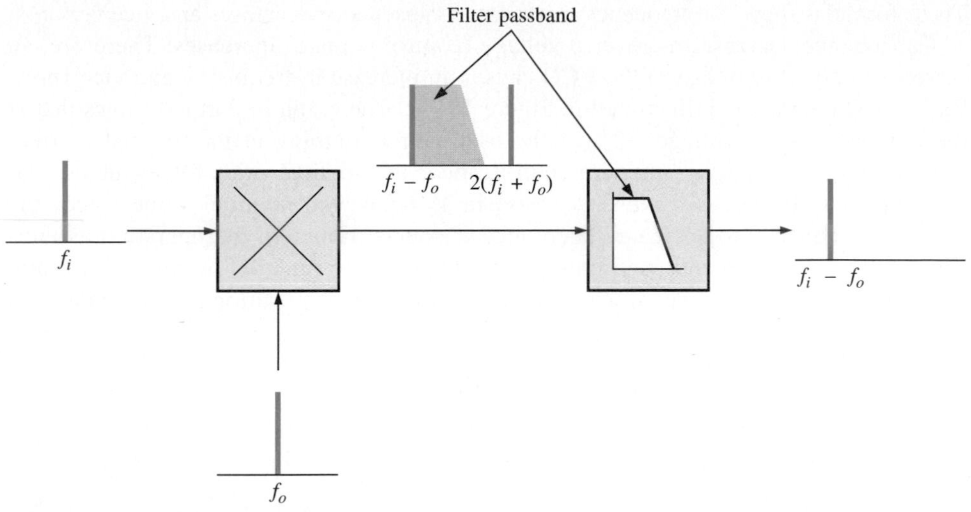

FIGURE 18–41
Basic phase detector/filter operation.

EXAMPLE 18–6

A PLL is locked onto an incoming signal with a frequency of 1 MHz at a phase angle of 50°. The VCO signal is at a phase angle of 20°. The peak amplitude of the incoming signal is 0.5 V and that of the VCO output signal is 0.7 V.

(a) What is the VCO frequency?

(b) What is the value of the control voltage being fed back to the VCO at this point?

Solution **(a)** Since the PLL is in lock, $f_i = f_o =$ **1 MHz.**

(b) $\theta_e = \theta_i - \theta_o = 50° - 20° = 30°$

$$V_c = \frac{V_i V_o}{2}\cos\theta_e = \frac{(0.5\text{ V})(0.7\text{ V})}{2}\cos 30° = (0.175\text{ V})\cos 30° = \mathbf{0.152\text{ V}}$$

Related Problem If the phase angle of the incoming signal changes instantaneously to 30°, indicating a change in frequency, what is the instantaneous VCO control voltage?

The Voltage-Controlled Oscillator (VCO)

Voltage-controlled oscillators can take many forms. A VCO can be some type of *LC* or crystal oscillator, or it can be some type of *RC* oscillator or multivibrator. No matter the exact type, most VCOs employed in PLLs operate on the principle of *variable reactance* using the varactor diode as a voltage-variable capacitor.

The capacitance of a varactor diode varies inversely with reverse-bias voltage. The capacitance decreases as reverse voltage increases and vice versa.

In a PLL, the control voltage fed back to the VCO is applied as a reverse-bias voltage to the varactor diode within the VCO. The frequency of oscillation is inversely related to capacitance for an *RC* type oscillator by the formula

$$f_o = \frac{1}{2\pi RC}$$

and for an *LC* type oscillator by the formula

$$f_o = \frac{1}{2\pi\sqrt{LC}}$$

These formulas show that frequency increases as capacitance decreases and vice versa.

Capacitance decreases as reverse voltage (control voltage) increases. Therefore, an increase in control voltage to the VCO causes an increase in frequency and vice versa. Basic VCO operation is illustrated in Figure 18–42. The graph in part (b) shows that at the nominal control voltage, $V_{c(nom)}$, the oscillator is running at its nominal or free-running frequency, $f_{o(nom)}$. An increase in V_c above the nominal value forces the oscillator frequency to increase, and a decrease in V_c below the nominal value forces the oscillator frequency to decrease. There are, of course, limits on the operation as indicated by the minimum and maximum points. The transfer function or conversion gain, K, of the VCO is normally expressed as a certain frequency deviation per unit change in control voltage.

$$K = \frac{\Delta f_o}{\Delta V_c}$$

▶ FIGURE 18–42

Basic VCO operation.

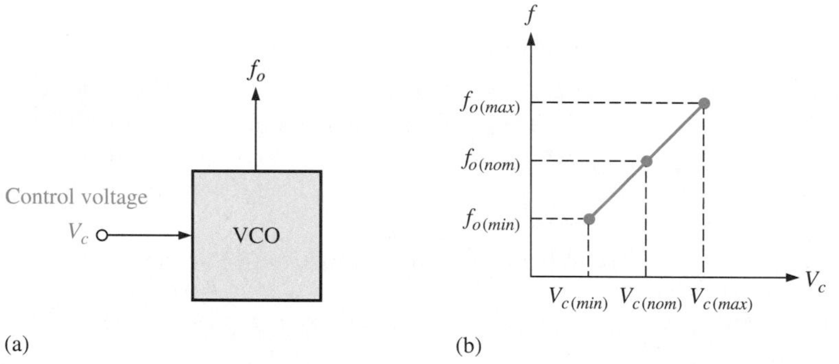

EXAMPLE 18–7

The output frequency of a certain VCO changes from 50 kHz to 65 kHz when the control voltage increases from 0.5 V to 1 V. What is the conversion gain, K?

Solution

$$K = \frac{\Delta f_o}{\Delta V_c} = \frac{65\text{ kHz} - 50\text{ kHz}}{1\text{ V} - 0.5\text{ V}} = \frac{15\text{ kHz}}{0.5\text{ V}} = \mathbf{30\ kHz/V}$$

Related Problem If the conversion gain of a certain VCO is 20 kHz/V, how much frequency deviation does a change in control voltage from 0.8 V to 0.5 V produce? If the VCO frequency is 250 kHz at 0.8 V, what is the frequency at 0.5 V?

Basic PLL Operation

When the PLL is locked, the incoming frequency, f_i, and the VCO frequency, f_o, are equal. However, there is always a phase difference between them called the *static phase error.* The phase error, θ_e, is the parameter that keeps the PLL locked in. As you have seen, the filtered voltage from the phase detector is proportional to θ_e (Equation 18–4). This voltage controls the VCO frequency and is always just enough to keep $f_o = f_i$.

Figure 18–43 shows the PLL and two sinusoidal signals of the same frequency but with a phase difference, θ_e. For this condition the PLL is in lock and the VCO control voltage is constant. If f_i decreases, θ_e increases to θ_{e1} as illustrated in Figure 18–44. This increase in θ_e is sensed by the phase detector causing the VCO control voltage to decrease, thus decreasing f_o until $f_o = f_i$ and keeping the PLL in lock. If f_i increases, θ_e decreases to θ_{e1} as

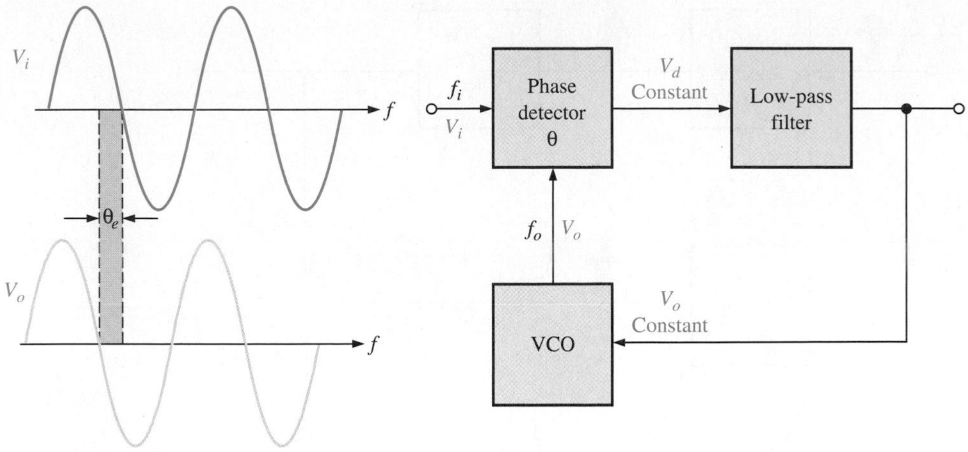

FIGURE 18–43

PLL in lock under static condition ($f_o = f_i$) and constant θ_e).

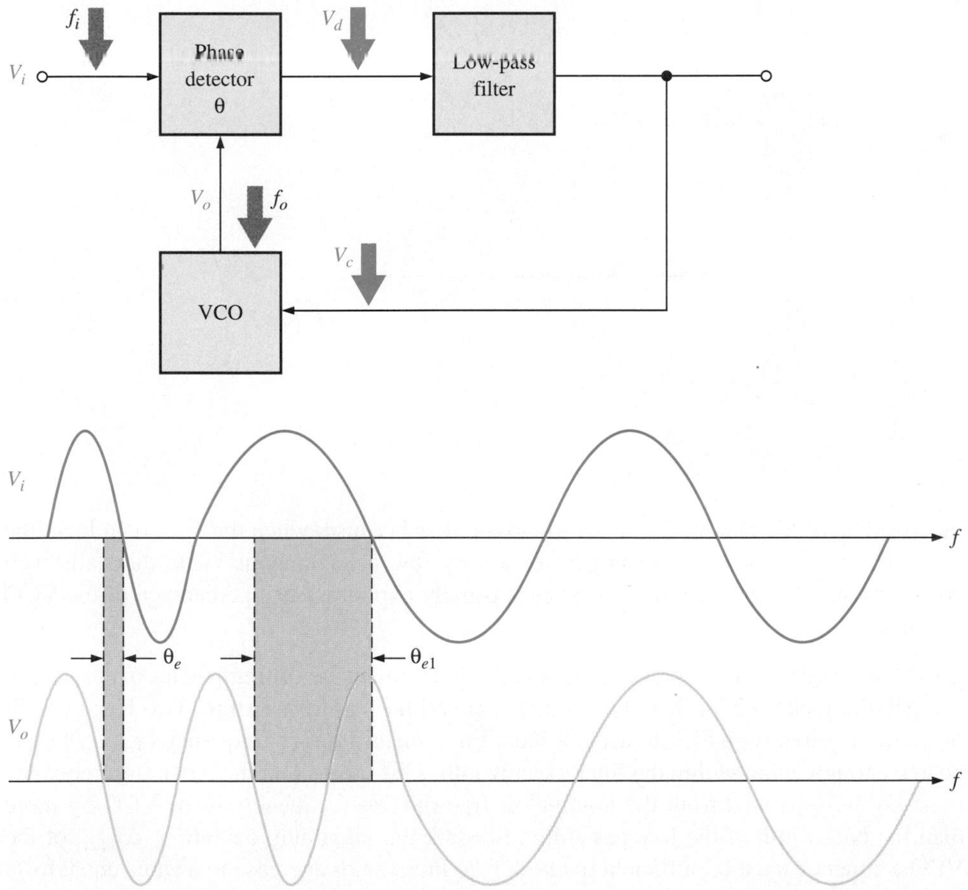

FIGURE 18–44

PLL action when f_i decreases.

illustrated in Figure 18–45. This decrease in θ_e causes the VCO control voltage to increase, thus increasing f_o until $f_o = f_i$ and keeping the PLL in lock.

Lock Range Once the PLL is locked, it will track frequency changes in the incoming signal. The range of frequencies over which the PLL can maintain lock is called the **lock range** or tracking range. Limitations on the hold-in range are the maximum frequency deviations of the VCO and the output limits of the phase detector. The hold-in range is

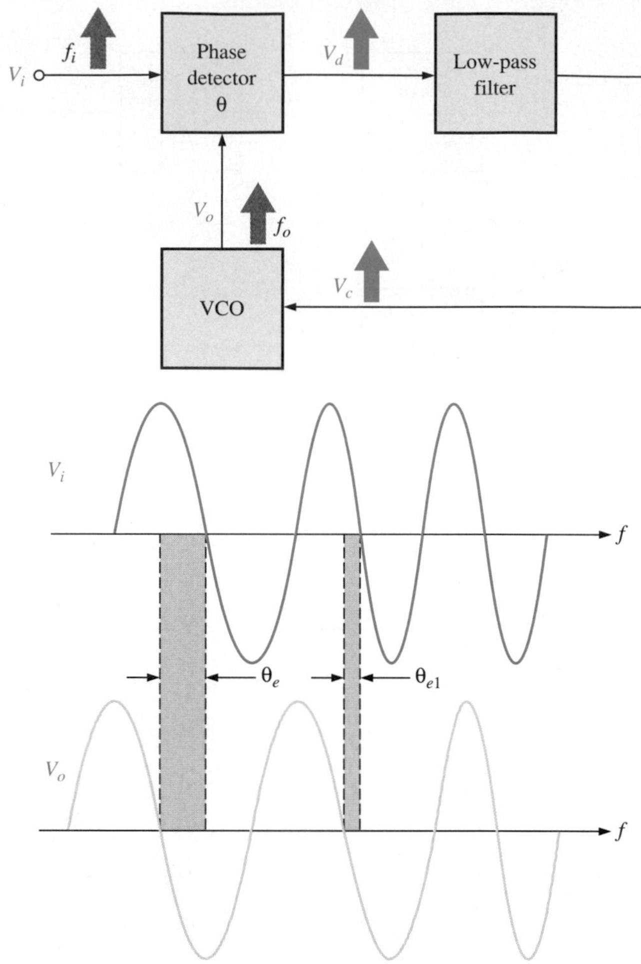

FIGURE 18–45

PLL action when f_i increases.

independent of the bandwidth of the low-pass filter because, when the PLL is in lock, the difference frequency ($f_i - f_o$) is zero or a very low instantaneous value that falls well within the bandwidth. The hold-in range is usually expressed as a percentage of the VCO frequency.

Capture Range Assuming the PLL is not in lock, the range of frequencies over which it can acquire lock with an incoming signal is called the **capture range**. Two basic conditions are required for a PLL to acquire lock. First, the difference frequency ($f_o - f_i$) must be low enough to fall within the filter's bandwidth. This means that the incoming frequency must not be separated from the nominal or free-running frequency of the VCO by more than the bandwidth of the low-pass filter. Second, the maximum deviation, Δf_{max}, of the VCO frequency must be sufficient to allow f_o to increase or decrease to a value equal to f_i. These conditions are illustrated in Figure 18–46; and when they exist, the PLL will "pull" the VCO frequency toward the incoming frequency until $f_o = f_i$.

The LM565 Phase-Locked Loop

The LM565 is a good example of an integrated circuit PLL. The circuit consists of a VCO, a phase detector, a low-pass filter formed by an internal resistor and an external capacitor, and an amplifier. The free-running VCO frequency can be set with external components. A block diagram is shown in Figure 18–47. The LM565 can be used for the frequency range from 0.001 Hz to 500 kHz.

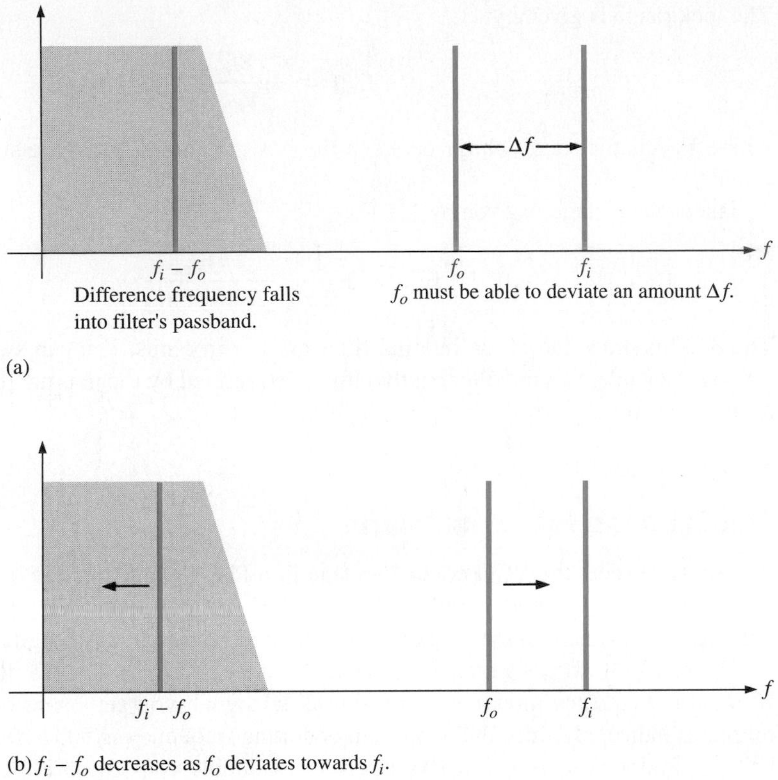

◄ **FIGURE 18–46**

Illustration of the conditions for a PLL to acquire lock.

The free-running frequency of the VCO is set by the values of R_1 and C_1 in Figure 18–47 according to the following formula. The frequency is in hertz when the resistance is in ohms and the capacitance is in farads.

$$f_o \cong \frac{1.2}{4R_1C_1}$$

Equation 18–5

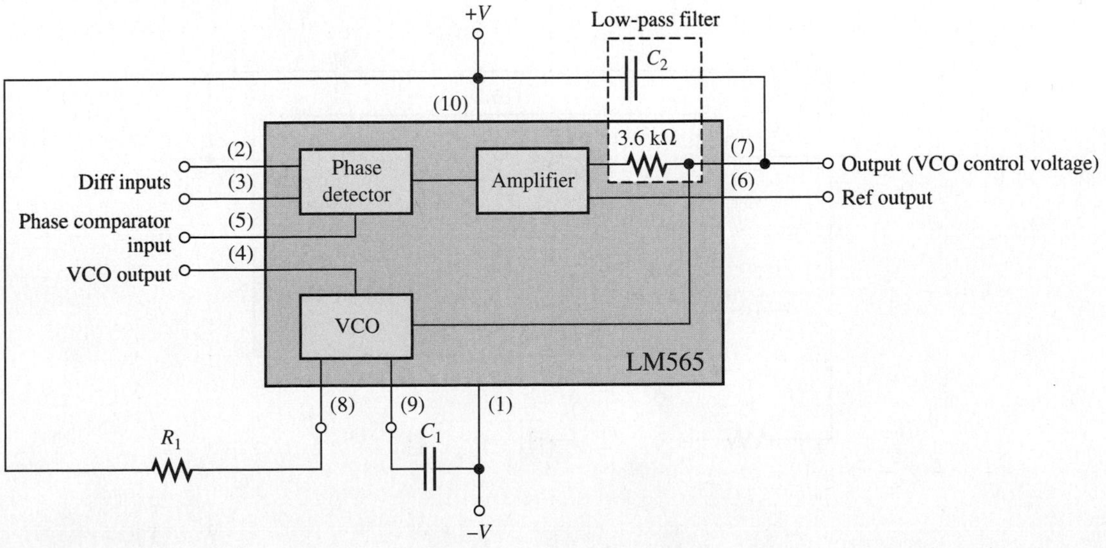

▲ **FIGURE 18–47**

Block diagram of the LM565 PLL.

The lock range is given by

Equation 18–6

$$f_{lock} = \pm\frac{8f_o}{V_{CC}}$$

where V_{CC} is the total voltage between the positive and negative dc supply voltage terminals.

The capture range is given by

Equation 18–7

$$f_{cap} \cong \pm\frac{1}{2\pi}\sqrt{\frac{2\pi f_{lock}}{(3600\ \Omega)C_2}}$$

The 3600 is the value of the internal filter resistor in ohms. You can see that the capture range is dependent on the filter bandwidth as determined by the internal resistor and the external capacitor C_2.

The PLL as an FM Demodulator

As you have seen, the VCO control voltage in a PLL depends on the deviation of the incoming frequency. The PLL will produce a voltage proportional to the frequency of the incoming signal which, in the case of FM, is the original modulating signal.

Figure 18–48 shows a typical connection for the LM565 as an FM demodulator. If the IF input is frequency modulated by a sinusoidal signal, you get a sinusoidal signal on the output as indicated. Since the maximum operating frequency is 500 kHz, this device must be used in double-conversion FM receivers. A double-conversion FM receiver is one in which essentially two mixers are used to first convert the RF to a 10.7 MHz IF and then convert this to a 455 kHz IF.

The free-running frequency of the VCO is adjusted to approximately 455 kHz, which is the center of the modulated IF range. C_1 can be any value, but R_1 should be in the range from 2 kΩ to 20 kΩ. The input can be directly coupled as long as there is no dc voltage difference between pins 2 and 3. The VCO is connected to the phase detector by an external wire between pins 4 and 5.

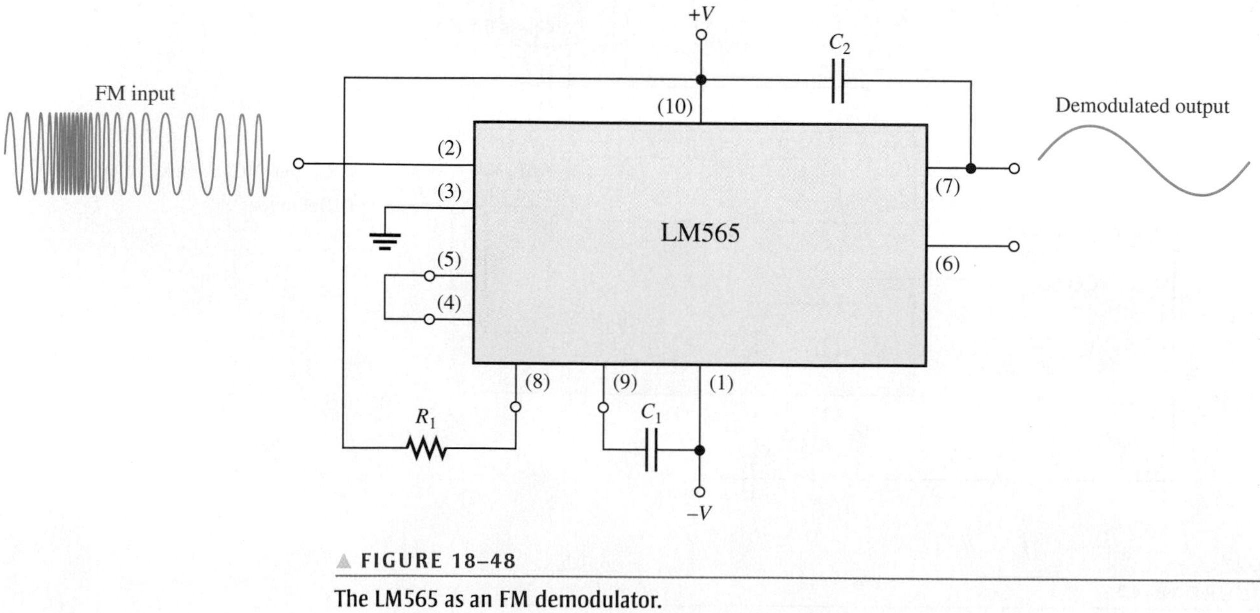

▲ FIGURE 18–48

The LM565 as an FM demodulator.

EXAMPLE 18–8

Determine the values for R_1, C_1, and C_2 for the LM565 in Figure 18–48 for a free-running frequency of 455 kHz and a capture range of ± 10 kHz. The dc supply voltages are ± 6 V.

Solution Use Equation 18–5 to calculate C_1. Choose R_1 = **4.7 kΩ**.

$$f_o \cong \frac{1.2}{4R_1C_1}$$

$$C_1 \cong \frac{1.2}{4R_1f_o} = \frac{1.2}{4(4700\ \Omega)(455 \times 10^3\ \text{Hz})} = 140 \times 10^{-12}\ \text{F} = \mathbf{140\ pF}$$

The lock range must be determined before you can calculate C_2.

$$f_{lock} = \pm\frac{8f_o}{V_{CC}} = \pm\frac{8(455\ \text{kHz})}{12\ \text{V}} = \pm 303\ \text{kHz}$$

Use Equation 18–7 to calculate C_2.

$$f_{cap} \cong \pm\frac{1}{2\pi}\sqrt{\frac{2\pi f_{lock}}{(3600\ \Omega)C_2}}$$

$$f_{cap}^2 \cong \left(\frac{1}{2\pi}\right)^2 \frac{2\pi f_{lock}}{(3600\ \Omega)C_2}$$

$$C_2 \cong \left(\frac{1}{2\pi}\right)^2 \frac{2\pi f_{lock}}{(3600\ \Omega)f_{cap}^2} = \left(\frac{1}{2\pi}\right)^2 \frac{2\pi(303 \times 10^3\ \text{Hz})}{(3600\ \Omega)(10 \times 10^3\ \text{Hz})^2}$$

$$= 0.134 \times 10^{-6}\ \text{F} = \mathbf{0.134\ \mu F}$$

Related Problem What can you do to increase the capture range from ± 10 kHz to ± 15 kHz?

SECTION 18–8 CHECKUP

1. List the three basic components in a phase-locked loop.
2. What is another circuit used in some PLLs other than the three listed in Question 1?
3. What is the basic function of a PLL?
4. What is the difference between the lock range and the capture range of a PLL?
5. Basically, how does a PLL track the incoming frequency?

18–9 FIBER OPTICS

Fiber-optic cables are replacing copper wire as a means of sending signals over long distances in many types of communications systems. Fiber optics is used by cable television, telephone, and electric utility companies, among others.

After completing this section, you should be able to

- **Discuss fiber-optic cables**
  - Describe how signals are sent through a fiber-optic cable
  - Define the basic types of fiber-optic cable

Instead of using electrical pulses to transmit information through copper lines, **fiber optics** uses light pulses to transmit information through fiber-optic cables about the diameter of a human hair, which is about 100 microns (one millionth of a meter). Fiber-optic systems have several advantages over systems using copper wire. These include faster speed, higher signal capacity, longer transmission distances without amplification, less susceptibility to interference, and they are more economical to maintain.

Basic Operation

When light is introduced into one end of a fiber-optic cable, it "bounces" along until it emerges from the other end. The fiber is generally made of pure glass or plastic that is surrounded by a highly reflective cladding that acts essentially as a mirrored surface but actually uses a physics phenomenon called *total internal reflection* to produce an almost lossless reflection. Think of a fiber-optic cable as a pipe lined inside with a mirror. As the light moves along the fiber, it is reflected off the cladding so that it can move around bends in the fiber with essentially no loss. A fiber-optic cable consists of the core, which is the glass fiber itself, the cladding that surrounds the fiber and provides the reflective surface, and the outer coating or jacket that provides protection. Other layers may be added for strengthening. The basic structure of a single fiber-optic cable is illustrated in Figure 18–49(a), and the propagation of light along a fiber with a bend is shown in part (b). It doesn't matter whether the fiber is straight or bent; the light still travels through it but with higher loss when the bend angle is high.

FIGURE 18–49

Simplified structure and operation of a fiber-optic cable.

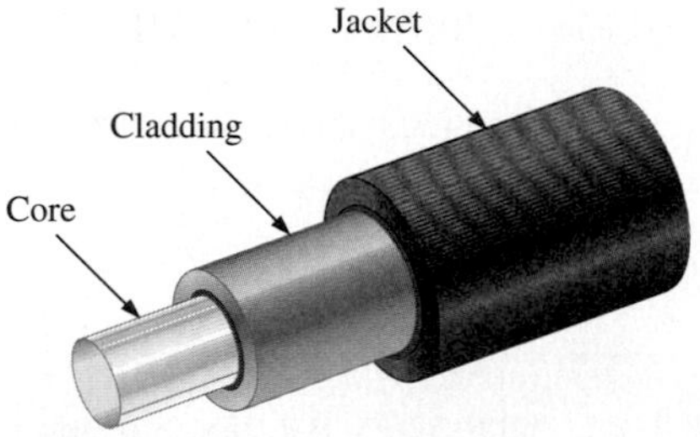

(a) Basic construction

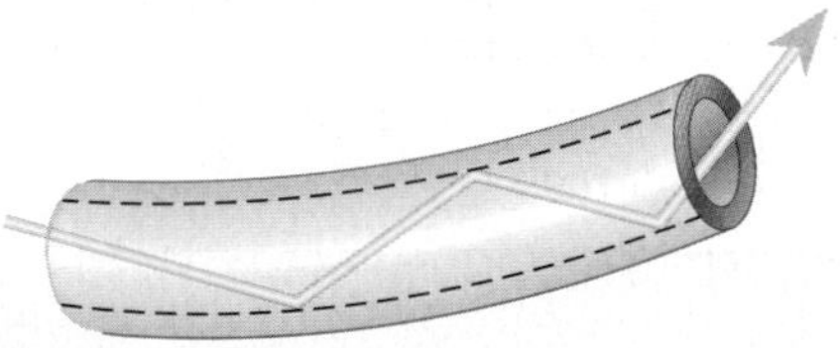

(b) Propagation of a light ray

When a light ray enters the fiber-optic cable, it strikes the reflective surface of the cladding at an angle called the **angle of incidence**, θ_i, If the angle of incidence is greater than a parameter known as the **critical angle**, θ_c, the light ray is then reflected back into the core at an angle called the **angle of reflection,** θ_r, as shown in Figure 18–50(a). The angle of incidence is always equal to the angle of reflection. If the angle of incidence is less than the critical angle, the light ray is refracted and passes into the cladding, causing energy to be lost, as shown in Figure 18–50(b). This is called *scattering* and any refracted light represents a loss or attenuation as a light ray is propagated through the fiber-optic cable. Another cause of attenuation of light in a fiber-optic cable is called *absorption,* which is caused by the interactions of the light photons and the molecules of the core.

FIGURE 18–50

Critical angle in a fiber-optic cable.

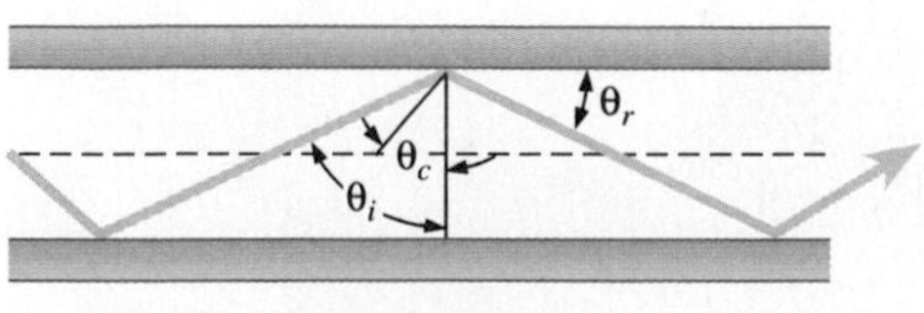

(a) Reflection of a light ray ($\theta_i > \theta_c$)

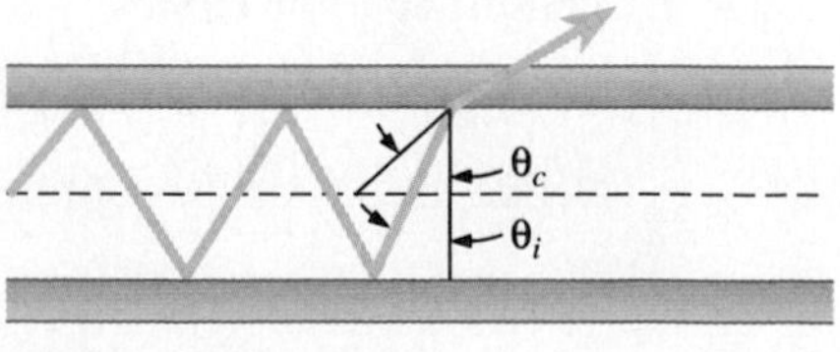

(b) Refraction of a light ray ($\theta_i < \theta_c$)

The core material and the cladding material each have a parameter known as the **index of refraction,** which determines the critical angle. The critical angle is defined by the formula

$$\theta_c = \cos^{-1}\left(\frac{n_2}{n_1}\right)$$

Equation 18–8

where n_1 is the index of refraction of the core and n_2 is the index of refraction of the cladding.

EXAMPLE 18–9

A certain fiber-optic cable has a core index of refraction of 1.35 and a cladding index of refraction of 1.30. Determine the critical angle.

Solution

$$\theta_c = \cos^{-1}\left(\frac{n_2}{n_1}\right) = \cos^{-1}\left(\frac{1.30}{1.35}\right) = \mathbf{15.6^\circ}$$

Related Problem Calculate the critical angle if $n_1 = 1.67$ and $n_2 = 1.59$.

Modes of Light Propagation

Three basic modes of light propagation in fiber-optic cables are multimode step index, single-mode step index, and multimode graded index.

Multimode Step Index Figure 18–51 shows a fiber-optic cable in which the diameter of the core is fairly large relative to the diameter of the cladding. As shown, there is a sharp transition in the index of refraction going from the core to the cladding, thus the term *step.* Light entering the cable will tend to propagate through the core in multiple rays or modes, as indicated. Some of the rays will go straight down the core while others will bounce back and forth as they propagate. Still others will scatter due to their small angle of incidence, causing attenuation in the light energy. As a result of the multiple modes, the light will encounter time dispersion; that is, all the light rays will not arrive at the end of the cable at exactly the same time.

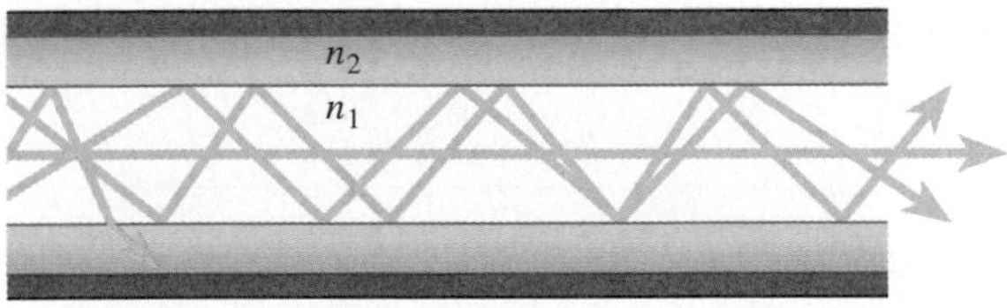

◀ **FIGURE 18–51**

Multimode step index fiber-optic cable.

Single-Mode Step Index Figure 18–52 shows a fiber-optic cable in which the diameter core is very small relative to the diameter of the cladding. There is a sharp transition in the index of refraction going from the core to the cladding. Light entering the cable tends to propagate through the core in a single ray or mode. This results in much less attenuation and, ideally, no time dispersion compared to the multimode cable.

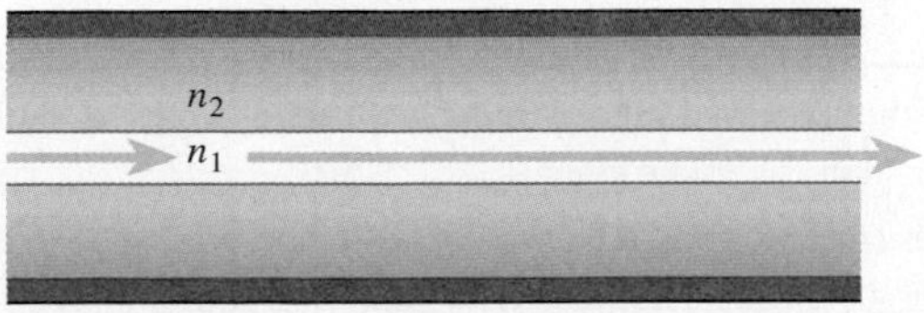

◀ **FIGURE 18–52**

Single-mode step index fiber-optic cable.

Multimode Graded Index Figure 18–53 shows a fiber-optic cable in which the diameter of the core is fairly large relative to the diameter of the cladding. There is a gradual or graded transition in the index of refraction going from the center of the core into the cladding. Light rays will be more curved as they bounce through the gradually changing indices of refraction resulting in less attenuation and time dispersion than in the multimode step index cable.

▶ FIGURE 18–53

Multimode graded index fiber-optic cable.

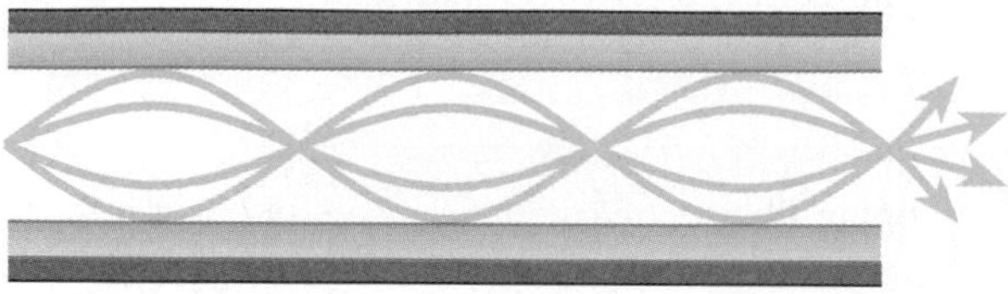

A Fiber-Optic Data Communications Link

A simplified block diagram of a fiber-optic data communications link is shown in Figure 18–54. The source provides the electrical signal that is to be transmitted. This electrical signal is converted to a light signal and coupled to the fiber-optic cable by the transmitter. At the receiving end, the light signal is coupled out of the cable into the receiver, which converts it to an electrical signal. This signal is then processed and connected to the end user.

▶ FIGURE 18–54

Basic block diagram of a fiber-optic data communication link.

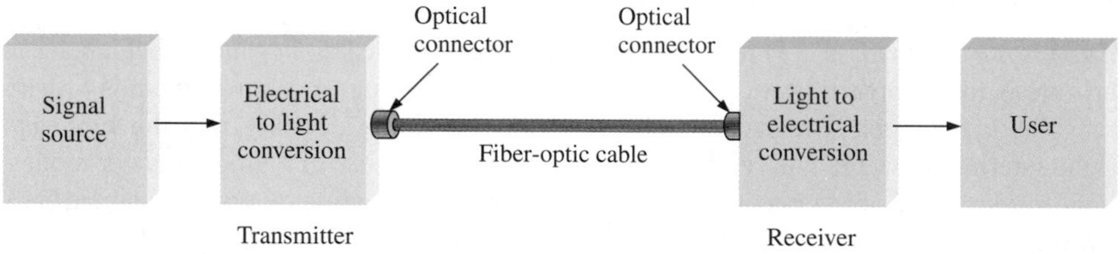

The electrical signal modulates the light intensity and produces a light signal that carries the same information as the electrical signal. A special connector then couples the light signal into the fiber-optic cable. At the other end the receiver demodulates the light signal and converts it back into the original electrical signal.

SECTION 18–9 CHECKUP

1. Generally, what is a fiber-optic cable made of?
2. Typically, what is the approximate diameter of a fiber-optic cable?
3. Name three basic parts of a fiber-optic cable.
4. What is the difference between the critical angle and the angle of incidence?
5. List three types of fiber-optic cables.

SUMMARY

Section 18–1

- In amplitude modulation (AM), the amplitude of a higher-frequency carrier signal is varied by a lower-frequency modulating signal (usually an audio signal).
- A basic superheterodyne AM receiver consists of an RF amplifier, a mixer, a local oscillator, an IF (intermediate frequency) amplifier, an AM detector, and audio and power amplifiers.
- The IF in a standard AM receiver is 455 kHz.
- The AGC (automatic gain control) in a receiver tends to keep the signal strength constant within the receiver to compensate for variations in the received signal.

- In frequency modulation (FM), the frequency of a carrier signal is varied by a modulating signal.
- A superheterodyne FM receiver is basically the same as an AM receiver except that it requires a limiter to keep the IF amplitude constant, a different kind of detector or discriminator, and a deemphasis network. The IF is 10.7 MHz.

Section 18–2

- A four-quadrant linear multiplier can handle any combination of voltage polarities on its inputs.
- The output of a linear multiplier is the product of the two inputs and a scale factor.

Section 18–3

- Amplitude modulation is basically a multiplication process.
- The multiplication of sinusoidal signals produces sum and difference frequencies.
- The output spectrum of a balanced modulator includes upper-side and lower-side frequencies, but no carrier frequency.
- The output spectrum of a standard amplitude modulator includes upper-side and lower-side frequencies and the carrier frequency.

Section 18–4

- A linear multiplier is used as the mixer in receiver systems.
- A mixer converts the RF signal down to the IF signal. The radio frequency varies over the AM or FM band. The intermediate frequency is constant.

Section 18–5

- One type of AM demodulator consists of a multiplier followed by a low-pass filter.

Section 18–6

- The audio and power amplifiers boost the output of the detector or discriminator and drive the speaker.

Section 18–7

- A voltage-controlled oscillator (VCO) produces an output frequency that can be varied by a control voltage. Its operation is based on a variable reactance.
- A VCO is a basic frequency modulator when the modulating signal is applied to the control voltage input.

Section 18–8

- A phase-locked loop (PLL) is a feedback circuit consisting of a phase detector, a low-pass filter, a VCO, and sometimes an amplifier.
- The purpose of a PLL is to lock onto and track incoming frequencies.
- A linear multiplier can be used as a phase detector.

Section 18–9

- Fiber optics provides a light path from a light-emitting device to a light-activated device.
- The three basic parts of a fiber-optic cable are the core, the cladding, and the jacket.
- Light rays must bounce off the core boundary at an angle (angle of incidence) greater than the critical angle in order to be reflected.
- Light rays that strike the core boundary at an angle less than the critical angle are refracted into the cladding, resulting in attenuation of the light.
- The angle of reflection is equal to the angle of incidence.
- Three types of fiber-optic cable are multimode step index, single-mode step index, and multimode graded index.

KEY TERMS

Key terms and other bold terms in the chapter are defined in the end-of-book glossary.

Amplitude modulation (AM) A communication method in which a lower-frequency signal modulates (varies) the amplitude of a higher-frequency signal (carrier).

Angle of incidence The angle at which a light ray strikes a surface.

Balanced modulation A form of amplitude modulation in which the carrier is suppressed; sometimes known as *suppressed-carrier modulation.*

Capture range The range of frequencies over which a PLL can acquire lock.

Critical angle The angle that defines whether a light ray will be reflected or refracted as it strikes a surface.

Fiber optics The use of light for the transmission of information through tiny fiber cables.

Four-quadrant multiplier A linear device that produces an output voltage proportional to the product of two input voltages.

Freqency modulation (FM) A communication method in which a lower-frequency intelligence-carrying signal modulates (varies) the frequency of a higher-frequency signal.

Index of refraction An optical characteristic of a material that determines the critical angle.

Lock range The range of frequencies over which a PLL can maintain lock.

Mixer A device for down-converting frequencies in a receiver system.

Phase-locked loop (PLL) A device for locking onto and tracking the frequency of an incoming signal.

KEY FORMULAS

18–1	$V_{OUT} = KV_XV_Y$	Multiplier output voltage
18–2	$v_1v_2 = \dfrac{V_{1(p)}V_{2(p)}}{2}\cos 2\pi(f_1 - f_2)t - \dfrac{V_{1(p)}V_{2(p)}}{2}\cos 2\pi(f_1 + f_2)t$	Sum and difference frequencies
18–3	$V_{out} = V_{c(p)}^2 \sin 2\pi f_c t + \dfrac{V_{c(p)}V_{m(p)}}{2}\cos 2\pi(f_c - f_m)t - \dfrac{V_{c(p)}V_{m(p)}}{2}\cos 2\pi(f_c + f_m)t$	Standard AM
18–4	$V_c = \dfrac{V_iV_o}{2}\cos\theta_e$	PLL control voltage
18–5	$f_o \cong \dfrac{1.2}{4R_1C_1}$	Output frequency LM565
18–6	$f_{lock} = \pm\dfrac{8f_o}{V_{CC}}$	Lock range LM565
18–7	$f_{cap} \cong \pm\dfrac{1}{2\pi}\sqrt{\dfrac{2\pi f_{lock}}{(3600\ \Omega)C_2}}$	Capture range LM565
18–8	$\theta_c = \cos^{-1}\left(\dfrac{n_2}{n_1}\right)$	Critical angle

TRUE/FALSE QUIZ

Answers are at the end of the chapter.

1. In amplitude modulation, the amplitude of a higher-frequency carrier signal varies according to a lower-frequency modulating signal.
2. The AM broadcast band is from 1 MHz to 540 MHz.
3. In frequency modulation, the frequency of a higher-frequency carrier signal varies according to a lower-frequency modulating signal.
4. In an AM receiver, the detector recovers the modulating signal from the RF carrier signal.
5. In an FM receiver, the descriminator recovers the modulating signal from the IF signal.
6. The linear multiplier is a key circuit in many types of communications systems.
7. Amplitude modulation is a division process.
8. There is no carrier signal in balanced modulation.
9. IF stands for interference frequency.
10. PLL stands for phase-locked loop.
11. A PLL consists of a phase detector, a low-pass filter, and a VCO.
12. A VCO is a variable capacitance oscillator.

CIRCUIT-ACTION QUIZ

Answers are at the end of the chapter.

1. If R_2 in Figure 18–12 is increased in value, the output voltage will
   (a) increase **(b)** decrease **(c)** not change
2. If R_1 in Figure 18–12 is increased in value, the output voltage will
   (a) increase **(b)** decrease **(c)** not change
3. Refer to Figure 18–23. If the amplitude modulating frequency is increased, the lower-side frequency will
   (a) increase **(b)** decrease **(c)** not change
4. Refer to Figure 18–23. If the carrier frequency is decreased, the upper-side frequency will
   (a) increase **(b)** decrease **(c)** not change
5. In amplitude modulation, if the amplitude of the modulating signal increases, the carrier frequency will
   (a) increase **(b)** decrease **(c)** not change
6. If a resistor is added in series with C_4 in Figure 18–36, the voltage gain will
   (a) increase **(b)** decrease **(c)** not change
7. If R_1 in Figure 18–48 is increased in value, the oscillation frequency will
   (a) increase **(b)** decrease **(c)** not change
8. If C_1 in Figure 18–48 is decreased in value, the lock range will
   (a) increase **(b)** decrease **(c)** not change

SELF-TEST

Answers are at the end of the chapter.

Section 18–1

1. In amplitude modulation, the pattern produced by the peaks of the carrier signal is called the
   (a) index **(b)** envelope
   (c) audio signal **(d)** upper-side frequency
2. Which of the following is not a part of an AM superheterodyne receiver?
   (a) Mixer **(b)** IF amplifier
   (c) DC restorer **(d)** Detector
   (e) Audio amplifier **(f)** Local oscillator
3. In an AM receiver, the local oscillator always produces a frequency that is above the incoming RF by
   (a) 10.7 kHz **(b)** 455 MHz **(c)** 10.7 MHz **(d)** 455 kHz
4. An FM receiver has an intermediate frequency that is
   (a) in the 88 MHz to 108 MHz range
   (b) in the 540 kHz to 1640 kHz range
   (c) 455 kHz
   (d) greater than the IF in an AM receiver
5. The detector or discriminator in an AM or an FM receiver
   (a) detects the difference frequency from the mixer
   (b) changes the RF to IF
   (c) recovers the audio signal
   (d) maintains a constant IF amplitude
6. The IF in a receiver is the
   (a) sum of the local oscillator frequency and the RF carrier frequency
   (b) local oscillator frequency
   (c) difference of the local oscillator frequency and the RF carrier frequency
   (d) difference of the carrier frequency and the audio frequency

7. When a receiver is tuned from one RF frequency to another,
   (a) the IF changes by an amount equal to the LO (local oscillator) frequency
   (b) the IF stays the same
   (c) the LO frequency changes by an amount equal to the audio frequency
   (d) both the LO and the IF frequencies change
8. The output of the AM detector goes directly to the
   (a) IF amplifier (b) mixer
   (c) audio amplifier (d) speaker

Section 18–2

9. In order to handle all combinations of input voltage polarities, a multiplier must have
   (a) four-quadrant capability
   (b) three-quadrant capability
   (c) four inputs
   (d) dual-supply voltages
10. The internal attenuation of a multiplier is called the
   (a) transconductance (b) scale factor (c) reduction factor
11. When the two inputs of a multiplier are connected together, the device operates as a
   (a) voltage doubler
   (b) square root circuit
   (c) squaring circuit
   (d) averaging circuit

Section 18–3

12. Amplitude modulation is basically a
   (a) summing of two signals
   (b) multiplication of two signals
   (c) subtraction of two signals
   (d) nonlinear process
13. The frequency spectrum of a balanced modulator contains
   (a) a sum frequency (b) a difference frequency
   (c) a carrier frequency (d) answers (a), (b), and (c)
   (e) answers (a) and (b) (f) answers (b) and (c)
14. Balanced modulation is sometimes known as
   (a) sum and difference modulation
   (b) carrier modulation
   (c) suppressed-carrier modulation
   (d) standard modulation

Section 18–4

15. The mixer in a receiver system
   (a) can be implemented with a linear multiplier
   (b) is basically a frequency converter
   (c) produces sum and difference frequencies
   (d) all of the above
16. The two inputs to a mixer are
   (a) the modulating signal and the carrier signal
   (b) the modulated carrier signal and the local oscillator signal
   (c) the IF signal and the RF signal
   (d) none of the above

Section 18–5

17. An AM demodulator is basically a
   (a) linear multiplier (b) low-pass filter
   (c) rectifier (d) linear multiplier followed by a low-pass filter

18. The final output of an AM demodulator is the
(a) IF signal (b) audio signal (c) carrier signal (d) side-band frequencies

Section 18–6

19. In an AM receiver, the IF amplifier is tuned to
(a) the carrier frequency (b) the local oscillator frequency
(c) 10.7 MHz (d) 455 kHz

20. In an AM receiver system, the audio amplifier
(a) drives the speaker(s)
(b) follows the detector
(c) has a typical bandwidth of 3 kHz to 15 kHz
(d) all of the above

Section 18–7

21. In FM, modulation is achieved by
(a) varying the frequency of an oscillator
(b) varying the amplitude of an oscillator
(c) mixing the local oscillator signal with the IF
(d) mixing the carrier signal with the modulating signal

22. A method for FM demodulation is
(a) slope detection (b) phase-shift discrimination (c) ratio detection
(d) quadrature detection (e) phase-locked loop demodulation (f) all of the above

Section 18–8

23. If the control voltage to a VCO increases, the output frequency
(a) decreases (b) does not change (c) increases

24. A PLL maintains lock by comparing
(a) the phase of two signals
(b) the frequency of two signals
(c) the amplitude of two signals

Section 18–9

25. In a fiber-optic cable, the light travels through the
(a) core (b) cladding (c) shell (d) jacket

26. If the angle of incidence of a light ray is greater than the critical angle, the light will be
(a) absorbed (b) reflected (c) amplified (d) refracted

27. The critical angle of a reflective material is determined by the
(a) absorption (b) amount of scattering (c) index of refraction (d) attenuation

PROBLEMS

Answers to all odd-numbered problems are at the end of the book.

BASIC PROBLEMS

Section 18–1 **Basic Receivers**

1. Label each block in the AM receiver in Figure 18–55.

FIGURE 18–55

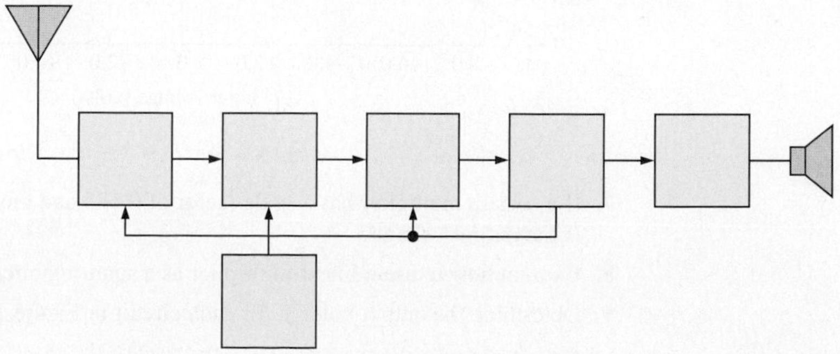

2. Label each block in the FM receiver in Figure 18–56.

FIGURE 18–56

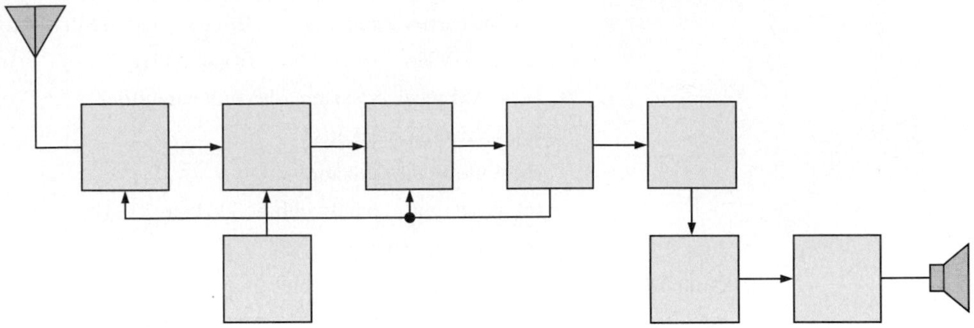

3. An AM receiver is tuned to a transmitted frequency of 680 kHz. What is the local oscillator (LO) frequency?
4. An FM receiver is tuned to a transmitted frequency of 97.2 MHz. What is the LO frequency?
5. The LO in an FM receiver is running at 101.9 MHz. What is the incoming RF? What is the IF?

Section 18–2 The Linear Multiplier

6. From the graph in Figure 18–57, determine the multiplier output voltage for each of the following pairs of input voltages.
   (a) $V_X = -4\text{ V}, V_Y = +6\text{ V}$
   (b) $V_X = +8\text{ V}, V_Y = -2\text{ V}$
   (c) $V_X = -5\text{ V}, V_Y = -2\text{ V}$
   (d) $V_X = +10\text{ V}, V_Y = +10\text{ V}$

FIGURE 18–57

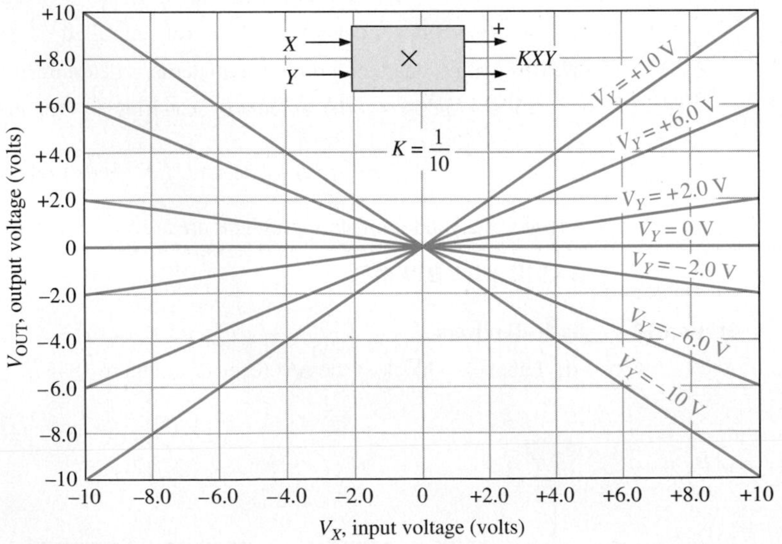

7. If a certain multiplier has a scale factor of 0.125 and the inputs are +3.5 V and −2.9 V, what is the output voltage?
8. Explain how to use a linear multiplier as a squaring circuit.
9. Determine the output voltage for each circuit in Figure 18–58.

▶ FIGURE 18–58

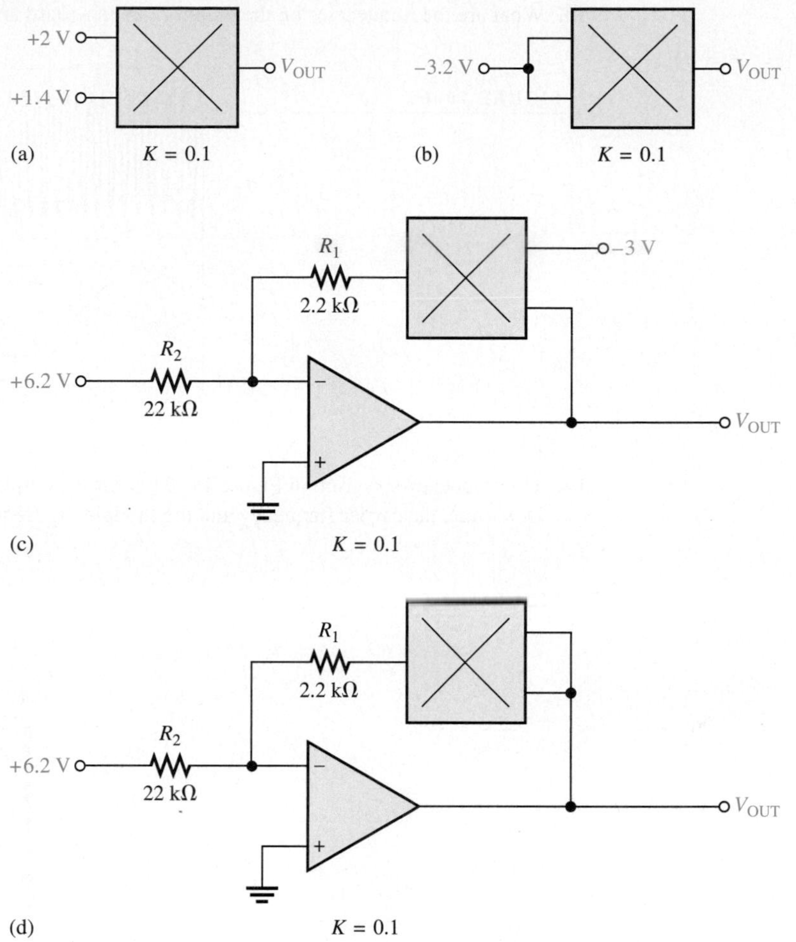

Section 18–3 **Amplitude Modulation**

10. If a 100 kHz signal and a 30 kHz signal are applied to a balanced modulator, what frequencies will appear on the output?

11. What are the frequencies on the output of the balanced modulator in Figure 18–59?

▶ FIGURE 18–59

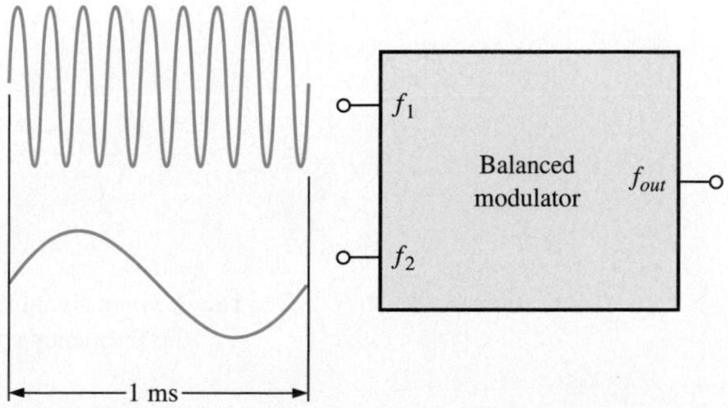

12. If a 1000 kHz signal and a 3 kHz signal are applied to a standard amplitude modulator, what frequencies will appear on the output?

13. What are the frequencies on the output of the standard amplitude modulator in Figure 18–60?

▶ FIGURE 18–60

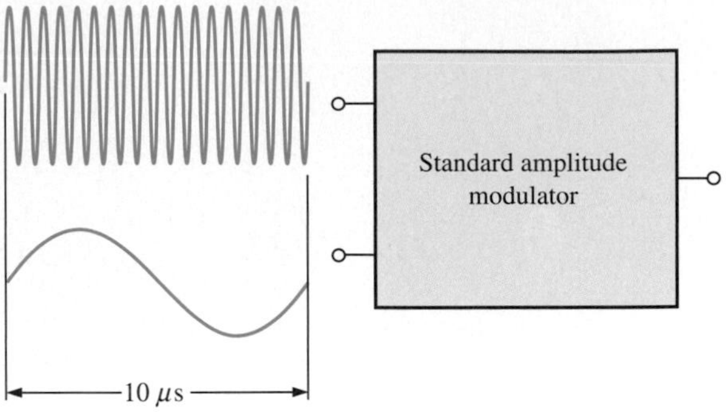

14. The frequency spectrum in Figure 18–61 is for the output of a standard amplitude modulator. Determine the carrier frequency and the modulating frequency.

▶ FIGURE 18–61

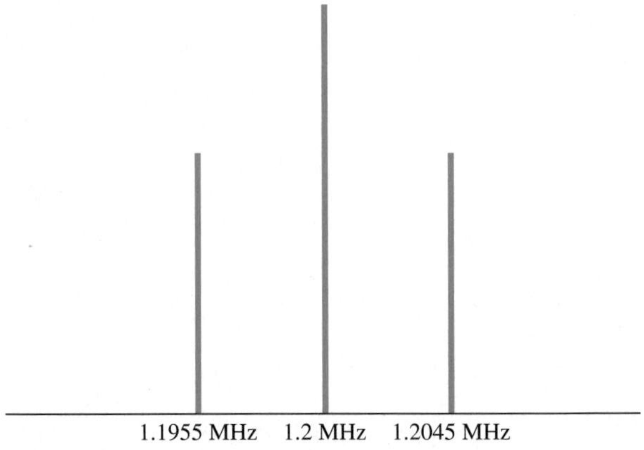

15. The frequency spectrum in Figure 18–62 is for the output of a balanced modulator. Determine the carrier frequency and the modulating frequency.

▶ FIGURE 18–62

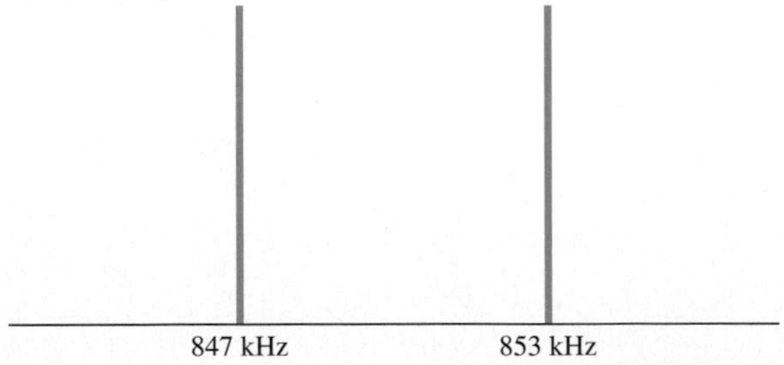

16. A voice signal ranging from 300 Hz to 3 kHz amplitude modulates a 600 kHz carrier. Develop the frequency spectrum.

Section 18–4 The Mixer

17. Determine the output expression for a multiplier with one sinusoidal input having a peak voltage of 0.2 V and a frequency of 2200 kHz and the other input having a peak voltage of 0.15 V and a frequency of 3300 kHz.

18. Determine the output frequency of the IF amplifier for the frequencies shown in Figure 18–63.

FIGURE 18–63

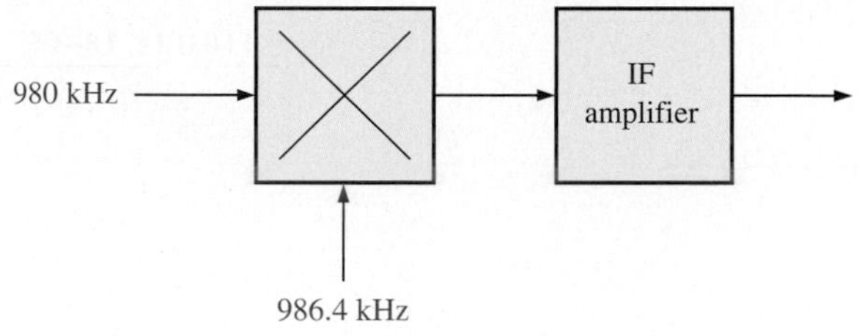

Section 18–5 **AM Demodulation**

19. The input to a certain AM receiver consists of a 1500 kHz carrier and two side frequencies separated from the carrier by 20 kHz. Determine the frequency spectrum at the output of the mixer amplifier.

20. For the same conditions stated in Problem 19, determine the frequency spectrum at the output of the IF amplifier.

21. For the same conditions stated in Problem 19, determine the frequency spectrum at the output of the AM detector (demodulator).

Section 18–6 **IF and Audio Amplifiers**

22. For a carrier frequency of 1.2 MHz and a modulating frequency of 8.5 kHz, list all of the frequencies on the output of the mixer in an AM receiver.

23. In a certain AM receiver, one amplifier has a passband from 450 kHz to 460 kHz and another has a passband from 10 Hz to 5 kHz. Identify these amplifiers.

24. Determine the maximum and minimum output voltages for the audio power amplifier in Figure 18–64.

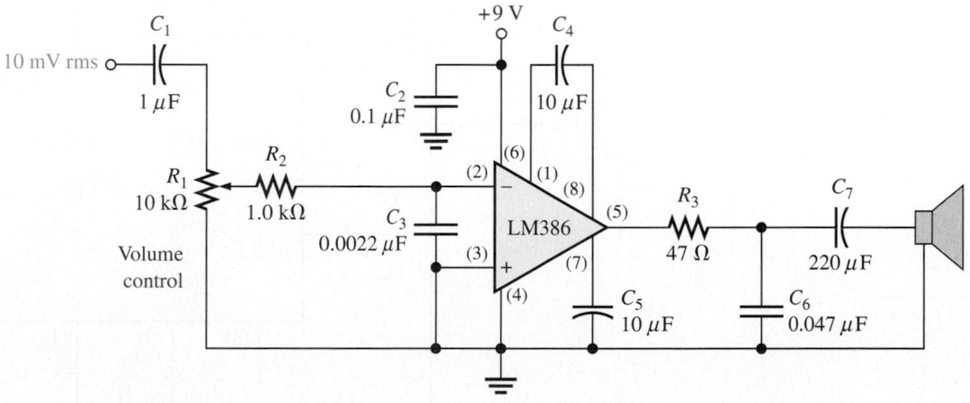

FIGURE 18–64

Section 18–7 **Frequency Modulation**

25. Explain how a VCO is used as a frequency modulator.

26. How does an FM signal differ from an AM signal?

27. What is the variable reactance element shown in Figure 18–39?

Section 18–8 The Phase-Locked Loop (PLL)

28. Label each block in the PLL diagram of Figure 18–65.

FIGURE 18–65

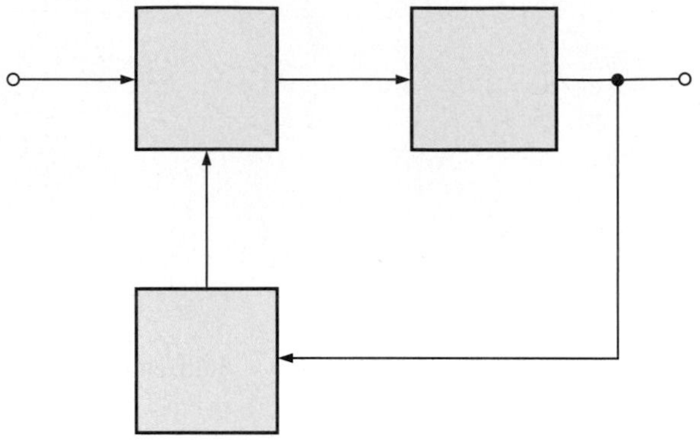

29. A PLL is locked onto an incoming signal with a peak amplitude of 250 mV and a frequency of 10 MHz at a phase angle of 30°. The 400 mV peak VCO signal is at a phase angle of 15°.

(a) What is the VCO frequency?

(b) What is the value of the control voltage being fed back to the VCO at this point?

30. What is the conversion gain of a VCO if a 0.5 V increase in the control voltage causes the output frequency to increase by 3.6 kHz?

31. If the conversion gain of a certain VCO is 1.5 kHz per volt, how much does the frequency change if the control voltage increases 0.67 V?

32. Name two conditions for a PLL to acquire lock.

33. Determine the free-running frequency, the lock range, and the capture range for the PLL in Figure 18–66.

FIGURE 18–66

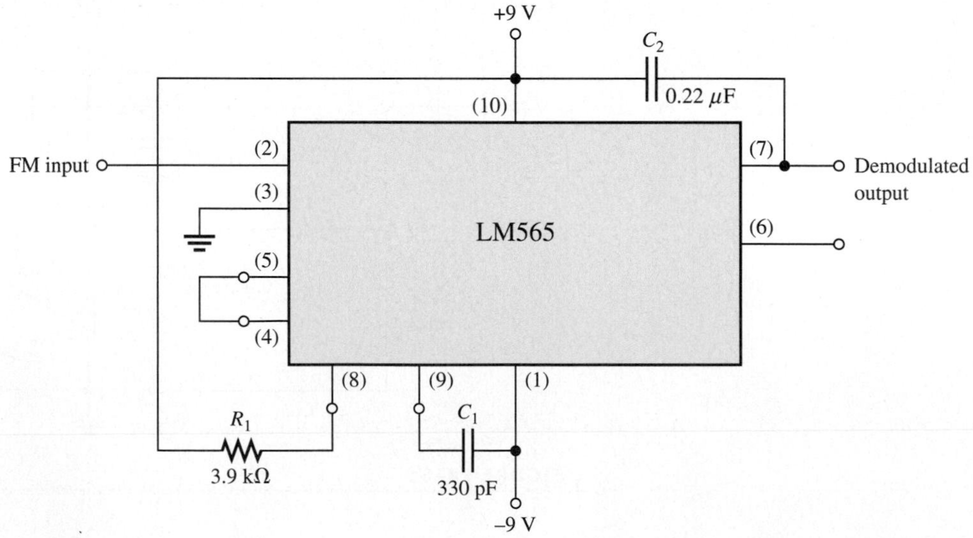

Section 18–9 Fiber Optics

34. A light ray strikes the core of a fiber-optic cable at 30° angle of incidence. If the critical angle of the core is 15°, will the light ray be reflected or refracted?

35. Determine the critical angle of a fiber-optic cable if the core has an index of refraction of 1.55 and the cladding has an index of refraction of 1.25.

ANSWERS

SECTION CHECKUPS

Section 18–1 Basic Receivers

1. AM is amplitude modulation. FM is frequency modulation.
2. In AM, the modulating signal varies the amplitude of a carrier. In FM, the modulating signal varies the frequency of a carrier.
3. AM: 540 kHz to 1640 kHz; FM: 88 MHz to 108 MHz

Section 18–2 The Linear Multiplier

1. A four-quadrant multiplier can handle any combination (4) of positive and negative inputs. A one-quadrant multiplier can only handle two positive inputs, for example.
2. Connect the two inputs together and apply a single input variable.

Section 18–3 Amplitude Modulation

1. Amplitude modulation is the process of varying the amplitude of a carrier signal with a modulating signal
2. Balanced modulation produces no carrier frequency on the output, whereas standard AM does.
3. The carrier signal is the modulated signal and has a sufficiently high frequency for transmission. The modulating signal is a lower-frequency signal that contains information and varies the carrier amplitude according to its waveshape.
4. The upper-side frequency is the sum of the carrier frequency and the modulating frequency. The lower-side frequency is the difference of the carrier frequency and the modulating frequency.
5. By summing the peak carrier voltage and the modulating signal before mixing with the carrier signal

Section 18–4 The Mixer

1. The mixer produces (among other frequencies) a signal representing the difference between the incoming carrier frequency and the local oscillator frequency. This is called the intermediate frequency.
2. The mixer multiplies the carrier and the local oscillator signals.
3. 1000 kHz + 350 kHz = 1350 kHz, 1000 kHz − 350 kHz = 650 kHz

Section 18–5 AM Demodulation

1. The filter removes all frequencies except the audio.
2. Only the 1 kHz

Section 18–6 IF and Audio Amplifiers

1. To amplify the 455 kHz amplitude modulated IF coming from the mixer
2. The IF center frequency is 455 kHz.
3. The 10 kHz bandwidth allows the upper-side and lower-side frequencies that contain the information to pass.
4. The audio amplifier follows the detector because the detector is the circuit that recovers the audio from the modulated IF.
5. The IF has a response of approximately 455 kHz ± 5 kHz. The typical audio amplifier has a maximum bandwidth from tens of hertz up to about 15 kHz although for many amplifiers, the bandwidth can be much less than this typical maximum.

Section 18–7 Frequency Modulation

1. The frequency variation of an FM signal bears the information.
2. VCOs are based on the principle of voltage-variable reactance.

Section 18–8 The Phase-Locked Loop (PLL)

1. Phase detector, low-pass filter, and VCO
2. Sometimes a PLL uses an amplifier in the loop.
3. A PLL locks onto and tracks a variable incoming frequency.
4. The lock range specifies how much a lock-on frequency can deviate without the PLL losing lock. The capture range specifies how close the incoming frequency must be from the free-running VCO frequency in order for the PLL to lock.
5. The PLL detects a change in the phase of the incoming signal compared to the VCO signal that indicates a change in frequency. The positive feedback then causes the VCO frequency to change along with the incoming frequency.

Section 18–9 Fiber Optics

1. A fiber-optic cable is made of glass.
2. A fiber-optic cable has the approximate diameter of a human hair.
3. Core, cladding, and jacket. There may also be other strengthening components.
4. The critical angle is fixed and determined by the indices of refractions of the core and cladding materials. The angle of incidence is the angle at which a light ray strikes the core/cladding boundary.
5. Multimode step index, single-mode step index, and multimode graded index

RELATED PROBLEMS FOR EXAMPLES

18–1 −3.6 V from the graph in Figure 18–9

18–2 Modulate the carrier with a higher-frequency signal.

18–3 The balanced modulator output has the same side frequencies but does not have a carrier frequency.

18–4 $V_p = 0.025$ mV, $f = 455$ kHz

18–5 455 kHz

18–6 0.172 V

18–7 A decrease of 6 kHz; 244 kHz

18–8 Decrease C_2 to 0.0595 μF.

18–9 17.8°

TRUE/FALSE QUIZ

1. T	**2.** F	**3.** T	**4.** F	**5.** T	**6.** T
7. F	**8.** T	**9.** F	**10.** T	**11.** T	**12.** F

CIRCUIT-ACTION QUIZ

1. (b)	**2.** (a)	**3.** (b)	**4.** (b)	**5.** (c)	**6.** (b)	**7.** (b)	**8.** (a)

SELF-TEST

1. (b)	**2.** (c)	**3.** (d)	**4.** (d)	**5.** (c)	**6.** (c)	**7.** (b)	**8.** (c)
9. (a)	**10.** (b)	**11.** (c)	**12.** (b)	**13.** (c)	**14.** (c)	**15.** (d)	**16.** (b)
17. (d)	**18.** (b)	**19.** (d)	**20.** (d)	**21.** (a)	**22.** (f)	**23.** (c)	**24.** (a)

TABLE OF STANDARD RESISTOR VALUES

Appendix A

Resistance Tolerance (±%)

0.1% 0.25% 0.5%	1%	2% 5%	10%	0.1% 0.25% 0.5%	1%	2% 5%	10%	0.1% 0.25% 0.5%	1%	2% 5%	10%	0.1% 0.25% 0.5%	1%	2% 5%	10%	0.1% 0.25% 0.5%	1%	2% 5%	10%	0.1% 0.25% 0.5%	1%	2% 5%	10%
10.0	10.0	10	10	14.7	14.7	—	—	21.5	21.5	—	—	31.6	31.6	—	—	46.4	46.4	—	—	68.1	68.1	68	68
10.1	—	—	—	14.9	—	—	—	21.8	—	—	—	32.0	—	—	—	47.0	—	47	47	69.0	—	—	—
10.2	10.2	—	—	15.0	15.0	15	15	22.1	22.1	22	22	32.4	32.4	—	—	47.5	47.5	—	—	69.8	69.8	—	—
10.4	—	—	—	15.2	—	—	—	22.3	—	—	—	32.8	—	—	—	48.1	—	—	—	70.6	—	—	
10.5	10.5	—	—	15.4	15.4	—	—	22.6	22.6	—	—	33.2	33.2	33	33	48.7	48.7	—	—	71.5	71.5	—	—
10.6	—	—	—	15.6	—	—	—	22.9	—	—	—	33.6	—	—	—	49.3	—	—	—	72.3	—	—	—
10.7	10.7	—	—	15.8	15.8	—	—	23.2	23.2	—	—	34.0	34.0	—	—	49.9	49.9	—	—	73.2	73.2	—	—
10.9	—	—	—	16.0	—	16	—	23.4	—	—	—	34.4	—	—	—	50.5	—	—	—	74.1	—	—	—
11.0	11.0	11	—	16.2	16.2	—	—	23.7	23.7	—	—	34.8	34.8	—	—	51.1	51.1	51	—	75.0	75.0	75	—
11.1	—	—	—	16.4	—	—	—	24.0	—	24	—	35.2	—	—	—	51.7	—	—	—	75.9	—	—	—
11.3	11.3	—	—	16.5	16.5	—	—	24.3	24.3	—	—	35.7	35.7	—	—	52.3	52.3	—	—	76.8	76.8	—	—
11.4	—	—	—	16.7	—	—	—	24.6	—	—	—	36.1	—	36	—	53.0	—	—	—	77.7	—	—	—
11.5	11.5	—	—	16.9	16.9	—	—	24.9	24.9	—	—	36.5	36.5	—	—	53.6	53.6	—	—	78.7	78.7	—	—
11.7	—	—	—	17.2	—	—	—	25.2	—	—	—	37.0	—	—	—	54.2	—	—	—	79.6	—	—	—
11.8	11.8	—	—	17.4	17.4	—	—	25.5	25.5	—	—	37.4	37.4	—	—	54.9	54.9	—	—	80.6	80.6	—	—
12.0	—	12	12	17.6	—	—	—	25.8	—	—	—	37.9	—	—	—	56.2	—	—	—	81.6	—	—	—
12.1	12.1	—	—	17.8	17.8	—	—	26.1	26.1	—	—	38.3	38.3	—	—	56.6	56.6	56	56	82.5	82.5	82	82
12.3	—	—	—	18.0	—	18	18	26.4	—	—	—	38.8	—	—	—	56.9	—	—	—	83.5	—	—	—
12.4	12.4	—	—	18.2	18.2	—	—	26.7	26.7	—	—	39.2	39.2	39	39	57.6	57.6	—	—	84.5	84.5	—	—
12.6	—	—	—	18.4	—	—	—	27.1	—	27	27	39.7	—	—	—	58.3	—	—	—	85.6	—	—	—
12.7	12.7	—	—	18.7	18.7	—	—	27.4	27.4	—	—	40.2	40.2	—	—	59.0	59.0	—	—	86.6	86.6	—	—
12.9	—	—	—	18.9	—	—	—	27.7	—	—	—	40.7	—	—	—	59.7	—	—	—	87.6	—	—	—
13.0	13.0	13	—	19.1	19.1	—	—	28.0	28.0	—	—	41.2	41.2	—	—	60.4	60.4	—	—	88.7	88.7	—	—
13.2	—	—	—	19.3	—	—	—	28.4	—	—	—	41.7	—	—	—	61.2	—	—	—	89.8	—	—	—
13.3	13.3	—	—	19.6	19.6	—	—	28.7	28.7	—	—	42.2	42.2	—	—	61.9	61.9	62	—	90.9	90.9	91	—
13.5	—	—	—	19.8	—	—	—	29.1	—	—	—	42.7	—	—	—	62.6	—	—	—	92.0	—	—	—
13.7	13.7	—	—	20.0	20.0	20	—	29.4	29.4	—	—	43.2	43.2	43	—	63.4	63.4	—	—	93.1	93.1	—	—
13.8	—	—	—	20.3	—	—	—	29.8	—	—	—	43.7	—	—	—	64.2	—	—	—	94.2	—	—	—
14.0	14.0	—	—	20.5	20.5	—	—	30.1	30.1	30	—	44.2	44.2	—	—	64.9	64.9	—	—	95.3	95.3	—	—
14.2	—	—	—	20.8	—	—	—	30.5	—	—	—	44.8	—	—	—	65.7	—	—	—	96.5	—	—	—
14.3	14.3	—	—	21.0	21.0	—	—	30.9	30.9	—	—	45.3	45.3	—	—	66.5	66.5	—	—	97.6	97.6	—	—
14.5	—	—	—	21.3	—	—	—	31.2	—	—	—	45.9	—	—	—	67.3	—	—	—	98.8	—	—	—

NOTE: These values are generally available in multiples of 0.1, 1, 10, 100, 1 k, and 1 M.

Appendix B

DERIVATIONS OF SELECTED EQUATIONS

Equation 2–1

The average value of a half-wave rectified sine wave is the area under the curve divided by the period (2π). The equation for a sine wave is

$$v = V_p \sin\theta$$

$$V_{AVG} = \frac{\text{area}}{2\pi} = \frac{1}{2\pi}\int_0^{\pi} V_p \sin\theta\, d\theta = \frac{V_p}{2\pi}(-\cos\theta)|_0^{\pi}$$

$$= \frac{V_p}{2\pi}[-\cos\pi - (-\cos 0)] = \frac{V_p}{2\pi}[-(-1) - (-1)] = \frac{V_p}{2\pi}(2)$$

$$V_{AVG} = \frac{V_p}{\pi}$$

Equation 2–10

Refer to Figure B–1.

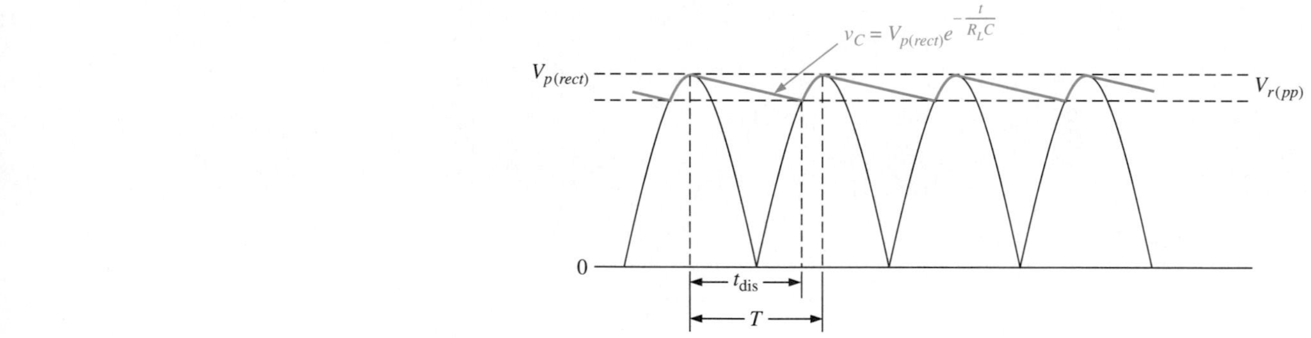

▲ FIGURE B–1

When the filter capacitor discharges through R_L, the voltage is

$$v_C = V_{p(rect)}e^{-t/R_LC}$$

Since the discharge time of the capacitor is from one peak to approximately the next peak, $t_{dis} \cong T$ when v_C reaches its minimum value.

$$v_{C(min)} = V_{p(rect)}e^{-T/R_LC}$$

Since $RC \gg T$, T/R_LC becomes much less than 1 (which is usually the case); e^{-T/R_LC} approaches 1 and can be expressed as

$$e^{-T/R_LC} \cong 1 - \frac{T}{R_LC}$$

Therefore,

$$v_{C(min)} = V_{p(rect)}\left(1 - \frac{T}{R_L C}\right)$$

The peak-to-peak ripple voltage is

$$V_{r(pp)} = V_{p(rect)} - V_{C(min)} = V_{p(rect)} - V_{p(rect)} + \frac{V_{p(rect)}T}{R_L C} = \frac{V_{p(rect)}T}{R_L C}$$

$$V_{r(pp)} \cong \left(\frac{1}{fR_L C}\right)V_{p(rect)}$$

Equation 2–11

To obtain the dc value, one-half of the peak-to-peak ripple is subtracted from the peak value.

$$V_{DC} = V_{p(rect)} - \frac{V_{r(pp)}}{2} = V_{p(rect)} - \left(\frac{1}{2fR_L C}\right)V_{p(rect)}$$

$$V_{DC} = \left(1 - \frac{1}{2fR_L C}\right)V_{p(rect)}$$

Equation 5–5

To develop a formula for the dc input resistance at the base of a transistor, we will use the diagram in Figure B–2. V_{IN} is applied between base and ground, and I_{IN} is the current at the base as shown.

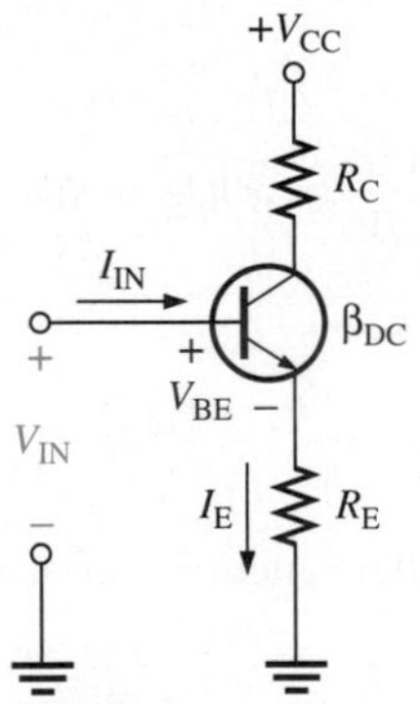

FIGURE B–2

By Ohm's law,

$$R_{IN(BASE)} = \frac{V_{IN}}{I_{IN}}$$

Kirchhoff's voltage law applied around the base-emitter circuit yields

$$V_{IN} = V_{BE} + I_E R_E$$

With the assumption that $V_{BE} << I_E R_E$, the equation reduces to

$$V_{IN} \cong I_E R_E$$

Now, since $I_E \cong I_C = \beta_{DC} I_B$,

$$V_{IN} \cong \beta_{DC} I_B R_E$$

The input current is the base current:

$$I_{IN} = I_B$$

By substitution,

$$R_{IN(BASE)} = \frac{V_{IN}}{I_{IN}} \cong \frac{\beta_{DC} I_B R_E}{I_B}$$

Cancelling the I_B terms gives

$$R_{IN(BASE)} = \beta_{DC} R_E$$

Equation 6–1

The Shockley equation for the base-emitter *pn* junction is

$$I_E = I_R(e^{VQ/kT} - 1)$$

where I_E = the total forward current across the base-emitter junction
I_R = the reverse saturation current
V = the voltage across the depletion layer
Q = the charge on an electron
k = a number known as Boltzmann's constant
T = the absolute temperature

At ambient temperature, $Q/kT \cong 40$, so

$$I_E = I_R(e^{V40} - 1)$$

Differentiating yields

$$\frac{dI_E}{dV} = 40 I_R e^{V40}$$

Since $I_R e^{V40} = I_E + I_R$,

$$\frac{dI_E}{dV} = 40(I_E + I_R)$$

Assuming $I_R << I_E$,

$$\frac{dI_E}{dV} \cong 40 I_E$$

The ac resistance r'_e of the base-emitter junction can be expressed as dV/dI_E.

$$r'_e = \frac{dV}{dI_E} \cong \frac{1}{40 I_E} \cong \frac{25\text{ mV}}{I_E}$$

Equation 6–14

The emitter-follower is represented by the *r* parameter ac equivalent circuit in Figure B–3(a).

By thevenizing from the base back to the source, the circuit is simplified to the form shown in Figure B–3(b). $V_{out} = V_e$, $I_{out} = I_e$, and $I_{in} = I_b$.

$$R_{out} = \frac{V_e}{I_e}$$

$$I_e \cong \beta_{ac} I_b$$

With $V_s = 0$ and with I_b produced by V_{out}, and neglecting the base-to-emitter voltage drop (and therefore r'_e),

$$I_b \cong \frac{V_e}{R_1 \| R_2 \| R_s}$$

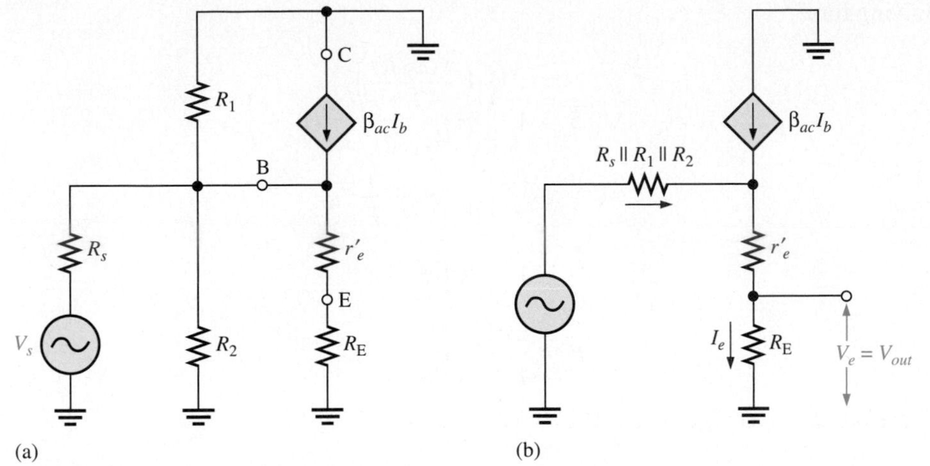

▲ FIGURE B–3

Assuming that $R_1 >> R_s$ and $R_2 >> R_s$,

$$I_b \cong \frac{V_e}{R_s}$$

$$I_{out} = I_e = \frac{\beta_{ac}V_e}{R_s}$$

$$\frac{V_{out}}{I_{out}} = \frac{V_e}{I_e} = \frac{V_e}{\beta_{ac}V_e/R_s} = \frac{R_s}{\beta_{ac}}$$

Looking into the emitter, R_E appears in parallel with R_s/β_{ac}. Therefore,

$$R_{out} = \left(\frac{R_s}{\beta_{ac}}\right) \parallel R_E$$

Midpoint Bias (Chapter 8)

The following proof is for the equation on page 384 that shows $I_D \cong 0.5I_{DSS}$ when $V_{GS} = V_{GS(off)}/3.4$.

Start with Equation 8–1:

$$I_D \cong I_{DSS}\left(1 - \frac{V_{GS}}{V_{GS(off)}}\right)^2$$

Let $I_D = 0.5I_{DSS}$.

$$0.5I_{DSS} = I_{DSS}\left(1 - \frac{V_{GS}}{V_{GS(off)}}\right)^2$$

Cancelling I_{DSS} on each side,

$$0.5 = \left(1 - \frac{V_{GS}}{V_{GS(off)}}\right)^2$$

We want a factor (call it F) by which $V_{GS(off)}$ can be divided to give a value of V_{GS} that will produce a drain current that is $0.5I_{DSS}$.

$$0.5 = \left[1 - \frac{\left(\frac{V_{GS(off)}}{F}\right)}{V_{GS(off)}}\right]^2$$

Solving for F,

$$\sqrt{0.5} = 1 - \frac{\left(\frac{V_{GS(off)}}{F}\right)}{V_{GS(off)}} = 1 - \frac{1}{F}$$

$$\sqrt{0.5} - 1 = -\frac{1}{F}$$

$$\frac{1}{F} = 1 - \sqrt{0.5}$$

$$F = \frac{1}{1 - \sqrt{0.5}} \cong 3.4$$

Therefore, $I_D \cong 0.5I_{DSS}$ when $V_{GS} = V_{GS(off)}/3.4$.

Equation 9–2

$$I_D \cong I_{DSS}\left(1 - \frac{I_D R_S}{V_{GS(off)}}\right)^2 = I_{DSS}\left(1 - \frac{I_D R_S}{V_{GS(off)}}\right)\left(1 - \frac{I_D R_S}{V_{GS(off)}}\right)$$

$$= I_{DSS}\left(1 - \frac{2I_D R_S}{V_{GS(off)}} + \frac{I_D^2 R_S^2}{V_{GS(off)}^2}\right) = I_{DSS} - \frac{2I_{DSS}R_S}{V_{GS(off)}}I_D + \frac{I_{DSS}R_S^2}{V_{GS(off)}^2}I_D^2$$

Rearranging into a standard quadratic equation form,

$$\left(\frac{I_{DSS}R_S^2}{V_{GS(off)}^2}\right)I_D^2 - \left(1 + \frac{2I_{DSS}R_S}{V_{GS(off)}}\right)I_D + I_{DSS} = 0$$

The coefficients and constant are

$$A = \frac{R_S^2 I_{DSS}}{V_{GS(off)}^2}$$

$$B = -\left(1 + \frac{2R_S I_{DSS}}{V_{GS(off)}}\right)$$

$$C = I_{DSS}$$

In simplified notation, the equation is

$$AI_D^2 + BI_D + C = 0$$

The solutions to this quadratic equation are

$$I_D = \frac{-B \pm \sqrt{B^2 - 4AC}}{2A}$$

Equation 9–10

A general model of a switched-capacitor circuit, as shown in Figure B–4(a), consists of a capacitor, two voltage sources, V_1 and V_2, and a two-pole switch. Let's examine this circuit for a specified period of time, T. Assume that V_1 and V_2 are constant during the time period T and $V_1 > V_2$. Of particular interest is the *average* current I_1 produced by the source V_1 during the time period T.

During the first half of the time period T, the switch is in position 1, as indicated in Figure B–4(b). The capacitor charges very rapidly to the source voltage V_1. Therefore, an average current I_1 due to V_1 is charging the capacitor during the interval from $t = 0$ to $t = T/2$. During the second half of the time period, the switch is in position 2, as indicated.

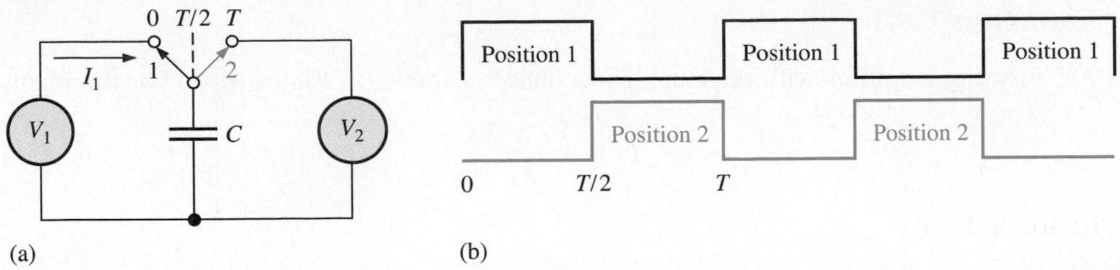

▲ FIGURE B–4

Because $V_1 > V_2$, the capacitor rapidly discharges to the voltage V_2. The average current produced by the source V_1 over the time period T is

$$I_{1(avg)} = \frac{Q_{1(T/2)} - Q_{1(0)}}{T}$$

$Q_{1(0)}$ is the charge at $t = 0$ and $Q_{1(T/2)}$ is the charge at $t = T/2$. Therefore, $Q_{1(T/2)} - Q_{1(0)}$ is the net charge transferred while the switch is in position 1.

The capacitor voltage at $T/2$ is equal to V_1, and the capacitor voltage at 0 or T is equal to V_2. By substituting CV for Q in the previous equation,

$$I_{1(avg)} = \frac{CV_{1(T/2)} - CV_{2(0)}}{T} = \frac{C(V_{1(T/2)} - V_{2(0)})}{T}$$

Since V_1 and V_2 are assumed to be constant during T, the average current can be expressed as

$$I_{1(avg)} = \frac{C(V_1 - V_2)}{T}$$

Figure B–5 shows a conventional resistive circuit with two voltage sources. From Ohm's law, the current is

$$I_1 = \frac{V_1 - V_2}{R}$$

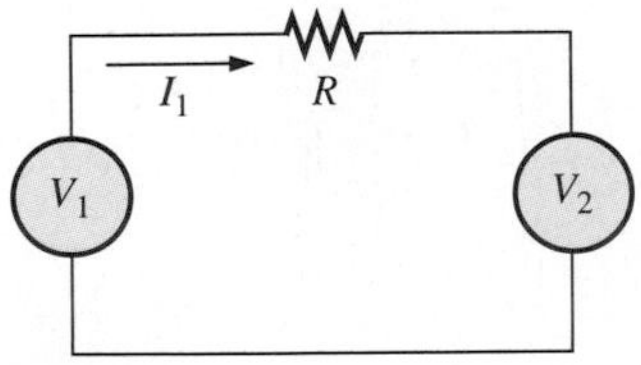

▲ FIGURE B–5

The current $I_{1(avg)}$ in the switched-capacitor circuit is equal to I_1 in the resistive circuit.

$$I_{1(avg)} = \frac{C(V_1 - V_2)}{T} = \frac{V_1 - V_2}{R}$$

By solving for R and canceling the $V_1 - V_2$ terms,

$$R = \frac{T(V_1 - V_2)}{C(V_1 - V_2)}$$

$$R = \frac{T}{C}$$

As you can see, a switched-capacitor circuit can emulate a resistor with a value determined by the time period T and the capacitance C. Remember that the two-pole switch is in each position for one-half of the time period T and that you can vary T by varying the frequency at which the switches are operated.

Since $T = 1/f$, the resistance in terms of frequency is

$$R = \frac{1}{fC}$$

Equation 10–1

An inverting amplifier with feedback capacitance is shown in Figure B–6. For the input,

$$I_1 = \frac{V_1 - V_2}{X_C}$$

Factoring V_1 out,

$$I_1 = \frac{V_1(1 - V_2/V_1)}{X_C}$$

The ratio V_2/V_1 is the voltage gain, $-A_v$.

$$I_1 = \frac{V_1(1 + A_v)}{X_C} = \frac{V_1}{X_C/(1 + A_v)}$$

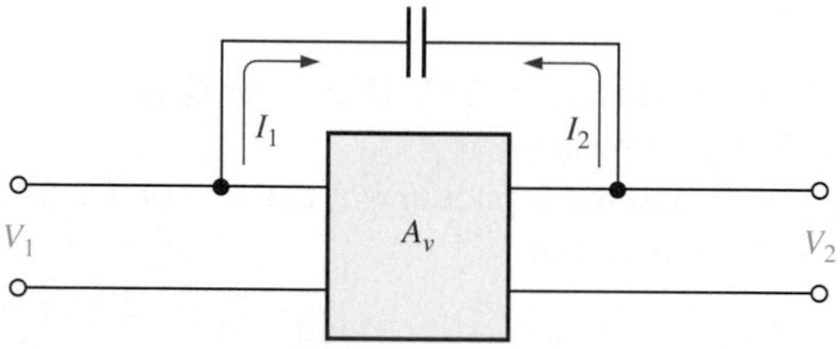

FIGURE B–6

The effective reactance as seen from the input terminals is

$$X_{C_{in(Miller)}} = \frac{X_C}{1 + A_v}$$

or

$$\frac{1}{2\pi f C_{in(Miller)}} = \frac{1}{2\pi f C(1 + A_v)}$$

Cancelling and inverting,

$$C_{in(Miller)} = C(A_v + 1)$$

Equation 10–2

For the output in Figure B–6,

$$I_2 = \frac{V_2 - V_1}{X_C} = \frac{V_2(1 - V_1/V_2)}{X_C}$$

Since $V_1/V_2 = -1/A_v$,

$$I_2 = \frac{V_2(1 + 1/A_v)}{X_C} = \frac{V_2}{X_C/(1 + 1/A_v)} = \frac{V_2}{X_C/[(A_v + 1)/A_v]}$$

The effective reactance as seen from the output is

$$X_{C_{out(Miller)}} = \frac{X_C}{(A_v + 1)/A_v}$$

$$\frac{1}{2\pi f C_{out(Miller)}} = \frac{1}{2\pi f C[(A_v + 1)/A_v]}$$

Cancelling and inverting yields

$$C_{out(Miller)} = C\left(\frac{A_v + 1}{A_v}\right)$$

Equations 10–29 and 10–30

The total gain, $A_{v(tot)}$, of an individual amplifier stage at the lower critical frequency equals the midrange gain, $A_{v(mid)}$, times the attenuation of the high-pass RC circuit.

$$A_{v(tot)} = A_{v(mid)}\left(\frac{R}{\sqrt{R^2 + X_C^2}}\right) = A_{v(mid)}\left(\frac{1}{\sqrt{1 + X_C^2/R^2}}\right)$$

$$f_{cl} = \frac{1}{2\pi RC}$$

Dividing both sides by any frequency f,

$$\frac{f_{cl}}{f} = \frac{1}{(2\pi fC)R}$$

Since $X_C = 1/2\pi fC$,

$$\frac{f_{cl}}{f} = \frac{X_C}{R}$$

Substitution in the gain formula gives

$$A_{v(tot)} = A_{v(mid)}\left(\frac{1}{\sqrt{1 + (f_{cl}/f)^2}}\right)$$

The gain ratio is

$$\frac{A_{v(tot)}}{A_{v(mid)}} = \frac{1}{\sqrt{1 + (f_{cl}/f)^2}}$$

For a multistage amplifier with n stages, each with the same f_{cl} and gain ratio, the product of the gain ratios is

$$\left(\frac{1}{\sqrt{1 + (f_{cl}/f)^2}}\right)^n$$

The critical frequency f'_{cl} of the multistage amplifier is the frequency at which $A_{v(tot)} = 0.707A_{v(mid)}$, so the gain ratio at f'_{cl} is

$$\frac{A_{v(tot)}}{A_{v(mid)}} = 0.707 = \frac{1}{1.414} = \frac{1}{\sqrt{2}}$$

Therefore, for a multistage amplifier,

$$\frac{1}{\sqrt{2}} = \left[\frac{1}{\sqrt{1 + (f_{cl}/f'_{cl})^2}}\right]^n = \frac{1}{(\sqrt{1 + (f_{cl}/f'_{cl})^2})^n}$$

So

$$2^{1/2} = (\sqrt{1 + (f_{cl}/f'_{cl})^2})^n$$

Squaring both sides,

$$2 = (1 + (f_{cl}/f'_{cl})^2)^n$$

Taking the nth root of both sides,

$$2^{1/n} = 1 + (f_{cl}/f'_{cl})^2$$
$$\left(\frac{f_{cl}}{f'_{cl}}\right)^2 = 2^{1/n} - 1$$
$$\left(\frac{f_{cl}}{f'_{cl}}\right) = \sqrt{2^{1/n} - 1}$$
$$f'_{cl} = \frac{f_{cl}}{\sqrt{2^{1/n} - 1}}$$

A similar process will give Equation 10–30:

$$f'_{cu} = f_{cu}\sqrt{2^{1/n} - 1}$$

Equations 10–31 and 10–32

The *rise time* is defined as the time required for the voltage to increase from 10 percent of its final value to 90 percent of its final value, as indicated in Figure B–7.

Expressing the curve in its exponential form gives

$$v = V_{final}(1 - e^{-t/RC})$$

When $v = 0.1V_{final}$,

$$0.1V_{final} = V_{final}(1 - e^{-t/RC}) = V_{final} - V_{final}e^{-t/RC}$$
$$V_{final}e^{-t/RC} = 0.9V_{final}$$
$$e^{-t/RC} = 0.9$$
$$\ln e^{-t/RC} = \ln (0.9)$$
$$-\frac{t}{RC} = -0.1$$
$$t = 0.1RC$$

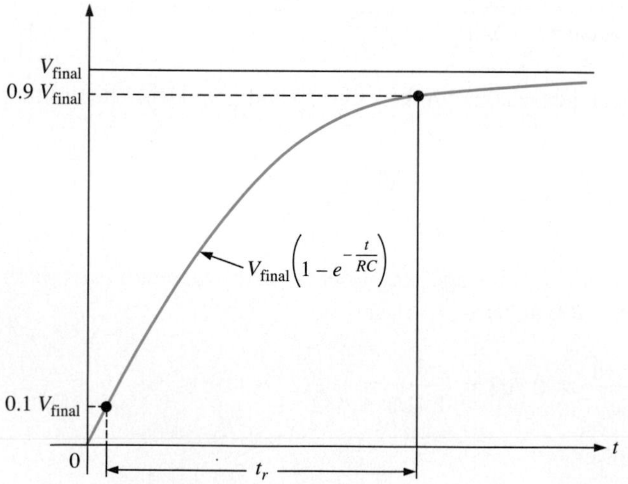

FIGURE B–7

When $v = 0.9\,V_{final}$,

$$0.9V_{final} = V_{final}(1 - e^{-t/RC}) = V_{final} - V_{final}e^{-t/RC}$$
$$V_{final}e^{-t/RC} = 0.1V_{final}$$
$$\ln e^{-t/RC} = \ln (0.1)$$
$$-\frac{t}{RC} = -2.3$$
$$t = 2.3RC$$

The difference is the rise time.

$$t_r = 2.3RC - 0.1RC = 2.2RC$$

The critical frequency of an RC circuit is

$$f_c = \frac{1}{2\pi RC}$$

$$RC = \frac{1}{2\pi f_c}$$

Substituting,

$$t_r = \frac{2.2}{2\pi f_{cu}} = \frac{0.35}{f_{cu}}$$

$$f_{cu} = \frac{0.35}{t_r}$$

In a similar way, it can be shown that

$$f_{cl} = \frac{0.35}{t_f}$$

Equation 12–21

The formula for open-loop gain in Equation 12–19 can be expressed in complex notation as

$$A_{ol} = \frac{A_{ol(mid)}}{1 + jf/f_{c(ol)}}$$

Substituting the above expression into the equation $A_{cl} = A_{ol}/(1 + BA_{ol})$ gives a formula for the total closed-loop gain.

$$A_{cl} = \frac{A_{ol(mid)}/(1 + jf/f_{c(ol)})}{1 + BA_{ol(mid)}/(1 + jf/f_{c(ol)})}$$

Multiplying the numerator and denominator by $1 + jf/f_{c(ol)}$ yields

$$A_{cl} = \frac{A_{ol(mid)}}{1 + BA_{ol(mid)} + jf/f_{c(ol)}}$$

Dividing the numerator and denominator by $1 + BA_{ol(mid)}$ gives

$$A_{cl} = \frac{A_{ol(mid)}/(1 + BA_{ol(mid)})}{1 + j[f/(f_{c(ol)}(1 + BA_{ol(mid)}))]}$$

The above expression is of the form of the first equation

$$A_{cl} = \frac{A_{cl(mid)}}{1 + jf/f_{c(cl)}}$$

where $f_{c(cl)}$ is the closed-loop critical frequency. Thus,

$$f_{c(cl)} = f_{c(ol)}(1 + BA_{ol(mid)})$$

Equation 14–1

In Figure B–8 the common-mode voltage, V_{cm}, on the noninverting input is amplified by the small common-mode gain of op-amp A1. (A_{cm} is typically less than 1.) The total output voltage of op-amp A1 is

$$V_{out1} = \left(1 + \frac{R_1}{R_G}\right)V_{in1} - \left(\frac{R_1}{R_G}\right)V_{in2} + V_{cm}$$

A similar analysis can be applied to op-amp A2 and results in the following output expression:

$$V_{out2} = \left(1 + \frac{R_2}{R_G}\right)V_{in2} - \left(\frac{R_2}{R_G}\right)V_{in1} + V_{cm}$$

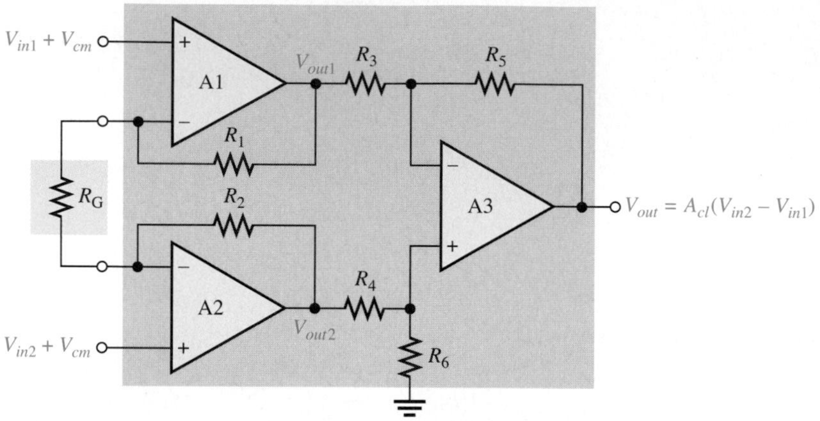

▲ FIGURE B–8

Op-amp A3 has V_{out1} on one of its inputs and V_{out2} on the other. Therefore, the differential input voltage to op-amp A3 is $V_{out2} - V_{out1}$.

$$V_{out2} - V_{out1} = \left(1 + \frac{R_2}{R_G} + \frac{R_1}{R_G}\right)V_{in2} - \left(1 + \frac{R_2}{R_G} + \frac{R_1}{R_G}\right)V_{in1} + V_{cm} - V_{cm}$$

For $R_1 = R_2 = R$,

$$V_{out2} - V_{out1} = \left(1 + \frac{2R}{R_G}\right)V_{in2} - \left(1 + \frac{2R}{R_G}\right)V_{in1} + V_{cm} - V_{cm}$$

Notice that, since the common-mode voltages (V_{cm}) are equal, they cancel each other. Factoring out the differential gain gives the following expression for the differential input to op-amp A3:

$$V_{out2} - V_{out1} = \left(1 + \frac{2R}{R_G}\right)(V_{in2} - V_{in1})$$

Op-amp A3 has unity gain because $R_3 = R_5 = R_4 = R_6$ and $A_v = R_5/R_3. = R_6/R_4.$ Therefore, the final output of the instrumentation amplifier (the output of op-amp A3) is

$$V_{out} = 1(V_{out2} - V_{out1}) = \left(1 + \frac{2R}{R_G}\right)(V_{in2} - V_{in1})$$

The closed-loop gain is

$$A_{cl} = \frac{V_{out}}{V_{in2} - V_{in1}}$$

$$A_{cl} = 1 + \frac{2R}{R_G}$$

Equation 16–1

$$\frac{V_{out}}{V_{in}} = \frac{R(-jX)/(R - jX)}{(R - jX) + R(-jX)/(R - jX)} = \frac{R(-jX)}{(R - jX)^2 - jRX}$$

Multiplying the numerator and denominator by j,

$$\frac{V_{out}}{V_{in}} = \frac{RX}{j(R - jX)^2 + RX} = \frac{RX}{RX + j(R^2 - j2RX - X^2)}$$

$$= \frac{RX}{RX + jR^2 + 2RX - jX^2} = \frac{RX}{3RX + j(R^2 - X^2)}$$

For a 0° phase angle there can be no j term. Recall from complex numbers in ac theory that a *nonzero* angle is associated with a complex number having a j term. Therefore, at f_r the j term is 0.

$$R^2 - X^2 = 0$$

Thus,

$$\frac{V_{out}}{V_{in}} = \frac{RX}{3RX}$$

Cancelling yields

$$\frac{V_{out}}{V_{in}} = \frac{1}{3}$$

Equation 16–2

From the derivation of Equation 16–1,

$$R^2 - X^2 = 0$$
$$R^2 = X^2$$
$$R = X$$

Since $X = \dfrac{1}{2\pi f_r C}$,

$$R = \frac{1}{2\pi f_r C}$$

$$f_r = \frac{1}{2\pi RC}$$

Equations 16–3 and 16–4

The feedback circuit in the phase-shift oscillator consists of three RC stages, as shown in Figure B–9. An expression for the attenuation is derived using the mesh analysis method for the loop assignment shown. All Rs are equal in value, and all Cs are equal in value.

$$(R - j1/2\pi fC)I_1 - RI_2 + 0I_3 = V_{in}$$
$$-RI_1 + (2R - j1/2\pi fC)I_2 - RI_3 = 0$$
$$0I_1 - RI_2 + (2R - j1/2\pi fC)I_3 = 0$$

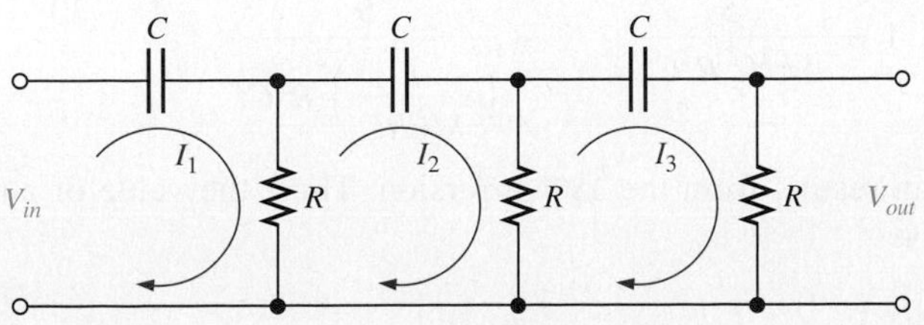

▲ FIGURE B–9

In order to get V_{out}, we must solve for I_3 using determinants:

$$I_3 = \frac{\begin{vmatrix} (R - j1/2\pi fC) & -R & V_{in} \\ -R & (2R - j1/2\pi fC) & 0 \\ 0 & -R & 0 \end{vmatrix}}{\begin{vmatrix} (R - j1/2\pi fC) & -R & 0 \\ -R & (2R - j1/2\pi fC) & -R \\ 0 & -R & (2R - j1/2\pi fC) \end{vmatrix}}$$

$$I_3 = \frac{R^2 V_{in}}{(R - j1/2\pi fC)(2R - j1/2\pi fC)^2 - R^2(2R - j1/2\pi fC) - R^2(R - 1/2\pi fC)}$$

$$\frac{V_{out}}{V_{in}} = \frac{RI_3}{V_{in}}$$

$$= \frac{R^3}{(R - j1/2\pi fC)(2R - j1/2\pi fC)^2 - R^3(2 - j1/2\pi fRC) - R^3(1 - 1/2\pi fRC)}$$

$$= \frac{R^3}{R^3(1 - j1/2\pi fRC)(2 - j1/2\pi fRC)^2 - R^3[(2 - j1/2\pi fRC) - (1 - j1/2\pi fRC)]}$$

$$= \frac{R^3}{R^3(1 - j1/2\pi fRC)(2 - j1/2\pi fRC)^2 - R^3(3 - j1/2\pi fRC)}$$

$$\frac{V_{out}}{V_{in}} = \frac{1}{(1 - j1/2\pi fRC)(2 - j1/2\pi fRC)^2 - (3 - j1/2\pi fRC)}$$

Expanding and combining the real terms and the j terms separately.

$$\frac{V_{out}}{V_{in}} = \frac{1}{\left(1 - \frac{5}{4\pi^2 f^2 R^2 C^2}\right) - j\left(\frac{6}{2\pi fRC} - \frac{1}{(2\pi f)^3 R^3 C^3}\right)}$$

For oscillation in the phase-shift amplifier, the phase shift through the RC circuit must equal 180°. For this condition to exist, the j term must be 0 at the frequency of oscillation f_r.

$$\frac{6}{2\pi f_r RC} - \frac{1}{(2\pi f_r)^3 R^3 C^3} = 0$$

$$\frac{6(2\pi)^2 f_r^2 R^2 C^2 - 1}{(2\pi)^3 f_r^3 R^3 C^3} = 0$$

$$6(2\pi)^2 f_r^2 R^2 C^2 - 1 = 0$$

$$f_r^2 = \frac{1}{6(2\pi)^2 R^2 C^2}$$

$$f_r = \frac{1}{2\pi\sqrt{6}RC}$$

Since the j term is 0,

$$\frac{V_{out}}{V_{in}} = \frac{1}{1 - \frac{5}{4\pi^2 f_r^2 R^2 C^2}} = \frac{1}{1 - \frac{5}{\left(\frac{1}{\sqrt{6}RC}\right)^2 R^2 C^2}} = \frac{1}{1 - 30} = -\frac{1}{29}$$

The negative sign results from the 180° inversion. Thus, the value of attenuation for the feedback circuit is

$$B = \frac{1}{29}$$

Circuit Simulation and Prototyping

Using Multisim and NI ELVIS

Appendix C

Design, Simulation, Prototyping, and Layout

Electronics has changed rapidly and has become a part of an increasing array of products. As circuits and systems become more advanced, circuit designers rely on computers to assist in the design process. The computer has become a vital part of this process. Producing a new circuit for a product can be broken into four main steps: design, simulation, prototyping, and layout. The last three steps usually involve a computer.

A circuit designer begins the development process with an idea for solving a problem. The idea is developed into a circuit, and the designer usually calculates the expected results by hand to get a good idea of circuit behavior. The second step is to enter the schematic into a computer (this is called "schematic capture") and test it with a circuit simulation program such as Multisim. The third step is to construct and test a prototype, which may reveal a hidden or unforeseen problem. Prototyping can also be computer-aided as will be discussed. Sometimes the simulation and prototyping steps are repeated to refine the design. Finally, the design is ready to be implemented. This fourth step is done by transferring the circuit design to a printed circuit board (PCB) using a graphical layout tool like Ultiboard to determine the optimum placement and interconnection of the components.

The focus here is for two of the steps in the design process—simulation and prototyping. As a design aid, Multisim is one of the most widely used circuit design and simulation tools in the industry. Multisim is also used in educational environments because it can simulate circuits quickly, including circuits with troubles. Many Multisim problems have been prepared for the text and the lab manual to introduce you to computer design and simulation tools and to give you troubleshooting practice in a simulated circuit. Although computer simulations are useful and allow you to test parameters that may be difficult, unsafe, or impossible to attain in the lab, they should not be considered a replacement for careful lab work.

The traditional circuit prototype in electronics classes is usually constructed on a solderless protoboard ("breadboard"). The prototype circuit can be tested with stand-alone instruments in the lab. Alternatively, the circuit can be tested with a complete prototyping system like National Instrument's Educational Laboratory and Virtual Instrumentation Suite (NI ELVIS). The NI ELVIS system has over 10 built-in instruments and an interface that can communicate with Multisim. Multisim can also simulate the NI ELVIS interface, as will be shown later.

More About Multisim

Electronics Workbench's Multisim provides an intuitive environment for schematic capture. You place "virtual" electrical components and wire them together into a schematic just as in the laboratory. The graphical circuit schematic capture tool within Multisim is built around a sophisticated industry standard SPICE simulator. Multisim provides built-in instruments that can be connected to schematic circuits in the same way they would connect to a real-world circuit. SPICE was developed at the University of California, Berkeley, and stands for "Simulation Program with Integrated Circuit Emphasis."

Multisim is available for the Windows operating system in a number of tiers for both students and educators. Once you have opened Multisim, you can create a new schematic such

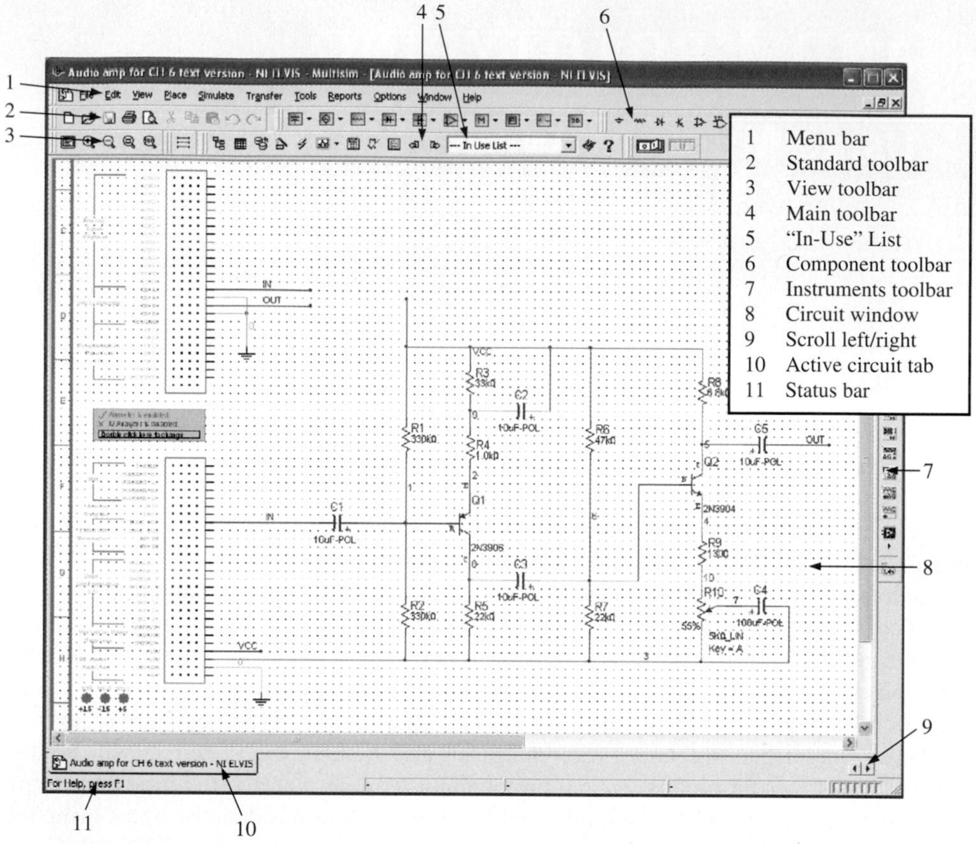

▲ **FIGURE C–1**

The Multisim environment.

as the one shown in Figure C–1. This figure outlines the various elements of the Multisim environment and is the amplifier used in the Application Activity for Chapter 6.

Create circuits in the circuit window by placing components from the Component toolbar. Clicking on the Component toolbar will open the component browser. Choose the family of components, and select an individual component to place on the circuit window by double-clicking on it. Once you have selected a component, it will attach itself and "ghost" the mouse cursor. Clicking again on the desired location in the schematic will place the component. New users to Multisim should use the BASIC_VIRTUAL family of components, which can be assigned any arbitrary value.

The next step is to wire components together. Simply left-click on the source terminal and then left-click on the destination terminal. Multisim will automatically choose the best path for the virtual wire between the two terminals. Always be sure your circuit has a source and a ground or reference, which are components found in the sources group. Once the circuit has been fully captured, it can then be simulated. Results of the simulation can then be used as a comparison with the physical circuit.

As an example of a simple Multisim circuit, the amplifier from Chapter 6 Application Activity is shown in Figure C–1. The amplifier is connected to a virtual interface that represents inputs and outputs on the NI ELVIS station. To create this type of simulated circuit, first create a new NI ELVIS schematic by clicking File >> New >> ELVIS Schematic. Place components by clicking the appropriate icon of the Components toolbar. Wire the NI ELVIS positive supply and ground reference to your circuit, and the circuit is ready to simulate.

Alternatively, you can create a traditional schematic by clicking File >> New >> Schematic. If you do this, you will need to manually place power and ground components. This is useful if you are using traditional stand-alone instruments to analyze your circuit.

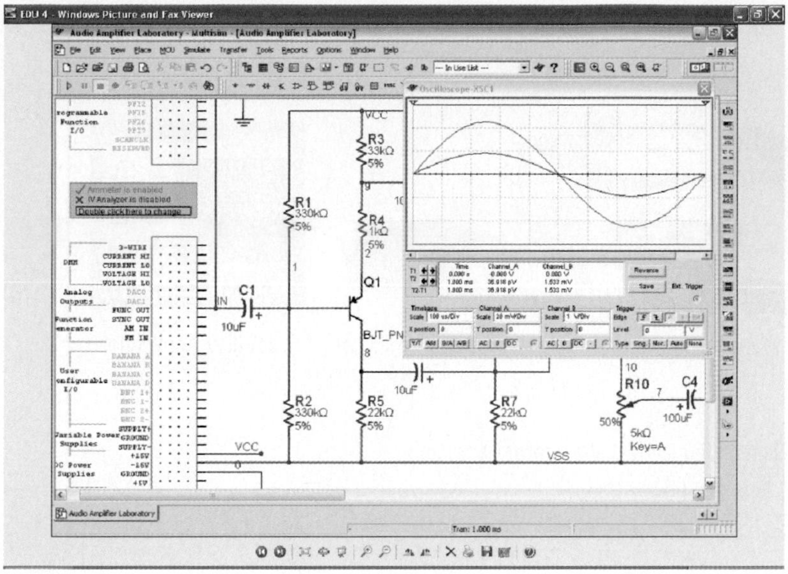

▲ FIGURE C–2

Once you have drawn the circuit, there are many ways to analyze your design using the simulation software.

The easiest way to interact with your circuit in Multisim is using the measurement probe. Use the measurement probe to measure voltages and other important characteristics from a circuit while the simulation is running. To use the measurement probe, click on the probe icon from the virtual Instruments toolbar. The probe becomes attached to the mouse cursor, and will display information about any node or wire that the mouse hovers above. To attach a probe to any particular wire, simply click on it.

You can also see waveforms in the circuit by using one of the oscilloscopes in the Instruments toolbar, as shown in Figure C–2 for a generic oscilloscope. A virtual Tektronix oscilloscope is used in many simulations in this text, and it can be controlled just as an actual oscilloscope by turning knobs using the mouse as a control.

More About NI ELVIS

NI ELVIS consists of LabVIEW virtual instruments, a multifunction data acquisition (DAQ) device, and a bench-top workstation. The combination of LabVIEW virtual instruments, a DAQ board, and prototyping workstation provides all the functions most commonly used in laboratories worldwide. Instructors and students can build customized instruments to suit their application using LabVIEW. Figure C–3 shows the components that make the NI ELVIS. The workstation can be customized by using different experiment boards such as the solderless protoboard shown.

After a circuit has been simulated successfully and you are satisfied with its operation, you can build a prototype. Prototyping hardware and instruments are designed to allow you to quickly construct circuits and test them in the real world. Prototyping is valuable for verifying the operation of the design and for uncovering real-world inconsistencies that simulation was not able to predict. For example, the simulation might not show the effect of unexpected spikes from the wall power or interference from a cell phone. It is critically important that there is a high correlation between real-world measurements and simulation results. A close match assures you that the design will effectively solve the problem.

Your lab may be equipped with separate instruments or you may have them provided as a functional package, such as the NI ELVIS prototyping hardware system. The advantage of the NI ELVIS system is that it can aid in the prototyping process by allowing you to construct

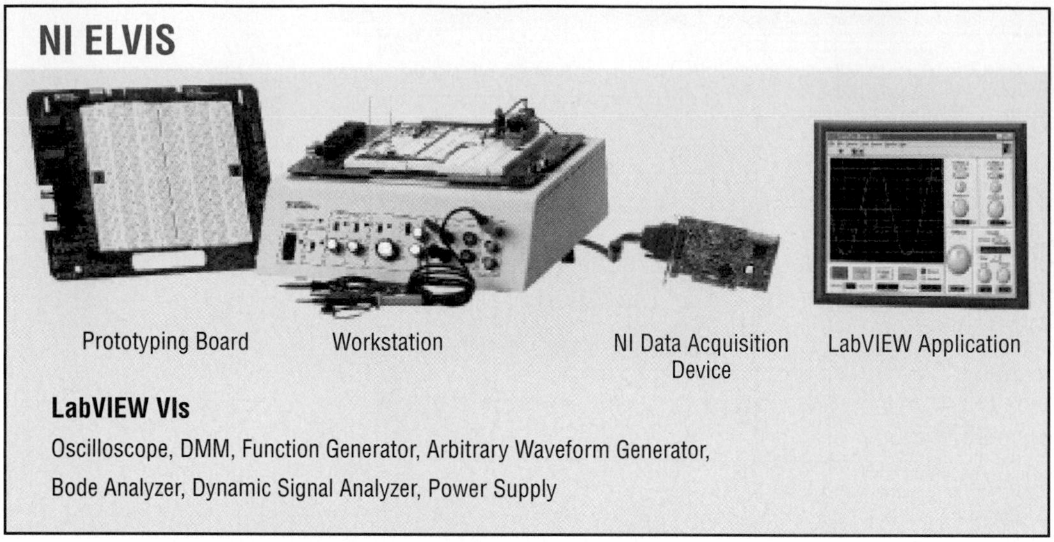

▲ **FIGURE C–3**
Components of NI ELVIS.

circuits first on a "virtual protoboard." After constructing the virtual circuit, you can transfer it to the hardware in the NI ELVIS system. The ideal electronics laboratory will provide an easy way for comparing measurements to simulations. The NI ELVIS system integrates naturally with Multisim to provide a seamless transition from design to prototyping. All the measured data is available in a single environment that allows for quick and easy comparison of simulations to real-world measurements.

Analyzing the Amplifier Using a Virtual Prototype Returning to the audio amplifier introduced with Multisim as an example, you can continue to the prototyping step. You can build the amplifier on a standard protoboard and test it with the laboratory instruments to complete the design.

Alternatively, you can repeat the simulation step but this time as a prototype using the 3D Virtual NI ELVIS from within Multisim. 3D prototyping allows you to learn more about using breadboards and experiment with your designs in a risk-free environment. To construct a 3D ELVIS prototype, open the 3D breadboard by clicking Tools >> Show Breadboard. Place components and wires to build up your circuit. The corresponding connection points and symbols on the NI ELVIS schematic will turn green, indicating the 3D connections are correct. If you created a traditional schematic, you will see a standard breadboard. Figure C–4 shows the virtual NI ELVIS. Figure C–5 shows the circuit on NI ELVIS prototyped in

▶ **FIGURE C–4**
3D virtual NI ELVIS.

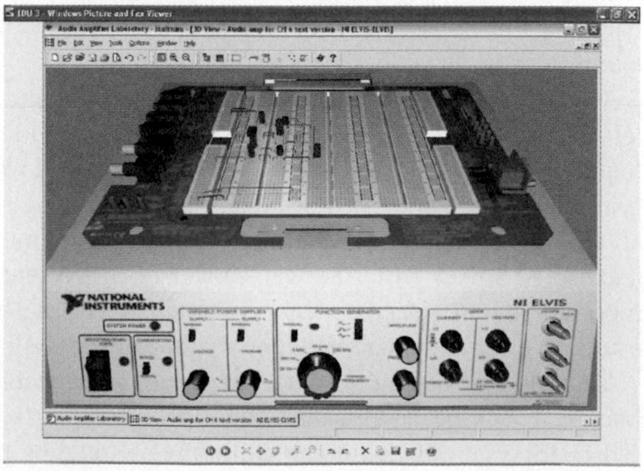

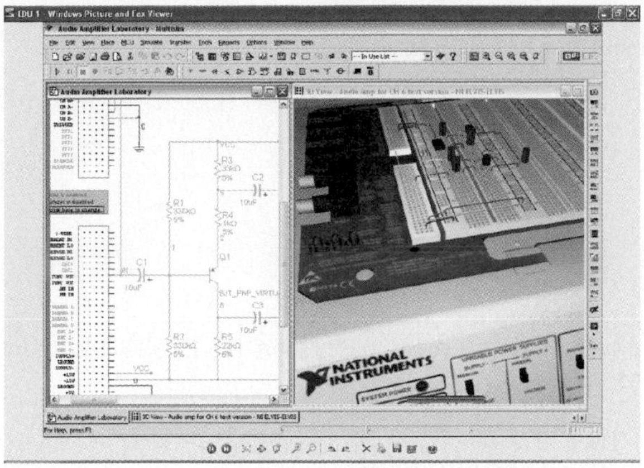

FIGURE C–5
Virtual amplifier prototyped in Multisim.

Multisim and ready for construction. Once the layout of your circuit has been verified using the 3D virtual environment, it can be physically built on the NI ELVIS.

Analyzing the Amplifier Using a Physical Prototype Figure C–6 shows the amplifier on the NI ELVIS actual prototyping board (real components, not virtual). Connections to various instruments such as the oscilloscope, DMM, and AM and FM modulator lines are also available on the breadboard for more advanced applications.

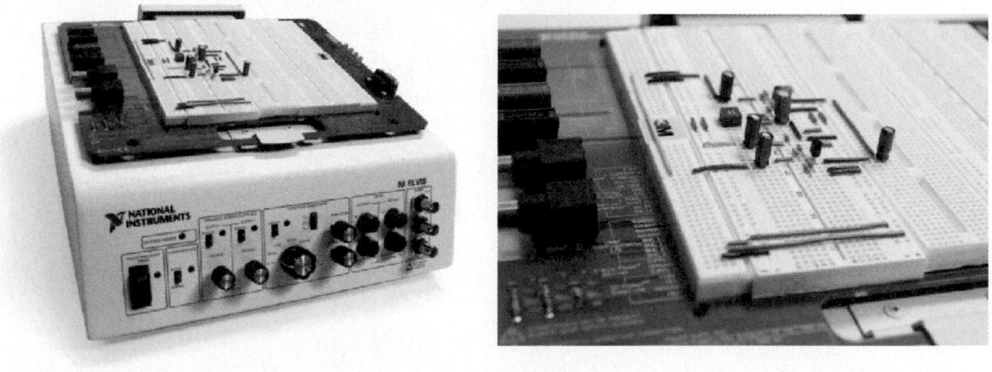

FIGURE C–6
The amplifier physically wired on an NI ELVIS prototyping board.

After you have wired the circuit, you can use the variable power supply provided by NI ELVIS to supply ±15V as needed by the experimental circuit and measure parameters on the board using the interactive graphical software.

The most important step in the laboratory procedure is to compare measurements of the actual circuit to the simulation. This will help you determine where potential errors exist in your design. For example, comparisons can help reveal inadequacies in the simulation model, or incorrect component values. After comparison with the theoretical values, you can revisit your design to improve it or prepare it for layout on a PC board.

To obtain your free Multisim Evaluation Version:
Visit the Electronics Workbench website at
http://www.electronicsworkbench.com/edu/eduprod_cd.html

To obtain your free 3-Hour tutorial:
Contact Electronics Workbench at

Toll Free	800.263.5552
Fax	416.977.1818
Email	sales@electronicsworkbench.com

To view the tutorial online:
Visit the Electronics Workbench website at
www.electronicsworkbench.com/edu/floyd

For more on the NI ELVIS:
Contact National Instruments at

Toll Free	888.280.7645
Fax	512.683.8411
Email	sales@ni.com
Web	www.ni.com

To download NI ELVIS courseware and read technical papers on NI ELVIS/Multisim integration, visit

http://www.ni.com/academic/ni_elvis/ltb.htm

To view conference papers on NI ELVIS in Academia, go to

http://www.ni.com/academic/ni_elvis/conference_papers_on_nielvis.htm

TI-89 Solution to Example 9-3

Appendix D

On the Applications screen select the Numeric Solver logo.

f(x) = 0

Numeric So …

Press ENTER to display the Numeric Solver screen.

Enter Equation

eqn:

Enter the equation. Each letter in the variables must be preceded by ALPHA.

Enter Equation

eqn: id=idss*(idss-id*rs/vgsoff)^2

Press ENTER to display the variables.

Enter Equation

eqn: id=idss*(idss-id*rs/vgsoff)^2

id=

idss=

rs=

vgsoff=

Enter the value of each variable except id.

```
Enter Equation

eqn: id=idss*(idss-id*rs/vgsoff)^2

  id=

  idss=.0043

  rs=1100

  vgsoff=7.7
```

Move the cursor to id and Press F2 to solve. The answer is **.0021037. . . . . .** (2.104 mA). This is in agreement with the results obtained with the TI-86 and with the graphical solution.

Answers to Odd-Numbered Problems

Chapter 1

1. 6 electrons; 6 protons
3. **(a)** insulator **(b)** semiconductor **(c)** conductor
5. Four
7. Conduction band and valence band
9. Antimony is a pentavalent material. Boron is a trivalent material. Both are used for doping.
11. No. The barrier potential is a voltage drop.
13. To prevent excessive forward current.
15. A temperature increase.
17. **(a)** −3 V **(b)** 0.7 V **(c)** 0.7 V **(d)** 0.7 V
19. **(a)** −3 V **(b)** 2.44 V **(c)** 0.731 V **(d)** 0.719 V
21. $V_A = 25$ V; $V_B = 24.3$ V; $V_C = 8.7$ V; $V_D = 8$ V
23. Diode open
25. Diode shorted
27. Diode shorted
29. Diode open

Chapter 2

1. See Figure ANS–1.

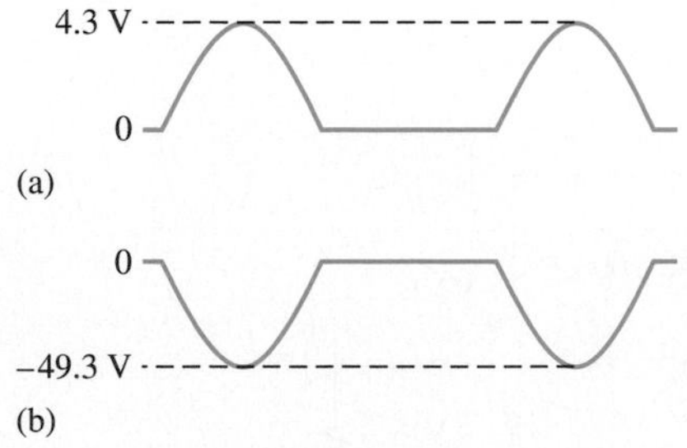

FIGURE ANS–1

3. 63.7 V
5. 24 V rms
7. **(a)** 1.59 V **(b)** 63.7 V **(c)** 16.4 V **(d)** 10.5 V
9. 186 V
11. 78.5 V
13. See Figure ANS–2.

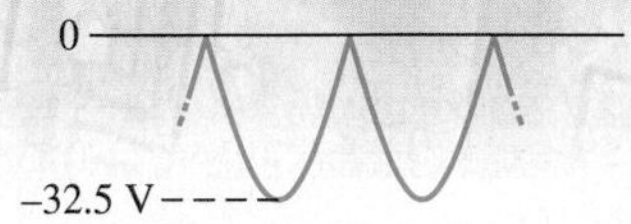

FIGURE ANS–2

15. $V_r = 8.33$ V; $V_{DC} = 25.8$ V
17. 556 μF
19. $V_{r(pp)} = 1.25$ V; $V_{DC} = 48.9$ V
21. 4%
23. See Figure ANS–3.

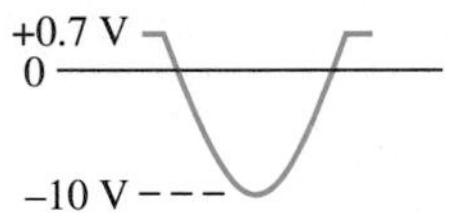

FIGURE ANS–3

25. See Figure ANS–4.

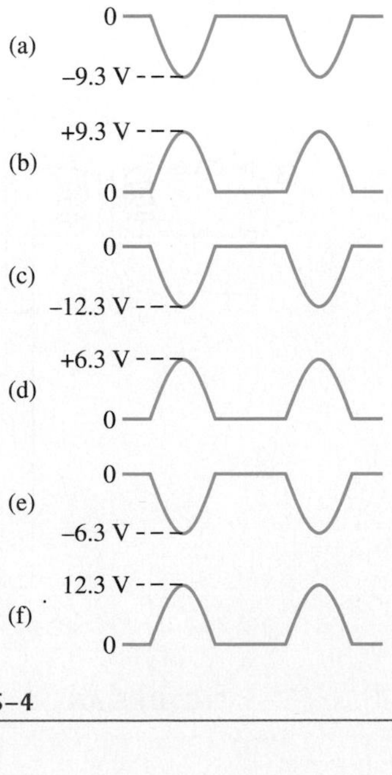

FIGURE ANS–4

27. See Figure ANS–5.

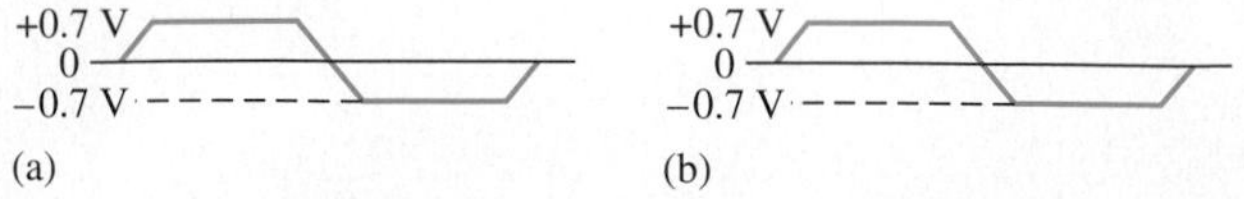

▲ FIGURE ANS–5

29. **(a)** 7.86 mA **(b)** 8.5 mA
(c) 18.8 mA **(d)** 19.4 mA

31. **(a)** A sine wave with a positive peak at +0.7 V, a negative peak at −7.3 V, and a dc value of −3.3 V.

(b) A sine wave with a positive peak at +29.3 V, a negative peak at −0.7 V, and a dc value of +14.3 V.

(c) A square wave varying from +0.7 V down to −15.3 V, with a dc value of −7.3 V.

(d) A square wave varying from +1.3 V down to −0.7 V, with a dc value of +0.3 V.

33. 56.6 V

35. 100 V

37. 50 Ω

39. R_{surge} is open. Capacitor is shorted.

41. The circuit should not fail because the diode ratings exceed the actual PIV and maximum current.

43. The rectifier must be connected backwards.

45. 177 μF

47. 651 mΩ (nearest standard 0.68 Ω)

49. See Figure ANS–6.

51. V_{C1} = 170 V; V_{C2} = 338 V

53. Diode leaky

55. Bottom diode open

57. Open filter capacitor

59. D_1 open

Chapter 3

1. See Figure ANS–7.

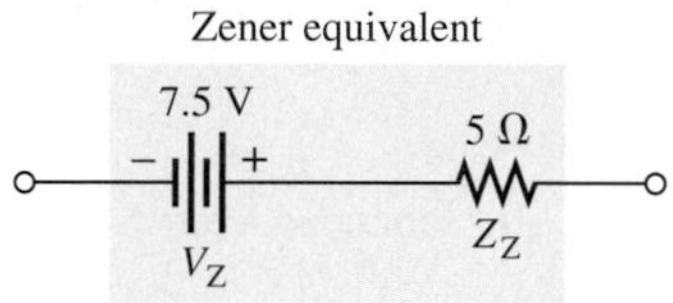

▲ FIGURE ANS–7

3. 5 Ω

5. 6.92 V

7. 14.3 V

9. See Figure ANS–8.

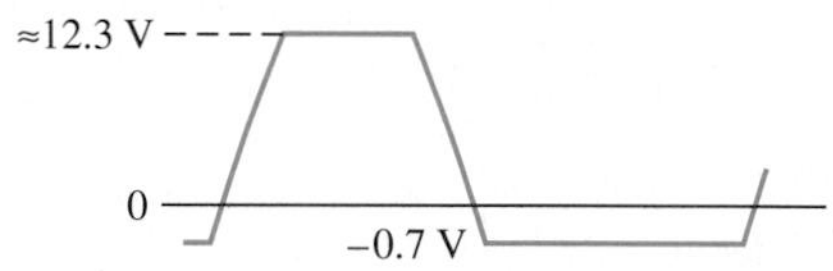

▲ FIGURE ANS–8

11. 10.3%

13. 3.13%

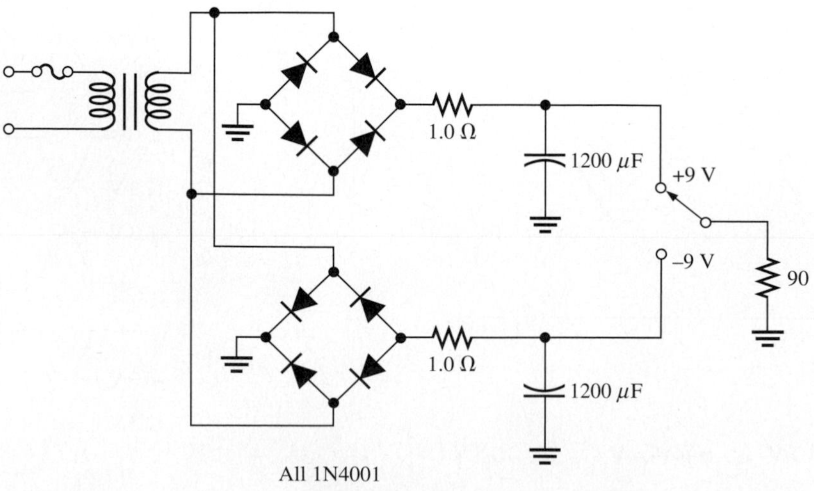

▲ FIG1URE ANS–6

15. 5.88%

17. 3 V

19. $V_R \cong 3$ V

21. See Figure ANS–9.

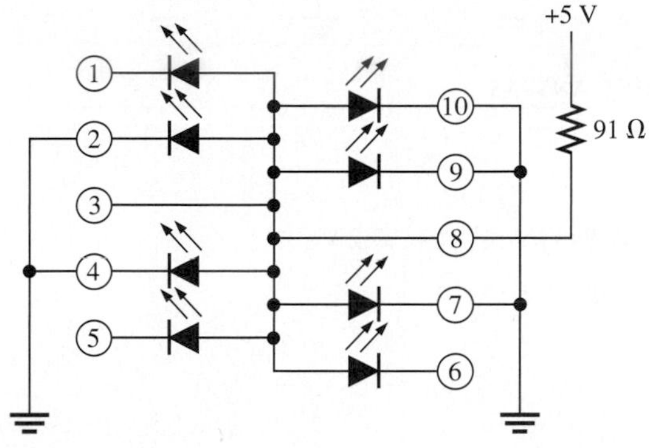

▲ FIGURE ANS–9

23. See Figure ANS–10.

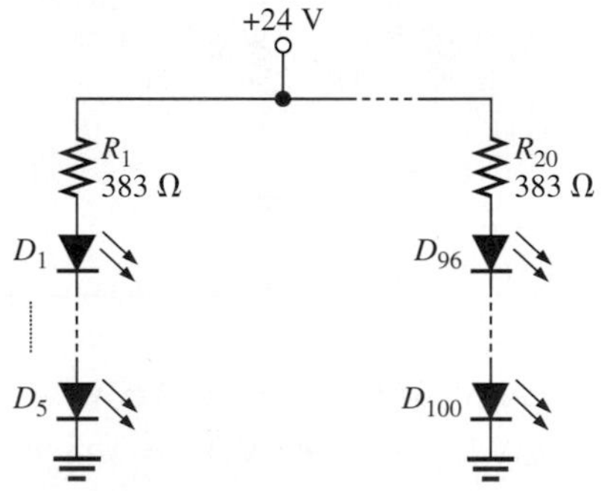

▲ FIGURE ANS–10

25. **(a)** 30 kΩ **(b)** 8.57 kΩ **(c)** 5.88 kΩ

27. −750 Ω

29. The reflective ends cause the light to bounce back and forth, thus increasing the intensity of the light. The partially reflective end allows a portion of the reflected light to be emitted.

31. **(a)** ≈30 V dc

(b) 0 V

(c) Excessive 120 Hz ripple limited to 12 V by zener

(d) Full-wave rectified waveform limited at 12 V by zener

(e) 60 Hz ripple limited to 12 V

(f) 60 Hz ripple limited to 12 V

(g) 0 V

(h) 0 V

33. Incorrect transformer secondary voltage

35. 48 mW

37. **(a)** 200 mA **(b)** 11 pF **(c)** 100 pF to 15.4 pF

39. **(a)** 1 μV **(b)** 940 nm **(c)** 830 nm

41. $V_{OUT(1)} = 6.8$ V; $V_{OUT(2)} = 24$ V

43. See Figure ANS–11.

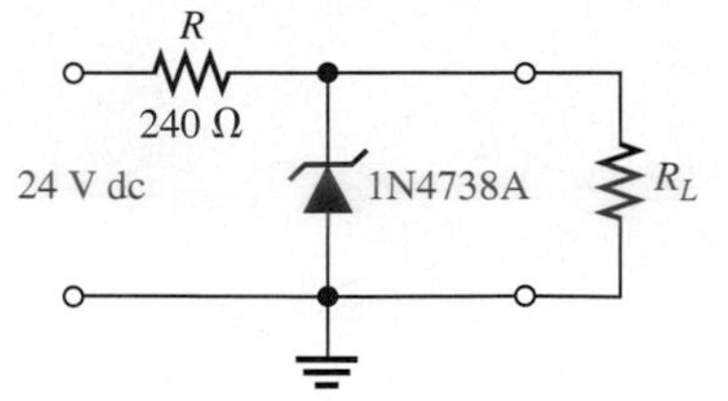

▲ FIGURE ANS–11

45. See Figure ANS–12.

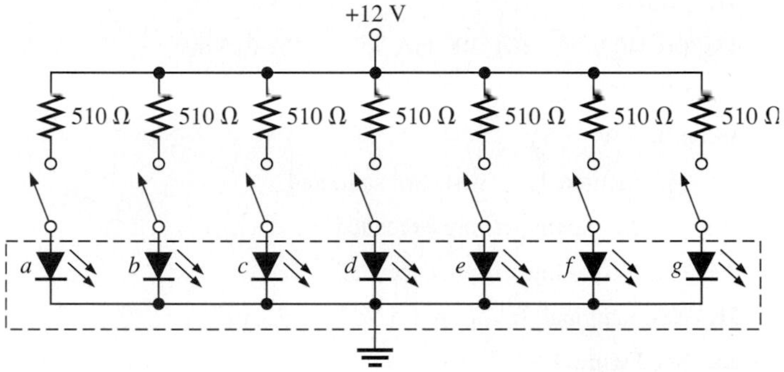

▲ FIGURE ANS–12

47. Zener diode open

49. Zener diode shorted

Chapter 4

1. Holes

3. The base is narrow and lightly doped so that a small recombination (base) current is generated compared to the collector current.

5. Negative, positive

7. 0.947

9. 101.5

11. 8.98 mA

13. 0.99

15. 5.3 V increase

17. **(a)** $V_{BE} = 0.7$ V, $V_{CE} = 5.10$ V, $V_{CB} = +4.40$ V

(b) $V_{BE} = -0.7$ V, $V_{CE} = -3.83$ V, $V_{CB} = -3.13$ V

19. $I_B = 30\ \mu$A, $I_E = 1.3$ mA, $I_C = 1.27$ mA

21. 3 μA

23. 425 mW

25. 33.3

27. 1.1 kΩ

29. 500 μA, 3.33 μA, 4.03 V

31. 20 mA

33. 30 mA

35. See Figure ANS–13.

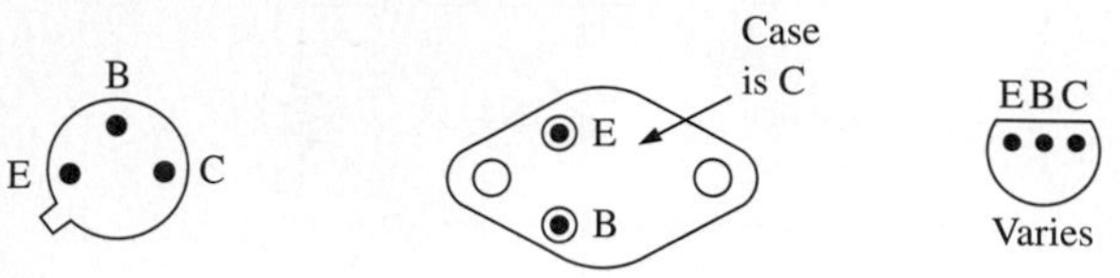

FIGURE ANS–13

37. Open, low resistance

39. **(a)** 27.8 **(b)** 109

41. 60 Ω

43. **(a)** 40 V **(b)** 200 mA dc **(c)** 625 mW
(d) 500 mW **(e)** 70

45. 840 mW

47. **(a)** Saturated **(b)** Not saturated

49. **(a)** No parameters are exceeded.
(b) No parameters are exceeded.

51. Yes, marginally; $V_{CE} = 1.5$ V; $I_C = 75$ mA

53. See Figure ANS–14.

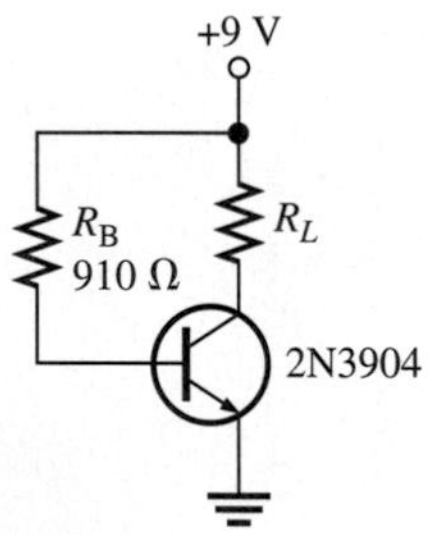

FIGURE ANS–14

55. R_B shorted

57. Collector-emitter shorted

59. R_E leaky

61. R_B open

Chapter 5

1. Saturation

3. 18 mA

5. $V_{CE} = 20$ V; $I_{C(sat)} = 2$ mA

7. See Figure ANS–15.

9. **(a)** $I_{C(sat)} = 50$ mA
(b) $V_{CE(CUTOFF)} = 10$ V
(c) $I_B = 250\ \mu$A; $I_C = 25$ mA; $V_{CE} = 5$ V

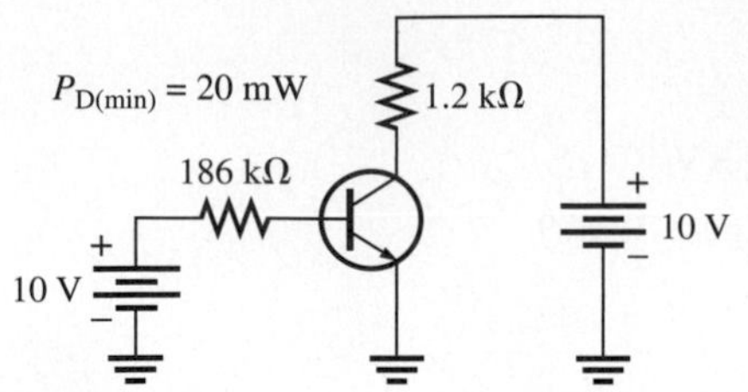

FIGURE ANS–15

11. 69.1

13. $I_C \cong 809\ \mu$A; $V_{CE} = 13.2$ V

15. See Figure ANS–16.

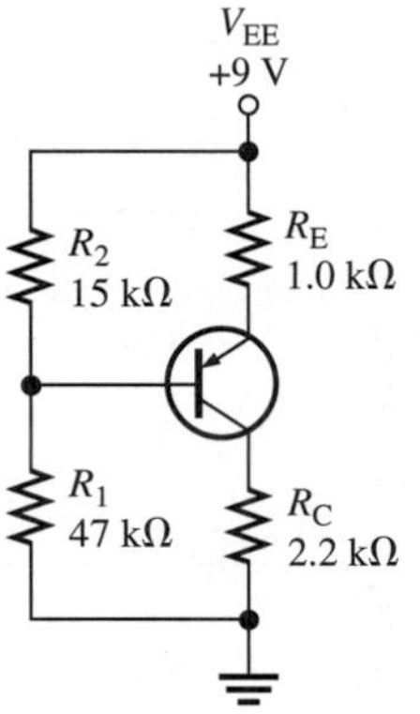

FIGURE ANS–16

17. **(a)** −1.41 mA, −8.67 V **(b)** 12.2 mW

19. $V_B = -186$ mV; $V_E = -0.886$ V; $V_C = 3.14$ V

21. 0.09 mA

23. $I_C = 16.3$ mA; $V_{CE} = -6.95$ V

25. 2.53 kΩ

27. 7.87 mA; 2.56 V

29. $I_{CQ} = 92.5$ mA; $V_{CEQ} = 2.75$ V

31. 27.7 mA to 69.2 mA; 6.23 V to 2.08 V; Yes

33. $V_1 = 0$ V, $V_2 = 0$ V, $V_3 = 8$ V

35. **(a)** Open collector **(b)** No problems
(c) Transistor shorted collector-to-emitter **(d)** Open emitter

37.

	1	2	3	4
(a)	1: 10 V,	2: float,	3: −3.59 V,	4: 10 V
(b)	1: 10 V,	2: 4.05 V,	3: 4.75 V,	4: 4.05 V
(c)	1: 10 V,	2: 0 V,	3: 0 V,	4: 10 V
(d)	1: 10 V,	2: 570 mV,	3: 1.27 V,	4: float
(e)	1: 10 V,	2: 0 V,	3: 0.7 V,	4: 0 V
(f)	1: 10 V,	2: 0 V,	3: 3.59 V,	4: 10 V

39. R_1 open, R_2 shorted, BE junction open

41. $V_C = V_{CC} = 9.1$ V, V_B normal, $V_E = 0$ V

43. None are exceeded.

45. 457 mW

47. See Figure ANS–17.

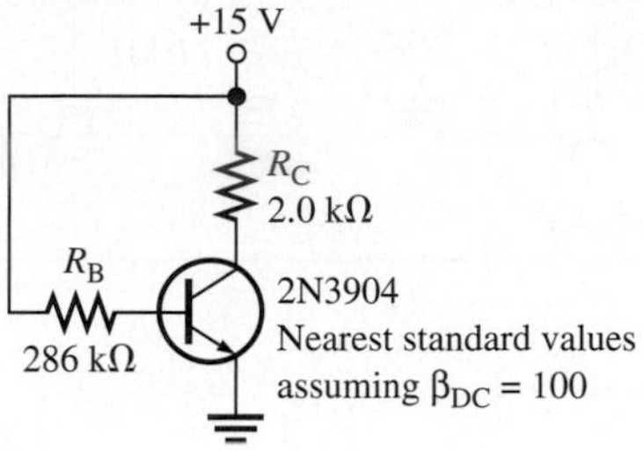

▲ **FIGURE ANS–17**

49. See Figure ANS–18.

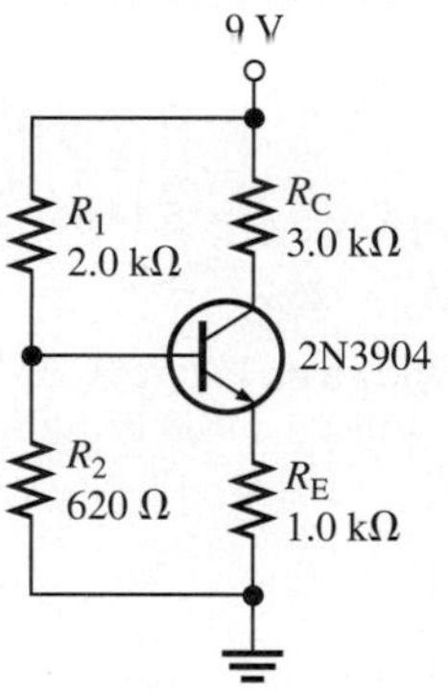

▲ **FIGURE ANS–18**

51. Yes

53. V_{CEQ} will be less, causing the transistor to saturate at a slightly higher temperature, thus limiting the low temperature response.

55. R_C open

57. R_2 open

59. R_C shorted

Chapter 6

1. Slightly greater than 1 mA min.

3. 8.33 Ω

5. $r'_e \cong 19\ \Omega$

7. See Figure ANS–19.

9. 37.5 mW

11. **(a)** 1.29 kΩ **(b)** 968 Ω **(c)** 171

13. **(a)** $V_B = 3.25$ V **(b)** $V_E = 2.55$ V
(c) $I_E = 2.55$ mA **(d)** $I_C \cong 2.55$ mA
(e) $V_C = 9.59$ V **(f)** $V_{CE} = 7.04$ V

15. $A'_v = 131;\ \theta = 180°$

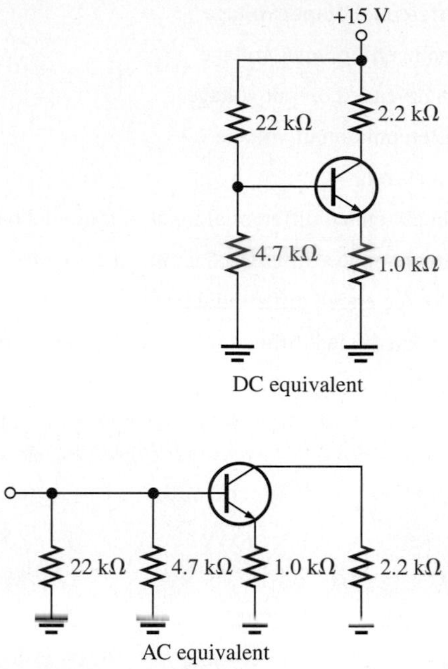

▲ **FIGURE ANS–19**

17. $A_{v(max)} = 65.5, A_{v(min)} = 2.06$

19. A_v is reduced to approximately 30. See Figure ANS–20.

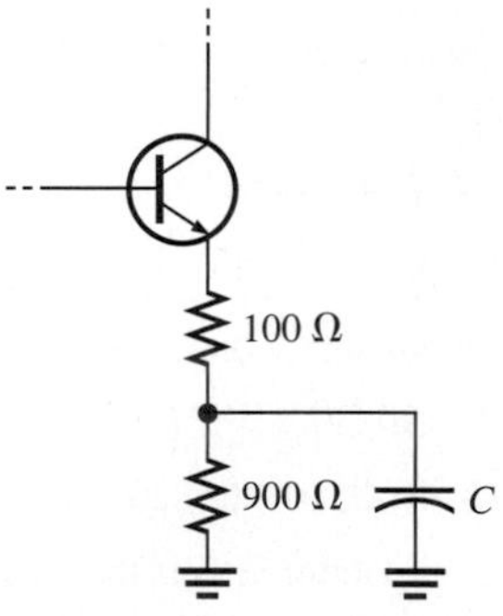

▲ **FIGURE ANS–20**

21. $R_{in(tot)} = 3.1\ \text{k}\Omega;\ V_{OUT} = 1.06$ V

23. 270 Ω

25. 8.8

27. $R_{in(emitter)} = 2.28\ \Omega;\ A_v = 526;\ A_i \cong 1;\ A_p = 526$

29. 400

31. **(a)** $A_{v1} = 93.6, A_{v2} = 303$
(b) $A'_v = 28{,}361$
(c) $A_{v1(dB)} = 39.4$ dB, $A_{v2(dB)} = 49.6$ dB, $A'_{v(dB)} = 89.1$ dB

33. $V_{B1} = 2.16$ V, $V_{E1} = 1.46$ V, $V_{C1} \cong 5.16$ V, $V_{B2} = 5.16$ V, $V_{E2} = 4.46$ V, $V_{C2} \cong 7.54$ V, $A_{v1} = 66, A_{v2} = 179$, $A'_v = 11{,}814$

35. **(a)** 1.41 **(b)** 2.00 **(c)** 3.16
(d) 10.0 **(e)** 100

37. V_1: differential output voltage
V_2: noninverting input voltage
V_3: single-ended output voltage
V_4: differential input voltage
I_1: bias current

39. **(a)** Single-ended differential input; differential output
(b) Single-ended differential input; single-ended output
(c) Double-ended differential input; single-ended output
(d) Double-ended differential input; differential output

41. Cutoff, 10 V

43.

TEST POINT	DC VOLTS	AC VOLTS (RMS)
Input	0 V	25 μV
Q_1 base	2.99 V	20.8 μV
Q_1 emitter	2.29 V	0 V
Q_1 collector	7.44 V	1.95 mV
Q_2 base	2.99 V	1.95 mV
Q_2 emitter	2.29 V	0 V
Q_2 collector	7.44 V	589 mV
Output	0 V	589 mV

45. **(a)** No output signal
(b) Reduced output signal
(c) No output signal
(d) Reduced output signal
(e) No output signal
(f) Increased output signal (perhaps with distortion)

47. **(a)** Q_1 is off **(b)** V_{EE} **(c)** Unchanged

49. **(a)** 700 **(b)** 40 Ω **(c)** 20 kΩ

51. A leaky coupling capacitor affects the bias voltages and attenuates the ac voltage.

53. Make $R_9 = 69.1\ \Omega$.

55. See Figure ANS–21.

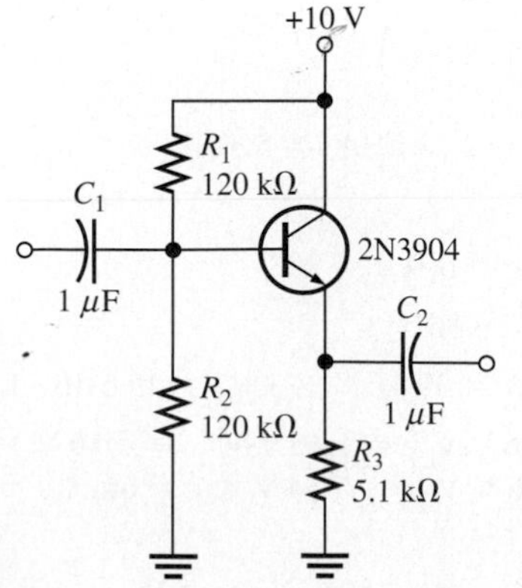

FIGURE ANS–21

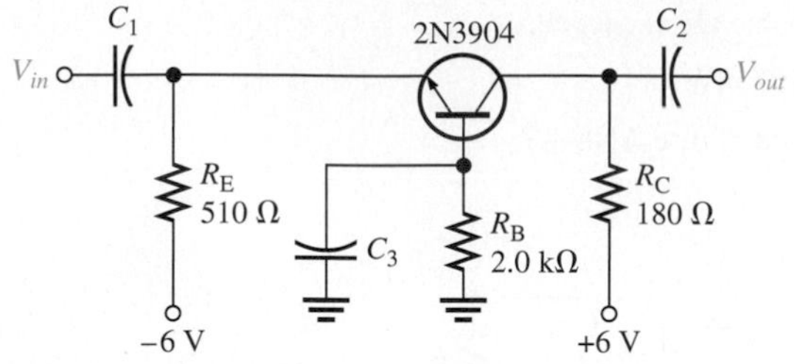

FIGURE ANS–22

57. See Figure ANS–22.

59. $A_v = R_C/r'_e$
$A_v \cong (V_{R_C}/I_C)/(0.025\text{ V}/I_C) = V_{R_C}/0.025 = 40V_{R_C}$

61. C_2 shorted

63. C_1 open

65. C_3 open

Chapter 7

1. **(a)** $I_{CQ} = 68.4$ mA; $V_{CEQ} = 5.14$ V
(b) $A_v = 11.7$; $A_p = 263$

3. The changes are shown on Figure ANS–23. The advantage of this arrangement is that the load resistor is referenced to ground.

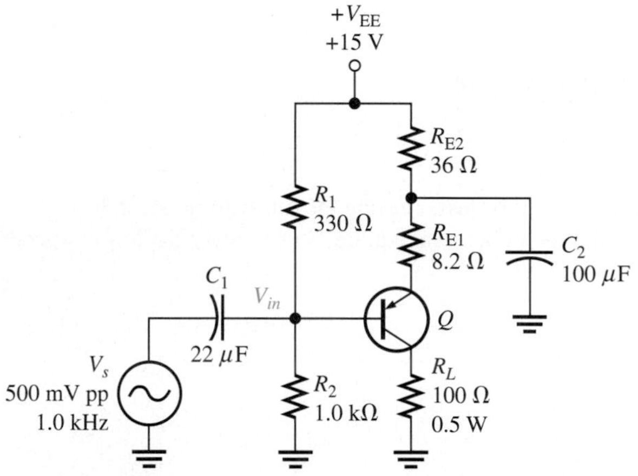

FIGURE ANS–23

5. **(a)** $I_C = 55.7$ mA; $V_{CE} = 1.99$ V
(b) $I_C = 14.6$ mA; $V_{CE} = 3.04$ V

7. For Figure 7–43(a): 39.8 mA, 1.99 V; For Figure 7–43(b): 12.3 mA, 2.88 V

9. 170 mW

11. **(a)** $V_{B(Q1)} = +0.7$ V; $V_{B(Q2)} = -0.7$ V; $V_E = 0$ V; $V_{CEQ(Q1)} = +9$ V; $V_{CEQ(Q2)} = -9$ V; $I_{CQ} = 8.3$ mA
(b) $P_L = 0.5$ W

13. 457 Ω

15. **(a)** $V_{B(Q1)} = +8.2\text{ V}$; $V_{B(Q2)} = +6.8\text{ V}$;
$V_E = +7.5\text{ V}$;
$I_{CQ} = 6.8\text{ mA}$; $V_{CEQ(Q1)} = +7.5\text{ V}$;
$V_{CEQ(Q2)} = -7.5\text{ V}$
(b) $P_L = 167\text{ mW}$

17. **(a)** C_2 open or Q_2 open
(b) Power supply off, open R_1, Q_1 base shorted to ground
(c) Q_1 has collector-to-emitter short
(d) One or both diodes shorted

19. 450 μW

21. 24 V

23. Negative half of input cycle

25. **(a)** No dc supply voltage or R_1 open
(b) D_2 open
(c) No fault
(d) Q_1 shorted C-to-E

27. −15 V dc, output signal same as input signal

29. D_3 is connected backwards.

31. 10 W

33. Gain increases, then decreases at a certain value of I_C.

35. T_C is much closer to the actual junction temperature than T_A. In a given operating environment, T_A is always less than T_C.

37. See Figure ANS–24.

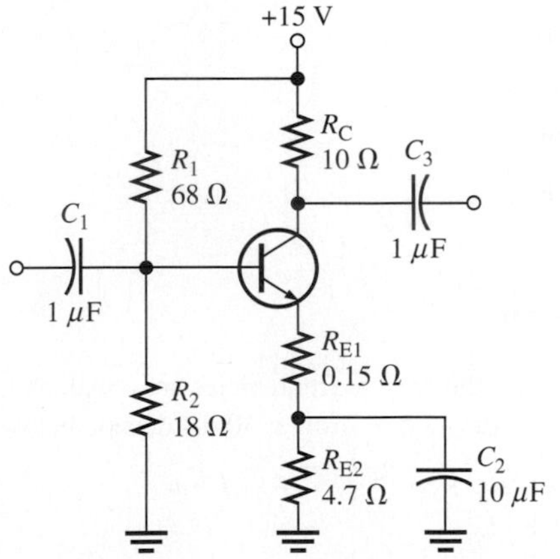

▲ FIGURE ANS–24

39. C_{in} open

41. Q_1 collector-emitter open

43. Q_2 drain-source open

Chapter 8

1. **(a)** Narrows **(b)** Increases

3. See Figure ANS–25.

5. 5 V

7. 10 mA

9. 4 V

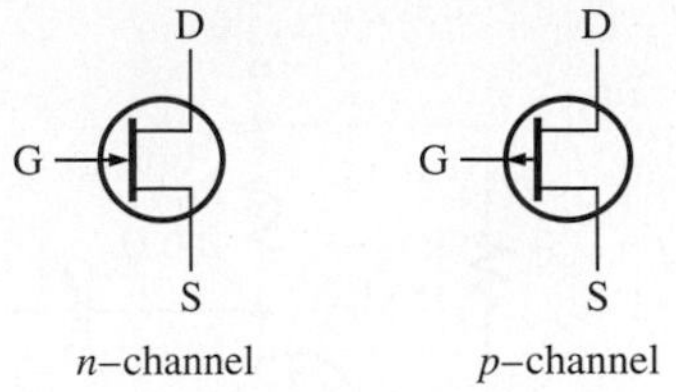

▲ FIGURE ANS–25

11. −2.63 V

13. $g_m = 1429\ \mu\text{S}$, $g_{fs} = 1429\ \mu\text{S}$

15. $V_{GS} = 0\text{ V}$, $I_D = 8\text{ mA}$
$V_{GS} = -1\text{ V}$, $I_D = 5.12\text{ mA}$
$V_{GS} = -2\text{ V}$, $I_D = 2.88\text{ mA}$
$V_{GS} = -3\text{ V}$, $I_D = 1.28\text{ mA}$
$V_{GS} = -4\text{ V}$, $I_D = 0.320\text{ mA}$
$V_{GS} = -5\text{ V}$, $I_D = 0\text{ mA}$

17. 800 Ω

19. **(a)** 20 mA **(b)** 0 A **(c)** Increases

21. 211 Ω

23. 9.80 MΩ

25. $I_D \cong 5.3\text{ mA}$, $V_{GS} \cong 2.1\text{ V}$

27. $I_D \cong 1.9\text{ mA}$, $V_{GS} \cong -1.5\text{ V}$

29. From 1.33 kΩ to 2.67 kΩ

31. 935 Ω

33. The enhancement mode

35. The gate is insulated from the channel.

37. 4.69 mA

39. **(a)** Depletion **(b)** Enhancement
(c) Zero bias **(d)** Depletion

41. **(a)** 4 V **(b)** 5.4 V **(c)** −4.52 V

43. **(a)** 5 V, 3.18 mA **(b)** 3.2 V, 1.02 mA

45. The input resistance of an IGBT is very high because of the insulated gate structure.

47. R_D or R_S open, JFET open D-to-S, $V_{DD} = 0\text{ V}$, or ground connection open.

49. Essentially no change

51. The 1.0 MΩ bias resistor is open.

53. $V_{G2S} = 6\text{ V}$, $I_D \cong 10\text{ mA}$; $V_{G2S} = 1\text{ V}$, $I_D \cong 5\text{ mA}$

55. 3.04 V

57. **(a)** −0.5 V **(b)** 25 V **(c)** 310 mW **(d)** −25 V

59. 2000 μS

61. $I_D \cong 1.4\text{ mA}$

63. $I_D \cong 13\text{ mA}$ when $V_{GS} = +3\text{ V}$, $I_D \cong 0.4\text{ mA}$ when $V_{GS} = -2\text{ V}$.

65. −3.0 V

67. $I_D = 3.58\text{ mA}$; $V_{GS} = -4.21\text{ V}$

69. 6.01 V

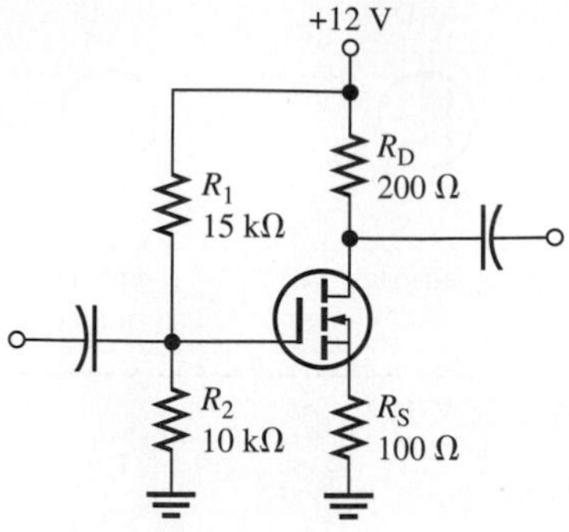

FIGURE ANS–26

71. See Figure ANS–26.

73. R_D shorted

75. R_1 open

77. R_D open

79. Drain-source shorted

Chapter 9

1. **(a)** 60 μA **(b)** 900 μA **(c)** 3.6 mA **(d)** 6 mA

3. 14.2

5. **(a)** n-channel D-MOSFET with zero-bias; $V_{GS} = 0$
(b) p-channel JFET with self-bias; $V_{GS} = -0.99$ V
(c) n-channel E-MOSFET with voltage-divider bias; $V_{GS} = 3.84$ V

7. **(a)** n-channel D-MOSFET
(b) n-channel JFET
(c) p-channel E-MOSFET

9. Figure 9–16(b): approximately 4 mA
Figure 9–16(c): approximately 3.2 mA

11. 920 mV

13. **(a)** 4.32 **(b)** 9.92

15. 7.5 mA

17. 2.54

19. 33.6 mV rms

21. 9.84 MΩ

23. $V_{GS} = 9$ V; $I_D = 3.13$ mA; $V_{DS} = 13.3$ V; $V_{ds} = 675$ mV

25. $R_{in} \cong 10$ MΩ; $A_v = 0.620$

27. **(a)** 0.906 **(b)** 0.299

29. 250 Ω

31. $A_v = 2640$; $R_{in} = 14.6$ MΩ

33. 0.95

35. 30 kHz

37. 40 kΩ

39. **(a)** 3.3 V **(b)** 3.3 V **(c)** 3.3 V **(d)** 0 V

41. The MOSFET has lower on-state resistance and can turn off faster.

43. **(a)** $V_{out} = 0$ V if C_1 is open
(b) $A_{v1} = 7.5, A_{v2} = 2.24, A_v = 16.8, V_{out} = 168$ mV
(c) $V_{GS} = 0$ V for Q_2. $I_D = I_{DSS}$. Output is clipped.
(d) No V_{out} because no signal at Q_2 gate.

45. **(a)** −3.0 V **(b)** 20 V dc
(c) 200 mW **(d)** ±10 V dc

47. 900 μS

49. 1.5 mA

51. 2.0; 6.82

53. See Figure ANS–27.

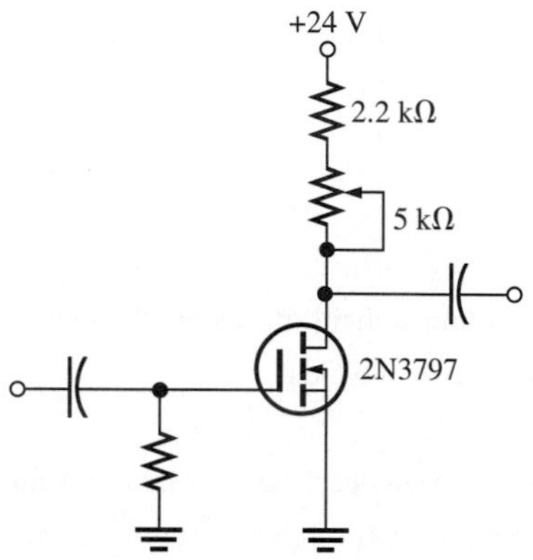

FIGURE ANS–27

55. C_2 open

57. R_S shorted

59. R_1 open

61. R_2 open

Chapter 10

1. If $C_1 = C_2$, the critical frequencies are equal, and they will both cause the gain to drop at 40 dB/decade below f_c.

3. BJT: C_{be}, C_{bc}, C_{ce}; FET: C_{gs}, C_{gd}, C_{ds}

5. 812 pF

7. $C_{in(miller)} = 6.95$ pF; $C_{out(miller)} \cong 5.28$ pF

9. 24 mV rms; 34 dB

11. **(a)** 3.01 dBm **(b)** 0 dBm
(c) 6.02 dBm **(d)** − 6.02 dBm

13. **(a)** 318 Hz
(b) 1.59 kHz

15. At $0.1f_c$: $A_v = 18.8$ dB
At f_c: $A_v = 35.8$ dB
At $10f_c$: $A_v = 38.8$ dB

17. Input RC circuit: $f_c = 3.34$ Hz
Output RC circuit: $f_c = 3.01$ kHz
Output f_c is dominant.

19. Input circuit: $f_c = 4.32$ MHz
Output circuit: $f_c = 94.9$ MHz
Input f_c is dominant.

21. Input circuit: $f_c = 12.9$ MHz
Output circuit: $f_c = 54.5$ MHz
Input f_c is dominant.

23. $f_{cl} = 136$ Hz, $f_{cu} = 8$ kHz

25. $BW = 5.26$ MHz, $f_{cu} \cong 5.26$ MHz

27. 230 Hz; 1.2 MHz

29. 514 kHz

31. $\approx$2.5 MHz

33. Increase the frequency until the output voltage drops to 3.54 V rms. This is f_{cu}.

35. 15.9 Hz

37. No effect

39. 112 pF

41. $C_{gd} = 1.3$ pF; $C_{gs} = 3.7$ pF; $C_{ds} = 3.7$ pF

43. $\approx$10.9 MHz

45. R_C open

47. R_2 open

Chapter 11

1. $I_A = 24.1$ mA

3. See "Turning the SCR On" in Section 11–2.

5. When the switch is closed, the battery V_2 causes illumination of the lamp. The light energy causes the LASCR to conduct and thus energize the relay. When the relay is energized, the contacts close and 115 V ac are applied to the motor.

7. Add a transistor to provide inversion of negative half-cycle in order to obtain a positive gate trigger.

9. See Figure ANS–28.

11. See Figure ANS–29.

13. Anode, cathode, anode gate, cathode gate

15. 6.48 V

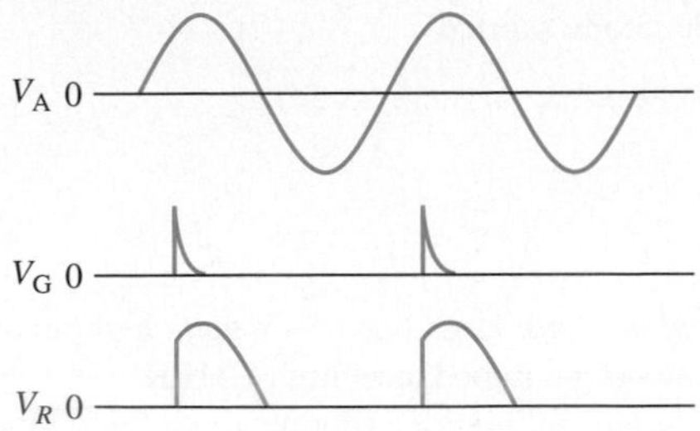

▲ FIGURE ANS–28

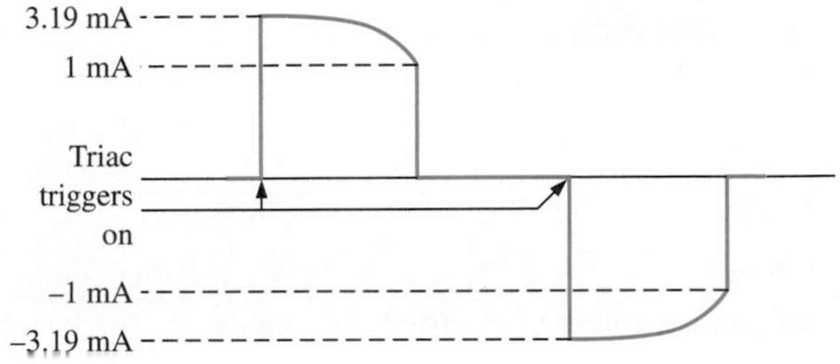

▲ FIGURE ANS–29

17. **(a)** 9.79 V **(b)** 5.2 V

19. See Figure ANS–30.

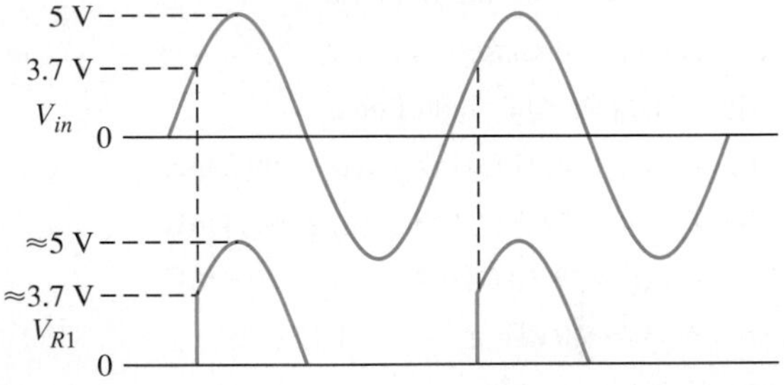

▲ FIGURE ANS–30

21. 0 V

23. As the PUT gate voltage increases, the PUT triggers on later to the ac cycle causing the SCR to fire later in the cycle, conduct for a shorter time, and decrease power to the motor.

25. See Figure ANS–31.

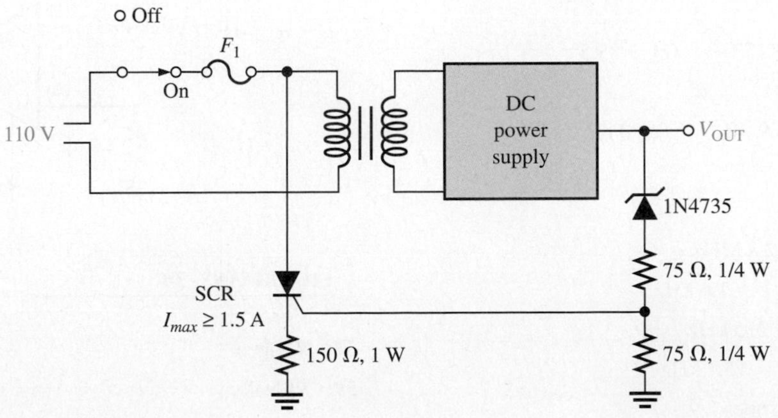

▲ FIGURE ANS–31

27. Cathode-anode shorted

29. R_1 shorted

Chapter 12

1. *Practical op-amp:* High open-loop gain, high input impedance, low output impedance, high CMRR.
Ideal op-amp: Infinite open-loop gain, infinite input impedance, zero output impedance, infinite CMRR.

3. **(a)** Single-ended differential input
(b) Double-ended differential input
(c) Common-mode input

5. 120 dB

7. 8.1 μA

9. 1.6 V/μs

11. **(a)** Voltage-follower **(b)** Noninverting **(c)** Inverting

13. **(a)** $A_{cl(NI)} = 374$ **(b)** $V_{out} = 3.74$ V rms
(c) $V_f = 9.99$ mV rms

15. **(a)** 49 kΩ **(b)** 3 MΩ **(c)** 84 kΩ **(d)** 165 kΩ

17. **(a)** 10 mV, in phase
(b) −10 mV, 180° out of phase
(c) 223 mV, in phase
(d) −100 mV, 180° out of phase

19. **(a)** $Z_{in(NI)} = 8.41$ GΩ; $Z_{out(NI)} = 89.2$ mΩ
(b) $Z_{in(NI)} = 6.20$ GΩ; $Z_{out(NI)} = 4.04$ mΩ
(c) $Z_{in(NI)} = 5.30$ GΩ; $Z_{out(NI)} = 19.0$ mΩ

21. **(a)** $Z_{in(I)} = 10$ kΩ; $Z_{out(I)} = 5.12$ mΩ
(b) $Z_{in(I)} = 100$ kΩ; $Z_{out(I)} = 7.32$ mΩ
(c) $Z_{in(I)} = 470$ Ω; $Z_{out(I)} = 6.22$ mΩ

23. **(a)** 2.69 kΩ **(b)** 1.45 kΩ **(c)** 53 kΩ
R_c is placed between V_{in} and the + input

25. 175 nV

27. $A_v = 125{,}892$; $BW_{ol} = 200$ Hz

29. **(a)** 0.997 **(b)** 0.923 **(c)** 0.707
(d) 0.515 **(e)** 0.119

31. **(a)** −51.5° **(b)** −7.17° **(c)** −85.5°

33. **(a)** 90 dB **(b)** −281°

35. **(a)** 29.8 dB **(b)** 23.9 dB **(c)** 0 dB
All are closed-loop gains.

37. 71.7 dB

39. **(a)** $A_{cl(VF)} = 1$; $BW = 2.8$ MHz
(b) $A_{cl(I)} = -45.5$; $BW = 61.6$ kHz
(c) $A_{cl(NI)} = 13$; $BW = 215$ kHz
(d) $A_{cl(I)} = -179$; $BW = 15.7$ kHz

41. **(a)** Faulty op-amp or R_1 open
(b) R_2 open, forcing open-loop operation

43. The gain becomes a fixed −100, with no effect as the gain potentiometer is adjusted.

45. Op-amp gain will be reduced by 100.

47. $Z_{in(NI)} = 3.96$ GΩ

49. 50,000

51. See Figure ANS–32.

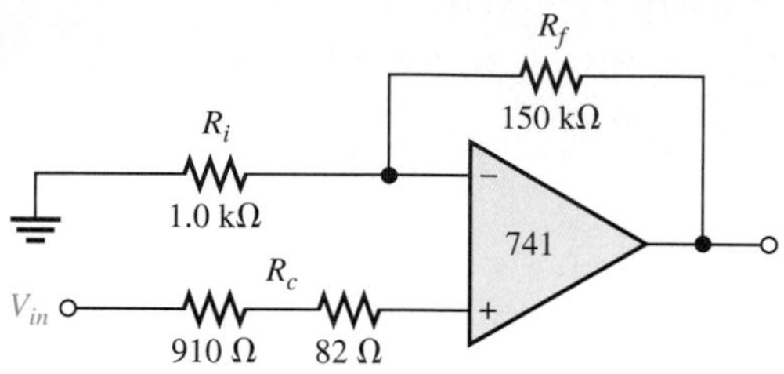

▲ FIGURE ANS–32

53. See Figure ANS–33.

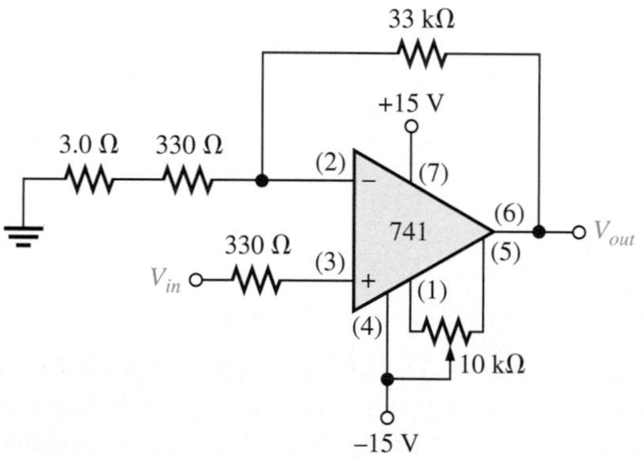

▲ FIGURE ANS–33

55. See Figure ANS–34.

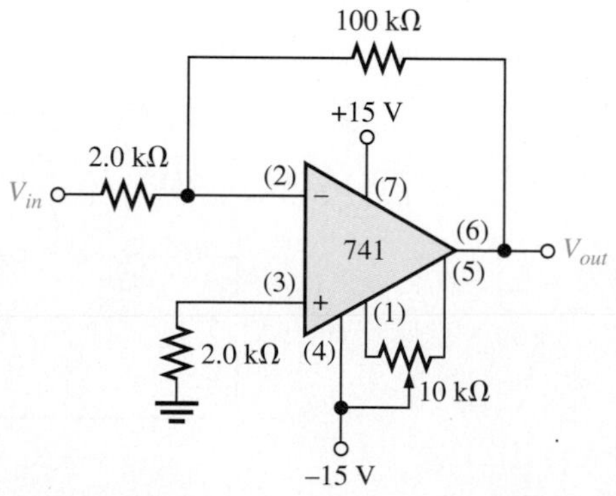

▲ FIGURE ANS–34

57. R_f open

59. R_f leaky

61. R_f shorted

63. R_f leaky

65. R_i shorted

67. R_f open

69. R_f open

71. R_i open

Chapter 13

1. 24 V, with distortion

3. $V_{UTP} = +2.77$ V, $V_{LTP} = -2.77$ V

5. See Figure ANS–35.

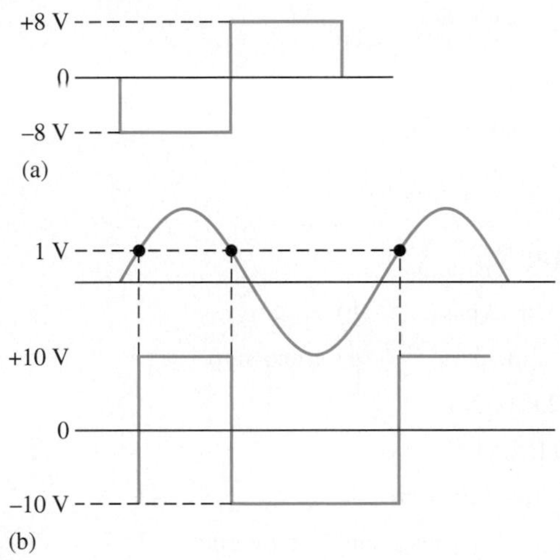

▲ FIGURE ANS–35

7. +8.57 V and −0.968 V

9. **(a)** −2.5 V **(b)** −3.52 V

11. 110 kΩ

13. $V_{OUT} = -3.57$ V, $I_f = 357\ \mu$A

15. −4.06 mV/μs

17. 1 mA

19. See Figure ANS–36.

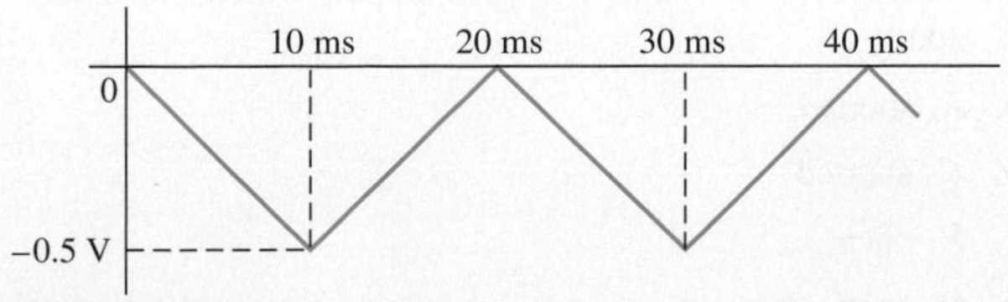

▲ FIGURE ANS–36

21. Output not correct; R_2 is open.

23. 50 kΩ resistor open

25. An increase in susceptibility to power line noise.

27. min. duty cycle = 6.39%
max. duty cycle = 93.6%

29. f_{in} = 100 kHz. See Figure ANS–37.

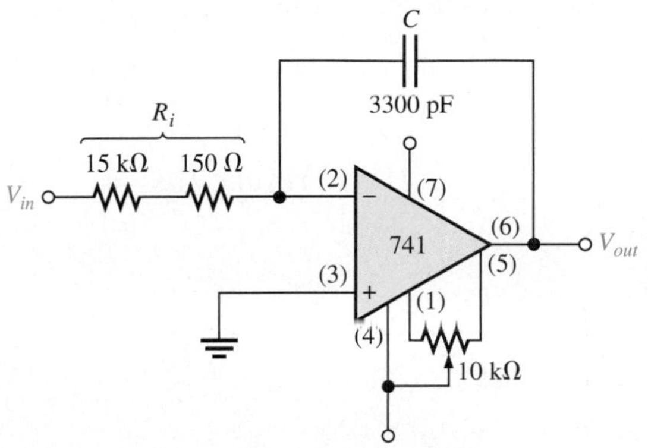

▲ FIGURE ANS–37

31. Op-amp inputs shorted together

33. D_1 shorted

35. Middle 10 kΩ resistor shorted

37. R_f open

39. C open

Chapter 14

1. $A_{v(1)} = A_{v(2)} = 101$

3. 1.005 V

5. 51.5

7. Change R_G to 2.2 kΩ.

9. 300

11. Change the 18 kΩ resistor to 68 kΩ.

13. Connect pin 6 directly to pin 10, and connect pin 14 directly to pin 15 to make $R_f = 0$.

15. 500 μA, 5 V

17. $A_v \cong 11.6$

19. See Figure ANS–38.

21. See Figure ANS–39.

23. **(a)** −0.301 **(b)** 0.301 **(c)** 1.70 **(d)** 2.11

25. The output of a log amplifier is limited to 0.7 V because of the transistor's *pn* junction.

27. −157 mV

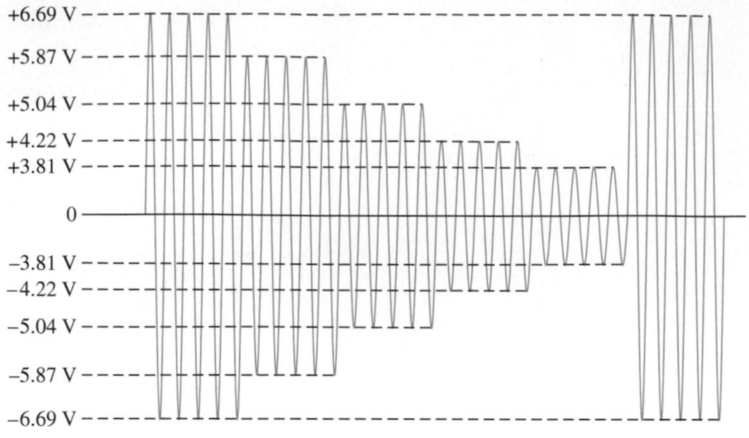

▲ FIGURE ANS–38

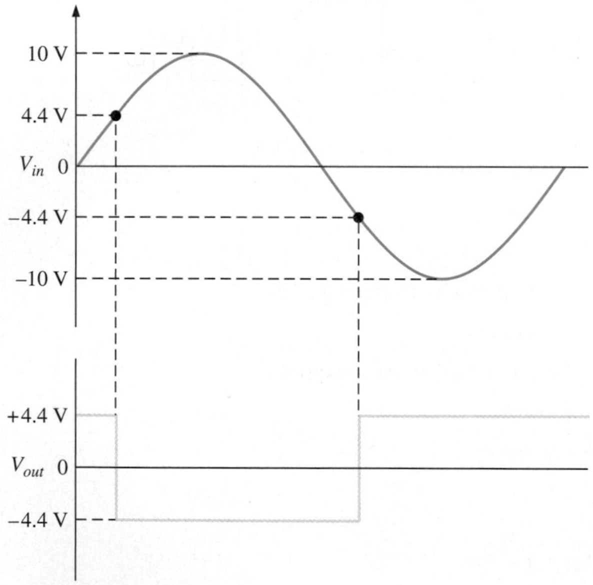

▲ FIGURE ANS–39

29. $V_{out(max)} = -147$ mV, $V_{out(min)} = -89.2$ mV; the 1 V input peak is reduced 85% whereas the 100 mV input peak is reduced only 10%.

31. See Figure ANS–40.

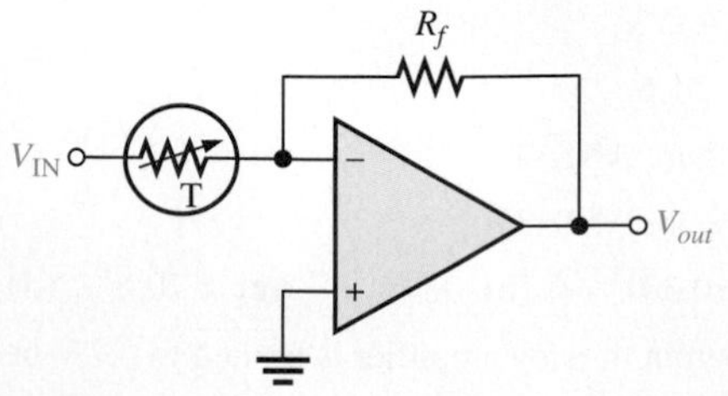

▲ FIGURE ANS–40

33. R open

35. Zener diode open

Chapter 15

1. **(a)** Band-pass **(b)** High-pass
(c) Low-pass **(d)** Band-stop

3. 48.2 kHz, No

5. 700 Hz, 5.04

7. **(a)** 1, not Butterworth
(b) 1.44, approximate Butterworth
(c) 1st stage: 1.67; 2nd stage: 1.67; Not Butterworth

9. **(a)** Chebyshev **(b)** Butterworth
(c) Bessel **(d)** Butterworth

11. 190 Hz

13. Add another identical stage and change the ratio of the feedback resistors to 0.068 for first stage, 0.586 for second stage, and 1.482 for third stage.

15. Exchange positions of resistors and capacitors.

17. **(a)** Decrease R_1 and R_2 or C_1 and C_2.
(b) Increase R_3 or decrease R_4.

19. **(a)** $f_0 = 4.95$ kHz, $BW = 3.84$ kHz
(b) $f_0 = 449$ Hz, $BW = 96.4$ Hz
(c) $f_0 = 15.9$ kHz, $BW = 838$ Hz

21. Sum the low-pass and high-pass outputs with a two-input adder.

23. R_4 shorted

25. C_3 shorted

27. R_1 open

29. R_1 open

31. R_7 open

Chapter 16

1. An oscillator requires no input (other than dc power).
3. $\frac{1}{75} = 0.0133$
5. 733 mV
7. 50 kΩ
9. 2.34 kΩ
11. 136 kΩ, 628 Hz
13. 10
15. Change R_1 to 3.54 kΩ
17. $R_4 = 65.8$ kΩ, $R_5 = 47$ kΩ
19. 3.33 V, 6.67 V
21. 0.0076 μF
23. Drain-to-source shorted
25. Collector-to-emitter shorted
27. R_2 open

Chapter 17

1. 0.0333%
3. 1.01%
5. A: Reference voltage, B: Control element, C: Error detector, D: Sampling circuit
7. 8.51 V
9. 9.57 V
11. 500 mA
13. 10 mA
15. $I_{L(max)} = 250$ mA, $P_{R1} = 6.25$ W
17. 40%
19. V_{OUT} decreases
21. 14.3 V
23. 1.3 mA
25. 2.8 Ω
27. $R_{limit} = 0.35$ Ω
29. See Figure ANS–41.

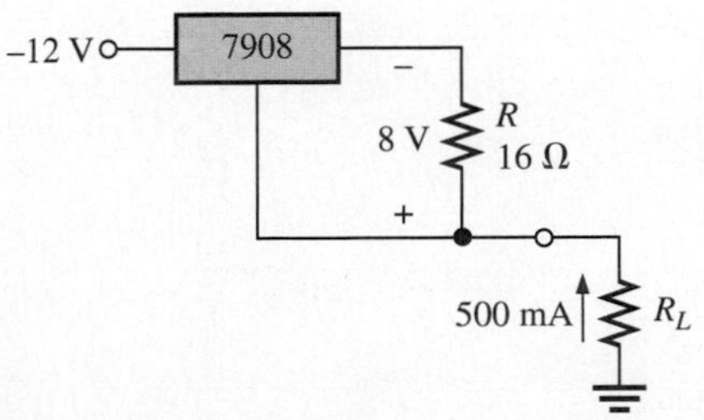

▲ FIGURE ANS–41

31. R_2 leaky
33. Q_2 collector-to-emitter open

Chapter 18

1. See Figure ANS–42.

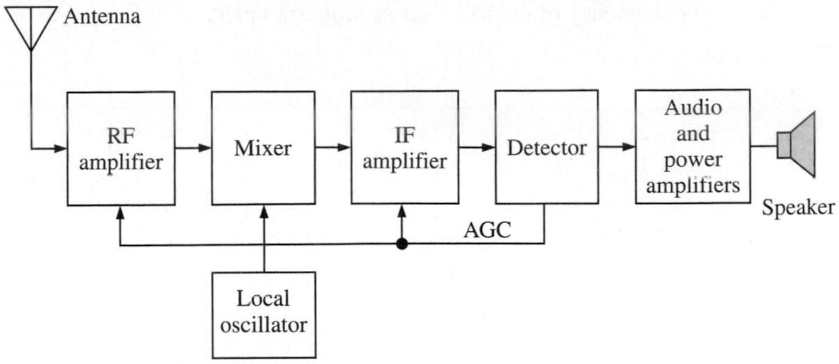

▲ FIGURE ANS–42

3. 1135 kHz
5. $f_{RF} = 91.2$ MHz; $f_{IF} = 10.7$ MHz
7. −1.27 V
9. **(a)** +0.28 V **(b)** +1.024 V
   (c) +2.07 V **(d)** +2.49 V
11. $f_{diff} = 8$ kHz; $f_{sum} = 10$ kHz
13. $f_{diff} = 1.7$ MHz; $f_{sum} = 1.9$ MHz; $f_c = 1.8$ MHz
15. $f_c = 850$ kHz; $f_m = 3$ kHz
17. $V_{out} = 15\text{ mV}\cos[2\pi(1100\text{ kHz})t] - 15\text{ mV}\cos[2\pi(5500\text{ kHz})t]$
19. See Figure ANS–43.

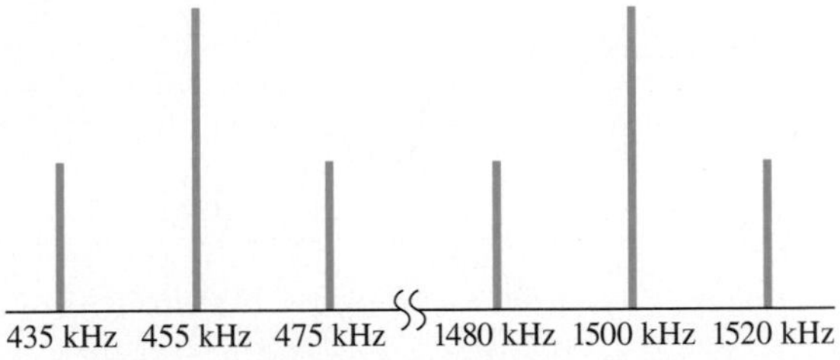

▲ FIGURE ANS–43

21. See Figure ANS–44.

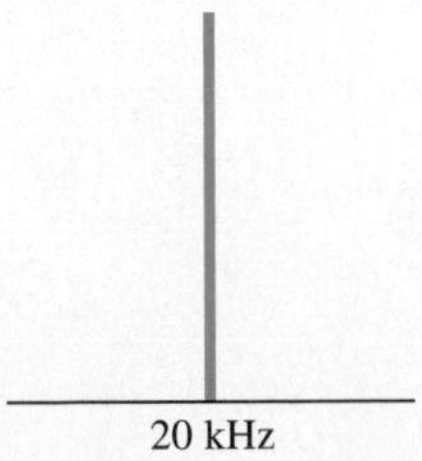

▲ FIGURE ANS–44

23. 450 kHz–460 kHz: IF amplifier; 10 Hz–5 kHz: Audio/Power amplifiers

25. The modulating input signal is applied to the control voltage terminal of the VCO. As the input signal amplitude varies, the output frequency of the VCO varies proportionately.

27. Varactor

29. **(a)** 10 MHz **(b)** 48.3 mV

31. 1005 Hz

33. $f_o = 233$ kHz; $f_{lock} = \pm 104$ kHz; $f_{cap} = \pm 4.56$ kHz

35. 36.2°

GLOSSARY

ac ground A point in a circuit that appears as ground to ac signals only.

active filter A frequency-selective circuit consisting of active devices such as transistors or op-amps coupled with reactive components.

A/D conversion A process whereby information in analog form is converted into digital form.

alpha (α) The ratio of dc collector current to dc emitter current in a bipolar junction transistor.

amplification The process of increasing the power, voltage, or current by electronic means.

amplifier An electronic circuit having the capability to amplify power, voltage, or current.

amplitude modulation (AM) A communication method in which a lower-frequency signal modulates (varies) the amplitude of a higher-frequency signal (carrier).

analog Characterized by a linear process in which a variable takes on a continuous set of values.

analog switch A device that switches an analog signal on and off.

angle of incidence The angle at which a light ray strikes a surface.

anode The *p* region of a diode.

antilogarithm The result obtained when the base of a number is raised to a power equal to the logarithm of that number.

astable Characterized by having no stable states.

atom The smallest particle of an element that possesses the unique characteristics of that element.

atomic number The number of protons in an atom.

attenuation The reduction in the level of power, current, or voltage.

audio Related to the frequency range of sound waves that can be heard by the human ear and generally considered to be in the 20 Hz to 20 kHz range.

avalanche breakdown The higher voltage breakdown in a zener diode.

avalanche effect The rapid buildup of conduction electrons due to excessive reverse-bias voltage.

balanced modulation A form of amplitude modulation in which the carrier is suppressed; sometimes known as *suppressed-carrier modulation.*

band-pass filter A type of filter that passes a range of frequencies lying between a certain lower frequency and a certain higher frequency.

band-stop filter A type of filter that blocks or rejects a range of frequencies lying between a certain lower frequency and a certain higher frequency.

bandwidth The characteristic of certain types of electronic circuits that specifies the usable range of frequencies that pass from input to output.

barrier potential The amount of energy required to produce full conduction across the *pn* junction in forward bias.

base One of the semiconductor regions in a BJT. The base is very thin and lightly doped compared to the other regions.

Bessel A type of filter response having a linear phase characteristic and less than −20 dB/decade/pole rolloff.

beta (β) The ratio of dc collector current to dc base current in a BJT; current gain from base to collector.

bias The application of a dc voltage to a diode, transistor, or other device to produce a desired mode of operation.

bipolar Characterized by both free electrons and holes as current carriers.

BJT Bipolar junction transistor; a transistor constructed with three doped semiconductor regions separated by two *pn* junctions.

Bode plot An idealized graph of the gain in dB versus frequency used to graphically illustrate the response of an amplifier or filter.

bounding The process of limiting the output range of an amplifier or other circuit.

breakdown The phenomenon of a sudden and drastic increase when a certain voltage is reached across a device.

bridge rectifier A type of full-wave rectifier consisting of diodes arranged in a four-cornered configuration.

Butterworth A type of filter response characterized by flatness in the passband and a −20 dB/decade/pole roll-off.

bypass capacitor A capacitor placed across the emitter resistor of an amplifier.

CAM Configurable analog module; a predesigned analog circuit used in an FPAA or dpASP for which some of its parameters can be selectively programmed.

capacitance ratio The ratio of varactor capacitances at minimum and at maximum reverse voltages.

capture range The range of frequencies over which a PLL can acquire lock.

carbon A semiconductive material.

cascade An arrangement of circuits in which the output of one circuit becomes the input to the next.

cascode A FET amplifier configuration in which a common-source amplifier and a common-gate amplifier are connected in a series arrangement.

cathode The *n* region of a diode.

center-tapped rectifier A type of full-wave rectifier consisting of a center-tapped transformer and two diodes.

channel The conductive path between the drain and source in a FET.

Chebyshev A type of filter response characterized by ripples in the passband and a greater than −20 dB/decade/pole roll-off.

clamper A circuit that adds a dc level to an ac voltage using a diode and a capacitor.

class A A type of amplifier that operates entirely in its linear (active) region.

class AB A type of amplifier that is biased into slight conduction.

class B A type of amplifier that operates in the linear region for 180° of the input cycle because it is biased at cutoff.

class C A type of amplifier that operates only for a small portion of the input cycle.

class D A nonlinear amplifier in which the transistors are operated as switches.

clipper See Limiter.

closed-loop An op-amp configuration in which the output is connected back to the input through a feedback circuit.

closed-loop voltage gain (A_{cl}) The voltage gain of an op-amp with external feedback.

CMOS Complementary MOS.

CMRR Common-mode rejection ratio; the ratio of open-loop gain to common-mode gain; a measure of an op-amp's ability to reject common-mode signals.

coherent light Light having only one wavelength.

collector The largest of the three semiconductor regions of a BJT.

common-base (CB) A BJT amplifier configuration in which the base is the common terminal to an ac signal or ground.

common-collector (CC) A BJT amplifier configuration in which the collector is the common terminal to an ac signal or ground.

common-drain (CD) A FET amplifier configuration in which the drain is the grounded terminal.

common-emitter (CE) A BJT amplifier configuration in which the emitter is the common terminal to an ac signal or ground.

common-gate (CG) A FET amplifier configuration in which the gate is the grounded terminal.

common mode A condition where two signals applied to differential inputs are of the same phase, frequency, and amplitude.

common-source (CS) A FET amplifier configuration in which the source is the grounded terminal.

comparator A circuit which compares two input voltages and produces an output in either of two states indicating the greater or less than relationship of the inputs.

complementary symmetry transistors Two transistors, one *npn,* and one *pnp,* having matched characteristics.

conduction electron A free electron.

conductor A material that conducts electrical current very well.

core The central part of an atom, includes the nucleus and all but the valence electrons.

covalent Related to the bonding of two or more atoms by the interaction of their valence electrons.

critical angle The angle that defines whether a light ray will be reflected or refracted when it strikes a surface.

critical frequency The frequency at which the response of an amplifier or filter is 3 dB less than at midrange.

crossover distortion Distortion in the output of a class B push-pull amplifier at the point where each transistor changes from the cutoff state to the *on* state.

crystal A solid material in which the atoms are arranged in a symmetrical pattern.

current The rate of flow of electrical charge.

current mirror A circuit that uses matching diode junctions to form a current source. The current in a diode junction is reflected as a matching current in the other junction (which is typically the base-emitter junction of a transistor). Current mirrors are commonly used to bias a push-pull amplifier.

cutoff The nonconducting state of a transistor.

cutoff frequency Another term for critical frequency.

cutoff voltage The value of the gate-to-source voltage that makes the drain current approximately zero.

D/A conversion The process of converting a sequence of digital codes to an analog form.

damping factor A filter characteristic that determines the type of response.

dark current The amount of thermally generated reverse current in a photodiode in the absence of light.

Darlington pair A configuration of two transistors in which the collectors are connected and the emitter of the first drives the base of the second to achieve beta multiplication.

dBm A unit for measuring power levels referenced to 1 mW.

dc load line A straight line plot of I_C and V_{CE} for a transistor circuit.

decade A ten-times increase or decrease in the value of a quantity such as frequency.

decibel (dB) A logarithmic measure of the ratio of one power to another or one voltage to another.

demodulation The process in which the information signal is recovered from the IF carrier signal; the reverse of modulation.

depletion In a MOSFET, the process of removing or depleting the channel of charge carriers and thus decreasing the channel conductivity.

depletion region The area near a *pn* junction on both sides that has no majority carriers.

diac A two-terminal four-layer semiconductor device (thyristor) that can conduct current in either direction when properly activated.

differential mode A mode of op-amp operation in which two opposite polarity signal voltages are applied to two inputs (double-ended) or in which a signal is applied to one input and ground to the other (single-ended).

differential amplifier (diff-amp) An amplifier in which the output is a function of the difference between two input voltages, used as the input stage of an op-amp.

differentiator A circuit that produces an output which approximates the instantaneous rate of change of the input function.

digital Characterized by a process in which a variable takes on either of two values.

diode A semiconductor device with a single *pn* junction that conducts current in only one direction.

diode drop The voltage across the diode when it is forward-biased; approximately the same as the barrier potential and typically 0.7 V for silicon.

discriminator A type of FM demodulator.

doping The process of imparting impurities to an intrinsic semiconductive material in order to control its conduction characteristics.

downloading The process of implementing the software description of a circuit in an FPAA.

drain One of the three terminals of a FET analogous to the collector of a BJT.

dynamic reconfiguration The process of downloading a design modification or new design in an FPAA while it is operating in a system without the need to power down or reset the system; also known as "on-the-fly" reprogramming.

dynamic resistance The nonlinear internal resistance of a semiconductive material.

efficiency The ratio of the signal power delivered to a load to the power from the power supply of an amplifier.

electroluminescence The process of releasing light energy by the recombination of electrons in a semiconductor.

electrostatic discharge (ESD) The discharge of a high voltage through an insulating path that can destroy an electronic device.

electron The basic particle of negative electrical charge.

electron-hole pair The conduction electron and the hole created when the electron leaves the valence band.

emitter The most heavily doped of the three semiconductor regions of a BJT.

emitter-follower A popular term for a common-collector amplifier.

enhancement In a MOSFET, the process of creating a channel or increasing the conductivity of the channel by the addition of charge carriers.

feedback The process of returning a portion of a circuit's output back to the input in such a way as to oppose or aid a change in the output.

feedback oscillator An electronic circuit that operates with positive feedback and produces a time-varying output signal without an external input signal.

FET Field-effect transistor; a type of unipolar, voltage-controlled transistor that uses an induced electric field to control current.

fiber optics The use of light for the transmission of information through tiny fiber cables.

filter In a power supply, a capacitor used to reduce the variation of the output voltage from a rectifier; a type of circuit that passes or blocks certain frequencies to the exclusion of all others.

floating point A point in the circuit that is not electrically connected to ground or a "solid" voltage.

fold-back current limiting A method of current limiting in voltage regulators.

forced commutation A method of turning off an SCR.

forward bias The condition in which a diode conducts current.

forward-breakover voltage ($V_{BR(F)}$) The voltage at which a device enters the forward-blocking region.

4-layer diode The type of two-terminal thyristor that conducts current when the anode-to-cathode voltage reaches a specified "breakover" value.

four-quadrant multiplier A linear device that produces an output voltage proportional to the product of two input voltages.

FPAA Field-programmable analog array; an integrated circuit that can be programmed for implementation of an analog circuit design.

free electron An electron that has acquired enough energy to break away from the valance band of the parent atom; also called a conduction electron.

frequency modulation (FM) A communication method in which a lower frequency intelligence-carrying signal modulates (varies) the frequency of a higher frequency signal.

frequency response The change in gain or phase shift over a specified range of input signal frequencies.

full-wave rectifier A circuit that converts an ac sinusoidal input voltage into a pulsating dc voltage with two output pulses occurring for each input cycle.

fuse A protective device that burns open when the current exceeds a rated limit.

gain The amount by which an electrical signal is increased or amplified.

gain-bandwidth product A constant parameter which is always equal to the frequency at which the op-amp's open-loop gain is unity (1).

gate One of the three terminals of a FET analogous to the base of a BJT.

germanium A semiconductive material.

half-wave rectifier A circuit that converts an ac sinusoidal input voltage into a pulsating dc voltage with one output pulse occurring for each input cycle.

high-pass filter A type of filter that passes frequencies above a certain frequency while rejecting lower frequencies.

holding current (I_H) The value of the anode current below which a device switches from the forward-conduction region to the forward-blocking region.

hole The absence of an electron in the valence band of an atom.

hysteresis Characteristic of a circuit in which two different trigger levels create an offset or lag in the switching action.

IGBT Insulated-gate bipolar transistor; a device that combines features of the MOSFET and the BJT and used mainly for high-voltage switching applications.

index of refraction An optical characteristic of a material that determines the critical angle.

infrared (IR) Light that has a range of wavelengths greater than visible light.

input resistance The resistance looking in at the transistor base.

instrumentation amplifier An amplifier used for amplifying small signals riding on large common-mode voltages.

insulator A material that does not conduct current.

integrated circuit (IC) A type of circuit in which all the components are constructed on a single tiny chip of silicon.

integrator A circuit that produces an output which approximates the area under the curve of the input function.

intrinsic The pure or natural state of a material.

inverting amplifier An op-amp closed-loop configuration in which the input signal is applied to the inverting input.

ionization The removal or addition of an electron from or to a neutral atom so that the resulting atom (called an ion) has a net positive or negative charge.

irradiance (*E*) The power per unit area at a specified distance for the LED; the light intensity.

isolation amplifier An amplifier with electrically isolated internal stages.

JFET Junction field-effect transistor; one of two major types of field-effect transistors.

large-signal A signal that operates an amplifier over a significant portion of its load line.

LASCR Light-activated silicon-controlled rectifier; a four-layer semiconductor device (thyristor) that conducts current in one direction when activated by a sufficient amount of light and continues to conduct until the current falls below a specified value.

laser *L*ight *a*mplification by *s*timulated *e*mission of *r*adiation.

light-emitting diode (LED) A type of diode that emits light when there is forward current.

limiter A diode circuit that clips off or removes part of a waveform above and/or below a specified level.

linear Characterized by a straight-line relationship.

linear region The region of operation along the load line between saturation and cutoff.

linear regulator A voltage regulator in which the control element operates in the linear region.

line regulation The change in output voltage for a given change in input (line) voltage, normally expressed as a percentage.

load The amount of current drawn from the output of a circuit through a load resistance.

load regulation The change in output voltage for a given change in load current, normally expressed as a percentage.

lock range The range of frequencies over which a PLL can maintain lock.

logarithm An exponent; the logarithm of a quantity is the exponent or power to which a given number called the base must be raised in order to equal the quantity.

loop gain An op-amp's open-loop gain times the attenuation.

lower-side frequency In balanced modulation, the difference of the carrier frequency and the modulation frequency.

low-pass filter A type of filter that passes frequencies below a certain frequency while rejecting higher frequencies.

majority carrier The most numerous charge carrier in a doped semiconductive material (either free electrons or holes).

midrange gain The gain that occurs for the range of frequencies between the lower and upper critical frequencies.

minority carrier The least numerous charge carrier in a doped semiconductive material (either free electrons or holes).

mixer A device for down-converting frequencies in a receiver system.

modulation The process in which a signal containing information is used to modify a characteristic of another signal such as amplitude, frequency, or pulse width so that the information on the first is also contained on the second.

monochromatic Related to light of a single frequency; one color.

MOSFET Metal oxide semiconductor field-effect transistor; one of two major types of FETs; sometimes called IGFET for insulated-gate FET.

multistage Characterized by having more than one stage; a cascaded arrangement of two or more amplifiers.

natural logarithm The exponent to which the base e ($e = 2.71828$) must be raised in order to equal a given quantity.

negative feedback The process of returning a portion of the output signal to the input of an amplifier such that it is out of phase with the input signal.

neutron An uncharged particle found in the nucleus of an atom.

noise An unwanted signal.

noninverting amplifier An op-amp closed-loop configuration in which the input signal is applied to the noninverting input.

nucleus The central part of an atom containing protons and neutrons.

octave A two-times increase or decrease in the value of a quantity such as frequency.

Ohmic region The portion of the FET characteristic curve lying below pinch-off in which Ohm's law applies.

OLED Organic light-emitting diode; a device that consists of two or three layers of materials composed of organic molecules or polymers that emit light with an application of voltage.

open-loop voltage gain (A_{ol}) The voltage gain of an op-amp without external feedback.

operational amplifier (op-amp) A type of amplifier that has a very high voltage gain, very high input impedance, very low output impedance, and good rejection of common-mode signals.

operational transconductance amplifier (OTA) A voltage-to-current amplifier.

optocoupler A device in which an LED is used to couple a photodiode or a phototransistor in a single package.

orbit The path an electron takes as it circles around the nucleus of an atom.

order The number of poles in a filter.

oscillator A circuit that produces a periodic waveform on its output with only the dc supply voltage as its input.

output resistance The resistance looking in at the transistor collector.

passband The range of frequencies that are allowed to pass through a filter with minimum attenuation.

peak inverse voltage (PIV) The maximum value of reverse voltage across a diode that occurs at the peak of the input cycle when the diode is reversed-biased.

pentavalent Describes an atom with five valence electrons.

phase-locked loop (PLL) A device for locking onto and tracking the frequency of an incoming signal.

phase shift The relative angular displacement of a time-varying function relative to a reference.

phase-shift oscillator A type of feedback oscillator that is characterized by three *RC* circuits in the positive feedback loop that produces a phase shift of 180°.

photodiode A diode in which the reverse current varies directly with the amount of light.

photon A particle of light energy.

phototransistor A transistor in which base current is produced when light strikes the photosensitive semiconductor base region.

piezoelectric effect The property of a crystal whereby a changing mechanical stress produces a voltage across the crystal.

pinch-off voltage The value of the drain-to-source voltage of a FET at which the drain current becomes constant when the gate-to-source voltage is zero.

pixel In an LED display screen, the basic unit for producing colored light and consisting of red, green, and blue LEDs.

***pn* junction** The boundary between two different types of semiconductive materials.

pole A circuit containing one resistor and one capacitor that contributes −20 dB/decade to a filter's roll-off.

positive feedback The return of a portion of the output signal to the input such that it reinforces and sustains the output. This output signal is in phase with the input signal.

power gain The ratio of output power to input power of an amplifier.

power supply A circuit that converts ac line voltage to dc voltage and supplies constant power to operate a circuit or system.

proton The basic particle of positive charge.

pulse width modulation A process in which a signal is converted to a series of pulses with widths that vary proportionally to the signal amplitude.

push-pull A type of class B amplifier with two transistors in which one transistor conducts for one half-cycle and the other conducts for the other half-cycle.

PUT Programmable unijunction transistor; a type of three-terminal thyristor (more like an SCR than a UJT) that is triggered into conduction when the voltage at the anode exceeds the voltage at the gate.

Q-point The dc operating (bias) point of an amplifier specified by voltage and current values.

quality factor (Q) For a reactive component, a figure of merit which is the ratio of energy stored and returned by the component to the energy dissipated; for a band-pass filter, the ratio of the center frequency to its bandwidth.

radiant intensity (I_θ) The output power of an LED per steradian in units of mW/sr.

radiation The process of emitting electromagnetic or light energy.

recombination The process of a free (conduction band) electron falling into a hole in the valence band of an atom.

rectifier An electronic circuit that converts ac into pulsating dc; one part of a power supply.

regulator An electronic device or circuit that maintains an essentially constant output voltage for a range of input voltage or load values; one part of a power supply.

relaxation oscillator An electronic circuit that uses an *RC* timing circuit to generate a nonsinusoidal waveform without an external input signal.

reverse bias The condition in which a diode prevents current.

ripple factor A measure of effectiveness of a power supply filter in reducing the ripple voltage; ratio of the ripple voltage to the dc output voltage.

ripple voltage The small variation in the dc output voltage of a filtered rectifier caused by the charging and discharging of the filter capacitor.

***r* parameter** One of a set of BJT characteristic parameters that include α_{DC}, β_{DC}, r'_e, r'_b, and r'_c.

roll-off The rate of decrease in the gain above or below the critical frequencies of a filter.

saturation The state of a BJT in which the collector current has reached a maximum and is independent of the base current.

schematic A symbolized diagram representing an electrical or electronic circuit.

Schmitt trigger A comparator with built-in hysteresis.

SCR Silicon-controlled rectifier; a type of three-terminal thyristor that conducts current when triggered on by a voltage at the single gate terminal and remains on until the anode current falls below a specified value.

SCS Silicon-controlled switch; a type of four-terminal thyristor that has two gate terminals that are used to trigger the device on and off.

semiconductor A material that lies between conductors and insulators in its conductive properties.

shell An energy band in which electrons orbit the nucleus of an atom.

signal compression The process of scaling down the amplitude of a signal voltage.

silicon A semiconductive material.

slew rate The rate of change of the output voltage of an op-amp in response to a step input.

source One of the three terminals of a FET analogous to the emitter of a BJT.

source-follower The common-drain amplifier.

spectral Pertaining to a range of frequencies.

stability A measure of how well an amplifier maintains its design values (Q-point, gain, etc.) over changes in beta and temperature.

stage One of the amplifier circuits in a multistage configuration.

standoff ratio The characteristic of a UJT that determines its turn-on point.

stiff voltage divider A voltage divider for which loading effects can be neglected.

summing amplifier An op-amp configuration with two or more inputs that produces an output voltage that is proportional to the negative of the algebraic sum of its input voltages.

suppressed-carrier modulation A form of amplitude modulation in which the carrier is suppressed; also called *balanced modulation.*

switched-capacitor circuit A combination of a capacitor and transistor switches used in programmable analog devices to emulate resistors.

switching current (I_S) The value of anode current at the point where the device switches from the forward-blocking region to the forward-conduction region.

switching regulator A voltage regulator in which the control element operates as a switch.

Sziklai pair A complementary Darlington arrangement.

thermal overload A condition in a rectifier where the internal power dissipation of the circuit exceeds a certain maximum due to excessive current.

thermistor A temperature-sensitive resistor with a negative temperature coefficient.

thyristor A class of four-layer (*pnpn*) semiconductor devices.

transconductance (g_m) The ratio of a change in drain current to a change in gate-to-source voltage in a FET; in general, the ratio of the output current to the input voltage.

transformer An electrical device constructed of two or more coils (windings) that are electromagnetically coupled to each other to provide a transfer of power from one coil to another.

transistor A semiconductive device used for amplification and switching applications.

triac A three-terminal thyristor that can conduct current in either direction when properly activated.

trigger The activating input of some electronic devices and circuits.

trivalent Describes an atom with three valence electrons.

troubleshooting A systematic process of isolating, identifying, and correcting a fault in a circuit or system.

turns ratio The number of turns in the secondary of a transformer divided by the number of turns in the primary.

UJT Unijunction transistor; a three-terminal single *pn* junction device that exhibits a negative resistance characteristic.

upper-side frequency In balanced modulation, the sum of the carrier frequency and the modulated frequency.

valence Related to the outer shell of an atom.

varactor A variable capacitance diode.

***V-I* characteristic** A curve showing the relationship of diode voltage and current.

voltage-controlled oscillator (VCO) A type of relaxation oscillator whose frequency can be varied by a dc control voltage; an oscillator for which the output frequency is dependent on a controlling input voltage.

voltage-follower A closed-loop, noninverting op-amp with a voltage gain of 1.

voltage multiplier A circuit using diodes and capacitors that increases the input voltage by two, three, or four times.

wavelength The distance in space occupied by one cycle of an electromagnetic or light wave.

Wien bridge oscillator A type of feedback oscillator that is characterized by an *RC* lead-lag circuit in the positive feedback loop.

zener breakdown The lower voltage breakdown in a zener diode.

zener diode A diode designed for limiting the voltage across its terminals in reverse bias.

INDEX